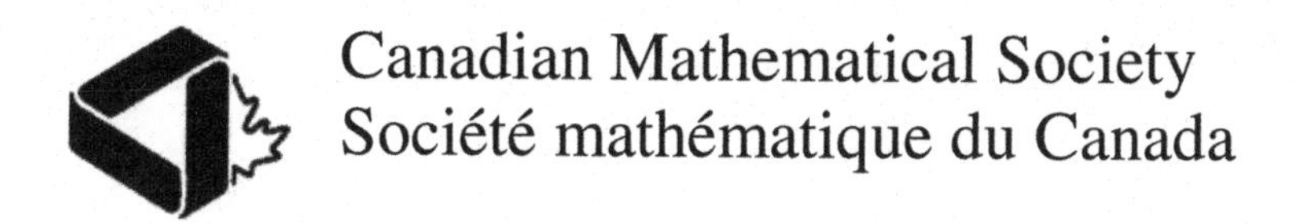

For other titles published in this series, go to
http://www.springer.com/series/4318

Marián Fabian · Petr Habala · Petr Hájek ·
Vicente Montesinos · Václav Zizler

Banach Space Theory

The Basis for Linear and Nonlinear Analysis

Marián Fabian
Mathematical Institute of the Academy
of Sciences of the Czech Republic
Žitná 25, Praha 1
11567 Prague, Czech Republic
fabian@math.cas.cz

Petr Habala
Czech Technical University in Prague
Department of Mathematics
Faculty of Electrical Engineering
Technická 2
16627 Prague, Czech Republic
habala@math.feld.cvut.cz

Petr Hájek
Mathematical Institute of the Academy
of Sciences of the Czech Republic
Žitná 25, Praha 1
11567 Prague, Czech Republic
hajek@math.cas.cz

Vicente Montesinos
Universidad Politécnica de Valencia
Departamento de Matematica Aplicada
Camino de Vera s/n
46022 Valencia, Spain
vmontesinos@mat.upv.es

Václav Zizler
University of Alberta
Department of Mathematical
and Statistical Sciences
Central Academic Building
Edmonton T6G 2G1
Alberta, Canada
zizler@math.cas.cz

ISSN 1613-5237
ISBN 978-1-4939-4114-8 ISBN 978-1-4419-7515-7 (eBook)
DOI 10.1007/978-1-4419-7515-7
Springer New York Dordrecht Heidelberg London

Mathematics Subject Classicication (2010):
Primary: 46Bxx
Secondary: 46A03, 46A20, 46A22, 46A25, 46A30, 46A32, 46A50, 46A55, 46B03, 46B04, 46B07, 46B10, 46B15, 46B20, 46B22, 46B25, 46B26, 46B28, 46B45, 46B50, 46B80, 46C05, 46C15, 46G05, 46G12, 47A10, 52A07, 52A21, 52A41, 58C20, 58C25

Printed on acid-free paper

Springer is part of Springer Science+Business Media (www.springer.com)

Preface

Many problems in modern linear and nonlinear analysis are of infinite-dimensional nature. The theory of Banach spaces provides a suitable framework for the study of these areas, as it blends classical analysis, geometry, topology, and linearity. This in turn makes Banach space theory a wonderful and active research area in Mathematics.

In infinite dimensions, neighborhoods of points are not relatively compact, continuous functions usually do not attain their extrema and linear operators are not automatically continuous. By introducing weak topologies, compactness can be obtained via Tychonoff's theorem. Similarly, functions often need to be perturbed so that the problem of finding extrema is solvable. To deal with problems in linear and nonlinear analysis, a good working knowledge of Banach space theory techniques is needed. It is the purpose of this introductory text to help the reader grasp the basic principles of Banach space theory and nonlinear geometric analysis.

The text presents the basic principles and techniques that form the core of the theory. It is organized to help the reader proceed from the elementary part of the subject to more recent developments. This task is not easy. Experience shows that working through a large number of exercises, provided with hints that direct the reader, is one of the most efficient ways to master the subject. Exercises are of several levels of difficulty, ranging from simple exercises to important results or examples. They illustrate delicate points in the theory and introduce the reader to additional lines of research. In this respect, they should be considered an integral part of the text. A list of remarks and open problems ends each chapter, presenting further developments and suggesting research paths.

An effort has been made to ensure that the book can serve experts in related fields such as Optimization, Partial Differential Equations, Fixed Point Theory, Real Analysis, Topology, and Applied Mathematics, among others.

As prerequisites, basic undergraduate courses in calculus, linear algebra, and general topology, should suffice.

The text is divided into 17 chapters.

In Chapter 1 we present basic notions in Banach space theory and introduce the classical Banach spaces, in particular sequence and function spaces.

In Chapter 2 we discuss two fundamental principles of Banach space theory, namely the Hahn–Banach Theorem on extension of bounded linear functionals and the Banach Open mapping Theorem, together with some of their applications.

In Chapter 3 we discuss weak topologies and their properties related to compactness. Then we prove the third fundamental principle, namely the Banach–Steinhaus Uniform Boundedness principle. Special attention is devoted to weak compactness, in particular to the theorems of Eberlein, Šmulyan, Grothendieck and James, and the theory of reflexive Banach spaces.

In Chapter 4 we introduce Schauder bases in Banach spaces. The possibility to represent each element of the space as the sequence of its coefficients in a given Schauder basis transfers the purely geometric techniques of the elementary Banach space theory to the analytic computations of the classical analysis. Although not every separable Banach space admits a Schauder basis, the use of basic sequences and Schauder bases with additional properties is one of the main tools in the investigation of the structural properties of Banach spaces.

In Chapter 5 we continue the study of the structure of Banach spaces by adding results on extensions of operators, injectivity, and weak injectivity. The core of the chapter is the theory of separable Banach spaces not containing isomorphic copies of ℓ_1.

Chapter 6 is an introduction to some basic results in the geometry of finite-dimensional Banach spaces and their connection to the structure of infinite-dimensional spaces. We do not discuss the deeper parts of the theory, which essentially depend on measure theoretical techniques. We introduce the notion of finite representability, and prove the principle of local reflexivity. We use the John ellipsoid to prove the Kadec–Snobar theorem and give a proof of Tzafriri's theorem. We indicate the connection of this result with Dvoretzky's theorem. Last part of the chapter is devoted to the Grothendieck inequality.

In Chapter 7 we present an introduction to nonlinear analysis, namely to variational principles and differentiability.

In Chapter 8 we study the interplay between differentiability of norms and the structure of separable Asplund spaces.

Chapter 9 introduces the subject of superreflexive spaces, whose structure is nicely described by the behavior of its finite-dimensional subspaces.

Chapter 10 studies the impact of the existence of higher order smooth norms on the structure of the underlying space. Special effort is devoted to countable compact spaces and ℓ_p spaces.

Chapter 11 deals with the property of dentability and results on differentiation of vector measures. We prove some basic results on Banach spaces with the Radon–Nikodým property.

Chapter 12 introduces the reader to the nonlinear geometric analysis of Banach spaces. Results on uniform and nonuniform homeomorphisms are presented, including Keller's theorem and basic fixed points theorems (Brouwer, Schauder, etc). We discuss a proof of the homeomorphisms of Banach spaces and results on uniform, in particular Lipschitz, homeomorphisms.

Chapter 13 contains a basic study of an important class of non-separable Banach spaces, the weakly compactly generated spaces. In particular, we discuss their decompositions and renormings. We also study weakly compact operators, absolutely summing operators, and the Dunford–Pettis property.

Chapter 14 deals with results on weak topologies, focusing on special types of compacta (scattered, Eberlein, Corson, etc.).

Chapter 15 presents basic results in the spectral theory of operators. We study compact and self-adjoint operators.

Chapter 16 deals with the basic theory of tensor products. We follow the Banach space approach, focusing on the Grothendieck duality theory of tensor products, Schauder bases, applications to spaces of compact operators, etc. We include Enflo's example of a Banach space without the approximation property.

A short appendix (Chapter 17) has been included collecting some very basic definitions and results that are used in the text, for the reader's immediate access.

In writing the text we strived to avoid excessive technicalities, keeping each subject as elementary as reasonably possible. Each chapter ends with a brief section of Remarks and Open Questions, containing further known results and some problems in the area that are—to our best knowledge—open.

Several more specialized books and survey articles appeared recently in Banach space theory, as [AlKa], [BeLi], [BoVa], [CasGon], [DJT], [HMVZ], [JoLi3], [Kalt4], [KaKuLP], [LPT], [MOTV2], [Wojt], among others. We hope that the present text can help both the student and the professional mathematician to get acquainted with the techniques needed in these directions. We also made an effort to make this text closer to a reference book in order to help researchers in Banach space theory.

We are grateful to many of our colleagues for suggestions, advice, and discussions on the subject of the book. We thank our Institutions: the Institute of Mathematics of the Czech Academy of Sciences, the Czech Technical University in Prague, the Department of Mathematical and Statistical Sciences at the University of Alberta, Edmonton, Canada, the Universidad Politécnica de Valencia, Spain, and its Instituto Universitario de Matemática Pura y Aplicada. This work has been supported by several Grant Agencies: The Czech National Grant Agency and the Institutional Research Plan of the Academy of Sciences (Czech Republic), NSERC Canada, the Ministerio de Educación (Spain) and the Generalitat Valenciana (Valencia, Spain). The grants involved are IAA 100 190 610, IAA 100 190 901, GAČR 201/07/0394, No. AVOZ 101 905 03 (Czech Republic), Proyecto MTM2008-03211 (Spain), BEST/2009/096 (Generalitat Valenciana) and PR2009-0267 (Ministerio de Educación), NSERC-7926 (Canada).

We would like to thank the Springer Team for their interest in this project. In particular, we are thankful to Keith F. Taylor, Karl Dilcher, Mark Spencer, Vaishali Damle, and Charlene C. Cerdas. We thank also Eulalia Noguera for her help with the tex file, and to Integra Software Services Pvt Ltd, in particular Sankara Narayanan, for their assistance in editing the final version of this book.

Above all, we are indebted to our families for their moral support and encouragement.

We would be glad if this book inspired some young mathematicians to choose Banach Space Theory and/or Nonlinear Geometric Analysis as their field of interest.

We wish the reader a pleasant time spent over this book.

Prague, Czech Republic — Marián Fabian
Prague, Czech Republic — Petr Habala
Prague, Czech Republic — Petr Hájek
Valencia, Spain — Vicente Montesinos
Edmonton, AB, Canada — Václav Zizler
Spring, 2010

Contents

Chapter 1
Basic Concepts in Banach Spaces

In this chapter we introduce basic notions and concepts in Banach space theory.

As a rule we will work with real scalars, only in a few instances, e.g., in spectral theory, we will use complex scalars. $\mathbb{K}$ denotes simultaneously the real ($\mathbb{R}$) or complex ($\mathbb{C}$) scalar field. We use $\mathbb{N}$ for the set $\{1, 2, \dots\}$.

All topologies are assumed to be Hausdorff, unless stated otherwise. In particular, by a *compact space* we mean a compact Hausdorff space. By a *neighborhood* of a point x in a topological space T we mean any subset of T that contains an open subset O of T such that $x \in O$.

If $(T, \mathcal{T})$ is a topological space, and S is a nonempty subset, we shall write $\mathcal{T}|_S$ for the restriction of the topology $\mathcal{T}$ to S (and so $(S, \mathcal{T})$ becomes a topological space). If there is no possibility of misunderstanding, the restricted topology will be called again $\mathcal{T}$. For a brief review on basic topological notions see, e.g., the Appendix.

1.1 Basic Definitions

Definition 1.1 *A non-negative function* $\|\cdot\|$ *on a vector (i.e., linear) space* X *is called a* norm *on* X *if*
(i) $\|x\| \geq 0$ *for every* $x \in X$,
(ii) $\|x\| = 0$ *if and only if* $x = 0$,
(iii) $\|\lambda x\| = |\lambda| \, \|x\|$ *for every* $x \in X$ *and every scalar* λ,
(iv) $\|x + y\| \leq \|x\| + \|y\|$ *for every* $x, y \in X$ *(the "triangle inequality").*
A vector space X *with a norm* $\|\cdot\|$ *is denoted by* $(X, \|\cdot\|)$, *and is called a* normed linear space *(or just a* normed space*).*

Note that the function $\rho(x, y) := \|x - y\|$, where $x, y \in X$, is indeed a metric on X. To check the triangle inequality we write

$$\rho(x, z) = \|x - z\| = \|x - y + y - z\| \leq \|x - y\| + \|y - z\| = \rho(x, y) + \rho(y, z).$$

By induction, $\left\|\sum_{i=1}^n x_i\right\| \leq \sum_{i=1}^n \|x_i\|$ for a finite number of vectors $x_1, \dots, x_n$ in X.

M. Fabian et al., *Banach Space Theory*, CMS Books in Mathematics,
DOI 10.1007/978-1-4419-7515-7_1,

All topological and uniform notions in normed spaces refer to the canonical metric given by the norm, unless stated otherwise. In situations when more than one normed space is considered, we will sometimes use $\|\cdot\|_X$ to denote the norm of X.

Definition 1.2 *A* Banach space *is a normed linear space $(X, \|\cdot\|)$ that is complete in the canonical metric defined by $\rho(x, y) = \|x - y\|$ for $x, y \in X$, i.e., every Cauchy sequence in X for the metric ρ converges to some point in X.*

Let $(X, \|\cdot\|)$ be a normed space. The set $B_X := \{x \in X : \|x\| \leq 1\}$ is said to be the *closed unit ball* of X, and $S_X := \{x \in X : \|x\| = 1\}$ the *unit sphere* of $(X, \|\cdot\|)$. Given $x_0 \in X$ and $r > 0$, the set $B(x_0, r) := \{x \in X : \|x - x_0\| \leq r\}$ is said to be the *closed ball centered at x_0 with radius r*. If $M \subset X$, then $\operatorname{span}(M)$ stands for the linear hull—or span—of M, that is, the intersection of all linear subspaces of X containing M. Equivalently, $\operatorname{span}(M)$ is the smallest (in the sense of inclusion) linear subspace of X containing M, or the set of all finite linear combinations of elements in M. Similarly, $\overline{\operatorname{span}}(M)$ stands for the closed linear hull of M, i.e., the smallest closed linear subspace of X containing M.

If no misunderstanding can arise, by a "subspace" of a vector space we will mean a linear subspace and, in case of normed spaces, a closed linear subspace.

Definition 1.3 *Let E be a vector space. Given $x, y \in E$, the set $[x, y] := \{\lambda x + (1 - \lambda)y : 0 \leq \lambda \leq 1\}$ is called the* closed segment defined by x and y. *If $x \neq y$, the set $(x, y) := \{\lambda x + (1 - \lambda)y : 0 < \lambda < 1\}$ is called the* open segment defined by x and y. *A subset C of a vector space E is called* convex *if $[x, y] \subset C$ whenever $x, y \in C$.*

If $M \subset X$, the *convex hull* of M is the smallest convex subset of X containing M, and will be denoted by $\operatorname{conv}(M)$; $\overline{\operatorname{conv}}(M)$ denotes the closed convex hull of M, i.e., the smallest closed convex subset of X containing M.

Definition 1.4 *Let U be a convex subset of a vector space V. We say that a function $f : U \to \mathbb{R}$ is* convex *if $f(\lambda x + (1-\lambda)y) \leq \lambda f(x) + (1-\lambda) f(y)$ for all $x, y \in U$ and $\lambda \in [0, 1]$. We say that f is* strictly convex *if $f(\lambda x + (1-\lambda)y) < \lambda f(x) + (1-\lambda) f(y)$ for all $x, y \in U$, $x \neq y$, and $\lambda \in (0, 1)$.*

For instance, every norm of a normed space X is a convex function on X. Observe that a function $f : U \to \mathbb{R}$ is convex if and only if the *epigraph* of f, i.e., the set $\operatorname{epi} f := \{(x, r) \in U \times \mathbb{R} : f(x) \leq r\} \subset X \times \mathbb{R}$, is convex (the linear structure of $X \times \mathbb{R}$ is defined coordinatewise).

For subsets A, B of a vector space X and a scalar α we also write $A + B := \{a + b : a \in A, b \in B\}$ and $\alpha A := \{\alpha a : a \in A\}$.

A set $M \subset X$ is called *symmetric* if $(-1)M \subset M$, and *balanced* if $\alpha M \subset M$ for all $\alpha \in \mathbb{K}$, $|\alpha| \leq 1$.

Let Y be a subspace of a normed space $(X, \|\cdot\|)$. By $(Y, \|\cdot\|)$ we denote Y endowed with the restriction of $\|\cdot\|$ to Y if there is no risk of misunderstanding.

Fact 1.5 *Let Y be a subspace of a Banach space X. Then Y is a Banach space if and only if Y is closed in X.*

Proof: Assume that Y is closed. Consider a Cauchy sequence $\{y_n\}_{n=1}^{\infty}$ in Y. Since the norm on Y is the restriction of the norm of X, the sequence is Cauchy in X and therefore converges to some $y \in X$. As Y is closed, $y \in Y$ and $y_n \to y$ in Y.

The other direction is proved by a similar argument. □

Definition 1.6 *A subset M of a normed space $(X, \|\cdot\|)$ is called* bounded *if there exists $r > 0$ such that $M \subset rB_X$. M is called* totally bounded *if for every $\varepsilon > 0$ the set M can be covered by a finite number of translates of εB_X. A sequence $\{x_n\}$ in X is called* bounded (totally bounded) *if the set $\{x_n : n \in \mathbb{N}\}$ is bounded (respectively, totally bounded).*

Note that every totally bounded set is already bounded. See also Exercises 1.47 and 1.48 for a description of total boundedness by using ε-nets, Definition 3.11, and Section 17.10.

1.2 Hölder and Minkowski Inequalities, Classical Spaces $C[0,1]$, ℓ_p, c_0, $L_p[0,1]$

We will now turn to some examples of Banach spaces.

Definition 1.7 *The symbol $C[0,1]$ denotes the vector space of all scalar valued continuous functions on the interval $[0,1]$ (the vector addition and the scalar multiplication being defined pointwise), endowed with the norm*

$$\|f\|_{\infty} := \sup\{|f(t)| : t \in [0,1]\} \ (= \max\{|f(t)| : t \in [0,1]\}).$$

Proposition 1.8 *The function $\|\cdot\|_{\infty}$ introduced in Definition 1.7 is indeed a norm, and $(C[0,1], \|\cdot\|_{\infty})$ is a Banach space.*

Proof: We easily check that $C[0,1]$ is a normed space. Consider a Cauchy sequence $\{f_n\}_{n=1}^{\infty}$ in $C[0,1]$. As $|f_k(t) - f_l(t)| \leq \|f_k - f_l\|_{\infty}$, the sequence $\{f_n(t)\}_{n=1}^{\infty}$ is a Cauchy sequence for every $t \in [0,1]$. Set $f(t) := \lim_{n\to\infty} f_n(t)$. This defines a scalar valued function f on $[0,1]$. It remains to show that f is continuous and $f_n \to f$ uniformly (i.e., in $\|\cdot\|_{\infty}$). Given $\varepsilon > 0$, there is n_0 such that $|f_n(t) - f_m(t)| \leq \varepsilon$ for every $t \in [0,1]$ and every $n, m \geq n_0$. By fixing $n \geq n_0$ and letting $m \to \infty$ we get $|f_n(t) - f(t)| \leq \varepsilon$ for every $n \geq n_0$ and every $t \in [0,1]$. Let $t_0 \in [0,1]$ and $\varepsilon > 0$ be fixed. Choose $\delta > 0$ so that $|f_{n_0}(t) - f_{n_0}(t_0)| < \varepsilon$ whenever $|t - t_0| < \delta$. Then, whenever $|t - t_0| < \delta$,

$$|f(t) - f(t_0)| \leq |f(t) - f_{n_0}(t)| + |f_{n_0}(t) - f_{n_0}(t_0)| + |f_{n_0}(t_0) - f(t_0)| < 3\varepsilon.$$

Therefore $f \in C[0,1]$. It has been shown above that, for every $n \geq n_0$, $\|f_n - f\|_{\infty} \leq \varepsilon$. This proves that $\|f_n - f\|_{\infty} \to 0$, so $C[0,1]$ is complete. □

Analogously, the space $C(K)$ of continuous scalar functions on a compact space K, endowed with the supremum norm, is a Banach space.

We note that $C[0, 1]$ is an infinite-dimensional Banach space. To see this, it is enough to produce, for any $n \in \mathbb{N}$, a linearly independent set of n elements in $C[0, 1]$. The set of functions $\{1, t, t^2, \ldots, t^{n-1}\}$ has this property. More generally, the space $C(K)$, where K is a compact topological space, is infinite-dimensional as soon as K is infinite; indeed, given a finite set of distinct points $S := \{k_i : i = 1, 2, \ldots, n\}$ in K, define the function δ_{k_i} on S for $i = 1, 2, \ldots, n$, where δ_k is the Kronecker delta function at k, i.e., $\delta_k(k) = 1$ and $\delta_k(k') = 0$ for all $k' \neq k$. Extend each δ_{k_i} to a continuous function on K by using the Tietze–Urysohn theorem (see Corollary 7.55). The resulting set of extended functions $\{\delta_{k_i} : i = 1, 2, \ldots, n\}$ is linearly independent in $C(K)$.

Definition 1.9 *The symbol ℓ_∞^n denotes the n-dimensional vector space of all n-tuples of scalars (that is, $\mathbb{R}^n$ or $\mathbb{C}^n$), the vector addition and the scalar multiplication being defined coordinatewise, endowed with the supremum norm $\|\cdot\|_\infty$ defined for $x = (x_1, \ldots, x_n) \in \ell_\infty^n$ by*

$$\|x\|_\infty = \max\{|x_i| : i = 1, \ldots, n\}.$$

Note that ℓ_∞^n is a special case of a $C(K)$ space, where $K := \{1, \ldots, k\}$, endowed with the discrete topology.

In order to introduce the class of ℓ_p spaces for $1 < p < \infty$ we need to prove the following classical inequalities.

Theorem 1.10 (Hölder inequality) *Let $p, q > 1$ be such that $\frac{1}{p} + \frac{1}{q} = 1$ and let $n \in \mathbb{N}$. Then for all $a_k, b_k \in \mathbb{K}$, $k = 1, \ldots, n$, we have*

$$\sum_{k=1}^{n} |a_k b_k| \leq \Big(\sum_{k=1}^{n} |a_k|^p\Big)^{\frac{1}{p}} \cdot \Big(\sum_{k=1}^{n} |b_k|^q\Big)^{\frac{1}{q}}. \tag{1.1}$$

For $p = 2, q = 2$, the inequality (1.1) is known as the *Cauchy–Schwarz inequality*.

In the proof of Theorem 1.10 we will use the following statement.

Lemma 1.11 *Let $p, q > 1$ be such that $\frac{1}{p} + \frac{1}{q} = 1$. Then $ab \leq \frac{a^p}{p} + \frac{b^q}{q}$ for all $a, b \geq 0$.*

Proof: Consider the graph of the function $y = x^{p-1}$, $x \geq 0$, and the areas A_1 of the region bounded by the curves $y = x^{p-1}$, $y = 0$, $x = a$, and A_2 of the region bounded by the curves $y = x^{p-1}$, $x = 0$, $y = b$ (see Fig. 1.1). Clearly, $A_1 = \int_0^a x^{p-1}\mathrm{d}x = \frac{a^p}{p}$. As $x = y^{1/(p-1)} = y^{q-1}$, we get $A_2 = \int_0^b y^{q-1}\mathrm{d}y = \frac{b^q}{q}$. It follows that $ab \leq A_1 + A_2 = \frac{a^p}{p} + \frac{b^q}{q}$. □

An alternative proof is by checking extrema of the function $\varphi(a) := \frac{a^p}{p} + \frac{b^q}{q} - ab$ for a fixed $b > 0$.

Proof of Theorem 1.10: We may assume that $a_i, b_i \geq 0$ and neither all a_i nor all b_i are zero. For $k = 1, \ldots, n$ define

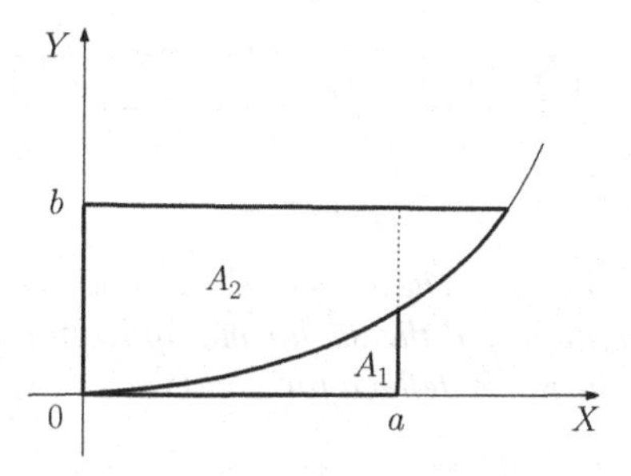

Fig. 1.1 Two areas and a rectangle in the proof of Lemma 1.11

$$A_k = a_k\Big(\sum_{j=1}^{n} a_j{}^p\Big)^{-\frac{1}{p}} \qquad \text{and} \qquad B_k = b_k\Big(\sum_{j=1}^{n} b_j{}^q\Big)^{-\frac{1}{q}}.$$

We note that $\sum_{k=1}^{n} A_k{}^p = \sum_{k=1}^{n} B_k{}^q = 1$. By Lemma 1.11, we have for $k = 1,\ldots,n$ that $A_k B_k \le \frac{1}{p}A_k{}^p + \frac{1}{q}B_k{}^q$. Summing up this inequality for $k = 1,\ldots,n$ we get

$$\sum_{k=1}^{n} A_k B_k \le \frac{1}{p}\sum_{k=1}^{n} A_k{}^p + \frac{1}{q}\sum_{k=1}^{n} B_k{}^q = \frac{1}{p} + \frac{1}{q} = 1,$$

which implies the desired inequality. □

Theorem 1.12 (Minkowski inequality) *Let $p \in [1,\infty)$ and $n \in \mathbb{N}$. Then for all $a_k, b_k \in \mathbb{K}$, $k = 1,\ldots,n$, we have*

$$\Big(\sum_{k=1}^{n} |a_k + b_k|^p\Big)^{\frac{1}{p}} \le \Big(\sum_{k=1}^{n} |a_k|^p\Big)^{\frac{1}{p}} + \Big(\sum_{k=1}^{n} |b_k|^p\Big)^{\frac{1}{p}}. \tag{1.2}$$

Proof: The statement is trivial for $p = 1$. If $p \in (1,\infty)$, let $q \in (1,\infty)$ be such that $\frac{1}{p} + \frac{1}{q} = 1$. We may assume that $a_i, b_i \ge 0$. Using the Hölder inequality (1.1) and the fact that $(p-1)q = p$ we obtain

$$\begin{aligned}
\sum (a_k + b_k)^p &= \sum (a_k + b_k)^{p-1}(a_k + b_k) = \sum (a_k + b_k)^{p-1} a_k + \sum (a_k + b_k)^{p-1} b_k \\
&\le \Big(\sum (a_k + b_k)^{(p-1)q}\Big)^{\frac{1}{q}}\Big(\sum a_k{}^p\Big)^{\frac{1}{p}} + \Big(\sum (a_k + b_k)^{(p-1)q}\Big)^{\frac{1}{q}}\Big(\sum b_k{}^p\Big)^{\frac{1}{p}} \\
&= \Big(\sum (a_k + b_k)^p\Big)^{\frac{1}{q}}\Big(\sum a_k{}^p\Big)^{\frac{1}{p}} + \Big(\sum (a_k + b_k)^p\Big)^{\frac{1}{q}}\Big(\sum b_k{}^p\Big)^{\frac{1}{p}}.
\end{aligned}$$

Dividing by $\big(\sum (a_k + b_k)^p\big)^{\frac{1}{q}}$ we get

$$\left(\sum (a_k + b_k)^p\right)^{\frac{1}{p}} = \left(\sum (a_k + b_k)^p\right)^{1-\frac{1}{q}} \leq \left(\sum a_k{}^p\right)^{\frac{1}{p}} + \left(\sum b_k{}^p\right)^{\frac{1}{p}}.$$

□

Definition 1.13 *Let $p \in [1, \infty)$. The symbol ℓ_p^n denotes the n-dimensional vector space $\mathbb{K}^n$, the vector addition and the scalar multiplication being defined coordinatewise, endowed with the norm defined for $x = (x_1, \dots, x_n) \in \ell_p^n$ by*

$$\|x\|_p = \left(\sum_{i=1}^{n} |x_i|^p\right)^{\frac{1}{p}}.$$

By Minkowski's inequality (1.2), $\|\cdot\|_p$ is indeed a norm on X.

The closed unit ball of ℓ_1^2 is the square with vertices $\pm e_1, \pm e_2$, where e_i are the standard unit vectors, $e_1 = (1, 0)$ and $e_2 = (0, 1)$. The unit ball of ℓ_∞^2 is the square with vertices $(\pm e_1 \pm e_2)$. The unit ball of ℓ_2^2 is the disk of radius 1 centered at the origin.

Balls for other ps are somehow in between (see Fig. 1.2).

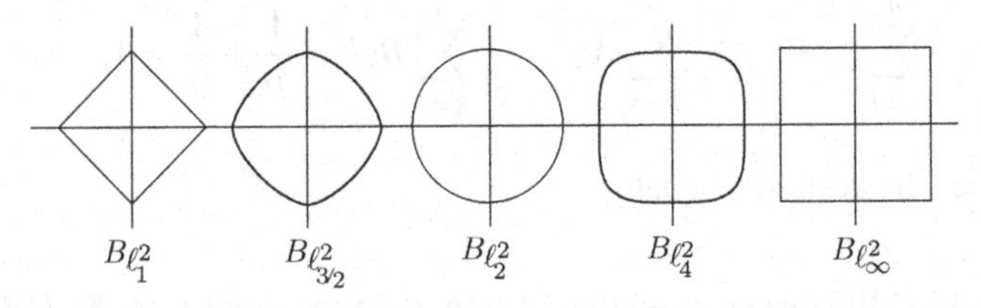

Fig. 1.2 Several balls in $\mathbb{R}^2$

The difference between ℓ_1^n and ℓ_∞^n becomes apparent once we increase the dimension. It is already apparent in three dimensions: The unit ball of ℓ_∞^3 is a cube, whereas the unit ball of ℓ_1^3 is an octahedron. The unit ball of ℓ_2^3 is a Euclidean ball (see Fig. 1.3).

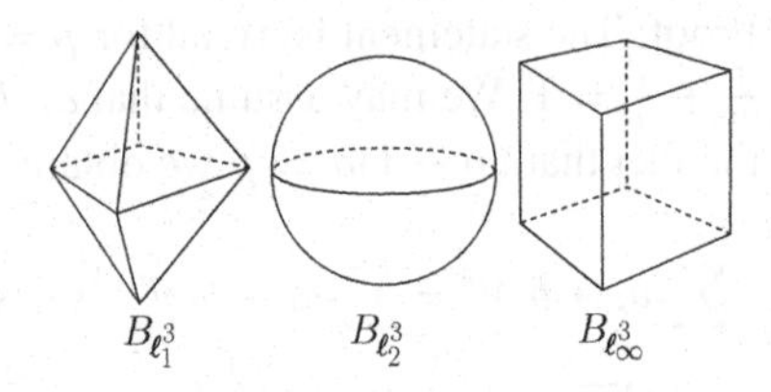

Fig. 1.3 Several balls in $\mathbb{R}^3$

Definition 1.14 *Let $p \in [1, \infty)$. The symbol $\ell_p = \ell_p(\mathbb{N})$ denotes the vector space of all scalar valued sequences $x = \{x_i\}_{i=1}^{\infty}$ satisfying $\sum |x_i|^p < \infty$, the vector addition and the scalar multiplication being defined coordinatewise, endowed with the norm*

$$\|x\|_p := \Big(\sum_{i=1}^{\infty} |x_i|^p\Big)^{\frac{1}{p}}.$$

When a scalar sequence $x = \{x_i\}_{i=1}^{\infty}$ is considered an element of ℓ_p, we use the notation $x = (x_i)$. This applies to other sequence spaces defined below.

To see that the definition is correct, we need to show that if $x = (x_i), y = (y_i) \in \ell_p$, then $x + y \in \ell_p$ and $\|x + y\|_p \leq \|x\|_p + \|y\|_p$.

For every $n \leq m \in \mathbb{N}$ we have by the Minkowski inequality (1.2):

$$\Big(\sum_{i=1}^{n} |x_i + y_i|^p\Big)^{\frac{1}{p}} \leq \Big(\sum_{i=1}^{n} |x_i|^p\Big)^{\frac{1}{p}} + \Big(\sum_{i=1}^{n} |y_i|^p\Big)^{\frac{1}{p}} \leq \Big(\sum_{i=1}^{m} |x_i|^p\Big)^{\frac{1}{p}} + \Big(\sum_{i=1}^{m} |y_i|^p\Big)^{\frac{1}{p}}.$$

By letting $m \to \infty$ we see that for every n:

$$\Big(\sum_{i=1}^{n} |x_i + y_i|^p\Big)^{\frac{1}{p}} \leq \Big(\sum_{i=1}^{\infty} |x_i|^p\Big)^{\frac{1}{p}} + \Big(\sum_{i=1}^{\infty} |y_i|^p\Big)^{\frac{1}{p}}$$

Letting $n \to \infty$ we get $x + y \in \ell_p$ and the triangle inequality follows.

Let $x = (x_i)_{i=1}^{\infty}$ be a sequence of scalars. We define the *support* of x by $\operatorname{supp}(x) = \{i \in \mathbb{N} : x_i \neq 0\}$.

Definition 1.15 *The symbol $\ell_\infty = \ell_\infty(\mathbb{N})$ denotes the vector space of all bounded scalar valued sequences endowed with the norm defined for $x = (x_i) \in \ell_\infty$ by*

$$\|x\|_\infty = \sup\{|x_i| : i \in \mathbb{N}\}.$$

The symbol $c_{00} = c_{00}(\mathbb{N})$ denotes the subspace of ℓ_∞ consisting of all $x = (x_i)$ such that $\operatorname{supp}(x)$ *is finite.*
The symbol $c = c(\mathbb{N})$ denotes the subspace of ℓ_∞ consisting of all $x = (x_i)$ such that $\lim\limits_{i\to\infty} x_i$ exists and is finite.
The symbol $c_0 = c_0(\mathbb{N})$ denotes the subspace of ℓ_∞ consisting of all $x = (x_i)$ such that $\lim\limits_{i\to\infty} x_i = 0$.
In all these cases, the vector addition and the scalar multiplication are defined coordinatewise.

Note that c_0 is the closure of c_{00} in ℓ_∞. Note also that if $x = (x_i)$ belongs to c_{00} or c_0, then $\|x\|_\infty = \max\{|x_i| : i \in \mathbb{N}\}$.

That the spaces $\ell_p(\mathbb{N})$, $\ell_\infty(\mathbb{N})$, $c_0(\mathbb{N})$, and $c_{00}(\mathbb{N})$ are infinite-dimensional follows from the fact that this is so for a vector space containing a linearly independent set of n vectors for each $n \in \mathbb{N}$. Vectors $\{e_i\}_{i=1}^{n}$ are linearly independent, where $e_i = (0, \dots, 0, 1, 0, \dots)$, and 1 is at the ith position. These vectors are called the *canonical unit vectors*.

Proposition 1.16 (i) *For $p \in [1, \infty]$, the space ℓ_p is a Banach space.*

(ii) *The spaces* c *and* c_0 *are closed subspaces of* ℓ_∞ *and thus they are Banach spaces.*
(iii) *The space* c_{00} *is not complete.*

In the proof we will use the following lemma.

Lemma 1.17 *Let* X *be a normed space. If a sequence* $\{x_n\}_{n=1}^\infty$ *in* X *is Cauchy, then it is bounded in* X.

Proof: Using the Cauchy property of $\{x_n\}$ we find $n_0 \in \mathbb{N}$ such that $\|x_n - x_{n_0}\| \leq 1$ for every $n \geq n_0$. Then $\|x_n\| \leq \|x_n - x_{n_0}\| + \|x_{n_0}\| \leq 1 + \|x_{n_0}\|$ for every $n \geq n_0$. Therefore for every $n \in \mathbb{N}$:

$$\|x_n\| \leq \max\{\|x_1\|, \|x_2\|, \dots, \|x_{n_0-1}\|, \|x_{n_0}\| + 1\}.$$

□

Proof of Proposition 1.16:

(i) It is easily checked, using a method similar to that for $C[0, 1]$, that ℓ_∞ is a Banach space. Now consider $p \in [1, \infty)$.

Let $\{x^k\}_{k=1}^\infty$ be a Cauchy sequence in ℓ_p, where $x^k = (x_i^k)$. Given $\varepsilon > 0$, find k_0 such that

$$\Big(\sum_{i=1}^{\infty} |x_i^k - x_i^l|^p\Big)^{\frac{1}{p}} \leq \varepsilon \tag{1.3}$$

for every $k, l \geq k_0$. In particular, $|x_i^k - x_i^l| \leq \varepsilon$ for every $k, l \geq k_0$ and $i \in \mathbb{N}$, hence the sequence $\{x_i^k\}_{k=1}^\infty$ converges to some x_i for every $i \in \mathbb{N}$.

Put $x = (x_i)$. We will show that $x \in \ell_p$. By Lemma 1.17, there is a constant $C > 0$ such that $\left(\sum_{i=1}^\infty |x_i^k|^p\right)^{\frac{1}{p}} \leq C$ for every k. Therefore $\left(\sum_{i=1}^n |x_i^k|^p\right)^{\frac{1}{p}} \leq C$ for all $n, k \in \mathbb{N}$. By letting $k \to \infty$ we get $\left(\sum_{i=1}^n |x_i|^p\right)^{\frac{1}{p}} \leq C$ for every $n \in \mathbb{N}$. Therefore $\left(\sum_{i=1}^\infty |x_i|^p\right)^{\frac{1}{p}} \leq C$ and $x \in \ell_p$.

We will now show that $x^k \to x$ in ℓ_p. Given $\varepsilon > 0$, we let $l \to \infty$ in (1.3) and get $\left(\sum_{i=1}^n |x_i^k - x_i|^p\right)^{\frac{1}{p}} \leq \varepsilon$ for every $n \in \mathbb{N}$ and every $k \geq k_0$. We let $n \to \infty$ to obtain

$$\Big(\sum_{i=1}^{\infty} |x_i^k - x_i|^p\Big)^{\frac{1}{p}} = \|x^k - x\|_p \leq \varepsilon$$

for every $k \geq k_0$. Therefore $x^k \to x$ in ℓ_p.

(ii) Let $x^k = (x_i^k) \in c$ and $x^k \to x$ in ℓ_∞. For $k \in \mathbb{N}$ we denote $l_k = \lim_{i\to\infty} x_i^k$. We will prove that $\lim_{k\to\infty} l_k$ exists and is finite by showing that $\{l_n\}_{k=1}^\infty$ is Cauchy. Given $\varepsilon > 0$ let n_0 be such that $\|x^n - x^m\|_\infty < \varepsilon$ for all $n, m \geq n_0$. Thus $|x_i^n - x_i^m| < \varepsilon$ for every $i \in \mathbb{N}$ and every $n, m \geq n_0$. Fixing $n, m \geq n_0$ and letting $i \to \infty$ we get $|l_m - l_n| < \varepsilon$. Therefore $l := \lim_{k\to\infty} l_k$ exists and is finite.

Since $x^k \to x = (x_i)$ in ℓ_∞, we have $\lim_{k\to\infty} x_i^k = x_i$ for all $i \in \mathbb{N}$. We will show that $\lim_{i\to\infty} x_i = l$, that is, $x \in c$. Given $\varepsilon > 0$, we find n_0 so that $\|x^k - x\|_\infty < \varepsilon$ and $|l_k - l| < \varepsilon$ for $k \geq n_0$, in particular $|x_i^k - x_i|_\infty < \varepsilon$. Then fix i_0 so that $|x_i^{n_0} - l_{n_0}| < \varepsilon$ for $i \geq i_0$. We have, for $i \geq i_0$,

$$|x_i - l| \leq |x_i - x_i^{n_0}| + |x_i^{n_0} - l_{n_0}| + |l_{n_0} - l| < 3\varepsilon.$$

Thus $\lim x_i = l$ and $x \in c$. This shows that c is closed in ℓ_∞.

Similarly we show that c_0 is a closed subspace of ℓ_∞. By Fact 1.5 the spaces c and c_0 are Banach spaces.

(iii) By Fact 1.5, it is enough to show that c_{00} is not closed in c_0. To this end, consider $x \in c_0$ defined by $x = (x_i)$, where $x_i = \frac{1}{i}$ for every i, and $x^n = (1, \frac{1}{2}, \ldots, \frac{1}{n}, 0, \ldots)$. Then $x^n \in c_{00}$ and $x^n \to x$ in c_0 since $\|x^n - x\|_\infty = \frac{1}{n+1} \to 0$ as $n \to \infty$. However, $x \notin c_{00}$. $\square$

Let $p \in [1, \infty)$. More generally, for an abstract nonempty set Γ we introduce spaces $\ell_p(\Gamma)$ and $c_0(\Gamma)$. The space $\ell_p(\Gamma)$ consists of all functions $f\colon \Gamma \to \mathbb{K}$ such that $\sum_{\gamma\in\Gamma} |f(\gamma)|^p < \infty$, with the norm $\|f\|_p := \left(\sum_{\gamma\in\Gamma} |f(\gamma)|^p\right)^{\frac{1}{p}}$, where the sum is defined by

$$\sum_{\gamma\in\Gamma} |f(\gamma)|^p = \sup\Big\{\sum_{\gamma\in F} |f(\gamma)|^p : \ F \text{ a finite subset of } \Gamma\Big\}.$$

The space $\ell_\infty(\Gamma)$ consists of all bounded functions $f\colon \Gamma \to \mathbb{K}$ and is endowed with the supremum norm $\|\cdot\|_\infty$. Its subspace $c_0(\Gamma)$ consists of all functions $f \in \ell_\infty(\Gamma)$ such that the set $\{\gamma \in \Gamma : \ |f(\gamma)| \geq \varepsilon\}$ is finite for every $\varepsilon > 0$. We also consider the space $c_{00}(\Gamma)$ of all functions $f \in \ell_\infty(\Gamma)$ whose *support* $\text{supp}(f) := \{\gamma \in \Gamma : \ f(\gamma) \neq 0\}$ is finite. Note that $c_0(\Gamma) = \overline{c_{00}(\Gamma)}$ (closure in $\ell_\infty(\Gamma)$).

Similarly as above we show that $\ell_p(\Gamma)$ and $c_0(\Gamma)$ are Banach spaces.

Note that every element f in $c_0(\Gamma)$ has countable support. Indeed, $\text{supp}(f) = \bigcup_{n=1}^\infty \{\gamma : \ |f(\gamma)| \geq 1/n\}$. Since $\ell_p(\Gamma) \subset c_0(\Gamma)$ for all $1 \leq p < \infty$, the same is true for every element in such an $\ell_p(\Gamma)$.

Definition 1.18 *Let $p \in [1, \infty)$. The symbol $L_p = L_p[0, 1]$ denotes the vector space of all classes of Lebesgue measurable scalar functions f defined almost everywhere on* $[0, 1]$ *(we identify functions that are equal almost everywhere) such that $\int_0^1 |f(t)|^p \, dt < \infty$, the vector addition and the scalar multiplication being defined pointwise, endowed with the norm*

$$\|f\|_p := \left(\int_0^1 |f(t)|^p \,dt\right)^{\frac{1}{p}}.$$

Note that L_p is a vector space. Indeed,

$$|f(t)+g(t)|^p \leq 2^p \max(|f(t)|, |g(t)|)^p = 2^p \max\left(|f(t)|^p, |g(t)|^p\right) \leq 2^p\left(|f(t)|^p + |g(t)|^p\right),$$

so $\int_0^1 |f+g|^p \,dt < \infty$ and $(f+g) \in L_p$ whenever $f, g \in L_p$. Similarly, $\alpha f \in L_p$ for every $\alpha \in \mathbb{K}$ and $f \in L_p$. These spaces are also infinite-dimensional; indeed, given $n \in \mathbb{N}$, the set $\{\chi_{[i-1/n, i/n]}\}_{i=1}^n$, where χ_S denotes the characteristic function of a set S, is linearly independent.

The triangle inequality for $\|\cdot\|_p$ follows from the following versions of the Hölder and Minkowski inequalities (Theorems 1.19 and 1.20).

Theorem 1.19 (Hölder inequality) *If $p > 1$, $(1/p) + (1/q) = 1$, $f \in L_p$, and $g \in L_q$, then $fg \in L_1$, and*

$$\int_0^1 |f(t)g(t)| \mathrm{d}t \leq \left(\int_0^1 |f(t)|^p \mathrm{d}t\right)^{1/p} \left(\int_0^1 |g(t)|^q \mathrm{d}t\right)^{1/q} \quad (= \|f\|_p \|g\|_q). \tag{1.4}$$

Proof: If $\|f\|_p = 0$ or $\|g\|_q = 0$, then the left-hand side of Equation (1.4) is also zero, so the inequality holds. Otherwise, put, for $t \in [0,1]$,

$$a = \frac{|f(t)|}{\|f\|_p}, \quad \text{and} \quad b = \frac{|g(t)|}{\|g\|_q},$$

and use Lemma 1.11 to obtain

$$\frac{|f(t)g(t)|}{\|f\|_p \|g\|_q} \leq \frac{1}{p}\frac{|f(t)|^p}{\|f\|_p^p} + \frac{1}{q}\frac{|g(t)|^q}{\|g\|_q^q}. \tag{1.5}$$

The function fg is measurable and Equation (1.5) shows that its absolute value is dominated by an integrable function, so fg is integrable, i.e., it is an element in L_1. Integration of both members in inequality (1.5) gives (1.4). □

Theorem 1.20 (Minkowski inequality) *If $p \geq 1$ and $f, g \in L_p$, then*

$$\|f+g\|_p \leq \|f\|_p + \|g\|_p. \tag{1.6}$$

Proof: For $p = 1$ the assertion is trivial. For $p > 1$ it follows from Hölder inequality (Theorem 1.19). Indeed, $f + g \in L_p$ and $|f+g|^{p-1} \in L_q$, so

$$\int_0^1 |f(t)+g(t)|^p \mathrm{d}t = \int_0^1 |f(t)+g(t)|.|f(t)+g(t)|^{p-1}\mathrm{d}t \leq \int_0^1 |f(t)|.|f(t)+g(t)|^{p-1}\mathrm{d}t$$
$$+\int_0^1 |g(t)|.|f(t)+g(t)|^{p-1}\mathrm{d}t \leq (\|f\|_p+\|g\|_p)\left(\int_0^1 |f(t)+g(t)|^p \mathrm{d}t\right)^{1/q},$$

so (1.6) holds. □

Theorem 1.21 *If $p \in [1,\infty)$, then L_p is a Banach space.*

In the proof we will use the following lemma. We say that a series $\sum_{i=1}^{\infty} x_i$ in a normed space X is *absolutely convergent* if $\sum_{i=1}^{\infty} \|x_i\| < \infty$. Recall that $\sum_{i=1}^{\infty} x_i$ is called *convergent* if the sequence $\{s_n\}$, where $s_n = \sum_{i=1}^{n} x_i$, $n \in \mathbb{N}$, is convergent in X.

Lemma 1.22 *A normed space X is a Banach space if and only if every absolutely convergent series in X is convergent.*

Proof: Assume that X is a Banach space and $\{x_n\}$ is a sequence in X such that $\sum \|x_n\| < \infty$. Then given $\varepsilon > 0$ there is $n_0 \in \mathbb{N}$ such that $\sum_{k+1}^{n} \|x_i\| < \varepsilon$ for every $n > k \geq n_0$; thus for $s_n := \sum_{i=1}^{n} x_i$ we have $\|s_n - s_k\| = \left\|\sum_{k+1}^{n} x_i\right\| \leq \sum_{k+1}^{n} \|x_i\| < \varepsilon$. Therefore the sequence $\{s_n\}_{n=1}^{\infty}$ is Cauchy and thus convergent in X. So every absolutely convergent series in a Banach space is convergent.

Now assume that the condition on absolute convergence is true and let $\{x_n\}_{n=1}^{\infty}$ be a Cauchy sequence in X. First we show that $\{x_n\}_{n=1}^{\infty}$ has a convergent subsequence. To this end, choose a subsequence $n_1 < n_2 < \dots$ such that $\|x_{n_k} - x_{n_l}\| < 2^{-k}$ for $l \geq k$. This is done by induction: first choose n_1 such that $\|x_{n_1} - x_l\| < 2^{-1}$ for every $l \geq n_1$. Then choose $n_2 > n_1$ such that $\|x_{n_2} - x_l\| < 2^{-2}$ for every $l \geq n_2$, etc. Put $x_{n_0} = 0$ and set $y_k = x_{n_k} - x_{n_{k-1}}$. We have $x_{n_k} = \sum_{i=1}^{k} y_i$, where $\|y_k\| \leq 2^{1-k}$ for every $k \geq 2$. As $\sum \|y_k\| < \infty$, by our assumption we get that $\sum y_k$ is convergent and hence its partial sums x_{n_k} form a convergent sequence.

To show that the whole sequence $\{x_n\}_{n=1}^{\infty}$ converges to the same element is standard. This concludes the proof of Lemma 1.22. □

For a slight improvement of Lemma 1.22, see Exercise 1.26.

Proof of Theorem 1.21: By Lemma 1.22 we need to show that $\sum f_n$ is a convergent series in L_p whenever $f_n \in L_p$ satisfy $\sum \|f_n\|_p < \infty$. Denote $M = \sum_{n=1}^{\infty} \|f_n\|_p$. For $n \in \mathbb{N}$ we define functions $g_n(t) = \sum_{k=1}^{n} |f_k(t)|$. Then $\|g_n\|_p \leq \sum_{k=1}^{n} \|\,|f_k|\,\|_p \leq M$, so $\int_0^1 |g_n|^p \,\mathrm{d}t = \int_0^1 g_n^p \,\mathrm{d}t \leq M^p$.

We have $g_{n+1}(t) \geq g_n(t)$ for every $n \in \mathbb{N}$ and $t \in [0,1]$. Thus the limit $g(t) := \lim_{n\to\infty} g_n(t)$ exists (finite or $+\infty$) for every $t \in [0,1]$. Since $g_n \geq 0$, by Fatou's lemma, $\int_0^1 g^p \,\mathrm{d}t \leq \liminf_{n\to\infty} \int_0^1 g_n^p \,\mathrm{d}t \leq M^p$.

Therefore $g^p \in L_1[0,1]$, so $g(t) < \infty$ on a set S of measure 1 in $[0,1]$. For $t \in S$ we have $\sum |f_k(t)| = g(t) < \infty$. Thus the sum $s(t) := \sum f_k(t)$ is finite. Define $s(t) = +\infty$ for $t \notin S$. We need to show that $s = \sum f_k$ in L_p. Denote

$s_n(t) = \sum_{k=1}^n f_k(t)$. Then $s = \lim_{n\to\infty} s_n$ almost everywhere (on S at least), so it is a measurable function on [0, 1]. We also have $|s_n(t)| \le g_n(t) \le g(t)$ for every $t \in S$. Therefore $|s(t)| \le g(t)$ almost everywhere on [0, 1], which implies that $\int |s(t)|^p \, dt < \infty$ and hence $s \in L_p$.

Finally, we have

$$|s_n(t) - s(t)|^p \le 2^p \max\bigl(|s_n(t)|^p, |s(t)|^p\bigr) \le 2^p g^p(t)$$

almost everywhere on [0, 1], $\int 2^p g^p \, dt < \infty$, and $|s_n(t) - s(t)| \to 0$ almost everywhere. By the Lebesgue dominated convergence theorem (see, e.g., [Rudi2, Theorem 1.34]),

$$\int_0^1 |s_n(t) - s(t)|^p \, dt \to 0.$$

This means that $\|s_n - s\|_p \to 0$, that is, $\sum f_k = s$ in L_p. □

Let f be a measurable function on [0, 1]. We define

$$\operatorname{ess\,sup}(f) = \inf\bigl(\sup\{f(t) : t \in N\}\bigr),$$

where the infimum is taken over all measurable subsets N of [0, 1] of Lebesgue measure 1. One can also use other equivalent definitions, for instance

$$\operatorname{ess\,sup}(f) = \inf\bigl\{\alpha : |\{t \in [0,1] : f(t) > \alpha\}| = 0\bigr\},$$

where $|A|$ denotes the Lebesgue measure of A. Clearly $f \le \operatorname{ess\,sup}(f)$ outside a set of measure zero.

Definition 1.23 *The space $L_\infty = L_\infty[0, 1]$ denotes the vector space of all classes of measurable functions f that are "essentially bounded," i.e., such that $\operatorname{ess\,sup}(|f|) < \infty$, endowed with the norm $\|f\|_\infty := \operatorname{ess\,sup}(|f|)$.*

Proposition 1.24 *The space L_∞ is a Banach space.*

Proof: It is easy to check that $\|\cdot\|_\infty$ is a norm on the space L_∞. We will show that it is a Banach space. First observe that $\{f_n\}_{n=1}^\infty$ being a Cauchy sequence in L_∞ implies that $\lim_{m,n\to\infty} (f_n - f_m) = 0$ uniformly on a set of measure 1. Indeed, for $k, l \in \mathbb{N}$ put

$$Z_{k,\ell} = \{t \in [0,1] : |f_k(t) - f_l(t)| \ge \|f_k - f_l\|_\infty\}.$$

On $[0,1]\setminus\bigcup_{k,l} Z_{k,l}$ the functions $\{f_n\}$ form a uniformly Cauchy and therefore uniformly convergent sequence. From this and from the completeness of the supremum norm, we obtain that L_∞ is a Banach space. □

The space L_∞ is infinite-dimensional; to see this follow the same argument as in the L_p case.

Given a measure space (Ω, μ) and $p \in [1, \infty]$, the space $L_p(\Omega, \mu)$ (also denoted $L_p(\mu)$) can be introduced in a similar fashion (see, e.g., [DiUh] or [Wojt]).

1.3 Operators, Quotients, Finite-Dimensional Spaces

We now begin an investigation of linear mappings between normed spaces. Recall that a mapping T from a vector space X over the field $\mathbb{K}$ into another vector space Y over $\mathbb{K}$ is called *linear* if $T(\alpha_1 x_1 + \alpha_2 x_2) = \alpha_1 T(x_1) + \alpha_2 T(x_2)$ for every $\alpha_1, \alpha_2 \in \mathbb{K}$ and $x_1, x_2 \in X$. The vector space of all linear mapping from X into Y will be denoted $\mathcal{L}(X, Y)$. A *linear functional* on X is a linear mapping from X into the field $\mathbb{K}$. Observe that a real linear functional (i.e., a linear functional from a vector space over $\mathbb{R}$ into $\mathbb{R}$) is a convex function. Recall, too, that a mapping $T: (P, \rho) \to (Q, \varrho)$ between metric spaces is called *Lipschitz*, more precisely, *C-Lipschitz*, if there is $C > 0$ such that $\varrho\big(T(x), T(y)\big) \le C\rho(x, y)$ for all $x, y \in P$. In case of normed spaces P, Q this inequality becomes $\|T(x) - T(y)\|_Q \le C\|x - y\|_P$. Note that every Lipschitz mapping is uniformly continuous.

Proposition 1.25 *Let $(X, \|\cdot\|_X)$ and $(Y, \|\cdot\|_Y)$ be normed spaces and let T be a linear mapping from X into Y. The following are equivalent:*
(i) *T is continuous on X.*
(ii) *T is continuous at the origin.*
(iii) *There is $C > 0$ such that $\|T(x)\|_Y \le C\|x\|_X$ for every $x \in X$.*
(iv) *T is Lipschitz.*
(v) *$T(B_X)$ is a bounded set in Y.*

Proof: (i) $\Longleftrightarrow$ (ii) follows from the linearity of T. (iii) $\Longleftrightarrow$ (iv) follows similarly as $\|T(x) - T(y)\|_Y = \|T(x - y)\|_Y \le C\|x - y\|_X$.

If (ii) is true, then given $\varepsilon > 0$, there is $\delta > 0$ such that $\|T(x)\|_Y \le \varepsilon$ whenever $\|x\|_X \le \delta$. If $x \in X$, $x \ne 0$, then $\|\delta \frac{x}{\|x\|_X}\|_X = \delta$ and thus $\|T\big(\delta \frac{x}{\|x\|_X}\big)\|_Y \le \varepsilon$. Therefore $\|T(x)\|_Y \le \varepsilon/\delta \|x\|_X$ for every $x \in X$, which shows (iii). On the other hand, (iii) clearly implies (ii) with $\delta := \varepsilon/C$.

Assuming (iii) we obtain that $\|T(x)\|_Y \le C$ whenever $x \in B_X$, so $T(B_X)$ is bounded. On the other hand, if $T(B_X)$ is bounded and, say, $\|T(x)\|_Y \le C$ for every $x \in B_X$, then for every $x \in X$, $x \ne 0$, we have $\|T\big(\frac{x}{\|x\|_X}\big)\|_Y \le C$, so $\|T(x)\|_Y \le C\|x\|_X$. This shows the equivalence of (iii) and (v). □

Definition 1.26 *Let X, Y be normed spaces. A linear mapping from X into Y is called an* operator. *An operator T from X into Y is called* bounded *if $T(B_X)$ is bounded in Y.*
We define the operator norm *of T by*

$$\|T\| = \sup\{\|T(x)\|_Y : x \in B_X\}.$$

$\mathcal{B}(X, Y)$ denotes the vector space of all bounded operators from X into Y, endowed with the operator norm. In the case of $X = Y$ we put $\mathcal{B}(X) = \mathcal{B}(X, X)$.

By Proposition 1.25, elements in $\mathcal{B}(X, Y)$ are precisely the continuous operators from X into Y. It is easy to check that $\mathcal{B}(X, Y)$ is indeed a normed linear space. Note that $\|T\|$ is the smallest number C that satisfies (iii) in Proposition 1.25 and if $T : X \to Y$ is a bounded operator, then the *kernel* $\text{Ker}(T) := \{x \in X : T(x) = 0\}$ is a closed subspace of X. That $\mathcal{B}(X, Y)$ is an infinite-dimensional space in the case that X is infinite-dimensional and Y is not reduced to $\{0\}$ follows from the Hahn–Banach Theorem 2.2, see Exercise 2.43.

Proposition 1.27 *Let X, Y be normed linear spaces. If Y is a Banach space then $\mathcal{B}(X, Y)$ is also a Banach space.*

Proof: The proof of completeness is similar to that for the space $C[0, 1]$. In particular, if $\{T_n\}$ is a sequence of linear mappings from X into Y and $T_n(x) \to T(x)$ for all $x \in X$, then T must be linear. □

Operators in $\mathcal{B}(X, \mathbb{K})$ are called *continuous linear functionals*. The operator norm introduced in Definition 1.26 appears now as $\|f\| = \sup\{|f(x)| : x \in B_X\}$ for $f \in \mathcal{B}(X, \mathbb{K})$. We formally introduce the following definition.

Definition 1.28 *Let $(X, \|\cdot\|)$ be a normed space. By X^* we denote the vector space $\mathcal{B}(X, \mathbb{K})$ of all continuous linear functionals on X, endowed with the operator norm $\|f\| = \sup\{|f(x)| : x \in B_X\}$, called the* canonical dual norm. *X^* is called the* dual space *of X.*

Remark: If $f \in X^*$ and $x \in X$, we shall use indistinctly the notation $f(x)$ or $\langle f, x\rangle$ for the action of f on x.

That the dual space does not reduce to $\{0\}$ if X is different from $\{0\}$ follows from the Hahn–Banach Theorem 2.2. More precisely, if $\dim(X) = n$ for some $n \in \mathbb{N}$, then $\dim(X^*) = n$, and if X is infinite-dimensional, so it is X^*, see Exercise 2.23.

Note that, by the definition, if $f \in X^*$ and $x \in X$, then $|f(x)| \leq \|f\| \cdot \|x\|$. The canonical dual norm $\|\cdot\|$ is denoted sometimes by $\|\cdot\|^*$ or $\|\cdot\|_{X^*}$ to emphasize that we work in the dual space.

Since $\mathbb{K}$ (that is, $\mathbb{R}$ or $\mathbb{C}$) is complete, using Proposition 1.27 we readily obtain

Proposition 1.29 *X^* is a Banach space for every normed space X.*

A mapping φ from a set D into a set R is called an *injection* (also called *one-to-one*) if $\varphi(d_1) \neq \varphi(d_2)$ whenever $d_1 \neq d_2$ in D. It is called a *surjection* (also called *onto*) if for every $r \in R$ there exists $d \in D$ such that $\varphi(d) = r$. The mapping φ is called a *bijection* if it is both an injection and a surjection. We use the terms *injective*, *surjective*, and *bijective* mapping, respectively.

An operator $T \in \mathcal{B}(X, Y)$ is called a *linear isomorphism from X onto Y*, (or just an *isomorphism*) if T is a bijection mapping from X to Y, and $T^{-1} \in \mathcal{B}(Y, X)$. If $T : X \to Y$ is an isomorphism, we also say that T is *invertible*. An isomorphism from X onto X is called an *automorphism*. Note that the inverse mapping of a linear bijection is also a linear mapping. An operator $T \in \mathcal{L}(X, Y)$ is an isomorphism from X onto Y if and only if it is surjective and there exist constants C_1 and C_2 such that, for all $x \in X$,

$$C_1\|x\| \leq \|T(x)\| \leq C_2\|x\|. \tag{1.7}$$

This follows from (iii) in Proposition 1.25, and the fact that Equation (1.7) implies that T is one-to-one. Note (Exercise 1.74) that (1.7) is equivalent to simultaneously $\|T\| \leq C_2$ and $\|T^{-1}\| \leq 1/C_1$. In geometrical terms, for an operator from X onto Y, Equation (1.7) is equivalent to $C_1 B_Y \subset T(B_X) \subset C_2 B_Y$.

Normed spaces X, Y are called *linearly isomorphic* (or just *isomorphic*) if there is a linear isomorphism T from X onto Y. It is easy to see that an isomorphism T carries Cauchy (convergent) sequences onto Cauchy (convergent) sequences. Therefore, if X, Y are isomorphic normed spaces and X is a Banach space, then Y is a Banach space as well.

An operator $T \in \mathcal{B}(X, Y)$ is called a *linear isomorphism from X into Y* (or just an *isomorphism from X into Y*, alternatively an *isomorphism into*) if T is an isomorphism from X onto a subset $T(X)$ of Y. Clearly $T(X)$ is a subspace of Y. If X is a Banach space then $T(X)$ is complete too, so in particular it must be closed in Y, by Fact 1.5.

An operator $T \in \mathcal{B}(X, Y)$ is called a *(linear) isometry from X into Y* if $\|T(x)\|_Y = \|x\|_X$ for every $x \in X$. Spaces X, Y are called *(linearly) isometric* if there exists a linear isometry from X onto Y.

Definition 1.30 *Let X, Y be isomorphic normed spaces. The* Banach–Mazur distance *between X and Y is defined by*

$$\mathrm{d}(X, Y) = \inf\{\|T\|\|T^{-1}\| : T \text{ an isomorphism of } X \text{ onto } Y\}.$$

Note that $\mathrm{d}(X, Y) \geq 1$ and we have $\mathrm{d}(X, Z) \leq \mathrm{d}(X, Y)\,\mathrm{d}(Y, Z)$. The fact that $\mathrm{d}(X, Y) < d$ means that there is an isomorphism T of X onto Y such that $\frac{1}{C_1} B_Y \subset T(B_X) \subset C_2 B_Y$ for some positive constants C_1 and C_2 satisfying $C_1 C_2 < \mathrm{d}$.

Let a vector space X be endowed with two norms, denote $X_1 = (X, \|\cdot\|_1)$ and $X_2 = (X, \|\cdot\|_2)$. Norms $\|\cdot\|_1$ and $\|\cdot\|_2$ are called *equivalent* if the formal identity mapping $I_X : X \to X$ defined by $I_X(x) = x$ is an isomorphism between the spaces X_1 and X_2, i.e., if there exist constants $c, C > 0$ such that $c\|x\|_2 \leq \|x\|_1 \leq C\|x\|_2$ for every $x \in X$ (see Equation (1.7)). According to Exercise 1.74, this is equivalent to $cB_1 \subset B_2 \subset CB_1$, where B_i denotes the closed unit ball of X_i, $i = 1, 2$.

We say that the space $(X, \|\cdot\|_2)$ is a *renorming* of the space $(X, \|\cdot\|_1)$. To *renorm* a normed space means to endow the space with an equivalent norm.

Definition 1.31 *Let X and Y be normed spaces.*
An operator $T \in \mathcal{B}(X, Y)$ is called a compact operator *if $\overline{T(B_X)}$ is compact in Y. The space of all compact operators from X into Y with the norm inherited from $\mathcal{B}(X, Y)$ is denoted by $\mathcal{K}(X, Y)$. If $X = Y$ then we write $\mathcal{K}(X)$ instead of $\mathcal{K}(X, X)$.*
An operator $T \in \mathcal{B}(X, Y)$ is called a finite rank operator *or a* finite-dimensional operator *if $\dim\big(T(X)\big) < \infty$. By $\mathcal{F}(X, Y)$ we denote the space of all finite rank operators from X into Y with the norm inherited from $\mathcal{B}(X, Y)$. If $X = Y$ then we write $\mathcal{F}(X)$ instead of $\mathcal{F}(X, X)$.*

If $f \in S_{X^*}$ and f does not attain its norm on B_X, then $f(B_X) = (-1, 1)$ (see Exercise 3.161). Thus we have $f \in \mathcal{K}(X, \mathbb{R})$, yet $f(B_X)$ is not compact. Therefore, the closure $\overline{T(B_X)}$ in the definition of a compact operator cannot be dropped.

Example: A compact operator that is not a finite rank operator, but it is the limit (in the operator norm) of a sequence of finite rank operators: Define $T \in \mathcal{B}(\ell_2)$ by $T(x) = (2^{-i}x_i)$ for $x = (x_i)$. Then $T \in \mathcal{K}(\ell_2)\backslash\mathcal{F}(\ell_2)$. Indeed, since $\overline{T(B_{\ell_2})}$ is a closed subset of the Hilbert cube (Exercise 1.51), it is compact. The sequence of finite rank operators we are looking for is $\{T_n\}$, where $T_n(x) := (2^{-1}x_1, \ldots, 2^{-n}x_n, 0, 0, \ldots)$, for all $n \in \mathbb{N}$.

The identity operator on ℓ_2 is an example of a bounded non-compact operator, since $\|e_i - e_j\| = \sqrt{2}$, if $i \neq j$.

Let $T\colon X \to Y$ be a finite rank operator. Let $n = \dim T(X)$. A simple result in Linear Algebra shows that we can find a biorthogonal system $\{e_i; g_i\}_{i=1}^n$ in $T(X) \times \big(T(X)\big)^*$ (i.e., $\langle e_i, g_j\rangle = \delta_{i,j}$ for $i, j \in \{1, 2, \ldots, n\}$). In particular, $\{e_i : i = 1, 2, \ldots, n\}$ is a Hamel basis (i.e., an algebraic basis) of $T(X)$. We define $f_i = g_i \circ T \in X^*$, $i = 1, 2, \ldots, n$. Then for every $x \in X$ we have $T(x) = \sum_{i=1}^n (g_i \circ T)(x)e_i = \sum_{i=1}^n f_i(x)e_i$. Conversely, given $f_1, \ldots, f_n \in X^*$ and $e_1, \ldots, e_n \in Y$, then $T\colon x \mapsto \sum_{i=1}^n f_i(x)e_i$ is a finite rank operator from X into Y. Such an operator is denoted by $T = \sum_{i=1}^n f_i \otimes e_i$.

The subject of compact operators will be treated extensively in Chapter 15.

Definition 1.32 *Let Z be a vector space. A* (linear) projection *P on Z is a linear mapping from Z into Z such that $P \circ P = P$. In this situation we say, too, that P* is a projection from Z onto $P(Z)$ parallel to $\operatorname{Ker} P$. *We shall often consider P as a mapping from the vector space Z onto the vector space $P(Z)$.*

Observe that associated to every linear projection on Z there exists an *algebraic direct sum decomposition* of Z into two subspaces X and Y, namely $X = PZ$ and $Y = \operatorname{Ker} P$, in the sense that each $z \in Z$ can be written in a unique way as $z = x+y$, with $x \in PZ$ and $y \in \operatorname{Ker} P$. Note that, in this case, $x = Pz$ and $y = z - Pz$. We shall write in this case $Z = X \oplus Y$, making sure that this is understood only in the algebraic sense, with no topologies involved. Conversely, given two subspaces X and Y of Z such that $Z = X \oplus Y$ (i.e., each $z \in Z$ can be written in a unique way as $x + y$, with $x \in X$ and $y \in Y$), then the mapping $P : Z \to Z$ defined as $Pz = x$ is a linear projection with range X and kernel Y.

More generally, given two vector spaces X and Y, the *algebraic direct sum* $X \oplus Y$ is the vector space of all ordered pairs (x, y), $x \in X$, $y \in Y$, with the vector operations defined coordinatewise. The spaces X and Y are algebraically isomorphic to the subspaces $\{(x, 0) : x \in X\}$ and $\{(0, y) : y \in Y\}$ of $X \oplus Y$, respectively.

Definition 1.33 *Let $(X, \|\cdot\|_X)$ and $(Y, \|\cdot\|_Y)$ be normed spaces. The algebraic direct sum $X \oplus Y$ of X and Y becomes a normed space, called the* topological direct sum of X and Y *and still denoted $X \oplus Y$, when it is endowed with the norm $\|(x, y)\| := \|x\|_X + \|y\|_Y$.*

The spaces X and Y are isometric to the subspaces $\{(x, 0) : x \in X\}$ and $\{(0, y) : y \in Y\}$ of $X \oplus Y$, respectively. If X and Y are Banach spaces, then so is $X \oplus Y$.

Let Y be a closed subspace of a normed space X. For $x \in X$ we consider the *coset* $\hat{x}$ relative to Y,

$$\hat{x} := \{z \in X : (x - z) \in Y\} = \{x + y : y \in Y\}.$$

The space $X/Y := \{\hat{x} : x \in X\}$ of all cosets, together with the addition and scalar multiplication defined by $\hat{x} + \hat{y} = \widehat{x+y}$ and $\lambda\hat{x} = \widehat{\lambda x}$, is clearly a vector space. It is easy to check that $\|\hat{x}\| := \inf\{\|y\| : y \in \hat{x}\}$ turns X/Y into a normed space. Indeed, for any $z \in \hat{x}$ we have $\|\hat{x}\| = \inf\{\|z - y\| : y \in Y\} = \operatorname{dist}(z, Y)$. Therefore $\hat{x} = \hat{0}$ if and only if $x \in Y$, as Y is closed.

If Y is a subspace of X, then $\operatorname{dist}(\alpha x, Y) = |\alpha| \operatorname{dist}(x, Y)$. Therefore $\|\lambda\hat{x}\| = |\lambda|\,\|\hat{x}\|$. The triangle inequality follows since if x_1, x_2 are in X and y_1, y_2 are in Y, then

$$\|x_1 + x_2 - (y_1 + y_2)\| \leq \|x_1 - y_1\| + \|x_2 - y_2\|.$$

Therefore $\operatorname{dist}(x_1 + x_2, Y) \leq \operatorname{dist}(x_1, Y) + \operatorname{dist}(x_2, Y)$.

Definition 1.34 *Let Y be a closed subspace of a normed space X. The space X/Y endowed with the canonical norm $\|\hat{x}\| := \inf\{\|x\| : x \in \hat{x}\}$, where $\hat{x} \in X/Y$, is called the* quotient space *of X with respect to Y.*
The canonical quotient mapping $q : X \to X/Y$ *associates to every $x \in X$ the coset $\hat{x}$ to which it belongs.*

Obviously, the mapping $q : X \to X/Y$ is linear and continuous. In fact, $\|q\| \leq 1$, as it follows from the definition of the norm in X/Y. We will see later, as a consequence of Lemma 1.37, that $\|q\| = 1$ if Y is a closed subspace of X such that $Y \neq X$.

Proposition 1.35 *Let Y be a closed subspace of a Banach space X. Then X/Y is a Banach space.*

Proof: Assume that $\sum_n \hat{x}_n$ is an absolutely convergent series in X/Y. For $n \in \mathbb{N}$, choose $x_n \in \hat{x}_n$ such that $\|x_n\| \leq \|\hat{x}_n\| + 2^{-n}$. Then $\sum x_n$ is an absolutely convergent series in X that, according to Lemma 1.22, converges. Clearly, the canonical quotient mapping $q : X \to X/Y$ is continuous, so the series $\sum \hat{x}_n$ converges, too. It is enough to apply again Lemma 1.22 to conclude that X/Y is a Banach space. □

It is easy to check that $(X \oplus Y)/X$ is isomorphic to Y and $(X \oplus Y)/Y$ is isomorphic to X. However, if Y is a closed subspace of X, then X may not be isomorphic to $Y \oplus (X/Y)$, see Exercise 12.50.

Proposition 1.36 *Let X be a vector space. If X is finite-dimensional, then any two norms on X are equivalent.*
In particular, all finite-dimensional normed spaces are Banach spaces and every normed space of dimension n is isomorphic to ℓ_2^n.

Consequently, if X is a Banach space and Y is a finite-dimensional subspace of X, then Y is closed in X by Fact 1.5.

Proof: Let $\{e_1, \ldots, e_n\}$ be an algebraic basis of X. We introduce a new norm $\|\cdot\|_1$ on X by $\|x\|_1 = \sum |\lambda_i|$ for $x = \sum \lambda_i e_i$. To check the triangle inequality, for $x = \sum \lambda_i e_i$ and $y = \sum \beta_i e_i$ we write

$$\|x + y\|_1 = \sum |\lambda_i + \beta_i| \leq \sum |\lambda_i| + \sum |\beta_i| = \|x\|_1 + \|y\|_1.$$

We will show that an arbitrary norm $\|\cdot\|$ on X is a Lipschitz function on $(X, \|\cdot\|_1)$. Indeed, if $x = \sum \lambda_i e_i$ and $y = \sum \beta_i e_i$, then

$$\begin{aligned}\|x - y\| &= \Big\|\sum (\lambda_i - \beta_i) e_i\Big\| \leq \sum |\lambda_i - \beta_i| \, \|e_i\| \\ &\leq \max\{\|e_i\|\} \sum |\lambda_i - \beta_i| = \max\{\|e_i\|\} \cdot \|x - y\|_1.\end{aligned}$$

Therefore $\big|\|x\| - \|y\|\big| \leq \|x - y\| \leq \max\{\|e_i\|\}\|x - y\|_1$.

We note that $S_1 := \{x \in X : \ \|x\|_1 = 1\}$ is compact in $(X, \|\cdot\|_1)$. Indeed, let $x^k \in S_1$, $x^k = \sum_{i=1}^n \lambda_i^k e_i$ for $k \in \mathbb{N}$. We have $\sum_{i=1}^n |\lambda_i^k| = 1$ for every k and thus $\{\lambda_i^k\}_{k=1}^\infty$ is bounded for every i. Let a subsequence $\{k_l\}_{l=1}^\infty$ be such that $\lambda_i^{k_l} \to \lambda_i$ for every i. Then $\sum_{i=1}^n |\lambda_i^{k_l} - \lambda_i| \to 0$ as $l \to \infty$, so we have $x^{k_l} \to x$ as $l \to \infty$, where $x = \sum \lambda_i e_i$. Since $\sum_{i=1}^n |\lambda_i^{k_l}| = 1$ for every l, we have $\sum |\lambda_i| = 1$ and thus $x \in S_1$.

Since $\|\cdot\|$ is continuous on the compact set S_1, there exist constants $c > 0$ and $d > 0$ such that $c \leq \|\frac{x}{\|x\|_1}\| < d$ for every non-zero $x \in X$. From the latter inequality we have $c\|x\|_1 \leq \|x\| \leq d\|x\|_1$ for every $x \in X$, so $\|\cdot\|$ is equivalent to $\|\cdot\|_1$. Consequently, all norms are equivalent.

If X is an n-dimensional vector space and T is a linear bijection from X onto ℓ_2^n, we can define a norm $\|\cdot\|_2$ on X by $\|x\|_2 = \|T(x)\|_{\ell_2^n}$. Then T is an isometry of $(X, \|\cdot\|_2)$ onto ℓ_2^n and $\|\cdot\|_2$ is an equivalent norm. □

We shall show later (see the paragraph after the proof of Theorem 4.49) that no space ℓ_p is isomorphic to ℓ_q for $p \neq q$.

Since $\|(x, y)\| = \|(\|x\|_X, \|y\|_Y)\|_1$ for every $(x, y) \in X \oplus Y$ (see Definition 1.33), it follows from Proposition 1.36 that $|||(x, y)||| := (\|x\|_X^p + \|y\|_Y^p)^{\frac{1}{p}}$ is an equivalent renorming of $X \oplus Y$ for every $p \geq 1$. Such a renormed space is denoted $(X \oplus Y)_p$.

To characterize finite-dimensional normed spaces, we need the following statement. Recall that a subspace Y of a vector space X is called *proper* if $Y \neq X$.

Lemma 1.37 (Riesz) *Let X be a normed space. If Y is a proper closed subspace of X then for every $\varepsilon > 0$ there is $x \in S_X$ such that* $\operatorname{dist}(x, Y) \geq 1 - \varepsilon$.

Proof: Choose an arbitrary element $\hat{z} \in X/Y$ satisfying $1 > \|\hat{z}\| > 1 - \varepsilon$. Now pick any $z \in \hat{z}$, $\|z\| \leq 1$, and set $x = z/\|z\|$ (see Fig. 1.4). We get

$$\operatorname{dist}(x, Y) = \operatorname{dist}(z, Y)/\|z\| = \|\hat{z}\|/\|z\| \geq \|\hat{z}\| > 1 - \varepsilon.$$

□

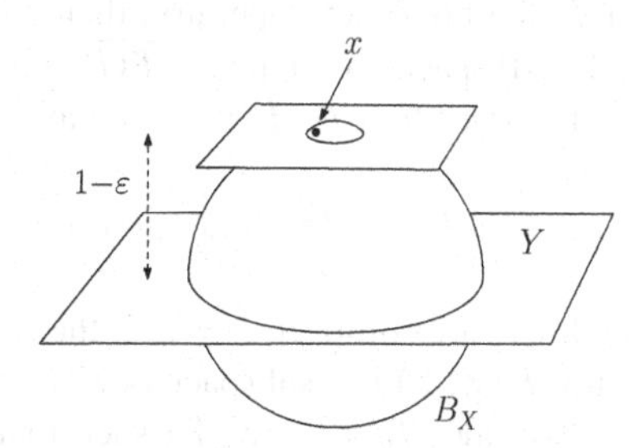

Fig. 1.4 Riesz's lemma

It follows from Lemma 1.37 that $\|q\| = 1$, where $q : X \to X/Y$ is the canonical quotient mapping.

Theorem 1.38 *Let X be a normed space. The space X is finite-dimensional if and only if the unit ball B_X of X is compact.*

Proof: If X is finite-dimensional, then B_X is compact by the proof of Proposition 1.36.

If X is infinite-dimensional, by Lemma 1.37 we find by induction an infinite sequence $\{x_n\}$ in S_X and such that $\operatorname{dist}(x_n, \operatorname{span}\{x_1, \ldots, x_{n-1}\}) > \frac{1}{2}$ for $n = 2, 3, \ldots$. Thus $\operatorname{dist}(x_n, x_m) > \frac{1}{2}$ for all $n \neq m$ and therefore $\{x_n\}$ does not have any convergent subsequence. Hence B_X is not compact. □

Proposition 1.39 *Every operator T from a finite-dimensional normed space X into a normed space Y is continuous.*

Proof: If $\dim X = n$, then X is isomorphic to ℓ_2^n (Proposition 1.36). It is enough, then, to prove that every operator T from ℓ_2^n into a normed space is continuous. If $\{e_i\}_{i=1}^n$ is the canonical basis of ℓ_2^n and $x = (x_i) \in \ell_2^n$, then $x = \sum_{i=1}^n x_i e_i$.

By linearity, $T(x) = \sum_{i=1}^{n} x_i T(e_i)$. For each i, $|x_i| \leq \|x\|_2$, so the ith coordinate mapping $x \to x_i$ from ℓ_2^n into $\mathbb{K}$ is continuous. It follows that T is continuous. □

A straightforward consequence of the previous result is that, for every $n \in \mathbb{N}$, two n-dimensional normed spaces X and Y are linearly isomorphic. Indeed, if $T : X \to Y$ is a linear isomorphism, then both T and T^{-1} are continuous. This result follows from Proposition 1.36, too.

The following statement is an application of the preceding result to finite rank operators. A complementary result on the class of compact operators is included.

Proposition 1.40 *Let X, Y be normed spaces. Then $\mathcal{F}(X, Y)$ is a linear subspace of $\mathcal{K}(X, Y)$.*
The space $\mathcal{K}(X, Y)$ is a closed subspace of $\mathcal{B}(X, Y)$; hence if Y is a Banach space, then $\mathcal{K}(X, Y)$ is also a Banach space.

Proof: Since $(T_1 + T_2)(X) \subset T_1(X) + T_2(X)$, $\mathcal{F}(X, Y)$ is a subspace of $\mathcal{B}(X, Y)$. If T is a finite rank operator, then $T(B_X)$ is a bounded set in a finite-dimensional (closed) space $T(X)$, hence $\overline{T(B_X)}$ is compact (see the proof of Proposition 1.36).

For operators T_1, T_2 we also have

$$(\alpha T_1 + \beta T_2)(B_X) \subset \alpha T_1(B_X) + \beta T_2(B_X) \subset \overline{\alpha T_1(B_X)} + \overline{\beta T_2(B_X)}$$

and if T_i are compact, $i = 1, 2$, the right hand side is a compact set (Exercise 1.61). Thus $\mathcal{K}(X, Y)$ is a subspace of $\mathcal{B}(X, Y)$. We will show that it is closed there.

Consider $T_n \in \mathcal{K}(X, Y)$ such that $\lim T_n = T$ in $\mathcal{B}(X, Y)$. To show that T is a compact operator, given $\varepsilon > 0$ we shall find a finite ε-net for $T(B_X)$ (see Exercises 1.47 and 1.48). First note that $T_n \to T$ in $\mathcal{B}(X, Y)$ means that $\lim T_n(x) = T(x)$ uniformly for $x \in B_X$. Thus there exists n_0 such that $\|T_n(x) - T(x)\| < \varepsilon/2$ for $x \in B_X$ and $n \geq n_0$. Since $T_{n_0}(B_X)$ is totally bounded in Y, there is a finite $\varepsilon/2$-net F for $T_{n_0}(B_X)$. We claim that F is a finite ε-net for $T(B_X)$. Indeed, given $x \in B_X$, we find $y \in F$ such that $\|T_{n_0}(x) - y\| < \varepsilon/2$. Then $\|T(x) - y\| \leq \|T(x) - T_{n_0}(x)\| + \|T_{n_0}(x) - y\| < \varepsilon$. Therefore T is a compact operator. The last statement follows from this and Fact 1.5. □

Remarks:

1. In general, $\mathcal{F}(X, Y)$ is not a closed subspace of $\mathcal{K}(X, Y)$. See the example after Definition 1.31.
2. Note that if X is infinite-dimensional, then no isomorphism from X into Y can be a compact operator by Theorem 1.38. In particular, the identity operator I_X on an infinite-dimensional normed space X is never compact.

The spaces $\mathcal{K}(X, Y)^*$ and $\mathcal{K}(X, Y)^{**}$ are discussed in Chapter 16.

Example: Let $X = L_2[0, 1]$ and $K \in L_2([0, 1] \times [0, 1])$. Define an operator T from $L_2[0, 1]$ into $L_2[0, 1]$ by

$$T(x): t \mapsto \int_0^1 K(t,s)x(s)\,\mathrm{d}s.$$

Note that, indeed, $T(x) \in L_2[0,1]$ whenever $x \in L_2[0,1]$:

$$\begin{aligned}
\|T(x)\|_{L_2} &= \left(\int_0^1 \left|\int_0^1 K(s,t)x(s)\,\mathrm{d}s\right|^2 \mathrm{d}t\right)^{\frac{1}{2}} \\
&\le \left(\int_0^1 \left(\int_0^1 |K(s,t)||x(s)|\,\mathrm{d}s\right)^2 \mathrm{d}t\right)^{\frac{1}{2}} \\
&\le \left(\int_0^1 \left(\int_0^1 |K(s,t)|^2\,\mathrm{d}s \int_0^1 |x(s)|^2\,\mathrm{d}s\right) \mathrm{d}t\right)^{\frac{1}{2}} \\
&= \left(\int_0^1 x^2(s)\,\mathrm{d}s\right)^{\frac{1}{2}} \left(\int_0^1\int_0^1 |K(s,t)|^2\,\mathrm{d}s\,\mathrm{d}t\right)^{\frac{1}{2}} < \infty.
\end{aligned}$$

Thus also $T \in \mathcal{B}(L_2[0,1])$ and $\|T\| \le \left(\int_0^1\int_0^1 |K(s,t)|^2\,\mathrm{d}s\,\mathrm{d}t\right)^{\frac{1}{2}}$.

We will show that T *is a compact operator*. First we will show that *if* K *is continuous on* $[0,1]\times[0,1]$, *then* T *maps* $L_2[0,1]$ *into* $C[0,1]$: By the continuity of K we have $M = \sup\{|K(s,t)| : (s,t) \in [0,1]\times[0,1]\} < \infty$ and hence for $x \in B_{L_2}$ we get

$$\begin{aligned}
|T(x)(t)| &= \left|\int_0^1 K(t,s)x(s)\,\mathrm{d}s\right| \le \int_0^1 |K(t,s)|\,|x(s)|\,\mathrm{d}s \\
&\le \left(\int_0^1 |K(t,s)|^2\,\mathrm{d}s\right)^{\frac{1}{2}} \|x\|_{L_2} \le \left(\int_0^1 M^2\,\mathrm{d}s\right)^{\frac{1}{2}} \|x\|_{L_2} \le M.
\end{aligned}$$

Given $\varepsilon > 0$, find $\delta > 0$ (from the uniform continuity of K on $[0,1]\times[0,1]$) such that if $t_1, t_2 \in [0,1]$, $|t_2 - t_2| < \delta$, then for every $s \in [0,1]$ we have $|K(t_1,s) - K(t_2,s)| < \varepsilon$. Consequently, for every $x \in B_{L_2}$ and $|t_1 - t_2| < \delta$ we have

$$\begin{aligned}
&|T(x)(t_1) - T(x)(t_2)| \\
&= \left|\int_0^1 K(t_1,s)x(s)\,\mathrm{d}s - \int_0^1 K(t_2,s)x(s)\,\mathrm{d}s\right| \le \int_0^1 |K(t_1,s) - K(t_2,s)|\,|x(s)|\,\mathrm{d}s \\
&\le \left(\int_0^1 |K(t_1,s) - K(t_2,s)|^2\,\mathrm{d}s\right)^{\frac{1}{2}} \|x\|_{L_2} \le \left(\int_0^1 \varepsilon^2\right)^{\frac{1}{2}} \|x\|_{L_2} \le \varepsilon.
\end{aligned}$$

Therefore if K is continuous on $[0,1]\times[0,1]$, then $T(x) \in C[0,1]$. In fact we also showed that $T(B_{L_2})$ is a uniformly bounded (by M) and equicontinuous subset of $C[0,1]$, hence by the Arzelà–Ascoli theorem, $T(B_{L_2})$ is relatively compact in $C[0,1]$. Since the norm topology of $L_2[0,1]$ is weaker than the topology of $C[0,1]$, we have that $\overline{T(B_{L_2})}$ is compact in $L_2[0,1]$ and T is a compact operator.

If $K \in L_2([0,1]\times[0,1])$ then choose a sequence K_n of continuous functions on $[0,1]\times[0,1]$ such that

$$\left(\int_0^1\int_0^1 \left(K(t,s)-K_n(t,s)\right)^2 \mathrm{d}s\,\mathrm{d}t\right)^{\frac{1}{2}} \to 0 \text{ as } n\to\infty.$$

For $n\in\mathbb{N}$ define a compact operator $T_n\colon L_2[0,1]\to L_2[0,1]$ by

$$T_n(x)\colon t\mapsto \int_0^1 K_n(t,s)x(s)\,\mathrm{d}s.$$

By the first estimate of this example,

$$\|T-T_n\|_{L_2}\le \left(\int_0^1\int_0^1 |K(t,s)-K_n(t,s)|^2\,\mathrm{d}s\,\mathrm{d}t\right)^{\frac{1}{2}} \to 0 \text{ as } n\to\infty.$$

Thus $T\in\overline{\mathcal{K}(L_2[0,1])}=\mathcal{K}(L_2[0,1])$.

A subset M of a topological space X is called *dense in* X if $\overline{M}$, the closure of M in X, is equal to X. A subset M of a normed space X is called *linearly dense* if $\overline{\operatorname{span}}(M)=X$.

Remark: In general, $\mathcal{F}(X,Y)$ is not a dense subspace of $\mathcal{K}(X,Y)$. See Chapter 16, in particular Theorems 16.35 and 16.54.

Definition 1.41 *Let X be a topological space. We say that X is* separable *if there exists a countable dense subset of X.*

Proposition 1.42 (i) *If $p\in[1,\infty)$, then the space ℓ_p is separable.*
(ii) *The spaces c and c_0 are separable.*
(iii) *The space ℓ_∞ is not separable.*

Proof: (i) Consider in ℓ_p the family $\mathcal{F}$ formed by all finitely supported vectors with rational coefficients. Then $\mathcal{F}$ is countable. We will show that $\mathcal{F}$ is dense in ℓ_p. Given $x\in\ell_p$ and $\varepsilon>0$, choose $n_0\in\mathbb{N}$ such that $\left(\sum_{i=n_0}^{\infty}|x_i|^p\right)\le\frac{\varepsilon^p}{2}$ and then find rational numbers $r_1,r_2,\dots,r_{n_0-1}$ such that $|x_i-r_i|^p\le\frac{\varepsilon^p}{2n_0}$ for $i=1,\dots,n_0-1$. Then $s:=\left(r_1,r_2,\dots,r_{n_0-1},0,\dots\right)$ is in $\mathcal{F}$ and

$$\|s-x\|_p^p=\sum_{i=1}^{n_0-1}|x_i-r_2|^p+\sum_{i=n_0}^{\infty}|x_i|^p\le\sum_{i=1}^{n_0-1}\frac{\varepsilon^p}{2n_0}+\frac{\varepsilon^p}{2}<\varepsilon^p.$$

Therefore $\mathcal{F}$ is dense in ℓ_p.

(ii) The proof of the separability of c_0 is similar to (i). The case of the space c needs only one adjustment, namely we define $\mathcal{F}$ as the family of vectors with rational coefficients such that the vectors are eventually constant.

(iii) Assume that ℓ_∞ is separable and let $\mathcal{D}$ be a dense countable set in ℓ_∞. The cardinal number of the family $\mathcal{F}$ of all subsets of $\mathbb{N}$ is the cardinality of the continuum c, in particular it is uncountable. For every $F\in\mathcal{F}$, let χ_F denote the characteristic function of F in $\mathbb{N}$. If $F_1, F_2 \in \mathcal{F}$ and $F_1 \neq F_2$, then $\|\chi_{F_1} - \chi_{F_2}\|_\infty \geq |\chi_{F_1}(n) - \chi_{F_2}(n)| = 1$ for some $n \in F_1 \setminus F_2$ or $n \in F_2 \setminus F_1$. For each $F \in \mathcal{F}$, let $d_F \in \mathcal{D}$ be chosen such that $\|\chi_F - d_F\| < \frac{1}{4}$. If $F_1 \neq F_2$, then $\|d_{F_1} - d_{F_2}\|_\infty > \frac{1}{4}$. Indeed, if we had $\|d_{F_1} - d_{F_2}\|_\infty \leq \frac{1}{4}$, then

$$\|\chi_{F_1} - \chi_{F_2}\|_\infty \leq \|\chi_{F_1} - d_{F_1}\|_\infty + \|d_{F_1} - d_{F_2}\|_\infty + \|d_{F_2} - \chi_{F_2}\| \leq \tfrac{3}{4} < 1,$$

a contradiction. Therefore the mapping $F \mapsto d_F$ is one-to-one and maps an uncountable set into a countable set, which is a contradiction. Thus ℓ_∞ is not separable. $\square$

Similarly we prove that $c_0(\Gamma)$ and $\ell_p(\Gamma)$, $p \in [1, \infty]$, are nonseparable for Γ uncountable.

Proposition 1.43 (i) *The space* $C[0, 1]$ *is separable.*
(ii) *If* $p \in [1, \infty)$, *then* L_p *is separable.*
(iii) *The space* L_∞ *is not separable.*

Proof: We only discuss the real case (real-valued functions). In the complex case, we consider real and imaginary parts of functions separately in order to reduce the problem to the real case.

(i) The collection $\mathcal{P}$ of polynomials on $[0, 1]$ forms an algebra in $C[0, 1]$ (i.e., a vector subspace closed for multiplication) that separates the points of $[0, 1]$ (i.e., given $s \neq t$ in $[0, 1]$, there exists $p \in \mathcal{P}$ such that $p(s) \neq p(t)$) and contains a constant function. Therefore the closure $\overline{\mathcal{P}}$ in $C[0, 1]$ is $C[0, 1]$ by the Stone–Weierstrass theorem. Since the countable set of polynomials on $[0, 1]$ with rational coefficients is dense in $\mathcal{P}$, we obtain that $C[0, 1]$ is separable.

(ii) It is proven in the theory of Lebesgue integral that $C[0, 1]$ is dense in L_p (see, e.g., [Royd]). Therefore the real case of (ii) follows from (i).

(iii) Consider the functions $f_t := \chi_{[0,t]}$, $t \in [0, 1]$, where $\chi_{[0,t]}$ is the characteristic function of $[0, t]$. Then $\|f_t - f_{t'}\| = 1$ if $t \neq t'$ and a similar argument as in the proof of Proposition 1.42 gives that L_∞ is not separable. $\square$

Proposition 1.44 $\mathcal{B}(\ell_2)$ *contains an isometric copy of* ℓ_∞ *and thus it is not separable.*

Proof: Define a mapping φ from ℓ_∞ into $\mathcal{B}(\ell_2)$ as follows: If $(a_i) \in \ell_\infty$, let $\varphi\big((a_i)\big)$ be the bounded operator from ℓ_2 into ℓ_2 defined for $(x_i)_i \in \ell_2$ by

$$\varphi\big((a_i)\big) \colon (x_i)_i \mapsto (a_i x_i)_i.$$

We claim that φ is a linear isometry from ℓ_∞ into $\mathcal{B}(\ell_2)$. It is enough to check that if $\|(a_i)\|_\infty = 1$, then the operator $\varphi\big((a_i)\big)$ has norm 1 in $\mathcal{B}(\ell_2)$. First note that if $x = (x_i) \in \ell_2$, then

$$\|(a_i x_i)\|_2 = \Big(\sum |a_i|^2 |x_i|^2\Big)^{\frac{1}{2}} \leq \|(a_i)\|_\infty \Big(\sum |x_i|^2\Big)^{\frac{1}{2}} = \|(a_i)\|_\infty \|x\|_2$$

and thus the operator $\varphi\big((a_i)\big)$ has norm at most 1. On the other hand, given $\varepsilon > 0$, choose n_0 such that $|a_{n_0}| > 1 - \varepsilon$. Then $\varphi\big((a_i)\big)(e_{n_0}) = a_{n_0} e_{n_0}$, which has norm $|a_{n_0}|$. Letting $\varepsilon \to 0$ we obtain $\|\varphi\big((a_i)\big)\| = 1$. □

1.4 Hilbert Spaces

An *inner product* (or a *scalar product* or a *dot product*) on a vector space X is a scalar valued function $(\cdot, \cdot)$ on $X \times X$ such that

(1) for every $y \in X$, the function $x \mapsto (x, y)$ is linear,
(2) $\overline{(x, y)} = (y, x)$, where the bar denotes the complex conjugation,
(3) $(x, x) \geq 0$ for every $x \in X$,
(4) $(x, x) = 0$ if and only if $x = 0$.

Note that by (1), $(0, y) = 0$ for any $y \in X$, hence also $(y, 0) = 0$ by (2).

Theorem 1.45 (Cauchy–Schwarz inequality) *Let (x, y) be an inner product on a vector space X.*
(i) *For $x, y \in X$ we have $|(x, y)| \leq \sqrt{(x, x)}\sqrt{(y, y)}$.*
(ii) *The function $\|x\| := \sqrt{(x, x)}$ is a norm on X.*

Proof: (i) If $(y, y) = 0$, we have $y = 0$ and the inequality is satisfied. Assume that $(y, y) > 0$. Then

$$0 \leq \Big(x - \frac{(x, y)}{(y, y)} y, x - \frac{(x, y)}{(y, y)} y\Big) = (x, x) - \frac{|(x, y)|^2}{(y, y)}$$

and the statement follows.

(ii) We will check the triangle inequality. For $x, y \in X$ we have

$$\begin{aligned}\|x + y\|^2 &= (x + y, x + y) = (x, x) + (y, y) + (x, y) + (y, x)\\ &= (x, x) + (y, y) + 2\operatorname{Re}(x, y) \leq (x, x) + (y, y) + 2|(x, y)|\\ &\leq (x, x) + (y, y) + 2\sqrt{(x, x)}\sqrt{(y, y)} = \big(\sqrt{(x, x)} + \sqrt{(y, y)}\big)^2 = \big(\|x\| + \|y\|\big)^2.\end{aligned}$$

□

One immediate consequence of Theorem 1.45 is that $(\cdot, \cdot)$ is a continuous mapping from $(X, \|\cdot\|) \times (X, \|\cdot\|)$ to the scalar field. In particular, it implies that for a fixed vector $y \in X$, $x \mapsto (x, y)$ is a continuous linear functional on X.

Definition 1.46 *A Banach space $(H, \|\cdot\|)$ is called a* Hilbert space *if there is an inner product $(\cdot, \cdot)$ on H such that $\|x\| = \sqrt{(x, x)}$ for every $x \in H$.*

It is straightforward to check that the norm $\|\cdot\|$ of a Hilbert space H satisfies the *parallelogram equality*, namely, for every $x, y \in H$ we have

$$\|x + y\|^2 + \|x - y\|^2 = 2\|x\|^2 + 2\|y\|^2. \tag{1.8}$$

We also have the *polarization identity*:

$$(x, y) = \frac{1}{4}\left(\|x + y\|^2 - \|x - y\|^2\right) \tag{1.9}$$

in the real case and

$$(x, y) = \frac{1}{4}\left(\|x + y\|^2 - \|x - y\|^2 + i\|x + iy\|^2 - i\|x - iy\|^2\right) \tag{1.10}$$

in the complex case.

On the other hand, assume that a norm $\|\cdot\|$ on a Banach space X satisfies the parallelogram equality. If we define (x, y) by the above equations, it turns out to be an inner product (Exercise 1.92) and $\|x\|^2 = (x, x)$. Thus X is a Hilbert space.

Therefore a Banach space X is a Hilbert space if and only if every two-dimensional subspace of X is a Hilbert space. The parallelogram equality (1.8) gives that ℓ_2^n, ℓ_2, and L_2 are Hilbert spaces. In Chapter 4 (see the paragraph after the proof of Theorem 4.53), we shall show that ℓ_p and L_p are not even isomorphic to a Hilbert space for $p \neq 2$.

Let H be a Hilbert space. Let $x, y \in H$. We say that x is *orthogonal* to y, denoted $x \perp y$, if $(x, y) = 0$. Let $M \subset H$. We say that x is *orthogonal* to M, denoted $x \perp M$, if x is orthogonal to every vector y from M.

Definition 1.47 *Let F be a subspace of a Hilbert space H. The set $F^\perp := \{h \in H : h \perp F\}$ is called the* orthogonal complement *of F in H.*

Fact 1.48 *If F is a subspace of a Hilbert space H, then $F^\perp$ is a closed subspace of H.*

Proof: Clearly, $F^\perp$ is a subspace. If $f, h, h_n \in H$, $(h_n, f) = 0$ and $h_n \to h$, then $(h, f) = 0$. This follows from the continuity of the linear functional $h \mapsto (h, f)$ discussed immediately before Definition 1.46. Hence $F^\perp$ is a closed subspace. □

Obviously $F \cap F^\perp = \{0\}$. Therefore every element $z \in F + F^\perp$ has a unique expression in the form $z = x + y$ with $x \in F$, $y \in F^\perp$ (i.e., $F + F^\perp$ is in fact an algebraic direct sum). We can also see that the orthogonality gives

$$\|z\|^2 = (x + y, x + y) = \|x\|^2 + \|y\|^2. \tag{1.11}$$

It follows that $T : F \oplus F^\perp \to H$ defined by $T(x, y) = x + y$ is an isomorphism of $F \oplus F^\perp$ onto $F + F^\perp \subset H$, so $F + F^\perp$ is a topological direct sum.

Theorem 1.49 (Riesz) *Let F be a subspace of a Hilbert space H. If F is closed, then $F + F^\perp = H$. Thus $T : F \oplus F^\perp \to H$ defined by $T(x, y) = x + y$ is an isomorphism of $F \oplus F^\perp$ onto H, and so H is the topological direct sum of F and $F^\perp$.*

Proof: Let $x \in H$. We claim there is a closest element x_1 to x in F and $(x - x_1) \perp F$. Let $y_n \in F$ be such that $\|x - y_n\|^2 < d^2 + \frac{1}{n}$, where $d := \operatorname{dist}(x, F) = \inf\{\|x - z\| :$

$z \in F\}$. We now prove that y_n is a Cauchy sequence. Consider $x - y_n$ and $x - y_m$ in the parallelogram equality (1.8):

$$\|2x - (y_n + y_m)\|^2 + \|y_m - y_m\|^2 = 2\|x - y_n\|^2 + 2\|x - y_m\|^2.$$

We can estimate

$$\begin{aligned}\|y_m - y_m\|^2 &= 2\|x - y_n\|^2 + 2\|x - y_m\|^2 - \|2x - (y_n + y_m)\|^2 \\ &= 2\|x - y_n\|^2 + 2\|x - y_m\|^2 - 4\left\|x - \frac{y_n + y_m}{2}\right\|^2 \\ &\leq 2(d^2 + \tfrac{1}{n}) + 2(d^2 + \tfrac{1}{m}) - 4d^2 = \tfrac{2}{n} + \tfrac{2}{m},\end{aligned}$$

where we used $\frac{y_n + y_m}{2} \in F$. Therefore $\{y_n\}_{n=1}^{\infty}$ converges to some point $x_1 \in F$. Then $\|x - x_1\| = \lim_{n\to\infty} \|x - y_n\| = d$. Put $x_2 = x - x_1$, note that $\|x_2\|^2 = d^2$. Assume that x_2 is not orthogonal to F, so there is $z \in F$ such that $(x_2, z) > 0$. Then for $\varepsilon > 0$ we have

$$\begin{aligned}\|x - (x_1 + \varepsilon z)\|^2 &= \|x_2 - \varepsilon z\|^2 = (x_2 - \varepsilon z, x_2 - \varepsilon z) \\ &= (x_2, x_2) - 2\varepsilon(x_2, z) + \varepsilon^2(z, z) = d^2 - \varepsilon\big(2(x_2, z) - \varepsilon\|z\|^2\big).\end{aligned}$$

Since that $(x_2, z) > 0$, for ε small enough we have $2(x_2, z) - \varepsilon\|z\|^2 > 0$ and therefore $\|x - (x_1 + \varepsilon z)\| < d$, a contradiction. Thus $x_2 \perp F$.

For any $x \in H$ we found $x_1 \in F$, $x_2 \in F^{\perp}$ such that $x = x_1 + x_2$, so $F + F^{\perp} = H$. □

Note that for a closed subspace F of a Hilbert space H we have $(F^{\perp})^{\perp} = F$. Indeed, write $H = F \oplus Z$, where $Z := F^{\perp}$. If $y \in F$ and $z \in Z$, then $(y, z) = 0$ and thus $F \subset Z^{\perp}$. If $x \in H$, $x \notin F$, write $x = y + z$ with $y \in F$, $z \in Z$. Then $z \neq 0$ and $(z, x) = (z, y + z) = \|z\|^2 \neq 0$. Thus $x \notin Z^{\perp}$. Therefore $F = Z^{\perp} = (F^{\perp})^{\perp}$.

Corollary 1.50 *If F is a closed subspace of a Hilbert space H, then F is one-complemented in H, i.e., there is a linear projection of norm 1 from H onto F.*

Proof: Any linear and continuous projection has, clearly, norm greater than or equal to 1. The other inequality follows from Equation (1.11). □

Proposition 1.51 *Let H be a Hilbert space and F be a subspace of H. Then F is linearly isometric to $H/F^{\perp}$.*

Proof: If $x \in F$, then $\inf\{\|x - y\|^2 : y \in F^{\perp}\} = \inf\{\|x\|^2 + \|y\|^2 : y \in F^{\perp}\} = \|x\|^2$, so the mapping $x \to \hat{x} \in H/F^{\perp}$ from F into $H/F^{\perp}$ is a linear isometry onto. □

Definition 1.52 *Let H be a Hilbert space and $S \subset H$. S is called an* orthonormal set *if $(s_1, s_2) = 0$ whenever $s_1 \neq s_2 \in S$ and $(s, s) = 1$ for every $s \in S$.*
A maximal orthonormal set (in the sense of inclusion) in H is called an orthonormal basis *of H.*

Theorem 1.53 *Every Hilbert space has an orthonormal basis.*

Proof: If $\{M_\alpha\}$ is a chain of orthonormal sets, i.e., $M_\alpha \subset M_\beta$ or $M_\beta \subset M_\alpha$ for all indices α, β, then $\bigcup M_\alpha$ is an orthonormal set. Indeed, whenever $x, y \in \bigcup M_\alpha$, $x \neq y$, there is M_β such that $x, y \in M_\beta$ and thus $(x, y) = 0$. By Zorn's lemma, given an orthonormal set S_0, there is a maximal orthonormal set S containing S_0. □

Theorem 1.54 *Let H be a Hilbert space, and let H_0 be a closed subspace of H. Then every orthonormal basis of H_0 can be extended to an orthonormal basis of H, i.e., if $\{e_\gamma\}_{\gamma\in\Gamma_0}$ is an orthonormal basis of H_0, then there exists an orthonormal set $\{e_\gamma\}_{\gamma\in\Gamma_1}$ in H such that $\{e_\gamma\}_{\gamma\in\Gamma_0\cup\Gamma_1}$ is an orthonormal basis of H.*

Proof: By Theorem 1.53 the space H_0 (a Hilbert space when endowed with the restriction of the inner product on H) has an orthonormal basis, say $\{e_\gamma\}_{\gamma\in\Gamma_0}$. Put $M_0 = \{e_\gamma\}_{\gamma\in\Gamma_0}$ and consider the family $\mathcal{M}$ of all orthonormal sets in H containing M_0. As in the proof of Theorem 1.53, every chain in $\mathcal{M}$ has an upper bound (the union of all elements in the chain), so by Zorn's lemma there exists a maximal orthonormal set (i.e., an orthonormal basis of H) in $\mathcal{M}$. This orthonormal basis extends $\{e_\gamma\}_{\gamma\in\Gamma_0}$. □

As an immediate consequence of Theorem 1.49, we obtain that if $\{e_\mu\}$ is an orthonormal basis of H, then $\overline{\operatorname{span}}(\{e_\mu\}) = H$. Indeed, otherwise we find a vector x from $S_H \cap \overline{\operatorname{span}}(\{e_\mu\})^\perp$, a contradiction with the maximality of $\{e_\mu\}$.

Theorem 1.55 *Every separable infinite-dimensional Hilbert space H has an orthonormal basis $\{e_i\}_{i=1}^\infty$.*
Moreover, if $\{e_i\}_{i=1}^\infty$ is an orthonormal basis of H, then for every $x \in H$,

$$x = \sum_{i=1}^{\infty} (x, e_i) e_i .$$

The numbers (x, e_i) are called *Fourier coefficients* and $\sum (x, e_i) e_i$ is the *Fourier expansion* of x or the *Fourier series* for x.

Proof: Let $\{e_\mu\}$ be an orthonormal basis of H by Theorem 1.53. Since $\|e_\mu - e_{\mu'}\| = \sqrt{2}$ for $\mu \neq \mu'$, by the proof of Proposition 1.43 (iii), $\{e_\mu\}$ is countable. Note that $\{e_\mu\}$ cannot be finite as H is infinite-dimensional. Choose an ordering of $\{e_\mu\}$ into a sequence $\{e_1, e_2, \dots\}$.

We claim that for $x \in H$ and $n \in \mathbb{N}$ we have

$$\operatorname{dist}\big(x, \operatorname{span}\{e_1, \dots, e_n\}\big) = \Big\| x - \sum_{i=1}^{n} (x, e_i) e_i \Big\| .$$

To this end, note that for all scalars $c_1, c_2, \dots, c_n$ we have

$$\left\| x - \sum_{i=1}^{n} c_i e_i \right\|^2 = (x, x) + \sum_{i=1}^{n} |c_i - (x, e_i)|^2 - \sum_{i=1}^{n} |(x, e_i)|^2.$$

Hence $\inf_{c_1,\dots,c_n \in \mathbb{K}} \left\| x - \sum_{i=1}^{n} c_i e_i \right\|^2 = (x, x) - \sum_{i=1}^{n} |(x, e_i)|^2$ and the infimum is attained when $c_i = (x, e_i)$. We now show that $x = \sum_{i=1}^{\infty} (x, e_i) e_i$. Given $\varepsilon > 0$, find n_0 such that $\operatorname{dist}\bigl(x, \operatorname{span}(\{e_i\}_{i=1}^{n_0})\bigr) < \varepsilon$ using $\overline{\operatorname{span}}(\{e_i\}_{i=1}^{\infty}) = H$. For $n \geq n_0$ we have

$$\left\| x - \sum_{i=1}^{n} (x, e_i) e_i \right\| = \operatorname{dist}\bigl(x, \operatorname{span}\{e_1, \dots, e_n\}\bigr) \leq \operatorname{dist}\bigl(x, \operatorname{span}\{e_1, \dots, e_{n_0}\}\bigr) < \varepsilon.$$

□

Proposition 1.56 *Let $\{e_i\}_{i=1}^{\infty}$ be an orthonormal set in a Hilbert space H and $x \in H$. Then*
(i)

$$\sum_{i=1}^{\infty} |(x, e_i)|^2 \leq \|x\|^2 \qquad \textit{(The Bessel inequality).} \tag{1.12}$$

(ii) *If $\{e_i\}_{i=1}^{\infty}$ is an orthonormal basis of H, then*

$$\|x\|^2 = \sum_{i=1}^{\infty} |(x, e_i)|^2 \qquad \textit{(The Parseval equality).} \tag{1.13}$$

(iii) *If the Parseval equality holds for every $x \in H$, then $\{e_i\}_{i=1}^{\infty}$ is an orthonormal basis of H.*
(iv) *If $\overline{\operatorname{span}}(\{e_i\}_{i=1}^{\infty}) = H$, then $\{e_i\}_{i=1}^{\infty}$ is an orthonormal basis of H.*

Proof: (i) For every $n \in \mathbb{N}$ we have

$$0 \leq \left\| x - \sum_{i=1}^{n} (x, e_i) e_i \right\|^2 = (x, x) - \sum_{i=1}^{n} |(x, e_i)|^2.$$

From this the Bessel inequality follows.

(ii) From the proof of Theorem 1.55 we have

$$\operatorname{dist}\bigl(x, \operatorname{span}\{e_1, \dots, e_n\}\bigr)^2 = (x, x) - \sum_{i=1}^{n} |(x, e_i)|^2$$

for every $n \in \mathbb{N}$ and $\operatorname{dist}(x, \operatorname{span}\{e_1, \ldots, e_n\}) \to 0$ as $n \to \infty$. Thus $(x, x) - \sum_{i=1}^{n} |(x_i e_i)|^2 \to 0$ as $n \to \infty$.

(iii) If $\{e_i\}$ is not an orthonormal basis of H, then $\overline{\operatorname{span}}\{e_i\} \neq H$. By Theorem 1.49 there is $e \in H \setminus \{0\}$ such that $(e, e_i) = 0$ for every $i \in \mathbb{N}$. Then $0 = \sum |(e, e_i)|^2 \neq \|e\|^2$, a contradiction.

(iv) Assume that $\overline{\operatorname{span}}\{e_i\} = H$ and $\{e_i\}$ is not an orthonormal basis. Choose e orthogonal to all e_i and $e \neq 0$. Let $y_n \in \operatorname{span}\{e_i\}$, $y_n \to e$. Then $(e, y_n) = 0$ for every n and therefore $(e, e) = \lim(e, y_n) = 0$, a contradiction. □

Theorem 1.57 (Riesz, Fischer) *Every separable infinite-dimensional Hilbert space H is linearly isometric to ℓ_2.*

Proof: Let $\{e_i\}_{i=1}^{\infty}$ be an orthonormal basis of H. For $x \in H$ put $T(x) = ((x, e_i))_{i=1}^{\infty}$. Parseval's equality (1.13) implies that $((x, e_i))_{i=1}^{\infty} \in \ell_2$ and that T is an isometry. It remains to prove that T maps H onto ℓ_2. Given $(c_i) \in \ell_2$, define $x \in H$ by $x = \sum c_i e_i$. Then clearly $\left\| \sum_{i=1}^{n} c_i e_i \right\|^2 = \sum_{i=1}^{n} |c_i|^2$, therefore $\sum c_i e_i$ is Cauchy and thus convergent in H. Next we will show that if $x = \sum_{i=1}^{\infty} c_i e_i$, then $c_i = (x, e_i)$ for every i, that is, $T(x) = (c_i)$. This follows from the continuity of the inner product:

$$\Big(\sum_{i=1}^{\infty} c_i e_i, e_j\Big) = \lim_n \Big(\sum_{i=1}^{n} c_i e_i, e_j\Big) = \lim_n c_j = c_j \text{ for every } j.$$

Thus T is an onto mapping. □

1.5 Remarks and Open Problems

Remarks

1. The space $L_2[0, 1]$ endowed with the inner product $(f, g) := \int_0^1 f \bar{g}\, dt$ is a separable Hilbert space. Similarly ℓ_2 is a Hilbert space, the inner product being $\big((a_n), (b_n)\big) := \sum_{n=1}^{\infty} a_n \bar{b}_n$. Therefore L_2 is linearly isometric to ℓ_2. Note that in Chapter 4 (see the paragraph after the proof of Theorem 4.53) we will show that ℓ_p is not isomorphic to L_p for $p \in [1, \infty)$, $p \neq 2$. The space ℓ_∞ is isomorphic (Exercise 4.42) but not isometric (Exercise 7.41) to L_∞.
2. We have shown that if $\{e_i\}_{i=1}^{\infty}$ is an orthonormal basis of a Hilbert space H, then $x = \sum_{i=1}^{\infty} (x, e_i) e_i$ for every $x \in H$ and this expression is unique, that is, if for some $c_i \in \mathbb{K}$ we have $x = \sum c_i e_i$, then $c_i = (x, e_i)$ for every i.
3. If $\{e_i\}_{i=1}^{\infty}$ is an orthonormal basis of H and $\{e_{\pi(i)}\}$ is a permutation of $\{e_i\}$, then $\sum (x, e_{\pi(i)}) e_{\pi i} = x$ and $\{e_{\pi(i)}\}$ are a basis of H. Indeed, given $\varepsilon > 0$, find $n_0 \in \mathbb{N}$ such that $\operatorname{dist}(x, \operatorname{span}\{e_1, \ldots, e_{n_0}\}) < \varepsilon$. Then find m_0 such that $\{1, \ldots, n_0\} \subset \{\pi(1), \ldots, \pi(m_0)\}$. For $m \geq m_0$ we have

$$\Big\|x - \sum_{i=1}^{m}(x, e_{\pi(i)})e_{\pi(i)}\Big\| = \operatorname{dist}\big(x, \operatorname{span}\{e_{\pi(i)}, \dots, e_{\pi(m_0)}\}\big)$$
$$\leq \operatorname{dist}\big(x, \operatorname{span}\{e_1, \dots, e_{n_0}\}\big) < \varepsilon.$$

4. If H is a non-separable Hilbert space and $\{e_\gamma\}_{\gamma\in\Gamma}$ is an orthonormal basis (see Theorem 1.53), every $x \in H$ has only a countable number of non-zero Fourier coefficients (x, e_γ). This follows from the Bessel inequality (1.12). Indeed, for every $\varepsilon > 0$ the set $\{\gamma \in \Gamma : |(x, e_\gamma)| > \varepsilon\}$ is finite. Then we still have $x = \sum_{\gamma\in\Gamma}(x, e_\gamma)e_\gamma$ (where the sum is in fact a countable one, and it is independent of the particular enumeration of the countable number of non-zero summands). Using this observation and proceeding similarly to how Theorem 1.57 was proven, we can show that every Hilbert space is isometric to some $\ell_2(\Gamma)$ (see Exercise 1.95).
5. The trigonometric system $\{1, e^{2\pi it}, e^{-2\pi it}, e^{4\pi it}, e^{-4\pi it}, \dots\}$ in the indicated order is an orthonormal basis in $L_2[0, 1]$, a separable Hilbert space when endowed with the inner product defined in (i) above. Indeed, it is an orthonormal system by inspection. Completeness of the trigonometric system, i.e., the fact that its closed linear span is $L_2[0, 1]$, follows from standard results: first, from the Lebesgue dominated convergence theorem, the set of all complex, measurable, simple functions s on $[0, 1]$ such that $\lambda(\{t \in [0, 1] : s(t) \neq 0\})$, is dense in $L_2[0, 1]$. Lusin's theorem then shows that the space $C[0, 1]$ is dense in $L_2[0, 1]$. The arithmetic means of the partial sums of the Fourier series of a continuous function in $[0, 1]$ converge uniformly to the function (Féjer). This finishes the proof of the completeness of the trigonometric system (see, e.g., [Katz, p. 15]).
6. We refer to [LiTz3] for more on the class of Orlicz and Lorentz Banach spaces defined in Exercises 1.100 and 1.101.

Open Problems

1. If the property (iv) in Definition 1.1 is replaced by: *There is a constant* $C \geq 1$ *so that* $\|x + y\| \leq C(\|x\| + \|y\|)$ *for every* $x, y \in X$, then we call $\|\cdot\|$ a *quasi-norm* on X. It is then possible to replace $\|\cdot\|$ by an equivalent quasi-norm $|||\cdot|||$ so that there is $0 < p \leq 1$ such that $|||x + y|||^p \leq |||x|||^p + |||y|||^p$ for all $x, y \in X$. Then $(X, |||\cdot|||)$ is called a *quasi-Banach* space if it is complete in the metric $d(x, y) := \|x - y\|^p$. We refer the reader to, e.g., [Kalt1, pp. 1009–1130] and references therein for the theory of quasi-Banach spaces. We just mention that the lack of convexity is behind the fact that the structure of quasi-Banach spaces often differs from that of Banach spaces. For example, there is a quasi-Banach space X so that X contains a vector $x \neq 0$ such that every closed infinite-dimensional subspace of X contains x. It is an open problem if there is a quasi-Banach space which has no proper closed infinite-dimensional subspaces. See also [Maur, pp. 1247–1297]. We will return to this topic in Remarks in Chapters 2 and 12.

Exercises for Chapter 1

1.1 Show that if A is a balanced subset of a vector space V, then $[0, a] \subset A$ for all $a \in A$.
Hint. It follows from the definition.

1.2 Prove that a convex set in a real vector space is symmetric if and only if it is balanced.
Hint. Balanced always implies symmetric. On the other hand, if S is symmetric and convex, $0 \in S$. To prove that $\alpha S \subset S$ for all $|\alpha| \leq 1$ split the argument into two parts: first for $0 \leq \alpha \leq 1$ and then for $-1 \leq \alpha < 0$, using that $(-1)S \subset S$.

1.3 Show that if V is a real vector space, $A \subset V$ is a balanced set and $f : V \to \mathbb{R}$ is a linear mapping, then, if $x_0 \in V$ satisfies $(x_0 + A) \cap \operatorname{Ker} f = \emptyset$ then f has constant sign on $x_0 + A$.
Hint. Assume that $f(x_0) > 0$ and that for some $x \in x_0 + A$ we have $f(x) < 0$. By Exercise 1.1, $[x_0, x] \subset x_0 + A$. Since $f\big|_{[x_0,x]}$is continuous, the intermediate value property gives some $x \in (x_0, x)$ with $f(x) = 0$, a contradiction.

1.4 Show that if A is a balanced subset of a vector space V, then $\operatorname{conv}(A)$ is balanced.
Hint. Use $\operatorname{conv}(A) = \{\lambda a + (1 - \lambda)b : a, b \in A, \lambda \in [0, 1]\}$.

1.5 Let V be a vector space. Recall that a mapping $\varphi \colon V \to V$ is called *affine* if there is a linear mapping $T \colon V \to V$ and a vector x_0 such that $\varphi(x) = x_0 + T(x)$ for all $x \in V$.

Prove that φ is an affine mapping of V into V if and only if $\varphi\Big(\sum_{i=1}^n \lambda_i x_i\Big) = \sum_{i=1}^n \lambda_i \varphi(x_i)$ for all $x_1, \ldots, x_n \in V$ and scalars λ_i such that $\sum_{i=1}^n \lambda_i = 1$.

In particular, if φ is an affine mapping and $K \subset V$ is convex, then $\varphi(K)$ is convex.
Hint. If φ is affine, we have $x_0 = \sum \lambda_i x_0$ and thus

$$\varphi\Big(\sum_{i=1}^n \lambda_i x_i\Big) = \sum_{i=1}^n \lambda_i x_0 + T\Big(\sum_{i=1}^n \lambda_i x_i\Big) = \sum_{i=1}^n \lambda_i (x_0 + T(x_i)) = \sum_{i=1}^n \lambda_i \varphi(x_i).$$

If φ has the stated property, it is enough to show that $\psi(x) = \varphi(x) - \varphi(0)$ is a linear mapping. For $x, y \in V$ and scalars α, β we have

$$\begin{aligned}
&\psi(\alpha x + \beta y) = \psi(\alpha x + \beta y + (1 - \alpha - \beta)0) \\
&= \varphi(\alpha x + \beta y + (1 - \alpha - \beta)0) - \varphi(0) = \alpha\varphi(x) + \beta\varphi(y) + (1 - \alpha - \beta)\varphi(0) - \varphi(0) \\
&= \alpha\varphi(x) + \beta\varphi(y) - \alpha\varphi(0) - \beta\varphi(0) = \alpha\psi(x) + \beta\psi(y).
\end{aligned}$$

1.6 Let X be a normed linear space. Prove that for any $x, y \in X$ we have $\big|\|x\| - \|y\|\big| \leq \|x - y\|$.
Hint. Triangle inequality, $\|x\| = \|(x - y) + y\|$.

1.7 Let X be a normed linear space. Assume that for $x, y \in X$ we have $\|x+y\| = \|x\| + \|y\|$. Show that then $\|\alpha x + \beta y\| = \alpha\|x\| + \beta\|y\|$ for every $\alpha, \beta \geq 0$.
Hint. Assume $\alpha \geq \beta$. Write

$$\begin{aligned}\|\alpha x + \beta y\| &= \|\alpha(x+y) - (\alpha-\beta)y\| \geq \alpha\|x+y\| - (\alpha-\beta)\|y\| \\ &= \alpha(\|x\| + \|y\|) - (\alpha-\beta)\|y\| = \alpha\|x\| + \beta\|y\|.\end{aligned}$$

1.8 Calculate the distance in $L_2[0,1]$, $L_4[0,1]$, and in $C[0,1]$, from the function $x(t) := t^2$ to the linear hull of the functions $y(t) := t$ and $z(t) = \sin t$.
Hint. Direct computation.

1.9 Prove that the closure of a subspace of a normed space is again a subspace.
Hint. A simple continuity argument.

1.10 Show that $\overline{\text{span}}(L) = \overline{\text{span}(L)}$ and $\overline{\text{conv}}(M) = \overline{\text{conv}(M)}$ (look at the given definitions of the closed linear hull and the closed convex hull, respectively).
Hint. By the definition, $\overline{\text{span}}(L)$ is the intersection of all closed subspaces containing L, and $\overline{\text{span}(L)}$ is one of such subspaces. From this one inclusion follows. The other inclusion follows similarly as $\text{span}(L)$ is the intersection of all subspaces containing L.

1.11 Show that C is a convex set in a vector space if and only if $\sum \lambda_i x_i \in C$ whenever $x_1, \ldots, x_n \in C$ and $\lambda_1, \ldots, \lambda_n \geq 0$ satisfy $\sum \lambda_i = 1$.
Hint. (a) $\lambda_1 x_1 + \lambda_2 x_2 + \lambda_3 x_3 = (\lambda_1 + \lambda_2)(\frac{\lambda_1}{\lambda_1+\lambda_2} x_1 + \frac{\lambda_2}{\lambda_1+\lambda_2} x_2) + \lambda_3 x_3$ and induction.

1.12 Let A and B be two convex sets in a normed space X. Show that $\text{conv}(A \cup B) = \{\lambda x + (1-\lambda)y : x \in A, y \in B, \lambda \in [0,1]\}$.
Hint. Show first that the set on the right hand side is convex.

1.13 Let A be a set in a real normed space X. Show that the *symmetric convex hull of* A, i.e., the intersection of all symmetric convex sets containing A, is equal to $\left\{\sum_{i=1}^n \lambda_i x_i : x_i \in A, \sum_{i=1}^n |\lambda_i| \leq 1, n \in \mathbb{N}\right\}$.
Hint. Show first that the above set is convex and symmetric.

1.14 Show that the linear subspace $\mathcal{R}$ of Riemann integrable functions in $L_1[0,1]$ is not closed in $L_1[0,1]$.
Hint. If C is the Cantor discontinuum of positive Lebesgue measure and $\chi(C)$ is its characteristic function, then there are Riemann integrable functions ψ_n that $\psi_n \to \chi(C)$ pointwise (use the step functions given by the definition of $\chi(C)$). Since the measure of C is positive, it can be shown that there is no Riemann integrable function that is equal to $\chi(C)$ almost everywhere (see, e.g., [Stro, p. 273]).

1.15 Let $1 \leq p \leq q \leq \infty$. Then $\|x\|_{\ell_q} \leq \|x\|_{\ell_p}$ for $x \in \ell_p$ and $\|f\|_{L_p} \leq \|f\|_{L_q}$ for $f \in L_q[0,1]$.

In particular, $\ell_p \subset \ell_q$ and, if $1 \le p < \infty$, then $\ell_p \subset c_0$. Moreover, $L_q[0,1] \subset L_p[0,1]$. All the corresponding identity operators have norm one.
Hint. Take $x = (x_i) \in \ell_p$ such that $\|x\|_{\ell_p} = 1$. Then $|x_i| \le 1$, so $|x_i|^q \le |x_i|^p$. Consequently $\|x\|_{\ell_q}^q \le \|x\|_{\ell_p}^p = 1$. If $1 \le p < \infty$ and $x \in \ell_p$ then obviously $x_i \to 0$. The inequality for function spaces follows from the Hölder inequality (1.1) used with $r = q/p$.

1.16 Let $f \in L_{p_0}[0,1]$ for some $p_0 > 1$. Show that $\lim\limits_{p\to 1^+} \|f\|_{L_p} = \|f\|_{L_1}$. If $f \in L_\infty[0,1]$, then $\lim\limits_{p\to\infty} \|f\|_{L_p} = \|f\|_{L_\infty}$.

Let $x \in \ell_q$ for some $q \ge 1$. Show that $\lim\limits_{p\to\infty} \|x\|_{\ell_p} = \|x\|_{\ell_\infty}$.
Hint. If f is bounded, we have that $f^p \to f$ pointwise as $p \to 1$. By the Lebesgue dominated convergence theorem, $\int_0^1 |f|^p \,\mathrm{d}t \to \int_0^1 |f| \,\mathrm{d}t$. We may assume that $|f| \le 1$, then $\int_0^1 |f|^p \,\mathrm{d}t \le 1$ and hence $\int_0^1 |f|^p \,\mathrm{d}t \le \|f\|_{L_p} \le \int_0^1 |f| \,\mathrm{d}t$, consequently $\|f\|_{L_p} \to \|f\|_{L_1}$ as $p \to 1$.

If f is not bounded, given ε, we first find a bounded function $F \in L_{p_0}$ such that $\|f - F\|_{L_{p_0}} < \varepsilon$. Then also $\|f - F\|_{L_p} < \varepsilon$ for every $p \in [1, p_0]$ and we can use the result above to find $\delta > 0$ such that $\big|\|f\|_{L_1} - \|f\|_{L_p}\big| < 3\varepsilon$ whenever $p \in (1, 1+\delta)$.

Let $f \in L_\infty[0,1]$. For $\varepsilon > 0$, set $M = \{t \in [0,1] : |f(t)| \ge \|f\|_{L_\infty} - \varepsilon\}$. By the definition, $\lambda(M) > 0$. We have $\|f\|_{L_p} \ge \left(\int_M |f|^p \,\mathrm{d}\lambda\right)^{1/p} \ge \lambda(M)^{1/p}(\|f\|_{L_\infty} - \varepsilon)$. Since $\lambda(M)^{1/p} \to 1$ as $p \to \infty$, there is p_0 such that $\big|\|f\|_{L_p} - \|f\|_{L_\infty}\big| < 2\varepsilon$ for $p \ge p_0$.

Now assume that $x \in \ell_q$ and $\|x\|_\infty = 1$. Then there is a finite set M of coordinates i such that $|x_i| = 1$, denote $K = |M|$. Note that $\sum_{i\in M} |x_i|^p = K$. Fix $n_0 > \max(M)$ such that $\sum_{i>n_0} |x_i|^q < K/2$. As $|x_i| < 1$ for $i \notin M$, we have $\sum_{i>n_0} |x_i|^p < K/2$ for every $p \ge q$, moreover, there is $p_0 \ge q$ such that $|x_i|^p < \frac{K}{2(n_0-K)}$ whenever $p \ge p_0$ and $i \ne n_0$, $i \notin M$. Consequently,

$$K \le \sum_{i=1}^{\infty} |x_i|^p \le K + \frac{K}{2} + \frac{K}{2} = 2K$$

for all $p \ge p_0$, that is, $K^{1/p} \le \|x\|_{\ell_p} \le (2K)^{1/p}$. Letting $p \to \infty$ we get $\lim\limits_{p\to\infty} \|x\|_{\ell_p} = 1 = \|x\|_{\ell_\infty}$.

1.17 Let $1 \le p_1 < p_2 < +\infty$ and let X be a subspace of $L_{p_2}[0,1]$ on which the $L_{p_1}[0,1]$ and $L_{p_2}[0,1]$ norms are equivalent. Show that then all $L_p[0,1]$ norms on X, for $1 < p \le p_2$, are equivalent.
Hint. Due to Exercise 1.16 we may assume that $p < p_1$. Suppose that, for some $C > 0$, we have $\|x\|_{p_2} \le C\|x\|_{p_1}$ for all $x \in X$. Let $1 < p < p_1$. Write $p_1 = \lambda p + (1-\lambda)p_2$, $0 < \lambda < 1$. By the Hölder inequality used for numbers $\frac{1}{\lambda}$ and $\frac{1}{1-\lambda}$, we have

$$\|x\|_{p_1}^{p_1} = \int_0^1 |x|^{\lambda p+(1-\lambda)p_2} = \int_0^1 |x|^{\lambda p} \cdot |x|^{(1-\lambda)p_2}$$
$$\leq \left(\int_0^1 (|x|^{\lambda p})^{\frac{1}{\lambda}}\right)^{\lambda} \cdot \left(\int_0^1 (|x|^{(1-\lambda)p_2})^{\frac{1}{1-\lambda}}\right)^{1-\lambda} = \|x\|_p^{p\lambda} \cdot \|x\|_{p_2}^{p_2(1-\lambda)}$$
$$\leq \|x\|_p^{p\lambda} \cdot (C\|x\|_{p_1})^{p_2(1-\lambda)} = \|x\|_p^{p\lambda} \cdot C^{p_2(1-\lambda)} \cdot \|x\|_{p_1}^{p_2(1-\lambda)} = \|x\|_p^{p\lambda} \cdot C^{p_2(1-\lambda)} \cdot \|x\|_{p_1}^{p_1-\lambda p}.$$

Thus

$$1 \leq \|x\|_p^{\lambda p} \cdot C^{p_2(1-\lambda)} \cdot \|x\|_{p_1}^{-\lambda p}.$$

Hence

$$\|x\|_{p_1}^{\lambda p} \leq C^{p_2(1-\lambda)} \cdot \|x\|_p^{\lambda p}.$$

Thus

$$\|x\|_{p_1} \leq C^{\frac{p_2(1-\lambda)}{\lambda p}} \cdot \|x\|_p \leq C^{\frac{p_2(1-\lambda)}{\lambda p}} \|x\|_{p_1}.$$

1.18 Show that for every $p \geq 1$, ℓ_p is linearly isometric to a subspace of $L_p[0, 1]$.
Hint. Consider span$\{f_n\}$, where $f_n := \big(n(n+1)\big)^{1/p} \chi_{[\frac{1}{n+1}, \frac{1}{n}]}$.

1.19 Let $1 \leq p \leq \infty$ and let μ be a σ-finite measure. Show that $L_p(\mu)$ is isometric to $L_p(\nu)$, where ν is a probability measure.
Hint. Take g an a.e. positive function such that $\int f \mathrm{d}\mu = 1$, and define a measure ν by $d\nu = g d\mu$. Then the mapping $f \to f \cdot g^{-1/p}$ defines an isometry from $L_p(\mu)$ onto $L_p(\nu)$. For $p = \infty$ the isometry is just the formal identity.

1.20 Show that $c_0(\Gamma)$ is the closure of $c_{00}(\Gamma)$ in $\ell_\infty(\Gamma)$.

1.21 Let Γ be a set and $p \in [1, \infty]$. Show that $c_0(\Gamma)$ and $\ell_p(\Gamma)$ are Banach spaces.
Hint. The proof of Proposition 1.16.

1.22 Show that $\ell_p(I)$ is linearly isometric to $\ell_p(J)$ whenever card(I) = card(J). Here card(I) denotes the cardinality of the set I.
Hint. If φ is a bijection from I onto J, consider the mapping $f \mapsto f \circ \varphi$.

1.23 Let $C^n[0, 1]$ be the space of all real-valued functions on $[0, 1]$ that have n continuous derivatives on $[0, 1]$, with the norm

$$\|f\| := \max_{0 \leq k \leq n} \big(\max\{|f^k(t)| : t \in [0, 1]\}\big).$$

Show that $C^n[0, 1]$ is a Banach space.
Hint. If $f_n \to f$ uniformly and $f_n' \to g$ uniformly, then $f' = g$.

1.24 Let $\mathcal{L}$ be the normed space of all Lipschitz functions on a Banach space X that are equal to 0 at the origin, under the norm

$$\|f\| := \sup\left\{\frac{|f(x)-f(y)|}{\|x-y\|} : x, y \in X\right\}.$$

Show that $\mathcal{L}$ is a Banach space.
Hint. Cauchy sequences are bounded.

1.25 Let $\mathcal{D}$ be the normed space of all bounded Lipschitz Fréchet differentiable functions (for definitions see Chapter 8) on a Banach space X under the norm $\|f\| := \sup\{|f(x)| : x \in X\} + \sup\{\|f'(x)\| : x \in X\}$, where $\|f'(x)\|$ is the norm of $f'(x)$ in X^*, i.e., $\|f'(x)\| := \sup\{f'(x)(h) : h \in B_X\}$. Show that $\mathcal{D}$ is a Banach space.
Hint. Classical rules of differentiation.

1.26 Show that a normed space Y is a Banach space if and only if $\sum y_n$ converges in Y whenever $\|y_n\| \leq 2^{-n}$ for every n.
Hint. Use Lemma 1.22. To prove the necessary condition, note that if $\|y_k\| \leq 2^{-k}$ for all $k \in \mathbb{N}$, then $\sum y_k$ is absolutely convergent. For sufficiency, observe that the sequence $\{y_k\}$ constructed in that lemma satisfies $\|y_k\| \leq 2^{1-k}$ for all $k \geq 2$.

1.27 Let Y be a closed subspace of a normed space X. Show that if Y and X/Y are both Banach spaces, then X is a Banach space.
Note: A property $\mathcal{P}$ is said to be a *three-space property* if the following holds: Let Y be a closed subspace of a space X. If Y and X/Y have $\mathcal{P}$ then X has $\mathcal{P}$.

Thus the property of being complete is a three-space property in the class of normed linear spaces.
Hint. If $\{x_n\}$ is Cauchy in X, there is $x \in X$ such that $\hat{x}_n \to \hat{x}$. There are $\{y_n\}$ in Y such that $\{x_n - x - y_n\} \to 0$. Thus $\{y_n\}$ is Cauchy, so $y_n \to y$ and $x_n \to x + y$.

1.28 Let Y, Z be subspaces of a Banach space X such that Y is isomorphic to Z. Are X/Y and X/Z isomorphic?
Hint. No. Let $X = \ell_2$, $Y = \{(0, x_2, x_3, \dots)\}$, and $Z = \{(0, 0, x_3, x_4, \dots)\}$.

1.29 Show that the distance $\mathrm{d}(x)$ of a point $x = (x_i) \in \ell_\infty$ to c_0 is equal to $\limsup_{i\to\infty} |x_i|$. Thus the norm in ℓ_∞/c_0 is $\|\hat{x}\| = \limsup_{i\to\infty} |x_i|$.
Hint. There is only finitely many i such that $|x_i| > \limsup |x_i| + \varepsilon$.

1.30 Let $\|\cdot\|_1, \|\cdot\|_2$ be two norms on a vector space X. Let B_1 and B_2 be the closed unit balls of $(X, \|\cdot\|_1)$ and $(X, \|\cdot\|_2)$, respectively. Prove that $\|\cdot\|_1 \leq C\,\|\cdot\|_2$ (that is, $\|x\|_1 \leq C\,\|x\|_2$ for all $x \in X$) if and only if $\frac{1}{C} B_2 \subset B_1$.

1.31 Let $\|\cdot\|_1$ and $\|\cdot\|_2$ be two equivalent norms on a vector space X. Let B_1 and B_2 be the closed unit balls of $(X, \|\cdot\|_1)$ and $(X, \|\cdot\|_2)$, respectively. Show that B_1 and B_2 are homeomorphic.

Recall that two topological spaces K and L are called *homeomorphic* if there exists a bijection φ from K onto L such that φ and φ^{-1} are continuous. Such φ is called a *homeomorphism*.
Hint. Define a mapping ϕ from B_1 onto B_2 by $\phi(0) = 0$ and $\phi(x) = \frac{\|x\|_1}{\|x\|_2} x$ for $x \in B_1 \setminus \{0\}$. Clearly $\|\phi(x)\|_2 = \|x\|_1$, continuity at 0 follows from the equivalence of the norms.

1.32 Let X be the normed space obtained by taking c_0 with the norm $\|x\|_0 := \sum 2^{-i}|x_i|$. Show that X is not a Banach space.
Note that this shows that $\|\cdot\|_0$ is not an equivalent norm on c_0.
Hint. The sequence $\{(1, 1, \ldots, \overset{n}{1}, 0, \ldots)\}_{n=1}^{\infty}$ is Cauchy and not convergent as the only candidate for the limit would be $(1, 1, \ldots) \notin c_0$.

1.33 Let M be a dense (not necessarily countable) subset of a Banach space X. Show that for every $x \in X \setminus \{0\}$ there are $x_k \in M$ such that $x = \sum x_k$ and $\|x_k\| \leq \frac{3\|x\|}{2^k}$.
Hint. Find $x_1 \in M$ such that $\|x - x_1\| \leq \frac{\|x\|}{2}$, then by induction $x_k \in M$ such that $\|x-(x_1+\cdots+x_{k-1})-x_k\| \leq \frac{\|x\|}{2^k}$. Then $x = \sum x_k$ and $\|x_k\| = \left\|\left(x - \sum_{n=1}^{k-1} x_n\right) - \left(x - \sum_{n=1}^{k} x_n\right)\right\| \leq \frac{\|x\|}{2^{k-1}} + \frac{\|x\|}{2^k} = \frac{3\|x\|}{2^k}$.

1.34 Show that a Banach space X is separable if and only if S_X is separable.
Hint. If $\{x_n\}$ is dense in S_X, consider $\{r_k x_n\}_{k,n}$ for some dense sequence $\{r_k\}$ in $\mathbb{K}$. If $(0 \notin)$ D is countable and dense in X, consider $\{\frac{x}{\|x\|} : x \in D\}$.

1.35 Let Y be a closed subspace of a Banach space X. Show that if X is separable, then Y and X/Y are separable.
Show that if Y and X/Y are separable, then X is then separable.
Thus separability is a three-space property.
Hint. If $\{\hat{x}_n\}$ is a dense set in X/Y and $\{x_n\}$ is dense in Y, choosing $y_n \in \hat{x}_n$ and considering $\{y_n + x_k : n, k \in \mathbb{N}\}$ we have a dense set in X.

1.36 Show that the space $BC(0, 1)$ of bounded continuous functions on $(0, 1)$ with the sup-norm is nonseparable.
Hint. Fix $\{f_n\}$ in $BC(0, 1)$ such that $\operatorname{supp} f_n \subset (\frac{1}{n+1}, \frac{1}{n})$ and $\|f_n\|_\infty = 1$ for all n and then define $T \colon \ell_\infty \to BC(0, 1)$ by $T((a_n)) = \sum_n a_n f_n$. Thus $BC(0, 1)$ contains an isometric copy of ℓ_∞. Note that this cannot be done in $C[0, 1]$ as $\sum_n a_n f_n$ is not necessarily continuous at 0.

1.37 If F is a finite set of positive integers and $\{x_j : j \in F\}$ is a finite set of elements in a Banach space X, then

$$\sup\left\{\left\|\sum_{i\in F}\varepsilon_i x_i\right\| : \varepsilon_i = \pm 1\right\} = \sup\left\{\sum_{i\in F}|\langle x^*, x_i\rangle| : x^* \in S_{X^*}\right\}$$
$$= \sup\left\{\left\|\sum_{i\in F}a_i x_i\right\| : \max|a_i| \leq 1\right\}. \quad (1.14)$$

Hint. Assume that the first listed supremum is ≤ 1 and let $x^* \in S_{X^*}$. For $i \in \mathbb{N}$ put $s_i = \operatorname{sign}\langle x^*, x_i\rangle$. Then

$$\sum_{i\in F}|\langle x^*, x_i\rangle| = \sum_{i\in F}s_i\langle x^*, x_i\rangle = \left\langle x^*, \sum_{i\in F}s_i x_i\right\rangle \leq \left\|\sum_{i\in F}s_i x_i\right\| \leq 1.$$

Assume that the second listed supremum is ≤ 1. Let $|a_i| \leq 1$ for $i \in F$. Pick $x^* \in S_{X^*}$ so that $\langle x^*, \sum_{i\in F}a_i x_i\rangle = \|\sum_{i\in F}a_i x_i\|$. Then

$$\left\|\sum_{i\in F}a_i x_i\right\| = \sum_{i\in F}a_i\langle x^*, x_i\rangle \leq \sum_{i\in F}|a_i||\langle x^*, x_i\rangle| \leq \sum_{i\in F}|\langle x^*, x_i\rangle|.$$

From these considerations and from homogeneity, it follows that the second supremum is less than or equal to the first supremum, and that the third supremum is less than or equal to the second one. Obviously, the first one is less than or equal to the third one.

1.38 Let X be a normed space. Let $\mathcal{P}_f(\mathbb{N})$ be the family of all finite subsets of $\mathbb{N}$ (including $\emptyset$). Let $\{x_n\}_{n=1}^{\infty}$ be a sequence in X. The series $\sum x_n$ in X is called

(i) *U-Cauchy* (for *unconditionally Cauchy*) if, given $\varepsilon > 0$ there exists $F_0 \in \mathcal{P}_f(\mathbb{N})$ such that $\|\sum_{n\in F}x_n\| < \varepsilon$ for every $F \in \mathcal{P}_f(\mathbb{N})$ with $F \cap F_0 = \emptyset$.

(ii) *S-Cauchy* (for *subseries Cauchy*) if, for every sequence $n_1 < n_2 < \ldots$ in $\mathbb{N}$, the series $\sum_k x_{n_k}$ is Cauchy.

(iii) *BM-Cauchy* (for *bounded-multiplier Cauchy*) if, for every bounded sequence (a_n) in $\mathbb{R}$, the series $\sum a_n x_n$ is Cauchy.

(iv) *R-Cauchy* (for *reordered Cauchy*) if, for every permutation π of $\mathbb{N}$, the series $\sum x_{\pi(n)}$ is Cauchy.

Prove that all these concepts coincide.

Hint. BM-Cauchy$\Longrightarrow$S-Cauchy: This is obvious.

S-Cauchy$\Longrightarrow$U-Cauchy: If $\sum x_n$ is S-Cauchy but not U-Cauchy, there exists $\varepsilon > 0$ and a sequence (F_k) in $\mathcal{P}_f(\mathbb{N})$ such that $\sup F_k < \inf F_{k+1}$ and $\|\sum_{n\in F_k}x_n\| \geq \varepsilon$ for all $k \in \mathbb{N}$. List the elements in $\bigcup_k F_k$ in increasing order to get a non-Cauchy subseries of $\sum x_n$, a contradiction.

U-Cauchy$\Longrightarrow$BM-Cauchy: Given $\varepsilon > 0$, we can find $F_0 \in \mathcal{P}_f(\mathbb{N})$ such that $\|\sum_{n\in F}x_n\| < \varepsilon$ for all $F \in \mathcal{P}_f(\mathbb{N})$ with $F \cap F_0 = \emptyset$. If $\varepsilon_n = \pm 1$ for all n then, for $F \in \mathcal{P}_f(\mathbb{N})$ with $F \cap F_0 = \emptyset$,

$$\left\|\sum_{n\in F}\varepsilon_n x_n\right\| = \left\|\sum_{n\in F^+} x_n - \sum_{n\in F^-} x_n\right\| \le \left\|\sum_{n\in F^+} x_n\right\| + \left\|\sum_{n\in F^-} x_n\right\| < 2\varepsilon,$$

where $F^+ := \{n \in F : \varepsilon_n = 1\}$ and $F^- := \{n \in F : \varepsilon_n = -1\}$. In view of equalities (1.14) in Exercise 1.37, we get $\|\sum_{n\in F} a_n x_n\| < 2\varepsilon$ whenever $|a_n| \le 1$ for all n. This proves that $\sum x_n$ is BM-Cauchy.

U-Cauchy$\Longrightarrow$R-Cauchy: Observe that if $\sum x_n$ is U-Cauchy, then every reordering $\sum x_{\pi(n)}$ is obviously U-Cauchy, too. Since U-Cauchy implies Cauchy, the series $\sum x_n$ is R-Cauchy.

R-Cauchy$\Longrightarrow$U-Cauchy: If $\sum x_n$ is R-Cauchy and not U-Cauchy, we can find (F_k) as in the second implication. Put $F_0 = \emptyset$, max $F_0 = 0$, and $D_n = \{\max F_{n-1}+1, \dots, \max F_n\} \setminus F_n$ for all $n \in \mathbb{N}$. Now define a reordering of $\mathbb{N}$ by listing in increasing order the elements of D_1, then F_1, then D_2, then F_2, and so on. The reordered sum is not Cauchy, a contradiction.

1.39 Let X be a normed space, $\{x_n\}_{n=1}^{\infty}$ a sequence in X and $x \in X$. The series $\sum x_n$ is said to be *U-convergent* (for *unconditionally convergent*) to x if, for every $\varepsilon > 0$ there exists $F_0 \in \mathcal{P}_f(\mathbb{N})$ such that $\|x - \sum_{n\in F} x_n\| < \varepsilon$ for all $F \in \mathcal{P}_f(\mathbb{N})$ with $F_0 \subset F$.

We can also say that the series is *S-convergent* (for *subseries convergent*), *BM-convergent* (for *bounded multiplier convergent*) *R-convergent* (for *reordered convergent*), when (ii), respectively (iii), respectively (iv) in Exercise 1.38, holds with the term "Cauchy" replaced by "convergent."

Prove that every U-Cauchy series in X that converges to some $x \in X$ is U-convergent to x. In particular, prove that in a Banach space, each of the Cauchy concepts introduced in Exercise 1.38 (and then any of them) implies the corresponding convergent concept introduced in this Exercise 1.39 (and then any of them).

Hint. Given $\varepsilon > 0$ find $n_0 \in \mathbb{N}$ such that $F_0 \subset \{1, 2, , \dots, n_0\}$ and $\|x - \sum_{n=1}^{n_0} x_n\| < \varepsilon$. Then, for $F \in \mathcal{P}_f(\mathbb{N})$ such that $F_0 \subset F$,

$$\left\|x - \sum_{n\in F} x_n\right\| \le \left\|x - \sum_{n=1}^{n_0} x_n\right\| + \left\|\sum_{n=1}^{n_0} x_n - \sum_{n\in F_0} x_n\right\| + \left\|\sum_{n\in F_0} x_n - \sum_{n\in F} x_n\right\| < 3\varepsilon.$$

This proves the assertion. For the second part, use Exercise 1.38 and the fact that every Cauchy series in a Banach space is convergent.

1.40 Let X be a normed space.

(i) Prove that, for a given series in X, BM-convergent$\Longrightarrow$S-convergent$\Longrightarrow$U-convergent.

(ii) Prove that every series in X that U-converges to some $x \in X$ is convergent (to x).

(iii) Prove that U-convergent$\Longrightarrow$R-convergent (and every reordered series has the same sum). Prove that R-convergent$\Longrightarrow$U-convergent (hence every reordered series has the same sum).

(iv) Give examples to prove that none of the reverse implications in (i) holds.

Hint. (i) The first implication is obvious. If $\sum x_n$ is S-convergent, then it is S-Cauchy, hence U-Cauchy (see Exercise 1.38). It is also convergent (say, to x). Use now Exercise 1.39 to conclude that $\sum x_n$ is U-convergent to x.

(ii) If $\sum x_n$ is U-convergent to x, given $\varepsilon > 0$ there is $F_0 \in \mathcal{P}_f(\mathbb{N})$ such that $\|x - \sum_{n\in F} x_n\| < \varepsilon$ for every $F \in \mathcal{P}_f(\mathbb{N})$ such that $F_0 \subset F$. Find $n_0 \in \mathbb{N}$ such that $F_0 \subset \{1, 2, \ldots, n_0\}$. Then, obviously, $\left\|x - \sum_{i=1}^{n} x_i\right\| \le \varepsilon$ for all $n \ge n_0$. This proves the assertion.

(iii) If $\sum x_n$ is U-convergent to x, and π is a permutation of $\mathbb{N}$, then obviously $\sum_n x_{\pi(n)}$ is U-convergent to x, hence convergent (to x) by (ii).

Assume now that $\sum x_n$ is R-convergent. Then it is R-Cauchy. Use Exercise 1.38 to obtain that it is U-Cauchy. Since in particular $\sum x_n = x$ for some $x \in X$, use Exercise 1.39 to conclude that $\sum x_n$ is U-convergent (to x).

(iv) An example of a U-convergent series that is not S-convergent in the space c_{00} of the eventually zero sequences in c_0 endowed with the supremum norm, is $\sum x_n$, where $x_n := n^{-1}e_n - (n+1)^{-1}e_{n+1}$, and e_n is the nth unit vector of the canonical basis of c_0. An example of a S-convergent series that is not BM-convergent in the space of real sequences that take only a finite number of values, endowed with the norm $\|x\| := \sup_n n^{-1}|x(n)|$, is $\sum e_n$, with e_n as above.

1.41 Let X be a Banach space. Prove that, for a series $\sum x_n$ in X, the concepts U-convergent, S-convergent, BM-convergent, and R-convergent coincide.

Hint. This is a consequence of Exercises 1.38, 1.39, and 1.40.

1.42 Prove that a series $\sum x_n$ is a normed space X is

(i) U-Cauchy if and only if the set $\{\sum_{n\in F} x_n,\ F \in \mathcal{P}_f(\mathbb{N})\}$ is totally bounded.

(ii) S-convergent if and only if the set $\{\sum_{n\in G} x_n\ :\ G \in \mathcal{P}(\mathbb{N})\}$ is relatively compact (in fact, S-convergent implies that this set is actually compact).

Hint. Put $s(G) = \sum_{n\in G} x_n$ for $G \in \mathcal{P}(\mathbb{N})$.

(i) If $\sum x_n$ is U-Cauchy, fix $\varepsilon > 0$ and find $F_0 \in \mathcal{P}_f(\mathbb{N})$ such that $\|s(F)\| < \varepsilon$ for all $F \in \mathcal{P}_f(\mathbb{N})$ with $F \cap F_0 = \emptyset$. Given $F \in \mathcal{P}_f(\mathbb{N})$, notice that $\|s(F) - s(F \cap F_0)\| = \|s(F \setminus F_0\| < \varepsilon$. Since $\mathcal{P}(F_0)$ is a finite set, it follows that $\{s(F)\ :\ F \in \mathcal{P}_f(\mathbb{N})\}$ is totally bounded.

Assume now that $\{s(F)\ :\ F \in \mathcal{P}_f(\mathbb{N})\}$ is totally bounded (in particular, bounded, say $\|s(F)\| \le M$ for all $F \in \mathcal{P}_f(\mathbb{N})$) and that $\sum x_n$ is not U-Cauchy. Then we can find $\varepsilon > 0$ and a sequence (F_k) in $\mathcal{P}_f(\mathbb{N})$ such that $\sup F_k < \inf F_{k+1}$ and $\|s(F_k)\| \ge \varepsilon$ for all $k \in \mathbb{N}$. The set $\{s(F)\ :\ F \in \mathcal{P}_f(\mathbb{N})\}$ can be covered by a finite number of sets with diameter not greater than $\varepsilon/2$, so at least one of those sets contains an infinite subsequence of $\{s(F_k)\}$, denoted by $\{z_0, z_1, \ldots\}$. Let k be an integer greater than $2M/\varepsilon$. Since $\|z_i - z_0\| \le \varepsilon/2$ for $i \in \mathbb{N}$, we have

$$\|\frac{1}{k}(z_1 + \ldots + z_k) - z_0\| \le \varepsilon/2.$$

However, $z_1+\ldots+z_k \in \{s(F) : F \in \mathcal{P}_f(\mathbb{N})\}$, hence $\|z_1+\ldots+z_k\| \le M$. It follows that $\|z_0\| \le \varepsilon/2 + M/k < \varepsilon$, a contradiction.

(ii) Let $2^{\mathbb{N}} = \{0,1\}^{\mathbb{N}}$ the Cantor space, endowed with the product topology. If $\sum x_n$ is S-convergent, we can define a mapping $\varphi : 2^{\mathbb{N}} \to X$ by $\varphi(a) = \sum a(n)x_n$ for $a = (a(n)) \in 2^N$. The mapping φ is continuous. Indeed, fix $a \in 2^{\mathbb{N}}$ and $\varepsilon > 0$. Then, since $\sum x_n$ is U-Cauchy (see Exercise 1.38), there exists $F_0 \in \mathcal{P}_f(\mathbb{N})$ such that $\|s(F)\| < \varepsilon$ for every $F \in \mathcal{P}_f(\mathbb{N})$ with $F \cap F_0 = \emptyset$. Hence, if $b \in 2^{\mathbb{N}}$ satisfies $b(n) = a(n)$ for $n \in F_0$, then $\|\varphi(b) - \varphi(a)\| < 2\varepsilon$. Since $2^{\mathbb{N}}$ is compact, $\varphi(2^{\mathbb{N}})$ $(= \{s(F) : F \in \mathcal{P}(\mathbb{N})\})$ is also compact.

Finally, assume that $\{s(G) : G \in \mathcal{P}(\mathbb{N})\}$ is relatively compact. This set is totally bounded, so, by (i), $\sum x_n$ is U-Cauchy and, by Exercise 1.38, S-Cauchy; moreover, all partial sums of any subseries lie in a relatively compact set, hence $\sum x_n$ is S-convergent.

1.43 Prove that every absolutely convergent series in a Banach space is U-convergent. Note that the series $\sum_i \frac{1}{i} e_i$ in ℓ_2 is U-convergent but not absolutely convergent.

Hint. Given $\varepsilon > 0$, find $n_0 \in \mathbb{N}$ such that $\sum_{n_0+1}^{\infty} \|x_n\| < \varepsilon$. Then, if $F_0 := \{1, 2, \ldots, n_0\}$ and $F \in \mathcal{P}_f(\mathbb{N})$ satisfies $F_0 \cap F = \emptyset$, we have $\|\sum_{n\in F} x_n\| \le \sum_{n\in F} \|x_n\| < \varepsilon$. To see that $\sum \frac{1}{i} e_i$ converges unconditionally in ℓ_2, note that $\left\|\sum_G \frac{1}{i} e_i\right\|$ is small if $\min(G)$ is large enough.

1.44 Let $\sum x_i$ be an unconditionally convergent series in a Banach space X. Show that for every $\varepsilon > 0$ there is n_0 such that $\left\|\sum_{i=n}^{m} \varepsilon_i x_i\right\| < \varepsilon$ for every $\varepsilon_i = \pm 1$ and $m \ge n \ge n_0$; in particular, $\left\|\sum_{i=n_0}^{\infty} \varepsilon_i x_i\right\| < \varepsilon$.

Hint. See the proof that U-Cauchy implies BM-Cauchy in Exercise 1.38.

1.45 Let $\sum_{n=1}^{\infty} x_n$ be a series in a Banach space such that the set $S := \{\sum_{i=1}^{n} \varepsilon_i x_i : \varepsilon_i = \pm 1, n \in \mathbb{N}\}$ is bounded. Assume that (a_n) is a sequence of real numbers such that $\lim a_n = 0$. Show that $\sum_{n=1}^{\infty} a_n x_n$ is unconditionally convergent.

Hint. Put $M = \sup\{\|s\| : s \in S\}$. Fix $n_0 \in \mathbb{N}$. Let F be a finite subset of $[n_0, +\infty)$. Put $m = \max_{i\in F} i$. If $\sup_{i\ge n_0} |a_i| \ne 0$ we get, due to Exercise 1.37,

$$\left\|\sum_{i\in F} a_i x_i\right\| = \sup_{i\ge n_0} |a_i| . \left\|\sum_{i\in F} \frac{a_i}{\sup_{i\ge n_0} |a_i|} x_i\right\|$$

$$\le \sup_{i\ge n_0} |a_i| . \sup_{|b_i|\le 1} \left\|\sum_{i=n_0}^{m} b_i x_i\right\| = \sup_{i\ge n_0} |a_i| . \sup_{\varepsilon_i=\pm 1} \left\|\sum_{i=n_0}^{m} \varepsilon_i x_i\right\| \le 2M \sup_{i\ge n_0} |a_i|$$

(if $\sup_{i\in F} |a_i| \ne 0$ we have obviously the same inequality). It follows that the series $\sum a_n x_n$ is unconditionally Cauchy (see the definition in Exercise 1.38) hence, by Exercise 1.39, unconditionally convergent.

1.46 Assume that $\sum_{n=1}^{\infty}$ is unconditionally convergent in a Banach space X and that (a_n) is a bounded sequence of real numbers. Show that $\sum_{n=1}^{\infty} a_n x_n$ is unconditionally convergent.
Hint. Similar to the hint in Exercise 1.45.

1.47 Let X be a normed space, $M \subset X$ and $\varepsilon > 0$. We say that $A \subset M$ is an *ε-net* in M if for every $x \in M$ there is $y \in A$ such that $\|x - y\| < \varepsilon$. We say that $A \subset M$ is *ε-separated* in M if $\|x - y\| \geq \varepsilon$ for all $x \neq y \in A$. Clearly, a maximal ε-separated subset of M in the sense of inclusion is an ε-net in M.

Let X be an n-dimensional real normed space. Show that if $\{x_j\}_{j=1}^{N}$ is an ε-net in B_X, then $N \geq \varepsilon^{-n}$. On the other hand, there is an ε-net $\{x_j\}_{j=1}^{N}$ for B_X with $N \leq (\frac{2}{\varepsilon} + 1)^n$.
Hint. Fix a linear isomorphism T of X onto $\mathbb{R}^n$. For a compact subset C of X we define the volume of C by $\text{vol}(C) = \lambda(T(C))$, where λ is the Lebesgue measure on $\mathbb{R}^n$. Let $B(x, r)$ is the closed ball centered at x with radius r. We have $B_X \subset \bigcup_{j=1}^{N} B(x_j, \varepsilon)$, so

$$\text{vol}(B_X) \leq \text{vol}\Big(\bigcup_{j=1}^{N} B(x_j, \varepsilon)\Big) \leq \sum_{j=1}^{N} \text{vol}\big(B(x_j, \varepsilon)\big) = N\cdot\text{vol}(\varepsilon B_X) = N\cdot\varepsilon^n\cdot\text{vol}(B_X).$$

Thus $N \geq \varepsilon^{-n}$.

On the other hand, if $\{x_j\}_{j=1}^{N}$ is a maximal ε-separated set in B_X, then $\{x_j\}_{j=1}^{N}$ is an ε-net in B_X. Since $B(x_j, \varepsilon/2) \cap B(x_i, \varepsilon/2) = \emptyset$ if $i \neq j$ and $B(x_j, \varepsilon/2) \subset (1 + \varepsilon/2)B_X$, we have

$$N\cdot(\varepsilon/2)^n\cdot\text{vol}(B_X) = \text{vol}\Big(\bigcup_{j=1}^{N} B(x_j, \varepsilon/2)\Big) \leq \text{vol}\big((1+\varepsilon/2)B_X\big) = (1+\varepsilon/2)^n\,\text{vol}(B_X).$$

Thus $N \leq \big(\frac{1+\varepsilon/2}{\varepsilon/2}\big)^n$.

1.48 Prove that a subset M of a normed space X is totally bounded if and only if for every $\varepsilon > 0$ there exists a finite ε-net in M.
Hint. It follows from the definition.

1.49 Prove that the closure of a totally bounded set in a normed space is totally bounded.
Hint. It follows from Exercise 1.48.

1.50 Show that a bounded set M in c_0 is totally bounded if and only if for every $\varepsilon > 0$ there is n_0 such that $|x_n| \leq \varepsilon$ for every $x \in M$ and $n \geq n_0$. Formulate and prove the analogous result for ℓ_p spaces.
Hint. Every bounded subset of $\mathbb{R}^{n_0}$ is totally bounded.

1.51 The Hilbert cube Q is defined as $\{x = (x_i) \in \ell_2 : \forall i, |x_i| \leq 2^{-i}\}$. Show that the Hilbert cube is a compact set in ℓ_2.
Hint. Given $\varepsilon > 0$ there is n_0 such that $\sum_{i=n_0+1}^{\infty} |x_i| \leq \varepsilon$ for every $x \in Q$. Then use finite ε-nets in $\mathbb{R}^{n_0}$.

1.52 Prove the following version of Riesz's Lemma 1.37: If Y is a finite-dimensional proper subspace of a normed space X, there is a point $x \in S_X$ whose distance from Y is 1.
Hint. See the proof of Lemma 1.37: for $\hat{z} \in X/Y$, the coset $\hat{z} := \{z + Y\}$ is the translate of a finite-dimensional subspace, so there is $y \in Y$ such that $\|z+y\| = \|\hat{z}\|$.

1.53 Let $X_1 \subsetneq X_2 \subsetneq \ldots$ be finite-dimensional subspaces of a normed space X. Then there are unit vectors $x_1, x_2, \ldots$ such that $x_n \in X_n$ for all $n \in \mathbb{N}$ and $\operatorname{dist}(x_n, X_{n-1}) = 1$ for all $n \geq 2$.
Hint. Use Exercise 1.52.

1.54 This exercise slightly improves Exercise 1.53. Let $X_1 \subset X_2 \subset \ldots$ subspaces of a real normed space X with $\dim X_n = n$ for all $n \in \mathbb{N}$. Then there is a sequence $x_1, x_2, \ldots$ of unit vectors such that $\|x_i - x_j\| > 1$ if $i \neq j$ and $\operatorname{span}\{x_i, \ldots, x_n\} = X_n$ for all $n \in \mathbb{N}$.
Hint. Given linearly independent vectors $\{x_1, \ldots, x_{n-1}\}$ in a n-dimensional real normed space X, let $f \in S_{X^*}$ such that $f(x_i) = 0$ for $i = 1, 2, \ldots, n-1$. Let $g \in X^*$ be such that $g(x_i) = 1$ for $i = 1, 2, \ldots, n-1$. The set $K := \{x \in S_X : f(x) = 1\}$ is compact and nonempty. Choose $x_n \in K$ such that $g(x_n) = \min\{g(x) : x \in K\}$ (see Fig. 1.5). Use f to prove that $\|x_n - x_i\| \geq 1$ for all $i \leq n-1$. If $\|x_n - x_i\| = 1$, then $x_n - x_i \in K$ but $g(x_n - x_i) = g(x_n) - 1 < g(x_n)$, a contradiction.

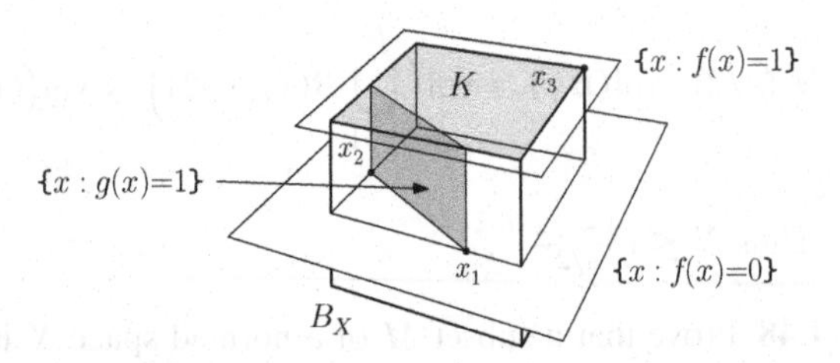

Fig. 1.5 The construction in Exercise 1.54

Let us remark that a result of Elton and Odell (see [Dies2, Chapter XIV]) says that for any infinite-dimensional Banach space X there is a constant $\varepsilon > 0$ and a sequence $\{x_n\}$ in S_X so that $\|x_n - x_m\| \geq 1 + \varepsilon$ whenever $n \neq m$.

1.55 Let C be a convex set in a normed space X and assume that $\operatorname{Int}(C) \neq \emptyset$ (recall that $\operatorname{Int}(C)$ denotes the *interior* of C). Prove that for every $x \in \overline{C}$ and $x_0 \in \operatorname{Int}(C)$, then $[x_0, x) \subset \operatorname{Int}(C)$.
Hint. Use the "cone" argument: there is an open ball B_δ^O such that $x_0 + B_\delta^O \subset C$; by the convexity of $\overline{C}$, the cone with vertex x and base $x_0 + B_\delta^O$ is a subset of $\overline{C}$ and all points in this cone but the vertex x are in $\operatorname{Int}(C)$ (draw a picture).

1.56 Let X be a normed space and C a nonempty convex subset of X. Prove that both $\text{Int}(C)$ and $\overline{C}$ are convex sets.
Hint. The first part follows from Exercise 1.55. For the second part use a simple continuity argument.

1.57 Let C be a convex set in a normed space X, assume $\text{Int}(C) \neq \emptyset$. Show that $\overline{\text{Int}(C)} = \overline{C}$ and $\text{Int}(\overline{C}) = \text{Int}(C)$.
Hint. The first part is a consequence of Exercise 1.55. To prove the second part, let x_0 be an element in $\text{Int}(C)$.

Given $c \in \text{Int}(\overline{C})$, there is $\varepsilon > 0$ such that $c + \varepsilon B_X \subset \overline{C}$. Let $d = c + \varepsilon(c - x_0)$. Then $d \in \overline{C}$, so for every $\nu > 0$ there is $y \in C$ with $\|d - y\| < \nu$. If ν is small (namely $\nu \leq \frac{\varepsilon\delta}{\sqrt{\delta^2 + \|x_0 - c\|^2}}$), c is in the cone with base $x_0 + B_\delta^O$ and vertex y, so $c \in \text{Int}(C)$. Draw a picture.

1.58 Let A be an open set in a normed space X. Show that $\text{conv}(A)$ is open.
Hint. Given $\{x_i\}_{i=1}^n \in A$ and $\{\lambda_i\}_{i=1}^n$ such that $\lambda_i \geq 0$ for every i and $\sum \lambda_i = 1$, note that if $\lambda_1 > 0$ and O_1 is an open set containing x_1 and contained in A, then $\text{conv}(A)$ contains the open set $\lambda_1 O_1 + \lambda_2 x_2 + \cdots + \lambda_n x_n$.

1.59 Is the convex hull of a closed set in $\mathbb{R}^2$ closed?
Hint. No in general, check $A := \{(x, \frac{1}{x}) : x > 0\} \cup \{(0, 0)\}$.

1.60 Let K, C be subsets of a normed space X.

(i) Show that if K, C are closed, $K + C$ need not be closed.

(ii) Show that if K is compact and C is closed, then $K + C$ is closed. Is $\text{conv}(K \cup C)$ closed?

(iii) Show that if K is compact and C is bounded and closed, then $\text{conv}(K \cup C)$ is closed.

Hint. (i) Consider $K := \{(x, 0) : x \in \mathbb{R}\}$ and $C := \{(x, \frac{1}{x}) : x > 0\}$.

(ii) If $x_n = k_n + c_n \to y$ for $k_n \in K$, $c_n \in C$, then by compactness assume $k_n \to k$, then also $c_n = x_n - k_n \to (y - k)$ and use that C is closed. For negative answer to $\text{conv}(K \cup C)$ see the previous exercise.

(iii) If $x_n = \lambda_n k_n + (1 - \lambda_n)c_n \to x$ for $k_n \in K$, $c_n \in C$, $\lambda \in [0, 1]$, find a subsequence n_i such that $k_{n_i} \to k$ and $\lambda_{n_i} \to \lambda$. If $\lambda = 1$, then by boundedness, $(1 - \lambda_{n_i})c_{n_i} \to 0$ and $x = k \in K$. If $\lambda \neq 1$, $c_{n_i} \to \frac{x - \lambda k}{1 - \lambda} \in C$ by closedness of C.

1.61 Let A, B be convex compact sets in a Banach space X. Show that $\text{conv}(A \cup B)$ and $A + B$ are compact. Generalize this statement to a finite number of sets.
Hint. Using Exercise 1.12, show that $\text{conv}(A \cup B)$ is a continuous image of the compact set $\{(\alpha, \beta) : \alpha, \beta \geq 0, \alpha + \beta = 1\} \times A \times B$, so it is compact. Similarly, $A + B$ is the image of $A \times B$ under the continuous mapping $(x, y) \mapsto x + y$.

1.62 Let A be a totally bounded set in a normed space X. Show that $\overline{\text{conv}}(A)$ is totally bounded. Likewise, prove that the closed convex and balanced hull of A is

totally bounded. In particular, if A is a compact subset of a Banach space, then the closed convex and balanced hull of A is compact.
Hint. Let B be a finite δ-net for A. It is straightforward to check that $\mathrm{conv}(B)$ is a δ-net for $\mathrm{conv}(A)$. Since $\mathrm{conv}(B)$ lies in a finite-dimensional space and is bounded, it is totally bounded, and its δ-net would produce a 2δ-net for $\mathrm{conv}(A)$. Note also that the closure of a totally bounded set is totally bounded (see Exercise 1.49). The argument for the convex and balanced hull of A is similar.

1.63 Show that the closed convex hull of a compact subset of an incomplete normed space need not to be compact.
Hint. Let $S = \{\frac{1}{n}e_n\} \cup \{0\} \subset c_{00}$. This set is compact in c_{00}. If $\overline{\mathrm{conv}}(S)$ in c_{00} were compact, it would be equal to $\overline{\mathrm{conv}}(S)$ in c_0 (compact spaces are closed in overspaces). However, the vector $(\frac{1}{2^n}\frac{1}{n})$ shows the contrary.

1.64 Let X be a Banach space and C be a compact set in X. Is it true that $\mathrm{conv}(C)$ is compact?
Hint. Not in general. Consider $C := \{\frac{1}{i}e_i\} \cup \{0\}$ in ℓ_2, where e_i are the standard unit vectors. Clearly C is compact. The vector $\left(2^{-i}\frac{1}{i}\right)$ is in $\overline{\mathrm{conv}(C)}$ and it is not in $\mathrm{conv}(C)$, since any point in $\mathrm{conv}(C)$ is finitely supported.

1.65 Let C be a compact set in a finite-dimensional Banach space X. Show that $\mathrm{conv}(C)$ is compact.
Hint. If a point x lies in $\mathrm{conv}(E) \subset \mathbb{R}^n$, then x lies in the convex hull of some subset of E that has at most $n+1$ points. Indeed, assume that $r > n$ and $x = \sum t_i x_i$ is a convex combination of some $r+1$ vectors $x_i \in E$. We will show that then x is actually a convex combination of some r of these vectors. Assume that $t_i > 0$ for $1 \leq i \leq r+1$. The r vectors $x_i - x_{r+1}$ for $1 \leq i \leq r$ are linearly dependent since $r > n$. Thus there are real numbers a_i not all zero such that $\sum_{i=1}^{r+1} a_i x_i = 0$ and $\sum_{i=1}^{r+1} a_i = 0$. Choose m so that $|a_i/t_i| \leq |a_m/t_m|$ for $1 \leq i \leq r+1$ and define $c_i = t_i - \frac{a_i t_m}{a_m}$ for $1 \leq i \leq r+1$. Then $c_i \geq 0$, $\sum c_i = \sum t_i = 1$, $x = \sum c_i x_i$ and $c_m = 0$. Having this, we can use the fact that in $\mathbb{R}^n$, $\mathrm{conv}(C)$ is the image of the compact set $S \times C^{n+1}$ in $\mathbb{R}^{n+1} \times C^{n+1}$, where S is formed by points $\{\lambda_i\}_1^{n+1}$ such that $\lambda_i \geq 0$ and $\sum \lambda_i = 1$, under the mapping $(\{\lambda\}, \{x_i\}) \mapsto \sum \lambda_i x_i$.

1.66 (See, e.g., [Jmsn, §22]) Let X be a normed space. Given a sequence $\{x_n\}$ in X, a *convex series* is a series (convergent or not) $\sum_{n=1}^{\infty} \lambda_n x_n$, where $\lambda_n \geq 0$ for all $n \in \mathbb{N}$ and $\sum_{n=1}^{\infty} \lambda_n = 1$. A subset A of X is called *CS-closed* (for *convex-series-closed*) if it contains the sum of every convergent convex series of its elements. A is called *CS-compact* if every convex series of its elements converges to an element in A. Prove the following statements. (i) Every closed convex set $A \subset X$ is CS-closed. (ii) Every open convex set $O \subset X$ is CS-closed. (iii) If $A \subset X$ is bounded, then every convex series of elements in A is Cauchy. (iv) Every CS-compact set is CS-closed and bounded; the converse is true in Banach spaces. (v) Suppose that A is CS-compact and B is CS-closed. Then $A + B$ and $\mathrm{conv}\,(A \cup B)$ are CS-closed. (vi)

The continuous linear image of a CS-compact set is CS-compact. (vii) The inverse image of a CS-closed set under a continuous linear mapping is CS-closed. (viii) If A is a CS-closed, then A and $\overline{A}$ have the same interior (compare with Exercise 1.57).
Hint. (i) We may assume that $0 \in A$. If $\lambda_k \geq 0$, $\sum \lambda_k = 1$, and $a_k \in A$ for all k, then $\sum_{k=1}^n \lambda_k a_k = \sum_{k=1}^n \lambda_k a_k + (1 - \sum_{k=1}^n \lambda_k)0 \in A$ for all n. (ii) If $x = \sum \lambda_k x_k$ is a convergent convex series of elements in O, put $x = \Lambda_n \sum_{k=1}^n \lambda_k \Lambda_n^{-1} x_k + (1 - \Lambda_n) \sum_{k=n+1}^{\infty} \lambda_k (1 - \Lambda_n)^{-1} x_k$, where $\Lambda_n := \sum_{k=1}^n \lambda_k$. Taking n such that $0 < \Lambda_n < 1$ we get $x \in O$ (use (i) and Exercise 1.55). (iii) should be clear. (iv) That a CS-compact set A is CS-closed is clear. If it is not bounded, choose $a_n \in A$ with $\|a_n\| > 2^n$ and consider the convex series $\sum 2^{-n} a_n$. The converse follows from (iii). (v) is simple; use Exercise 1.12. (vi) and (vii) are easy. (viii) is certainly the most interesting feature of CS-closed sets. It is enough to prove that $\mathrm{Int}(\overline{A}) \subset A$ and, by translating, that if $0 \in \mathrm{Int}(\overline{A})$ then $0 \in A$. Let $\delta > 0$ such that $\delta B_X \subset \overline{A}$. Find $a_1 \in A$, $\|a_1\| \leq \delta/2$, and put $s_1 = a_1/2$. Since $\|-4s_1\| \leq \delta$, find $a_2 \in A$ with $\|-4s_1 - a_2\| \leq \delta/2$. Put $s_2 = a_1/2 + a_2/4$. Then $\|-8s_2\| \leq \delta$, hence $a_3 \in A$ exists with $\|-8s_2 - a_3\| \leq \delta/2$. In this way, $0 = \sum 2^{-n} a_n$ $(\in A)$.

Observe that the argument in (viii) is in the core of the proof of the Banach Open Mapping Theorem 2.25.

1.67 Let A be a subset of a Banach space X. Denote by sconv (A)—the *superconvex hull of A*—the set of all $x \in X$ that can be written as $x = \sum_{i=1}^{\infty} \lambda_i x_i$, where $x_i \in A$, $\lambda_i \geq 0$ and $\sum \lambda_i = 1$. Show that (i) A is CS-closed (see Exercise 1.66) if and only if $A = \mathrm{sconv}(A)$. (ii) $\mathrm{sconv}(A) \subset \overline{\mathrm{conv}}(A)$. (iii) Let A be the set of all standard unit vectors e_i in ℓ_2; then $0 \in \overline{\mathrm{conv}}(A)$ and $0 \notin \mathrm{sconv}(A)$.
Hint. (i) is easy. (ii) Obviously, $A \subset C$, where $C := \overline{\mathrm{conv}}(A)$. Then $\mathrm{sconv}(A) \subset \mathrm{sconv}(C)$. Since C is closed and convex, it is CS-closed (see (i) in Exercise 1.66), hence $\mathrm{sconv}(C) = C$ by (i) here. Finally we get $\mathrm{sconv}(A) \subset C$. (iii) If $\sum \lambda_i e_i = 0$, then $\lambda_i = 0$ for every i.

1.68 Let $\{K_i\}$ be a finite family of convex compact sets in $\mathbb{R}^n$ such that every subfamily of them consisting of $n+1$ members has a nonempty intersection. Then $\bigcap K_i \neq \emptyset$. Prove this Helly's theorem for $n = 1$.

Show an example for $n = 2$ that $n + 1$ is necessary.
Hint. Consider the interval between the maximum of the left endpoints of K_i and the minimum of the right endpoints of K_i. For general case see, e.g., [DGK]. Example: three lines forming a triangle.

1.69 Let X be a Banach space. Show that if $A \subset X$ is totally bounded, then there is a sequence $\{x_n\} \in X$ such that $x_n \to 0$ in X and $A \subset \mathrm{sconv}\{x_n\}$ (see Exercise 1.67).

In particular, for every compact subset A of X there exists a sequence $\{x_n\}$ such that $x_n \to 0$ and $A \subset \overline{\mathrm{conv}}\{x_n\}$ (Grothendieck).
Hint. We set $A_1 = A$, let B_1 be a finite 2^{-2}-net in A_1. If A_i and B_i were defined for $i \leq n$, let $A_{n+1} = (A_n - B_n) \cap 2^{-2n} B_X$, note that every $a_n \in A_n$ is of the form $a_n = a_{n+1} + b_n$, where $a_{n+1} \in A_{n+1}$, $b_n \in B_n$. Let B_{n+1} be a finite 2^{-2n}-net in A_{n+1}. Therefore every element $a \in A$ $(= A_1)$ is of the form $a = b_1 + a_2 = b_1 + b_2 + a_3 =$

$\dots = \sum_1^n b_n + a_{n+1}$. Since $a_{n+1} \in 2^{-2n} B_X$, we have $a = \sum_{n=1}^{\infty} b_i = \sum 2^{-i}(2^i b_i)$. Then it suffices to take for $\{x_n\}$ the sequence that contains first all vectors from $2^1 B_1$, then from $2^2 B_2$ and so on. Since $2^i b_i \in 2^i B_X \subset 2^i A_i$, we have $\|2^i b_i\| \le 2^{2-i}$ and $\|x_n\| \to 0$.

Note that we in fact proved that if $x_n \to 0$, then the closed convex symmetric hull of $\{x_n\}$ is $\left\{\sum_{n=1}^{\infty} \lambda_n x_n : \sum_{n=1}^{\infty} |\lambda_n| \le 1\right\}$.

1.70 Let X, Y be non-zero normed spaces and $T \in \mathcal{B}(X, Y)$. Show that

$$\|T\| = \sup\{\|T(x)\|_Y : \|x\|_X < 1\} = \sup\{\|T(x)\|_Y : \|x\|_X = 1\}.$$

Hint. Clearly, both suprema are not greater than $\|T\|$. Given $\varepsilon > 0$, find $x \in B_X$ such that $\|T(x)\|_Y \ge \sqrt{1-\varepsilon}\,\|T\|$. Then $\|\sqrt{1-\varepsilon}\,x\| < 1$ and $\frac{x}{\|x\|_X} \in S_X$, both vectors give $\|T(y)\|_Y \ge (1-\varepsilon)\|T\|$.

1.71 Assume that T is an operator from a normed space X into a normed space Y such that $\{T(x_n)\}$ is bounded for every sequence $\{x_n\} \subset X$ satisfying $\|x_n\| \to 0$. Is T necessarily continuous?
Hint. Yes. Assuming the contrary, consider a sequence $\{x_n/\sqrt{\|x_n\|}\}$ for $\{x_n\}$ such that $x_n \to 0$ and $T(x_n) \not\to 0$.

1.72 Let T be a one-to-one bounded operator from a normed space X into a normed space Y. Show that T is an isometry onto Y if and only if $T(B_X) = B_Y$ if and only if $T(S_X) = S_Y$ if and only if $T(B_X^O) = B_Y^O$, where B_X^O is the open unit ball in X.
Hint. By homogeneity, T is an isometry onto Y if and only if $T(S_X) = S_Y$. Assume that $T(B_X) = B_Y$. If there is $x \in S_X$ such that $\|T(x)\| = C < 1$, then $\|x/C\| > 1$ and $\|T(x/C)\| = 1$. But there must be $y \in B_X$ such that $T(y) = T(x/C)$, a contradiction with T being one-to-one. A slight modification of this argument gives the last equivalence.

1.73 Let X, Y be Banach spaces and $T \in \mathcal{B}(X, Y)$. If there is $\delta > 0$ such that $\|T(x)\| \ge \delta\|x\|$ for all $x \in X$, then $T(X)$ is closed in Y. Moreover, T is an isomorphism from X into Y.
Hint. The inequality we assume implies that T is one-to-one, so the inverse $T^{-1}\colon T(X) \to X$ is well defined and $\|T^{-1}\| \le \frac{1}{\delta}$. The mapping $T : X \to TX$ is an isomorphism, so TX is a Banach subspace of Y, hence closed by Fact 1.5.

1.74 Let T be an operator from a normed space X onto another normed space Y. Prove that the following are equivalent: (i) There exist two positive constants C_1, C_2 such that, for every $x \in X$, $C_1\|x\| \le \|Tx\| \le C_2\|x\|$. (ii) $C_1 B_Y \subset T(B_X) \subset C_2 B_Y$. (iii) T is an isomorphism and $\|T\| \le C_2$, $\|T^{-1}\| \le 1/C_1$.
Hint. Direct computation.

1.75 Let X be a normed space isomorphic to another normed space Y and let $\mathrm{d}(\cdot, \cdot)$ be the Banach–Mazur distance. Let C be a positive constant. Prove that the following are equivalent: (i) $\mathrm{d}(X, Y) < C$. (ii) There exist positive constants C_1 and C_2 and an linear isomorphism from X onto Y such that $(1/C_1)B_Y \subset T(B_X) \subset C_2 B_Y$, and $C_1 C_2 < C$. (iii) There exists a linear isomorphism T from X onto Y, and a positive constant $\widetilde{C} < C$ such that $B_Y \subset T(B_X) \subset \widetilde{C} B_Y$.
Hint. Use Exercise 1.74 and the definition of the Banach–Mazur distance.

1.76 Let X be a finite-dimensional normed space and let Y be a normed space. Let T be a (bounded) operator from X onto Y. Assume that there is $\delta \in (0, \frac{1}{4})$ and a finite δ-net $M := \{x_i\}$ in S_X (see Exercise 1.47) such that $(1+\delta)^{-1} \leq \|T(x_i)\| \leq (1+\delta)$ for every i. Then T^{-1} exists and $\|T\|\|T^{-1}\| \leq \theta(\delta) = (\frac{1+\delta}{1-\delta})\big(\frac{1}{1+\delta} - \frac{\delta(1+\delta)}{1-\delta}\big)^{-1}$.
Hint. Given $x \in S_X$, there is $x_i \in M$ with $\|x - x_i\|_X < \delta$. Then $\|Tx - Tx_i\|_Y \leq \|T\|.\|x - x_i\|_X < \|T\|\delta$, hence

$$\|Tx\|_Y \leq \|Tx_i\|_Y + \|Tx - Tx_i\|_Y \leq (1+\delta) + \|T\|\delta.$$

By Exercise 1.70, $\|T\| \leq (1+\delta)(1-\delta)^{-1}$.
On the other hand,

$$\begin{aligned}\|Tx\|_Y &= \|Tx - Tx_i + Tx_i\|_Y \geq \|Tx_i\|_Y - \|Tx - Tx_i\|_Y \\ &\geq (1+\delta)^{-1} - \|T\|\delta \geq \frac{1}{1+\delta} - \frac{1+\delta}{1-\delta}\delta \ (> 0).\end{aligned}$$

In particular, T is one-to-one (and onto), so an isomorphism. It follows that $\|T^{-1}\| \leq \big(\frac{1}{1+\delta} - \frac{\delta(1+\delta)}{1-\delta}\big)^{-1}$. Both estimates give the conclusion.

1.77 Let X, Y be Banach spaces and $T \in \mathcal{K}(X, Y)$. Prove that if $\{x_n\}_{n=1}^{\infty}$ is a sequence in X, $x \in X$, and $x_n \xrightarrow{w} x$ in X, then $T(x_n) \xrightarrow{\|\cdot\|} T(x)$ in Y.
An operator satisfying the conclusion of the statement is called a *completely continuous operator*. Thus every compact operator is completely continuous. The example of the identity operator on ℓ_1 shows that the converse implication is not valid in general.
Hint. If $x_n \xrightarrow{w} x$, then $\{x_n\}$ is weakly (and hence norm-) bounded, and we may assume that $x, x_n \in B_X$. We have $T(x_n) \xrightarrow{w} T(x)$ by the w-w-continuity of T. Since $\overline{T(B_X)}$ is a compact space in the norm topology and the w-topology is weaker than the norm topology and Hausdorff, these two topologies coincide on $\overline{T(B_X)}$. Consequently, $T(x_n) \xrightarrow{\|\cdot\|} T(x)$.

1.78 Let X be a Banach space and G be a subspace of X that is a G_δ set in X. Show that G is closed in X.
Hint. Let $S = \overline{G} \setminus G$, where $\overline{G}$ denotes the closure of G. Since G is a G_δ set in X, G is G_δ in $\overline{G}$. Hence $G = \bigcap G_n$, where G_n are open subsets in $\overline{G}$. Therefore

$S = \bigcup(\overline{G}\setminus G_n)$ and each G_n is dense in $\overline{G}$ as it contains G. Thus each $\overline{G}\setminus G_n$ is nowhere dense in $\overline{G}$ and so S is of first category in $\overline{G}$. We will show that S is an empty set. If $x_0 \in S$, consider the set $G^* := \{x_0 + x : x \in G\}$. Note that $G^* \subset S$. Indeed, any point $x_0 + x$, $x \in G$, is in $\overline{G}$ since $\overline{G}$ is a linear set. If for some $x \in G$ we have $x_0 + x \in G$ then by linearity of G we have $x_0 \in G$, a contradiction. Therefore $G^* \subset S$ and G^* is of first category in $\overline{G}$. Thus the shift G of G^* is also of first category in $\overline{G}$. Hence the whole $\overline{G}$ as a union of G and S, which are both of first category in $\overline{G}$, is of first category in itself. This is a contradiction, since $\overline{G}$ is a complete metric space.

1.79 Let X be a normed linear space X. Show that if X is *topologically complete* in its norm topology (that is, X is homeomorphic to a complete metric space), then X is a Banach space.
Hint. If X is topologically complete, then by the Alexandrov–Hausdorff–Lavrentieff–Mazurkiewicz theorem, see, e.g., [Enge, Theorems 4.3.23 and 4.3.24], X is G_δ in the completion of X, which is a Banach space. Note that this means that a non-complete normed space cannot be homeomorphic to a Banach space.

1.80 Let X be an infinite-dimensional Banach space. Show that there is no translation invariant Borel measure μ on X such that $\mu(U) > 0$ for every open set U and such that $\mu(U_1) < \infty$ for some open U_1.
Hint. Every open ball contains an infinite number of disjoint open balls of equal radii (see Lemma 1.37).

1.81 Let X be an infinite-dimensional Banach space. Show that X admits no countable Hamel basis.

Therefore c_{00} cannot be normed to become a Banach space.
Hint. If $\{e_i\}$ is a countable infinite Hamel basis of a Banach space X, put $F_n = \operatorname{span}\{e_1, \dots, e_n\}$. F_n are closed and thus by the Baire category theorem, at least one F_{n_0} has a nonempty interior, that is, there is $x \in X$ and a ball $B = \delta B_X$ such that $x + B \subset F_{n_0}$. Using linearity of F_{n_0} we have that $-x + B \subset F_{n_0}$, so $B \subset (x + B) + (-x + B) \subset F_{n_0}$. Thus 0 is an interior point of F_{n_0}. This would mean that $F_{n_0} = X$, a contradiction.

1.82 Let X be an infinite-dimensional separable Banach space and $\{e_\gamma\}$ be a Hamel basis for X. Define a norm $\|\cdot\|$ on X by $|||x||| = \sum |x_\gamma|$ for $x = \sum x_\gamma e_\gamma$. Show that $|||\cdot|||$ is indeed a norm on X. Prove that it is not equivalent to the original norm of X.
Hint. Consider the question of separability of X in $|||\cdot|||$ using the previous exercise.

1.83 Prove that on every infinite-dimensional Banach space $(X, \|\cdot\|)$ there is a norm $|||\cdot|||$ that is not equivalent to $\|\cdot\|$.
Hint. If X is separable, use Exercise 1.82. In general, choose an infinite-dimensional separable subspace Y of X and define a non-equivalent norm $|||\cdot|||$ on Y using the

first part. Then, if Z is an algebraic complement of Y in X, put $|||x||| = |||y||| + \|z\|$, whenever $x = y + z$, $y \in Y$, $z \in Z$. This defines a norm on X that cannot be equivalent to $\|\cdot\|$ (otherwise the restriction $|||\cdot|||$ to Y will be also an equivalent norm).

1.84 Show that the norm $|||x||| := \sum_{i=1}^{\infty} 2^{-i}|x_i|$ in ℓ_2 is not equivalent to the $\|\cdot\|_2$-norm.
Hint. Check this norm on $\{e_i\}_{i=1}^{\infty}$.

1.85 Show that the linear dimension (the cardinality of a Hamel basis) of the space ℓ_p, $p \in [1, \infty)$, is the continuum c.
Hint. First note that $\operatorname{card}(\ell_p) = c$. Therefore the linear dimension of ℓ_p is less than or equal to c. On the other hand, note that if $\lambda < 1$, then the vector $(\lambda, \lambda^2, \dots) \in \ell_p$, and these vectors form a linearly independent set.

1.86 Find a vector space X with two norms on it such that both of them are complete norms and they are not equivalent.
Hint. Take a vector space V of linear dimension c and let T_1 and T_2 be linear bijections of V onto ℓ_2 and ℓ_4, respectively. Define norms on V by $\|x\|_1 = \|T_1(x)\|_2$ and $\|x\|_2 = \|T_2(x)\|_4$. Then $(V, \|\cdot\|_1)$ is isomorphic to ℓ_2 and $(V, \|\cdot\|_2)$ is isomorphic to ℓ_4. Since ℓ_2 is not isomorphic to ℓ_4 (see Exercise 1.99), $\|\cdot\|_1$ and $\|\cdot\|_2$ are not equivalent.

1.87 Let X and Y be Banach spaces and T be a bounded operator from X into Y. Show that $|||x||| := (\|x\|^2 + \|Tx\|^2)^{\frac{1}{2}}$ is an equivalent norm on X.
Hint. Hölder inequality.

1.88 Show that the following function f defined on ℓ_1 by $f(x) = \big((\sum |x_i|)^2 + \sum_i \frac{1}{2^i}|x_i|^2\big)^{\frac{1}{2}}$ is an equivalent norm on ℓ_1.
Hint. Hölder inequality.

1.89 If $0 < p < 1$, show that the function $f(x, y) := (|x|^p + |y|^p)^{\frac{1}{p}}$ is not a norm on $\mathbb{R}^2$.
Hint. Look at Fig. 1.6, where the "unit ball" for $p = 1/2$ is shown.

1.90 A family $\{x_\alpha\}_{\alpha \in \Gamma}$ in a Banach space X is said to be ω_0-*independent* if, for every sequence $(\alpha_n)_{n=1}^{\infty}$ in Γ of distinct indices and every sequence of real numbers $(\lambda_n)_{n=1}^{\infty}$, the series $\sum_{n=1}^{\infty} \lambda_n x_{\alpha_n}$ converges to zero in X if and only if all λ_n are zero. Observe that, since ω_0-independence clearly implies linear independence, no finite-dimensional space can contain an infinite ω_0-independent family. The following result is due to Kalton [Kalt3] (see also [HMVZ, Theorem 1.58]): *Let X be a Banach space, and let G be a subset of X. Let H be the set of accumulation points of G, and suppose that X is the closed linear span of H. Then, given any $x \in X$ and any*

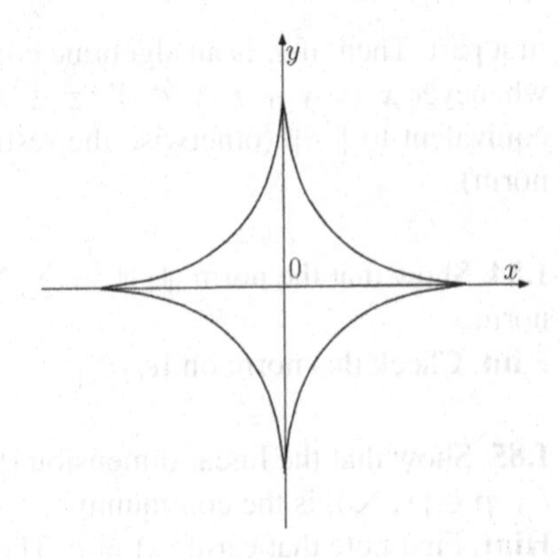

Fig. 1.6 The ball for $p = 1/2$ in Exercise 1.89

sequence of numbers (a_n) *with* $\sum |a_n| = \infty$ *and* $\lim a_n = 0$, *there is a sequence of signs* ϵ_n *and distinct elements* $g_n \in G$ *so that* $x = \sum_{n=1}^{\infty} \epsilon_n a_n g_n$.

Use this theorem to prove the following result of Fremlin and Sersouri [FrSe], that answers a question of Z. Lipecki: *Suppose X is a separable Banach space. Then every* ω_0*-independent family in X is countable.*

Hint. By contradiction, let G be an uncountable ω_0-independent family in X. Because G is a subset of the separable (metric) space X, it contains an uncountable set H that is dense in itself. Let Y be the closed linear span of H. According to Kalton's theorem quoted above, given any sequence (a_n) with $\sum_n |a_n| = \infty$ and $\lim a_n = 0$, we can find signs ϵ_n and distinct $g_n \in G$ so that $0 = \sum_{n=1}^{\infty} \epsilon_n a_n g_n$, a contradiction.

1.91 Let H be a Hilbert space. Prove the *generalized parallelogram equality*, i.e., if $x_1, \ldots, x_n \in H$, then

$$\sum_{\varepsilon_i=\pm 1} \left\| \sum_{i=1}^{n} \varepsilon_i x_i \right\|^2 = 2^n \sum_{i=1}^{n} \|x_i\|^2.$$

Hint. Induction on n.

1.92 Let X be a Banach space whose norm $\|\cdot\|$ satisfies the parallelogram equality (1.8). Define (x, y) by the polarization identity (1.9), (1.10) and prove that (x, y) is an inner product.

Hint. Clearly $(\cdot,\cdot)$ is continuous in both coordinates, $(x, y) = \overline{(y, x)}$ and $(-x, y) = -(x, y)$. Using the parallelogram equality show that $(x + y, z) = (x, z) + (y, z)$. Then by induction $(nx, y) = n(x, y)$ for all $n \in \mathbb{N}$, hence also for all integers n.

Given $\frac{n}{m}$, write $(\frac{n}{m}x, y) = n(\frac{1}{m}x, y) = \frac{n}{m}m(\frac{1}{m}x, y) = \frac{n}{m}(\frac{m}{m}x, y) = \frac{n}{m}(x, y)$. By continuity we get $(\alpha x, y) = \alpha(x, y)$ for all $\alpha \in \mathbb{R}$.

1.93 Show that ℓ_4^n is not a Hilbert space.

Hint. Parallelogram equality.

1.94 Find a Hilbert space H and a linear (not closed) subspace F of it such that $H \neq F + F^{\perp}$. This shows that the assumption of closedness in Theorem 1.49 is crucial.
Hint. Consider the subspace F of finitely supported vectors in ℓ_2. Then $F^{\perp} = \{0\}$ as for every $x \in H \setminus \{0\}$, $(x, e_i) \neq 0$ for $i \in \operatorname{supp}(x)$.

1.95 Let H be a Hilbert space. Show that there exists an abstract set Γ such that H is isometric to $\ell_2(\Gamma)$. The density character of a normed space will be introduced in Definition 13.2. Prove that the cardinality of Γ is the same as the density character of H.
Hint. Take an orthonormal basis $\{e_\gamma\}_{\gamma \in \Gamma}$ and follow the proof of Theorem 1.57. For the second part, observe that $\|e_\gamma - e_\delta\| = \sqrt{2}$, so $B(e_\gamma, \sqrt{2}/2) \cap B(e_\delta, \sqrt{2}/2) = \emptyset$ for all $\gamma \neq \delta$. Let $D \subset H$ be a dense subset of H such that $\operatorname{card}(D)$, the cardinality of D, is the density of H. The mapping that to each $\gamma \in \Gamma$ associates a single element in $D \cap B(e_\gamma, \sqrt{2}/2)$ is one-to-one, so $\operatorname{card}(\Gamma) \leq \operatorname{card}(D)$. On the other hand, the set of all finite rational combinations of elements in $\{e_\gamma : \gamma \in \Gamma\}$ is dense in H. This gives the other inequality.

1.96 (Orlicz) Show that an unconditionally convergent series $\sum x_i$ in a Hilbert space satisfies $\sum \|x_i\|^2 < \infty$.
Hint. Use Exercises 1.44 and 1.91.

1.97 Suppose $\{x^k\}_{k=1}^{\infty}$ is an orthonormal sequence in ℓ_2, where $x^k := (x_i^k)$. Show that $\lim\limits_{k \to \infty} x_i^k = 0$ for every $i \in \mathbb{N}$.
Hint. Use the Bessel inequality (1.12) to show that $(e_i, x^k) \to 0$ as $k \to \infty$.

1.98 A Banach space X is said to have *cotype 2* if there exists a constant $C > 0$ such that for all vectors $x_1, \ldots, x_n \in X$ we have

$$\frac{1}{2^n} \sum_{\varepsilon_i = \pm 1} \Big\| \sum_{i=1}^{n} \varepsilon_i x_i \Big\| \geq \frac{1}{C} \Big(\sum_{i=1}^{n} \|x_i\|^2 \Big)^{1/2}.$$

We say that X has *type 2* if there exists a constant $C > 0$ such that for all vectors $x_1, \ldots, x_n \in X$ we have

$$\frac{1}{2^n} \sum_{\varepsilon_i = \pm 1} \Big\| \sum_{i=1}^{n} \varepsilon_i x_i \Big\| \leq C \Big(\sum_{i=1}^{n} \|x_i\|^2 \Big)^{1/2}.$$

Similarly we define type/cotype q.

Taking for granted the Kahane–Khintchine inequality (see, e.g., [AlKa, p. 134]) that the average on the left hand side can be replaced by the expression

$\left(\frac{1}{2^n}\sum_{\varepsilon_i=\pm1}\left\|\sum_{i=1}^n \varepsilon_i x_i\right\|^2\right)^{1/2}$, show that if X is a Banach space isomorphic to a Hilbert space, then X is both of type 2 and cotype 2.

A theorem of Kwapień says that every space that is of type 2 and cotype 2 is isomorphic to a Hilbert space ([Kwap], see, e.g., [AlKa, p. 187]).

Hint. Square both sides of the inequality and use Exercise 1.91.

1.99 Show that ℓ_4 is not isomorphic to a subspace of ℓ_2.

Hint. Show that ℓ_4 is not of cotype 2 by considering the standard unit vectors.

1.100 An *Orlicz function* M is a continuous nondecreasing convex function defined for $t \geq 0$ such that $M(0) = 0$ and $\lim_{t\to\infty} M(t) = \infty$. We assume that $M(t) > 0$ for all $t > 0$ and that M satisfies the *Δ_2-condition at zero*, i.e., that $\limsup_{t\to 0} M(2t)/M(t) < \infty$.

Let h_M be the space of all sequences of real numbers $x = (a_1, a_2, \dots)$ such that $\sum_{n=1}^{\infty} M(|a_n|/\rho) < \infty$ for all $\rho > 0$. (Due to the Δ_2 condition for M this is the same as to require that $\sum_{n=1}^{\infty} M(|a_n|/\rho) < \infty$ for some $\rho > 0$.)

Define a norm $\|\cdot\|$ on h_M for $x = (a_1, a_2, \dots)$ by

$$\|x\| = \inf\{\rho > 0 : \sum_{n=1}^{\infty} M(|a_n|/\rho) \leq 1\}.$$

Show that h_M equipped with the norm $\|\cdot\|$ is a Banach space. It is called an *Orlicz sequence space.*

Hint. Fatou lemma.

1.101 For every $1 \leq p < \infty$ and every nonincreasing sequence of positive numbers $w = (w_n)_{n=1}^{\infty}$ we consider the space $d(w, p)$ of all sequences of real numbers $x = (a_n)_{n=1}^{\infty}$ for which

$$\|x\| := \sup_{\pi}\left(\sum_{n=1}^{\infty} |a_{\pi(n)}|^p w_n\right)^{1/p} < \infty.$$

where π ranges over all permutations of $\mathbb{N}$. Show that $d(w, p)$ with the norm $\|\cdot\|$ is a Banach space. It is called a *Lorentz sequence space.*

1. Show that $d(w, p)$ is isomorphic to ℓ_p if $\inf_n w_n > 0$.
2. Show that $d(w, p)$ is isomorphic to ℓ_∞ if $\sum_{n=1}^{\infty} w_n < \infty$.

Therefore it is usually assumed that $w_1 = 1$, that $\lim_{n\to\infty} w_n = 0$, and that $\sum_{n=1}^{\infty} w_n = \infty$.

Hint. Fatou lemma.

Chapter 2
Hahn–Banach and Banach Open Mapping Theorems

The Hahn-Banach theorem, in the geometrical form, states that a closed and convex set can be separated from any external point by means of a hyperplane. This intuitively appealing principle underlines the role of convexity in the theory. It is the first, and most important, of the fundamental principles of functional analysis. The rich duality theory of Banach spaces is one of its direct consequences. The second fundamental principle, the Banach open mapping theorem, is studied in the rest of the chapter.

A real-valued function p on a vector space X is called a *subadditive* if $p(x+y) \leq p(x) + p(y)$ for all $x, y \in X$. It is called *positively homogeneous* if for all $x \in X$ and $\alpha \geq 0$ it satisfies $p(\alpha x) = \alpha p(x)$. If p is subadditive and, moreover, $p(\alpha x) = |\alpha| p(x)$ for all $x \in X$ and all scalars α, then p is called a *seminorm* on X. Note that every norm is a seminorm. Note, too, that every positively homogeneous subadditive function is a convex function.

By a *linear functional* on a vector space X, we mean a linear mapping from X into $\mathbb{K}$.

Theorem 2.1 (Hahn, Banach) *Let Y be subspace of a real linear space X, and let p be a positively homogeneous subadditive functional on X. If f is a linear functional on Y such that $f(x) \leq p(x)$ for every $x \in Y$, then there is a linear functional F on X such that $F = f$ on Y and $F(x) \leq p(x)$ for every $x \in X$.*

Proof: Let $\mathcal{P}$ be the collection of all ordered pairs (M', f'), where M' is a subspace of X containing Y and f' is a linear functional on M' that coincides with f on Y and satisfies $f' \leq p$ on M'. $\mathcal{P}$ is nonempty as it contains the pair (Y, f). We partially order $\mathcal{P}$ by $(M', f') \prec (M'', f'')$ if $M' \subset M''$ and $f''\big|_{M'} = f'$. If $\{M_\alpha, f_\alpha\}$ is a chain, then $M' := \bigcup M_\alpha$ and a linear functional f' on M' defined by $f'(x) = f_\alpha(x)$ for $x \in M_\alpha$ satisfy $(M_\alpha, f_\alpha) \prec (M', f')$ for all α. By Zorn's lemma, $\mathcal{P}$ has a maximal element (M, F). We need to show that $M = X$.

Assume $M \neq X$, pick $x_1 \in X \backslash M$ and put $M_1 = \operatorname{span}\{M, x_1\}$. We will find $(M_1, F_1) \in \mathcal{P}$ such that $(M, F) \prec (M_1, F_1)$, a contradiction. For a fixed $\alpha \in \mathbb{R}$ we define $F_1(x + t x_1) = F(x) + t\alpha$ for $x \in M, t \in \mathbb{R}$. Then F is linear. It remains to show that we can choose α so that $F_1 \leq p$.

M. Fabian et al., *Banach Space Theory*, CMS Books in Mathematics,
DOI 10.1007/978-1-4419-7515-7_2, © Springer Science+Business Media, LLC 2011

Due to the positive homogeneity of p and F, it is enough to choose α such that

$$\begin{array}{l} F_1(x+x_1) \leq p(x+x_1) \\ F_1(x-x_1) \leq p(x-x_1) \end{array} \text{ for every } x \in M. \tag{2.1}$$

Indeed, for $t > 0$ we then have

$$F_1(x+tx_1) = tF_1\left(\tfrac{x}{t}+x_1\right) \leq tp\left(\tfrac{x}{t}+x_1\right) = p(x+tx_1)$$

and for $t = -\eta < 0$ we have

$$\begin{aligned} F_1(x+tx_1) &= F_1(x-\eta x_1) = \eta F_1\left(\tfrac{x}{\eta}-x_1\right) \\ &\leq \eta p\left(\tfrac{x}{\eta}-x_1\right) = p(x-\eta x_1) = p(x+tx_1). \end{aligned}$$

But (2.1) is equivalent to ($\alpha :=$) $F_1(x_1) \leq p(x+x_1) - F(x)$ and ($-\alpha =$) $-F_1(x_1) \leq p(x-x_1) - F(x)$ for every $x \in M$. This in turn is equivalent to

$$F(y) - p(y-x_1) \leq \alpha \leq p(x+x_1) - F(x)$$

for every $x, y \in M$. Thus to find a suitable $\alpha \in \mathbb{R}$ we need to show that $\sup\{F(y) - p(y-x_1) :\ y \in M\} \leq \inf\{p(x+x_1) - F(x) :\ x \in M\}$. This is in turn equivalent to the statement that for every $x, y \in M$ we have

$$F(y) - p(y-x_1) \leq p(x+x_1) - F(x).$$

The latter reads $F(x+y) \leq p(x+x_1) + p(y-x_1)$, which is true as

$$F(x+y) \leq p(x+y) = p(x+x_1+y-x_1) \leq p(x+x_1) + p(y-x_1).$$

This completes the proof of Theorem 2.1. □

2.1 Hahn–Banach Extension and Separation Theorems

Before we pass to normed space versions of the Hahn–Banach theorem, we need to establish the relationship between the real and the complex normed spaces.

Let X be a complex normed space. The space X is also a real normed space. We will denote this real version of X by $X_{\mathbb{R}}$.

On the other hand, if X is a real normed space, then $X \times X$ becomes a complex normed space $X_{\mathbb{C}}$ when its linear structure and norm are defined for $x, y, u, v \in X$ and $a, b \in \mathbb{R}$ by

$$\begin{aligned} (x, y) + (u, v) &:= (x+u, y+v) \\ (a+ib)(x, y) &:= (ax-by, bx+ay) \\ \|(x, y)\|_{\mathbb{C}} &:= \sup\{\|\cos(\theta)x + \sin(\theta)y\| :\ 0 \leq \theta \leq 2\pi\}. \end{aligned}$$

The set $X \times \{0\} := \{(x, 0) : x \in X\}$ is a closed $\mathbb{R}$-linear subspace of $X_{\mathbb{C}}$ which is—as a real space—isometric to X under the mapping $(x, 0) \mapsto x$. Conversely, $X_{\mathbb{C}} = \{h + ik : h, k \in X \times \{0\}\}$.

We will verify that $\|\cdot\|_{\mathbb{C}}$ is actually a norm on $X_{\mathbb{C}}$. It is clear that $\|\cdot\|_{\mathbb{C}}$ is non-negative, satisfies the triangle inequality, and factors real constants to their absolute value. If α is real and $z := (x, y) \in X_{\mathbb{C}}$, then

$$
\begin{aligned}
\|e^{-i\alpha}z\|_{\mathbb{C}} &= \|\big(\cos(\alpha)x + \sin(\alpha)y, -\sin(\alpha)x + \cos(\alpha)y\big)\|_{\mathbb{C}} \\
&= \sup\{\|\cos(\theta)[\cos(\alpha)x + \sin(\alpha)y] + \sin(\theta)[-\sin(\alpha)x + \cos(\alpha)y]\| : 0 \le \theta \le 2\pi\} \\
&= \sup\{\|\cos(\theta + \alpha)x + \sin(\theta + \alpha)y\| : 0 \le \theta \le 2\pi\} \\
&= \sup\{\|\cos(\eta)x + \sin(\eta)y\| : 0 \le \eta \le 2\pi\} = \|z\|_{\mathbb{C}}.
\end{aligned}
$$

Therefore $\|\cdot\|_{\mathbb{C}}$ is a norm on $X_{\mathbb{C}}$. Since $\max\{\|x\|, \|y\|\} \le \|(x, y)\|_{\mathbb{C}} \le \|x\| + \|y\|$, we have that the topology induced on $X_{\mathbb{C}} = X \times X$ by $\|\cdot\|_{\mathbb{C}}$ is equivalent to the product topology induced on $X \times X$ by $\|\cdot\|$.

We will now relate duals of X and $X_{\mathbb{R}}$. Consider the mapping $R\colon X^* \to X_{\mathbb{R}}{}^*$ defined by $R(f)(x) = \mathrm{Re}\big(f(x)\big)$ for $x \in X$, where $\mathrm{Re}(f(x))$ is the real part of $f(x)$. We claim that it is a norm-preserving mapping from X^* onto $X_{\mathbb{R}}{}^*$ and is linear as a mapping $(X^*)_{\mathbb{R}} \to X_{\mathbb{R}}{}^*$.

To see this claim, note that if X is a complex Banach space and $f \in X^*$, then $\sup_{z \in B_X} |f(z)| = \sup_{z \in B_X} |\mathrm{Re}\big(f(z)\big)|$. Indeed, for all z we have $|f(z)| \ge |\mathrm{Re}\big(f(z)\big)|$, so one inequality is clear. On the other hand, for $z \in B_X$ we write $f(z) = e^{i\alpha}|f(z)|$ and have $f(e^{-i\alpha}z) = e^{-i\alpha}f(z) = |f(z)|$. Thus $|\mathrm{Re}\big(f(e^{-i\alpha}z)\big)| = |f(z)|$ and $\|e^{-i\alpha}z\| = \|z\|$.

Now we show that R is onto $X_{\mathbb{R}}{}^*$. To $g \in X_{\mathbb{R}}{}^*$ we assign the functional defined on X by $G(x) = g(x) - ig(ix)$. Then G is linear over $\mathbb{R}$, but also

$$
G(ix) = g(ix) - ig(-x) = g(ix) + ig(x) = i\big(g(x) - ig(ix)\big) = iG(x).
$$

Therefore G is linear over $\mathbb{C}$ and hence $G \in X^*$. Moreover, $R(G) = g$.

Theorem 2.2 (Hahn, Banach) *Let Y be a subspace of a normed space X. If $f \in Y^*$ then there exists $F \in X^*$ such that $F\big|_Y = f$ and $\|F\|_{X^*} = \|f\|_{Y^*}$.*

Proof: First assume that X is a real normed space. Define a new norm $|||\cdot|||$ on X by $|||x||| = \|f\|_{Y^*}\|x\|$, where $\|\cdot\|$ is the original norm of X. We have $|f(y)| \le |||y|||$ for all $y \in Y$, so by Theorem 2.1 there is a linear functional F on X that extends f and $|F(x)| \le |||x|||$ $(= \|f\|_{Y^*}\|x\|)$ for every $x \in X$. Therefore $\|F\|_{X^*} := \sup\{|F(x)| : \|x\| \le 1\} \le \|f\|_{Y^*}$. Since F extends f, we obviously have $\|F\|_{X^*} \ge \|f\|_{Y^*}$ as well. Consequently $\|F\|_{X^*} = \|f\|_{Y^*}$.

Now assume that X is a complex normed space. Consider the linear functional $R(f)$ on $Y_{\mathbb{R}}$, where R is the isometry defined above. By the first part of this proof, we extend $R(f)$ to a linear functional $g \in X_{\mathbb{R}}{}^*$ that satisfies $\|g\|_{X_{\mathbb{R}}{}^*} = \|R(f)\|_{Y_{\mathbb{R}}{}^*} = \|f\|_{Y^*}$. Then the norm of the linear functional $F(x) := g(x) - ig(ix) \in X^*$ is equal to $\|g\|_{X_{\mathbb{R}}{}^*}$ $(= \|f\|_{Y^*})$.

The real part of F is g and thus $\mathrm{Re}(F\big|_Y) = \mathrm{Re}(f)$, that is, $R(F\big|_Y) = R(f)$. Since R is a bijection of Y^* onto $Y_{\mathbb{R}}{}^*$, we get $F\big|_Y = f$. □

Corollary 2.3 (Hahn, Banach) *Let X be a normed space. For every $x \in X$ there is $f \in S_{X^*}$ such that $f(x) = \|x\|$. In particular, $\|x\| = \max\{|f(x)| : f \in B_{X^*}\}$ for every $x \in X$.*

As a consequence, if $X \neq \{0\}$ then $X^* \neq \{0\}$ as well (see Corollary 3.33).

Proof: Put $Y = \mathrm{span}\{x\}$ and define $f \in Y^*$ by $f(tx) = t\|x\|$. Clearly $\|f\|_{Y^*} = 1$ and $f(x) = \|x\|$. Using Theorem 2.2 we extend f to a linear functional from X^* with the same norm. From $|f(x)| \leq \|f\| \|x\|$ we have $\sup_{f \in B_{X^*}} |f(x)| \leq \|x\|$. On the other hand, the linear functional constructed above shows that the supremum is attained and equal to $\|x\|$. □

Corollary 2.4 *Let $\{x_i\}_{i=1}^n$ be a linearly independent set of vectors in a normed space X and $\{\alpha_i\}_{i=1}^n$ be a set of real numbers. Then there is $f \in X^*$ such that $f(x_i) = \alpha_i$ for $i = 1, \ldots, n$.*

Proof: Define a linear functional f on $\mathrm{span}\{x_i\}$ by $f(x_i) = \alpha_i$ for $i = 1, \ldots, n$. Proposition 1.39 shows that f is continuous. The result follows from Theorem 2.2. □

Definition 2.5 *Let C be a convex subset of a normed space X and let $x \in C$. A non-zero linear functional $f \in X^*$ is called a* supporting functional *of C at x if $f(x) = \sup\{f(y) : y \in C\}$. The point x is said to be a* support point *of C (supported by f).*

By Corollary 2.3, for every $x \in S_X$ there is a supporting functional of B_X at x, and so x is a support point of B_X.

There exists a closed convex and bounded set C in a Banach space, having empty interior, and a point in C that is not a support point (see Exercise 2.17). However, every closed convex and bounded subset of a Banach space must have support points. This follows from Theorem 7.41.

Consider a Banach space X. If Y is a subset of X, we define its *annihilator* by $Y^\perp = \{f \in X^* : f(y) = 0 \text{ for all } y \in Y\}$. Note that $Y^\perp$ is a closed subspace of X^*. Similarly, for a subset Y of X^* we define $Y_\perp = \{x \in X : f(x) = 0 \text{ for every } f \in Y\}$, which is a closed subspace of X.

Note that if F is a subset of a Hilbert space H, then the orthogonal complement $F^\perp$ when considered a subspace of the dual H^* under the canonical duality (see Theorem 2.22) coincides with the annihilator $F^\perp$.

Proposition 2.6 *Let Y be a closed subspace of a Banach space X. Then $(X/Y)^*$ is isometric to $Y^\perp$ and Y^* is isometric to $X^*/Y^\perp$.*

Proof: Consider the mapping $\delta : Y^\perp \to (X/Y)^*$ defined by $\delta(x^*) : \hat{x} \mapsto x^*(x)$, where $x \in \hat{x}$. This definition is correct, since $x^*(x_1) = x^*(x_2)$ whenever $x_1, x_2 \in \hat{x}$

as $x^* \in Y^\perp$. To see that δ maps $Y^\perp$ onto $(X/Y)^*$, given $f \in (X/Y)^*$, define $x^* \in X^*$ by $x^*(x) = f(\hat{x})$, where $x \in \hat{x}$. Then $x^* \in Y^\perp$ and $\delta(x^*)(\hat{x}) = x^*(x) = f(\hat{x})$. To check that δ is an isometry, write

$$\|\delta(x^*)\| = \sup_{\|\hat{x}\|<1} |\delta(x^*)(\hat{x})| = \sup_{\|x\|<1} |x^*(x)| = \|x^*\|.$$

The middle equality follows since given $\|\hat{x}\| < 1$, there is $x \in \hat{x}$ such that $\|x\| < 1$. On the other hand, given $\|x\| < 1$, we have $\|\hat{x}\| < 1$.

To prove the second part of this proposition, define a mapping σ from Y^* into $X^*/Y^\perp$ by $\sigma(y^*) = \{\text{all extensions of } y^* \text{ on } X\}$. It is easy to see that $\sigma(y^*)$ is a coset in $X^*/Y^\perp$ and the Hahn–Banach theorem gives that $\|\sigma(y^*)\| = \|y^*\|_{Y^*}$. It follows that σ is a linear and onto mapping. □

We will now establish several separation results.

Proposition 2.7 *Let Y be a closed subspace of a normed space X. If $x \notin Y$ then there is $f \in S_{X^*}$ such that $f(y) = 0$ for all $y \in Y$ and $f(x) = \operatorname{dist}(x, Y)$.*

Proof: Let $(0 <)\ d = \operatorname{dist}(x, Y)$. Put $Z = \operatorname{span}\{Y, x\}$ and define a linear functional f on Z by $f(y + tx) = td$ for $y \in Y$ and $t \in \mathbb{K}$. Clearly $f\big|_Y = 0$ and $f(x) = d$. For $u := y + tx$, where $y \in Y$ and t is a scalar such that $u \neq 0$, we have

$$\begin{aligned} |f(u)| = |t|\,d = \frac{|t|.\|u\|}{\|u\|}d = \frac{|t|.\|u\|}{\|y+tx\|}d = \frac{\|u\|}{\|(y/t)+x\|}d \\ = \frac{\|u\|d}{\|x - (-(y/t))\|} \leq \frac{\|u\|d}{\operatorname{dist}(x, Y)} = \|u\|. \end{aligned}$$

Therefore $\|f\| \leq 1$.

On the other hand, there is a sequence $y_n \in Y$ such that $\|y_n - x\| \to d$. We have $d = |f(y_n) - f(x)| \leq \|f\| \cdot \|y_n - x\|$, so by passing to the limit when $n \to \infty$ we obtain $d \leq \|f\|d$.

Thus $\|f\| = 1$, and $f(x) = \operatorname{dist}(x, Y)$. Extending f on X with the same norm we obtain the desired functional. □

Proposition 2.8 *Let X be a normed space. If X^* is separable, then X is separable.*

Proof: Choose a dense subset $\{f_n\}$ of S_{X^*}. For every $n \in \mathbb{N}$, pick $x_n \in S_X$ such that $f_n(x_n) > \frac{1}{2}$. Let $Y = \overline{\operatorname{span}}\{x_n\}$. As Y is separable (finite rational combinations of $\{x_n\}$ are dense in Y), it is enough to show that $X = Y$. If $Y \neq X$, then there is $f \in X^*$, $\|f\| = 1$ such that $f(x) = 0$ for every $x \in Y$. Let n be such that $\|f_n - f\| < \frac{1}{4}$. Then

$$\begin{aligned} |f(x_n)| &= |f_n(x_n) - (f_n(x_n) - f(x_n))| \geq |f_n(x_n)| - |f_n(x_n) - f(x_n)| \\ &\geq |f_n(x_n)| - \|f - f_n\| \cdot \|x_n\| > \tfrac{1}{2} - \tfrac{1}{4} = \tfrac{1}{4}, \end{aligned}$$

a contradiction. □

To prove separation results for sets we need a new notion.

Definition 2.9 *Let C be a set in a normed space X. We define the* Minkowski functional *of C, $\mu_C\colon X \to [0, +\infty]$, by*

$$\mu_C(x) = \begin{cases} \inf\{\lambda > 0 : x \in \lambda C\}, & \text{if } \{\lambda > 0 : x \in \lambda C\} \neq \emptyset, \\ +\infty, & \text{if } \{\lambda > 0 : x \in \lambda C\} = \emptyset. \end{cases}$$

Lemma 2.10 *Let μ be a subadditive real function on a real normed space X. Then*
(i) μ is continuous if and only if it is continuous at 0.
(ii) If μ is continuous, every linear functional $f : X \to \mathbb{R}$ such that $f \leq \mu$ is also continuous.

Proof: From the subadditivity of μ it follows easily that, for $x, y \in X$, $-\mu(y-x) \leq \mu(x) - \mu(y) \leq \mu(x-y)$, so μ is continuous if (and only if) it is continuous at 0. This proves (i). In order to prove (ii), use Exercise 2.1. □

Lemma 2.11 *Let C be a convex neighborhood of 0 in a normed space X. Then its Minkowski functional μ_C is a finite non-negative positively homogeneous subadditive continuous functional. Moreover, $\{x : \mu_C(x) < 1\} = \text{Int}(C) \subset C \subset \overline{C} = \{x : \mu_C(x) \leq 1\}$.*

Proof: Let $B_\delta = \{x : \|x\| \leq \delta\} \subset C$ for some $\delta > 0$. Since $0 \in C$, the point 0 is in λC for every $\lambda > 0$ and thus $\mu_C(0) = 0$. Given $x \in X\backslash\{0\}$, we get $\delta \frac{x}{\|x\|} \in B_\delta \subset C$, so $x \in \frac{\|x\|}{\delta} C$. Thus

$$(0 \leq)\ \mu_C(x) \leq \frac{\|x\|}{\delta} < \infty. \tag{2.2}$$

Given $\alpha, \lambda > 0$, clearly $x \in \lambda C$ if and only if $\alpha x \in \lambda\alpha C$. Therefore $\mu_C(\alpha x) = \alpha\mu_C(x)$ and thus μ_C is positively homogeneous. We claim that $\mu_C(x) < \lambda$ implies that $x \in \lambda C$. Indeed, there exists λ_0 such that $\mu_C(x) \leq \lambda_0 < \lambda$ and $x \in \lambda_0 C$. Then we can find $c \in C$ such that $x = \lambda_0 c$. Therefore

$$(x =)\ \lambda_0 c = \lambda \left(\frac{\lambda_0}{\lambda} c + \frac{(1-\lambda_0)}{\lambda} 0 \right). \tag{2.3}$$

Since C is convex and $0 \in C$ we get $x \in \lambda C$ as claimed.

To prove subadditivity, let $x, y \in X$ and s, t such that $\mu_C(x) < s$, $\mu_C(y) < t$. By the former claim, $x \in sC$ and $y \in tC$. Then $x + y \in sC + tC$, and thus by the convexity

$$x + y \in (t+s)\Big(\frac{s}{t+s} C + \frac{t}{t+s} C\Big) \subset (t+s)C.$$

Therefore $\mu_C(x+y) \leq t + s$, so by the choice of s and t we have $\mu_C(x+y) \leq \mu_C(x) + \mu_C(y)$.

The continuity of μ_C at 0 follows from (2.2), so μ_C is continuous by Lemma 2.10. By the continuity of μ_C, the set $\{x \in X : \mu_C(x) < 1\}$ is open, and, by the claim, a subset of C, hence a subset of $\mathrm{Int}(C)$. It follows that, if $\mu_C(x) = 1$ and $0 < s < 1 < t$, then $sx \in \mathrm{Int}(C)$ and $tx \notin C$. Therefore, if $\mu_C(x) = 1$ then x belongs to the boundary of C and if $\mu_C(x) > 1$ then, again by the continuity of μ_C, $x \in \mathrm{Int}(X \setminus C)$. This proves the statement. □

Theorem 2.12 (Hahn, Banach) *Let C be a closed convex set in a normed space X. If $x_0 \notin C$ then there is $f \in X^*$ such that* $\mathrm{Re}\big(f(x_0)\big) > \sup\{\mathrm{Re}\big(f(x)\big) : x \in C\}$.

Proof: First, let X be a real space. We may assume without loss of generality that $0 \in C$, otherwise we consider $(C - x)$ and $x_0 - x$ for some $x \in C$. Let $\delta = \mathrm{dist}(x_0, C)$. Then δ is positive as C is closed. Set $D = \{x \in X : \mathrm{dist}(x, C) \leq \delta/2\}$. Since $0 \in C$, we have $\frac{\delta}{4} B_X \subset D$ and so D contains 0 as an interior point. D is also closed, convex, and $x_0 \notin D$. Let μ_D be the Minkowski functional of D. Since D is closed and $x_0 \notin D$, we have $\mu_D(x_0) > 1$ (Exercise 2.21).

Define a linear functional on $\mathrm{span}\{x_0\}$ by $f(\lambda x_0) = \lambda \mu_D(x_0)$. Then on $\mathrm{span}\{x_0\}$ we have $f(\lambda x_0) \leq \mu_D(\lambda x_0)$. For $\lambda \geq 0$ it is clear from the definition of f, for $\lambda < 0$ we have $f(\lambda x_0) = \lambda \mu_D(x_0) < 0$ while $\mu_D(\lambda x_0) \geq 0$. Extend f onto X by Theorem 2.1 and denote this extension by f again. Then $f(x) \leq \mu_D(x)$ for every $x \in X$. The continuity of f follows from Lemma 2.10.

Since $\mu_D(x_0) > 1$ and $f(x_0) = \mu_D(x_0)$, we get $f(x_0) > 1$, so $f(x_0) > \sup\{f(x) : x \in C\}$.

If X is a complex space, we construct g from $X^*_{\mathbb{R}}$ as in the real case and then define $f(x) = g(x) - ig(ix)$. □

For simplicity we will state the following result only for the real case.

Proposition 2.13 *Let X be a real normed space.*
(i) *Let C be an open convex set in X. If $x_0 \notin C$ then there is $f \in X^*$ such that $f(x) < f(x_0)$ for all $x \in C$.*
(ii) *Let A, B be disjoint convex sets in X. If A is open then there is $f \in X^*$ such that $f(a) < \inf_{f(b):\, b \in B}$ for all $a \in A$.*

Proof: (i) We pick some $y \in C$ and consider $D := C - y$, $y_0 := x_0 - y$. Then define μ_D and f on $\mathrm{span}\{y_0\}$ as in the proof of Theorem 2.12. Let f denote the extended functional as well. We have $f(y_0) = \mu_D(y_0) \geq 1$ as $y_0 \notin D$. Also $f(x) < 1$ for $x \in D$ as D is open (Exercise 2.21), and the statement follows.

(ii) Applying (i) to the open convex set $C := A - B$ and to $x_0 := 0$ we obtain f such that $f(x) < f(0)$ $(= 0)$ for $x \in A - B$. Thus $f(a) < f(b)$ for every $a \in A$, $b \in B$. It follows that $f(a) \leq \inf\{f(b) : b \in B\}$ for $a \in A$. If $a \in A$ is such that $f(a) = \inf\{f(b) : b \in B\}$, then, from the openness of A we get $f(a + h) > \inf\{f(b) : b \in B\}$ for some $a + h \in A$, a contradiction. Therefore $f(a) < \inf\{f(b) : b \in B\}$ for all $\alpha \in A$. □

Proposition 2.14 *Let $(X, \|\cdot\|)$ be a normed space. Let Y be a subspace of X and let $|||\cdot|||$ be an equivalent norm on Y. Then there is an equivalent norm $|\cdot|$ on X inducing on Y the norm $|||\cdot|||$.*

Proof: Without loss of generality, we may assume that $B_{(Y,\|\cdot\|)} \subset B_{(Y,|||\cdot|||)}$. The set $B := \operatorname{conv}\{B_{(Y,|||\cdot|||)} \cup B_{(X,\|\cdot\|)}\}$ is convex and balanced. Obviously, B is bounded and contains $B_{(X,\|\cdot\|)}$, so its Minkowski functional is an equivalent norm $|\cdot|$ on X (see Fig. 2.1).

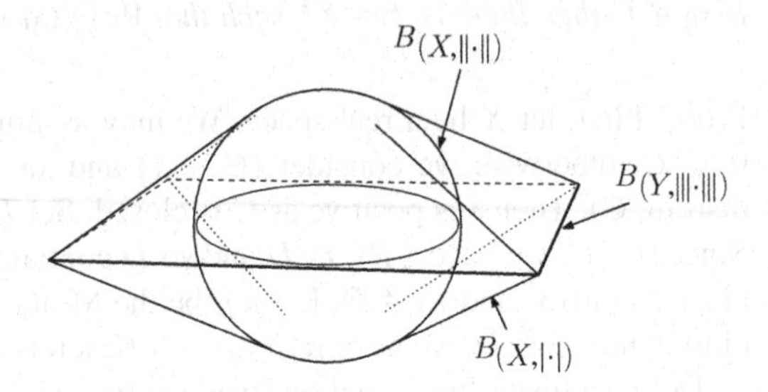

Fig. 2.1 Extending a norm

This norm certainly induces on Y the norm $|||\cdot|||$, since $B \cap Y = B_{(Y,|||\cdot|||)}$. Indeed, $B_{(Y,|||\cdot|||)} \subset B$. On the other hand, if $y \in B \cap B_{(Y,|||\cdot|||)}$, then $y = \lambda y_1 + (1-\lambda)x$, where $0 \leq \lambda \leq 1$, $y_1 \in B_{(Y,|||\cdot|||)}$, and $x \in B_{(X,\|\cdot\|)}$ (see Exercise 1.12). We get $(1-\lambda)x = y - \lambda y_1 \in Y$. If $\lambda \neq 1$ then $x \in Y$, so $x \in B_{(X,\|\cdot\|)} \cap Y = B_{(Y,\|\cdot\|)} \subset B_{(Y,|||\cdot|||)}$. By convexity, $y \in B_{(Y,|||\cdot|||)}$. If, on the contrary, $\lambda = 1$, we obtain again $y \in B_{(Y,|||\cdot|||)}$. □

We refer to Exercise 5.95 for an alternative proof of the Hahn–Banach theorem.

2.2 Duals of Classical Spaces

In Propositions 2.15, 2.16, 2.17, 2.18, 2.19, and 2.20, we assume the scalar field to be $\mathbb{R}$.

Proposition 2.15 (Riesz) *$c_0^* = \ell_1$ in the sense that for every $f \in c_0^*$ there is a unique $(a_i) \in \ell_1$ such that $f(x) = \sum a_i x_i$ for all $x = (x_i) \in c_0$, and the mapping $f \mapsto (a_i)$ is a linear isometry from c_0^* onto ℓ_1.*

Proof: Given $f \in c_0^*$, define $a_i = f(e_i)$, where $e_i := (0, \ldots, 0, \overset{i}{1}, 0, \ldots)$ are the standard unit vectors in c_0. For $n \in \mathbb{N}$ we set

$$x^n = (\operatorname{sign}(a_1), \ldots, \operatorname{sign}(a_n), 0, \ldots) \in c_0.$$

Then $\|x^n\|_\infty = 1$ and $f(x^n) = \sum_{i=1}^n |a_i| \leq \|f\| \cdot \|x^n\|_\infty = \|f\|$. Therefore $\sum_{i=1}^\infty |a_i| \leq \|f\| < \infty$, that is, the mapping $f \mapsto \big(f(e_i)\big)$ is a continuous mapping into ℓ_1. It is obviously linear.

On the other hand, if $\sum_{i=1}^{\infty} |a_i| < \infty$ then $\sum |a_i x_i| < \infty$ for every $x = (x_i) \in c_0$. Indeed, we have $\sum |a_i x_i| \leq \sup |x_i| \cdot \sum |a_i| = \|(a_i)\|_1 \|(x_i)\|_\infty$. Consider the linear functional h defined on c_0 by $h(x) = \sum a_i x_i$. Then from the above estimate we have $\|h\| \leq \|(a_i)\|_1$ and also $h(e_i) = a_i$, so $h \in c_0^*$ and the mapping $f \mapsto \big(f(e_i)\big)$ is thus onto. We also obtain that $\|f\| = \|\big(f(a_i)\big)\|_1$, hence the considered mapping is an isometry onto ℓ_1. □

Proposition 2.16 (Riesz) $\ell_1^* = \ell_\infty$ *in the sense that for every* $f \in \ell_1^*$ *there is a unique* $(a_i) \in \ell_\infty$ *such that* $f(x) = \sum a_i x_i$ *for all* $x = (x_0) \in \ell_1$*, and the mapping* $f \mapsto (a_i)$ *is a linear isometry from* ℓ_1^* *onto* ℓ_∞.

Proof: Given $f \in \ell_1^*$, put $a_i = f(e_i)$ for $i \in \mathbb{N}$, where e_i are the standard unit vectors in ℓ_1. Then $|a_i| \leq \|f\|$, so $\|(a_i)\|_\infty \leq \|f\|$. Conversely, for $(a_i) \in \ell_\infty$ consider the functional h defined on ℓ_1 by $h(x) = \sum a_i x_i$. Again, $|h(x)| \leq \|(a_i)\|_\infty \|x\|_1$, hence $h \in \ell_1^*$ and $\|h\| \leq \|(a_i)\|_\infty$. Similarly as in the proof of Proposition 2.15, we conclude that the mapping is a linear isometry onto ℓ_∞. □

Proposition 2.17 (Riesz) *Let* $p, q \in (1, \infty)$ *be such that* $\frac{1}{p} + \frac{1}{q} = 1$. *Then* $\ell_p^* = \ell_q$ *in the sense that for every* $f \in \ell_p^*$ *there exists a unique element* $(a_i) \in \ell_q$ *such that* $f(x) = \sum a_i x_i$ *for all* $x =: (x_i) \in \ell_p$*, and the mapping* $f \mapsto (a_i)$ *is a linear isometry from* ℓ_p^* *onto* ℓ_q.

Proof: For $f \in \ell_p^*$, put $a_i = f(e_i)$. Considering

$$x^n := \big(|a_1|^{q-1} \operatorname{sign}(a_1), \ldots, |a_n|^{q-1} \operatorname{sign}(a_n), 0, \ldots\big)$$

we see that

$$\sum_{i=1}^{n} |a_i|^q = f(x^n) \leq \|f\| \cdot \|x^n\|_p = \|f\| \Big(\sum_{i=1}^{n} \big(|a_i|^{q-1}\big)^p\Big)^{\frac{1}{p}} = \|f\| \cdot \Big(\sum_{i=1}^{n} |a_i|^q\Big)^{\frac{1}{p}}.$$

This reads $\Big(\sum_{i=1}^{n} |a_i|^q\Big)^{\frac{1}{q}} \leq \|f\|$. Hence $\|(a_i)\|_q \leq \|f\| < \infty$.

If $(a_i) \in \ell_q$ and $(x_i) \in \ell_p$, then the series $\sum x_i a_i$ is convergent by the Hölder inequality (1.1) as $\sum |x_i a_i| \leq \|(x_i)\|_p \|(a_i)\|_q$. Therefore the functional h defined on ℓ_p by $h(x) = \sum x_i a_i$ is well defined and $\|h\| \leq \|(a_i)\|_q$. The rest of the proof is analogous to those above. □

Similarly we show that for a set Γ and $p \in [1, \infty)$ we have $c_0(\Gamma)^* = \ell_1(\Gamma)$ and $\ell_p(\Gamma)^* = \ell_q(\Gamma)$, where $\frac{1}{p} + \frac{1}{q} = 1$. This applies, in particular, for a finite set Γ.

Proposition 2.18 (Riesz) *Let* $p, q \in (1, \infty)$ *be such that* $\frac{1}{p} + \frac{1}{q} = 1$. *Then* $L_p[0, 1]^* = L_q[0, 1]$ *in the sense that for every* $F \in L_p^*$ *there is a unique* $f \in L_q$ *such that* $F(g) = \int_0^1 g f \, \mathrm{d}x$ *for all* $g \in L_p$*, and the mapping* $F \mapsto f$ *is a linear isometry of* L_p^* *onto* L_q.

Proof: Let $F \in L_p^*$. For $t \in [0,1]$, let $u_t = \chi_{[0,t)}$ be the characteristic function of $[0,t)$. Define $\alpha(t) = F(u_t)$. We claim that α is absolutely continuous (see the definition right before Proposition 11.13). Indeed, if $[\tau_i, t_i]$, $i = 1, \dots, n$, is a collection of non-overlapping intervals, that is, their interiors are pairwise disjoint, put $\varepsilon_i = \operatorname{sign}(\alpha(t_i) - \alpha(\tau_i))$ and estimate:

$$\begin{aligned}
\sum_{i=1}^{n} |\alpha(t_i) - \alpha(\tau_i)| &= \sum_{i=1}^{n} \varepsilon_i(\alpha(t_i) - \alpha(\tau_i)) = F\Big(\sum_{i=1}^{n} \varepsilon_i(u_{t_i} - u_{\tau_i})\Big) \\
&\le \|F\|_{L_p^*} \cdot \Big\|\sum_{i=1}^{n} \varepsilon_i(u_{t_i} - u_{\tau_i})\Big\|_{L_p} = \|F\|_{L_p^*}\Big(\int_0^1 \Big|\sum_{i=1}^{n} \varepsilon_i(u_{t_i} - u_{\tau_i})\Big|^p \mathrm{d}x\Big)^{\frac{1}{p}} \\
&= \|F\|\Big(\sum_{i=1}^{n} \int_{\tau_i}^{t_i} 1 \,\mathrm{d}x\Big)^{\frac{1}{p}} = \|F\| \cdot \Big(\sum_{i=1}^{n} (t_i - \tau_i)\Big)^{\frac{1}{p}}.
\end{aligned}$$

Therefore α is an absolutely continuous function on $[0,1]$. By the Lebesgue fundamental theorem of calculus, we have $\alpha(t) - \alpha(0) = \int_0^t \alpha' \,\mathrm{d}x$ for every $t \in [0,1]$. Setting $f = \alpha'$ and using $\alpha(0) = F(u_0) = 0$ we get

$$F(u_t) = \alpha(t) = \int_0^t f \,\mathrm{d}x = \int_0^1 u_t f \,\mathrm{d}x.$$

Since F is linear, we also have $F(g_n) = \int_0^1 g_n f \,\mathrm{d}x$ for all step functions $g_n := \sum_{k=1}^{n} c_k\big(u_{\frac{k}{n}} - u_{\frac{k-1}{n}}\big)$.

Let g be a bounded measurable function on $[0,1]$. Then there is a sequence of step functions g_n such that $g_n \to g$ a.e. and $\{g_n\}$ is uniformly bounded. By the Lebesgue dominated convergence theorem, we get

$$\lim_{n\to\infty} F(g_n) = \lim_{n\to\infty} \int_0^1 g_n f \,\mathrm{d}x = \int_0^1 \lim_{n\to\infty} g_n f \,\mathrm{d}x = \int_0^1 g f \,\mathrm{d}x.$$

On the other hand, since $g_n \to g$ a.e. and g_n are uniformly bounded, the same theorem implies $\|g_n - g\|_{L_p} \to 0$ as $n \to \infty$. By the continuity of F on L_p, we thus have $F(g) = \lim\limits_{n\to\infty} F(g_n) = \int_0^1 g f \,\mathrm{d}x$. Hence $F(g) = \int_0^1 g f \,\mathrm{d}x$ for every bounded measurable function g on $[0,1]$.

We will show that $f \in L_q$ and $\|f\|_q \le \|F\|$. Consider a family of functions g_n defined by

$$g_n(x) = \begin{cases} |f(x)|^{q-1} \operatorname{sign}\big(f(x)\big) & \text{if } |f(x)| \le n, \\ 0 & \text{if } |f(x)| > n. \end{cases}$$

The functions g_n are bounded and measurable. Thus we have $F(g_n) = \int_0^1 g_n f \,\mathrm{d}x$. Note also that $|F(g_n)| \le \|F\| \, \|g_n\|_p$. On the other hand,

$$\int_0^1 |g_n|^p\,dx = \int_0^1 |g_n|^{\frac{q}{q-1}}\,dx = \int_0^1 |g_n(t)|\,|g_n(t)|^{\frac{1}{q-1}}\,dx$$
$$\leq \int_0^1 |g_n|\,|f|\,dx = \int_0^1 g_n f\,dx = F(g_n) = |F(g_n)|.$$

Hence $\int_0^1 |g_n|^p\,dx \leq \|F\|\cdot\|g_n\|_p = \|F\|\Big(\int_0^1 |g_n|^p\,dx\Big)^{\frac{1}{p}}$, so $\Big(\int_0^1 |g_n|^p\,dx\Big)^{\frac{1}{q}} \leq \|F\|$.

Since f is integrable, we have $|g_n| \to |f|^{q-1}$ a.e. By Fatou's lemma, the last inequality implies that

$$\Big(\int_0^1 |f|^q\,dx\Big)^{\frac{1}{q}} = \Big(\int_0^1 |f|^{(q-1)p}\,dx\Big)^{\frac{1}{q}} = \Big(\int_0^1 |g_n|^p\,dx\Big)^{\frac{1}{q}} \leq \|F\|.$$

This shows that $f \in L_q$. Finally, let $g \in L_p$. There exists a sequence $\{g_n\}$ of bounded measurable functions that converges to g in L_p. Then $F(g_n) \to F(g)$ and by Hölder's inequality (1.1) we have $\int_0^1 g_n f\,dx \to \int_0^1 gf\,dx$. We have shown that $F(g_n) = \int_0^1 g_n f\,dx$ for bounded measurable functions, so $F(g) = \int_0^1 gf\,dx$ as claimed.

On the other hand, given a function $f \in L_q$, we can define a linear functional on L_p by $F(g) = \int_0^1 gf\,dx$. It follows from the Hölder inequality (1.1) that F is continuous and $\|F\| \leq \|f\|_q$. □

Using similar methods, we obtain an analogous result for the space L_1.

Proposition 2.19 (Riesz) $L_1[0,1]^* = L_\infty[0,1]$ *in the sense that for every* $F \in L_1^*$ *there exists a unique* $f \in L_\infty$ *such that* $F(g) = \int_0^1 gf\,dx$ *for all* $g \in L_1$, *and the mapping* $F \mapsto f$ *is a linear isometry of* L_1^* *onto* L_∞.

Proposition 2.20 (Riesz) *For every* $F \in C[0,1]^*$ *there exists a function* f *on* $[0,1]$ *with bounded variation such that* $F(g) = \int_0^1 g\,\mathrm{d}f$ *(Stieltjes integral) for all* $g \in C[0,1]$ *and* $\|F\| = \bigvee_0^1 f$, *where* $\bigvee_0^1 f$ *denotes the* variation *of* f *on* $[0,1]$.
On the other hand, if f *is a function of bounded variation on* $[0,1]$, *then* $F(g) := \int_0^1 g\,\mathrm{d}f$ *is a continuous linear functional on* $C[0,1]$.

Proof: Consider the space $\ell_\infty[0,1]$ of bounded functions on $[0,1]$ with the supremum norm denoted by $\|\cdot\|_\infty$. If $F \in C[0,1]^*$, we have that $|F(g)| \leq \|F\|\cdot\|g\|_\infty$ for every $g \in C[0,1]$. Since $C[0,1]$ is a subspace of $\ell_\infty[0,1]$, by the Hahn–Banach theorem we can extend F to a functional $\widetilde{F}$ on $\ell_\infty[0,1]$ such that $|\widetilde{F}(g)| \leq \|F\|\cdot\|g\|_\infty$. We will represent $\widetilde{F}$ similarly to the L_p setting above. For $t \in [0,1]$, let $u_t = \chi_{[0,t)}$, the characteristic function of $[0,t)$. Put $f(t) = \widetilde{F}(u_t)$ for $t \in [0,1]$ (note that F is not defined on u_t as u_t is not continuous). We will prove that f has bounded variation on $[0,1]$. To this end, consider $t_0 = 0 < t_1 < \cdots < t_{n-1} < t_n = 1$ and put $\varepsilon_i = \mathrm{sign}(f(t_i) - f(t_{i-1}))$. We have

$$\sum_{i=1}^{n} |f(t_i) - f(t_{i-1})| = \sum_{i=1}^{n} \varepsilon_i (f(t_i) - f(t_{i-1})) = \widetilde{F}\Big(\sum_{i=1}^{n} \varepsilon_i (u_{t_i} - u_{t_{i-1}})\Big)$$
$$\le \|\widetilde{F}\| \cdot \Big\|\sum_{i=1}^{n} \varepsilon_i (u_{t_i} - u_{t_{i-1}})\Big\|_\infty = \|F\| \cdot 1.$$

Hence f has bounded variation on $[0, 1]$ which is bounded by $\|F\|$.

For $g \in C[0, 1]$ and $g_n := \sum_{i=1}^{n} g(\frac{k}{n})(u_{\frac{k}{n}} - u_{\frac{k-1}{n}})$ we have

$$\widetilde{F}(g_n) = \sum_{i=1}^{n} g(\tfrac{k}{n})\big(f(\tfrac{k}{n}) - f(\tfrac{k-1}{n})\big) = \int_0^1 g_n \,\mathrm{d}f.$$

Therefore $\lim\limits_{n\to\infty} \widetilde{F}(g_n) = \lim\limits_{n\to\infty} \sum_{k=1}^{n} g(\frac{k}{n})(f(\frac{k}{n}) - f(\frac{k-1}{n})) = \int_0^1 g \,\mathrm{d}f$. Since $\widetilde{F} \in \ell_\infty[0, 1]^*$ and $g_n \to g$ in $\|\cdot\|_\infty$, we have $\lim \widetilde{F}(g_n) = \widetilde{F}(g)$, so $\widetilde{F}(g) = \int_0^1 g \,\mathrm{d}f$. However, for $g \in C[0, 1]$ we have $\widetilde{F}(g) = F(g)$, hence $F(g) = \int_0^1 g(t) \,\mathrm{d}f(t)$.

We have already shown that $\bigvee_0^1 f \le \|F\|$. On the other hand, from the theory of Riemann–Stieltjes integral we have that given a function f of bounded variation, $F \colon g \mapsto \int_0^1 g \,\mathrm{d}f$ is a linear mapping and $\int_0^1 g(t) \,\mathrm{d}f(t) \le \|g\|_\infty \bigvee_0^1 f$. Therefore F is continuous and $\|F\| \le \bigvee_0^1 f$. □

In general, if K is a compact set, the space $C(K)^*$ can be identified with the space of all regular Borel measures on K of bounded variation. Every such measure μ defines a functional $F_\mu(f) := \int_K f \,\mathrm{d}\mu$, the correspondence $\mu \mapsto F_\mu$ is a linear isometry ([Rudi2, Theorem 2.14]).

Let $k \in K$. We define the corresponding *Dirac functional* (or *Dirac measure*) by $\delta_k(f) = f(k)$ for every $f \in C(K)$. Observe that δ_k is a continuous linear functional of norm one. Indeed, on one hand, $\|\delta_k\| = \sup\limits_{\|f\|\le 1} (\delta_k(f)) = \sup\limits_{\|f\|\le 1} (f(k)) \le 1$. By considering the constant function $f = 1$, we obtain $\|\delta_k\| = 1$.

Proposition 2.21 *The space $C[0, 1]^*$ is not separable.*

Proof: Consider the Dirac measures δ_t for $t \in [0, 1]$. We claim that if $t_1 \ne t_2$ then $\|\delta_{t_1} - \delta_{t_2}\| = 2$. Indeed, $\|\delta_{t_1} - \delta_{t_2}\| \le \|\delta_{t_1}\| + \|\delta_{t_2}\| = 2$. On the other hand, choose $f_0 \in C[0, 1]$ such that $f_0(t_1) = 1$, $f_0(t_2) = -1$, and $\|f_0\|_\infty = 1$. Then $\|\delta_{t_1} - \delta_{t_2}\| \ge |f_0(t_1) - f_0(t_2)| = 2$. Similarly to the case of ℓ_∞ we find that $C[0, 1]^*$ is not separable. □

Recall that the inner product on a complex Hilbert space ℓ_2, respectively L_2, is defined by $\big((x_i), (y_i)\big) = \sum x_i \bar{y}_i$, respectively $(g, f) = \int_0^1 g \bar{f} \,\mathrm{d}x$. This motivates the following identification of the dual space in case of *complex scalars*. Recall that a mapping Φ is called *conjugate linear* if $\Phi(\alpha x + y) = \bar{\alpha}\Phi(x) + \Phi(y)$ for all vectors x, y and scalars α.

Theorem 2.22 (Riesz) *Let H be a Hilbert space. For every $f \in H^*$ there is a unique $a \in H$ such that $f(x) = (x, a)$ for all $x \in H$. The mapping $f \mapsto a$ is a conjugate-linear isometry of H^* onto H.*

Proof: The uniqueness of such a is clear. Indeed, if $f(x) = (x, a_1) = (x, a_2)$ then using $x = a_1 - a_2$ we get $(a_1 - a_2, a_1) = (a_1 - a_2, a_2)$. Thus $(a_1 - a_2, a_1 - a_2) = 0$, so $a_1 = a_2$.

By the Cauchy–Schwarz inequality,

$$\|f\| = \sup_{\|x\|\leq 1} |f(x)| = \sup_{\|x\|\leq 1} |(x, a)| \leq \sup_{\|x\|\leq 1} \big(\|a\| \cdot \|x\|\big) \leq \|a\|.$$

On the other hand, $\|f\| = \sup\{|f(x)| : \|x\| \leq 1\} \geq (a/\|a\|, a) = \|a\|$. Hence $\|f\| = \|a\|$.

To obtain the representation of $0 \neq f \in H^*$, consider $N := \mathrm{Ker}(f)$. It is a proper closed subspace of H. Choose $z_0 \in N^{\perp}$ and assume without loss of generality that $f(z_0) = 1$.

We claim that $H = N \oplus \mathrm{span}\{z_0\}$. Indeed, given $h \in H$, it suffices to find a scalar α such that $h - \alpha z_0 \in N$, that is, $f(h - \alpha z_0) = 0$. This is satisfied for $\alpha = f(h)$.

We now show that $f(x) = \big(x, \frac{z_0}{\|z_0\|^2}\big)$ for every $x \in H$. Given $x := y + \alpha z_0$, where $y \in N$ and α is a scalar, we have

$$\begin{aligned} f(x) &= \alpha f(z_0) = \alpha = \alpha(z_0, z_0)/\|z_0\|^2 \\ &= (y, z_0)/\|z_0\|^2 + (\alpha z_0, z_0)/\|z_0\|^2 = \Big(x, \frac{z_0}{\|z_0\|^2}\Big). \end{aligned}$$

□

2.3 Banach Open Mapping Theorem, Closed Graph Theorem, Dual Operators

Definition 2.23 *Let φ be a mapping from a topological space X into a topological space Y. We say that φ is an* open mapping *if it maps open sets in X onto open sets in Y.*

Let T be an operator from a normed space X into a normed space Y. Observe that if T is an open mapping, then T is necessarily onto. Indeed, by Exercise 2.37, $\delta B_Y \subset T(B_X)$ for some $\delta > 0$ and hence by linearity, $Y \subset T(X)$. We will now establish the converse for bounded operators.

By $B_X^O(r)$ we denote the open ball with radius r centered at the origin of a Banach space X.

Lemma 2.24 (Banach) *Let X be a Banach space, Y a normed space and $T \in \mathcal{B}(X, Y)$. If $r, s > 0$ satisfy $B_Y^O(s) \subset \overline{T(B_X^O(r))}$, then $B_Y^O(s) \subset T(B_X^O(r))$.*

Proof: By considering $\frac{r}{s}T$ if necessary, we may assume that $r = s = 1$. Denote $B_X^O = B_X^O(1)$ and $B_Y^O = B_Y^O(1)$. Let $z \in B_Y^O$ be given. Choose $\delta > 0$ such that $\|z\|_Y < 1 - \delta < 1$ and put $y = (1 - \delta)^{-1}z$. Note that $\|y\|_Y < 1$. We will show that $y \in (1 - \delta)^{-1}T(B_X^O)$, which implies that $z \in T(B_X^O)$.

We start with $y_0 = 0$ and inductively find a sequence $y_n \in Y$ such that $\|y - y_n\|_Y < \delta^n$ and $(y_n - y_{n-1}) \in T(\delta^{n-1}B_X^O)$. Indeed, having chosen $y_0, y_1, \ldots, y_{n-1} \in Y$, we have $(y - y_{n-1}) \in \delta^{n-1}B_Y^O \subset \overline{T(\delta^{n-1}B_X^O)}$, hence there is $w \in T(\delta^{n-1}B_X^O)$ such that $\|w - (y - y_{n-1})\|_Y < \delta^n$. Setting $y_n = y_{n-1} + w$ we complete the construction.

Next we find a sequence $\{x_n\}_{n=1}^{\infty} \subset X$ such that $\|x_n\|_X < \delta^{n-1}$ and $T(x_n) = y_n - y_{n-1}$ for $n \in \mathbb{N}$. Since the series $\sum x_i$ is absolutely convergent, we put $x = \sum_{n=1}^{\infty} x_n$. Then $\|x\|_X \leq \sum_{n=1}^{\infty} \|x_n\|_X < \sum_{n=1}^{\infty} \delta^{n-1} = \frac{1}{1-\delta}$ and by the continuity and linearity of T,

$$T(x) = \lim_{N\to\infty} \sum_{n=1}^{N} T(x_n) = \lim_{N\to\infty} \sum_{n=1}^{N} (y_n - y_{n-1}) = \lim_{N\to\infty} y_N = y.$$

□

Note that $\overline{T(B_X^O(r))} = \overline{T(B_X(r))}$, so the conclusion of the lemma is true if we assume for instance $\delta B_Y \subset \overline{T(B_X)}$.

Theorem 2.25 (Banach open mapping principle) *Let X, Y be Banach spaces and $T \in \mathcal{B}(X, Y)$. If T is onto Y then T is an open mapping.*

Proof: Put $G = T(B_X^O)$. Since T is linear, we only need to prove that G contains a neighborhood of the origin. Note that we have $T(B_X^O(r)) = rG$ and $\overline{rG} = r\overline{G}$ for every $r > 0$. Therefore $\overline{T(B_X^O(r))} = r\overline{G}$ for every $r > 0$. This implies that $Y = T(X) = \bigcup_{n=1}^{\infty} n\overline{G}$. By the Baire category theorem, there is $n \in \mathbb{N}$ such that $n\overline{G}$ contains an interior point, so there is $x_0 \in \overline{G}$ and $\delta > 0$ such that $\left(x_0 + B_Y^O(\delta)\right) \subset n\overline{G}$. Since $n\overline{G}$ is symmetric, we have $\left(-x_0 + B_Y^O(\delta)\right) \subset n\overline{G}$. If $x \in B_Y^O(\delta)$ then from the convexity of $n\overline{G}$ we have $x = \frac{1}{2}(x_0 + x) + \frac{1}{2}(-x_0 + x) \in n\overline{G}$. Therefore $B_Y^O(\delta) \subset \overline{T(B_X^O(n))}$ and consequently $B_Y^O\left(\frac{\delta}{n}\right) \subset \frac{1}{n}\overline{T(B_X^O(n))} = \overline{T(B_X^O)}$. By Lemma 2.24, we have $B_Y^O\left(\frac{\delta}{n}\right) \subset T(B_X^O)$ as claimed. □

It follows from the proof that if $T : X \to Y$ is onto, then there is $\delta > 0$ such that $\delta B_Y \subset T(B_X)$.

Note that even if $T \in \mathcal{B}(X, Y)$ is open, it does not imply that $T(M)$ is closed in Y whenever M is closed in X (Exercise 15.11).

In Exercise 2.33, a rewording of the proof of the Banach open mapping principle in the language of convex series is presented.

Corollary 2.26 *Let X, Y be Banach spaces and let $T \in \mathcal{B}(X, Y)$ be onto Y.*
(i) *If T is one-to-one, then T^{-1} is a bounded operator.*
(ii) *There is a constant $M > 0$ such that for every $y \in Y$ there is $x \in T^{-1}(y)$ satisfying $\|x\|_X \le M\|y\|_Y$.*
(iii) *Y is isomorphic to $X/\operatorname{Ker}(T)$.*

Proof: (i) If O is open in X, then $(T^{-1})^{-1}(O) = T(O)$ is open in Y showing that T^{-1} is continuous.

(ii) By the open mapping theorem, there is $\delta > 0$ such that $\delta B_Y \subset T(B_X)$. Therefore for every $y \in Y$ such that $\|y\|_Y = \delta$, there is $x \in B_X$ such that $T(x) = y$. Thus it is enough to put $M = 1/\delta$.

(iii) Define a linear mapping $\widehat{T}$ from $X/\operatorname{Ker}(T)$ onto Y by $\widehat{T}(\hat{x}) = T(x)$ for $x \in \hat{x}$. The mapping $\widehat{T}$ is well defined. Moreover $\widehat{T}$ is one-to-one and onto Y. Let $\hat{x}_n \to 0$. Then there is $x_n \in \hat{x}_n$ such that $\|x_n\|_X < \|\hat{x}_n\| + 1/n$ and therefore $x_n \to 0$. Since T is continuous, we have $T(x_n) \to 0$ and thus $\widehat{T}(\hat{x}_n) \to 0$. Hence $\widehat{T}$ is continuous and one-to-one, so by (i) it is an isomorphism of $X/\operatorname{Ker}(T)$ onto Y. □

Theorem 2.27 (Banach closed graph theorem) *Let X, Y be Banach spaces and let T be an operator from X into Y. T is a bounded operator if and only if its graph $G := \{(x, T(x)) : x \in X\}$ is closed in $X \oplus Y$.*

Recall that the norm on $X \oplus Y$ is defined by $\|(x, y)\| = \|x\|_X + \|y\|_Y$. In particular, $(x_n, y_n) \to (x, y)$ if and only if $x_n \to x$ and $y_n \to y$ (see Definition 1.33).

Note that G, the graph of T, is a subspace of $X \oplus Y$.

Proof: If T is continuous and $(x_n, T(x_n)) \to (x_0, y_0)$, then $y_0 = T(x_0)$. Indeed, we have $x_n \to x_0$ and $T(x_n) \to y_0$, while the continuity of T implies that $T(x_n) \to T(x_0)$. This means that (x_0, y_0) $(= (x_0, T(x_0)))$ is in the graph of T, showing that G is closed.

If G is closed in $X \oplus Y$, then G is a Banach space in the norm induced from $X \oplus Y$. Consider the mapping $p \colon G \to X$ defined by $p(x, T(x)) = x$. By the definition of the norm in $X \oplus Y$ we see that p is continuous, maps G onto X, and is one-to-one. By Corollary 2.26, $p^{-1} \colon x \mapsto (x, T(x))$ is a continuous mapping from X onto G. Since also $q \colon X \oplus Y \to Y, q(x, y) := y$, is continuous and $T = q \circ p^{-1}$, T must be continuous. □

Definition 2.28 *Let X, Y be Banach spaces and $T \in \mathcal{B}(X, Y)$. We define the* dual *(also called* adjoint*) operator $T^* \in \mathcal{B}(Y^*, X^*)$ for $f \in Y^*$ by $(T^*(f))(x) = f(T(x))$, for all $x \in X$.*

It is easy to observe that $x \mapsto f(T(x))$ is a linear mapping. If $\|x\| \le 1$ then $|T^*(f)(x)| = |f(T(x))| \le \|f\|\,\|T\|$. Thus $T^*(f)$ is also bounded, so $T^*(f) \in Y^*$ and T^* is well defined. Also the mapping $f \mapsto T^*(f)$ is linear and the above estimate shows that $\|T^*(f)\| \le \|T\|\,\|f\|$. Consequently, T^* is a bounded operator from X^* into Y^*.

Proposition 2.29 *Let X, Y be Banach spaces. If $T \in \mathcal{B}(X, Y)$ then $\|T^*\| = \|T\|$.*

Proof: We have

$$\|T^*\| = \sup_{f \in B_{Y^*}} \|T^*(f)\|_{X^*} = \sup_{f \in B_{Y^*}} \Big\{ \sup_{x \in B_X} |T^*(f)(x)| \Big\}$$
$$= \sup_{f \in B_{Y^*}} \{ \sup_{x \in B_X} |f\big(T(x)\big)|\} = \sup_{x \in B_X} \{ \sup_{f \in B_{Y^*}} |f\big(T(x)\big)|\} = \sup_{x \in B_X} \{\|T(x)\|_Y\} = \|T\|.$$

□

Let X, Y, Z be Banach spaces and let $T \in \mathcal{B}(X, Y)$, $S \in \mathcal{B}(Y, Z)$. Then $(ST)^* = T^*S^*$. Indeed, consider $f \in Z^*$. Then for every $x \in X$ we get $(ST)^*(f)(x) = f\big(ST(x)\big) = (S^*f)\big(T(x)\big) = \big(T^*S^*(f)\big)(x)$, so $(ST)^*(f) = (T^*S^*)(f)$.

2.4 Remarks and Open Problems

Remarks

1. We mentioned in Open Problem 1 in Chapter 1 that quasi-Banach spaces behave differently from Banach spaces. This is mainly due to the fact that the Hahn–Banach theorem fails in that context, see [Kalt1].

Open Problems

1. It is an open problem if every infinite-dimensional Banach space has a separable infinite-dimensional quotient, i.e., if for every Banach space X there is an infinite-dimensional separable Banach space Y and a bounded operator from X onto Y. This problem is equivalent to the problem whether in every Banach space X there is an increasing sequence $\{E_n\}_{n=1}^{\infty}$ of distinct closed subspaces such that $\overline{\bigcup_n E_n} = X$ (see, e.g., [Muji] and [HMVZ, Chapter 4]).

Exercises for Chapter 2

2.1 Let X be a real normed space. If f is a linear functional on E that is dominated by a function $p : E \to \mathbb{R}$ (i.e., $f \leq p$), and p is continuous at 0, then f is continuous.
Hint. $-f(x) = f(-x) \leq p(-x)$, hence $-p(-x) \leq f(x) \leq p(x)$ for all $x \in E$. Since this implies that f is continuous at 0, the conclusion follows from Proposition 1.25.

2.2 Let C be a convex symmetric set in a Banach space X. Assume that a linear functional f on X is continuous at 0 when restricted to C. Show that the restriction of f to C is uniformly continuous.
Hint. Given $\varepsilon > 0$, we look for a neighborhood U of the origin in X such that $x, y \in C$ and $x - y \in U$ imply $|f(x - y)| < \varepsilon$. We have $\frac{1}{2}(x - y) \in C$, so by homogeneity of f we only need to find an open ball U centered at 0 such that $|f(w)| < \varepsilon/2$ for point $w \in C \cap U$. Such U exists by the continuity of $f|_C$ at 0.

2.3 Show that if X is a finite-dimensional Banach space, then every linear functional f on X is continuous on X.
Hint. Use Proposition 1.39.

2.4 Show that if X is an infinite-dimensional normed space, then X admits a discontinuous linear functional.
Hint. Let $\{e_\gamma\}$ be a Hamel basis formed by vectors of norm 1. Define a linear functional f on $\{e_\gamma\}$ so that the set $\{f(e_\gamma)\}$ is unbounded, and extend f on X linearly. Then f is not bounded on the unit ball.

2.5 Show that if $f \neq 0$ is a linear functional on a normed space X, then the codimension of $f^{-1}(0)$ in X is 1.
Hint. For $x \in X$ write $x = (x - (f(x)/f(x_0))x_0) + (f(x)/f(x_0))x_0$, where x_0 is some fixed element in X with $f(x_0) \neq 0$.

2.6 Recall that by a *hyperplane* of a normed space X we mean any subspace Y of codimension 1 (that is, $\dim(X/Y) = 1$).

Let Y be a subspace of a normed space X. Show that Y is a hyperplane if and only if there is a linear functional f, $f \neq 0$, such that $Y = f^{-1}(0)$. Show that Y is a closed hyperplane if and only if there is $f \in X^*$, $f \neq 0$, such that $Y = f^{-1}(0)$.
Hint. One direction: Exercise 2.5. Given a closed hyperplane Y, take $e \notin Y$, use Proposition 2.7 to find f. Then $Y \subset f^{-1}(0)$ and since $\operatorname{codim}(Y) = 1$, equality follows. For a general hyperplane, the proof is similar.

2.7 Let H be a hyperplane in a normed space X, and let F be a two-dimensional subspace of X. Show that $\dim(F \cap H) \geq 1$.
Hint. Use algebraic complementability of H in X.

2.8 Let H be a closed hyperplane of a Banach space X. Let $x_0 \in X \setminus H$. Prove that there is a linear and continuous projection P from X onto H *parallel* to x_0, i.e., such that $Px_0 = 0$ (for a more precise statement, see Exercise 5.7).
Hint. Exercise 2.6 gives $f \in X^*$ such that $\operatorname{Ker} f = H$. By scaling we may assume that $f(x_0) = 1$. Let $P : X \to X$ be defined by $P(x) = x - f(x)x_0$. This is the sought projection.

2.9 Let X be a Banach space. Show that all closed hyperplanes of X are mutually isomorphic. By induction we get that given $k \in \mathbb{N}$, all closed subspaces of X of codimension k are mutually isomorphic. In fact, Zippin proved that the Banach–Mazur distance of two hyperplanes of the same infinite-dimensional Banach space is less than or equal to 25, see [AlKa, p. 238].

Hint. Let F and G to distinct closed hyperplanes of X. According to Exercise 2.6, there exists $f, g \in X^*$ such that $F := f^{-1}(0)$ and $G := g^{-1}(0)$. Find $e \in X$ such that $f(e) = g(e) = 1$. Let P_f (resp., P_g) be the (linear and continuous) projection of X onto F (resp., onto G) parallel to e (see Exercise 2.8). Clearly, $P_f \circ P_g(x) = x$ for every $x \in F$, and $P_g \circ P_f(y) = y$ for every $y \in G$. From this it follows that $P_g|_F : F \to G$ is an isomorphism. See Fig. 2.2.

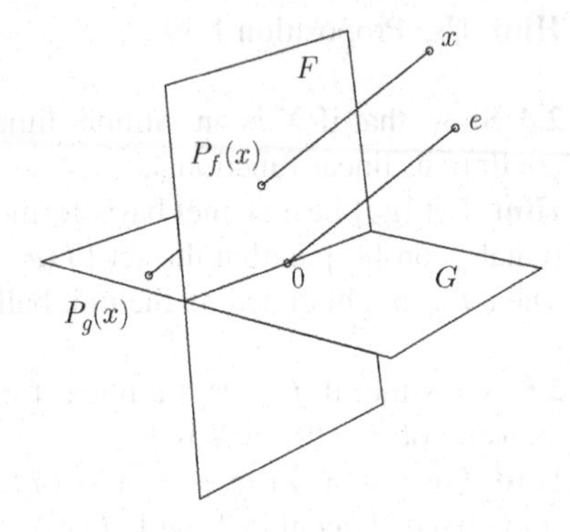

Fig. 2.2 All hyperplanes are mutually isomorphic

2.10 Let f be a linear functional on a Banach space X. Show that if f is not identically 0, the following are equivalent (see also Proposition 3.19):
(i) f is continuous.
(ii) f is continuous at 0.
(iii) $f^{-1}(0)$ is closed.
(iv) $f^{-1}(0)$ is not dense in X.

Hint. (i)$\Longrightarrow$(ii) is obvious, and (ii)$\Longrightarrow$(i) is clear from the linearity of f. (i)$\Longrightarrow$(iii) is clear. (iii)$\Longrightarrow$(iv) is obvious. (iv)$\Longrightarrow$(iii) follows from the fact that if $f^{-1}(0)$ is not closed, then $f^{-1}(0) \subsetneqq \overline{f^{-1}(0)} \subset X$ and $\overline{f^{-1}(0)}$ is a linear subspace. It is enough to use now Exercise 2.5. (iii)$\Longrightarrow$(i): Since $f^{-1}(\mathbb{R}\setminus\{0\}) \neq \emptyset$ is open, there is some ball $B = x_0 + \delta B_X$ such that $f|_B \neq 0$. Assume $f(x_0) > 0$, then also $f|_B > 0$ (connect x_0 with points of B, $f \neq 0$ on the connecting segments and f is continuous on each of those segments). Then $f|_{B_X} \geq -\frac{1}{\delta} f(x_0)$, so by symmetry of B_X we get $|f(x)| \leq \frac{1}{\delta} f(x_0)$ for $x \in B_X$ and f is continuous.

Another related approach is the following: two subspaces A and B of a normed space X form an (algebraic) *direct sum decomposition* of X (written $X = A \oplus B$) if $A \cap B = \{0\}$ and $A + B = X$ (see the paragraph prior to Definition 1.33). Prove first that if f is a non-zero linear functional and $x_0 \in X$ such that $f(x_0) \neq 0$, and $K := f^{-1}(0)$, then $X = K \oplus \text{span}\{x_0\}$ (see Exercise 2.5). As a consequence, no subspace S of X exists such that $K \subsetneqq S \subsetneqq X$. Since $K \subset \overline{K} \subset X$, and the closure of a subspace is also a subspace (see Exercise 1.9) it follows that K is dense if it is not closed (the converse being trivially true). To finish the exercise, notice that if K is closed, then X/K is then a one-dimensional space. Let $\hat{f} : X/K \to \mathbb{K}$ a linear functional such that $\hat{f} \circ q = f$, where $q : X \to X/K$ is the canonical quotient mapping (see Exercise 2.35). Apply now Exercise 2.3.

2.11 Find a discontinuous linear mapping T from some Banach space X into X such that $\mathrm{Ker}(T)$ is closed.
Hint. Let $X = c_0$ and $T(x) = (f(x), x_1, x_2, \dots)$ for $x = (x_i)$, where f is a discontinuous linear functional on X.

2.12 Let X be a Banach space, $f \in S_{X^*}$. Show that for every $x \in X$ we have $\mathrm{dist}\big(x, f^{-1}(0)\big) = |f(x)|$.
Hint. The result is obviously true if $f = 0$. If not, put $K = f^{-1}(0)$. There exists, by Proposition 2.7, $g \in S_{X^*}$ that vanishes on K and $g(x) = \mathrm{dist}(x, K)$. Since $g^{-1}(0) = f^{-1}(0)$, we get $g = \lambda f$ for some scalar λ (see Exercise 2.5), and $|\lambda| = 1$ since both f and g belong to S_{X^*}. This proves the assertion.

2.13 (The "parallel-hyperplane lemma".) Let X be a real Banach space, $f, g \in S_{X^*}$ and $\varepsilon > 0$ be such that $|f(x)| \le \varepsilon$ for every $x \in g^{-1}(0) \cap B_X$. Prove that either $\|f - g\| \le 2\varepsilon$ or $\|f + g\| \le 2\varepsilon$ (see Fig. 2.3).

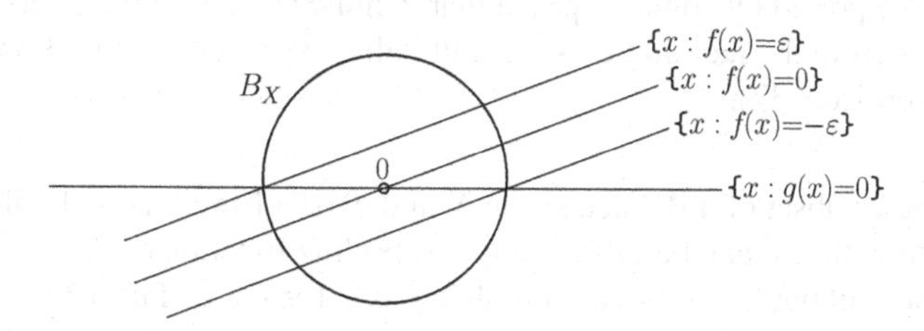

Fig. 2.3 The "parallel-hyperplane lemma"

Hint. Consider f on $g^{-1}(0)$ and extend it with the same norm (at most ε) on X, calling this extension $\tilde{f}$. Then $\tilde{f} - f = 0$ on $g^{-1}(0)$ and thus $\tilde{f} - f = \alpha g$ for some α by Lemma 3.21. Note that $|1 - |\alpha|| = \big|\|f\| - \|f - \tilde{f}\|\big| \le \|\tilde{f}\| \le \varepsilon$. Thus if $\alpha \ge 0$, then $\|g + f\| = \|(1-\alpha)g + \tilde{f}\| \le |1 - \alpha| + \|\tilde{f}\| \le 2\varepsilon$. If $\alpha < 0$, calculate $\|g - f\|$.

2.14 If X is an infinite-dimensional Banach space, show that there are convex sets C_1 and C_2 such that $C_1 \cup C_2 = X$, $C_1 \cap C_2 = \emptyset$, and both C_1 and C_2 are dense in X.
Hint. Take a discontinuous functional f on X (Exercise 2.4), define $C_1 = \{x : f(x) \ge 0\}$ and $C_2 = \{x : f(x) < 0\}$, use Exercise 2.10.

2.15 Let X be a finite-dimensional Banach space. Let C be a convex subset of X that is dense in X. Prove that $C = X$.

Hint. We may assume that $0 \in C$. Let $\{e_1, e_2, \ldots, e_n\}$ be an algebraic basis of X consisting of unit vectors. Fix $\varepsilon > 0$. For each $i \in \{1, 2, \ldots, n\}$ we can find $v_i \in C$ such that $\|e_i - v_i\| < \varepsilon$. If $\varepsilon > 0$ is small enough, $\{v_1, \ldots, v_n\}$ is a linearly independent set in C (look at the determinant of the matrix with columns $v_i, i = 1, 2, \ldots .n$). The set $\operatorname{conv}(\{v_i : i = 1, 2, \ldots, n\} \cup \{0\})$ has a nonempty interior and is contained in C, so C has a nonempty interior. If $x_0 \in X \setminus C$, then there exists a closed hyperplane H that separates $\{x_0\}$ and $\operatorname{Int}(C)$, a contradiction with the denseness of C.

2.16 Let N be a maximal ε-separated set in the unit sphere of a Banach space X (see Exercise 1.47). Show that $(1 - \varepsilon)B_X \subset \overline{\operatorname{conv}}(N)$.
Hint. Otherwise, by the separation theorem, we find $x \in X$ and $f \in S_{X^*}$ with $\|x\| \leq 1 - \varepsilon$ and $f(x) > \sup_{\overline{\operatorname{conv}}(N)}(f) = \sup_N(f)$. For $\delta > 0$ choose $y \in S_X$ such that $f(y) > 1 - \delta$. By the maximality of N, there exists $z \in N$ with $\varepsilon > \|y - z\| \geq f(y) - f(z)$. Thus $\sup_N(f) \geq f(z) > f(y) - \varepsilon > 1 - \delta - \varepsilon$. This holds for any $\delta > 0$, so we have $1 - \varepsilon \leq \sup_N(f) < f(x) \leq \|x\| \leq 1 - \varepsilon$, a contradiction.

2.17 Let $D = \{\pm e_i : i \in \mathbb{N}\} \subset \ell_2$, where e_i is the ith unit vector. The set $C := \overline{\operatorname{conv}}(D)$ has empty interior, so it coincides with its boundary. Show that 0 is not a support point of C.
Hint. If 0 is supported by some f, prove that f must be 0. That the interior of C is empty follows from the fact that C is the unit ball of ℓ_1 (use Exercise 3.36 or, more generally, Exercise 3.37).

2.18 Let C be a subset of a Banach space X and f be a Lipschitz real-valued function on C. Show that f can be extended to a Lipschitz function on X.
Hint. Assume without loss of generality that f is 1-Lipschitz. Put for $x \in X$,

$$F(x) = \inf\{f(z) + \|z - x\|; z \in C\}.$$

To see that F is finite for every $x \in X$, pick an arbitrary $z_0 \in C$. Then for any $z \in C$,

$$f(z) + \|x - z\| \geq f(z_0) - \|z - z_0\| + \|x - z\| \geq f(z_0) - \|x - z_0\|.$$

Thus

$$F(x) \geq f(z_0) - \|x - z_0\|.$$

If $x \in C$, then for every $z \in C$, $f(x) \leq f(z) + \|z - x\|$. Thus $F(x) = f(x)$. To show that F is 1-Lipschitz, pick $x, y \in X$, $\varepsilon > 0$ and choose $z_0 \in C$ so that

$$f(z_0) + \|z_0 - x\| < F(x) + \varepsilon.$$

Then

$$\begin{aligned}F(y) - F(x) &\le F(y) - f(z_0) - \|z_0 - x\| + \varepsilon \\ &\le f(z_0) + \|z_0 - y\| - f(z_0) - \|z_0 - x\| + \varepsilon \\ &\le \|x - y\| + \varepsilon.\end{aligned}$$

In Exercises 2.19, 2.20, 2.21, and 2.22, μ_C denotes the Minkowski functional of a set C.

2.19 Let $(X, \|\cdot\|)$ be a Banach space. Show that $\mu_{B_X}(x) = \|x\|$.
Hint. Use continuity of the norm.

2.20 Let A, B be convex sets in a Banach space X. Show that if $A \subset B$ then $\mu_B \le \mu_A$. Show that $\mu_{cA}(x) = \frac{1}{c}\mu_A(x)$ for $c > 0$.
Hint. Follows from the definition.

2.21 Let C be a convex neighborhood of 0 in a real Banach space X (then μ_C is a non-negative positive homogeneous subadditive continuous functional on X, see Lemma 2.11). Prove the following:

(i) If C is also open, then $C = \{x : \mu_C(x) < 1\}$. If C is closed instead, then $C = \{x : \mu_C(x) \le 1\}$.

(ii) There is $c > 0$ such that $\mu_C(x) \le c\|x\|$.

(iii) If C is moreover symmetric, then μ_C is a continuous seminorm, that is, it is a continuous homogeneous subadditive functional.

(iv) If C is moreover symmetric and bounded, then μ_C is a norm that is equivalent to $\|\cdot\|_X$. In particular, it is complete, that is, (X, μ_C) is a Banach space.

Note that the symmetry condition is good only for the real case. In a complex normed space X we have to replace it by C being *balanced*.

Hint. (i) It follows from Lemma 2.11.

(ii) See Equation (2.2).

(iii) Observing that $\mu_C(-x) = \mu_C(x)$ and positive homogeneity are enough to prove $\mu_C(\lambda x) = |\lambda|\mu_C(x)$ for all $\lambda \in \mathbb{R}$, $x \in X$.

(iv) From (iii) we already have the homogeneity and the triangle inequality. We need to show that $\mu_C(x) = 0$ implies $x = 0$ (the other direction is obvious). Indeed, $\mu_C(x) = 0$ implies that $x \in \lambda C$ for all $\lambda > 0$, which by the boundedness of C only allows for $x = 0$.

In (ii) we proved $\mu_C(x) \le c\|x\|$, an upper estimate follows from $C \subset dB_X$.

The equivalence then implies completeness of the new norm.

2.22 Let K be a bounded closed convex and symmetric set in a Banach space X. Denote by Y the linear hull of K. Let $|||\cdot|||$ on Y be defined as the Minkowski functional of K. Show that $(Y, |||\cdot|||)$ is a Banach space, i.e., K is a *Banach disc*. For an extension of this result to the setting of locally convex spaces and for some of its consequences see Exercises 3.71, 3.72, 3.73, and 3.74.

Hint. If (x_n) is a Cauchy sequence in $(Y, |||\cdot|||)$ it is Cauchy in X and converges, say to x_0, in X. Given a closed ball U in $(Y, |||\cdot|||)$, note that U is closed in X. As (x_n) is Cauchy in $(Y, |||\cdot|||)$, there is $n_0 \in \mathbb{N}$ such that $x_n - x_m \in U$ for all $n, m \geq n_0$. As U is closed in X, $x_n - x_0 \in U$ for $n \geq n_o$ (in particular, $x_0 \in Y$). It follows that $x_n \to x_0$ in $(Y, |||\cdot|||)$.

2.23 Prove that, if $n \in \mathbb{N}$, the dual space of a n-dimensional Banach space is again n-dimensional. Prove that the dual space of an infinite-dimensional normed space is again infinite-dimensional.
Hint. Use Propositions 1.36 and 2.17—this last one for a finite index set. The infinite-dimensional assertion follows from this.

2.24 Show that if Y is a subspace of a Banach space X and X^* is separable then so is Y^*.
Hint. Y^* is isomorphic to the separable space $X^*/Y^\perp$.

2.25 Show that ℓ_1 is not isomorphic to a subspace of c_0.
Hint. The dual of ℓ_1 is nonseparable. Use now Exercise 2.24.

2.26 Show that c_0 is not isomorphic to $C[0, 1]$.
Hint. Check the separability of their duals—Proposition 2.21.

2.27 Let X be a Banach space.
(i) Show that in X^* we have $X^\perp = \{0\}$ and $\{0\}^\perp = X^*$. Show that in X we have $(X^*)_\perp = \{0\}$ and $\{0\}_\perp = X$.
(ii) Let $A \subset B$ be subsets of X. Show that $B^\perp$ is a subspace of $A^\perp$.
Hint. Follows from the definition.

2.28 Let X be a Banach space. Show that:
(i) $\overline{\text{span}}(A) = (A^\perp)_\perp$ for $A \subset X$.
(ii) $\overline{\text{span}}(B) \subset (B_\perp)^\perp$ for $B \subset X^*$. Note that in general we cannot put equality.
(iii) $A^\perp = \left((A^\perp)_\perp\right)^\perp$ for $A \subset X$ and $B_\perp = \left((B_\perp)^\perp\right)_\perp$ for $B \subset X^*$.
Hint. (i) Using definition, show that $A \subset (A^\perp)_\perp$. Then use that $B_\perp$ is a closed subspace for any $B \subset X^*$, proving that $\overline{\text{span}}(A) \subset (A^\perp)_\perp$. Take any $x \notin \overline{\text{span}}(A)$. Since $\overline{\text{span}}(A)$ is a closed subspace, by the separation theorem there is $f \in X^*$ such that $f(x) > 0$ and $f|_{\overline{\text{span}}(A)} = 0$. But then $f\big|_A = 0$, hence $f \in A^\perp$, also $f(x) > 0$, so $x \notin (A^\perp)_\perp$.
(ii) Similar to (i).
(iii) Applying (i) to $A^\perp$ we get $A^\perp \subset \left((A^\perp)_\perp\right)^\perp$. On the other hand, using $A \subset (A^\perp)_\perp$ and the previous exercise we get $\left((A^\perp)_\perp\right)^\perp \subset A^\perp$. The dual statement is proved in the same way.

2.29 Let $X = \mathbb{R}^2$ with the norm $\|x\| = (|x_1|^4 + |x_2|^4)^{1/4}$. Calculate directly the dual norm on X^* using the Lagrange multipliers.

Hint. The dual norm of $(a,b) \in X^*$ is $\sup\{ax_1 + bx_2 : x_1^4 + x_2^4 = 1\}$. Define $F(x_1, x_2, \lambda) = ax_1 + bx_2 - \lambda(x_1^4 + x_2^4 - 1)$ and multiply by x_1 and x_2, respectively, the equations you get from $\frac{\partial F}{\partial x_1} = 0$ and $\frac{\partial F}{\partial x_2} = 0$.

2.30 Let Γ be a set and let $p \in [1,\infty)$, $q \in (1,\infty]$ be such that $\frac{1}{p} + \frac{1}{q} = 1$. Show that $c_0(\Gamma)^* = \ell_1(\Gamma)$ and $\ell_p(\Gamma)^* = \ell_q(\Gamma)$.
Hint. See the proofs of Propositions 2.15, and 2.16, 2.17.

2.31 Show that c^* is linearly isometric to ℓ_1.
Hint. We observe that $c = c_0 \oplus \operatorname{span}\{e\}$, where $e := (1,1,\ldots)$ (express $x := (\xi_i) \in c$ in the form $x = \xi_0 e + x_0$ with $\xi_0 := \lim\limits_{i\to\infty} \xi_i$ and $x_0 \in c_0$). If $u \in c^*$, put $v_0' = u(e)$ and $v_i = u(e_i)$ for $i \geq 1$. Then we have $u(x) = u(\xi_0 e) + u(x_0) = \xi_0 v_0' + \sum_{i=1}^{\infty} v_i(\xi_i - \xi_0)$ and $(v_1, v_2, \ldots) \in \ell_1$ as in Proposition 2.15. Put $\tilde{u} = (v_0, v_1, \ldots)$, where $v_0 := v_0' - \sum_{i=1}^{\infty} v_i$, and write $\tilde{x} := (\xi_0, \xi_1, \ldots)$. We have $u(x) = \xi_0 v_0 + \sum_{i=1}^{\infty} v_i \xi_i = \tilde{u}(\tilde{x})$.

Conversely, if $\tilde{u} \in \ell_1$ then the above rule gives a continuous linear functional u on c with $\|u\| \leq \|\tilde{u}\|$, as $|\tilde{u}(\tilde{x})| \leq \left(\sum_{i=0}^{\infty} |v_i|\right) \sup\limits_{i\geq 0} |\xi_i| = \|\tilde{u}\| \sup\limits_{i\geq 0} |\xi_i| = \|\tilde{u}\|_1 \|x\|_\infty$.
The inequality $\|\tilde{u}\| \leq \|u\|$ follows like this: Let ξ_i be such that $|v_i| = \xi_i v_i$ if $v_i \neq 0$ and $\xi_i = 1$ otherwise, $i = 0, 1, \ldots$. Set $x^n = (\xi_1, \ldots, \xi_n, \xi_0, \xi_0, \ldots)$. Then $\|x^n\|_\infty = 1$ and $|u(x^n)| = |\tilde{u}(\tilde{x}^n)| \geq |v_0| + \sum_{i=1}^{n} |v_i| - \sum_{i=n+1}^{\infty} |v_i|$. Since $|u(x^n)| \leq \|u\|$, we have $\|u\| \geq |v_0| + \sum_{i=1}^{n} |v_i| - \sum_{i=n+1}^{\infty} |v_i|$. By letting $n \to \infty$ we get $\|\tilde{u}\| \leq \|u\|$.

2.32 Let $p \in (1,\infty)$ and X_n be Banach spaces for $n \in \mathbb{N}$. By $X := \left(\sum X_n\right)_p$ we denote the normed linear space of all sequences $x = \{x_i\}_{i=1}^{\infty}$, $x_i \in X_i$, such that $\sum \|x_i\|_{X_i}^p < \infty$, with the norm $\|x\| := \left(\sum \|x_i\|_{X_i}^p\right)^{\frac{1}{p}}$.

Show that X is a Banach space and that X^* is isometric to $\left(\sum X_i^*\right)_q$ (where $\frac{1}{p} + \frac{1}{q} = 1$) in the following sense: to $f \in X^*$ we assign $\{f_i\}_{i=1}^{\infty}$ such that $f_i \in X_i^*$ and $f\left(\{x_i\}_{i=1}^{\infty}\right) = \sum f_i(x_i)$.

Remark: Sometimes the notation $\sum_{\ell_p} X_n$ will be used instead of $\left(\sum X_n\right)_p$.
Hint. Follow the proof for ℓ_p, which is the case of $X_i = \mathbb{R}$.

2.33 Prove the open mapping theorem by using the concept of convex series (Exercise 1.66) and the Baire category theorem.
Hint. Let $T : X \to Y$ be a bounded linear and onto mapping between Banach spaces. According to Exercise 1.66, B_X is a CS-compact set, so TB_X is again CS-compact, hence CS-closed. Since $Y = \bigcup_{n=1}^{\infty} n\overline{TB_X}$, the Baire category theorem ensures that $\emptyset \neq \operatorname{Int}(\overline{TB_X})$. According to Exercise 1.66, $\operatorname{Int}(TB_X) = \operatorname{Int}(\overline{TB_X})$. Again a "cone argument" (see Exercise 1.55) concludes that $0 \in \operatorname{Int}(TB_X)$, so T is an open mapping.

2.34 We proved the closed graph theorem using the open mapping theorem. Now prove the open mapping principle using the closed graph theorem.

Hint. First prove it for one-to-one mappings using the fact that $\{(y, T^{-1}(y))\}$ is closed. For the general case, note that the quotient mapping is an open mapping by the definition of the quotient topology.

2.35 Let X, Y be normed spaces, $T \in \mathcal{B}(X, Y)$. Show that $\widehat{T}: X/\operatorname{Ker}(T) \to Y$ defined by $\widehat{T}(\hat{x}) = T(x)$ is a bounded operator onto $T(X)$.

2.36 (i) Prove directly that if X is a Banach space and f is a non-zero linear functional on X, then f is an open mapping from X onto the scalars.

(ii) Let the operator T from c_0 into c_0 be defined by $T\big((x_i)\big) = \left(\frac{1}{i}x_i\right)$. Is T a bounded operator? Is T an open map? Does T map c_0 onto a dense subset in c_0?
Hint. (i) If $f(x) = \delta > 0$ for some $x \in B_X^O$, then $(-\delta, \delta) \subset f(B_X^O)$.

(ii) Yes. No. Yes (use finitely supported vectors).

2.37 Let T be an operator (not necessarily bounded) from a normed space X into a normed space Y. Show that the following are equivalent:

(i) T is an open mapping.

(ii) There is $\delta > 0$ such that $\delta B_Y \subset T(B_X)$.

(iii) There is $M > 0$ such that for every $y \in Y$ there is $x \in T^{-1}(y)$ satisfying $\|x\|_X \le M\|y\|_Y$.
Hint. (i)$\Longrightarrow$(ii): $T(B_X^O)$ is open and contains 0; hence it contains a closed ball centered at 0.

(ii)$\Longrightarrow$(iii): Let $0 \neq y \in Y$. We have $\delta\|y\|_Y^{-1}y \in \delta B_Y$ $(\subset T(B_X))$. We can find then $u \in B_X$ such that $\delta\|y\|_Y^{-1}y = Tu$, so $y = T(x)$, where $x := \|y\|_Y\delta^{-1}u$. Certainly $\|x\|_X \le M\|y\|_Y$, where $M := \delta^{-1}$.

(iii)$\Longrightarrow$(i): If $y \in M^{-1}B_Y$ there exists $x \in X$ such that $Tx = y$ and $\|x\|_X \le M\|y\|_Y$ (≤ 1), so $y \in T(B_X)$. This proves that $M^{-1}B_Y \subset T(B_X)$. By linearity, T is open.

2.38 Let X, Y be normed spaces, $T \in \mathcal{B}(X, Y)$. Show that if X is complete and T is an open mapping, then Y is complete.
Hint. Use (iii) in the previous exercise and Exercise 1.26.

2.39 Let X, Y be Banach spaces, $T \in \mathcal{B}(X, Y)$. Show that if T is one-to-one and $B_Y^O \subset T(B_X) \subset B_Y$, then T is an isometry onto Y.
Hint. Since $B_Y^O \subset T(B_X)$, T is onto (Exercise 2.37) and hence invertible. From $T(B_X) \subset B_Y$ we get $\|T\| \le 1$. Assume that there is $x \in S_X$ such that $\|T(x)\| < \|x\|$. Pick $\delta > 1$ such that $\delta\|T(x)\| < 1$. Then $T(\delta x) \in B_Y^O \subset T(B_X)$. Thus there must be $z \in B_X$ such that $T(z) = T(\delta x)$ but it cannot be $\delta x \notin B_X$, a contradiction with T being one-to-one.

2.40 Let X, Y be Banach spaces and $T \in \mathcal{B}(X, Y)$. Show that the following are equivalent:

(i) $T(X)$ is closed.

(ii) T is an open mapping when considered as a mapping from X onto $T(X)$.

(iii) There is $M > 0$ such that for every $y \in T(X)$ there is $x \in T^{-1}(y)$ satisfying $\|x\|_X \le M\|y\|_Y$.

Hint. (i)$\Longrightarrow$(ii): Theorem 2.25. (ii)$\Longrightarrow$(iii): Exercise 2.37. (iii)$\Longrightarrow$(i): By Exercise 2.37, $T : X \to T(X)$ is an open mapping. Now use Exercise 2.38 and Fact 1.5.

2.41 Let X, Y be Banach spaces and $T \in \mathcal{B}(X, Y)$. Show that if T maps bounded closed sets in X onto closed sets in Y, then $T(X)$ is closed in Y.

Hint. Assume $T(x_n) \to y \notin T(X)$. Put $M = \text{Ker}(T)$, set $d_n = \text{dist}(x_n, M)$ and find $w_n \in M$ such that $d_n \le \|x_n - w_n\| \le 2d_n$. If $\{x_n - w_n\}$ is bounded then $T(x_n - w_n) \to y \in T(X)$, since the closure of $\{x_n - w_n\}$ is mapped onto a closed set containing y, a contradiction. Therefore we may assume that $\|x_n - w_n\| \to \infty$. Since $T(x_n - w_n) \to y$, we have $T(\frac{x_n - w_n}{\|x_n - w_n\|}) \to 0$. By the hypothesis, M must contain a point w from the closure of $\{\frac{x_n - w_n}{\|x_n - w_n\|}\}$ as 0 lies in the closure of the image of this sequence. Fix n so that $\|\frac{x_n - w_n}{\|x_n - w_n\|} - w\| < 1/3$. Then $\big\|x_n - w_n - \|x_n - w_n\|w\big\| \le \frac{1}{3}\|x_n - w_n\| < (2/3)d_n$ and $w_n + \|x_n - w_n\|w \in M$, a contradiction.

2.42 Let X and Y be Banach spaces. Then $\mathcal{K}(X, Y)$ contains isomorphic copies of Y and X^*.

Hint. $T(x) = f^*(x)y$.

2.43 Let X and Y be normed spaces. Prove that $\mathcal{B}(X, Y)$ is an infinite-dimensional space if X is infinite-dimensional and Y is not reduced to $\{0\}$.

Hint. The space $\mathcal{B}(X, Y)$ contains an isometric copy of X^*. Use now Exercise 2.23.

2.44 Let $T \in \mathcal{B}(X, Y)$. Prove the following:

(i) $\text{Ker}(T) = T^*(Y^*)_\perp$ and $\text{Ker}(T^*) = T(X)^\perp$.

(ii) $\overline{T(X)} = \text{Ker}(T^*)_\perp$ and $\overline{T^*(Y^*)} \subset \text{Ker}(T)^\perp$.

Hint. (i) Assume $x \in T^*(Y^*)_\perp$. Then for any $g \in Y^*$ we have $g\big(T(x)\big) = T^*(g)(x) = 0$, hence $T(x) = 0$. Thus $x \in \text{Ker}(T)$.

(ii) $\overline{T(X)} = \overline{\text{span}}\big(T(X)\big) = (T(X)^\perp)_\perp = \text{Ker}(T^*)_\perp$.

2.45 Let X, Y be normed spaces, $T \in \mathcal{B}(X, Y)$. Consider $\widehat{T}(\hat{x}) := T(x)$, where $x \in \hat{x}$, as an operator from $X/\text{Ker}(T)$ into $\overline{T(X)}$. Then we get $\widehat{T}^* : \overline{T(X)}^* \to (X/\text{Ker}(T))^*$. Using Proposition 2.6 and $\overline{T(X)}^\perp = T(X)^\perp = \text{Ker}(T^*)$ we may assume that $\widehat{T}^*$ is a bounded operator from $Y^*/\text{Ker}(T^*)$ into $\text{Ker}(T)^\perp \subset X^*$. On the other hand, for $T^* : Y^* \to X^*$ we may consider $\widehat{T^*} : Y^*/\text{Ker}(T^*) \to X^*$. Show that $\widehat{T}^* = \widehat{T^*}$.

Hint. Take any $\hat{y} \in Y^*/\text{Ker}(T^*)$ and $x \in X$. Then using the above identifications we obtain

$$\widehat{T}^*(\widehat{y^*})(\hat{x}) = \widehat{y^*}\big(\widehat{T}(\hat{x})\big) = y^*\big(T(x)\big) = T^*(y^*)(x) = \widehat{T^*}(\widehat{y^*})(\hat{x}).$$

2.46 Let X, Y be Banach spaces and $T \in \mathcal{B}(X, Y)$. Show that T maps X onto a dense set in Y if and only if T^* maps Y^* one-to-one into X^*.

Also, if T^* maps onto a dense set, then T is one-to-one.

Hint. If $\overline{T(X)} \neq Y$, let $f \in Y^* \setminus \{0\}$ be such that $f = 0$ on $T(X)$. Then $T^*(f) = 0$. The other implications are straightforward.

2.47 Let X, Y be Banach spaces and $T \in \mathcal{B}(X, Y)$. If T is one-to-one, is T^* necessarily onto?

Hint. No, consider the identity mapping from ℓ_1 into ℓ_2.

2.48 Let X, Y be Banach spaces and $T \in \mathcal{B}(X, Y)$. If T is an isomorphism into Y, is T^* necessarily an isomorphism into X^*?

Hint. No, embed $\mathbb{R}$ into $\mathbb{R}^2$.

2.49 Let X, Y be Banach spaces and $T \in \mathcal{B}(X, Y)$. Show that:

(i) T^* is onto if and only if T is an isomorphism into Y.

(ii) T is onto if and only if T^* is an isomorphism into X^*.

(iii) $T(X)$ is closed in Y if and only if $T^*(Y^*)$ is closed in X^*.

Hint. (i) If T^* is onto, it is an open mapping (Theorem 2.25) and by Exercise 2.37 there is $\delta > 0$ so that $\delta B_{X^*} \subset T^*(B_{Y^*})$. Then

$$\|T(x)\|_Y = \sup_{y^* \in B_{Y^*}} y^*\big(T(x)\big) = \sup_{y^* \in B_{Y^*}} T^*(y^*)(x) = \sup_{x^* \in T^*(B_{Y^*})} \big(x^*(x)\big)$$
$$\geq \sup_{x^* \in \delta B_{X^*}} \big(x^*(x)\big) = \delta \|x\|_X$$

and use Exercise 1.73.

If T is an isomorphism into, then T^{-1} is a bounded operator from $T(X)$ into X. Given $x^* \in X^*$, define y^* on $T(X)$ by $y^*(y) = x^*\big(T^{-1}(y)\big)$. Clearly $y^* \in T(X)^*$, extend it to a functional in Y^*. Then $T^*(y^*) = x^*$.

(ii) If T is onto, as in (i) we find $\delta > 0$ such that $\delta B_Y \subset T(B_X)$, then $\|T^*(y^*)\|_{X^*} \geq \delta \|y^*\|_{Y^*}$ and use Exercise 1.73.

Assume T^* is an isomorphism into. By Exercise 2.37 and Lemma 2.24, it is enough to find $\delta > 0$ so that $\delta B_Y \subset \overline{T(B_X)}$. Assume by contradiction that no such δ exists. Then find $y_n \to 0$ such that $y_n \notin \overline{T(B_X)}$. The set is closed, so $d_n := \operatorname{dist}(y_n, \overline{T(B_X)}) > 0$.

Fix n, set $V_n = \bigcup_{y \in T(B_X)} \big(y + B_Y^O(\frac{d_n}{2})\big)$. Then V_n is an open convex set and $y_n \notin V_n$, so by Proposition 2.13 there is $y^* \in Y^*$ such that $|y^*| < 1$ on V_n and $y^*(y_n) = 1$. Since $T(B_X) \subset V_n$, we get

$$\|T^*(y^*)\| = \sup_{x \in B_X} T^*(y^*)(x) = \sup_{x \in B_X} y^*\big(T(x)\big) = \sup_{y \in T(B_X)} \big(y^*(y)\big) \leq 1,$$

so $\|y^*\| \leq \|(T^*)^{-1}\| \, \|T^*(y^*)\| \leq \|(T^*)^{-1}\|$, $1 = y^*(y_n) \leq \|(T^*)^{-1}\| \, \|y_n\|$. This shows that $\|y_n\| \geq 1/\|(T^*)^{-1}\|$ for every n, contradicting $y_n \to 0$.

(iii) If $T(X)$ is closed and $q : X \to X/\operatorname{Ker}(T)$ is the canonical quotient mapping, then $\widehat{T}$ such that $\widehat{T} \circ q$, is an operator from $X/\operatorname{Ker}(T)$ onto a Banach space $T(X)$, hence by (ii) above, $\widehat{T}^*$ is an isomorphism into, in particular $\widehat{T}^*(Y^*/\operatorname{Ker}(T^*))$ is closed. By Exercise 2.45, $\widehat{T^*}(Y^*/\operatorname{Ker}(T^*)) = T^*(Y^*)$ is closed.

If $T^*(Y^*)$ is closed, consider $\widehat{T}\colon X \to \overline{T(X)}$. Then $\widehat{T}^*(Y^*/\operatorname{Ker}(T^*)) = \widehat{T^*}(Y^*/\operatorname{Ker}(T^*)) = T^*(Y^*)$ is closed and $\widehat{T}^*$ is one-to-one, hence it is an isomorphism into. By (ii), $\widehat{T}$ must be onto, that is, $T(X) = \overline{T(X)}$.

2.50 Show that there is no $T \in \mathcal{B}(\ell_2, \ell_1)$ such that T is an onto mapping.
Hint. By Exercise 2.49, T^* would be an isomorphism of ℓ_∞ into ℓ_2, which is impossible as ℓ_∞ is nonseparable and ℓ_2 is separable.

2.51 Let X, Y be Banach spaces, $T \in \mathcal{B}(X, Y)$. Show that:

(i) T is an isomorphism of X onto Y if and only if T^* is an isomorphism of Y^* onto X^*.

(ii) T is an isometry of X onto Y if and only if T^* is an isometry of Y^* onto X^*.
Hint. (i) Follows from Exercise 2.49.

(ii) If T is an isometry, then by (i), T^* is an isomorphism. Also $T(B_X) = B_Y$, so $\|T^*(y^*)\| = \sup\limits_{x \in B_X} T^*(y^*)(x) = \|y^*\|$. The other direction is similar.

2.52 We have $\|T\| = \|T^*\|$ for a bounded operator on a Banach space. So, if for a sequence of bounded operators T_n we have $\|T_n\| \to 0$, then $\|T_n^*\| \to 0$. Find an example of a sequence of bounded operators T_n on a Banach space X such that $\|T_n(x)\| \to 0$ for every $x \in X$ but it is not true that $\|T_n^*(x^*)\| \to 0$ for every $x^* \in X^*$.
Hint. Let $T_n(x) = (x_n, x_{n+1}, \dots)$ in ℓ_2. Then $T_n^*(x) = (0, \dots, 0, x_1, x_2, \dots)$, where x_1 is on the nth place.

2.53 Let X be a normed space with two norms $\|\cdot\|_1$ and $\|\cdot\|_2$ such that X in both of them is a complete space. Assume that $\|\cdot\|_1$ is not equivalent to $\|\cdot\|_2$. Let I_1 be the identity mapping from $(X, \|\cdot\|_1)$ onto $(X, \|\cdot\|_2)$ and I_2 be the identity mapping from $(X, \|\cdot\|_2)$ onto $(X, \|\cdot\|_1)$. Show that neither I_1 nor I_2 are continuous.
Hint. The Banach open mapping theorem.

2.54 Let L be a closed subset of a compact space K. Show that $C(L)$ is isomorphic to a quotient of $C(K)$.
Hint. Let $T\colon C(K) \to C(L)$ be defined for $f \in C(K)$ by $T(f) = f\big|_L$. Then T is onto by Tietze's theorem, use Corollary 2.26.

2.55 Let X, Y be Banach spaces and $T \in \mathcal{B}(X, Y)$. Show that if Y is separable and T is onto Y, then there is a separable closed subspace Z of X such that $T(Z) = Y$.
Hint. Let $\{y_n\}$ be dense in B_Y and take $x_n \in X$ such that $T(x_n) = y_n$ and $\|x_n\| < K$ for some $K > 0$ (Corollary 2.26). Set $Z = \overline{\operatorname{span}}\{x_n\}$, clearly $T(Z) \subset Y$. By density

of $\{y_n\}$, $B_Y^O \subset \overline{T(KB_Z^O)}$, hence by Lemma 2.24 we have $B_Y^O \subset T(KB_Z^O)$. Thus $Y \subset T(Z)$.

2.56 Let Y be a closed subspace of a Banach space X. Assume that X/Y is separable. Denote by q the canonical quotient mapping of X onto X/Y. Show that there is a separable closed subspace $Z \subset X$ such that $q(Z) = X/Y$.
Hint. Apply the previous exercise.

2.57 Let X be a Banach space and let Y be a separable closed subspace of X^*. Then there is a separable closed subspace $Z \subset X$ such that Y is isometric to a subspace of Z^*.
Hint. Let $\{f_n\}$ be dense in S_{Y^*}. For every n, let $\{x_n^k\}_k \subset S_X$ be such that $f_n(x_n^k) \to 1$ as $k \to \infty$. Put $Z = \overline{\text{span}}\{x_n^k : n, k \in \mathbb{N}\}$.

2.58 Let X be the normed space of all real-valued functions on $[0, 1]$ with continuous derivative, endowed with the supremum norm. Define a linear mapping T from X into $C[0, 1]$ by $T(f) = f'$. Show that T has closed graph. Prove that T is not bounded. Explain why the closed graph theorem cannot be used here.
Hint. The graph of T is closed: let $(f_n, f_n') \to (f, g)$ in $X \oplus C[0, 1]$. Then $f_n \to f$ uniformly on $[0, 1]$ and $f_n' \to g$ uniformly. Hence by a standard result of real analysis, $f' = g$.

T is not bounded: use $\{f_n\}$ bounded with $\{f_n'\}$ unbounded. The space in question is not complete.

2.59 Let X be a closed subspace of $C[0, 1]$ such that every element of X is a continuously differentiable function on $[0, 1]$. Show that X is finite-dimensional.
Hint. Let $T \colon X \to C[0, 1]$ be defined for $f \in X$ by $T(f) = f'$. The graph of T is closed (see the previous exercise). Therefore T is continuous by the closed graph theorem.

Thus for some $n \in \mathbb{N}$ we have $\|f'\|_\infty \leq n$ whenever $f \in X$ satisfies $\|f\|_\infty \leq 1$. Let $x_i = \frac{i}{4n}$ for $i = 0, 1, \ldots, 4n$. Define an operator $S \colon X \to \mathbb{R}^{4n+1}$ by $S(f) = \{f(x_i)\}$. We claim that S is one-to-one. It is enough to show that if $\|f\|_\infty = 1$, then for some i, $S(f)(x_i) \neq 0$. Assume that this is not true. If $f(x) = 1$ and $x \in (\frac{i}{4n}, \frac{i+1}{4n})$, then by the Lagrange mean value theorem we have $|f(x) - f(\frac{i}{4n})| = |f'(\xi)||x - \frac{i}{4n}| \leq n \cdot \frac{1}{4n}$, a contradiction. Therefore $\dim(X) \leq 4n + 1$.

2.60 (Grothendieck) Let X be a closed subspace of $L_2[0, 1]$ whose every element belongs also to $L_\infty[0, 1]$. Show that $\dim(X) < \infty$.
Hint. The identity mapping from X to $(L_\infty[0, 1], \|\cdot\|_\infty)$ has a closed graph, so for some α we get $\|f\|_\infty \leq \alpha\|f\|_2$ for every $f \in X$. Let $\{f_1, \ldots, f_n\}$ be an orthonormal set in X. For every $x := \{x_1, \ldots x_n\} \in \mathbb{C}^n$ we put $f_x = \sum x_k f_k$. Then $|f_x(t)| \leq \alpha\|f_x\|_2 \leq \alpha\|x\|_2$ for almost all $t \in [0, 1]$ and so if Λ is a countable dense set in $\mathbb{C}^n$, there exists a set of measure zero N such that $|f_x(t)| \leq \alpha\|x\|_2$ for every $x \in \Lambda$ and every $t \in [0, 1] \backslash N$. Each mapping $x \mapsto f_x(t)$ from $\mathbb{C}^n$ to $\mathbb{C}$ is linear

and continuous, so $|f_x(t)| \leq \alpha\|x\|_2$ for all $x \in \mathbb{C}^n$ and $t \in [0,1]\backslash N$. In particular, $|f_x(t)| \leq \alpha$ for $x \in B_{\mathbb{C}^n}$ and $t \in [0,1]\backslash N$. The choice $x := (f_1(t), \dots, f_n(t))$ gives us $\sum |f_k(t)|^2 \leq \alpha^2$. Integration then gives $n = \|\sum f_k\|_2^2 = \sum \int |f_k(t)|^2 \, dt \leq \alpha^2$.

2.61 Show that the bounded linear one-to-one mapping ϕ from $L_1[0, 2\pi]$ into c_0 defined by $T(f) = \hat{f}(n)$, where $\hat{f}(n)$ are Fourier coefficients of f, is not onto c_0.
Hint. If T were onto c_0, then by the Banach open mapping theorem, T^{-1} would be bounded, which is not the case as the sequence $\{\chi_{\{1,\dots,n\}}\}$ shows (note that we have $\|D_n\|_1 \to \infty$, where D_n is the Dirichlet kernel).

2.62 Show that there is a linear functional L on ℓ_∞ with the following properties:

(1) $\|L\| = 1$,
(2) if $x := (x_i) \in c$, then $L(x) = \lim\limits_{i\to\infty} x_i$,
(3) if $x := (x_i) \in \ell_\infty$ and $x_i \geq 0$ for all i, then $L(x) \geq 0$,
(4) if $x := (x_i) \in \ell_\infty$ and $x' = (x_2, x_3, \dots)$, then $L(x) = L(x')$.

This functional is called a *Banach limit* or a *generalized limit*.
Hint. We propose several approaches.

(a) For simplicity we consider only the real scalars setting. Let M be the subspace of ℓ_∞ formed by elements $x - x'$ for $x \in \ell_\infty$ and x' as above. Let 1 denote the vector $(1, 1, \dots)$. We claim that $\text{dist}(1, M) = 1$. Note that $0 \in M$ and thus $\text{dist}(1, M) \leq 1$. Let $x \in \ell_\infty$. If $(x - x')_i \leq 0$ for any of i then $\|1 - (x - x')\|_\infty \geq 1$. If $(x - x')_i \geq 0$ for all i, then $x_i \geq x_{i+1}$ for all i, meaning that $\lim x_i$ exists. Therefore $\lim(x_i - x_i') = 0$ and thus $\|1 - (x - x')\| \geq 1$.

By the Hahn–Banach theorem, there is $L \in \ell_\infty^*$ with $\|L\| = 1$, $L(1) = 1$, and $L(m) = 0$ for all $m \in M$. This functional satisfies (1) and (4). To prove (2), it is enough to show that $c_0 \subset L^{-1}(0)$. To see this, for $x \in \ell_\infty$ we inductively define $x^{(1)} = x'$ and $x^{(n+1)} = (x^{(n)})'$ and note that by telescopic argument we have $x^{(n)} - x \in M$. Hence $L(x) = L(x^{(n)})$ for every $x \in \ell_\infty$ and every n. If $x \in c_0$ then $\|x^{(n)}\| \to 0$ and thus $L(x) = 0$. To show (3), assume that for some $x = (x_n)$ we have $x_i \geq 0$ for all i and $L(x) < 0$. By scaling, we may assume that $1 \geq x_i \geq 0$ for all i. Then $\|1 - x\|_\infty \leq 1$ and $L(1 - x) = 1 - L(x) > 1$, a contradiction with $\|L\| = 1$.

(b) For $x := (x_i) \in \ell_\infty$, $k \in \mathbb{N}$ and $n_1 < \dots < n_k$ in $\mathbb{N}$, put $\pi(x; n_1, \dots, n_k) = \limsup_n \frac{1}{k} \sum_{i=1}^k x_{n+n_i}$ and $p(x) = \inf\{\pi(x; n_1, \dots, n_k) : n_1 < \dots < n_k,\ k \in \mathbb{N}\}$. Then p is a convex function on ℓ_∞. Deduce the existence of a continuous linear functional $L : \ell_\infty \to \mathbb{R}$ such that $L \leq p$ and check the sought properties of L.

and continuous, so $|f_i(t)| \le c\|x\|_2$ for all $x \in \mathbb{C}^n$ and $t \in [0,1] \setminus N$. In particular, $|f_i(t)| \le c$ for $x \in B_{\ell_2^n}$ and $t \in [0,1] \setminus N$. The choice $x = (f_1(t), \dots, f_n(t))$ gives $\sum_i |f_i(t)|^2 \le c^2$. Integration then gives $n = \|\sum f_i\|_2^2 = \sum \int |f_i(t)|^2 \, dt \le c^2$.

2.61 Show that the bounded linear one-to-one mapping $\mathcal{F}$ from $L_1[0, 2\pi]$ into c_0 defined by $\mathcal{F}(f) = \hat{f}(n)$, where $\hat{f}(n)$ are Fourier coefficients of f, is not onto c_0.

Hint. If $\mathcal{F}$ were onto c_0, then by the Banach open mapping theorem, $\mathcal{F}^{-1}$ would be bounded, which is not the case as the sequence $\{f(n)\}_{n \in \mathbb{Z}}$ shows (note that we have $\|D_n\|_1 \to \infty$, where D_n is the Dirichlet kernel).

2.62 Show that there is a linear functional L on ℓ_∞ with the following properties:

(1) $\|L\| = 1$.

(2) If $x = (x_i) \in c$, then $L(x) = \lim_{i \to \infty} x_i$.

(3) If $x = (x_i) \in \ell_\infty$ and $x_i \ge 0$ for all i, then $L(x) \ge 0$.

(4) If $x = (x_i) \in \ell_\infty$ and $x' = (x_2, x_3, \dots)$, then $L(x) = L(x')$.

This functional is called a *Banach limit* or a *generalized limit*.

Hint. We propose several approaches.

(a) For simplicity, we consider only the real scalars setting. Let M be the subspace of ℓ_∞ formed by elements $x - x'$ for $x \in \ell_\infty$ and x' as above. Let 1 denote the vector $(1, 1, \dots)$. We claim that $\operatorname{dist}(1, M) = 1$. Note that $0 \in M$ and thus $\operatorname{dist}(1, M) \le 1$. Let $x \in \ell_\infty$. If $(x - x')_i \le 0$ for any i, then $\|1 - (x - x')\|_\infty \ge 1$. If $(x - x')_i > 0$ for all i, then $x_i \ge x_{i+1}$ for all i, meaning that $\lim x_i$ exists. Therefore $\lim(x_i - x_{i+1}) = 0$ and thus $\|1 - (x - x')\|_\infty \ge 1$.

By the Hahn–Banach theorem, there is $L \in \ell_\infty^*$ with $\|L\| = 1$, $L(1) = 1$, and $L(m) = 0$ for all $m \in M$. This functional satisfies (1) and (4). To prove (2), it is enough to show that $c_0 \subset L^{-1}(0)$. To see this, for $x \in \ell_\infty$ we inductively define $x^{(1)} = x$ and $x^{(n+1)} = (x^{(n)})'$ and note that by telescopic argument we have $x^{(n)} - x \in M$. Hence $L(x) = L(x^{(n)})$ for every $x \in \ell_\infty$ and every n. If $x \in c_0$, then $\|x^{(n)}\|_\infty \to 0$ and thus $L(x) = 0$. To show (3), assume that for some $x = (x_n)$ we have $x_i \ge 0$ for all i and $L(x) < 0$. By scaling, we may assume that $1 \ge x_i \ge 0$ for all i. Then $\|1 - x\|_\infty \le 1$ and $L(1 - x) = 1 - L(x) > 1$, a contradiction with $\|L\| = 1$.

(b) For $x = (x_i) \in \ell_\infty$, $k \in \mathbb{N}$ and $n_1 < \dots < n_k$ in $\mathbb{N}$, put $\pi(x; n_1, \dots, n_k) = \limsup_j \frac{1}{k} \sum_{i=1}^{k} x_{n_i + j}$ and $p(x) = \inf\{\pi(x; n_1, \dots, n_k) : n_1 < \dots < n_k, k \in \mathbb{N}\}$. Then p is a convex function on ℓ_∞. Deduce the existence of a continuous linear functional $L : \ell_\infty \to \mathbb{R}$ such that $L \le p$ and check the sought properties of L.

Chapter 3
Weak Topologies and Banach Spaces

An indispensable tool in the study of deeper structural properties of a Banach space X is its *weak topology*, i.e., the topology on X of the pointwise convergence on elements of the dual space X^*, or the *weak* topology* on X^*, i.e., the topology on X^* of the pointwise convergence on elements of X. The topology on X^* of the uniform convergence on the family of all convex balanced and weakly compact subsets of X plays also an important role. All those topologies can be efficiently studied in the general framework of topological vector spaces.

This allows the use of Tychonoff's compactness theorem for weak* topologies in duals to Banach spaces and the results of Banach–Dieudonné type. We also study extreme points, the Choquet representation theorem, properties of James boundaries, and characterizations of weakly compact sets. We briefly discuss nonlocally convex spaces and the space of distributions.

We then study the third fundamental principle of Functional Analysis—the Hahn–Banach theorem and the Banach open mapping principle being the subject of the previous chapter—, namely the Banach–Steinhaus uniform boundedness principle.

Finally we introduce and study reflexive Banach spaces.

3.1 Dual Pairs, Weak Topologies

Definition 3.1 *Let E be a vector space over the field $\mathbb{K}$ ($\mathbb{R}$ or $\mathbb{C}$). By $E^{\#}$ we denote the vector space of all linear functionals on E, i.e., $\mathcal{L}(E, \mathbb{K})$. This space is called the* algebraic dual of E.

Let E and F be vector spaces over the field $\mathbb{K}$ of real or complex numbers. A *bilinear form* on $E \times F$ is a mapping $\langle \cdot, \cdot \rangle : E \times F \to \mathbb{K}$ that is linear in each variable separately, i.e., for every $e_1, e_2, e \in E$, $f_1, f_2, f \in F$, $\alpha_1, \alpha_2 \in \mathbb{K}$,

(i) $\langle \alpha_1 e_1 + \alpha_2 e_2, f \rangle = \alpha_1 \langle e_1, f \rangle + \alpha_2 \langle e_2, f \rangle$, and

(ii) $\langle e, \alpha_1 f_1 + \alpha_2 f_2 \rangle = \alpha_1 \langle e, f_1 \rangle + \alpha_2 \langle e, f_2 \rangle$.

The choice of the notation is done in order to emphasize the symmetry between the roles played by E and F.

M. Fabian et al., *Banach Space Theory*, CMS Books in Mathematics,
DOI 10.1007/978-1-4419-7515-7_3, © Springer Science+Business Media, LLC 2011

A subset S of F *separates points of* E under the bilinear form $\langle\cdot,\cdot\rangle$ if, given $e \in E$ such that $\langle e, s\rangle = 0$ for all $s \in S$, then $e = 0$. If the space E and the bilinear form are understood, we also say that S is *separating*.

Definition 3.2 *A couple of vector spaces E, F over the field $\mathbb{K}$, together with a bilinear form $\langle\cdot,\cdot\rangle : E \times F \to \mathbb{K}$, is said to form a* dual pair *if, under the bilinear form, E separates points of F and F separates points of E. For simplicity, we shall write $\langle E, F\rangle$ to denote a dual pair.*

Remark: Given a dual pair $\langle E, F\rangle$, the mapping ϕ that to an element $e \in E$ associates the element $\phi(e) \in F^{\#}$ given by $\phi(e)(f) = \langle e, f\rangle$ for every $f \in F$ is linear and one-to-one thanks to the fact that F separates points of E. In this way, E is identified with a (linear) subspace of $F^{\#}$. In the same way, F is identified with a (linear) subspace of $E^{\#}$. Indeed, the roles of E and F can be reversed. These natural identifications will be made in the sequel and we shall use indistinctly the notation $\langle e, f\rangle$, $\langle f, e\rangle$, $e(f)$, or $f(e)$, where $e \in E$ and $f \in F$, when dealing with a dual pair $\langle E, F\rangle$, if there is no risk of misunderstanding. Usually, the notation $f(e)$ or, equivalently, $\langle f, e\rangle$, will be adopted when it is important to stress that f is understood as a function of the variable e.

Recall that, given a nonempty set Γ and a family $\{S_\gamma : \gamma \in \Gamma\}$ of nonempty sets S_γ, the *product* $\prod_{\gamma\in\Gamma} S_\gamma$ is the set of all elements $(s_\gamma)_{\gamma\in\Gamma}$, where $s_\gamma \in S_\gamma$ for all $\gamma \in \Gamma$. A particular case is when all S_γ coincide with a certain set S. In this case the product is called the *power set* S^Γ, and it can be identified with the set of all functions $f : \Gamma \to S$. Dealing with elements in the power set, we shall adopt the functional notation $f(\gamma)$, $\gamma \in \Gamma$ or, alternatively, the vector notation $(f_\gamma)_{\gamma\in\Gamma}$, where $f_\gamma := f(\gamma)$ for all $\gamma \in \Gamma$. If S is a vector space, the *support* of an element $f \in S^\Gamma$ is the set $\operatorname{supp}(f) := \{\gamma \in \Gamma : f(\gamma) \neq 0\}$.

Some examples of dual pairs follow:

1. Let E be a vector space. Then $\langle E, E^{\#}\rangle$ is a dual pair under the bilinear form $\langle f, e\rangle := f(e)$ for $e \in E$ and $f \in E^{\#}$. Obviously E separates points of $E^{\#}$. That $E^{\#}$ separates points of E follows from a simple linear algebra argument.
2. Let X be a normed space and let X^* be its dual space (see Definition 1.28). Then $\langle X, X^*\rangle$ is a dual pair under the bilinear form $\langle x^*, x\rangle := x^*(x)$, where $(x, x^*) \in X \times X^*$. Obviously X separates points of X^*; the fact that X^* separates points of X is a consequence of the Hahn–Banach theorem.
3. Let K be a compact space, and let $C(K)$ be the Banach space of all continuous scalar-valued functions on K, equipped with the supremum norm (see Definition 1.7, Proposition 1.8, and the subsequent note). For $k \in K$ define the element δ_k in $C(K)^*$ by $\delta_k(f) = f(k)$ for all $f \in C(K)$. Then $\langle C(K), \operatorname{span}(\{\delta_k : k \in K\})\rangle$, where the bilinear form is the one induced by the bilinear form on $C(K) \times C(K)^*$ defined in Example 1 above, is a dual pair. The fact that $C(K)$ separates points of $\operatorname{span}(\{\delta_k : k \in K\})$ is a consequence of the Tietze–Urysohn theorem. Obviously, $\operatorname{span}(\{\delta_k : k \in K\})$ separates points of $C(K)$.

4. Let Γ be a nonempty set and $\varphi(\Gamma)$ be the vector space of all the finitely supported vectors in $\mathbb{K}^\Gamma$. Then $\langle \mathbb{K}^\Gamma, \varphi(\Gamma)\rangle$ is a dual pair under the bilinear form $\langle x, y\rangle := \sum_{\gamma\in\Gamma} x_\gamma y_\gamma$, where $x = (x_\gamma) \in \mathbb{K}^\Gamma$ and $y = (y_\gamma) \in \varphi(\Gamma)$.

We introduce now a natural topology associated to a dual pair on one of the vector spaces involved. We stress that the symmetric character of the dual pair allows us to work either in one or the other of the two vector spaces involved. Since it is a particular case of a product topology (also called topology of the pointwise convergence), let us recall here first this well-known concept.

Definition 3.3 *Let $\{(S_\gamma, \mathcal{T}_\gamma)\}_{\gamma\in\Gamma}$ be a family of Hausdorff topological spaces. The product set $\prod_{\gamma\in\Gamma} S_\gamma$ becomes a (Hausdorff) topological space when equipped with the* product topology *$\mathcal{T}_p$ (sometimes called the* topology of the pointwise convergence*). A base for this topology is given by the family of sets (called* elementary open sets*) of the form $\prod_{\gamma\in\Gamma} O_\gamma$, where O_γ is an open set in S_γ and each O_γ is equal to the corresponding S_γ but for a finite number of $\gamma \in \Gamma$. In particular, if all $(S_\gamma, \mathcal{T}_\gamma)$ are equal (to some $(S, \mathcal{T})$) we speak of the* power topological space *$(S^\Gamma, \mathcal{T}_p)$.*

That the topology $\mathcal{T}_p$ on $\prod_{\gamma\in\Gamma} S_\gamma$ is Hausdorff as soon as the topology $\mathcal{T}_\gamma$ on S_γ is Hausdorff for every $\gamma \in \Gamma$ can be easily proved by looking at the way the elementary open sets are defined.

It is standard to check that a net $\{f_i\}_{i\in I}$ in $\prod_{\gamma\in\Gamma} S_\gamma$ $\mathcal{T}_p$-converges to an element $f \in \prod_{\gamma\in\Gamma} S_\gamma$ if and only if $\{f_i(\gamma)\}_{i\in I}$ $\mathcal{T}_\gamma$-converges to $f(\gamma)$ for every $\gamma \in \Gamma$.

If S is a topological space and nothing is said on the contrary, the power set S^Γ will be always considered as a topological space in the pointwise topology $\mathcal{T}_p$.

Tychonoff's theorem (see, e.g., [Enge, Theorem 3.2.4]) states that an arbitrary product of compact spaces is itself compact. (This statement is equivalent to Zorn's lemma, or to the Axiom of Choice, see, e.g., [Kell].)

The following simple example justifies the use of nets in convergence (an introduction to the concept of net is done in Section 17.2), since sequences are not enough in this case for describing the topology. Let Γ be an uncountable set. Let $\psi(\Gamma)$ be the subspace of $\mathbb{R}^\Gamma$ of all the countably supported elements in $\mathbb{R}^\Gamma$. Then $\overline{\psi(\Gamma)} = \mathbb{R}^\Gamma$, while there is no sequence in $\psi(\Gamma)$ that converges to the element $f \in \mathbb{R}^\Gamma$ defined by $f(\gamma) = 1$ for every $\gamma \in \Gamma$.

If $\langle E, F\rangle$ is a dual pair, we agreed to identify E with a (linear) subspace of $F^\#$, in turn a (linear) subspace of $\mathbb{K}^F$ (see the paragraph after Definition 3.2). In this way, we have the following definition.

Definition 3.4 *Let $\langle E, F\rangle$ be a dual pair. The* weak topology *on E associated to the dual pair, denoted $w(E, F)$, is the restriction to E of the topology on $\mathbb{K}^F$ of the pointwise convergence.*

The following proposition gives two more descriptions of the topology $w(E, F)$, and it is a straightforward consequence of the nature of the product, i.e., pointwise, topology on $\mathbb{K}^F$.

Proposition 3.5 *Let $\langle E, F\rangle$ be a dual pair. Then*

(i) *A base of neighborhoods of an element* $e_0 \in E$ *in the topology* $w(E, F)$ *is given by*

$$\{U(e_0; f_1, \ldots, f_n; \varepsilon) : f_1, \ldots, f_n \in F, \ \varepsilon > 0, \ n \in \mathbb{N}\}, \tag{3.1}$$

where $U(e_0; f_1, \ldots, f_n; \varepsilon) := \{e \in E : |\langle e - e_0, f_i \rangle| < \varepsilon, \ i = 1, 2, \ldots, n\}$.
(ii) *A net* $\{e_i\}$ *in* E *is* $w(E, F)$*-convergent to an element* $e \in E$ *if and only if* $\langle e_i, f \rangle \to \langle e, f \rangle$ *for every* $f \in F$.

Since $\mathbb{K}^F$ is Hausdorff, the topology $w(E, F)$ is Hausdorff, too, whenever $\langle E, F \rangle$ is a dual pair.

If E is a vector space, we saw in Example 1 above that $\langle E, E^\# \rangle$ is a dual pair, so we can endow $E^\#$ with the topology $w(E^\#, E)$. The following proposition gives a precise description of the sort of space we obtain.

Proposition 3.6 *Let* E *be a vector space not reduced to* $\{0\}$*. Then,* $\big(E^\#, w(E^\#, E)\big)$ *is linearly isomorphic to some* $(\mathbb{K}^B, \mathcal{T}_p)$*, where* $\operatorname{card}(B)$ *is the algebraic dimension of* E*.*

Proof: Let B be an algebraic basis of E. We claim that the mapping $\phi : (E^\#, w(E^\#, E)) \to (\mathbb{K}^B, \mathcal{T}_p)$ given by $\phi(f) = f|_B$ for $f \in E^\#$ is a linear isomorphism. Linearity is clear, and the fact that B is an algebraic basis gives that ϕ is one-to-one and onto. If a net $\{f_\alpha\}$ in $E^\#$ is $w(E^\#, E)$-convergent to some $f \in E^\#$, then $\{\phi(f_\alpha)\}$ is $\mathcal{T}_p$-convergent to $\phi(f)$, since $f_\alpha(b) \to_\alpha f(b)$ for all $b \in B$ (see Proposition 3.5). On the other hand, if $f_\alpha(b) \to_\alpha f(b)$ for all $b \in B$, then linearity gives $f_\alpha(e) \to_\alpha f(e)$ for all $e \in E$. Again Proposition 3.5 implies that $f_\alpha \to f$ in the topology $w(E^\#, E)$. □

3.2 Topological Vector Spaces

Definition 3.7 *Let* E *be a vector space and let* $\mathcal{T}$ *be a Hausdorff topology on* E *such that the operations* $(x, y) \in E \times E \mapsto x + y$ *and* $(x, \alpha) \in E \times \mathbb{K} \mapsto \alpha x$ *are continuous on* $E \times E$ *and* $E \times \mathbb{K}$*, respectively. Then* $(E, \mathcal{T})$ *is called a* topological vector space.
If the topology $\mathcal{T}$ *is clear from the context, we shall write briefly* E *for the topological vector space* $(E, \mathcal{T})$*.*

A normed space in its norm topology is a particular instance of a topological vector space. Many of the concepts introduced in the context of normed spaces extend naturally to the more general setting of topological vector spaces. For example, a subset M of a topological vector space $(E, \mathcal{T})$ is said to be *linearly dense* if $\overline{\operatorname{span}}(M) = E$. Two topological vector spaces E and F are called *linearly isomorphic* (or just *isomorphic*) if there exists a continuous and one-to-one operator from E onto F such that its inverse mapping is also continuous (if *isomorphism* is meant only in the topological sense, with no linearity involved, we will mention this explicitly).

If E is a topological vector space, for every $x_1, x_2 \in E$ and a neighborhood W of $x_1 + x_2$ there are neighborhoods V_1 and V_2 of x_1 and x_2 respectively such that $V_1 + V_2 \subset W$, where $V_1 + V_2 := \{x + y : x \in V_1, y \in V_2\}$. Also, for every $x \in E$, $\alpha \in \mathbb{K}$ and a neighborhood W of αx there is a neighborhood V of x in E and $\delta > 0$ such that $\beta V \subset W$ for every $|\beta - \alpha| < \delta$, where $\beta V := \{\beta v : v \in V\}$.

We have the following statement.

Fact 3.8 *Let X be a topological vector space.*
(i) *For every $a \in E$, the translation operator T_a defined for $x \in E$ by $T_a(x) = x + a$ is a homeomorphism of E onto E.*
(ii) *For every $\alpha \in \mathbb{K}$, $\alpha \neq 0$, the multiplication operator M_α defined for $x \in E$ by $M_\alpha(x) = \alpha x$ is a homeomorphism of E onto E.*

By a *local base* for a topological vector space $(E, \mathcal{T})$ we mean a base of neighborhoods of the origin 0, that is, a collection $\mathcal{B}$ of neighborhoods of 0 such that every neighborhood of 0 contains an element of $\mathcal{B}$.

Lemma 3.9 *If $\mathcal{B}$ is a local base of a topological vector space E, then every set from $\mathcal{B}$ contains the closure of some set from $\mathcal{B}$.*
In particular, E has a local base consisting of closed sets.

Proof: Let $U \in \mathcal{B}$. Since $(x, y) \mapsto x - y$ is continuous, we find an open neighborhood V of zero in E such that $V - V \subset U$, that is, $V \cap \big((E \setminus U) + V\big) = \emptyset$. Since $(E \setminus U) + V$ is an open set, we have $\overline{V} \cap \big((E \setminus U) + V\big) = \emptyset$, in particular $\overline{V} \cap (E \setminus U) = \emptyset$ as $0 \in V$. Thus $\overline{V} \subset U$, and it is enough to choose an element $B \in \mathcal{B}$ such that $B \subset V$ to obtain the conclusion. $\square$

The fact that for every $a \in E$ the translation operator T_a is a homeomorphism implies that, if $\mathcal{B}$ is a local base for the topological vector space $(E, \mathcal{T})$, then $\{a + B : B \in \mathcal{B}\}$ is a base of neighborhoods of a.

Symmetric and balanced sets were introduced in Chapter 1 in the context of normed spaces. Those definitions make sense in any vector space over the field $\mathbb{K}$.

Note that a convex set in a real vector space is symmetric (see Exercise 1.2) if and only if it is balanced. The two concepts do not agree in complex vector spaces.

Proposition 3.10 *Every topological vector space E has a local base consisting of balanced sets.*

Proof: Let U be a neighborhood of 0 in E. By the continuity of scalar multiplication, there is $\delta > 0$ and a neighborhood V of 0 in E such that $\alpha V \subset U$ for every $|\alpha| < \delta$. Put $W = \bigcup_{|\alpha| < \delta} \alpha V$. Then W is a balanced neighborhood of 0 in E and $W \subset U$. $\square$

A topological vector space $(E, \mathcal{T})$ is an example of a uniform space (see Section 17.10). Henceforth, all concepts associated to uniformities (like system of vicinities, Cauchy nets, completeness, and so on) are applicable to topological vector spaces. For example, a net $\{x_\alpha\}_{\alpha\in I}$ in E is *Cauchy* if for every neighborhood U of 0 in E there is $\alpha_0 \in I$ such that $(x_\alpha - x_\beta) \in U$ whenever $\alpha_0 \le \alpha, \beta$. E is *complete* if every Cauchy net is convergent. In the case of normed spaces, completeness can be checked just by using sequences (see Exercise 3.11). Thus, a normed space X is complete, i.e., it is a Banach space, if every Cauchy sequence in X converges to some point in X. Similarly to the case of normed spaces, every topological vector space $(E, \mathcal{T})$ can be embedded in a smallest complete topological vector space, its *completion* $(\widetilde{E}, \widetilde{\mathcal{T}})$. It is unique up to linear isomorphisms. The closures in $(\widetilde{E}, \widetilde{\mathcal{T}})$ of the elements of a local base of $(E, \mathcal{T})$ form a local base of $(\widetilde{E}, \widetilde{\mathcal{T}})$.

In the same way that in the case of normed spaces (see the discussion after Proposition 1.29), to every topological vector space $(E, \mathcal{T})$ over the field $\mathbb{C}$ we associate a topological vector space $(E_{\mathbb{R}}, \mathcal{T})$ over the field $\mathbb{R}$. There is a one-to-one correspondence between complex and real linear functionals that to $f : E \to \mathbb{C}$ associates $\operatorname{Re} f$. In this way, to a real linear functional $u : E_{\mathbb{R}} \to \mathbb{R}$ it corresponds the complex linear functional $f : E \to \mathbb{C}$ defined by $f(x) = u(x) - iu(ix)$ for $x \in E$ (so $\operatorname{Re} f = u$). Obviously, f is continuous if and only if $\operatorname{Re} f$ is continuous.

Given a topological vector space $(E, \mathcal{T})$ and a vector subspace G of E, clearly G becomes a topological vector space when endowed with the restriction of the topology $\mathcal{T}$. Instead of introducing a new symbol for this restricted topology, we shall denote it again $\mathcal{T}$, and so the topological vector *subspace* will be denoted $(G, \mathcal{T})$.

Let $\{(E_\gamma, \mathcal{T}_\gamma) : \gamma \in \Gamma\}$ be a collection of topological vector spaces. The vector space $\prod_{\gamma\in\Gamma} E_\gamma$ becomes a topological vector space when endowed with the product topology introduced in Definition 3.3. Checking this fact is a simple exercise. In particular, if $(E, \mathcal{T})$ is a topological vector space and Γ is a nonempty set, then $(E^\Gamma, \mathcal{T}_p)$ is a topological vector space, where $\mathcal{T}_p$ denotes the product topology.

Let $(E, \mathcal{T})$ be a topological vector space, and let F be a closed (linear) subspace. The set E/F, called the *quotient* of E and F, is the vector space of all the *cosets* $x + F$ of E, $x \in E$. The mapping $q : E \to E/F$ that maps $x \in E$ to the coset $x + F$ is called the *canonical quotient mapping*. When the vector space E/F is endowed with the *quotient topology* $\mathcal{T}_q$, i.e., the (Hausdorff) topology on E/F consisting on all the sets $\widehat{U}$ such that $q^{-1}(\widehat{U}) \in \mathcal{T}$, it becomes a topological vector space, the *quotient space* $(E/F, \mathcal{T}_q)$. (Observe that the elements of $\mathcal{T}_q$ are precisely the images by q of the elements of $\mathcal{T}$.) That the topology $\mathcal{T}_q$ is indeed a topology and that, moreover, it makes E/F a topological vector space can be checked easily. The mapping $q : E \to E/F$ is onto, continuous and open.

We provide now several natural examples of topological vector spaces.

1. A normed space in the topology defined by the norm is an example of a topological vector space.
2. Let X be a normed space, Γ a nonempty set. Then, the space $(X^\Gamma, \mathcal{T}_p)$ is a topological vector space.
3. Given a dual pair $\langle E, F\rangle$, the space $\big(E, w(E, F)\big)$ is a topological vector space. It is enough to recall that $\big(E, w(E, F)\big)$ is just a (linear) subspace of the topo-

logical vector space $(\mathbb{K}^F, \mathcal{T}_p)$. In particular, every normed space X in its weak topology, or its dual in its weak* topology (see Definition 3.20), is topological vector spaces.

4. Given $0 < p < 1$, put $\ell_p = \{(x_n) \in \mathbb{K}^{\mathbb{N}} : \sum_{i=1}^{\infty} |x_n|^p < \infty\}$. The space ℓ_p, with the topology induced by q_p, where $q_p(x) := (\sum_{n=1}^{\infty} |x_n|^p)^{1/p}$, $x = (x_n) \in \ell_p$, is a topological vector space. However, it is not a normed space, see Exercise 3.6.

Bounded and totally bounded sets were already introduced in the setting of normed spaces. The corresponding notions for topological vector spaces are defined below. See also Section 17.10.

Definition 3.11 *Let A be a subset of a topological vector space E.*
A is called bounded *if for every neighborhood U of 0 in E there is $\alpha > 0$ such that $\alpha A \subset U$.*
A is called totally bounded *if for every neighborhood U of zero in E there is a finite set $F \subset E$ such that $A \subset F + U$.*

Note that every totally bounded set A in a topological vector space E is bounded. Indeed, given a balanced neighborhood W of zero in E, we use the continuity of the addition to obtain a balanced neighborhood V of zero in E such that $V + V \subset W$. Since A is totally bounded, there is a finite set $F \subset E$ such that $A \subset F + V$. Let $F = \{x_1, \ldots, x_n\}$. From the continuity of the scalar multiplication, we find $1 > \delta_i > 0$ such that $\delta x_i \in V$ whenever $0 < \delta \leq \delta_i$. Set $\delta = \min\{\delta_i\}$, then $\delta x_i \in V$ for every i. From this and the balancedness of V we obtain $\delta A \subset \delta F + \delta V \subset V + V \subset W$. This shows that A is bounded in E.

Obviously, every compact subset of a topological vector space is totally bounded.

From the description of a local base for the topology w (see Proposition 3.5), we get easily the following result.

Proposition 3.12 *Let $\langle E, F\rangle$ be a dual pair. A subset A of E is $w(E, F)$-bounded if and only if $\sup_{x \in A} |\langle x, f\rangle| < +\infty$ for every $f \in F$.*

The following result extends Proposition 1.36 to the class of topological vector spaces.

Proposition 3.13 *Let E be a topological vector space. If $\dim(E) = n$ then E is linearly homeomorphic to ℓ_2^n.*

Proof: Let $\{e_1, \ldots, e_n\}$ be a basis of E. We define a mapping u from ℓ_2^n onto E for $x := (x_i)$ by $u(x) = \sum_{i=1}^{n} x_i e_i$.

From the continuity of vector operations in E, it follows that u is a continuous linear bijection of ℓ_2^n onto E. To complete the proof we need to show that u^{-1} is continuous. Let B_1 denote the unit ball of ℓ_2^n, we will show that $u(B_1)$ contains a neighborhood of 0 in E. By the linearity of u and of the topologies of E and ℓ_2^n, this will prove continuity at all points of E.

Let S_1 be the unit sphere in ℓ_2^n. Then S_1 is compact in ℓ_2^n and since u is continuous, $u(S_1)$ is compact in E. As u is one-to-one, we have that $0 \notin u(S_1)$. Since E is a Hausdorff space, for every $s \in u(S_1)$ there are neighborhoods U_s and V_s of s and 0 respectively such that $U_s \cap V_s \neq \emptyset$.

By compactness, there are $s_1, \dots, s_p \in u(S_1)$ be such that $u(S_1) \subset \bigcup_{i=1}^{p} U_{s_i}$. Put $V = \bigcap_{i=1}^{p} V_{s_i}$. Then V is a neighborhood of zero in E such that $V \cap u(S_1) = \emptyset$. From Proposition 3.10 it follows that there is a balanced neighborhood W of zero in E such that $W \subset V$ and thus $W \cap u(S_1) = \emptyset$. We claim that $W \subset u(B_1)$. Indeed, if for some $w \in W$ we have $w \notin u(B_1)$, then $w = u(v)$ for some $v \notin B_1$. Then $\frac{v}{\|v\|} \in S_1$ and $u\left(\frac{v}{\|v\|}\right) = \frac{w}{\|v\|} \in W$ since W is balanced, $w \in W$ and $\|v\| > 1$. The proof is complete. □

Corollary 3.14 *Let F be a subspace of a topological vector space E. If F is finite-dimensional then F is closed in E.*

Proof: Since ℓ_2^n is a complete normed space, by Exercise 3.11 it is also complete as a topological vector space. By Proposition 3.13 the same is true for F. Let $y \in \overline{F}$. Then there is a net $\{x_\alpha\} \subset F$ such that $\lim x_\alpha = y$. Clearly $\{x_\alpha\}$ is a Cauchy net in F, so it converges to some element in F. Hence $y \in F$. □

Corollary 3.15 *Let E be a finite-dimensional vector space. Then, all the topological vector space topologies on E coincide (i.e., they have the same family of open sets).*

Corollary 3.16 *Let E be a finite-dimensional vector space and let F be a topological vector space. Then, every linear mapping from E into F is continuous.*

Proof: Let $T : E \to F$ be a linear mapping. It is enough to prove that $T : E \to T(E)$ is continuous. The result follows from Propositions 3.13 and 1.39, since $T(E)$ is a finite-dimensional topological vector space. □

Proposition 3.17 *Let E be a topological vector space. If E has a totally bounded neighborhood of zero, then E is finite-dimensional (and conversely).*

Proof: Let U be a totally bounded neighborhood of 0. It is also bounded, hence $\{2^{-n}U\}_{n=1}^{\infty}$ forms a local base of E. By the total boundedness of U, there is a finite set $A \subset E$ such that $U \subset A + \frac{1}{2}U$. Let $F = \operatorname{span}(A)$. Then $U \subset F + \frac{1}{2}U$. Multiplying this inclusion by $\frac{1}{2}$ we have $\frac{1}{2}U \subset F + \frac{1}{4}U$. Combining these inclusions we have $U \subset F + F + \frac{1}{4}U = F + \frac{1}{4}U$. By induction we obtain $U \subset F + 2^{-n}U$ for every n. Therefore $U \subset \bigcap(F + 2^{-n}U) \subset \overline{F}$. The last inclusion follows from the fact that $\{2^{-n}U\}$ forms a local base in E. Since F is closed in E by Corollary 3.14, we have $U \subset F$. Given $x \in E$, we have $0 \cdot x \in U$; thus, by continuity there is $\delta > 0$ such that $\delta x \subset U \subset F$ and hence $x \in F$. Therefore $F = E$ and E is finite-dimensional.

If E is finite-dimensional, then E is linearly isomorphic to ℓ_2^n for some $n \in \mathbb{N}$ (see Proposition 3.13). Use now Theorem 1.38. □

The following notion was introduced in Definition 1.28 for the case of a normed space.

Definition 3.18 *Let E be a topological vector space. The space E^* denotes the set of all continuous linear functionals on E, and it is called the* (topological) dual *space of E.*

Proposition 3.19 *Let E be a topological vector space and let f be a linear functional on E. The following are equivalent:*
(i) *f is continuous on E, that is, $f \in E^*$.*
(ii) *f is continuous at some point of E.*
(iii) *There is a neighborhood $U(x_0)$ of some $x_0 \in E$ such that $|f|$ is bounded on $U(x_0)$.*
(iv) Ker *f is a closed subspace of E.*
(v) Ker *f is not dense in E.*

Proof: The implications (i)$\Longrightarrow$(ii)$\Longrightarrow$(iii) are trivial.

(iii)$\Longrightarrow$(iv): By translating we get that $|f|$ is bounded (say by M) on some neighborhood $U(0)$ of 0. Let (x_i) be a net in Ker f that converges to some $x \in E$. Fix $n \in \mathbb{N}$. Since $(nx_i) \to nx$, there exists i_0 such that $(nx_i - nx) \in U(0)$ for all $i \geq i_0$. In particular, $|f(nx_i) - f(nx)| \leq M$, hence $|f(x)| = |f(x_i) - f(x)| \leq M/n$ for all $i \geq i_0$. Since $n \in \mathbb{N}$ is arbitrary, $f(x) = 0$, so $x \in \text{Ker } f$, hence Ker f is closed.

(iv)$\Longrightarrow$(v) is obvious.

(v)$\Longrightarrow$(iv): We have $\text{Ker } f \subset \overline{\text{Ker } f} \subsetneq E$. Since $\overline{\text{Ker } f}$ is a subspace of E, and Ker f is a one-codimensional subspace of E, we necessarily have $\text{Ker } f = \overline{\text{Ker } f}$.

(iv)$\Longrightarrow$(i): Without loss of generality we may assume that E is a real topological vector space. If Ker f is closed, it follows from Exercise 3.1 that for every $\alpha \in \mathbb{R}$, $\{x \in E : f(x) \leq \alpha\}$ and $\{x \in E : f(x) \geq \alpha\}$ are both closed. We shall prove that f is continuous at 0 (and this will obviously suffice, by linearity). Assume, on the contrary, that (x_i) is a net in E that converges to 0 and that the net $(f(x_i))$ does not converge to 0. By passing to a subnet if necessary, we may assume that , for some $\varepsilon > 0$, $|f(x_i)| \geq \varepsilon$ for all i. Since $\{x \in E : f(x) \leq -\varepsilon\} \cup \{x \in E : f(x) \geq \varepsilon\}$ is closed, we get $|f(0)| \geq \varepsilon$, a contradiction. □

In particular, a closed hyperplane in E is the kernel of a continuous linear functional on E and conversely. Indeed, if H is a closed hyperplane of E, then (algebraically) $E = H \oplus [x_0]$ for some $0 \neq x_0 \in E$, so the mapping $f : E \to \mathbb{K}$ defined by $f(x) = \lambda$ for $x := h + \lambda x_0$, $h \in H$, $\lambda \in \mathbb{K}$, is linear and $\text{Ker} f = H$. By the previous proposition, f is continuous. The converse is trivial.

Definition 3.20 *If E is a topological vector space, the topology $w(E, E^*)$ is called the* weak topology of *E, and will be denoted briefly by w. The topology $w(E^*, E)$ is called the* weak* topology of *E^*, and will be denoted briefly by w^*.*

If $(E, \mathcal{T})$ is a topological vector space, the weak topology of E is weaker that the topology $\mathcal{T}$. This follows directly from the description of a local base for the topology w (see Proposition 3.5, all w-neighborhoods of the form (3.1) are clearly $\mathcal{T}$-open sets) or, alternatively, from the fact that if a net $\{x_i\}$ in E $\mathcal{T}$-converges to $x \in E$ then $f(x_i) \to f(x)$ for every $f \in E^*$ (see Proposition 3.5 again).

If E is a vector space, we already mentioned that $\langle E, E^{\#}\rangle$ is a dual pair. On the other hand, if $\langle E, F\rangle$ is a dual pair, we can always consider F as a linear subspace of $E^{\#}$ (see Section 3.1). From the very definition of a dual pair, the subspace F of $E^{\#}$ separates points of E, so the topology $w(E, F)$ on E is Hausdorff.

By a slight abuse of the notation, we can also consider the case of a nonempty (otherwise arbitrary) subset F of $E^{\#}$ (although F would not separate points of E), and the topology $w(E, F)$ on E of the pointwise convergence on all points in F (which may not be Hausdorff).

In Proposition 3.2 we shall describe the dual space of a space endowed with a weak topology of this type. We shall need the following lemma.

Lemma 3.21 *Let X be a vector space and let $f, f_1, f_2, \dots, f_n$ be linear functionals on X. If $\bigcap_{i=1}^{n} f_i^{-1}(0) \subset f^{-1}(0)$, then f is a linear combination of $f_1, \dots, f_n$ (and conversely).*

We use the following fact from linear algebra.

Claim:
Let E_1, E_2, E_3 be vector spaces, let $f\colon E_1 \to E_3$ and $g\colon E_1 \to E_2$ be linear mappings. There is a linear mapping $h\colon E_2 \to E_3$ such that $f = h \circ g$ if and only if $g^{-1}(0) \subset f^{-1}(0)$.

Proof of the Claim: Suppose that $g^{-1}(0) \subset f^{-1}(0)$. Define $h\colon g[E_1] \to E_3$ by $h(g(x)) = f(x)$ for $x \in E$. To check the consistence, assume $g(x_1) = g(x_2)$. Then $(x_1 - x_2) \in g^{-1}(0) \subset f^{-1}(0)$, so $f(x_1) = f(x_2)$. Extend h to a linear mapping on E_2, clearly $h(g) = f$. The other implication is clear. $\triangle$

Proof of Lemma 3.21: Using the Claim with $E_1 := X$, $E_2 := \mathbb{R}^n$, $E_3 := \mathbb{R}$, $f := f$ and $g(x) := \big(f_1(x), \dots, f_n(x)\big)$ we get that there is a linear mapping $h\colon \mathbb{R}^n \to \mathbb{R}$ such that $f(x) = h(g(x))$. The mapping h can be written as $h(y) = \sum \alpha_i y_i$ for some $\alpha_1, \dots, \alpha_n$ and every $y = (y_i) \in \mathbb{R}^n$. Therefore $f(x) = \sum \alpha_i f_i(x)$ for every $x \in X$. $\square$

Proposition 3.22 *Let E be a vector space. For every subset F of $E^{\#}$, $\big(E, w(E, F)\big)^* = \operatorname{span}(F)$.*

Proof: By the definition of the $w(E, F)$-topology, $F \subset \big(E, w(E, F)\big)^*$. We need to show that given $f \in \big(E, w(E, F)\big)^*$ there are $f_1, \dots, f_n$ such that $f = \sum \alpha_i f_i$.

Since f is continuous in the $w(E, F)$-topology, there is an open neighborhood V of 0 such that $|f(x)| < 1$ for all $x \in V$. We may assume that there are $f_1, \dots, f_n \in F$ and $\varepsilon > 0$ such that $V := \{x \in E : |f_i(x)| < \varepsilon\}$.

Take any z such that $f_i(z) = 0$ for all i. Then also $|f_i(nz)| < \varepsilon$ for all $n \in \mathbb{N}$. By the linearity this means that $|f(z)| < \frac{1}{n}$ for all $n \in \mathbb{N}$, that is, $f(z) = 0$. This shows that $\bigcap_i \operatorname{Ker}(f_i) \subset \operatorname{Ker}(f)$ and the proposition follows from Lemma 3.21. $\square$

Proposition 3.23 *Let X be an infinite-dimensional Banach space. Then the w-topology of X is not first countable, in particular, it is not metrizable. The same is true for the w^*-topology of X^*.*

Proof: If $\{V_n\}$ were a countable base of neighborhoods at 0, assume without loss of generality that $V_n = \{x \in X : \max_{i=1,\dots,N_n} |f_i^n(x)| < \varepsilon_n\}$. Since X^* has no countable Hamel basis, there is $f \in X^*$ such that f is not in the span of all f_i^n. We claim that $U = f^{-1}(-1, 1)$ (which is a weak neighborhood of 0) does not contain any of these V_n. Indeed, if $V_n = \{x : \max |f_i(x)| < \varepsilon_n\} \subset U$, then $\bigcap f_i^{-1}(0) \subset f^{-1}(0)$ and f would be a linear combination of $\{f_i\}$. The proof for X^* is similar. □

A simple consequence of the Hahn–Banach theorem is the following.

Proposition 3.24 *There exists a non-zero continuous functional on a topological vector space E if and only if E has a convex neighborhood of the origin which is different of the whole space E.*

Proof: Without loss of generality, we may assume that E is a real space. If f is a non-zero continuous functional on E, the set $\{x \in E : |f(x)| \leq 1\}$ is a convex neighborhood of 0 which differs from E. Conversely, assume that U is a convex neighborhood of 0 that differs from E. Its Minkowski functional p_U is a continuous positively homogeneous sublinear function on E (see Exercise 3.2). Let $x_0 \in E$ such that $p_U(x_0) \neq 0$. Put $f(\lambda x_0) = \lambda p_U(x_0)$ for all $\lambda \in \mathbb{R}$. If $\lambda \geq 0$, $f(\lambda x_0) = p_U(\lambda x_0)$. Otherwise, $f(\lambda x_0) < 0 \leq p_U(\lambda x_0)$. It follows that $f(\lambda x_0) \leq p_U(\lambda x_0)$ for all $\lambda \in \mathbb{R}$. Use the Hahn–Banach theorem 2.1 to extend f to a linear functional $\widetilde{f}$ on E such that $\widetilde{f} \leq p_U$. Then $\widetilde{f}$ is continuous (see Exercise 3.3). □

The following example shows that there are topological vector spaces where Proposition 3.24 does not apply, i.e., topological vector spaces E for which the space E^* may be degenerate.

Example: Fix $p \in (0, 1)$. Let L_p denote the vector space of Lebesgue measurable real-valued functions on $[0, 1]$ for which

$$q(f) := \int_0^1 |f(t)|^p \,\mathrm{d}t < \infty.$$

Since for $a \geq 0$ and $b \geq 0$ we have $(a+b)^p \leq a^p + b^p$, it follows that $q(f + g) \leq q(f) + q(g)$. Therefore the formula $\mathrm{d}(f, g) = q(f - g)$ defines a translation invariant metric on L_p in which L_p is complete. This follows in the same way as for L_p, $p \geq 1$ (see Theorem 1.21). We claim that if O is a nonempty open convex set in L_p, then $O = L_p$.

To prove this claim, assume that $V \neq \emptyset$ is an open convex set in L_p such that $0 \in V$. Choose $f \in L_p$. We will show that $f \in V$. Choose $r > 0$ such that $B_r \subset V$, where $B_r := \{f \in L_p : q(f) < r\}$, then choose $n \in \mathbb{N}$ such that $n^{p-1} q(f) < r$. By the continuity of the indefinite integral $\int_0^x |f|^p \,\mathrm{d}t$, there are points $0 =: x_0 < x_1 < \cdots < x_n := 1$ such that $\int_{x_{i-1}}^{x_i} |f|^p \, dt = n^{-1} q(f)$ for $i = 1, \ldots, n$.

Define for $i = 1, \ldots, n$ functions g_i on $[0, 1]$ by

$$g_i(t) = \begin{cases} nf(t) & \text{for } x_{i-1} < t \le x_i \\ 0 & \text{otherwise.} \end{cases}$$

Then $g_i \in V$ as $q(g_i) = \int_{x_{i-1}}^{x_i} |nf|^p \, dt = n^{p-1} q(f) < r$ and $B_r \subset V$.

Since V is convex and $f = \frac{1}{n}(g_1 + \cdots + g_n)$, it follows that $f \in V$ and the claim is proved.

Consequently, the only open convex neighborhood of 0 is the whole L_p, so the topology is not locally convex (see Definition 3.25).

Also, let $F \in L_p^*$. As $F^{-1}\big((-\alpha, \alpha)\big)$ is a convex neighborhood of zero in L_p for every $\alpha > 0$, we have $F^{-1}\big((-\alpha, \alpha)\big) = L_p$. Thus $F = 0$, and $L_p^* = \{0\}$.

3.3 Locally Convex Spaces

We saw in the example after Proposition 3.24 that there are topological vector spaces with no continuous linear functional but 0. In order to avoid this "pathological" behavior, and motivated by Proposition 3.24, we introduce now a particular subclass.

Definition 3.25 *A topological vector space E is called a* locally convex space *if E has a local base consisting of convex sets.*
A topology $\mathcal{T}$ on E is called a locally convex topology *if $(E, \mathcal{T})$ is a locally convex space.*

If $(E, \mathcal{T})$ is a locally convex space and F is a vector subspace of E, then F becomes a locally convex space when endowed with the restriction of the topology $\mathcal{T}$.

If $(E_\gamma, \mathcal{T}_\gamma)_{\gamma \in \Gamma}$ is a family of locally convex space, then $(\prod_{\gamma \in \Gamma} E_\gamma, \mathcal{T}_p)$ is a locally convex space.

If $(E, \mathcal{T})$ is a locally convex space and F is a closed (linear) subspace, then $(E/F, \mathcal{T}_q)$ is a locally convex space.

These three assertions are easy to verify.

The following are natural examples of locally convex spaces.

1. Every normed space E is a locally convex space. Indeed, every ball is a convex subset of E.
2. If Γ is a nonempty set, the space $(\mathbb{K}^\Gamma, \mathcal{T}_p)$ is a locally convex space. Indeed, elementary open sets of the form $\prod_{\gamma \in \Gamma} O_\gamma$, where $O_\gamma := \mathbb{K}$ but for a finite number of them, all equal to $(-\varepsilon, \varepsilon)$, where $\varepsilon > 0$, form a local base of convex subsets of $\mathbb{K}^\Gamma$.
3. Let $\langle E, F \rangle$ be a dual pair. Then $(E, w(E, F))$ is a locally convex space, since it is a linear subspace of the locally convex space $(\mathbb{K}^F, \mathcal{T}_p)$. In other words, every

element $U(0; f_1, \ldots, f_n; \varepsilon)$ (see Equation (3.1)) of the local base described in Proposition 3.5 is convex.

4. In particular, every normed space X in its weak topology is a locally convex space, and so it is X^* endowed with its weak* topology (see Definition 3.20).

Some more examples of locally convex spaces will be given later. See in particular Section 3.10. The class of spaces in Example 4, Section 3.2, consists of non-locally convex topological vector spaces, see Exercise 3.6. Another class of non-locally convex topological vector spaces was given in the example after Proposition 3.24.

Proposition 3.26 *Every locally convex space E has a local base consisting of convex balanced sets.*

Proof: Let U be a convex neighborhood of 0 in E. First we construct W as in the proof of Proposition 3.10. Since W is balanced, then $V := \mathrm{conv}(W)$ is a balanced (see Exercise 1.4) and convex neighborhood of 0 and $V \subset U$ as U is convex. □

A topological space $(S, \mathcal{T})$ is *completely regular* if it is Haussdorf and, given any $x \in S$ and any nonempty closed subset F of S such that $x \notin F$, there exists a continuous function $f : S \to [0, 1]$ such that $f(x) = 0$ and $f(y) = 1$ for all $y \in F$. It is known that every uniform space is, as a topological space, completely regular, see, e.g., [Enge, Theorem 8.1.20]. Since every topological vector space $(E, \mathcal{T})$ has a uniformity on it whose associated topology is $\mathcal{T}$ (see Section 17.10), we get that every topological vector space is completely regular. We can prove this result for the class of locally convex spaces without relying on the result about uniformities.

Proposition 3.27 *Every locally convex space is completely regular.*

Proof: Let $(E, \mathcal{T})$ be a locally convex space. Certainly, the topology $\mathcal{T}$ is Haussdorf. It is enough to prove that, given a closed and convex neighborhood U of 0, there is a continuous function $f : E \to [0, 1]$ such that $f(0) = 0$ and $f(x) = 1$ for all $x \notin U$. Let p_U be the Minkowski functional of U (a continuous function according to Lemma 2.11). The sought function is $f(x) := \min\{p_U(x), 1\}$, $x \in E$. □

A metric $\mathrm{d}(\cdot, \cdot)$ on a vector space E is called *(translation) invariant* if $\mathrm{d}(x + z, y + z) = \mathrm{d}(x, y)$ for all $x, y, z \in E$.

Definition 3.28 *A locally convex space E is called a* Fréchet space *if its topology is induced by a complete invariant metric.*
A locally convex space E is called normable *if its topology is induced by a norm on E.*

In particular, if X is a Banach space then the metric $\mathrm{d}(x, y) := \|x - y\|$ makes it a Fréchet space.

Proposition 3.29 *Let E be a locally convex space. If E has a countable local base, then the topology of E is induced by a translation invariant metric (and conversely).*

Proof: Let $\{U_n\}$ be a countable local base in E. By Proposition 3.26 we can assume that every U_n is a convex and balanced set. For every n, let p_n be the Minkowski functional of U_n. Then p_n is a seminorm, i.e., $p_n(x+y) \leq p_n(x) + p_n(y)$ for $x, y \in E$ and $p_n(\alpha x) = |\alpha| p(x)$ for $x \in E$ and $\alpha \in \mathbb{R}$ (see Exercise 2.21 for the Banach space case).

Define a metric d on E by

$$\mathrm{d}(x, y) = \sum_{n=1}^{\infty} 2^{-n} \frac{p_n(x-y)}{1 + p_n(x-y)}. \tag{3.2}$$

To verify that d is indeed a metric, note that $\frac{t}{1+t}$ is increasing. Thus for $x, y, z \in E$,

$$\begin{aligned} \frac{p_n(x-z)}{1+p_n(x-z)} &\leq \frac{p_n(x-y)+p_n(y-z)}{1+p_n(x-y)+p_n(y-z)} \\ &= \frac{p_n(x-y)}{1+p_n(x-y)+p_n(y-z)} + \frac{p_n(y-z)}{1+p_n(x-y)+p_n(y-z)} \\ &\leq \frac{p_n(x-y)}{1+p_n(x-y)} + \frac{p_n(y-z)}{1+p_n(y-z)}. \end{aligned}$$

Since the series in (3.2) converges uniformly and every p_n is continuous (see Exercise 3.2), d is also continuous. As the metric d is translation invariant, to prove that d induces the topology of E it is enough to show that every U_n contains the ball $\{x \in E : \mathrm{d}(x, 0) < 2^{-(n+1)}\}$. If $\mathrm{d}(x, 0) < 2^{-(n+1)}$ then $\frac{p_n(x)}{1+p_n(x)} < \frac{1}{2}$, and since the function $\frac{t}{1+t}$ is increasing, we have $p_n(x) < 1$ and thus $x \in U_n$. □

Proposition 3.30 *Let E be a locally convex space. E is normable if and only if E has a bounded neighborhood of zero.*

Proof: Clearly, if E is normable then it has a bounded neighborhood of zero, namely the open unit ball centered at the origin. Assume that U is a bounded balanced convex neighborhood of zero in E. Then the Minkowski functional p of U is a continuous norm on E which induces the topology of E. (See Exercise 3.2.) □

Example: The space $\mathbb{R}^{\mathbb{N}}$ (denoted also by s), i.e., the vector space of all sequences of real numbers with the product topology, is a separable Fréchet space that is not normable, see Exercise 3.12 (for another example, see Fact 3.76). In Section 12.4 we will discuss the result that all infinite-dimensional separable Banach spaces are (non-linearly) homeomorphic to the space $\mathbb{R}^{\mathbb{N}}$.

Theorem 3.31 *Let E be a locally convex space. Let F be a (not necessarily closed) subspace of E. Let $f : F \to \mathbb{K}$ be a continuous linear functional. Then there exists a continuous linear functional $\widetilde{f} : E \to \mathbb{K}$ that extends f.*

Proof: Assume first that E is real. Put $B = \{r \in \mathbb{R} : r \leq 1\}$. The set $V := f^{-1}(B)$ is a neighborhood of 0 in F. There exists a closed convex and balanced neighborhood U of 0 in E such that $U \cap F \subset V$. Let $p_V : F \to \mathbb{R}$ be the Minkowski functional of V on F. If $y \in \lambda V$ for some $\lambda > 0$, then $y = \lambda v$ for some $v \in V$, hence $f(y) = \lambda f(v) \leq \lambda$. It follows that $f(y) \leq p_V(y)$ ($\leq p_U(y)$) for all $y \in F$, where $p_U : E \to \mathbb{R}$ is the Minkowski functional of U on E. It is enough to use Theorem 2.1 to conclude that there exists a linear functional $\tilde{f} : E \to \mathbb{R}$ that extends f and such that $\tilde{f}(x) \leq p_U(x)$ for all $x \in E$ (so $\tilde{f}$ is continuous by Lemma 2.10, that holds in this context, too).

If E is complex, the previous argument gives an extension of $\operatorname{Re} f$ to a continuous real functional u on $E_{\mathbb{R}}$, and u is the real part of a (continuous) functional $\tilde{f}$ that extends f (see the paragraph in page 88). □

Note that this extension theorem does not work in general for topological vector spaces. Indeed, we can define a non-zero functional on a given one-dimensional subspace of any non-zero topological vector space. By Corollary 3.16, this functional is continuous. If it would be always possible to extend such a functional to the whole space, then the dual space will never be reduced to 0. The example in Section 3.2 proves that in general this is not the case.

The following separation results are the topological vector space version of those that were formulated in the context of Banach spaces (Theorem 2.12 and Proposition 2.13).

Theorem 3.32 *Let E be a topological vector space.*
(i) *If A is a nonempty open convex set in E and $x_0 \in E \setminus A$, then there is $f \in E^*$ such that* $\operatorname{Re} f(a) < \operatorname{Re} f(x_0)$ *for all $a \in A$.*
(ii) *If E is a locally convex space, C a nonempty closed convex set in E, and $x_0 \in E \setminus C$, then there exists $f \in E^*$ such that* $\sup_{c\in C} \operatorname{Re} f(c) < \operatorname{Re} f(x_0)$.

Proof: Without loss of generality, we may assume that E is a real space.

(i) By considering a shift of A we may also assume that $0 \in A$. The Minkowski functional μ_A of A is a non-negative continuous positively homogeneous and subadditive functional on E, and $A = \{x \in E : \mu_A(x) < 1\}$ (see Exercise 3.2). Define f on $\operatorname{span}\{x_0\}$ by $f(\alpha x_0) = \alpha\mu_A(x_0)$ for $\alpha \in \mathbb{R}$ and proceed like in the proof of Proposition 3.24 to obtain a linear functional on E (denoted by f again) that agrees with f on $\operatorname{span}\{x_0\}$ and such that $f \leq \mu_A$. Then $f \in E^*$ (see Exercise 3.3) and we have $f(a) \leq \mu_A(a) < 1 \leq \mu_A(x_0) = f(x_0)$ for every $a \in A$.

(ii) Assume now that E is locally convex. Let U be an open convex and balanced neighborhood of 0 such that $(x_0 + U) \cap C = \emptyset$. Since $x_0 \notin C + U$, and $C + U$ is an open and convex set, by (i) we can find $f \in E^*$ such that $f(c + u) < f(x_0)$ for all $c \in C$ and $u \in U$. Then $\sup_{c\in C} \leq f(x_0) - f(u)$ for all $u \in U$. It is enough to choose $u \in U$ such that $f(u) > 0$ to obtain the conclusion. □

The following corollary is immediate.

Corollary 3.33 *Let E be a locally convex space and $x, y \in E$. If $x \neq y$ then there is $f \in E^*$ such that $f(x) \neq f(y)$. In other words, $\langle E, E^*\rangle$ is a dual pair. In particular, $E^* \neq \emptyset$.*

We introduced in Definition 3.20 the weak* topology (in short, w^*) in the dual E^* of a locally convex space E. The space (E^*, w^*) is a topological vector space (even more, a locally convex space) whose dual space is E, and all the separation theorems formulated above in the context of topological (and locally convex) vector spaces hold true. For example, we have the following result.

Corollary 3.34 *Let E be a locally convex space. Then, given a nonempty w^*-closed convex subset C of E^* and an element $x_0^* \in E^* \setminus C$, there exists $x \in E$ such that $\sup_{x^* \in C} \mathrm{Re}\langle x^*, x\rangle < \langle x_0^*, x\rangle$.*

Another useful separation result is the following.

Theorem 3.35 *Let E be a topological vector space. Let A, B be two convex subsets of E such that $A \cap B = \emptyset$. Then*
(i) *If A is open, then there exist $f \in E^*$ such that* $\mathrm{Re}\, f(a) < \inf_{b\in B} \mathrm{Re}\, f(b)$ *for all $a \in A$, $b \in B$.*
(ii) *If E is a locally convex space, A is closed and B is compact, then there exists $f \in E^*$ and $\lambda \in \mathbb{R}$ such that* $\mathrm{Re}\, f(a) < \lambda < \inf_{b\in B} \mathrm{Re}\, f(b)$ *for all $a \in A$.*

Proof: Without loss of generality, we may assume that E is a real space.

(i) The set $A - B$ is an open convex subset of E and $0 \notin A - B$. Use (i) in Theorem 3.32 to obtain $f \in E^*$ such that $f(a - b) < f(0)$ $(= 0)$ for all $a \in A$, $b \in B$. Then $f(a) < f(b)$ for all $a \in A$, $b \in B$, and this implies $f(a) \le \lambda := \inf_{b\in B} f(b)$ for all $a \in A$. If $f(a_0) = \lambda$ for some $a_0 \in A$, by the fact that A is open we can find $a \in A$ such that $f(a) > \lambda$, a contradiction.

(ii) Assume now that E is a locally convex space. Given $b \in B$ there exists an open convex and balanced neighborhood U_b of 0 such that $(b + U_b) \cap A = \emptyset$. Since $\{b + U_b : b \in B\}$ is an open cover of B and B is compact, there exists a finite set $B_0 \subset B$ such that $B \subset \bigcup_{b\in B_0} b + U_b$. Set $U = \bigcup_{b\in B_0} U_b$. Then U is an open convex and balanced neighborhood of 0 and $(A + U) \cap B = \emptyset$. By (ii) in Theorem 3.32 applied to $(A + U)$ and B, there exists $f \in E^*$ such that $f(a + u) < \inf_{b\in B} f(b)$ for all $a \in A$, $u \in U$. It is enough to choose some $u \in U$ with $f(u) > 0$ to obtain the conclusion. □

For further separation result we refer, e.g., to the remark following Theorem 3.122.

3.4 Polarity

Definition 3.36 *Let $\langle E, F\rangle$ be a dual pair. Let A be a subset of E. By the* polar *of A in F we mean $A^\circ := \{f \in F : \mathrm{Re}\langle x, f\rangle \le 1$ for all $x \in A\}$. The set $\{f \in F : |\langle x, f\rangle| \le 1$, for all $x \in A\}$ is called the* absolute polar *of A.*

It easily follows that, for every $A \subset E$, A° is a convex and $w(F, E)$-closed subset of F that contains 0. If A is balanced, so is the corresponding polar set; we then have that the polar and the absolute polar sets of A coincide.

If E is a locally convex space, and $A \subset E^*$, it is customary to write $A_\circ$ for the polar set of A in E, i.e., the polar set of A with respect to the dual pair $\langle E, E^* \rangle$. If we are dealing with a dual pair, say $\langle E, F \rangle$, we shall not do this distinction, so, for example, the polar of a set $A \subset G$ will be denoted $A^{\circ\circ}$. Note, too, that the polar set of a linear subspace G of a vector space E with respect to a dual pair $\langle E, F \rangle$ is the *annihilator* $G^\perp$ of G in F (introduced for Banach spaces right before Proposition 2.6), i.e., $G^\circ = G^\perp := \{f \in F : \langle g, f \rangle = 0 \text{ for all } g \in G\}$.

Observe that if X is a normed space, then the polar set $(B_X)^\circ$ of B_X in X^* is just B_{X^*}, and the polar set $(B_{X^*})_\circ$ of B_{X^*} in X is simply B_X.

Theorem 3.37 (Alaoglu, Bourbaki) *Let E be a locally convex space. If $A \subset E$ is a neighborhood of 0, then A° is w^*-compact.*

Proof: Assume first that A is an convex balanced neighborhood of 0. Consider the dual pair $\langle E, E^\# \rangle$ (see Example 1 in Section 3.1). The set A° ($\subset E^\#$) is bounded (and closed) in $(E^\#, w(E^\#, E))$. Indeed, given $x \in E$ there exists $\lambda > 0$ such that $\lambda x \in A$. Then, $\sup_{f \in A^\circ} |f(x)| \leq \lambda^{-1}$ and we can use Proposition 3.12. Since $(E^\#, w(E^\#, E))$ is linearly homeomorphic to $(\mathbb{K}^B, \mathcal{T}_p)$ for some B (see Proposition 3.6), the Tychonoff theorem implies that A° is compact in $(E^\#, w(E^\#, E))$. Observe now that A° ($\subset E^\#$) is indeed a subset of E^* (hence a ($w^* =$) $w(E^*, E)$-compact set). This follows from Proposition 3.19, since every $f \in A^\circ$ is bounded on A, so $f \in E^*$. This finishes the proof for the case of an convex balanced set A.

For a general neighborhood U of 0 we find a convex balanced neighborhood A of 0 such that $A \subset U$ (Proposition 3.26). Then U° in E^* is a w^*-closed subset of the w^*-compact set A° and the theorem is proved. □

In particular, consider a dual pair $\langle E, F \rangle$ and $A \subset E$ a $w(E, F)$-neighborhood of 0; then A° is $w(F, E)$-compact in F. If X is a Banach space, then B_X°, that is, B_{X^*}, is a w^*-compact subset of X^*.

Theorem 3.38 (Bipolar theorem) *Let $\langle E, F \rangle$ be a dual pair, and denote $w := w(E, F)$. Then, for every subset A of E we have $A^{\circ\circ}$ ($:= (A^\circ)^\circ$) $= \overline{\text{conv}}^w(A \cup \{0\})$. In particular, let E be a locally convex space. If $A \subset E$ is w-closed, convex and $0 \in A$, then $A = (A^\circ)_\circ$. Similarly, if $B \subset E^*$ is w^*-closed, convex and $0 \in B$, then $B = (B_\circ)^\circ$.*

Proof: Let $C = \overline{\text{conv}}^w(A \cup \{0\})$. Since $A \subset A^{\circ\circ}$, $0 \in A^{\circ\circ}$ and $A^{\circ\circ}$ is w-closed and convex, we have $C \subset A^{\circ\circ}$. Assume that there is $x \in A^{\circ\circ} \setminus C$. By Theorem 3.32 there is $f \in F$ such that $\text{Re}\, f(x) > \sup_C(\text{Re}\, f)$. Since $0 \in C$, we have $\sup_C(\text{Re}\, f) \geq 0$, so by scaling we may assume that $\sup_C(\text{Re}\, f) \leq 1$ and $\text{Re}\, f(x) > 1$. Then $\sup_A(\text{Re}\, f) \leq 1$, hence $f \in A^\circ$, also $x \in A^{\circ\circ}$, so we have $\text{Re}\, f(x) \leq 1$, a contradiction. □

We now show an application of the bipolar Theorem 3.38.

Proposition 3.39 *Let $\langle E, F \rangle$ be a dual pair. Let G be a linear subspace of F. Then G separates points of E (and so $\langle E, G \rangle$ is a dual pair) if and only if G is $w(F, E)$-dense in F.*

Proof: Recall that, with respect to a dual pair, the polar set of a linear subspace is its annihilator. If G separates points of E then $G^\circ = \{0\}$, so $G^{\circ\circ} = F$. By Theorem 3.38, $F = \overline{G}^{w(F,E)}$. If G is $w(F,E)$-dense in F and for some $e \in E$ we have $\langle e, g \rangle = 0$ for all $g \in G$, then $\langle e, f \rangle = 0$ for all $f \in F$, since e is, by Proposition 3.2, $w(F,E)$-continuous on F. This implies that $e = 0$ and so G separates points of E. □

3.5 Topologies Compatible with a Dual Pair

Let $\langle E, F \rangle$ be a dual pair. Let $\mathcal{M}$ be a family of $w(F,E)$-bounded subsets of F that separates points of E, i.e., if $\langle e, f \rangle = 0$ for all $f \in \bigcup_{\mathcal{M}} M$, then $e = 0$. A locally convex topology on E associated to the family $\mathcal{M}$ can be defined by giving a local base. It consists of all sets of the form M°, where $M \in \mathcal{M}$, then their scalar multiples, and finally taking the finite intersections of these. It is simple to prove that the topology so generated is indeed a (Hausdorff) locally convex topology on E. It is called the *topology on E of the uniform convergence on the sets of* $\mathcal{M}$ or, briefly, the *topology on E of the uniform convergence on* $\mathcal{M}$.

Let $\langle E, F \rangle$ be a dual pair. A family $\mathcal{G}$ of $w(F,E)$-bounded sets in F is called *saturated* if $\mathcal{G}$ is closed with respect to taking subsets, scalar multiples and balanced convex $w(F,E)$-closed hulls of the union of any two of its members. If a saturated family $\mathcal{G}$ separates points of E, then a local base for the topology $\mathcal{T}_{\mathcal{G}}$ is given by the polars of the $w(F,E)$-closed convex and balanced elements of $\mathcal{G}$.

The *saturation* of a family $\mathcal{M}$ of $w(F,E)$-bounded subsets of F is, by definition, the smallest saturated family of subsets of F that contains $\mathcal{M}$. It is simple to prove that if $\mathcal{S}(\mathcal{M})$ is the saturation of a family $\mathcal{M}$ of $w(F,E)$-bounded subsets of F that separates points of E, then the topologies $\mathcal{T}_{\mathcal{M}}$ and $\mathcal{T}_{\mathcal{S}(\mathcal{M})}$ on E coincide.

Examples:

1. Let $\langle E, F \rangle$ be a dual pair. The topology $w(E,F)$ on E coincides with the topology $\mathcal{T}_{\mathcal{G}}$, where $\mathcal{G}$ is the family of all $w(F,E)$-closed convex balanced hulls of finite subsets of F and their subsets. This family is certainly saturated and covers F.
2. Let X be a normed space. Then, the norm topology on X is the topology $\mathcal{T}_{\mathcal{G}}$, where $\mathcal{G}$ is the family $\{rB_{X^*} : r \in \mathbb{R},\ r > 0\}$ and their subsets, i.e., the family of all bounded subsets of X^*. This is again a saturated family of subsets of X^* that covers X^*.
3. Let $\langle E, F \rangle$ be a dual pair. The family of all convex balanced $\sigma(F,E)$-compact subsets of F and their subsets is saturated and covers F (see Exercise 3.14 and Definition 3.43).

Definition 3.40 *Let* $\langle E, F \rangle$ *be a dual pair. A topology* $\mathcal{T}$ *on* E *that makes* E *a topological vector space is called* compatible with the dual pair *if* $(E, \mathcal{T})^* = F$.

Note that Proposition 3.2 implies that if $(E, \mathcal{T})$ is a locally convex space, then the weak topology w on E is compatible with the dual pair $\langle E, E^* \rangle$, and so it is the topology w^* on E^*.

The following result characterizes the topologies compatible with a dual pair.

Theorem 3.41 (Mackey, Arens) *Let $\langle E, F \rangle$ be a dual pair. A locally convex topology $\mathcal{T}$ on E is compatible with the dual pair $\langle E, F \rangle$ if and only if $\mathcal{T}$ is the topology of uniform convergence on some saturated family $\mathcal{G}$ that covers F and consists of $w(F, E)$-compact convex balanced sets in F and their subsets.*

Proof: Let $\mathcal{G}$ be a saturated family satisfying the assumptions. If $f \in F$, then $f \in G$ for some $G \in \mathcal{G}$ convex, balanced, and $w(F, E)$-closed. Then $|f| \leq 1$ on G°, hence by Proposition 3.19, $f \in (E, \mathcal{T}_\mathcal{G})^*$. Conversely, if $f \in (E, \mathcal{T}_\mathcal{G})^*$, then there is $G \in \mathcal{G}$, convex, balanced, and $w(F, E)$-closed such that $|f| \leq 1$ on G°, i.e., $f \in G^{\circ\circ}$. By the bipolar theorem, $f \in G^{\circ\circ} = G \subset F$.

Now assume that a locally convex topology $\mathcal{T}$ on E is such that $(E, \mathcal{T})^* = F$. Define $\mathcal{G} = \{G : G \subset U^\circ, U \in \mathcal{U}\}$, where $\mathcal{U}$ is the family of all closed convex balanced $\mathcal{T}$-neighborhood of 0. Note that U° is a $w(F, E)$-compact set in F by Alaoglu's theorem and it is also balanced and convex. The separation theorem shows that the $w(F, E)$-closed convex and balanced hull of $U_1^\circ \cap U_2^\circ$ is $(U_1 \cap U_2)^\circ$ for any U_1 and U_2 in $\mathcal{U}$. This proves that $\mathcal{G}$ is a saturated family. If $f \in F$, then $|f| \leq 1$ on some $U \in \mathcal{U}$, so $f \in U^\circ \in \mathcal{G}$, hence $\mathcal{G}$ covers F.

We will show that $\mathcal{T} = \mathcal{T}_\mathcal{G}$. Let U be a $\mathcal{T}$-neighborhood of 0. By Lemma 3.9 and Proposition 3.26 we may assume that U is $\mathcal{T}$-closed, convex, and balanced. Thus $U^\circ \in \mathcal{G}$. Since $U = U^{\circ\circ}$ by the bipolar theorem, U is a $\mathcal{T}_\mathcal{G}$-neighborhood of 0.

Let U be a $\mathcal{T}_\mathcal{G}$-neighborhood of 0. We may assume that $U = V^{\circ\circ}$, where V is some closed convex balanced $\mathcal{T}$-neighborhood of 0. Then $U = V$ (so U is a $\mathcal{T}$-neighborhood of 0). □

In the proof of Theorem 3.41 is shown the following result. Recall the definition of an equicontinuous set of functions (see Section 17.10). It is obvious that a subset M of the dual E^* of a locally convex space $(E, \mathcal{T})$ is uniformly equicontinuous if and only if it equicontinuous, if and only if it is equicontinuous at 0, if and only if it is a subset of the polar of a closed convex and balanced neighborhood of 0 in $(E, \mathcal{T})$.

Proposition 3.42 *The topology of a locally convex space coincides with the topology $\mathcal{T}_\mathcal{G}$, where $\mathcal{G}$ is the (saturated) family of all the equicontinuous subsets of the dual space E^*.*

A consequence of Theorem 3.41 is that every locally convex topology on E compatible with a dual pair $\langle E, F \rangle$ is stronger than $w(E, F)$.

Definition 3.43 *Let $\langle E, F \rangle$ be a dual pair. The locally convex topology on E of uniform convergence on all $w(F, E)$-compact convex balanced sets in F is called the* Mackey topology *on E associated to the dual pair $\langle E, F \rangle$, and is denoted $\mu(E, F)$.*

Observe that the family of all $w(F, E)$-compact convex balanced sets in F and their subsets is saturated (see Example 3 after Definition 3.40).

From Proposition 3.2, Theorem 3.41 and the note after this theorem we obtain the following corollary.

Corollary 3.44 *Let $\langle E, F\rangle$ be a dual pair. Then the strongest locally convex topology $\mathcal{T}$ on E compatible with the dual pair $\langle E, F\rangle$ is the Mackey topology $\mu(E, F)$, and the weakest locally convex topology w on E compatible with this same dual pair is the weak topology $w(E, F)$ on E.*

If E is a locally convex space, a *closed* (respectively, *open*) *half-space* in E is a set of the form $\{x \in E : \operatorname{Re} f(x) \le \alpha\}$ (respectively, $\{x \in E : \operatorname{Re} f(x) < \alpha\}$), where $f \in E^*$ and $\alpha \in \mathbb{R}$.

As a consequence of Theorem 3.32, the following result holds.

Theorem 3.45 (Mazur) *The closure $\overline{C}$ of a convex set C in a locally convex space E is the intersection of all closed half-spaces that contain C. In particular, $\overline{C}^{\mathcal{T}_1} = \overline{C}^{\mathcal{T}_2}$ if $\mathcal{T}_1$ and $\mathcal{T}_2$ are two compatible locally convex topologies topology! compatible with a dual pair on E. If $\langle E, F\rangle$ is a dual pair, all compatible locally convex topologies on E define the same family of closed convex sets.*

Proof: Let $x_0 \in E\backslash\overline{C}$. By Theorem 3.32 we can find $f \in X^*$ such that $\operatorname{Re} f(x_0) > \sup\{\operatorname{Re} f(x) : x \in \overline{C}\}$. If $\operatorname{Re} f(x_0) > \alpha > \sup\{\operatorname{Re} f(x) : x \in \overline{C}\}$, the closed half-space $\{x \in E : \operatorname{Re} f(x) \le \alpha\}$ contains $\overline{C}$ and excludes x_0. This proves one inclusion. For the other, just recall that an arbitrary intersection of closed sets is itself closed. The particular case and the last statement follow from the fact that all compatible topologies give, by the very definition, the same dual space. □

Remark: In particular, if X is a normed space, every $\|\cdot\|$-closed convex subset of X is w-closed. so the family of all w-closed convex subsets of X and the family of all $\|\cdot\|$-closed convex subsets of X coincide. Another way to formulate this is to say that the w-closure and the $\|\cdot\|$-closure of any convex subset of a normed space coincide. For non-convex sets in normed spaces, the norm and weak closure may differ, see Exercise 3.46.

In an infinite-dimensional locally convex space $(E, \mathcal{T})$, $x_i \stackrel{w}{\to} x$ does not in general imply $x_i \stackrel{\mathcal{T}}{\to} x$ (for instance, $e_n \stackrel{w}{\to} 0$ in c_0 or ℓ_p, $p > 1$, and $\|e_n\| = 1$ for all $n \in \mathbb{N}$). However, we have the following.

Corollary 3.46 *Let $(E, \mathcal{T})$ be a locally convex space. If a net $\{x_i\}_{i\in I}$ in E is w-convergent to some $x \in E$, there exists a net $\{y_j\}_{j\in J}$ in* $\operatorname{conv}\{x_\alpha\}$ *that $\mathcal{T}$-converges to x. If E is a normed space, we can find a sequence in* $\operatorname{conv}\{x_\alpha\}$ *that $\|\cdot\|$-converges to x.*

Proof: Define $C = \overline{\operatorname{conv}}^{\mathcal{T}}\{x_i : i \in I\}$. Since $x_i \stackrel{w}{\to} x$, we have $x \in \overline{C}^w = C$, since C is convex and $\mathcal{T}$-closed. The statement follows. □

Lemma 3.47 *Let E be a locally convex space with a countable local base. Assume that A is a balanced convex set in E such that for every bounded set $B \subset E$ there is $\alpha > 0$ satisfying $\alpha B \subset A$. Then A is a neighborhood of zero in E.*

Proof: Let $\{U_n\}$ be a countable local base for X. We may assume that $U_{n+1} \subset U_n$ for every n. By contradiction, assume that there is no $n \in \mathbb{N}$ for which $U_n \subset nA$. Then there are $x_n \in U_n$ such that $x_n \notin nA$. Clearly $\lim x_n = 0$. Since every convergent sequence is bounded, there is $K > 0$ such that $x_n \in KA$ for every n and thus $x_n \in nA$ for $n \geq K$. This is a contradiction with $x_n \notin nA$ for all n. □

Let $(E, \mathcal{T})$ be a complex locally convex space. Let $E^*_{\mathbb{R}}$ be the dual space of its associated real locally convex space. If A is a nonempty subset of E, the set $\operatorname{Re} A^\circ := \{\operatorname{Re} f : f \in A^\circ\}$ is the polar set of A in $E^*_{\mathbb{R}}$ (see Definition 3.36). This implies, in particular, that the topology $\mathcal{T}$ can be obtained as a $\mathcal{T}_{\mathcal{G}}$-topology by using the saturated family $\mathcal{G}$ in E^* as in the proof of Theorem 3.41 or the family $\operatorname{Re}\mathcal{G}$ of all the sets $\operatorname{Re} G$, $G \in \mathcal{G}$, in $E^*_{\mathbb{R}}$ (a saturated family in $E^*_{\mathbb{R}}$). In particular, the topologies $w(E, E^*)$ and $\mu(E, E^*)$ coincide, respectively, with the topologies $w(E, E^*_{\mathbb{R}})$ and $\mu(E, E^*_{\mathbb{R}})$. The same happens in the case of a normed complex space and its norm topology.

3.6 Topologies of Subspaces and Quotients

Let $\langle E_1, E_2\rangle$ be a dual pair. Let G be a linear subspace of E_1. The couple $(G, E_2/G^\perp)$ forms a dual pair under the bilinear form $\langle\cdot,\cdot\rangle$ defined for all $g \in G$, $\hat{e}_2 \in E_2/G^\perp$, by $\langle g, \hat{e}_2\rangle = \langle g, e_2\rangle$, where $e_2 \in \hat{e}_2$. This bilinear form is well defined and, clearly, under $\langle\cdot,\cdot\rangle$ the space G separates points of $E_2/G^\perp$ and the space $E_2/G^\perp$ separates points of G. Let $q : E_2 \to E_2/G^\perp$ be the canonical quotient mapping.

Let $\mathcal{M}$ be a saturated family consisting of $w(E_2, E_1)$-bounded subsets of E_2 and such that $\bigcup_{\mathcal{M}} M$ separates points of E_1. This family defines a locally convex topology $\mathcal{T}_{\mathcal{M}}$ on E_1. Observe that the family $\widetilde{M} := \{q(M) : M \in \mathcal{M}\}$ consists of $w(E_2/G^\perp, G)$-bounded subsets of $E_2/G^\perp$ and $\bigcup_{M\in\mathcal{M}} q(M)$ separates points of G. On G we may, then, consider two natural locally convex topologies associated to the family $\mathcal{M}$: (i) the restriction $\widetilde{\mathcal{T}}_{\mathcal{M}}$ of $\mathcal{T}_{\mathcal{M}}$ to G, and (ii) the topology $\mathcal{T}_{\widetilde{M}}$ on G.

Proposition 3.48 *Let $(E, \mathcal{T})$ be a locally convex spaces and let G be a (not necessarily closed) subspace of E. Then, in the situation described in the two paragraphs above, we have $\widetilde{\mathcal{T}}_{\mathcal{M}} = \mathcal{T}_{\widetilde{M}}$.*

Proof: It is enough to check that $M^\circ \cap G = (q(M))^\circ$ for every $M \in \mathcal{M}$. Then, if $\widetilde{U}$ is a $\widetilde{\mathcal{T}}_{\mathcal{M}}$-neighborhood of 0 in G, there exists a $\mathcal{T}_{\mathcal{M}}$-neighborhood of 0 in E_1 such that $U \cap G = \widetilde{U}$. We can find a closed convex and balanced set $M \in M$ such that $M^\circ \subset U$. It follows that $(q(M))^\circ = M^\circ \cap G \subset U \cap G = \widetilde{U}$, so $\widetilde{U}$ is a $\mathcal{T}_{\widetilde{M}}$-neighborhood of 0. On the other hand, given $M \in \mathcal{M}$, again by the fact that $M^\circ \cap G = (q(M))^\circ$ we get that $(q(M))^\circ$ is a $\widetilde{\mathcal{T}}_{\mathcal{M}}$-neighborhood of 0. This proves that, conversely, every $\mathcal{T}_{\widetilde{M}}$ neighborhood of 0 is a $\widetilde{\mathcal{T}}_{\mathcal{M}}$-neighborhood of 0. □

Corollary 3.49 *Let E be a locally convex space and let G be a (non necessarily closed) subspace. Then, the topology $w(E, E^*)$ on E induces the topology $w(G, G^*)$ on G.*

Let $(E, \mathcal{T})$ be a locally convex space and let $\mathcal{M}$ be the family of all the equicontinuous subsets of E^*. This family is saturated and covers E^* (see the paragraph preceding Proposition 3.42). We know from the same Proposition 3.42 that $\mathcal{T} = \mathcal{T}_{\mathcal{M}}$. Let G be a closed subspace of E. The natural embedding I from $(E/G)^*$ into E^* is an algebraic isomorphism from $(E/G)^*$ onto $G^\perp$ $(\subset E^*)$. That I is onto follows from the fact that, given $e^* \in G^\perp$, a mapping $(\hat{e})^*$ from E/G into $\mathbb{K}$ can be defined by

$$\langle (\hat{e})^*, \hat{e} \rangle = \langle e^*, e \rangle, \quad \text{for all } e \in \hat{e} \in E/G.$$

The mapping $(\hat{e})^*$ is continuous: indeed, given $\varepsilon > 0$ there exists a neighborhood of 0 in E such that $|\langle e^*, u \rangle| \leq 1$ for all $u \in U$. Then $|\langle (\hat{e})^*, q(u) \rangle| \leq 1$ for all $u \in U$. Continuity follows from the fact that $q(U)$ is a neighborhood of 0 in (E/G). From now on we shall identify $(E/G)^*$ and $G^\perp$ in the sense described.

On E/G we may consider two natural topologies: the quotient topology of $\mathcal{T}$, denoted by $\widehat{\mathcal{T}}$, and the topology $\mathcal{T}_{\widehat{\mathcal{M}}}$, where $\widehat{\mathcal{M}}$ is the family of all elements in $\mathcal{M}$ that are in $G^\perp$. This family $\widehat{\mathcal{M}}$ consists of $w(G^\perp, E/G)$-bounded sets, is also saturated with respect to the pair $\langle E/G, G^\perp \rangle$, and covers $G^\perp$, so it defines a locally convex topology $\mathcal{T}_{\widehat{\mathcal{M}}}$ on E/G.

Proposition 3.50 *Let $(E, \mathcal{T})$ be a locally convex space. Let G be a closed subspace of E. Then, in the situation described in the two paragraphs above, the quotient topology $\widehat{\mathcal{T}}$ of $\mathcal{T}$ on E/G coincides with the topology $\mathcal{T}_{\widehat{\mathcal{M}}}$.*

Proof: Let $\widehat{M}$ be a $\widehat{\mathcal{T}}$-equicontinuous subset of $(E/G)^*$. There exists an open convex and balanced neighborhood U of 0 in $(X, \mathcal{T}))$ such that $\widehat{M} \subset \big(q(U)\big)^\circ$. Therefore $M \subset U^\circ \cap G^\perp$, that is, $\widehat{M} \in \widehat{\mathcal{M}}$. Conversely, if $\widehat{M} \in \widehat{\mathcal{M}}$, then there is a $\mathcal{T}$-neighborhood U of 0 such that $\widehat{M} \subset (U^\circ \cap G^\perp)$ $(= (U+G)^\perp)$, so $\widehat{M} \subset \big(q(U)\big)^\circ$; thus, $\widehat{M}$ is $\widehat{\mathcal{T}}$-equicontinuous. □

Corollary 3.51 *Let E be a locally convex space, and let G be a closed subspace. Then, the topology $w(E, E^*)$ induces on G the topology $w(G, G^*)$. Likewise, the topology $\mu(E, E^*)$ induces on G the topology $\mu(G, G^*)$.*

3.7 Weak Compactness

A subset A of a topological space T is said to be *relatively compact* if $\overline{A}$ is compact. In the case of topological vector spaces, this is equivalent to the fact that every net in A has a subnet that converges to some point in $\overline{A}$, see Exercise 3.10. A is said to be *(relatively) countably compact* if every sequence in A has a subnet that converges to some point in A (in $\overline{A}$). A is said to be *(relatively) sequentially compact* if every sequence in A has a subsequence that converges to some point in A (in $\overline{A}$). Obviously, every (relatively) compact set is (relatively) countably compact, and every (relatively) sequentially compact set is (relatively) countably compact.

Let K be a compact topological space. Let $C_p(K)$ denote the space of all real continuous functions on K endowed with the restriction of the topology $\mathcal{T}_p$ on $\mathbb{R}^K$ of the pointwise convergence on K (see Definition 3.3). The restriction of this topology to $C(K)$ will be denoted also by $\mathcal{T}_p$.

Theorem 3.52 (Eberlein, Grothendieck, see, e.g., [Todo1]) *Let K be a compact topological space. In $C_p(K)$, every relatively countably compact set is relatively compact (and conversely).*

Proof: The space $C_p(K)$ is a subspace of $(\mathbb{R}^K, \mathcal{T}_p)$. Let M be a (nonempty) relatively countably compact subset of $C_p(K)$. Note that $\{f(k) : f \in M\}$ is bounded for every $k \in K$; then, by Tychonoff's theorem, $\overline{M}^{\mathbb{R}^K}$ is compact in $(\mathbb{R}^K, \mathcal{T}_p)$.
Claim: $\overline{M}^{\mathbb{R}^K} \subset C(K)$.

In order to prove the claim we proceed by contradiction. Assume that there exists $f \in \overline{M}^{\mathbb{R}^K}$ that is discontinuous (at some point $k_0 \in K$). We can find then a net $\{t_i\}_{i\in I}$ in K and some $\varepsilon > 0$ such that

$$t_i \to k_0, \text{ and } |f(t_i) - f(k_0)| \geq \varepsilon \text{ for all } i \in I. \tag{3.3}$$

Step 1:

$$\begin{cases} \exists\, f_1 \in M \text{ so that } |f_1(k_0) - f(k_0)| < \varepsilon/2 \text{ (since } f \in \overline{M}^{\mathbb{R}^K}). \\ \text{Put } U_1 = \{k \in K : |f_1(k) - f_1(k_0)| \leq \varepsilon/2\}, \\ \qquad \text{a closed neighborhood of } k_0 \text{ (since } f_1 \text{ is continuous).} \\ \exists\, k_1 \in U_1 \text{ such that } |f(k_1) - f(k_0)| \geq \varepsilon \text{ (since (3.3) holds).} \end{cases}$$

Step 2:

$$\begin{cases} \exists\, f_2 \in M \text{ so that } |f_2(k_i) - f(k_i)| < \varepsilon/2^2,\ i = 0, 1 \text{ (since } f \in \overline{M}^{\mathbb{R}^K}). \\ \text{Put } U_2 = \{k \in K : |f_i(k) - f_i(k_0)| \leq \varepsilon/2^2,\ i = 1, 2\}, \\ \qquad \text{a closed neighborhood of } k_0 \text{ (since } f_i,\ i = 1, 2, \text{ are continuous).} \\ \exists\, k_2 \in U_2 \text{ so that } |f(k_2) - f(k_0)| \geq \varepsilon \text{ (since (3.3) holds).} \end{cases}$$

Step n:

$$\begin{cases} \exists\, f_n \in M \text{ so that } |f_n(k_i) - f(k_i)| < \varepsilon/2^n,\ 0 \leq i < n \text{ (since } f \in \overline{M}^{\mathbb{R}^K}). & (n1) \\ \text{Put } U_n = \{k \in K : |f_i(k) - f_i(k_0)| \leq \varepsilon/2^n,\ 1 \leq i \leq n\}, & \\ \qquad \text{a closed neighborhood of } k_0 \text{ (since } f_i,\ 1 \leq i \leq n, \text{ are continuous).} & (n2) \\ \exists\, k_n \in U_n \text{ so that } |f(k_n) - f(k_0)| \geq \varepsilon \text{ (since (3.3) holds).} & (n3) \end{cases}$$

Note that $U_{n+1} \subset U_n$ for all $n \in \mathbb{N}$. Also, for $i = 1, 2, \ldots, n$ and $n \in \mathbb{N}$,

$$\begin{aligned} &|f_{n+1}(k_i) - f(k_0)| \\ &\quad \geq |f(k_i) - f(k_0)| - |f_{n+1}(k_i) - f(k_i)| \geq \varepsilon - \varepsilon/2^{n+1} > \varepsilon/2. \end{aligned} \tag{n4}$$

The sequence $\{k_n\}$ has a cluster point $k_\infty \in K$. Observe that $k_\infty \in \bigcap_{n\in\mathbb{N}} U_n$, hence

$$f_n(k_\infty) = f_n(k_0), \text{ for all } n \in \mathbb{N}. \tag{3.4}$$

The sequence $\{f_n\}$ has a cluster point g in $C_p(K)$. In particular, $g(k_0)$ is a cluster point of $\{f_n(k_0)\}$. There exists a subsequence of $\{f_n\}$ (denoted again $\{f_n\}$) such that $f_n(k_0) \to g(k_0)$. In view of (3.4), $f_n(k_\infty) \to g(k_0)$. At the same time, $g(k_\infty)$ is a cluster point of the sequence $\{f_n(k_\infty)\}$, and we get

$$g(k_0) = g(k_\infty). \tag{3.5}$$

Fixing $i \in \{0, 1, 2, \ldots\}$ and letting $n \to \infty$ in $(n1)$ we get

$$g(k_i) = f(k_i), \text{ for all } i = 0, 1, 2, \ldots \tag{3.6}$$

Fixing $i \in \mathbb{N}$ and letting $n \to \infty$ in (n4) we get $|g(k_i) - f(k_0)| \geq \varepsilon/2$. By (3.6) we get $|g(k_i) - g(k_0)| \geq \varepsilon/2$ for all $i \in \mathbb{N}$. By (3.5) we get $|g(k_i) - g(k_\infty)| \geq \varepsilon/2$ for all $i \in \mathbb{N}$. This contradicts the continuity of g at x_∞ and proves the Claim. △

Since $\overline{M}^{\mathbb{R}^K}$ is $\mathcal{T}_p$-compact, the conclusion follows. □

Definition 3.53 *A topological space T is said to have* countable tightness *if for every $A \subset T$ and $x \in \overline{A}$ there is a countable set $S \subset A$ such that $x \in \overline{S}$.*
A topological space T is called a Fréchet–Urysohn space *if for every $A \subset T$ and $x \in \overline{A}$ there is a sequence $\{x_n\} \subset A$ such that $x_n \to x$.*
A topological space T is called angelic *if for every relatively countably compact subset A of T, the two following properties hold:* (i) *A is relatively compact, and* (ii) *given $x \in \overline{A}$ there is a sequence $\{x_n\} \subset A$ such that $x_n \to x$.*

Obviously, if K is a compact topological space, K is a Fréchet–Urysohn space if and only if it is angelic.

Example: Let $x_n = \sqrt{n}e_n \in \ell_2$, where e_n is the standard nth unit vector in ℓ_2. Then $0 \in \overline{\{\sqrt{n}e_n\}}^w$ and there is no subsequence of x_n that weakly converges to 0 (Exercise 3.55). Thus ℓ_2 in its weak topology is not a Fréchet–Urysohn space.

It is simple to prove that in angelic spaces the classes of relatively compact, relatively countably compact and relatively sequentially compact sets coincide; the same is true if the word "relatively" is dropped (Exercise 3.9).

Example: Let Γ be uncountable. The element $a = (1, 1, \ldots) \in \ell_\infty(\Gamma)$ is in the closure of $B_{c_0(\Gamma)}$ in the w^*-topology by Goldstine's theorem. However, there is no countable $S \subset B_{c_0(\Gamma)}$ such that $a \in \overline{S}^{w^*}$. Indeed, assume that $a \in \overline{S}^{w^*}$ for some countable $S \subset B_{c_0(\Gamma)}$. Every element of $c_0(\Gamma)$ has a countable support and therefore there is a countable set $M \subset \Gamma$ such that $s(\gamma) = 0$ for every $\gamma \notin M$ and every $s \in S$. Then for $a \in \overline{S}^{w^*}$ we have $a(\gamma) = 0$ for $\gamma \notin M$, a contradiction. Therefore $B_{\ell_\infty(\Gamma)}$ $(= B_{\ell_1^*(\Gamma)})$ in its w^*-topology does not have countable tightness.

Theorem 3.54 (Kaplansky, see, e.g., [Koth, §24.1(6)]) *Let $(E, \mathcal{T})$ be a locally convex space such that $E^* = \bigcup_{n=1}^{\infty} M_n$, where each M_n is a $w(E^*, E)$-compact subset of E^*. Then, (E, w) has countable tightness.*

Proof: Let A be an arbitrary nonempty subset of E. Take $\overline{a} \in \overline{A}$. We shall prove that there exists a countable subset A_0 of A such that $\overline{a} \in \overline{A_0}$. Without loss of generality, we may assume that the sequence $\{M_n\}_{n=1}^{\infty}$ is increasing. Fix $n, m, k \in \mathbb{N}$. Given $x_1^*, \ldots, x_k^* \in M_n$, define a w-neighborhood of $\overline{a}$ of the form $\{x \in E : |\langle x^* - i, x - \overline{a}\rangle| < 1/m, \ i = 1, 2, \ldots, k\}$. Associate to this an element $a \in A$ in this neighborhood. The set $V := \{x^* \in M_n : |\langle x^*, a - \overline{a}\rangle| < 1/m\}$ is open in (M_n, w^*) and contains x_i^*, $i = 1, 2, \ldots, k$, so $V \times \overset{(k)}{\ldots} \times V$ is a neighborhood of $(x_1^*, \ldots, x_k^*)$ in M_n^k. This is done for every $(x_1^*, \ldots, x_k^*) \in M_n^k$; we obtain an open covering of the set M_n^k (a compact set in the product topology of w^*, by Tychonoff's theorem). Therefore we can find a finite subcovering, which leads to a finite subset $A_{n,m,k}$ of A. The sought set is $A_0 := \bigcup_{n,m,k} A_{n,m.k}$. It is clear that this set satisfies the requirement. □

Let K be a compact topological space. Put $\widehat{K} = \{\delta_k : k \in K\}$ ($\subset C(K)^*$), where δ_k is the Dirac functional defined by the element $k \in K$.

Lemma 3.55 *Let K be a compact topological space. Then K is homeomorphic to the set $\widehat{K}$ of all Dirac functionals endowed with the topology induced by w^* in $C(K)^*$.*

Proof: The mapping $\delta : K \to C(K)^*$ that associates δ_k to each $k \in K$ is continuous for the topology of K and the w^*-topology of $C(K)^*$. Indeed, if (k_i) is a net in K that converges to some $k \in K$, then $f(k_i) \to f(k)$ for every $f \in C(K)$. It is one-to-one thanks to Urysohn's lemma. Since K is compact, δ is a homeomorphism onto $\delta(K)$ ($= \widehat{K}$). □

Let $L = \text{span}(\widehat{K})$ ($\subset C(K)^*$). The space L separates points of $C(K)$ (in fact, it is a 1-norming subspace of $C(K)^*$), hence we may consider the dual pair $\langle C(K), L\rangle$, see Example 3 after Definition 3.2 and the paragraph preceding Exercise 3.88.

The following lemma is almost obvious.

Lemma 3.56 *The topology $w(C(K), L)$ on $C(K)$ coincides with the topology $\mathcal{T}_p$ of the pointwise convergence on $C(K)$.*

Proof: Let $\{f_i\}$ be a net in $C(K)$ that $\mathcal{T}_p$-converges to some $f \in C(K)$. Fix $n \in \mathbb{N}$, scalars λ_j, and $k_j \in K$, for $j = 1, 2, \ldots, n$. Then

$$\left\langle \sum_{j=1}^{n} \lambda_j \delta_{k_j}, f_i \right\rangle = \sum_{j=1}^{n} \lambda_j \langle \delta_{k_j}, f_i\rangle \xrightarrow{i} \sum_{j=1}^{n} \lambda_j \langle \delta_{k_j}, f\rangle = \left\langle \sum_{j=1}^{n} \lambda_j \delta_{k_j}, f \right\rangle.$$

On the other hand, if $\{f_i\}$ is a net in $C(K)$ that $w(C(K), L)$-converges to some $f \in C(K)$, certainly it $\mathcal{T}_p$-converges to f. □

Corollary 3.57 *Let K be a compact topological space. Then, the space $C_p(K)$ has countable tightness.*

Proof: The space L is the topological dual of the locally convex space $\big(C(K), w(C(K), L)\big)$. Observe that, for $n, m \in \mathbb{N}$, the sets

$$K_{n,m} := \left\{ \sum_{j=1}^{n} \lambda_j \delta_{k_j} : \sum_{j=1}^{n} |\lambda_j| \le m,\ k_1, \dots, k_n \in K \right\}$$

are $w\big(L, C(K)\big)$-compact, due to Lemma 3.55, and that $L = \bigcup_{n,m\in\mathbb{N}} K_{n,m}$. It is enough to apply Theorem 3.54 and to note that Lemma 3.56 ensures that $\big(C(K), w(C(K), L)\big) = C_p(K)$. □

Theorem 3.58 (Eberlein, Šmulyan, Grothendieck, see, e.g., [Todo1]) *For a given nonempty compact topological space K, the space $C_p(K)$ is angelic. In particular, the following classes of subsets of $C_p(K)$ coincide: (relatively) countably compact, (relatively) sequentially compact, and (relatively) compact.*

Proof: Let $A \subset C(K)$ be $\mathcal{T}_p$-relatively countably compact, and let $\overline{a} \in \overline{A}^{\mathcal{T}_p}$. By Corollary 3.57 there exists a countable set $A_0 \subset A$ such that $\overline{a} \in \overline{A_0}^{\mathcal{T}_p}$. Let $F = \overline{\text{span}}^{\mathcal{T}_p}(A_0)$, a $\mathcal{T}_p$-separable subspace of $C(K)$. We consider the dual pair $\langle C(K), L\rangle$, where $L := \text{span}(\widehat{K})$, see the paragraph before Lemma 3.56. Recall that $\mathcal{T}_p$ on $C(K)$ is just the topology $w\big(C(K), L\big)$ (Lemma 3.56). The space F carries the topology induced by $\mathcal{T}_p$, denoted again $\mathcal{T}_p$. The dual space of $(F, \mathcal{T}_p)$ is $L/F^{\perp}$. The canonical quotient mapping

$$q : \big(L, w(L, C(K))\big) \to \big(L/F^{\perp}, w(L/F^{\perp}, F)\big)$$

is continuous. Therefore $q(\widehat{K})$ is compact in the space $\big(L/F^{\perp}, w(L/F^{\perp}, F)\big)$, where $\widehat{K} := \{\delta_k : k \in \mathbb{K}\}$, see Lemma 3.55. Since the space $\big(F, w(F, L/F^{\perp})\big)$ is separable, the set $q(\widehat{K})$ is $w(L/F^{\perp}, F)$-metrizable, hence $w(L/F^{\perp}, F)$-separable. Certainly, $q(\widehat{K})$ separates points of F, hence it is $w(L/F^{\perp}, F)$-linearly dense in $L/F^{\perp}$, so $\big(L/F^{\perp}, w(L/F^{\perp}, F)\big)$ is separable. The set $\overline{A_0}^{\mathcal{T}_p}$ ($\subset F$) is $\mathcal{T}_p$-compact, according to Theorem 3.52. It follows that it is $\mathcal{T}_p$-metrizable. Henceforth there exists a sequence $\{a_n\}$ in A_0 that $\mathcal{T}_p$-converges to $\overline{a}$.

The angelicity of $C_p(K)$ follows from this and from Theorem 3.52.

The identity of the classes of $\mathcal{T}_p$-compacta mentioned in the statement follows from Exercise 3.9. □

For the case that the sets involved in Theorem 3.58 would be uniformly bounded, see Theorem 3.139.

3.8 Extreme Points, Krein–Milman Theorem

We will now study extreme points in real locally convex spaces. This subject, in the case of Banach spaces, will be considered again in Subsection 3.11.8. Some stronger results in this direction will be discussed in Chapters 8 and 11.

Definition 3.59 *Let E be a topological vector space and C be a nonempty convex subset of E. We say that a point $x_0 \in C$ is an* extreme point *of C if $x_1 = x_2 = x_0$ whenever $x_1 \in C$, $x_2 \in C$ and $x_0 = (1/2)(x_1 + x_2)$.*

This is a particular case of the following concept.

Definition 3.60 *Let E be a locally convex space and S a nonempty subset of E. A real linear manifold $H \subset E$ (i.e., the translate of a real linear subspace of E) is called a* supporting manifold *of S if $H \cap S \neq \emptyset$ and if every open interval $(x, y) \subset S$ which contains a point of H satisfies $(x, y) \subset H$. An* extreme point *of S is a supporting manifold of dimension 0 of S.*

A simple example of a supporting manifold is given by the following lemma.

Lemma 3.61 *Let S be a nonempty subset of a locally convex space E. Let f be a real linear functional $f : E \to \mathbb{R}$ that attains the supremum on S. Then the set $H := \{y \in E : f(y) = s\}$, where $s := \sup_{x \in S} f(x)$, is a supporting manifold of S. In particular, if $H \cap S$ is a single point, this point is an extreme point of S.*

Proof: Certainly, $H \cap S \neq \emptyset$. Let $(x, y) \subset S$ be an open interval such that $(x, y) \cap H \neq \emptyset$. Choose $x_0, y_0 \in (x, y)$ such that $x_0 \neq y_0$ and $[x_0, y_0] \cap H \neq \emptyset$. If $[x_0, y_0] \not\subset H$ then either $f(x_0) > s$ or $f(y_0) > s$, in both cases a contradiction. It follows that $(x, y) \subset H$. □

The supporting manifolds of a set S have a certain "transitive" property.

Lemma 3.62 *Let H be a supporting manifold of a nonempty set S in a locally convex space E. A real linear manifold $H_1 \subset H$ is a supporting manifold of S if and only in it is a supporting manifold of $H \cap S$.*

Proof: Assume first that H_1 is a supporting manifold of S. Then $H_1 \cap (H \cap S) = H_1 \cap S \neq \emptyset$. Let $(x, y) \subset H \cap S$ ($\subset S$) be an open interval such that $(x, y) \cap H_1 \neq \emptyset$. Then $(x, y) \subset H_1$ and so H_1 is a supporting manifold of $H \cap S$. Conversely, assume that H_1 is a supporting manifold of $H \cap S$. Let $(x, y) \subset S$ be an open interval such that $(x, y) \cap H_1 \neq \emptyset$. Then $(x, y) \cap H \neq \emptyset$, hence $(x, y) \subset H$, so $(x, y) \subset H \cap S$. It follows that $(x, y) \subset H_1$ and so H_1 is a supporting manifold of S. □

Lemma 3.63 *Let K be a nonempty compact subset of a locally convex space E. Let $\mathcal{H} = \{H_\gamma\}_{\gamma \in \Gamma}$ be a family of closed supporting manifolds of K directed by inclusion, i.e., given H_{γ_1} and H_{γ_2} in $\mathcal{H}$, there exists $H_{\gamma_3} \in \mathcal{H}$ such that $H_{\gamma_3} \subset H_{\gamma_1} \cap H_{\gamma_2}$. Then $H := \bigcap_{\gamma \in \Gamma} H_\gamma$ is a closed supporting manifold of K.*

Proof: Let $(x, y) \subset K$ be an open interval such that $(x, y) \cap H \neq \emptyset$. Then, for every $\gamma \in \Gamma$ we have $(x, y) \cap H_\gamma \neq \emptyset$. Since H_γ is a supporting manifold of K we get $(x, y) \subset H_\gamma$. This is true for all $\gamma \in \Gamma$, so $(x, y) \subset H$. We claim that $H \cap K \neq \emptyset$. In order to prove the claim, define a partial order in Γ by $\gamma_1 \leq \gamma_2$ whenever $H_{\gamma_1} \supset H_{\gamma_2}$. With this order, Γ is directed upward. By choosing, for every $\gamma \in \Gamma$, some $x_\gamma \in H_\gamma \cap K$, we obtain a net $\{x_\gamma\}_\Gamma$ in K. It has a cluster point in K (that belongs also to H), and this proves the claim. It follows that H is a supporting manifold of K. □

Proposition 3.64 *Let K be a nonempty compact subset of a locally convex space E. Let H be a closed supporting manifold of K. Then H contains at least one extreme point of K.*

Proof: By Lemmas 3.62 and 3.63, the (nonempty) family of closed supporting manifolds of K contained in H is directed by inclusion and inductive (i.e., every decreasing chain has a least element). By Zorn lemma, it has a minimal element, say H_1. Assume that H_1 does not reduces to a point. By translating, we may always assume $0 \in H_1$. The set $H_1 \cap K$ is a compact nonempty subset of the locally convex space H_1. Let $H_0 \subset H_1$ be a closed hyperplane of H_1 that supports $H_1 \cap K$ (use the compactness). Then, Lemma 3.62 proves that H_0 ($\subsetneq H_1$) is a closed supporting manifold of K. This violates the minimality of H_1. Since H_1 reduces to a point and it is a supporting manifold of K, this point is an extreme point of K. □

Theorem 3.65 (Krein–Milman) *Let E be a locally convex space and let K be a nonempty convex and compact subset of E. Then*
(i) *K has at least an extreme point.*
(ii) $K = \overline{\text{conv}}\,(\text{Ext}(K))$, *where* $\text{Ext}(K)$ *is the set of all extreme points of K.*

Proof: (i) Every real continuous linear functional $f : E \to \mathbb{R}$ attains its supremum on K. By Lemma 3.61, K has a supporting manifold. By Proposition 3.64, K has an extreme point.

(ii) Assume that there exists $x_0 \in K \setminus \overline{\text{conv}}\,(\text{Ext}(K))$. By Theorem 3.35 (ii), we can find a real continuous linear functional f such that $f(x_0) > \sup_{x \in \overline{\text{conv}}\,(\text{Ext}(K))} f(x)$. The set $\{y \in E : f(y) = \sup_{x \in K} f(x)\}$ is, according to Lemma 3.61, a supporting manifold of K; by Proposition 3.64, it has at least one extreme point of K. This is a contradiction. □

Remark: If $0 < p < 1$, then, in the L_p-space introduced in the example after Proposition 3.24, there is a convex compact set with no extreme point ([Robe1] and [Robe2]). This is not in contradiction with Theorem 3.65, since, in this case, L_p is not a locally convex space.

The following result is a kind of converse of Theorem 3.65.

Theorem 3.66 (Milman) *Let E be a locally convex space and B be a nonempty subset of E such that $C := \overline{\text{conv}}\,(B)$ is compact. Then* $\text{Ext}(C) \subset \overline{B}$ *(and so every extreme point of C is also an extreme point of $\overline{B}$).*

Proof: It is enough to prove that every extreme point of C lies in $B + U$, where U is an arbitrary closed convex and balanced neighborhood of 0 in E. Fix such a set U. Since B is totally bounded, it can be covered by a finite number of sets $x_i + U$, where $x_i \in B$ for $i = 1, 2, \dots, n$. The sets $B_i := \overline{\text{conv}}\,(B \cap (x_i + U))$ are convex and compact, hence conv $(\bigcup_{i=1}^n B_i)$ is also convex and compact. It follows easily that $C = \text{conv}\,(\bigcup_{i=1}^n B_i)$. If $e \in \text{Ext}(C)$, then $e = \sum_{i=1}^n \lambda_i b_i$, where $b_i \in B_i$, $\lambda_i \geq 0$ for all $i = 1, 2, \dots, n$, and $\sum_{i=1}^n \lambda_i = 1$. Necessarily $e = b_i$ for some b_i. This proves that $\text{Ext}(C) \subset B + U$. □

For a proof of Theorem 3.66 for the weak topology of a locally convex space, based on Lemma 3.69, see Exercise 3.145.

Let E and F be two locally convex spaces, and $T : E \to F$ be a continuous linear mapping. If S is a nonempty subset of E and $x_0 \in \text{Ext}(S)$, it is not always true that $Tx_0 \in \text{Ext}(T(S))$ (see Exercise 3.21). It is easy to prove that if H is a closed supporting manifold of $T(S)$ then $T^{-1}(H)$ is a (closed) supporting manifold of S (see Exercise 3.22). The following result is a consequence of this observation and Lemma 3.64.

Proposition 3.67 *Let E and F be two locally convex spaces, $T : E \to F$ be a continuous linear mapping and K be a nonempty compact subset of E. Then, every extreme point of $T(K)$ is the image of an extreme point of K.*

Definition 3.68 *Let C be a set in a locally convex space X. A* slice *of C is a nonempty intersection of C with an open half-space of X.*

We observed that finite intersections of slices form a basis of the weak topology. Due to the remark at the end of Section 3.5, this is true both in the real or complex case. The definition of the concept of supporting manifold (in particular, extreme point) (see Definition 3.60) uses only real linear manifolds and segments (x, y) defined by using only real scalars. In this section we will assume that all spaces considered are real.

Lemma 3.69 (Choquet) *Let C be a w-compact convex set in a locally convex space X. For every $x \in$ Ext(C), the slices of C containing x form a neighborhood base of x in the relative weak topology of C.*

Proof: Let V be a neighborhood of x in the relative weak topology of C of the form $V = \tilde{V}_1 \cap \dots \cap \tilde{V}_k$, where $\tilde{V}_i$ are slices of C, i.e., $\tilde{V}_i := V_i \cap C$ and V_i are open half-spaces in X (see Fig. 3.1). Then $x \notin \bigcup_{i=1}^k \big((X \setminus V_i) \cap C\big)$. Consequently $x \notin \text{conv} \bigcup_{i=1}^k \big((X \setminus V_i) \cap C\big)$ as x is an extreme point of C (Exercise 3.127). As the convex hull of a finite number of weakly compact convex sets is weakly compact (Exercise 3.14), we have that $x \notin \overline{\text{conv}} \bigcup_{i=1}^k \big((X \setminus V_k) \cap C\big)$. By Theorem 2.12, there is $f \in X^*$ and $\alpha \in \mathbb{R}$ such that $f(x) > \alpha > \sup \left\{ f(x) : x \in \bigcup_{i=1}^k (X \setminus V_i) \cap C \right\}$. Then the slice $C \cap \{x \in X : f(x) > \alpha\}$ contains x and is contained in V. □

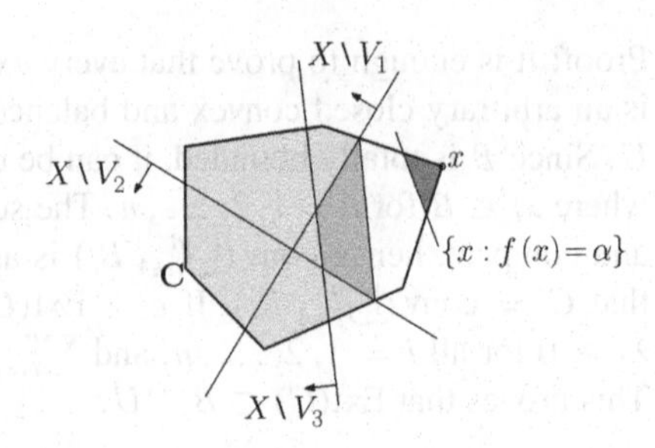

Fig. 3.1 The proof of Choquet's Lemma 3.69

3.9 Representation and Compactness

In the case of finite-dimensional topological vector space, Krein–Milman Theorem 3.65 can be strengthened. This is the content of the following result.

Proposition 3.70 (Carathéodory, see, e.g., [Rudi3]) *Let D be a compact convex subset of an n-dimensional topological vector space E. Every $x \in D$ is a convex combination of at most $n+1$ extreme points of D.*

Proof: By induction on n. For $n = 1$ we have $D = [a, b]$ and the statement follows. Assume that the result holds for $n - 1$. Consider $D \subset E$ with $\dim(E) = n$ and $\text{Int}(D) \neq \emptyset$ (Exercise 3.23). If x is a boundary point of D, let H be a supporting hyperplane of D such that $x \in H \cap D$. By the induction hypothesis, x is a convex combination of at most n extreme points of $H \cap D$ which are extreme points of D (see the proof of the Krein–Milman theorem). If $x \in \text{Int}(D)$, choose an extreme point $z_0 \in D$ and let y be another boundary point of D that lies on the line going through x and z_0. By the same argument as above, y is a convex combination of at most n extreme points $z_1, \ldots, z_n$ of D. Thus $y = \sum_{i=1}^n \alpha_i z_i$ with $\alpha_i \geq 0$ and $\sum_{i=1}^n \alpha_i = 1$.

Now $x = \alpha y + (1-\alpha)z_0$ for some $\alpha \in (0, 1)$, so $x = \sum_{i=1}^n \alpha\alpha_i z_i + (1-\alpha)z_0$, which is a convex combination of $n+1$ extreme points $z_0, \ldots, z_n$. □

A related argument was suggested in the Hint of Exercise 1.65.

We will now reformulate Carathéodory's theorem in the language of integral representation.

Let D be a compact convex subset of an n-dimensional topological vector space E. For $y \in D$, let δ_y be the Dirac measure at y.

Let $x \in D$ be given. By Carathéodory's theorem, there are extreme points $x_1, \ldots, x_k$ of D, $k \leq n+1$, and $\alpha_i \geq 0$, $\sum_{i=1}^k \alpha_i = 1$ such that $x = \sum_{i=1}^k \alpha_i x_i$. Put $\mu = \sum_{i=1}^k \alpha_i \delta_{x_i}$. Then μ is a probability measure on D. If f is a continuous function on D, we have

$$\int_D f \,\mathrm{d}\mu = \int_D f \,\mathrm{d}\Big(\sum \alpha_i \delta_{x_i}\Big) = \sum \alpha_i \int_D f \,\mathrm{d}\delta_{x_i} = \sum \alpha_i f(x_i).$$

If f is linear, we have $f(x) = f(\sum \alpha_i x_i) = \sum \alpha_i f(x_i) = \int_D f \,\mathrm{d}\mu$.

This motivates the following definition.

Definition 3.71 *Let D be a compact convex subset of a locally convex space E. We say that a probability measure μ on D* represents $x \in D$ *if for all $f \in E^*$,*

$$f(x) = \int_D f \, \mathrm{d}\mu.$$

If S is a Borel subset of D, we say that a measure μ is supported by S if $\mu(D \backslash S) = 0$.

Proposition 3.72 *Let D be a convex compact set in a locally convex space E. For every $x \in D$ there is a probability measure μ on D supported by $\overline{\mathrm{Ext}(D)}$ that represents x.*

Proof: By the Krein–Milman theorem, $\overline{\mathrm{conv}}(\mathrm{Ext}(D)) = D$. Hence there is a net $\{y_\alpha\}$ such that $y_\alpha \to x$ and $y_\alpha = \sum_{i=1}^{n_\alpha} \alpha_i^\alpha x_i^\alpha$, where $\alpha_i^\alpha \geq 0$, $\sum \alpha_i^\alpha = 1$, and $x_i^\alpha \in \mathrm{Ext}(D)$. We represent each y_α by a probability measure $\mu_\alpha = \sum \alpha_i^\alpha \delta_{x_i^\alpha}$. By the Riesz representation theorem, the set of all probability measures on the compact set $\overline{\mathrm{Ext}(D)}$ is identified with a w^*-compact set in $B_{C(\overline{\mathrm{Ext}(D)})^*}$. Therefore there is a subnet $\{y_\beta\}$ of $\{y_\alpha\}$ convergent in the w^*-topology to a probability measure μ on $\overline{\mathrm{Ext}(D)}$. If f is a continuous linear functional on E, then its restriction to $\overline{\mathrm{Ext}(D)}$ is in $C(\overline{\mathrm{Ext}(D)})$ and thus we have

$$f(x) = \lim f(y_\beta) = \lim \int\limits_{\overline{\mathrm{Ext}(D)}} f \, \mathrm{d}\mu_\beta = \int\limits_{\overline{\mathrm{Ext}(D)}} f \, \mathrm{d}\mu.$$

To finish the proof, we extend μ to $\tilde{\mu}$ on D by $\tilde{\mu}(B) = \mu(B \cap \overline{\mathrm{Ext}(D)})$ for every Borel set B in D. □

A natural question is whether a representing measure is supported by $\mathrm{Ext}(D)$ alone, as in Carathéodory's theorem. One of the difficulties is that $\mathrm{Ext}(D)$ may not in general be a Borel set ([Phel1]). However, if D is metrizable, we have the following result.

Lemma 3.73 *Let D be a compact convex set in a locally convex space E. If D is metrizable then the set of extreme points of D is a G_δ set in D.*

Proof: Let d be a metric on D that induces the topology of D. For $n \in \mathbb{N}$, put

$$F_n = \left\{x \in D : \, x = \tfrac{y+z}{2}, y, z \in D, \mathrm{d}(y, z) \geq \tfrac{1}{n}\right\}.$$

Then F_n is a closed set in D. Indeed, let $x_k \in F_n$ and $\lim x_k = x$, where $x_k = \frac{y_k + z_k}{2}$ for $y_k, z_k \in D$, $\mathrm{d}(y_k, z_k) \geq \frac{1}{n}$. Since D is metrizable and compact, we may assume that $\lim y_k = y$ and $\lim z_k = z$. Thus we have $x = \frac{y+z}{2}$. Moreover, $\mathrm{d}(y_k, z_k) \to \mathrm{d}(y, z)$, so $\mathrm{d}(y, z) \geq \frac{1}{n}$. This shows that $x \in F_n$, proving that F_n is closed. To conclude the proof, observe that $x \in D$ is not an extreme point of D if and only if $x \in F_n$ for some n. □

Theorem 3.74 (Choquet representation theorem, see, e.g., [Phel1]) *Let D be a metrizable compact convex set in a locally convex space E. For every $x_0 \in D$ there is a probability measure μ on D supported by* Ext(D) *that represents x_0.*

Proof: Note that Ext(D) is a Borel set by Lemma 3.73. Since D is a metrizable compact set, the space $C(D)$ of all continuous functions on D with the supremum norm $\|\cdot\|$ is separable (Lemma 3.102). Let A denote the subspace of $C(D)$ formed by all continuous functions f on D such that $f(\alpha x + \beta y) = \alpha f(x) + \beta f(y)$ for every $x, y \in D$ and $\alpha, \beta \geq 0$ with $\alpha + \beta = 1$ and $\alpha x + \beta y \in D$. Note that $f \in A$ need not, in general, be a restriction onto D of some $F(x) = k + \varphi(x)$ for $\varphi \in E^*$ (Exercise 3.26). Let $S_A = \{f \in A : \ \|f\| = 1\}$, let $\{h_n\}$ be a dense set in S_A. Put $h = \sum_{n=1}^{\infty} 2^{-n} h_n^2$.

We claim that h is a strictly convex continuous function on D. Let $x \neq y \in D$ be given. Choose n such that $h_n(x) \neq h_n(y)$. Using the affine property of h_n and the fact that the real function $t \mapsto t^2$ is strictly convex, for $\alpha \in (0, 1)$ we have

$$h_n^2(\alpha x + (1-\alpha)y) < \alpha h_n^2(x) + (1-\alpha)h_n^2(y).$$

Therefore

$$\begin{aligned}
h(\alpha x + (1-\alpha)y) &= 2^{-n}h_n^2(\alpha x + (1-\alpha)y) + \sum_{j\neq n} 2^{-j}h_j^2(\alpha x + (1-\alpha)y) \\
&\leq 2^{-n}h_n^2(\alpha x + (1-\alpha)y) + \sum_{j\neq n} 2^{-j}\left(\alpha h_j^2(x) + (1-\alpha)h_j^2(y)\right) \\
&< \alpha 2^{-n}h_n^2(x) + (1-\alpha)2^{-n}h_n^2(y) + \sum_{j\neq n} \alpha 2^{-j}h_j^2(x) + \sum_{j\neq n}(1-\alpha)2^{-j}h_j^2(y) \\
&= \alpha h(x) + (1-\alpha)h(y).
\end{aligned}$$

Let Y be the subspace of $C(D)$ spanned by A and h, i.e., $Y = A \oplus \mathbb{R}h$. For $f \in C(D)$ the *concave envelope* $\bar{f} : D \to \mathbb{R}$ is defined by

$$\bar{f}(x) = \inf\{g(x) : \ g \in A, \ g \geq f\}.$$

We list a few basic properties of $\bar{f}$:

(1) $\bar{f}$ is concave, bounded, upper semicontinuous and thus Borel measurable,
(2) $f \leq \bar{f}$, and $f = \bar{f}$ if f is concave,
(3) $\overline{f+g} \leq \bar{f} + \bar{g}$; $\overline{f+g} = \bar{f} + g$ if $g \in A$; $\overline{rf} = r\bar{f}$ if $r > 0$,
(4) $|\bar{f} - \bar{g}| \leq \|f - g\|$.
(to see (4), write $\bar{f} = \overline{f - g + g} \leq \overline{f - g} + \bar{g} \leq \|f - g\| + \bar{g}$).

Define a positively homogeneous subadditive functional p on $C(D)$ by $p(g) = \bar{g}(x_0)$ and define $\varphi \in Y^*$ by $\varphi(g + rh) = g(x_0) + r\bar{h}(x_0)$ for $g \in A$ and $r \in \mathbb{R}$. Note that φ is dominated by p on Y. We only need to verify that $g(x_0) + r\bar{h}(x_0) \leq \overline{g + rh}(x_0)$, which follows from (3) if $r > 0$. If $r < 0$, then $\overline{g + rh} = g + rh \geq g + r\bar{h}$ by the concavity of $g + rh$.

By the Hahn–Banach theorem, there is an extension m of φ onto the whole $C(D)$ such that $m(g) \leq \bar{g}(x_0)$ for every $g \in C(D)$.

If $g \in C(D)$ and $g \leq 0$, then $0 \geq \bar{g}(x_0) \geq m(g)$. Thus if $\|g\| \leq 1$, then $g - 1 \leq 0$ and $m(g) \leq m(1)$. Since $m(1) = \varphi(1) = 1$, we obtain that m is a norm-one functional on $C(D)$. By Riesz' representation theorem, there is a probability measure μ on D such that $m(g) = \int_D g \,\mathrm{d}\mu$ for every $g \in C(D)$. Note that $\int_D h \,\mathrm{d}\mu = \bar{h}(x_0)$ and $g(x_0) = m(x_0) = \int_D g \,\mathrm{d}\mu$ for every $g \in A$, in particular for every $g \in E^*$. To conclude the proof, it is enough to show that:

(a) $\int_D h \,\mathrm{d}\mu = \int_D \bar{h} \,\mathrm{d}\mu$,

(b) $\{x \in D : h(x) = \bar{h}(x)\} \subset \operatorname{Ext}(D)$.

Indeed, since $\bar{h} \geq h$ on D, from (a) we get $\mu\{x \in D : \bar{h}(x) < h(x)\} = 0$. From (b) it follows that $D \setminus \operatorname{Ext}(D) \subset \{x \in D : \bar{h}(x) > h(x)\}$ and thus $\mu\big(D \setminus \operatorname{Ext}(D)\big) = 0$. To verify (a), write

$$\int_D h \,\mathrm{d}\mu \leq \int_D \bar{h} \,\mathrm{d}\mu \leq \inf\Big\{\int_D g \,\mathrm{d}\mu : g \in A, g \geq \bar{h}\Big\} = \inf\Big\{\int_D g \,\mathrm{d}\mu : g \in A, g \geq h\Big\}$$

$$= \inf\{g(x_0) : g \in A, g \geq h\} = \bar{h}(x_0) = \varphi(h) = m(h) = \int_D h \,\mathrm{d}\mu.$$

To prove (b), assume that $x \in D \setminus \operatorname{Ext}(D)$. Then $x = \frac{y+z}{2}$, where $y, z \in D$, $y \neq z$. Hence using concavity of $\bar{h}$

$$h(x) < \tfrac{1}{2}h(y) + \tfrac{1}{2}h(z) \leq \tfrac{1}{2}\bar{h}(y) + \tfrac{1}{2}\bar{h}(z) \leq \bar{h}\big(\tfrac{y+z}{2}\big) = \bar{h}(x).$$

□

3.10 The Space of Distributions

Let Ω be a nonempty open set in $\mathbb{R}^n$. We will now construct the space of distributions and establish some of its properties.

Choose any sequence $\{K_n\}_{n\in\mathbb{N}}$ of compact subsets of Ω such that $K_n \subset \operatorname{Int}(K_{n+1})$ and $\bigcup K_n = \Omega$; for example $K_n = \{x : \operatorname{dist}(x, \mathbb{R}^n \setminus \Omega) \geq \frac{1}{n}, \|x\|_2 \leq n\}$, where $\|\cdot\|_2$ is the canonical norm in ℓ_2^n.

Definition 3.75 *Let $C(\Omega)$ denote the vector space of all real-valued continuous functions on Ω with the topology κ determined by the local base $\{U_n\}$, where*

$$U_n := \Big\{f \in C(\Omega) : \sup_{x\in K_n} |f(x)| < \tfrac{1}{n}\Big\}.$$

Fact 3.76 *$\big(\mathcal{C}(\Omega), \kappa\big)$ is a locally convex space whose topology does not depend on the choice of $\{K_n\}$.*
The topology κ is the topology of uniform convergence on every compact subset of Ω. That is, $f_\alpha \to f$ in κ if and only if $f_\alpha \to f$ uniformly on every compact subset of Ω.
$\big(\mathcal{C}(\Omega), \kappa\big)$ is a Fréchet space but the topology κ is not normable.

Proof: Local convexity of κ is routine to verify.

Let $\{K_n'\}$ be another family of compact sets with the required properties. Then for $n \in \mathbb{N}$, $\{\text{Int}(K_m)\}_m$ is an open cover of K_n', hence using $K_n \subset \text{Int}(K_{n+1})$, there is K_m such that $K_n' \subset \text{Int}(K_m) \subset K_m$.

This shows the independence of κ on the choice of $\{K_n\}$ and also that κ is the topology of uniform convergence on every compact set in Ω.

By a standard argument we see that $\big(\mathcal{C}(\Omega), \kappa\big)$ is a Fréchet space. Since every U_n contains functions f for which $\sup_{K_{n+1}} |f|$ can be arbitrary large, we have that no U_n is bounded. Therefore $(\mathcal{C}(\Omega), \kappa)$ is not normable. □

In particular, any set of the form $\Big\{f \in \mathcal{C}(\Omega) : \sup_{x\in K} |f(x)| < \varepsilon\Big\}$ with $\varepsilon > 0$ and K a compact set in Ω is an open neighborhood of 0 in $\mathcal{C}(\Omega)$.

Recall that for every multi-index $\alpha := (\alpha_1, \dots, \alpha_n)$, where α_i are non-negative integers, the corresponding derivative is defined by

$$D^\alpha = \left(\tfrac{\partial}{\partial x_1}\right)^{\alpha_1} \cdots \left(\tfrac{\partial}{\partial x_n}\right)^{\alpha_n}.$$

The degree of the derivative is defined as $|\alpha| = \sum \alpha_i$.

Definition 3.77 *By $\mathcal{C}^\infty(\Omega)$ we denote the set of all real-valued functions f on Ω such that $D^\alpha f \in \mathcal{C}(\Omega)$ for every α. A topology on $\mathcal{C}^\infty(\Omega)$ is defined by its local base $\{U_{n,\alpha}\}$*

$$U_{n,\alpha} := \Big\{f \in \mathcal{C}^\infty(\Omega) : \sup_{x\in K_n} |D^\alpha f(x)| < \frac{1}{n}\Big\}.$$

Note that the base $\{U_{n,\alpha}\}$ is countable.

Fact 3.78 *$\mathcal{C}^\infty(\Omega)$ is a Fréchet space whose topology does not depend on the choice of $\{K_n\}$.*
$f_\beta \to f$ in $\mathcal{C}^\infty(\Omega)$ if and only if $D^\alpha f_\beta \to D^\alpha f$ uniformly on K for every compact set K in Ω and every multi-index α.
$\mathcal{C}^\infty(\Omega)$ has the Heine–Borel property.

Recall that a space has the *Heine–Borel property* if all of its closed bounded subsets are compact.

Proof: We will show the Heine–Borel property. Let A be a bounded and closed set in $\mathcal{C}^\infty(\Omega)$. Then for every n there is an $M_n > 0$ such that $|D^\alpha f(x)| \le M_n$ for all $|\alpha| \le n$, $f \in A$ and $x \in K_n$. Therefore $\{D^\alpha f : f \in A\}$ is equicontinuous on K_{n-1} for $|\alpha| \le n-1$. From the Arzelà–Ascoli theorem and the Cantor diagonal process, it follows that every sequence $\{f_i\} \subset A$ contains a subsequence f_{i_j} for which all $D^\alpha f_{i_j}$ converge uniformly on every compact subset of Ω. Hence f_{i_j} is convergent in $\mathcal{C}^\infty(\Omega)$. This proves that A is compact. □

The differentiation mapping $D^\alpha : \mathcal{C}^\infty(\Omega) \to \mathcal{C}^\infty(\Omega)$ is continuous and linear.

In this section we shall depart from the previously defined concept of *support* and use the symbol supp(f) and the word *support* of a real-valued function on Ω to denote the set $\overline{\{x \in \Omega : f(x) \neq 0\}}$.

Definition 3.79 *Let K be a compact set in Ω such that* $\mathrm{Int}(K) \neq \emptyset$. *We define $\mathcal{D}_K$ as the closed subspace of $\mathcal{C}^\infty(\Omega)$ formed by functions f such that $supp(f) \subset K$. We define $\mathcal{D}(\Omega)$ as the union of all $\mathcal{D}_K(\Omega)$ for compact sets K as above.*

Note that $\mathcal{D}(\Omega)$ is not a complete subspace of $\mathcal{C}^\infty(\Omega)$. For $\Omega = \mathbb{R}$ this is seen as follows: Choose $\varphi \in \mathcal{C}^\infty(\mathbb{R})$ with supp$(\varphi) \subset [0, 1]$ and $\varphi > 0$ on $(0, 1)$. For $m \in \mathbb{N}$ put

$$\varphi_m(x) = \varphi(x-1) + \tfrac{1}{2}\varphi(x-2) + \cdots + \tfrac{1}{m}\varphi(x-m).$$

Then $\{\varphi_m\}$ is a Cauchy sequence in $\mathcal{D}(\mathbb{R}) \subset \mathcal{C}^\infty(\Omega)$ and $\lim \varphi_m$ does not have compact support.

We now introduce a locally convex topology on $\mathcal{D}(\Omega)$.

For every compact set $K \subset \Omega$ with nonempty interior, let τ_K denote the topology of $\mathcal{D}_K$ inherited from $\mathcal{C}^\infty(\Omega)$. Let $\mathcal{B}$ denote the collection of all convex balanced sets $W \subset \mathcal{D}(\Omega)$ such that $\mathcal{D}_K \cap W$ is τ_K-open in $\mathcal{D}_K$ for every such compact set $K \subset \Omega$. Then $\mathcal{B}$ is a local base for a locally convex topology on $\mathcal{D}(\Omega)$, which will be denoted by τ.

Proposition 3.80 (i) *The topology on $\mathcal{D}_K$ inherited from $\mathcal{D}(\Omega)$ coincides with τ_K for every compact set $K \subset \Omega$ with* $\mathrm{Int}(K) \neq \emptyset$.
(ii) *If A is a bounded subset of $\mathcal{D}(\Omega)$, then there is a compact set $K \subset \Omega$ such that $A \subset \mathcal{D}_K$.*
(iii) *$\mathcal{D}(\Omega)$ has the Heine–Borel property.*
(iv) *Every Cauchy sequence in $\mathcal{D}(\Omega)$ is convergent.*
(v) *$\mathcal{D}(\Omega)$ is not metrizable.*

Proof: (Sketch) To prove (i) it is enough to show that given an open set $\mathcal{O}$ in $\mathcal{D}_K$, there exists an open set V in $\mathcal{D}(\Omega)$ such that $\mathcal{O} = V \cap \mathcal{D}_K$. Let $\{U_n\}$ be the local base in $\mathcal{D}_K$ as above and p_n be the Minkowski functional of U_n. For every $\varphi \in \mathcal{O}$ there is n and $\delta > 0$ such that

$$\{\psi \in \mathcal{D}_K : p_n(\psi - \varphi) < \delta\} \subset \mathcal{O}.$$

Put $W_\varphi = \{\psi \in \mathcal{D}(\Omega) : p_n(\psi) < \delta\}$. Then $W_\varphi \in \mathcal{B}$ and

$$\mathcal{D}_K \cap (\varphi + W_\varphi) = \varphi + (\mathcal{D}_K \cap W_\varphi) \subset \mathcal{O}.$$

Let $V = \bigcup_{\varphi \in \mathcal{O}} (\varphi + W_\varphi)$. Then V has the desired property.

(ii) Assume that A is bounded in $\mathcal{D}(\Omega)$ and A lies in no $\mathcal{D}_K$. Then there are $\varphi_m \in A$ and distinct points $x_m \in \Omega$ without a limit point in Ω such that $\varphi_m(x_m) \neq 0$, $m \in \mathbb{N}$. Let W be the set of all $\varphi \in \mathcal{D}(\Omega)$ that satisfy $|\varphi(x_m)| < \frac{1}{m}|\varphi_m(x_m)|$ for all $m \in \mathbb{N}$.

Note that W is convex and balanced. Since every compact set K contains only finitely many x_m, it is easy to see that $\mathcal{D}_K \cap W$ is open in $\mathcal{D}_K$. Therefore $W \in \mathcal{B}$. Since $\varphi_m \notin mW$, no multiple of A is contained in W. Therefore A is not bounded in $\mathcal{D}(\Omega)$.

(iii) follows from (ii) and the fact that $\mathcal{D}_K$ has the Heine–Borel property.

(iv) follows from the fact that every Cauchy sequence is bounded, from (ii) and from the completeness of $\mathcal{D}_K$.

(v) It is easy to see that $\mathcal{D}_K$ has no interior point in $\mathcal{D}(\Omega)$. Assume that $\mathcal{D}(\Omega)$ is metrizable. Then $\mathcal{D}(\Omega)$ is a complete metrizable space as any Cauchy sequence in $\mathcal{D}(\Omega)$ is convergent. Note that $\mathcal{D}_k$ are closed subspaces of $\mathcal{D}(\Omega)$. This leads to a contradiction with the Baire category theorem. □

We now characterize continuity of linear mappings on $\mathcal{D}(\Omega)$.

Proposition 3.81 *A linear functional f on $\mathcal{D}(\Omega)$ is continuous if and only if f is sequentially continuous if and only if the restriction of f to every $\mathcal{D}_K$ is continuous. Similarly, a linear mapping $T\colon \mathcal{D}(\Omega) \to E$, where E is a locally convex space, is continuous if and only if $T\big|_{\mathcal{D}_K}$ is continuous for every $\mathcal{D}_K$.*

Proof: Assume that the restrictions of f to all $\mathcal{D}_K$ are continuous. Consider the set $V := f^{-1}\big((-\delta, \delta)\big)$. Then V is balanced and convex. By Proposition 3.80, V is open in $\mathcal{D}(\Omega)$ if and only if $\mathcal{D}_K \cap V$ is open in $\mathcal{D}_K$ for every K. But this is the case as the restriction of f to each $\mathcal{D}_K$ is continuous.

If f is sequentially continuous on $\mathcal{D}(\Omega)$, then the restriction of f to every $\mathcal{D}_K$ is continuous as $\mathcal{D}_K$ is metrizable for every K. □

Definition 3.82 *Continuous linear functionals on $\mathcal{D}(\Omega)$ are called* distributions. *It is traditional to denote $\mathcal{D}(\Omega)^*$ by $\mathcal{D}'(\Omega)$.*

We will show some important examples of distributions:

Recall that a measurable function f in Ω is *locally integrable* if $\int_K |f| < \infty$ for every compact subset K.

The space of locally integrable functions $L^1_{\mathrm{loc}}(\Omega)$ can be embedded into $\mathcal{D}'(\Omega)$. Indeed, let $f \in L^1_{\mathrm{loc}}(\Omega)$. We define $T_f(\varphi) = \int_{\mathbb{R}^n} f\varphi \,\mathrm{d}\omega$. Then T_f is a distribution. The continuity on every $\mathcal{D}_K$ follows from

$$\left|\int_K f\varphi \right| \mathrm{d}\omega \leq \int_K |f| \max_K |\varphi| \,\mathrm{d}\omega.$$

In particular, we may consider $C(\Omega)$ a subset of $\mathcal{D}'(\Omega)$.

The space of all Borel measures on Ω can be also embedded into $\mathcal{D}'(\Omega)$. It is easy to see that the functional $T_\mu(\varphi) := \int_{\mathbb{R}^n} \varphi \, d\mu$ is a distribution. In particular, Dirac measures are distributions.

Let α be a multi-index. For $F \in \mathcal{D}'(\Omega)$ we define the *derivative* (in the distributional sense) by

$$D^\alpha F : \varphi \mapsto (-1)^{|\alpha|} F(D^\alpha(\varphi))$$

It is easy to see that $D^\alpha F \in \mathcal{D}'(\Omega)$.

The definition is motivated by the following. Let T be the distribution corresponding to some function f, that is, $T := T_f$. Assume that f has continuous derivatives of order $|\alpha|$. Using integration by parts and the fact that $\varphi \in \mathcal{D}(\Omega)$ vanish on the boundary of Ω we obtain $D^\alpha T_f(\varphi) = T_{D^\alpha f}(\varphi)$ for all $\varphi \in \mathcal{D}(\Omega)$. So for differentiable functions, the distributional derivative D^α agrees with the usual partial derivative.

Fact 3.83 $D^\alpha : \mathcal{D}'(\Omega) \to \mathcal{D}'(\Omega)$ *is a continuous operator.*

Proof: Linearity is obvious.

Let K be a compact set in Ω and let $F \in \mathcal{D}'(\Omega)$ be bounded by $C > 0$ on a neighborhood U_n from the local base in $\mathcal{D}_K$. Let p_n be the Minkowski functional of U_n. Then $|F(\varphi)| \leq C p_n(\varphi)$ for every $\varphi \in \mathcal{D}_K$. Consequently

$$|(D^\alpha F)(\varphi)| \leq C p_n(D^\alpha \varphi) \leq C p_{n+|\alpha|}(\varphi).$$

Therefore $D^\alpha F$ is a continuous linear functional when restricted to an arbitrary $\mathcal{D}_K$, which by Proposition 3.81 means that $D^\alpha F \in \mathcal{D}'(\Omega)$. □

The space $\mathcal{D}'(\Omega)$ is important in the theory of generalized solutions to differential equations. Note that every continuous function on Ω has all partial derivatives in the sense of distributions.

3.11 Banach Spaces

3.11.1 Banach–Steinhaus Theorem

Definition 3.84 *Let X, Y be normed spaces and $\mathcal{A} \subset \mathcal{B}(X, Y)$. We say that $\mathcal{A}$ is* pointwise bounded *if* $\sup\{\|T(x)\|_Y : T \in \mathcal{A}\} < \infty$ *for all $x \in X$.*

Observe that a set $\mathcal{A} \subset \mathcal{B}(X, Y)$ is pointwise bounded precisely when it is bounded in $\mathcal{B}(X, Y)$ equipped with the topology $\mathcal{T}_p$ of the pointwise convergence, see Definitions 3.3 and 3.11.

If $\mathcal{A}$ is bounded in $\mathcal{B}(X, Y)$ equipped with the operator norm, that is, there is $C > 0$ so that $\|T\| \leq C$ for all $T \in \mathcal{A}$, then $\mathcal{A}$ is pointwise bounded. Indeed, for $x \in X$ we have $\|T(x)\|_Y \leq \|T\| \, \|x\|_X \leq C\|x\|_X$, so $\sup\{\|T(x)\|_Y : T \in \mathcal{A}\} \leq C\|x\|_X$. The opposite direction is often called the Banach–Steinhaus Uniform Boundedness principle.

Theorem 3.85 (Banach, Steinhaus) *Let X, Y be Banach spaces and $\mathcal{A} \subset \mathcal{B}(X, Y)$. If $\mathcal{A}$ is pointwise bounded then $\mathcal{A}$ is bounded in $\mathcal{B}(X, Y)$.*

Proof: For $n \in \mathbb{N}$ set $N_n = \big\{x \in X : \sup_{T \in \mathcal{A}} \|T(x)\|_Y \leq n\big\}$.

We claim that N_n is closed, convex, and balanced in X. The balancedness of N_n is obvious. To check closedness, let $x_k \in N_n$ and $x_k \to x \in X$. Given $T \in \mathcal{A}$, we have $\|T(x_k)\|_Y \leq n$, so $\|T(x)\|_Y \leq n$ by continuity. To see that N_n is convex, let $x_1, x_2 \in N_n$ and $\lambda \in [0, 1]$. Then for every $T \in \mathcal{A}$ we have

$$\|T(\lambda x_1 + (1-\lambda)x_2)\|_Y \leq \lambda\|T(x_1)\|_Y + (1-\lambda)\|T(x_2)\|_Y \leq \lambda n + (1-\lambda)n = n.$$

Since for every $x \in X$ we have $\sup_{T \in \mathcal{A}} \|T(x)\|_Y < \infty$, there is some $n \in \mathbb{N}$ greater than the supremum, hence $x \in N_n$. So $\bigcup_{n=1}^{\infty} N_n = X$. By the Baire category theorem, there is n_0 such that the set N_{n_0} contains an interior point x_0. Thus there is $\delta > 0$ such that $x_0 + \delta B_X \subset N_{n_0}$. Because of the symmetry of N_{n_0} we have $-x_0 + \delta B_X \subset N_{n_0}$. If $b \in B_X$, then by the convexity of N_{n_0} we have $b = \frac{1}{2}(x_0 + b) + \frac{1}{2}(-x_0 + b) \in N_{n_0}$. Hence $\delta B_X \subset N_{n_0}$. Consequently, given $T \in \mathcal{A}$, for every $x \in B_X$ we have $\|T(\delta x)\|_Y \leq n_0$, that is, $\|T\| \leq \frac{n_0}{\delta}$. This means that $\sup\limits_{T \in \mathcal{A}} \|T\| \leq \frac{n_0}{\delta}$. □

For an alternative proof of Theorem 3.85 that makes no use of the Baire category argument see Exercise 3.68.

Corollary 3.86 *Let X, Y be Banach spaces and $T_n \in \mathcal{B}(X, Y)$ for $n \in \mathbb{N}$. Assume that for every $x \in X$ there exists the limit $T(x) := \lim\limits_{n \to \infty} T_n(x)$. Then $T \in \mathcal{B}(X, Y)$ and $\|T\| \leq \liminf\limits_{n \to \infty} \|T_n\|$ $(< \infty)$.*

Proof: The linearity of T is easy:

$$T(x+y) = \lim T_n(x+y) = \lim\big(T_n(x)+T_n(y)\big) = \lim T_n(x) + \lim T_n(y) = T(x)+T(y).$$

From $T_n(x) \to T(x)$ we have $\|T_n(x)\| \to \|T(x)\|$, in particular $\{\|T_n(x)\| : n \in \mathbb{N}\}$ is bounded for all $x \in X$. By the Banach–Steinhaus theorem, $\{T_n\}$ is bounded in $\mathcal{B}(X, Y)$. Denote $C = \liminf \|T_n\|$. Then

$$\|T(x)\| = \lim \|T_n(x)\| = \liminf \|T_n(x)\| \leq \liminf(\|x\| \, \|T_n\|).$$

This shows that $\|T(x)\| \leq C\,\|x\|$, that is, T is bounded and $\|T\| \leq C$. □

Corollary 3.87 *Let X, Y be Banach spaces and $T_1, T_2, \dots \in \mathcal{B}(X, Y)$. If $\lim T_n(x) = T(x)$ for every $x \in X$, then for every compact set K in X we have $T_n(x) \to T(x)$ uniformly on K.*

Proof: By contradiction, assume that there is a compact set K in X, $\varepsilon > 0$, a subsequence of $\{T_n\}$ (denoted by $\{T_n\}$ again) and $x_n \in K$ such that $\|T_n(x_n) - T(x_n)\| \geq \varepsilon$. Since $\{x_n\} \subset K$, we may assume that $x_n \to x$ for some $x \in K$. Note that by the Banach–Steinhaus theorem we have $M := \sup\{\|T\|, \|T_1\|, \|T_2\|, \dots\} < \infty$. Thus

$$\|(T_n - T)(x_n)\| \leq \|(T_n - T)(x)\| + \|(T_n - T)(x_n - x)\|$$
$$\leq \|(T_n - T)(x)\| + \|T_n - T\| \cdot \|x_n - x\| \leq \|(T_n - T)(x)\| + 2M \cdot \|x_n - x\| \to 0,$$

a contradiction with $\|T_n(x_n) - T(x_n)\| \geq \varepsilon$. □

Theorem 3.88 (Banach, Steinhaus) *Let X be a Banach space. If $M \subset X^*$ is w^*-bounded then M is bounded.*
If $M \subset X$ is w-bounded then M is bounded.

Proof: The space (X^*, w^*) is $\mathcal{B}(X, \mathbb{K})$ endowed with the topology of the pointwise convergence. The result follows from Theorem 3.85.

The mapping $\pi : X \to \mathcal{B}(X^*, \mathbb{K})$ given by $\pi(x)(x^*) = \langle x^*, x\rangle$ for every $x^* \in X^*$, $x \in X$, is an isometry from $(X, \|\cdot\|)$ into $\mathcal{B}(X^*, \mathbb{K})$ endowed with the operator norm. Moreover, the topology $\mathcal{T}_p$ on $\mathcal{B}(X^*, \mathbb{K})$ of the pointwise convergence induces on $\pi(X)$ a topology, denoted again $\mathcal{T}_p$, that makes $\pi : (X, w) \to (\pi(X), \mathcal{T}_p)$ an isomorphism. The result follows from Theorem 3.85. □

For an extension of Theorem 3.88 to the setting of locally convex spaces, see Exercises 3.71, 3.72, 3.73, and 3.74.

We saw in Corollary 3.15 that all topological vector space topologies on a finite-dimensional vector space coincide. In particular, if X is a finite-dimensional normed space, then on X the weak topology coincides with the norm topology. However, for infinite-dimensional spaces X, the weak topology *never* coincides with the norm topology of X. This is proved in the next result.

Proposition 3.89 *Let X be an infinite-dimensional normed space and $\mathcal{O} \subset X$. If $\mathcal{O} \neq \emptyset$ is w-open, then $\mathcal{O}$ is not bounded.*
In particular, the weak and norm topology on X do not coincide.

Proof: We can assume that $0 \in \mathcal{O}$. Then there is $\varepsilon > 0$, $f_1, \dots, f_n \in X^*$ such that $\{x : |f_i(x)| < \varepsilon\} \subset \mathcal{O}$. Clearly, the set

$$N := \{x \in X : f_i(x) = 0 \text{ for } i = 1, \dots, n\} = \bigcap f_i^{-1}(0)$$

is contained in $\mathcal{O}$. We claim that $N \neq \{0\}$. Indeed, suppose that $N = \{0\}$. Then for every $f \in X^*$ we have $N \subset f^{-1}(0)$, so by Lemma 3.21, f is a linear combination of $f_1, \dots, f_n$. Thus $X^* = \operatorname{span}\{f_i\}$, a contradiction.

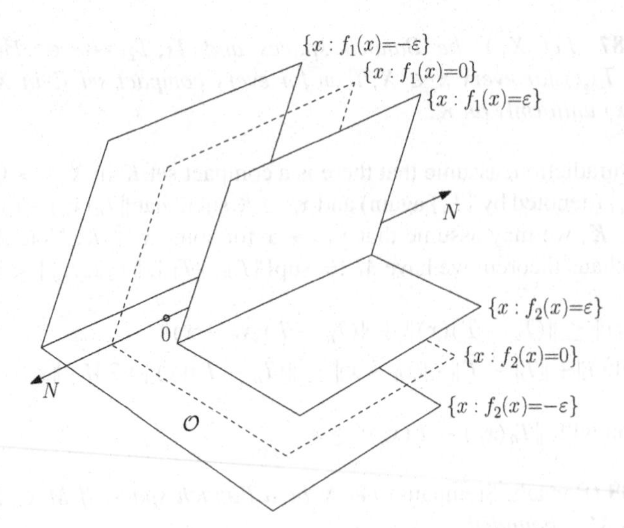

Fig. 3.2 Non-empty w-open sets in infinite dimension are unbounded

Therefore we can find $0 \neq x \in N$. Then for every scalar λ we have $\lambda x \in N$, so $\mathcal{O}$ contains a line through x. Hence $\mathcal{O}$ cannot be bounded in X (see Fig. 3.2). □

Proposition 3.90 *Let X be a Banach space. Then the original norm topology on X coincides with the Mackey topology on X.*

Proof: By Theorem 3.41, $\mu(X, X^*)$ is stronger than the norm topology. Conversely, let V be a $\mu(X, X^*)$-neighborhood of zero. Then there is a w^*-compact, convex, and balanced set S in X^* such that $S_\circ \subset V$. Since S is w^*-bounded, it is also norm-bounded in X^* and there is $\alpha > 0$ such that $S \subset \alpha B_{X^*}$. Then $V \supset S_\circ \supset \frac{1}{\alpha}(B_{X^*})_\circ = \frac{1}{\alpha} B_X$. □

Proposition 3.91 *Let X be a Banach space and $\mathcal{T}$ be a locally convex topology on X such that $(X, \mathcal{T})^* = X^*$. If $\mathcal{T}$ has a countable local base, then $\mathcal{T}$ coincides with the original norm topology of X.*

Proof: By the Banach–Steinhaus theorem, the systems of all bounded sets in $(X, \|\cdot\|)$ and (X, w) coincide. By Corollary 3.44 and Proposition 3.90, this system is the same for all locally convex topologies $\mathcal{T}$ on X for which $(X, \mathcal{T})^* = X^*$. Applying Lemma 3.47 for the set $A = B_X$ completes the proof. □

3.11.2 Banach–Dieudonné Theorem

Theorem 3.92 (Banach, Dieudonné) *Let X be a Banach space and let A be a convex set in X^*. If $A \cap nB_{X^*}$ is w^*-closed for every $n \in \mathbb{N}$, then A is w^*-closed.*

Proof: First note that under our assumption, $A \cap B$ is w^*-closed for every closed ball B in X^* (centered at a general point of X^*). Indeed, if n is large enough so that $B \subset nB_{X^*}$, then $A \cap B = (A \cap nB_{X^*}) \cap B$, where both B and $A \cap nB_{X^*}$ are w^*-closed sets.

Also note that A is norm closed. Indeed, if $f_n \in A$, $f_n \to f$ in X^*, then f_n is bounded and thus for some j we have that $f_n \in jB_{X^*}$ for every n. Since $A \cap jB_{X^*}$ is w^*-closed, $f_n \overset{w^*}{\to} f$, and $f_n \in A \cap jB_{X^*}$ for every n, we have $f \in A \cap jB_{X^*} \subset A$.

We may assume without loss of generality that $0 \in A$. For $n \in \mathbb{N}$, put $A_n = A \cap 2^n B_{X^*}$. Then $A = \bigcup A_n$ and each A_n is w^*-closed, convex and contains 0, therefore $A_n = ((A_n)_\circ)^\circ$ by the bipolar theorem. Put $G = \bigcap (A_n)_\circ$.

We will show that $A = G^\circ$ and this will finish the proof, as every polar set is w^*-closed. First observe that $G \subset (A_n)_\circ$ for every n and thus $G^\circ \supset ((A_n)_\circ)^\circ = A_n$ for every n. Consequently, $A = \bigcup A_n \subset G^\circ$.

It remains to show that $G^\circ \subset A$. This will be done in several steps.

Claim A
$(A_n)_\circ \subset (A_{n+1})_\circ + 2^{-n} B_X$ for every $n \in \mathbb{N}$.

Assume $x \notin (A_{n+1})_\circ + 2^{-n} B_X$. By the separation theorem, we obtain $f \in X^*$ such that $f(x) \geq 1 \geq \sup\{f(x) : x \in (A_{n+1})_\circ + 2^{-n} B_X\}$. By the definition of A_n we get $(A_{n+1})_\circ \supset 2^{-(n+1)} B_X$ and $f\big(3 \cdot 2^{-(n+1)} B_X\big) = f\big(2^{-(n+1)} B_X + 2^{-n} B_X\big) \subset f\big((A_{n+1})_\circ + 2^{-n} B_X\big)$. Therefore $\|f\| \leq \frac{2}{3} 2^n$.

Also, $\sup\{f(x) : x \in (A_{n+1})_\circ + 2^{-n} B_X\} = \sup\limits_{(A_{n+1})_\circ} (f) + \sup\limits_{2^{-n} B_X} (f) \leq 1$ and thus $\sup\limits_{(A_{n+1})_\circ} (f) \leq 1 - 2^{-n} \|f\|$.

Choose $\varepsilon \in \big(0, \min(\frac{1}{3}, 2^{-n} \|f\|)\big)$ and put $g = \frac{1}{1-\varepsilon} f$. From $\varepsilon < 2^{-n} \|f\|$ and $\sup\limits_{(A_{n+1})_\circ} (f) \leq 1 - 2^{-n} \|f\|$ we get $\sup\limits_{(A_{n+1})_\circ} (g) \leq 1$, that is, $g \in ((A_{n+1})_\circ)^\circ = A_{n+1}$. In particular, $g \in A$. Since $\|g\| \leq 2^n$ (as $\|f\| \leq \frac{2}{3} 2^n$ and $\varepsilon < \frac{1}{3}$) and $g \in A$, we have $g \in A_n$. Moreover, $g(x) > 1$ and thus $x \notin (A_n)_\circ$. This proves Claim A. △

Claim B
$(A_n)_\circ \subset G + 2^{-(n-1)} B_X$ for every $n \in \mathbb{N}$.

Given $n \in \mathbb{N}$ and $x_n \in (A_n)_\circ$, define inductively, using Claim A, vectors $x_{n+1}, x_{n+2}, \ldots$ in such a way that for $m = n, n+1, \ldots$ we have $x_{m+1} \in (A_{m+1})_\circ$ and $\|x_m - x_{m+1}\| \leq 2^{-m}$. Then $\{x_m\}$ is norm Cauchy and thus convergent to some $x \in X$. Since $(A_m)_\circ$ is norm closed for all m, we have that $x \in \bigcap (A_m)_\circ = G$. Moreover, $\|x_n - x\| \leq \sum_{j=0}^{\infty} \dfrac{1}{2^{n+j}} \leq \dfrac{1}{2^{n-1}}$. △

Claim C
$A = \bigcap\limits_{\varepsilon > 0} (1 + \varepsilon) A$.

Since A is convex and $0 \in A$, write $a = \frac{1}{1+\varepsilon}(1+\varepsilon)a + \big(1 - \frac{1}{1+\varepsilon}\big)0$ for $a \in A$ and $\varepsilon > 0$ to see that $A \subset (1+\varepsilon)A$. Therefore $A \subset \bigcap\limits_{\varepsilon > 0} (1+\varepsilon) A$.

If $x \in \bigcap(1+\varepsilon)A$ and $x = \left(1+\frac{1}{n}\right)x_n$ with $x_n \in A$ for every n, then $x - x_n = \frac{1}{n}x_n$ and $\|x_n\| \leq \|x\|$. Thus $x_n \to x$ and since A is closed, we have $x \in A$. Claim C is proved. $\triangle$

Claim D
$G^\circ \subset \bigcap_{\varepsilon>0}(1+\varepsilon)A.$

Let $\varepsilon > 0$ and $n \in \mathbb{N}$ be given. For every x and y in X,

$$x + y = (1+\varepsilon)\left(\frac{x}{1+\varepsilon} + \frac{\varepsilon}{1+\varepsilon}\frac{y}{\varepsilon}\right).$$

From this and Claim B we have $(A_n)_\circ \subset (1+\varepsilon)\,\mathrm{conv}\left\{G \cup \frac{1}{2^{n-1}\varepsilon}B_X\right\}$. Therefore by taking polars we have $\frac{1}{1+\varepsilon}\left(G^\circ \cap 2^{n-1}\varepsilon B_{X^*}\right) \subset A_n \subset A$ for every $n \in \mathbb{N}$

Keeping $\varepsilon > 0$ and taking the union over n we have $\frac{1}{1+\varepsilon}G^\circ \subset A$, which means $G^\circ \subset (1+\varepsilon)A$ for every $\varepsilon > 0$. Thus $G^\circ \subset \bigcap_{\varepsilon>0}(1+\varepsilon)A$. This proves Claim D, $\triangle$ and Theorem 3.92 follows. $\square$

Let X be a Banach space. A set $C \subset X^*$ is called *w^*-sequentially closed* if $\{f_n\} \subset C$, $f_n \overset{w^*}{\to} f$ implies that $f \in C$.

Corollary 3.93 *Let X be a separable Banach space and let C be a convex set in X^*. If C is w^*-sequentially closed then it is w^*-closed.*

Proof: The w^*-topology of B_{X^*} is metrizable (see Proposition 3.101). In metrizable spaces the closures and sequential closures coincide, so since $C \cap nB_{X^*}$ is w^*-sequentially closed (in nB_{X^*}), $C \cap nB_{X^*}$ is w^*-closed in nB_{X^*} and thus also in X^* as nB_{X^*} itself is w^*-closed. It is enough now to apply Theorem 3.92 to finish the proof. $\square$

Remark: If $X = \ell_2$ and $x_n := \sqrt{n}e_n$, where e_n is the standard unit vector in ℓ_2, then $0 \in \overline{\{x_n\}}^w$ (Exercise 3.55). Thus the set $\{x_n\}$ is not w-closed in ℓ_2 and yet $\{x_n\} \cap jB_X$ is finite and thus weakly closed in X. Since on ℓ_2 the w- and w^*-topology coincide as ℓ_2 is reflexive (see Definition 3.110 and examples after it), we see that the assumption of convexity of the set A in the Banach–Dieudonné theorem cannot be dropped.

Corollary 3.94 *Let X be a Banach space and let F be a linear functional on X^*. The following are equivalent:*
(i) *F is w^*-continuous.*
(ii) *$F \in X$, i.e., there is $x \in X$ such that $F(f) = f(x)$ for every $f \in X^*$.*
(iii) *The restriction of F to B_{X^*} is w^*-continuous.*
(iv) *$F^{-1}(0) \cap B_{X^*}$ is w^*-closed.*

Proof: (i) is equivalent to (ii) by Proposition 3.2 and the note after Definition 3.40.

(i)$\Longrightarrow$(iii) and (iii)$\Longrightarrow$(iv) are trivial.

(iv)$\Longrightarrow$(i): If (iv) holds, then by Theorem 3.92 and linearity of F, $F^{-1}(0)$ is w^*-closed. It is enough now to use Proposition 3.19. $\square$

3.11.3 The Bidual Space

Given a normed space $(X, \|\cdot\|)$, we endow X^* with its dual norm. This space becomes a Banach space (see Proposition 1.29). By X^{**} we denote the space $(X^*)^*$ with the norm $\|F\| := \sup_{f \in B_{X^*}} |F(f)|$. We define higher duals by induction as $X^{***} = (X^{**})^*$, etc.

Definition 3.95 *Let X be a normed space. The* canonical embedding π *of X into X^{**} is defined for $x \in X$ by*

$$\pi(x) \colon f \mapsto f(x),$$

for all $f \in X^*$.

Note that π is a linear isometry into. Indeed,

$$\pi(\alpha x + \beta y)(f) = f(\alpha x + \beta y) = \alpha f(x) + \beta f(y) = [\alpha\pi(x) + \beta\pi(y)](f).$$

Also, for $x \in X$ we have $\|\pi(x)\| = \sup_{f \in B_{X^*}} |f(x)| \le \sup_{f \in B_{X^*}} \|f\| \cdot \|x\| \le \|x\|$. Considering $f_0 \in S_{X^*}$ such that $f_0(x) = \|x\|$ we have $\|\pi(x)\| \ge |f_0(x)| = \|x\|$. Therefore $\|\pi(x)\| = \|x\|$.

Using this natural identification, for $x \in X$ we often write $x \in X^{**}$ instead of $\pi(x) \in X^{**}$, and we identify X with $\pi(X) \subset X^{**}$. In particular, $\pi(x)(f) = x(f)$ for $x \in X$ and $f \in X^*$.

Note too that if $F \in X^{**}$ is w^*-continuous on X^*, then there exists $x \in X$ such that $\pi(x) = F$ (see Corollary 3.94).

We can consider then two dual pairs, namely $\langle X, X^*\rangle$ and $\langle X^*, X^{**}\rangle$. By the very definition, the topology $w(X^{**}, X^*)$ (i.e., the w^* topology on X^{**}) induces on X the topology $w(X, X^*)$ (i.e., the w topology on X). A consequence of Corollary 3.49 is that the weak topology of X^{**} (i.e., the topology $w(X^{**}, X^{***})$, a topology stronger than $w(X^{**}, X^*)$) induces again the topology w on X.

Proposition 3.39 implies that X is $w(X^{**}, X^*)$-dense in X^{**}. The following result proves something more precise, namely that the closed unit ball of X is w^*-dense in the closed unit ball of X^{**}.

Theorem 3.96 (Goldstine) *Let X be a Banach space. The w^*-closure of B_X in X^{**} is $B_{X^{**}}$.*

Proof: We consider the dual pair $\langle X^*, X^{**}\rangle$. According to the bipolar theorem 3.38, $(B_X)^{\circ\circ} = \overline{B_X}^{w(X^{**},X^*)}$. It is elementary that $(B_X)^\circ = B_{X^*}$ and, similarly, $(B_{X^*})^\circ = B_{X^{**}}$. □

The following result will be used later.

Lemma 3.97 *Let $(X, \|\cdot\|)$ be a Banach space and let $|||\cdot|||$ be an equivalent norm on X^*. $|||\cdot|||$ is a dual norm to some equivalent norm on X if and only if $|||\cdot|||$ is w^*-lower semicontinuous.*

Recall that a norm $|\cdot|^*$ on X^* is a *dual norm* to some norm $|\cdot|$ on X if for every $f \in X^*$ we have $|f|^* = \sup\{f(x) : |x| \leq 1\}$.

Proof: As a supremum of w^*-continuous functions $x : f \mapsto f(x)$ for $x \in S_X$, every dual norm is w^*-lower semicontinuous (see Exercise 3.31).

Assume that $|||\cdot|||$ is a w^*-lower semicontinuous equivalent norm on X^* and denote by B its closed unit ball. Then B is w^*-closed and, by the bipolar theorem, $B = (B_0)^0$. It follows that $|||\cdot|||$ is the dual norm to the (equivalent) norm given by the Minkowski functional of B_0. □

Given normed spaces X, Y and $T \in \mathcal{B}(X, Y)$, we introduced in Definition 2.28 the adjoint operator $T^* \in \mathcal{B}(Y^*, X^*)$. If the procedure is applied to T^*, we obtain $T^{**} := (T^*)^*$, an element in $\mathcal{B}(X^{**}, Y^{**})$. Note that for $x \in X$ we have $T^{**}(\pi(x)) = \pi(T(x))$, in particular $T^{**}(\pi(X)) \subset \pi(Y)$. By using the natural identification between X and $\pi(X)$ (and between Y and $\pi(Y)$) described above, this amounts to say that $T^{**}(X) \subset Y$ and $T^{**}\big|_X = T$.

3.11.4 The Completion of a Normed Space

Definition 3.98 *Let X be a normed space. We say that a Banach space Z is a* completion *of X if X is linearly isometric to a dense subspace of Z.*

Given a completion $\widetilde{X}$ of a normed space X, we shall always assume that X is a (linear) subspace of $\widetilde{X}$.

Fact 3.99 *Every normed space has a completion.*

Proof: It is enough to recall that X^{**} is a Banach space (see Proposition 1.29) and that X is isometric to a subspace of X^{**} via the canonical mapping $\pi : X \to X^{**}$ (see Definition 3.95). As $\widetilde{X}$ we can then use $\overline{\pi(X)}^{X^{**}}$. □

The following simple result will be used later.

Lemma 3.100 *Let X be a normed space. Then every element $\widetilde{x}$ of its completion $\widetilde{X}$ can be written $\widetilde{x} = \sum_{n=1}^{\infty} x_n$, where $x_n \in X$, $\sum_{n=1}^{\infty} \|x_n\| < \infty$. Moreover, $\|\widetilde{x}\| = \inf \sum_{n=1}^{\infty} \|x_n\|$, where the infimum is taken over all series in X summing up to $\widetilde{x}$.*

Proof: Obviously, $\overline{B_X} = B_{\widetilde{X}}$. We may assume, without loss of generality, that $\widetilde{x} \in B_{\widetilde{X}}$. There exists $x_1 \in B_X$ such that $\|\widetilde{x} - x_1\| \leq 1/2$. Since $2(\widetilde{x} - x_1) \in B_{\widetilde{X}}$, we can find $x_2 \in B_X$ such that $\|2(\widetilde{x} - x_1) - x_2\| \leq 1/2$, i.e., $\|\widetilde{x} - x_1 - (1/2)x_2\| \leq 1/2^2$. Find $x_3 \in B_X$ such that $\|2^2(\widetilde{x} - x_1 - (1/2)x_2) - x_3\| \leq 1/2$. Proceed recursively to obtain a sequence $\{x_n\}$ in B_X such that

$$\|\widetilde{x} - x_1 - (1/2)x_2 - \ldots - (1/2^n)x_{n+1}\| \leq 1/2^{n+1}, \text{ for all } n \in \mathbb{N}.$$

This proves that $\widetilde{x} = \sum_{n=1}^{\infty}(1/2^{n-1})x_n$. To prove the second assertion, modify the construction in the following way: Fix $m \in \mathbb{N}$, and start choosing $x_1 \in B_X$ such that $\|\widetilde{x} - x_1\| \leq 1/2^m$. Then choose $x_2 \in B_X$ such that $\|\widetilde{x} - x_1 - (1/2^m)x_2\| \leq 1/2^{m+1}$. Proceed recursively to obtain a sequence $\{x_n\}$. Since $\|x_1\| \leq \|\widetilde{x}\| + 1/2^m$ and $\sum_{n=1}^{\infty} 1/2^m = 1/2^{m-1}$, we get the conclusion. □

3.11.5 Separability and Metrizability

Proposition 3.101 *Let X be a separable Banach space. Then (B_{X^*}, w^*) is a metrizable compact space. A metric on B_{X^*} compatible with the topology w^* is given by*

$$\rho(f, g) := \sum_{i=1}^{\infty} 2^{-i}|(f - g)(x_i)|, \tag{3.7}$$

where $\{x_n\}_{n=1}^{\infty} \subset S_X$ is dense in S_X.

Proof: Let $\{x_n\}_{n=1}^{\infty} \subset S_X$ be a dense subset of S_X. Since $D := \operatorname{span}\{x_n : n \in \mathbb{N}\}$ is a dense subspace of X, $\langle X^*, D\rangle$ is a dual pair (see Proposition 3.39). The topology $w(X^*, D)$ is metrizable (see Proposition 3.29), Hausdorff, and coarser than w^*. Therefore it agrees with w^* on B_{X^*}, since (B_{X^*}, w^*) is compact (Theorem 3.37). That ρ defined by (3.7) is a metric is clear. In order to prove that the associated topology is $w(X^*, D)$, let $f, f_1, f_2, \ldots$ elements in B_{X^*}. Assume first that (f_n) is $w(X^*, D)$-convergent to f. Fix $\varepsilon > 0$. There exists $i_0 \in \mathbb{N}$ such that $\sum_{i=i_0+1}^{\infty} 2^{-i} < \varepsilon/4$. Find then $n_0 \in \mathbb{N}$ such that $\sum_{i=1}^{i_0} 2^{-i}|(f - f_n)(x_i)| < \varepsilon/2$ for every $n \geq n_0$. Since $|(f - f_n)(x_i)| \leq 2$ for all n and i, we get $\rho(f, f_n) < \varepsilon$ for every $n \geq n_0$. Conversely, assume that $\rho(f - f_n) \to 0$. Then, for each $i \in \mathbb{N}$ we have $(f - f_n)(x_i) \to_n 0$. This proves that $f_n \to f$ in the topology $w(X^*, D)$. □

Thus if X is separable, then the topological space (B_{X^*}, w^*) is metrizable. In fact, the converse is also true.

This will be proved in Proposition 3.103. In its proof we shall need the following general result concerning the space $C(K)$ of continuous functions on a compact topological space K. Recall that the set of all Dirac functionals $\widehat{K} := \{\delta_k : k \in K\}$ is a (w^*-compact) subset of $S_{C(K)^*}$ (see the paragraph right before Proposition 2.21 and Lemma 3.55).

Lemma 3.102 *Let K be a compact metric space. The Banach space $C(K)$ is separable if and only if K is metrizable.*

Proof: If $C(K)$ is separable then $B_{C(K)^*}$ in the w^*-topology is metrizable by Proposition 3.101. Since $\widehat{K} \subset (B_{C(K)^*}, w^*)$ is metrizable, so is K, see Lemma 3.55.

Assume now that K is a compact metric space. Let $\{x_n\}_{n=1}^{\infty}$ be a dense sequence in K. Let $f_{n,m}(x) = \frac{1}{m} - \text{dist}(x, x_n)$ if $\text{dist}(x, x_n) \leq \frac{1}{m}$ and $f_{n,m} = 0$ if $\text{dist}(x, x_n) > \frac{1}{m}$. Then the family $\{f_{n,m}\}$ together with a constant function generates a separable algebra that separates the points of K and thus its closure is $C(K)$ by the Stone–Weierstrass theorem. □

Proposition 3.103 *Let X be a Banach space. (B_{X^*}, w^*) is metrizable if and only if X is separable.*

Proof: Assume that (B_{X^*}, w^*) is metrizable. Lemma 3.102 implies that $C(B_{X^*}, w^*)$ is separable in its supremum norm. The mapping $I : X \to C(B_{X^*}, w^*)$, $I(x) : f \mapsto f(x)$, is an isometry, hence X is separable as a subspace of a separable metric space.

The other implication is contained in Proposition 3.101. □

Since $X^* = \bigcup_{n \in \mathbb{N}} nB_{X^*}$, and compact metrizable space are separable, from Proposition 3.101 we also have

Corollary 3.104 *Let X be a Banach space. If X is separable then B_{X^*}, and so X^*, are w^*-separable.*

We note that (X^*, w^*) may be separable although (B_{X^*}, w^*) may be not, [JoLi1]. Also, (B_{X^*}, w^*) may be separable although X may be not, [JoLi1]. For more in this direction, we refer to [HMVZ, Chapter 7].

Proposition 3.105 *Let X be a Banach space. If X is w-separable then X is separable.*

Proof: Let S be a countable set that is weakly dense in X, denote $D = \text{span}(S)$. The countable set C of rational linear combinations of S is dense in D, hence D is norm-separable. Since D is convex and dense in X, by Theorem 3.45, $\overline{D} = \overline{D}^w = X$, so C is dense in X (see also Exercise 3.100). □

We will now investigate (B_X, w).

Proposition 3.106 *Let X be a Banach space. (B_X, w) is metrizable if and only if X^* is separable.*

If X^* is separable, then $(B_{X^{**}}, w^*)$ is metrizable by Proposition 3.101. Since (B_X, w) is a topological subspace of $(B_{X^{**}}, w^*)$ (see Corollary 3.49 and the paragraph after Definition 3.95), (B_X, w) is also metrizable.

For the other direction, let $\mathcal{U} = \{U_n : n \in \mathbb{N}\}$ be a countable base of neighborhoods of 0 in (B_X, w). Without loss of generality we may assume that $U_n := \{x \in B_X : |f(x)| < 1, \ f \in F_n\}$, where F_n is a finite subset of X^*, $n \in \mathbb{N}$.

To finish the proof it will be enough to ensure that $\overline{\text{span}}\{\bigcup_{n\in\mathbb{N}} F_n\} = X^*$. To this end, let $x^{**} \in X^{**}$ that vanishes on $\bigcup_{n\in\mathbb{N}} F_n$. By Theorem 3.96 there exists a net $\{x_\alpha\}_{\alpha\in I}$ in B_X that w^*-converges to x^{**}. Fix $n \in \mathbb{N}$. Then, there exists $\alpha_0 \in I$ such that $|f(x_\alpha)| = |f(x_\alpha - x^{**})| < 1$ for $\alpha \geq \alpha_0$ and for all $f \in F_n$. Then $x_\alpha \in U_n$ for $\alpha \geq \alpha_0$. This holds for all $n \in \mathbb{N}$, so $w\text{-}\lim x_\alpha = 0$, hence $x^{**} = 0$. □

Proposition 3.107 *Let C be a weakly compact set in a Banach space X. If X^* is w^*-separable, then (C, w) is metrizable.*
In particular, if X is separable then (C, w) is metrizable.

Proof: If D is a countable separating set in X^*, then the topology $w(X, D)$ is metrizable and coincides with the weak topology on C as C is weakly compact. Such a separating set exists if X^* is w^*-separable (take any w^*-dense countable set). □

3.11.6 Weak Compactness

Note that a subset A of a Banach space is relatively weakly compact if and only if A is bounded and its w^*-closure in X^{**} is a subset of X. Indeed, if A is relatively weakly compact, then certainly A is bounded and $\overline{A}^w$ is w-compact in X, hence $\overline{A}^{w^*} = \overline{A}^w$ $(\subset X)$.

On the other hand, if A is bounded, then A° is a neighborhood of 0 in X^* and so, with respect to the dual pair $\langle X^*, X^{**}\rangle$, $A^{\circ\circ}$ is w^*-compact by Alaoglu's theorem. Since $A \subset A^{\circ\circ}$ we get that $\overline{A}^{w^*}$ is w^*-compact; if it is a subset of X, then $\overline{A}^w$ $(= \overline{A}^{w^*})$ is w-compact.

For the concept of angelic space see Definition 3.53.

Proposition 3.108 *Let X be a Banach space. Then (X, w) is an angelic space.*

Proof: It is enough to observe that subspaces of angelic spaces are themselves angelic, and to identify (X, w) with a subset of the topological space $C_p(B_{X^*})$ where B_{X^*} carries the topology induced by w^*. Use then Theorem 3.58. □

From the angelicity of a Banach space in its weak topology, we obtain the following result (see Exercise 3.9).

Theorem 3.109 (Eberlein, Šmulyan) *Let A be a subset of a Banach space X. The following are equivalent:*
(i) *A is (relatively) w-compact.*
(ii) *A is (relatively) w-sequentially compact.*
(iii) *A is (relatively) w-countably compact.*

3.11.7 Reflexivity

Definition 3.110 *A Banach space X is said to be* reflexive *if the mapping π in Definition* 3.95 *maps X onto X^{**}.*

In particular, a reflexive space X is isometric to X^{**}. On the other hand, there are examples of nonreflexive spaces X such that X is linearly isometric to X^{**}—of course, not by means of π (James' space J, see Definition 4.43).

We will see in Proposition 3.113 that X is reflexive if and only if the w-topology and the w^*-topology agree on X^*.

Examples:

(i) ℓ_p and L_p spaces for $1 < p < \infty$ are reflexive. Indeed, if $\frac{1}{p} + \frac{1}{q} = 1$ then $\ell_p^{**} = \ell_q^* = \ell_p$ in the sense that that for every $f \in \ell_q^*$ there is $g \in \ell_p$ such that $f(x) = \sum x_i \bar{g}_i$ for every $x \in \ell_q$. So the action of f on x is the same as the action of g on x and π is thus a mapping from ℓ_p onto ℓ_p^{**}. Similarly it is shown for L_p.

(ii) c_0 is not reflexive since $c_0^{**} = \ell_\infty$, c_0 is separable and ℓ_∞ is not.

(iii) $C[0, 1]$ is not reflexive. Indeed, $C[0, 1]^{**}$ would then be separable and thus $C[0, 1]^*$ would be separable by Proposition 2.8, which is not true (Proposition 2.21).

Theorem 3.111 *A Banach space X is reflexive if and only if B_X is weakly compact.*

Proof: Let X be reflexive. Then $X = X^{**}$ and thus $B_X = B_{X^{**}}$ is w^*-compact in X^{**} by Alaoglu's theorem, so it is weakly compact in X.

If B_X is weakly compact, then it is w^*-closed in X^{**}. Since the w^*-closure of B_X in X^{**} is $B_{X^{**}}$ by Theorem 3.96, we get that $B_X = B_{X^{**}}$ and the space X is reflexive. □

Proposition 3.112 *A Banach space X is reflexive if and only if X^* is reflexive.*

Proof: Let X be reflexive. Then $X = X^{**}$ and thus the w^*-topology and w-topology on X^* coincide. Therefore, by the Alaoglu's theorem, B_{X^*} is weakly compact and X^* is reflexive by Theorem 3.111.

Let X^* be reflexive. By the first part of this proof, we have that X^{**} is reflexive. Therefore $B_{X^{**}}$ is w-compact. B_X is a closed and therefore w-closed (Theorem 3.45) subset of $B_{X^{**}}$ and thus w-compact in X^{**}.

Since the weak topology on X^{**} induces on X its weak topology (see Corollary 3.49), B_X is w-compact, and we are done by Theorem 3.111. □

Proposition 3.113 *A Banach space is reflexive if and only if the w-topology and the w^*-topology agree on X^*.*

Proof: If X is reflexive, then X^* is also reflexive, by Proposition 3.112, so on X^* the w- and w^*-topologies agree. Conversely, if these two topologies agree on X^*, then, by Alaoglu's theorem (Theorem 3.37), (B_{X^*}, w^*) $(= (B_{X^*}, w))$ is compact. By Theorem 3.111, the space X^* is reflexive. The result follows from Proposition 3.112. □

Observe that, if X is not reflexive, there is a convex set in X^*, namely $f^{-1}(0)$ for $f \in X^* \backslash X$, that is w-closed but not w^*-closed—it is w^*-dense and different from X^*.

Proposition 3.114 *Let X be a reflexive Banach space. If Y is a closed subspace of X, then Y is a reflexive Banach space.*

Proof: By Corollary 3.49, the weak topology on Y coincides with the restriction to Y of the weak topology on X. To complete the proof, it suffices to observe that B_Y is a weakly closed subset of the weakly compact set B_X. Therefore B_Y is weakly compact in Y and thus Y is reflexive by Theorem 3.111. $\square$

Proposition 3.115 *Let X be a Banach space. If X is separable and reflexive, then (B_X, w) is a compact metrizable space.*

Proof: The result follows from Proposition 3.107 and Theorem 3.111. $\square$

3.11.8 Boundaries

3.11.8.1 Dirac Deltas and Extreme Points of $B_{C(K)^*}$

Lemma 3.116 *Let K be a compact topological space. Then $\mathrm{Ext}\big(B_{C(K)^*}\big) = \{\pm\delta_k\}_{k\in K}$, where δ_k are the Dirac functionals in $C(K)^*$.*

Proof: We will first show that all extreme points of $B_{C(K)^*}$ are of the form $\pm\delta_k$. Let $A = \overline{\mathrm{conv}}^{w^*}\{\pm\delta_k\}_{k\in K}$. We claim that $A = B_{C(K)^*}$. Indeed, assuming $F \in B_{C(K)^*}\setminus A$, by Corollary 3.34 there is $f \in C(K)$ such that $\sup_A(f) \leq 1$ and $F(f) > 1$. Since A is symmetric, $\sup_A |f| \leq 1$, which implies $\sup_{k\in K} |\delta_k(f)| = \|f\| \leq 1$. Thus $F(f) \leq 1$, a contradiction. Now by Theorem 3.66 we obtain that $\mathrm{Ext}\big(B_{C(K)^*}\big) \subset \overline{\{\pm\delta_k\}_{k\in K}}^{w^*}$.

In Lemma 3.55 we showed that $(\{\delta_k\}_{k\in K}, w^*)$ is homeomorphic to K, so $\{\delta_k\}_{k\in K}$ is w^*-compact. Thus also $\{\pm\delta_k\}_{k\in K} = \{\delta_k\}_{k\in K} \cup \{-\delta_k\}_{k\in K}$ is w^*-compact, hence w^*-closed, so $\mathrm{Ext}(B_{C(K)^*}) \subset \{\pm\delta_k\}_{k\in K}$.

It remains to show that every Dirac measure δ_k is an extreme point of $B_{C(K)^*}$. This will complete the proof as then $-\delta_k$ are extreme by the symmetry of $B_{C(K)^*}$.

Given $k_0 \in K$, consider the family $\mathcal{U}$ of all open neighborhoods of k_0 in K. Given $U \in \mathcal{U}$, by Urysohn's lemma there is $f_U \in B_{C(K)}$ such that $f_U(k_0) = 1$ and $f_U = 0$ on $K\setminus U$. Then $H_U := \{F \in C(K)^* : F(f_U) = 1\}$ is a w^*-closed supporting manifold for $B_{C(K)^*}$. Define $H = \bigcap_{U\in\mathcal{U}} H_U$.

Since $\delta_{k_0} \in H_U$ for all $U \in \mathcal{U}$, we have that $H \neq \emptyset$, and it is a w^*-closed supporting manifold for $B_{C(K)^*}$. Since $H \cap B_{C(K)^*}$ is w^*-compact, it contains an extreme point, which is also an extreme point of $B_{C(K)^*}$ (see Lemma 3.62). However, the only candidates are then $\pm\delta_k$ for $k \in K$ and we easily check that $\delta_k \notin H$ for $k \neq k_0$, as the topology of K is Hausdorff. So δ_{k_0} must be an extreme point. $\square$

Theorem 3.117 (Banach, Stone) *Let K, L be compact spaces. $C(K)$ is linearly isometric to $C(L)$ if and only if K and L are homeomorphic.*

Proof: Let φ be a homeomorphism of K onto L. Define a mapping $T: C(K) \to C(L)$ by $T(f) = f \circ \varphi^{-1}$. It is standard to check that T is an isometry of $C(K)$ onto $C(L)$.

Assume now that T is an isometry of $C(K)$ onto $C(L)$. Then T^* is an isometry (Exercise 2.51). Thus $T(B_{C(K)^*}) = B_{C(L)^*}$ and by Lemma 3.116 and Exercise 3.126, for $l \in L$ we have $T^*(\delta_l) = \varepsilon(l)\delta_{k_l}$, where $k_l \in K$ and $\varepsilon(l) \in \{-1, 1\}$. The mapping $\varrho: l \mapsto \varepsilon(l)k_l$ is continuous as T^* is w^*-w^*-continuous.

We now show that the function $\varepsilon : L \to \{-1, 1\}$ is continuous. Indeed, for the constant function 1, denoted x, we can write $\varepsilon(l) = \varepsilon(l)\delta_{k_l}(x) = T^*(\delta_l)(x) = T(x)(l)$, so $\varepsilon = T(x)$, hence ε is continuous. Then the mapping $\rho: L \to K$ defined by $\rho = \varrho/\varepsilon$, i.e., $\rho(l) = k_l$ for $l \in L$, is a continuous one-to-one mapping from L onto K. □

Remark: The space $C(K)$ is isomorphic to $C(L)$ whenever K and L are metrizable uncountable compact spaces (Milyutin, see, e.g., [Rose10] and [AlKa]).

3.11.8.2 James Boundaries

Definition 3.118 *Let $(X, \|\cdot\|)$ be a Banach space. Let C be a bounded subset of X^*. A set $B \subset C$ is called a* James boundary *for C if for every $x \in X$ there is $g \in B$ such that $g(x) = \sup\{f(x) : f \in C\}$.*
A set $B \subset B_{X^*}$ is called a James boundary *of X if it is a James boundary for B_{X^*}.*

Note that the condition is homogeneous in x. For instance, $B \subset B_{X^*}$ is a James boundary of X if for every $x \in S_X$ there is $f \in B$ such that $f(x) = 1$.

Fact 3.119 *Let X be a Banach space. Let C be a bounded w^*-closed subset of X^*. Then* Ext(C) *is a James boundary for C. In particular,* Ext(B_{X^*}) *is a James boundary of X.*

Proof: Given $x \in S_X$, consider $H := \{f \in X^* : f(x) = \sup_{g \in C} g(x)\}$. Then H is a w^*-closed supporting manifold for C which contains an extreme point of C by Proposition 3.64. □

Note that there are James boundaries B of X such that $B \cap \text{Ext}(B_{X^*}) = \emptyset$ (see Exercise 3.151).

Another simple observation is contained in the following result.

Fact 3.120 *Let X be a Banach space. Let C be a convex w^*-compact subset of X^* and $B \subset C$ be a James boundary for C. Then $\overline{\text{conv}}^{w^*}(B) = C$.*

Proof: Assume that there exists $g \in C$ such that $g \notin \overline{\text{conv}}^{w^*}(B)$ $(\subset C)$. By the separation theorem, there exists $x \in X$ such that $\sup_{f \in B} f(x) < g(x)$. It follows that there is no point in B where x attains its supremum on C, a contradiction with the definition of a James boundary. □

It is important to be able to check weak compactness of a bounded subset S of a Banach space X with a minimum of requirements. An example of the kind of result we are looking for is the following: *A uniformly bounded subset S of $C[0, 1]$ is weakly compact if (and only if) it is pointwise sequentially compact.* This is a consequence of the Lebesgue's dominated convergence theorem. Indeed, let $\{f_n\}$ be a sequence in S. By hypothesis, it has a subsequence $\{f_{n_k}\}$ that pointwise converges to some $f \in S$. Since there is a constant $M \geq 0$ such that $|f_{n_k}| \leq M$ for all $k \in \mathbb{N}$, it happens that $\int_0^1 f_{n_k}(t)\, d\mu(t) \to 0$ as $k \to \infty$, for every regular Borel measure μ on $[0, 1]$. This can be stated, by the Riesz's representation theorem, as $\langle \mu, f_{n_k} \rangle \to 0$ as $k \to \infty$ for every $\mu \in C[0, 1]^*$ or, in other words, $f_{n_k} \stackrel{w}{\to} 0$. Since this is true for every sequence $\{f_n\}$ in S, the set S is w-sequentially compact. By the Eberlein–Šmulyan theorem, S is w-compact.

Observe that, in the former example, the set $B := \{\pm\delta_t : t \in [0, 1]\}$ is a James boundary of $X := C[0, 1]$, due to Lemma 3.116 and Fact 3.119. We just proved, in this case, that a bounded subset S of X is w-compact whenever S is $w(X, B)$-sequentially compact. This is an instance of a theorem of Pfitzner.

Theorem 3.121 (Pfitzner, [Pfi]) *Let B be a James boundary of a Banach space X. Then a bounded subset S of a Banach space is w-compact whenever it is $w(X, B)$-compact.*

We are not going to prove Pfitzner's theorem in full generality, but we will present some important special cases (obtained earlier by various authors) in this section and in Exercises 3.155, 3.156, and 3.157.

A closely related problem is to provide sufficient conditions on a Banach space X, on a convex w^*-compact subset C of X^*, or on a James boundary B for C, so that B would be a *strong James boundary for C*, i.e., $\overline{\text{conv}}^{\|\cdot\|}(B) = C$. Fact 3.120 says that $\overline{\text{conv}}^{w^*}(B) = C$ always hold. However, not every James boundary is strong. For example, consider $X := C[0, 1]$ and $B = \{\pm\delta_t : t \in [0, 1]\} \subset B_{X^*}$. Then $\overline{\text{conv}}^{\|\cdot\|}(B) \neq B_{X^*}$ (Exercise 3.130) and yet, B is a James boundary of X.

Note that, *for a strong James boundary B of a Banach space X, the conclusion of Pfitzner's theorem holds.* This is a consequence of the following trivial fact: if A is a bounded subset of a normed space, B is a subset of X^*, and a net in A converges to some point $x \in X$ pointwise on elements in B, then it converges to x pointwise on elements in $\overline{B}^{\|\cdot\|}$.

3.11.8.3 Strong James Boundaries

The following result gives a sufficient condition for a James boundary to be strong. Variants of it were obtained independently by Rodé and Godefroy using different methods. Here we will follow Godefroy's approach using Simons' inequality (Lemma 3.123), which, in turn, originated in the work of James, [Jame2] and [Jame5].

Theorem 3.122 (Godefroy [Gode2]) *Let X be a Banach space and let C be a bounded closed convex subset of X^*. If B is a norm-separable James boundary for C, then $C = \overline{\operatorname{conv}}^{\|\cdot\|}(B)$.*

In particular, if a norm-separable subset B of B_{X^*} is such that $\overline{\operatorname{conv}}(B) \neq B_{X^*}$, then there is $x \in S_X$ such that $x(f) < 1$ for all $f \in B$.

Theorem 3.122 does not hold in general if the assumption on the separability of B is dropped, as we noted above (see again Exercise 3.130).

Let B be a nonempty set. Consider the space $\ell_\infty(B)$ with its canonical sup-norm. If $x = (x_b) \in \ell_\infty(B)$, we denote $\sup_B x = \sup\{x_b : b \in B\}$. If there is $b_0 \in B$ such that $x_{b_0} = \sup_B x$, we say that x *attains its supremum over* B. In the proof of Theorem 3.122 we will use the following results.

Lemma 3.123 (Simons' inequality [Simo]) *Let B be a nonempty set. Let $\{x_n\}$ be a bounded sequence in $\ell_\infty(B)$. Assume that for all $\Lambda_n \geq 0$, $n \in \mathbb{N}$, satisfying $\sum_{n=1}^{\infty} \Lambda_n = 1$, the vector $\sum_{n=1}^{\infty} \Lambda_n x_n$ attains its supremum over B (i.e., every element in the superconvex hull of $\{x_n : n \in \mathbb{N}\}$ attains its supremum over B). Then, if $u(b) := \limsup_n x_n(b)$ for all $b \in B$,*

$$\sup_B u \geq \inf\bigl\{\sup_B x : x \in \operatorname{conv}\{x_n\}\bigr\}.$$

Proof: (Oja [Oja1]) Set $C_k = \left\{\sum_{n=k}^{\infty} \Lambda_n x_n : \Lambda_n \geq 0, \sum_{n=k}^{\infty} \Lambda_n = 1\right\}$ for $k \in \mathbb{N}$. It will be enough to prove that

$$\inf_{x \in C_1} \sup_B x \leq \sup_B u. \tag{3.8}$$

Let $\varepsilon > 0$. Choose inductively $z_k \in C_k$ such that for $k = 0, 1, \ldots,$

$$\sup_B (2^k v_k + z_{k+1}) \leq \inf_{z \in C_{k+1}} \sup_B (2^k v_k + z) + \frac{\varepsilon}{2^{k+1}},$$

where $v_0 = 0$ and $v_k := \sum_{n=1}^{k} \frac{z_n}{2^n}$ for $k \in \mathbb{N}$. Put $v = \sum_{n=1}^{\infty} \frac{z_n}{2^n}$.

Since $z_{k+1} = 2^{k+1} v_{k+1} - 2^{k+1} v_k$, we get $2^{k+1} v_{k+1} - 2^k v_k = 2^k v_k + z_{k+1}$, so using $2^k v - 2^k v_k = 2^k \sum_{n=k+1}^{\infty} \frac{z_n}{2^n} \in C_{k+1}$ we get for all $k = 0, 1, \ldots,$

$$\begin{aligned}
\sup_B (2^{k+1} v_{k+1} - 2^k v_k) &\leq \sup_B \bigl(2^k v_k + (2^k v - 2^k v_k)\bigr) + \frac{\varepsilon}{2^{k+1}} \\
&= \sup_B 2^k v + \frac{\varepsilon}{2^{k+1}} = 2^k \sup_B v + \frac{\varepsilon}{2^{k+1}}.
\end{aligned}$$

As $v \in C_1$, we can choose $t \in B$ such that $v(t) = \sup_B v$. Since $\sum_{k=0}^{m-1} 2^k = 2^m - 1$, from the last inequality we have for $m \in \mathbb{N}$:

$$
\begin{aligned}
2^m v_m(t) &= \sum_{k=0}^{m-1} (2^{k+1} v_{k+1} - 2^k v_k)(t) \\
&\leq (2^m - 1) \sup_B v + \varepsilon = 2^m v(t) + \varepsilon - \sup_B v.
\end{aligned}
$$

Thus $\sup\limits_B v \leq 2^m v(t) - 2^m v_m(t) + \varepsilon$, hence

$$
\begin{aligned}
\inf_{x \in C_1} \sup_B x \leq \sup_B v &\leq \limsup_m (2^m v - 2^m v_m)(t) + \varepsilon \\
&\leq \limsup_m x_m(t) + \varepsilon = u(t) + \varepsilon.
\end{aligned}
$$

as $2^m v - 2^m v_m \in C_{m+1}$. Inequality (3.8) follows as $\varepsilon > 0$ was arbitrary. □

If X is a Banach space and $B \subset X^*$ is bounded, we can consider X a subset of $\ell_\infty(B)$ by identifying $x \in X$ with $\big(f(x)\big)_{f \in B}$. Note that if B is a James boundary of X, then $\sup\limits_B x = \|x\|$ and the supremum is attained. Thus we get

Theorem 3.124 (Simons [Simo]) *Let B be a James boundary for a bounded subset M of the dual of a Banach space X. If $\{x_n\}$ is a bounded sequence in X, and $u(x^*) := \limsup_n x^*(x_n)$ for every $x^* \in X^*$, then*

$$
\sup_B u = \sup_M u.
$$

Proof: Assume $\sup_B u < \alpha < \sup_M u$. Let $f \in M$ be such that $u(f) > \alpha$. Assume without loss of generality that $f(x_n) > \alpha$ for all $n \in \mathbb{N}$. If $x \in \operatorname{conv}\{x_n\}$, from a standard convexity argument we have $f(x) > \alpha$. Moreover, every element $x := \sum_{n=1}^{\infty} \lambda_n x_n$, where $\lambda_n \geq 0$ for all n and $\sum_{n=1}^{\infty} \lambda_n = 1$, belongs to X (in other language, $\operatorname{sconv}(x_n) \subset X$, see Exercise 1.67), so x attains its supremum on B (and $\sup_B x = \sup_M x$). From Simons'inequality,

$$
\begin{aligned}
(\alpha >) \sup_B u &\geq \inf_{x \in \operatorname{conv}(x_n)} \sup_B x \\
&= \inf_{x \in \operatorname{conv}(x_n)} \sup_M x \geq \inf_{x \in \operatorname{conv}(x_n)} f(x) \geq \alpha,
\end{aligned}
$$

a contradiction. □

We are now ready to prove Godefroy's Theorem 3.122.

Proof of Theorem 3.122: [Gode2] By contradiction, assume that $\overline{\operatorname{conv}}(B) \neq C$ (see Fig. 3.3). By the separation theorem, we can find $F \in S_{X^{**}}$, $\alpha < \beta$ and $y_0^* \in C \setminus \overline{\operatorname{conv}}(B)$ such that $F(f) \leq \alpha$ for all $f \in B$ and $F(y_0^*) > \beta$. Let $S = \{x \in B_X : y_0^*(x) > \beta\}$. Using Goldstine's theorem we get that $F \in \overline{S}^{w^*}$. As B is separable, the topology on bounded sets in X^{**} of pointwise convergence on B is metrizable.

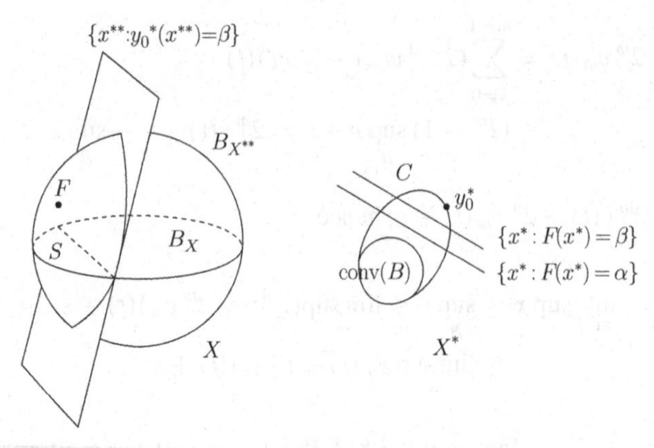

Fig. 3.3 The proof of Theorem 3.122

Thus there is a sequence $\{x_n\}$ in S which converges to F on points of $B \cup \{y_0^*\}$. Put $u(x^*) = \limsup x^*(x_n)$ for $x^* \in X^*$. Then, by Theorem 3.124, we get

$$\alpha \geq \sup_B F = \sup_B u = \sup_C u \geq u(y_0^*) = F(y_0^*) > \beta,$$

a contradiction. □

Corollary 3.125 (Godefroy) *Let X be a Banach space X. If X has a separable James boundary, then X^* is separable.*

Proof: If B is a separable James boundary of X, then by Theorem 3.122, $\overline{\text{conv}}(B) = B_{X^*}$. Thus B_{X^*} and also X^* are separable. □

By Fact 3.119 we have the following result, which was obtained independently by several authors (for references see, e.g., [LoMo]).

Corollary 3.126 *Let X be a Banach space. If* $\text{Ext}(B_{X^*})$ *is separable, then X^* is separable.*

3.11.8.4 James Boundaries and James Theorem

Using Simons' lemma we also see how James boundaries interact with second duals.

Corollary 3.127 *Let B be a James boundary for a bounded subset M of the dual of a Banach space X. If $x^{**} \in X^{**}$ is the w^*-limit of a bounded sequence in X, then* $\sup_{x^* \in B} x^{**}(x^*) = \sup_{x^* \in M} x^{**}(x^*)$.

Proof: It is enough to observe that $x^{**} = \limsup x_n$ when we consider x^{**} and x_n, $n \in \mathbb{N}$, functions on X^*, and apply then Theorem 3.124. □

In particular we obtain.

Corollary 3.128 *Let B be a James boundary of a Banach space X. If $x^{**} \in X^{**}$ is the w^*-limit of a bounded sequence in X, then $\|x^{**}\| = \sup_{x^* \in B} x^{**}(x^*)$.*

Corollary 3.129 *Let X be a Banach space and let C be a closed convex and bounded subset of X^*. If every $x^{**} \in S_{X^{**}}$ is the w^*-limit of a sequence in B_X, then $C = \overline{\text{conv}}^{\|\cdot\|}(B)$ for every James boundary B for C.*
In particular, if X is a separable Banach space that does not contain an isomorphic copy of ℓ_1, then $C = \overline{\text{conv}}^{\|\cdot\|}(B)$ for every closed convex and bounded subset of X^* and every James boundary B for C.

Proof: Assume that there exists $x^* \in C \setminus \overline{\text{conv}}^{\|\cdot\|}(B)$. The separation theorem provides $x^{**} \in S_{X^{**}}$ and $\alpha \in \mathbb{R}$ such that $\sup_B x^{**} < \alpha < x^{**}(x^*)$. This contradicts Corollary 3.127. The particular case follows from the first part by using Theorem 5.40. $\square$

The fundamental James' theorem below (Theorem 3.130) can be seen as a boundary problem: If every $f^* \in X^*$ attains its supremum over a set $A \subset X$ at some point of A, then the set A is certainly bounded, and it is a James boundary for the w^*-compact and convex set $\overline{\text{conv}}^{w^*}(A) \subset X^{**}$.

Theorem 3.130 (James [Jame2]) *Let A be a w-closed subset of a Banach space X. A is w-compact if and only if every $f \in X^*$ attains its supremum over A at some point of A.*

Proof: Assume that A is w-compact. Since every $f \in X^*$ is w-continuous, the image $f(A)$ of the w-compact set A must be compact in $\mathbb{R}$, hence closed, and the supremum is attained.

We present the proof of the other direction only for a separable set A (in particular, if the space X is separable). Let $C = \overline{\text{conv}}(A)$. Observe that, for $f \in X^*$, $\sup_A f = \sup_C f = \sup_{\overline{C}^{w^*}} f$. By our assumption, A is a separable James boundary for $\overline{C}^{w^*}$ in X^{**}, hence by Theorem 3.122, $(X \supset) \overline{\text{conv}}(A) = \overline{C}^{w^*}$. Note that if every $f \in X^*$ attains its supremum over C, then C is bounded by Theorem 3.88. It follows that C (a convex and w-closed subset of X) is w-compact (see the paragraph before Proposition 3.108). In particular, A is w-compact. $\square$

The proof in the nonseparable case is much more difficult (see, e.g., [Dies2] or [Flor]).

Corollary 3.131 (James [Jame2]) *A Banach space X is reflexive if and only if every $f \in X^*$ attains its norm.*

We will now show another James' characterization of reflexivity.

Theorem 3.132 (James [Jame1]) *A Banach space X is reflexive if and only if there exists $\theta \in (0, 1)$ such that whenever $\{x_n\}$ is a sequence in S_X with $\inf\{\|u\| : u \in \text{conv}\{x_n\}\} \geq \theta$, then there are $n_0 \in \mathbb{N}$, $u \in \text{conv}\{x_1, \dots, x_{n_0}\}$ and $v \in \text{conv}\{x_{n_0+1}, x_{n_0+2}, \dots\}$ such that $\|u - v\| \leq \theta$.*

Proof: (Oja [Oja2]) Assume that X is reflexive. Fix $\theta \in (0,1)$, let $\{x_n\}$ be a sequence in S_X. For an integer $n \geq 0$ let $K_n = \overline{\text{conv}}\{x_{n+1}, x_{n+2}, \ldots\}$.

Since $\{K_n\}$ is a nested sequence of nonempty weakly compact sets, there is $x \in \bigcap K_n$. As $x \in K_0$, there is n_0 and $u \in \text{conv}\{x_1, \ldots, x_{n_0}\}$ such that $\|x - u\| < \theta/2$. As $x \in K_{n_0}$, there is $v \in \text{conv}\{x_{n_0+1}, x_{n_0+2}, \ldots,\}$ such that $\|x - v\| < \theta/2$. Then $\|v - u\| < \theta$.

Let $X \neq X^{**}$ and any $\theta \in (0,1)$ be given. We will construct a sequence $\{x_n\}$ failing the above condition. Denote $B_\theta = \{F \in X^{**} : \|F\| \leq \theta\}$. By Lemma 1.37, there is $F_\theta \in S_{X^{**}} \setminus \bigcup_{x \in X} (x + B_\theta)$ (see Fig. 3.4).

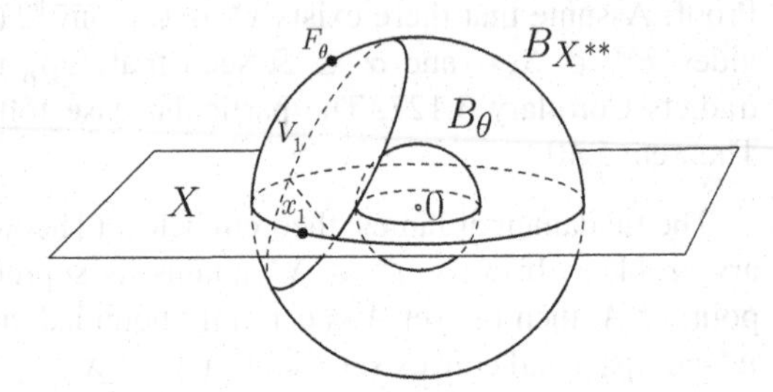

Fig. 3.4 The first step in the proof of Theorem 3.132

As $F_\theta \notin B_\theta$, there exists a convex w^*-neighborhood V_1 of F_θ in X^{**} such that $V_1 \cap B_\theta = \emptyset$. Pick $x_1 \in V_1 \cap S_X$. As $F_\theta \notin x_1 + B_\theta$, there is a convex w^*-neighborhood $V_2 \subset V_1$ of F_θ which is disjoint with $x_1 + B_\theta$. Choose $x_2 \in V_2 \cap S_X$. As $F_\theta \notin (\text{conv}\{x_1, x_2\} + B_\theta)$, there is a convex w^*-neighborhood $V_3 \subset V_2$ of F_θ such that $V_3 \cap (\text{conv}\{x_1, x_2\} + B_\theta) = \emptyset$. Pick $x_3 \in V_3 \cap S_X$ and continue by induction to get a decreasing sequence $\{V_n\}$ of convex sets and a sequence $\{x_n\}$ of vectors with the following properties:

$\text{conv}\{x_n\} \subset V_1$, hence $\inf\{\|u\| : u \in \text{conv}\{x_n\}\} \geq \theta$.

For all $n \in \mathbb{N}$, $\text{conv}\{x_{n+1}, x_{n+2}, \ldots\} \subset V_{n+1}$, and $V_{n+1} \cap (\text{conv}\{x_1, \ldots, x_n\} + B_\theta) = \emptyset$, so given $u \in \text{conv}\{x_1, \ldots, x_n\}$ and $v \in \text{conv}\{x_{n+1}, x_{n+2}, \ldots\}$, $\|u - v\| > \theta$. □

We will now show several applications of Theorem 3.130.

In Exercise 1.62 we mentioned that the closed convex hull of a $\|\cdot\|$-compact subset of a Banach space is $\|\cdot\|$-compact. The following theorem extends this result in a significant way.

Theorem 3.133 (Krein) *Let X be a Banach space. If A is a weakly compact set in X, then $C := \overline{\text{conv}}(A)$ is weakly compact in X.*

Proof: If $f \in X^*$, then $\sup_A(f) = \sup_C(f)$, so f attains its supremum over C (a w-closed set) at some point of C. It is enough now to apply Theorem 3.130. □

It is possible to prove Theorem 3.133 in full generality (see Exercise 3.177) by using just the separable version of James Theorem 3.130, whose proof, presented above, is much easier than the general case.

Theorem 3.130 can be used to give another proof of (iii)$\Longrightarrow$(i) in Theorem 3.109 (however, observe that the proof of Theorem 3.130 was only completed in the separable case). See Exercise 3.178.

Note the similarity of the following result with the Lebesgue dominated convergence theorem for continuous functions.

3.11.8.5 The Rainwater–Simons Theorem

Theorem 3.134 (Rainwater [Rain1], Simons [Simo]) *Let B be a James boundary for a bounded subset M of the dual of a Banach space X. Let $\{x_n\}$ be a bounded sequence in X and $x \in X$. If $f(x_n) \to f(x)$ for all $f \in B$, then $f(x_n) \to f(x)$ for all $f \in M$. In particular, if B is a James boundary of a Banach space X and $f(x_n) \to f(x)$ for all $f \in B$, then $x_n \overset{w}{\to} x$.*

Proof: Without loss of generality, we may assume that $\{x_n\}$ converges to 0 on elements in B. Let $u = \limsup x_n$, where each x_n is considered as a map from X^* into $\mathbb{R}$. According to Theorem 3.124, $\sup_M u = 0$. If there exists $x^* \in M$ such that $x^*(x_n) \not\to 0$, we can find $\varepsilon > 0$ and a subsequence $\{x_{n_k}\}$ such that $|x^*(x_{n_k})| > \varepsilon$ for all $k \in \mathbb{N}$. If $x^*(x_{n_k}) > \varepsilon$ for infinitely many k's, we get a contradiction. Otherwise, use the same argument for the sequence $\{-x_n\}$ to get again a contradiction. □

Corollary 3.135 *Let B be a James boundary of a Banach space X. Let $A \subset X$ be a bounded and (relatively) $w(X, B)$-sequentially compact subset of X. Then A is (relatively) w-compact.*

Proof: Every sequence $\{a_n\}$ in A has a subsequence $\{a_{n_k}\}$ that $w(X, B)$-converges to some point $a \in A$ ($a \in X$). By Theorem 3.134, $\{a_{n_k}\}$ is w-convergent to a. The set A is then (relatively) w-sequentially compact. The Eberlein–Šmulyan theorem gives the conclusion. □

Corollary 3.136 *Let X be a separable Banach space. Let B be a James boundary of X. Let A be a bounded and (relatively) $w(X, B)$-compact subset of X. Then A is (relatively) w-compact.*

Proof: The set (B_{X^*}, w^*) is compact and metrizable, hence it is separable, and so it is its subset B. Let B_0 be a countable and w^*-dense subset of B. Given a sequence $\{a_n\}$ in A, a diagonal argument gives a subsequence $\{a_{n_k}\}$ of $\{a_n\}$ such that $\{\langle b_0, a_n\rangle\}_n$ converges for every $b_0 \in B_0$. The set B_0 separates points of X, and $(A, w(X, B))$ $\big(\big(\overline{A}^{w(X,B)}, w(X, B)\big)\big)$ is compact, so the restrictions of the topologies $w(X, B)$ and $w(X, B_0)$ to A (to $\overline{A}^{w(X,B)}$) coincide. The sequence $\{a_n\}$ has a $w(X, B)$-cluster point $a \in A$ ($a \in \overline{A}^{w(X,B)}$); hence the sequence $\{a_{n_k}\}$ is $w(X, B)$-convergent to a. This proves that A is (relatively) $w(X, B)$-sequentially compact. Corollary 3.135 gives the conclusion. □

Corollary 3.137 (Rainwater [Rain1]) *Let X be a Banach space, let $\{x_n\}$ be a bounded sequence in X and $x \in X$. If $f(x_n) \to f(x)$ for every $f \in \operatorname{Ext}(B_{X^*})$, then $x_n \overset{w}{\to} x$.*

Corollary 3.138 *Let K be a compact topological space. Let $\{f_n\}$ be a bounded sequence in $C(K)$ and $f \in C(K)$. Then, if $f_n \to f$ pointwise, we have $f_n \overset{w}{\to} f$.*

Proof: Use Corollary 3.137 and Lemma 3.116. □

The following result contains the conclusion of Pfitzner's theorem in the particular setting of $C(K)$ spaces and the James boundary consisting of the extreme points of the closed dual unit ball.

Theorem 3.139 (Eberlein, Šmulyan, Grothendieck) *Let K be a compact space. Then, if A is a uniformly bounded subset of $C(K)$, the following are equivalent: The set A is*
(i-p) *(relatively) $\mathcal{T}_p$-countably compact,*
(ii-p) *(relatively) $\mathcal{T}_p$-sequentially compact,*
(iii-p) *(relatively) $\mathcal{T}_p$-compact,*
(i-w) *(relatively) w-countably compact,*
(ii-w) *(relatively) w-sequentially compact, and*
(iii-w) *(relatively) w-compact.*

Proof: The equivalence between (i-p), (ii-p), and (iii-p) holds for any subset of $C(K)$, see Theorem 3.52. Assume now that A is uniformly bounded. If A is (relatively) $\mathcal{T}_p$-sequentially compact and $\{f_n\}$ is a sequence in A, there exists a subsequence $\{f_{n_k}\}$ of $\{f_n\}$ that $\mathcal{T}_p$-converges to some $f \in A$ ($f \in C(K)$). By Corollary 3.138, the sequence $\{f_{n_k}\}$ is w-convergent to f. This proves that A is (relatively) w-sequentially compact. It is enough now to apply Theorem 3.109. □

Remark: The version of Theorem 3.139 for an arbitrary James boundary of $C(K)$ is given in Exercises 3.155, 3.156, and 3.157.

Another important special instance of Pfitzner's theorem is given in the next result, that covers the case of convex sets and arbitrary James boundaries.

Theorem 3.140 (See [Gode3] and references therein) *Let X be a Banach space. Let B be a James boundary of X. Then, a convex and $w(X, B)$-compact subset of X is w-compact.*

Proof: Let A be a convex and $w(X, B)$-compact subset of X. Consider the locally convex space $\big(X, w(X, B)\big)$ and the fact that B is bounded in the space $\big(\operatorname{span}(B), w(\operatorname{span}(B), X)\big)$; then, according to Exercises 3.73 and 3.74, the set A is uniformly bounded on the set B. Since $\overline{\operatorname{conv}}^{w^*}(B) = B_{X^*}$, A is uniformly bounded on B_{X^*}, i.e., A is bounded in X.

In order to prove that A is w-compact, it is enough to show, by the Eberlein–Šmulyan theorem, that every countable subset $\{a_n : n \in \mathbb{N}\}$ of A is w-relatively compact. The boundedness of A allows us to define a linear and continuous mapping $S : \ell_1 \to X$ by

$$S((\lambda_n)) = \sum_{n=1}^{\infty} \lambda_n a_n, \quad \text{for all } (\lambda_n) \in \ell_1.$$

We shall show that S is a w-compact mapping; this will finish the proof. By Theorem 13.34, it is enough to show that $S^* : X^* \to \ell_\infty$ is w-compact. Let $Y = \overline{\operatorname{span}}^{\|\cdot\|}(B)$, a closed linear subspace of $(X^*, \|\cdot\|)$. By Fact 3.120, $\overline{B_Y}^{w^*} = B_{X^*}$. Since S^* is w^*-w^*-continuous, it suffices to show that $S^*(B_Y)$ is w-relatively compact. To that end, we shall prove that every $\mu \in B_{\ell_\infty^*}$ attains its supremum on $S^*(B_Y)$ at some point of $S^*(B_Y)$ (or, equivalently, that every $z := S^{**}(\mu)$ attains its supremum on B_Y at some point of B_Y). James' Theorem 3.130 will give then the conclusion. There is a net $\{l_i\}_{i \in I}$ in B_{ℓ_1} that w^*-converges to μ. For $i \in I$, put $x_i = S(l_i)$. The net $\{x_i\}_{i \in I}$ is in A, since A is convex and w-closed, so it has a $w(X, B)$-cluster point $a \in A$. Notice that, moreover, $\{x_i\}_{i \in I}$ is w^*-convergent to z. Obviously, $\langle b, a \rangle = \langle b, z \rangle$ for every $b \in B$, so a and z coincide on Y. Since B is a James boundary, a attains its supremum on B_Y at some point of B_Y, and so does z. This concludes the proof. □

We refer to [FLP] for an approach to James boundaries via the Choquet representation theory (see Theorem 3.74).

3.12 Remarks and Open Problems

Remarks

1. The *dual Mackey topology*, i.e., the topology on X^* of the uniform convergence on the family of all convex, balanced and w-compact subsets of X, is discussed, e.g., in [HMVZ, Section 3.1]. See also Exercises 3.40 and 3.41.
2. We refer, e.g., to the article [PeWo, pp. 1361–1423] for a discussion on Sobolev spaces.
3. For a proof of the Eberlein–Šmulyan theorem by using Schauder bases' techniques we refer to [Pelc3d], see [AlKa, p. 23].
4. Bourgain and Talagrand [BoTa] gave the positive solution to the boundary problem for an arbitrary Banach space and its James boundary consisting of all extreme points of the closed dual unit ball. Godefroy's boundary problem appears in [Gode2], see also [Gode3]. For a convex set (and an arbitrary James boundary) the result was implicit in Simons [Simo], see also [Gode3, p. 44] and [Flor]. This is Theorem 3.140. The answer was also known to be positive for Banach spaces X such that (B_{X^*}, w^*) is angelic [CasVe]. For the case

of $C(K)$ spaces, the positive solution was in [CasGod]. This is presented in Exercise 3.157. Finally, a complete positive solution has been given by Pfitzner in [Pfi].

Open Problems

1. Assume that for a Banach space X, $\|f - f_n\| \to 0$ whenever $f, f_n \in S_{X^*}$ satisfy $f_n \overset{w^*}{\to} f$. It is apparently unknown whether it is then true that Y^* is separable for all separable subspaces Y of X.
2. A separable Banach space X is reflexive if and only if it admits an equivalent $2R$-norm, that is, a norm $\|\cdot\|$ such that a sequence $\{x_n\}$ in S_X is convergent whenever $\lim_{n,m\to\infty} \|x_n + x_m\| = 2$, [OdSc3]. This is unknown for general nonseparable spaces.

Exercises for Chapter 3

3.1 Let $f : E \to \mathbb{R}$ be a linear functional on the real topological vector space E such that $\operatorname{Ker} f$ is closed and $f \neq 0$. Show, without using Proposition 3.19, that, for every $\alpha \in R$, the set $\{x \in E : f(x) < 0\}$ is open, and that we have $\overline{\{x \in E : f(x) < 0\}} = \{x \in E : f(x) \leq 0\}$. Conclude that the same is true whenever 0 is replaced by any $\alpha \in \mathbb{R}$ and the $<$ and $\leq$ symbols are replaced, respectively, by $>$ and $\geq$.
Hint. Let $x_0 \in E$ such that $f(x_0) < 0$. Since $\operatorname{Ker} f$ is closed, we can find a balanced neighborhood U of 0 such that $(x_0+U)\cap\operatorname{Ker} f = \emptyset$. By Exercise 1.3 we get $f(x) < 0$ for all $x \in x_0 + U$. This proves that $\{x \in E : f(x) < 0\}$ is open. Its complement, i.e., the set $\{x \in E : f(x) \geq 0\}$, is closed. It follows that $\overline{\{x \in E : f(x) > 0\}} \subset \{x \in E : f(x) \geq 0\}$. Let $x_0 \in E$ such that $f(x_0) = 0$, and let $y \in E$ such that $f(y) > 0$. Then $[y, x_0) \subset \{x \in E : f(x) > 0\}$, hence $x_0 \in \overline{\{x \in E : f(x) > 0\}}$; this proves that $\{x \in E : f(x) \geq 0\} \subset \overline{\{x \in E : f(x) > 0\}}$.

3.2 Let $(E, \mathcal{T})$ be a topological vector space. Assume that there exists a convex neighborhood C of 0. Let p_C be the Minkowski functional of C (use Definition 2.9 now in the setting of topological vector spaces). Prove the following assertions. (i) p_C is a finite non-negative positively homogeneous and subadditive continuous functional on E. (ii) $\operatorname{Int}(C) = \{x \in E : p_C(x) < 1\} \subset C \subset \{x \in E : p_C(x) \leq 1\} = \overline{C}$. (iii) If C is moreover balanced then p_C is a continuous seminorm (a continuous norm if C is also bounded; in this case, this norm induces the topology $\mathcal{T}$).
Hint. (i) for the positive homogeneity and the subadditivity proceed as in the proof of Lemma 2.11; to prove that $p_C(0) = 0$, follow the argument there, too. If $\{x_i\}$ is a net converging to 0, then for every $\lambda > 0$ we have $\lambda x_i \to 0$, hence there exists i_0 such that $\lambda x_i \in C$ for $i \geq i_0$. Then $p_C(\lambda x_i) \leq 1$, i.e., $p_C(x_i) \leq 1/\lambda$. This proves the continuity of p_C at 0. Lemma 2.10 holds in this context with the same proof, so

p_C is continuous. (ii) Follow again the proof of Lemma 2.11. (iii) Follow the hint in Exercise 2.21. If C is bounded and $p_C(x) = 0$ it follows that $x \in U$ for every neighborhood U of 0, hence $x = 0$.

3.3 Let E be a real topological vector space. Let $f : E \to \mathbb{R}$ be a linear functional and $p : E \to \mathbb{R}$ a function that is continuous at 0, and such that $f \leq p$. Then f is continuous.
Hint. Since $-f(x) = f(-x) \leq p(-x)$, we get $-p(-x) \leq f(x) \leq p(x)$ for all $x \in E$, hence f is continuous at 0. The conclusion follows from Proposition 3.19.

3.4 Let L_0 be the vector space of all Lebesgue measurable functions on $[0, 1]$ with the metric $\rho(x, y) = \int_0^1 \frac{|x(t)-y(t)|}{1+|x(t)-y(t)|}\,dt$. Show that $\rho(x_n, x) \to 0$ if and only if $x_n \to x$ in measure, and that L_0 is a topological vector space. For the definition of convergence in measure, see the proof of Theorem 4.55.
Hint. Let λ be the Lebesgue measure in $[0, 1]$. Denote $A_\varepsilon = \{t : |(x_n - x)(t)| \geq \varepsilon\}$, $B_\varepsilon = \{t : |(x_n - x)(t)| \leq \varepsilon\}$, then

$$\int_0^1 \frac{|x_n - x|}{1 + |x_n - x|}\,dt = \int_{A_\varepsilon} \frac{|x_n - x|}{1 + |x_n - x|}\,dt + \int_{B_\varepsilon} \frac{|x_n - x|}{1 + |x_n - x|}\,dt$$
$$\leq \lambda\{t : |(x_n - x)(t)| \geq \varepsilon\} + \varepsilon.$$

If x_n does not converge to 0 in measure, there exists $\delta > 0$ such that $\lambda(A_\varepsilon) \geq \delta$ for all $\varepsilon > 0$. Then

$$\int_0^1 \frac{|x_n - x|}{1 + |x_n - x|}\,dt \geq \int_{A_\varepsilon} \frac{|x_n - x|}{1 + |x_n - x|}\,dt \geq \int_{A_\varepsilon} \frac{\varepsilon}{1 + \varepsilon}\,dt = \frac{\varepsilon\delta}{1 + \varepsilon},$$

and the latter does not converge to 0. We used here the monotonicity of the function $\frac{t}{1+t}$. For the reverse implication, use a similar argument.

To observe that the convergence in measure generates a topology in which both operations are continuous, write

$$|\alpha_n x_n - \alpha_0 x_0| \leq |\alpha_n|\,|x_n - x_0| + |\alpha_n - \alpha_0|\,|x_0|$$

and note that

$$\{t : |\alpha_n x_n - \alpha_0 x_0| \geq \varepsilon\} \subset \{t : |\alpha_n|\,|x_m - x_0| \geq \varepsilon/2\} \cup \{t : |\alpha_n - \alpha_0|\,|x_0| \geq \varepsilon/2\}.$$

Measures of the sets on the right hand side go to zero (for the second summand use the Chebyshev inequality).

3.5 Let X be the vector space of all continuous functions on $(0, 1)$. For $f \in X$ and $r > 0$, let $V(f, r) = \{g \in X : |g(x) - f(x)| < r \text{ for all } x \in (0, 1)\}$. Let T be the

topology on X that these sets generate. Show that the addition is $\mathcal{T}$-continuous but scalar multiplication is not.
Hint. $\{\frac{1}{n}\frac{1}{x}\} \not\to 0$.

3.6 Let $0 < p < 1$. Let ℓ_p be the space of all sequences of real numbers $x = (x_i)$ such that $\sum |x_i|^p < \infty$, with the metric $\mathrm{d}(x, y) = \sum |x_i - y_i|^p$. Show that d is a complete metric and ℓ_p is a topological vector space but not a locally convex space. Show that $\ell_p^* = \ell_\infty$ and the set $U = \{(x_i) : \sum |x_i| \leq 1\}$ has the property that $f(U)$ is bounded for every $f \in \ell_p^*$ but U is unbounded in ℓ_p.
Hint. To show that $\ell_p^* = \ell_\infty$, note that the topology of ℓ_p is stronger than that of ℓ_1 and thus every element of ℓ_∞ is a continuous linear functional on ℓ_p. If $f \notin \ell_\infty$, we find $x \in \ell_p$ such that $\sum f_i x_i$ diverges. To see that ℓ_p is not locally convex, consider the set $A = \{x^n\} \subset \ell_p$, where $x^n \in \ell_p$ is defined by $x_i^n = n^{p-1}$ if $i = n$ and $x_i^n = 0$ otherwise. This is a relatively compact set in ℓ_p whose convex hull is not bounded as the sequence $\{y^n\}$ defined by $y^n = n^{-1}\sum_{k=1}^n x^k$ is unbounded in ℓ_p. The latter is seen by an elementary calculation, as $d(0, y^{2n})$ can be estimated from below by

$$(2n)^{-p}\sum_{k=n}^{2n} k^{p^2-p} \geq (2n)^{-p} n\, n^{p^2-p} = \frac{n^{(1-p)^2}}{2^p} \to \infty.$$

3.7 Show that if $\{x_n\}$ is a sequence in a metrizable topological vector space X and $x_n \to 0$, then there are positive numbers γ_n such that $\gamma_n \to \infty$ and $\gamma_n x_n \to 0$.

However, if X is the vector space of all real-valued functions on $[0, 1]$ with the topology of pointwise convergence on $[0, 1]$, then there is a sequence $\{f_n\} \subset X$ that pointwise converges to zero in X and if γ_n is any sequence of numbers such that $\gamma_n \to \infty$, then $\{\gamma_n f_n\}$ does not converge to zero.
Hint. First part is standard, $\gamma_n = \frac{1}{\sqrt{\rho(x_n,0)}}$. To see the second part, note that the cardinality of the family of all sequences converging to zero is c.

3.8 Let X, Y be topological vector spaces. Assume that T is a linear mapping from X onto Y such that $T^{-1}(0)$ is closed in X and Y is finite-dimensional. Show that T is a continuous and open mapping.
Hint. If U is a balanced neighborhood of zero in X and $\{e_i\}$ is a basis for Y, then for some n, $T(nU) \supset \{e_i\}$. The convex hull of $\{\pm e_i\}$ contains 0 as an interior point. This shows that T is an open mapping. T is an algebraic isomorphism of $X/T^{-1}(0)$ and Y. Since all linear vector topologies on finite-dimensional spaces coincide, and since the mapping $x \mapsto \hat{x}$ of X onto $X/T^{-1}(0)$ is continuous, T continuous.

3.9 Prove that in an angelic topological space, the following classes of sets coincide: (relatively) compact, (relatively) countably compact, (relatively) sequentially compact.
Hint. (1) (Relative) compactness implies (relative) countable compactness. (2) Relative countable compactness implies relative compactness by definition of angelicity. (3) Let A be a relatively compact set. Given a sequence $\{a_n\}$ in A, if the set

$A_0 := \{a_n : n \in \mathbb{N}\}$ has no accumulation point, then A_0 is finite, and so the sequence $\{a_n\}$ has a convergent subsequence. Otherwise, let z be an accumulation point of A_0. If $z = a_n$ infinitely often, then $\{a_n\}$ has a convergent subsequence. If not, we may assume $z \neq a_n$ for all $n \in \mathbb{N}$. The angelicity gives a sequence in $\{a_n : n \in \mathbb{N}\}$ that converges to z, hence a subsequence of $\{a_n\}$ that also converges to z. This proves that A is relatively sequentially compact. (4) Obviously, (relative) sequential compactness implies (relative) countable compactness. (5) If A is countably compact then it is relatively compact; an element z in $\overline{A}$ is the limit of a sequence in A, so $z \in A$. We get that A is closed, hence compact. (5) If A is compact it is, by (3), relatively sequentially compact; it is also closed, hence sequentially compact.

3.10 Let $(E, \mathcal{T})$ be a topological vector space. Prove that a set $A \subset E$ is relatively compact if and only if every net in A has a convergent subnet.

Hint. The necessary condition is clear. To prove sufficiency, let $\{\overline{a}_i : i \in I, \leq\}$ be a net in $\overline{A}$. We shall prove that it has a subnet that converges to some point of $\overline{A}$. Let $\mathcal{U}$ be a local base of balanced neighborhoods in E. Given $i \in I$ and $U \in \mathcal{U}$, find $a_{i,U} \in (\overline{a}_i + U) \cap A$. Define in $I \times \mathcal{U}$ a partial order by $(i_1, U_1) \prec (i_2, U_2)$ whenever $i_1 \leq i_2$ and $U_1 \supset U_2$. Prove that $(I \times \mathcal{U}, \prec)$ is directed upwards. Since $\{a_{i,U} : (i, U) \in I \times \mathcal{U}, \prec\}$ is a net in A, there exists a subnet $\{a_{\phi(j)} : j \in J, \lhd\}$ that converges (to an element $\overline{a} \in \overline{A}$). If $p_1 : I \times \mathcal{U} \to I$ denotes the first coordinate mapping, $p_1 \circ \phi$ defines a subnet of $\{\overline{a}_i : i \in I, \leq\}$ that converges to $\overline{a}$. Indeed, the mapping $p_1 \circ \phi$ satisfies the condition for defining a subnet; moreover, given $U \in \mathcal{U}$ there exists $j_0 \in J$ such that for all $j_0 \lhd j$, $a_{\phi(j)} \in \overline{a} + U$. Recall that $a_{\phi(j)} \in \overline{a}_{p_1 \circ \phi(j)} + U$, hence $\overline{a}_{p_1 \circ \phi(j)} \in a_{\phi(j)} + U \in \overline{a} + 2U$ for all $j_0 \lhd j$. This proves the assertion.

3.11 Let X be a normed space. Prove that X is complete in the canonical metric given by its norm if and only if it is complete in its norm topology as a topological vector space.

Hint. If every Cauchy net in X converges, then also every Cauchy sequence (which is also a net) converges.

Assume that X is a complete normed space and $\{x_\alpha\}_{\alpha \in I}$ is a Cauchy net in X. Given $n \in \mathbb{N}$, find $i_n \in I$, $i_n > i_{n-1}$, such that $(x_\alpha - x_\beta) \in \frac{1}{n} B_X$ for all $\alpha, \beta \geq i_n$, in particular, $x_\alpha \in B(x_{i_n}, \frac{1}{n})$ for all $\alpha \geq i_n$. Then $\{x_{i_n}\}$ is a Cauchy sequence which must converge to some $x \in X$. We claim that x is the limit of the net $\{x_\alpha\}_{\alpha \in I}$: Choose $n \in \mathbb{N}$. There exists $n_0 \in \mathbb{N}$ such that $n_0 \geq n$ and $x_{i_m} \in B(x, \frac{1}{n})$ for all $m \geq n_0$, in particular, $x_{i_{n_0}} \in B(x, \frac{1}{n})$. Then for all $\alpha \in I$ such that $\alpha \geq i_{n_0}$ we get $x_\alpha \in B(x_{i_{n_0}}, \frac{1}{n_0})$, hence $x_\alpha \in B(x, \frac{2}{n})$. Consequently, $x_\alpha \to x$.

3.12 Prove that the space $\mathbb{R}^{\mathbb{N}}$ (equipped with the product topology) is a separable Fréchet space that is not normable.

Hint. The separability follows from the fact that it is a countable product of separable spaces. To show the completeness is also standard. The fact that it is not normable follows from the non-boundedness of the elements in the local base.

3.13 (Klee) Let $K_1, \ldots, K_n$ be open convex sets in a locally convex space X. Show that if $\bigcap K_i = \emptyset$ then there is a bounded operator L from X into $\mathbb{R}^{n-1}$ such that $\bigcap L(K_i) = \emptyset$.
Hint. Let $K = \{(x_1 - x_2, x_1 - x_3, \ldots, x_1 - x_n) : x_i \in K_i, i = 1, \ldots, n\}$. Then $K \subset X^{n-1}$ is an open convex set that does not contain 0. Thus by the Hahn–Banach theorem, there exist $x_1^*, x_2^*, \ldots, x_{n-1}^* \in X^*$ such that $\sum_{i=1}^{n-1} x_i^*(x_1 - x_i) > 0$ for all $x_i \in K_i, i = 1, \ldots, n$. Put $L(x) = \big(x_1^*(x), \ldots, x_{n-1}^*(x)\big)$ for $x \in X$.

3.14 Let $(E, \mathcal{T})$ be a locally convex space. Let $K_1, \ldots, K_n$ be convex (respectively convex and balanced) compact subsets of E. Then the convex hull (respectively the convex and balanced hull) of $\bigcup_{i=1}^n K_i$ is again compact.
Hint. For the convex case, observe that

$$\operatorname{conv}\left(\bigcup_{i=1}^{n} K_i\right) = \left\{\sum_{i=1}^{n} \alpha_i k_i : \alpha_i \geq 0,\ k_i \in K,\ \text{for } i = 1, 2, \ldots, n,\ \sum_{i=1}^{n} \alpha_i = 1\right\}.$$

Note then that $\{(\alpha_1, \ldots, \alpha_n) : \alpha_i \geq 0 \text{ for } i = 1, 2, \ldots, n,\ \sum_{i=1}^{n} \alpha_i = 1\}$ is a compact subset of $\mathbb{K}^n$. It is enough now to use a convergence argument on nets in $\operatorname{conv}\big(\bigcup_{i=1}^n K_i\big)$. For the convex and balanced case change $\alpha_i \geq 0$ and $\sum_{i=1}^n \alpha_i = 1$ to $(\alpha_1, \ldots, \alpha_n) \in \mathbb{K}^n$ with $\sum_{i=1}^n |\alpha_i| \leq 1$.

3.15 Given a family $\{E_\gamma\}_{\gamma \in \Gamma}$ of vector spaces, its *algebraic direct sum* $\bigoplus_\Gamma E_\gamma$ is the subspace of $\prod_\Gamma E_\gamma$ consisting of all elements $x = (x_\gamma)$ having finitely many non-zero coordinates. For $\gamma \in \Gamma$, let $I_\gamma : E_\gamma \to \bigoplus_\Gamma E_\gamma$ be the natural embedding. If each E_γ is a locally convex space and $\{B_\beta^\gamma\}$ is a base of neighborhoods of 0 in E_γ, endow $\bigoplus_\Gamma E_\gamma$ with the topology having as a base of neighborhoods of 0 the collection of sets that are convex and balanced hulls $\bigcap_{\gamma \in \Gamma} I_\gamma(B_\beta^\gamma)$. Prove that $\bigoplus_\Gamma E_\gamma$ endowed with this topology is a locally convex space (called the *locally convex direct sum of the family* $\{E_\gamma\}_\Gamma$.

Prove that each E_γ is a subspace of $\bigoplus_\Gamma E_\gamma$. Prove that $\bigoplus_\Gamma E_\gamma$ is complete if each E_γ is complete. Prove that the locally convex direct sum and the topological product of a *finite* family of locally convex spaces agree.
Hint. Standard computation.

3.16 Let E be a locally convex space and let $\{E_\gamma\}_{\gamma \in \Gamma}$ be a family of subspaces of E. We say that E is the *locally convex direct sum* of the family $\{E_\gamma\}_{\gamma \in \Gamma}$ if E is isomorphic to the locally convex direct sum $\bigoplus_\Gamma E_\gamma$ defined in Exercise 3.15. Prove that if E is a locally convex space and F is a finite-dimensional subspace of E, there exists a subspace G of E such that $E = F \oplus G$ (in this case, E/F is isomorphic

to G and E/G is isomorphic to F). Observe that the result is no longer true in the class of topological vector spaces.
Hint. Since F is isomorphic to ℓ_2^n if $\dim(F) = n$ (see Proposition 3.13), we can find $\{f_i : i = 1, 2, \dots, n\}$ in F and $\{f_i^* : i = 1, 2, \dots, n\}$ in F^* such that $\langle f_j^*, f_i\rangle = \delta_{ij}$ for all i, j. Put $G = \bigcap_{i=1}^n \operatorname{Ker}(f_i^*)$. It is easy to prove that $E = F \oplus G$. If the result should be true in the class of topological vector spaces, by taking a non-zero (continuous) linear functional on F and extending this to be 0 on G we shall always obtain non-zero continuous linear functionals on every topological vector space, against the example in Section 3.2.

3.17 Let E be a locally convex space and A a nonempty subset of E. Prove directly that A° is a w^*-closed set.
Hint. $A^\circ = \bigcap_{x\in A} \{f : x(f) \leq 1\}$ and the sets are w^*-closed, since vectors from X are w^*-continuous as elements of $(X^*, w^*)^*$.

3.18 Let E be a locally convex space, and $\{A_\alpha\}$ a nonempty family of nonempty subsets of E. Show that $\bigcup (A_\alpha)^\circ \subset \left(\bigcap A_\alpha\right)^\circ$, $\left(\bigcup A_\alpha\right)^\circ = \bigcap (A_\alpha)^\circ$ and, if $\lambda \neq 0$ and A is a nonempty subset of E, $(\lambda A)^\circ = \frac{1}{\lambda} A^\circ$. Show that if A_1 and A_2 are two nonempty subsets of E such that $A_1 \subset A_2$ then $(A_2)^\circ \subset (A_1)^\circ$.
Hint. Use the definition.

3.19 Show that if X is a normed space, then $(B_X)^\circ = B_{X^*}$, and $(B_{X^*})_\circ = B_X$.
Hint. Use the definition.

3.20 (Corson) Let K be a convex compact set in a locally convex space E. Show that if $\operatorname{Ext}(K)$ is countable then K is metrizable.
Hint. Let X be the Banach space of all continuous affine functions on K with the sup-norm. Identify K with its image in X^*. By the bipolar theorem, B_{X^*} is the w^*-closed convex hull of $K \cup (-K)$. Hence $\operatorname{Ext}(B_{X^*})$ is countable by Theorem 3.66 and thus X^* is separable by Corollary 3.126. Thus X is separable and hence (B_{X^*}, w^*) is metrizable, so K is metrizable.

3.21 Give an example of two locally convex spaces E and F, a continuous linear mapping $T : E \to F$, a compact convex subset K of E and an extreme point e_0 of K such that $T(x_0)$ is not an extreme point of $T(K)$.
Hint. $E := \mathbb{R}^2$, $F := \mathbb{R}$, T the orthogonal projection.

3.22 Let E, F two locally convex spaces, $T : E \to F$ a continuous linear mapping, S a nonempty subset of E, $H \subset F$ a closed supporting manifold of $T(S)$. Prove that $T^{-1}(H)$ is a supporting manifold of S.
Hint. Obviously, $T^{-1}(H) \cap S \neq \emptyset$. Let $(x, y) \subset S$ such that $(x, y) \cap T^{-1}(H) \neq \emptyset$. Consider first the case $T(x) = T(y)$ and then the case $T(x) \neq T(y)$. Use the definitions.

3.23 Let D be a convex set in a $\mathbb{R}^n$. Show that either $\mathrm{Int}(D) \neq 0$ or D lies in a proper subspace of $\mathbb{R}^n$.
Hint. Assume that $0 \in D$, let $\{e_i\}_{i=1}^m$ be a maximal linearly independent set in D. Since $D \subset \mathrm{span}\{e_i\}_{i=1}^m$, it remains to prove that $\mathrm{Int}(D) \neq \emptyset$ if $n = m$. Consider $C = \mathrm{conv}\{\{e_i\} \cup \{0\}\} \subset D$, note that $\{\sum \alpha_i e_i : 0 \leq \alpha_i \leq \frac{1}{m}\} \subset C$. Let $x_0 = \sum \frac{1}{2m} e_i$. Then $B = \{\sum \alpha_i e_i : |\alpha_i - \frac{1}{2m}| < \frac{1}{2m}\}$ is an open neighborhood of x_0 and $B \subset C \subset D$.

3.24 Let K be a nonempty convex compact subset of a finite-dimensional topological vector space, and let $f : K \to \mathbb{R}$ be a continuous convex function. Prove that f attains its supremum at an extreme point of K.
Hint. That f is bounded and attains its supremum is clear. Use now Carathéodory's result 3.70.

3.25 Assume that C is a convex compact set in a locally convex space X and D is a closed subset of C. Show that if μ is a probability measure on D, then there is a unique point $x \in C$ represented by μ. Moreover, the mapping $\mu \mapsto x$ is continuous from the w^*-topology in $C(D)^*$ into the topology of X. Prove this result.
Hint. For $f \in X^*$ put $H_f = \{x \in D : f(x) = \mu(f)\}$. We will show that $\bigcap_{f \in X^*} H_f \cap C \neq \emptyset$. Since C is compact, by the finite intersection property it is enough to show that $(\bigcap H_{fi}) \cap C \neq \emptyset$ for every finite set $f_1, \ldots, f_n \in X^*$. Define a mapping T from X into $\mathbb{R}^n$ by $T(x) = (f_1(x), \ldots, f_n(x))$. Then $T(C)$ is compact and convex. It suffices to show that $p \in T(C)$, where $p = (\int_D f_1 \,\mathrm{d}\mu, \ldots, \int_D f_n \,\mathrm{d}\mu)$. Assume that $p \notin T(C)$. Then there is a linear functional on $\mathbb{R}^n$ which strictly separates p from the compact convex set $T(C)$, i.e., there is $a = (a_1, \ldots, a_n)$ such that $(a, p) > \sup\{(a, T(x)) : x \in C\}$, where $(.,.)$ is the inner product on $\mathbb{R}^n$. Define $g \in X^*$ by $g = \sum a_i f_i$. Then we have

$$\begin{aligned}\int_D g \,\mathrm{d}\mu &= \int_D \Big(\sum a_i f_i\Big) \,\mathrm{d}\mu = \sum a_i \int_D f_i \,\mathrm{d}\mu \\ &= \sum a_i \mu(f_i) = (a, p) > \sup_C \sum a_i f_i(x) = \sup_C(g).\end{aligned}$$

This is a contradiction as $\mu(D) = 1$.

If the mapping in question is not continuous as stated, then there are measures μ_α and μ such that $\mu_\alpha \to \mu$ and their corresponding points x_α and x do not converge in X. Since C is compact, assume without loss of generality that $x_\alpha \to y \neq x$. Then for every $f \in X^*$ we have $f(x_\alpha) = \mu_\alpha(f) \to \mu(f) = f(x)$. Because of continuity of f we also have $f(y) = \lim f(x_\alpha)$. Thus $f(x) = f(y)$. Since X^* separates points, we have a contradiction.

3.26 Give an example of an affine function, in the sense as in the proof of Choquet's theorem, defined on a metrizable compact subset D of a Banach space X, that is not a restriction to D of a function of the form $f(x) = k + \varphi(x)$ for $k \in \mathbb{R}$, $\varphi \in X^*$.

Hint. $X = \ell_1$, $D = \{x \in \ell_1 : |x_i| \leq 4^{-i}\}$, $f(x) = \sum 2^i x_i$.

3.27 Let X be a separable Banach space. Assume that $(B_{X^{**}}, w^*)$ is angelic. Show that then every w^*-compact convex set C in X^* satisfies $C = \overline{\text{conv}}(\text{Ext}(C))$.
Hint. Use Choquet's theorem in (X, w), then the representation also holds for $f \in X^{**}$ by the Lebesgue dominated convergence theorem. Then use the separation theorem. Note that the result follows directly from Corollary 3.129.

3.28 Assume that $f_j \in \mathcal{D}'(\Omega)$ and $f(\varphi) = \lim_j f_j(\varphi)$ exists finite for every $\varphi \in \mathcal{D}(\Omega)$. Show that $f \in \mathcal{D}'(\Omega)$.
Hint. By the Banach–Steinhaus theorem, f is continuous on every $\mathcal{D}_K$.

3.29 Let Ω be an open set in $\mathbb{R}^n$ and $\{f_j\} \subset \mathcal{D}'(\Omega)$ be such that $\{f_j(\varphi)\}$ is bounded for every $\varphi \in \mathcal{D}(\Omega)$. Show that there is a subsequence $\{f_{j_k}\}$ and $f \in \mathcal{D}'(\Omega)$ such that $\lim f_{j_k}(\varphi) = f(\varphi)$ uniformly for φ on every bounded subset of $\mathcal{D}(\Omega)$.
Hint. By the Banach–Steinhaus theorem, the restrictions of f_j to $\mathcal{D}_K$ are equicontinuous. Then use the Arzelà–Ascoli theorem.

3.30 Let $C^\infty[0, 1]$ be the vector space of all real-valued functions on $[0, 1]$ whose all derivatives are continuous on $[0, 1]$. Let $k \in \mathbb{N}$ and $p \in [1, \infty)$. Define the norm $\|\cdot\|_p^k$ on C^∞ by $\|f\|_p^k = \left(\sum_{0 \leq l \leq k} \|f^{(l)}\|_p^p\right)^{\frac{1}{p}}$, where $f^{(l)}$ denotes the lth derivative of f and $\|\cdot\|_p$ is the norm from L_p. The *Sobolev space* $W_p^k[0, 1]$ is the completion of the normed space $(C^\infty[0, 1], \|\cdot\|_p^k)$. Show that $W_p^k[0, 1]$ is isomorphic to $L_p[0, 1]$.

3.31 Let $\{x_n\}$ be a sequence in a Banach space X. Prove that if $x_n \xrightarrow{w} x$ then $\{x_n\}$ is bounded and $\|x\| \leq \liminf_n \|x_n\|$ (the second statement has a more general formulation: the norm is a w-lower semicontinuous function, see Definition 17.6).

Let $\{f_n\}$ be a sequence in X^*. Prove that if $f_n \xrightarrow{w^*} f$ then $\{f_n\}$ is bounded and $\|f\| \leq \liminf_n \|f_n\|$ (the second statement has a more general formulation: the dual norm in X^* is a w^*-lower semicontinuous function).
Hint. Proof is similar to that of Corollary 3.86. To prove lower semicontinuity in full generality consider the w-closedness (w^*-closedness) of the level sets, i.e., the balls centered at 0 in X (in X^*), see the paragraph after Definition 17.6. For X use Mazur's Theorem 3.45, for X^* the fact that the polar of a subset of X in X^* is w^*-closed.

3.32 Let X be a normed space. (i) Let A be a nonempty bounded subset of X. Prove that $\text{diam}\, A = \text{diam}\, \overline{A}^w$. (ii) Let B be a nonempty bounded subset of X^*. Prove that $\text{diam}\, B = \text{diam}\, \overline{B}^{w^*}$.
Hint. (i) The w-lower semicontinuity of the norm. (ii) The w^*-lower semicontinuity of the dual norm.

3.33 Let $\{x_n\}$ be a sequence in a Banach space X. Prove that $x_n \overset{w}{\to} x$ if and only if $\{x_n\}$ is bounded and the set $\{f \in X^* : f(x_n) \to f(x)\}$ is dense in X^*.

Similarly, let $\{f_n\}$ be a sequence in X^*. Prove that $f_n \overset{w^*}{\to} f$ if and only if $\{f_n\}$ is bounded and the set $\{x \in X : f_n(x) \to f(x)\}$ is dense in X.

In particular, we have the following corollary. Assume that $\{x_n\}$ is bounded and $M \subset X^*$ is such that $\overline{\text{span}}(M) = X^*$ and $f(x_n) \to f(x)$ for all $f \in M$. Then $x_n \overset{w}{\to} x$. Analogous statement is true for w^*-convergence.

Hint. One direction is a direct consequence of the Banach–Steinhaus Theorem 3.85. The other one is standard: Let $f \in X^*$. Given $\varepsilon > 0$, approximate f by g such that $\|g - f\| < \varepsilon$ and $g(x_n) \to g(x)$. In the estimate

$$|f(x_n) - f(x)| \le |f(x_n) - g(x_n)| + |g(x_n) - g(x)| + |g(x) - f(x)|$$

we pass to the limit and obtain

$$\limsup |f(x_n) - f(x)| \le \varepsilon \sup \|x_n\| + 0 + \varepsilon \|x\|.$$

Since $\varepsilon > 0$ was arbitrary, we get $f(x_n) \to f(x)$ as needed.

3.34 Let $\{x_n\}$ be a sequence in a Banach space X. Assume that $f(x_n) \to 0$ for every $f \in A \subset X^*$, where A is a set of second Baire category in X^*. Show that $x_n \overset{w}{\to} 0$.

Hint. See the proof of the Banach–Steinhaus theorem.

3.35 Let X be a normed space. Show that if $\{x_n\}$ is Cauchy and $x_n \overset{w}{\to} 0$, then $x_n \to 0$ in norm.

Hint. $x_n \in x_m + \varepsilon B_X$ and $x_m + \varepsilon B_X$ is weakly closed.

3.36 Let X be a Banach space and let $\{x_n\}$ be a sequence in X that w-converges to 0. Prove that the closed convex and balanced hull M of $\{x_n : n \in \mathbb{N}\}$ is equal to the set $B := \{\sum_{n=1}^{\infty} \lambda_n x_n : \sum_{n=1}^{\infty} |\lambda_n| \le 1\}$, and that B is w-compact ([Koth, 20.9(6)]).

Hint. The w-compactness of M follows from Krein's theorem. Given $\lambda = (\lambda_n) \in B_{\ell_1}$, the sequence $\{\sum_{i=1}^{n} \lambda_i x_i\}_{n=1}^{\infty}$ is $\|\cdot\|$-Cauchy, so it converges in X (to the element $\sum_{i=1}^{\infty} \lambda_i x_i$) and we get $B \subset M$. Let $\varphi : \ell_1 \to X$ be defined as $\varphi(\lambda) = \sum_{i=1}^{\infty} \lambda_i x_i$ for $\lambda = (\lambda_i) \in \ell_1$; it is well defined, $\|\cdot\|_1$-$\|\cdot\|$-continuous and linear; moreover, $M = \varphi(B_{\ell_1})$. Let $\varphi^* : X^* \to \ell_\infty$ be its adjoint mapping. Since, for every $x^* \in X^*$, we have $\langle \varphi^*(x^*), e_n \rangle = \langle x^*, \varphi(e_n) \rangle = \langle x^*, x_n \rangle \to_n 0$, we get $\varphi^*(x^*) \in c_0$, hence φ is $w(\ell_1, c_0)$-w-continuous. Thus, by Alaoglu's theorem, M is w-compact, in particular w-closed. This proves that $M = B$.

3.37 Assume that $\{x_n\}$ is a weakly convergent sequence in an infinite-dimensional Banach space X. Show that $\overline{\text{conv}}\{x_n\}$ does not have any interior point. This is not the case for the w^*-convergence as the standard unit vectors of ℓ_1 show.

Hint. (Veselý, Zanco) Assume that $x_n \xrightarrow{w} 0$ and that $K = \overline{\text{conv}}\{x_n\}$ contains an interior point 0. Let x be an arbitrary point of the topological boundary of K. Using the Minkowski functional of K, we construct a functional f supporting K at x. Let N be such that $f(x_n) < f(x)/2$ for $n > N$. Put $A = \text{conv}\{x_n : n \leq N\}$ and $B = \text{conv}\{x_n : n > N\}$. Then there are $\lambda_k \in [0, 1]$, $a_k \in A$ a $b_k \in B$ such that the points $y_k = (1 - \lambda_k)a_k + \lambda_k b_k$ tend to x. By applying f we can see that $\lambda_k \to 0$ as otherwise f could not have a maximum at x. Hence $a_k \to x$ and A is closed. Thus $x \in A \subset \text{conv}\{x_n\}$. Hence X would have a countable Hamel basis, a contradiction.

3.38 Assume that B_{X^*} is angelic in its w^*-topology and assume that M is a linearly dense set in X. Assume that for a w^*-linearly dense set $D \subset X^*$, D has a countable support on M. Then all elements of X^* have countable support on M.
Hint. Let S be the collection of all elements of X^* that have countable support on M. Then observe that $S \cap B_{X^*}$ is w^*-closed and use Theorem 3.92 to see that S is w^*-closed in X^*. The set S is assumed to be w^*-dense in X^*.

3.39 The Josefson–Nissenzweig theorem asserts that for every Banach space X, there is a sequence $\{f_n\} \subset S_{X^*}$ such that $f_n \xrightarrow{w^*} 0$ ([Dies2]).

Show that given $f \in B_{X^*}$, there is a sequence $\{f_n\} \subset S_{X^*}$ such that $f_n \xrightarrow{w^*} f$.
Hint. Consider the sphere S centered at f with radius 1. Then there are $g_n \in S$ such that $(g_n - f) \xrightarrow{w^*} 0$. Project all g_n onto S_X in a radial way; namely, given g_n, let f_n be the intersection of S_X and the ray emanating from f and going through g_n. Then $f_n = f + c_n(g_n - f)$ for some $c_n \geq 0$ (use the fact that $f \in B_{X^*}$, draw a picture). We also have $c_n \leq 2$ and hence $(f_n - f) \xrightarrow{w^*} 0$ as needed.

3.40 Let X be a Banach space and let $\mathcal{T}_{\mathcal{K}}$ be the topology on X^* of the uniform convergence on the family $\mathcal{K}$ of all $\|\cdot\|$-compact subsets of X. Prove that $(X^*, \mathcal{T}_{\mathbb{K}})$ is complete.
Hint. Let $\{x^*_\alpha\}$ be a $\mathcal{T}_{\mathbb{K}}$-Cauchy net in X^*; it is also w^*-Cauchy. Since $(X^\#, w(X^\#, X))$ is complete (see Proposition 3.6), the net $\{x_\alpha\}$ is $w(X^\#, X)$-convergent to some $x^\#_0 \in X^\#$ (here $X^\#$ denotes the space of all linear forms on X). Fix a $\|\cdot\|$-compact convex and balanced subset K of X and some $\varepsilon > 0$. There exists α_0 such that $\sup_{x\in K} |\langle x^*_\alpha - x^*_\beta, x\rangle| < \varepsilon/2$ for all $\alpha, \beta \geq \alpha_0$. Then $\sup_{x\in K} |\langle x^*_\alpha - x^\#_0, x\rangle| \leq \varepsilon/2$ for all $\alpha \geq \alpha_0$. Since $x^*_{\alpha_0}$ is continuous, there exists a neighborhood U of 0 in X such that $\sup_{x\in K\cap U} |\langle x^*_{\alpha_0}, x\rangle| \leq \varepsilon/2$. It follows that $\sup_{x\in K\cap U} |\langle x^\#_0, x\rangle| \leq \varepsilon$. This proves that $x^\#|_K$ is continuous at 0. Then, by a standard convexity argument, it is (uniformly) continuous on K. Assume now that $x^\#_0$ is not continuous on X. We can then find a sequence $\{x_n\}$ in X such that $x_n \to 0$ and $\langle x^\#, x_n\rangle \not\to 0$. This is a contradiction, since the closed convex and balanced hull of $\{x_n : n \in \mathbb{N}\}$ is a $\|\cdot\|$-compact subset of X.

3.41 Prove that, for a Banach space X, the space $(X^*, \mu(X^*, X))$ is complete, where $\mu(X^*, X)$ denotes the Mackey topology on X^* associated to the dual pair $\langle X^*, X\rangle$.

Hint. Use Exercise 3.40 and Proposition 17.17.

3.42 [Groth3] A Banach space is said to have the *Grothendieck property* if every w^*-convergent sequence in X^* is w-convergent.

(i) Show that if X has the Grothendieck property and T is a bounded operator onto a Banach space Y, then Y has the Grothendieck property.

(ii) Show that a quotient of a Banach space with the Grothendieck property has the Grothendieck property.

(iii) Show that every separable Banach space with the Grothendieck property is reflexive.

(iv) Show that every separable quotient space of ℓ_∞ is reflexive.

Hint. (i) Direct proof, use the fact that T^* is w^*-w^*-continuous and T^{**} is onto (Exercise 2.49).

(ii) It follows from (i).

(iii) The Eberlein–Šmulyan Theorem 3.109 for the dual unit ball.

(iv) ℓ_∞ has the Grothendieck property ([Dies2, Chapter VII], [HMVZ, Chapter VII]).

3.43 Prove that if $F \in \ell_\infty^{**} \backslash \ell_\infty$, then $F^{-1}(0)$ $(\subset \ell_\infty^*)$ is w^*-sequentially closed but not w^*-closed.

Hint. The w^*-sequential closedness follows from the fact that in ℓ_∞^*, every w^*-convergent sequence is w-convergent (the Grothendieck property of ℓ_∞, see Exercise 3.42).

3.44 Does there exist a bounded operator from ℓ_∞ onto c_0?

Does there exist a bounded operator from ℓ_∞ onto ℓ_1?

Does there exist a bounded operator from ℓ_∞ onto $C[0, 1]$?

Hint. No. It is known that ℓ_∞ has the Grothendieck property ([Dies2]). Use Exercise 3.42.

3.45 Is c_0, ℓ_1 or $C[0, 1]$ isomorphic to a quotient of ℓ_∞?

Hint. No, see Exercise 3.44.

3.46 Show that if X is infinite-dimensional, then $0 \in \overline{S_X}^w$.

Using this, as in the previous exercise we can show that $\overline{S_X}^w = B_X$ for infinite-dimensional spaces.

Hint. Consider a neighborhood $V = \{x : |f_i(x)| < \varepsilon \text{ for } i = 1, \dots, n\}$. Since all $f_i^{-1}(0)$ are 1-codimensional (Exercise 2.5), $\bigcap f_i^{-1}(0)$ is finite-codimensional. As X is infinite-dimensional, there is $x \in S_X \cap V$.

3.47 If $p > 1$ show that a sequence $\{f_n\}$ in $L_p[0, 1]$ converges to zero weakly if and only if it is bounded in $L_p[0, 1]$ and $\int_0^x f_n \to_n 0$ for every $x \in [0, 1]$.

Hint. Linear denseness of step functions.

3.48 In c_0, let $x_n = ne_n$, where e_n is the standard nth unit vector, $n \in \mathbb{N}$. Show that $x_n \to 0$ pointwise, but not weakly by finding a concrete $f \in \ell_1$ so that $\{f(x_n)\}$ is unbounded.
Hint. Put $f = \sum_n \frac{n}{2^n} e_{2^n}$.

3.49 Let $\{e_i\}$ be an orthonormal basis of a Hilbert space H. Show that $e_i \xrightarrow{w} 0$.
Hint. $(e_i, e_j) \to 0$ as $i \to \infty$ for each j, and $\overline{\text{span}}\{e_j\} = H = H^*$.

3.50 Show that in the space ℓ_1, 0 is not in $\overline{\text{conv}}\{e_i\}$.
Hint. Consider $(1, 1, 1, \dots) \in \ell_\infty$.

3.51 Let $X = \ell_1 \cong c_0^*$. Show that the standard unit vectors e_i satisfy $e_i \xrightarrow{w^*} 0$ but not $e_i \xrightarrow{w} 0$.
Hint. Show that $e_i(x) \to 0$ for $x \in c_0$.
For the second part, use the previous exercise.

3.52 Show that in the space ℓ_∞, the set $\{e_i\} \cup \{0\}$ is weakly compact but not norm compact.
Hint. The set is weakly compact in c_0 as $e_i \xrightarrow{w} 0$ in c_0.

3.53 Show that the set $\{e_i\}$ of the standard unit vectors in ℓ_2 is norm closed but not w-closed in ℓ_2.
Hint. If $x^n \in \{e_i\}$ and $x^n \to x$ in ℓ_2, then $\{x^n\}$ is eventually constant. Recall (Exercise 3.49) that the origin is in the w-closure of $\{e_i\}$.

3.54 Let $\{e_i\}_{i=1}^\infty$ be the standard unit vectors in ℓ_2. Show that $A = \{e_i\}_{i=1}^\infty \cup \{0\}$ is a weakly compact set in ℓ_2.
Hint. By Exercise 3.49, $e_i \xrightarrow{w} 0$.

3.55 Let $\{e_i\}$ be the sequence of the standard unit vectors in ℓ_2. Show that $0 \in \overline{\{\sqrt{n}e_n\}}^w$ and that no subsequence of $\{\sqrt{n}e_n\}$ converges weakly to 0. Note that this fact also shows that the weak topology of ℓ_2 is not metrizable.
Hint. Let V be the neighborhood of 0 given by vectors $x^1, x^2, \dots, x^n$ in ℓ_2 and $\varepsilon > 0$. Consider the element $z \in \ell_2$ defined by $z_i = \sum_{k=1}^n |x_i^k|$. Note that for an infinite number of indexes i we have $|z_i|^2 < \frac{\varepsilon}{i}$ since otherwise $z \notin \ell_2$. Therefore for an infinite number of indexes i we have $|\sqrt{i}e_i(x^k)| < \varepsilon$ for $k = 1, \dots, n$, in particular $V \cap \{\sqrt{i}e_i\} \neq \emptyset$. The second part follows from Exercise 3.31.

3.56 Let $f_n \in C[0, 1]$, $\|f_n\| \leq 1$. Show that $f_n \xrightarrow{w} 0$ in $C[0, 1]$ if and only if $f_n(x) \to 0$ for every $x \in [0, 1]$.
This shows that for bounded sequences in $C[0, 1]$, weak convergence is equivalent to pointwise convergence. An analogous statement is true for c_0.
Hint. Use Corollary 3.137 and Lemma 3.116.

3.57 Let $x, x_n \in \ell_2$ be such that $x_n \xrightarrow{w} x$ in ℓ_2. Show that there is a subsequence $\{x_{n_k}\}$ such that the Cesàro means

$$\frac{x_{n_1} + x_{n_2} + \cdots + x_{n_k}}{k}$$

converge to x in ℓ_2 (the Banach–Saks theorem).

Show that an analogous statement is true for c_0.

Spaces that have this property are called spaces with the *weak Banach–Saks property*. On the other hand, a Banach space X is said to have the *Banach–Saks property* if the former statement holds for every *bounded* sequence in X. Since every bounded sequence in ℓ_2 has a w-convergent subsequence, we are proving in fact that ℓ_2 has the Banach–Saks property.

Hint. ℓ_2 case: Assume $x = 0$. Let $\|x_n\| \leq M$ for all n. Put $n_1 = 1$ and if $n_1, \ldots, n_k$ were chosen, pick n_{k+1} such that $|(x_{n_j}, x_{n_{k+1}})| \leq \frac{1}{k}$ for every $j \leq k$. Then

$$\frac{\|x_{n_1} + x_{n_2} + \cdots + x_{n_{k+1}}\|^2}{(k+1)^2} \leq \frac{(k+1)M^2 + 2 \cdot 1 + 2 \cdot \frac{2}{2} + \cdots + \frac{2k}{k}}{(k+1)^2} \to 0.$$

c_0 case: The sliding hump argument.

3.58 Define a functional f on ℓ_1 by $f(x) = \sum x_i$ for $x = (x_i)$. Show that f is norm continuous on ℓ_1 and that f is not w^*-continuous on $\ell_1 \cong c_0^*$ (that is, $f \in \ell_\infty \backslash c_0$).

Hint. Consider the standard basis $\{e_n\}$ of ℓ_1. Note that $e_n \xrightarrow{w^*} 0$ in ℓ_1.

3.59 Let X and Y be Banach spaces. Show that if an operator T from X into Y is w–w-continuous then $T \in \mathcal{B}(X, Y)$. On the other hand, every bounded operator is w–w-continuous.

Hint. Given $f \in Y^*$, consider $V = f^{-1}(-1, 1)$. Then V is a w-neighborhood of zero in Y. By our assumption, $T^{-1}(V)$ is a w-neighborhood of zero in X. Therefore $\lambda B_X \subset T^{-1}(V)$ for some λ, that is, $|f(T(B_X))| \leq \frac{1}{\lambda}$. Thus $T(B_X)$ is w-bounded and also bounded in Y. On the other hand, if $x_\alpha \xrightarrow{w} 0$ in X and $f \in Y^*$, then $f \circ T \in X^*$ and thus $T(x_\alpha) \xrightarrow{w} 0$ in Y.

3.60 Let T be a bounded operator from Y^* into X^*. Show that T is a dual operator if and only if T is w^*-w^*-continuous.

Hint. Assume T is w^*-w^*-continuous. Given $x \in X$, the function $\pi(x) \circ T$ is w^*-continuous on Y^*, so $\pi(x) \circ T = x(T) \in Y$. The assignment $x \mapsto \pi(x) \circ T$ defines the predual operator. The other implication is straightforward.

3.61 Let X, Y be Banach spaces and $T \in \mathcal{B}(X, Y)$. Show that

(i) T is an isomorphism into if and only if T^{**} is an isomorphism into.

(ii) T is an isometry into if and only if T^{**} is an isometry into.

Hint. (i) If T is an isomorphism into, it follows directly from Exercise 2.49 using (i) and then (ii) applied to T^*. If T^{**} is an isomorphism into, we have $\|T^{**}(x^{**})\|_{Y^{**}} \geq \delta\|x^{**}\|_{X^{**}}$ for some δ, hence the same holds for T, then use Exercise 1.73.

(ii) Going through the proof of Exercise 2.49 we see that $T^*(B_{Y^*}) = B_{X^*}$, which implies $\|T^{**}(F)\|_{Y^{**}} = \|F\|_{X^{**}}$.

3.62 Let X, Y be Banach spaces, $T \in \mathcal{B}(X, Y)$. Show that if T is an isomorphism of X onto Y, then $T^{**}(X) = Y$ and $(T^{**})^{-1}(Y) = X$.
Hint. T is onto Y, hence $T^{**}(X) = T(X) = Y$. Also T^{**} is an isomorphism, in particular T^{**} is one-to-one, so $(T^{**})^{-1}(Y) \subset X$.

3.63 Let X, Y be Banach spaces and $T \in \mathcal{B}(X, Y)$. If $T: X \to Y$ is one-to one, is T^{**} necessarily one-to-one?
Hint. No. Consider the identity mapping from ℓ_1 into ℓ_2. Its dual operator cannot be onto a dense set in ℓ_∞ as ℓ_∞ is not separable. Since T^* is not onto a dense set, T^{**} is not one-to-one by Exercise 2.46.

3.64 Let X be a separable Banach space. Find $T \in \mathcal{B}(\ell_2, X)$ such that $T(\ell_2)$ is dense in X. Note that then $T^*: X^* \to \ell_2$ is a bounded w^*-w-continuous one-to-one operator.
Hint. Let $\{y_i\}$ be dense in B_X and T be defined for $x = (x_i) \in \ell_2$ by $T(x) = \sum 2^{-i} x_i y_i$. Show that $y_i \in T(\ell_2)$ for every i.

3.65 Assume that X and Y are Banach spaces. Assume that T is a bounded operator from X^* into Y^* that is w^*-w^*-continuous when restricted to B_{X^*}. Show that then T is w^*-w^*-continuous, thus T is a dual operator to a bounded operator T_0 from Y into X.
Hint. We need to show that if $y \in Y$, then $y(T)$ is a w^*-continuous linear functional on X^*, i.e., that its kernel, i.e., $y(T)^{-1}(0)$ is w^*-closed (Corollary 3.94). This fact follows from Banach–Dieudonné Theorem 3.92. The operator T_0 is defined as follows: For every $f \in X^*$ and for every $y \in Y$, $f(T_0(y)) = f(y(T)$.

3.66 Assume that X and Y are Banach spaces, that M be a linearly dense set in Y, i.e., that its norm-closed linear hull is Y. Assume that T is a bounded operator T from X^* into Y^* such that $y(T)$ is a w^*-continuous linear functional on X^* for every $y \in M$. Prove that then T is a w^*-w^*-continuous operator from X^* into Y^*.
Hint. Observe that $y(T) = T^*(y)$ for all $y \in M$. Then $T^*(Y) = T^*\big(\overline{\text{span}}(M)\big) \subset \overline{T^*\big(\text{span}(M)\big)} \subset \overline{X} = X$, and the conclusion follows.

3.67 Let X be a separable Banach space, $\{x_i\}$ be a sequence in B_X and T be a bounded operator from X^* into ℓ_2 defined by $(Tx^*)_i = \frac{1}{2^i} x^*(x_i)$. Show that T is a w^*-w^*-continuous compact operator.
Hint. Use Exercise 3.66.

3.68 (Hahn, see [Megg, p. 49]) Prove the Banach–Steinhaus Theorem 3.85 by providing the details of the following "gliding hump" argument (the notation refers to the aforesaid theorem): Assume that $\mathcal{A}$ is unbounded. Then there exists a sequence $\{T_n\}$ in $\mathcal{A}$ and a sequence $\{x_n\}$ in X such that

$$\begin{cases} \|T_1x_1\| \geq 1, \ \|T_nx_n\| \geq n + \sum_{j=1}^{n-1} \|T_nx_j\|, \text{ for } n = 2, 3, \ldots, \\ \|x_1\| \leq 2^{-1}, \ \|x_n\| \leq 2^{-n} \min_{j<n} \|T_j\|^{-1}, \quad \text{for } n = 2, 3, \ldots \end{cases} \tag{3.9}$$

It follows that (a) the series $\sum_{n=1}^{\infty} x_n$ converges (to some element $x \in X$), that (b) $\sum_{j=n+1}^{\infty} \|T_nx_j\| \leq 1$ for all $n \in \mathbb{N}$, and that (c) $\|T_nx\| \geq n - 1$ for all $n \in \mathbb{N}$ (so conclude that $\sup_{T \in \mathcal{A}} \|Tx\| = \infty$).
Hint. Assume that $\sup_{T\in\mathcal{A}} \|T\| = \infty$ and that, at the same time, we have $\sup_{T\in\mathcal{A}} \|Tz\| < \infty$ for every $z \in X$. Choose first $T_1 \in \mathcal{A}$ such that $\|T_1\| > 2$. Let $s_1 \in S_X$ so that $\|T_1s_1\| > 2$ and put $x_1 = s_1/2$. Since $M_1 := \sup_{T\in\mathcal{A}} \|Tx_1\| < \infty$, we can choose $T_2 \in \mathcal{A}$ so that $\|T_2\| > (2 + M_1)2^2\|T_1\|$. Find $s_2 \in S_X$ such that $\|T_2s_2\| > (2 + M_1)2^2\|T_1\|$ and put $x_2 = 2^{-2}\|T_1\|^{-1}s_2$. Since $M_2 := \sup_{T\in\mathcal{A}} \|Tx_2\| < \infty$, we can choose $T_3 \in \mathcal{A}$ so that $\|T_3\| > (3 + M_1 + M_2)2^3 \max\{\|T_1\|, \|T_2\|\}$. Find $s_3 \in S_X$ such that $\|T_3s_3\| > (3 + M_1 + M_2)2^3 \max\{\|T_1\|, \|T_2\|\}$ and put $x_3 = 2^{-3} \min\{\|T_1\|^{-1}, \|T_2\|^{-1}\}s_2$. Continue in this way to define the two sequences $\{T_n\}$ and $\{x_n\}$ verifying (3.9). Now (a) should be clear. To prove (b) observe that $\|T_nx_j\| \leq \|T_n\|.\|x_j\| \leq \|T_n\|2^{-j}\|T_n\|^{-1} = 2^{-j}$ for $j > n$ and $n \in \mathbb{N}$. (c) follows from $\|T_nx\| = \|T_n \sum_{j=1}^{\infty} x_j\| \geq \|T_nx_n\| - \sum_{j=1}^{n-1} \|T_nx_j\| - \sum_{j=n+1}^{\infty} \|T_nx_j\|$ and the estimates above.

3.69 Show that there exists a sequence $\{f_n\}_{n=1}^{\infty}$ in the dual of some normed linear space X such that $\{f_n(x)\}_{n=1}^{\infty}$ is bounded for every $x \in X$ and yet $\{\|f_n\|\}$ is unbounded. This shows that the assumption of completeness of X cannot be dropped in the Banach–Steinhaus theorem.
Hint. In c_{00}^* put $f_n = ne_n$, where e_n is the standard unit vector. Then check that for every $x \in c_{00}$ we have $f_n(x) = 0$ for n large enough.

3.70 Show that every weakly compact set C in a Banach space is a bounded set.
Hint. If $f \in X^*$, then $f(C)$ is compact in $\mathbb{K}$ since every $f \in X^*$ is w-continuous. Then use the Banach–Steinhaus theorem.

3.71 (Banach discs, I) Let $(E, \mathcal{T})$ be a locally convex space. A bounded convex and balanced subset of E is called a *disc*. The Minkowski functional μ_D of a disc $D \subset E$ in the linear space span(D) is a norm. A disc D in E such that (span$(D), \mu_D)$ is a Banach space is called a *Banach disc*. This concept was introduced in the context of Banach spaces in Exercise 2.22. Prove that a Banach disc D is uniformly bounded on every w^*-bounded subset of E^*.
Hint. Given $x^* \in E^*$, the set $\{x^*(d) : d \in D\}$ is bounded, so $x^*|_{\text{span}(D)}$ belongs to (span$(D), \mu_D)^*$. Let B be a w^*-bounded subset of E^*. The family

$\{x^*|_{\mathrm{span}(M)} : x^* \in B\}$ is pointwise bounded, and so μ_D-bounded by the Banach–Steinhaus theorem. This gives the conclusion.

3.72 (Banach discs, II, see, e.g., [Koth, §20.11(2)]) Let D be a closed disc in a locally convex space $(E, \mathcal{T})$. Prove that if D is sequentially complete (i.e., every Cauchy sequence in D converges to some point in D), then D is a Banach disc; moreover, D is the closed unit ball of $(\mathrm{span}(D), \mu_D)$.
Hint. Obviously, on $\mathrm{span}(D)$ the topology induced by $\mathcal{T}$ is coarser than the norm topology. It follows that if $\{x_n\}$ is a μ_D-Cauchy sequence in D, then it is $\mathcal{T}$-Cauchy, hence $\mathcal{T}$-convergent (to some element $d \in D$). Since D is $\mathcal{T}$-closed, it follows that $\{x_n\}$ is also μ_D-convergent (to d). This gives the conclusion.

3.73 (Banach discs, III) Let $(E, \mathcal{T})$ be a locally convex space. Prove that every w-bounded subset of E is $\mathcal{T}$-bounded.
Hint. Let A be a w-bounded subset of E. Let $U \subset E$ be a neighborhood of 0. Then U° is a w^*-compact convex balanced subset of E^*. Exercise 3.72 gives that U° is a Banach disc. Now Exercise 3.71 ensures that U° is uniformly bounded on A. This gives the conclusion.

3.74 (Banach discs, IV, see, e.g., [Flor, p. 17]) Let $(E, \mathcal{T})$ be a locally convex space. Prove that every convex and relatively countably compact subset A of E is contained in a Banach disc.
Hint. A is clearly $\mathcal{T}$-bounded. If $\widetilde{E}$ denotes the completion of E, the mapping $T : \ell_1(A) \to \widetilde{E}$ given by $T(\lambda) = \sum_{a\in A} \lambda(a)a$—observe that this sum has only a countable number of non-zero summands—is well defined, continuous, and therefore $T(B_{\ell_1(A)})$ is a Banach disc in $\widetilde{E}$ that contains A. Actually, $T(B_{\ell_1(A)}) \subset E$. In order to show this, take first $\lambda_n \geq 0$ for all $n \in \mathbb{N}$ such that $s_N := \sum_{n=1}^N \lambda_n \to s$, and $a_n \in A$ for all $n \in \mathbb{N}$. Then

$$\sum_{n=1}^{\infty} \lambda_n a_n = \lim_N s_N \sum_{n=1}^{N} \frac{\lambda_n}{s_N} a_n = s . \lim_N \sum_{n=1}^{N} \frac{\lambda_n}{s_N} a_n \in E,$$

since A is convex and relatively countably compact. Now, for arbitrary signs of the coefficients, split the sum collecting the positive or negative ones. This proves that $T(B_{\ell_1(A)}) \subset E$.

3.75 Let $\{f_n\} \subset X^*$. Show that if there exists $\varepsilon_n \to 0$ such that for every $x \in X$ there is $K_x > 0$ with $|f_n(x)| \leq K_x \varepsilon_n$, then $\|f_n\| \to 0$.
Hint. The Baire category.

3.76 Let X be a Banach space. Assume that $x_n \to x$ in X and $f_n \overset{w^*}{\to} f$ in X^*. Show that $f_n(x_n) \to f(x)$.

Hint. Write $f_n(x_n) - f(x) = f_n(x_n) - f_n(x) + f_n(x) - f(x)$ and note that $\{f_n\}$ is bounded by the Banach–Steinhaus theorem.

3.77 Assume that $\|\cdot\|_1$ and $\|\cdot\|_2$ are norms on a normed space X that are not equivalent. Show that then there is a linear functional on X that is continuous in one of the norms and is not continuous in the other.
Hint. One of the unit balls in these norms is not a bounded set in the normed space whose unit ball is the other unit ball. This unbounded ball is not weakly bounded in the topology given by the other norm. Thus there is a functional continuous in the other norm and not in this one.

3.78 As an application of the Banach–Steinhaus uniform boundedness principle show that there is a continuous 2π-periodic function whose Fourier series diverges at 0.
Hint. Let C be the Banach space of 2π-periodic continuous functions on $\mathbb{R}$ with the supremum norm. For $f \in C$ put $\sigma_n(f) = \sum_{k=-n}^{n} c_k$, where c_k is the kth Fourier coefficient of f ($\sigma_n(f)$ is the value at 0 of the nth symmetric partial sum of the Fourier series of f). As $c_k = \frac{1}{2\pi}\int_{-\pi}^{\pi} f(t)e^{-ikt}\,dt$, we have $\sigma_n(f) = \frac{1}{2\pi}\int_{-\pi}^{\pi} f(t)D_n(t)\,dt$, where $D_n(t) = \sum_{k=-n}^{n} e^{-ikt} = \dfrac{\sin\big((n+\frac{1}{2})t\big)}{\sin(\frac{1}{2}t)}$. Thus σ_n is a continuous linear functional on C and $\|\sigma_n\| = \frac{1}{2\pi}\int_{-\pi}^{\pi} |D_n(t)|\,dt$. We estimate

$$\|\sigma_n\| = \frac{1}{\pi}\int_0^{\pi}\left|\frac{\sin\big((n+\frac{1}{2})t\big)}{\sin(\frac{1}{2}t)}\right| dt = \frac{1}{\pi}\int_0^{\pi}\left|\frac{\sin\big((n+\frac{1}{2})t\big)}{t}\right|\cdot\left|\frac{t}{\sin(\frac{1}{2}t)}\right| dt$$
$$\geq c\int_0^{\pi}\left|\frac{\sin\big((n+\frac{1}{2})t\big)}{t}\right| dt = c\int_0^{(n+\frac{1}{2})}\left|\frac{\sin(u)}{u}\right| du$$

and this expression tends to infinity as $n \to \infty$ since $\int_0^{\infty} |\frac{\sin u}{u}|\,du$ diverges. Thus $\sup_n \|\sigma_n\| = \infty$. By the Banach–Steinhaus theorem, there is $f \in C$ such that $\{\sigma_n(f)\}_{n=1}^{\infty}$ is unbounded, so the Fourier series for f diverges at 0.

3.79 Find a Banach space X and $f_n \in X^*$ such that $f_n \overset{w^*}{\to} 0$ and yet for every convex combination h of $\{f_n\}$ we have $\|h\| = 1$.

This shows that Mazur's theorem does not hold for the w^*-topology.

Hint. Put $X = c_0$, $f_n = (0, \dots, 0, \overset{n}{1}, 0, \dots) \in \ell_1$.

3.80 Let C be a bounded set in a Banach space X. Show that C is relatively w-compact if and only if $\overline{C}^{w^*} \subset X$.
Hint. $\overline{C}^{w^*}$ is w^*-compact (Alaoglu).

3.81 Let X be a Banach space. Assume that C is a w^*-compact set in X^*. Is $\overline{\operatorname{conv}}^{w^*}(C)$ also w^*-compact?
Hint. Yes, as $\overline{\operatorname{conv}}^{w^*}(C)$ is a w^*-closed subset of some dual ball, which is w^*-compact by Alaoglu's theorem.

3.82 Can we replace B_X and $B_{X^{**}}$ in the Goldstine's theorem by S_X and $S_{X^{**}}$?
Hint. Yes. Use the w^*-lower semicontinuity of the second dual norm (see Exercise 3.31).

3.83 Show that if X is an infinite-dimensional Banach space, then there is a set S in X^* such that the set of all w^*-sequential limits of S is not norm closed. Compare this with the situation with Baire 1 functions and find the reason for the difference.
Hint. Let $f_n \in X^*$ with $\|f_n\| = 1/n$ for each n. Put $f_n = w^*\text{-}\lim\limits_k f_{n,k}$ so that $\|f_{n,k}\| = n$ for all n and k. By the Banach–Steinhaus principle, no sequence in $\{f_{n,k} : n, k \in \mathbb{N}\}$ can w^*-converge to 0. Concerning Baire 1 functions: a lattice structure.

3.84 Show the following Helly's theorem: Let X be a Banach space. If $x^{**} \in S_{X^{**}}$, F is a finite set in X^* and $\varepsilon > 0$, then there is $x \in X$ with $\|x\| = (1+\varepsilon)$ such that $f(x) = x^{**}(f)$ for every $f \in F$. Note the difference between Helly's and Goldstine's theorems.
Hint. Put $Z = F_\perp$. Let $q : X \to X/Z$ be the canonical quotient mapping. Consider $q^{**}x^{**} \in (X/Z)^{**}$ ($= (X/Z)$, since X/Z is finite-dimensional). Then find in the coset $q^{**}x^{**}$ an element $x \in X$ of norm $1+\varepsilon$.

Note that ε cannot in general be taken 0. Let $f \in S_{X^*}$ be such that it does not attain its norm and take $x^{**} \in S_{X^{**}}$ such that $x^{**}(f) = 1$. Then there does not exist any $x \in B_X$ such that $f(x) = x^{**}(f) = 1$.

3.85 Let X be a Banach space. In Exercise 2.28 we proved that for a subset Y of X we have $(Y^\perp)_\perp = \overline{\operatorname{span}}(Y)$. Since $\operatorname{span}(Y)$ is convex, by Mazur's theorem it also reads $(Y^\perp)_\perp = \overline{\operatorname{span}}^{w}(Y)$.

Prove that for a subset Y of X^* we have $(Y_\perp)^\perp = \overline{\operatorname{span}}^{w^*}(Y)$.

For $T \in \mathcal{B}(X, Y)$ we then have $\overline{T(X)}^{w} = \operatorname{Ker}(T^*)_\perp$ and $\overline{T^*(Y^*)}^{w^*} = \operatorname{Ker}(T)^\perp$.

Hint. Show that $(Y_\perp)^\perp$ is w^*-closed, so $\overline{\operatorname{span}}^{w^*}(Y) \subset (Y_\perp)^\perp$. Assuming that the inclusion is strict, derive a contradiction using Corollary 3.34.

3.86 Let X, Y be Banach spaces and $T \in \mathcal{B}(X, Y)$. Consider $\widehat{T}$ as an operator from $X/\operatorname{Ker}(T)$ into $T(X)$. Show that then $\widehat{T}^{**}$ can be considered an operator from $X^{**}/\operatorname{Ker}(T^{**})$ into $T(X)^{**}$; in particular, $\widehat{T}^{**}(\hat{x}) = T(x)$ for all $x \in X$ (see Exercise 2.45).

Show also that under the isometry of $(X/\operatorname{Ker}(T))^{**}$ onto $X^{**}/\operatorname{Ker}(T^{**})$, the subspace $X/\operatorname{Ker}(T)$ of $(X/\operatorname{Ker}(T))^{**}$ corresponds to the subspace $\{x + \operatorname{Ker}(T^{**}) : x \in X\}$ of $X^{**}/\operatorname{Ker}(T^{**})$.

Hint. We have $\widehat{T}^*: T(X)^* \to (X/\operatorname{Ker}(T))^*$, so $\widehat{T}^{**}: (X/\operatorname{Ker}(T))^{**} \to T(X)^{**}$. But

$$(X/\operatorname{Ker}(T))^{**} = \left(\operatorname{Ker}(T)^{\perp}\right)^* = \left(\overline{T^*(X^*)}^{w^*}\right)^* = X^{**}/\overline{T^*(X^*)}^{w^*\perp}.$$

But in general, $\overline{Z}^{w^*\perp} = Z^{\perp}$ for any subspace Z of X^*, so $(X/\operatorname{Ker}(T))^{**} = X^{**}/T^*(X^*)^{\perp} = X^{**}/\operatorname{Ker}(T^{**})$.

3.87 Let X, Y be Banach spaces, $T \in \mathcal{B}(X, Y)$. Assume that $T(X)$ is closed, let $x^{**} \in X^{**}$. If $T^{**}(x^{**}) \in T(X)$, then for every $\delta > 0$ there exists $x \in X$ such that $T(x) = T^{**}(x^{**})$ and $\|x\| \leq (1+\delta)\|x^{**}\|$.
Hint. Assume that $T(X) = Y$, then $\widehat{T}: X/\operatorname{Ker}(T) \to Y$ is an isomorphism and thus $\widehat{T}^{**}: X^{**}/\operatorname{Ker}(T^{**}) \to Y^{**}$ is an isomorphism (Exercises 3.86 and 2.51). Since $\widehat{T}^{**}(\widehat{x^{**}}) \in Y$, we have $\widehat{x^{**}} = x + \operatorname{Ker}(T^{**})$ for some $x \in X$. Thus $\widehat{x^{**}}$ corresponds to some $\hat{x} \in X/\operatorname{Ker}(T)$ (Exercise 3.86), hence there is $x \in \hat{x}$ such that $\frac{1}{1+\delta}\|x\| \leq \|\hat{x}\| = \|\widehat{x^{**}}\| \leq \|x^{**}\|$.

Let X be a Banach space and $A \subset X^*$. Let λ be a positive number. We say that A is *λ-norming* if $\sup\{f(x) : f \in A \cap B_{X^*}\} \geq \frac{1}{\lambda}\|x\|$ for every $x \in X$ (if this is the case, certainly $\lambda \geq 1$). We say that A is *norming* if it is λ-norming for some $\lambda \geq 1$.

By the Hahn–Banach theorem, both X^* and B_{X^*} are 1-norming. Also, X is 1-norming for X^*. Observe that a norming subset of X^* separates points of X (see Exercise 3.92).

3.88 Let X be a Banach space and let $F \in X^{**}\setminus X$. Show that $F^{-1}(0)$ is a norming subspace of X^*. If $x \in X\setminus\{0\} \subset X^{**}$, can $x^{-1}(0) \subset X^*$ be norming?
Hint. Let the distance of F to X be $2\delta > 0$. Assume that for some $x \in S_X$ we have $\sup\{f(x) : f \in F^{-1}(0), \|f\| \leq 1\} \leq \delta/2$. Then by Exercise 2.13 we have $\|x \pm F\| < \delta$, a contradiction with the distance of F to X.

If $x \in X\setminus\{0\}$, $x^{-1}(0)$ cannot be norming as it is w^*-closed. If it were norming, it would be separating, hence w^*-dense in X^* and thus $x^{-1}(0) = X^*$, a contradiction.

3.89 Let $(X, \|\cdot\|)$ be a Banach space, let Y be a closed subspace of X^*. For $x \in X$ define $\|x\|^Y = \sup\{|y^*(x)| : y^* \in B_Y\}$. Clearly $\|\cdot\|^Y$ is a seminorm on X and $\|\cdot\|^Y \leq \|\cdot\|$. Show that:

(i) $\|x\|^Y = \operatorname{dist}(x, Y^{\perp})$ (distance in X^{**}).

(ii) Y is separating if and only if $\|\cdot\|^Y$ is a norm on X if and only if $X \cap Y^{\perp} = \{0\}$ in X^{**}.

(iii) Y is norming if and only if $\|\cdot\|^Y$ is an equivalent norm on X if and only if $\operatorname{dist}(S_X, Y^{\perp}) > 0$ (distance in X^{**}).

(iv) Y is 1-norming if and only if $\|\cdot\|^Y = \|\cdot\|_X$ if and only if $\operatorname{dist}(S_X, Y^{\perp}) = 1$ (distance in X^{**}).
Hint. (i) $\|x\|^Y = \|x\|_{Y^*} = \|x\|_{X^{**}/Y^{\perp}} = \operatorname{dist}(x, Y^{\perp})$. (ii) through (iv) are routine.

3.90 Let X be a Banach space, let Y be a separating closed subspace of X^*. The topology $w(X, Y)$ on X is locally convex and Hausdorff. Denote $B^Y = \{x \in X : \|x\|^Y \leq 1\}$ (see Exercise 3.89). Clearly $B_X \subset B^Y$. Show that:

(i) $B^Y = \overline{B_X}^{w(X,Y)}$.

(ii) Y is norming if and only if $\overline{B_X}^{w(X,Y)}$ is bounded.

(iii) Y is 1-norming if and only if B_X is $w(X, Y)$-closed if and only if the norm of X is $w(X, Y)$-lower semicontinuous.

Hint. (i) Consider $(X, w(X, Y))$, then $(X, w(X, Y))^* = Y$. We have $B_Y = (B_X)^\circ$ (polar in Y) and $B^Y = ((B_X)^\circ)_\circ = \overline{B_X}^{w(X,Y)}$. (ii) and the first two equivalences in (iii) then follow using Exercise 3.89. The latter two conditions are equivalent since $\alpha B_X = \|\cdot\|^{-1}((-\infty, \alpha])$ for $\alpha \geq 0$.

3.91 Assume that B_X is $w(X, Y)$-closed for every norming subspace Y of X^*. Show that X is then reflexive.

Hint. By contradiction. Let $G \in S_{X^{**}} \setminus X$ be such that $\text{dist}(G, X) < \frac{1}{4}$. Put $Y = G^{-1}(0)$. Then Y is norming (Exercise 3.88), hence B_X is $w(X, Y)$-closed and by the previous exercise, Y is 1-norming. This is a contradiction: Since $\text{dist}(G, X) < \frac{1}{4}$, we have $f(x) - G(f) < \frac{1}{4}$ for all $f \in B_{X^*}$. Thus $f(x) < \frac{1}{4}$ for all $f \in B_Y$ and Y cannot be 1-norming.

3.92 Show that every norming subspace of X^* is a separating set for X.

Find an example of a separating subspace that is not norming.

Hint. For the second statement, take a partition $I, I_1, I_2, \ldots$ of $\mathbb{N}$ into disjoint infinite sets. Let $S = \left\{y \in \ell_1 : \text{ for each } k \in I,\ y_k = \frac{1}{k} \sum_{n \in I_k} y_n\right\}$. Then S is separating for c_0 but not norming (consider e_k for $k \in I$).

Note that if X is separable and X^{**}/X is infinite-dimensional, then there is a separating subspace of X^* that is not norming (see [DaLi]).

3.93 Let X be a Banach space and Y be a Banach space such that $X^* \subset Y$. Let $F \subset Y^*$ be the closed linear subset formed by all functionals whose restriction to X^* are w^*-continuous. Show that F is a norming subspace in Y^*.

Hint. Observe first that F is closed due to the Banach–Dieudonné theorem (more precisely, to Corollary 3.94). Consider $y \in S_Y$. If $\text{dist}(y, X^*) < \frac{1}{4}$, choose $y_1 \in X^*$ such that $\|y - y_1\| < \frac{1}{4}$ and then choose $f \in X$ such that $\|f\| = 1$ and $f(y_1) > \|y\| - \frac{1}{4}$. Extend f to Y with norm 1 and call this extension again f. Then $f(y) \geq f(y_1) - |f(y - y_1)| \geq \|y\| - \frac{1}{4} - \|y - y_1\| \geq \frac{1}{2} \geq \frac{1}{4}$. If, on the other side, $\text{dist}(y, X^*) \geq \frac{1}{4}$, then choose $f \in Y^*$ such that $f = 0$ on X^*, $\|f\| = \text{dist}(y, X^*)^{-1}$, and $f(y) = 1$. Then $\text{dist}(y, X^*) f$ is a norm-1 functional on Y which belong to F, and $\text{dist}(y, X^*) f(y) \geq \frac{1}{4}$.

3.94 Let X be a Banach space and $A \subset X^*$. Show that A separates the points of X if and only if $A_\perp = \{0\}$ if and only if $\overline{\text{span}}^{w^*}(A) = X^*$.

Hint. Easy from definition and Exercises 3.85 and 2.27.

3.95 Let X be a separable Banach space. Show that there is a sequence $\{f_n\} \subset S_{X^*}$ such that $\{f_n\}$ is separating in X.
Hint. Let $\{x_n\}$ be dense in S_X. For each n choose $f_n \in S_{X^*}$ such that $f_n(x_n) = 1$. Given $x \in S_X$, take n such that $\|x - x_n\| < \frac{1}{2}$, then $f_n(x) \geq \frac{1}{2}$.

3.96 Let X be a reflexive space and let $Y \subset X^*$ be a closed subspace of X^*. Show that if Y separates points of X then $Y = X^*$.
Hint. We have $\overline{Y}^{w^*} = X^*$. By reflexivity $\overline{Y}^{w} = X^*$, then use Theorem 3.45.

3.97 Let T be an operator from a Banach space X into a Banach space Y. Suppose that F is a subset of Y^* that separates points of Y. Assume that $f(T(x_n)) \to 0$ whenever $f \in F$ and $\{x_n\} \subset X$ is such that $\|x_n\| \to 0$. Show that T is a bounded operator.
Hint. The closed graph theorem.

3.98 Let C be a weakly separable subset of a Banach space. Show that C is norm separable.
Hint. The norm closed linear hull is weakly and thus norm-separable, thus C is norm separable.

3.99 Prove directly that if X is a separable Banach space, then X^* is w^*-separable.
Hint. Take the sequence $\{f_n\} \subset S_{X^*}$ from Exercise 3.95. Then $\operatorname{span}\{f_n\}$ is w^*-dense. Therefore the rational combinations form a w^*-dense set in X^*.

3.100 Let X be a Banach space. Show that the following are equivalent:
(i) B_X is w-separable,
(ii) S_X is w-separable,
(iii) X is separable.
Hint. (i)$\Longrightarrow$(ii): If a countable $C \subset B_X$ is w-dense in B_X and $x \in S_X$, then there is a net $\{x_\alpha\} \subset C$ such that $x_\alpha \overset{w}{\to} x$. We can assume that $\|x_\alpha\| \to 1$ as the norm is weakly lower semicontinuous (see Exercise 3.31), hence $\frac{x_\alpha}{\|x_\alpha\|} \overset{w}{\to} x$. $\{\frac{x}{\|x\|} : x \in C\}$ is w-dense in S_X.

Every weak neighborhood of a point from B_X intersect S_X if X is infinite-dimensional. Thus (ii)$\Longrightarrow$(i). Clearly, (i) implies that X is weakly and thus norm-separable (Proposition 3.105). (iii) implies that the ball and the sphere are separable and hence w-separable.

3.101 Let X be a Banach space. Show that B_{X^*} is w^*-separable if and only if S_{X^*} is w^*-separable.

Note that the w^*-separability of B_{X^*} is not preserved when passing to an equivalent norm (Exercise 14.42). Also, there is a space X for which X^* is w^*-separable and no equivalent norm on X exists so that B_{X^*} is w^*-separable ([JoLi1]).
Hint. Similar to the previous exercise.

3.102 Show that ℓ_∞^* is w^*-separable.
Hint. The Goldstine's theorem and $\ell_1^{**} = \ell_\infty^*$.

3.103 Find a Banach space X such that X^* is w^*-separable and a subspace Y of X such that Y^* is a w^*-dense subspace of X^* that is not w^*-separable.
Hint. Take $X := \ell_1(\Gamma)$ with $\text{card}(\Gamma) = c$, and $Y := c_0(\Gamma)$. The w^*-separability of X^* follows from the fact that $\ell_1(\Gamma) \subset C[0, 1]^* \subset \ell_\infty$ and the Goldstine's theorem (see Exercise 3.102). The restriction of the w^*-topology of X^* to $c_0(\Gamma)$ is the w-topology on $c_0(\Gamma)$, and $c_0(\Gamma)$ is not norm-separable.

The same proof gives that $[0, 1]^{[0,1]}$ is pointwise separable.

3.104 Show that every weakly compact set in ℓ_∞ is norm-separable.
Hint. Let C be w-compact in ℓ_∞. Since ℓ_∞^* is w^*-separable, C is w-metrizable by Proposition 3.107, hence also w-separable. Thus $\text{span}(C)$ is weakly and thus norm-separable.

3.105 Use the following hint to show that the weak topology of an infinite-dimensional Banach space is never metrizable.
Hint. Assume that the weak topology of a Banach space X is metrizable by a metric ρ. We know that 0 is in the weak closure of nS_X for every n. Therefore we can find a point $x_n \in nS_X$ for which $\rho(x_n, 0) < \frac{1}{n}$. Then $\rho(x_n, 0) \to 0$, which means that $x_n \to 0$ in the weak topology. Thus by Banach–Steinhaus, $\{x_n\}$ must be bounded, which is not the case.

3.106 Let Y be a closed subspace of a Banach space X. Show that the weak topology of X/Y is the factor topology of the weak topology of X, that is, U is weakly open in X/Y if and only if $q^{-1}(U)$ is weakly open in X, where q is the canonical quotient mapping.

Show that if T is a bounded operator from a Banach space X onto a Banach space Y, then T is an open mapping in the respective weak topologies of X and Y.
Hint. $(X/Y)^* = Y^\perp$ and every bounded linear mapping is w–w-continuous.

3.107 Let X be an infinite-dimensional Banach space. Show that (X^*, w^*) is not complete in the uniformity of the topological vector space (see the paragraph preceding Proposition 3.27, and Section 17.10). Similar result holds for the w-topology. Recall that a sequence $\{x_i\}$ is w-Cauchy if $\{f(x_i)\}$ is Cauchy for every $f \in X^*$.
Hint. Consider a linear discontinuous functional f on X (it always exists, see Exercise 2.4). For every finite-dimensional subspace of X with a basis B and $\varepsilon > 0$ consider a continuous linear functional g on X such that f and g differ by at most ε on vectors in B. By ordering B by inclusion one can get a net of continuous linear functionals converging to f. This net is thus w^*-Cauchy and not w^*-convergent in X^*.

3.108 Let X be an infinite-dimensional Banach space. Show that S_X is a dense G_δ set in (B_X, w). Deduce from this that the norm of an infinite-dimensional space is never a weakly continuous function on X.
Hint. Given $\|x_0\| < 1$, let $U = \{x : |f_i(x - x_0)| < \varepsilon\}$. Take $x \neq 0$ such that $x \in \bigcap f_i^{-1}(0)$ and note that the continuous function $\|\cdot\|$ considered on the set $\{x_0 + tx : t \in \mathbb{R}\}$ attains the value 1 somewhere, showing that $U \cap S_X \neq \emptyset$. Therefore S_X is w-dense in B_X. Put $G_n = \{x \in B_X : \|x\| > 1 - \frac{1}{n}\}$. Then G_n are open in (B_X, w) and $S_X = \bigcap G_n$.

3.109 Construct an equivalent norm on $\ell_1 \cong c_0^*$ that is not w^*-lower semicontinuous.
Hint. Consider the norm $\|(x_i)\| = \sum |x_i| + 2|\sum x_i|$. Then take $a_i < 0$ such that $\sum a_i = -1$ and consider the vectors $x_n = (1, 0, \ldots, 0, a_1, a_2, \ldots)$, where a_1 is at the nth place. Show that $x_n \overset{w^*}{\to} (1, 0, \ldots)$ and compare the norms.

3.110 Let X be an infinite-dimensional Banach space. Can X be a Baire space in its weak topology? Recall that a topological space T is called a *Baire space* if the intersection of a countable family of open dense subsets of T is dense in T.
Hint. No. nB_X are nowhere dense in (X, w).

3.111 Show that 0 is not a G_δ-point of $B_{\ell_2(\Gamma)}$ in the weak topology if Γ is uncountable. Recall that $p \in W$ is a G_δ-point of W if $\{p\} = \bigcap O_n$, where O_n are open in W.
Hint. If $\{p\} = \bigcap O_n$ and O_n are basic neighborhoods in the weak topology, then each O_n restricts only a countable number of coordinates, so $\bigcap O_n$ restrict only a countable number of coordinates and thus cannot be a singleton if Γ is uncountable.

3.112 Let X be a reflexive Banach space. Show that if Y is isomorphic to X then Y is reflexive.
Hint. B_Y is w-compact as a w-closed (see Theorem 3.45) subset of $T(\|T^{-1}\|B_X)$, which is w-compact by the w-w-continuity of T.

3.113 Show that c_0 is not isomorphic to a subspace of ℓ_p, $p \in (1, \infty)$.
Hint. c_0 is not reflexive.

3.114 Let Y be a closed subspace of a reflexive Banach space X. Show that X/Y is reflexive.
Hint. $(X/Y)^* = Y^\perp \subset X^*$ and X^* is reflexive.

3.115 Let X be a Banach space. Show that if every separable subspace of X is reflexive, then X is reflexive.
Hint. The Eberlein–Šmulyan theorem for the weak compactness of B_X.

3.116 Let Y be a closed subspace of a Banach space X. Show that if Y and X/Y are reflexive, then X is reflexive. Thus reflexivity is a three-space property.
Hint. If Y is reflexive, then Y is w^*-closed in X^{**}. Thus $Y = (Y^{\perp})^{\perp}$. Then $X^{**}/Y^{\perp\perp} = (Y^{\perp})^*$ and $Y^* = (X/Y)^*$, so $(Y^{\perp})^* = (X/Y)^{**} = X/Y$. Consequently $X = X^{**}$.

Note that this is a delicate proof although it looks easy. We have to prove not $X = X^{**}$ but actually $\pi(X) = X^{**}$, that is, we have to check that the mappings in the above equalities work as supposed to.

3.117 Let X be a reflexive Banach space and let T be a bounded operator from X onto a Banach space Y. Show that Y is reflexive.
Hint. Y is isomorphic to $X/\operatorname{Ker}(T)$, which is a reflexive Banach space.

3.118 Show that if T is a bounded one-to-one operator from ℓ_1 into ℓ_2 then $T(\ell_1)$ is not closed in ℓ_2.
Hint. Otherwise T would be an isomorphism of ℓ_1 onto a reflexive space $T(\ell_1)$, hence ℓ_1 would be reflexive.

3.119 Prove that a Banach space is reflexive if and only if the weak and weak* topologies on B_{X^*} coincide. Compare with Proposition 3.113 and with Exercise 3.121.
Hint. One direction is obvious. For the other, use Corollary 3.94.

3.120 ([DaJo]) Show that a Banach space X is reflexive if and only if the closed unit ball of every equivalent norm on X^* is w^*-closed.
Hint. If the condition on w^*-closedness holds and $F \in X^{**}$ is given, consider the set $V_n = \{f \in B_{X^*} : |F(f)| \le \frac{1}{n}\}$. The Minkowski functional of V_n is an equivalent norm on X^* and V_n is its closed unit ball. By our assumption, V_n is w^*-closed. Therefore $F^{-1}(0) \cap B_{X^*} = \bigcap V_n$ is w^*-closed and thus $F \in X$, meaning that X is reflexive. If X is reflexive, then the closed unit ball of any equivalent norm is w-compact and thus w^*-compact.

3.121 Let X be a Banach space. Show that if the w-topology and the w^*-topology coincide on X^{**}, then X is reflexive. (See also Exercise 3.119.)
Hint. Apply Proposition 3.113 and then Proposition 3.112.

3.122 Recall that, given a Banach space X, a *biorthogonal system* $\{x_n; x_n^*\}_{n=1}^{\infty}$ in $X \times X^*$ consists of two sequences, $\{x_n\}_{n=1}^{\infty}$ in X and $\{x_n^*\}_{n=1}^{\infty}$ in X^* such that $\langle x_m^*, x_n\rangle = \delta_{n,m}$ for all $n, m \in \mathbb{N}$. It is called *total* if $\{x_n^* : n \in \mathbb{N}\}$ is w^*-linearly dense in X^*. Prove that if X is a Banach space X and $\{x_n; x_n^*\}_{n=1}^{\infty}$ is a biorthogonal system in $X \times X^*$ such that the set $\{x_n : n \in \mathbb{N}\}$ has the property that for every $x^* \in X^*$ and every $c > 0$, the set $\{n \in \mathbb{N} : |\langle x^*, x_n\rangle| > c\}$ is finite, then $\|\sum_{i=1}^{n} x_i^*\| \to_n \infty$.

Hint. If not, there exists an increasing sequence $\{n_p\}$ in $\mathbb{N}$ such that $\{\|\sum_{i=1}^{n_p} x_i^*\| : p \in \mathbb{N}\}$ is bounded in X^*. Let x^* be a w^*-cluster point of this sequence. Note that $\langle \sum_{i=1}^{n_p} x_i^*, x_j \rangle = 1$ for every $j \leq n_p$. Then $\langle x^*, x_j \rangle = 1$ for all $j \in \mathbb{N}$. This is a contradiction.

3.123 Using Exercise 3.122 and Theorem 3.132, prove the following result: (Pták [Ptak2]) *Let X be a Banach space. Then the following assertions are equivalent:* (i) *X is reflexive.* (ii) *For every biorthogonal system $\{x_n; x_n^*\}_{n=1}^\infty$ in $X \times X^*$ such that $\{x_n^* : n \in \mathbb{N}\}$ is bounded, the sequence $\{\sum_{k=1}^n x_k\}_{n=1}^\infty$ is unbounded.* (iii) *For every biorthogonal system $\{x_n; x_n^*\}_{n=1}^\infty$ in $X \times X^*$ such that $\{x_n : n \in \mathbb{N}\}$ is bounded, the sequence $\{\sum_{k=1}^n x_k^*\}_{n=1}^\infty$ is unbounded.*

Hint. [FGM] (i)$\Longrightarrow$(ii): Assume that the space X is reflexive, and let $\{x_n; x_n^*\}_{n\in\mathbb{N}}$ be a biorthogonal system in $X \times X^*$ such that $\{x_n^* : n \in \mathbb{N}\}$ is bounded. Let $Y = \overline{\operatorname{span}}\{x_n : n \in \mathbb{N}\}$; this is a reflexive space. Let $q : X^* \to X^*/Y^\perp$ the canonical quotient mapping. Since $(X^*/Y^\perp)^* = Y$, the system $\{q(x_n^*); x_n\}_{n\in\mathbb{N}}$ is total and biorthogonal in $(X^*/Y^\perp) \times Y$.

We claim that the set $\{q(x_n^*) : |\langle q(x_n^*), y\rangle| > c\}$ is finite for every $y \in Y$ and every $c > 0$. Otherwise, since $\{q(x_n^*) : n \in \mathbb{N}\}$ is a weakly relatively compact subset of $X^*/Y^\perp$, the set $\{q(x_n^*) : |\langle q(x_n^*), y\rangle| > c\}$ would have a w-cluster point $\hat{x}^*$ in X^*/Y; by the orthogonality and the fact that $\{x_n : n \in \mathbb{N}\}$ is linearly dense in Y, $\hat{x}^* = 0$, a contradiction with $|\langle \hat{x}^*, y\rangle| \geq c$. The claim holds, and it follows from Lemma 3.122 that $\left\|\sum_{k=1}^n x_k\right\| \to_n \infty$.

(i)$\Longrightarrow$(iii): If X is reflexive, so is X^*. Given a biorthogonal system $\{x_n; x_n^*\}_{n\in\mathbb{N}}$ in $X \times X^*$, it can also be seen as a biorthogonal system $\{x_n^*; x_n\}_{n\in\mathbb{N}}$ in $X^* \times X^{**}$. If the set $\{x_n : n \in \mathbb{N}\}$ is bounded, it follows from (i)$\Longrightarrow$(ii) that $\left\|\sum_{k=1}^n x_k^*\right\| \to_n \infty$.

(ii)$\Longrightarrow$(i): Assume that X is not reflexive. Theorem 3.132 says, in particular, that there exist two sequences (x_n) in S_X and (x_n^*) in S_{X^*} such that $\langle x_m^*, x_n\rangle = \frac{1}{2}$ if $n \geq m$, and $\langle x_m^*, x_n\rangle = 0$ if $n < m$. Let $d_1 = 2x_1$, $d_n = 2(x_n - x_{n-1})$, $n = 2, 3, \ldots$ Then, it is clear that the family $\left\{d_n : x_n^*\right\}_{n\in\mathbb{N}}$ is a biorthogonal system in $X \times X^*$. Moreover, $\{x_n^* : n \in \mathbb{N}\}$ is bounded. Observe, too, that $\sum_{k=1}^n d_k = 2x_n$ for all $n \in \mathbb{N}$. We obtain thus a contradiction with (ii).

(iii)$\Longrightarrow$(i): Starting from the assumption that X is not reflexive, we proceed as in the proof of (iii)$\Longrightarrow$(i). Once we have the two sequences (x_n) and (x_n^*), put $d_n^* = 2(x_n^* - x_{n+1}^*)$ for $n \in \mathbb{N}$. The system $\{x_n; d_n^*\}_{n\in\mathbb{N}}$ is again a biorthogonal system and the set $\{x_n : n \in \mathbb{N}\}$ is bounded. Now $\sum_{k=1}^n d_n^* = 2(x_1^* - x_{n+1}^*)$ for all $n \in \mathbb{N}$. We obtain again a contradiction, this time with (iii).

3.124 Let X, Y be Banach spaces, $T \in \mathcal{B}(X, Y)$. Show that if X is reflexive then $T(B_X)$ is closed in Y.

Hint. B_X is weakly compact and T is weakly continuous.

3.125 Assume that the unit sphere of a given Banach space X contains an infinite sequence of points such that the mutual distance of two different members of the sequence is 2. Does it imply that X is nonreflexive?

Hint. No. Consider the equivalent norm on ℓ_2 defined for $x = \{x_i\}$ by

$$\|x\| = \max\big\{\|x\|_2, \max\{|x_i| + |x_j| : i \neq j\}\big\}.$$

Check the standard unit vectors.

3.126 (i) Let X, Y be Banach spaces and let T be a one-to-one mapping from X onto Y. Show that if C is a convex subset of X, then $T\big(\mathrm{Ext}(C)\big) = \mathrm{Ext}\big(T(C)\big)$. In particular, if T is an isometry of X into Y then $T\big(\mathrm{Ext}(B_X)\big) = \mathrm{Ext}\big(T(B_X)\big)$.

(ii) Let T be a bounded operator from X into Y. Show that if C is w-compact in X, then $\mathrm{Ext}\big(T(C)\big) \subset T\big(\mathrm{Ext}(C)\big)$.

Hint. (i) If $x \in C \setminus \mathrm{Ext}(C)$, then $x = \frac{1}{2}(y + z)$ for some $y, z \in C$. But then also $T(x) = \frac{1}{2}\big(T(y) + T(z)\big)$. Since $T\big|_C$ is invertible, the other direction follows.

(ii) Let $d \in \mathrm{Ext}\big(T(C)\big)$. Denote $K = C \cap T^{-1}(d)$. Since T is w–w-continuous, K is w-compact and hence there is $c \in \mathrm{Ext}(K)$ such that $T(c) = d$. Check that $c \in \mathrm{Ext}(C)$: If $c = \frac{1}{2}(x + y)$, $x, y \in C$, then $d = \frac{1}{2}\big(T(x) + T(y)\big)$. Since d is extreme, $T(x) = T(y) = T(d)$, so $x, y \in K$. Since c is extreme in K, $x = y$.

3.127 Let C be a convex set in a Banach space X and let c be an extreme point in C. Show that if $c = \sum_{i=1}^{n} \lambda_i x_i$ with $\sum \lambda_i = 1$, $\lambda_i \geq 0$, and $x_i \in C$ for $i = 1, \dots, n$ then $c = x_i$ for some i.

Hint. $\lambda_1 x_1 + \lambda_2 x_2 + \lambda_3 x_3 = (\lambda_1 + \lambda_2)(\frac{\lambda_1}{\lambda_1+\lambda_2} x_1 + \frac{\lambda_2}{\lambda_1+\lambda_2} x_2) + \lambda_3 x_3$, use induction on n.

3.128 Let m_0 be the set of all sequences assuming only finitely many different values. Show that m_0 is dense in ℓ_∞. A more precise statement is given in Exercise 3.129.

Hint. Let $\alpha = (a_n)_{n=1}^{\infty}$ be a sequence such that $\|a\|_\infty \leq 1$. For any $\varepsilon > 0$, pick $N \in \mathbb{N}$ such that $1/N < \varepsilon$. Define the sequence $b = (b_n)_{n=1}^{\infty} \in m_0$ by $b_n = (\operatorname{sign} a_n) j/N$, if $j/N \leq |a_n| < (j+1)/N$, $j = 1, 2, \dots, N$. Then $\|a - b\|_\infty \leq 1/N < \varepsilon$.

3.129 Show directly that $B_{\ell_\infty} = \overline{\mathrm{conv}}\big(\mathrm{Ext}(B_{\ell_\infty})\big)$.

Hint. Note that (x_n) is extreme in B_{ℓ_∞} if and only if $|x_n| = 1$ for each n. Given $x \in B_{\ell_\infty}$ and $\varepsilon > 0$, find a simple function y on $\mathbb{N}$ such that $\|x - y\| < \varepsilon$. If $\{a_i\}_{i=1}^{k}$ are the values of y, write $\{a_i\}_{i=1}^{k}$ as a convex combination of extreme points in $B_{\ell_\infty^k}$. Then express y as the same combination of the characteristic functions corresponding to $\{a_i\}_1^k$, which are extreme points in B_{ℓ_∞}.

3.130 Show that $\overline{\mathrm{conv}}\big(\mathrm{Ext}(B_{C[0,1]^*})\big) \neq B_{C[0,1]^*}$.

Hint. Show that the Lebesgue measure λ is not in $\overline{\mathrm{conv}}\big(\mathrm{Ext}(B_{C[0,1]^*})\big)$. Indeed, if λ is norm-close to $\sum \alpha_i \delta_{t_i}$, find $f \in B_{C[0,1]}$ such that $f(t_i) = 0$ but $\lambda(f)$ is close to 1.

3.131 Show that B_{c_0} has no extreme point.
Hint. If $x = (x_i) \in B_{c_0}$, let i_0 be such that $|x_{i_0}| < \frac{1}{4}$. Then x is the center of a line segment with endpoints $(x_i) \pm \frac{1}{4} e_{i_0}$.

3.132 Show that extreme points of B_c are vectors $x = (x_n) \in c$ such that $|x_n| = 1$ for all $n \in \mathbb{N}$.

Show that c and c_0 are not isometric.
Hint. Isometry preserves extreme points, the previous exercise.

3.133 Show that the extreme points of $B_{C[0,1]}$ are exactly two functions: constant functions 1 and -1.

Show that $B_{C(K)}$, for an arbitrary compact set K, has some extreme points.
Hint. Use the definition.

3.134 Show that c_0 is not isometric to any $C(K)$ space.
Hint. Lack of extreme points in B_{c_0} unlike in $B_{C(K)}$.

3.135 Show that the extreme points of B_{ℓ_1} are exactly the vectors $\pm e_i$.
Hint. Do it first in ℓ_1^2.

3.136 Show that $B_{L_1[0,1]}$ has no extreme points.
Hint. For c with $\int_0^c |f| \, \mathrm{d}t = 1/2$ put $f_1 = 2f \cdot \chi_{[0,c]}$ and $f_2 = 2f \cdot \chi_{[c,1]}$.

3.137 Find a norm-closed convex and bounded subset of $C[0,1]^*$ that has no extreme point.
Hint. First note that $L_1[0,1] \subset C[0,1]^*$ by using the mapping $f \to \int_0^1 f(x)g(x)\mathrm{d}x$, $g \in C[0,1]$. Then use Exercise 3.136.

3.138 Show that the extreme points of $B_{L_\infty[0,1]}$ are exactly functions for which $|f|$=1 almost everywhere.
Hint. Use the definition.

3.139 Show that none of $C[0,1]$, c_0, $L_1[0,1]$ is isometric to a dual space.
Hint. Isometry preserves extreme points, the Krein–Milman theorem.

3.140 Show that $C([0,1] \cup [2,3])$ is not isometric to $C[0,1]$.
Hint. Use the Banach–Stone theorem and the connectedness.

3.141 Use the Banach–Stone theorem to show that $c \oplus c$ is not isometric to c. Recall that these spaces are isomorphic.
Hint. $c \oplus c$ is isometric to the space of continuous functions on the disjoint union of two copies of one-point-compactifications of $\mathbb{N}$, see Section 17.1. This set has two limit points. c is isometric to the space of continuous functions on the one-point-

compactification of natural numbers. This set has only one limit point. (For another approach to the same result see Exercise 5.21.)

3.142 Let K be a compact metric space such that the identity I_K is the only homeomorphism of K onto K. Show that all linear isometries of $C(K)$ onto $C(K)$ are $\pm I_{C(K)}$.

Such K does exist. In fact, there is a connected compact set K in $\mathbb{R}^2$ such that the only continuous mappings $K \to K$ are constants and the identity.
Hint. The Banach–Stone theorem.

3.143 Show that $\ell_1(c)$ is isometric to a subspace of $C[0, 1]^*$, where c denotes a set of cardinality continuum.
Hint. Given $(a_t) \in \ell_1(c)$, consider the functional $\sum a_t \delta(t)$.

3.144 Find an example of a compact convex set C in $\mathbb{R}^3$ such that extreme points of C do not form a closed set. Can that happen in $\mathbb{R}^2$?
Hint. Let D denote the disk of radius 1 in the (x, y)-plane centered at $(1, 0, 0)$. Let L be the line segment on the z-axis from the point $(0, 0, -1)$ to the point $(0, 0, 1)$. For the set C take $\operatorname{conv}\{D \cup L\}$. This cannot happen in $\mathbb{R}^2$, since if x is not an extreme point then points close to x are not extreme as well.

3.145 (Milman) Use Lemma 3.69 to prove that if C is a weakly compact convex set in a locally convex space X and $B \subset C$ is such that $\overline{\operatorname{conv}}(B) = C$, then $\operatorname{Ext}(C) \subset \overline{B}^w$.
Hint. Assume that for some $x \in \operatorname{Ext}(C)$ we have $x \notin \overline{B}^w$. By Lemma 3.69, there is a slice S of C that contains $x \in C$ and such that $S \cap B = \emptyset$. Thus there are $f \in X^*$ and $\alpha \in \mathbb{R}$ such that $f(x) > \alpha \geq \sup_B(f)$. By the linearity of f, $\sup_B(f) = \sup_{\overline{\operatorname{conv}}(B)}(f)$, so $f(x) > \alpha \geq \sup_{\overline{\operatorname{conv}}(B)}(f)$. This contradicts $C = \overline{\operatorname{conv}}(B)$.

3.146 (Troyanski–Lin) Let $x \in \operatorname{Ext}(B_X)$ and assume that at x the relative norm and weak topology of B_X coincide. Show that the slices form a neighborhood base of the norm topology at x.
Hint. (Rosenthal) Show that x is an extreme point of $B_{X^{**}}$. To this end, assume that $x = \frac{1}{2}(x_1 + x_2)$ with $x_i \in B_{X^{**}}$. By a geometric argument, show that the w^*-topology and the norm topology of $B_{X^{**}}$ coincide at x_i. By Goldstine, $x_i \in B_X$, a contradiction. Having shown that x is an extreme point of the second dual ball, use Choquet's lemma.

3.147 Show that B_{l_∞} does not have arbitrarily small slices.
Hint. Given $F \in S_{\ell_\infty^*}$ and $\varepsilon > 0$, consider the slice $S_\varepsilon = \{x \in B_{\ell_\infty} : F(x) > 1-\varepsilon\}$. We claim that $\operatorname{diam} S_\varepsilon \geq 2$. To this end, it is enough to prove that S_ε contains at least two extreme points of B_{ℓ_∞}. Note that functionals not attaining their norm are dense in $B_{\ell_\infty^*}$ (if there was a ball of norm-attaining, all would attain and ℓ_∞ would be reflexive). So assume that F does not attain its norm.

By Exercise 3.129, S_ε contains at least one extreme point x. If it was the only one, it would be in all slices S_δ for $\delta \in (0, \varepsilon)$. Then $F(x) = 1$, so F attains its norm, a contradiction.

3.148 Let X be a Banach space and A (respectively B) be a bounded subset of X (respectively, of X^*). Show that A has arbitrary small slices if and only if $\overline{\text{conv}}(A)$ does. Show that B has arbitrarily small w^*-slices if and only if $\overline{\text{conv}}^{w^*}(B)$ does. .
Hint. Assume that A has small slices. Pick a small ε-slice A_1 of A that results from a cut of A by a hyperplane H and put $D_1 = \overline{\text{conv}}(A \setminus A_1)$, $A_2 = \overline{\text{conv}}(A_1)$. Then $\text{conv}\, A \subset \overline{\text{conv}}(A_2 \cup D_1)$. Show that if you move H in the parallel way, the cuts of $\overline{\text{conv}}(A_2 \cup D_1)$ have eventually diameter smaller than 2ε. The argument for B is similar. Details can be found in [Bour], see also [Dies2, pp. 157–158].

3.149 Find a compact convex set C in a Banach space such that C is not the convex hull of its extreme points.
Hint. Put $C = \overline{\text{conv}}(\{2^{-i}e_i\} \cup \{0\})$, where e_i are the standard unit vectors in ℓ_2. Then $c = \sum 4^{-i}e_i \in C$. By Theorem 3.66, $\text{Ext}(C) \subset \{2^{-i}e_i\} \cup \{0\}$. Therefore $\text{conv}\big(\text{Ext}(C)\big)$ is formed by points that are all finitely supported and thus it does not contain c.

3.150 Let Y be a closed subspace of a Banach space X. We know that Y^* is isometric to $X^*/Y^\perp$. Show that for every $\hat{x} \in \text{Ext}\big(B_{X^*/Y^\perp}\big)$ there exists $x \in \hat{x}$ such that $x \in \text{Ext}(B_{X^*})$.
Hint. Exercise 3.126.

3.151 Find a James boundary B for a Banach space X with the property that $B \cap \text{Ext}(B_{X^*}) = \emptyset$.
Hint. Take $X := \ell_1(\Gamma)$, where Γ is uncountable. Let $B = \{x^* \in X^* : x^*(\gamma) \in \{-1, 0, 1\}$ for all $\gamma \in \Gamma$, $\text{supp}(x^*)$ countable$\}$.

3.152 Consider $X = \mathbb{R}^2$ with the dual norm on X^* being the norm whose unit ball is

$$D = B((1,0), 10) \cap B((-1,0), 10)) \cap \big\{(x, y) \in \mathbb{R}^2 : -1 \leq y \leq 1\big\},$$

where $B(x, r)$ is a closed ball centered at x with radius r.

Prove that $B = \text{Ext}(D \setminus \{\text{upper left corner}\})$ is a James boundary of X which is not closed.

3.153 Let X be a Banach space, let B be a separable subset of X^* such that $\overline{B}^{w^*} \neq B_{X^*}$. Can we conclude as in Theorem 3.122 that B is not a James boundary of X?
Hint. No. Let X be $\mathbb{R}^2$ with the maximum norm and let $B = \{\pm e_1, \pm e_2\}$, where e_i are the standard unit vectors in $\mathbb{R}^2$. Then for each $x \in S_X$ we have $\max_B(x) = 1$.

3.154 Show that if B is a James boundary, then $\mathrm{Ext}(B_{X^*}) \subset \overline{B}^{w^*}$.
Hint. By separation, $\overline{\mathrm{conv}}^{w^*}(B) = B_{X^*}$. Then use Theorem 3.66.

3.155 (Cascales, Godefroy [CasGod]) Prove the following result: *Let K be a compact space and $B \subset S_{C(K)^*}$ a James boundary of $C(K)$. If $\{f_n\}$ is an arbitrary sequence in $C(K)$ and $x \in K$, then there exists $\mu \in B$ such that $f_n(x) = f_n(\mu)$ for every $n \in \mathbb{N}$.*
Hint. The set $F := \{y \in K : f_n(y) = f_n(x) \text{ for all } n \in \mathbb{N}\}$ is a closed G_δ-subset of K. It follows from Urysohn's lemma (see [Dugu2, p. 148]) that there exists a continuous function h from K to $[0, 1]$ such that $F = \{x \in K : h(x) = 1\}$. Since B is a James boundary, there exists $\mu \in B$ such that $\mu(h) = 1$. But this clearly implies that μ is a probability measure supported by F, and thus that μ works.

3.156 (Cascales, Godefroy [CasGod]) Prove the following result: *Let K be a compact space and $B \subset S_{C(K)^*}$ a James boundary of $C(K)$.*

(i) If $\{f_n\}$ is an arbitrary sequence of $C(K)$, then any $w(C(K), B)$-cluster point of $\{f_n\}$ in $C(K)$ is also a $\mathcal{T}_p$-cluster point.

(ii) If L is a $w(C(K), B)$-relatively countably compact subset of $C(K)$, then it is $\mathcal{T}_p$- relatively compact in $C(K)$.

(iii) If L is a $w(C(K), B)$-compact subset of $C(K)$, then it is $\mathcal{T}_p$-compact.

Hint. (i) is a straightforward consequence of Lemma 3.155.

(ii) follows from (i) together with the fact that the $\mathcal{T}_p$-relatively countably compact subsets of $C(K)$ are actually $\mathcal{T}_p$-relatively compact, see Theorem 3.139.

To prove (iii) let us fix a $\mathcal{T}_p$-compact subset L of $C(K)$. From (ii) we get that L is $\mathcal{T}_p$-relatively compact in $C(K)$. To finish the proof it will be enough to show that L is $\mathcal{T}_p$-closed. Pick g in the $\mathcal{T}_p$-closure of L. Using that $(C(K), \mathcal{T}_p)$ is angelic (see Theorem 3.58), there is a sequence $\{g_n\}$ in L with

$$g = \mathcal{T}_p\text{ - }\lim_{n\to\infty} g_n. \tag{3.10}$$

By compactness, there is $h \in L$ which is a $w(C(K), B)$-cluster point of $\{g_n\}$. Statement (i) implies that h is a $\mathcal{T}_p$-cluster point of $\{g_n\}$, hence, by (3.10), $h = g$ and thus $g \in L$. This shows that L is $\mathcal{T}_p$-compact and the proof is complete.

3.157 (Cascales, Godefroy [CasGod]) Prove the following result, the case of Pfitzner's theorem for the case of $C(K)$ spaces: *Let K be a compact space and $B \subset S_{C(K)^*}$ a James boundary of $C(K)$. Then a subset A of $C(K)$ is weakly compact if, and only if, it is norm bounded and $w(C(K), B)$-compact.*
Hint. Obviously any weakly compact subset of $C(K)$ is norm bounded and $w(C(K), B)$-compact. Conversely, let A be a bounded and $w(C(K), B)$-compact subset of $C(K)$. Statement (iii) in Proposition 3.156 ensures that A is $\mathcal{T}_p$-compact. Since A is bounded, Theorem 3.139 implies that A is weakly compact.

3.158 Let C be a bounded convex set in a Banach space X such that every continuous linear functional on X attains its supremum over C. Is C necessarily closed?

Hint. No. Consider in the plane ℓ_2^2 the convex hull of two circles of radius one centered at $(2, 0)$ and $(-2, 0)$, remove the points $(\pm 2, \pm 1)$. The resulting set is bounded and not closed, one can show that every bounded linear functional attains its supremum over this set (level sets of linear functionals are lines in $\mathbb{R}^2$).

3.159 Find a Banach space X and a convex continuous bounded below function f on X such that $\lim\limits_{\|x\|\to\infty} f(x) = \infty$ and f does not attain its infimum on X. Can X be reflexive? Can X be finite-dimensional if we drop the requirement on the limit of f?

Hint. Let $X = c_0$ and let f be in S_{X^*} such that f does not attain its norm. Define ϕ for $x \in X$ by $\phi(x) = \|x\|^2 - f(x)$. Then ϕ is bounded below. Indeed, if $\|x\| \geq 2$, then $\phi(x) \geq \|x\|^2 - \|x\| \geq 2$. If $\|x\| \leq 2$, then $|\phi(x)| \leq \|x\|^2 + |f(x)| \leq \|x\|^2 + 2\|f\| \leq 6$. Let $\|x\| < \frac{1}{2}$ be such that $\frac{1}{2} \geq f(x) > \frac{1}{4}$. Then $\phi(x) = \|x\|^2 - f(x) < \frac{1}{4} - f(x) < 0$. Hence ϕ attains negative values at some points of X. Assume that ϕ attains its minimum on X at x_0. As $\phi(0) = 0$, we get $x_0 \neq 0$. $\phi(x_0) \leq \phi(x)$ reads $\|x_0\|^2 - f(x_0) \leq \|x\|^2 - f(x)$ for all $x \in X$, hence $f(x) \leq f(x_0) + \|x\|^2 - \|x_0\|^2$ for all $x \in X$. If $x \in X$ is such that $\|x\| \leq \|x_0\|$, then $f(x) \leq f(x_0)$. Thus f attains its maximum on the set $\{x \in X : \ \|x\| \leq \|x_0\|\}$. Hence f attains its maximum on B_X, a contradiction. The function e^x on $\mathbb{R}$ can be used to finish the solution of this exercise. The space X cannot be reflexive, as on such space the weakly lower semicontinuity of ϕ can be used to produce the minimum of ϕ, by using the limit condition on ϕ.

3.160 Find an example of an incomplete infinite-dimensional normed space X such that every $f \in X^*$ attains its norm ([Jame4]).

Hint. Let $X = \left(\sum_{n=1}^{\infty} \ell_\infty^n\right)_2$. Note that if $x = (x_n) \in X$ is an extreme point of B_X, then x_n is an extreme point of $B_{\ell_\infty^n}$ for every n, so $x_n = (\pm 1, \pm 1, \dots, \pm 1)$. Let $Z = \operatorname{span}\big(\operatorname{Ext}(B_X)\big)$. X is reflexive, so by the Krein–Milman theorem we have $B_X = \overline{\operatorname{conv}}\big(\operatorname{Ext}(B_X)\big)$ and thus $\overline{Z} = X$.

Note that $Z \neq X$. To see this, consider the following $x \in X$: $x_1 = 1$, $x_2 = (\frac{1}{2}, \frac{1}{3})$, $x_3 = (\frac{1}{4}, \frac{1}{5}, \frac{1}{6})$, ... Assume that $x \in Z$, that is, $x = \sum_{l=1}^{m} \alpha_l y^l$, where $y^1, y^2, \dots, y^m \in \operatorname{Ext}(B_X)$. Let $n > 2^m$. Then $x_n = \sum_{l=1}^{m} \alpha_l y_n^l$. Since every coordinate of each vector y_n^l is ± 1, we have at most 2^m different coordinates of the vector x_n. As $n > 2^m$, at least two coordinates of x_n should be equal, which is not the case. Hence $x \notin Z$.

Now pick any $f \in Z^*$. Extend it on X and call it f again. As B_X is weakly compact, f attains its norm on B_X. From the proof of Krein–Milman theorem, we have that f attains its norm at an extreme point of B_X. Using the fact that the norm of f on X and Z is the same we get that f attains its norm on B_Z.

3.161 Show using Theorem 3.130 that B_{c_0} is not weakly compact.
Hint. $(2^{-i}) \in \ell_1$ does not attain its norm on B_{c_0}. Indeed, if $x \in B_{c_0}$ then eventually $|x_i| < 1$, so $|\sum 2^{-i} x_i| < \sum 2^{-i} = 1$.

Direct proof of the exercise: $\left\{\sum_{i=1}^{n} e_i\right\}$ has no w-cluster point.

Note that exactly finitely supported elements in ℓ_1 attain their norms.

3.162 Show that the functionals in c_0^* that attain their norm do not form a residual set in c_0^*.
Hint. Recall that it is exactly finitely supported functionals that attain their suprema over B_{c_0} (see the previous exercise). Then find the Baire category of the set of finitely supported vectors. Use the fact that finite-dimensional subspaces of ℓ_1 are closed and have empty interiors in ℓ_1.

3.163 Let X be a Banach space and $f \in S_{X^*}$. Does there exist $f_n \in X^*$ that attain their norm such that $\|f_n - f\| \to 0$ and f_n lie on one line through f?
Hint. Not in general. Consider in $c_0^* = \ell_1$ an element with infinite support and use the fact that precisely finitely supported vectors in ℓ_1 attain their norm.

3.164 Assume that C is a separable closed convex subset of X^* such that each $x \in X$ attains its supremum over C. Is C w^*-compact?
Hint. Yes, by the proof of Theorem 3.122, C is equal to its w^*-closure.

3.165 Show that a Banach space X is reflexive if and only if the distance of any point to a given closed convex subset of X is attained.
Hint. Assume there always is a closest point. Let $f \in S_{X^*}$ and consider the closed convex set $F := \{x \in X : f(x) = 1\}$. The distance of F to 0 is 1. A closest point in F to 0 is then an element in B_{X^*} where f attains its norm. It is enough to apply Corollary 3.131. Assume now that X is reflexive. Let C be a nonempty closed convex subset of X, $x_0 \in X \backslash C$ and $d := \operatorname{dist}(x_0, C)$ (> 0). Fix $\varepsilon > 0$. The set $B(x_0, d+\varepsilon) \cap C$ is nonempty, closed, convex and bounded (so w-compact). The function $x \to \|x - x_0\|$ is w-lower semicontinuous, hence it attains its minimum on $B(x_0, d+\varepsilon) \cap C$ at some x, also where the distance from x_0 to C is attained.

3.166 Show the following Šmulyan's theorem ([Smul1]): *A Banach space X is reflexive if each nested sequence $C_n \supset C_{n+1}$ of closed convex subsets of B_X has a nonempty intersection.*
Hint. Show that then every $f \in X^*$ attains its norm by considering the sets $C_n := \{x \in B_X : f(x) \geq \sup_{B_X} f - \frac{1}{n}\}$.

3.167 Let X be a Banach space, let C be a separable set in X^*. If C is w^*-compact, is $D = \overline{\operatorname{conv}}(C)$ also w^*-compact?

What if the separability assumption is dropped?
Hint. Yes for the first part. Each $x \in X$ attains its supremum over C which is the supremum of x over D. Then use Exercise 3.164.

If the separability is dropped the situation is different. Consider the Dirac measures in $B_{C[0,1]^*}$, see Exercise 3.130.

3.168 Let X be a separable Banach space and let B be a James boundary for X. Show that if $A \subset X$ is a bounded set in X compact in the topology of pointwise convergence on B, then A is w-compact.
Hint. We may assume that B is convex. Then $\overline{B}^{w^*} = B_{X^*}$. As B_{X^*} is metrizable in the w^*-topology, B is w^*-separable and there is a countable w^*-dense subset D of B, in particular $\overline{D}^{w^*} = B_{X^*}$. The topology of the pointwise convergence on D is a metrizable topology on A which, by the compactness assumption, coincides with the pointwise topology on B. Thus every sequence in A has a subsequence that is convergent in the topology of the pointwise convergence on B, which by Theorem 3.134 gives that A is weakly compact.

3.169 (i) Show that Theorem 3.134 does not in general hold for nets.
(ii) Show that Theorem 3.134 does not in general hold for unbounded sequences.
(iii) Show that Theorem 3.134 does not in general hold for $x \in X^{**}$.
Hint. (i) For finite $F \subset [0, 1]$ consider the following function $f_F \in S_{C[0,1]}$: Assume $F = \{x_i\}, 0 \leq x_1 < \cdots < x_n \leq 1$. Define f_F as the piece-wise linear function with nodes in x_i, where $f_F(x_i) = 1$, and nodes in 0,1, and midpoints of $[x_i, x_{i+1}]$, where $f_F = 0$. Then for the Lebesgue measure λ we have $\lambda(f_F) = \frac{1}{2}$, but considering the net $\{f_F\}$, F partially ordered by the inclusion, we see that $\delta_t(f_F) \to 1$ for all Dirac measures.
(ii) Put $x^i = ie_i$, where e_i denotes the standard unit vector in c_0. Then $x^i \to 0$ pointwise in c_0, but x^i does not converge weakly to 0 in c_0 as it is unbounded.
(iii) Let $F \in C[0, 1]^{**}$ be zero on all Dirac measures but $F \neq 0$. Take $x_n = 0 \in C[0, 1]$.

3.170 Show that Simons's Lemma 3.123 does not hold if the requirement that elements in the superconvex hull of a sequence attains their supremum is dropped.
Hint. Take $B := [0, +\infty)$ and $x_n := \chi_{[n,+\infty)}$, $n \in \mathbb{N}$.

3.171 Show that Lemma 3.123 does not hold in general if the assumption on the completeness of X is dropped.
Hint. Consider the subspace X of ℓ_∞ formed by all sequences that are eventually constant. Then the standard unit vectors $\{e_i\}$ form a James boundary B of X. Consider $x^i = \chi_{\{i,i+1,\dots\}}$. Then $\sup_B(\limsup_i x^i) = 0$. If $x \in \operatorname{conv}\{x^i\}$, then for large n, the nth coordinate of x equals 1. Note that this is caused by the fact that infinite convex combinations are used in the proof of the Simons' inequality.

3.172 Show the following Lyapunoff theorem: Suppose that $\mu_1, \dots, \mu_n$ are real-valued nonatomic measures on a σ-algebra $\mathcal{M}$. Define

$$\mu(E) = (\mu_1(E), \dots, \mu_n(E)) \qquad (E \in \mathcal{M}).$$

Then μ is a function with domain $\mathcal{M}$ whose range is a compact convex set in $\mathbb{R}^n$.

Note that μ is called *nonatomic* if every set $E \in \mathcal{M}$ with $|\lambda|(E) > 0$ contains a set $A \in \mathcal{M}$ with $0 < |\lambda|(A) < |\lambda|(E)$. Here $|\lambda|$ denotes the variation measure of λ.

Hint. (Lindenstrauss, see [Rudi3, p. 113]) Put $|\sigma| = |\mu_1| + \cdots + |\mu_n|$ and $K = \{g \in L_\infty(\sigma) : 0 \leq g \leq 1\}$. Define $\Lambda : L_\infty(\sigma) \to \mathbb{R}^n$ by

$$\Lambda(g) = \Big(\int g\mathrm{d}\mu_1, \ldots, \int g\mathrm{d}\mu_n\Big).$$

The mapping Λ is a weak*-continuous linear mapping from $L_\infty(\sigma)$ into $\mathbb{R}^n$. Moreover, K is a weak*-compact convex set in $L_\infty(\sigma)$ $(= L_1(\sigma)^*)$. Therefore $\Lambda(K)$ is a compact convex set in $\mathbb{R}^n$. So, it remains to show that $\mu(\mathcal{M}) = \Lambda(K)$.

It is easy to show that $\mu(\mathcal{M}) \subset \Lambda(K)$. To obtain the opposite inclusion, pick a point $p \in \Lambda(K)$ and define $K_p = \{g \in K : \Lambda g = p\}$. We have to show that K_p contains some χ_E, a characteristic function of $E \in \mathcal{M}$, for then $p = \mu(E)$.

The set K_p is a weak*-compact convex set. By the Krein–Milman theorem, K_p has an extreme point g_0. Assume that g_0 is not a characteristic function in $L_\infty(\sigma)$. Then there is a set $E \in \mathcal{M}$ and a constant $0 < r < 1/2$ such that $\sigma(E) > 0$ and $r \leq g_0 \leq 1 - r$ on E. Put $Y = \chi_E \cdot L_\infty(\sigma)$. Since $\sigma(E) > 0$ and σ is nonatomic, $\dim Y > n$. Hence there exists $g \in Y$, not the zero element in $L_\infty(\sigma)$, such that $\Lambda g = 0$ and such that $-r < g < r$. Therefore $g_0 + g$ and $g_0 - g$ are both in K_p. Thus g_0 is not an extreme point of K_p. This shows that all extreme points of K_p are characteristic functions. This finishes the proof.

3.173 Prove that an operator T from a Banach space X into a Banach space Y is completely continuous if and only if $T(K)$ is $\|\cdot\|$-compact for every w-compact subset K of X.

Hint. Assume that T is completely continuous. Note first that T is continuous, since it maps $\|\cdot\|$-null sequences into $\|\cdot\|$-null sequences. Let K be a w-compact subset of X. Given a sequence $\{Tx_n\}$ in TK, the sequence $\{x_n\}$ has a subsequence $\{x_{n_k}\}$ that w-converges to some point $x \in K$, due to the Eberlein–Šmulyan Theorem 3.109. Then $\{Tx_{n_k}\}$ is $\|\cdot\|$-convergent to Tx, and this proves that $T(K)$ is $\|\cdot\|$-compact. Conversely, assume that T maps w-compact subsets of X onto $\|\cdot\|$-compact subsets of Y. Note that T is continuous: otherwise there would exist a sequence $\{x_n\}$ in B_X such that $\|Tx_n\| \geq n^2$ for all $n \in \mathbb{N}$. The set $\{x_n/n : n \in \mathbb{N}\} \cup \{0\}$ is $\|\cdot\|$-compact, and $\{Tx_n : n \in \mathbb{N}\}$ is not even bounded, a contradiction. Let $\{x_n\}$ be a w-null sequence. The set $\{x_n\} \cup \{0\}$ is w-compact, hence $\{Tx_n\} \cup \{0\}$ is $\|\cdot\|$-compact. Let $\{Tx_{n_k}\}$ be an arbitrary subsequence of $\{Tx_n\}$. There exists a subsequence $\{Tx_{n_{k_m}}\}$ of $\{Tx_{n_k}\}$ that $\|\cdot\|$-converges to some Tx, where $x \in K$. Since $x_{n_{k_m}} \stackrel{w}{\to} 0$, we get $Tx_{n_{k_m}} \stackrel{w}{\to} 0$, and so $Tx = 0$. Since this happens for every subsequence $\{Tx_{n_k}\}$ of $\{Tx_n\}$, we get $Tx_n \stackrel{\|\cdot\|}{\to} 0$.

3.174 Show that $B_{\ell_\infty^*}$ is not w^*-sequentially compact.

Hint. Define a sequence of elements of $B_{\ell_\infty^*}$ by $f_n\big((x_i)\big) = x_n$. This sequence has no w^*-convergent subsequence. To see it, assume that $\{f_{n_k}\}$ is a w^*-convergent subsequence. Define a vector $x = (x_i) \in \ell_\infty$ by $x_{n_k} = (-1)^k$ and $x_i = 0$ if $i \notin \{n_k\}$. Then $f_{n_k}(x)$ is not convergent.

3.175 Show that the sequence $\{\sin nx\}_{n=1}^\infty$ in $[-1,1]^{[0,2\pi]}$ has no pointwise convergent subsequence (this proves, in particular, that $[-1,1]^{[0,2\pi]}$, endowed with the topology of the pointwise convergence, is not sequentially compact).
Hint. Otherwise, $\{\sin_{n_{k+1}} - \sin_{n_k}\}_k \to 0$ pointwise, hence $\int_0^{2\pi}(\sin_{n_{k+1}}(x) - \sin_{n_k}(x))^2\,\mathrm{d}x \to 0$ because of the Lebesgue dominated convergence theorem. However, $\int_0^{2\pi}(\sin_{n_{k+1}}(x) - \sin_{n_k}(x))^2\,\mathrm{d}x = 2\pi$ for all k.

3.176 (i) Let $\{f_n\}$ be a sequence in the dual X^* of a Banach space X. Show directly that if X is separable and $\{f_n\}$ is bounded, then there exists a w^*-convergent subsequence of $\{f_n\}$.
(ii) Let $\{x_n\}$ be a sequence in a Banach space X. Show directly that if X is reflexive and $\{x_n\}$ is bounded, then there exists a w-convergent subsequence of $\{x_n\}$.

Hint. (i) Let $\{x_i\}$ be a dense countable subset of X. Since $\{f_n\}$ is bounded, so is $\{f_n(x_1)\}$ and there exists a convergent subsequence $\{f_n^1(x_1)\}$. Applying this trick to $\{f_n^1\}$ and x_2 we get a subsequence $\{f_n^2\}$ such that $\{f_n^2(x_i)\}$ converges for $i = 1, 2$. Continue in this manner and set $f_{n_k} = f_k^k$. Then for every x_i the limit $\lim\limits_k f_{n_k}(x_i)$ exists.
Using the boundedness of $\{f_n\}$ and density of $\{x_i\}$, show that $\lim\limits_k f_{n_k}(x)$ exists for all $x \in X$, call it $f(x)$. f is linear and assuming $\|f_n\| \le C$ we show that $|f(x_i)| \le C\|x_i\|$, so f is continuous.
(ii) Define $Y = \overline{\operatorname{span}}\{x_n\}$. Then Y is reflexive and separable, hence also $Y^{**} = Y$ and consequently Y^* are separable. Since $\{x_n\}$ is bounded in Y^{**}, by the first part it has a subsequence $\{x_{n_k}\}$ that w^*-converges in Y^{**} to some $x \in Y^{**} = Y$. So $f(x_{n_k}) \to f(x)$ for all $f \in Y^*$, hence also for all $f \in X^*$ and $x_{n_k} \xrightarrow{w} x$.

3.177 Prove Krein's Theorem 3.133 by using only the separable version of James' Theorem 3.130.
Hint. Let A be a weakly compact subset of a Banach space X and let $C = \overline{\operatorname{conv}}(A)$. Take a sequence $\{c_n\}$ in C. There exists a countable subset S of A such that $\{c_n : n \in \mathbb{N}\} \subset \overline{\operatorname{conv}}(S)$. The set $\overline{\operatorname{conv}}(S)$ is $\|\cdot\|$-separable and every $f \in X^*$ attains its supremum over $\overline{\operatorname{conv}}(S)$ at some point in the weakly compact set $\overline{S}^w$ $(\subset \overline{\operatorname{conv}}(S))$. By the separable version of Theorem 3.130 (a complete proof of this was provided), $\overline{\operatorname{conv}}(S)$ is w-compact. Therefore, C is weakly countably compact. By the Eberlein–Šmulyan Theorem 3.109, C is weakly compact.

3.178 Prove (iii)$\Longrightarrow$(i) in Theorem 3.109 by using Theorem 3.130.

Hint. Let A be a relatively weakly countably compact subset of a Banach space X. Certainly A is bounded. Given $f \in X^*$, there is a sequence $\{a_n\}$ in A such that $f(a_n) \to \sup_A f$ $(= \sup_{\overline{A}^w} f)$. The sequence $\{a_n\}$ has a w-cluster point in $\overline{A}^w$. This proves that f attains its supremum over $\overline{A}^w$ at some point in $\overline{A}^w$. Use Theorem 3.130 to conclude that $\overline{A}^w$ is w-compact.

3.179 Show that the convexity assumption in Theorem 3.130 cannot be dropped. This shows that any characterization of general weakly compact sets via attainment of maxima (for instance Theorem 14.49) must consider more functions than just elements of X^*.
Hint. Consider $B_{\ell_2} \setminus \frac{1}{2} B_{\ell_2}$.

Chapter 4
Schauder Bases

In this chapter we shall introduce Schauder bases, an important concept in Banach space theory. Elements of a Banach space with a Schauder basis may be represented as infinite sequences of "coordinates," which is very natural and useful for analytical work. Although not every separable Banach space admits a Schauder basis (this is a deep and difficult result of Enflo), basic sequences exist in every infinite-dimensional Banach space (Mazur) and are ideal for the study of linear subspaces and quotients. In many respects, the assumption of having a Schauder basis is not very restrictive, and in fact, for naturally defined separable Banach spaces, it is usually easy to find their Schauder basis. This notion has proved to be an extremely useful tool in the study of the structure of classical as well as abstract Banach spaces. Among the results in this chapter let us mention the Pełczyński decomposition principle, leading to a complete characterization of complemented subspaces of the classical sequence spaces, James' structural theory of spaces with an unconditional Schauder basis, Bessaga–Pełczyński characterization of spaces containing c_0, etc.

4.1 Projections and Complementability, Auerbach Bases

Let V be a vector space. In Definition 1.32 we introduced the notion of *linear projection* in V, and we mentioned, right after it, that associated to a projection P in V there is an algebraic direct sum (or *algebraic direct decomposition*) $V = PX \oplus \operatorname{Ker} P$. Conversely, if $V = M_1 \oplus M_2$ is an algebraic direct decomposition of V, the mappings $P_i : V \to V$ given by $P_i(v) = v_i$, where $v = v_1 + v_2$, $v_i \in M_i$, $i = 1, 2$, are linear projection onto M_i (parallel to the other subspace) for $i = 1, 2$, and $P_1 + P_2 = I_V$ (the identity mapping on V). If T is a mapping from a vector space into itself, we denote $T^2 = T \circ T$ and we define by induction T^n for $n \in \mathbb{N}$.

Given a subspace M_1 of a vector space V, there is another subspace M_2 such that $V = M_1 \oplus M_2$. Such a subspace M_2 is called an *algebraic complement* of M_1 in V.

Definition 4.1 *A closed subspace Y of a Banach space X is said to be* complemented *in X if there is a bounded linear projection P of X onto Y. If λ is a real number (necessarily $\lambda \geq 1$) such that $\|P\| \leq \lambda$ we say that Y is* λ-complemented *in X.*

M. Fabian et al., *Banach Space Theory*, CMS Books in Mathematics,
DOI 10.1007/978-1-4419-7515-7_4, © Springer Science+Business Media, LLC 2011

Let Y_1 be a closed subspace of X. We say that Y_2 is a (topological) complement *of Y_1 in X if $X = Y_1 \oplus Y_2$ and Y_2 is a closed subspace of X.*

Note that if Y is a complemented subspace of a Banach space X, then Y is closed. Indeed, if $Y = P(X)$ for some projection P on X, then $Y = (I - P)^{-1}(0)$. On the other hand, not every closed subspace of X is complemented. However, finite-dimensional and finite-codimensional closed subspaces are complemented (Theorem 4.5 and Exercise 5.8, respectively).

Proposition 4.2 *Let Y be a closed subspace of a Banach space X. Y is complemented in X if and only if there exists a topological complement of Y in X.*

Proof: Assume that there is a closed subspace Z of X such that $X = Y \oplus Z$. Let P be the linear projection onto Y such that $Z = \text{Ker}(P)$. We will show that the graph $\mathcal{G} = \{(x, P(x)) : x \in X\}$ is closed.

Let $x_n \in X$ be such that $(x_n, P(x_n)) \to (x, y)$. Since Y is closed, $y \in Y$. Note that $(x_n - P(x_n)) \in Z = \text{Ker}(P)$. Indeed, $P(x_n - P(x_n)) = P(x_n) - P^2(x_n) = 0$. Since Z is closed, we get $x - y \in Z$. By the uniqueness of the decomposition we get $y = P(x)$. Thus $(x, y) \in \mathcal{G}$ and by the closed graph theorem, P is a bounded projection onto Y.

To prove the opposite implication, assume $P^2 = P$, $P(X) = Y$. We will show that $\text{Ker}(P)$ is a topological complement of Y.

If $x \in \text{Ker}(P) \cap P(X)$, then $x = P(x) = 0$, so $Y \cap \text{Ker}(P) = \{0\}$. Consider any $x \in X$. Since P is a projection, $x - P(x) \in \text{Ker}(P)$ and we get a decomposition $x = P(x) + (x - P(x))$. This shows that $X = Z \oplus \text{Ker}(P)$.

Since P is continuous, $\text{Ker}(P)$ is a closed subspace of X. $\square$

Observe that if H is a Hilbert space and F its closed subspace, the orthogonal complement $F^{\perp}$ of F is a complement of F in the sense of Definition 4.1, and the orthogonal projection of H to F is a norm-one projection onto F (see Corollary 1.50).

Fact 4.3 *Let X, Y be Banach spaces, let T be an isomorphism of X onto Y. If X_1 is a complemented subspace of X with topological complement X_2, then $T(X_1)$ is a complemented subspace of Y with topological complement $T(X_2)$.*

Proof: Clearly $Y = T(X_1) \oplus T(X_2)$ and both subspaces are closed. $\square$

Note that if P is a projection of X onto X_1 with $\text{Ker}(P) = X_2$, then $Q = TPT^{-1}$ is a projection of Y onto $T(X_1)$ with $\text{Ker}(Q) = T(X_2)$.

Let X be a normed space such that $X = Y \oplus Z$ for complemented subspaces Y, Z. Then X is isomorphic to the direct sum $(Y \oplus Z)_\infty$ of spaces Y, Z with the norm $\|(y, z)\| = \max(\|y\|, \|z\|)$.

Proposition 4.4 *Let Y be a closed subspace of a Banach space X. If Y is complemented and Z is a complement of Y in X, then X/Y is isomorphic to Z. Moreover,*

if $Q : X \to X$ is the projection onto Z parallel to Y and $\|Q\| = 1$, then X/Y is isometric to Z
The dual X^* is isomorphic to $Y^* \oplus Z^*$; in short, $(Y \oplus Z)^* = Y^* \oplus Z^*$. If $P : X \to X$ is the projection onto Y parallel to Z, then Y^* is isomorphic to $P^*(X^*)$. Moreover, if $\|P\| = 1$, then Y^* is isometric to $P^*(X^*)$.
If Y is a closed subspace of a Hilbert space H and Z is the orthogonal complement of Y in H, then H/Y is isometric to Z.

Proof: Consider the operator $I : Z \to X/Y$ defined by $I(z) = \hat{z}$ ($\in X/Y$), where $\hat{z}$ denotes the equivalence class containing z. From the definition of the norm of X/Y, it follows that $\|I(z)\| \leq \|z\|$ for every $z \in Z$, so I is a continuous operator. We claim that I is one-to-one. To see this, assume that for some $z \in Z$ we have $\hat{z} = \hat{0}$ $(= Y)$. Since $z \in \hat{z}$, we get $z \in Z \cap Y$ and thus $z = 0$. I is also an onto mapping. Indeed, let $\hat{x} \in X/Y$. Pick an arbitrary $x \in \hat{x}$. Write $x = y + z$, where $y \in Y$ and $z \in Z$. Then $\hat{x} = \hat{z}$ $(= I(z))$. Therefore I is an isomorphism of Z onto X/Y by the open mapping theorem.

Assume now that $\|Q\| = 1$. Take $z \in Z$. If $x \in X$ is an arbitrary element in $\hat{z}$, then $x = y + z$ for some $y \in Y$. Thus $\|z\| = \|Q(x)\| \leq \|x\|$. By the definition of the norm in X/Y we get $\|z\| \leq \|\hat{z}\|$. This proves the reverse inequality and so I is an isometry.

Let P, respectively Q, be projections of X onto Y, respectively Z. By Exercise 4.10, Y^* is isomorphic to $P^*(X^*)$ and Z^* is isomorphic to $Q^*(X^*)$. Moreover, from $I_X = P + Q$ we get $I_{X^*} = P^* + Q^*$, that is, $P^*(X^*) + Q^*(X^*) = X^*$. It remains to show that $P^*(X^*) \cap Q^*(X^*) = \{0\}$. Take $f \in P^*(X^*) \cap Q^*(X^*)$. For $y \in Y$ we have $f(y) = Q^*(f)(y) = f\big(Q(y)\big) = 0$, and for $z \in Z$, $f(z) = P^*(f)(z) = f\big(P(z)\big) = 0$, hence $f = 0$. Since $P^*(X^*)$ and $Q^*(X^*)$ are closed subspaces, it follows from Proposition 4.2 that $X^* = P^*(X^*) \oplus Q^*(X^*)$ is a topological direct sum. If $\|P\| = 1$, use again Exercise 4.10.

If H is a Hilbert space, then $\|Q\| = 1$ by Corollary 1.50, and the conclusion follows from the first part of the statement. □

Recall the definition of the Kronecker delta: $\delta_{ij} = 0$ if $i \neq j$ and $\delta_{ij} = 1$ if $i = j$.

Let X be a Banach space, consider vectors $\{e_1, \dots, e_n\} \subset X$. Functionals $\{f_1, \dots, f_n\} \subset X^*$ are called *biorthogonal* to $\{e_i\}_{i=1}^n$ if $f_i(e_j) = \delta_{ij}$ for $i, j = 1, \dots, n$. In other words, the set $\{e_i; f_i\}_{i=1}^n$ is a *biorthogonal system* in $X \times X^*$.

A biorthogonal system $\{e_i; f_i\}_{i=1}^n$ is called an *Auerbach basis* of X if $\{e_i\}_{i=1}^n$ is a basis of X and $\|e_i\| = \|f_i\| = 1$ for every i.

Theorem 4.5 (Auerbach, see, e.g., [LiTz3]) (i) *Let X be a Banach space. If $\dim(X) = n$ then there exists an Auerbach basis $\{e_i; f_i\}_{i=1}^n$ of X.*
(ii) *Let Y be a subspace of a Banach space X. If $\dim(Y) = n$ then there exists a projection P of X onto Y such that $\|P\| \leq n$.*

Part (ii) in Theorem 4.5 gives that the so called *projection constant* $\lambda(Y)$ of an n-dimensional Banach space Y (see Definition 5.15) is less or equal than n. This will be substantially improved in Theorem 6.28.

Proof of Theorem 4.5: (i) Let $\{x_1, \ldots, x_n\}$ be an algebraic basis of X. For $u_1, \ldots, u_n \in B_X$ let $v(u_1, \ldots, u_n)$ be the determinant of the matrix whose jth row is formed by the coordinates of u_j in the basis $\{x_1, \ldots, x_n\}$. The function $|v|$ is continuous on the compact set $B_X \times \cdots \times B_X$, therefore there is $(e_1, \ldots, e_n) \in B_X \times \cdots \times B_X$ such that

$$v(e_1, \ldots, e_n) = \max\{|v(x_1, \ldots, x_n)| : (x_1, \ldots, x_n) \in B_X \times \cdots \times B_X\}.$$

Since determinants are homogeneous in each coordinate, we have $e_i \in S_X$. As $v(e_1, \ldots, e_n) \neq 0$, the vectors $\{e_i\}$ are linearly independent, so they form a basis of X. For $i = 1, \ldots, n$, define $f_i \in X^*$ by

$$f_i(x) = \frac{v(e_1, \ldots, e_{i-1}, x, e_{i+1}, \ldots, e_n)}{v(e_1, e_2, \ldots, e_n)}.$$

Then $f_i(e_j) = \delta_{ij}$, so $\big\{\{e_1, \ldots, e_n\} : \{f_1, \ldots, f_n\}\big\}$ is a biorthogonal system. Moreover, $\sup\{|f_j(x)| : x \in B_X\} \leq 1$ for each j, therefore $\|f_j\| = 1$. Thus $\{e_i; f_i\}_{i=1}^n$ is an Auerbach basis of X.

(ii) Let $\{e_i; f_i\}_{i=1}^n$ be an Auerbach basis of Y. We extend f_i to norm-one functionals on X. Then we define an operator $P\colon X \to Y$ by $P(x) = \sum_{i=1}^n f_i(x)e_i$ for $x \in X$.

We claim that P is a projection onto Y. Indeed, for every $y = \sum \alpha_i e_i \in Y$ we have $\alpha_i = f_i(y)$. Therefore $P(y) = \sum_{i=1}^n f_i(y)e_i = y$. Finally, if $x \in X$ and $\|x\| \leq 1$, then $\|P(x)\| \leq \sum_{i=1}^n |f_i(x)| \, \|e_i\| \leq \sum_{i=1}^n 1 = n$. □

4.2 Basics on Schauder Bases

Definition 4.6 *Let X be an infinite-dimensional normed linear space. A sequence $\{e_i\}_{i=1}^\infty$ in X is called a* Schauder basis *of X if for every $x \in X$ there is a unique sequence of scalars $\{a_i\}_{i=1}^\infty$, called the* coordinates *of x, such that $x = \sum_{i=1}^\infty a_i e_i$.*

Clearly $\{e_i\}_{i=1}^\infty$ is a linearly independent set in X.

If a Banach space X has a Schauder basis, then it is separable as finite rational combinations of elements of the basis form a countable dense set.

If X is n-dimensional, the notion of Schauder basis coincides with the notion of algebraic basis. We will write $\{e_i\}$, $\operatorname{span}\{e_i\}$, etc. to cover both the finite- and infinite-dimensional case.

The condition in the definition can be weakened. If every vector $x \in X$ has a unique decomposition $\sum a_i e_i$ such that $\sum_{i=1}^n a_i e_i \overset{w}{\to} x$ as $n \to \infty$, it already implies that $\{e_i\}$ is a Schauder basis (see, e.g., [Karl]).

If $\{e_i\}$ is a Schauder basis of a normed space X, then the *canonical projections* $P_n\colon X \to X$ are defined for $n \in \mathbb{N}$ by $P_n\Big(\sum_{i=1}^\infty a_i e_i\Big) = \sum_{i=1}^n a_i e_i$. Each P_n

is a linear projections from X onto the linear subspace spanned by $\{e_i : i = 1, 2, \ldots, n\}$.

Lemma 4.7 *Let $\{e_i\}$ be a Schauder basis of a normed space X. The canonical projections P_n satisfy:*
(i) $\dim\big(P_n(X)\big) = n$,
(ii) $P_n P_m = P_m P_n = P_{\min(m,n)}$,
(iii) $P_n(x) \to x$ *in X for every $x \in X$.*
Conversely, if bounded linear projections $\{P_n\}_{n=1}^{\infty}$ in a normed space X satisfy (i)–(iii), *then P_n are canonical projections associated with some Schauder basis of X.*

Proof: The set $\{e_i : n \in \mathbb{N}\}$ is a linearly independent set in X. Thus (i) follows, (ii) is obvious. The property (iii) follows directly from the definition of the Schauder basis.

Conversely, if projections P_n satisfy (i)–(iii) put $P_0 = 0$ and choose $0 \neq e_i \in P_i(X) \cap \mathrm{Ker}(P_{i-1})$ for $i \in \mathbb{N}$. Then

$$x = \lim_{n\to\infty} P_n(x) = \lim_{n\to\infty} \big(P_n(x) - P_0(x)\big) = \lim_{n\to\infty} \sum_{i=1}^{n} \big(P_i(x) - P_{i-1}(x)\big)$$
$$= \sum_{i=1}^{\infty} \big(P_i(x) - P_{i-1}(x)\big) = \sum_{i=1}^{\infty} \alpha_i e_i$$

for some scalars α_i as $\dim\big(P_n(X)/P_{n-1}(X)\big) = 1$, $i \in \mathbb{N}$.

The uniqueness of $\{\alpha_i : i \in \mathbb{N}\}$ follows from the fact that if $x = \sum_{i=1}^{\infty} \beta_i e_i$, then by the continuity of P_n we get $P_n(x) = \sum_{i=1}^{n} \beta_i e_i$, hence $\beta_i e_i = P_i(x) - P_{i-1}(x) = \alpha_i e_i$, for all $n \in \mathbb{N}$.

Thus $\{e_i\}$ is a Schauder basis of X and P_n are the projections associated with $\{e_i\}$. □

Fact 4.8 *Let $\{e_i\}$ be a Schauder basis of a normed linear space X with canonical projections P_n. If* $\sup_n \|P_n\| < \infty$ *(we say that P_n are* uniformly bounded*), then $\{e_i\}$ is also a Schauder basis of the completion $\widetilde{X}$ of X.*

Proof: We will show that the extensions $\widetilde{P}_n$ of P_n on $\widetilde{X}$ satisfy (i)–(iii) of Lemma 4.7. Since $P_n(X)$ is finite-dimensional, it is closed in $\widetilde{X}$ and thus $\widetilde{P}_n\big(\widetilde{X}\big) = P_n(X)$, so (i) follows. (ii) is extended from P_n to $\widetilde{P}_n$ by the continuity of P_n. Since $P_n(x) \to x$ for all x in a dense set X and P_n are uniformly bounded, we have $\widetilde{P}_n(\tilde{x}) \to \tilde{x}$ in $\widetilde{X}$, so (iii) is also true. Since $e_n \in P_n(X) \cap \mathrm{Ker}(P_{n-1})$, we get $e_n \in \widetilde{P}_n\big(\widetilde{X}\big) \cap \mathrm{Ker}\big(\widetilde{P}_{n-1}\big)$ for every n. Therefore $\widetilde{P}_n$ are canonical projections associated with the Schauder basis $\{e_i\}$ of $\widetilde{X}$. □

Lemma 4.9 *Let $\{e_i\}$ be a Schauder basis of a Banach space $(X, \|\cdot\|)$. Define $|||\cdot|||$ on X by $|||x||| = \sup_n \left\|\sum_{i=1}^n a_i e_i\right\|$ for $x = \sum_{i=1}^\infty a_i e_i$. Then:*
(i) $|||\cdot|||$ is a norm on X, $\{e_i\}$ is a Schauder basis of $(X, |||\cdot|||)$ and the canonical projections P_n are uniformly bounded by 1 in $|||\cdot|||$,
(ii) $|||\cdot|||$ is an equivalent norm on X.

Proof: (i) The triangle inequality and homogeneity of $|||\cdot|||$ are simple to check. Since for every $x \in X$ we have $\|x\| = \lim_{n\to\infty} \left\|\sum_{i=1}^n a_i e_i\right\|$ by Lemma 4.7, we obtain that $|||x||| \geq \|x\|$ for every $x \in X$. This in particular means that $|||\cdot|||$ is a norm on X.

To show that $\{e_i\}$ is a Schauder basis of $(X, |||\cdot|||)$, we use Lemma 4.7. The properties (i) and (ii) are straightforward. To check (iii), we note that for $x \in X$ we have

$$|||x - P_m(x)||| = \sup_n \|P_n(x) - P_n P_m(x)\| = \sup_{n\geq m} \|P_n(x) - P_m(x)\| \to 0$$

as $m \to \infty$. Finally, for $m \in \mathbb{N}$ we estimate

$$\begin{aligned} |||P_m||| &= \sup_{|||x|||\leq 1} |||P_m(x)||| = \sup_{|||x|||\leq 1} \sup_n \|P_n P_m(x)\| = \sup_n \sup_{|||x|||\leq 1} \|P_n P_m(x)\| \\ &= \sup_n \left\{ \sup \left\{ \|P_n P_m(x)\| : \ x \text{ with } \sup_i \|P_i(x)\| \leq 1 \right\} \right\} \leq 1. \end{aligned}$$

(ii) We will show that X is complete in the norm $|||\cdot|||$, i.e., that $\widetilde{X} \subset X$, where $\widetilde{X}$ is the completion of X in $|||\cdot|||$. By (i) we already know that $\{e_i\}$ is a Schauder basis of $\widetilde{X}$. Given $x \in \widetilde{X}$, there is a unique sequence of scalars α_i such that $x = \sum \alpha_i e_i$, where the convergence is in the norm $|||\cdot|||$. Since $|||\cdot||| \geq \|\cdot\|$ on X, we get that $\sum \alpha_i e_i$ is Cauchy in $\|\cdot\|$ and thus convergent to some element $x' \in X$ in the norm $\|\cdot\|$. As shown in part (i), $\sum \alpha_i e_i$ then converges to x' in the norm $|||\cdot|||$. Thus $x = x' \in X$. This means that X is complete in $|||\cdot|||$.

The formal identity mapping $I_X \colon \big(X, |||\cdot|||\big) \to \big(X, \|\cdot\|\big)$ is a linear bijection of a Banach space $\big(X, |||\cdot|||\big)$ onto a Banach space $\big(X, \|\cdot\|\big)$ which is continuous as $|||\cdot||| \geq \|\cdot\|$. From the Banach open mapping principle it follows that I_X^{-1} is continuous, which means that $|||\cdot|||$ is an equivalent norm on X. □

Theorem 4.10 (Banach) *Let $\{e_i\}$ be a Schauder basis of a Banach space X. The canonical projections P_n associated with $\{e_i\}$ are uniformly bounded.*

The value $\mathrm{bc}\{e_i\} = \sup_n \|P_n\|$ is called the *basis constant* of $\{e_i\}$.

Proof: Define $|||\cdot|||$ as in Lemma 4.9. Then $|||P_n||| \leq 1$ for every n and since $|||\cdot|||$ is an equivalent norm, the result follows. □

Considering the vectors e_n we see that $\|P_n\| \geq 1$, in particular $\mathrm{bc}\{e_i\} \geq 1$. A Schauder basis $\{e_i\}$ is called *normalized* if $\|e_i\| = 1$ for every n. A Schauder basis $\{x_i\}$ of a Banach space X is called *seminormalized* if $0 < \inf \|x_i\| \leq \sup \|x_i\| <$

∞. It is called *monotone* if $\mathrm{bc}\{e_i\} = 1$, that is, its associated projections satisfy $\|P_n\| = 1$ for every n. We proved in Lemma 4.9 that every Banach space with a Schauder basis can be renormed in such a way that, in the new norm the basis becomes monotone.

It follows from Theorem 4.10 that, if $\{e_i\}$ is a Schauder basis of a Banach space X and $n \in \mathbb{N}$, then $X = \mathrm{span}\{e_i : i = 1, 2, \ldots, n\} \oplus \overline{\mathrm{span}}\{e_i : i = n+1, n+2, \ldots\}$ is a topological direct decomposition. Indeed, P_n is continuous, and certainly $(I - P_n)X = \mathrm{Ker}\, P_n$.

Let $\{e_i\}$ be a Schauder basis of a Banach space X. For $j \in \mathbb{N}$ and $x = \sum_{i=1}^{\infty} a_i e_i$ denote $f_j(x) = a_j$. Then $\|P_j(x) - P_{j-1}(x)\| = \|f_j(x)e_j\| = |f_j(x)| \cdot \|e_j\|$ and thus

$$\|f_j\| = \sup_{x \in B_X} |f_j(x)| = \|e_j\|^{-1} \sup_{x \in B_X} \|f_j(x)e_j\| \leq 2\|e_j\|^{-1} \sup_n \|P_n\|.$$

Therefore $f_j \in X^*$ for every j. The functionals $\{f_i\}$ are called the *associated biorthogonal functionals* (or *coordinate functionals*) to $\{e_i\}$ and we have $x = \sum_{i=1}^{\infty} f_i(x)e_i$ for every $x \in X$.

We will denote the biorthogonal functionals f_i by e_i^* and say that $\{e_i; e_i^*\}$ *is a Schauder basis of the Banach space* X. It is a biorthogonal system and we have just proved that $\|e_i\| \, \|e_i^*\| \leq 2\,\mathrm{bc}\{e_i\}$. Note that $\{e_i^*\}$ is separating for X.

Examples:

(i) Any linear basis of a finite-dimensional Banach space X is a Schauder basis of X. In particular, any Auerbach basis (see the definition right before Theorem 4.5) is a Schauder basis.

(ii) Any orthogonal basis of a Hilbert space H is a Schauder basis of H.

(iii) If $X = c_0$ or $X = \ell_p$ for $p \in [1, \infty)$, then the sequence $\{e_i\}$ of the standard unit vectors $e_i = (0, \ldots, 0, \overset{i}{1}, 0, \ldots)$ is a Schauder basis of X.

All the above statements are easy to verify.

(iv) Let $\{t_j\}_{j=1}^{\infty}$ be a sequence of distinct points in $[0, 1]$ such that $t_1 = 0, t_2 = 1$, and $\overline{\{t_j\}} = [0, 1]$. Define projections P_n from $C[0, 1]$ into $C[0, 1]$ by $P_1(f) = f(0)$ and $P_n(f)$ to be the piecewise linear function with nodes at t_j, $j = 1, \ldots, n$ and such that $P_n(f)(t_j) = f(t_j)$ for $j = 1, \ldots, n$.

By Lemma 4.7 and the uniform continuity of continuous functions on $[0, 1]$, it follows that the projections P_n determine a monotone Schauder basis of $C[0, 1]$. This basis is called the *Faber–Schauder basis of* $C[0, 1]$. Figure 4.1 shows the first elements of this basis for the particular election of the dyadic points.

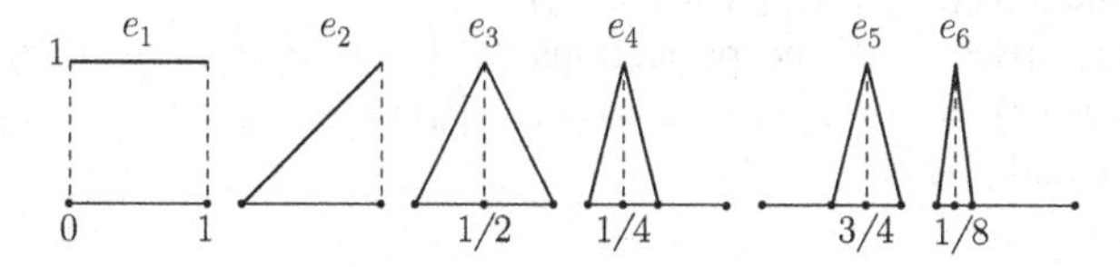

Fig. 4.1 The first elements of the Faber–Schauder basis for dyadic nodes

(v) Let the functions h_i be defined on $[0, 1]$ as follows (see Fig. 4.2):

$h_0(x) = 1$ for $x \in [0, 1]$; $h_1(x) = 1$ for $x \in \left[0, \frac{1}{2}\right)$ and $h_1(x) = -1$ for $x \in \left[\frac{1}{2}, 1\right]$; $h_2(x) = 1$ for $x \in \left[0, \frac{1}{4}\right)$, $h_2(x) = -1$ for $x \in \left[\frac{1}{4}, \frac{1}{2}\right]$, $h_2(x) = 0$ for $x \in \left(\frac{1}{2}, 1\right]$; $h_3(x) = 1$ for $x \in \left[\frac{1}{2}, \frac{3}{4}\right)$, $h_3(x) = -1$ for $x \in \left(\frac{3}{4}, 1\right]$ and $h_3(x) = 0$ elsewhere, etc. Fix $p \in [1, \infty)$. The set $\{h_i\}$ is linearly independent. Since $H = \operatorname{span}\{h_i\}$ contains the characteristic functions of dyadic intervals, we get $\overline{H}^{L_p} = L_p[0, 1]$.

For $x = \sum_{i=0}^{m} \alpha_i h_i \in H$ define $P_n(x) = \sum_{i=0}^{n} \alpha_i h_i$ (we can always assume that $m \geq n$ by adding $0h_j$ to the sum). These projections satisfy (i)–(iii) of Lemma 4.7, so by Fact 4.8, to show that $\{h_i\}$ is a Schauder basis (called the *Haar basis*) of $L_p[0, 1]$ we only need to prove that P_n are uniformly bounded on $(H, \|\cdot\|_p)$.

Assume that $f = \sum_{i=1}^{n} a_i h_i$ and $g = \sum_{i=1}^{n+1} a_i h_i$ for some real numbers a_i. Then f and g differ only on some dyadic interval I where f has a constant value, say b, and g has value $b + a_{n+1}$ on the first half of I and the value $b - a_{n+1}$ on the second half of I. Since for every $p \in [1, \infty)$ we have $|b|^p = |\frac{1}{2}(b+a_{n+1})+\frac{1}{2}(b-a_{n-1})|^p \leq \frac{1}{2}|b + a_{n+1}|^p + \frac{1}{2}|b - a_{n+1}|^p$ by convexity of $|x|^p$, we get $\|f\| \leq \|g\|$ in L_p and it follows that the projections P_n have norm at most 1. Thus $\{h_i\}$ is a monotone Schauder basis of $L_p[0, 1]$. By inspection, we see that $\{h_i\}$ is an orthonormal basis of $L_2[0, 1]$.

Note that, for the Faber–Schauder basis, $f_0 = 1$ and f_n, for $n \geq 1$, is the indefinite integral of h_{n-1}, where $\{h_n\}$ is the Haar basis.

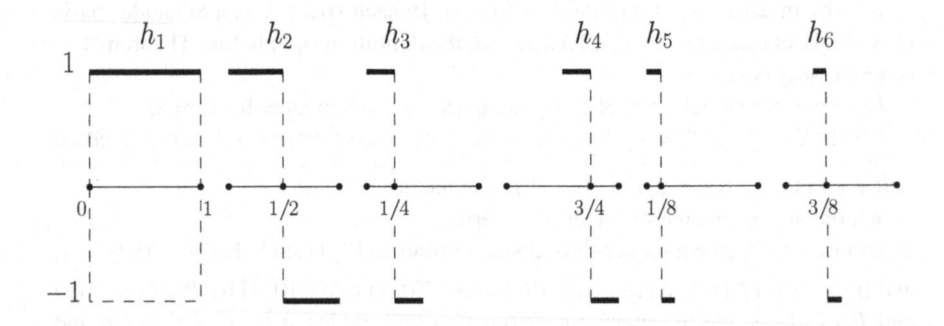

Fig. 4.2 The first elements of the Haar basis

(vi) The trigonometric system defined in Note (v) after Theorem 1.57 is a Schauder basis in $L_p[0, 1]$, $p > 1$ (see, e.g., [Katz, p. 50]) but it is not a Schauder basis in $L_1[0, 1]$ (see, e.g., [Katz, p. 47]). It is not a Schauder basis in the space of continuous functions (see, e.g., [AlKa, p. 9]).

(vii) On the space c, define projections P_n for $x = (x_i) \in c$ by $P_1(x) = (x_1, x_1, \ldots)$, $P_2(x) = (x_1, x_2, x_2, \ldots)$, $\ldots$, $P_n(x) = (x_1, x_2, \ldots, x_n, x_n, \ldots)$. Then for $x \in c$ and $n \in \mathbb{N}$,

$$\|x - P_n(x)\| = \|(0, \ldots, 0, x_{n+1} - x_n, x_{n+2} - x_n, \ldots)\| = \max_{j>n} |x_j - x_n| \to 0$$

as $n \to \infty$. By Lemma 4.7, P_n generate a Schauder basis $x_1 = (1, 1, \ldots)$, $x_2 = (0, 1, 1, \ldots)$, ..., $x_n = (0, \ldots, 0, \overset{n}{1}, 1, \ldots)$. Given $z = (z_1, z_2, \ldots) \in c$ and $(\alpha_1, \alpha_2, \ldots, \alpha_n)$ such that $(z_1, z_2, \ldots, z_n, z_n, \ldots) = \alpha_1 x_1 + \cdots + \alpha_n x_n$, we calculate that $z_1 = \alpha_1$, $z_2 = \alpha_1 + \alpha_2$, ..., $z_n = \alpha_1 + \cdots + \alpha_n$. Therefore $P_n(z) = \alpha_1 x_1 + \cdots + \alpha_n x_n$ and $\|P_n(z)\| = \max_{j \le n} |z_j| = \max\{|\alpha_1| : |\alpha_1 + \alpha_2|, \ldots, |\alpha_1 + \cdots + \alpha_n|\}$. The basis $\{x_i\}$ is called the *summing basis* of c.

4.3 Shrinking and Boundedly Complete Bases, Perturbation

Let $\{e_i\}$ be a Schauder basis of X. Considering $e_i \in X^{**}$ we see that $\{e_i^*; e_i\}$ is a biorthogonal system in X^*.

Fact 4.11 *Let $\{e_i; e_i^*\}$ be a Schauder basis of a Banach space X with the canonical projections P_n.*
(i) *For $n \in \mathbb{N}$, $P_n^*(f) = \sum_{i=1}^n f(e_i)e_i^* = \sum_{i=1}^n e_i(f)e_i^*$ for every $f \in X^*$.*
(ii) *$P_n^*(f) \overset{w^*}{\to} f$ in X^* for every $f \in X^*$.*
(iii) *$\{e_i^*; e_i\}$ is a Schauder basis of $\overline{\operatorname{span}}\{e_i^*\}$ with the canonical projections P_n^*. In particular, $P_n^*(f) \to f$ for every $f \in \overline{\operatorname{span}}\{e_i^*\}$.*

Proof: (i) For $n \in \mathbb{N}$, $f \in X^*$ and $x = \sum_{i=1}^\infty e_i^*(x)e_i$ we have

$$P_n^*(f)(x) = f\big(P_n(x)\big) = f\Big(\sum_{i=1}^n e_i^*(x)e_i\Big) = \sum_{i=1}^n f(e_i)e_i^*(x).$$

(ii) By the continuity of $f \in X^*$,

$$\lim_{n\to\infty} P_n^*(f)(x) = \lim_{n\to\infty} \sum_{i=1}^n e_i^*(x)f(e_i) = f\Big(\lim_{n\to\infty} \sum_{i=1}^n e_i^*(x)e_i\Big) = f(x).$$

(iii) We easily check that $P_n^* P_m^* = P_{\min(n,m)}^*$. If $f \in \operatorname{span}\{e_i^*\}$, then $P_n^*(f) = f$ for n large enough and thus $\lim \|P_n^*(f) - f\| = 0$. Since $\|P_n\| = \|P_n^*\|$, $\{P_n^*\}$ are uniformly bounded and we can apply Lemma 4.7 and Fact 4.8. □

Definition 4.12 *Let $\{e_i; e_i^*\}$ be a Schauder basis of a Banach space X.*
It is called shrinking *if* $\overline{\operatorname{span}}\{e_i^*\} = X^*$.
It is called boundedly complete *if $\sum_{i=1}^\infty a_i e_i$ converges whenever the scalars a_i are such that* $\sup_n \Big\| \sum_{i=1}^n a_i e_i \Big\| < \infty$.

Every normalized shrinking basis $\{e_i\}$ has the property that $e_i \overset{w}{\to} 0$ because $\lim_{i\to\infty} e_k^*(e_i) = 0$ for every $k \in \mathbb{N}$. However, there exists a Banach space with a normalized non-shrinking Schauder basis $\{e_i\}$ such that $e_i \overset{w}{\to} 0$ in X ([PeSz]).

Fact 4.13 *Let $\{e_i; e_i^*\}$ be a Schauder basis of a Banach space X with the canonical projections P_n. The following are equivalent:*
(i) $\{e_i; e_i^*\}$ is shrinking.
(ii) $\{e_i^*\}$ is a Schauder basis of X^*.
(iii) $\lim_{n\to\infty} \|f|_{\overline{\text{span}}\{e_i\}_{i>n}}\| = 0$ for every $f \in X^*$.

Proof: (i)$\Longrightarrow$(ii): Fact 4.11.

(ii)$\Longrightarrow$(i): If projections $\{P_n^*\}$ generate a Schauder basis of X^*, then $P_n^*(f) \to f$ for every f and thus $X^* = \overline{\text{span}}\{e_i^*\}$.

(i) $\Longleftrightarrow$ (iii): Note that if P is a bounded linear projection of a Banach space X onto $P(X)$, then $\sup_{P(B_X)} (f) = \sup_{B_X}\big(P^*(f)\big) = \|P^*(f)\|$ and

$$B_{P(X)} \subset P(B_X) \subset \|P\| B_X \cap P(X) \subset \|P\| B_{P(X)}.$$

Thus we have for every $f \in X^*$ and n,

$$\begin{aligned} \|f|_{(I_X-P_n)(X)}\| &= \sup\{f(x) : x \in B_{(I_X-P_n)(X)}\} \\ &\le \sup\{f(x) : x \in (I_X - P_n)(B_X)\} \le \sup\{f(x) : x \in (\|P_n\| + 1)B_{(I_X-P_n)(X)}\}. \end{aligned}$$

Hence $\|f|_{(I_X-P_n)(X)}\| \le \|f - P_n^*(f)\| \le (\|P_n\| + 1)\|f|_{(I_X-P_n)(X)}\|$. Thus $\{e_i\}$ is shrinking if and only if $\|f|_{(I_X-P_n)(X)}\| \to 0$ for every $f \in X^*$. $\square$

For a sufficient condition for shrinkingness in terms of the smoothness of the norm, see Proposition 8.22.

The standard unit vector basis of c_0 and ℓ_p, $p \in (1, \infty)$, is shrinking, while the standard unit vector basis of ℓ_1 is not. The standard unit vector basis of ℓ_p is boundedly complete for $p \in [1, \infty)$ while the standard unit vector basis of c_0 is not, as the vectors $(1, \dots, \overset{n}{1}, 0, \dots)$ show.

Proposition 4.14 *Let $\{e_i; e_i^*\}$ be a Schauder basis of a Banach space X. If $\{e_i\}$ is shrinking, then the mapping $T(x^{**}) = \big(x^{**}(e_i^*)\big)$ is an isomorphism of X^{**} onto to the space of all sequences (a_i) such that $|||(a_i)||| = \sup_n \left\|\sum_{i=1}^n a_i e_i\right\| < \infty$. Moreover, if $\{e_i\}$ is monotone, then T is an isometry.*

Proof: It is routine to check that $|||\cdot|||$ defines a norm on the vector space of (a_i) such that $|||(a_i)||| < \infty$. Denote $K = \text{bc}\{e_i\}$, let P_n be the canonical projections associated with $\{e_i\}$. For $x \in X$, $x^* \in X^*$ and $x^{**} \in X^{**}$ we have $P_n^*(x^*) = \sum_{i=1}^n x^*(e_i)e_i^*$ and $P_n^{**}(x^{**})(x^*) = \sum_{i=1}^n x^{**}(e_i^*)x^*(e_i)$, so we can write $P_n^{**}(x^{**}) = \sum_{i=1}^n x^{**}(e_i^*)e_i$. Thus

$$|||T(x^{**})||| = \sup_n \left\| \sum_{i=1}^{n} x^{**}(e_i^*)e_i \right\| = \sup_n \|P_n^{**}(x^{**})\| \leq K\|x^{**}\|$$

and T is a bounded operator with $\|T\| \leq K$.

Now consider $(a_i)_{i=1}^{\infty}$ such that $|||(a_i)||| < \infty$. Since X^* is separable (as $\{e_i\}$ is shrinking) and $\left\{\sum_{i=1}^{n} a_i e_i\right\}$ is bounded in X^{**}, there exists a w^*-cluster point x^{**} of $\left\{\sum_{i=1}^{n} a_i e_i\right\}$ which satisfies $x^{**}(e_i^*) = a_i$. Moreover, $\|x^{**}\| \leq \limsup \left\| \sum_{i=1}^{n} a_i e_i \right\| \leq |||(a_i)|||$. Thus $T(x^{**}) = (a_i)$ and $\|T(x^{**})\| \geq \|x^{**}\|$, which completes the proof. □

Theorem 4.15 *Let $\{e_i; e_i^*\}$ be a Schauder basis of a Banach space X. If $\{e_i\}$ is boundedly complete, then X is isomorphic to $\left(\overline{\text{span}}\{e_i^*\}\right)^*$.*

Proof: Let P_n be the canonical projections associated with $\{e_i\}$. Denote $Z = \overline{\text{span}}\{e_i^*\}$ and define $J: X \to Z^*$ by $J(x): z \mapsto z(x)$, then J is a bounded operator. We will show that J is an isomorphism of X onto Z^*.

Let $x \in X$. Then for every $z \in Z$ we have $|J(x)(z)| = |z(x)| \leq \|z\|\,\|x\|$, so $\|J(x)\| \leq \|x\|$. On the other hand, for $n \in \mathbb{N}$ find $x^* \in S_{X^*}$ such that $x^*\big(P_n(x)\big) = \|P_n(x)\|$. Since $P_n^*(X^*) = \text{span}\{e_i^*\}_{i=1}^{n}$, we have $P_n^*(x^*) \in Z$, and $\|P_n^*(x^*)\| \leq K = \text{bc}\{e_i\}$. By definition, $J\big(P_n(x)\big)\big(P_n^*(x^*)\big) = \big(P_n^*(x^*)\big)\big(P_n(x)\big) = x^*\big(P_n^2(x)\big) = x^*\big(P_n(x)\big) = \|P_n(x)\|$ and therefore $\|J\big(P_n(x)\big)\|_{Z^*} \geq J\big(P_n(x)\big)\big(\frac{P_n^*(x^*)}{\|P_n^*(x^*)\|}\big) = \frac{1}{\|P_n^*(x^*)\|}\|P_n(x)\| \geq \frac{1}{K}\|P_n(x)\|$. By the continuity of J, we have $\frac{1}{K}\|x\| \leq \|J(x)\|_{Z^*} \leq \|x\|$ for every $x \in X$.

We will now show that J maps X onto Z^*. To this end, observe first that $\{e_i^*, J(e_i)\}$ is a Schauder basis of Z, let $\widetilde{P}_n$ denote its canonical projections. Then $\widetilde{P}_n^*(z^*) \overset{w^*}{\to} z^*$ in Z^* and $\sup_n \|\widetilde{P}_n^*\| = \sup_n \|\widetilde{P}_n\| \leq K < \infty$, hence for every $z^* \in Z^*$ and $n \in \mathbb{N}$ we have

$$\left\| J\Big(\sum_{i=1}^{n} z^*(e_i^*)e_i\Big) \right\|_{Z^*} = \left\| \sum_{i=1}^{n} z^*(e_i^*)J(e_i) \right\|_{Z^*} = \|\widetilde{P}_n^*(z^*)\| \leq K \cdot \|z^*\|.$$

Thus we have $\left\| \sum_{i=1}^{n} z^*(e_i^*)e_i \right\| \leq K \cdot \left\| J\Big(\sum_{i=1}^{n} z^*(e_i^*)e_i\Big) \right\|_{Z^*} \leq K^2 \cdot \|z^*\|$. Since the basis $\{e_i\}$ is boundedly complete, the series $\sum_{i=1}^{\infty} z^*(e_i^*)e_i$ is convergent in X to some $x \in X$. J is a continuous mapping, therefore

$$J(x) = \lim_{n\to\infty} J\Big(\sum_{i=1}^{n} z^*(e_i^*)e_i\Big) = \lim_{n\to\infty} \sum_{i=1}^{n} z^*(e_i^*)J(e_i) = \lim_{n\to\infty} \widetilde{P}_n^*(z^*)$$

in the norm topology of Z^*. But $\widetilde{P}_n^*(z^*) \overset{w^*}{\to} z^*$ in Z^*, so $z^* = J(x)$ and J is an onto mapping. This concludes the proof. □

Theorem 4.16 (James, see, e.g., [LiTz3]) *Let X be a Banach space with a Schauder basis $\{e_i\}$. The space X is reflexive if and only if $\{e_i\}$ is both shrinking and boundedly complete.*

Proof: Let X be reflexive. By Fact 4.11, for every $f \in X^*$ we have $P_n^*(f) \overset{w^*}{\to} f$ and thus $P_n^*(f) \overset{w}{\to} f$ in X^* as X is reflexive. Therefore $X^* = \overline{\text{span}}^w\{e_i^*\} = \overline{\text{span}}\{e_i^*\}$ by Mazur's theorem and $\{e_i\}$ is a shrinking basis of X.

By Proposition 4.14, the space X^{**} is isomorphic to the (Banach) space $Y := \left\{(a_i) : |||(a_i)||| := \max_n \left\|\sum_{i=1}^n a_i e_i\right\| < \infty\right\}$. Under this correspondence, $X \subset X^{**}$ corresponds to $Y_1 = \left\{(a_i) : \sum_{i=1}^\infty a_i e_i \text{ converges}\right\}$. Since X is reflexive, we get $Y = Y_1$ and $\{e_i\}$ is boundedly complete.

Conversely, if $\{e_i\}$ is shrinking and boundedly complete, then we have the above identification and $Y_1 = Y$, thus $X = X^{**}$ and X is reflexive. □

Definition 4.17 *A sequence $\{e_i\}$ in a Banach space X is called a* basic sequence *if $\{e_i\}$ is a Schauder basis of $\overline{\text{span}}\{e_i\}$.*
A basic sequence $\{e_i\}$ is called shrinking *(respectively,* boundedly complete*) if it is a shrinking (respectively boundedly complete) basis of $\overline{\text{span}}\{e_i\}$.*

Proposition 4.18 (Banach) *Let $\{e_i\}$ be a sequence of non-zero vectors in a Banach space X. The sequence $\{e_i\}$ is a basic sequence if and only if there is $K > 0$ such that for all $n < m$ and scalars $a_1, \dots, a_m$ we have*

$$\left\|\sum_{i=1}^n a_i e_i\right\| \le K \left\|\sum_{i=1}^m a_i e_i\right\|. \tag{4.1}$$

Moreover, the smallest such K is equal to $\text{bc}\{e_i\}$.

Proof: One implication is clear from

$$\left\|\sum_{i=1}^n a_i e_i\right\| = \left\|P_n\Big(\sum_{i=1}^m a_i e_i\Big)\right\| \le \|P_n\| \left\|\sum_{i=1}^m a_i e_i\right\| \le \text{bc}\{e_i\} \left\|\sum_{i=1}^m a_i e_i\right\|.$$

On the other hand, suppose that K satisfies $\left\|\sum_{i=1}^n a_i e_i\right\| \le K \left\|\sum_{i=1}^m a_i e_i\right\|$ for all a_i and $n \le m$. We define projections P_n on $\text{span}\{e_i\}$ by $P_n\left(\sum_{i=1}^m a_i e_i\right) := \sum_{i=1}^n a_i e_i$ for $m > n$ and scalars a_i and observe that P_n have norm at most K. We check that P_n satisfy (i)–(iii) of Lemma 4.7 on $\text{span}\{e_i\}$, so by Fact 4.8, $\{e_i\}$ is a Schauder basis of $\overline{\text{span}}\{e_i\}$ and $\text{bc}\{e_i\} \le K$. □

Not every separable Banach space admits a Schauder basis (Enflo, see Section 16.5 and Theorem 16.54). But we have the following.

Theorem 4.19 (Mazur) *Every infinite-dimensional Banach space contains a basic sequence. If X^* is separable, then X contains a shrinking basic sequence.*

In the proof, we use the following lemma.

Lemma 4.20 *Let Y be a finite-dimensional subspace of an infinite-dimensional Banach space X. For every $\varepsilon > 0$ there is $x \in S_X$ such that $\|y\| \leq (1+\varepsilon)\|y+\lambda x\|$ for every $y \in Y$ and every scalar λ.*

Proof: (see Fig. 4.3) Let $\varepsilon \in (0, 1)$. Let $\{y_i\}_{i=1}^m$ be an $\frac{\varepsilon}{2}$-net in S_Y. For $i \in \{1, \ldots, m\}$ choose $y_i^* \in S_{X^*}$ with $y_i^*(y_i) = 1$. Since X is infinite-dimensional, there is $x \in S_X$ such that $y_i^*(x) = 0$ for every $i = 1, \ldots, m$. We claim that x has the desired property. Indeed, let $y \in S_Y$. Choose $i \in \{1, \ldots, m\}$ such that $\|y_i - y\| < \frac{\varepsilon}{2}$. Let λ be a scalar. Then

$$\|y + \lambda x\| \geq \|y_i + \lambda x\| - \tfrac{\varepsilon}{2} \geq y_i^*(y_i + \lambda x) - \tfrac{\varepsilon}{2} = 1 - \tfrac{\varepsilon}{2} \geq \tfrac{1}{1+\varepsilon}.$$

Thus, given $y \in Y\backslash\{0\}$ and a scalar λ, we have $\|\frac{y}{\|y\|} + \frac{\lambda}{\|y\|}x\| \geq \frac{1}{1+\varepsilon}$. □

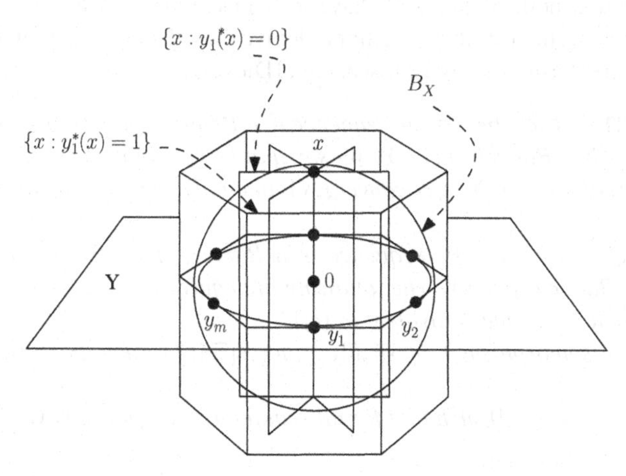

Fig. 4.3 Proof of Lemma 4.20

Proof of Theorem 4.19: Given $\varepsilon > 0$, find $\varepsilon_n > 0$ with $\prod_{n=1}^{\infty}(1 + \varepsilon_n) \leq 1 + \varepsilon$. Let $x_1 \in S_X$ be an arbitrary element. Using Lemma 4.20, construct inductively a sequence $\{x_n\}_{n=2}^{\infty}$ in S_X such that for every $n \geq 1$,

$$\|y\| \leq (1 + \varepsilon_n)\|y + \lambda x_{n+1}\| \text{ for all } y \in \operatorname{span}\{x_1, \ldots, x_n\}.$$

By induction, for $n < m$ and scalars $a_1, \ldots, a_m$ we have

$$\Big\|\sum_{i=1}^{n} a_i x_i\Big\| \leq (1+\varepsilon_n)\cdots(1+\varepsilon_{m-1})\Big\|\sum_{i=1}^{m} a_i x_i\Big\| \leq (1+\varepsilon)\Big\|\sum_{i=1}^{m} a_i x_i\Big\|.$$

By Proposition 4.18, $\{x_i\}$ is a basic sequence and $\mathrm{bc}\{x_i\} \leq 1+\varepsilon$.

Note that $\|P_n\| \leq \prod_{i=n}^{\infty}(1+\varepsilon_i) \to 1$ as $n \to \infty$.

An easy modification of Mazur's construction in Lemma 4.20 and then the proof of Theorem 4.19 gives the last part: let $D = \{d_n^* : n \in \mathbb{N}\}$ be a $\|\cdot\|$-dense countable subset of X^*. At the nth step in the construction of the basic sequence $\{x_n\}$ and for ε_n, consider in X^* not only the (finite) set $\{y_i^* : i = 1, 2, \ldots, m\}$ (see the proof of Lemma 4.20) but also the set $\{d_i^* : i = 1, 2, \ldots, n\}$ in order to find $x_{n+1} \in \bigcap_{i=1}^{m} \mathrm{Ker}\, y_i^* \cap \bigcap_{i=1}^{n} \mathrm{Ker}\, d_i^* \cap S_X$. Now, an appeal to Fact 4.13 gives the conclusion: indeed, given $x^* \in X^*$ and $\varepsilon > 0$, find $n_0 \in \mathbb{N}$ such that $\|d_{n_0}^* - x^*\| < \varepsilon$. Then, $\|d_{n_0}^*|_{\overline{\mathrm{span}}\{x_n:\, n\geq n_1\}}\| = 0$, so $\|x^*|_{\overline{\mathrm{span}}\{x_n:\, n\geq n_1\}}\| < \varepsilon$ for every $n_1 > n_0$. This proves that the basic sequence is shrinking. $\square$

For another proof of the last part in the statement of Theorem 4.19, see the paragraph after Proposition 8.22.

It is not known whether every separable Banach space X contains a closed subspace Y such that both Y and X/Y have a Schauder basis. We note that if X is separable and nonreflexive, there is a nonreflexive closed subspace Y of X such that Y has a Schauder basis (Pełczyński, see, e.g., [Dies2]).

Definition 4.21 *Let $\{e_i\}$ be a basic sequence in a Banach space X and let $\{f_i\}$ be a basic sequence in a Banach space Y. We say that $\{e_i\}$ is* equivalent *to $\{f_i\}$ if for all sequences of scalars (a_i), $\sum a_i e_i$ converges if and only if $\sum a_i f_i$ converges.*

Fact 4.22 *Let $\{e_i\}$ be a basic sequence in a Banach space X and let $\{f_i\}$ be a sequence in a Banach space Y. The following are equivalent:*
(i) *$\{f_i\}$ is a basic sequence equivalent to $\{e_i\}$.*
(ii) *There is an isomorphism T of $\overline{\mathrm{span}}\{e_i\}$ onto $\overline{\mathrm{span}}\{f_i\}$ such that $T(e_i) = f_i$ for every i.*
(iii) *There are $C_1, C_2 > 0$ such that for all scalars $a_1, \ldots, a_n$ we have*

$$\frac{1}{C_1}\Big\|\sum_{i=1}^{n} a_i e_i\Big\|_X \leq \Big\|\sum_{i=1}^{n} a_i f_i\Big\|_Y \leq C_2\Big\|\sum_{i=1}^{n} a_i e_i\Big\|_X.$$

Proof: (i)$\Longrightarrow$(ii): Define a mapping T from $\overline{\mathrm{span}}\{e_i\}$ into $\overline{\mathrm{span}}\{f_i\}$ by $T\Big(\sum_{i=1}^{\infty} a_i e_i\Big) = \sum_{i=1}^{\infty} a_i f_i$. From the equivalence of $\{e_i\}$ and $\{x_i\}$ we have that T is well defined, one-to-one and onto $\overline{\mathrm{span}}\{f_i\}$.

We will now show that T has a closed graph. Indeed, if $x^k = \sum_{i=1}^{\infty} a_i^k e_i$ converge to $x = \sum_{i=1}^{\infty} a_i e_i$ and $T(x^k) = \sum_{i=1}^{\infty} a_i^k f_i$ converge to $\sum_{i=1}^{\infty} c_i f_i$, then by the continuity of the coordinate functionals we have $a_i^k \to a_i$ and in the same way $a_i^k \to c_i$ for every i. Hence $a_i = c_i$ for every i and thus $T(x^k) \to \sum a_i f_i = T(x)$. By the closed graph theorem, T is continuous, and by the open mapping theorem, T^{-1} is continuous as well.

(ii)$\Longrightarrow$(iii): This follows easily with $C_2 = \|T\|$, $C_1 = \|T^{-1}\|$.

(iii)$\Longrightarrow$(i): For $n < m$ and scalars $a_1, \ldots, a_m$ we have $\left\|\sum_{i=1}^n a_i f_i\right\| \leq C_1 C_2 \operatorname{bc}\{e_i\}\left\|\sum_{i=1}^m a_i f_i\right\|$, so by Proposition 4.18, $\{f_i\}$ is a basic sequence. From (iii) we also have that $\sum a_i f_i$ is Cauchy if and only if $\sum a_i e_i$ is Cauchy. □

The following result is sometimes called the *small perturbation lemma.*

Theorem 4.23 (Krein, Milman, Rutman [LiTz3], Valdivia [Vald5]) *Let $\{e_i\}$ be a basic sequence in a Banach space X and let $\{e_i^*\}$ be the coefficient functionals of the basis $\{e_i\}$ of $\overline{\operatorname{span}}\{e_i\}$. Assume that $\{f_i\}$ is a sequence in X such that $\sum_{i=1}^\infty \|e_i - f_i\|\,\|e_i^*\| = C < 1$. Then:*
(i) $\{f_i\}$ is a basic sequence in X equivalent to $\{e_i\}$.
(ii) If $\overline{\operatorname{span}}\{e_i\}$ is complemented in X, then so is $\overline{\operatorname{span}}\{f_i\}$.
(iii) If $\{e_i\}$ is a Schauder basis of X, then so is $\{f_i\}$. Moreover, let $\{f_i^*\}$ be the coefficient functionals of the basis $\{f_i\}$ of X. Then $\overline{\operatorname{span}}\{e_i^*\} = \overline{\operatorname{span}}\{f_i^*\}$.

Proof: (i) Extend e_i^* to functionals on X of the same norm. For $x \in X$ we have $\sum_{i=1}^\infty \|e_i^*(x)(e_i - f_i)\| \leq \|x\| \sum_{i=1}^\infty \|e_i^*\|\,\|e_i - f_i\| = C\|x\|$, so $S(x) = \sum_{i=1}^\infty e_i^*(x)(e_i - f_i)$ defines a bounded operator from X into X with $\|S\| \leq C < 1$. Let $T = I_X - S$. We have $\|x - T(x)\| = \|S(x)\| \leq C\|x\|$, hence $\|T(x)\| \geq (1 - C)\|x\|$. Since $1 - C > 0$, by Exercise 1.73, T is an isomorphism into, in particular $T(X)$ is closed.

We claim that $T(X) = X$. Assume $T(X) \neq X$. Since $C < 1$, by Lemma 1.37 there is $x \in S_X$ such that $\operatorname{dist}(x, T(X)) > C$, which contradicts $\|x - T(x)\| \leq C$. So T is an isomorphism of X onto X. Then using $T(e_i) = f_i$ we obtain that T maps $\overline{\operatorname{span}}\{e_i\}$ onto $\overline{\operatorname{span}}\{f_i\}$ and (i) is proved.

(ii) follows from the proof of (i) using T, Fact 4.3 and the note following this fact.

(iii) Let T be the isomorphism of X onto X from (i). Since $\{e_i\}$ is a Schauder basis of X, we get $X = T(X) = T(\overline{\operatorname{span}}\{e_i\}) = \overline{\operatorname{span}}\{f_i\}$, so $\{f_i\}$ is a Schauder basis of X.

Fix some $i \in \mathbb{N}$ and denote $x_k^* = \sum_{j=1}^k f_i^*(e_j)e_j^*$ for $k \in \mathbb{N}$. Then $x_k^* \in \overline{\operatorname{span}}\{e_j^*\}$ and by Fact 4.11, $x_k^* \overset{w^*}{\to} f_i^*$. For $x \in B_X$ we then have $f_i^*(x) = \sum_{j=1}^\infty f_i^*(e_j)e_j^*(x)$, so for $k \geq i$ we estimate

$$\begin{aligned}|(f_i^* - x_k^*)(x)| &= \Big|\sum_{j=k+1}^\infty f_i^*(e_j)e_j^*(x)\Big| = \Big|\sum_{j=k+1}^\infty f_i^*(e_j - f_j)e_j^*(x)\Big| \\ &\leq \|f_i^*\| \sum_{j=k+1}^\infty \|e_j - f_j\|\,\|e_j^*\| \to 0 \text{ as } k \to \infty.\end{aligned}$$

Since the estimate is independent of $x \in B_X$, we get $\|f_i^* - x_k^*\| \to 0$ as $k \to \infty$ and thus $f_i^* \in \overline{\operatorname{span}}\{e_i^*\}$. Therefore $\overline{\operatorname{span}}\{f_i^*\} \subset \overline{\operatorname{span}}\{e_i^*\}$.

Note that $(T^{-1})^*(e_i^*) = f_i^*$, so $\|f_i^*\| \leq \|(T^{-1})^*\| \|e_i^*\|$ for $i \in \mathbb{N}$. Thus the series $\sum \|f_i^*\| \|e_i - f_i\|$ converges and we can reverse the roles of e_i and f_i in the last paragraph, obtaining $\overline{\operatorname{span}}\{e_i^*\} \subset \overline{\operatorname{span}}\{f_i^*\}$. □

The following corollary will be used in Chapter 5.

Corollary 4.24 *Assume that a finite-dimensional subspace E of a Banach space X is complemented in X by a projection P. If $\{e_i : i = 1, 2, \dots, n\}$ is a basis in E, then there is $\varepsilon > 0$ so that if $f_i \in X$, $i = 1, 2, \dots, n$, satisfy $\|e_i - f_i\| < \varepsilon$ for all $i = 1, 2, \dots, n$, then* span$\{f_i : i = 1, 2, \dots, n\}$ *is complemented in X with a projection of norm less or equal than $2\|P\|$.*

4.4 Block Bases, Bessaga–Pełczyński Selection Principle

Definition 4.25 *Let $\{e_i\}$ be a basic sequence in a Banach space X. A sequence of non-zero vectors $\{u_j\}$ in X of the form $u_j = \sum_{i=p_j+1}^{p_{j+1}} a_i e_i$ with scalars a_i and $p_1 < p_2 < \dots$ is called a* block basic sequence *of $\{e_i\}$.*

Note that a block basic sequence of $\{e_i\}$ is a basic sequence with basis constant not greater than bc$\{e_i\}$. Indeed, for $k \leq l$ we have

$$\Big\|\sum_{j=1}^{k} \alpha_j u_j\Big\| = \Big\|\sum_{j=1}^{k} \alpha_j \sum_{i=p_j+1}^{p_{j+1}} a_i e_i\Big\| = \Big\|\sum_{j=1}^{k} \sum_{i=p_j+1}^{p_{j+1}} \alpha_j a_i e_i\Big\|$$

$$\leq \operatorname{bc}\{e_i\}\Big\|\sum_{j=1}^{l} \sum_{i=p_j+1}^{p_{j+1}} \alpha_j a_i e_i\Big\| = \operatorname{bc}\{e_i\}\Big\|\sum_{j=1}^{l} \alpha_j u_j\Big\|.$$

The following result is often used when investigating subspaces of a given space.

Theorem 4.26 (Pełczyński [Pelc3]) *Let X be a Banach space with a Schauder basis $\{e_i\}$. If Y is an infinite-dimensional closed subspace of X, then Y contains an infinite-dimensional closed subspace Z with a Schauder basis that is equivalent to a block basic sequence of $\{e_i\}$.*

Proof: Let $K = \operatorname{bc}\{e_i\}$. Given $p \in \mathbb{N}$, let W_p be the finite-codimensional subspace of X defined by

$$W_p = \Big\{x \in X : x = \sum_{i=p+1}^{\infty} a_i e_i\Big\} = \overline{\operatorname{span}}\{e_i\}_{i>p}.$$

Then $W_p \cap Y$ is infinite-dimensional, so there is $y \in S_Y \cap W_p$. We will inductively construct the two equivalent basic sequences.

Choose an arbitrary $y_1 = \sum_{i=1}^{\infty} a_i^1 e_i \in Y$ with $\|y_1\| = 1$. Find $p_1 \in \mathbb{N}$ such that for $u_1 = \sum_{i=1}^{p_1} a_i^1 e_i \in X$ we have $\|y_1 - u_1\| < \frac{1}{4K}$. Choose $y_2 = \sum_{i=p_1+1}^{\infty} a_i^2 e_i \in S_Y \cap W_{p_1}$ and fix $p_2 \in \mathbb{N}$ such that for $u_2 = \sum_{i=p_1+1}^{p_2} a_i^2 e_i$ we have $\|y_2 - u_2\| < \dfrac{1}{2 \cdot 2^2 K}$. Continue in this manner. Then $\{u_j\}$ is a block basic sequence of $\{e_i\}$. Since $\sum_{j=1}^{\infty} \|y_j - u_j\| < \frac{1}{2K}$ and $\|u_j^*\| \leq 2\,\mathrm{bc}\{u_j\} \leq 2K$, by Theorem 4.23, $\{y_j\}$ is a basic sequence in Y equivalent to $\{u_j\}$, so the subspace $Z = \overline{\mathrm{span}}\{y_j\}$ has the desired property. □

The above procedure of finding vectors with almost successive supports is called the "sliding hump argument." An easy modification provides the following.

Corollary 4.27 (Bessaga–Pełczyński selection principle, see, e.g., [LiTz3]) *Let X be a Banach space with a Schauder basis $\{e_i\}$. If a sequence $\{x_n\}$ satisfies $\inf \|x_n\| > 0$ and $x_n \overset{w}{\to} 0$, then some subsequence $\{x_{n_k}\}$ of $\{x_n\}$ is a basic sequence equivalent to a block basic sequence of $\{e_i\}$.*

Theorem 4.28 (Johnson–Rosenthal, [JoRo], see also [LiTz3, p. 10]) *Let X be a separable Banach space. Then X has a quotient with a Schauder basis.*

Proof: Here $[L]$ will denote the norm-closed linear hull of a set L. Let $\{y_k\} \subset S_{X^*}$, $y_k \overset{w^*}{\to} 0$. Let $\{\varepsilon_n\}$ be a sequence of positive numbers less than 1 such that $\sum_{n=1}^{\infty} \varepsilon_n < \infty$ and, consequently, $\prod_{n=1}^{\infty}(1 - \varepsilon_n)^{-1} < \infty$.

Using Helly's theorem, the compactness of the unit ball of a finite-dimensional space and the separability of X, we may choose an increasing sequence $k_1 < k_2 < \ldots$ of positive integers and finite subsets $F_1 \subset F_2 \subset \ldots$ of S_X such that the linear span of $\bigcup_{i=1}^{\infty} F_i$ is dense in X and that for each $n \in \mathbb{N}$,

(i) for each $f \in [\{y_{k_i}\}_{i=1}^n]^*$ with $\|f\| = 1$, there is $x \in F_n$ such that

$$|y(x) - f(y)| \leq (\varepsilon_n/3)\|y\|, \text{ for all } y \in [(y_{k_i})_{i=1}^n],$$

(ii) $|y_{k_{n+1}}(x)| < \varepsilon_n/3$, for all $x \in F_n$.

We will show that (y_{k_i}) is a basic sequence. For it, fix $n \in \mathbb{N}$ and let numbers $a_1, a_2, \ldots, a_n$ be given such that $\|\sum_{i=1}^n a_i y_{k_i}\| = 1$. Pick $f \in [\{y_{k_i}\}_{i=1}^n]^*$ such that $f(\sum_{i=1}^n a_i y_{k_i}) = 1 = \|f\|$ and choose $x \in F_n$ that satisfies (i) for this f.

It follows that

$$\left| \left(\sum_{i=1}^{n} a_i y_{k_i} \right)(x) \right| \geq 1 - \frac{\varepsilon_n}{3}.$$

Then, for any number λ,

$$\left\|\sum_{i=1}^{n} a_i y_{k_i} + \lambda y_{k_{n+1}}\right\| \geq \left|\sum_{i=1}^{n} a_i y_{k_i}(x) + \lambda y_{k_{n+1}}(x)\right|$$
$$\begin{cases} \geq 1 - \frac{\varepsilon_n}{3} - \lambda\frac{\varepsilon_n}{3} \geq 1 - \frac{\varepsilon_n}{3} - 2\frac{\varepsilon_n}{3} = 1 - \varepsilon_n, & \text{if } \lambda \leq 2, \\ \geq 1, & \text{otherwise.} \end{cases}$$

Thus

$$\left\|\sum_{i=1}^{n} a_i y_{k_i}\right\| \leq \frac{1}{1-\varepsilon_n}\left\|\sum_{i=1}^{n+1} a_i y_{k_i}\right\|$$

holds for any numbers $a_1, a_2, \dots, a_{n+1}$. Then, by induction, we get that for any k and for any numbers $a_1, \dots, a_{n+k}$,

$$\left\|\sum_{i=1}^{n} a_i y_{k_i}\right\| \leq \left(\prod_{j=1}^{n+k-1} \frac{1}{1-\varepsilon_j}\right)\left\|\sum_{i=1}^{n+k} a_i y_{k_i}\right\|$$

and (y_{k_i}) is thus a basic sequence. Let $(f_i)_{i=1}^{\infty}$ denote the functionals in $[(y_{k_i})_{i=1}^{\infty}]$ biorthogonal to $(y_{k_i})_{i=1}^{\infty}$.

Define the projections $[(f_i)_{i=1}^{\infty}] \to [(f_i)_{i=1}^{\infty}]$ by

$$P_m f = \sum_{i=1}^{m} f(y_{k_i}) f_i, \text{ for } f \in [(f_i)_{i=1}^{\infty}].$$

We have $\|P_m\| \leq \prod_{n=m}^{\infty}(1-\varepsilon_n)^{-1}$, and thus $\|P_m\| \to 1$.

Therefore $(f_i)_{i=1}^{\infty}$ is a Schauder basis for $[(f_i)_{i=1}^{\infty}]$. We will show that $[(f_i)_{i=1}^{\infty}]$ is a quotient of X. This will finish the proof.

In order to show that $[(f_i)_{i=1}^{\infty}]$ is a quotient of X, we need to find a bounded operator from X onto $[(f_i)_{i=1}^{\infty}]$. For it, let $T : X \to [(y_{k_i})_{i=1}^{\infty}]^*$ be defined for $x \in X$ by

$$Tx(y) = y(x), \text{ for } y \in [(y_{k_i})_{i=1}^{\infty}].$$

We need to show that $T(X) = [(f_i)_{i=1}^{\infty}]$. We will first show that $T(X) \subset [(f_i)_{i=1}^{\infty}]$. For this, let $x \in F_n$ for some $n \in \mathbb{N}$. Then, by (ii), $\sum_{i=1}^{\infty} |y_{k_i}(x)| < \infty$ and thus

$$Tx(y) = \sum_{i=1}^{\infty} y_{k_i}(x) f_i \in [(f_i)_{i=1}^{\infty}].$$

Since the linear hull of $\bigcup_{i=1}^{\infty} F_i$ is dense in X and T is a bounded operator, we get $T(X) \subset [(f_i)_{i=1}^{\infty}]$. We will now show the following.

Claim:
For every g in the linear hull of $(f_i)_{i=1}^\infty$ with $\|g\| = 1$ and for every $\varepsilon > 0$, there is $x \in X$ with $\|x\| = 1$ such that $\|Tx - g\| < 4\varepsilon$.

Having this claim proved, the open mapping principle gives that $T(X) = [(f_i)_{i=1}^\infty]$ and finishes the proof.

To show the Claim, let $0 < \varepsilon < 1$ and choose N such that $\sum_{j=n}^\infty \varepsilon_j < \varepsilon$ and $\|P_n\| \leq 1 + \varepsilon$ for all $n > N$. Fix $n > N$. Define $\|f\|_1 = \|f|_{[(y_{k_j})_{j=1}^n]}\|$ for $f \in [(f_j)_{j=1}^n]$. Observe that $\|f\|_1 \leq \|f\| \leq \|P_n\|.\|f\|_1 \leq 2\|f\|_1$ for all $f \in [(f_j)_{j=1}^\infty]$.

Now fix $g \in [(f_j)_{j=1}^n]$ with $\|g\| = 1$ and put $f = g/\|g\|_1$. Choose $x \in F_n$ satisfying (i) for f. We have, by (i),

$$\left\| \sum_{j=1}^n y_{k_j}(x) f_j - f \right\|_1 \leq \frac{\varepsilon_n}{3}.$$

Hence $\| \sum_{j=1}^n y_{k_i}(x) f_j - f\| \leq 2\varepsilon_n/3 < (2/3)\varepsilon$.

Moreover, $\|f_j\| = \|P_j - P_{j-1}\| \leq 4$ for all $j > n$ and hence, by (ii),

$$\left\| \sum_{j=n+1}^\infty y_{k_j}(x) f_j \right\| \leq 4 \sum_{j=n}^\infty \frac{\varepsilon_j}{3} < \frac{4}{3}\varepsilon.$$

Thus

$$\|Tx - f\| \leq \frac{4}{3}\varepsilon + \frac{2}{3}\varepsilon = 2\varepsilon.$$

Moreover, since as above,

$$1 = \|g\| \leq \|P_n\|.\|g\|_1 \leq (1 + \varepsilon)\|g\|_1.$$

Thus $\|g\|_1 \geq (1+\varepsilon)^{-1}$ and

$$\|f - g\| = \left\| \frac{g}{\|g\|_1} - g \right\| = \|g\| \left(\frac{1}{\|g\|_1} - 1 \right) \leq \varepsilon.$$

Therefore

$$\|Tx - g\| \leq \|Tx - f\| + \|f - g\| \leq 2\varepsilon + \varepsilon = 3\varepsilon.$$

This completes the proof of the theorem. □

Corollary 4.29 (Johnson–Rosenthal, [JoRo], see also [LiTz3, p. 14]) *Let X be a Banach space whose dual is separable. Assume that Y is an infinite-dimensional subspace of X^* with separable dual Y^*. Then Y has an infinite-dimensional reflexive subspace.*

Proof: By Theorem 4.19, Y contains a shrinking basic sequence $\{y_k\}_{k=1}^{\infty} \subset S_Y$. Then $y_k \overset{w}{\to} 0$ in Y and thus $y_k \overset{w}{\to} 0$ in X^*; hence $y_k \overset{w^*}{\to} 0$ in X^*. By the proof of Theorem 4.28, there is a subsequence $\{y_{k_n}\}_{n=1}^{\infty}$ of $\{y_k\}_{k=1}^{\infty}$ which is a boundedly complete basic sequence. Since $\{y_k\}_{k=1}^{\infty}$ is shrinking, $\{y_{k_n}\}_{n=1}^{\infty}$ is shrinking as well. Therefore, the norm-closed linear hull of $\{y_{k_n}\}_{n=1}^{\infty}$ is reflexive by Theorem 4.16. □

Corollary 4.30 (Johnson–Rosenthal, [JoRo], see also [LiTz3, p. 14]) *Assume that X is an infinite-dimensional Banach space such that X^{**} is separable. Then every infinite-dimensional subspace of X or of X^* contains an infinite-dimensional reflexive subspace.*

Proof: The second statement follows directly from Corollary 4.29. To prove the first statement, assume that Y is an infinite-dimensional subspace of X. Then Y is also an infinite-dimensional subspace of the separable space X^{**}. Moreover Y^* is separable as X^* is separable. Therefore, by Corollary 4.29, Y contains an infinite-dimensional reflexive subspace. □

Definition 4.31 *Let X be a Banach space. A sequence $\{X_n\}_{n=1}^{\infty}$ of finite-dimensional subspaces of X is called a* finite-dimensional decomposition of X (FDD, *in short) if every $x \in X$ has a unique representation of the form $x = \sum_{n=1}^{\infty} x_n$ with $x_n \in X_n$ for every $n \in \mathbb{N}$.*

If $\dim X_n = 1$ for each $n \in \mathbb{N}$, i.e., if $X_n = \operatorname{span}\{x_n\}$ for some $x_n \in X_n$, then $\{X_n\}_{n=1}^{\infty}$ is an FDD for X if and only if $\{x_n\}_{n=1}^{\infty}$ is a Schauder basis of X.

If $\{X_n\}_{n=1}^{\infty}$ is an FDD for a Banach space X, define projections P_n on X by $P_n(\sum_{i=1}^{\infty} x_i) = \sum_{i=1}^{n} x_i$.

The following result is the finite-dimensional-decomposition version of Lemma 4.7.

Proposition 4.32 *If $\{X_n\}_{n=1}^{\infty}$ is an FDD for a Banach space X, then $\sup_n \|P_n\| < \infty$. Conversely, If $\{P_n\}$ is a sequence of finite rank projections on X such that $P_n P_m = P_{\min(m,n)}$ and $\lim_n P_n x = x$ for every $x \in X$, then $\{X_n\}_{n=1}^{\infty}$ determines a unique FDD on X by putting $X_1 = P_1 X$ and $X_n = (P_n - P_{n-1})X$ for $n > 1$.*

Definition 4.33 *An FDD $\{X_n\}_{n=1}^{\infty}$ of a Banach space X is called* shrinking *if the associated projections satisfy $\lim_n \|P_n^* x^* - x^*\| = 0$ for every $x^* \in X^*$.*

Theorem 4.34 (Johnson–Rosenthal, [JoRo], see also [LiTz3, p. 480]) *If X is a separable Banach space, then there is a subspace Y of X such that both Y and X/Y have FDD. If X^* is separable, Y may be chosen so that both Y and X/Y have a shrinking FDD.*

Proof: Let $\{x_n; x_n^*\}$ be a 1-norming Markushevich basis for X, i.e., a biorthogonal system with $\overline{\operatorname{span}}\{x_n : n \in \mathbb{N}\} = X$ and $\overline{\operatorname{span}}\{x_n^* : n \in \mathbb{N}\}$ a 1-norming subspace of X^* (see Definition 4.58). Choose finite sets $\sigma_1 \subset \sigma_2 \subset \ldots$ and $\Delta_1 \subset \Delta_2 \subset \ldots$ so that $\sigma := \bigcup \sigma_n$ and $\Delta := \bigcup \Delta_n$ are complementary infinite subsets of $\mathbb{N}$ and that, for $n \in \mathbb{N}$,

(i) If $x^* \in \overline{\text{span}}\{x_i^* : i \in \Delta_n\}$ there is $x \in \overline{\text{span}}\{x_i : i \in \Delta_n \cup \sigma_{n+1}\}$ so that $\|x\| = 1$ and $|x^*(x)| > (1 - 1/(n+1))\|x^*\|$;

(ii) If $x \in \overline{\text{span}}\{x_i : i \in \sigma_n\}$ there is $x^* \in \overline{\text{span}}\{x_i^* : i \in \sigma_n \cup \Delta_n\}$ so that $\|x^*\| = 1$ and $|x^*(x)| > (1 - 1/(n+1))\|x\|$.

For $n \in \mathbb{N}$ define $S_n : X \to X$ and $T_n : X \to X$ by $S_n x = \sum_{i \in \sigma_n} x_i^*(x) x_i$ and $T_n x = \sum_{i = \Delta_n} x_i^*(x) x_i$. Then, for $n \in \mathbb{N}$,

$$\|T_n^*|_{\{x_i :\, i \in \sigma_{n+1}\}^\perp}\| \leq 1 + \frac{1}{n}$$

and

$$\|S_n^*|_{\{x_i^* :\, i \in \Delta_n\}}\| \leq 1 + \frac{1}{n}.$$

Indeed, to see the first inequality, suppose $y \in \{x_i : i \in \sigma_{n+1}\}^\perp$. Pick $x \in \overline{\text{span}}\{x_i : i \in \Delta_n \cup \sigma_{n+1}\}$ so that $\|x\| = 1$ and $\|T_n^* y(x)\| \geq (1 - 1/(n+1))\|T_n^* y\|$. Since $y \in \{x_i : i \in \sigma_{n+1}\}^\perp$, $y(x) = T_n^* y(x)$ and hence

$$|y(x)| \geq \left(1 - \frac{1}{n+1}\right) \|T_n^* y\|,$$

which gives

$$\|y\| \geq \left(1 - \frac{1}{n+1}\right) \|T_n^* y\|,$$

so

$$\|T_n^* y\| \leq \left(1 - \frac{1}{n}\right) \|y\|.$$

The other inequality follows similarly. Put $Y = \overline{\text{span}}\{x_i : i \in \sigma\}$. We will show that $Y^\perp = \overline{\text{span}}^{w^*}\{x_i^* : i \in \Delta\}$. The only nontrivial inclusion is $Y^\perp \subset \overline{\text{span}}^{w^*}\{x_i^* : i \in \Delta\}$. For it , let $y \in \{x_i : i \in \sigma\}^\perp$. It is enough to show that $T_n^* y \xrightarrow{w^*} y$. Since $\{T_n^* y : n \in \mathbb{N}\}$ is bounded, assume that $T_n^* y \xrightarrow{w^*} x^*$ for some $x^* \in X^*$. Then $T_n^* x^* = T_n^* y$ for every n and hence $x^* - y \in \{x_i : i \in \Delta_n\}^\perp$ for each n. Thus $x^* - y \in \{x_i : i \in \Delta\}^\perp$. However, obviously, $x^* - y \in \{x_i : i \in \sigma\}^\perp$. Hence, $y = x^*$ and $T_n^* y \xrightarrow{w^*} y$. Hence $Y \subset \overline{\text{span}}^{w^*}\{x_i^* : i \in \Delta\}$. Now,

$$\overline{\text{span}}\{x_i : i \in \sigma\} = \{x_i : i \in \sigma\}^{\perp\top} = (\overline{\text{span}}^{w^*}\{x_i^* : i \in \Delta\})^\top = \{x_i^* : i \in \Delta\}^\top.$$

We have

$$(X/Y)^* = Y^\perp = \overline{\operatorname{span}}^{w^*}\{x_i^* : i \in \Delta\}$$

and (T_n^*) produce by duality an FDD for X/Y. The sequence of projections $(S_n|_Y)$ then in turn produce an FDD for Y. If X^* is separable and the dual norm is locally uniformly rotund (see Definition 7.9 and Theorem 8.7), which is shared by subspaces and quotients of X, we get that there is a shrinking FDD both in Y and X/Y (see the situation with Schauder bases). □

4.5 Unconditional Bases

Recall that a series $\sum x_i$ in a Banach space X is unconditionally convergent if $\sum \varepsilon_i x_i$ converges for all choices of signs $\varepsilon_i = \pm 1$ (see Exercise 1.39).

Definition 4.35 *A Schauder basis $\{e_i\}$ of a Banach space X is said to be* unconditional *if for every $x \in X$, its expansion $x = \sum a_i e_i$ converges unconditionally. A sequence $\{e_i\}$ in a Banach space X is called an* unconditional basic sequence *if it is an unconditional basis of* $\overline{\operatorname{span}}\{e_i\}$.

It is straightforward to check that the canonical basis $\{e_i\}$ of c_0 or ℓ_p, $p \in [1, \infty)$, is an unconditional basis.

Also, as we saw in the notes following Theorem 1.57, every orthonormal basis of a separable Hilbert space is unconditional.

The Haar basis is an unconditional basis for $L_p[0, 1]$ for $p > 1$ (Paley and Marcinkiewicz, see, e.g., [AlKa, p. 130]), but not for $L_1[0, 1]$, as the latter space does not admit any unconditional basis, see Corollary 4.39.

The trigonometric system is not an unconditional basis for $L_p[0, 1]$ if $p \neq 2$ (see, e.g., [LiTz2, p. 143]).

Note that every basis equivalent to an unconditional basis is also unconditional.

Example: Let $\{e_i\}$ be the standard unit vector basis of c_0. For $n \in \mathbb{N}$ set $x_n = \sum_{i=1}^n e_i$. We check that if $x = \sum_{i=1}^\infty \alpha_i e_i$, then $x = \sum_{n=1}^\infty \beta_n x_n$, where $\beta_n = \alpha_n - \alpha_{n+1}$. It follows that $\{\beta_n\}_{n=1}^\infty$ forms a convergent series with $\sum_{n=1}^\infty \beta_n = \alpha_1$ and $\|x\|_\infty = \sup_k \left|\sum_{n=k}^\infty \beta_n\right|$. On the other hand, for every convergent series $\sum \beta_k$ we have

$$x = \sum_{k=1}^\infty \beta_k x_k = \sum_{i=1}^\infty \Big(\sum_{k=i}^\infty \beta_k\Big) e_i \in c_0.$$

We just showed that $(c_0, \|\cdot\|_\infty)$ is isometric to the space of convergent series $\{\beta_n\}$ with the norm $\|(\beta_n)\| = \sup_k \left|\sum_{n=k}^\infty \beta_k\right|$. The uniqueness of expansion $x = \sum \beta_n x_n$ follows by induction, using the fact that $\sum_{i=i}^\infty \beta_k = \alpha_i$, where α_i is the standard ith coordinate of x (β_n is then necessarily $\alpha_n - \alpha_{n+1}$). It follows that $\{x_n\}$ is a Schauder basis of c_0 which is not unconditional. Indeed, $\sum \frac{(-1)^n}{n} x_n \in c_0$ while $\sum \frac{1}{n} x_n \notin c_0$. The basis $\{x_n\}$ is called the *summing basis of* c_0.

Proposition 4.36 *Let $\{e_i\}$ be a sequence in a Banach space X. The following are equivalent:*
(i) *$\{e_i\}$ is an unconditional basic sequence.*
(ii) *There is a constant K such that for all scalars $a_1, \dots, a_m$ and signs $\varepsilon_i = \pm 1$ we have*

$$\Big\| \sum_{i=1}^{m} \varepsilon_i a_i e_i \Big\| \leq K \Big\| \sum_{i=1}^{m} a_i e_i \Big\|.$$

(iii) *There is a constant L such that for all scalars $a_1, \dots, a_m$ and every subset σ of $\{1, \dots, m\}$ we have*

$$\Big\| \sum_{i \in \sigma} a_i e_i \Big\| \leq L \Big\| \sum_{i=1}^{m} a_i e_i \Big\|.$$

We claim that the condition (iii) can be equivalently stated as $\left\| \sum_{i\in\sigma} a_i e_i \right\| \leq L \left\| \sum_{i=1}^{\infty} a_i e_i \right\|$ whenever $\sigma \subset \mathbb{N}$.

Indeed, let σ be any (even infinite) subset of $\mathbb{N}$. Assuming (iii) we show that $\sum_{i\in\sigma} a_i e_i$ is Cauchy: Given $\varepsilon > 0$, there is n_0 such that for $m > n > n_0$ we have $\left\| \sum_{i=n}^{m} a_i e_i \right\| < \frac{\varepsilon}{L}$. Considering $\sigma' = \sigma \cap \{n, \dots, m\}$ and $b_i = a_i$ for $n \leq i \leq m$, $b_i = 0$ otherwise, we use the condition (iii) to see that $\left\| \sum_{i\in\sigma, n\leq i\leq m} a_i e_i \right\| < \varepsilon$. Thus $\sum_{i\in\sigma} a_i e_i$ is convergent. Passing to the limit we get the claim.

However, in most applications the finite-sum statement is easier to handle as one does not need to discuss the convergence. Similar observation can be made about the condition (ii).

Proof: (i)$\Longrightarrow$(iii): Let $Y = \overline{\text{span}}\{e_i\}$. Given $\sigma \subset \mathbb{N}$, define an operator P_σ from Y into Y by $P_\sigma(x) = \sum_{i\in\sigma} a_i e_i$ for $x = \sum a_i e_i$. The operator P_σ is well defined as $\sum_{i\in\sigma} a_i e_i$ converges whenever $\sum a_i e_i$ converges (use (i) and Exercise 1.41). We now check that P_σ has a closed graph. Indeed, let $x^k \to x$ in Y for $x^k = \sum_i a_i^k e_i$, $x = \sum a_i e_i$, and $P_\sigma(x^k) = \sum_{i\in\sigma} a_i^k e_i \to y = \sum b_i e_i$. From the continuity of the biorthogonal functionals in Schauder bases, we have $a_i^k \to a_i$ for every i and for the same reason, $a_i^k \to b_i$ for every i. Thus $b_i = a_i$ for every i and hence $P_\sigma(x) = y$, meaning that P_σ has a closed graph and is thus continuous.

Consider now the family of operators P_σ, σ running through all subsets of $\mathbb{N}$. We claim that for every fixed $x = \sum a_i e_i \in X$, the family $\{P_\sigma(x)\}$ is bounded. Indeed, from the unconditionality we get that given $\varepsilon > 0$, there is a finite set $F \subset \mathbb{N}$ such that $\left\| \sum_{i\in A} a_i e_i \right\| < \varepsilon$ whenever $A \cap F = \emptyset$ (see exercises in Chapter 1). From this the boundedness of $\{P_\sigma(x)\}$ for every $x \in X$ follows. Now the Banach–Steinhaus uniform boundedness principle gives that the operators P_σ are uniformly bounded by some L.

(iii)$\Longrightarrow$(ii): Given scalars $a_1, \dots, a_m$ and signs $\varepsilon_i = \pm 1$, we define $\sigma = \{i : \varepsilon_i = 1\}$ and $\sigma' = \{1, \dots, m\} \backslash \sigma$. Then

$$\Big\|\sum_{i=1}^{m} \varepsilon_i a_i e_i\Big\| = \Big\|\sum_{i\in\sigma} a_i e_i - \sum_{i\in\sigma'} a_i e_i\Big\|$$
$$\le \Big\|\sum_{i\in\sigma} a_i e_i\Big\| + \Big\|\sum_{i\in\sigma'} a_i e_i\Big\| \le 2L\Big\|\sum_{i=1}^{m} a_i e_i\Big\|.$$

(ii)$\Longrightarrow$(iii): Given $a_1, \dots, a_m$ and $\sigma \subset \{1, \dots, m\}$, we define $\varepsilon_i = 1$ if $i \in \sigma$ and $\varepsilon_i = -1$ for $i \in \{1, \dots, m\} \backslash \sigma$. Then

$$\Big\|\sum_{i\in\sigma} a_i e_i\Big\| = \frac{1}{2}\Big\|\sum_{i=1}^{m} (\varepsilon_i a_i e_i + a_i e_i)\Big\|$$
$$\le \frac{1}{2}\Big\|\sum_{i=1}^{m} \varepsilon_i a_i e_i\Big\| + \frac{1}{2}\Big\|\sum_{i=1}^{m} a_i e_i\Big\| \le K\Big\|\sum_{i=1}^{m} a_i e_i\Big\|.$$

(ii) and (iii)$\Longrightarrow$(i): Given $n < m$ and scalars $a_1, \dots, a_m$, we use (iii) with $\sigma = \{1, \dots, n\}$ to see that by Proposition 4.18, $\{e_i\}$ is a basic sequence with $\mathrm{bc}\{e_i\} \le L$. Now let $\sum_{i=1}^{\infty} a_i e_i$ be a convergent series. Given $\varepsilon_i = \pm 1$, using (ii) we show that $\Big\|\sum_{i=n}^{m} \varepsilon_i a_i e_i\Big\| \le 2K\Big\|\sum_{i=n}^{m} a_i e_i\Big\|$. Thus $\sum \varepsilon_i a_i e_i$ is Cauchy, hence convergent. This shows that $\sum a_i e_i$ converges unconditionally. □

The best possible constant K from the condition (ii) in Proposition 4.36 is called the *unconditional basis constant* of $\{e_i\}$ and is denoted by $\mathrm{ubc}\{e_i\}$. In the the proof above, we have shown that $L \le K$ and $\mathrm{bc}\{e_i\} \le \mathrm{ubc}\{e_i\}$.

A natural question is whether every Banach space contains an unconditional basic sequence (see Theorem 4.19). This long-standing problem was answered in the negative by Gowers and Maurey ([GoMa]).

Theorem 4.37 (James, see, e.g., [LiTz3]) *Let X be a separable Banach space. If X has an unconditional Schauder basis that is not boundedly complete, then X contains an isomorphic copy of c_0.*

In the proof, we will use the following statement.

Lemma 4.38 *Let $\{e_i\}$ be an unconditional basic sequence in a Banach space X. Then for all scalars (a_i) such that $\sum a_i e_i$ converges and all bounded sequences of scalars $\{\lambda_i\}$ we have*

$$\Big\|\sum_{i=1}^{\infty} \lambda_i a_i e_i\Big\| \le \mathrm{ubc}\{e_i\}(\sup_i |\lambda_i|)\Big\|\sum_{i=1}^{\infty} a_i e_i\Big\|.$$

Proof: Given $m \in \mathbb{N}$, pick $x^* \in S_{X^*}$ so that $x^*\left(\sum_{i=1}^m \lambda_i a_i e_i\right) = \left\|\sum_{i=1}^m \lambda_i a_i e_i\right\|$ and define ε_i by $\varepsilon_i = 1$ if $a_i x^*(e_i) \geq 0$ and $\varepsilon_i = -1$ if $a_i x^*(e_i) < 0$. Then

$$\left\|\sum_{i=1}^m \lambda_i a_i e_i\right\| \leq \sum_{i=1}^m |\lambda_i|\,|a_i x^*(e_i)| \leq \big(\sup_{1\leq i\leq m} |\lambda_i|\big) \sum_{i=1}^m \varepsilon_i a_i x^*(e_i)$$
$$\leq \big(\sup_{1\leq i\leq m} |\lambda_i|\big)\|x^*\| \left\|\sum_{i=1}^m \varepsilon_i a_i e_i\right\| \leq \big(\sup_{1\leq i\leq m} |\lambda_i|\big)\,\mathrm{ubc}\{e_i\} \cdot \left\|\sum_{i=1}^m a_i e_i\right\|.$$

□

Proof of Theorem 4.37: Let $\{e_i\}$ be an unconditional basis of X that is not boundedly complete. Then there are scalars (a_i) such that $\left\|\sum_{i=1}^n a_i e_i\right\| \leq 1$ for every n and $\sum_{i=1}^\infty a_i e_i$ does not converge. By the Cauchy criterion, there are $\varepsilon > 0$ and natural numbers $p_1 < q_1 < p_2 < q_2 \cdots$ such that for $u_j = \sum_{j=p_j}^{q_i} a_j e_j$ we have $\|u_j\| \geq \varepsilon$ for every j, yet $\left\|\sum_{i=1}^m u_i\right\| \leq K \left\|\sum_{i=1}^\infty a_i e_i\right\| \leq K$, where $K = \mathrm{ubc}\{e_i\}$.

By the previous lemma, for every sequence $\{\lambda_j\}_{j=1}^m$ of scalars we have

$$\left\|\sum_{j=1}^m \lambda_j u_j\right\| \leq K\big(\sup_j |\lambda_j|\big)\left\|\sum_{j=1}^m u_j\right\| \leq K^2\big(\sup_j |\lambda_j|\big) = K^2\|(\lambda_j)\|_\infty.$$

On the other hand, from the unconditionality of $\{e_i\}$ we have for each $i \in \{1, \ldots, m\}$: $\left\|\sum_{j=1}^m \lambda_j u_j\right\| \geq \frac{1}{K}\|\lambda_i u_i\| \geq \frac{\varepsilon}{K}|\lambda_i|$, that is, $\frac{\varepsilon}{K}\|(\lambda_j)\|_\infty \leq \left\|\sum_{j=1}^m \lambda_j u_j\right\|$. Thus $\{u_j\}$ is equivalent to the canonical basis of c_0. □

Corollary 4.39 *The space $L_1[0,1]$ does not have an unconditional Schauder basis.*

Proof: The space $L_1[0,1]$ is weakly sequentially complete (a classical Steinhaus theorem, see Definition 5.35 and Exercises 13.48 and 13.49 for a proof). Let $\{e_i\}$ be the standard unit vector basis of c_0. Then the sequence $x_n = \sum_{i=1}^n e_i$ is weakly Cauchy but not weakly convergent, therefore c_0 is not isomorphic to a subspace of $L_1[0,1]$. If $L_1[0,1]$ had an unconditional basis $\{x_i\}$, $\{x_i\}$ would be boundedly complete by Theorem 4.37 and thus $L_1[0,1]$ would be isomorphic to a dual space by Theorem 4.15. But this is not possible: In a separable dual space every closed convex bounded set is the closed convex hull of its extreme points (see Theorem 3.122), while the unit ball of $L_1[0,1]$ has no extreme points (Exercise 3.136). For another approach, see Exercise 8.23. □

Theorem 4.40 *The space $C[0,1]$ does not have an unconditional basis.*

Proof: (Kadec) Assume $C[0,1]$ has an unconditional basis. Let $\alpha = \sup \|P_A\|$, where P_A are the canonical projections onto $\{e_i\}_{i \in A}$ for finite $A \subset \mathbb{N}$. Pick A such that $\|P_A\| > \alpha - \frac{1}{2}$. Then by the Daugavet result in Exercise 9.27,

$$\|I - P_A\| = \|I\| + \|P_A\| = 1 + \|P_A\| > 1 + \alpha - \tfrac{1}{2} = \alpha + \tfrac{1}{2}.$$

On the other hand, $\|I - P_A\| = \|P_{\mathbb{N} \setminus A_0}\| \leq \alpha$. Hence $\alpha \geq \alpha + \frac{1}{2}$, a contradiction. □

Theorem 4.41 (James, see, e.g., [LiTz3]) *Let X be a separable Banach space. If X has an unconditional Schauder basis that is not shrinking, then X contains an isomorphic copy of ℓ_1.*
In particular, a separable Banach space with an unconditional Schauder basis contains an isomorphic copy of ℓ_1 if X^* is nonseparable.

Proof: Let $\{e_i\}$ be the assumed basis, let $K = \mathrm{bc}\{e_i\}$. Since $\{e_i\}$ is not shrinking, there is $f \in S_{X^*}$ such that $\sup\{f(z) : z \in B_X \cap \overline{\mathrm{span}}\{e_i\}_{i \geq n}\} \not\to 0$. This means that there is $\varepsilon > 0$ and a sequence $n_k < n_{k+1}$ such that

$$\sup\{f(z) : z \in B_X \cap \mathrm{span}\{e_{n_k}, e_{n_k+1}, \ldots\}\} \geq \varepsilon$$

for every k. Using the sliding hump technique, we can construct a normalized block basic sequence u_j of $\{e_i\}$ such that $f(u_j) \geq \varepsilon/2$ for every j. Let $m \in \mathbb{N}$ and $a_1, \ldots, a_m$ be scalars. Assume $\sum_{i \leq m, a_i \geq 0} a_i \geq \sum_{i \leq m, a_i < 0} -a_i$. Then

$$4\Big\|\sum_{i=1}^{m} a_i u_i\Big\| \geq \frac{4}{K}\Big\|\sum_{i \leq m, a_i \geq 0} a_i u_i\Big\| \geq \frac{4}{K} \cdot f\Big(\sum_{i \leq m, a_i \geq 0} a_i u_i\Big)$$

$$\geq 2\frac{\varepsilon}{K} \sum_{i \leq m, a_i \geq 0} a_i \geq \frac{\varepsilon}{K} \sum_{i \leq m, a_i \geq 0} a_i + \frac{\varepsilon}{K} \sum_{i \leq m, a_i < 0} -a_i \geq \frac{\varepsilon}{K} \sum_{i=1}^{m} |a_i|.$$

If $\sum_{i \leq m, a_i \geq 0} a_i < \sum_{i \leq m, a_i < 0} -a_i$, we consider $-a_i$ to obtain the same inequality. On the other hand, since $\{u_j\}$ is normalized, we have $\left\|\sum_{i=1}^{m} a_i u_i\right\| \leq \sum_{i=1}^{m} |a_i|$. Therefore $\frac{\varepsilon}{4K}\|(a_i)\|_{\ell_1} \leq \left\|\sum_{i=1}^{m} a_i u_i\right\| \leq \|(a_i)\|_{\ell_1}$, that is, $\{u_j\}$ is equivalent to the canonical basis of ℓ_1.

To see the second statement, note that if $\{e_i; e_i^*\}$ is shrinking, then $X^* = \overline{\mathrm{span}}\{e_i\}$ and thus X^* is separable. □

Corollary 4.42 (James, see, e.g., [LiTz3]) *Let X be a Banach space with an unconditional basis. Then X is reflexive if and only if X contains no isomorphic copy of ℓ_1 or c_0.*

Proof: It follows directly from Theorems 4.16, 4.41, and 4.37. □

The following definition is due to James ([Jame1] and [Jame1b]).

Definition 4.43 *The* James space J *consists of all sequences* (α_i) *of real numbers such that* $\lim \alpha_i = 0$ *and* $\|(a_i)\| < \infty$, *where the norm* $\|(a_i)\|$ *is defined by*

$$\|(a_i)\| = \sup_{n_1 < \cdots < n_k} \left((\alpha_{n_1} - \alpha_{n_2})^2 + (\alpha_{n_2} - \alpha_{n_3})^2 + \cdots + (\alpha_{n_{k-1}} - \alpha_{n_k})^2 \right)^{\frac{1}{2}}.$$

It is standard to verify that $(J, \|\cdot\|)$ is a Banach space. Moreover, the sequence $\{e_i\}$ of the standard unit vectors $e_n = (0, \ldots, 0, \overset{n}{1}, 0, \ldots)$ is a monotone Schauder basis of J.

Claim:
The sequence $\{e_i\}$ is a shrinking Schauder basis of J and $J^{**} = J \oplus \operatorname{span}\{(1, 1, 1, \ldots)\}$, in particular, J is not reflexive.

Proof: Assume that the basis is not shrinking. As in the proof of Proposition 4.41 we find $x^* \in J^*$, $\varepsilon > 0$, and a normalized block basic sequence $\{u_j\}$ of $\{e_i\}$ such that $x^*(u_j) > \varepsilon$ for every j. Consider the vector $u = \sum_{j=1}^{\infty} \frac{1}{j} u_j$. By Hölder's inequality (1.1) applied to the norm of J and by considering the division points of u_js we find that $\|u\|^2 \leq 2 \sum \frac{1}{j^2} \|u_j\|^2 \leq 2 \sum \frac{1}{j^2} < \infty$. On the other hand, $x^*(u) = \sum x^*(u_j)/j$ is not finite, a contradiction. $\triangle$

By Proposition 4.14, J^{**} is identified with (β_i) such that $\sup_k \left\| \sum_{i=1}^k \beta_i e_i \right\| < \infty$ via the mapping $T(x^{**}) = \left(x^{**}(e_i^*) \right)$. From the definition of the norm $\|\cdot\|$ of J and from the fact that $\sup_k \left\| \sum_{i=1}^k \beta_i e_i \right\| < \infty$ it follows that $\lim_{i \to \infty} \beta_i = \beta$ exists. Thus $(\beta_i - \beta)_i$ corresponds to an element of J as $\lim_{i \to \infty} (\beta_i - \beta) = 0$. Therefore $J^{**} = J \oplus \operatorname{span}\{(1, 1, \ldots)\}$. $\square$

Since J^{**} is separable, J does not contain an isomorphic copy of c_0 or ℓ_1. Indeed, if J contained an isomorphic copy of c_0, then J^* would have a quotient isomorphic to c_0^* and J^{**} would have a subspace isomorphic to $c_0^{**} = \ell_\infty$. Similar argument shows that J does not contain an isomorphic copy of ℓ_1.

By Corollary 4.42, J does not have an unconditional basis. However, J contains an isomorphic copy of ℓ_2, in particular it contains an unconditional basic sequence.

Bessaga and Pełczyński proved (see, e.g., [LiTz3]) that if a Banach space X has an unconditional basis and Y is a nonreflexive closed subspace of X, then Y contains an isomorphic copy of c_0 or ℓ_1. Using this result and the fact that J (as every separable Banach space) is isometric to a subspace of $C[0, 1]$ (Theorem 5.8), we see again that $C[0, 1]$ does not admit any unconditional basis.

4.6 Bases in Classical Spaces

It was a long-standing problem whether every Banach space contains either a reflexive subspace or an isomorphic copy of c_0 or ℓ_1. This problem was recently answered in the negative by Gowers. In fact, Gowers showed ([Gowe4]) that there is a separable Banach space X such that every infinite-dimensional closed subspace of X has nonseparable dual and yet X contains no isomorphic copy of ℓ_1.

Theorem 4.44 (Bessaga, Pełczyński, see, e.g., [LiTz3]) *Let X be a Banach space. If X^* has a subspace isomorphic to c_0, then X has a complemented subspace isomorphic to l_1. In particular, X^* has a subspace isomorphic to ℓ_∞.*

Proof: Let T be an isomorphism from c_0 into X^*. Then T^* maps X^{**} onto ℓ_1 (Exercise 2.49). Since B_X is w^*-dense in $B_{X^{**}}$ by Goldstine's theorem, using Corollary 2.26 we find $K > 0$ and $x_n \in X$ for $n \in \mathbb{N}$ such that $\|x_n\| \le K, T^*(x_n)(e_n) = 1$ and $\sum_{i=1}^{n-1} |T^*(x_n)(e_i)| < \frac{1}{n}$, where e_i are the standard unit vectors in c_0. From Corollary 4.52 and Proposition 4.45, it follows that $\{T^*(x_n)\}$ has a subsequence $\{T^*(x_{n_k})\}$ that is equivalent to the canonical basis of ℓ_1 and whose span is complemented in ℓ_1 by a projection P. Therefore for some constant $M > 0$, and every choice of scalars $\{a_k\}_{k=1}^{\infty}$ satisfying $\sum |a_k| < \infty$ we have

$$\left\|\sum_{k=1}^{\infty} a_k x_{n_k}\right\| \le K \sum_{k=1}^{\infty} |a_k| \le KM \left\|\sum_{k=1}^{\infty} a_k T^*(x_{n_k})\right\| \le KM\|T^*\| \left\|\sum_{k=1}^{\infty} a_k x_{n_k}\right\|.$$

Hence T^* is an isomorphism of $Y = \overline{\operatorname{span}}\{x_{n_k}\}$ onto $\overline{\operatorname{span}}\{T^*(x_{n_k})\}$. Thus Y is isomorphic to ℓ_1 and $\widetilde{P} = (T^*)^{-1} P T^*$ is a projection of X onto Y. Finally, $\widetilde{P}^*(X^*)$ is isomorphic to $\widetilde{P}(X)^* = \ell_1^* = \ell_\infty$ (Exercise 4.10). □

Given a bounded sequence in a Banach space with a separable dual, we can extract a weakly Cauchy subsequence by a standard Cantor diagonal procedure. Rosenthal characterized spaces share this property as spaces not containing an isomorphic copy of ℓ_1. Precisely, he proved that if $\{x_n\}$ is a bounded sequence in a Banach space X, then either $\{x_n\}$ has a weak Cauchy subsequence or contains a subsequence that is equivalent to the standard unit vector basis of ℓ_1 (see Theorem 5.37). There are separable spaces whose dual is nonseparable and yet they do not contain an isomorphic copy of ℓ_1 (e.g., the space JT discussed in exercises).

We will now investigate the structure of subspaces of ℓ_p spaces.

Proposition 4.45 *Let X be c_0 or ℓ_p with $p \in [1, \infty)$. If $\{u_j\}$ is a normalized block basic sequence of the standard unit vector basis $\{e_i\}$, then $\{u_i\}$ is equivalent to $\{e_i\}$, $\overline{\operatorname{span}}\{u_i\}$ is isometric to X, and there is a projection of norm one of X onto $\overline{\operatorname{span}}\{u_i\}$.*

Proof: We present the proof for ℓ_p. Let $u_j = \sum_{i=p_j+1}^{p_{j+1}} \lambda_i e_i$ with $\sum_{i=p_j+1}^{p_{j+1}} |\lambda_i|^p = 1$ for $j \in \mathbb{N}$. Then

$$\Big\|\sum_{j=1}^{m} a_j u_j\Big\| = \Big(\sum_{j=1}^{m}\sum_{i=p_j+1}^{p_{j+1}} |a_j|^p|\lambda_i|^p\Big)^{\frac{1}{p}} = \Big(\sum_{j=1}^{m}|a_j|^p\sum_{i=p_j+1}^{p_{j+1}}|\lambda_i|^p\Big)^{\frac{1}{p}}$$
$$= \Big(\sum_{j=1}^{m}|a_j|^p\Big)^{\frac{1}{p}} = \Big\|\sum_{j=1}^{m} a_j e_j\Big\|.$$

So $\{e_i\}$ and $\{u_i\}$ are equivalent and $T(\sum a_j u_j) = \sum a_j e_j$ is the isometry.

To find a projection, for every $j \in \mathbb{N}$ choose $u_j^* \in \operatorname{span}\{e_i\}_{i=p_j+1}^{p_{j+1}} \subset \ell_p^*$ for which $\|u_j^*\| = u_j^*(u_j) = 1$. Then $u_j^*(u_k) = 0$ for $k \neq j$ and the operator P from X into $\overline{\operatorname{span}}\{u_j\}$ defined by $P(x) = \sum_{j=1}^{\infty} u_j^*(x)u_j$ is a linear projection of X onto $\overline{\operatorname{span}}\{u_j\}$. For $x = \sum a_i e_i \in X$ we have $|u_j^*(x)|^p \leq \sum_{i=p_j+1}^{p_{j+1}} |a_i|^p$ for every j as $\|u_j^*\| = 1$ and thus

$$\|P(x)\|^p = \sum_{j=1}^{\infty}\sum_{i=p_j+1}^{p_{j+1}} |u_j^*(x)|^p \cdot |\lambda_i|^p = \sum_{j=1}^{\infty}|u_j^*(x)|^p$$
$$\leq \sum_{j=1}^{\infty}\sum_{i=p_j+1}^{p_{j+1}} |a_i|^p = \|x\|^p.$$

This shows that $\|P\| = 1$. □

Combining Theorem 4.26 and Proposition 4.45 we have

Theorem 4.46 *Let X be c_0 or ℓ_p with $p \in [1, \infty)$. If Y is an infinite-dimensional closed subspace of X, then Y contains a subspace Z which is isomorphic to X and complemented in X.*

The following result is called the *Pełczyński decomposition method.*

Theorem 4.47 (Pełczyński, see, e.g., [LiTz3]) *Let X and Y be Banach spaces so that X is isomorphic to a complemented subspace of Y and Y is isomorphic to a complemented subspace of X. Assume that either:*
(i) *X is isomorphic to $X \oplus X$ and Y is isomorphic to $Y \oplus Y$, or*
(ii) *X is isomorphic to $(\sum X)_{c_0}$ or X is isomorphic to $(\sum X)_{\ell_\infty}$ or X is isomorphic to $(\sum X)_{\ell_p}$ for some $1 \leq p < \infty$.*
Then X is isomorphic to Y.

Proof: Put $Y \approx X \oplus E$ and $X \approx Y \oplus F$, where $\approx$ denotes isomorphism. If (i) holds, then we have $Y \approx X \oplus E \approx X \oplus X \oplus E \approx X \oplus Y$. Similarly, $X \approx Y \oplus F \approx Y \oplus Y \oplus F \approx Y \oplus X$. Thus $X \approx Y$.

If (ii) holds, we have in particular $X \approx X \oplus X$ and $Y \approx X \oplus E \oplus X \approx X \oplus X \oplus E \approx X \oplus Y$. Furthermore,

$$\Big(\sum X\Big)_{\ell_p} \approx \Big(\sum Y \oplus F\Big)_{\ell_p} \approx \Big(\sum Y\Big)_{\ell_p} \oplus \Big(\sum F\Big)_{\ell_p}.$$

Hence if $X \approx (\sum X)_{\ell_p}$, we have

$$X \approx \Big(\sum X\Big)_{\ell_p} \approx \Big(\sum Y\Big)_{\ell_p} \oplus \Big(\sum F\Big)_{\ell_p} \approx Y \oplus \Big(\sum Y\Big)_{\ell_p} \oplus \Big(\sum F\Big)_{\ell_p}$$
$$\approx Y \oplus \Big(\sum Y \oplus F\Big)_{\ell_p} \approx Y \oplus \Big(\sum X\Big)_{\ell_p} \approx Y \oplus X \approx Y.$$

A similar argument is used for the c_0 or for the ℓ_∞ sums. □

Corollary 4.48 (Pełczyński) *Let X be c_0 or ℓ_p, $p \in [1, \infty)$. If Y is an infinite-dimensional complemented subspace of X, then Y is isomorphic to X.*

Proof: By Theorem 4.46, Y has a subspace Z that is isomorphic to X and complemented in X (hence complemented in Y). So X is isomorphic to a complemented subspace of Y (and Y *is* a complemented subspace of X). If $X := c_0$ then X is isomorphic to $(\sum X)_{c_0}$. Similarly, if $X := \ell_p$ $(1 \leq p < \infty)$ then X is isomorphic to $(\sum X)_{\ell_p}$ (Exercise 4.39). Using (ii) in Theorem 4.47, we get the conclusion. □

We note that for $p \in (1, \infty)$, $p \neq 2$, there are closed subspaces of ℓ_p that are isomorphic to ℓ_p and not complemented in ℓ_p [BDGJN], [Rose4], and [Rose1]. Also, if $p \in [1, \infty)\backslash\{2\}$ then there are infinite-dimensional closed subspaces of ℓ_p that are not isomorphic to ℓ_p [Pelc3].

It seems to be unknown whether every isomorphic copy of ℓ_1 in ℓ_1 has to be complemented in ℓ_1, [BDGJN].

When comparing structure of Banach spaces, an important role is played by certain classes of operators. Let X, Y be Banach spaces and $T \in \mathcal{B}(X, Y)$. We say that T is *strictly singular* if there is no infinite-dimensional subspace Z of X such that $T\big|_Z$ is an isomorphism into Y. We recall that T is a *compact operator* if $\overline{T(B_X)}$ is compact in Y. Compact operators will be treated extensively in Chapter 15. Note that every compact operator is strictly singular (Exercise 4.45).

Proposition 4.49 (Pitt, see, e.g., [LiTz3, p. 76]) *Let $1 \leq p < r < \infty$. Every bounded operator T from ℓ_r into ℓ_p or from c_0 into ℓ_p is compact.*

Proof: Assume that $T\colon \ell_r \to \ell_p$ is not compact. Since ℓ_r is reflexive, there is a sequence $\{x_n\}$ in ℓ_r such that $x_n \overset{w}{\to} 0$ in ℓ_r and $\|T(x_n)\| \geq \varepsilon$ for some $\varepsilon > 0$ (Exercise 15.8). Then $\inf \|x_n\| > 0$, so using Corollary 4.27 and Proposition 4.45 first in ℓ_r and then in ℓ_p we find a subsequence $\{x_{n_k}\}$ of $\{x_n\}$ such that $\{x_{n_k}\}$ is equivalent to the canonical basis of ℓ_r and $\{T(x_{n_k})\}$ is equivalent to the canonical basis of ℓ_p.

If $(\alpha_k) \in \ell_r\backslash\ell_p$, then $\sum \alpha_k x_{n_k}$ is convergent, hence

$$T(\sum a_k x_{n_k}) = \sum a_k T(x_{nk}) \in \ell_p.$$

Thus $\sum |\alpha_k|^p < \infty$, which is a contradiction.

The c_0 case follows from the first part by using Theorem 15.3. □

We thus obtain that ℓ_p, ℓ_q are not isomorphic for $p \neq q$. Using Theorems 4.46 and 4.49 we deduce that if $p, q \in [1, \infty)$ and $p \neq q$, then the spaces ℓ_p and ℓ_q do not have isomorphic infinite-dimensional closed subspaces. Such spaces are called *totally incomparable.*

Proposition 4.50 *Assume that T is a bounded operator from c_0 into a Banach space X. Then the following are equivalent.*

(i) T is compact.

(ii) $\sum_{i=1}^{\infty} Te_i$ converges unconditionally in X, where $\{e_i\}_{i=1}^{\infty}$ is the canonical basis of c_0.

Proof: Assume that $\sum_i Te_i$ is unconditionally convergent. Then given $\varepsilon > 0$, there is $n_0 \in \mathbb{N}$ such that $\| \sum_{i\in F} a_i Te_i \| \leq \varepsilon$ whenever F is a finite set in $\{n_0, n_0+1, \ldots\}$ and $|a_i| \leq 1$ for $i \in F$. Let P_{n_0} denote the canonical projection onto the span of the first n_0 coordinates in c_0. Then it follows that the compact set $TP_{n_0}B_{c_0}$ provides a 2ε-net for TB_{c_0}. Therefore T is a compact operator. Assume, conversely, that T is compact. Consider

$$T^{**} : c_0^{**} \ (= \ell_\infty) \to X \ (\subset X^{**}).$$

The restriction of T^{**} to B_{ℓ_∞} is w^*-$\|\cdot\|$-continuous because on the $\|\cdot\|$-compact set $\overline{TB_{c_0}}$ the w^*- and the norm-topologies coincide. Since $\sum e_{\pi(n)}$ is w^*-convergent in ℓ_∞, $\sum_{n=1}^{\infty} Te_{\pi(n)}$ is $\|\cdot\|$-convergent for any permutation π, which gives that $\sum Te_i$ is unconditionally convergent in X. □

Theorem 4.51 (Pełczyński, [Pelc3b]) *Let T be a bounded noncompact operator from c_0 into a Banach space X. Then there is a subspace Z of c_0 isomorphic to c_0 such that the restriction $T|_Z$ of T to Z is an isomorphism of Z into X.*

Proof: By Proposition 4.50, $\sum Te_i$ is not unconditionally convergent in X, where $\{e_i\}$ denotes the canonical basis of c_0. By Exercises 1.38 and 1.39, there is $\varepsilon > 0$ and a sequence $(F_n)_{n=1}^{\infty}$ of mutually disjoint finite subsets of $\mathbb{N}$ so that $\| \sum_{k\in F_n} Te_k \| \geq \varepsilon$ for every n. Put $x_n = \sum_{k\in F_n} Te_k$ for $n \in \mathbb{N}$. Since the sequence $\{\sum_{k\in F_n} e_k\}_{n=1}^{\infty}$ in c_0 is a weakly null sequence, and T is a bounded operator, the sequence $\{x_n\}$ is weakly null in X.

By passing possibly to a subsequence we can assume, by Corollary 4.27, that the sequence $\{x_n\}$ is basic in X, with a basis constant K, say. Then, for $\xi = (\xi(n))_{n=1}^{\infty} \in c_{00}$, we have

$$\left\| \sum_{n=1}^{\infty} \xi(n) x_n \right\| = \left\| T \left(\sum_{n=1}^{\infty} \xi(n) \sum_{k\in F_n} e_k \right) \right\| \leq \|T\| . \max\{|\xi(n)| : n \in \mathbb{N}\}.$$

On the other hand,

$$\max\{|\xi(n)| : n \in \mathbb{N}\} \leq 2K \left\| \sum_{n=1}^{\infty} \xi(n) x_n \right\|,$$

as $\{x_n\}$ is basic with basis constant K, and $\|P_{n+1} - P_n\| \leq 2K$, where P_n denotes the nth canonical projection for the basis $\{x_n\}$ of $\overline{\operatorname{span}}\{x_n : n \in \mathbb{N}\}$. Thus $\{x_n\}$ is equivalent to the canonical basis of c_0 and therefore also to the basis $\{\sum_{k \in F_n} e_k\}_{n=1}^{\infty}$. Thus to finish the proof, we put $Z = \overline{\operatorname{span}}\{\sum_{k \in F_n} e_k : n \in \mathbb{N}\}$. □

Corollary 4.52 (Bessaga, Pełczyński [BePe0]) *Let X be a Banach space. Then the following are equivalent.*
(i) *X does not contain an isomorphic copy of c_0.*
(ii) *If $\{x_n\}$ is a sequence in X such that the set $S := \{\sum_{i=1}^{n} \varepsilon_i x_i : \varepsilon_i = \pm 1, n \in \mathbb{N}\}$ is bounded, then S is $\|\cdot\|$-relatively compact.*
(iii) *If $\{x_n\}$ is a sequence in X such that the set $S := \{\sum_{i=1}^{n} \varepsilon_i x_i : \varepsilon_i = \pm 1, n \in \mathbb{N}\}$ is bounded, then $\sum_{i=1}^{\infty} x_i$ is unconditionally convergent.*

Proof: (i)$\Longrightarrow$(iii): By Exercise 1.37, the operator $T : c_{00} \to X$ defined by $Te_i = x_i$, where e_i is the ith unit vector, is bounded and thus has a bounded extension on c_0. If $\sum x_i$ is not unconditionally convergent, then T is not compact (see Exercise 1.42), and thus, by Theorem 4.51, X contains an isomorphic copy of c_0.

(iii)$\Longrightarrow$(ii): Use Exercise 1.42.

(ii)$\Longrightarrow$(i): Assume that X contains an isomorphic copy of c_0, and let T be an isomorphism of c_0 into X. If $\{e_i\}_{i=1}^{\infty}$ denotes the canonical basis of c_0, the set

$$S := \left\{ \sum_{i=1}^{n} \varepsilon_i e_i : \varepsilon_i = \pm 1, \ n \in \mathbb{N} \right\} \ (\subset c_0)$$

is bounded and not $\|\cdot\|$-relatively compact (it contains for instance $x_n := \sum_{i=1}^{n} e_i$, where $\operatorname{dist}(x_n, x_m) = 1$ if $n \neq m$). So the set TS is not $\|\cdot\|$-relatively compact but bounded in X. □

Note that the canonical basis of c_0 satisfies the condition (iii) in Corollary 4.52 and is not unconditionally convergent.

We will now show that the structure of subspaces of L_p is different from that of ℓ_p.

Theorem 4.53 (Khintchine, see, e.g., [LiTz3]) *If $p \in (1, \infty)$ then $L_p[0, 1]$ contains a complemented subspace isomorphic to ℓ_2.*
$L_1[0, 1]$ contains a subspace isomorphic to ℓ_2.

Note that $L_1[0, 1]$ has no complemented subspace isomorphic to ℓ_2 (Exercise 13.43).

In the proof of Theorem 4.53, we will use Rademacher's functions defined by $r_n(t) = \operatorname{sign}\bigl(\sin(2^n \pi t)\bigr)$ for $t \in [0, 1]$ and $n \in \mathbb{N}$. Clearly $r_n \in S_{L_p[0,1]}$ for every $p \geq 1$ (see Fig. 4.4).

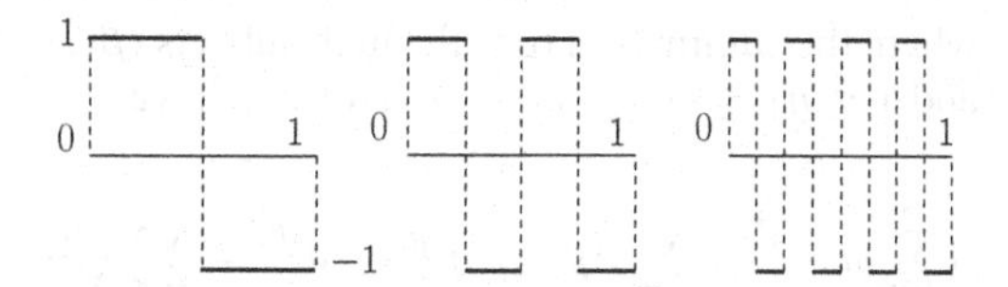

Fig. 4.4 The first Rademacher functions

It is easy to observe that $\int_0^1 r_l(t)r_k(t)\,\mathrm{d}t = \delta_{kl}$. Consequently, $\{r_n\}$ is an orthonormal set in $L_2[0,1]$. In particular, $\left\|\sum_{n=1}^m a_n r_n\right\|_{L_2} = \|(a_n)\|_{\ell_2}$, that is, $\{r_n\}$ considered in L_2 is a basic sequence that is equivalent (with constant 1) to the canonical basis of ℓ_2. The behavior of $\{r_n\}$ in L_p spaces is described in the following result.

Lemma 4.54 (Khintchine inequality) *Let r_n be the Rademacher functions on* $[0,1]$. *For every $p \in [1,\infty)$ there exist positive constants A_p and B_p such that for every $a_1,\ldots,a_m$,*

$$A_p\Big(\sum_{n=1}^m |a_n|^2\Big)^{\frac{1}{2}} \le \Big(\int_0^1 \Big|\sum_{n=1}^m a_n r_n(t)\Big|^p \,\mathrm{d}t\Big)^{\frac{1}{p}} \le B_p\Big(\sum_{n=1}^m |a_n|^2\Big)^{\frac{1}{2}}. \tag{4.2}$$

By A_p and B_p, we denote the best possible constants in this inequality. They are called Khintchine's constants and their values are known. We observed that $A_2 = B_2 = 1$. From the Hölder inequality (1.1) it follows that if $p > r$, then $\left(\int_0^1 |f|^p \,\mathrm{d}t\right)^{\frac{1}{p}} \ge \left(\int_0^1 |f|^r \,\mathrm{d}t\right)^{\frac{1}{r}}$. Consequently $A_r \le A_p$ and $B_r \le B_p$.

Proof: By the last remark, it is enough to show that there exist $A_1 > 0$ and $B_{2k} < \infty$ for all $k \in \mathbb{N}$. We start with B_{2k}.

$$\begin{aligned}\int_0^1 \Big|\sum_{n=1}^m a_n r_n(t)\Big|^{2k} \mathrm{d}t &= \int_0^1 \Big(\sum_{n=1}^m a_n r_n(t)\Big)^{2k} \mathrm{d}t \\ &= \sum A_{\alpha_1,\ldots,\alpha_j} a_{n_1}^{\alpha_1}\cdots a_{n_j}^{\alpha_j} \int_0^1 r_{n_1}^{\alpha_1}(t)\cdots r_{n_j}^{\alpha_j}(t)\,\mathrm{d}t,\end{aligned}$$

here the summation runs through all multi-indices $(\alpha_1,\ldots,\alpha_j)$ with $\sum_{i=1}^j \alpha_i = 2k$ and $1 \le n_1 \le \cdots \le n_j \le m$. By elementary combinatorics, $A_{\alpha_1,\ldots,\alpha_j} = \frac{(\sum \alpha_j)!}{(\alpha_1)!\cdots(\alpha_j)!}$. Observe that $\int_0^1 \prod_{n_1<\cdots<n_k} r_{n_k}^{\alpha_i}(t)\,\mathrm{d}t = 1$ if all α_i are even and it is equal to 0 otherwise. Using $\beta_i = \alpha_i/2$ we therefore write

$$\int_0^1 \Big|\sum_{n=1}^n a_n r_n(t)\Big|^{2k} \mathrm{d}t = \sum A_{2\beta_1,\ldots,2\beta_j} a_{n_1}^{2\beta_1}\cdots a_{n_j}^{2\beta_j},$$

where the summation runs through subsets $(\beta_1, \dots, \beta_j)$ of $\mathbb{N}$ such that $\sum_{i=1}^{j} \beta_i = k$ and $1 \leq n_1 \leq \cdots \leq n_j \leq m$. Thus we have

$$\Big(\sum_{n=1}^{m} |a_n|^2\Big)^k = \sum A_{\beta_1,\dots,\beta_j} a_{n_1}^{2\beta_1} \cdots a_{n_j}^{2\beta_j} = \sum \frac{A_{\beta_1,\dots,\beta_j}}{A_{2\beta_1,\dots,2\beta_j}} A_{2\beta_1,\dots,2\beta_j} a_{n_1}^{2\beta_1} \cdots a_{n_j}^{2\beta_j}$$

$$\geq \min\Big\{\frac{A_{\beta_1,\dots,\beta_j}}{A_{2\beta_1,\dots,2\beta_j}}\Big\} \sum A_{2\beta_1,\dots,2\beta_j} a_{n_1}^{2\beta} \cdots a_{n_j}^{2\beta_j} = \min\Big\{\frac{A_{\beta_1,\dots,\beta_j}}{A_{2\beta_1,\dots,2\beta_j}}\Big\} \int_0^1 \Big|\sum_{n=1}^{m} a_n r_n(t)\Big|^{2k} \,dt.$$

Since $\sum_{i=1}^{j} \beta_i = k$, the minimum above is a minimum of a finite set of positive numbers and so it is positive. Hence B_{2k} exists finite.

To prove the existence of A_1, put $f(t) = \sum_{n=1}^{m} a_n r_n(t)$. By Hölder's inequality (1.1) used for $p = \frac{3}{2}$ and $q = 3$, by the first part of the proof and our observation before this lemma, we have

$$\int_0^1 |f(t)|^2 \,\mathrm{d}t = \int_0^1 |f(t)|^{\frac{2}{3}} |f(t)|^{\frac{4}{3}} \,\mathrm{d}t \leq \Big(\int_0^1 |f(t)| \,\mathrm{d}t\Big)^{\frac{2}{3}} \Big(\int_0^1 |f(t)|^4 \,\mathrm{d}t\Big)^{\frac{1}{3}}$$

$$\leq \Big(\int_0^1 |f(t)| \,\mathrm{d}t\Big)^{\frac{2}{3}} B_4^{\frac{4}{3}} \Big(\sum_{n=1}^{m} |a_n|^2\Big)^{\frac{2}{3}} = \Big(\int_0^1 |f(t)| \,\mathrm{d}t\Big)^{\frac{2}{3}} B_4^{\frac{4}{3}} \Big(\int_0^1 |f(t)|^2 \,\mathrm{d}t\Big)^{\frac{2}{3}}.$$

Therefore $\Big(\int_0^1 |f(t)| \,\mathrm{d}t\Big)^{\frac{2}{3}} \geq B_4^{-\frac{4}{3}} \Big(\int_0^1 |f(t)|^2 \,\mathrm{d}t\Big)^{\frac{1}{3}}$, that is,

$$\int_0^1 |f(t)| \,\mathrm{d}t \geq B_4^{-2} \Big(\int_0^1 |f(t)|^2 \,\mathrm{d}t\Big)^{\frac{1}{2}} = B_4^{-2} \Big(\sum_{n=1}^{m} |a_n|^2\Big)^{\frac{1}{2}}.$$

Hence $A_1 \geq B_4^{-2}$. □

Proof of Theorem 4.53: Define a mapping T from ℓ_2 into $L_p[0, 1]$ by $T\big((a_n)\big) = \sum_{n=1}^{\infty} a_n r_n$, where $\{r_n\}$ are the Rademacher functions. From Lemma 4.54, it follows that T is an isomorphism from ℓ_2 into $L_p[0, 1]$. If $p \geq 2$, then $L_p[0, 1]$ is a subspace of $L_2[0, 1]$ (Exercise 1.15). Let P be the restriction to $L_p[0, 1]$ of the orthogonal projection of $L_2[0, 1]$ onto $\overline{\operatorname{span}}\{r_n\}$. We have $P(f)(t) = \sum_{n=1}^{\infty} (\int_0^1 f(s) r_n(s) \,\mathrm{d}s) r_n(t)$. This projection is bounded in $L_p[0, 1]$, since $\|P(f)\|_{L_p} \leq B_p \|P(f)\|_{L_2} \leq B_p \|f\|_{L_2} \leq B_p \|f\|_{L_p}$ by the Khintchine and Hölder inequalities.

If $p \in (1, 2)$ and $q > 2$ satisfies $\frac{1}{p} + \frac{1}{q} = 1$, there is a subspace Y of $L_q[0, 1]$ isomorphic to ℓ_2 complemented by a projection P. Then $P^*(Y^*)$ is a complemented subspace of $L_p[0, 1]$ isomorphic to ℓ_2 (Exercise 4.10). □

One of the consequences is that ℓ_p is not isomorphic to $L_p[0, 1]$ for $p \neq 2, \infty$. Indeed, by duality it is enough to show that ℓ_p is not isomorphic to $L_p[0, 1]$ for $p < 2$. Suppose that ℓ_p is isomorphic to $L_p[0, 1]$ for some $p < 2$. By Theorem 4.53, ℓ_2

must be isomorphic to a subspace of ℓ_p, which is a contradiction with Pitt's theorem. The situation is different if $p = \infty$ as ℓ_∞ is isomorphic to L_∞ (Exercise 4.42). Also, $L_2[0, 1]$ is isometric to ℓ_2 by Theorem 1.57.

The spaces L_p and L_q are not isomorphic if $p \neq q$, $p, q \in [1, \infty)$. This follows for instance using the notion of type and cotype. If $1 \leq p < q \leq 2$ then ℓ_p is not isomorphic to a subspace of L_q (this can be proved for instance using the notion of type). However, L_p contains a subspace isometric to L_q (see, e.g., [AlKa, p. 140]).

4.7 Subspaces of L_p Spaces

Theorem 4.55 (Kadec and Pełczyński, [KaPe], see, e.g., [AlKa, p. 148]) *Let $2 < p < \infty$ and X be an infinite-dimensional subspace of L_p. Then either X is isomorphic to ℓ_2 (in which case is complemented in L_p) or X contains a subspace that is isomorphic to ℓ_p and complemented in L_p.*

Before starting on the proof, we remark that if $1 < p < 2$, then L_p contains an uncomplemented isomorphic copy of ℓ_2, [BDGJN].

Proof of Theorem 4.55: Assume that X is isomorphic to ℓ_2. Since $p > 2$, $L_p \subset L_2$. Denote the continuous inclusion mapping from L_p into L_2 by T. The restriction of T to X is, by our assumption, an isomorphism of X onto $TX \subset L_2 \cap X$, which is of course complemented in L_2 by a projection P. Then the mapping $T^{-1}PT$ is a bounded linear projection of L_p onto X. Assume that X is not isomorphic to ℓ_2. Then there is a sequence (f_n) of vectors in X such that $\|f_n\|_p = 1$ and $\|f_n\|_2 \to 0$. Then, obviously, $f_n \to 0$ in measure (a sequence $\{f_n\}$ of measurable functions is said to converge *in measure* to a function f if, for every $\varepsilon > 0$, $\mu\{x : |f_n(x) - f(x)| \geq \varepsilon\} \to 0$; for the result, see, e.g., [Rudi2, Ex. 3.18], or [WhZy, Theorem 4.21]). By passing to a subsequence if needed, we can assume that $f_n \to 0$ almost everywhere, see, e.g., [Stro, p. 315].

We will prove the following.

Claim 1: *There is a sequence (f_{n_k}) of (f_n) and a sequence of disjoint measurable sets $\{A_k\}$ such that $\|f_n.\chi_{A_k}\|_p \to 1$, where χ_{A_k} is the characteristic function of A_k.*

Assuming Claim 1 proved, by normalizing we can see that it is enough to prove the following

Claim 2: *Let (f_n) be a sequence of norm-one disjointly supported functions in L_p. Then the closed linear hull of (f_n) is a 1-complemented subspace of L_p that is isomorphic to ℓ_p.*

Proof of Claim 2: By the disjointness of the supports of f_is, for any sequence $(a_i)_{i=1}^\infty \subset c_{00}$ we have

$$\left\|\sum_{i=1}^{\infty} a_i f_i\right\|_p^p = \int \left|\sum_{i=1}^{\infty} a_i f_i(t)\right|^p \mathrm{d}t$$

$$= \int \sum_{i=1}^{\infty} |a_i f_i(t)|^p \mathrm{d}t = \sum_{i=1}^{\infty} |a_i|^p \int |f_i(t)|^p \mathrm{d}t = \sum_{i=1}^{\infty} |a_i|^p.$$

Therefore the closed linear hull of $\{f_n : n \in \mathbb{N}\}$ is even isometric to ℓ_p.

We will now prove that the closed linear hull of $\{f_n : n \in \mathbb{N}\}$ is complemented in L_p. Let $p^{-1} + q^{-1} = 1$. By the Hahn–Banach theorem, for $i \in \mathbb{N}$, there is $g_i \in L_q$ with $\|g_i\|_q = 1$ and $1 = \|f_i\|_p = \int f_i(t) g_i(t) \mathrm{d}t$. We can assume that the support of g_i is the same as that of f_i for all i. Define an operator P from L_p into the closed linear hull of $\{f_n : n \in \mathbb{N}\}$ for $f \in L_p$ by

$$P(f) = \sum_{i=1}^{\infty} \left(\int f(t) g_i(t) \mathrm{d}t \right) f_i.$$

It is clear that P maps L_p into the closed linear hull of $\{f_n : n \in \mathbb{N}\}$ and that P is an operator. Moreover, $Pf_i = f_i$ for all i. Thus P is a linear projection. Its norm can be estimated as follows:

$$\|Pf\|_p = \left(\sum_{i=1}^{\infty} \left| \int f(t) g_i(t) \mathrm{d}t \right|^p \right)^{1/p}$$

$$= \left(\sum_{i=1}^{\infty} \left| \int_{\operatorname{supp} f_i} f(t) g_i(t) \mathrm{d}t \right|^p \right)^{1/p} \le \left(\sum_{i=1}^{\infty} \int_{\operatorname{supp} f_i} |f(t)|^p \mathrm{d}t \right)^{1/p} \le \left(\int |f(t)|^p \mathrm{d}t \right)^{1/p}.$$

△

Therefore, it remains to prove Claim 1.

In doing so, let $f_n \in L_p$, $\|f_n\| = 1$ for all n and $f_n \to 0$ almost everywhere. Note that then $f_n \to 0$ in measure μ. We use now a variant of a classical "gliding hump" argument: Put $f_{n_1} = f_1$ and take $F_1 := \{w : |f_{n_1}(w)|^p > 1/2\}$. Since $f_{n_1} \in L_p$, there is $\delta_1 > 0$ such that $\int_E |f_{n_1}|^p < 1/2$ whenever $\mu(E) < \delta_1$. Then pick $n_2 > n_1$ such that $\mu\{w : |f_{n_2}(w)| > (1/2^2)\} < \delta_1$ and put $F_2 = \{w : |f_{n_2}(w)|^p > 1/2^2\}$. In the same way, there is $\delta_2 > 0$ such that $\int_E |f_{n_i}|^p < 1/2^2$ for $i = 1, 2$ whenever $\mu(E) < \delta_2$. Then pick $n_3 > n_2$ such that $\mu\{w : |f_{n_3}(w)|^p > 1/2^3\} < \delta_2$ and put $F_3 = \{w : |f_{n_3}(w)|^p > 1/2^3\}$. By induction, we find a subsequence $\{f_{n_k}\}$ of $\{f_n\}$ and a sequence of sets $\{F_k\}$ such that

$$\|f_{n_k} - f_{n_k} \chi_{F_k}\| < \frac{1}{2^k}, \text{ for all } k.$$

Put $A_1 = F_1 \setminus \bigcup_{k>1} F_k$, $A_2 = F_2 \setminus \bigcup_{k>2} F_k$, $\ldots$, $A_j = F_j \setminus \bigcup_{k>j} F_k$. Then the sets A_j, $j \in \mathbb{N}$, are mutually disjoint. We have

$$\int_{F_k} |f_{n_k}|^p - \int_{A_k} |f_{n_k}|^p \leq \sum_{j>k} \int_{F_j} |f_{n_k}|^p \leq \sum_{j>k} \frac{1}{2^{j-1}} = \frac{1}{2^{k-1}}.$$

This means

$$\|f_{n_k}\chi_{F_k} - f_{n_k}\chi_{A_k}\|_p^p < \frac{1}{2^{k-1}}$$

for all k.

Hence

$$\|f_{n_k} - f_{n_k}\chi_{A_k}\|_p \leq \|f_{n_k} - f_{n_k}\chi_{F_k}\|_p + \|f_{n_k}\chi_{F_{n_k}} - f_{n_k}\chi_{A_k}\|_p \leq \left(\frac{1}{2^k}\right)^{1/p} + \left(\frac{1}{2^{k-1}}\right)^{1/p}.$$

Thus

$$\|f_{n_k}\chi_{A_k}\|_p \geq \|f_{n_k}\| - \|f_{n_k} - f_{n_k}\chi_{A_k}\| \to 1.$$

□

Theorem 4.56 *If $1 < p, q < \infty$, then ℓ_p is isomorphic to a complemented subspace of L_q if and only if either $p = 2$ or $p = q$.*

Proof: First of all, ℓ_2 is isomorphic to a complemented subspace of L_q for all $q > 1$ (Theorem 4.53). If ℓ_p is isomorphic to a subspace of L_q for $q > 2$, then $p = 2$ or $p = q$ by Theorem 4.55. If ℓ_p, $p \neq 2$, is isomorphic to a complemented subspace of L_q, $q < 2$, then ℓ_p^* (isomorphic to ℓ_{p^*}) is isomorphic to a subspace of L_{q^*}, where p^* is the dual index of p, i.e., $p^{-1} + (p^*)^{-1} = 1$. By Theorem 4.55, $p^* = q^*$, i.e., $p = q$. □

Theorem 4.57 *L_1 does not contain any complemented subspace isomorphic to ℓ_p, $p > 1$. For any $1 < q \leq 2$, L_1 contains a subspace isomorphic to ℓ_q. For no $q > 2$, L_1 contains a subspace isomorphic to ℓ_q. Of course, L_1 contains a complemented subspace isomorphic to ℓ_1.*

Proof: The first statement is a consequence of the fact that L_1 has the Dunford–Pettis property (since its dual space does) and thus L_1 does not contain any infinite-dimensional reflexive complemented subspace (Proposition 13.44 and the remark after Proposition 13.42). That L_1 contains subspaces isomorphic to ℓ_q for all $1 < q < 2$ is a result of Kadec [Kade2]. For $q = 2$ use Theorem 4.53. That L_1 cannot contain any subspace isomorphic to ℓ_q for $q > 2$ follows from the fact that L_1 has cotype 2 and ℓ_q, $q > 2$, not (see [AlKa, p. 140]). That L_1 contains a complemented subspace isomorphic to ℓ_1 is shown in [AlKa, p. 121]. □

It is an open problem whether every infinite-dimensional complemented subspace of L_1 is isomorphic either to L_1 or to ℓ_1, see, e.g., [AlKa, p. 122].

If $1 \leq p < q \leq 2$, then L_p contains an isomorphic copy of L_q, [LiPe].

4.8 Markushevich Bases

Definition 4.58 *Let X be a Banach space. A biorthogonal system $\{x_\alpha; f_\alpha\}_{\alpha\in\Gamma}$ in X is called a* Markushevich basis *of X if $\overline{\operatorname{span}}\{x_\alpha\}_{\alpha\in\Gamma} = X$ and $\{f_\alpha\}_{\alpha\in\Gamma}$ separates the points of X.*
A Markushevich basis $\{x_\alpha; f_\alpha\}_{\alpha\in\Gamma}$ is called shrinking *if $\overline{\operatorname{span}}\{f_\alpha\} = X^*$.*

Sometimes, a Markushevich basis $\{x_\alpha; f_\alpha\}_{\alpha\in\Gamma}$ will be denoted by $\{x_\alpha\}_{\alpha\in\Gamma}$ if the set of functional coefficients is understood.

Clearly, every Schauder basis of a Banach space X is a Markushevich basis of X. An example of a Markushevich basis that is not a Schauder basis is the sequence of trigonometric polynomials $\{e^{i2\pi nt} : n = 0, \pm1, \pm2, \dots\}$ in the space $\widetilde{C}[0, 1]$ of complex continuous functions on $[0, 1]$ whose values at 0 and 1 are equal, with the sup-norm.

Theorem 4.59 (Markushevich, see, e.g., [LiTz3, Section 1.f] or [HMVZ, p. 8]) *Let X be a separable Banach space. If $\{z_i\}_i \subset X$ satisfies $\overline{\operatorname{span}}\{z_i\}_i = X$ and $\{g_i\}_i \subset X^*$ separates points of X, then there is a Markushevich basis $\{x_i; f_i\}$ of X such that* $\operatorname{span}\{x_i\} = \operatorname{span}\{z_i\}$ *and* $\operatorname{span}\{f_i\} = \operatorname{span}\{g_i\}$.

Proof: Define $x_1 = z_1$ and $f_1 = g_{k_1}/g_{k_1}(z_1)$, where $k_1 \in \mathbb{N}$ is such that $g_{k_1}(z_1) \neq 0$. Then find the smallest integer h_2 such that $g_{h_2} \notin \operatorname{span}\{f_1\}$. Define $f_2 = g_{h_2} - g_{h_2}(x_1)f_1$. Find an index k_2 such that $f_2(z_{k_2}) \neq 0$, and set $x_2 = (z_{k_2} - f_1(z_{k_2})x_1)/f_2(z_{k_2})$. Let h_3 be the smallest integer such that $z_{h_3} \notin \operatorname{span}\{x_1, x_2\}$. Put $x_3 = z_{h_3} - f_1(z_{h_3})x_1 - f_2(z_{h_3})x_2$ and $f_3 = (g_{k_3} - g_{k_3}(x_1)f_1 - g_{k_3}(x_2)f_2)/g_{k_3}(x_3)$, where k_3 is an index such that $g_{k_3}(x_3) \neq 0$. Continue by induction. At the step $2n$ we construct f_{2n} first, at the step $2n + 1$ we start by constructing x_{2n+1}. It follows that $\operatorname{span}\{z_i\}_1^n \subset \operatorname{span}\{x_i\}_1^{2n}$ and $\operatorname{span}\{g_i\}_1^n \subset \operatorname{span}\{f_i\}_1^{2n}$. Clearly $f_i(x_j) = \delta_{ij}$, $\operatorname{span}\{x_i\} \subset \operatorname{span}\{z_i\}$ and $\operatorname{span}\{f_i\} \subset \operatorname{span}\{g_i\}$. □

It is an open problem whether every separable Banach space X admits a Markushevich basis $\{x_i; f_i\}_{i=1}^\infty$ with $\|x_i\| = \|f_i\| = 1$ for all i. It is known that given a separable space X and $\varepsilon > 0$, a Markushevich basis of X exists so that $\sup \|x_i\|\,\|f_i\| < 1 + \varepsilon$ (Ovsepian–Pełczyński, see, e.g., [HMVZ, p. 14]).

Theorem 4.60 (Gurarii, Kadec [GuKa], see, e.g., [HMVZ, p. 30]) *Let Z be a closed subspace of a separable Banach space X. Any Markushevich basis $\{x_i; f_i\}$ of Z can be extended to a Markushevich basis of X.*
Precisely, there are $z_j \in X$, $g_j \in X^*$, and extensions of f_i to functionals on X such that $\big\{\{x_i\} \cup \{z_j\}; \{f_i\} \cup \{g_j\}\big\}$ is a Markushevich basis of X.

Proof: Extend all f_i onto X and denote these extensions by $\tilde{f}_i$. Let $\{\hat{y}_j, \phi_j\}$ be a Markushevich basis of X/Z (it is separable). For all j choose $y_j \in \hat{y}_j$ and define $\varphi_j(x) = \phi_j(\hat{x})$ for $x \in X$, note that $\varphi_j(x_i) = 0$ for all i. We have $\overline{\operatorname{span}}\big\{\{x_i\} \cup \{y_j\}\big\} = X$ and $\{\tilde{f}_i\} \cup \{\varphi_j\}$ is a family separating points of X.

Put $z_j = y_j - \sum_{i=1}^{j} \lambda_{ij} x_i$ and $\psi_i = \tilde{f}_i - \sum_{j=1}^{i} \lambda_{ij}\varphi_j$, where $\lambda_{ij} = \tilde{f}_i(y_j)$ for $i \neq j$ and $\lambda_{ii} = \frac{1}{2}(\tilde{f}_i(y_i) - 1)$. Then $\big\{\{x_i\} \cup \{z_j\}; \{\psi_i\} \cup \{\varphi_j\}\big\}$ is a Markushevich basis of X that extends $\{x_i, f_i\}$. Indeed, from the definition of z_j and ψ_i it is clear that $\overline{\text{span}}\{x_i, z_j\} = X$, $\{\psi_i\} \cup \{\varphi_j\}$ is separating points of X, and ψ_i extend f_i onto X. It is routine to check that the system is biorthogonal. □

Theorem 4.61 (Johnson, [Johns1]) *The space ℓ_∞ does not admit a Markushevich basis.*

We will need the following.

Lemma 4.62 (Rosenthal) *Let X be a Banach space. Every reflexive subspace of X^* is w^*-closed in X^*.*

Proof: Let Y be a reflexive subspace of X^*. By the Banach–Dieudonné theorem, we only need to show that B_Y is w^*-closed in X^*. Since B_Y is w-compact in Y, it is w-compact as a subset of X^*. Therefore B_Y is w^*-compact in X^* and thus w^*-closed in X^*. □

Proof of Theorem 4.61: Assume that $\left\{x_\alpha;\, f_\alpha\right\}_{\alpha\in\Gamma}$ is a Markushevich basis of ℓ_∞. Put $Y = \overline{\text{span}}\left\{f_\alpha\right\}_{\alpha\in\Gamma}$. We claim that Y is reflexive. It is enough to prove that B_Y is weakly sequentially compact.

Let $\{y_n\}$ be a sequence in B_Y. Since every element of $\overline{\text{span}}\{f_\alpha\}$ has a countable support over $\{x_\alpha\}$ (as a limit in norm of a sequence of points in $\text{span}\{f_\alpha\}$ which have finite support), there is a countable subset N of Γ such that for $\alpha \in \Gamma\setminus N$ and $n \in \mathbb{N}$ we have $y_n(x_\alpha) = 0$.

By the Cantor diagonal argument, let $\{y_{n_k}\}$ be a subsequence of $\{y_n\}$ such that $\lim_{k\to\infty} y_{n_k}(e_\alpha)$ exists for every $\alpha \in N$. Since $y_n(e_\alpha) = 0$ for every $\alpha \in \Gamma\setminus N$ and $\{y_n\}$ is bounded, we have that $\lim_{k\to\infty} y_{n_k}(x)$ exists and is finite for every $x \in X$. Denote this limit by $y(x)$. Then y is a bounded linear functional on ℓ_∞ and $y_{n_k} \overset{w^*}{\to} y$ in ℓ_∞^*. By the Grothendieck property of ℓ_∞ (see Exercises 3.42 and 3.44), this means that $y_{n_k} \overset{w}{\to} y$ in X^* and $y \in B_Y$ as B_Y is w-closed in X^*. Therefore B_Y is w-compact in X^* and thus also w-compact in Y. Consequently, Y is reflexive.

By Lemma 4.62, Y is w^*-closed in X^*, and since Y is separating for X, it is w^*-dense in X^*. Thus $Y = \ell_\infty^*$. Consequently ℓ_∞^* and thus ℓ_∞ is reflexive, a contradiction. □

Theorem 4.63 (Plichko) *Let X be a Banach space. If X is separable then X^* is a complemented subspace of a Banach space with Markushevich basis.*
In particular, ℓ_∞ is a complemented subspace of a Banach space with Markushevich basis.

Proof: Let I be a set with $\text{card}(I) = \text{dens}(X^*)$. Define $U = \left(\sum_I c_0\right)_{\ell_1(I)}$ and $Z = \left(X^* \oplus U\right)_1$. Let $\{e_n;\, f_n\}$ be a Markushevich basis of X scaled so that $\|e_n\| \le 1$ for every n. Pick $\{y^i\}_{i\in I}$ dense in B_{X^*}. For $i \in I$ and $n \in \mathbb{N}$, let

$$y_n^i = y^i - \sum_{j=1}^{n} y^i(e_j) f_j.$$

For a fixed $n \in \mathbb{N}$ define $h_n = \left\{\{y_k^i(e_n)\}_k\right\}_i$. We have $y_k^i(e_n) = 0$ for $k \geq n$ and $|y_k^i(e_n)| \leq 1$ for $k < n$, hence

$$h_n \in \left(\sum_I \ell_1\right)_{\ell_\infty(\Gamma)} = U^*.$$

Let u_n^i be the standard unit vector in U and g_n^i be the standard unit vector in the space $\left(\sum_I \ell_1\right)_{\ell_\infty(I)} = U^*$. For $n \in \mathbb{N}$ and $i \in I$ put $x_n^i = (y_n^i, u_n^i) \in Z$ and $e_n' = (e_n, -h_n) \in Z^*$. We claim that $\left\{\{x_n^i\}_{n\in\mathbb{N}, i\in I} \cup \{(f_m, 0)\}_{n,m\in\mathbb{N}}; \{e_m'\}_{m\in\mathbb{N}} \cup \{(0, g_n^i)\}_{n\in\mathbb{N}, i\in I}\right\}$ is a Markushevich basis of Z.

It is easy to see that this is a biorthogonal system and that the functionals separate points of Z ($e_n' = e_n$ on X^*).

It remains to show that the span of vectors $\{x_n^i, f_j\}_{n,j\in\mathbb{N}, i\in I}$ is dense in Z. To see this, we note that for $n \in \mathbb{N}$ and $i \in I$ we have

$$u_n^i + y^i = u_n^i + y_n^i + \sum_{j=1}^{n} y^i(e_j) f_j = x_n^i + \sum_{j=1}^{n} y^i(e_j) f_j \in \operatorname{span}\{x_n^i, f_j\}_{j=1}^{\infty}.$$

Also, for every $i \in I$ we have

$$\left\| \sum_{n=1}^{k} \frac{u_n^i + y^i}{k} - y^i \right\| = \frac{1}{k} \to 0 \text{ as } k \to \infty.$$

Therefore $y^i \in \overline{\operatorname{span}}\{x_n^i, f_j\}$. Furthermore, for every $n \in \mathbb{N}$ and $i \in I$ we have $u_n^i = x_n^i - y_n^i = x_n^i - y^i + \sum_{j=1}^{n} y^i(e_j) f_j \in \overline{\operatorname{span}}\{x_n^i, f_j\}$.

Thus $\overline{\operatorname{span}}\{x_n^i, x_j\}_{n,j\in\mathbb{N}, i\in I} = Z$ and the proof is complete. □

4.9 Remarks and Open Problems

Remarks

1. In [Zipp3], M. Zippin proved that every Banach space with separable dual is isomorphic to a subspace of a Banach space with a shrinking Schauder basis. It follows that every separable reflexive Banach space is isomorphic to a subspace of a separable reflexive space with a Schauder basis. By a dual argument, it follows that every separable reflexive space is a quotient of a reflexive space with a Schauder basis.

2. In [DFJP] it is shown that every Banach space with a separable dual is a quotient space of a space with a shrinking basis.
3. The "convexified" Tsirelson space T_2 (see, e.g., [Casa, p. 276]) is an example of a space that, in particular, does not contain an isomorphic copy of ℓ_2 and every subspace of it has a Schauder basis.
4. If K is a compact metric space, then $C(K)$ has a Schauder basis (Vakher [Vakh]). This at present time follows since, if K is uncountable, then $C(K)$ is isomorphic to $C[0, 1]$ by Miljutin's theorem (see, e.g., [AlKa, p. 94]) and, if K is countable, then K is homeomorphic to an ordinal segment $[0, \alpha]$ for some countable α by Mazurkiewicz–Sierpiński theorem (see, e.g., [HMVZ, p. 73]).
5. Pełczyński and Singer proved in [PeSi] that if an infinite-dimensional Banach space X has a basis, then it has a basis that is not unconditional, see, e.g., [AlKa, Theorem 9.5.6].
6. Let us remark that we did not discuss symmetric bases in this text. We refer to [LiTz3] for this topic.
7. Let us remark that there is a Banach space X with a Schauder basis whose dual is separable and X does not have any shrinking Schauder basis (see [LiTz3, p. 10] and Theorem 16.56).
8. Let us remark that if a Banach space X has the property that X^* has a Schauder basis, then X has a shrinking Schauder basis, [JRZ], see, e.g., [LiTz3, p. 10].
9. Zippin showed in [Zipp0] that if a nonreflexive space X has a basis, then it has a non-shrinking basis, see, e.g., [AlKa, p. 59].
10. There is a separable Banach space with a normalized non-shrinking unconditional basis $\{e_n\}_{n=1}^{\infty}$ such that $e_n \overset{w}{\to} 0$, [PeSz].
11. Pełczyński showed in [Pelc3c] that a Banach space X is reflexive if every subspace with a Schauder basis is reflexive.
12. We showed in Theorem 4.40 that $C[0, 1]$ does not admit any unconditional basis. More generally, let us refer to [AlKa, p. 96], where it is shown that if for a compact metric space K, $C(K)$ is a subspace of a Banach space with unconditional basis, then $C(K)$ is isomorphic to c_0. Let us also refer to [Pisi3, p. 112] for the proof that the space $\mathcal{K}(\ell_2)$ of compact operators on ℓ_2 does not admit any unconditional basis. One way to check whether a Banach space fails to have an unconditional basis is to use the so-called property (u), introduced by Pełczyński in [Pelc1b]. A Banach space X is said to have *property (u)* if whenever $\{x_n\}$ is a weakly Cauchy sequence in X, there is a weakly unconditionally Cauchy series $\sum_{k=1}^{\infty} u_k$ in X so that $x_n - \sum_{k=1}^{n} u_k \to 0$ in the weak topology. In [Pelc1b], it was proved that every Banach space with unconditional basis has property (u). We refer to [AlKa, Section 3.5] for more on Pełczyński's property (u).
13. Let $L_p(\mu)$ for $1 \leq p < \infty$ be a separable space and let μ be a purely nonatomic probability measure (i.e., a probability measure such that for every measurable set of positive measure there is a measurable subset with strictly smaller positive measure). Then $L_p(\mu)$ is isometric to $L_p(0, 1)$ (see, e.g., [JoLi3, p. 15]). If the measure μ is purely atomic with atoms $\{A_\gamma\}_{\gamma \in \Gamma}$, then the function which maps for each γ the unit vector e_γ to $\mu(A_\gamma)^{-1/p}$ times the indicator function of A_γ, extends to an isometry from $\ell_p(\Gamma)$ onto $L_p(\mu)$. From these facts, it follows (see

[JoLi3, p. 15]) that for $1 \le p < \infty$, ℓ_p^n, ℓ_p, $L_p(0,1)$, $\ell_p \oplus_p L_p(0,1)$, $\ell_p^n \oplus_p L_p(0,1)$, $n = 1, 2, ...$, is a complete listing, up to isometry, of the separable $L_p(\mu)$-spaces when $p \neq 2$ and these are all mutually nonisometric. Of course, in the Hilbertian case $p = 2$, ℓ_p^n, $n = 1, 2, \ldots, \ell_2$, is the appropriate listing. By the Pełczyński decomposition method, we thus get that ℓ_p^n, ℓ_p, $L_p(0,1)$ is a complete listing, up to isomorphism, of the separable L_p-spaces. Moreover, all these spaces are mutually nonisomorphic.
In nonseparable cases, the situation is much more complicated. For example, we will show in Exercise 13.52 that $\ell_1(\omega_1)$ is not a subspace of any $L_1(\mu)$, where μ is a σ-finite measure.
14. We refer to, e.g., [HMVZ] for more on Markushevich bases and their applications in the geometry of Banach spaces.

Open Problems

1. It is an open problem whether every infinite-dimensional complemented subspace of L_1 is isomorphic either to L_1 or to ℓ_1, see, e.g., [AlKa, p. 122].
2. It is an open problem whether every complemented subspace of a space with unconditional basis has an unconditional basis, see [Casa, p. 279].
3. It is an open problem whether every separable Banach space contains a Markushevich basis $\{x_i; f_i\}_{i=1}^{\infty}$ with $\|x_i\| = \|f_i\| = 1$ for all $i \in \mathbb{N}$ (i.e., an Auerbach basis). It is known that given a separable space X and $\varepsilon > 0$, a Markushevich basis of X exists so that $\sup \|x_i\| \, \|f_i\| < 1 + \varepsilon$ (Ovsepian–Pełczyński, see, e.g., [HMVZ, p. 14]). See also, e.g., [HMVZ, Section 1.2].
4. It is apparently an open problem whether every Banach space contains a monotone subbasis.
5. It is not known whether every separable Banach space X contains a closed subspace Y such that both Y and X/Y have a Schauder basis, see, e.g., [LiTz3, p. 12]. See Remark 4.11.

Exercises for Chapter 4

4.1 (i) Let M be a subspace of a vector space V. Show that there is a linear projection of V onto M, i.e., $P\big|_M = I_M$ and $P(V) = M$.

(ii) Show that if a linear mapping $P\colon V \to V$ satisfies $P^2 = P$, then $V = P(V) \oplus \operatorname{Ker}(P)$. Moreover, $Q = I_V - P$ is a projection such that $Q(V) = \operatorname{Ker}(P)$ and $\operatorname{Ker}(Q) = P(V)$.

Hint. (i) Extend an algebraic basis $\{x_\alpha\}$ of M to an algebraic basis $\{x_\alpha, y_\beta\}$ of V and define P by $P(x_\alpha) = x_\alpha$, $P(y_\beta) = 0$.

(ii) Any $x \in V$ can be written as $x = P(x) + \big(x - P(x)\big)$.

4.2 Let P be a bounded linear projection in a Banach space X. Show that P^* is a (bounded) projection in X^*.

Hint. $(P^*)^2 = (P^2)^* = P^*$.

4.3 Let P, Q be projections in a Banach space X. Show that the following are equivalent:

(i) $P(X) \subset Q(X)$ and $P^*(X^*) \subset Q^*(X^*)$.

(ii) $PQ = QP = P$.

(iii) $P(X) \subset Q(X)$ and $\text{Ker}(Q) \subset \text{Ker}(P)$.

Hint. $P(X) \subset Q(X)$ if and only if $QP = P$. Using dual projections, (i) and (ii) are therefore equivalent. Assume (ii) holds, let $Q(x) = 0$. Then $P(x) = PQ(x) = P(0) = 0$. If (iii) holds, then $QP = P$. Moreover, given $x \in X$, $PQ(x) - P(x) = P\big(Q(x) - x\big) = 0$ as $Q\big(Q(x) - x\big) = 0$. Thus (ii) holds.

4.4 Let P be a bounded linear projection of a Banach space X onto $P(X)$. Show that for every $x \in X$,

$$\text{dist}\big(x, P(X)\big) \leq \|x - P(x)\| \leq (\|P\| + 1)\,\text{dist}\big(x, P(X)\big).$$

Hint. Obviously, $\|x - P(x)\| \geq \text{dist}\big(x, P(X)\big)$. On the other hand, if $y \in P(X)$, write $x - P(x) = x - y + y - P(x) = x - y + P(y) - P(x)$. Thus $\|x - P(x)\| \leq \|x - y\| + \|P(x) - P(y)\| \leq (1 + \|P\|)\|x - y\|$.

4.5 Let Y be a subspace of a Banach space X. Show that if there is a bounded projection P onto Y then Y is closed.

Hint. Take $y_n \in Y$ such that $y_n \to y$. Then $P(y_n) \to P(y)$, so also $y_n \to P(y)$. By the uniqueness of limit, $y = P(y)$ and $y \in Y$.

4.6 Assume that Y_i are subspaces of a Banach space X such that $X = Y_1 \oplus Y_2$ (algebraic sum). Let P_i be the associated linear projections onto Y_i (so $P_1 + P_2 = I_X$). Show that:

(i) $P_1(X) = \text{Ker}(P_2)$.

(ii) P_1 is bounded if and only if P_2 is bounded.

(iii) Both Y_1 and Y_2 are closed if and only if both P_i are bounded.

Find an example when Y_1 is closed but Y_2 is not.

The complement of a subspace Y is sometimes defined as a subspace Z such that $X = Y \oplus Z$ and the corresponding projections are bounded. (iii) shows that it is an equivalent definition, the previous exercise shows that closedness of Y is necessary for the existence of a complement.

Hint. (i) $x \in \text{Ker}(P_2)$ satisfies $x = (P_1 + P_2)(x) = P_1(x)$, similarly the other inclusion.

(ii) $P_2 = I_X - P_1$. (iii) Proposition 4.2.

For the example, $Y_1 = \mathbb{R}$, $Y_2 = \text{Ker}(f)$ for some discontinuous linear functional f.

4.7 Let X be a Banach space. Show that X^* is complemented in X^{***}.

Hint. Define $P\colon X^{***} \to X^*$ by $P(f) = f|_{X^*}$. This is called *Dixmier's projection.* Note that $\|P\| = 1$.

4.8 Let Y be a complemented subspace of a Banach space X. Let P be a projection of X onto Y. Show that the dual operator P^* is a mapping that extends elements of Y^* to elements in X^*. If $\|P\| = 1$, we get a linear Hahn–Banach extension.

Let Y be a closed subspace of a reflexive Banach space X. Assume that there is a bounded operator $E\colon Y^* \to X^*$ such that that $E(f)$ is an extension of f on X with the same norm. Show that Y is then complemented in X.
Hint. Show that E^* is a projection of X onto Y.

4.9 Let X be a Banach space. Show that if Y is a complemented subspace of X (by a projection P), then $\mathrm{Ker}(P)^\perp$ is a complemented subspace of X^* (by the projection P^*) and $Y^\perp = \mathrm{Ker}(P^*)$ is a complemented subspace of X^*.
Hint. Clearly $P^*(X^*) \subset \mathrm{Ker}(P)^\perp$. On the other hand, fix $y^* \in \mathrm{Ker}(P)^\perp$. For $x \in X$ write $x = y+z$, $y \in Y$ and $z \in \mathrm{Ker}(P)$, then $P^*(y^*)(x) = y^*(x)$, so $P^*(y^*) = y^*$. $\mathrm{Ker}(P^*) = Y^\perp$ is straightforward.

4.10 Let X be a Banach space and P be a bounded linear projection of X onto $P(X)$. Show that $P(X)^*$ is isomorphic to $P^*(X^*)$. If, moreover, $\|P\| = 1$ then $P(X)^*$ is isometric to $P^*(X^*)$.
Hint. Note that P can be considered as a mapping from X onto $P(X)$, so P^* maps $P(X)^*$ into X^*. Check this operator for the isomorphism. If $\|P\| = 1$, then $\|P^*\| = 1$, so $\|P^*(g)\| \leq \|g\|$ for every $g \in P(X)^*$. Moreover, $P^*(g)$ $(= g \circ P)$ extends g to X (see Exercise 4.8).

4.11 Show that if Y is isomorphic to a complemented subspace of X, then Y^* is isomorphic to a complemented subspace of X^*.
Hint. Let Y be isomorphic to Z and P be a projection of X onto Z. By Exercise 4.10, $P^*(Z^*)$ is closed in X^* and isomorphic to Z^*, hence to Y^*. By Exercise 4.2, $P^*(Z^*)$ is complemented in X^*.

4.12 Let $P : X \to X$ be a norm-one projection onto a subspace E of a Banach space X. Is PB_X closed in E?
Hint. Yes: $PB_X = B_X \cap E$.

4.13 Let $P : X \to X$ be a bounded projection onto a subspace E of a Banach space X. Is PB_X necessarily closed in E?
Hint. No. Let $f \in 2S_{X^*}$, $e \in S_X$, f not attaining its norm. Put $Px = f(x)e$.

4.14 Calculate the norm of the projection on the line $y = \frac{1}{2}x$ that has as kernel the y-axis, in spaces ℓ_p^2 for $p \in [1, \infty)$, and in ℓ_∞^2.
Hint. Draw a picture. The results: $(1 + \frac{1}{2^p})^{\frac{1}{p}}$, and 1 in the case ℓ_∞^2.

4.15 Let X be a normed space, Y, Z subspaces of X such that $X = Y \oplus Z$. Assume that there is a bounded linear projection P of X onto Y with $\mathrm{Ker}(P) = Z$. Show that:

(i) X is isomorphic to $(Y \oplus Z)_\infty$ (hence to $(Y \oplus Z)_p$ for $p \geq 1$).

(ii) If Y, Z are complete spaces, then so is X.

Hint. (i) Define $T(y, z) = y + z$. Since $X = Y + Z$ and $Y \cap Z = \{0\}$, this is a bijection. We have $\|T(y, z)\| = \|y + z\| \leq \|y\| + \|z\| \leq 2\max(\|y\|, \|z\|)$. On the other hand, $\|y\| \leq \|P\|\,\|y+z\|$ and $\|z\| \leq \|I_X - P\|\,\|y+z\|$, so $\max(\|y\|, \|z\|) \leq (1 + \|P\|)\|T(y, z)\|$.

(ii) If Y, Z are complete then so is $(Y \oplus Z)_\infty$

4.16 Let Y, Z be closed subspaces of a Banach space X. Show that $X = Y \oplus Z$ (topological sum) if and only if $X^* = Y^\perp \oplus Z^\perp$ (topological sum).

Hint. If $X = Y \oplus Z$, the statement follows from Exercise 4.9. Assume $X^* = Y^\perp \oplus Z^\perp$. Then $f(x) = 0$ for all $f \in X^*$ and $x \in Y \cap Z$, hence $Y \cap Z = \{0\}$. Using the separation theorem and $Y^\perp \cap Z^\perp = \{0\}$ we find that $Y + Z$ is dense in X. We claim $Y + Z$ is closed. Let Q be the linear projection of $Y \oplus Z$ (algebraic sum) onto Y. We claim that it is bounded. Let P be the bounded projection of $Y^\perp \oplus Z^\perp$ onto $Y^\perp$. Fix $y \in Y$, $z \in Z$. Then for $x^* \in X^*$ write $x^* = y^* + z^*$ and

$$x^*(y) = y^*(y) = y^*(y + z) \leq \|y^*\|\,\|y + z\| \leq \|P\|\,\|x^*\|\,\|y + z\|,$$

so $\|y\| \leq \|P\|\,\|y + z\|$. Thus $\|Q\| \leq \|P\|$. By Exercise 4.15, $Y \oplus Z$ is closed.

4.17 Find two closed subspaces Y, Z of a Banach space X so that $Y + Z$ is not closed in X.

Hint. Consider $Y = \overline{\mathrm{span}}\{e_{2i-1}\}$, $Z = \overline{\mathrm{span}}\{e_{2i-1} + 2^{-i}e_{2i}\}$ in $X = \ell_1$. Clearly both are closed subspaces, but $Y + Z$ is not closed. Indeed, $z_N = \sum_{i=1}^N 2^{-i}e_{2i} \in Y + Z$ as $z_N = \sum_{i=1}^N (e_{2_i-1} + 2^{-i}e_{2i}) - \sum_{i=1}^N e_{2_i-1}$. But $z_N \to z = \sum_{i=1}^\infty 2^{-i}e_{2i}$, which cannot be in $Y + Z$ as $(1, 1, \dots) \notin \ell_1$.

4.18 Let Y, Z be closed subspaces of a Banach space X such that $Y \cap Z = \{0\}$. Denote $d = \mathrm{dist}(S_Y, S_Z)$. Show that the subspace $Y + Z$ is closed in X if and only if $d > 0$.

Hint. Assume that $Y + Z$ is closed in X. Then Y is complemented in the Banach space $Y + Z$, so there is a bounded linear projection of $Y + Z$ onto Z. For every $y \in S_Y$, $z \in S_Z$ we have $\|P\|\|y - z\| \geq \|P(y - z)\| = \|y\| = 1$, hence $\mathrm{dist}(S_Y, S_Z) \geq \|P\|^{-1}$.

Assume that $Y + Z$ is not closed in X. By the first part of this proof, there is no $C > 0$ such that $C\|y - z\| \geq \|y\|$ for every $y \in Y$, $z \in Z$. By scaling we have the existence of $y_n \in Y$ and $z_n \in Z$, $n \in \mathbb{N}$ such that $\|y_n\| = 1$ and $\|y_n + z_n\| \leq \frac{1}{n}$ for all n. Hence $\big|1 - \|z_n\|\big| = \big|\|y_n\| - \| - z_n\|\big| \leq \|y_n + z_n\| \leq \frac{1}{n}$. Therefore by the triangle inequality, $\big\|y_n + \|z_n\|^{-1}z_n\big\| \to 0$, so $d = 0$.

4.19 Let X be a Banach space and Y be a subspace of X. Show that Y is 1-complemented in X if and only if there is a w^*-closed subspace Q of X^* such that Q 1-norms Y and whenever for some $q \in Q$ we have $q(y) = 0$ for all $y \in Y$, then $q = 0$.
Hint. For the necessary condition, let $X = Y \oplus Z$, with $P : X \to Y$ the associated projection onto Y, and $\|P\| = 1$. Put $Q = P^* X^*$. Check that $Q = Z^{\perp}$. This proves that Q is w^*-closed. The other properties of Q are also easy to check. For the sufficient condition, let $Z = Q_{\perp}$. Prove first that $Y \cap Z = \{0\}$. This shows that $Y + Z$ is an algebraic direct sum. Let $P : Y + Z \to Y$ be the associated projection. Given $y \in Y, z \in Z$, note that

$$\begin{aligned}\|P(y+z)\| &= \|y\| = \sup\{|\langle q, y\rangle| : q \in Q,\ \|q\| \le 1\} \\ &= \sup\{|\langle q, y+z\rangle| : q \in Q,\ \|q\| \le 1\} \le \|y+z\|,\end{aligned}$$

hence P is continuous and $\|P\| = 1$, since P is a projection. This proves, in particular, that $Y \oplus Z$ is closed. The other condition shows that $Y \oplus Z$ is dense in X, hence $X = Y \oplus Z$.

4.20 Let Y and Z be closed subspaces of a Banach space X such that $Y \cap Z = \{0\}$. Define a norm $|||\cdot|||$ on $Y \oplus Z$ by $|||y + z||| = \|y\| + \|z\|$.
(i) Show that $|||\cdot|||$ is a complete norm on $Y \oplus Z$.
(ii) Show that the following are equivalent:
(1) $|||\cdot|||$ is equivalent to the original norm on $Y \oplus Z$.
(2) $Y + Z$ is closed.
(3) Y is complemented in $Y + Z$, that is, $Y \oplus Z$ is a topological sum.
(4) There is $k > 0$ such that $\|y\| \le k\|y + z\|$ for every $y \in Y$.
Hint. (i) $\{y_n + z_n\}$ is Cauchy (respectively convergent) if and only if both $\{y_n\}$ and $\{z_n\}$ are Cauchy (respectively convergent).
(ii) If $|||\cdot|||$ is an equivalent norm, then $(Y \oplus Z, \|\cdot\|)$ is complete, hence $Y + Z$ is closed. But then $Y \oplus Z$ is an algebraic sum of two closed subspaces of a Banach space, hence it is a topological sum. In particular, Y is complemented by projection P such that $\|y\| \le \|P\|\,\|y+z\|$. Finally, the inequality implies $\|z\| \le (1+k)\|y+z\|$ and so $|||y + z||| \le (2k + 1)\|y + z\|$, clearly $\|y + z\| \le |||y + z|||$.

4.21 Show that there are two closed subspaces M_1 and M_2 of a Hilbert space such that $M_1 \cap M_2 = \{0\}$ and $M_1 + M_2$ is not closed, and thus there is no projection on M_1 that has kernel M_2.
Hint. Define $T \in \mathcal{B}(\ell_2)$ by $T(x) = (2^{-i}x_i)$ for $x = (x_i)$. Show that T maps ℓ_2 onto a dense subset of ℓ_2 but does not map it onto ℓ_2.
Denote by G the graph of T in $\ell_2 \oplus \ell_2$ and let $M = \ell_2 \oplus \{0\} \subset \ell_2 \oplus \ell_2$. Show that $G \cap M = \{0\}$, $G + M$ is dense in $\ell_2 \oplus \ell_2$ but that $G + M \neq \ell_2 \oplus \ell_2$.

4.22 Let X be a Banach space. Assume that Y, Z are w^*-closed subspaces of X^* such that $Z \oplus Y = X^*$ (algebraic sum). Show that the corresponding (algebraic) projection Q of X^* onto Y is w^*–w^*-continuous.

There is a Banach space X such that an isomorphic copy Y of a Hilbert space is a complemented subspace of X^*; however, there is no w^*-w^*-continuous projection of X^* onto Y (see [JoLi1]).

Hint. Note that $Y_\perp \oplus Z_\perp = X$ (algebraic sum). Let P be the corresponding (algebraic) projection of X onto $Z_\perp$. Then $P^*(y+z)(y_\perp + z_\perp) = Q(y+z)(y_\perp + z_\perp) = y(z_\perp)$.

4.23 Let $\{e_\gamma\}$ be a Hamel basis of an infinite-dimensional Banach space X. Show that some of the coordinate functionals associated with this basis are not continuous.

Hint. Pick an infinite sequence $\{e_{n_i}\}$ in $\{e_\gamma\}$. Consider the vector $x = \sum_{i=1}^{\infty} 2^{-i} \frac{e_{n_i}}{\|e_{n_i}\|}$. Since $\{e_\gamma\}$ is a Hamel basis of X, we have $x = \sum_F x_j e_j$, where F is a finite set. Let n_p be such that $n_p \notin F$, i.e., $x_{n_p} = 0$. For every m, $\sum_{i=1}^{m} 2^{-i} \frac{e_{n_i}}{\|e_{n_i}\|}$ has the n_p-coordinate equal to $2^{-n_p}/\|e_{n_p}\|$. If the n_p-coordinate functional were continuous, we would have $x_{n_p} = 2^{-n_p}/\|e_{n_p}\| \neq 0$, a contradiction.

4.24 Why do we not use (iii) in Lemma 4.7 and the Banach–Steinhaus theorem to conclude that P_n are uniformly bounded?

Hint. Do we know that they are bounded operators?

4.25 Show that the canonical projections of a Schauder basis of a normed space X need not be uniformly bounded if X is not a Banach space.

Hint. Consider the trigonometric polynomials in the space of continuous functions on $[0, 2\pi]$.

4.26 Use the notion of basic sequence to prove that a Hamel basis of an infinite-dimensional Banach space has cardinality at least the continuum.

Hint. Take any basic sequence $\{x_n\}$ in X. Let $\{N_\alpha\}_\alpha \in \Gamma$ be a collection of infinite subsets of $\mathbb{N}$ such that $N_\alpha \cap N_\beta$ is finite if $\alpha \neq \beta$ and $\mathrm{card}(\Gamma) \geq c$ (Lemma 5.7). Define $y_\alpha = \sum_{i \in N_\alpha} 2^{-i} x_i$. Then $\{y_\alpha\}$ is a linearly independent set of cardinality at least c.

4.27 Let $\{e_i\}$ be a Schauder basis of a Banach space X. Prove that there is an equivalent norm on X in which $\{e_i\}$ is monotone.

Hint. Put $|||x||| = \sup_n \|P_n(x)\|$. Then

$$|||P_m(x)||| = \sup_n \|P_n P_m(x)\| = \sup_{n \leq m} \|P_n(x)\| \leq \sup_n \|P_n(x)\|.$$

4.28 Let $\{e_i\}$ be a Schauder basis. For $n \leq m \in \mathbb{N}$ define $T_{n,m}\left(\sum_{i=1}^{\infty} a_i e_i\right) = \sum_{i=n}^{m} a_i e_i$. We say that $\{e_i\}$ is *bimonotone* if $\|T_{n,m}\| = 1$ for all n, m.

(i) Show that $\|T_{n,m}\| \leq 2\,\mathrm{bc}\{e_i\}$.

(ii) Show that there is an equivalent norm $|||\cdot|||$ on X such that $\{e_i\}$ is a bimonotone basis of $(X, |||\cdot|||)$.

Hint. (i) $T_{n,m} = P_m - P_{n-1}$. (ii) $|||x||| = \sup\limits_{n \leq m} \|T_{n,m}(x)\|$.

4.29 Let $\{e_i\}$ be a Schauder basis of a Banach space X. Show that $\{e_i\}$ is monotone if and only if $\|P_{n+1}(x)\| \geq \|P_n(x)\|$ for every $x \in X$ and $n \in \mathbb{N}$.

Hint. If $\|P_n\| = 1$ for all n, use $P_n(x) = P_n P_{n+1}(x)$. If the condition holds, prove first $\|P_k(x)\| \geq \|P_n(x)\|$ for all $k \geq n$ and then take the limit for $k \to \infty$.

4.30 Show that $C[0, 1]$ has a Schauder basis consisting of polynomials.

Hint. Consider any Schauder basis $\{e_i\}$ of $C[0, 1]$ and approximate $\{e_i\}$ by polynomials using the Stone–Weierstrass theorem. Use the stability theorem for Schauder bases.

4.31 Find two dense linear subspaces of a Banach space having intersection $\{0\}$.

Hint. Let $X = \{f \in C[0, 1] : \ f(0) = f(1) = 0\}$. This space, endowed with the supremum norm, is a Banach space. The set $\{e_n : n = 3, 4, \ldots\}$ is a Schauder basis of X, where $\{e_n\}_{n\in\mathbb{N}}$ is the Faber–Schauder basis of $C[0, 1]$ defined in Example (iv) after Theorem 4.10 by using the dyadic points in $[0, 1]$. Indeed, $\{e_n : n = 3, 4, \ldots\} \subset X$, and, if $f \in X$ is written as $f = \sum_{n=1}^{\infty} a_n e_n$, then $0 = f(0) = a_1$, and $0 = f(1) = a_1 + a_2$, hence $f = \sum_{n=3}^{\infty} a_n e_n$. Choose a C^{∞}-function g_n in $[0, 1]$ such that $g_n(0) = g_n(1) = 0$ and $\|g_n - e_n\| < 1/2^n$, $n = 3, 4, \ldots$. Then $\{g_n : n = 3, 4, \ldots\}$ is a Schauder basis (see Theorem 4.23). Every element in $\mathrm{span}\{g_n : n = 3, 4, \ldots\}$ is in the class C^{∞}, but no element in $\mathrm{span}\{e_n : n = 3, 4, \ldots\} \setminus \{0\}$ is.

4.32 Let X, Y be closed subspaces of a separable Banach space Z such that $X \cap Y = \{0\}$. Assume that the algebraic sum $X + Y$ is not closed in Z. Show that there exist a basic sequence $\{x_i\} \subset X$ and a basic sequence $\{y_i\} \subset Y$ such that $\|x_i\| = \|y_i\| = 1$ and $\|x_i - y_i\| < 4^{-i}$ for $n \in \mathbb{N}$. In particular, X and Y have infinite-dimensional closed subspaces that are isomorphic.

Thus if X and Y are totally incomparable spaces (for example ℓ_p, ℓ_q for $p \neq q$), then $X + Y$ is closed in every overspace.

Hint. If $X + Y$ is not closed, then (Exercise 4.18) there exists $x \in X$, $y \in Y$, $\|x\| = \|y\| = 1$ and $\|x - y\|$ arbitrarily small. Then use the proof of Propositions 4.19 and 4.23.

4.33 Let $\{e_i; e_i^*\}$ be a Schauder basis of a Banach space X. Show that if $\{e_i\}$ is shrinking, then $\{e_i^*\}$ is a boundedly complete basis of X^*.

Hint. $\{e_i^*\}$ is a Schauder basis of X^*. Consider a_i such that $\sup\left\|\sum_{i=1}^n a_i e_i^*\right\| < \infty$. X is separable, so there is a sequence $\{n_k\} \subset \mathbb{N}$ and $x^* \in X^*$ such that $x_{n_k}^* = \sum_{i=1}^{n_k} a_i e_i^* \overset{w^*}{\to} x^*$ (Exercise 3.176). Since $\{e_i^*\}$ is a basis of X^*, we have $x^* = \sum_{i=1}^{\infty} \beta_i e_i^*$. Fix $j \in \mathbb{N}$. Then $\lim_k x_{n_k}^*(e_j) = x^*(e_j) = \beta_j$ and $x_{n_k}^*(e_j) = a_j$ for $n_k \geq j$, hence $a_j = \beta_j$. Thus $\sum_{i=1}^{\infty} a_i e_i^* = x^*$.

4.34 Let $\{e_i; e_i^*\}$ be a Schauder basis of a Banach space X. Show that $\overline{\operatorname{span}}\{e_i^*\}$ is a norming subspace of X^*.
Hint. Let $|||x||| = \sup \|P_n(x)\|$. Then $|||P_n||| = 1$ for every n, so $|||P_n^*||| = 1$. Given $f \in S_{X^*}$, we have $|||P_n^*(f)||| \leq 1$, $P_n^*(f) \in \operatorname{span}\{e_1^*, \dots, e_n^*\}$ and $P_n^*(f) \overset{w^*}{\to} f$. Hence $\overline{\operatorname{span}\{e_i^*\} \cap B'_{X^*}}^{w^*} = B'_{X^*}$, where B'_{X^*} is the dual unit ball in the dual norm to $|||\cdot|||$. Thus $\overline{\operatorname{span}}\{e_i^*\}$ is 1-norming for $(X, |||\cdot|||)$. Since $|||\cdot|||$ is an equivalent norm on X, $\overline{\operatorname{span}}\{e_i^*\}$ is a norming set.

4.35 Let $X = \left(\sum \ell_\infty^n\right)_2$. Show that every infinite-dimensional closed subspace of X contains an isomorphic copy of ℓ_2.
Hint. Given a closed subspace Z of X, there is a subsequence in Z which is equivalent to a block basic sequence of the canonical basis of X such that the "nods" are at the ends of the canonical blocks. This block basis is equivalent to the standard unit vector basis of ℓ_2.

4.36 Let X be a Banach space. Assume that there is $\lambda \geq 1$ so that for every $\varepsilon > 0$ and every finite set $F \subset X$, there is a finite rank operator $T : X \to X$ so that $\|Tx - x\| < \varepsilon$ for every $x \in F$ and $\|T\| \leq \lambda$. Must X have the bounded approximation property?
Hint. Yes, given a compact set K and $\varepsilon > 0$ consider a finite set $F \subset X$ such that $K \subset \bigcup_{x \in F} B(x, \varepsilon/(3\lambda))$. If T is a finite rank operator on X such that $\|Tx - x\| < \varepsilon/3$ for $x \in F$, then $\|Tx - x\| < \varepsilon$ for every $x \in K$.

4.37 Show that every Banach space X has the following property: Given $\varepsilon > 0$ and given a finite set $F \subset X$, there is a finite rank operator T on X such that $\|x - Tx\| < \varepsilon$ for every $x \in F$.
Hint. Any projection from X onto $\operatorname{span}(F)$. Note that ε can be taken 0.

4.38 Show that $\left(\sum L_p[0, 1]\right)_{\ell_p}$ is isometric to $L_p[0, 1]$.
Hint. Put $A_n = [\frac{1}{n+1}, \frac{1}{n}]$ and identify $f \in L_p[0, 1]$ with the sequence $\{f|_{A_n}\}$. Use the isometry of $L_p[0, 1]$ and $L_p(A_n)$.

4.39 Show that $\left(\sum \ell_p\right)_{\ell_p}$ is isometric to ℓ_p and $\left(\sum c_0\right)_{c_0}$ is isometric to c_0.
Hint. Direct computation.

4.40 Show that $\left(\sum C[0, 1]\right)_{c_0}$ is isomorphic to $C[0, 1]$.

Hint. Use the Pełczyński's decomposition method. Represent $\left(\sum C[0,1]\right)_{c_0}$ as a complemented subspace of $C[0,1]$ using an infinite number of nods in $[0,1]$ and the closed subspace of $C[0,1]$ of functions that vanish at these nodes.

4.41 Let $p \in (1,\infty)$. Show that $L_p[0,1] \oplus \ell_2$ is isomorphic to $L_p[0,1]$.
Hint. Use that ℓ_2 is isomorphic to a complemented subspace of $L_p[0,1]$.

4.42 Use that $L_\infty[0,1]$ is complemented in every overspace (see Exercise 5.91) to show that ℓ_∞ is isomorphic to $L_\infty[0,1]$ ([Pelc2]). As ℓ_1 is not isomorphic to $L_1[0,1]$, this provides an example of two non-isomorphic spaces whose duals are isomorphic. This also shows that no isomorphism of ℓ_∞ and $L_\infty[0,1]$ can be w^*-w^*-continuous (Exercises 2.51 and 3.60).
Hint. As the dual of a separable space, L_∞ is isomorphic to a subspace of ℓ_∞. ℓ_∞ is complemented in every overspace, so we use the Pełczyński decomposition method.

4.43 Show that ℓ_2 is isomorphic to a quotient of ℓ_∞.
Compare this with Exercise 13.40.
Hint. The space ℓ_2 is isomorphic to a subspace of $L_1[0,1]$ by Theorem 4.53. Therefore $\ell_2^* \cong \ell_2$ is isomorphic to a quotient of $L_1^* = L_\infty$ and then use the previous exercise.

4.44 Assume that T is a bounded operator from a Banach space X into X such that $T(X)$ is not closed. Show that on no finite-codimensional subspace Y of X, T is an isomorphism from Y into X.
Hint. Assume that T is an isomorphism from a finite-codimensional Y into X. We have $X = Y \oplus Z$ as Y is complemented. Then $T(X) = T(Y) + T(Z)$. If T is an isomorphism on Y then $T(Y)$ is closed. Since Z is finite-dimensional, we have that $T(Y) + T(Z)$ is closed.

4.45 Let X, Y be Banach spaces. Show that every compact operator from X into Y is strictly singular.
Give an example of a strictly singular operator that is not compact.
Hint. Let Z be a subspace of X such that T is an isomorphism of Z onto $T(Z)$. Then $T(B_Z)$ is a closed subset of $\overline{T(B_X)}$, hence it is compact. As $T\big|_Z$ is an isomorphism, B_Z must be compact, hence Z is finite-dimensional.
For the second question, consider the formal identity mapping from ℓ_1 into ℓ_2.

4.46 (KaTo) Let X be an infinite-dimensional Banach space and let T be a bounded operator from X into X such that the restriction to every infinite-dimensional closed subspace of X is not compact. Show that there is a finite-codimensional subspace Z of X such that the restriction of T on Z is an isomorphism.
Hint. Assume that the restriction of T to every finite-codimensional subspace is not an isomorphism. Then for every $\delta > 0$ and $\{x_i^*\}_{i=1}^m \subset X^*$ there is $x \in S_X$ with $\|T(x)\| \le \delta$ and $x_i^*(x) = 0$ for $i = 1, \ldots, m$. By the proof of Mazur's theorem,

there is a basic sequence $\{x_i\} \subset X$ with $\text{bc}\{x_i\} \leq 2$ so that $\|T(x_i)\| < 8^{-i}$. Let $Z = \overline{\text{span}}\{x_i\}$. Then $T(B_Z)$ is contained in the set $\{\sum a_i T(x_i) : |a_i| \leq 2\}$, which is totally bounded due to $\|T(x_i)\| \leq 8^{-i}$ (see Exercise 1.50).

4.47 Show that the sum of two strictly singular operators T, S from a Banach space X into a Banach space Y is strictly singular.
Hint. By repeated use of the previous exercise, first for T and then for S get an infinite-dimensional closed subspace Z of X on which both T and S are compact. Then $T + S$ is compact on Z and thus not an isomorphism on Z.

4.48 Let T be a strictly singular operator from a Banach space X into a Banach space Y. Is T^* necessarily strictly singular?
Hint. No. Consider a bounded operator T from ℓ_1 onto ℓ_2 (Theorem 5.1). T is strictly singular as ℓ_2 is not isomorphic to any subspace of ℓ_1 (Pitt's theorem). However, T^* is an isomorphism into (Exercise 2.49) and thus T^* is not strictly singular.

4.49 Let Γ be uncountable and $1 \leq p < q < \infty$. Show that there is no bounded one-to-one operator from $\ell_q(\Gamma)$ into $\ell_p(\Gamma)$. Similarly, there is no bounded one-to-one operator from $c_0(\Gamma)$ into $\ell_p(\Gamma)$, $p \in [1, \infty)$.
Hint. Assume that such an operator $T : \ell_q(\Gamma) \to \ell_p(\Gamma)$ exists. Let $\varepsilon > 0$ and an uncountable set $\Gamma_1 \subset \Gamma$ be such that $\|T(e_\gamma)\| \geq \varepsilon$ for $\gamma \in \Gamma_1$. An infinite sequence in $\{e_\gamma\}_{\gamma \in \Gamma_1}$ tends weakly to zero and we use Pitt's theorem for its closed span.

4.50 Show that every strictly singular operator T from ℓ_p into ℓ_p is necessarily compact for $p \in [1, \infty)$.
Hint. Examine the proof of Pitt's theorem.

4.51 Does there exist a bounded operator from $C[0, 1]$ onto ℓ_2?
Note that there is no isomorphic copy of ℓ_2 complemented in $C[0, 1]$ (this follows using the Dunford–Pettis property, see Chapter 13).
Hint. Yes. ℓ_2^* is isomorphic to a subspace of $L_1[0, 1]$, which is isomorphic to a subspace of $C[0, 1]^*$ by the Riesz representation theorem. By Lemma 4.62, ℓ_2^* is w^*-closed in $C[0, 1]^*$.

4.52 Does there exist a bounded operator from ℓ_p onto ℓ_q, $p \neq q \in [1, \infty)$?
Hint. No. Pitt's theorem, duality.

4.53 Does there exist a bounded operator from ℓ_p onto c_0?
Hint. If and only if $p = 1$. For $p = 1$ see Theorem 5.1. For $p > 1$ use the reflexivity of ℓ_p.

4.54 Show that a block basic sequence of an unconditional basis is unconditional.

4.55 Show that every infinite-dimensional subspace of a Banach space with unconditional basis has an infinite-dimensional unconditional basic sequence.
Hint. A block basis of an unconditional basis is unconditional. Then use the Bessaga–Pełczyński selection principle.

4.56 Show that the vectors $e_1, e_2 - e_1, e_3 - e_2, \dots$ form a basis of ℓ_1 that is not unconditional (e_i denotes the standard ith unit vector).
Hint. Standard.

4.57 (Köthe, Lorch) Show that every normalized unconditional basis $\{x_i\}$ of a Hilbert space is equivalent to the canonical basis of ℓ_2.
Hint. We need to prove that for a sequence $\{\lambda_i\}$ of scalars, the series $\sum \lambda_i x_i$ converges if and only if $\sum |\lambda_i|^2 < \infty$. Assume that $\sum \lambda_i x_i$ converges. It converges unconditionally so, by Exercise 1.96, ($\sum |\lambda_i|^2 =$) $\sum \|\lambda_i x_i\|^2$ converges. On the other hand

$$\left\|\sum_F \lambda_i x_i\right\| \leq K \left\| \frac{1}{2^n} \sum_{\varepsilon_i = \pm 1} \sum_F \varepsilon_i \lambda_i x_i \right\| = K \left(\sum_F |\lambda_i|^2 \right)^{1/2}$$

for every nonempty finite set $F \subset \mathbb{N}$, where K is the unconditional basis constant (the inequality follows from the unconditionality, the equality from the parallelogram identity).

Note that an analogous result holds for c_0 and ℓ_1. However, in these cases the proof is more difficult (see, e.g., [LiTz3, paraghaph preceeding Proposition 2.b.9]).

4.58 Assume that an X admits an unconditional basis and that X^* is separable. Is it true that then X^* has an unconditional basis?
Hint. Yes, James' characterization of spaces with unconditional bases not containing ℓ_1.

4.59 Assume that X^* has an unconditional basis. Does X necessarily admit an unconditional basis?
Hint. No. $C(\omega_0^{\omega_0})$, see [AlKa, p. 96].

4.60 Let X be a Banach space (not necessarily separable). A family $\{e_\gamma\}_{\gamma \in \Gamma}$ in X is called an *unconditional Schauder basis of* X if for every $x \in X$ there is a unique family of real numbers $\{a_\gamma\}_{\gamma \in \Gamma}$ such that $x = \sum a_\gamma e_\gamma$, where the summation is meant in the sense that for every $\varepsilon > 0$ there is a finite set $F \subset \Gamma$ such that $\left\|x - \sum_{\gamma \in F'} a_\gamma e_\gamma\right\| \leq \varepsilon$ for every $F' \supset F$. Note that for every $x \in X$ only countably many coordinates a_γ are nonzero. Indeed, given $n \in \mathbb{N}$, there is a finite set $F \subset \Gamma$ such that $\left\|\sum_{\gamma \in F'} a_\gamma e_\gamma\right\| \leq \frac{1}{n}$ for every finite F' disjoint from F. Applying this to $F' = \{\gamma\}$ and assuming $\|e_\gamma\| = 1$ we get $\{\gamma : |a_\gamma| \geq \frac{1}{n}\} \subset F$.

Similarly as for the case of countable Schauder bases it can be shown that the coordinate functionals a_γ^* are in fact in X^*.

Show that the closed linear span of the coordinate functionals of an unconditional basis is a norming subspace in X^*.

Hint. Use $(P_F)^*$.

4.61 Let X be a Banach space. Show that if X^* is separable, then X admits a shrinking Markushevich basis.

Hint. The proof of Theorem 4.59.

4.62 Let X, Y be separable Banach spaces. Show that there is a bounded operator mapping X onto a linearly dense subset in Y.

Hint. Let $\{x_i; f_i\}$ and $\{z_i; g_i\}$ be Markushevich bases of X and Y respectively, with $\{f_i\}$ and $\{z_i\}$ bounded. Put $T(x) = \sum 2^{-i} f_i(x) z_i$.

4.63 Let X be an infinite-dimensional separable Banach space. Show that there is a biorthogonal system $\{x_i; f_i\}$ such that $\overline{\operatorname{span}}\{x_i\} = X$ and $\{f_i\}$ is not separating.

Note that this cannot be done in finite-dimensional spaces.

Hint. Let $\{y_i\}$ be a linearly independent sequence such that $\overline{\operatorname{span}}\{y_i\} = X$. Pick $x \in X \setminus \operatorname{span}\{y_i\}$ (infinite-dimensional spaces cannot have a countable Hamel basis, Exercise 1.81). Since $\{x, y_1, \ldots\}$ is a linearly independent set, using Hahn–Banach we find $g_i \in S_{X^*}$ such that $g_i(x) = 0$, $g_i(y_i) = 1$, and $g_i(y_j) = 0$ for $j = 1, \ldots, i-1$. Then $\{g_i\}$ separates points of $\operatorname{span}\{y_i\}$ and as in Theorem 4.59 we find a biorthogonal system $\{x_i; f_i\}$ so that $\operatorname{span}\{x_i\} = \operatorname{span}\{y_i\}$ and $\operatorname{span}\{f_i\} = \operatorname{span}\{g_i\}$. Then $\overline{\operatorname{span}}\{x_i\} = X$ and $\{f_i\}$ does not separate points. Indeed, $g_i(x) = 0$ for all i, hence also $f(x) = 0$ for all $f \in \operatorname{span}\{g_i\}$.

4.64 Let X be a Banach space and $\{x_\gamma; x_\gamma^*\}_{\gamma \in \Gamma}$ be a biorthogonal system in $X \times X^*$ such that $\{x_\gamma : \gamma \in \Gamma\}$ is linearly dense in X. Prove that $\operatorname{card}(\Gamma) = \operatorname{dens} X$.

Hint. Let $D \subset X$ be a dense subset of X with $\operatorname{card}(D) = \operatorname{dens} X$. Given $d \in D$ there exists a countable set $\Gamma_d \subset \Gamma$ such that $d \in \overline{\operatorname{span}}\{x_\gamma : \gamma \in \Gamma_d\}$. Then $D \subset \overline{\operatorname{span}}\{x_\gamma : \gamma \in \bigcup_{d \in D} \Gamma_d\}$, hence $X = \overline{\operatorname{span}}\{x_\gamma : \gamma \in \bigcup_{d \in D} \Gamma_d\}$. If $\gamma_0 \in \Gamma \setminus \bigcup_{d \in D} \Gamma_d$, note that $\langle x_{\gamma_0}^*, x_\gamma \rangle = 0$ for all $\gamma \in \bigcup_{d \in D} \Gamma_d$, hence $x_{\gamma_0} \notin \overline{\operatorname{span}}\{x_\gamma : \gamma \in \bigcup_{d \in D} \Gamma_d\}$ $(= X)$, a contradiction. Thus $\bigcup_{d \in D} \Gamma_d = \Gamma$ and we get $\operatorname{card}(\Gamma) = \operatorname{card}(D) = \operatorname{dens} X$.

4.65 Show that James' space J (Definition 4.43) contains an isomorphic copy of ℓ_2.

Hint. Let H denote the subspace of J consisting of vectors with even coordinates equal to zero. Show that on H, James' norm is equivalent to the ℓ_2 norm.

4.66 Let $u_i = e_i - e_{i+1}$, where $\{e_i\}$ is the shrinking basis of J. Show that $\{u_i\}$ is a boundedly complete basis of J which is not shrinking, and

$$\left\|\sum \xi_i u_i\right\| = \sup\Big\{\Big(\sum_{j=1}^{m-1}\Big(\sum_{i=n_j}^{n_{j+1}-1} \xi_i\Big)^2\Big)^{\frac{1}{2}} : 1 \leq n_1 < \cdots < n_m\Big\}.$$

4.67 Let $\{u_i\}$ be the boundedly complete basis of J from the previous exercise. Let $\{v_i\}$ be the biorthogonal functionals to $\{u_i\}$, which span the predual space J_* to J by Theorem 4.15. Define $g \in J^*$ by $g(\sum \xi_i u_i) = \sum \xi_i$. Show that $J^* = \overline{\operatorname{span}}(\{v_i\} \cup \{g\}) = \overline{\operatorname{span}}(J_* \cup \{g\})$.
Hint. $J_* = \overline{\operatorname{span}}\{v_i\}$. Clearly, $g \in J^* \setminus J_*$, then use $\dim(J^*/J_*) = 1$.

4.68 Assume that we replace the exponent 2 in the definition of James' space J by 1. Is the resulting space isomorphic to ℓ_1? Can c_0 be obtained in a similar way?
Hint. Yes. Yes.

4.69 Is the following an equivalent norm on James' space J?

$$\|z\|_0^2 = \sup_{n_1 < \ldots < n_{2m}} \Big\{\sum_{i=1}^{m} |z_{n_{2i-1}} - z_{n_{2i}}|^2\Big\}$$

Hint. Yes.

4.70 Define a norm $|||\cdot|||$ on James' space J by

$$|||x||| = \sup \frac{1}{\sqrt{2}}\big((x_{n_1} - x_{n_2})^2 + \ldots + (x_{n_{m-1}} - x_{n_m})^2 + (x_{n_m} - x_{n_1})^2\big)^{\frac{1}{2}}.$$

Show that $|||\cdot|||$ is an equivalent norm on J and in this norm, J^{**} is isometric to J.
Hint. Consider the mapping $U\colon J^{**} \to J$ defined by

$$U(x^{**}) = (-\lambda, x^{**}(e_1) - \lambda, x^{**}(e_2) - \lambda, \ldots),$$

where $\lambda = \lim x^{**}(e_i^*)$.

4.71 Find a Schauder basis $\{e_i\}$ of a Banach space X with separable dual such that $\{e_i\}$ is not shrinking.
Hint. Consider the biorthogonal functionals $\{f_i\}$ to the standard unit vector basis $\{e_i\}$ in James' space J. Since the standard basis is shrinking, $\{f_i\}$ is a basis of J^*. However, $\{f_i\}$ cannot be shrinking as its biorthogonal functionals are $\{e_i\}$ in J and $\overline{\operatorname{span}}\{e_i\} = J \neq J^{**}$.

4.72 Does there exist a Banach space X not isomorphic to a Hilbert space and such that X^* is isomorphic to X?

Hint. Consider $(\ell_p \oplus \ell_q)_2$, where $p \in (1, \infty)$ and $\frac{1}{p} + \frac{1}{q} = 1$. To obtain a nonreflexive example, consider $J \oplus J^*$, where J is James' space.

4.73 Does there exist a Banach space X such that $X \oplus X$ is not isomorphic to X?
Hint. Consider James' space. What is the codimension of $J \oplus J$ in $(J \oplus J)^{**}$? Note that the first reflexive space X that is not isomorphic to $X \oplus X$ was constructed by T. Figiel [Figi2].

In Exercises 4.74, 4.75, 4.76, 4.77, 4.78, and 4.79 we define and investigate the James tree space JT, [Jame2] and [LiSt]. All notation and definitions made in one exercise carry to the following ones.

4.74 Let $T = \{(n, i) : 0 \leq i \leq 2^n - 1, n \in \mathbb{N}_0\}$ be equipped with a partial ordering $<$ determined by the relation $(n, i) < (n + 1, j)$ if, and only if, $j \in \{2i, 2i + 1\}$. Then $(T, <)$ is a *dyadic tree*. By a *segment* we mean a subset $S = \{t \in T : (n, i) \leq t \leq (m, j)\}$ for some $(n, i), (m, j) \in T$. A maximal linearly ordered subset of T is called a *branch*. Let Γ denote the set of all branches in T. Show that $\mathrm{card}(\Gamma) = c$, the cardinality of the continuum.

4.75 The James tree space JT consists of all real functions defined on T with the norm

$$\|x\| = \sup\Big(\sum_{j=1}^{k}\Big(\sum_{(n,i)\in S_j} x(n, i)\Big)^2\Big)^{\frac{1}{2}},$$

where the supremum is taken over all finite sets of pairwise disjoint segments in T. Verify that JT is a Banach space with a boundedly complete basis $\{e_{(n,i)}\}$ ordered lexicographically, that is,

$$(0, 0) \leq (1, 0) \leq (1, 1) \leq \cdots \leq (n, i) \leq (n, i + 1) \leq \cdots \leq (n, 2^n - 1) \leq (n + 1, 0) \leq \ldots$$

and the functions $e_{(n,i)}$ are defined by $e_{(n,i)}(m, j) = 1$ if $(n, i) = (m, j)$ and zero otherwise (i.e., $e_{(n,i)}(m, j) = \delta_{(m,j)(n,i)}$).

In particular, there exists a predual JT_* to JT with a shrinking basis $\{f_{(n,i)}\}$ with $e_{(n,i)}(f_{(m,j)}) = \delta_{(m,j),(n,i)}$. Thus JT^* can be represented as a space of real functions F on T satisfying $F = w^*\text{-}\lim \sum_{n=0}^{\infty} \sum_{i=0}^{2^n-1} F(n, i) f_{(n,i)}$.
Hint. By contradiction. Consider $x \in JT$ and $\varepsilon > 0$ such that for every $n \in \mathbb{N}$ there exists a finite set $\{S_j^n\}_{j=1}^k$ of pairwise disjoint segments satisfying $\Big(\sum_{j=1}^k \Big(\sum_{(n,i)\in S_j^n} x(n, i)\Big)^2\Big)^{\frac{1}{2}} \geq \varepsilon$. Then there exists a fast growing sequence $\{n_l\}$ of integers such that $S_{l_1}^{n_{l_1}} \cap S_{l_2}^{n_{l_2}} = \emptyset$ for $l_1 \neq l_2$. Thus $\sum_l \Big(\sum_{(n,i)\in S_l^{n_l}} x(n, i)\Big)^2 = \infty$, a contradiction.

4.76 For every branch $\gamma \in \Gamma$, let $Y_\gamma = \{x \in JT : \operatorname{supp}(x) \subset \gamma\}$. Show that Y_γ is a subspace of JT isomorphic to James' space J.

Show that the operator $P_\gamma : JT \to JT$ defined by

$$P_\gamma\Big(\sum_{n=0}^{\infty}\sum_{i=0}^{2^n-1} \alpha_{(n,i)} e_{(n,i)}\Big) = \sum_{t\in\gamma} \alpha_t e_t$$

is a norm-one projection of JT onto Y.
Hint. See Exercise 4.66

4.77 For $F \in JT^*$ define $S(F) \in \ell_2(\Gamma)$ by $S(F) = \big(\lim_{n\to\infty,(n,i)\in\gamma} F(n,i)\big)_\gamma$ for $\gamma \in \Gamma$. Show that $S : JT^* \to \ell_2(\Gamma)$ is a bounded linear mapping.
Hint. $\|F\| = \sup\{F(x) : x \in B_{JT}\}$. Using this one can get $\|S\| = 1$. To show that S is onto, given $\{\gamma_i\}_{i=1}^n \subset \Gamma$ and scalars $a_1, \dots, a_n$, find $F \in JT^*$ such that $S(F) = \sum_{i=1}^n a_i \gamma_i$ and $\|F\| = \sqrt{\sum a_i^2}$.

4.78 Show that $\operatorname{Ker}(S) = JT_*$.
Hint. For $m = 0, 1, 2, \dots$, define for $j \in \{0, \dots, 2^m - 1\}$

$$P_{m,j}\Big(\sum_{n=0}^{\infty}\sum_{i=0}^{2^n-1} t_{(n,i)} e_{(n,i)}\Big) = \sum_{n=m}^{\infty}\sum_{(m,j)\le(n,i))} t_{(n,i)} e_{(n,i)}$$

and $P_m = \sum_{j=0}^{2^m-1} P_{m,j}$. Then P_m are projections on JT. First show that for $x^* \in \operatorname{Ker}(S)$, $\lim_{n\to\infty} \max_{0\le i\le 2^n-1} \|P^*_{n,i}(x^*)\| = 0$. If it were not the case, there would be a sequence $\{(n_k, i_k)\}_{k\in\mathbb{N}}$ and $\varepsilon > 0$ such that $\|P^*_{(n_k,i_k)}(x^*)\| > \varepsilon$. One can show that there exists only a limited number of incomparable elements in $\{(n_k, i_k)\}_{k\in\mathbb{N}}$, thus there exists a branch $\gamma \in \Gamma$ containing infinitely many elements from $\{(n_k, i_k)\}_{k\in\mathbb{N}}$. We may assume that $\{(n_k, i_k)\}_{k\in\mathbb{N}} \subset \gamma$ and $\|P^*_{(n_k,i_k)}(x^*) - P^*_{(n_{k+1},i_{k+1})}(x^*)\| > \varepsilon$ for all k.

By the previous exercise, we have $P_\gamma(x^*) \in \overline{\operatorname{span}}\{f_{(n,i)} : (n,i) \in \gamma\}$ and for k large enough, $\|(P^*_{n_k} - P^*_{n_{k+1}})P_\gamma(x^*)\| < \frac{1}{2}\varepsilon$. Define for $k \in \mathbb{N}$ mappings

$$U_k^* = P^*_{(n_k,i_k)} - P^*_{(n_{k+1},i_{k+1})} - (P^*_{n_k} - P^*_{n_{k+1}})P^*_\gamma.$$

Then U_k^* is a dual projection to some U_k and $\|U_k^*(x^*)\| > \frac{1}{2}\varepsilon$. The supports of the subspaces $U_k(JT)$ are mutually disjoint and no branch $\gamma' \in \Gamma$ intersects more than one of them. Thus $\left\|\sum_{k=1}^j U_k^*(x^*)\right\|^2 = \sum_{k=1}^j \|U_k^*(x^*)\|^2$ and for j large enough we obtain a contradiction.

Now assume that there is $v^* \in \operatorname{Ker}(S)\setminus JT_*$, let $\operatorname{dist}(v^*, JT_*) > (1-\delta)\|v^*\|$ for some δ. To get a contradiction, we find $x, y \in S_{JT}$ such that $v(x) > 1 - \delta$,

$v(y) > 1 - \delta$, and $P_{m_1}(x) = 0$, $P_{m_2}(y) = y$ for some m_2 much larger than m_1. For δ small enough we use our limit formula and obtain by calculating the norm of $x + y$ from definition that $\|x + y\| < 3.5$. Details may be found in [LiSt].

4.79 Show that JT is a separable space with non-separable dual containing no isomorphic copy of ℓ_1.
Hint. Using Dixmier's projection (see Exercise 4.7), from the above exercise we obtain that $JT^{**} \sim JT \oplus \ell_2(\Gamma)$. Thus $\operatorname{card}(JT^{**}) = c$. If there was an isomorphic copy of ℓ_1 in JT, we would find an isomorphic copy of $\ell_1(c)$ in JT^* by Exercise 5.35 and thus $\ell_\infty(c)$ would be isomorphic to some quotient of JT^{**}. This would imply that $\operatorname{card}(JT^{**}) > 2^c$, which is a contradiction.

Remark: The space JT has the property of being saturated by copies of ℓ_2, see Remark 3 in Chapter 5. Hagler constructed a modification of JT: a separable Banach space X with non-separable dual such that X is saturated by copies of c_0 and X^* is saturated by copies of ℓ_1, [Hag].

We mentioned in the introduction to Section 4.6 a stronger result of Gowers in [Gowe4]. The space X constructed there contains no c_0 or ℓ_1 or reflexive subspace. This was a solution to a long-standing problem.

4.80 (Lindenstrauss, [Lind12b]) Let X be a separable space. Then there exists a separable Banach space Z, such that $X = Z^{**}/Z$, and Z^{**} has a boundedly complete basis.
Hint. Let $\{x_i\}_{i=1}^\infty$ be a norm dense sequence from S_X. We let

$$Z = \Big\{(a_i)_{i=1}^\infty : \sum_{i=1}^\infty a_i x_i = 0\Big\}$$

to be a space of scalar sequences with the norm

$$\|z\| = \sup_{0=p_0<p_1<p_2<\cdots<p_m} \left(\sum_{j=1}^m \left\| \sum_{i=p_{j-1}+1}^{p_j} a_i x_i \right\|^2 \right)^{\frac{1}{2}} < \infty. \tag{4.3}$$

Show that Z^{**} consists of all scalar sequences satisfying (4.3), and such that $\sum_{i=1}^\infty a_i x_i$ converges. Next show that $T : Z^{**} \to X$, $T(a_1, a_2, \dots) = \sum_{i=1}^\infty a_i x_i$ is a quotient mapping. Finally, show that the canonical sequence $e_n = (0, \dots, 1_n, 0, \dots)$ is a boundedly complete basis of Z^{**}.

Chapter 5
Structure of Banach Spaces

The focus of the study in the present chapter is on Banach spaces containing subspaces isomorphic to c_0 or ℓ_1. We prove Sobczyk's theorem on complementability of c_0 in separable overspaces, lifting property of ℓ_1 and Pełczyński's characterization of separable Banach spaces containing ℓ_1. We present Rosenthal's ℓ_1 theorem, Odell–Rosenthal theorem and the Rosenthal–Bourgain–Fremlin–Talagrand theory of Baire-1 functions on Polish spaces.

5.1 Extension of Operators and Lifting

Theorem 5.1 *Every separable Banach space is linearly isometric to a quotient space of* ℓ_1.

Proof: Let X be a separable Banach space. Let $\{x_i : i \in \mathbb{N}\}$ be a dense subset of B_X. Define a mapping T from ℓ_1 into X by

$$T(\xi) = \sum_{i=1}^{\infty} \xi_i x_i, \text{ for } \xi = (\xi)_i \in \ell_1.$$

Note that, for every $\xi \in \ell_1$, the series $\sum_i \xi_i x_i$ is absolutely convergent since $\sum_i \|\xi_i x_i\| \leq \sum_i |\xi_i| = \|\xi\|_{\ell_1}$. We will show that $TB_{\ell_1} = B_X$. To that end, given $x \in B_X$ choose a sequence $\{x_{n_k}\}$ as follows: First choose n_1 such that $\|x - x_{n_1}\| < 2^{-1}$. Since $2(x - x_{n_1}) \in B_X$, by the density of $\{x_n : n \in \mathbb{N}\}$ in B_X we can choose $n_2 > n_1$ such that $\|2(x - x_{n_1}) - x_{n_2}\| < 2^{-1}$, i.e., $\|x - x_{n_1} - (1/2)x_{n_2}\| < 2^{-2}$. Then pick $n_3 > n_2$ such that $\|2^2(x - x_{n_1} - 2^{-1}x_{n_2}) - x_{n_3}\| < 2^{-1}$, i.e., $\|x - x_{n_1} - 2^{-1}x_{n_2} - 2^{-2}x_{n_3}\| < 2^{-3}$, etc. We have $\sum_k 2^{-k} x_{n_k} = x$. The element $\xi = (\xi_i)_{i=1}^{\infty}$ such that $\xi_i = 0$ for $i \notin \{n_k : k \in \mathbb{N}\}$ and $\xi_{n_k} = 2^{-k}$ for $k \in \mathbb{N}$ is in B_{ℓ_1}, and $T\xi = x$. Therefore $T_{B_{\ell_1}} = B_X$ and hence X is linearly isometric to $\ell_1 / T^{-1}(0)$. $\square$

Similarly we can prove that given a Banach space X, there is a set Γ such that X is isometric to a quotient of $\ell_1(\Gamma)$.

M. Fabian et al., *Banach Space Theory*, CMS Books in Mathematics,
DOI 10.1007/978-1-4419-7515-7_5, © Springer Science+Business Media, LLC 2011

Note that if X is not isomorphic to ℓ_1 and T_1, T_2 are bounded operators from ℓ_1 onto X, then $\mathrm{Ker}(T_1)$ is isomorphic to $\mathrm{Ker}(T_2)$ ([LiRo1], see also [LiTz2]).

We say that a Banach space X has the *lifting property* if for all Banach spaces Y, Z such that there is an onto operator $S \in \mathcal{B}(Y, Z)$ and for all $T \in \mathcal{B}(X, Z)$ there is $\widetilde{T} \in \mathcal{B}(X, Y)$ such that $T = S \circ \widetilde{T}$.

Proposition 5.2 *ℓ_1 has the lifting property.*

Proof: Let Y and Z be two Banach spaces and $S : Y \to Z$ an onto continuous linear mapping. If $T : \ell_1 \to Z$ is a continuous linear mapping, put $x_i = T(e_i)$ for $i \in \mathbb{N}$, where e_i denote the standard unit vectors of ℓ_1. By the open mapping theorem, there is a constant C such that for every i there is $y_i \in Y$ with $S(y_i) = x_i$ and $\|y_i\| \leq C$. We define $\widetilde{T}(e_i) = y_i$ for $i \in \mathbb{N}$. Then $S(\widetilde{T}(e_i)) = S(y_i) = T(e_i)$ for all $i \in \mathbb{N}$, and by linearity, $S \circ \widetilde{T} = T$. From $B_{\ell_1} = \overline{\mathrm{conv}}\{e_i : i \in \mathbb{N}\}$ we then get $\|\widetilde{T}\| \leq C$. □

Corollary 5.3 *If there is a bounded operator from a Banach space X onto ℓ_1 then ℓ_1 is isomorphic to a subspace of X.*

Proof: It follows from Proposition 5.2, for $Z = X$, $T = I_X$, and S being the canonical quotient mapping. □

Proposition 5.4 *Every separable Banach space is linearly isometric to a subspace of ℓ_∞.*

Proof: By Proposition 3.101 there is a sequence $\{f_i\}$ that is w^*-dense in B_{X^*}. Define $T\colon X \to \ell_\infty$ by $T(x) = \big(f_i(x)\big)_i$. Then $\|x\| = \sup |f_i(x)| = \|T(x)\|_\infty$ and T is a linear isometry into ℓ_∞. □

Similarly we prove that *every Banach space is linearly isometric to a subspace of $\ell_\infty(\Gamma)$ for some Γ*. Note (Exercise 5.33) that a dual space to an arbitrary separable Banach space is isometric to a subspace of ℓ_∞.

Proposition 5.5 *Every separable Banach space is linearly isometric to a subspace of ℓ_∞/c_0.*

Proof: Let $\{x_i^*\}$ be a w^*-dense sequence in B_{X^*}. Define $T\colon X \to \ell_\infty$ by

$$T(x) = (x_1^*(x), x_2^*(x), x_1^*(x), x_2^*(x), x_3^*(x), x_1^*(x), \ldots).$$

As in Proposition 5.4, we prove that T is a linear isometry into ℓ_∞. Let $q : \ell_\infty \to \ell_\infty/c_0$ be the canonical quotient operator. We claim that $q \circ T$ is an isometry from X into ℓ_∞/c_0. This follows from the fact that $\|q(x_i)\| = \|(x_i)\|_{\ell_\infty/c_0} = \limsup |x_i|$ (Exercise 1.29). □

Theorem 5.6 (Phillips [Phil], Sobczyk [Sobc]) *The space c_0 is not complemented in ℓ_∞.*

In the proof of Theorem 5.6, we will use the following lemma.

Lemma 5.7 (Sierpiński) *There exists a family $\mathcal{F}$ of cardinality continuum formed by infinite subsets of $\mathbb{N}$ such that if $F_1, F_2 \in \mathcal{F}$, $F_1 \neq F_2$, then* $\operatorname{card}(F_1 \cap F_2) < \infty$.

Proof: Let $i \colon \mathbb{N} \to \mathbb{Q}$ be a one-to-one and onto mapping. For every real number r choose a sequence of distinct rationals $\{g_n^r\}$ convergent to r. Define a mapping $\varphi \colon \mathbb{R} \to \mathbb{N}^{\mathbb{N}}$ by $\varphi(r) = \{i^{-1}(g_n^r)\}_{n=1}^{\infty}$.

For $r_1, r_2 \in \mathbb{R}$, $r_1 \neq r_2$, we have $\operatorname{card}\big(\{i^{-1}(g_n^{r_1})\}_{n\in\mathbb{N}} \cap \{i^{-1}(g_n^{r_2})\}_{n\in\mathbb{N}}\big) < \infty$. Putting $\mathcal{F} = \{\varphi(r) : r \in \mathbb{R}\}$ concludes the proof of the lemma. □

Proof of Theorem 5.6: (Whitley [Whitl], see, e.g., [Jmsn]) Assume, by contradiction, that P is a bounded linear projection of ℓ_∞ onto c_0, and put $Q = I - P$, where I is the identity mapping of ℓ_∞. For $\gamma \in \mathbb{R}$ put $x_\gamma = \chi_{F_\gamma}$ (the characteristic function of F_γ in $\mathbb{N}$), where $\{F_\gamma\}_{\gamma\in\mathbb{R}}$ is the family constructed above.

Claim:
For every $\varepsilon > 0$ and $n \in \mathbb{N}$ we have $\operatorname{card}\{\gamma \in \mathbb{R} : |Q(x_\gamma)_n| > \varepsilon\} < \infty$.

Indeed, assume that $|Q(x_{\gamma_i})_n| \geq \varepsilon$ for $i = 1, \ldots, k$. Define

$$x'_{\gamma_i} = \chi_{\left(F_{\gamma_i} \setminus \bigcup\{F_{\gamma_j} \colon j \neq i, j=1,\ldots,k\}\right)}.$$

Then $x_{\gamma_i} - x'_{\gamma_i} = \chi_{\left(F_{\gamma_i} \cap \bigcup\{F_{\gamma_j} \colon j \neq i, j=1,\ldots,k\}\right)} \in c_0$ as the intersection is finite. Thus we obtain that $Q(x_{\gamma_i}) = Q(x'_{\gamma_i})$ for $i = 1, \ldots, k$. We have

$$\begin{aligned} Q\Big(\sum_{i=1}^{k} \operatorname{sign}[Q(x'_{\gamma_i})_n] x'_{\gamma_i}\Big)_n &= \sum_{i=1}^{k} \operatorname{sign}[Q(x'_{\gamma_i})_n] Q(x'_{\gamma_i})_n \\ &= \sum_{i=1}^{k} |Q(x'_{\gamma_i})_n| = \sum_{i=1}^{k} |Q(x_{\gamma_i})_n| \geq k\varepsilon. \end{aligned}$$

On the other hand, $\Big\|\sum_{i=1}^{k} \operatorname{sign}[Q(x'_{\gamma_i})_n] x'_{\gamma_i}\Big\|_\infty \leq 1$ due to the disjoint supports of x'_{γ_i}, $i = 1, \ldots, k$. We thus have

$$k\varepsilon \leq Q\Big(\sum_{i=1}^{k} \operatorname{sign}[Q(x'_{\gamma_i})_n] x'_{\gamma_i}\Big)_n \leq \|Q\| \Big\|\sum_{i=1}^{k} \operatorname{sign}[(Qx'_{\gamma_i})_n] x'_{\gamma_i}\Big\|_\infty \leq \|Q\|.$$

Therefore $k \leq \frac{\|Q\|}{\varepsilon}$. △

Using the Claim, we see that $A = \bigcup_{n=1}^{\infty}\{\gamma \in \mathbb{R} : Q(x_\gamma)_n \neq 0\}$ is countable. Therefore there is $\gamma \in \mathbb{R}\setminus A$ for which we have $Q(x_\gamma)_n = 0$ for every n. Hence

$Q(x_\gamma) = 0$ and thus $x_\gamma = P(x_\gamma)$. This implies that $x_\gamma \in c_0$, a contradiction with the fact that x_γ is a characteristic function of an infinite set. □

In fact, we saw in Exercise 3.44 that c_0 is not even isomorphic to a quotient of ℓ_∞. However, there exists a Lipschitz mapping from ℓ_∞ onto c_0 whose restriction to c_0 is the identity ([Lind1], Exercise 5.29).

Theorem 5.8 (Banach, Mazur) *Every separable Banach space is linearly isometric to a subspace of* $C[0, 1]$.

Proof: Let X be a separable Banach space. Then B_{X^*} is a compact metric space in the w^*-topology, therefore there exists a continuous mapping φ from the Cantor set $\Delta \subset [0, 1]$ onto (B_{X^*}, w^*) (Theorem 17.11). Define a mapping $\widetilde{T}$ from X into $C(\Delta)$, the space of continuous functions on Δ, by $\widetilde{T}(x) = x \circ \varphi$, where $x \in X$ is considered as a (continuous) mapping acting on (X^*, w^*). Then $\widetilde{T}(x)$ is a continuous function on Δ. Now we define a mapping T from X into $C[0, 1]$. Given $x \in X$, extend the function $\widetilde{T}(x)$ defined on Δ onto $[0, 1]$ as follows: for $r \notin \Delta$ put $T(x)(r) = \alpha \widetilde{T}(x)(r_1) + \beta \widetilde{T}(x)(r_2)$, where $r_1 = \max\{p \in \Delta : p < r\}$, $r_2 = \min\{p \in \Delta : r < p\}$ and $r = \alpha r_1 + \beta r_2$.

T is a linear mapping from X into $C[0, 1]$ and

$$\|T(x)\| := \sup_{r \in [0,1]} |T(x)(r)| = \sup_{d \in \Delta} |\widetilde{T}(x)(d)| = \sup_{d \in \Delta} |(x \circ \phi)(d)| = \sup_{f \in B_{X^*}} |x(f)| = \|x\|.$$

Therefore T is a linear isometry from X into $C[0, 1]$. □

We say that $C[0, 1]$ is *isometrically universal for all separable Banach spaces*. Note that there is no finite-dimensional Banach space isometrically universal for all two-dimensional Banach spaces ([Bess1]).

It is conjectured that every complemented subspace of $C[0, 1]$ is isomorphic to $C[0, 1]$ or to $C(K)$ for some countable metric space K.

Corollary 5.9 (Banach, Fréchet, Mazur) *Every separable metric space* (P, ρ) *is isometric to a subset of* $C[0, 1]$.

Proof: Let $\{x_n\}_{n=1}^\infty$ be a dense sequence in (P, ρ). Choose $x_0 \in P$ and define a mapping φ from P into ℓ_∞ by $\varphi(x) = \big(\rho(x, x_n) - \rho(x_0, x_n)\big)_n$.

Note that given $x \in P$, by the triangle inequality

$$|\rho(x, x_n) - \rho(x_0, x_n)| \le \rho(x, x_0)$$

and therefore indeed $\varphi(x) \in \ell_\infty$. We now show that φ is an isometry from P into ℓ_∞. First, if x and x' are in P then

$$\begin{aligned}\|\varphi(x) - \varphi(x')\|_{\ell_\infty} &= \sup_n \left|\big(\rho(x, x_n) - \rho(x_0, x_n)\big) - \big(\rho(x', x_n) - \rho(x_0, x_n)\big)\right| \\ &= \sup_n |\rho(x, x_n) - \rho(x', x_n)| \le \rho(x, x').\end{aligned}$$

We claim that $\|\varphi(x) - \varphi(x')\|_{\ell_\infty} \geq \rho(x, x')$ for every x, x' in P. Given $\varepsilon > 0$, by density choose $n \in \mathbb{N}$ such that $\rho(x, x_n) < \varepsilon/2$. Then

$$\rho(x', x_n) \geq \rho(x', x) - \rho(x, x_n) \geq \rho(x', x) - \varepsilon/2.$$

Therefore

$$\begin{aligned}\|\varphi(x) - \varphi(x')\|_{\ell_\infty} &\geq |\rho(x', x_n) - \rho(x, x_n)| \geq \rho(x', x_n) - \rho(x, x_n)\\ &\geq \rho(x', x) - \varepsilon/2 - \varepsilon/2 = \rho(x', x) - \varepsilon.\end{aligned}$$

Thus $\|\varphi(x) - \varphi(x')\|_{\ell_\infty} = \rho(x', x)$ as $\varepsilon > 0$ was arbitrary.

We see that φ is an isometric bijection of P onto a separable subset $\varphi(P)$ of ℓ_∞. Let $X = \overline{\operatorname{span}}(\varphi(P))$ in ℓ_∞. Then X is separable and there is a linear isometry T of X onto a separable subspace of $C[0, 1]$. Then the composition $T \circ \varphi$ is an isometric bijection of P onto a subset of $C[0, 1]$. □

Proposition 5.10 *Let T be a bounded operator from a Banach space X into a Banach space $\ell_\infty(\Gamma)$ for some Γ. Let Y be a Banach space such that $Y \supset X$. Then T can be extended to a bounded operator from Y into $\ell_\infty(\Gamma)$ with the same norm as T.*

Proof: Assume without loss of generality that $\|T\| = 1$. For $\gamma \in \Gamma$, let $f_\gamma \in B_{X^*}$ be defined by $f_\gamma(x) = (Tx)_\gamma$, where $(Tx)_\gamma$ is the γ-coordinate of Tx in $\ell_\infty(\Gamma)$. Let $\widetilde{f_\gamma} \in B_{Y^*}$ be a Hahn–Banach extension of f_γ on Y. Define an operator $\widetilde{T} : Y \to \ell_\infty(\Gamma)$ by

$$\widetilde{T}y = \big(\tilde{f}_\gamma(y)\big)_\gamma, \quad \text{for } y \in Y.$$

Then $\|\widetilde{T}\| \leq 1$ and $\widetilde{T}$ extends T (so $\|\widetilde{T}\| = \|T\|$). □

Theorem 5.11 (Sobczyk [Sobc], see also [LiTz3, p. 106]) *Let $c_0 \subset X$ where X is a separable Banach space. Then there is a projection P from X onto c_0 with $\|P\| \leq 2$.*

Proof: (Veech [Veec]) Let ρ be a translation-invariant metric that induces the w^*-topology on B_{X^*} and let $F = B_{X^*} \cap c_0^\perp$. For $n \in \mathbb{N}$, let x_n^* be a Hahn–Banach extension of the n-th standard unit vector of $c_0^* = \ell_1$. Note that every w^*-limit point of (x_n^*) belongs to F. Since (B_{X^*}, ρ) is a compact space, it follows that $\rho(x_n^*, F) \to 0$ as $n \to \infty$. Thus we can find $w_n^* \in F$ such that $\rho(x_n^*, w_n^*) \to 0$ as $n \to \infty$. This means that $x_n^* - w_n^* \to 0$ in the w^*-topology of X^*.

Define an operator $P : X \to c_0$ by

$$Px = \big((x_n^* - w_n^*)(x)\big)$$

for $x \in X$. Then $\|P\| \le 2$ and, if $x = (x_n) \in c_0$, then $w_n^*(x) = 0$ for all n, and thus $Px = (x_n^*(x)) = (x_n)$. Hence P is a projection from X onto c_0. □

For the complementability of $c_0(\Gamma)$ in overspaces, see, e.g., [GKL1] and [ACGJM].

Definition 5.12 *A Banach space X is said to be* injective *(respectively* separably injective*) if for every Banach space Y containing X (respectively for every separable Banach space Y containing X) there is a bounded projection from Y onto X. If there is $\lambda \ge 1$ so that such projection exists with norm $\le \lambda$, then X is called* λ-injective *(respectively* λ-separably injective*).*

Proposition 5.13 *Let X be a Banach space and $\lambda \ge 1$. Then the following are equivalent.*

(i) *X is λ-injective.*
(ii) *For every pair of Banach spaces $Z \supset Y$, and every bounded operator $T : Y \to X$ there is a bounded operator $\widetilde{T} : Z \to X$ that extends T and $\|\widetilde{T}\| \le \lambda \|T\|$.*
(iii) *For every Banach space $Y \supset X$, for every Banach space Z and for every bounded operator $T : X \to Z$ there is a bounded operator $\widetilde{T} : Y \to Z$ that extends T and $\|\widetilde{T}\| \le \lambda \|T\|$.*

Proof: (ii)$\Longrightarrow$(i) and (iii)$\Longrightarrow$(i): If Y is a Banach space such that $X \subset Y$, extend the identity operator $I : X \to X$ to Y to obtain a projection $\widetilde{I} : Y \to X$ with $\|\widetilde{I}\| \le \lambda$.

(i)$\Longrightarrow$(iii): Let Y, Z and T be as in (iii). Let P be a bounded linear projection from Y onto X such that $\|P\| \le \lambda$. Then the operator $\widetilde{T} := TP$ has the desired property.

(i)$\Longrightarrow$(ii): Let Y, Z and T as in (ii). Let $\{x_\gamma^*\}_{\gamma \in \Gamma}$ be a subset of S_{X^*} such that $\|x\| = \sup_\gamma |x_\gamma^*(x)|$ for $x \in X$. Define an isometry S from X into $\ell_\infty(\Gamma)$ by

$$Sx(\gamma) = x_\gamma^*(x), \text{ for } \gamma \in \Gamma,\ x \in X.$$

Use Proposition 5.10 to extend ST to $\widetilde{ST} : Z \to \ell_\infty(\Gamma)$ with $\|\widetilde{ST}\| = \|ST\|$ $(= \|T\|)$. Let $P : \ell_\infty(\Gamma) \to X$ be a projection such that $\|P\| \le \lambda$. Then $P\widetilde{ST}$ is the desired extension. □

Note that, for every nonempty set Γ, the space $\ell_\infty(\Gamma)$ is 1-injective. This follows from Proposition 5.10 and (ii) in Proposition 5.13.

Proposition 5.14 *Let X be an injective space. Then there is $\lambda \ge 1$ such that X is λ-injective.*

Proof: Suppose that no such λ exists. Then, by Proposition 5.13, for every $n \in \mathbb{N}$ there are spaces $Z_n \supset Y_n$ and operators $T_n : Y_n \to X$ with $\|T_n\| = 1$ such that any bounded linear extension $\widetilde{T}_n$ of T_n from Z_n to X satisfies $\|\widetilde{T}_n\| \ge n^3$. Let $Y = (\sum Y_n)_{c_0}$ and T be a bounded operator from Y into X defined by

$$T(y_1, y_2, \ldots, y_n, \ldots) = \sum_{n=1}^{\infty} \frac{T_n y_n}{n^2}.$$

Let $\widetilde{T}$ be a bounded linear extension of T from $(\sum Z_n)_{c_0}$ to X. The restriction of $n^2\widetilde{T}$ to Z_n (i.e., to elements $(0, 0, \ldots, z_n, 0, \ldots)$) is, by the definition of T, an extension of T_n, since for $y_n \in Y_n$, $\widetilde{T}(0, 0, \ldots, y_n, 0, \ldots) = T(0, 0, \ldots, y_n, 0, \ldots) = T_n(y_n)/n^2$. The norm of this restriction of $n^2\widetilde{T}$ to Z_n is $\leq n^2\|\widetilde{T}\| < n^3$ for $n > \|\widetilde{T}\|$. This contradiction finishes the proof. □

Definition 5.15 *Let X be a Banach space. Let Z be a Banach space such that $X \subset Z$. If X is not complemented in Z put $\lambda(X, Z) = \infty$. Otherwise,*

$$\lambda(X, Z) := \inf\{\|P\| : \ P : Z \to X \text{ a linear and bounded projection}\}. \tag{5.1}$$

Define the projection constant *of X as*

$$\lambda(X) = \sup\{\lambda(X, Z) : \ Z \text{ an overspace of } X\}. \tag{5.2}$$

The following result proves that in order to compute $\lambda(X)$ for any Banach space X, it is enough to restrict the class of overspaces of X to *any* (in fact, to *a single*) overspace of type $\ell_\infty(\Gamma)$. Recall that every Banach space is isometric to a subspace of $\ell_\infty(B_{X^*})$.

Proposition 5.16 *Let X be a subspace of $\ell_\infty(\Gamma)$ for some nonempty index set Γ. Then either* (a) *X is not complemented in $\ell_\infty(\Gamma)$ (and then $\lambda(X) = \infty$), or* (b) *X is complemented in $\ell_\infty(\Gamma)$ (and, in such a case, $\lambda(X) = \lambda(X, \ell_\infty(\Gamma))$).*

Proof: If (a) holds, certainly $\lambda(X) = \infty$. Assume that, on the contrary, X is complemented in $\ell_\infty(\Gamma)$ (and let $P : \ell_\infty(\Gamma) \to X$ be a projection). Then X is injective. Indeed, let Y and Z be two Banach spaces such that $Y \subset Z$, and let $T : Y \to X$ be a bounded operator. Let $J : X \hookrightarrow \ell_\infty(\Gamma)$ be the inclusion mapping. By Theorem 5.10, we obtain a bounded linear extension $\widetilde{S}$ to Z of the mapping $S := J \circ T$. Then $\widetilde{T} := P \circ \widetilde{S}$ (: $Z \to X$) is an extension of T to Z. Moreover,

$$\|\widetilde{T}\| \leq \|P\|.\|\widetilde{S}\| = \|P\|.\|S\| = \|P\|.\|T\|,$$

hence, according to Definition 5.12, X is $\|P\|$-injective. In particular, $\lambda(X, Z) \leq \|P\|$ for every overspace Z of X, so $\lambda(X) \leq \|P\|$. This holds for every projection $P : \ell_\infty(\Gamma) \to X$, so $\lambda(X) \leq \lambda(X, \ell_\infty(\Gamma))$. On the other side, $\lambda(X, \ell_\infty(\Gamma)) \leq \lambda(X)$. This proves the assertion. □

Remarks:

1. From Proposition 5.16 we get that a Banach space is injective if and only if it is complemented in any (and then in all) overspace of the form $\ell_\infty(\Gamma)$ (and that

$\lambda(X)$ can be computed by using just one of those overspaces). A space of the form $\ell_\infty(\Gamma)$ is a particular case of a space of type $C(K)$ for a compact space K (just endow Γ with the discrete topology and consider its Stone–Čech compactification $\beta\Gamma$; then $\ell_\infty(\Gamma) = C(\beta\Gamma)$, see Section 17.1). For finite-dimensional spaces, the computation of $\lambda(X)$ will be reduced to checking $\lambda(X, C(K))$ for any overspace $C(K)$ (see Proposition 5.26).

2. Assume that $\dim(X) = n$. Let Z be an overspace of X. We can use the fact that an Auerbach basis in X exists to prove that we can always find a projection $P_0 : Z \to X$ such that $\|P_0\| \le n$ (see Theorem 4.5). It follows that $\lambda(X, Z) \le n$. Since this is true for every overspace Z, we get $\lambda(X) \le n$. This result will be substantially improved in Theorem 6.28.

The three following lemmas are simple. We collect them here for future references.

Lemma 5.17 *Let X be an injective Banach space.*
(i) *If, for some $\lambda \ge 1$, X is λ-injective, then $\lambda(X) \le \lambda$.*
(ii) *The space X is, for every $\varepsilon > 0$, $(\lambda(X) + \varepsilon)$-injective.*
(iii) *If X is finite-dimensional, then X is $\lambda(X)$-injective.*

Proof: (i) Assume that, for some $\lambda \ge 1$, X is λ-injective. Let Z be a Banach space with $X \subset Z$. There exists a projection P from Z onto X such that $\|P\| \le \lambda$. Then, $\lambda(X, Z) \le \lambda$. Since this is true for all such Z, we get $\lambda(X) \le \lambda$.

(ii) Given a Banach space Z with $X \subset Z$ we have $\lambda(X, Z) \le \lambda(X)$. It follows that for each $\varepsilon > 0$ we can find a projection $P : Z \to X$ such that $\|P\| < \lambda(X) + \varepsilon$. This shows that X is $(\lambda(X) + \varepsilon)$-injective.

(iii) Assume now that X is finite-dimensional. Let Z be a Banach space such that $X \subset Z$. According to (ii), for each $n \in \mathbb{N}$ there exists a projection P_n from Z onto X such that $\|P_n\| \le \lambda(X) + (1/n)$. Choose an algebraic basis $\{e_i : i = 1, 2, \ldots, n\}$ of X and let $\pi_i : X \to \mathbb{K}$ be the mapping that associates α_i to an element $x = \sum_{i=1}^n \alpha_i e_i$, for $i = 1, 2, \ldots, n$. Fix $i \in \{1, 2, \ldots, n\}$. The mapping $\pi_i \circ P_n$ is an element of Z^* for each $n \in \mathbb{N}$. Moreover, $\{\pi_i \circ P_n : n \in \mathbb{N}\}$ is a bounded subset of Z^*, hence we can extract a w^*-convergent subnet. If this is done consecutively for $i = 1, 2, \ldots, n$, it follows that there exists a single subnet $\{P_{n_j}\}_j$ of $\{P_n\}$ such that $\{\pi_i \circ P_{n_j}\}_j$ is w^*-convergent for all $i \in \{1, 2, \ldots, n\}$. This means that $\{P_{n_j}\}_j$ is pointwise-to-norm convergent to an element P that is, certainly, a projection from Z onto X, and $\|P\| \le \lambda(X)$. This proves that X is $\lambda(X)$-injective. □

Lemma 5.18 *Assume that, for some $\varepsilon > 0$, the Banach–Mazur distance $\mathrm{d}(X, X')$ between two Banach spaces X and X' satisfy $\mathrm{d}(X, X') < 1 + \varepsilon$. Then, if X is λ-injective, then X' is $(1 + \varepsilon)\lambda$-injective.*

Proof: Let Y and Z be two Banach spaces such that $Y \subset Z$. Let $T : Y \to X'$ be a bounded operator, and let $S : X' \to X$ be an isomorphism such that $\|S\|.\|S^{-1}\| \le (1 + \varepsilon)$. The mapping $S \circ T : Y \to X$ has an extension $\widetilde{S \circ T} : Z \to X$ such that $\|\widetilde{S \circ T}\| \le \lambda \|S \circ T\|$ $(\le \lambda \|S\|.\|T\|)$. Then, $\widetilde{T} := S^{-1} \circ \widetilde{S \circ T}$, a bounded linear mapping from Z into X', is an extension of T, and $\|\widetilde{T}\| \leq (1+\varepsilon)\lambda\|T\|$. This proves that X' is $(1+\varepsilon)\lambda$-injective. □

Lemma 5.19 *For some* $\varepsilon > 0$, *let* X *and* X' *be two finite-dimensional Banach spaces such that* $\mathrm{d}(X, X') < 1+\varepsilon$. *Then,*

$$(1+\varepsilon)^{-1}\lambda(X') \leq \lambda(X) \leq (1+\varepsilon)\lambda(X'). \tag{5.3}$$

Proof: According to (iii) in Lemma 5.17, X is $\lambda(X)$-injective. By Lemma 5.18, X' is $(1+\varepsilon)\lambda(X)$-injective. Use now (i) in Lemma 5.17 to conclude that $\lambda(X') \leq (1+\varepsilon)\lambda(X)$. Since the Banach–Mazur distance is symmetric we obtain, by reversing the roles of X and X', the second inequality in (5.3). □

Theorem 5.20 *Let* Y *and* Z *be Banach spaces,* $Y \subset Z$ *and let* T *be a bounded operator from* Y *into* c_0. *Assume that* Z/Y *is separable. Then* T *extends to a bounded operator* $\hat{T}$ *from* Z *into* c_0 *with* $\|\hat{T}\| \leq 2\|T\|$.

Proof: Since ℓ_∞ is 1-injective and $c_0 \subset \ell_\infty$, T extends to a bounded operator $\widetilde{T}$: $Z \to \ell_\infty$ such that $\|\widetilde{T}\| = \|T\|$. Since moreover Z/Y is separable, $\widetilde{T}Z$ is separable. Put $E = \overline{\mathrm{span}}\{\widetilde{T}(Z) \cup c_0\}$ $(\subset \ell_\infty)$. This is a separable Banach space that contains c_0. By Sobczyk's Theorem 5.11, there is a projection P from E onto c_0 such that $\|P\| \leq 2$. Then the operator $\hat{T} := P\widetilde{T}$ maps Z into c_0, $\|\hat{T}\| \leq \|P\| \cdot \|\widetilde{T}\| \leq 2\|\widetilde{T}\|$ and if $y \in Y$, then $\hat{T}y = PTy = Ty$ since $Ty \in c_0$. □

Corollary 5.21 *If* X *is isomorphic to* c_0, *then* X *is separably injective.*

Proof: Let $T : X \to c_0$ be an isomorphism from X onto c_0. Let Z be a separable Banach space containing X. By Theorem 5.20, T extends to a bounded operator $\widetilde{T} : Z \to c_0$. Put $P = T^{-1}\widetilde{T} : Z \to X$. If $x \in X$, then $\widetilde{T}x = Tx$ and thus $Px = T^{-1}Tx = x$. Hence P is a bounded linear projection from Z onto X. □

Definition 5.22 *A bounded operator* T *from a Banach space* X *into a Banach space* Y *is called* weakly compact *if* $\overline{T(B_X)}$ *is a weakly compact subset of* Y.

We shall study weakly compact operators in Chapter 13.

Note: The space ℓ_2 is, as every separable Banach space, linearly isometric to a subspace of $C[0, 1]$ by the Banach–Mazur Theorem 5.8, and certainly $C[0, 1]/\ell_2$ is separable. The formal identity operator $T : \ell_2 \to c_0$, which is weakly compact as ℓ_2 is reflexive, cannot be extended to a weakly compact operator from $C[0, 1]$ into c_0. Indeed, if such extension $\widetilde{T}$ existed, then $\widetilde{T}$ would carry weakly compact sets to norm compact sets, as $C[0, 1]$ has the Dunford–Pettis property (see Theorem 13.43). This is not true as $e_i \xrightarrow{w} 0$ in $C[0, 1]$, where e_i are the standard unit vectors in ℓ_2, and $\|Te_i\| = 1 \not\to 0$.

For compact operators, the situation is different, as the next theorem shows. However, note that if X is a Banach space that is not isomorphic to a Hilbert space, then there is a subspace Y of X and a compact operator T from Y into Y such that there is no bounded operator $\widetilde{T}$ from X into Y that extends T. To see this, modify accordingly the proof of Theorem 6.16 (see [LiTz1]).

Theorem 5.23 *Let $X \subset Y$ be Banach spaces, K be a compact space and let T be a compact operator from X into $C(K)$. Then T extends to a compact operator $\widetilde{T}$ from Y into $C(K)$.*

In the proof, the following proposition will be used. The technique used in its proof is sometimes called the "small perturbation principle."

Proposition 5.24 *Let K be a compact space. Let F be a finite-dimensional subspace of $C(K)$ and $\eta > 0$. Then there is a finite-dimensional subspace $G \supset F$ of $C(K)$ such that $\mathrm{d}(G, \ell_\infty^n) < 1 + \eta$, where $\mathrm{d}(\cdot,\cdot)$ denotes the Banach–Mazur distance. Moreover, there is a projection P from $C(K)$ onto G with $\|P\| < 1 + \eta$. Therefore, for all compact spaces K, the space $C(K)$ has the bounded approximation property.*

Proof: For simplicity, we will prove the statement only for metrizable spaces K. Let $f_1, f_2, \ldots, f_n$ be a basis of F with $\|f_i\| = 1$ for all i. Given $\varepsilon > 0$, choose for every $y \in K$ a neighborhood U_y of y such that the oscillation of all $f_1, f_2, \ldots, f_n$ on U_y is smaller than ε. Let $U_1, U_2, \ldots, U_m$ be a finite subcollection of $\{U_y : y \in K\}$ that covers K. Let $x_i \in U_i$ with $x_i \neq x_j$ if $i \neq j$. We claim that there is a *peaked partition of unity* $\Phi_1, \Phi_2, \ldots, \Phi_m$ on K (i.e., nonnegative continuous functions on K such that $\sum_i \Phi_i(x) = 1$ for all $x \in K$, Φ_i vanishes outside U_i, $\|\Phi_i\| = 1$, and $\Phi_i(x_i) = 1$ for all $i = 1, 2, \ldots, m$). Indeed, put $V_i = U_i \setminus \{x_j : j \neq i\}$, $i = 1, 2, \ldots, m$. Then $V_1, V_2, \ldots, V_m$ is an open cover of K. Put $\Phi_i(x) = \rho(x, X \setminus V_i)/\sum_{j=1}^m \rho(x, X \setminus V_j)$, where ρ is a metric on K. Then $\Phi_j(x_i) = 0$ for $j \neq i$, so $\Phi_i(x_i) = 1$ for all $i = 1, 2, \ldots, m$. We note that $\|\sum_{i=1}^m \alpha_i \Phi_i\| = \sup\{|\alpha_i| : i = 1, 2, \ldots, m\}$ for all real numbers $\alpha_1, \ldots, \alpha_m$. Indeed, if $x \in K$ then, as $\Phi_i \geq 0$,

$$\left|\sum_{i=1}^{m} \alpha_i \Phi_i(x)\right| \leq \sum_{i=1}^{m} |\alpha_i| \Phi_i(x)$$

$$\leq \sup\{|\alpha_i| : i = 1, 2, \ldots, m\} \sum_{i=1}^{m} \Phi_i(x) = \sup\{|\alpha_i| : i = 1, 2, \ldots, m\}.$$

On the other hand,

$$\left\|\sum_{i=1}^{n} \alpha_i \Phi_i\right\| \geq \left|\sum_{i=1}^{m} \alpha_i \Phi_i(x_j)\right| = |\alpha_j|$$

for all j.

Let $G = \operatorname{span}\{\Phi_i : i = 1, 2, \ldots, m\}$ $\big(\subset C(K)\big)$. Moreover, it easily follows that the mapping $P : C(K) \to G$ defined by

$$Pf = \sum_{i=1}^{n} f(x_i)\Phi_i, \quad f \in C(K)$$

ia a norm-one projection from $C(K)$ onto G. Given f_i, an element of the basis $\{f_1, f_2, \ldots, f_m\}$ for F, let $g_i \in G$ be defined by

$$g_i(x) = \sum_{j=1}^{m} \Phi_j(x) f_i(x_j).$$

Then, for all $x \in K$,

$$|f_i(x) - g_i(x)| = \left|\sum_{j=1}^{m} \Phi_j(x)\big(f_i(x) - f_i(x_j)\big)\right| \le \sum_{j=1}^{m} |f_i(x) - f_i(x_j)|\Phi_j(x) \le \varepsilon,$$

since a term in the sum is not zero only if x and x_j belong to the same U_y where $|f_i(x) - f_i(x_j)| \le \varepsilon$.

We will now show how to "shift" G to an overspace of F. This method is called "the principle of small perturbations."

Since all norms on a finite-dimensional space F are equivalent, there is $K \ge 1$ such that

$$K^{-1} \max |\lambda_i| \le \left\|\sum_{i=1}^{m} \lambda_i f_i\right\| \le K \max |\lambda_i|$$

for all real numbers $\lambda_1, \lambda_2, \ldots, \lambda_m$. Let $1 > \delta > 0$ such that $(1+\delta)/(1-\delta) < 1+\eta$, and put $\varepsilon = \delta/(2nK)$. For $i = 1, 2, \ldots, n$, let $g_i \in G$ be such that $\|g_i\| = 1$ and $\|f_i - g_i\| < \varepsilon$. Then

$$(2K)^{-1} \max |\lambda_i| \le \left\|\sum_{i=1}^{m} \lambda_i g_i\right\| \le 2K \max |\lambda_i|.$$

By the Hahn–Banach theorem, there exists $x_i^* \in C(K)^*$ with $\|x_i^*\| \le 2K$ and such that $x_i^*(g_j) = \delta_{i,j}$, the Kronecker delta, $i = 1, 2, \ldots, n$. Let an operator $T : C(K) \to C(K)$ be defined by

$$Th = h + \sum_{i=1}^{n} x_i^*(h)(f_i - g_i)$$

for $h \in C(K)$. Then $Tg_i = f_i$ for $i = 1, 2, \ldots, n$ and

$$(1-\delta)\|h\| \leq \|Th\| \leq (1+\delta)\|h\|, \quad h \in C(K).$$

Indeed, to see for example the right-hand-side inequality, estimate

$$\|Th - h\| \leq \sum_{i=1}^{n} \|x_i^*\|.\|h\|.\|f_i - g_i\| \leq \|h\|2Kn\varepsilon \leq \delta\|h\|, \quad \text{for } h \in C(K).$$

Thus $TG \supset F$ and $\mathrm{d}(TG, \ell_\infty^n) \leq (1+\delta)/(1-\delta) < 1+\eta$. If $P : C(K) \to G$ is the norm-one projection defined above in this proof, then $P_1 := TPT^{-1}$ is a projection from $C(K)$ onto $T(G)$ with norm less or equal than $1+\eta$. This implies that $C(K)$ has the bounded approximation property. Indeed, if C is a compact set in $C(K)$ and $\eta > 0$ is given, find a finite set $\{f_i : i = 1, 2, \dots, m\}$ so that

$$C \subset \bigcup_i B\big(f_i, \eta/(1+\varepsilon)\big).$$

By the above, let P be a projection from $C(K)$ onto $\operatorname{span}\{f_i : i = 1, 2, \dots m\}$ with $\|P\| \leq 1+\varepsilon$. If $f \in C$ then, for some i,

$$\begin{aligned} \|Pf - f\| &= \|(I-P)f\|\| \\ &\leq \|(I-P)f_i\| + \|I-P\|.\|f - f_i\| \leq (2+\varepsilon)\|f - f_i\| \leq \frac{2+\varepsilon}{2+\varepsilon}\eta = \eta. \end{aligned}$$

□

Proof of Theorem 5.23: Given a compact operator T from X into $C(K)$, by using the bounded approximation property of $C(K)$ (see Proposition 5.24) we can write $T = \sum_{n=0}^{\infty} T_n$ in the operator-norm topology, where T_n are finite rank operators and $\|T_n\| \leq 2^{-n}$ for $n \geq 1$.

By Proposition 5.24, each T_n extends to a finite rank operator $\widetilde{T}_n$ from Y to $C(K)$ with $\|\widetilde{T}_n\| \leq (1+\varepsilon)/2^n$ for $n = 1, 2, \dots$ Thus the operator $\widetilde{T} := \sum_{n=0}^{\infty} \widetilde{T}_n$ is a compact operator from Y into $C(K)$ that extends T. □

Corollary 5.25 *If $Y \supset X$ are Banach spaces and T is a compact operator from X into c_0, then T can be extended to a compact operator from Y into c_0.*

Proof: By Theorem 5.23, T extends to a compact operator $\widetilde{T}$ from Y into $C[0,1]$ ($\supset c_0$). If P is a projection from $C[0,1]$ onto c_0 (see Sobczyk Theorem 5.11), then the compact operator $\widehat{T} := P\widetilde{T}$ extends the operator T. □

Another consequence of Proposition 5.24 is the following result. Although it is of interest on its own, we present it here since it will be used in the proof of Theorem 6.28.

Proposition 5.26 *Let X be a finite-dimensional Banach space. Let K be a compact space such that $X \subset C(K)$. Then $\lambda(X) = \lambda(X, C(K))$.*

Proof: Fix $\varepsilon > 0$. By Proposition 5.24, there exists a finite-dimensional Banach space G such that $X \subset G \subset C(K)$ and $\mathrm{d}(G, \ell_\infty^n) < 1+\varepsilon$. Let $S : G \to \ell_\infty^n$ be an isomorphism such that $\|S\|.\|S^{-1}\| \leq 1+\varepsilon$. Given an arbitrary projection P from G onto X, the mapping $P' := S \circ P \circ S^{-1}$ is a projection from ℓ_∞^n onto $X' := S(X)$, and $\|P'\| \leq \|S\|.\|P\|.\|S^{-1}\| \leq (1+\varepsilon)\|P\|$. Then

$$\lambda(X', \ell_\infty^n) \leq \|P'\| \leq (1+\varepsilon)\|P\|.$$

This holds for all projections P from G onto X, so

$$(1+\varepsilon)^{-1}\lambda(X', \ell_\infty^n) \leq \lambda(X, G).$$

By Proposition 5.16 we know that $\lambda(X') = \lambda(X', \ell_\infty^n)$. Moreover, if P is a projection from $C(K)$ onto X, then $P\big|_G$ is a projection from G onto X, and $\|P\big|_G\| \leq \|P\|$; this proves that $\lambda(X, G) \leq \lambda(X, C(K))$.

Finally, we get

$$\lambda(X, C(K)) \geq \lambda(X, G) \geq (1+\varepsilon)^{-1}\lambda(X', \ell_\infty^n) = (1+\varepsilon)^{-1}\lambda(X') \geq (1+\varepsilon)^{-2}\lambda(X),$$

where the last inequality follows from Lemma 5.19.

Since $\varepsilon > 0$ was arbitrary and, certainly, $\lambda(X, C(K)) \leq \lambda(X)$, we obtain the conclusion. □

Theorem 5.27 (Michael [Mich]) *Assume that C is a closed convex subset of a Banach space X. Then there is a continuous mapping from X into C that is the identity on C.*

Proof: Define a set-valued mapping Φ from X into the subsets of C by

$$\Phi(x) = \begin{cases} x, & \text{if } x \in C, \\ C, & \text{if } x \notin C. \end{cases}$$

Since C is closed, the mapping Φ is lower semicontinuous (see Definition 17.6). Indeed, if $x_0 \in X$ and G is an open subset of X that intersects $\Phi(x_0)$, the set $\mathcal{O} := \{x \in X : \Phi(x) \cap G \neq \emptyset\}$ is open, since $\mathcal{O} = (X \setminus C) \cup G$. By Michael's selection theorem (Theorem 7.53), Φ admits a continuous selection, and this is the desired mapping. □

Corollary 5.28 *For every closed subspace Y of a Banach space X there is a (non-linear) continuous projection from X onto Y, i.e., there is a continuous mapping $\varphi : X \to Y$ such that φ in the identity mapping on Y.*

Lindenstrauss' result on uniformly continuous (non-linear) projections will be discussed in Chapter 12.

Let $S_1 \subset S_2$ be two topological spaces. A function (in general, non-linear) $T : C(S_1) \to C(S_2)$ is called an *extension operator* if, for every $f \in C(S_1)$, $T(f)$ is an extension of f to S_2, i.e., $T(f)(s_1) = f(s_1)$ for all $s_1 \in S_1$.

Theorem 5.29 (Arens, Borsuk [Bors], Dugundji [Dugu1], Kakutani [Kaku2]) *Let K be a compact metric space and H be a closed subset of K. Then there is a linear isometry extension operator T from $C(H)$ onto a norm-one complemented subspace of $C(K)$ such that $T1 = 1$. In particular, $C(H)$ is linearly isometric to a 1-complemented subspace of $C(K)$.*

Proof: The space $C(H)$ is separable and thus $B_{C(H)^*}$ in its w^*-topology is affinely homeomorphic to a compact convex subset of ℓ_2 in its norm topology (see the remark after Definition 12.24 and Exercise 12.20), and so is the set $P \subset B_{C(H)^*}$ of probability measures on H. Define a set-valued mapping Φ from K into P for every $k \in K$ by

$$\Phi(k) = \begin{cases} \delta_k, & \text{if } k \in H, \\ P, & \text{if } k \in K \setminus H, \end{cases}$$

where $\delta_k \in C(H)^*$ is the Dirac measure at k in the compact set H. Then Φ is lower semicontinuous and $\Phi(k)$ is closed and convex for every $k \in K$.

Thus, by Michael's selection Theorem 7.53, the mapping Φ has a continuous selection φ from K into P.

Given $f \in C(H)$, define $Tf(k) = f(\varphi(k))$ for $k \in K$. If $k_n \in K$, and $k_n \to k$ in K, then $k_n \to k$ in the w^*-topology of $C(K)^*$, hence $\varphi(k_n) \to \varphi(k)$ in the w^*-topology of $C(H)^*$, and thus $f(\varphi(k_n)) \to f(\varphi(k))$ as $f \in C(H)$. This proves that T is a (linear) mapping from $C(H)$ into $C(K)$. Moreover, $Tf(h) = f(h)$ for all $h \in H$, $T1 = 1$ and T is a linear isometry.

If R denotes the operator of restriction from K to H, then the operator $Q = T(R)$ is a norm-one linear operator from $C(K)$ into $C(K)$. Moreover, if $f \in C(K)$ and $g = R(f)$ then

$$Q(Q(f)) = TR(TR(f)) = TR(Tg) = T(g) = T(Rf) = Q(f).$$

Therefore Q is a linear projection. □

Remark: We refer to [Pelc5] for more in this area.

5.2 Weak Injectivity

We say that a separable topological space $(X, \mathcal{T})$ is a *Polish space*, if its topology $\mathcal{T}$ is generated by a complete metric. So, by a *Polish space* we will simply mean a separable complete metric space (see Section 17.9). We denote by Δ the Cantor set.

Lemma 5.30 *Let (P, ρ) be a Polish space and $\phi : P \to \Delta$ a continuous and onto mapping. Then there exists a subset S of P such that S is homeomorphic to Δ and $\phi|_S$ is a homeomorphism onto $\phi(S)$.*

Proof: We will construct by induction a sequence $\{\Delta_k\}_{k\in\mathbb{N}}$ of relatively open subsets of P, with the following properties.

(i) For $k \in \mathbb{N}$ we have $\Delta_k = \bigcup_{i=1}^{2^k} \Delta_k^i$, where $\{\Delta_k^i\}_{i=1}^{2^k}$ is a family of pairwise disjoint relatively open and nonempty subsets of P.

(ii) $\overline{\Delta_{k+1}^{2i-1} \cup \Delta_{k+1}^{2i}} \subset \Delta_k^i$ and ρ-diam $\Delta_k^i < \frac{1}{k}$, for $1 \leq i \leq 2^k$ and $k \in \mathbb{N}$.

(iii) $\phi(\overline{\Delta_{k+1}^{2i-1}}) \cap \phi(\overline{\Delta_{k+1}^{2i}}) = \emptyset$ and all $\phi(\Delta_{k+1}^i)$ are uncountable, for $1 \leq i \leq 2^k$ and $k \in \mathbb{N}$.

Let us describe the inductive step (the initial step is similar). Having constructed $\{\Delta_l\}_{l\leq k}$, and given i, choose a countable base $\{B_j\}_{j\in\mathbb{N}}$ of the open set Δ_k^i consisting of sets of ρ-diameter at most $\frac{1}{k+1}$. Clearly, $\phi(\Delta_k^i) = \bigcup_{j\in\mathbb{N}} \phi(B_j)$ so there exists some $j \in \mathbb{N}$ such that $\phi(B_j)$ is uncountable. There exist points $t \neq r, t, r \in \phi(B_j)$, such that every open neighborhood of either t, r in $\phi(B_j)$ is uncountable. Fix two disjoint open neighborhoods U of t and V of r, and set $\Delta_{k+1}^{2i-1} = B_j \cap \phi^{-1}(U)$ and $\Delta_{k+1}^{2i} = B_j \cap \phi^{-1}(V)$. The conditions (i)–(iii) guarantee that $S := \bigcap_{k=1}^{\infty} \Delta_k$ is homeomorphic to the Cantor set Δ , and moreover $\phi|_S$ is injective. □

The following result is sometimes called the *weak injectivity theorem.*

Theorem 5.31 (Pełczyński [Pelc6]) *Let X be a separable Banach space containing a subspace Y that is isomorphic to $C[0, 1]$. Then Y has a subspace Z that is isomorphic to $C[0, 1]$ and complemented in X.*

Proof: Let Δ be the Cantor set. By Theorem 5.29, the subspace Y contains a further subspace Y_0 that is isomorphic to $C(\Delta)$. There is an equivalent norm $|||\cdot|||_{Y_0}$ on Y_0 such that $(Y_0, |||\cdot|||_{Y_0})$ is isometric to $C(\Delta)$. Use Proposition 2.14 to define an equivalent norm $|||\cdot|||_X$ on X that induces $|||\cdot|||_{Y_0}$ on Y_0. The (separable) space $(X, |||\cdot|||_X)$ is, by Theorem 5.8, isometric to a subspace of $C[0, 1]$, so there is no loss of generality assuming that X is actually isometric to $C[0, 1]$ (and that the subspace Y_0 is isometric to $C(\Delta)$). So $T^*(B_{X^*}) \to B_{Y_0^*}$ is onto, and by Proposition 3.67, $\text{ext}B_{Y_0^*} \subset T^*(\text{ext}B_{X^*})$. We have $\text{ext}B_{X^*} = \{\pm\delta_t : t \in [0, 1]\}$, $\text{ext}B_{Y_0^*} = \{\pm\delta_t : t \in \Delta\}$, where δ_t is the Dirac functional at a point t. Use a simple argument and Lemma 5.30 to obtain some $S \subset [0, 1]$, homeomorphic to Δ, such that $T^*|_S \to \Delta$ is a homeomorphism onto the image $\widetilde{S} := T^*(S) \subset \Delta$. Let $\Phi : C(S) \to C(\widetilde{S})$ be an isometry onto resulting from the action of T^*. By the Borsuk–Dugundji Theorem 5.29 there exists an isometric extension operator $E : C(\widetilde{S}) \to C(\Delta)$, with the image $\widetilde{Z} := E(C(\widetilde{S})) \hookrightarrow Y_0$. Let $Z = T(\widetilde{Z})$. It is standard to check that $P : C[0, 1] \to Z$ defined as $P(f) = T \circ E \circ \Phi(f|_S)$, is the sought projection onto Z. □

Corollary 5.32 *Let X be a separable Banach space. If $C[0, 1]$ is isomorphic to a quotient of a subspace of X, then it is isomorphic to a quotient of X.*

Proof: Let Y be a subspace of X such that $C[0, 1]$ is isomorphic to a quotient Z of Y. Let $T : Y \to Z$ be a quotient mapping. The space Z is isometric to a subspace of

ℓ_∞. Since ℓ_∞ is an injective space (see Proposition 5.10), there is an extension $\widetilde{T}: X \to \ell_\infty$ of T, and $E := \overline{\widetilde{T}(X)}$ is a separable space with $Z \subset E$. By Theorem 5.31 there is $Z_1 \hookrightarrow Z$, $Z_1 \cong C[0,1]$, that is complemented in E, say by a projection P. Then $P \circ \widetilde{T} : X \to Z_1$ is the sought quotient mapping. □

Corollary 5.33 (Pełczyński [Pelc6]) *Let X be a separable Banach space. The following are equivalent.*
(i) $C[0,1]$ *is isomorphic to a quotient of* X.
(ii) $\ell_1 \hookrightarrow X$.

Proof: (i)⟹(ii): By Theorem 5.8, ℓ_1 is isometric to a subspace of $C[0,1]$. By the lifting property of ℓ_1 (see Proposition 5.2) we obtain (ii).

(ii)⟹(i): By Theorem 5.1, the space $C[0,1]$ is isometric to a quotient of ℓ_1. By Corollary 5.32, $C[0,1]$ is isomorphic to a quotient of X. □

The following result is a particular case of Milyutin theorem (see, e.g., [Rose10]).

Corollary 5.34 *If Δ denotes the Cantor ternary set, the spaces $C[0,1]$ and $C(\Delta)$ are linearly isomorphic.*

Proof: Since Δ is a compact subset of $[0,1]$, Theorem 5.29 shows that $C(\Delta)$ is isomorphic to a complemented subspace of $C[0,1]$. On the other hand, there is a continuous onto function $\phi : \Delta \to [0,1]$. This implies that $C[0,1]$ is linearly isometric to a subspace Y of the separable Banach space $C(\Delta)$. Theorem 5.31 implies that Y has a further subspace Z that is isomorphic to $C[0,1]$ and complemented in $C(\Delta)$. It is enough now to apply Pełczyński's decomposition method (Theorem 4.47) to get the conclusion. □

5.2.1 Schur Property

Recall that a sequence $\{x_n\} \subset X$ is weakly Cauchy if and only if $\{f(x_n)\}_{i=1}^{\infty}$ is Cauchy (that is, convergent) for every $f \in X^*$.

Definition 5.35 *A Banach space X is called* weakly sequentially complete *if every weakly Cauchy sequence in X is weakly convergent to some point in X.*

Note that there are Banach spaces which are not weakly sequentially complete. An example is c_0; indeed, the sequence $\{\sum_{i=1}^{n} e_i\}_{n=1}^{\infty}$ is weakly Cauchy, although is not weakly convergent in c_0.

The space ℓ_1 has a stronger property than being weakly sequentially complete. This is the content of the following result.

Theorem 5.36 (Schur) *Let $\{x^n\}$ be a sequence in ℓ_1. If $\{x^n\}$ is weakly Cauchy then $\{x^n\}$ is norm convergent in ℓ_1.*

Proof: First we show that if $x^n \overset{w}{\to} 0$ in ℓ_1, then $x^n \to 0$. By contradiction, assume that for some $\varepsilon > 0$ there is an increasing sequence $n_1 < n_2 < \dots$ such that $\|x^{n_j}\| = \sum_{i=1}^{\infty} |x_i^{n_j}| > \varepsilon$. Choose N_1 so that $\sum_{i=N_1+1}^{\infty} |x_i^{n_1}| < \frac{\varepsilon}{5}$.

Then $\sum_{i=1}^{N_1} |x_i^{n_1}| \ge \frac{4}{5}\varepsilon$, which is equivalent to $\sum_{i=1}^{N_1} \varepsilon_i^{n_1} x_i^{n_1} \ge \frac{4}{5}\varepsilon$, where $\varepsilon_i^{n_1} = \operatorname{sign}(x_i^{n_1})$ for $i = 1, \dots, N_1$. Note that if we choose an arbitrary sequence of signs $\{\varepsilon_i = \pm 1\}$ such that $\varepsilon_i = \varepsilon_i^{n_1}$ for $i \le N_1$, we have

$$\left|\sum_{i=1}^{\infty} \varepsilon_i x_i^{n_1}\right| = \left|\sum_{i=1}^{N_1} \varepsilon_i^{n_1} x_i^{n_1} + \sum_{i=N_1+1}^{\infty} \varepsilon_i x_i\right| \ge \left|\sum_{i=1}^{N_1} \varepsilon_i^{n_1} x_i^{n_1}\right| - \sum_{i=N_1+1}^{\infty} |x_i| \ge \frac{4}{5}\varepsilon - \frac{1}{5}\varepsilon = \frac{3}{5}\varepsilon.$$

Next, find n_{j_2} such that $\sum_{i=1}^{N_1} |x_i^{n_{j_2}}| < \frac{\varepsilon}{5}$. This is possible since $x_i^n \to 0$. Then we choose $N_2 > N_1$ such that $\sum_{i=N_2+1}^{\infty} |x_i^{n_{j_2}}| \le \frac{\varepsilon}{5}$ and consequently $\sum_{i=1}^{N_2} |x_i^{n_{j_2}}| \ge \frac{4}{5}\varepsilon$. Then for arbitrary choice of signs $\{\varepsilon_i = \pm 1\}$ satisfying $\varepsilon_i = \varepsilon_i^{n_1}$ for $i \le N_1$ and $\varepsilon_i = \varepsilon_i^{n_2}$ for $N_1 < i \le N_2$ we have

$$\begin{aligned}\left|\sum_{i=1}^{\infty} \varepsilon_i x_i^{n_{j_2}}\right| &\ge \left|\sum_{i=N_1+1}^{N_2} \varepsilon_i x_i^{n_{j_2}}\right| - \sum_{i=1}^{N_1} |x_i^{n_{j_2}}| - \sum_{i=N_2+1}^{\infty} |x_i^{n_{j_2}}| \\ &\ge \left|\sum_{i=N_1+1}^{N_2} \varepsilon_i x_i^{n_{j_2}}\right| - \frac{2}{5}\varepsilon = \sum_{i=1}^{N_2} |x_i^{n_{j_2}}| - \sum_{i=1}^{N_1} |x_i^{n_{j_2}}| - \frac{2}{5}\varepsilon \ge \frac{4}{5}\varepsilon - \frac{\varepsilon}{5} - \frac{2}{5}\varepsilon = \frac{\varepsilon}{5}.\end{aligned}$$

Repeating this process, we obtain a vector $u = (u_i) \in \ell_\infty$ defined by $u_i = \varepsilon_i^{n_k}$ for $N_{k-1} < i \le N_k$, which has the property that $u(x^{n_{j_k}}) \ge \frac{\varepsilon}{5}$ for all k. This is a contradiction with $x^n \overset{w}{\to} 0$.

For a weakly Cauchy sequence we proceed in an analogous way: If $\{x^n\}$ is weakly Cauchy and not norm Cauchy, there exists $\varepsilon > 0$ and indices $n_j, m_j \to \infty$ such that $\|x^{n_j} - x^{m_j}\| \ge \varepsilon$. Then consider the sequence $\{x^{n_j} - x^{m_j}\}_j$, which converges weakly to zero. □

We say that a Banach space X has the *Schur property* if every weakly convergent sequence is norm convergent. We have just shown that ℓ_1 has this property. The same is true for $\ell_1(\Gamma)$ for any nonempty set Γ. Observe, too, that the technique of proof of Theorem 5.36 is another instance of what has been called the "sliding hump argument."

5.3 Rosenthal's ℓ_1 Theorem

In the proof of Theorem 5.37, we shall use the following notation: The family of all infinite subsets of a given infinite set S is denoted $\mathcal{P}_\infty(S)$, while the family of all finite subsets (including the empty set) of a set M is denoted $\mathcal{P}_f(M)$. Given $Z \in \mathcal{P}_\infty(\mathbb{N})$, we will always consider Z as an increasing sequence $Z := \{s_1, s_2, \dots\}$.

Put then $Z_1 = \{s_1, s_3, s_5, \ldots\}$ and $Z_2 = \{s_2, s_4, s_6, \ldots\}$. Given a sequence $\{x_n\}$ in a Banach space, and $\varepsilon > 0$, let S_ε denote the set of all infinite subsets A of $\mathbb{N}$ such that there exists $f \in B_{X^*}$ with the property

$$\inf_{n\in A_2} f(x_n) \geq \sup_{n\in A_1} f(x_n) + 2\varepsilon.$$

Theorem 5.37 (Rosenthal [Rose6]) *Let $\{x_n\}_{n=1}^{\infty}$ be a bounded sequence in an infinite-dimensional Banach space X. Then there is a subsequence $\{x_{n_k}\}_{k=1}^{\infty}$ that is either weakly Cauchy or it is equivalent to the unit basis of ℓ_1.*

Proof: For simplicity, we give the proof for the case of real scalars. Let us start with a simple but important observation.

The ℓ_1-criterion: *Let X be a Banach space, Γ be a set, $\{x_i\}_{i\in\Gamma} \subset B_X$. Then $\{x_i\}_{i\in\Gamma}$ is equivalent to the unit basis of $\ell_1(\Gamma)$ if and only if there is $\varepsilon > 0$ such that, for every disjoint sets $A, B \subset \Gamma$, there exists $f \in B_{X^*}$ such that $\inf_{i\in A} f(x_i) \geq \varepsilon$ and $\sup_{i\in B} f(x_i) \leq -\varepsilon$.*

Necessity is clear. To prove sufficiency, let $x = \sum_{i\in\Gamma} a_i x_i$, $\sum_{i\in\Gamma} |a_i| = 1$, and let $A = \{i : a_i \geq 0\}$, $B = \Gamma \setminus A$. Choose $f \in B_{X^*}$ such that $\inf_{i\in A} f(x_i) \geq \varepsilon$, $\sup_{i\in B} f(x_i) \leq -\varepsilon$. Then

$$\|x\| \geq f(x) = \sum_{i\in\Gamma} a_i f(x_i) = \sum_{i\in A} a_i f(x_i) + \sum_{i\in B} a_i f(x_i) \geq \varepsilon.$$

This proves the criterion.

Suppose that $\sup_n \|x_n\| < K$, for some $K \geq 1$.

Claim 1: *A bounded sequence $\{x_n\}_{n=1}^{\infty}$ contains a subsequence equivalent to the ℓ_1-basis if and only if there exists some infinite set $M \subset \mathbb{N}$ and $\varepsilon > 0$ such that $B \in S_\varepsilon$ for every infinite set $B \subset M$. Formally*

$$\big(\exists M \in \mathcal{P}_\infty(\mathbb{N})\big)(\exists \varepsilon > 0)\big(\forall B \in \mathcal{P}_\infty(M)\big)(B \in S_\varepsilon) \tag{5.4}$$

The proof of necessity is immediate. To prove sufficiency, apply (5.4) first to $B := M$. We get $f_0 \in B_{X^*}$ such that $\inf_{m\in M_2} f_0(x_m) \geq \sup_{m\in M_1} f_0(x_m) + 2\varepsilon$. By changing f_0 to $-f_0$ if necessary, we may assume that for some $M' \in \mathcal{P}_\infty(M)$ we have $\inf_{m\in M'} f_0(x_m) \geq \varepsilon$. Let $l \in [\varepsilon, K]$ be a cluster point of $\{f_0(x_m) : m \in M'\}$. Put $g = (\varepsilon/2l) f_0$. It follows that $\{g(x_m) : m \in M'\}$ clusters to $\varepsilon/2$, and we can take $M'' \in \mathcal{P}_\infty(M')$ such that $|g(x_m) - \varepsilon/2| < \varepsilon^2/(4K)$ for all $m \in M''$. Put $M''' = M''_1$. We shall prove, by using the criterion above, that $\{x_n : n \in M'''\}$ is equivalent to the ℓ_1-basis.

To this end, let A and B be two mutually disjoint elements in $\mathcal{P}(M''')$. We can always construct an infinite subset C of M such that $A \subset C_1$ and $B \subset C_2$. By using (5.4) again, we can find $f \in B_{X^*}$ such that $\inf_{m\in C_2} f(x_m) \geq \sup_{m\in C_1} f(x_m) + 2\varepsilon$, so, in particular, $\beta := \inf_{m\in B} f(x_m) \geq \sup_{m\in A} f(x_m) + 2\varepsilon =: \alpha + 2\varepsilon$. Take $\lambda := -(\alpha + \beta)/\varepsilon$. It is easy to verify that the function $h_0 := f + \lambda g$ satisfies the following properties.

(i) $\|h_0\| \le 1 + K/(2\varepsilon)$.
(ii) $\inf_{m\in B} h_0(x_m) \ge \varepsilon/2, \quad \sup_{m\in A} h_0(x_m) < -\varepsilon/2$.
Hence, the function $h := \big(1 + K/(2\varepsilon)\big)^{-1} h_0$ satisfies
(i) $h \in B_{X^*}$,
(ii) $\inf_{m\in B} h(x_m) \ge \varepsilon^2/(2\varepsilon + K), \quad \sup_{m\in A} h(x_m) < -\varepsilon^2/(2\varepsilon + K)$.
The conclusion follows by using the ℓ_1-criterion above. $\triangle$

Claim 2: *A bounded sequence $\{x_n\}_{n=1}^\infty$ contains a weakly Cauchy subsequence if and only if there exists some infinite set $M \subset \mathbb{N}$ such that for every infinite set $L \subset M$ and for every $\varepsilon > 0$ there exists an infinite set $A_{\varepsilon,L} \subset L$, such that $B \notin S_\varepsilon$ for every infinite set $B \subset A_{\varepsilon,L}$. Using the logical symbols, this means that*

$$\begin{aligned}&(\exists M \in \mathcal{P}_\infty(\mathbb{N}))(\forall \varepsilon > 0)(\forall L \in \mathcal{P}_\infty(M))\\ &\quad(\exists A_{\varepsilon,L} \in \mathcal{P}_\infty(L))(\forall B \in \mathcal{P}_\infty(A_{\varepsilon,L}))(B \notin S_\varepsilon).\end{aligned} \tag{5.5}$$

Necessity is immediate as the "natural" formulation of the existence of a weakly Cauchy subsequence is

$$(\exists M \in \mathcal{P}_\infty(\mathbb{N}))(\forall \varepsilon > 0)(\forall L \in \mathcal{P}_\infty(M))(L \notin S_\varepsilon). \tag{5.6}$$

To prove sufficiency, construct a nested sequence of infinite sets $\{A_n\}_{n=1}^\infty$, $A_0 := M$, $A_{n+1} := A_{\frac{1}{n+1}, A_n}$ for $n \in \mathbb{N}$ by using (5.5) repeatedly. Then, for $n \in \mathbb{N}$ and every $B \in \mathcal{P}_\infty(A_n)$, $B \notin S_{1/n}$. Let $\{m_i\}_{i=1}^\infty$ be a diagonal increasing sequence, $m_i \in A_i$ for $i \in \mathbb{N}$. It is clear (use (5.6) that $\{x_{m_i}\}_{i=1}^\infty$ is weakly Cauchy, which establishes the claim. $\triangle$

We proceed with the proof of the theorem. If $\{x_n\}_{n=1}^\infty$ contains no weakly Cauchy subsequence, then by negating (5.5) we obtain

$$\begin{aligned}&(\forall M \in \mathcal{P}_\infty(N))(\exists \varepsilon > 0)(\exists L \in \mathcal{P}_\infty(M))\\ &\quad(\forall A \in \mathcal{P}_\infty(L))(\exists B \in \mathcal{P}_\infty(A))(B \in S_\varepsilon).\end{aligned} \tag{5.7}$$

For notational convenience assume that $L = \mathbb{N}$ in (5.7). So there exists $\varepsilon > 0$, to be fixed from now on, such that for every infinite $A \subset \mathbb{N}$ there exists an infinite $B \subset A$, $B \in S := S_\varepsilon$, i.e., (note that we have achieved a favorable change of the order of quantifiers compared to the negation of (5.6))

$$(\exists \varepsilon > 0)(\forall A \in \mathcal{P}_\infty(N))(\exists B \in \mathcal{P}_\infty(A))(B \in S_\varepsilon). \tag{5.8}$$

The strategy of the proof is to build inductively the "initial finite segments" of a sequence M that is ultimately going to satisfy (5.4). Heuristically, we are building an almost basis of ℓ_1 (although our condition is weaker and to get an ℓ_1 basis we need to pass to a subsequence, at the end). The main point here is to make sure that the initial finite sequence can be extended arbitrarily many times. This is achieved by controlling simultaneously a candidate set of tails for the future extensions. More precisely, we shall prove that as soon as we have a finite set $F \subset \mathbb{N}$ and an infinite set $I \subset \mathbb{N}$ with $\max F < \min I$ and satisfying

$$(\forall J \in \mathcal{P}_\infty(I))(\forall G \in \mathcal{P}(F))(\exists L \in \mathcal{P}_\infty(J))(G \cup L \in S), \tag{5.9}$$

then we can enlarge F to a finite superset and reduce I to an infinite subset in such a way that (5.9) still holds for those two new sets.

Note that the couple $(F := \emptyset, I := \mathbb{N})$ satisfies (5.9), due to (5.8). Assume now that we have a couple (F, I) that satisfies (5.9). To produce the new couple, we shall prove the following
Claim:

$$\begin{gathered}(\exists n \in I)(\exists J \in \mathcal{P}_\infty(I), n < \min J)\\ (\forall L \in \mathcal{P}_\infty(J))(\forall G \in \mathcal{P}(F \cup \{n\}))(\exists M \in \mathcal{P}_\infty(L))(G \cup M \in S)\end{gathered} \tag{5.10}$$

Proceeding by contradiction, suppose that the Claim does not hold. This means

$$\begin{gathered}(\forall n \in I)(\forall J \in \mathcal{P}_\infty(I), n < \min J)\\ (\exists L_{n,J} \in \mathcal{P}_\infty(J))\ (\exists G_{n,J} \in \mathcal{P}(F \cup \{n\}))\ (\forall M \in \mathcal{P}_\infty(L_{n,J}))(G_{n,J} \cup M \notin S)\end{gathered} \tag{5.11}$$

By applying (5.9), note that $n \in G_{n,J}$ for all $n \in I$. Using a simple inductive argument, there exists an increasing sequence $J := \{n_i\}_{i=1}^\infty$, a sequence $\{G_{n_i}\}_{i=1}^\infty$ of finite sets, and a sequence of infinite sets $\{L_i\}_{i=1}^\infty$ such that

$$n_1 = \min I,\ G_{n_1} = G_{n_1, I\setminus\{n_1\}},\ L_1 = L_{n_1, I\setminus\{n_1\}}, \tag{5.12}$$

and, for $i = 2, 3, \ldots$,

$$n_i = \min L_{i-1},\ G_{n_i} = G_{n_i,\, L_{i-1}\setminus\{n_i\}},\ L_i = L_{n_i, L_{i-1}\setminus\{n_i\}}, \tag{5.13}$$

such that for $i \in \mathbb{N}$ and for all $M \in \mathcal{P}_\infty(L_i)$, we have $G_{n_i} \cup M \notin S$. By passing to an infinite subsequence of J, we may, without loss of generality, assume that $G_n\setminus\{n\} = G$ is independent of the choice of $n \in J$. According to (5.9), there exists an infinite subset $L \subset J$ such that $G \cup L \in S$. Put $n_l = \min L$. On the other hand, $G \cup L = (G \cup \{n_l\}) \cup (L\setminus\{n_l\}) = G_{n_l} \cup (L\setminus\{n_l\})$, and $L\setminus\{n_l\} \in \mathcal{P}_\infty(L_l)$, so $G \cup L \notin S$. This is a contradiction and proves (5.10). $\triangle$

The inductive procedure leads to an increasing sequence $M := \{n_i\}_{i=1}^\infty$, and an auxiliary decreasing sequence of infinite sets $\{I_{n_i}\}_{i=1}^\infty$, such that, for all $k \in \mathbb{N}$, the sets $F := \{n_1, \ldots, n_k\}$ and $I := I_{n_k}$ satisfy (5.9), and $n_{k+1} \in I_{n_k}$. Given any finite set $F \subset M$ and an infinite set $I \subset M$ such that $\max F < \min I$, we have that there exists some infinite set $L \subset I$ such that $F \cup L \in S$, i.e.,

$$\begin{gathered}(\exists M \in \mathcal{P}_\infty(\mathbb{N}))(\forall F \in \mathcal{P}_f(M))\\ (\forall I \in \mathcal{P}_\infty(M), \max F < \min I)(\exists L \in \mathcal{P}_\infty(I))(F \cup L \in S)\end{gathered} \tag{5.14}$$

Let $A = \{a_i\}_{i=1}^{\infty} \subset M$ be any infinite subset written as an increasing sequence. For each $n \in \mathbb{N}$ there exists some $L_n \in \mathcal{P}_\infty(\mathbb{N})$ such that $\max\{a_1, \dots, a_{2n}\} < \min L_n$ and $\{a_1, \dots, a_{2n}\} \cup L_n \in S$; in particular, there exists $g_n \in B_{X^*}$ such that $\inf_{i\in\{1,\dots,n\}} g_n(x_{a_{2i}}) \geq \sup_{i\in\{1,\dots,n\}} g_n(x_{a_{2i-1}}) + 2\varepsilon$. Let g be a w^*-cluster point of $\{g_n\}_{n=1}^{\infty}$. Clearly, $\inf_{i\in\mathbb{N}} g(x_{a_{2i}}) \geq \sup_{i\in\mathbb{N}} g(x_{a_{2i-1}}) + 2\varepsilon$, so $A \in S$. The proof is finished by an appeal to Claim 1. □

We are going now to investigate further properties of Banach spaces related to ℓ_1. As we shall see later, it will be natural to work in the setting of Baire 1 functions on Polish spaces.

If X is a topological space, we denote by $\mathcal{T}_p$ the topology of pointwise convergence in the space $\mathbb{R}^X$ of all real-valued functions on X.

Definition 5.38 *Let X be a metric space, f be a real function on X. We say that f is a* Baire 1 class function *if f is a $\mathcal{T}_p$-limit of a sequence of elements from $C(X)$. By $B_1(X)$ we denote the space of all Baire 1 real functions, equipped with the topology $\mathcal{T}_p$.*

Remark: The usual definition of a Baire 1 mapping $f : X \to Y$ is that the inverse image of an open set is an F_σ set. This turns out to be equivalent to our definition whenever X is a metric space and $Y = \mathbb{R}$ (see, e.g., [DGZ3, p. 18]).

If X is a separable Banach space, it is natural to consider the Polish space $P := (B_{X^*}, w^*)$, and the embedding $X^{**} \subset \mathbb{R}^P$, where the $\mathcal{T}_p$-topology corresponds to the w^*-topology on X^{**}.

Lemma 5.39 *Let (P, ρ) be a Polish space, F be a subset of $\mathbb{R}^P$, and $f \in \overline{F}^{\mathcal{T}_p}$. If f is not a Baire 1 function, then there is a sequence $\{f_k\}_{k=1}^{\infty}$ in F that has no $\mathcal{T}_p$-convergent subsequence.*

Proof: By Baire's Great Theorem 17.12, there exists a closed and nonempty set $K \subset P$ such that $f\big|_K$ has no point of continuity.

For each $n \in \mathbb{N}$ and $m \in \mathbb{Z}$, we define $K_{n,m} \subset K$ by putting $r \in K_{n,m}$ if and only if for every neighborhood U of r in K, there exist $t, s \in U$ such that $f(t) \geq \frac{m+3}{n}$, $f(s) \leq \frac{m}{n}$. Clearly, $K_{n,m}$ are closed sets and $K = \bigcup_{n\in\mathbb{N}, m\in\mathbb{Z}} K_{n,m}$. By the Baire category theorem some $K_{n,m}$ has a nonempty interior. For simplicity of notation let $K = K_{n,m}$. We are going to construct inductively a pair of sequences $\{f_k\}_{k=1}^{\infty}$ from F and $\{D_k\}_{k=1}^{\infty}$ of relatively open subsets of K, with the following properties.

(i) $D_k = \bigcup_{i=1}^{2^k} D_k^i$, where $\{D_k^i\}_i$ is a collection of pairwise disjoint relatively open and nonempty subsets of K. For convenience we also denote $\widetilde{D}_k^0 = \bigcup_{i=1}^{2^{k-1}} D_k^{2i}$, $\widetilde{D}_k^1 = \bigcup_{i=1}^{2^{k-1}} D_k^{2i-1}$.

(ii) $\overline{D_{k+1}^{2i-1} \cup D_{k+1}^{2i}} \subset D_k^i$, ρ-diam $D_k^i < \frac{1}{k}$.

(iii) $\sup_{\widetilde{D}_k^0} f_k \leq \frac{m+1}{n}$ and $\inf_{\widetilde{D}_k^1} f_k \geq \frac{m+2}{n}$.

Let us describe the inductive step (the initial step is similar). Having constructed $\{f_k\}_{k\leq l}$ and $\{D_k\}_{k\leq l}$, we find points $\{t_l^i : 1 \leq i \leq 2^{l+1}\}$, such that $t_l^{2i} \in D_l^i$, $t_l^{2i-1} \in$

D^i_l, and $f(t^i_l) \leq \frac{m}{n}$ for i even and $f(t^i_l) \geq \frac{m+3}{n}$ for i odd. There exists $f_{l+1} \in F$ such that $f_{l+1}(t^i_l) < \frac{m+1}{n}$ for i even and $f_{l+1}(t^i_l) > \frac{m+2}{n}$ for i odd. It remains to fix some small enough pairwise disjoint neighborhoods D^i_{l+1} of the points t^i_l, $i = 1, \ldots 2^{l+1}$, so that (i)–(iii) will be satisfied for $D_{l+1} = \bigcup_{i=1}^{2^{l+1}} D^i_{l+1}$. Note that condition (ii) implies that $\bigcap_{k=1}^{\infty} D^{i(k)}_k \neq \emptyset$ if and only if $i(k+1) \in \{2i(k), 2i(k)-1\}$ for all $k \in \mathbb{N}$. To finish the proof, let $\{f_{k(j)}\}_{j=1}^{\infty}$ be any subsequence of $\{f_k\}_{k=1}^{\infty}$. Choose a sequence $\{i(k)\}_{k=1}^{\infty}$, satisfying $i(k+1) \in \{2i(k), 2i(k)-1\}$ for all $k \in \mathbb{N}$, and moreover such that $i(k(j))$ is even if and only if j is even. There exists some $t \in \bigcap_{k=1}^{\infty} D^{i(k)}_k$. We have now $f_{k(j)}(t) \leq \frac{m+1}{n}$ for all j even, while $f_{k(j)}(t) \geq \frac{m+2}{n}$ for all j odd. This implies that $\{f_{k(j)}\}_{j=1}^{\infty}$ is not $\mathcal{T}_p$-convergent and the proof is finished. □

Theorem 5.40 (Odell and Rosenthal [OdRo], Rosenthal [Rose7]) *Let X be a separable infinite-dimensional Banach space. The following are equivalent.*
(i) *X does not contain an isomorphic copy of ℓ_1.*
(ii) *Every element of $B_{X^{**}}$ is the w^*-limit of a sequence in B_X.*
(iii) $\mathrm{card}(X^{**}) = \mathrm{card}(X) = c$.

Proof: (ii)$\Longrightarrow$(iii): Choose a norm-dense sequence $\{x_n\}_{n=1}^{\infty}$ in B_X. Since each element of $B_{X^{**}}$ is the w^*-limit of a subsequence of $\{x_n\}_{n=1}^{\infty}$, the conclusion follows.

(iii)$\Longrightarrow$(i): By contradiction, assume that $\ell_1 \hookrightarrow X$. Then $\ell_\infty^* \subset X^{**}$. It is known that $\ell_1(c) \hookrightarrow \ell_\infty$ (Exercise 5.34), and thus ℓ_∞^* has $\ell_\infty(c)$ as a quotient. This gives the conclusion.

(i)$\Longrightarrow$(ii): We proceed by contradiction, assuming that $x^{**} \in B_{X^{**}}$ is not a w^*-limit of any sequence from B_X. By [DGZ3, Lemma 3.4, p. 112], this implies that x^{**} is not a Baire 1 function. We apply Lemma 5.39 to $P = (B_{X^*}, w^*)$, $F = B_X \subset B_{X^*}$, $f = x^{**}$. Then the sequence $\{f_i\}_{i=1}^{\infty}$ contains no $\mathcal{T}_p$-convergent subsequence, which is equivalent to saying that it has no weakly Cauchy subsequence. By Rosenthal's Theorem 5.37, we obtain the conclusion. □

Theorem 5.41 (Rosenthal [Rose7]) *Let (P, ρ) be a Polish space, and F be a subset of $B_1(P)$. The following are equivalent in the $\mathcal{T}_p$-topology.*
(i) *F is relatively compact.*
(ii) *F is relatively countably compact.*
(iii) *F is relatively sequentially compact.*

Proof: (i)$\Longrightarrow$(ii) is immediate.

We proceed with the proof that (ii)$\Longrightarrow$(iii).

Lemma 5.42 *Let (P, ρ) be a Polish space and $\{f_k\}_{k=1}^{\infty}$ a pointwise bounded sequence from $\mathbb{R}^P$ that has no $\mathcal{T}_p$-convergent subsequence. Then there is an infinite subset M of $\mathbb{N}$, $r \in \mathbb{R}$, $\delta > 0$ such that for every infinite set $A \subset M$, there exists $t_A \in P$ such that*

$$\liminf_{i \in \mathbb{N}} f_{a_i}(t_A) < r, \quad \limsup_{i \in \mathbb{N}} f_{a_i}(t_A) > r + \delta. \tag{5.15}$$

Proof: Let $\{(r_i, \delta_i)\}_{i=1}^\infty$ be an enumeration of all rational pairs with $\delta_i > 0$. Proceed by contradiction. Using induction, we are going to construct a sequence $\{M_i\}_{i=1}^\infty$ of infinite subsets of the integers. $M_0 = \mathbb{N}$, and $M_{i+1} \subset M_i$ is chosen so that letting $M = M_i$, $A = M_{i+1}$, $r = r_i$ and $\delta = \delta_i$, (5.15) fails for every $t \in P$. Let $M = \{m_i\}_{i=1}^\infty$ be the diagonal set for the sequence $\{M_i\}_{i=1}^\infty$. We obtain that for every $t \in P$, $\{f_{m_i}(t)\}_{i=1}^\infty$ is convergent, a contradiction. □

Lemma 5.43 *Let (P, ρ) be a Polish space and $\{f_k\}_{k=1}^\infty$ be a pointwise bounded sequence from $\mathbb{R}^P$, that has no $\mathcal{T}_p$-convergent subsequence. Then there is an infinite $M \subset \mathbb{N}$ such that $\{f_k\}_{k\in M}$ has no cluster point in $B_1(P)$.*

Proof: By Lemma 5.42 there exists $M \subset \mathbb{N}$, r, δ such that for every $A \subset M$ there exists $t \in P$ such that $\overline{f_m(t) > r + \delta \text{ and } f_m(t) < r \text{ happens infinitely often for}}$ $\overline{m \in A}$. For every $A \subset M$, denote $K(A) = \overline{\{t_A : \text{ satisfies (5.15)}\}}$.

Lemma 5.44 *There exists some infinite set N such that $K(N) = K(L)$ $(=: K)$ for all infinite $L \subset N$, where $K(\cdot)$ is defined in the proof of Lemma 5.43.*

Proof: Clearly, $L \subset M$ implies $\emptyset \neq K(L) \subset K(M)$. If for every L there exists $N \subset L$ with $K(N) \subsetneq K(L)$, then we can construct an ordinal sequence $\{L_\alpha\}_{\alpha<\omega_1}$, such that $\beta > \alpha$ implies that $L_\beta \backslash L_\alpha$ is finite and $K(L_\beta) \subsetneq K(L_\alpha)$. Indeed, having obtained L_α for all $\alpha < \beta < \omega_1$, reindex the set $\{L_\alpha\}_{\alpha<\beta}$ as $\{\widetilde{L}_i\}_{i=1}^\infty$, and put $L_\beta = \{a_i\}_{i=1}^\infty$, where $a_i \in \bigcap_{j\leq i} \widetilde{L}_j$. This contradicts the Lindelöf property of the separable metric space $P \setminus \bigcap_{\alpha<\omega_1} K(L_\alpha)$. □

We proceed by constructing sequences $\{t_i\}_{i=1}^\infty$ from K and infinite sets $\{M_i\}_{i=1}^\infty$, with the properties:

1. $N = M_1 \supset M_2 \ldots$
2. $f_i(t_j) > r + \rho$ for all j even and $i \in M_j$, $f_i(t_j) < r$ for all j odd and $i \in M_j$,
3. $\{t_{2i}\}_{i=1}^\infty$ and $\{t_{2i+1}\}_{i=1}^\infty$ are both dense in K.

Denote by M the diagonal set of the system $\{M_i\}_{i=1}^\infty$. It remains to see that $\{f_i\}_{i\in M}$ has no Baire 1 $\mathcal{T}_p$-cluster point. Assuming the contrary, let $f \in \overline{\{f_i\}_{i\in M}}^{\mathcal{T}_p}$. By Baire's Great Theorem 17.12, f has a point of continuity $t \in K$. We have $f(t_j) \geq r + \rho$ for every j even, and $f(t_j) \leq r$ for all j odd. We have obtained a contradiction. This finishes the proof of Lemma 5.43 and the proof of (ii)$\Longrightarrow$(iii). □

(iii)$\Longrightarrow$(i) follows from Lemma 5.39. This finishes the proof of Theorem 5.41. □

The next abstract lemma will be useful in what follows.

Lemma 5.45 *Let $(\mathcal{V}, \subseteq)$ be a system of nonempty closed subsets of a Polish space (P, ρ), partially ordered by inclusion, with the following properties:*

1. For every pair $A, B \in \mathcal{V}$ there exists some $C \in \mathcal{V}$, such that $C \subset A \cap B$.

2. For every sequence $\{V_n\}_{n=1}^\infty \subset \mathcal{V}$, $V_{n+1} \subset V_n$ there exists a $V \in \mathcal{V}$, $V \subset \bigcap_n V_n$.

Then $\mathcal{V}$ has a minimal element.

Proof: P has a countable base $\{U_n\}_{n\in\mathbb{N}}$. If for all $V \in \mathcal{V}$ we have $V = P$ there is nothing to prove. We may assume then that $V \neq P$ for all $V \in \mathcal{V}$. Let $M = \{n \in \mathbb{N} : U_n \cap V = \emptyset \text{ for some } V \in \mathcal{V}\}$. Given $n \in M$, let $V_n \in \mathcal{V}$ such that $U_n \cap V_n = \emptyset$. By the assumption, there exists $\widetilde{V} \in \mathcal{V}$ such that $\widetilde{V} \subset \bigcap_{n\in M} V_n$. We claim that $\widetilde{V}$ is the sought minimal element. Indeed, assume that, on the contrary, there is $V \in \mathcal{V}$ such that $V \subsetneq \widetilde{V}$. Certainly, there exists $n \in \mathbb{N}$ such that $U_n \cap (\widetilde{V} \setminus V) \neq \emptyset$. This implies that $n \in M$. Then we have $U_n \cap (\widetilde{V} \setminus V) \subset U_n$, and $U_n \cap (\widetilde{V} \setminus V) \subset \widetilde{V}$ $(\subset V_n)$, so $U_n \cap V_n \neq \emptyset$, a contradiction.

□

Theorem 5.46 (Rosenthal [Rose7]) *Let (P, ρ) be a Polish space, F be a $\mathcal{T}_p$-relatively compact subset of $B_1(P)$. Then F has the property that every $f \in \overline{F}^{\mathcal{T}_p}$ lies in the $\mathcal{T}_p$-closure of a countable subset of F.*

Proof: We start with an auxiliary lemma.

Lemma 5.47 *Let S be a $\mathcal{T}_p$-relatively compact subset of $B_1(P)$, so that $0 \in \overline{S}^{\mathcal{T}_p}$. Then for every $\delta > 0$ there is a countable subset $H \subset S$ so that $\inf_{h\in H} h(t) < \delta$ for all $t \in P$.*

Proof: By contradiction, there exists $\delta > 0$ such that for every countable subset $H \subset S$ it holds $\inf_{h\in H} h(t) \geq \delta$ for some $t \in P$. We define $V(H) = \{t \in P : \inf_{f\in H} f(t) \geq \delta\}$, where H is a countable subset of S. Define $\mathcal{V} = \{V(H) : H \subset S, H \text{ countable}\}$ a system of closed subsets of P. By Lemma 5.45, there exists in $\mathcal{V}$ a nonempty minimal element $V_P = V(H_P)$. Choose a dense set $\{t_i\}_{i=1}^{\infty} \subset V_P$, and find a sequence $\{f_k\}_{k=1}^{\infty} \in S$ such that $\lim\limits_{k\to\infty} f_k(t_i) = 0$ for all $i \in \mathbb{N}$. By Theorem 5.41, S is $\mathcal{T}_p$-sequentially compact, so we may, without loss of generality, assume that $\{f_k\}_{k=1}^{\infty}$ is $\mathcal{T}_p$-convergent to a Baire 1 function f. Thus f restricted to V_P has a point of continuity $t \in V_P$ by Baire's Great Theorem 17.12, and necessarily $f(t) = 0$. This however implies that $V(H_P \cup \{f_k\}_{k=1}^{\infty}) \subsetneq V_P$, a contradiction. □

To finish the proof of the theorem, suppose that F is $\mathcal{T}_p$-relatively compact, and for each $m \in \mathbb{N}$ define the mapping $\phi_m : B_1(P) \to B_1(P^m)$ by $\phi_m(f)(t_1, \ldots, t_m) = \sum_{i=1}^{m} |f(t_i)|$. Let $g \in \overline{F}^{\mathcal{T}_p}$. We may, without loss of generality, assume that $g = 0$. Hence $S_m = \phi_m(F)$ are $\mathcal{T}_p$-relatively compact in $B_1(P^m)$, with $0 \in \overline{S_m}^{\mathcal{T}_p}$. Using Lemma 5.47, there is a countable subset H_m of F so that $\frac{1}{m} > \inf\{\phi_m(h)(y) : h \in H_m\}$ for all $y \in P^m$. It follows that $0 \in \overline{H}^{\mathcal{T}_p}$, where $H = \bigcup_{m=1}^{\infty} H_m$ is a countable subset of F. □

We need the following topological lemma.

Lemma 5.48 *Let $(\mathcal{X}, \mathcal{T})$ be a regular topological space, that is sequentially compact and countably tight. Let $x \in \mathcal{X}$, $\{x_i\}_{i=1}^{\infty}$ be a sequence in $\mathcal{X}$ and $\{I_n\}_{n=1}^{\infty}$ be a decreasing sequence of infinite subsets of $\mathbb{N}$, such that $x \in \overline{\{x_i\}_{i\in I_n}}^{\mathcal{T}}$ for all $n \in \mathbb{N}$.*

Then there is an infinite set $I \subset \mathbb{N}$ such that $I \setminus I_n$ is finite for all $n \in \mathbb{N}$, and $x \in \overline{\{x_i\}_{i\in I}}^{\mathcal{T}}$.

Proof: Let

$$F = \{\lim_{i\in I} x_i \,:\, I \text{ infinite, } (\forall n \in \mathbb{N}) \text{ card}(I \setminus I_n) < \infty,\ \lim_{i\in I} x_i \text{ exists}\}. \tag{5.16}$$

Let U be an open neighborhood of x. Then $J := \{i \,:\, x_i \in U\}$ intersects every I_n in an infinite set. As I_n is decreasing, there is some $K \subset J$ such that $K \setminus I_n$ is finite for all $n \in \mathbb{N}$. As $\mathcal{X}$ is sequentially compact, there is some $I \subset K$ such that $z_I = \lim_{i\in I} x_i \in \overline{U}$ exists. Since $\mathcal{X}$ is regular and U was arbitrary, we have $x \in \overline{F}$. Since $\mathcal{X}$ is countably tight, there exists a sequence $\{z_m\}_{m=1}^{\infty} \in F$ such that $x \in \overline{\{z_m\}_{m\in\mathbb{N}}}^{\mathcal{T}}$. Using that $z_m = z_{J_m} = \lim_{i\in J_m} x_i$, where $J_m \setminus I_n$ is finite for all $n \in \mathbb{N}$, let $I = \bigcup_{n=1}^{\infty}(I_n \cap J_n)$. Because I_n is decreasing, we have that $I \setminus I_n$ and $J_n \setminus I$ are finite for every $n \in \mathbb{N}$. It follows that $z_m \in \overline{\{x_i\}_{i\in I}}^{\mathcal{T}}$ for all $m \in \mathbb{N}$. Therefore $x \in \overline{\{x_i\}_{i\in I}}^{\mathcal{T}}$, as requested. □

Theorem 5.49 (Bourgain, Fremlin, Talagrand [BFT]) *Let (P, ρ) be a Polish space. Then $B_1(P)$ is $\mathcal{T}_p$-angelic.*

Proof: In view of Theorems 5.41 and 5.46, it remains to show that the closure of a countable and relatively compact set in $B_1(P)$ coincides with its sequential closure. We begin with a variant of Lemma 5.39.

Lemma 5.50 *Let (P, ρ) be a Polish space, $\{f_n\}_{n=1}^{\infty}$ be a $\mathcal{T}_p$-relatively compact sequence in $B_1(P)$ of continuous functions, $0 \in \overline{\{f_n\}_{n\in\mathbb{N}}}^{\mathcal{T}_p}$. Let $W \subset P$ be a nonempty closed subset, $\varepsilon > 0$. Then there is a nonempty relatively open $U \subset W$ and an infinite $I \subset \mathbb{N}$ such that $0 \in \overline{\{f_i\}_{i\in I}}^{\mathcal{T}_p}$, and moreover $\limsup_{i\in I} |f_i(t)| \leq \varepsilon$ for all $t \in U$.*

Proof: Proceeding by contradiction, there exists a closed $W \subset P$ and $\varepsilon > 0$ so that for every relatively open $U \subset W$ and every infinite $I \subset \mathbb{N}$, $\limsup_{i\in I} |f_i(t)| > \varepsilon$ for all $t \in U$. We are going to construct $I \subset \mathbb{N}$ so that $\{f_i\}_{i\in I}$ has no $\mathcal{T}_p$-convergent subsequence. By Theorem 5.41 this will be a contradiction. We start by making a simple observation. Suppose that E_i, $i = 1, \ldots, n$ be nonempty relatively open subsets of W. Denote $I_i = \{j \,:\, |f_j(t)| < \varepsilon, \text{ for all } t \in E_i\}$. We have that $0 \notin \overline{\{f_j\}_{j\in I_i}}^{\mathcal{T}_p}$ for all $i = 1, \ldots, n$. Consequently, we have

$$0 \in \overline{\{f_j\}_{j\in\mathbb{N}\setminus\bigcup_{i=1}^{n} I_i}}^{\mathcal{T}_p}. \tag{5.17}$$

We construct a pair of sequences consisting of $\{g_k\}_{k=1}^{\infty}$, a subsequence of the original one $\{f_n\}_{n=1}^{\infty}$, and $\{D_k\}_{k=1}^{\infty}$, consisting of relatively open subsets of W, with the following properties.

(i) $D_k = \bigcup_{i=1}^{2^k} D_k^i$, D_k^i pairwise disjoint relatively open and nonempty subsets of K. For convenience, we also denote $\widetilde{D}_k^0 = \bigcup_{i=1}^{2^{k-1}} D_k^{2i}$, $\widetilde{D}_k^1 = \bigcup_{i=1}^{2^{k-1}} D_k^{2i-1}$.

(ii) $\overline{D_{k+1}^{2i-1} \cup D_{k+1}^{2i}} \subset D_k^i$, ρ-diam $D_k^i < \frac{1}{k}$.

(iii) $\sup_{\widetilde{D}_k^0} g_k \le \frac{\varepsilon}{2}$ and $\inf_{\widetilde{D}_k^1} g_k \ge \varepsilon$.

Let us describe the inductive step (the initial step is similar). Assume that we have constructed $\{g_k\}_{k\le l}$ and $\{D_k\}_{k\le l}$. We use (5.17) to find points $\{t_l^i : 1 \le i \le 2^{l+1}\}$, such that $t_l^{2i} \in D_l^i$, $t_l^{2i-1} \in D_l^i$, and a function $g_{l+1} \in \{f_n\}_{n=1}^\infty$, so that $g_{l+1}(t_l^i) \le \frac{\varepsilon}{2}$ for i even and $g_{l+1}(t_l^i) > \varepsilon$ for i odd. Using the continuity of g_{l+1} at this point of the proof, it remains to fix some small enough pairwise disjoint neighborhoods D_{l+1}^i of the points t_l^i, $i = 1, \dots 2^{l+1}$, so that (i)–(iii) will be satisfied for $D_{l+1} = \bigcup_{i=1}^{2^{l+1}} D_{l+1}^i$. Note that condition (ii) implies that $\bigcap_{k=1}^\infty D_k^{i(k)} \ne \emptyset$ if and only if $i(k+1) \in \{2i(k), 2i(k)-1\}$ for all $k \in \mathbb{N}$. To finish the proof, it suffices to show that $\{g_k\}_{k=1}^\infty$ has no $\mathcal{T}_p$-convergent subsequence. Let $\{g_{k(j)}\}_{j=1}^\infty$ be any subsequence of $\{g_k\}_{k=1}^\infty$. Choose a sequence $\{i(k)\}_{k=1}^\infty$, satisfying $i(k+1) \in \{2i(k), 2i(k)-1\}$ for all $k \in \mathbb{N}$, and moreover such that $i(k(j))$ is even if and only if j is even. There exists some $t \in \bigcap_{k=1}^\infty D_k^{i(k)}$. We have now $g_{k(j)}(t) \le \frac{\varepsilon}{2}$ for all j even, while $g_{k(j)}(t) \ge \varepsilon$ for all j odd. This implies that $\{g_{k(j)}\}_{j=1}^\infty$ is not $\mathcal{T}_p$-convergent and the proof is finished. □

Lemma 5.51 *Under the assumptions of Lemma* 5.50, *for every* $\varepsilon > 0$ *there is an infinite set* $I \subset \mathbb{N}$ *such that* $0 \in \overline{\{f_i\}_{i\in I}}^{\mathcal{T}_p}$, *and moreover* $\limsup_{i\in I} |f_i(t)| \le \varepsilon$ *for all* $t \in P$.

Proof: For each infinite set $I \subset \mathbb{N}$ put $U(I) = \operatorname{int}\{t : \limsup_{i\in I} |f_i(t)| \le \varepsilon\}$, and let $A(I) = \overline{\{f_i\}_{i\in I}}^{\mathcal{T}_p}$. Note that $I\backslash J$ finite implies that $U(J) \subset U(I)$. Let $\{V_k\}_{k=1}^\infty$ be a base of the topology of P, and choose $\{I_k\}_{k=1}^\infty$, $0 \in A(I_k)$ as follows. $I_0 := \mathbb{N}$. Given I_k, we choose $I_{k+1} \subset I_k$ such that either $0 \in A(I_{k+1})$ and $V_k \subset U(I_{k+1})$, if such a set I_{k+1} exists, or else we put $I_{k+1} = I_k$. We obtain a decreasing sequence $\{I_k\}_{k=1}^\infty$ to which Lemma 5.48 applies. Thus for some $I \subset \mathbb{N}$, such that $I\backslash I_k$ is finite, we have $0 \in \overline{\{f_i\}_{i\in I}}^{\mathcal{T}_p}$. Let $J \subset I$ be infinite and such that $0 \in \overline{\{f_i\}_{i\in J}}^{\mathcal{T}_p}$. We claim that $U(J) = U(I)$. Indeed, assuming the contrary there is some $V_k \subset U(J)$ but $V_k \not\subset U(I)$. Since $J\backslash I_k$ is finite, $J \cap I_k$ is an infinite subset of I_k such that $0 \in A(J \cap I_k)$ and $V_k \subset U(J \cap I_k)$. Accordingly, we must have chosen I_{k+1} such that $V_k \subset U(I_{k+1}) = U(I)$. As $I\backslash I_{k+1}$ is finite, we reach a contradiction. It remains to show that $U(I) = P$. Assuming the contrary again, let $W = X\backslash U(I)$ be a nonempty closed set. By Theorems 5.41 and 5.46, the assumptions of Lemma 5.48 are satisfied. Thus there is a nonempty relatively open $U \subset W$ and an infinite $J \subset I$, such that $0 \in \overline{\{f_i\}_{i\in J}}^{\mathcal{T}_p}$ and moreover $\limsup_{i\in J} |x_i(t)| \le \varepsilon$ for all $t \in U$. Thus $U \subset U(J)\backslash U(I) \ne \emptyset$, a contradiction which finishes the proof. □

Repeating the lemma above for a decreasing sequence $\varepsilon_n \to 0$, and choosing a diagonal sequence $I \subset \mathbb{N}$, we obtain that $\lim_{i\in I} f_i(t) = 0$ holds for every $t \in P$. So far we have proved that if the countable set $\{f_n\}_{n=1}^{\infty}$ consists of continuous functions, and a cluster point f is also a continuous function (passing from a zero function to a general continuous function is trivial), then f can be reached by a $\mathcal{T}_p$-convergent subsequence of $\{f_n\}_{n=1}^{\infty}$. In order to dispose of the continuity assumption, we use Theorem 17.16.

As a consequence, if $F \subset B_1(P)$ is a countable $\mathcal{T}_p$-relatively compact set with a cluster point f (without loss of generality, $f = 0$), then there exists a metric $\widetilde{\rho}$ on P, stronger than the original ρ, which turns P into a Polish space $(P, \widetilde{\rho})$, and moreover all $g \in F$ are continuous in the topology coming from $\widetilde{\rho}$. It is immediate that $F \subset B_1((P, \widetilde{\rho}))$ is $\mathcal{T}_p$-relatively compact. This finishes the proof. $\square$

We finish this section by collecting, in Theorem 5.52, some previous results that characterized the absence of copies of ℓ_1 in the space in terms of the extreme points of weak*-compact convex sets in the dual space. Another characterization, this time in terms of convergence of sequences in the dual space, is given in Theorem 5.53.

Theorem 5.52 (Odell and Rosenthal [OdRo], Rosenthal [Rose7]) *Let X be a separable Banach space. Then, the following statements are equivalent:*
(i) *X contains no isomorphic copy of ℓ_1.*
(ii) *Every $x^{**} \in B_{X^{**}}$ is the w^*-limit of a sequence in B_X.*
(iii) *Every weak*-compact convex set in X^* is the norm-closed convex hull of its extreme points.*

Proof: (i)$\Longrightarrow$(ii) is in Theorem 5.40.

(ii)$\Longrightarrow$(iii): This is Corollary 3.129.

(iii)$\Longrightarrow$(i): If $\ell_1 \hookrightarrow X$, by Theorem 5.44 $C[0, 1]$ is isomorphic to a quotient of X, hence $(B_{C[0,1]^*}, w^*)$ is a subspace of (B_{X^*}, w^*). Therefore it suffices to show that $\lambda \notin \overline{\text{conv}}^{\|\cdot\|}(\text{Ext}(B_{C[0,1]^*}))$, where λ denotes the Lebesgue measure on $[0, 1]$. To see this, note that if λ is norm-close to a finite convex sum $\sum a_i \delta_i$ of Dirac's measures, we can find $f \in B_{C[0,1]}$ such that $f(t_i) = 0$ and $\lambda(f)$ is close to 1. $\square$

Recall that the Mackey topology $\mu(X^*, X)$ on X^* associated to the dual pair $\langle X, X^* \rangle$ is the topology on X^* of the uniform convergence on the family of all convex balanced and w-compact subsets of X (see Definition 3.43). The following theorem was motivated by a result of Borwein [Bor].

Theorem 5.53 (Orno [Orn91], Valdivia [Vald6], see, e.g., [HMVZ, p. 94]) *A Banach space X does not contain an isomorphic copy of ℓ_1 if and only if the two topologies $\mu(X^*, X)$ and $\|\cdot\|$ agree sequentially on X^*.*

Proof: Assume that X does not contain an isomorphic copy of ℓ_1 and $\{f_n\}$ is a sequence in X^* that converges to 0 uniformly on every weakly compact set but not in norm of X^*. Then there is a subsequence, which we will denote again by $\{f_n\}$, vectors $x_n \in B_X$ and $\varepsilon > 0$ such that $f_n(x_n) \geq \varepsilon$ for every n. By Rosenthal Theorem 5.37, there is a subsequence of $\{x_n\}$, which we will again denote by $\{x_n\}$, such that $\{x_n\}$ is weakly Cauchy. Since $f_n \xrightarrow{w^*} 0$, it is easy to see that there is a subsequence of $\{f_n\}$, which we will again denote by $\{f_n\}$, such that $f_n(x_{2n}) - f_n(x_{2n-1})$

stays bounded away from zero. The sequence $\{x_{2n} - x_{2n-1}\}$ is weakly null. This is a contradiction with the fact that f_n converge to 0 uniformly on weak compact sets.

To prove that the condition suffices, suppose that Y is a subspace of X which is isomorphic to ℓ_1 and let $\{e_n\}_{n=1}^{\infty}$ be the image of the unit vector basis under some isomorphism from ℓ_1 onto Y. Define a bounded operator from Y into $L_\infty[0, 1]$ by mapping e_n to the nth Rademacher function r_n. By the injective property of $L_\infty[0, 1]$ (see Exercise 5.91), this operator extends to a bounded linear operator T from X into $L_\infty[0, 1]$. Let r_n^* be the nth Rademacher function in $L_1[0, 1]$ considered as a subspace of $(L_\infty[0, 1])^*$. Thus the sequence $\{r_n^*\}$, being equivalent to an orthonormal sequence in a Hilbert space, converges weakly to zero. Since $L_\infty[0, 1]$ has the Dunford–Pettis property (see Exercise 4.42 and Theorem 13.43, and observe that ℓ_∞ is a $C(K)$-space), $\{r_n^*\}$ converges in the topology $\mu(X^*, X)$ to zero, a fortiori $\{T^* r_n^*\}$ is $\mu(X^*, X)$-convergent to zero. But $\langle T^* r_n^*, e_n\rangle = \langle r_n^*, r_n\rangle = 1$, so $\{T^* r_n^*\}$ does not converge to zero in norm. □

5.4 Remarks and Open Problems

Remarks

1. If $0 < p < 1$, then there is in L_p a subspace such that all bounded operators on it are multiples of the identity [KaRo]. Shelah and Steprāns proved that there is a nonseparable Banach space X for which every bounded operator from X into X has the form $S + \rho I$, where S is an operator with separable range, I is the identity operator and ρ is a real number [ShSt1].
2. An infinite-dimensional Banach space X is called *hereditarily indecomposable* (HI in short) if no infinite-dimensional subspace Y of X can be written as a topological direct sum of two infinite-dimensional closed subspaces Y_1 and Y_2 of Y. Gowers and Maurey [GoMa] proved that there exists a reflexive HI space. In this direction see the subsequent paper [Fere]. Note that an HI space cannot contain an infinite-dimensional subspace with unconditional basis (the existence of such space was a longstanding famous open problem). Also, if X is an HI space, then X is not isomorphic to any proper subspace of X, in particular it is not isomorphic to any of its hyperplanes (see [Maur, p. 1265]). This solved another longstanding open problem. Gowers showed in [Gowe6] that if X is an arbitrary infinite-dimensional Banach space, then either X contains an infinite-dimensional subspace with unconditional basis or X contains an HI subspace. We refer also to [ArTo], where it is shown that there is a separable Banach space Y with separable dual Y^* and non-separable Y^{**} such that all Y, Y^* and Y^{**} are hereditarily indecomposable and any bounded operator on Y^{**} is of the form $\lambda I + S$, where S has a separable range. These results are beyond the scope of this text and we refer to the article [Maur, pp. 1247–1297] for information on these topics.

3. A Banach space $(X, \|\cdot\|)$ is called *distortable* if there is $\lambda > 1$ and an equivalent norm $|\cdot|$ on X so that for all infinite-dimensional subspaces Y of X,

$$\sup\left\{\frac{|y_1|}{|y_2|} : y_1, y_2 \in S_{(X,\|\cdot\|)}\right\} \geq \lambda.$$

We will see, in Exercise 5.42, James' classical result that if X is isomorphic to c_0 or to ℓ_1, then X does not contain a distortable subspace.
Milman showed in [Milm2] that if X does not contain isomorphic copies of c_0 or some ℓ_p, $p \in [1, \infty)$, then X contains a distortable subspace (for the proof see [OdSc4]).
So, the existence of a distortable space was clear after Tsirelson space appeared in 1974 [Tsir], see Exercises 9.29, 9.30, and 9.31. Now it is known that any ℓ_p, $1 < p < \infty$ is distortable, and moreover, if X does not contain a distortable subspace, then X is saturated by isomorphic copies of c_0 or ℓ_1 (we say that a Banach space is *saturated by isomorphic copies of a Banach space* Y if every infinite-dimensional subspace of X contains an isomorphic copy of Y), see [OdSc4].
If X is a Banach space, we will say that a function $f : S_X \to \mathbb{R}$ *stabilizes* if for every infinite-dimensional subspace $Y \hookrightarrow X$ and every $\varepsilon > 0$, there exists an infinite-dimensional subspace $Z \hookrightarrow Y$ so that

$$\operatorname{osc}(f, S_Z) := \sup\{f(z_1) - f(z_2) : z_1, z_2 \in S_Z\} < \varepsilon.$$

Note that X does not contain a distortable subspace if and only if every equivalent norm on X stabilizes.
Gowers showed in [Gowe1] that every Lipschitz function $f : S_{c_0} \to \mathbb{R}$ stabilizes.
These topics are beyond the scope of this text and we refer to [OdSc4] for information in this direction.
4. Talagrand proved in [Tala4] the following result: *if* $|||\cdot|||$ *is an equivalent norm on* ℓ_∞*, then there is* $\delta > 0$ *such that for every* $n \in \mathbb{N}$*, there is a subspace* X_n *of* ℓ_∞ *isomorphic to* ℓ_∞ *and such that on* X_n*,* $(\delta - 2^{-n})\|\cdot\|_\infty \leq |||\cdot||| \leq (\delta + 2^{-n})\|\cdot\|_\infty$.
Partington showed in [Part] that *if* Γ *is uncountable, then* $\ell_\infty(\Gamma)$ *in any equivalent norm contains an isometric copy of* ℓ_∞ *in the supremum norm.* Thus, for Γ uncountable, $\ell_\infty(\Gamma)$ admits no equivalent strictly convex norm.
5. Gowers showed in [Gowe5] that there is a Banach space Z that is isomorphic to $Z \oplus Z \oplus Z$ but not isomorphic to $Z \oplus Z$. Then Z is isomorphic to a complemented subspace of $Z \oplus Z$ and $Z \oplus Z$ is isomorphic to a complemented subspace of Z, since the latter is isomorphic to $Z \oplus Z \oplus Z$. This is a negative solution to the Schroeder–Bernstein problem for Banach spaces and should be compared with the Pełczyński decomposition method. Note that if we do not insist on the complementability, then for a negative solution of this problem, we can just consider an infinite-dimensional subspace of c_0 that is not isomorphic to c_0

(see Exercise 5.67), as this subspace has to contain an isomorphic copy of c_0 by Theorem 4.46.
Let us remark in passing that Lindenstrauss solved Köthe's problem in [Lind2] by showing that there are two nonisomorphic spaces that have isomorphically the same family of all subspaces and isomorphically the same family of all quotients, see Exercise 5.70.

6. Concerning a problem on isometrically universal spaces, let us remark that Szankowski showed in [Szan1] that there is a reflexive separable space X such that every finite-dimensional space W is isometric to a one-complemented subspace of X.
   Lindenstrauss proved in [Lind6] that any two-dimensional space is isometric to a subspace of L_1, and Bessaga showed in [Bess1] that there is no finite-dimensional space that would be isometrically universal for all 2-dimensional spaces.
   Godefroy and Kalton showed in [GoKa2] that if a separable Banach space Z contains isometric copies of all strictly convex separable Banach spaces, then Z contains an isometric copy of all separable Banach spaces.
7. Maurey showed in [Maur0] that if X is a Banach space of type 2 and Y is a subspace of X which is isomorphic to a Hilbert space, then Y is complemented in X.
8. Lindenstrauss and Pełczyński, in [LiPe2], proved that if E is a subspace of c_0 and K is a compact space, then any bounded operator $T : E \to C(K)$ extends to a bounded operator from c_0 into $C(K)$. For generalizations, see [JoZi].
9. For a characterization of spaces that do not contain a copy of ℓ_1 in terms of renorming, we refer to [DGZ3, p. 106].
10. Stegall proved in [Steg1] the following result: *If X is a separable Banach space such that X^* is not separable and $\varepsilon > 0$ is given, then there is a subset $\Delta \subset S_{X^*}$ w^*-homeomorphic to the Cantor set and a sequence $\{x_n\}$ in X with $\|x_n\| < 1 + \varepsilon$ for all $n \in \mathbb{N}$ such that $\sum_{n=1}^{\infty} \|Tx_n - \chi_{A_n}\| < \varepsilon$, where $T : X \to C(\Delta)$ is the evaluation operator (i.e., $Tx(x^*) = x^*(x)$) and the sets A_n are the (homeomorphic images of) the dyadic intervals of the Cantor set.* (For more in this direction see, e.g., [LoMo].)
11. We refer to, e.g., [HMVZ, Ch. 7] for more on fixing c_0 by operators.
12. There is a Banach space X not isomorphic to c_0 such that X^* is isometric to ℓ_1 [BeLi0].

Open Problems

1. Is it true that every complemented subspace in a $C(K)$ space, for K a compact topological space, is isomorphic to a $C(L)$ space, L a compact topological space? See, e.g., [Rose10, p. 1593].
2. Assume that a Banach space X is complemented in all its overspaces. Is X isomorphic to $C(K)$ where K is a extremely disconnected compact space (i.e., the closure of any open subset is open)? See, e.g., [Zipp2, p. 1714].

3. Assume that a separable Banach space X has the property that all of its subspaces with Schauder basis are complemented in X. Is X isomorphic to ℓ_2? [Pel8, Problem 2.3].
4. Assume that a separable Banach space X has the property that all of its infinite-dimensional subspaces with basis are isomorphic. Is X isomorphic to ℓ_2? [Pel8, Problem 2.4]. It is known that a separable infinite-dimensional Banach space all of whose infinite-dimensional subspaces are isomorphic is necessarily isomorphic to a Hilbert space, [Gowe6] and [KoTo].
5. Assume that every subspace of a separable Banach space X has an unconditional basis. Is X isomorphic to a Hilbert space? See, e.g., [Casa, p. 279].
6. Does the space H_∞ of all bounded analytical functions on $\{z \in \mathbb{C} : |z| < 1\}$ with the supremum norm have the approximation property? (See Definition 16.34.) See, e.g., [Casa, p. 285].

Exercises for Chapter 5

5.1 Consider the space $L_p(\Omega, \Sigma, \mu)$ for $1 \leq p < \infty$ and fix $\varepsilon > 0$. Let F be a finite-dimensional subspace of $L_p(\Omega, \Sigma, \mu)$. Prove that there is a finite-dimensional space $\widetilde{F} \subset L_p(\Omega, \Sigma, \mu)$ such that $F \subset \widetilde{F}$ and $d(\widetilde{F}, \ell_p^m) < 1 + \varepsilon$ for some m and that there is a norm-2 projection from $L_p(\Omega, \Sigma, \mu)$ onto $\widetilde{F}$. Therefore $L_p(\Omega, \Sigma, \mu)$ has the bounded approximation property.
Hint. Let $\{f_i\}_{i=1}^n$ be a Hamel basis for F. Since every function in L_p is the limit of simple functions, we can find a partition of Ω into sets $\{A_j\}_{j=1}^m$ so that $0 < \mu(A_j) < \infty$ for each j and so that there are $\{g_i\}_{i=1}^n$ in $G := \operatorname{span}\{\chi_{A_j} : j = 1, 2, \dots, m\}$ such that $\|f_i - g_i\| < \varepsilon$ for all j. Thus F is "almost" contained in G and G is isometric to ℓ_p^m.

We will now show that G can be replaced by a subspace $\widetilde{F}$ of L_p which is closed to G, contains F and, moreover, there is a projection of L_p onto $\widetilde{F}$ of norm less or equal than 2. (See also Proposition 5.24.)

Since all norms on a finite-dimensional space F are equivalent, there is $K \geq 1$ such that

$$K^{-1} \max |\lambda_i| \leq \left\| \sum_{i=1}^{m} \lambda_i f_i \right\| \leq K \max |\lambda_i|$$

for all real numbers $\lambda_1, \lambda_2, \dots, \lambda_m$. Let $1 > \delta > 0$ such that $(1+\delta)/(1-\delta) < 1+\eta$, and put $\varepsilon = \delta/(2nK)$. For $i = 1, 2, \dots, n$, let $g_i \in G$ be such that $\|g_i\| = 1$ and $\|f_i - g_i\| < \varepsilon$. Then

$$(2K)^{-1} \max |\lambda_i| \leq \left\| \sum_{i=1}^{m} \lambda_i g_i \right\| \leq 2K \max |\lambda_i|.$$

By the Hahn–Banach theorem, there exists $x_i^* \in L_p^*$ with $\|x_i^*\| \leq 2K$ and such that $x_i^*(g_j) = \delta_{i,j}$, the Kronecker delta, $i = 1, 2, \dots, n$. Let an operator $T : L_p \to L_p$

be defined by

$$Th = h + \sum_{i=1}^{n} x_i^*(h)(f_i - g_i)$$

for $h \in L_p$. Then $Tg_i = f_i$ for $i = 1, 2, \ldots, n$ and

$$(1 - \delta)\|h\| \leq \|Th\| \leq (1 + \delta)\|h\|.$$

Indeed, to see for example the right-hand-side inequality, estimate

$$\|Th - h\| \leq \sum_{i=1}^{n} \|x_i^*\|.\|h\|.\|f_i - g_i\| \leq \|h\|2Kn\varepsilon \leq \delta\|h\|.$$

Thus $TG \supset F$ and $\rho(TG, \ell_\infty^m) \leq (1+\delta)/(1-\delta) < 1 + \varepsilon$.

Put $\widetilde{F} = TG$. Let P be a norm-one projection from L_p onto G. Then $P_1 := TPT^{-1}$ is the sought projection onto $\widetilde{F}$.

5.2 Show that a Banach space X is isomorphic to a Hilbert space if every separable subspace of X is isomorphic to a Hilbert space.
Hint. For separable $Y \subset X$, put

$$\begin{aligned} C_Y = \sup\{C > 0 :\ &\text{there exists a bilinear form } b_Y \\ &\text{on } Y \times Y \text{ such that } C\|y\|^2 \leq b_Y(y, y) \leq \|y\|^2, \text{ for all } y \in Y\}. \end{aligned}$$

Show that $C := \inf\{C_Y : Y \text{ separable subspace of } X\} > 0$ (otherwise there are separable subspaces Y_n, $n \in \mathbb{N}$, such that $C_{Y_n} \to 0$ and consider the closed linear span of $\bigcup_n Y_n$). Given a separable subspace Y of X define on $X \times X$ a bilinear form that is b_Y on $Y \times Y$ and such that $(C/2)\|y\|^2 \leq b_Y(y, y) \leq \|y\|^2$ for $y \in Y$. Consider $[-1, 1]^{B_{X \times X}}$ and Tychonoff's theorem to get a limit point, obtaining the desired bilinear form on X to give a dot product on X.

The assertion follows also from Kwapień's theorem (see Exercise 1.98).

5.3 Let X be a Banach space, let Y be a closed separating subspace of X^*. For $x \in X$ define $\|x\|^Y = \sup\{|y^*(x)| :\ y^* \in B_Y\}$ (see Exercise 3.89). Show that:

(i) Y is norming if and only if $X + Y^\perp$ is closed in X^{**} (that is, it is a topological sum).

(ii) Y is 1-norming if and only if $X + Y^\perp$ is closed in X^{**} and the projection P of $X + Y^\perp$ onto X has $\|P\| = 1$.

Hint. (i) Assume Y is c-norming. Then for $x \in X$, $y \in Y^\perp$ we have

$$\|x\| \leq c\|x\|^Y = c\|x + y\|^Y \leq c\|x + y\|$$

and the claim follows by Exercise 4.20. If the sum $X \oplus Y^{\perp}$ is topological, then X is isomorphic to $(X + Y^{\perp})/Y^{\perp}$. Thus for some $c > 0$, we have $\frac{1}{c}\|x\| \leq \|\hat{x}\| = \operatorname{dist}(x, Y^{\perp}) = \|x\|^{Y}$.

(ii) If Y is 1-norming, by (i) we have $\|x\| \leq \|x + y\|$. Now assume there is a norm-one projection P of $Z = X + Y^{\perp}$ onto X. Consider $\widehat{P}\colon Z/\operatorname{Ker}(P) \to X$. We have $\widehat{P}(B_{Z/\operatorname{Ker}(P)}) = P(B_Z) = B_X$ and $\widehat{P}$ is one-to-one, hence it is an isometry of $Z/\operatorname{Ker}(P) = Z/Y^{\perp}$ onto X. We get $\|x\| = \|\hat{x}\| = \operatorname{dist}(x, Y^{\perp}) = \|x\|^{Y}$.

5.4 Let $x^{**} \in X^{**} \setminus X$. Show that $\operatorname{Ker}(x^{**})$ is 1-norming if and only if $\|x^{**} + x\| \geq \|x\|$ for all $x \in X$ (see Exercise 3.88).

Hint. If $\operatorname{Ker}(x^{**})$ is 1-norming, then the inequality follows by the previous exercise. On the other hand, having the inequality we obtain a norm-one projection of $X + \operatorname{Ker}(x^{**})^{\perp} = X + \operatorname{span}\{x^{**}\}$ onto X (note that $\operatorname{span}\{x^{**}\}$ is w^*-closed, so $\operatorname{Ker}(x^{**})^{\perp} = (\operatorname{span}\{x^{**}\}_{\perp})^{\perp} = \operatorname{span}\{x^{**}\}$).

5.5 Assume that the codimension of X in X^{**} is finite. Show that then every w^*-dense closed subspace in X^* is norming.

Hint. Since X is finite-codimensional in $X + Y^{\perp}$, $X + Y^{\perp}$ is closed in X^{**}. Use the preceding exercises.

5.6 Let Y be a closed subspace of a Banach space, recall that Y^* is isometric to $X^*/Y^{\perp}$. Let q be the canonical quotient mapping $X^* \to X^*/Y^{\perp}$. Show that if Z a closed norming subspace of $Y^* = X^*/Y^{\perp}$, then $W = q^{-1}(Z)$ is a norming subspace of X^*.

Hint. Assume without loss of generality that Z is a 1-norming subspace of Y^* (consider $\|\cdot\|^{Z}$ on Y, see Exercise 5.3, and extend it to an equivalent norm in X, see Proposition 2.14). Take $x \in S_X$. If $\operatorname{dist}(x, Y) < \frac{1}{4}$, take $y \in Y$ such that $\|x - y\| < \frac{1}{4}$, so $\|y\| > \frac{3}{4}$. Find $y^* \in B_Z$ such that $y^*(y) > \frac{3}{4}$. Thus there is $w^* \in \frac{3}{2}B_W$ such that $q(w^*) = y^*$, that is, $w^*(y) > \frac{3}{4}$. Then $x^* := \frac{2}{3}w^* \in B_W$ and $x^*(x) = \frac{2}{3}[w^*(y) + w^*(x - y)] > \frac{1}{4}$. Now assume that $\operatorname{dist}(x, Y) \geq \frac{1}{4}$. Then $\|\hat{x}\| \geq \frac{1}{4}$ ($\hat{x}$ is the coset containing x and the norm is calculated in X/Y), so there is $x^* \in Y^{\perp}$ such that $x^*(x) \geq \frac{1}{4}$ and $\|x^*\| \leq 1$. Since $q(x^*) = 0 \in Z$, we get that $x^* \in B_W$. So in any case, for every $x \in S_X$ we get $\sup\{x^*(x) : x^* \in B_W\} \geq \frac{1}{4}$.

5.7 Let Y be a closed hyperplane of a Banach space X. Show that for any $\varepsilon > 0$ there is a projection P of X onto Y with $\|P\| \leq 2 + \varepsilon$ (see Exercise 5.8 for an extension).

In particular, X is isomorphic to $Y \oplus \mathbb{R}$.

Hint. Write $X = Y \oplus Z$ (algebraic sum), where $Z = \operatorname{span}(e)$. Using $\hat{f} \in (X/Y)^* \cong Y^{\perp}$ such that $\hat{f}(\hat{e}) = 1$, find $f \in (1 + \varepsilon)B_{X^*}$ such that $f(e) = 1$ and $f(y) = 0$ for $y \in Y$. Let $P(x) = x - f(x)e$.

The last part follows as Y is complemented.

5.8 Let Y be a closed subspace of codimension n in a Banach space X. Show that for every $\varepsilon > 0$ there is a projection P of X onto Y such that $\|P\| < n+1+\varepsilon$. In particular, Y is complemented.
Hint. Let $\{f_i; F_i\}_1^n \subset (Y^\perp \times (Y^\perp)^*)$ be an Auerbach basis. Identify $(Y^\perp)^*$ with $(X/Y)^{**}$. Take $x_i \in \mathcal{F}_i$ with $\|x_i\| < 1+\varepsilon/n$. Let $P(x) = x - \sum f_i(x)x_i$.

5.9 Let Z be a hyperplane of a Banach space X. Show that if there exists a bounded operator T from some Banach space Y onto Z then Z is closed in X.
Hint. By taking quotients assume that $T\colon Y \to Z$ is one-to-one. Consider the mapping T_1 from $Y \oplus \mathbb{R}$ onto X defined by $T_1(y,r) = T(y) + rz_0$, where z_0 is in the algebraic complement of Z in X. Then $T_1 \in \mathcal{B}(X \oplus \mathbb{R}, X)$ and is onto, hence it is an isomorphism and so $T(Y)$ must be closed in X.

5.10 Let Y be a finite-codimensional subspace of a Banach space X. Show that if there is a bounded operator T from some Banach space Z onto Y, then Y is closed.
Hint. Induction on the previous exercise.

5.11 Let Y be a subspace of a Banach space X, let $x_0 \notin Y$. Show that if Y is closed then $\operatorname{span}\big(Y \cup \{x_0\}\big)$ is closed in X.

This gives by induction that if Y is a closed subspace of X and F is a finite-dimensional subspace of X, then $\operatorname{span}(Y \cup F)$ is closed in X.
Hint. Let $f \in X^*$ be such that $f(z) = 0$ for all $y \in Y$ and $f(x_0) = 1$. Let $y_n + \lambda_n x_0 \to x \in X$, $y_n \in Y$. Then $\lambda_n = f(y_n + \lambda_n x_0) \to f(x)$. Therefore $y_n \to x - f(x)x_0$ and $x - f(x)x_0 \in Y$ since Y is closed. Then $x = x - f(x)x_0 + f(x)x_0 \in \operatorname{span}\big(Y \cup \{x_0\}\big)$.

5.12 Let X, Y, Z be Banach spaces, let T be a bounded operator from X into Y such that $T(X)$ is closed in Y, and let S be a *finite rank* operator from X into Z (that is, $\dim\big(S(X)\big)$ is finite). Define $U\colon X \to Y \oplus Z$ by $U(x) = \big(T(x), S(x)\big)$. Show that $U(X)$ is closed in $Y \oplus Z$.
Hint. If $y^* \in Y^*$, $z^* \in Z^*$ and $x \in X$, we have

$$\begin{aligned} U^*(y^*, z^*)(x) &= (y^*, z^*)\big(U(x)\big) = (y^*, z^*)\big(T(x), S(x)\big) \\ &= y^*\big(T(x)\big) + z^*\big(S(x)\big) - \big(T^*(y^*) + S^*(z^*)\big)(x), \end{aligned}$$

so $U^*(y^*, z^*) = T^*(y^*) + S^*(z^*)$. Since S^* is a finite rank operator and T^* has a closed range by Exercise 2.49, we get that U^* has a closed range $T^*(Y^*) + S^*(Z^*)$ by Exercise 5.11. By Exercise 2.49, we obtain that U has a closed range as well.

5.13 Show that the intersection of a finite-codimensional subspace and an infinite-dimensional subspace in a Banach space is an infinite-dimensional subspace.
Hint. The intersection of a subspace with a hyperplane has codimension at most one in the subspace. Then use induction.

5.14 Let X be a Banach space. Show that $X \oplus \mathbb{R}$ is isomorphic to X if and only if X is isomorphic to all its closed hyperplanes.
Hint. Assume that X is isomorphic to a closed hyperplane H. Then $H \oplus \mathbb{R}$ is isomorphic to $X \oplus \mathbb{R}$ and to X as well (Exercise 5.7).

If X is isomorphic to $X \oplus \mathbb{R}$ by an isomorphism $T: X \oplus \mathbb{R} \to X$, put $H = T(X \oplus \{0\})$ and observe that X is isomorphic to H, which is a closed hyperplane in X. Then use Exercise 2.9.

The question whether every Banach space is isomorphic to its closed hyperplanes was answered in the negative by Gowers ([Gowe3]).

5.15 Let $f \in X^* \setminus \{0\}$, where X is c_0, c, or the space ℓ_p for $1 < p < \infty$. Show that $f^{-1}(0)$ is isomorphic to the original space. Equivalently, all closed hyperplanes in c_0 or ℓ_p are isomorphic to the whole space.
Hint. The closed hyperplane $\{(0, x_1, x_2, \dots) : (x_i) \in X\}$ in X is isomorphic to X by the mapping $(x_1, x_2, \dots) \mapsto (0, x_1, x_2, \dots)$, then use Exercise 2.9.

5.16 Show that the spaces c_0 and c are isomorphic. Recall that they are not isometric (Exercise 3.132)
Hint. c_0 is a hyperplane in c given by the functional $f(x) = \lim x_n$. Alternatively, consider $T(x) = (\lim x_n, x_1 - \lim x_n, x_2 - \lim x_n, \dots)$.

5.17 Show that all closed hyperplanes of $C[0, 1]$ are isomorphic to $C[0, 1]$ and $C[0, 1] \oplus \mathbb{R}$ is isomorphic to $C[0, 1]$.

Consider the subspace $(C[0, 1])_0$ formed by all functions in $C[0, 1]$ that vanish at 0. Use it to show directly that $C[0, 1] \oplus \mathbb{R}$ is isomorphic to $C[0, 1]$.
Hint. By the Sobczyk and Banach–Mazur theorems, there is an isometric copy of c_0 complemented in $C[0, 1]$, therefore $C[0, 1] \sim c_0 \oplus Z$ for some Z. Show that the hyperplane $\{(0, x_1, x_2, \dots) : (x_i) \in c_0\} \oplus Z$ is isomorphic to $c_0 \oplus Z$. Then use Exercise 5.14.

For the second part: $(C[0, 1])_0 \oplus \mathbb{R}$ is isomorphic to $C[0, 1]$ by the mapping $(f, r) \mapsto f + r$ for $f \in (C[0, 1])_0$, $r \in \mathbb{R}$.

5.18 Show that $C[0, 1] \oplus C[0, 1]$ is isomorphic to $C[0, 1]$.
Hint. Let C_0 be a subspace of $C[0, 1]$ formed by functions equal to 0 at 0 and let C_1 be a subspace of $C[0, 1]$ formed by functions equal to 0 at 1. By the previous exercise, C_0, C_1 are isomorphic to $C[0, 1]$. Putting together graphs of functions, we can see that $C_1 \oplus C_0$ is isomorphic to a subspace of $C[0, 2]$ consisting of functions equal to 0 at 1. By scaling, this is in turn isometric to a subspace C_2 of $C[0, 1]$ consisting of functions that are 0 at $\frac{1}{2}$. This is a hyperplane in $C[0, 1]$, hence isomorphic to $C[0, 1]$ itself.

5.19 Show that $C[0, 1] \oplus c_0$ is isomorphic to $C[0, 1]$.
Hint. Use previous exercises and the Pełczyński decomposition method.

5.20 Show that c_0 is isometric to $c_0 \oplus c_0$ with max-norm.

Hint. Map $\big((x_1, x_2, \dots), (y_1, y_2, \dots)\big)$ to $(x_1, y_1, x_2, y_2, \dots)$.

5.21 Show that $c \oplus c$ is not isometric to c.
Hint. Assume that T is an isometry of $c \oplus c$ onto c. Consider $x, y \in (c \oplus c)$ defined by $x = \big((1, 1, \dots), (1, 1, \dots)\big)$ and $y = \big((1, 1, \dots), (-1, -1, \dots)\big)$. Then x, y are extreme points of $B_{c \oplus c}$. Both $h_1 = \frac{x+y}{2}$, $h_2 = \frac{x-y}{2}$ have norm one and for each of them there exists an infinite-dimensional closed subspace Y_i of $c \oplus c$ such that $\|h_i + y_i\| = 1$ for every $y_i \in Y_i \cap B_{c \oplus c}$. Isometry would have to carry these properties to c (in particular, it carries extreme points to extreme points), but such behavior is impossible in c. (For another approach to the same result see Exercise 3.141.)

5.22 Show that $C([0, 1] \cup [2, 3])$ is isomorphic to $C[0, 1]$. Note that $[0, 1] \cup [2, 3]$ is not homeomorphic to $[0, 1]$ (connectedness).
Hint. Similar to the previous exercises.

5.23 Show that $L_p[0, 1] \oplus \ell_p$ is isomorphic to L_p.
Hint. Similar to the previous exercises.

5.24 Show that $C^1[0, 1]$ is isomorphic to $C[0, 1]$.
Hint. Consider the mapping $f \mapsto (f', f(0))$ and use that $C[0, 1] \oplus \mathbb{R}$ is isomorphic to $C[0, 1]$.

5.25 (Klee, [Klee1]) Let Γ be an infinite set. Prove, by a suitable modification of the proof of Theorem 5.1, that $\ell_1(\Gamma)$ has an equivalent rotund norm $|||\cdot|||$ such that every Banach space of density character less or equal than card(Γ) is isometric to a quotient of $(\ell_1, |||\cdot|||)$.
Hint. For each $n \in \mathbb{N}$, let $g_n : [0, 1] \to [0, 1]$ be a continuous strictly convex function such that $nt/(n+1) \le g_n(t) \le t$, for all $t \in [0, 1]$. Put $\Gamma = \bigcup_{n=1}^{\infty} \Gamma_n$, where the sets Γ_n are pairwise disjoint and $\text{card}(\Gamma_n) = \text{card}(\Gamma)$ for all $n \in \mathbb{N}$. For $n \in \mathbb{N}$ and $\gamma \in \Gamma_n$, put $g_\gamma = g_n$. Let

$$B = \{x \in \ell_1(\Gamma) : |x_\gamma| \le 1, \sum_{\gamma \in \Gamma} g_\gamma(|x_\gamma|) < 1\}.$$

The set B is absolutely convex, and $B_{\ell_1(\Gamma)} \subset B \subset 2B_{\ell_1(\Gamma)}$, hence $\|\cdot\|_B$, the Minkowski functional of B, is an equivalent norm on $\ell_1(\Gamma)$. It is easy to show that $\|\cdot\|_B$ is rotund.

Now consider a Banach space Y of density character less than or equal to card(Γ). It is easy to define a mapping φ from Γ onto a dense subset of the open unit ball B_Y^O of Y and such that $((n+1)/n)\varphi(\Gamma_n) \subset B_Y^O$ for each $n \in \mathbb{N}$. For $x \in \ell_1(\Gamma)$ put $T(x) = \sum_{\gamma \in \Gamma} x_\gamma \varphi(\gamma)$. Then T is a linear mapping from $\ell_1(\Gamma)$ into Y. It is easy to see that $B_Y^O \subset T(B)$. Conversely, if $x \in B$,

$$\|Tx\| \le \sum_{n\in\mathbb{N}} \left(\sum_{\gamma\in\Gamma_n} |x_\gamma|.\|\varphi(\gamma)\| \right) \le \sum_{n\in\mathbb{N}} \sum_{\gamma\in\Gamma_n} \frac{n}{n+1}|x_\gamma| \le \sum_{\gamma\in\Gamma} g_\gamma(|x_\gamma|) < 1,$$

hence $T(B) \subset B_Y^O$, and the conclusion follows now as in the proof of Theorem 5.1.

5.26 Show that for any infinite compact space K, $C(K)$ contains a subspace isometric to c_0. If K is metrizable, then this subspace is complemented in $C(K)$.
Hint. Let $\{U_n\}$ be a sequence of nonempty disjoint open subsets of K. Such a sequence can be found by induction: Pick U_1 such that $K_1 := K\backslash\overline{U}_1$ is infinite, and take $U_2 \subset K_1$ such that $K_2 := K_1\backslash\overline{U}_2$ is infinite, and so on. Then pick a sequence $\{\varphi_n\}$ of continuous functions form K into $[0, 1]$ such that $\max_{s\in K} \varphi_n(s) = 1$ and $\{s \in K : \varphi_n(s) > 0\} \subset U_n$ for all n. Then, for any $(a_n) \in c_{00}$, we have $\|\sum_{n=1}^{\infty} a_n\varphi_n\|_\infty = \max_n |a_n|$. Then (φ_n) is a basic sequence in $C(K)$ isometrically equivalent to the unit vector basis in c_0. For the second statement, note that $C(K)$ is then separable; use now Sobczyk's Theorem 5.11.

5.27 Prove that a Banach space is injective if and only if it is isomorphic to a complemented subspace of some $\ell_\infty(\Gamma)$.
Hint. Every Banach space is isometric to a subspace of some $\ell_\infty(\Gamma)$ (see the comments after Proposition 5.4 and Exercise 5.30). If it is injective, then it must be complemented there. Conversely, use (iii) in Proposition 5.13.

5.28 Let c be the subspace of ℓ_∞ of all convergent sequences. Show that every bounded projection $P : c \to c_0$ satisfies $\|P\| \ge 2$. This proves that 2 is the best possible constant in Sobczyk's theorem. Compare with Exercise 5.7.
Hint. The space c_0 is the kernel of the bounded functional $f : C \to \mathbb{K}$ given by $f(x) = \lim x_n$, for $x = (x_n) \in c$. Every bounded projection $P : c \to c_0$ is written $P(x) = x - f(x)a$, for some $a \in c$ such that $f(a) = 1$. It is easy to find $x \in c$ such that $f(x) = 1$, $\|x\| = 1$ and $\|x - a\|$ $(= \|P(x)\|)$ is almost 2.

5.29 Recall that the distance $\mathrm{d}(x)$ of a point $x = (x_i) \in \ell_\infty$ to c_0 is equal to $\limsup |x_i|$. Define a mapping ϕ from ℓ_∞ onto c_0 by $\phi(x)_i = 0$ if $|x_i| \le \mathrm{d}(x)$ and $\phi(x)_i = \mathrm{sign}(x_i)(|x_i| - \mathrm{d}(x))$ if $|x_i| > \mathrm{d}(x)$. Show that ϕ is a Lipschitz retraction of ℓ_∞ onto c_0, that is, a Lipschitz mapping from ℓ_∞ onto c_0 that is the identity on c_0.

Note that ϕ cannot be linear as c_0 is not complemented in ℓ_∞.
Hint. Direct calculation.

5.30 Let X be a Banach space. Show that there is a set Γ such that X is isometric to a quotient of $\ell_1(\Gamma)$. Show that there is a set Γ such that X is isometric to a subspace of $\ell_\infty(\Gamma)$.
Hint. The proofs of Theorem 5.1 and Proposition 5.4.

5.31 We have $\overline{\text{conv}}(\text{Ext}(B_{\ell_\infty})) = B_{\ell_\infty}$ (see Exercise 3.129). Show that $\ell_\infty = \ell_1^*$ contains a w^*-compact convex subset C that is not equal to $\overline{\text{conv}}(\text{Ext}(C))$.
Hint. As any separable space, $C[0, 1]$ is a quotient of ℓ_1 and we thus can carry the unit ball of $C[0, 1]^*$ into ℓ_1^*. Then use Exercise 3.130.

5.32 Show that ℓ_2^n is isometric to a subspace of ℓ_∞ but not to any subspace of c_0.
Hint. Any separable Banach space is isometric to a subspace of ℓ_∞ (Proposition 5.4). ℓ_2^n has uncountably many extreme points, while finite-dimensional subspaces of c_0 have only finitely many of them. To see the latter, use the uniform convergence to zero of elements of compact sets in c_0.

5.33 Let X be a Banach space. Show that if X is separable then X^* is isometric to a subspace of ℓ_∞.
Hint. The proof of Theorem 5.1.

5.34 Recall that $\ell_1(c)$ is isometric to a subspace of $C[0, 1]^*$ (Exercise 3.143). Use it to show that $\ell_1(c)$ is isometric to a subspace of ℓ_∞. Show that this is not the case for $c_0(c)$.

Note that ℓ_∞^* contains an isometric copy of $\ell_2(c)$ ([Rose3]).
Hint. Use Exercise 5.33. For the second part, there is no countable separating family $\mathcal{F}$ in the dual of $c_0(c)$. Indeed, if such an $\mathcal{F}$ exists, every functional in $\mathcal{F}$ has a countable support, hence there is $r \in c$ which is not in $\bigcup_{f\in\mathcal{F}} \text{supp}(f)$. Clearly $f(e_r) = 0$ for every $f \in \mathcal{F}$.

5.35 Suppose a Banach space X contains an isomorphic copy of ℓ_1. Show that then X^* contains an isomorphic copy of $\ell_1(2^{\mathbb{N}})$.
Hint. We have ℓ_∞ as a quotient of X^*. The space ℓ_∞ has a subspace isometric to $\ell_1(c) = \ell_1(2^{\mathbb{N}})$. So X^* has a subspace Y such that $\ell_1(2^{\mathbb{N}})$ is a quotient of it. By the lifting property, $\ell_1(2^{\mathbb{N}})$ is a subspace of Y, hence of X^*.

5.36 Show that ℓ_∞/c_0 contains a subspace isometric to $c_0(c)$.
Hint. Consider a system of characteristic functions of subsets A_λ of $\mathbb{N}$ with the property $\text{card}(A_\lambda) = \infty$ and $\text{card}(A_{\lambda_1} \cap A_{\lambda_2}) < \infty$ whenever $\lambda_1 \neq \lambda_2$ (Lemma 5.7). Then the cosets containing χ_{A_λ} form the standard unit vector basis of $c_0(c)$ in the space ℓ_∞/c_0. Note that this implies that ℓ_∞/c_0 is not isomorphic to a subspace of ℓ_∞ (since ℓ_∞^* is w^*-separable and $c_0(c)$ is not).

5.37 Show that ℓ_1 is of first Baire category in c_0.
Hint. B_{ℓ_1} is closed in c_0: If $\{a^k\} \subset B_{\ell_1}$, $a^k = (a_i^k)$, and $a^k \to a$ in c_0, then $a^k \to a$ pointwise. Given n, we have $\sum_{i=1}^n |a_i^k| \leq 1$. Passing $k \to \infty$ we get $\sum_{i=1}^n |a_i| \leq 1$, so $a \in B_{\ell_1}$.

Then observe that B_{ℓ_1} does not contain any open ball of c_0.

5.38 Show that $L_2[0, 1]$ is a set of the first Baire category in $L_1[0, 1]$.

Hint. B_{L_2} is a weakly compact set in L_1, so it is closed. Show by constructing an appropriate function that B_{L_2} contains no interior point as a set in L_1.

5.39 (Lindenstrauss [Lind4]) Let X be a Banach space and assume that $\left(X \oplus \mathbb{R}\right)_\infty$ is isometric to a subspace of X. Show that X has a subspace isometric to c_0.
Hint. X contains a vector $x_1 \in S_X$ and a subspace Y_1 that is isometric to X such that $\|x_1 + y\| = \max(1, \|y\|)$. Continuing by induction we get for every $n \in \mathbb{N}$ a vector $x_n \in S_{Y_n}$ and a subspace Y_{n-1} of Y_n isometric to X such that $\|x_n + y\| = \max(1, \|y\|)$ for every $y \in Y_n$. It follows that for all real $\lambda_1, \ldots, \lambda_n$ we have $\left\|\sum_{i=1}^n \lambda_i x_i\right\| = \max |\lambda_i|$. Hence the closed subspace of X spanned by $\{x_i\}$ is isometric to c_0.

5.40 (Lindenstrauss [Lind8]) Show that there is no separable reflexive Banach space X_0 such that every separable reflexive Banach space is isometric to a subspace of X.

Szlenk proved that there is no separable reflexive Banach space X so that every separable reflexive Banach space is isomorphic to a subspace of X ([Szle]).
Hint. Use the previous exercise for X_0 and the fact that c_0 is not reflexive. Note that this method works for many other classes of Banach spaces.

5.41 Numbers C_1, C_2 in the definition of equivalence of basic sequences may serve as a measure of "closedness" of basic sequences.

Show that if basic sequences $\{e_i\}$, $\{f_i\}$ are equivalent with constants C_1, C_2, then the spaces $\overline{\text{span}}\{e_i\}$, $\overline{\text{span}}\{f_i\}$ are isomorphic and we can estimate $\mathrm{d}(\overline{\text{span}}\{e_i\}, \overline{\text{span}}\{f_i\}) \leq C_1 C_2$, where $\mathrm{d}(\cdot, \cdot)$ denotes the Banach–Mazur distance.
Hint. Standard.

5.42 (James [Jame3], see also [LiTz3, p. 97]) Let $(X, \|\cdot\|)$ be the space ℓ_1 or c_0 with its usual norm. Let $|||\cdot|||$ be an equivalent norm on X. Show that then, for every $\varepsilon > 0$ there is a subspace Y of X with $\mathrm{d}((Y, |||\cdot|||), (X, \|\cdot\|)) < 1+\varepsilon$, where $\mathrm{d}(\cdot, \cdot)$ is the Banach–Mazur distance.
Hint. Assume $X = \ell_1$ and that $\alpha|||x||| \leq \|x\| \leq |||x|||$ for some α and all $x \in X$. Let $\varepsilon > 0$ and let (P_n) be the sequence of natural projections in X. For every n put $\lambda_n = \sup\{\|x\| : |||x||| = 1, \ P_n x = 0\}$. Clearly, $\lambda_n \downarrow \lambda$ for some $1 \geq \lambda \geq \alpha$. Let n_0 be such that $\lambda_{n_0} < \lambda(1+\varepsilon)$. By the definition of λ_n, there is a block basis (y_k) of the unit vector basis of X so that for all k, $|||y_k||| = 1$, $P_{n_0} y_k = 0$ and $\|y_k\| > \lambda/(1+\varepsilon)$. For every choice of numbers (a_k), we have $P_{n_0}(\sum_{k=1}^\infty a_k y_k) = 0$ and thus

$$\left|\left|\left|\sum_{k=1}^\infty a_k y_k\right|\right|\right| \geq \lambda_{n_0}^{-1}\left\|\sum_{k=1}^\infty a_k y_k\right\|$$

$$= \lambda_{n_0}^{-1}\sum_{k=1}^\infty |a_k|.\|y_k\| \geq \lambda_{n_0}^{-1}\frac{\lambda}{1+\varepsilon}\sum_{k=1}^\infty |a_k| \geq \frac{1}{(1+\varepsilon)^2}\sum_{k=1}^\infty |a_k|.$$

On the other hand, by the triangle inequality,

$$\Big\| \sum_{k=1}^{\infty} a_k y_k \Big\| \leq \sum_{k=1}^{\infty} |a_k|.|||y_k||| \leq \sum_{k=1}^{\infty} |a_k|.$$

So,

$$\mathrm{d}\Big((\overline{\mathrm{span}}\{y_k : k \in \mathbb{N}\}, ||| \cdot |||), (\ell_1, \| \cdot \|) \Big) < (1+\varepsilon)^2,$$

(consider the mapping $T\big(\sum_i a_i e_i\big) := \sum_i a_i y_i$ from ℓ_1 onto $\overline{\mathrm{span}}\{y_k : k \in \mathbb{N}\}$).

The proof for c_0 is similar, we only replace the sup by the inf in the definition of λ_n.

5.43 Assume X and Y are two Banach spaces so that there is an isomorphism T of X onto Y such that $\|T\| \cdot \|T^{-1}\| = 1$. Show that X is isometric to Y.
Hint. Consider $\widetilde{T} := T/\|T\|$. Note that T itself need not be an isometry: $X = Y$ and $Tx = 2x$ for $x \in X$.

5.44 Find an example of two Banach spaces whose Banach–Mazur distance is 1 but these spaces are not isometric (compare with Exercise 5.43).
Hint. (see [PeBe]) For X take c_0 with the norm

$$\|x\|_1 = \Big(\max |x_i|^2 + \sum_{i=1}^{\infty} 2^{-i} |x_i|^2 \Big)^{\frac{1}{2}}$$

and for Y take c_0 with the norm

$$\|x\|_2 = \Big(\max |x_i|^2 + \sum_{i=2}^{\infty} 2^{-i+1} |x_i|^2 \Big)^{\frac{1}{2}}.$$

Define $T_n \in \mathcal{B}(X, Y)$ by $T_n(x) = (x_n, x_1, \ldots, x_{n-1}, x_{n+1}, \ldots)$. It is clearly onto, $\|T_n\| \to 1$ and $\|T_n^{-1}\| \to 1$. We find that X is strictly convex (see the proof of Theorem 8.2). However, Y is not strictly convex (consider $(-1, 1, 0, 0, \ldots)$ and $(1, 1, 0, 0, \ldots)$). Therefore X and Y cannot be isometric.

5.45 Let Γ be an abstract set. Show that $\ell_1(\Gamma)$ has the Schur property.
Hint. Let $x_n \xrightarrow{w} x$ in $\ell_1(\Gamma)$. Since $x_n \in \ell_1(\Gamma)$, we have that $\sum_{\gamma \in \Gamma} |x_n(\gamma)| < \infty$. Therefore $\mathrm{supp}(x_n)$ and $\mathrm{supp}(x)$ is countable. Let $\Gamma' = \bigcup \mathrm{supp}(x_n) \cup \mathrm{supp}(x)$. Then Γ' is countable, hence $\ell_1(\Gamma')$ is isometric to ℓ_1. Since $x_n, x \in \ell_1(\Gamma')$, Theorem 5.36 implies that $x_n \to x$.

5.46 Show that on S_{ℓ_1} the w^*- and the norm topology coincide.

Hint. By metrizability, it is enough to check sequences. Let $x, x_n \in S_{\ell_1}$ and $x_n \overset{w^*}{\to} x$. Given $\varepsilon > 0$, fix k_0 such that $\sum_{i=k_0+1}^{\infty} |x_i| < \varepsilon$. Using $x_n \overset{w^*}{\to} x$, find n_0 such that $\sum_{i=1}^{k_0} |x_i^n - x_i| < \varepsilon$ for every $n \geq n_0$. Since $\sum_{i=1}^{k_0} |x_i| + \sum_{k_0=1}^{\infty} |x_i| = 1$, we also have $\sum_{i=1}^{k_0} |x_i| > 1 - \varepsilon$. Moreover, for $n \geq n_0$, $\left|\sum_{i=1}^{k_0} |x_i^n| - \sum_{i=1}^{k_0} |x_i|\right| \leq \sum_{i=1}^{k_0} |x_i^n - x_i| < \varepsilon$ and hence $\sum_{i=1}^{k_0} |x_i^n| \geq \sum_{i=1}^{k_0} |x_i| - \varepsilon > 1-2\varepsilon$ and also because $\sum_{i=1}^{\infty} |x_i^n| = 1$ for every n, that $\sum_{i=k_0+1}^{\infty} |x_i^n| = 1-\sum_{i=1}^{k_0} |x_i^n| \leq 1-(1-2\varepsilon) = 2\varepsilon$.

Therefore, for $n \geq n_0$, $\sum_{i=k_0+1}^{\infty} |x_i^n - x_i| \leq \sum_{i=k_0+1}^{\infty}\big(|x_i^n| + |x_i|\big) \leq 3\varepsilon$ and thus $\sum_{i=1}^{\infty} |x_i^n - x_i| = \sum_{i=1}^{k_0} |x_i^n - x_i| + \sum_{i=k_0+1}^{\infty} |x_i^n - x_i| \leq 4\varepsilon$.

5.47 Let C be a weakly compact set in $\ell_1(\Gamma)$. Show that C is compact.
Hint. By the Eberlein–Šmulyan theorem, C is weakly sequentially compact. If $\{x_n\}$ is a sequence in C, there is a subsequence $\{x_{n_k}\}$ that is weakly convergent, hence—by the Schur property—norm convergent.

5.48 Let X be an infinite-dimensional closed subspace of ℓ_1. Show that X^* is non-separable.
Hint. $0 \in \overline{S_X}^{w}$. If X^* were separable, then (B_X, w) would be metrizable. Thus there would be $\{x_n\} \subset S_X$ such that $x_n \overset{w}{\to} 0$ in X, therefore $x_n \overset{w}{\to} 0$ in ℓ_1. By Schur's theorem, $x_n \to 0$, a contradiction.

5.49 Let $f, f_1, f_2, \cdots \in L_1[0, 1]$. Show that if $f_n \to f$ almost everywhere and $\|f_n\|_1 \to \|f\|_1$, then $f_n \to f$ in $L_1[0, 1]$ (Vitali).
Hint. $|f_n| + |f| - |f_n - f| \to 2|f|$ almost everywhere. Thus by Fatou's lemma,

$$\begin{aligned} 2\int_0^1 |f(t)|\,\mathrm{d}t &\leq \liminf \int_0^1 |f_n(t)| + |f(t)| - |(f_n - f)(t)|\,\mathrm{d}t \\ &= 2\int_0^1 |f(t)|\,\mathrm{d}t - \limsup \int_0^1 |(f_n - f)(t)|\,\mathrm{d}t. \end{aligned}$$

Hence $\limsup \int_0^1 |(f_n - f)(t)|\,\mathrm{d}t = 0$.

5.50 Assume that X is an infinite-dimensional Schur space. Then it contains an isomorphic copy of ℓ_1.
Hint. If not then from any bounded sequence can be extracted a subsequence that is weak- and thus norm-Cauchy, so X is finite-dimensional.

5.51 Let X be c_0 or ℓ_p for $p \in [1, \infty]\backslash\{2\}$. Let T be an isometry of X onto X. Show that there exists a sequence of signs $\varepsilon_i = \pm 1$ and a permutation π of natural numbers such that $T\big((x_i)\big) = (\varepsilon_i x_{\pi(i)})$.
Hint. ℓ_p case: Note that if vectors x, y satisfy $\|x + y\|^p = \|x - y\|^p = \|x\|^p + \|y\|^p$, then they have disjoint supports. This is easy to see for $p = 1$ and requires

elementary calculations for $p > 1$. Note also that this fails for $p = 2$, since then any two orthogonal vectors satisfy the condition.

Using this, we have that $T(e_i)$ and $T(e_j)$ have disjoint supports for $i \neq j$. Since T is onto, we readily obtain that supports of $T(e_i)$ must be singletons and $\operatorname{span}\{T(e_i)\} = \ell_p$.

If $X = c_0$, we consider T^{**}, which is an isometry of ℓ_∞ onto ℓ_∞, so we use the previous result and $T = T^{**}\big|_X$.

5.52 Let E and F be vector spaces and $B(E \times F)$ be the vector space of all bilinear forms on $E \times F$. Let the mapping $\chi : E \times F \to B(E \times F)^{\#}$ be defined by

$$\chi(e, f)(b) = b(e, f)$$

for all $b \in B(E \times F)$ and all $e \in E$, $f \in F$. Then χ is a bilinear mapping from $E \times F$ into $B(E \times F)^{\#}$.

The linear hull of $\chi(E \times F)$ in $B(E \times F)^{\#}$ is called the *tensor product of E and F* and is denoted by $E \otimes F$. The mapping χ is called the *canonical bilinear mapping from $E \times F$ into $E \otimes F$* and $\chi(e, f)$ is denoted by $e \otimes f$. The tensor product $E \otimes F$ is thus the vector space of finite sums $\sum \lambda_i (e_i \otimes f_i)$ for $e_i \in E$, $f_i \in F$, and λ_i real numbers.

Show that, for $e \in E$ and $f \in F$,

$$\lambda(e \otimes f) = (\lambda e) \otimes f = e \otimes (\lambda f)$$
$$(e_1 + e_2) \otimes f = e_1 \otimes f + e_2 \otimes f$$
$$e \otimes (f_1 + f_2) = e \otimes f_1 + e \otimes f_2.$$

Note that thus every element $u \in E \otimes F$ is of the form $u = \sum e_i \otimes f_i$. If we assume that both sets $\{e_i\}$ and $\{f_i\}$ are linearly independent, then show that the number of terms in the sum for u is uniquely determined. This number is called the *rank of u*. Show, too, that the $\otimes$-product of two linearly independent sets is linearly independent. This means that, in the finite-dimensional case, $\dim(E \otimes F) = \dim E \cdot \dim F$.

Show that the mapping $U \to U \circ \chi$ is a linear isomorphism from $(E \otimes F)^{\#}$ onto $B(E \times F)$.

Hint. Easy computation. We will study tensor products in Chapter 16.

5.53 Let X, Y be Banach spaces and let b be a bilinear form b on $X \times Y$. Show that the following are equivalent:

(i) b is continuous at the origin $(0, 0)$ of $X \times Y$.

(ii) b is uniformly continuous on bounded sets of $X \times Y$.

(iii) There is $K > 0$ such that $|b(x, y)| \leq K\|x\|\,\|y\|$ for all $x \in X$ and $y \in Y$.

(iv) b is separately continuous on $X \times Y$, i.e., $b(x, y_0)$ is continuous in x for every $y_0 \in Y$ and $b(x_0, y)$ is continuous in y for every $x_0 \in X$.

Hint. (i)$\Longrightarrow$(iii): Assume that there is $\delta > 0$ such that $|b(x, y)| \leq 1$ whenever $x \in \delta B_X$ and $y \in \delta B_Y$. By a homogeneity argument, we get that $|b(x, y)| \leq \delta^{-2}\|x\|\,\|y\|$ for every $x \in X$ and $y \in Y$.

(iii)$\Longrightarrow$(ii): If $x, x' \in B_X$ and $y, y' \in B_Y$ are such that $\|x - x'\| < \delta$ and $\|y - y'\| < \delta$, then

$$\begin{aligned}|b(x, y) - b(x', y')| &\leq |b(x - x', y - y') + b(x', y - y') + b(x - x', y')| \\ &\leq K\|x - x'\|\,\|y - y'\| + K\|x'\|\,\|y - y'\| + K\|y'\|\,\|x - x'\| \leq K\delta^2 + 2K\delta.\end{aligned}$$

(ii)$\Longrightarrow$(i) and (ii)$\Longrightarrow$(iv) are trivial.

(iv)$\Longrightarrow$(iii): Consider the family $\mathcal{F} = \{b(x_0, \cdot) : x_0 \in B_X\}$ of continuous linear functionals on Y. For every $y_0 \in Y$ we have $\sup_{x \in B_X} |b(x, y_0)| < \infty$ as this functional is continuous. So $\{b(x_0, y) : x_0 \in B_X\}$ is bounded for every $y \in Y$ and the uniform boundedness principle, $\sup\{|b(x, y)| : x \in B_X, y \in B_Y\} < \infty$, which gives (iii).

5.54 For Banach spaces X and Y, we say that a bilinear form b on $X \times Y$ is *bounded* if it satisfies (iii) (and so all the other equivalent conditions) in Exercise 5.53. In this case, its *norm* $\|b\|$ is defined by $\|b\| = \sup\{|b(x, y)| : x \in B_X,\ y \in B_Y\}$. Show that the normed space of all bounded bilinear forms on $X \times Y$ with this norm is a Banach space. We will denote it by $(\mathcal{B}il(X \times Y), \|\cdot\|)$.

Show that $(\mathcal{B}il(X \times Y), \|\cdot\|)$ is isometric to the Banach space $\mathcal{B}(X, Y^*)$ via the mapping $b \to T$ defined by

$$\big(T(x), \cdot\big) = b(x, \cdot)$$

Note that from these facts it follows that $\mathcal{B}il(\ell_2 \times \ell_2)$ is nonseparable.

Hint. Note that the space of bounded operators on a Hilbert space is nonseparable (Proposition 1.44). The rest is standard.

5.55 Find a basis for the linear space $\mathcal{B}il(\ell_2^n \times \ell_2^m)$.

Hint. If $\{e_i\}_{i=1}^n$ is a basis for ℓ_2^n and $\{f_j\}_{j=1}^m$ is a basis for ℓ_2^m, let b_{i_0, j_0} be a bilinear form defined by $b_{i_0, j_0}(\sum \alpha_i e_i, \sum \beta_j f_j) = \alpha_{i_0}\beta_{j_0}$ for $1 \leq i_0 \leq n,\ 1 \leq j_0 \leq m$. Then if $b \in \mathcal{B}il(\ell_2^n \times \ell_2^m)$, and $x = \sum_i \alpha_i e_i$ and $y = \sum_j \beta_j f_j$, then $b(x, y) = b\big(\sum \alpha_i e_i, \sum \beta_j f_j\big) = \sum_{i,j} \alpha_i \beta_j b(e_i, f_j) = \sum_{i,j} b_{i,j}(x, y) b(e_i, f_j)$. Moreover, $\{b_{i,j} : i = 1, 2, \ldots, n,\ j = 1, 2, \ldots, m\}$ is a linearly independent set, as for given (i_0, j_0), $b_{i_0, j_0}(i_0, j_0) = 1$ and $b_{i,j}(i_0, j_0) = 0$ if $(i, j) \neq (i_0, j_0)$.

5.56 Let $p \in [1, \infty)$, $x \in \ell_p$. Let $a > 0$ and $h_j \in \ell_p$, $j \in \mathbb{N}$, be such that $\|h_j\| = a$ and $h_j \xrightarrow{w} 0$ in ℓ_p. Show that then $\|x + h_j\|_p^p \to \|x\|_p^p + a^p$.

Hint. For $z = (z_i) \in \ell_p$ and $n \in \mathbb{N}$, denote $\tilde{z}^n = (z_1, \ldots, z_n, 0, \ldots)$ and $\tilde{z}_n = (0, \ldots, 0, z_{n+1}, z_{n+2}, \ldots)$.

Given $\varepsilon > 0$, from the uniform continuity of $\|\cdot\|_p^p$ on bounded sets there is n such that $\big|\|x + h_j\|_p^p - \|\tilde{x}^n + h_j\|_p^p\big| < \varepsilon$ for every j. Since $h_j \xrightarrow{w} 0$, there is

j_0 such that for every $j \geq j_0$ we have $\left|\|\tilde{x}^n + h_j\|_p^p - \|\tilde{x}^n + \widetilde{(h_j)}_n\|_p^p\right| < \varepsilon$ and $\left|\|\widetilde{(h_j)}_n\|_p^p - \|h_j\|_p^p\right| < \varepsilon$. Finally note that $\|\tilde{x}^n + \widetilde{(h_j)}_n\|_p^p = \|\tilde{x}^n\|_p^p + \|\widetilde{(h_j)}_n\|_p^p$ for every j.

5.57 A function Q on a Banach space X is called a continuous *quadratic form* if there is a continuous bilinear form b on $X \times X$ such that $Q(x) = b(x, x)$ for $x \in X$ and b is symmetric, that is, $b(x, y) = b(y, x)$ for every $x, y \in X$.

Let $p \in (2, \infty)$, let Q be a continuous quadratic form on ℓ_p. Show that Q is w-sequentially continuous, i.e., $Q(x_j) \to Q(x)$ whenever $x_j \overset{w}{\to} x$ in ℓ_p.
Hint. Fixing $x \in \ell_p$, write $Q(x + h) = Q(x) + Q(h) + A(x, h)$ for $h \in X$. Since $A(x, h)$ is a continuous linear functional and thus weakly continuous, we have that $A(x, h_j) \to 0$ for every x. We need to prove that $Q(h_j) \to 0$ whenever $h_j \overset{w}{\to} 0$. Assume that this is not true and for some $\varepsilon > 0$ and $\|h_j\| = 1$, $j \in \mathbb{N}$, we have $h_j \overset{w}{\to} 0$ and $Q(h_j) \geq \varepsilon$ for all j. Put $n_1 = 1$ and $x_{n_1} = x_1 = h_1$. Find n_2 such that $|A(x_{n_1}, h_{n_2})| < \frac{\varepsilon}{2}$ and $\|x_{n_1} + h_{n_2}\|_p^p \leq \|x_{n_1}\|_p^p + 2$ (use Exercise 5.56). Put $x_2 = x_{n_1} + h_{n_2}$, find n_3 such that $|A(x_2, h_{n_3})| < \frac{\varepsilon}{2}$ and $\|x_2 + h_{n_3}\|_p^p \leq \|x_2\|_p^p + 2$, etc.

Then we have $\|x_j\|_p^p \leq 2j$ for every j and

$$Q(x_j) = Q(x_{j-1}) + A(x_{j-1}, h_{n_j}) + Q(h_{n_j}) \geq Q(x_{j-1}) + \varepsilon - \tfrac{\varepsilon}{2} \geq Q(x_{j-1}) + \tfrac{\varepsilon}{2}.$$

Since $Q(x_1) = Q(h_1) > \varepsilon$, we get by induction that $Q(x_j) > j\frac{\varepsilon}{2}$ for every j. Then $\frac{Q(x_j)}{\|x_j\|_p^2} \geq j\frac{\varepsilon}{2}(2j)^{-2/p} = \left(\frac{\varepsilon}{2^{1+2/p}}\right)j^{1-2/p} \to \infty$, which contradicts the continuity of Q using Exercise 5.53.

Similarly, all polynomials of degree smaller than p on ℓ_p and all polynomials on c_0 are weakly sequentially continuous ([BoFr] and Exercise 10.5).

5.58 Show that the property to be isomorphic to a subspace of c_0 is a three-space property, i.e., if a subspace Y of X and X/Y both have it, then X has it.
Hint. Let $Y \subset X$ and let both Y and X/Y be isomorphic to subspaces of c_0. By Theorem 5.20, the isomorphism of Y into c_0 extends to a bounded operator T from X into c_0. If $j : X/Y \to c_0$ denotes the isomorphism into c_0 and $q : X \to X/Y$ is the quotient mapping, then $T \oplus (j \circ q) : X \to c_0 \oplus c_0$ is an isomorphism into c_0.

5.59 Let T be a bounded operator from ℓ_1 onto c_0. Show that $T^{-1}(0)$ is not complemented in ℓ_1.
Hint. If $\ell_1 = T^{-1}(0) \oplus Z$, then, by the open mapping theorem, Z would be isomorphic to c_0, a contradiction with Theorem 4.46.

5.60 Show that the Dirac measures $\delta_{1/n}$ do not converge to δ_0 in the weak topology of $C[0, 1]^*$. Note that this gives another proof of the fact that $C[0, 1]$ does not have the Grothendieck property.
Hint. They are in $\ell_1[0, 1] \subset C[0, 1]^*$; use then the Schur property of ℓ_1 spaces.

5.61 Show that the sequence $(\chi_{[n,\infty)})$ of characteristic functions in $\mathbb{N}$ converge to 0 in the w^*-topology of ℓ_∞ $(= \ell_1^*)$, but not in the w-topology of ℓ_∞.
Hint. $0 \notin \overline{\text{conv}}^{\|\cdot\|}\{\chi_{[n,\infty)} : n \in \mathbb{N}\}$.

5.62 Show that if X is an infinite-dimensional Banach space and Y is a finite-dimensional Banach space, then there is a bounded operator from X onto Y.
Hint. If $\dim Y = n$, choose an n-dimensional subspace E of X. Then E is isomorphic to Y by an isomorphism T. Moreover, E is complemented in X by a projection P. Then $T \circ P$ maps X onto Y.

5.63 If X is an infinite-dimensional Banach space, show that there is a closed infinite-dimensional subspace Y of X such that X/Y is infinite-dimensional.
Hint. If X is separable, take a Markushevich basis $\{x_n; x_n^*\}$ for X and put $Y = \overline{\text{span}}\{x_{2n} : n \in \mathbb{N}\}$.

The following few exercises are taken from [Lind2].

5.64 Assume that (X_n) is a strictly increasing sequence of finite-dimensional subspaces of ℓ_∞ such that there is $\lambda > 0$ so that, for every $n \in \mathbb{N}$, X_n is complemented in ℓ_∞ by a projection of norm less or equal than λ. Show that $\overline{\bigcup_n X_n}$ is not reflexive..
Hint. Assume that it is reflexive. If (P_n) is the sequence of the associated uniformly bounded projections, consider its limit in the weak operator topology, i.e., $T_n x \to Tx$ in the weak topology of the target space, for every x in the domain space (see the definition right before Exercise 16.17). Then, for $x \in \bigcup_n X_n$, $P_n x = x$ for sufficiently large n, and thus $Px = x$ for $x \in \overline{\bigcup_n X_n}$. Thus P is a projection onto $\overline{\bigcup_n X_n}$. This is impossible by Proposition 13.45.

5.65 Show that the space $\left(\sum \ell_2^n\right)_0$ is not isomorphic to c_0.
Hint. If it were, put $X_n = \ell_2^n \subset \ell_2 \subset \ell_\infty$ and note that then $\left(\sum \ell_2^n\right)_\infty$ is isomorphic to ℓ_∞ and so ℓ_2 would be isomorphic to a complemented subspace of ℓ_∞, which is impossible as ℓ_∞ has the Dunford–Pettis property.

5.66 If X_n is a finite-dimensional Banach space for each $n \in \mathbb{N}$, show that the canonical norm of $(\sum_n X_n)_0$ is a Lipschitz Kadec–Klee smooth norm.
Hint. For the notion of Lipschitz Kadec–Klee smooth norm, see Definition 12.60. Follow the proof of Theorem 12.64.

5.67 Show that c_0 and $(\sum_n \ell_2^n)_0$ have the same linear dimension, i.e., both these spaces have, up to a isomorphism, the same family of closed subspaces.
Hint. $(\sum_n \ell_2^n)_0$ is isomorphic to a subspace of c_0 by Exercise 5.66 and c_0 is clearly isomorphic to a subspace of $(\sum_n \ell_2^n)_0$ (take from each ℓ_2^n a one-dimensional subspace).

5.68 Show that $(\sum_n \ell_2^n)_1$ is isomorphic to a subspace of ℓ_1.

Hint. The space ℓ_2^n is 2-isomorphic to an n-dimensional subspace M_n of ℓ_1 by Dvoretzky's Theorem 6.15. Then $(\sum_n \ell_2^n)_1$ is isomorphic to $(\sum_n M_n)_1$, which is isomorphic to a subspace of ℓ_1.

5.69 Show that $(\sum_n \ell_2^n)_1$ is not isomorphic to ℓ_1.
Hint. Otherwise, its dual space $(\sum_n \ell_2^n)_\infty$ would be isomorphic to ℓ_∞, which is impossible by Exercise 5.64.

5.70 Show that $(\sum \ell_2^n)_1$ and ℓ_1 are two non-isomorphic spaces that have, up to isomorphisms, the same family of subspaces and quotient spaces.
Hint. As any separable space, $(\sum_n \ell_2^n)_1$ is a quotient of ℓ_1. If we take in each ℓ_2^n a one-dimensional subspace X_n, we get $(\sum_n X_n)$, a complemented subspace of $(\sum_n \ell_2^n)_1$, that is thus a quotient of $(\sum_n \ell_2^n)_1$ and is isomorphic to ℓ_1.

5.71 (Lindenstrauss [Lind5b]) Let T be an operator from ℓ_1 onto $L_1[0,1]$ and $X := T^{-1}(0) \hookrightarrow \ell_1$. Show that X is not isomorphic to a dual space—and thus X does not have an unconditional basis, although ℓ_1 does have an unconditional basis. Compare with Exercise 4.55.
Hint. The rough idea behind the proof is the following. Under the negation of the conclusion, we shall define an isomorphism from $L_1[0,1]$ into ℓ_1. Since $L_1[0,1]$ contains a subspace isomorphic to ℓ_2 (see Theorem 4.53), the space ℓ_2 will be isomorphic to a subspace of ℓ_1, hence it will contain a subspace isomorphic to ℓ_1 (see Theorem 4.46), something impossible due to the reflexivity of ℓ_2.

In order to construct the isomorphism into under the negation of the conclusion, define a family of finite-dimensional subspaces of $L_1[0,1]$ and, for each one of them, a "linear local inverse" of T. Define, too, a "global bounded inverse" (not even linear or continuous) of T. Then, the difference of this two mappings has range in X. Were X isomorphic to a dual space, it would carry a locally convex topology $\mathcal{T}$ making the (original) ball of X $\mathcal{T}$-relatively compact. A pointwise $\mathcal{T}$-limit of those differences will exist, and the sought linear mapping will be obtained by adding the "non-linear part" that was previously subtracted.

To be precise, let $\mathcal{A}$ be the set of all partitions of $[0,1]$ into a finite number of disjoint measurable sets, and for every $\alpha \in \mathcal{A}$, let Y_α be the subspace of $L_1[0,1]$ spanned by the characteristic functions of the sets in the decomposition. Then each Y_α is isometric to an $\ell_1^{n_\alpha}$ for some positive integer n_α—and we may identify this two spaces. For each $\alpha \in \mathcal{A}$, define an operator S_α from $\ell_1^{n_\alpha}$ into ℓ_1 by $S_\alpha(\sum_{i=1}^{n_\alpha} \beta_i e_i) = \sum_{i=1}^{n_\alpha} \beta_i x_i$, for $\beta_1, \ldots, \beta_{n_\alpha} \in \mathbb{R}$, where $e_1, \ldots, e_{n_\alpha}$ are the standard unit vectors in $\ell_1^{n_\alpha}$ and $x_i \in \ell_1$ satisfy $\|x_i\| \le 2$ and $Tx_i = e_i$, $i = 1, 2, \ldots, n_\alpha$. Then $\|S_\alpha\| \le 2$ and $TS_\alpha = I_{Y_\alpha}$, the identity on Y_α, for all $\alpha \in \mathcal{A}$. Let φ be a mapping (not necessarily linear or continuous) from $L_1[0,1]$ to ℓ_1 such that $T\varphi = I\big|_{L_1[0,1]}$ and, for all $y \in L_1[0,1]$, $\|\varphi(y)\| \le \eta\|y\|$ for some $\eta < \infty$ (use the Banach Open Mapping theorem). Then for each $\alpha \in \mathcal{A}$ and each $y \in Y_\alpha$, we have $S_\alpha(y) - \varphi(y) \in X$.

Put, for $a \in A$ and $y \in \bigcup_{\alpha \in \mathcal{A}} Y_\alpha$,

$$\pi_a(y) = \begin{cases} S_\alpha(y) - \varphi(y) & \text{if } y \in Y_\alpha, \\ 0 & \text{otherwise.} \end{cases}$$

This defines an element $\pi_\alpha \in \prod_{y \in \bigcup Y_\alpha} (2+\eta)\|y\| B_X$. If $\mathcal{A}$ is partially ordered by $\preceq$, where $\alpha \preceq \beta$ means that every element in α is a union of elements in β, we obtain a net $\{\pi_\alpha;\ \alpha \in \mathcal{A},\ \preceq\}$. Assume that X is isomorphic to a dual Banach space. Then a locally convex topology $\mathcal{T}$ exists on X such that B_X is relatively compact in $(X, \mathcal{T})$. The set $\prod_{y \in \bigcup Y_\alpha} (2+\eta)\|y\| B_X$ is then relatively compact in $\prod_{\alpha \in \mathcal{A}} (X, \mathcal{T})$ by Tychonov's theorem, hence the net $\{\pi_\alpha\}$ has a cluster point π in $\prod_{\alpha \in \mathcal{A}} (X, \mathcal{T})$. Fix $y \in \bigcup_{\alpha \in \mathcal{A}} Y_\alpha$. Then $y \in Y_{\alpha_0}$ for some $\alpha_0 \in \mathcal{A}$ and, for every $\alpha \succeq \alpha_0$ we have $S_\alpha = \pi_\alpha + \varphi$ (a linear mapping). It follows that $S := \pi + \varphi$ is a linear mapping and, obviously, $\|S(y)\| \leq (2+\eta)K\|y\|$ for some $K > 0$ independent of y. The mapping S can be extended to a continuous operator $\widetilde{S}$ from $(L_1[0,1] =) \overline{\bigcup_{\alpha \in \mathcal{A}} Y_\alpha}$ into ℓ_1.

Moreover, $TS(y) = T\pi(y) + T\varphi(y) = T\varphi(y) = y$ for all $y \in \bigcup_{\alpha \in \mathcal{A}} Y_\alpha$, since $\pi(y) \in X$. It follows that the mapping $\widetilde{S}$ is an isomorphism from $L_1[0,1]$ into ℓ_1. This finishes the proof.

Had the space X an unconditional basis, it will certainly be not boundedly complete, since X is not isomorphic to a dual space, see Theorem 4.15. Then, the space X would have an isomorphic copy of c_0, see Theorem 4.37, and this is impossible.

5.72 Let X be a subspace of a Banach space Y, Z a finite-dimensional Banach space and T a bounded operator from X into Z. Show that T can be extended to a bounded operator from Y into Z.

Hint. Let $(e_i, e_i^*)_{i=1}^n$ be a Hamel basis for Z and define elements $f_i \in X^*$ by $f_i = T^* e_i^*$, $i = 1, 2, \ldots, n$. Let $\widetilde{f_i}$ be a Hahn–Banach extension of f_i to Y, $i = 1, 2, \ldots, n$. Define an operator $\widetilde{T} : Y \to Z$ by $\widetilde{T}y = \sum_{i=1}^n \widetilde{f_i}(y) e_i$, for $y \in Y$. Then for $x \in X$ we have

$$\widetilde{T}x = \sum_{i=1}^n f_i(x) e_i = \sum_{i=1}^n (T^* e_i^*)(x) e_i = \sum_{i=1}^n e_i^*(Tx) e_i = Tx.$$

5.73 Does there exist an absolute constant $K > 0$ so that $\widetilde{T}$ can be taken in Exercise 5.72 so that $\|\widetilde{T}\| \leq K\|T\|$?

Hint. No. There is an infinite-dimensional Banach space X and a number $\delta > 0$ such that any finite rank projection P from X into X satisfies $\|P\| \geq \delta (\text{rank } P)^{1/2}$ (Pisier, see Remarks in Chapter 6). So if X_n are n-dimensional subspaces of this X, then the identity mappings on them cannot be extended to mappings from X into X_n with norms less than or equal to K.

5.74 Let X be isomorphic to a Hilbert space, $Y \subset X$ be a subspace of X and T a bounded operator from Y into a Banach space Z. Show that there is an operator $\widetilde{T}$

from X into Z that extends T. Note that the Hahn–Banach theorem follows from this in the case of Hilbert spaces.
Hint. Let P be a projection from X into Y. Consider the operator $T \circ P$.

5.75 Let X and Y be infinite-dimensional separable Banach spaces. Show that there is a one-to-one compact operator T from X into Y such that $T(X)$ is linearly dense in Y.
Hint. Let $\{x_i; f_i\}_{i=1}^{\infty}$ and $\{y_i; g_i\}_{i=1}^{\infty}$ be Markushevich bases for X and Y, respectively. Assume that $\{f_i : i \in \mathbb{N}\}$ and $\{y_i : i \in \mathbb{N}\}$ are bounded. Put for $x \in X$, $Tx = \sum_i (1/2^i) f_i(x) y_i$. Then T is compact and one-to-one (if $x \neq 0$ there is i such that $f_i(x) \neq 0$ and then

$$g_i(Tx) = \frac{1}{2^i} f_i(x) g_i(y_i) = \frac{1}{2^i} f_i(x) \neq 0$$

by the biorthogonality of $\{y_i\}$ and $\{g_i\}$). Moreover,

$$Tx_i = \sum \frac{1}{2^j} f_j(x_i) y_j = \frac{1}{2^i} y_i$$

by the biorthogonality of $\{x_i\}$ and $\{f_i\}$.

5.76 Let X and Y be infinite-dimensional separable Banach spaces. Then there is a compact one-to-one operator T from X^* into Y such that $T(X^*)$ is linearly dense in Y.
Hint. Follow the hint in Exercise 5.75.

5.77 Assume that Y is a subspace of X and W is a subspace of Z. Assume that Y is isomorphic to W and X/Y is isomorphic to Z/W. Is X necessarily isomorphic to Z?
Hint. No. There is a space Z that is not isomorphic to a Hilbert space and a subspace W such that both W and Z/W are isomorphic to a Hilbert space (Enflo, Lindenstrauss, and Pisier [ELP]). Alternatively one can use the Ciesielski–Pol space in [DGZ3, p. 260].

5.78 Find a non-separable space X with a non-complemented subspace isomorphic to c_0.
Hint. Use Theorem 14.54.

5.79 Find an example of an element of ℓ_∞ that does not attain its norm.
Hint. $x_n = (1 - 1/n)$.

5.80 Show that $(\sum_n \ell_2^n)_0$ is isomorphic to a quotient of c_0.
Hint. Use the following dual version of the Dvoretzky Theorem 6.15: Let k be an integer. Then every infinite-dimensional Banach space has a k-dimensional quotient

space M_k with the quotient norm $\|\cdot\|$ and an Euclidean norm $|||\cdot|||$ in it such that for every $x \in M_k$ we have $\|x\| \leq |||x||| \leq 2\|x\|$. Then $(\sum_n \ell_2^n)_0$ is 2-isomorphic to $(\sum_n M_n)_0$ which is a quotient of c_0 $(= (\sum_n c_0)_0)$. Note that c_0 is isomorphic to a complemented subspace of $(\sum_n \ell_2^n)_0$ (consider the first coordinates in ℓ_2^n). Therefore the spaces $(\sum_n \ell_2^n)_0$ and c_0 have, up to isomorphisms, the same quotient spaces.

5.81 Let T be a completely regular topological space and $BC(T)$ be the space of bounded continuous real-valued functions on T with the supremum norm. Then $BC(T)$ is a Banach space. Show that the mapping $\varphi : t \to \delta_t$, where δ_t is the Dirac measure at t, is a homeomorphic mapping from T into $(B_{BC(T)^*}, w^*)$ and that the Čech–Stone compactification of T is just $\overline{\varphi(T)}^{w^*} \subset B_{BC(T)^*}$.
Hint. Definition of complete regularity and Alaoglu's theorem.

5.82 Show that $(\sum_n \ell_2^n)_p$ is isomorphic to ℓ_p if $1 < p < \infty$. Note that this gives that the image of the standard basis in $(\sum \ell_2^n)_p$ by this isomorphism provides an unconditional basis in ℓ_p that is not equivalent to the unit vector basis in ℓ_p if $p \neq 2$.
Hint. Let $p > 2$. Let F_n be the subspace of $L_p[0, 1]$ spanned by the characteristic functions of the intervals $[k/2^n, (k+1)/2^n]$, $k = 0, 1, \ldots, 2^{n-1}$. Then F_n is isometric to $\ell_p^{2^n}$ and $PF_n = \overline{\operatorname{span}}\{r_i : i = 1, \ldots, n\}$, where P is the projection $Pf := \sum_{n=1}^{\infty}(\int_0^1 f(s)r_n(s)\, ds)r_n$, and r_n are the Rademacher functions (see the proof of Khintchine's inequality, Lemma 4.54). There is a constant K_p so that for every n, there is a subspace C_n of $\ell_p^{2^n}$ with $\operatorname{dist}(C_n, \ell_2^n) \leq K$ and the projection from $\ell_p^{2^n}$ onto C_n is of norm $\leq K_p$. Thus the space $(\sum_n \ell_2^n)_p$ is isomorphic to a complemented subspace of $(\sum_n \ell_p^{2^n})_p \approx \ell_p$. Therefore $(\sum_n \ell_2^n)_p$ is isomorphic to ℓ_p by Corollary 4.48. By duality the result holds for $1 < p < 2$.

5.83 Assume that T is a noncompact operator from X into ℓ_1. Show that X contains an isomorphic copy of ℓ_1. If T is a nonweakly compact operator from X into L_1 then X contains an isomorphic copy of ℓ_1.
Hint. Since $T(B_X)$ is not $\|\cdot\|$-relatively compact, there exists a sequence $\{x_n\}$ in B_X and some $\varepsilon > 0$ such that $\|T(x_n) - T(x_m)\| \geq \varepsilon$ for all $n \neq m$. Schur's Theorem 5.36 and Rosenthal's Theorem 5.37 give the conclusion. For the second part, if $T(B_X)$ is not w-relatively compact, there exists, by the Eberlein–Šmulyan theorem, a sequence $\{x_n\}$ in B_X such that $\{T(x_n)\}$ has no w-convergent subsequence. The sequence $\{x_n\}$ cannot be w-Cauchy due to the weak sequential completeness of L_1. Use again Rosenthal's Theorem 5.37.

5.84 Let X be a Banach space. Let $\{x_n\}$ be a bounded sequence in X such that $\{x_n\}$ has no weakly Cauchy subsequence. It follows from Rosenthal's Theorem 5.37 that X contains an isomorphic copy of ℓ_1. Prove this result for a Banach space having an unconditional basis without using Rosenthal's theorem.

Hint. X^* must be non-separable, otherwise we use the Cantor diagonal procedure to produce a weak Cauchy subsequence. If X^* is non-separable, use Theorem 4.42.

5.85 An example of a weakly sequentially complete space is any reflexive space (show this) and any general $L_1(\mu)$ space (this is a classical Steinhaus theorem, see the proof in Exercises 13.48 and 13.49). Show that if X is weakly sequentially complete and nonreflexive, then X must contain an isomorphic copy of ℓ_1.
Hint. The first problem: use the Eberlein–Šmulyan theorem. The second problem: If X is not reflexive, by the Eberlein–Šmulyan theorem there exists a bounded sequence $\{x_n\}$ without w-convergent subsequence. Since X is weakly sequentially complete, $\{x_n\}$ has no w-Cauchy subsequence. Use now Rosenthal's Theorem 5.37.

5.86 A set D in a Banach space X is called *limited in* X if $f_n(x) \to 0$ uniformly on $x \in D$ whenever $f_n \in X^*$ and $f_n \overset{w^*}{\to} 0$. The space X is called a *Gelfand–Phillips* space if every limited set in X is $\|\cdot\|$-relatively compact. (Note that any $\|\cdot\|$-relatively compact subset of X is limited in X by the Banach–Steinhaus uniform boundedness principle). Show that

(i) Every limited set in X is bounded.

(ii) Every separable Banach space is a Gelfand–Phillips space.

(iii) The space ℓ_∞ is not a Gelfand–Phillips space.

Hint. (i) If not, there are $d_i \in D$ with $\|d_i\| > i$ for all $i \in \mathbb{N}$. Pick $f_i \in S_{X^*}$ with $f_i(d_i) = \|d_i\|$. Then $(1/i)f_i \overset{w^*}{\to} 0$ and $(1/i)f_i(d_i) > 1$ for all $i \in \mathbb{N}$.

(ii) If D is not $\|\cdot\|$-relatively compact, then there are $x_n \in D$, $n \in \mathbb{N}$, so that $\liminf_n \operatorname{dist}(x_n, E_{n-1}) > 0$, where $E_{n-1} := \operatorname{span}\{x_1, x_2, \ldots, x_{n-1}\}$. This follows from the compactness of balls in finite-dimensional spaces. Choose $\varepsilon > 0$ and $f_n \in S_{X^*}$ so that $f_n(x_n) > \varepsilon$ and $f_n(x_k) = 0$ for $k < n$. Since B_{X^*} is w^*-sequentially compact, let $\lim\limits_k f_{n_k} = f$ in the w^*-topology of X^*, where n_k is a subsequence of the integers. Then $f_{n_k}(x_{n_k}) > \varepsilon$ and $f(x_{n_k}) = \lim\limits_j f_{n_j}(x_{n_k}) = 0$ for all k.

(iii) Put $D = B_{c_0} \subset \ell_\infty$. If $f_n \in \ell_\infty^*$ are such that $f_n \overset{w^*}{\to} 0$, then, by the Grothendieck property of ℓ_∞, $f_n \overset{w}{\to} 0$, and the restrictions $\tilde{f}_n$ of f_n to c_0 converge weakly to 0 in ℓ_1. By the Schur property of ℓ_1 ($= c_0^*$), $\tilde{f}_n \to 0$ in the norm topology of c_0^* and thus $f_n(x) \to 0$ uniformly on $x \in D$ ($= B_{c_0}$).

5.87 Let X be an infinite-dimensional Banach space. Show that B_X is not limited in X.
Hint. This is a reformulation of the Josefson–Nissenzweig theorem (see Exercise 3.39).

5.88 Assume that K is a compact space and X is a separable subspace of $C(K)$. Show that there is a separable subspace $\tilde{X}$ of $C(K)$ such that $\tilde{X} \supset X$ and $\tilde{X}$ is isometric to a $C(S)$ space for some compact metric space S.

Hint. Let $\widetilde{X}$ be a closed separable subalgebra in $C(K)$ that contains X and the constant functions on K. Let S be formed by all nonzero continuous linear functionals x^* in the closed unit ball of $\widetilde{X}^*$ for which $x^*(fg) = (x^*(f))(x^*(g))$ for all $f, g \in \widetilde{X}$. Then it is shoed in Dunford–Schwartz, p. 275 that it follows from the Stone–Weierstrass theorem that $\widetilde{X}$ is isometric to $C(S)$.

5.89 Assume that $Y \hookrightarrow X$ is such that both Y and X/Y are isomorphic to subspaces of ℓ_∞. Show that X is isomorphic to a subspace of ℓ_∞.
Hint. Let $q : X \to X/Y$ be the quotient mapping, T be an isomorphism of Y into ℓ_∞, and S be an isomorphism of X/Y into ℓ_∞.

Then $X \ni x \mapsto (Sqx, \widetilde{T}x)$ is an isomorphism from X into $\ell_\infty \oplus \ell_\infty$, where $\widetilde{T}$ is an extension of T to X.

5.90 There exists a Banach space X, $\varepsilon > 0$ and a bounded sequence $\{x_i\}_{i=1}^\infty \subset X$, such that for every disjoint $A, B \subset \mathbb{N}$, there exists $f \in B_{X^*}$ with $\inf_{i \in A} f(x_i) \geq \sup_{i \in B} f(x_i) + 2\varepsilon$, yet the sequence $\{x_i\}$ is not equivalent to the unit basis of ℓ_1.
Hint. The idea is to define the norm on the space c_{00} by setting the norming set consisting of $a\chi_A + b\chi_B$ choosing a and b conveniently in order that $(\frac{1}{n}, \frac{1}{n}, \dots, \frac{1}{n}, 0 \dots)$ will evaluate almost zero.

5.91 Show that $L_\infty(\Omega, \mu) = L_i^*(\Omega, \mu)$ is 1-injective, if μ is σ-finite.
Hint. We use a well-known compactness argument:

Let $\omega = (\Omega_1, \Omega_2, \cdots, \Omega_m)$ denote a finite partition of Ω into mutually disjoint measurable sets of positive measure. Let $\mathcal{A}$ denote the collection of all such partitions, partially ordered as follows: For $\omega, \gamma \in \mathcal{A}$, $\omega = (\Omega_1, \Omega_2, \cdots, \Omega_m) < \gamma = (\Gamma_1, \Gamma_2, \cdots, \Gamma_n)$ if $n > m$ and each Ω_i is a union of members of γ. Then $\mathcal{A}$ is directed by $<$ and each subspace $E_\omega = [\chi_{\Omega_i}]_{i=1}^m$ of $L_\infty(\Omega, \mu)$ is isometric to $\ell_\infty(1, 2, \cdots, m)$. Hence E_ω is 1-injective and thus, whenever $X \supset L_\infty(\Omega, \mu)$, there is a projection P_ω from X onto E_ω with $\|P_\omega\| = 1$. It is clear that $\bigcup_\omega E_\omega$ is dense in $L_\infty(\Omega, \mu)$. The unit ball $U := B_{\mathcal{B}(X, L_\infty(\Omega,\mu))}$ of the space of bounded operators from X into $L_\infty(\Omega, \mu)$, under the pointwise w^*-topology, is compact. Since $\{P_\omega\}$ is a net in U directed by $<$, it contains a convergent subnet with limit $P \in U$. Then P is a projection from X onto $L_\infty(\Omega, \mu)$ with $\|P\| = 1$, and hence $L_\infty(\Omega, \mu)$ is a 1-injective space.

Note that this gives, by Pełczyński Decomposition Method, that L_∞ is isomorphic to ℓ_∞.

5.92 Show the following result [Zipp2]: Let X be a Banach space, let E be a subspace of X and let $\lambda \geq 1$. Let K be a compact space and let $T : E \to C(K)$ be a bounded operator. Then T can be extended to a bounded operator $\hat{T} : X \to C(K)$ so that $\|\hat{T}\| \leq \lambda \|T\|$ if there is a w^*-w^*-continuous function $\varphi : B_{E^*} \to \lambda B_{X^*}$ which extends functionals (i.e., $\varphi(e^*)(e) = e^*(e)$ for all $e \in E$ and $e^* \in B_{E^*}$).
Hint. Let $T : E \to C(K)$ be an operator of $\|T\| = 1$. Let the function $\psi_T : K \to B_{E^*}$ be defined by $\psi_T(k)(e) = (Te)(k)$, $k \in K$. Put $\psi = \varphi(\psi_T) : K \to \lambda B_{X^*}$. Then ψ is w^*-continuous. Define $\hat{T} : X \to C(K)$ by $\hat{T}x(k) = \psi(k)(x)$. Then $\hat{T}$ is

linear as for each $k \in K$, ψ is a linear functional; $\|\hat{T}\| \leq \lambda$ because $\psi(k) \in \lambda B_{X^*}$ and so $\|\psi(k)\| \leq \lambda$ and $\hat{T}$ extends T because φ extends functionals from E^*: if $e \in E$ then

$$(\hat{T}e)(k) = \psi(k)(e) = (\varphi(\psi_T(k))(e) = \psi_T(k)(e) = (Te)(k)$$

for all $k \in K$.

5.93 Show that if T is a bounded operator from a subspace of ℓ_p into $C(K)$ space for $p > 1$, then T can be extended to ℓ_p with the same norm.
Hint. Exercise 5.92.

5.94 A Banach space X is said to have the *binary intersection property* if every family of mutually intersecting closed balls has a common point.

Assume that K is an extremely disconnected compact space (i.e., the closure of every open subset of K is open). This is known to be equivalent to saying that every nonempty subset of $C(K)$ that has an upper bound (in the natural order of $C(K)$) has a least upper bound (see, e.g., [GiJe]). Show that then $C(K)$ has the binary intersection property.
Hint. Let u denote the constant function 1 on K. A closed ball $B(x, r)$ in $C(K)$ with center x and radius r is exactly the order segment $[x - ru, x + ru] = \{y \in C(K) : x - ru \leq y \leq x + ru\}$. If $\mathcal{A} = [x_\alpha, y_\alpha]_{\alpha \in \mathcal{A}}$ is a collection of mutually intersecting segments, then, for every $\alpha, \beta \in \mathcal{A}$, there is a $z_{\alpha,\beta}$ such that $x_\alpha, x_\beta \leq z_{\alpha,\beta} \leq y_\alpha, y_\beta$. Hence $\{x_\alpha\}_{\alpha \in \mathcal{A}}$ is order bounded from above and thus $x = \sup\{x_\alpha : \alpha \in \mathcal{A}\}$ exists and is a common point of all the segments in $\mathcal{A}$.

5.95 Assume that X has the binary intersection property. Show that X is a 1-injective space. In particular, by taking $X = \mathbb{R}$, this gives an alternative proof of the Hahn–Banach theorem.
Hint. By Zorn's lemma, it suffices to show that if $Z \supset Y$, $\dim(Z/Y) = 1$ and $T : Y \to X$ is an operator with norm $\|T\| = 1$, then T admits an extension $\hat{T} : Z \to X$ with $\|\hat{T}\| = 1$. Let $z \in Z/Y$ and consider the family $\{B(Ty, \|z - y\|) : y \in Y\}$ of balls in X. Any two of these balls intersect because for $y_1, y_2 \in Y$, $\|Ty_1 - Ty_2\| \leq \|y_1 - y_2\| \leq \|z - y_1\| + \|z - y_2\|$. Therefore, there is a point e common to all the balls of this family. Define $\hat{T} : Z \to X$ by $\hat{T}(az + y) = ae + Ty$ for all $az + y \in Z$. It is routine to check that $\hat{T}$ extends T and $\|\hat{T}\| = \|T\| = 1$.

5.96 Let Γ be a set with $\text{card}(\Gamma) = 2^{\aleph_0}$. Show that the duals of the spaces $c_0(\Gamma) \oplus_\infty C([0, 1])$ and $C([0, 1])$ are isometric, although one of these spaces is separable and the other is not.
Hint. Duality of classical spaces.

5.97 Let X be a hereditarily indecomposable Banach space. Show that X does not contain any infinite-dimensional subspace with an unconditional basis; a fortiori it contains no subspace isomorphic to either c_0 or ℓ_p.
Hint. Projections on spans of infinite subsets of the unconditional basis.

5.98 Show that a Banach space X is an HI space if and only if for all subspaces Y and Z of X, we have

$$\inf\{\|y - z\| : y \in S_Y, z \in S_Z\} = 0.$$

Hint. Exercise 4.18.

5.99 Show that X is an HI space if and only if for every subspace $Y \hookrightarrow X$, the quotient mapping $\pi_Y : X \to X/Y$ is strictly singular.
Hint. See Exercise 4.32.

Chapter 6
Finite-Dimensional Spaces

The interplay between the structure of an infinite-dimensional Banach space and properties of its finite-dimensional subspaces belongs to the subject of the local theory of Banach spaces. It is a vast and deep part of Banach space theory intimately related to probability and combinatorics. Our goal is to familiarize the reader with some of its basic notions and results that are accessible without the use of deep probabilistic tools.

We begin with the notion of finite representability, the principle of local reflexivity and more finite representability results such as the Brunel–Sucheston spreading models technique. We prove Tzafriri's theorem and combine it with previous results to get the crude version of Dvoretzky's Theorem 6.15, which is sufficient for the proof of the Lindenstrauss–Tzafriri theorem on complemented subspaces. We introduce the John ellipsoid of maximal volume, and apply it to obtain the Kadec–Snobar estimate on norms of projections onto finite-dimensional subspaces. Finally we prove the Grothendieck inequality.

6.1 Finite Representability

Definition 6.1 *Let X, Y be Banach spaces. We say that Y is* crudely finitely representable *in X if there is $K > 0$ such that for every finite-dimensional subspace F of Y there is a linear isomorphism T of F onto $T(F) \subset X$ so that $\|T\| \cdot \|T^{-1}\| < K$. We say that Y is* finitely representable *in X if for every $\varepsilon > 0$, Y is crudely finitely representable in X with constant $K = 1 + \varepsilon$.*

We observe that if Y is finitely representable in X and W is finitely representable in Y, then W is finitely representable in X. The concept of finite representability was first studied by James ([Jame7], [Jame8]). We begin with some simple results.

Theorem 6.2 (i) *Every Banach space is finitely representable in c_0.*
(ii) *Every Banach space is finitely representable in $\left(\sum \ell_\infty^n\right)_2$.*
(iii) *Let $p \geq 1$. Then $L_p[0, 1]$ is finitely representable in ℓ_p.*
(iv) *If X is crudely finitely representable in a Hilbert space, then X is isomorphic to a Hilbert space.*

M. Fabian et al., *Banach Space Theory*, CMS Books in Mathematics,
DOI 10.1007/978-1-4419-7515-7_6, © Springer Science+Business Media, LLC 2011

Proof: (i) and (ii) Let Y be a Banach space, $\varepsilon > 0$, and F be a finite-dimensional subspace of Y. Let $\{y_i\}_{i=1}^n$ be an ε-net in S_F. Find $f_i \in S_{Y^*}$ such that $f_i(y_i) = 1$, $i = 1, \dots, n$. Consider the mapping $T: F \to \ell_\infty^n$ defined for $y \in F$ by $T(y) = \big(f_i(y)\big)_{i=1}^n$. Then

$$\|T(y)\| = \max_i\{|f_i(y)|\} \leq \max_i\big\{\|f_i\| \cdot \|y\|\big\} \leq \|y\|.$$

Given $y \in S_F$, we find $i \in \{1, \dots, n\}$ such that $\|y - y_i\| < \varepsilon$. Then

$$\|T(y)\| \geq |f_i(y)| = |f_i(y_i) + f_i(y - y_0)| \geq |f_i(y_i)| - \|f_i\| \cdot \|y - y_i\| \geq 1 - \varepsilon.$$

Therefore $\|T(y)\| \geq (1 - \varepsilon)\|y\|$ for every $y \in F$, which means that $\|T^{-1}\| \leq (1 - \varepsilon)^{-1}$.

To prove (i) it is enough to consider the mapping ST of F into c_0, where S is a linear isometry from ℓ_∞^n into c_0 defined by $S\big((x_i)\big) = (x_1, \dots, x_n, 0, 0, \dots)$.

For (ii) we consider an isometry $\widetilde{S}$ from ℓ_∞^n into $\big(\sum_{n=1}^\infty \ell_\infty^n\big)_2$ defined by $\widetilde{S}\big((x_i)\big) = \big(0, \dots, 0, (x_i), 0, \dots\big)$, where (x_i) is in the nth block in $\Big(\sum_{n=1}^\infty \ell_\infty^n\Big)_2$.

For (iii) assume that E is a finite-dimensional subspace of $L_p[0, 1]$, with a linear basis $\{b_1, \dots, b_n\}$. By perturbing (as closely as we wish) the basis functions into a system of simple functions $\{f_1, \dots, f_n\}$, we see that those form a linear basis of a subspace F of $L_p[0, 1]$ which is (as closely as we wish) almost isometric to E (i.e., $d(E, F)$ is as small as we wish). There exists a system of disjoint measurable subsets E_i, $1 \leq i \leq N$, of $[0, 1]$, such that, for $k = 1, 2, \dots, n$, $f_k \in \operatorname{span}\{\chi_{E_i}; 1 \leq i \leq N\} := Y$. It is easy to see that Y is isometric to ℓ_p^N.

For (iv) assume for simplicity that X is separable, $X_1 \subset \ldots \subset X_n \subset X_{n+1} \subset \ldots$ are subspaces of X such that $X = \overline{\bigcup_{n=1}^\infty X_n}$ and, for $n \in \mathbb{N}$, the space X_n is n-dimensional. By assumption, there exists $K > 0$ and operators $T_n : X_n \to \ell_2^n$ such that $\|T_n^{-1}\| > K$, $\|T_n\| \leq 1$. For $n \in \mathbb{N}$, let $G_n : X_n \times X_n \to \mathbb{R}$ be defined by $G_n(x, y) = \langle T_n x, T_n y\rangle$. The mapping G_n defines an inner product on the respective space X_n, such that $G_n(x, x) \leq \|x\|^2 \leq K^2 G_n(x, x)$. To finish the proof, it suffices to pass to a subsequence $\{n_k\}_{k=1}^\infty$, so that $G(x, y) := \lim\limits_{k\to\infty} G_{n_k}(x, y)$ exists for every $x, y \in \bigcup_{n=1}^\infty X_n$. Indeed, it is clear that G is an inner product extendible to the whole X, whose corresponding norm is equivalent to the original norm on X.

□

The following fundamental principle due to Lindenstrauss and Rosenthal plays an important role in the theory. It implies, of course, that for every Banach space X, X^{**} is finitely representable in X, but the statement contains much more information.

Theorem 6.3 (principle of local reflexivity, [LiRo2], [JRZ]) *Let X be a Banach space. For every finite-dimensional subspace E of X^{**}, finite-dimensional subspace F of X^* and $\varepsilon > 0$, there exists a linear isomorphism T of E onto $T(E) \subset X$ such that $\|T\| \cdot \|T^{-1}\| \leq 1 + \varepsilon$, $x^*\big(T(x^{**})\big) = x^{**}(x^*)$ for $x^* \in F$ and $x^{**} \in E$, and T is the identity on $E \cap X$.*

Proof: (Stegall [Steg3]) Choose $\delta > 0$ so that $\theta(\delta) < (1+\varepsilon)$, where $\theta(\delta)$ is the function in Exercise 1.76. Choose $a_1^*, \dots, a_m^* \in S_{X^*}$ containing a basis of F and such that $\|x^{**}\| \le (1+\delta)\sup_j |x^{**}(a_j^*)|$ for all $x^{**} \in E$. Finally, choose a δ-net $\{b_1^{**}, \dots, b_n^{**}\}$ in S_E such that $\{b_1^{**}, \dots, b_k^{**}\}$ is a basis of $E \cap X$ and $\{b_1^{**}, \dots, b_r^{**}\}$, $r \ge k$, is a basis of E. Let $q = n - r$. Then for every $p \in \{1, \dots, q\}$ we have unique scalars $t_{p,i}$, $1 \le i \le r$ such that $b_{r+p}^{**} = \sum_{i=1}^{r} t_{p,i} b_i^{**}$. Define

$$s_{p,i} = \begin{cases} t_{p,i} \ i \le r, \\ -1 \ i = r+p, \\ 0, \ r < i \le n \text{ and } i \ne r+p. \end{cases}$$

Denote $X^m = \left(\bigoplus_{i=1}^{m} X\right)_\infty$ for $m \in \mathbb{N}$ and define $A_0 \colon X^n \to X^{k+q}$ by

$$A_0(x_1, \dots, x_n) = \left(x_1, \dots, x_k, \sum_{i=1}^{n} s_{1,i} x_i, \dots, \sum_{i=1}^{n} s_{q,i} x_i\right).$$

The operator A_0 is onto, since the matrix $(s_{p,i})$ has rank q. Note that $\sum_{i=1}^{n} s_{j,i} b_i^{**} = 0$ for $j = 1, \dots, q$, so $A_0^{**}(b_1^{**}, \dots, b_n^{**}) \in X^{k+q}$. Define $U \colon X^n \to \mathbb{R}^{nm}$ by $U(x_1, \dots, x_n) = \{a_j^*(x_i)\}_{i,j=1}^{n,m}$. Let $Z = X^{k+q} \times \mathbb{R}^{nm}$, define $A \colon X^n \to Z$ by

$$A(x_1, \dots, x_n) = \Big(A_0(x_1, \dots, x_n), U(x_1, \dots, x_n)\Big).$$

By Exercise 5.12, A is an operator with a closed range. Since A is onto and $A_0^{**}(b_1^{**}, \dots, b_n^{**}) \in X^{k+q}$, there exist $d_1, \dots, d_n$ such that $A_0(d_1, \dots, d_n) = A_0^{**}(b_1^{**}, \dots, b_n^{**})$. We claim that in fact there exist $e_1, \dots, e_n$ such that $A(e_1, \dots, e_n) = A^{**}(b_1^{**}, \dots, b_n^{**})$.

Indeed, since U is a finite rank operator, $U(\mathrm{Ker}(A_0)) = U\big(\mathrm{Ker}(A_0)^{**}\big) = U(\mathrm{Ker}(A_0^{**}))$. Because $(b_1^{**}, \dots, b_n^{**}) \in (d_1, \dots, d_n) + \mathrm{Ker}(A_0^{**})$, there exists $(e_1, \dots, e_n) \in (d_1, \dots, d_n) + \mathrm{Ker}(A_0)$ so that $U(e_1, \dots, e_n) = U(b_1^{**}, \dots, b_n^{**})$ and the claim follows. In particular, $A^{**}(b_1^{**}, \dots, b_n^{**}) \in A(X^n)$. By Exercise 3.87, there is $(b_1, \dots, b_n) \in X$ such that $A(b_1, \dots, b_n) = A^{**}(b_1^{**}, \dots, b_n^{**})$ and moreover $\sup \|b_i\| \le (1+\delta)\sup \|b_i^{**}\| = 1+\delta$.

Define an operator $T \colon E \to X$ by $T(b_i^{**}) = b_i$ for $1 \le i \le r$. Note that for $1 \le p \le q$ we have $\sum_{i=1}^{r} s_{p,i} b_i^{**} = 0$ and $\sum_{i=1}^{r} s_{p,i} b_i = 0$. Thus we have also that $T(b_i^{**}) = b_i$ for $r < i \le n$. Note that for each i we have

$$\|T(b_i^{**})\| \ge \sup_j |a_j^*\big(T(b_i^{**})\big)| = \sup_j |b_i^{**}(a_j^*)| \ge (1+\delta)^{-1}$$

Thus $\|T\| \cdot \|T^{-1}\| \le 1+\varepsilon$ by the choice of δ. □

6.2 Spreading Models

Given a set X, we let $X^{(n)}$ to be the set of all subsets of X of cardinality n. We say that a system of k disjoint sets $\{S_i\}_{i=1}^k$ forms a *partitioning of* $X^{(n)}$ whenever $X^{(n)} = \bigcup_{i=1}^k S_i$.

Proposition 6.4 (Ramsey [Rams]) *Let $k, n \in \mathbb{N}$. Then for every partitioning $\{S_i\}_{i=1}^k$ of $\mathbb{N}^{(n)}$ there exists $i \in \{1, \ldots, k\}$ and an infinite set $M \subset \mathbb{N}$, such that $M^{(n)} \subset S_i$.*
An equivalent formulation is the following.
Let n be a natural number. Let ψ be a mapping from $\mathbb{N}^{(n)}$ to some finite set C. Then there is an infinite subset M of $\mathbb{N}$ such that ψ is constant on $M^{(n)}$.

Still a paraphrasis of the previous statement is the following: *If a coloring (with finite number of colors) of sets of natural numbers of a given length n is defined, then there is an infinite subset M of $\mathbb{N}$ such that all subsets of M of length n have the same color.*

Proof of Proposition 6.4: By induction on n. For $n = 1$ the result is obvious. Assume that for some $n > 1$ the statement has been proved for $1, 2, \ldots, n-1$. We shall prove it for n.

The argument is based on the following observation: if we fix some $j \in \mathbb{N}$, we may consider all elements in $\mathbb{N}^{(n)}$ that contain j. Define a mapping ψ' from $(\mathbb{N}\setminus\{j\})^{(n-1)}$ into C as $\psi'(F) = \psi(F \cup \{j\})$ for all $F \in (\mathbb{N}\setminus\{j\})^{(n-1)}$. This is a coloring of all finite subsets of length $n-1$ in $(\mathbb{N}\setminus\{j\})^{(n-1)}$, hence, by the induction hypothesis, there exists an infinite subset M_1 of $\mathbb{N}\setminus\{j\}$ such that all subsets of M_1 of length $n-1$ get the same color (i.e., ψ' is constant on them). This means that ψ is constant (the same constant) on all sets $F \cup \{j\}$, where $F \in (\mathbb{N}\setminus\{j\})^{(n-1)}$.

To prove the assertion for n, we iterate the construction above: let us start with $n_0 := 1$, and find an infinite subset M_1 of $\mathbb{N}\setminus\{n_0\}$ such that ψ is constant on all sets of the form $\{n_0\} \cup F$, for F a subset of length $n-1$ of M_1. Let $n_1 = \min M_1$ ($> n_0$). Find an infinite subset M_2 of $M_1\setminus\{n_1\}$ such that ψ is constant (maybe a different constant) on all sets of the form $\{n_1\} \cup F$, for F a subset of length $n-1$ of M_2, and put $n_2 = \min M_2$ ($> n_1$). Continue in this way to obtain a sequence $M_1 \supset M_2 \supset \ldots$ of infinite sets (and the sequence $n_0 < n_1 < n_2 < \ldots$). By passing to a subsequence if necessary (denoted again $\{M_i\}$), we may assume that the same constant is associated to all M_is. The sought set is then $\{n_i\}_{i=1}^\infty$. □

Lemma 6.5 *Let $\{x_n\}_{n=1}^\infty$ be a seminormalized basic sequence in a Banach space X, $k \in \mathbb{N}$, $\varepsilon > 0$. Then there exists a subsequence $\{y_n\}_{n=1}^\infty$ of $\{x_n\}_{n=1}^\infty$ such that for all scalars a_i, $1 \le i \le k$,*

$$(1-\varepsilon)\left\|\sum_{i=1}^k a_i y_i\right\| \le \left\|\sum_{i=1}^k a_i y_{n_i}\right\| \le (1+\varepsilon)\left\|\sum_{i=1}^k a_i y_i\right\|, \quad \textit{whenever } n_1 < \cdots < n_k. \tag{6.1}$$

Proof: Since the sequence $\{x_n\}_{n=1}^\infty$ is basic and normalized, we can find $C > 0$ that satisfies (4.1) in Proposition 4.18, and $R > 0$ such that $R^{-1} \le \|x_n\| \le R$ for all

$n \in \mathbb{N}$. In particular, given scalars $a_1, \ldots, a_k$, we have, for every $i \in \{1, 2, \ldots, k\}$,

$$C^{-1}R^{-1}|a_i| \le C^{-1}|a_i|.\|x_{n_i}\| = C^{-1}\|a_i x_{n_i}\| \le \left\| \sum_{i=1}^{k} a_i x_{n_i} \right\| \le R \sum_{i=1}^{k} |a_i| \le Rk \max_{1\le i\le k} |a_i|.$$

This implies the existence of $K > 0$ such that, for all scalars $a_1, \ldots, a_k$,

$$K^{-1} \max_{1\le i\le k} |a_i| \le \left\| \sum_{i=1}^{k} a_i x_{n_i} \right\| \le K \max_{1\le i\le k} |a_i|, \quad \text{whenever } n_1 < \cdots < n_k. \tag{6.2}$$

Fix $\delta > 0$. Consider a δ-net $\{(a_1^j, \ldots, a_k^j)\}_{j=1}^{M}$ in the unit ball of ℓ_∞^k. Let $N_0 = \mathbb{N}$ and proceed by finite induction in $j = 1, \ldots, M$ as follows.

Assume that for some $j \in \{1, 2, \ldots, M\}$, $N_0, N_1, \ldots, N_{j-1}$ have been already defined. Partition $N_{j-1}^{(k)}$ into $\{S_m\}_{1\le m\le p}$, for $p := [\delta^{-1}(K - K^{-1}) \max_{1\le i\le k} |a_i|]$, where $[\cdot]$ denotes here the integer part of a number, and

$$\begin{aligned} S_m := \{&(n_1, \ldots, n_k) : n_1 < n_2 < \ldots < n_k, \\ &K^{-1} \max_{1\le i\le k} |a_i| + (m-1)\delta \le \left\| \sum_{i=1}^{k} a_i^j x_{n_i} \right\| < K^{-1} \max_{1\le i\le k} |a_i| + m\delta \right\}. \end{aligned}$$

Then, Proposition 6.4 gives $m_j \in \{1, 2, \ldots, p\}$ and an infinite subset N_j of N_{j-1} such that $N_j^{(k)} \subset S_{m_j}$. This defines N_j and finishes the finite induction process. Denote $N_M = \{m_i\}_{i=1}^{\infty}$ and let $y_i = x_{m_i}$ for $i \in \mathbb{N}$. It follows that

$$\left| \left\| \sum_{i=1}^{k} a_i^j y_{n_i} \right\| - \left\| \sum_{i=1}^{k} a_i^j y_i \right\| \right| < \delta, \text{ for every } n_1 < \ldots < n_k. \tag{6.3}$$

Observe that

$$\begin{aligned} &\left| \left\| \sum_{i=1}^{k} a_i y_{n_i} \right\| - \left\| \sum_{i=1}^{k} a_i^j y_{n_i} \right\| \right| \\ &\le \left\| \sum_{i=1}^{k} (a_i - a_i^j) y_{n_i} \right\| \le Rk\delta, \text{ whenever } \|(a_i)_1^k - (a_i^j)_1^k\| < \delta. \end{aligned} \tag{6.4}$$

A similar estimate holds replacing $\{y_{n_i}\}_{i=1}^k$ by $\{y_i\}_{i=1}^k$. It follows from (6.3) and (6.4) that

$$\left| \left\| \sum_{i=1}^{k} a_i y_{n_i} \right\| - \left\| \sum_{i=1}^{k} a_i y_i \right\| \right| < \delta(1 + 2Rk). \tag{6.5}$$

Take now an arbitrary $(a_1, \ldots, a_k) \in \ell_\infty^k$. From (6.5) we get

$$\left\| \frac{\sum_{i=1}^k a_i y_i}{\max_{1 \le i \le k} |a_i|} \right\| - \delta(1+2Rk) \le \left\| \frac{\sum_{i=1}^k a_i y_{n_i}}{\max_{1 \le i \le k} |a_i|} \right\| \le \left\| \frac{\sum_{i=1}^k a_i y_i}{\max_{1 \le i \le k} |a_i|} \right\| + \delta(1+2Rk),$$

so

$$\left\| \sum_{i=1}^k a_i y_i \right\| - \delta(1+2Rk) \max_{1 \le i \le k} |a_i| \le \left\| \sum_{i=1}^k a_i y_{n_i} \right\| \le \left\| \sum_{i=1}^k a_i y_i \right\| + \delta(1+2Rk) \max_{1 \le i \le k} |a_i|.$$

Taking in account (6.2), we get

$$\big(1 - K\delta(1+2Rk)\big) \left\| \sum_{i=1}^k a_i y_i \right\| \le \left\| \sum_{i=1}^k a_i y_{n_i} \right\| \le \big(1 + K\delta(1+2Rk)\big) \left\| \sum_{i=1}^k a_i y_i \right\|.$$

Choosing δ small enough leads to the estimate (6.1). □

Recall that a basis $\{e_n\}_{n=1}^\infty$ of a Banach space is called *subsymmetric* if $\|\sum_{i=1}^\infty a_i e_i\| = \|\sum_{i=1}^\infty a_i e_{k_i}\|$ for every infinite increasing sequence of integers $(k_i)_{i=1}^\infty$.

Theorem 6.6 (Brunel, Sucheston [BrSu2]) *Let $\varepsilon_n \searrow 0$, $\{x_n\}_{n=1}^\infty$ be a normalized basic sequence in a Banach space X. Then there exists a subsequence $\{y_n\}_{n=1}^\infty$ of $\{x_n\}_{n=1}^\infty$ and a Banach space $(Y, |||\cdot|||)$ with a subsymmetric basis $\{e_n\}_{n=1}^\infty$, such that for all $k \in \mathbb{N}$ and all scalars $a_1, a_2, \ldots, a_k$,*

$$(1-\varepsilon_k) \left|\left|\left| \sum_{i=1}^k a_i e_i \right|\right|\right| \le \left\| \sum_{i=1}^k a_i y_{n_i} \right\| \le (1+\varepsilon_k) \left|\left|\left| \sum_{i=1}^k a_i e_i \right|\right|\right|, \quad \textit{whenever } k \le n_1 < \cdots < n_k. \tag{6.6}$$

Proof: A repeated application of Lemma 6.5 leads to a decreasing sequence of infinite subsets $M_1 \supset M_2 \supset \ldots$ of the integers, $M_k := (m_i^k)_{i=1}^\infty$, such that

$$(1-\varepsilon_k) \left\| \sum_{i=1}^k a_i x_{m_i^k} \right\| \le \left\| \sum_{i=1}^k a_i x_{n_i} \right\| \le (1+\varepsilon_k) \left\| \sum_{i=1}^k a_i x_{m_i^k} \right\| \tag{6.7}$$

whenever $n_1 < \cdots < n_k$ belong to M_k. Pass to the diagonal sequence $M = (m_i^i)_{i=1}^\infty$ of the system $\{M_n\}_{n=1}^\infty$, and let $y_i = x_{m_i^i}$ for $i \in \mathbb{N}$. Finally, let Y be the completion of the normed space of all finitely supported elements of c_0 equipped with the norm

$$\left|\left|\left| \sum_{i=1}^k a_i e_i \right|\right|\right| := \lim_{n_1 < \cdots < n_k, n_1 \to \infty} \left\| \sum_{i=1}^k a_i y_{n_i} \right\|. \tag{6.8}$$

It is clear from (6.7) that the limit exists and (6.6) is satisfied. □

The Banach space Y constructed above is called the *spreading model built from the sequence* $\{x_n\}_{n=1}^{\infty}$. Typically, a given basic sequence leads to many mutually non-isomorphic spreading models. We also say that Y is a spreading model of X provided that Y results as a spreading model built on some basic sequence in X.

Theorem 6.7 *Every spreading model Y of a Banach space X is finitely representable in X. Moreover, if $\{x_n\}_{n=1}^{\infty}$ is weakly null then $\{e_i\}_{n=1}^{\infty}$ is an unconditional basis of Y, where $\{e_i\}_{n=1}^{\infty}$ is the basis constructed in Theorem 6.6.*

Proof: The finite representability is obvious from (6.6). Suppose that $x_n \overset{w}{\to} 0$. Given a finite set $S = (s_i)_{i=1}^{k} \subset \mathbb{N}$, we introduce the projection $P_S(\sum_{i=1}^{\infty} a_i e_i) := \sum_{i\in S} a_i e_i$. By Proposition 4.36, it suffices to prove that given any finitely supported vector $v = \sum_{i=1}^{N} a_i e_i$, $P_S(\sum_{i=1}^{\infty} a_i e_i)$ is bounded by $|||v|||$. Let $\{y_n\}_{n=1}^{\infty}$ be the subsequence of $\{x_n\}_{n=1}^{\infty}$ constructed in Theorem 6.6. Given a finite increasing sequence of integers $n_1 < n_2 < \ldots < n_N$, put $M = (n_i)_{i=1}^{N}$ and consider $v_M = \sum_{i=1}^{N} a_i y_{n_i}$ and $v_M^S = \sum_{i\in S} a_i y_{n_i}$. By (6.6),

$$\lim_{n_1\to\infty} \|v_M^S\| = |||P_S(v)|||, \quad \lim_{n_1\to\infty} \|v_M\| = |||v|||. \tag{6.9}$$

Choose $f_M^S \in B_{X^*}$ such that $f_M^S(v_M^S) = \|v_M^S\|$. Now fix $\delta > 0$ and observe that for a fixed n_1, we have that for n_2 large enough it holds that $\operatorname{card}\{i : n_1 < i < n_2, |f_M^S(y_i)| < \delta\} > N$. Indeed, assuming the contrary there exist for arbitrarily large $n_2 \in \mathbb{N}$ some $f_{n_2} \in B_{X^*}$ such that $\operatorname{card}\{i : n_1 < i < n_2, |f_{n_2}(y_i)| < \delta\} \leq N$. Taking $f \in \overline{\{f_{n_2}\}}^{w^*}$, we reach a contradiction with the fact that $y_i \overset{w}{\to} 0$. A similar argument applies to all $n_i \in M$. Consequently, there exist an $M = (n_i)_{i=1}^{N}$ and $L = (m_i)_{i=1}^{N}$, such that $m_i = n_i$ if, and only if, $i \in S$, and such that $|f_M^S(y_{m_i})| < \delta$ for all m_i, $i \notin S$. So,

$$f_M^S(v_L) \geq f_M^S(v_L^S) - N\delta. \tag{6.10}$$

Since δ can be chosen arbitrarily small, and N is fixed, using (6.9) leads to the desired estimate $|||P_S(v)||| \leq |||v|||$. □

Corollary 6.8 *Let X be an infinite-dimensional Banach space. Then there exists an infinite-dimensional Banach space with an unconditional Schauder basis which is crudely finitely representable in X.*

Proof: Mazur's technique (see Theorem 4.19 and Lemma 4.20) gives a subspace Y of X with a Schauder basis. Applying Rosenthal's ℓ_1 Theorem 5.37 to a separate sequence on the unit sphere of Y (see the proof of Theorem 1.38), Y contains either an ℓ_1 basic sequence, or a weakly Cauchy sequence. By taking consecutive differences and applying Corollary 4.27, we get a normalized basic sequence. Theorem 6.6 gives then the assertion. □

6.3 Complemented Subspaces in Spaces with an Unconditional Schauder Basis

In this section, we prove a theorem of Tzafriri on sequences of complemented finitely-dimensional subspaces—behaving like ℓ_p spaces—of Banach spaces with an unconditional basis. In subsequent sections, this theorem will be referred to as the "crude Dvoretzky theorem," in view of the theorem of Dvoretzky—Theorem 6.15 below.

Theorem 6.9 (Tzafriri [Tzaf]) *Let X be an infinite-dimensional Banach space with an unconditional basis. Then there exist $p \in \{1, 2, \infty\}$, a constant $M > 0$, and a sequence $\{P_n\}_{n=1}^{\infty}$ of projections in X such that $\|P_n\| \leq M$ and $\mathrm{d}(P_n X, \ell_p^n) \leq M$ for all $n \in \mathbb{N}$.*

The proof will be carried out for real Banach spaces. The complex version follows with some minor adjustments.

Proposition 6.10 *Let X be a Banach space with a normalized monotone Schauder basis $\{e_n\}$. Then, for every $\varepsilon > 0$ there exists a sequence $\{\lambda(j)\}$ of positive real numbers and a subsequence $\{z_k := e_{n_k}\}$ of $\{e_n\}$ such that*

$$0 < \lambda(j) - \|z_{k_1} + z_{k_2} + \ldots + z_{k_j}\| < \varepsilon \tag{6.11}$$

for every set of indices $j < k_1 < k_2 < \ldots < k_j$, $j \in \mathbb{N}$. In addition,

$$1 \leq \lambda(j) < \lambda(m) + \varepsilon \tag{6.12}$$

for every $j, m \in \mathbb{N}$ such that $j < m$.

Proof: Fix $\varepsilon > 0$. For every integer $k > 1$, we consider a fixed partition of the interval $[1, k+1]$

$$\lambda_0^{(k)} := 1 < \lambda_1^{(k)} < \lambda_2^{(k)} < \ldots < \lambda_{s(k)}^{(k)} := k+1$$

with the property that $\lambda_j^{(k)} - \lambda_{j-1}^{(k)} < \varepsilon$ for $j = 1, 2, \ldots, s(k)$. Define now a function φ_k from all the unordered k-tuples of different positive integers into the set $\{1, 2, \ldots, s(k)\}$ by setting $\varphi_k\{n_1, n_2, \ldots, n_k\} = j$ if $\lambda_{j-1}^{(k)} \leq \|e_{n_1} + e_{n_2} + \ldots + e_{n_k}\| < \lambda_j^{(k)}$.

Now proceed inductively: put $N^{(1)} = \mathbb{N}$ and $\lambda(1) = 1$. Apply Proposition 6.4 to the function φ_2 to obtain an infinite set $N^{(2)} \subset N^{(1)}$ such that φ_2 restricted to all unordered couples from $N^{(2)}$ is constant, say j, for some $j \in \{1, 2, \ldots, s(2)\}$. In this case, put $\lambda(2) = \lambda_j^{(2)}$. Apply again Proposition 6.4 to the set $N^{(2)}$ and the function φ_3 restricted to unordered 3-tuples from elements in $N^{(2)}$: we obtain an infinite set $N^{(3)} \subset N^{(2)}$ such that φ_3 restricted to all unordered 3-tuples from $N^{(3)}$ is constant, say k, for some $k \in \{1, 2, \ldots, s(3)\}$. In this case, put $\lambda(3) = \lambda_k^{(3)}$.

Continue in this way to obtain a sequence $N^{(1)} := \mathbb{N} \supset N^{(2)} \supset N^{(3)} \ldots$ of infinite sets and a sequence $\{\lambda(j)\}$. A standard diagonal argument gives the sequence $n_1 < n_2 < n_3, \ldots$ and the result. The last estimate follows clearly from the monotonicity of the basis. □

Two estimates follow trivially from inequality (6.11), the monotonicity of the basis and the fact that all of its vectors are normalized. We single them out for future references.

$$1 = \|z_{k_1}\| \leq \|z_{k_1} + z_{k_2} + \ldots + z_{k_j}\| \leq \lambda(j), \tag{6.13}$$

and, if $\varepsilon \leq 1$,

$$\lambda(j) \leq \|z_{k_1} + z_{k_2} + \ldots + z_{k_j}\| + \varepsilon \leq 2\|z_{k_1} + z_{k_2} + \ldots + z_{k_j}\|, \tag{6.14}$$

for all $j < k_1 < k_2 < \ldots < k_j$, $j \in \mathbb{N}$.

Proposition 6.11 *Fix $0 < \varepsilon < 1/2$ in Proposition 6.10 and let $\{\lambda(j)\}$ and $\{z_k\}$ be the sequences of numbers and vectors, respectively, obtained there. Assume that there exists an integer $h > 1$ such that*

$$\frac{\lambda(hn)}{\lambda(n)} \geq 1 + \varepsilon, \quad n \in \mathbb{N}. \tag{6.15}$$

Then there exists a constant A and a number $q > 2$ such that

$$\frac{1}{\lambda(n)} \left\| \sum_{j=1}^{n} a_j z_{m+k_j} \right\| \leq n^{-1/q} A \left(\sum_{j=1}^{n} |a_j|^q \right)^{1/q} \tag{6.16}$$

for every $n \in \mathbb{N}$, $m \in \mathbb{N}$ such that $m > n$, integers $0 < k_1 < k_2 < \ldots < k_n$, and every sequence of scalars $\{a_1, a_2, \ldots, a_n\}$.

Proof: Find $r > 2$ such that $1 < h^{1/r} < 1 + \varepsilon$. Then $\lambda(hn)/\lambda(n) > h^{1/r}$ for all $n \in \mathbb{N}$. It follows that, for $n, s \in \mathbb{N}$,

$$\lambda(h^s n) > h^{1/r}\lambda(h^{s-1}n) > (h^2)^{1/r}\lambda(h^{s-2}n) \ldots > (h^{s-1})^{1/r}\lambda(hn) > (h^s)^{1/r}\lambda(n),$$

i.e.,

$$\frac{\lambda(h^s n)}{\lambda(n)} > (h^s)^{1/r}. \tag{6.17}$$

Let α and β be two integers such that $h^{j-1} < \beta \leq h^j$ and $h^{i-1} < \alpha \leq h^i$ for some $i \leq j$ in $\mathbb{N}$. We consider two cases.

(1) If $i < j$ we get $h^{j-1} < \beta \leq h^j \leq h^{i-1} < \alpha \leq h^i$. Then we have $\lambda(\beta) < \lambda(h^j) + \varepsilon$ and $\lambda(h^{i-1}) < \lambda(\alpha) + \varepsilon$. Hence, putting $\rho = \lambda(h^{i-1})/\lambda(h^j)$,

$$\begin{aligned}\frac{\lambda(\alpha)}{\lambda(\beta)} &> \frac{\lambda(h^{i-1}) - \varepsilon}{\lambda(h^j) + \varepsilon} = \frac{\rho - \varepsilon/\lambda(h^j)}{1 + \varepsilon/\lambda(h^j)} \geq \frac{\rho - \varepsilon}{1 + \varepsilon} \\ &> \frac{\rho - 1/2}{1 + 1/2} = \frac{2\rho - 1}{3} > \frac{2\rho - \rho}{3} = \frac{\rho}{3} > \frac{\rho}{4} \\ &= \frac{1}{4} \frac{\lambda(h^{i-j-1}h^j)}{\lambda(h^j)} \geq \frac{1}{4}\left(h^{i-j-1}\right)^{1/r} \geq \frac{1}{4h^{2/r}}\left(\frac{\alpha}{\beta}\right)^{1/r},\end{aligned}$$

where the last but one inequality comes from (6.17) and the last one from the fact that $\alpha/\beta \leq h^i/h^{j-1}$.

(2) Assume now that $i = j$. Then $h^{j-1} < \beta \leq \alpha \leq h^j$ and we get

$$\frac{\lambda(\alpha)}{\lambda(\beta)} \geq \frac{\lambda(\beta) - \varepsilon}{\lambda(\beta)} = 1 - \frac{\varepsilon}{\lambda(\beta)} \geq 1 - \varepsilon \geq \frac{1}{2} \geq \frac{1}{2h^{1/r}}\left(\frac{h^j}{h^{j-1}}\right)^{1/r} \geq \frac{1}{2h^{1/r}}\left(\frac{\alpha}{\beta}\right)^{1/r},$$

where the first inequality follows from (6.12) and the last one from the fact that $\alpha/\beta \leq h^j/h^{j-1}$.

Thus, in both cases we have

$$\frac{\lambda(\alpha)}{\lambda(\beta)} \geq \frac{1}{4h^{2/r}}\left(\frac{\alpha}{\beta}\right)^{1/r}. \tag{6.18}$$

Fix positive integers $n < m$ and $k_1 < k_2 < \ldots < k_n$, and a sequence of scalars $\{a_1, a_2, \ldots, a_n\}$. Put $a_j = b_j^{(1)} - b_j^{(2)}$, where $0 \leq b_j^{(s)} \leq |a_j|$ for $s = 1, 2$, and $j = 1, 2, \ldots, n$. Then

$$\left\| \sum_{j=1}^n a_j z_{m+k_j} \right\| \leq \sum_{s=1}^2 \left\| \sum_{j=1}^n b_j^{(s)} z_{m+k_j} \right\|. \tag{6.19}$$

Let $\sum_{j=1}^n b_j z_{m+k_j}$ be one of the two sums in the right hand side of (6.19). Let π be a permutation of the integers $\{1, 2, \ldots, n\}$ such that $b_{\pi(1)} \geq b_{\pi(2)} \geq \ldots \geq b_{\pi(n)} \geq 0$. Then, by using (6.13),

$$\left\|\sum_{j=1}^{n} b_j z_{m+k_j}\right\| = \left\|\sum_{j=1}^{n} b_{\pi(j)} z_{m+k_{\pi(j)}}\right\|$$
$$= \left\|(b_{\pi(1)} - b_{\pi(2)}) \sum_{s=1}^{1} z_{m+k_{\pi(s)}} + (b_{\pi(2)} - b_{\pi(3)}) \sum_{s=1}^{2} z_{m+k_{\pi(s)}} + \dots\right.$$
$$\left. + (b_{\pi(n-1)} - b_{\pi(n)}) \sum_{s=1}^{n-1} z_{m+k_{\pi(s)}} + b_{\pi(n)} \sum_{s=1}^{n} z_{m+k_{\pi(s)}}\right\|$$
$$\leq (b_{\pi(1)} - b_{\pi(2)})\lambda(1) + (b_{\pi(2)} - b_{\pi(3)})\lambda(2) + \dots$$
$$+ (b_{\pi(n-1)} - b_{\pi(n)})\lambda(n-1) + b_{\pi(n)}\lambda(n).$$

Thus, using (6.18),

$$\frac{1}{\lambda(n)}\left\|\sum_{j=1}^{n} b_j z_{m+k_j}\right\|$$
$$\leq 4h^{2/r}\left\{(b_{\pi(1)} - b_{\pi(2)})\left(\frac{1}{n}\right)^{1/r} + (b_{\pi(2)} - b_{\pi(3)})\left(\frac{2}{n}\right)^{1/r} + \dots\right.$$
$$\left. + (b_{\pi(n-1)} - b_{\pi(n)})\left(\frac{n-1}{n}\right)^{1/r} + b_{\pi(n)}\right\}$$
$$= 4h^{2/r}\sum_{j=1}^{n} b_{\pi(j)}\left[\left(\frac{j}{n}\right)^{1/r} - \left(\frac{j-1}{n}\right)^{1/r}\right]. \tag{6.20}$$

Set $q > r$, and let q' and r' be the conjugate indices to q and r, respectively, i.e., $1/q + 1/q' = 1$ and $1/r + 1/r' = 1$. Observe that, for $j \in \mathbb{N}$, we have $(j-1)/j \leq \left((j-1)/j\right)^{1/r}$. This implies $1 - \left((j-1)/j\right)^{1/r} \leq 1/j$, hence

$$j^{1/r} - (j-1)^{1/r} \leq j^{-1/r'}. \tag{6.21}$$

Carry (6.21) to (6.20) and use Hölder inequality for q and q' to get

$$\frac{1}{\lambda(n)}\left\|\sum_{j=1}^{n} b_j z_{m+k_j}\right\|$$
$$\leq \frac{4h^{2/r}}{n^{1/r}}\sum_{j=1}^{n}\frac{b_{\pi(j)}}{j^{1/r'}} \leq \frac{4h^{2/r}}{n^{1/r}}\left(\sum_{j=1}^{n}|b_{\pi(j)}|^q\right)^{1/q}\left(\sum_{j=1}^{n}\frac{1}{j^{q'/r'}}\right)^{1/q'}$$
$$\leq \frac{4h^{2/r}}{n^{1/r}}\left(\sum_{j=1}^{n}|a_j|^q\right)^{1/q}\left(\frac{1}{1-q'/r'}\right)^{1/q'} n^{(1/q'-1/r')}, \tag{6.22}$$

where the last inequality is obtained by integrating the function $1/x^{q'/r'}$ between 0 and n. This amounts to

$$\left\| \sum_{j=1}^{n} b_j z_{m+k_j} \right\| \Big/ \lambda(n) \leq \frac{4h^{2/r}}{(1-q'/r')^{1/q'}} \left(\sum_{j=1}^{n} |a_j|^q \right)^{1/q} n^{-1/q}. \tag{6.23}$$

Since this is true for any of the sums at the right hand side of (6.19), by putting $A = 8h^{2/r}/(1-q'/r')^{1/q'}$ we obtain inequality (6.16). □

Proposition 6.12 *Let V be a 2^n-dimensional Banach space generated by a system of vectors $\{v_1, v_2, \dots, v_{2^n}\}$. Suppose that there exist constants $K > 1$ and $p > 2$ such that*

$$K^{-1} \left(\sum_{j=1}^{2^n} |a_j|^{p'} \right)^{1/p'} (2^{-n})^{1/p'} \leq \left\| \sum_{j=1}^{2^n} a_j v_j \right\| \left\| \sum_{j=1}^{2^n} v_j \right\|^{-1} \leq K \left(\sum_{j=1}^{2^n} |a_j|^p \right)^{1/p} (2^{-n})^{1/p} \tag{6.24}$$

for every set of scalars $\{a_1, a_2, \dots, a_{2^n}\}$, where p' is the conjugate index to p, i.e., $1/p + 1/p' = 1$. Then there is a constant $M := M(K, p)$ (and so independent of n and V while $\|\cdot\|$ in V satisfies (6.24)*), and a projection $P : V \to V$ such that $\|P\| \leq M$ and $\mathrm{d}(PV, \ell_2^n) \leq M$.*

Proof: For $j = 1, 2, \dots, 2^n$, $k = 1, 2, \dots, n$ and $h = 1, 2, \dots, 2^{k-1}$, put

$$\varepsilon_{kj} = \begin{cases} 1 & \text{if } (2h-2)2^{n-k} + 1 \leq j \leq (2h-1)2^{n-k}, \\ -1 & \text{if } (2h-1)2^{n-k} + 1 \leq j \leq 2h2^{n-k}. \end{cases} \tag{6.25}$$

Let χ_S denote the characteristic function of a set $S \subset [0, 1]$. Observe that the functions

$$r_k := \sum_{j=1}^{2^n} \varepsilon_{kj} \chi_{[(j-1)/2^n, j/2_n]}, \qquad k = 1, 2, \dots, n \tag{6.26}$$

are the first n Rademacher functions (see the proof of Lemma 4.54). Put

$$w_k = \sum_{j=1}^{2^n} \varepsilon_{kj} v_j, \qquad k = 1, 2, \dots, n.$$

Vectors $w_1, w_2, \dots, w_n$ are the n first elements of the so-called *Rademacher system in V associated to the basis $\{v_1, v_2, \dots, v_n\}$*.

We shall prove that the space $W := \mathrm{span}\{w_1, w_2, \dots, w_n\}$ is the range of a projection in V that satisfies the requirements.

Khintchine inequalities (Lemma 4.54) for this index p say that there exists $K_p > 0$ such that

$$K_p^{-1}\left(\sum_{k=1}^n |a_k|^2\right)^{1/2} \leq \left\|\sum_{k=1}^n a_k r_k\right\|_p \leq K_p\left(\sum_{k=1}^n |a_k|^2\right)^{1/2}. \tag{6.27}$$

Observe that

$$\sum_{k=1}^n a_k w_k = \sum_{k=1}^n a_k \sum_{j=1}^{2^n} \varepsilon_{kj} v_j = \sum_{j=1}^{2^n}\left(\sum_{k=1}^n \varepsilon_{kj} a_k\right) v_j. \tag{6.28}$$

Hence, putting together (6.24), (6.27), and (6.28), we get

$$\begin{aligned}\left\|\sum_{k=1}^n a_k w_k\right\| \left\|\sum_{j=1}^{2^n} v_j\right\|^{-1} &= \left\|\sum_{j=1}^{2^n}\left(\sum_{k=1}^n \varepsilon_{kj} a_k\right) v_j\right\| \left\|\sum_{j=1}^{2^n} v_j\right\|^{-1} \\ &\leq K\left(\sum_{j=1}^{2^n}\left|\sum_{k=1}^n \varepsilon_{kj} a_k\right|^p\right)^{1/p} (2^{-n})^{1/p} = K\left\|\sum_{k=1}^n a_k r_k\right\|_p \\ &\leq K K_p\left(\sum_{k=1}^n |a_k|^2\right)^{1/2},\end{aligned} \tag{6.29}$$

where the computation of the norm $\left\|\sum_{k=1}^n a_k r_k\right\|_p$ is done in Exercise 6.15.

Analogously,

$$\left\|\sum_{k=1}^n a_k w_k\right\| \left\|\sum_{j=1}^{2^n} v_j\right\|^{-1} \geq K^{-1} K_{p'}^{-1}\left(\sum_{k=1}^n |a_k|^2\right)^{1/2}. \tag{6.30}$$

Inequalities (6.29) and (6.30) together show that $\mathrm{d}(W, \ell_2^n) \leq K^2 K_p K_{p'}$.

Let Q be the orthogonal projection in $L_2[0,1]$ whose range is $\operatorname{span}\{r_1, r_2, \dots, r_n\}$. In the proof of Theorem 4.53, we showed that Q acts as a bounded linear projection in every $L_r[0,1]$, $r > 1$, and that the norm $\|Q\|_r$ of Q in $L_r[0,1]$ is independent of n. If

$$Q\left(\sum_{j=1}^{2^n} a_j \chi_{[(j-1)/2^n, j/2^n]}\right) = \sum_{k=1}^n b_k r_k, \tag{6.31}$$

then we set

$$P\left(\sum_{j=1}^{2^n} a_j v_j\right) = \sum_{k=1}^{n} b_k w_k.$$

The mapping P so defined is a linear projection in V whose range is W. Moreover, by using consecutively (6.29), (6.27), (6.31) and, finally, (6.24),

$$\begin{aligned}\left\| P\left(\sum_{j=1}^{2^n} a_j v_j\right)\right\| &= \left\|\sum_{k=1}^{n} b_k w_k\right\| \le K K_p \left\|\sum_{j=1}^{2^n} v_j\right\| \left(\sum_{k=1}^{n} |b_k|^2\right)^{1/2} \\ &\le K K_p K_{p'} \left\|\sum_{j=1}^{2^n} v_j\right\| \cdot \left\|\sum_{k=1}^{n} b_k r_k\right\|_{p'} \\ &\le \|Q\|_{p'} K K_p K_{p'} \left\|\sum_{j=1}^{2^n} v_j\right\| \cdot \left(\sum_{j=1}^{2^n} |a_j|^{p'}\right)^{1/p'} (2^{-n})^{1/p'} \\ &\le \|Q\|_{p'} K^2 K_p K_{p'} \left\|\sum_{j=1}^{2^n} a_j v_j\right\|.\end{aligned}$$

In conclusion, we proved that, for $M := \|Q\|_{p'} K^2 K_p K_{p'}$, we have $\|P\| \le M$ and $d(PV, \ell_2^n) \le M$. □

Proof of Theorem 6.9: Assume, without loss of generality, that the unconditional constant of the basis is 1. Fix $0 < \varepsilon < 1/2$. The basic sequence $\{e_n^*\}_{n=1}^\infty$ formed by the functional coefficients is also unconditional with the same unconditional constant. Apply Proposition 6.10 to obtain two sequences of real numbers $\{\lambda(j)\}_{j=1}^\infty$ and $\{\mu(j)\}_{j=1}^\infty$, and two subsequences $\{z_i := e_{n_i}\}_{i=1}^\infty$ and $\{z_i^* := e_{n_i}^*\}_{i=1}^\infty$ so that

$$0 < \lambda(j) - \|z_{k_1} + z_{k_2} + \ldots + z_{k_j}\| < \varepsilon,$$
$$0 < \mu(j) - \|z_{k_1}^* + z_{k_2}^* + \ldots + z_{k_j}^*\| < \varepsilon,$$

for every set of indices $j < k_1 < k_2 < \ldots < k_j$, $j \in \mathbb{N}$.

We shall distinguish three cases. Each of them will produce one of the three situations in the statement of the theorem: (I) the case $p = \infty$, (II) the case $p = 1$ and, finally, (III) the case $p = 2$.

Case (I): Assume first that for every integer $h > 1$ there exists an integer $n := n(h) \ge 1$ such that $\lambda(hn)/\lambda(n) < 1 + \varepsilon$. Fix h and let $n = n(h)$. Consider the following vectors:

$$
\begin{aligned}
u_1 &:= (z_{hn+1} + \ldots + z_{hn+n})/\lambda(n) \\
u_2 &:= (z_{hn+n+1} + \ldots + z_{hn+2n})/\lambda(n) \\
&\vdots \\
u_h &:= (z_{hn+(h-1)n+1} + \ldots + z_{hn+hn})/\lambda(n).
\end{aligned}
$$

From Equation (6.11) we get $0 < 1 - \|u_n\| < \varepsilon/\lambda(n) < 1/2$, hence $\|u_n\| \geq 1/2$ for $n = 1, 2, \ldots, h$. The basis $\{x_n\}$ is unconditional (with unconditional constant 1). Use the monotonicity of the basis, then Lemma 4.38, inequality (6.13) and finally the assumption on λ to get, for every sequence $\{a_1, a_2, \ldots, a_h\}$ of scalars,

$$
\begin{aligned}
\frac{1}{2} \max_{1\leq i\leq h} |a_i| &\leq \left\| \sum_{i=1}^{h} a_i u_i \right\| \\
&\leq \left(\max_{1\leq i\leq h} |a_i| \right) \left\| \sum_{i=1}^{h} u_i \right\| = \left(\max_{1\leq i\leq h} |a_i| \right) \left\| \sum_{j=1}^{hn} z_{hn+j} \right\| \frac{1}{\lambda(n)} \\
&\leq \left(\max_{1\leq i\leq h} |a_i| \right) \frac{\lambda(hn)}{\lambda(n)} \leq 2 \max_{1\leq i\leq h} |a_i|. \qquad (6.32)
\end{aligned}
$$

This shows that $\mathrm{d}(X_h, \ell_\infty^h) \leq 4$, where $X_h := \mathrm{span}\{u_1, u_2, \ldots, u_h\}$. Since ℓ_∞ is an injective space, it follows immediately that there are projections P_h in X such that $P_h X = X_h$ and $\|P_h\| \leq 4$ for all $h \in \mathbb{N}$.

Case (II): Assume now that for every integer $h > 1$ there exists an integer $n := n(h) \geq 1$ such that $\mu(hn)/\mu(n) < 1 + \varepsilon$. The basic sequence $\{x_n^*\}$ is also unconditional, monotone, and it has unconditional constant 1. Case (I) applied to $\{z_i^*\}$ gives, for every $h > 1$, vectors $\{y_1^*, y_2^*, \ldots, y_h^*\}$ in X^* which have disjoint supports relatively to the basis $\{x_n\}$ and such that

$$
\frac{1}{2} \max_{1\leq i\leq h} |a_i| \leq \left\| \sum_{i=1}^{h} a_i y_i^* \right\| \leq 2 \max_{1\leq i\leq h} |a_i| \qquad (6.33)
$$

for every sequence $\{a_1, a_2, \ldots, a_h\}$ of scalars (see Equation 6.32). Due again to the unconditionality of the basis, we can find, for every $i = 1, 2, \ldots, h$, a vector $y_i \in S_X$ such that its support (with respect to $\{x_n\}$) is contained in the support of y_i^* and such that $1/2 \leq \langle y_i^*, y_i \rangle \leq 2$. We get, for every sequence $\{b_i : i = 1, 2, \ldots, h\}$ of scalars,

$$
\left\| \sum_{i=1}^{h} b_i y_i \right\| \leq \sum_{i=1}^{h} |b_i| \leq 2 \left\langle \sum_{i=1}^{h} (\mathrm{sign}\, b_i) y_i^*, \sum_{i=1}^{h} b_i y_i \right\rangle \leq 4 \left\| \sum_{i=1}^{h} b_i y_i \right\|
$$

(here we used the last inequality in (6.33)). This shows that $\mathrm{d}(Y_h, \ell_1^h) \leq 4$, where $Y_h := \mathrm{span}\{y_1, y_2, \ldots, y_h\}$.

For $x \in X$ put

$$P_h x = \sum_{i=1}^{h} \frac{\langle y_i^*, x\rangle}{\langle y_i^*, y_i\rangle} y_i.$$

Then P_h is a projection from X onto Y_h for which

$$\|P_h x\| \leq \sum_{i=1}^{h} \frac{|\langle y_i^*, x\rangle|}{\langle y_i^*, y_i\rangle} = \left(\sum_{i=1}^{h} \frac{\text{sign } \langle y_i^*, x\rangle}{\langle y_i^*, y_i\rangle} y_i^*\right) x \leq 4\|x\|,$$

i.e., again $\|P_h\| \leq 4$ for $h \in \mathbb{N}$.

Case (III): Finally, assume that cases (I) and (II) do not hold. We can then use Proposition 6.11 for $\{z_k\}$ in X and then for $\{z_k^*\}$ in X^* to obtain constants $B > 1$ and $p > 2$ such that, simultaneously,

$$\frac{1}{\lambda(n)}\left\|\sum_{j=1}^{n} a_j z_{n+j}\right\| \leq B\left(\sum_{j=1}^{n} |a_j|^p\right)^{1/p} n^{-1/p}, \tag{6.34}$$

$$\frac{1}{\mu(n)}\left\|\sum_{j=1}^{n} a_j z_{n+j}^*\right\| \leq B\left(\sum_{j=1}^{n} |a_j|^p\right)^{1/p} n^{-1/p}, \tag{6.35}$$

for every $n \in \mathbb{N}$ and every sequence $\{a_1, a_2, \ldots, a_n\}$ of scalars.

Fix $n \in \mathbb{N}$. To obtain $\|\sum_{j=1}^n z_{n+j}^*\|$ we compute $\sup_{x \in S_X} \langle \sum_{j=1}^n z_{n+j}^*, x\rangle$. It is certainly enough to take the supremum on all $x \in S_X$ having a support (with respect to $\{z_n\}$) contained in the support of $\sum_{j=1}^n z_{n+j}^*$. Those vectors are in a finite-dimensional subspace of X, hence there exists $x := \sum_{j=1}^n b_j z_{n+j} \in S_X$ such that $\|\sum_{j=1}^n z_{n+j}^*\| = \langle \sum_{j=1}^n z_{n+j}^*, x\rangle$ $(= \sum_{j=1}^n b_j)$.

Choose a positive integer C such that $C > (8B)^p$. We Claim that if

$$\eta := \left\{j : 1 \leq j \leq n,\ |b_j| \geq \frac{8C}{\lambda(n)}\right\}$$

and s denotes the cardinal of η, then $n > Cs$. Indeed, if $n \leq Cs$ then

$$\lambda(n) \leq 2\left\|\sum_{j=1}^{n} z_{Cs+j}\right\| \leq 2\left\|\sum_{j=1}^{Cs} z_{Cs+j}\right\| \leq 2C\lambda(s), \tag{6.36}$$

where the first inequality follows from (6.14), the second from the monotonicity of the basis, and the third observing that $Cs \geq n \geq s$, that $\left\|\sum_{j=1}^{Cs} z_{Cs+j}\right\| \leq$

$\left\|\sum_{j=1}^{s} z_{Cs+j}\right\| + \left\|\sum_{j=s+1}^{2s} z_{Cs+j}\right\| + \ldots, + \left\|\sum_{j=(C-1)s+1}^{Cs} z_{Cs+j}\right\|$ and then using inequality (6.13).

On the other hand,

$$1 = \|x\| \geq \left\|\sum_{j\in\eta} |b_j| z_{n+j}\right\| \geq \frac{8C}{\lambda(n)} \left\|\sum_{j\in\eta} z_{n+j}\right\| \geq 4C\frac{\lambda(s)}{\lambda(n)}, \tag{6.37}$$

where the first and second inequalities follow from the unconditionality of the basis, and the third from (6.11) and the fact that $\lambda(s) - \varepsilon \geq \lambda(s)/2$. Certainly, (6.36) and (6.37) are in contradiction, so the Claim holds. $\triangle$

We have

$$\frac{\mu(s)}{\mu(n)} \leq \frac{2}{\mu(n)} \left\|\sum_{j=1}^{s} z_{n+j}^*\right\| \leq 2B\left(\frac{s}{n}\right)^{1/p} \leq \frac{2B}{C^{1/p}} < \frac{1}{4}, \tag{6.38}$$

where the first inequality follows from 6.14, the second one from (6.35), the third one from the Claim and the last one from the condition on C above.

From (6.14) again, $\mu(n) \leq 2\|\sum_{j=1}^{n} z_{n+j}^*\| = 2\sum_{j=1}^{n} b_j$. Hence,

$$\begin{aligned}
\mu(n) &\leq 2\sum_{j=1}^{n} b_j \leq 16C\frac{n}{\lambda(n)} + 2\sum_{j\in\eta} b_j \\
&\leq 16C\frac{n}{\lambda(n)} + 2\left\langle \sum_{j\in\eta} z_{n+j}^*, x\right\rangle \leq 16C\frac{n}{\lambda(n)} + 2\mu(s) \leq 16C\frac{n}{\lambda(n)} + \frac{\mu(n)}{2},
\end{aligned}$$

that is

$$\mu(n) \leq 32C\frac{n}{\lambda(n)}, \quad \text{for } n \in \mathbb{N}. \tag{6.39}$$

Observe, too, that

$$\begin{aligned}
\sum_{j=1}^{n} |a_j|^{p'} &\leq \left\langle \sum_{j=1}^{n} |a_j|^{p'-1} (\operatorname{sign} a_j) z_{n+j}^*, \sum_{j=1}^{n} a_j z_{n+j} \right\rangle \\
&\leq \left\|\sum_{j=1}^{n} a_j z_{n+j}\right\| \cdot \left\|\sum_{j=1}^{n} |a_j|^{p'-1} (\operatorname{sign} a_j) z_{n+j}^*\right\|.
\end{aligned}$$

Consequently, for $1/p + 1/p' = 1$, we have, by using (6.35) and (6.39),

$$\begin{aligned}
\left\|\sum_{j=1}^{n} a_j z_{n+j}\right\| &\geq \sum_{j=1}^{n} |a_j|^{p'} \Bigg/ \left\|\sum_{j=1}^{n} |a_j|^{p'-1}(\text{sign } a_j) z_{n+j}^*\right\| \\
&\geq \frac{1}{B} n^{1/p} \left(\sum_{j=1}^{n} |a_j|^{p'}\right) \frac{1}{\mu(n)} \left(\sum_{j=1}^{n} |a_j|^{(p'-1)p}\right)^{-1/p} \\
&\geq \frac{1}{32BC} \frac{\lambda(n)}{n^{1/p'}} \left(\sum_{j=1}^{n} |a_j|^{p'}\right)^{1/p'}.
\end{aligned}$$

To finish the proof, it is enough to apply Proposition 6.12, since B, C and p are independent of n. □

Let us state without proof a result of Krivine [Kri] (see, e.g., [MiSc] or [AlKa]) going in the same direction. Before the actual statement, let us highlight the differences between Theorems 6.9 and 6.13 below. In Tzafriri's result (Theorem 6.9), the finite representability is only crude, but we have uniform complementability and the candidates for p are just $1, 2, \infty$. On the other hand, in Krivine's result the unit bases of the embedded ℓ_p^n consists of vectors whose supports are disjoint blocks (we say, then, that the space ℓ_p is *block finitely representable in* X).

Theorem 6.13 (Krivine, [Kri]) *Let X be a Banach space with a Schauder basis. Then there exists a $p \in [1, \infty]$ such that ℓ_p is block finitely representable in X.*

Theorems 6.7, 6.9, and comments after the statement of Theorem 6.9, give the following

Corollary 6.14 *The Hilbert space is crudely finitely representable in every infinite-dimensional Banach space X.*

Proof: By Corollary 6.8 there exists a Banach space Y with an unconditional basis that is crudely finitely representable in X. By Tzafriri's Theorem 6.9, one of the Banach spaces $Z = c_0, \ell_2, \ell_1$ is crudely finitely representable in Y, and thus also in X. By (i) in Theorem 6.2, ℓ_2 is finitely representable in $Z = c_0$, so it remains to check the case $Z = \ell_1$. By Khintchine's inequality (4.2) in Lemma 4.54, ℓ_2 is isomorphic to a subspace of $L_1[0, 1]$, and the later space is finitely representable in ℓ_1. □

Although the above corollary is powerful enough to be yield the proof of the complemented subspace theorem in the next section, it is far from optimal. By using powerful probabilistic arguments which are beyond the scope of this book, Dvoretzky proved the following fundamental theorem (see also, e.g., [BeLi], [AlKa], [MiSc], [Pisi4], and [Tomc]).

Theorem 6.15 (Dvoretzky [Dvor]) *For every $\varepsilon > 0$ there exists $c(\varepsilon) > 0$ with the following property. Let X be an n-dimensional Banach space. There exists a k-dimensional subspace Y of X, such that $d(Y, \ell_2^k) \leq 1+\varepsilon$, where $k := [c(\varepsilon)\log(n)]$ and $[\cdot]$ denotes the integer part of a number.*

6.4 The Complemented-Subspace Result

The main result in this section illustrates how finite-dimensional arguments yield a powerful result whose formulation appears to be purely infinite-dimensional.

Theorem 6.16 (Lindenstrauss, Tzafriri, [LiTz1]) *Let X be a Banach space. Every closed subspace of X is complemented in X if and only if X is isomorphic to a Hilbert space.*

In the proof, we will use the following lemma.

Lemma 6.17 *Under the assumption of the theorem, there is $\lambda \geq 1$ so that every finite-dimensional subspace E is λ-complemented in X.*

Proof of Lemma 6.17: For a finite-dimensional subspace E of X, denote by $\lambda(E)$ the minimum of the norms of projections from X onto E (use Alaoglu–Bourbaki theorem for a sequence of projections composed with elements in an algebraic basis of E^*). We are to show that $\sup\{\lambda(E) : \dim E < \infty\} < \infty$.

Assume that this is not the case. Then we claim that for every subspace X_0 of X of finite codimension, we have

$$\sup\{\lambda(E) : \dim E < \infty, \ E \subset X_0\} = \infty. \tag{6.40}$$

Indeed, assume that for some subspace X_0 of X of codimension k we have

$$M := \sup\{\lambda(E) : \dim E < \infty, \ E \subset X_0\} < \infty.$$

Let E be a finite-dimensional subspace of X. Put $E_0 = E \cap X_0$. Let P_0 be a projection from X onto E_0 with $\|P_0\| \leq M$. Put $F = \{x \in E : P_0 x = 0\}$. Since $\dim F \leq k$, by using Theorem 4.5 there is a projection P_1 from X onto F with $\|P_1\| \leq k$. Let $P = P_0 + P_1 - P_1 P_0$, then it can be checked that P is a projection from X onto E with $\|P\| \leq (M + 1)(k + 1)$, a contradiction. This proves (6.40).

If E is a finite-dimensional subspace of X and $\varepsilon > 0$, then there is a finite-codimensional subspace X_0 of X such that

$$\|e + x\| \geq (1 - \varepsilon)\|e\|, \ e \in E, \ x \in X_0$$

(see the proof of Lemma 4.20).

We will now proceed by induction to construct a sequence $\{E_n\}_{n=1}^{\infty}$ of finite-dimensional subspaces and a sequence of $\{X_n\}_{n=1}^{\infty}$ finite-codimensional subspaces so that, for all $n \in \mathbb{N}$,

(i) $\lambda(E_n) > n$,
(ii) $\|e + x\| \geq \frac{1}{2}\|e\|, \ e \in E_n, \ x \in X_n$,
(iii) $E_{n+1} \subset X_n$,
(iv) $X_{n+1} \subset X_n$.

Let $Y = \overline{\operatorname{span}} \bigcup_{n=1}^{\infty} E_n$. Fix $N \in \mathbb{N}$. If $e_j \in E_j$, $j = 1, 2, ..., N$, and $1 \leq m \leq N$, we have

$$\|e_1 + \cdots + e_m\| \le 2\|e_1 + \cdots + e_N\|.$$

Thus

$$\|e_m\| \le 4\|e_1 + \cdots + e_N\|.$$

Therefore each E_m is 4-complemented in Y. Since, by assumption, Y is complemented in X, we get that $\sup_n \lambda(E_n) < \infty$, which is a contradiction. This finishes the proof of the lemma. □

Proof of Theorem 6.16: Assume that each subspace of X is complemented. By the crude Dvoretzky Theorem 6.9, there is a $K > 0$ such that the Hilbert space is K-finitely representable in X. It is easy to see that the same value of K may be used for all subspaces of finite codimension of X. Let λ be the constant from Lemma 6.17. We will show that X is crudely finitely represented in a Hilbert space. In view of (iv) in Theorem 6.2, this will give the conclusion.

For this, let E be an n-dimensional subspace of X and let

$$\alpha = \mathrm{d}(E, \ell_2^n) \tag{6.41}$$

where $\mathrm{d}(\cdot, \cdot)$ denotes the Banach–Mazur distance.

We are going to show that $\alpha \le \lambda^4 2^6 K^3$. For this, let Q be a projection from X onto E with $\|Q\| \le \lambda$. There is a subspace F of $(I - Q)X$ such that

$$\mathrm{d}(F, \ell_2^n) \le K. \tag{6.42}$$

By the multiplicative triangle inequality for the Banach–Mazur distance it follows from (6.41) and (6.42) that there is an operator T from E onto F such that

$$\|x\|/(K\alpha) \le \|Tx\| \le \|x\|, \text{ for all } x \in E. \tag{6.43}$$

Let P be a projection of norm less than or equal to λ from X onto the subspace $G := \{x + \mu Tx : x \in E\}$, where $\mu := 2^4\lambda^2 K^2$. Observe that each element $d \in G$ can be written in a unique way as $x + \mu Tx$, where $x \in E$.

Let $V : E \to E$ be the operator defined by

$$PTx = Vx + \mu TVx, \ x \in E. \tag{6.44}$$

Then

$$Vx = QPTx, \text{ for } x \in E.$$

Therefore

$$\|Vx\| \le \lambda^2 \|Tx\|. \tag{6.45}$$

Note that for each $x \in E$ we have

$$Px = P(x+\mu Tx) - \mu PTx = x + \mu Tx - \mu(Vx + \mu TVx) = (x - \mu Vx) + \mu(Tx - \mu TVx),$$

where $(x - \mu Vx) \in E$ $(= QX)$ and $\mu(Tx - \mu TVx) \in F$ $\big(\subset (I - Q)X\big)$. Hence

$$(I - Q)Px = \mu(Tx - \mu TVx), \ x \in E. \tag{6.46}$$

Combining (6.43), (6.45), and (6.46) we obtain, for $x \in E$,

$$\begin{aligned} \|(I - Q)Px\| &= \mu\|T(x - \mu Vx)\| \geq \frac{\mu}{K\alpha}\|x - \mu Vx\| \\ &\geq \frac{\mu}{K\alpha}(\|x\| - \mu\|Vx\|) \geq \frac{\mu}{K\alpha}(\|x\| - \mu\lambda^2\|Tx\|) \end{aligned} \tag{6.47}$$

Using (6.42), we can find an operator U from F onto ℓ_2^n such that

$$\|y\|/K \leq \|Uy\| \leq \|y\|, \ y \in F. \tag{6.48}$$

Let $\widehat{T}$ be the operator from E into the $2n$-dimensional Hilbert space $\ell_2^n \oplus \ell_2^n$ defined by

$$\widehat{T}x = \left(\frac{1}{2}UTx, (4\lambda^2)^{-1}U(I - Q)Px\right) \tag{6.49}$$

Then, for $x \in E$,

$$\begin{aligned} \|\widehat{T}x\| &\leq \|U\| \cdot \|T\| \cdot \|x\|/2 + \|U\| \cdot \|I - Q\| \cdot \|P\| \cdot \|x\|/(4\lambda^2) \\ &\leq \|x\|/2 + (\lambda + 1)\lambda\|x\|/(4\lambda^2) \leq \|x\| \end{aligned} \tag{6.50}$$

On the other hand, by (6.47), (6.48), and (6.49), we have, for $x \in E$,

$$\begin{aligned} \|\widehat{T}x\| &\geq \max\{\|UTx\|/2, \|U(I - Q)Px\|/(4\lambda^2)\} \\ &\geq \max\{\|Tx\|/(2K), \|(I - Q)Px\|/(4K\lambda^2)\} \\ &\geq \max\left\{\|Tx\|/(2K), \frac{\mu}{4K^2\alpha\lambda^2}(\|x\| - \mu\lambda^2\|Tx\|)\right\} \end{aligned} \tag{6.51}$$

We will distinguish between two cases.

(a) If $\lambda^4 2^5 K^2\|Tx\| \geq \|x\|$, then

$$\|\widehat{T}x\| \geq \|Tx\|/(2K) \geq \|x\|/(\lambda^4 2^6 K^3). \tag{6.52}$$

(b) Otherwise, i.e., if $\lambda^4 2^5 K^2\|Tx\| < \|x\|$, then, since $\mu = 2^4\lambda^2K^2$, we have

$$\|\widehat{T}x\| \geq \frac{\mu}{4\alpha\lambda^2K^2}(\|x\| - \lambda^4 2^4 K^2\|Tx\|) \geq \frac{\mu}{8\alpha\lambda^2K^2}\|x\| = \frac{2}{\alpha}\|x\| \tag{6.53}$$

Thus, in either case,

$$\|\widehat{T}x\| \geq \|x\| \min\left\{\frac{2}{\alpha}, \frac{1}{\lambda^4 2^6 K^3}\right\}, \quad x \in E, \tag{6.54}$$

i.e., $\widehat{T}$ is an isomorphism from E onto ℓ_2^n. Since α is minimal, it follows from (6.41), (6.50), and (6.54), that $\alpha \leq \lambda^4 2^6 K^3$. This finishes the proof of one of the implications.

Assume now that X is isomorphic to a Hilbert space. The conclusion follows from Corollary 1.50 and Fact 4.3. □

6.5 The John Ellipsoid

Definition 6.18 *Let n be a positive integer. An n-*dimensional ellipsoid *in a normed space X is a translate of the image of $B_{\ell_2^n}$ under a one-to-one bounded operator from ℓ_2^n into X.*

Let $T : \ell_2^n \to X$ be a one-to-one bounded operator into a normed space X. Since T is an isomorphism onto $T\ell_2^n$ and T preserves middle points, it is easy to see that the ellipsoid $TB_{\ell_2^n}$ is a strictly convex body in the space $T\ell_2^n$ (in the sense that it generates a strictly convex norm in $T\ell_2^n$). If $e_0 \in X$, the ellipsoid $TB_{\ell_2^n} + e_0$ is said to be *centered at* e_0.

Lemma 6.19 *Let X be an n-dimensional normed space and $E \subset X$ be an n-dimensional ellipsoid centered at 0. Then, given x_0 and y_0 in E, the segment*

$$[x_0, y_0] - \left(\frac{x_0 + y_0}{2}\right)$$

lies in Int(E).

Proof: Note that (see Fig. 6.1)

$$x_0 - \left(\frac{x_0 + y_0}{2}\right) = \frac{x_0 - y_0}{2} = \frac{x_0 + (-y_0)}{2}.$$

Since $-y_0 \in E$ and E is convex, we get $x_0 - (1/2)(x_0 + y_0) \in E$. Moreover, E is strictly convex (see the preceding observation). Then $x_0 - (1/2)(x_0 + y_0)$ does not belong to the boundary of E. The argument applies to y_0 as well. Finally, it is enough to observe that the interior set of a convex set is itself convex. □

In this section, an n-dimensional real normed space X is identified algebraically with $\mathbb{R}^n$ by fixing a Hamel basis of X. The corresponding Euclidean product on X is denoted by "$\cdot$", the Euclidean norm by $\|\cdot\|_2$, and the Euclidean volume by "vol." We put v for the Euclidean volume of $B_{(X,\|\cdot\|_2)}$.

Fig. 6.1 Lemma 6.19

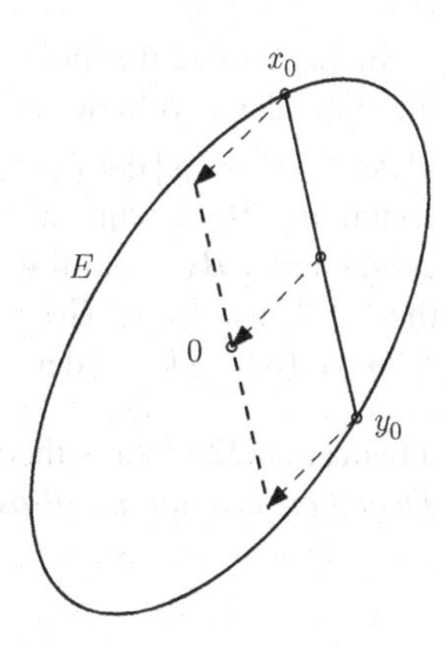

Corollary 6.20 *Let X be an n-dimensional normed space. Assume that an ellipsoid E in X of minimum volume containing B_X exists. Then E is centered at* 0.

Proof: Assume that E is an ellipsoid of minimum volume in X containing B_X and that E is centered at some e_0 ($\neq 0$). Lemma 6.19 ensures that $B_X + e_0$ in contained in Int(E). A compactness argument gives $\varepsilon > 0$ such that $\operatorname{dist}(x, \partial E) \geq \varepsilon$ for all $x \in B_X + e_0$, where ∂E denotes the boundary of E. Therefore, a suitable homothetic image of E from e_0 (using a positive factor strictly less than 1) still contains $B_X + e_0$. By adding $-e_0$ we obtain an ellipsoid containing B_X having a volume strictly smaller than the volume of E, a contradiction (see Fig. 6.2). □

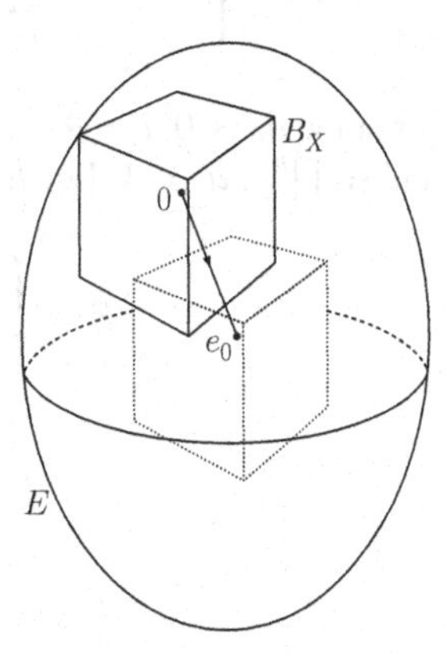

Fig. 6.2 Corollary 6.20

Lemma 6.21 *Let $(X, \|\cdot\|)$ be an n-dimensional normed space. Then, there exists at least one ellipsoid of minimum volume containing B_X.*

Proof: By Corollary 6.20, it is enough to consider the family of all ellipsoids in X containing B_X and centered at 0. This family corresponds to the set of all isomorphisms $T \in \mathcal{B}(\ell_2^n, X)$ such that $TB_{\ell_2^n} \supset B_X$, or, equivalently,

$$\left((T^*)^{-1}(B_{(\ell_2^n)^*}) =\right) (TB_{\ell_2^n})^\circ \subset (B_X)^\circ \ (= B_{X^*}).$$

In particular, the polar set of an ellipsoid in X is an ellipsoid in X^*, and the Euclidean volume of the ellipsoid $\big(T(B_{\ell_2^n})\big)^\circ$ in X^* is $v|\det(T^*)^{-1}|$ ($= v|\det T^*|^{-1} = v|\det T|^{-1}$). So, the existence of an ellipsoid of minimum volume containing B_X is equivalent to the existence of an ellipsoid of maximal volume contained in B_{X^*}, and this is proved by a simple compactness argument involving the set $\mathcal{I} := \{S \in \mathcal{B}(\ell_2^n, X^*) : \|S\| \leq 1\}$ and the continuity of the function $S \to \mathrm{vol}(SB_{\ell_2^n})$ ($= v|\det S|$, where $v := \mathrm{vol}(B_{(X,\|\cdot\|_2)})$). □

Theorem 6.22 (John's theorem [JohnF]) *Let $X = (\mathbb{R}^n, \|\cdot\|)$ be a normed space. Then there is a unique ellipsoid D of minimum volume containing B_X. Furthermore,*

$$n^{-1/2}D \subset B_X \subset D. \tag{6.55}$$

In particular, $\mathrm{d}(X, \ell_2^n) \leq n^{1/2}$, where $\mathrm{d}(\cdot,\cdot)$ denotes the Banach–Mazur distance.

Proof: Lemma 6.21 ensures the existence of such ellipsoid. By Corollary 6.20, every such ellipsoid is centered at 0.

Let us first prove the uniqueness. Suppose that D and D' are ellipsoids of minimum volume containing B_X (see Fig. 6.3). If $T \in \mathcal{B}(\mathbb{R}^n)$ is invertible then $T(D)$ and $T(D')$ are ellipsoids of minimum volume containing $T(B_X)$. Hence we may assume that

$$D = \left\{x \in \mathbb{R}^n : \sum_{i=1}^n x_i^2 \leq 1\right\} \text{ and } D' = \left\{x \in \mathbb{R}^n : \sum_{i=1}^n \frac{x_i^2}{a_i^2} \leq 1\right\},$$

for some $a_i > 0$, $i = 1, 2, \ldots, n$. Putting $v = \mathrm{vol}\, D$, we have $\mathrm{vol}\, D' = v\prod_{i=1}^n a_i$, and so $\prod_{i=1}^n a_i = 1$. Let E be the ellipsoid

$$E = \left\{x \in \mathbb{R}^n : \sum_{i=1}^n \frac{1}{2}x_i^2\left(1 + \frac{1}{a_i^2}\right) \leq 1\right\}.$$

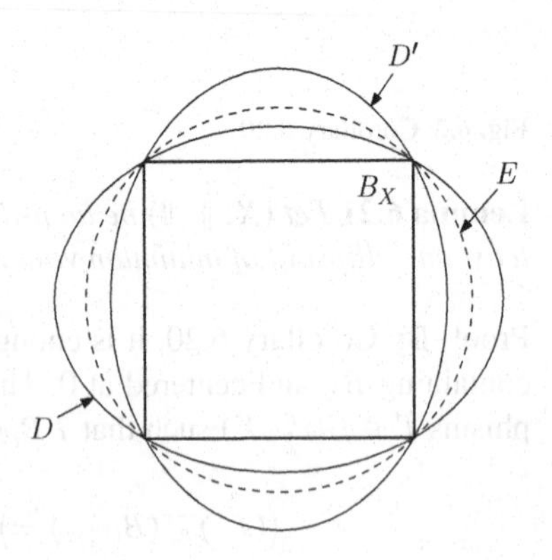

Fig. 6.3 The minimum volume ellipsoid E

Then

$$B_X \subset D \cap D' \subset E.$$

Indeed, given $x \in D \cap D'$, we have

$$\sum_{i=1}^{n} \frac{1}{2} x_i^2 \left(1 + \frac{1}{a_i^2}\right) = \frac{1}{2}\left(\sum_{i=1}^{n} x_i^2 + \sum_{i=1}^{n} \frac{x_i^2}{a_i^2}\right) \le 1.$$

Moreover,

$$\frac{(\mathrm{vol}\, E)^2}{(\mathrm{vol}\, D)^2} = \prod_{i=1}^{n} \frac{2}{1 + 1/a_i^2} = \prod_{i=1}^{n} \frac{2a_i^2}{1 + a_i^2} = \prod_{i=1}^{n} \frac{2a_i}{1 + a_i^2} < 1,$$

since $\prod_{i=1}^{n} a_i = 1$, we have $1 + a_1^2 \ge 2a_i$ for all $i = 1, 2, \ldots, n$, and not every a_i is 1. This contradicts the assumption that D was an ellipsoid of minimum volume containing B_X.

Let us turn to the proof of $n^{-1/2} D \subset B_X$. Suppose that this is not the case, so S_X has a point in the interior of $n^{-1/2} D$. By taking a support plane of B_X at such a point and rotating B_X to make this support plane parallel to the plane of the axes $x_2, x_3, \ldots, x_n$, we may assume that

$$B_X \subset P := \left\{ x \in \mathbb{R}^n : |x_1| \le \frac{1}{c} \right\}$$

for some $c > n^{1/2}$.

For $a > b > 0$ define an ellipsoid $E_{a,b}$ by

$$E_{a,b} = \left\{ x \in \mathbb{R}^n : a^2 x_1^2 + b^2 \sum_{i=2}^{n} x_i^2 \le 1 \right\},$$

so that $(\mathrm{vol}\, D)/(\mathrm{vol}\, E_{a,b}) = ab^{n-1}$ (see Fig. 6.4). If $x \in B_X$ then $x \in D \cap P$, and so

$$a^2 x_1^2 + b^2 \sum_{i=2}^{n} x_i^2 = (a^2 - b^2) x_1^2 + b^2 \sum_{i=1}^{n} x_i^2 \le \frac{a^2 - b^2}{c^2} + b^2.$$

It follows that $B_X \subset E_{a,b}$ and $\mathrm{vol}\, E_{a,b} < \mathrm{vol}\, D$ whenever

$$\frac{a^2 - b^2}{c^2} + b^2 \le 1 \text{ and } ab^{n-1} > 1.$$

Thus, to complete the proof, it suffices to show that these inequalities are satisfied for some choice of $0 < b < a$. To show this, take $0 < \varepsilon < (1/2)$, $a = (1 + \varepsilon + 2\varepsilon^2)^{n-1}$

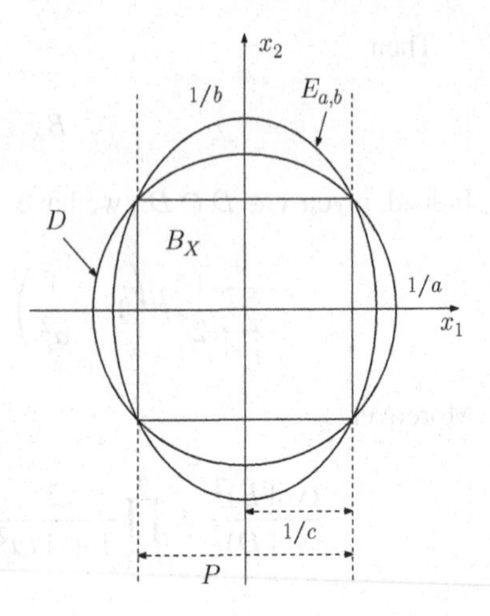

Fig. 6.4 The ellipsoid $E_{a,b}$

and $b = 1 - \varepsilon$. Then $ab^{n-1} > 1$. Also

$$\frac{a^2 - b^2}{c^2} + b^2 = \frac{(1 + \varepsilon + 2\varepsilon^2)^{2(n-1)}}{c^2} + (1 - \varepsilon)^2 \left(1 - \frac{1}{c^2}\right)$$
$$= 1 + 2\varepsilon \left(\frac{n-1}{c^2} - 1 + \frac{1}{c^2}\right) + O(\varepsilon^2) = 1 + 2\varepsilon \left(\frac{n}{c^2} - 1\right) + O(\varepsilon^2) < 1$$

if $\varepsilon > 0$ is sufficiently small. □

Corollary 6.23 *Let X and Y be n-dimensional normed spaces. Then* $\mathrm{d}(X, Y) \leq n$.

Proof: $\mathrm{d}(X, Y) \leq \mathrm{d}(X, \ell_2^n)\mathrm{d}(\ell_2^n, Y) \leq n^{1/2}n^{1/2} = n$. □

Remark: Theorem 6.22 can be formulated in terms of *maximum volume ellipsoids inside the unit ball of an n-dimensional normed space X* (see the argument in the proof of Lemma 6.21). The "inside" version of the theorem is thus a consequence of the "outside" formulation.

Let n be a positive integer. John's Theorem 6.22, in the version of the previous remark, ensures that, if $(X, \|\cdot\|)$ is an n-dimensional real space, there exists a unique n-ellipsoid of maximum volume contained in B_X. This amounts to say that an isomorphism $T : \ell_2^n \to X$ exists such that $TB_{\ell_2^n}$ is the unique ellipsoid of maximum volume contained in B_X or, in other words, that $B_{\ell_2^n}$ is the maximum volume ellipsoid contained in the compact convex and symmetric set $T^{-1}B_X$. Theorem 6.27 below gives a characterization of sets of the form $T^{-1}B_X \subset \ell_2^n$ for such a mapping T.

The rest of this section is an elaboration of the proof of John's ellipsoid theorem given in [GrSc].

Before we formulate the statement, we need to fix some notation. Let us recall that ∂C denotes the boundary of a subset C of a normed space.

Fix $n \in \mathbb{N}$. As it is customary, an element $u \in \mathbb{R}^n$ has a representation as a $n \times 1$-matrix. Let $\mathcal{M}$ be the (finite-dimensional) vector space of all $n \times n$-real matrices. Given $u, v \in \mathbb{R}^n$, let $u \otimes v = uv^t \in \mathcal{M}$. This is consistent with the general definition of the tensor product of a vector and a linear functional. Given $A = (a_{i,j})$, $B = (b_{i,j})$ in $\mathcal{M}$, put $A \cdot B = \sum_{i,j=1}^n a_{i,j} b_{i,j}$. This is a scalar product in $\mathcal{M}$. The same symbol $\cdot$ stands also for the Euclidean scalar product in $\mathbb{R}^n$.

Let $M \in \mathcal{M}$, $u, v, w \in \ell_2^n$. It is easy to prove the following:

$$(u \otimes v)w = (v \cdot w)u. \tag{6.56}$$

$$(u \otimes v) \cdot M = (Mv) \cdot u. \tag{6.57}$$

If $\det M \neq 0$, *then* $M = SQ$,
where S *is a symmetric, positive-definite* $n \times n$*-matrix*
and Q *is a orthogonal* $n \times n$*-matrix.* (6.58)

Indeed, put $S = (MM^t)^{1/2}$, $Q = S^{-1}M$. It is simple to prove that this two matrices satisfy the requirements in (6.58).

In dealing with all possible n-ellipsoids in an n-dimensional normed space, we consider then $MB_{\ell_2}^n$, where $M \in \mathcal{M}$ and $\det M \neq 0$. By (6.58), it is enough to consider the n-ellipsoids $SB_{\ell_2^n}$, where S is a symmetric positive-definite $n \times n$-matrix. So we are driven to consider $\mathcal{S}$, the $(1/2)n(n+1)$-dimensional vector subspace of $\mathcal{M}$ of all symmetric $n \times n$-matrices. Put $\mathcal{P}$ for the open cone with vertex at 0 of all positive-definite matrices in $\mathcal{S}$. Let $\mathcal{D} = \{S \in \mathcal{P} : \det(S) \geq 1\}$.

Proposition 6.24 *The set* $\mathcal{D}$ *is a strictly convex and smooth subset of* $\mathcal{P}$.

Proof: Given two distinct elements S_1 and S_2 in $\mathcal{D}$, a standard result in linear algebra says that we can find a single orthogonal $n \times n$-matrix that reduces both S_1 and S_2 to a diagonal form, say D_1 and D_2, respectively. Obviously, $\det S_i = \prod_{k=1}^n \lambda_k^{(i)}$, where the diagonal of D_i is $(\lambda_k^{(i)})_{k=1}^n$, $i = 1, 2$ (and so $\det(\delta S_1 + (1-\delta)S_2) = \prod_{i=1}^n (\delta\lambda_i^{(1)} + (1-\delta)\lambda_i^{(2)})$, for $0 \leq \delta \leq 1$). The function $f : \mathbb{R}^n \to \mathbb{R}$ given by $f(\lambda_1, \ldots, \lambda_n) := \prod_{k=1}^n \lambda_k$ is strictly convex on the domain $\{(\lambda_1, \ldots, \lambda_n) \in \mathbb{R}^n : \lambda_k > 0,\ k = 1, 2, \ldots, n\}$. This proves the strict convexity of the set $\mathcal{D}$. The smoothness follows from the implicit mapping theorem. □

Each $v \in \ell_2^n$ defines an element in $(\ell_2^n)^*$, namely the mapping $u \mapsto u \cdot v$, $u \in \ell_2^n$. Let C be a compact convex and symmetric subset of ℓ_2^n such that $B_{\ell_2^n} \subset C$. The set C is the intersection of all the closed half-spaces containing C determined by its support hyperplanes, see Theorem 3.45. It follows that

$$C = \{u \in \ell_2^n : u \cdot v \leq \sup_C v,\ v \in S_{\ell_2^n}\},$$

hence the family of all n-ellipsoids in ℓ_2^n contained in C is represented by

$$\mathcal{E} := \mathcal{P} \cap \bigcap_{u,v \in S_{\ell_2^n}} H_{u,v}, \tag{6.59}$$

where $H_{u,v} := \{S \in \mathcal{S} : \ Su \cdot v \le \sup_C v\}$, a closed half-space in $\mathcal{S}$ for $u, v \in S_{\ell_2^n}$. In particular, $\mathcal{E}$ is a convex subset of $\mathcal{P}$. Since $B_{\ell_2^n} \subset C$, we get $I \in \mathcal{E}$, where I is the $n \times n$-identity matrix.

Lemma 6.25 *Let $u, v \in S_{\ell_2^n}$. Then, $I \in \partial H_{u,v}$ if and only if $u = v \in S_{\ell_2^n} \cap \partial C$ (and, if this is the case, $\sup_C u = \sup_{B_{\ell_2^n}} u$).*

Proof: Assume first that $I \in \partial H_{u,v}$. Then $u \cdot v = \sup_C v$ and we get

$$u \cdot v = \sup_C v \ge \sup_{B_{\ell_2^n}} v \ge u \cdot v,$$

so $u \cdot v = \sup_C v = \sup_{B_{\ell_2^n}} v = v \cdot v$. Due to the strict convexity of $B_{\ell_2^n}$ we get $u = v$, and $u \in \partial C \cap S_{\ell_2^n}$. The converse is obvious. □

Lemma 6.26 *The cone generated by $\mathcal{E}$ in $\mathcal{S}$ with vertex at I is the set $\mathcal{K} := \bigcap_{u \in S_{\ell_2^n} \cap \partial C} H_{u,u}$.*

Proof: Assume that $M \in \mathcal{S}$ belongs to the cone generated by $\mathcal{E}$ with vertex at I, and that $M \ne I$. Then, there exists $\lambda > 0$ and $E \in \mathcal{E}$ such that $M = I + \lambda(E - I)$. We get, for $u \in S_{\ell_2^n} \cap \partial C$,

$$Mu \cdot u = \big(u + \lambda(Eu - u)\big) \cdot u = 1 + \lambda(Eu \cdot u - 1) \le 1 + \lambda(\sup_C u - 1) = 1,$$

since $Eu \cdot u \le \sup_C u = 1$. This shows that $M \in H_{u,u}$.

Assume now that $M \in \bigcap_{u \in S_{\ell_2^n} \cap \partial C} H_{u,u}$. Let $E_\lambda = (1 - \lambda)I + \lambda M$, where $\lambda > 0$. Since I is positive-definite and M is symmetric, there exists an orthogonal $n \times n$-matrix that reduces I and M simultaneously to a diagonal form. This shows that we can choose $\lambda > 0$ small enough to ensure that $E_\lambda \in \mathcal{P}$. Moreover, for every $v \in S_{\ell_2^n}$ we have $\sup_C v \ge 1$. Since $E_\lambda u \cdot v = \big((1-\lambda)u + \lambda Mu\big) \cdot v$, we may choose $0 < \lambda$ small enough to have, too, $E_\lambda u \cdot v \le 1$ ($\le \sup_C v$) for all $u, v \in S_{\ell_2^n}$. This proves that $E_\lambda \in \mathcal{E}$ for this λ. □

Given a nonempty convex set C in a Euclidean space E and a point $c_0 \in C$, the *normal cone to C at c_0* is the set

$$\mathcal{N}_C(c_0) := \{x \in E : \ x \cdot (c - c_0) \le 0, \ \text{for all } c \in C\}. \tag{6.60}$$

The normal cone has, by definition, vertex at 0.

In the next result, (i)$\Longrightarrow$(ii) is due to John, (ii)$\Longrightarrow$(i) to Ball and Pełczyński. The proof is an elaboration of the argument in [GrSc].

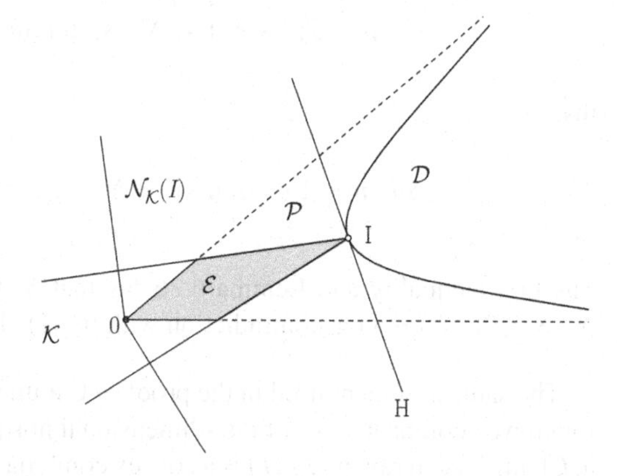

Fig. 6.5 The proof of Theorem 6.27

Theorem 6.27 (John [JohnF], Ball [Ball], Pełczyński [Pelc7]) *Let n be a positive integer. Let C be a compact convex and symmetric subset of ℓ_2^n such that $B_{\ell_2^n} \subset C$. Then the two following statements are equivalent.*
(i) *$B_{\ell_2^n}$ is the unique ellipsoid of maximum volume contained in C.*
(ii) *If I denotes the $n \times n$-identity matrix, there are $u_k \in B_{\ell_2^n} \cap \partial C$, $\lambda_k > 0$, $k = 1, 2, \dots m$, where $n \leq m \leq \frac{1}{2}n(n+1)$, such that*

$$I = \sum_{k=1}^{m} \lambda_k u_k \otimes u_k. \tag{6.61}$$

Proof: ([GrSc]) (i)$\Longrightarrow$(ii): Observe that if $S \in \mathcal{P}$, then $v \det(S)$ is the Euclidean volume of the ellipsoid $SB_{\ell_2^n}$, where $v := \mathrm{vol}(B_{\ell_2^n})$. From (i) it follows that $\det(S) < 1$ for $S \in \mathcal{E}\backslash\{I\}$. In particular, $\mathcal{E} \cap \mathcal{D} = \{I\}$ (see Fig. 6.5).

Claim 1: $I \in \mathcal{N}_{\mathcal{K}}(I)$.

Indeed, the separation theorem gives a hyperplane H ($:= \{M \in \mathcal{S} : f(M) = \alpha\}$ for some $f \in \mathcal{S}^*$ and some $\alpha > 0$) containing I and separating $\mathcal{E}$ and $\mathcal{D}$, say $f(K) \leq \alpha$ for every $K \in \mathcal{K}$ and $f(D) \geq \alpha$ (> 0) for all $D \in \mathcal{D}$. The smoothness of the set $\mathcal{D}$ (see Proposition 6.24) ensures that this separating hyperplane is unique. Since the gradient of the mapping $M \to \det M$ at I is again I, we get that I is orthogonal to H. It is enough to use Exercise 6.9 to conclude the proof of Claim 1. $\triangle$

Claim 2: $\mathcal{N}_{\mathcal{K}}(I) = \mathrm{conv}\,\{\lambda(u \otimes u) : u \in B_{\ell_2^n} \cap \partial C, \lambda \geq 0\}$.

Observe that, because of (6.57), $(u \otimes u) \cdot I = (Iu) \cdot u = 1$ for all $u \in B_{\ell_2^n} \cap \partial C$, so $\{u \otimes u : B_{\ell_2^n} \cap \partial C\} \subset G$, where $G := \{S \in \mathcal{S} : S \cdot I = 1\}$, a hyperplane in $\mathcal{S}$ that does not contain the origin. To prove the Claim is clearly enough to show that $\mathcal{N}_{\mathcal{K}}(I) \cap G = \mathrm{conv}\,\{u \otimes u : u \in B_{\ell_2^n} \cap \partial C\}$. Assume not. Observe that $B_{\ell_2^n} \cap \partial C$ is a compact set, so it is $\{u \otimes u : u \in B_{\ell_2^n} \cap \partial C\}$ and then $\mathrm{conv}\,\{u \otimes u : u \in B_{\ell_2^n} \cap \partial C\}$, too. By the separation theorem, we can find $S \in \mathcal{S}$ and $N \in \mathcal{N}_{\mathcal{K}}(I) \cap G$ such that

$$(u \otimes u) \cdot S \leq 1 < N \cdot S, \text{ for all } u \in B_{\ell_2^n} \cap \partial C,$$

that is

$$Su \cdot u \leq 1 \; (= \sup_C u) \; < N \cdot S, \text{ for all } u \in B_{\ell_2^n} \cap \partial C. \tag{6.62}$$

The first inequality and Lemma 6.26 say that $S \in \mathcal{K}$. Since $N \in \mathcal{N}_{\mathcal{K}}(I)$, we get $N \cdot S \leq N \cdot I \; (= 1)$, a contradiction with (6.62). This proves Claim 2. $\triangle$

The same argument used in the proof of Carathéodory's result (Proposition 3.70) for convex compact sets in finite-dimensional normed spaces shows that, according to Claim 2, each ray in $\mathcal{N}_K(I)$ is a convex combination of a number $m \leq (1/2)n(n+1)$ of rays of the form $\{\lambda(u \otimes u) : \lambda \geq 0\}$, where $u \in B_{\ell_2^n} \cap \partial C$. This proves Equation (6.61).

For showing that $m \geq n$, it suffices to prove that $\operatorname{span}\{u_1, \ldots, u_m\} = \ell_2^n$. Assume that this is not the case; we can find then $u \neq 0$ such that $u \perp u_i$ for $i = 1, 2, \ldots, m$. Then, by (6.61) and (6.56),

$$0 \neq u \cdot u = Iu \cdot u = \left(\sum_{k=1}^{m} \lambda_k (u_k \otimes u_k) u \right) \cdot u = \left(\sum_{k=1}^{m} \lambda_k (u_k \cdot u) u_k \right) \cdot u = 0,$$

a contradiction.

(ii)$\Longrightarrow$(i): Since $B_{\ell_2^n} \subset C$, we get $I \in \mathcal{E}$. Observe that, for $k \in \{1, 2, \ldots, n\}$, $Iu_k \cdot u_k = 1 \; (= \sup_C u_k)$. It follows that $I \in \partial H_{u_k, u_k}$. This implies that, in fact, $I \in \partial \mathcal{E}$. Certainly, $I \in \partial \mathcal{D}$, and the gradient argument used before implies that I is orthogonal to $\mathcal{D}$ at I. This shows that the hyperplane $H := \{S \in \mathcal{S} : S \cdot I = 1\}$ supports $\mathcal{D}$ at I, in the sense of Definition. By (6.61) and Claim 2 we get $I \in \mathcal{N}_{\mathcal{K}}(I)$, so $I \cdot (K - I) \leq 0$ for all $K \in \mathcal{K}$, that is, $I \cdot K \leq 1$ for all $K \in \mathcal{K}$. This implies that $\mathcal{K}$ is in one of the half-spaces defined by H, so H separates $\mathcal{K}$ and $\mathcal{D}$ and, a fortiori, $\mathcal{E}$ and $\mathcal{D}$. Since $\mathcal{D}$ is smooth (see Proposition 6.24), $\mathcal{D} \cap \mathcal{E} = \{I\}$, so $B_{\ell_2^n}$ is the unique ellipsoid of maximal volume contained in C. $\square$

6.6 Kadec–Snobar Theorem

In Theorem 4.5 we proved, as a consequence of the existence in every finite-dimensional Banach space of an Auerbach basis, that the projection constant $\lambda(X)$ of an n-dimensional Banach space is less than or equal to n. Theorem 6.28 provides a substantial improvement. In order to prove it, we need some preliminary results.

Theorem 6.28 (Kadec, Snobar [KaSn]) *Let $(X, \|\cdot\|)$ be an n-dimensional Banach space. Then $\lambda(X) \leq \sqrt{n}$.*

Proof: The space X is isometric to a subspace of $\big(C(B_{X^*}), \|\cdot\|_\infty\big)$, where $\|\cdot\|_\infty$ denotes the supremum norm. According to Proposition 5.26, $\lambda(X) = \lambda\big(X, C(B_{X^*})\big)$.

By Fritz John's Theorem 6.22, there exists an ellipsoid E of maximum volume in B_X. This is the image of $B_{\ell_2^n}$ under an isomorphism from ℓ_2^n onto X. Let us denote this isomorphism T^{-1} (hence T is an isomorphism from X onto ℓ_2^n). The absolutely convex and compact subset TB_X has the property that $B_{\ell_2^n}$ is the maximum volume ellipsoid contained in it. Apply Theorem 6.27 to ensure that T enjoys, additionally, the following properties (as usual, we denote by $\|\cdot\|$ both the norm in X and the dual norm in X^*, and $\|\cdot\|_2$ the canonical norm in ℓ_2^n. Moreover, if $u \in \ell_2^n$, we may consider, as customary, u as an element in $(\ell_2^n)^*$, in such a way that $\langle u, v\rangle = (u, v)$ for $v \in \ell_2^n$, where $\langle\cdot,\cdot\rangle$ denotes the duality bilinear mapping on $\ell_2^n \times (\ell_2^n)^*$, and $(\cdot,\cdot)$ the scalar product in ℓ_2^n):

(i) $\|T^{-1}\| \le 1$.
(ii) There exist a set of vectors $\{y_1, y_2, \dots, y_s\}$ in ℓ_2 and a set of positive numbers $\{\lambda_1, \lambda_2, \dots, \lambda_s\}$ such that

(a) $n \le s \le n(n+1)/2$.
(b)

$$\|y_r\|_2 = \|T^{-1}y_r\| = \|T^*y_r\| = 1, \; r = 1, 2, \dots, s. \tag{6.63}$$

(c) Given $u, v \in \ell_2^n$, we have

$$\sum_{r=1}^{s} \lambda_r (y_r, u)(y_r, v) = (u, v). \tag{6.64}$$

Observe that, if e_i is the ith vector of the canonical basis of ℓ_2^n, we have, for $u = v := e_i$, $\sum_{r=1}^{s} \lambda_r (y_r, e_i)^2 = (e_i, e_i) = 1$. This holds for $i = 1, 2, \dots, n$, hence

$$n = \sum_{i=1}^{n}\sum_{r=1}^{s} \lambda_r (y_r, e_i)^2 = \sum_{r=1}^{s} \lambda_r \sum_{i=1}^{n} (y_r, e_i)^2 = \sum_{r=1}^{n} \lambda_r. \tag{6.65}$$

From a geometric point of view, (i) means that $B_{\ell_2} \subset TB_X$, (ii.b) that $\{T^{-1}y_r : r = 1, 2, \dots, s\}$ are contact points between S_X and the boundary of the maximal volume ellipsoid E inscribed in B_X (or that $\{y_r : r = 1, 2, \dots, s\}$ are contact points between $B_{\ell_2^n}$ and the boundary of TB_X), and that the tangents to $S_{\ell_2^n}$ at points y_r are also tangents to the boundary of TB_X at those points (see Fig. 6.6), and, finally, (iii) that although $\{y_r : r = 1, 2, \dots, s\}$ is not an orthonormal set, it behaves, after weighting, as such.

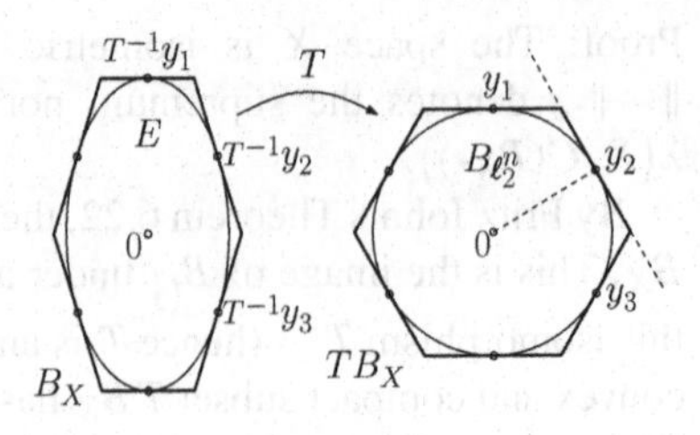

Fig. 6.6 The effect of the mapping T on B_X and the ellipsoid E

The required projection is defined by the formula

$$P(F) = \sum_{r=1}^{s} \lambda_r F(T^* y_r) T^{-1} y_r, \quad F \in C(B_{X^*}). \tag{6.66}$$

Observe that P is well defined, since $T^* y_r \in B_{X^*}$ for all $1 \leq r \leq s$, and P maps clearly $C(B_{X^*})$ into X. In order to see that it is a projection onto X, let $x \in X$. Then,

$$P(x) = \sum_{r=1}^{s} \lambda_r \langle T^* y_r, x\rangle T^{-1}(y_r) = \sum_{r=1}^{s} \lambda_r \langle y_r, Tx\rangle T^{-1} y_r = \sum_{r=1}^{s} \lambda_r (y_r, Tx) T^{-1} y_r.$$

Take $x^* \in X^*$ arbitrary. Then

$$\begin{aligned}\langle x^*, P(x)\rangle &= \sum_{r=1}^{s} \lambda_r (y_r, Tx) \langle x^*, T^{-1} y_r\rangle \\ &= \sum_{r=1}^{s} \lambda_r (y_r, Tx) \langle (T^{-1})^* x^*, y_r\rangle = (Tx, (T^{-1})^* x^*) = (x, x^*).\end{aligned}$$

This implies that $P(x) = x$ for all $x \in X$ and so P is a projection from $C(B_{X^*})$ onto X. We will estimate now $\|P\|$.

$$\begin{aligned}
\|P\| &= \sup_{F\in C(B_{X^*}),\ \|F\|_\infty=1} \|P(F)\| \\
&= \sup_{F\in C(B_{X^*}),\ \|F\|_\infty=1}\ \sup_{x^*\in B_{X^*}} \sum_{r=1}^{s} \lambda_r F(T^*y_r)\langle x^*, T^{-1}y_r\rangle \\
&\le \sup_{x^*\in B_{X^*}} \sum_{r=1}^{s} \lambda_r |\langle x^*, T^{-1}y_r\rangle| = \sup_{x^*\in B_{X^*}} \sum_{r=1}^{s} \lambda_r |\langle (T^{-1})^*x^*, y_r\rangle| \\
&= \sup_{x^*\in B_{X^*}} \sum_{r=1}^{s} \sqrt{\lambda_r}\sqrt{\lambda_r} |\langle (T^{-1})^*x^*, y_r\rangle| \\
&\le \sup_{x^*\in B_{X^*}} \left(\sum_{r=1}^{s}\lambda_r\right)^{1/2} \left(\sum_{r=1}^{s} \lambda_r |\langle (T^{-1})^*x^*, y_r\rangle|^2\right)^{1/2} \\
&= \sqrt{n} \sup_{x^*\in B_{X^*}} \|(T^{-1})^*x^*\| = \sqrt{n}\|T^{-1}\| \le \sqrt{n}.
\end{aligned}$$

For the last inequalities we used both (6.64) and (6.65). □

It is known that

$$\lambda(\ell_2^n) = \frac{n\Gamma(\frac{n}{2})}{\sqrt{\pi}\,\Gamma(\frac{n+1}{2})} \sim \sqrt{\frac{2n}{\pi}},$$

(see [KaSn]), hence the estimate given in Theorem 6.28 is almost optimal.

6.7 Grothendieck's Inequality

The following fundamental inequality will be used in Chapter 13 for the study of absolutely summing operators.

Theorem 6.29 (Grothendieck, [Groth2]) *Let $(\alpha_{i,j})_{i,j=1}^n$ be a matrix of scalars such that $\left|\sum_{i,j=1}^n \alpha_{i,j}s_i t_j\right| \le 1$ for every choice of scalars $\{s_i\}_{i=1}^n$ and $\{t_j\}_{j=1}^n$ such that $|s_i| \le 1$, $i = 1, 2, \dots, n$ and $|t_j| \le 1$, $j = 1, 2, \dots, n$. Then, there exists a universal constant K_G (called* Grothendieck's constant*) such that, for any choice of vectors $\{x_i\}_{i=1}^n$ and $\{y_j\}_{j=1}^n$ in a Hilbert space,*

$$\left|\sum_{i,j=1}^{n} \alpha_{i,j}(x_i, y_j)\right| \le K_G \max_i \|x_i\| \max_j \|y_j\|. \tag{6.67}$$

Proof: It is enough to prove the result for real scalars and for norm-one vectors. Since every finite-dimensional subspace of a Hilbert space is isometric to ℓ_2^k for some positive integer k, we may assume that vectors $\{x_i\}_{i=1}^n$ and $\{y_j\}_{j=1}^n$ are in a Hilbert space $X := \ell_2^k$ for some positive integer k. Let μ be the unique probability measure on S_X that is rotation-invariant. Take two vectors x and y in S_X. A simple two-dimensional computation gives

$$\int_{S_X} \operatorname{sign}(x, u).\operatorname{sign}(y, u)\, d\mu(u) = 1 - \frac{2\angle[x, y]}{\pi}, \tag{6.68}$$

where $\angle[x, y]$ is the angle between x and y, i.e., $0 \le \angle[x, y] \le \pi$ and $(x, y) = \cos \angle[x, y]$.

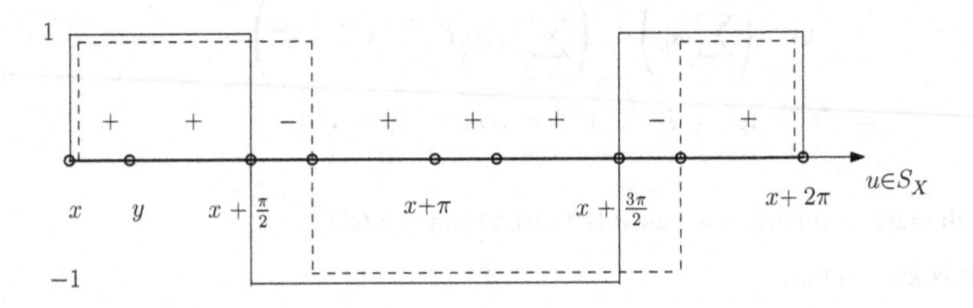

Fig. 6.7 The graphs of the sign functions

(Observe Fig. 6.7: the graph in full corresponds to the function $u \to \operatorname{sign}(x, u)$, while the dashed graph corresponds to the function $u \to \operatorname{sign}(y, u)$. Signs show where the product $\operatorname{sign}(x, u).\operatorname{sign}(y, u)$ has value $+1$ or -1. Equation (6.68) should now be clear). In view of the assumption on $(\alpha_{i,j})$ we have, for every $u \in S_X$ and for every $\{s_i\}_{i=1}^n$ and $\{t_j\}_{j=1}^n$ with $|s_i| \le 1$ and $|t_j| \le 1$, $i, j = 1, 2, \dots, n$, that

$$-1 \le \sum_{i,j=1}^n \alpha_{i,j} s_i t_j \operatorname{sign}(x_i, u)\operatorname{sign}(y_j, u) \le 1.$$

By integrating on $u \in S_X$ with respect to μ we get that

$$-1 \le \sum_{i,j=1}^n \alpha_{i,j} s_i t_j \left(1 - \frac{2\angle[x_i, y_j]}{\pi}\right) \le 1.$$

Hence, the matrix $\left(\alpha_{i,j}\left(1 - \frac{2\angle[x_i, y_j]}{\pi}\right)\right)_{i,j=1}^n$ also satisfies the assumptions made on $(\alpha_{i,j})_{i,j=1}^n$. By iterating this argument we get, for every positive integer m, that the matrix

$$\left(\alpha_{i,j}\left(1 - \frac{2\angle[x_i, y_j]}{\pi}\right)^m\right)_{i,j=1}^n$$

satisfies the assumption made on $(\alpha_{i,j})_{i,j=1}^n$. In particular, by using $s_i = t_j = 1$, $i, j = 1, 2, \ldots, n$, we get

$$\left| \sum_{i,j=1}^{n} \alpha_{i,j} \left(1 - \frac{2\angle[x_i, y_j]}{\pi}\right)^m \right| \leq 1. \tag{6.69}$$

Since, for $1 \leq i, j \leq n$,

$$\begin{aligned}
(x_i, y_j) &= \cos \angle[x_i, y_j] = \sin\left(\frac{\pi}{2} - \angle[x_i, y_j]\right) \\
&= \sum_{m=0}^{\infty} (-1)^m \left[\frac{\pi}{2} - \angle[x_i, y_j]\right]^{2m+1} \frac{1}{(2m+1)!} \\
&= \sum_{m=0}^{\infty} (-1)^m \left(\frac{\pi}{2}\right)^{2m+1} \frac{1}{(2m+1)!} \left[1 - \frac{2\angle[x_i, y_j]}{\pi}\right]^{2m+1},
\end{aligned}$$

we get, using (6.69),

$$\begin{aligned}
&\left| \sum_{i,j=1}^{n} \alpha_{i,j}(x_i, y_j) \right| \\
&= \left| \sum_{i,j=1}^{n} \alpha_{i,j} \sum_{m=0}^{\infty} (-1)^m \left(\frac{\pi}{2}\right)^{2m+1} \frac{1}{(2m+1)!} \left[1 - \frac{2\angle[x_i, y_j]}{\pi}\right]^{2m+1} \right| \\
&= \left| \sum_{m=0}^{\infty} (-1)^m \left(\frac{\pi}{2}\right)^{2m+1} \frac{1}{(2m+1)!} \sum_{i,j=1}^{n} \alpha_{i,j} \left[1 - \frac{2\angle[x_i, y_j]}{\pi}\right]^{2m+1} \right| \\
&\leq \sum_{m=0}^{\infty} \left(\frac{\pi}{2}\right)^{2m+1} \frac{1}{(2m+1)!} = \frac{1}{2}(e^{\pi/2} - e^{-\pi/2}).
\end{aligned}$$

□

Observe that, in the real case, this computation gives the estimate $K_G \leq (1/2)(e^{\pi/2} - e^{-\pi/2})$.

6.8 Remarks

The evaluation of the Banach–Mazur distance $\mathrm{d}(\cdot, \cdot)$ between ℓ_p^n spaces is as follows (see [GKM1] and [GKM2], see also [PeBe, p. 231] and Exercises 6.11 and 6.12):

If either $1 \leq p < q \leq 2$ or $2 \leq p < q \leq \infty$, then

$$\mathrm{d}(\ell_p^n, \ell_q^n) = n^{\frac{1}{p} - \frac{1}{q}}, \qquad (n = 1, 2, \ldots).$$

If $1 \leq p < 2 < q \leq \infty$, then

$$(\sqrt{2}-1)\mathrm{d}(\ell_p^n, \ell_q^n) \leq \max(n^{\frac{1}{p}-\frac{1}{2}}, n^{\frac{1}{2}-\frac{1}{q}}) \leq \sqrt{2}\mathrm{d}(\ell_p^n, \ell_q^n), \quad (n = 1, 2, \ldots).$$

Bohnenblust proved in his pioneering work [Bohn], in particular, that there is a three-dimensional space that admits no monotone Schauder basis. The method of the proof is probabilistic. Gordon and Lewis showed in [GoLe] that some well-known n-dimensional spaces of operators have unconditional basis constant of order greater than or equal to $\sqrt[4]{n}$. Later, Szarek showed in [Szar] that there is a constant $C > 0$ so that, for every n, there is a $2n$-dimensional Banach space X such that, for every projection P on X of rank n, $\|P\| \geq C\sqrt{n}$. Independently, Pisier proved in [Pisi3, p. 145] that there is an infinite-dimensional Banach space X and a number $\delta > 0$ such that any finite rank projection P from X into X satisfies $\|P\| \geq \delta\sqrt{\operatorname{rank} P}$. By probabilistic methods, Gluskin showed in [Glus] that for every n there are two n-dimensional spaces X_n and Y_n such that $\inf_n \mathrm{d}(X_n, Y_n)/n > 0$.

Johnson and Odell showed in [JoOd] that if X is an infinite-dimensional separable Banach space then, given a number $C > 0$, there are two equivalent norms $\|\cdot\|_1$ and $\|\cdot\|_2$ on X such that the Banach–Mazur distance from $(X, \|\cdot\|_1)$ to $(X, \|\cdot\|_2)$ is greater than C. For a result in this direction in a nonseparable setting see [Gode6].

Exercises for Chapter 6

6.1 Show that ℓ_2 is finitely representable in every ℓ_p, $1 \leq p < \infty$.
Hint. Use the fact that ℓ_2 is isometric to a subspace of every L_p for all $1 \leq p < \infty$ (see, e.g., [AlKa, p. 155]), together with Theorem 6.2.

6.2 Let X be a finite-dimensional Banach space. Show that if a Banach space Y is finitely representable in X, then Y is isometric to a subspace of X.
Hint. First note that $\dim(Y) \leq \dim(X)$ and that the unit ball of the space of operators on an n-dimensional Banach space is compact. If $T_n \colon Y \to X$ have the property that both $\|T_n\|$ and $\|T_n^{-1}\|$ tend to 1, take their limit point.

6.3 Show that if Y is crudely finitely representable in X and X has type p (respectively cotype p), then Y has type p (respectively cotype p).
Hint. Direct examination of definition.

6.4 Show that ℓ_q is not crudely finitely representable in ℓ_p for $q < p \leq 2$ or $2 \leq p < q$.

However, the Dvoretzky Theorem 6.15 gives that ℓ_2 is finitely representable in every Banach space.
Hint. Use Exercise 6.3 and the best types/cotypes of ℓ_p

6.5 It is known that the space ℓ_1 has cotype 2 (see, e.g., [LiTz4] or [AlKa]). By looking at the standard unit vectors in c_0 we see that c_0 does not have cotype 2. This gives that c_0 is not crudely finitely representable in ℓ_1. Find an elementary proof that c_0 is not finitely representable in ℓ_1.
Hint. (James) First show an auxiliary elementary statement that if $a, b, c \geq 0$, then the sum of four terms $|a \pm b \pm c|$ is at least $\frac{5}{3}(a+b+c)$ (consider $a \geq b \geq c \geq 0$). Consider now $x, y, z \in S_{\ell_1}$ and use the above on each coordinate of the four vectors $x \pm y \pm z$ to see that the sum of the norms of these four vectors is at least 5. Therefore at least one of the vectors $x \pm y \pm z$ has norm at least $\frac{5}{4}$. This is cannot happen in c_0 (use the standard unit vectors).

6.6 Consider the set $\mathcal{BM}_n$ of all equivalence classes of n-dimensional Banach spaces $X = (\mathbb{R}^n, \|\cdot\|)$, where X is said to be *equivalent* to Y if X and Y are isometric. Show that $\mathcal{BM}_n$ becomes a compact metric space with the metric $\ln \mathrm{d}$, where d is the Banach–Mazur distance. The metric space $\mathcal{BM}_n$ is called the *Banach–Mazur compactum.*
Hint. Arzelà–Ascoli. Use the following observation: If T is an isomorphism from a Banach space X onto a Banach space Y, with $\|T\| \cdot \|T^{-1}\| < C$, then put $\tilde{T} = T/\|T\|$ and note that $\|\tilde{T}\| = 1$ and $\|\tilde{T}^{-1}\| = \|T^{-1}\| \cdot \|T\| < C$.

6.7 For $n \in \mathbb{N}$, let $\mathcal{K}(n)$ be the set of all Banach spaces of dimension n equipped with the distance $\ln\big(d(X, Y)\big)$ for $X, Y \in \mathcal{K}(n)$. Show that $\mathcal{K}(n)$ is a compact metric space. It is called the *Minkowski compactum.*
Hint. The Arzelà–Ascoli theorem.

6.8 Let $X = \mathbb{R}^2$ with the norm whose unit ball is given by $\{(x, y) : 2x^2 + y^2 \leq 1\}$. Calculate $\mathrm{d}(X, \ell_2^2)$.
Hint. Find an *isometry* between X and ℓ_2^2.

6.9 Let C be a convex subset of an Euclidean space E and c_0 an element in C. Let f be a support functional of C at c_0, say $f(c) \leq \alpha = f(c_0)$ for all $c \in C$. Let e_0 be an element in E orthogonal to $H_0 := \operatorname{Ker} f$ such that $f(e_0) > 0$. Prove that e_0 belongs to $\mathcal{N}_C(c_0)$, the normal cone to C at c_0, see (6.60).
Hint. Put $E = H_0 \oplus \operatorname{span}\{e_0\}$ and do a simple calculation.

6.10 What is the Banach–Mazur distance between the space $(\mathbb{R}^2, \|\cdot\|_2)$ and the space $\mathbb{R}^2$ having as its closed unit ball an ellipsoid E centered at 0?
Hint. There is, by definition, a continuous linear mapping $T : \mathbb{R}^2 \to \mathbb{R}^2$ such that $TB_{(\mathbb{R}^2, \|\cdot\|_2)} = E$. In particular, this means that $\|T\| = \|T^{-1}\| = 1$.

6.11 Let $\mathrm{d}(X, Y)$ denotes the Banach–Mazur distance between Banach spaces X and Y. If $p > 2$, show that $d(\ell_p^n, \ell_2^n) = n^{\frac{1}{2} - \frac{1}{p}}$. Compare with Exercise 6.10.

Hint. Let I denote the formal identity mapping from ℓ_2^n into ℓ_p^n. Then, clearly, $\|I\| = 1$ and by using the Lagrange multipliers technique for computing extrema with constraints, we directly get that $\|I^{-1}\| = n^{\frac{1}{2}-\frac{1}{p}}$. Thus $d(\ell_2^n, \ell_p^n) \leq n^{\frac{1}{2}-\frac{1}{p}}$.

To obtain the reverse inequality, let T be an isomorphism from ℓ_p^n into ℓ_2^n normalized so that $\|T^{-1}\| = 1$, so that $\|Tx\|_2 \geq \|x\|_p$ for all $x \in \ell_p^n$. If e_i are the standard unit vectors, we have from the parallelogram equality in the Hilbert space,

$$2^n n \leq 2^n \sum_{i=1}^{n} \|Te_i\|_2^2 = \sum_{\varepsilon_i=\pm 1} \left\| \sum_{i=1}^{n} \varepsilon_i T e_i \right\|_2^2$$

$$\leq 2^n \max_{\varepsilon_i=\pm 1} \left\{ \left\| T\left(\sum_{i=1}^{n} \pm e_i\right) \right\|_2^2 \right\} \leq 2^n \|T\|^2 n^{\frac{2}{p}},$$

which gives that $\|T\| \geq n^{\frac{1}{2}-\frac{1}{p}}$.

Therefore $\|T\| \cdot \|T^{-1}\| \geq n^{\frac{1}{2}-\frac{1}{p}}$. Summing up with the above, we get that $\mathrm{d}(\ell_2^n, \ell_p^n) = n^{\frac{1}{2}-\frac{1}{p}}$.

6.12 Show that if $p < q$ are both greater than or equal to 2 or both smaller than or equal to 2, then $\mathrm{d}(\ell_p^n, \ell_q^n) = n^{\frac{1}{p}-\frac{1}{q}}$.
Hint. First of all, note that the second case follows by duality. So assume that $2 \leq p < q$. The same technique used in the first part of the hint in Exercise 6.11 gives $\mathrm{d}(\ell_p^n, \ell_q^n) \leq n^{\frac{1}{p}-\frac{1}{q}}$.

To get the reverse inequality, we use the multiplicative triangle inequality for the Banach–Mazur distance, i.e., $\mathrm{d}(\ell_q^n, \ell_2^n) \leq \mathrm{d}(\ell_q^n, \ell_p^n) \cdot \mathrm{d}(\ell_p^n, \ell_2^n)$, and the result of Exercise 6.11. This gives

$$\mathrm{d}(\ell_p^n, \ell_q^n) \geq \frac{\mathrm{d}(\ell_q^n, \ell_2^n)}{\mathrm{d}(\ell_p^n, \ell_2^n)} = \frac{n^{\frac{1}{2}-\frac{1}{q}}}{n^{\frac{1}{2}-\frac{1}{p}}} = n^{\frac{1}{p}-\frac{1}{q}}.$$

Thus $\mathrm{d}(\ell_p^n, \ell_q^n) = n^{\frac{1}{p}-\frac{1}{q}}$.

6.13 Show that ℓ_2^2 is isometric to a subspace of ℓ_4^3. Note that it is not isometric to any subspace of ℓ_p if p is not an even integer by [DJP] and mentioned in [KoKo, p. 914]. Note in passing: Every two dimensional subspace of L_4 is sometric to a subspace of ℓ_4^3 [KoKo, p. 914].
Hint. See [LyVa, p. 330]. Let $c = (\frac{8}{9})^{\frac{1}{4}}$. Put $u = c(\frac{1}{2}, -\frac{1}{2}, -1)$, and $v = c(\frac{\sqrt{3}}{2}, \frac{\sqrt{3}}{2}, 0)$. Then for any real numbers a and b, $\|au + bv\|_4 = (a^2 + b^2)$.

6.14 Let X be a hyperplane in ℓ_∞^3 defined by

$$X = \left\{ x \in \ell_\infty^3 : \sum_{i=1}^{3} x_i = 0 \right\}$$

where $x = (x_i)$.

Show that the orthogonal projection from ℓ_∞^3 onto X (in the sense of ℓ_2^3) has norm $\frac{4}{3}$ and that this is the minimal norm projection on X in ℓ_3^∞. Therefore $\lambda(X) = \frac{4}{3}$ by Proposition 5.16.

Hint. Elementary geometry.

6.15 Fix $p > 1$ and $n \in \mathbb{N}$. Let $\{r_1, r_2, \dots, r_n\}$ be the sequence of the first n Rademacher functions in $L_p[0, 1]$. Prove that, for any set of scalars $\{a_1, a_2, \dots, a_n\}$, we have $\left\| \sum_{k=1}^n a_k r_k \right\|_p^p = \frac{1}{2^n} \sum_{j=1}^{2^n} \left| \sum_{k=1}^n \varepsilon_{kj} a_k \right|^p$, where ε_{kj} are given in Equation (6.25).

Hint. Use the description of r_k given in (6.26). Then,

$$\begin{aligned}
\left\| \sum_{k=1}^n a_k r_k \right\|_p^p &= \left\| \sum_{k=1}^n a_k \sum_{j=1}^{2^n} \varepsilon_{kj} \chi_{[(j-1)/2^n, j/2_n]} \right\|_p^p \\
&= \left\| \sum_{j=1}^{2^n} \left(\sum_{k=1}^n \varepsilon_{kj} a_k \right) \chi_{[(j-1)/2^n, j/2_n]} \right\|_p^p \\
&= \int_0^1 \left| \sum_{j=1}^{2^n} \left(\sum_{k=1}^n \varepsilon_{kj} a_k \right) \chi_{[(j-1)/2^n, j/2_n]}(t) \right|^p \mathrm{d}t \\
&= \sum_{j=1}^{2^n} \int_{(j-1)/2^n}^{j/2^n} \left| \sum_{j=1}^{2^n} \left(\sum_{k=1}^n \varepsilon_{kj} a_k \right) \chi_{[(j-1)/2^n, j/2_n]}(t) \right|^p \mathrm{d}t \\
&= \frac{1}{2^n} \sum_{j=1}^{2^n} \left| \sum_{k=1}^n \varepsilon_{kj} a_k \right|^p .
\end{aligned}$$

6.16 Show that if X is an n-dimensional Banach space, then $\mathrm{d}(X, \ell_1^n) \leq n$. Therefore if X and Y are n-dimensional spaces, then $\mathrm{d}(X, Y) \leq n^2$. Note that we showed in John's theorem that it is actually $\leq n$.

Hint. Let $\{x_i; f_i\}$ be an Auerbach basis of X. Let X_1 be the space X renormed by the norm $\|x\|_1 = \sum_{i=1}^n |x_i|$, and let X_∞ be the space X renormed with the norm $\|x\|_\infty = \sup_{1 \leq i \leq n} |x_i|$, where $x = (x_i)$. Then $B_{X_1} \subset B_X \subset B_{X_\infty}$. Check that $B_{X_\infty} \subset n B_{X_1}$. Therefore $B_{X_1} \subset B_X \subset n B_{X_1}$.

6.17 Let X and Y be n-dimensional spaces such that $\mathrm{d}(X, Y) = 1$. Show that X and Y are isometric.

Hint. Use a compactness argument. If T is an isomorphism of X onto Y with $\|T\|\|T^{-1}\| = 1$, then the mapping $\tilde{T} = \|T^{-1}\|T$ is an isomorphism of X onto Y that satisfies $\|\tilde{T}\| = \|\tilde{T}^{-1}\| = 1$, so $\tilde{T}$ is an isometry.

6.18 If X and Y are n-dimensional Banach spaces, show that there is an isomorphism T from X onto Y such that $\mathrm{d}(X, Y) = \|T\|.\|T^{-1}\|$.
Hint. Compactness.

Chapter 7
Optimization

In this chapter we discuss some basic techniques in nonlinear analysis on Banach spaces that are frequently used in applications in related fields. The classical approach uses differentiability, and we discuss this concept in infinite-dimensional Banach spaces.

The lack of compactness in infinite-dimensional Banach spaces results in functions not attaining their suprema or infima, and some perturbations of them are thus needed to establish optimization results. One convenient approach to these problems is through variational principles. We prove Ekeland's and Bishop–Phelps theorems, and Lindenstrauss' norm attaining operators theorem. We discuss some topological tools used in this area, such as Michael's selection theorem.

7.1 Introduction

Definition 7.1 *Let $(X, \|\cdot\|)$, $(Y, \|\cdot\|)$ be Banach spaces, let $U \subset X$ be an open set, let $f : U \longrightarrow Y$ be a mapping and let $x \in U$. We say that f is* Gâteaux differentiable *at x if there exists $L \in \mathcal{B}(X, Y)$ such that*

$$\left\|\tfrac{1}{t}\big(f(x+th) - f(x)\big) - Lh\right\| \longrightarrow 0 \quad \text{as} \quad 0 \neq t \to 0 \quad \textit{for every} \quad h \in X. \tag{7.1}$$

If, moreover, the limit above is uniform for $h \in B_X$, then we say that f is Fréchet differentiable *at x. The (uniquely determined) operator L is then denoted by $f'(x)$ and called the* Gâteaux *(respectively* Fréchet*)* derivative of f at x.
We say that f is Gâteaux (Fréchet) differentiable *on U if it is Gâteaux (Fréchet) differentiable at every point $x \in U$. Sometimes we replace "differentiable" by* "smooth." *We say that f is* C^1-smooth *on U if it is Fréchet smooth on U and moreover the mapping $U \ni x \longmapsto f'(x)$ is continuous.*

Definition 7.2 *Let f be a real-valued function on a Banach space X. Given $x \in X$ and $h \in X$, if the limit*

$$\lim_{t \to 0} \frac{f(x+th) - f(x)}{t}$$

M. Fabian et al., *Banach Space Theory*, CMS Books in Mathematics,
DOI 10.1007/978-1-4419-7515-7_7,

exists and is finite, we call it the directional derivative of f at x in the direction h *and we denote it by* $Df(x)h$.

A norm $\|\cdot\|$ on a Banach space X is called *Fréchet* (respectively *Gâteaux) differentiable* if $\|\cdot\|$ is Fréchet (respectively Gâteaux) differentiable on the open set $X\setminus\{0\}$. Note that a norm is never differentiable at 0.

Since differentiability conditions for a norm are homogeneous, a norm is differentiable at x if it is differentiable at λx for some scalar λ. Consequently, it is enough to check the differentiability at points of S_X.

Consider a norm $\|\cdot\|$ of a Banach space X. If it is Gâteaux differentiable at $x \in S_X$, then from $\left|\frac{\|x+th\|-\|x\|}{t}\right| \leq \frac{\|th\|}{t} = 1$ for $h \in S_X$ we get that $F = \|x\|'$ (the derivative of $\|\cdot\|$ at x) satisfies $\|F\| \leq 1$. On the other hand, considering $h = x$ we get $F(x) = 1$. Thus F is a supporting functional of B_X at x.

For convex functions, continuity implies a stronger property. This is the content of the following result.

Lemma 7.3 *Let f be a convex function on an open convex set U in a Banach space X. Then, given an element $x_0 \in G$, the following are equivalent.*
(i) *f is continuous at x_0.*
(ii) *f is* locally bounded above *at x_0, i.e., bounded above on a neighborhood of x_0.*
(iii) *f is* locally Lipschitz *at x_0, i.e., Lipschitz on a neighborhood of x_0.*

Proof: (i)$\Longrightarrow$(ii) is obvious.

(ii)$\Longrightarrow$(iii): We will first show that if f is bounded above on $B(x_0, r)$, then it is bounded on $B(x_0, r)$. Assume without loss of generality that $x_0 = 0$ and that f is bounded above by K on the closed unit ball B_X. From the convexity of f, we have that $f(0) = f\left(\frac{1}{2}(x + (-x))\right) \leq \frac{1}{2}f(x) + \frac{1}{2}f(-x)$ for every $x \in B_X$. Thus $f(x) \geq 2f(0) - f(-x) \geq 2f(0) - K$. Then to finish the proof of this implication it is enough to prove the following:

Claim: *If a convex function f is bounded by 1 on B_X, then f is 5-Lipschitz on $\frac{1}{2}B_X$.*

Indeed, assume that $x, y \in \frac{1}{2}B_X$ and $f(y) - f(x) > 5\|y - x\|$. Consider the point $z := y + \dfrac{y-x}{2\|y-x\|}$. Clearly $z \in B_X$. On the other hand, since the points x, y, z lie on a line in this order (see Fig. 7.1), by the convexity of f we have

$$\frac{f(z) - f(y)}{\|z - y\|} \geq \frac{f(y) - f(x)}{\|y - x\|} > 5.$$

Also, $\|z - y\| = \frac{1}{2}$. Therefore $f(z) \geq f(y) + \frac{5}{2} \geq -1 + \frac{5}{2} = \frac{3}{2}$, a contradiction. This proves the Claim and the implication (ii)$\Longrightarrow$(iii).

(iii)$\Longrightarrow$(i) is trivial. $\square$

Note that if U is a convex subset of a vector space V, a function $f: U \to \mathbb{R}$ is convex if and only if the function $t \mapsto \frac{f(x+ty)-f(x)}{t}$ is increasing in t for all $x \in U$, $y \in V$ and all $t \neq 0$ for which $x + ty \in U$ (see Fig. 7.2 and Exercise 7.4).

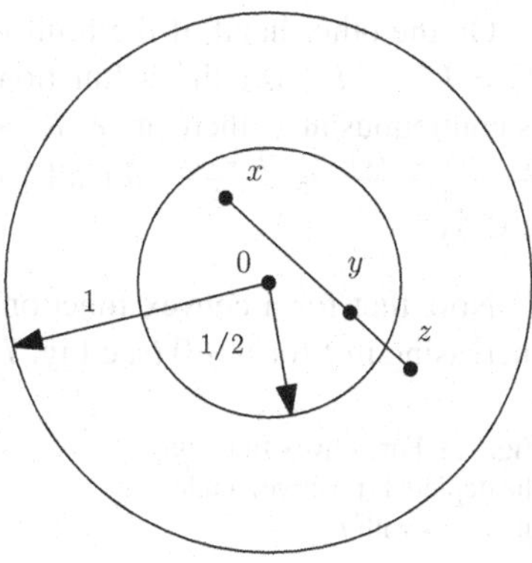

Fig. 7.1 The three points in the proof of Lemma 7.3

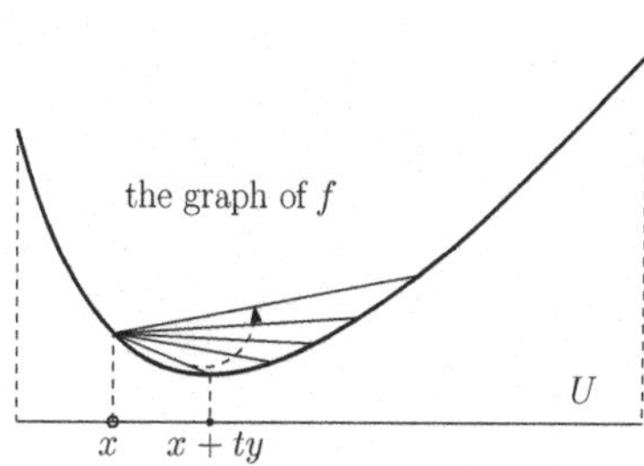

Fig. 7.2 For convex functions the angle increases with t

Now assume that f is a convex function on an open convex subset U of a Banach space X. It follows from the preceding observation that for all $x \in U$ and $h \in S_X$, the one-sided limits $F^+(h) = \lim\limits_{t \to 0^+} \frac{f(x+th)-f(x)}{t}$ and $F^-(h) = \lim\limits_{t \to 0^-} \frac{f(x+th)-f(x)}{t}$ exist finite. Moreover, we have $F^+(h) \geq F^-(h)$ and it is easy to check that F^+ is a subadditive function in h, $F^-(h)$ is a superadditive function in X and both are positively homogeneous.

Lemma 7.4 *Let f be a convex function defined on an open convex subset U of a Banach space X that is continuous at $x \in U$. Then f is Fréchet differentiable at x if and only if*

$$\lim_{t \to 0} \frac{f(x+th) + f(x-th) - 2f(x)}{t} = 0$$

uniformly for $h \in S_X$.

An analogous result is true for the Gâteaux differentiability. Note that by the triangle inequality and convexity of f we always have $f(x+th) + f(x-th) - 2f(x) \geq 0$.

Proof: If f is Fréchet differentiable at x, then $F^+(h) = F^-(h)$ for all h and the claim follows from the following equality for $t > 0$:

$$\frac{f(x+th) - f(x)}{t} - \frac{f(x-th) - f(x)}{-t} = \frac{f(x+th) + f(x-th) - 2f(x)}{t}.$$

On the other hand, if the limit is 0, by this equation we have $F^+ = F^-$. Then $F = F^+ = F^-$ is a linear functional. We need to show that it is bounded. Since f is continuous at x, there are $\delta, K > 0$ such that $f \le K$ on $x + \delta B_X$. By convexity, $\frac{f(x+th)-f(x)}{t} \le \frac{K-f(x)}{\delta}$ for all $|t| \le \delta$, $h \in S_X$. Thus $F(h) \le \frac{K-f(x)}{\delta}$ for all $h \in S_X$. □

Note that for a convex function f, the function $t \mapsto \frac{f(x+th)+f(x-th)-2f(x)}{t}$ is increasing in t for $t > 0$ (see Fig. 7.3).

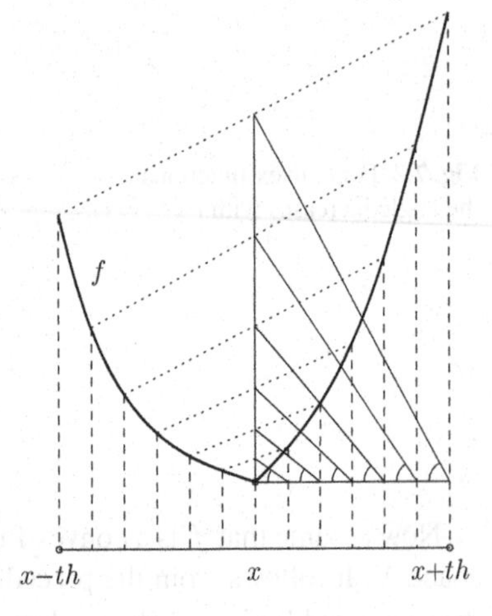

Fig. 7.3 For convex functions the depicted (positive) angle increases with t

The uniformity in $h \in S_X$ in the case of Fréchet differentiability allows us to write the definition as $\lim_{y\to 0} \frac{f(x+y)-f(x)-F(y)}{\|y\|} = 0$; similarly for the limit in Lemma 7.4.

Definition 7.5 *Let f be a real-valued function on an open subset U of a Banach space X and let S be a subset of U.*
We say that f is uniformly Gâteaux differentiable (UG) *on S if it is Gâteaux differentiable at each $x \in S$ and for every $h \in S_X$ the limit* (7.1) *is uniform for $x \in S$; more precisely, given any $\varepsilon > 0$, there exists $\delta := \delta(\varepsilon, h) > 0$ with the following property: for every $x \in S$ and $0 < |t| < \delta$ with $x + th \in U$, we have $\left|\frac{1}{t}\big(f(x+th) - f(x)\big) - f'(x)h\right| < \varepsilon$.*
We say that f is uniformly Fréchet differentiable (UF) *if it is Fréchet differentiable at each $x \in S$ and the limit* (7.1) *is uniform in $x \in S$ and also in $h \in S_X$; more precisely, given any $\varepsilon > 0$, there exists $\delta := \delta(\varepsilon) > 0$, with the following property: for every $x \in S$, $h \in S_X$, and $0 < |t| < \delta$ with $x + th \in U$, we have $\left|\frac{1}{t}\big(f(x + th) - f(x)\big) - f'(x)h\right| < \varepsilon$.*

Note that f is uniformly Fréchet differentiable on an open *convex* set U if and only if it is Fréchet differentiable at every point of U and the mapping $x \mapsto f'(x)$ is uniformly continuous as a mapping $U \to X^*$. Note, too, that for convex continuous functions the uniformly Fréchet differentiability can be formulated in terms of the uniform convergence to 0 of the "symmetric" quotient given in Lemma 7.4.

The norm $\|\cdot\|$ of a Banach space X is called *uniformly Fréchet differentiable* (or *UF-smooth*), respectively *uniformly Gâteaux differentiable* (or *UG-smooth*) if $\|\cdot\|$ is uniformly Fréchet (respectively uniformly Gâteaux) differentiable on S_X (note that if the limit (7.1) exists for the function $\|\cdot\|$ at some $x \in S_X$ then, by homogeneity, it exists at every point λx for $\lambda \neq 0$. So Fréchet (Gâteaux) differentiability on S_X implies Fréchet (Gâteaux) differentiability on the open set $X\backslash\{0\}$).

Definition 7.6 *The norm of a Banach space X is called* strictly convex *(or* rotund*) if* $\mathrm{Ext}(B_X) = S_X$. *If this is the case, we also say that X is a* strictly convex *(or* rotund*) Banach space.*

Fact 7.7 *Let $(X, \|\cdot\|)$ be a normed space. The following are equivalent:*
(i) *$\|\cdot\|$ is strictly convex.*
(ii) *If $x, y \in S_X$ satisfy $\|x + y\| = 2$, then $x = y$.*
(iii) *If $x, y \in X$ satisfy $2\|x\|^2 + 2\|y\|^2 - \|x + y\|^2 = 0$, then $x = y$.*
(iv) *If $x, y \neq 0$ satisfy $\|x + y\| = \|x\| + \|y\|$, then $x = \lambda y$ for some $\lambda > 0$.*

Proof: (i) $\Longleftrightarrow$ (ii) is easy, $\|x + y\| = 2$ means that $\frac{1}{2}(x + y) \in S_X$.

(ii)$\Longrightarrow$(iii): Note that

$$\begin{aligned} 2\|x\|^2 + 2\|y\|^2 - \|x + y\|^2 &\geq 2\|x\|^2 + 2\|y\|^2 - (\|x\| + \|y\|)^2 \\ &= (\|x\| - \|y\|)^2 \geq 0. \end{aligned}$$

Thus if $2\|x\|^2 + 2\|y\|^2 - \|x + y\|^2 = 0$, then $\|x\| = \|y\|$. Hence we may assume that $x, y \in S_X$; we get $\|x + y\| = 2$ and (ii) implies $x = y$.

(iii)$\Longrightarrow$(ii) is immediate.

(iv)$\Longrightarrow$(ii): If $x, y \in S_X$ and $\|x + y\| = 2$, then $\|x\| + \|y\| = \|x + y\|$ and hence $x = y$.

(ii)$\Longrightarrow$(iv): Let $\|x + y\| = \|x\| + \|y\|$ for some $x, y \neq 0$. We may assume that $0 < \|x\| \leq \|y\|$. Then

$$\begin{aligned} 2 &\geq \big\|x/\|x\| + y/\|y\|\big\| \geq \big\|x/\|x\| + y/\|x\|\big\| - \big\|y/\|x\| - y/\|y\|\big\| \\ &= (1/\|x\|)\|x + y\| - \|y\|(1/\|x\| - 1/\|y\|) \\ &= (1/\|x\|)(\|x\| + \|y\|) - \|y\|(1/\|x\| - 1/\|y\|) = 2. \end{aligned}$$

Thus $\big\|x/\|x\| + y/\|y\|\big\| = 2$ and then $x/\|x\| = y/\|y\|$. □

Definition 7.8 *The norm $\|\cdot\|$ of a Banach space X is called* weakly uniformly rotund (WUR) *if $(x_n - y_n) \xrightarrow{w} 0$ whenever $x_n \in X$ and $y_n \in X$ are such that $\{x_n\}_{n=1}^{\infty}$ is bounded and $\lim\big(2\|x_n\|^2 + 2\|y_n\|^2 - \|x_n + y_n\|^2\big) = 0$. If this is the case, we also say that X is a* weakly uniformly rotund *Banach space. The norm of the* *dual space is said to be w^*-uniformly rotund* (W*UR) *if* $(f_n - g_n) \overset{w^*}{\to} 0$ *whenever* $f_n, g_n \in X^*$, $\{f_n\}_{n=1}^{\infty}$ *is bounded, and* $\lim(2\|f_n\|^2 + 2\|g_n\|^2 - \|f_n + g_n\|^2) = 0$.

Definition 7.9 *The norm* $\|\cdot\|$ *of a Banach space* X *is called* locally uniformly rotund (LUR) *if for all* $x, x_n \in X$ *satisfying* $\lim(2\|x\|^2 + 2\|x_n\|^2 - \|x + x_n\|^2) = 0$ *we have* $\lim_{n\to\infty} \|x_n - x\| = 0$. *If this is the case, we also say that* X *is a* locally uniformly rotund *Banach space* (see Fig. 7.4).

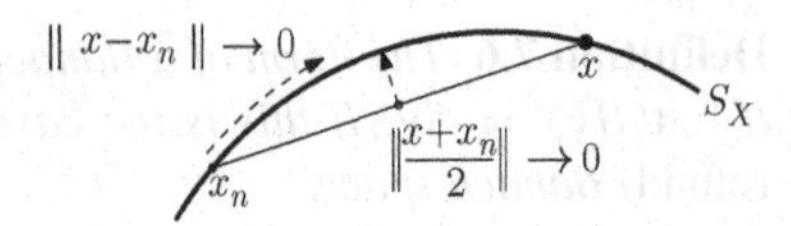

Fig. 7.4 An LUR norm at x

Note that a norm $\|\cdot\|$ on X is LUR if and only if $\lim \|x_n - x\| = 0$ whenever $x_n, x \in S_X$ are such that $\lim \|x_n + x\| = 2$ (Exercise 7.11). A similar remark holds for WUR and W*UR. It is also obvious that LUR implies strict convexity (rotundity), but it is a strictly stronger notion (Exercise 7.12). The same is true for WUR. Note that it follows from the parallelogram law that the norm of a Hilbert space is LUR and WUR.

Definition 7.10 *Let* S *be a nonempty subset of a Banach space* X. *A point* $x \in S$ *is called an* exposed point *of* S *if there is* $f \in X^*$ *such that* $f(x) = \sup_S f$ *and* $\{s \in S : f(s) = f(x)\} = \{x\}$. *We say in this case that* f exposes S at x.
A point $x \in S$ *is called a* strongly exposed point *of* S *if there exists* $f \in X^*$ *such that* $f(x) = \sup_S f$ *and* $x_n \to x$ *for all sequences* $\{x_n\} \subset S$ *such that* $\lim f(x_n) = \sup_S f$. *We say in this case that* f strongly exposes S at x.

Observe that every exposed point of a subset S of a Banach space is an extreme point of S (see Exercise 7.72), and that there are closed convex sets having exposed points that are not strongly exposed (see Exercise 7.73). From the definition, it follows that if S is a nonempty subset of a Banach space X, a point $x_0 \in S$ is strongly exposed if and only if there exists $f \in X^*$ such that $\operatorname{diam} S(f, \delta) \to 0$ whenever $\delta \downarrow 0$, where $S(f, \delta) := \{s \in S : f(s) > M - \delta\}$ and $M := \sup_S f$.

7.2 Subdifferentials: Šmulyan's Lemma

The standard optimization problem consists in finding a point where a real function defined on a certain subset, say S, of a Banach space, say X, attains its infimum. Sometimes it is cumbersome to carry on the computations keeping in mind that the function is defined only on S, and the trick of extending the function by keeping its values on S and attributing the "value" $+\infty$ to f on $X \setminus S$ is useful and hence widespread. In fact, the arithmetic on $[0, +\infty]$ is simple (since the function to be minimized should be bounded below, we may always assume that it takes only non-negative values), and certainly this extension of f does not introduce undesirable points where the function may attain its minimum. Moreover, if f was defined and continuous on a nonempty open subset of a Banach space, this extension keeps some continuity property, as it becomes lower semicontinuous; this follows from the very definition of lower semicontinuity.

To avoid trivialities, we shall be dealing with "proper" functions. We say that a function $f : X \to (-\infty, +\infty]$ is *proper* if the *domain* $\operatorname{dom} f := \{x \in X : f(x) < +\infty\}$ is nonempty.

We observe that functions with values in $[0, +\infty]$ already appeared (for example, the Minkowski functional of a subset of a Banach space is such a function, see Definition 2.9).

Quite often we shall be considering real-valued (or extended real-valued) functions defined on open subsets of Banach spaces that lack differentiability at some points. In order to deal with such situations, it is customary to introduce some more relaxed differentiability concepts.

Definition 7.11 *Let X be a Banach space, let $f : X \to (-\infty, +\infty]$ be a proper function. For $x_0 \in \operatorname{dom} f$ and $\varepsilon \geq 0$ we define the* ε-subdifferential of f at x_0 *as the subset of X^* given by*

$$\partial_\varepsilon f(x_0) := \{x^* \in X^* : f(x) \geq f(x_0) + \langle x^*, x - x_0 \rangle - \varepsilon, \text{ for all } x \in X\}. \quad (7.2)$$

The 0-subdifferential of f at x_0, also denoted by $\partial f(x_0)$, is called the subdifferential *of f at x_0; i.e.,*

$$\partial f(x_0) := \{x^* \in X^* : f(x) \geq f(x_0) + \langle x^*, x - x_0 \rangle, \text{ for all } x \in X\}. \quad (7.3)$$

Remark: A simple convexity argument shows that if $f : X \to (-\infty, +\infty]$ is a *convex* function, then $x^* \in \partial_\varepsilon f(x_0)$ whenever there exists $\delta > 0$ such that $f(x) - f(x_0) \geq \langle x^*, x - x_0 \rangle - \varepsilon$ for every $x \in B(x_0, \delta)$.

Figure 7.5 depicts the ε-subdifferential of a convex and continuous function on $\mathbb{R}$. For a description of the ε-subdifferential of the Minkowski functional of a closed convex neighborhood of 0 (in particular, of the norm function), see Lemma 7.19.

The epigraph of an extended-valued function $f : X \to (-\infty, +\infty]$ has the same definition as for the case of real-valued functions: precisely, $\operatorname{epi} f := \{(x, r) : x \in X,\ r \in \mathbb{R},\ \ f(x) \leq r\} \subset X \times \mathbb{R}$. The definition of an extended-valued *convex* function is the same as for the case of real-valued functions, see Definition 1.4. If f is convex and lower semicontinuous, the epigraph of f is a closed convex subset of $X \times \mathbb{R}$, and from the separation Theorem 3.32 we get that for every $\varepsilon > 0$ the ε-subdifferential of f at any point of $\operatorname{dom} f$ is nonempty. If f is, moreover, continuous, ε can be taken even to be 0, since the epigraph has then a nonempty interior. For details and for an analytic approach, see Exercises 7.14, 7.15, and 7.16.

The following observation will be used often.

Fact 7.12 *Let $C > 0$ and let f be a C-Lipschitz real-valued function defined on an open subset U of a Banach space. Then, if $x \in U$ and $x^* \in \partial f(x)$, we have $\|x^*\| \leq C$.*

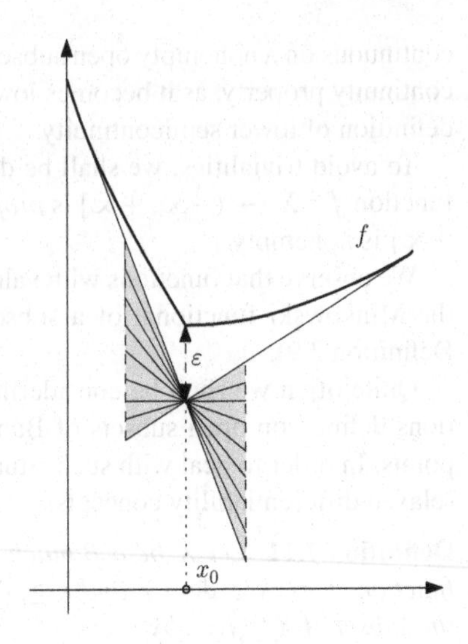

Fig. 7.5 The graphs of the elements in $\partial_\varepsilon f(x_0)$ (translated to $(x_0, f(x_0) - \varepsilon)$)

Proof: Take $y \in S_X$ and $t > 0$ such that $x + ty \in U$. Then $\langle x^*, ty\rangle \leq f(x+ty) - f(x) \leq |f(x+ty) - f(x)| \leq Ct$, hence $\langle x^*, y\rangle \leq C$. Since this happens for every $y \in S_X$, we get $\|x^*\| \leq C$. □

The same argument proves that if $x \in U$ and $r > 0$ is such that $B(x, r) \subset U$, then $\|x^*\| \leq C + \varepsilon/r$ for all $x^* \in \partial f_\varepsilon(x)$ and for all $\varepsilon > 0$.

There is a simple connection between the ε-subdifferential of a Lipschitz convex function at a point, and its subdifferential at close points. It is given in the next result for latter reference.

Lemma 7.13 *Let $C > 0$ and let f be a C-Lipschitz convex function defined on a nonempty open and convex subset U of a Banach space X. Let $x_0 \in U$ and $\varepsilon > 0$ be such that $B(x_0, \varepsilon) \subset U$. Then, $\partial f\big(B(x_0, \varepsilon)\big) \subset \partial_{2C\varepsilon} f(x_0)$.*

Proof: Let $x \in B(x_0, \varepsilon)$. If $x^* \in \partial f(x)$ and $y \in B(x_0, \varepsilon)$, we have

$$\begin{aligned} f(y) &\geq f(x) + \langle x^*, y - x\rangle \\ &= f(x_0) + \langle x^*, y - x_0\rangle + \big(f(x) - f(x_0)\big) + \langle x^*, x - x_0\rangle \\ &\geq f(x_0) + \langle x^*, y - x_0\rangle - 2C\|x - x_0\| \geq f(x_0) + \langle x^*, y - x_0\rangle - 2C\varepsilon, \end{aligned}$$

since, by Fact 7.12, $\|x^*\| \leq C$. This concludes the proof, in view of the Remark following Definition 7.11. □

Remark: There is a kind of reverse inclusion, a consequence of the following result of Brøndsted and Rockafellar, see, e.g., [Phelps, Theorem 3.17] (the version for norms is proved in Exercise 7.53): *Assume that f is a convex proper lower semi-continuous function on a Banach space X. Then, given any point $x_0 \in \operatorname{dom} f$, $\varepsilon > 0$, $\lambda > 0$ and any $x_0^* \in \partial_\varepsilon f(x_0)$, there exists $x \in \operatorname{dom} f$ and $x^* \in X^*$ such that*

$$x^* \in \partial f(x), \quad \|x - x_0\| \leq \varepsilon/\lambda, \quad \textit{and} \quad \|x^* - x_0^*\| \leq \lambda.$$

In particular, the domain of ∂f *is dense in* $\operatorname{dom} f$ See also Theorem 7.41 and Exercise 7.53.

As a consequence, we get that for a convex proper lower semicontinuous function f on a Banach space X, for $x_0 \in \operatorname{dom}(f)$, and for $\varepsilon > 0$ and $\lambda > 0$, $\partial_\varepsilon f(x_0) \subset \partial f(B(x_0, \varepsilon/\lambda) + \lambda B_{X^*}$.

Proposition 7.14 *Let X be a Banach space and f a continuous convex function defined on a nonempty open convex subset U of X. Then, the multi-valued mapping $x \mapsto \partial f(x)$ from U into 2^{X^*} is $\|\cdot\|$-w^*-upper semicontinuous.*

Proof: Fix $x_0 \in U$, and let W be a w^*-open subset of X^* that contains $\partial f(x_0)$. We shall prove that there exists $\delta > 0$ such that $B(x_0, \delta) \subset U$ and $\partial f(x) \in W$ for all $x \in B(x_0, \delta)$. If not, we can find a sequence $\{x_n\}$ in U such that $x_n \to x_0$ and a sequence $\{x_n^*\}$ in X^* such that $x_n^* \in \partial f(x_n) \setminus W$ for all $n \in \mathbb{N}$. By Lemma 7.3 and Fact 7.12, we may assume that $\|x_n^*\| \leq C$ for some constant $C > 0$ and for all $n \in \mathbb{N}$, so the sequence $\{x_n^*\}$ has a w^*-cluster point x_0^* ($\notin W$). It is simple to prove that $x_0^* \in \partial f(x_0)$, and this is a contradiction. □

Theorems 7.15 and 7.17 below are nonhomogeneous generalizations of what is usually called the *Šmulyan lemma*—Corollary 7.20 gives a slight extension of Corollary 7.22, the classical version—,which characterizes Fréchet (respectively Gâteaux) differentiability of the norm of a Banach space by looking at w^*-sections of the closed dual unit ball.

Theorem 7.15 *Let f be a convex continuous function defined on a nonempty open convex subset U of a Banach space X, and let $x_0 \in U$. Then the following are equivalent:*

(i) *f is Fréchet differentiable at x_0;*
(ii) $\|x_n^* - x_0^*\| \to 0$ *whenever* $x_0^* \in \partial f(x_0)$ *and* $x_n^* \in \partial_{\varepsilon_n} f(x_0)$, *where* $\varepsilon_n \downarrow 0$.
(iii) $\|x_n^* - x_0^*\| \to 0$ *whenever* $\{x_n\}$ *is a sequence in* U *such that* $\|x_n - x_0\| \to 0$ *and* $x_n^* \in \partial f(x_n)$ *for every* $n \in \mathbb{N} \cup \{0\}$.

Proof: (i)$\Longrightarrow$(ii): Let f be Fréchet differentiable at x_0 and let $\{x_0^*\} = \partial f(x_0)$. Suppose that (ii) does not hold. Then there exists $\varepsilon_n \downarrow 0$, $x_n^* \in \partial_{\varepsilon_n} f(x_0)$, and $\eta > 0$ such that

$$\|x_n^* - x_0^*\| \geq \eta \quad \text{for all } n \in \mathbb{N}.$$

Let $t_n = \frac{4\varepsilon_n}{\eta}$ and choose $h_n \in S_X$ so that

$$\langle x_n^* - x_0^*, h_n \rangle \geq \eta/2, \ n \in \mathbb{N}.$$

Thus, for n big enough to have $x_0 + t_n h_n \in U$,

$$f(x_0 + t_n h_n) - f(x_0) - \langle x_0^*, t_n h_n \rangle$$
$$\geq \langle x_n^*, t_n h_n \rangle - \langle x_0^*, t_n h_n \rangle - \varepsilon_n \geq t_n \frac{\eta}{2} - \varepsilon_n = t_n \frac{\eta}{4},$$

a contradiction with the Fréchet differentiability of f at x_0.

(ii)$\Longrightarrow$(iii) is a consequence of Lemma 7.13

(iii)$\Longrightarrow$(i): Observe, first, that (iii) implies that $\partial f(x_0)$ is a singleton, say $\{x_0^*\}$ (from previous observations, the set $\partial f(x_0)$ is nonempty). Suppose f is not Fréchet differentiable at x_0. Then there exist sequences $t_n \downarrow 0$, $\{h_n\}$ in S_X, and $\varepsilon > 0$ such that

$$f(x_0 + t_n h_n) - f(x_0) - \langle x_0^*, t_n h_n \rangle > \varepsilon t_n, \text{ for all } n \in \mathbb{N}.$$

Pick $x_n^* \in \partial f(x_0 + t_n h_n)$ for $n \in \mathbb{N}$. Thus

$$\langle x_n^*, -t_n h_n \rangle \leq f(x_0) - f(x_0 + t_n h_n), \text{ for all } n \in \mathbb{N},$$

i.e.,

$$\langle x_n^*, t_n h_n \rangle \geq f(x_0 + t_n h_n) - f(x_0) \left(> \langle x_0^*, t_n h_n \rangle + \varepsilon t_n \right)$$

for every $n \in \mathbb{N}$. Therefore $\|x_0^* - x_n^*\| > \varepsilon$ for all $n \in \mathbb{N}$, a contradiction with (iii). □

Remarks: (a) Observe that (i)$\Longrightarrow$(ii) in Theorem 7.15 does not depend on the convexity of the mapping f, so it applies to every function f defined on an open subset U of a Banach space that is Fréchet differentiable at some point $x_0 \in U$.

(b) The statement (ii) in Theorem 7.15 can be formulated by saying that $\operatorname{diam} \partial_\varepsilon f(x_0) \to 0$ whenever $\varepsilon \downarrow 0$.

Corollary 7.16 *Let f be a convex continuous function defined on a nonempty open convex subset U of a Banach space X. Let $x_0 \in U$. If f is Fréchet differentiable at x_0, then the set-valued mapping $\partial f : U \to 2^{X^*}$ is $\|\cdot\|$-$\|\cdot\|$-upper semicontinuous at x_0. Conversely, if $\partial f(x_0)$ is a singleton and the set-valued mapping $\partial f : U \to 2^{X^*}$ is $\|\cdot\|$-$\|\cdot\|$-upper semicontinuous at x_0, then f is Fréchet differentiable at x_0.*

Proof: This follows directly from the equivalence (i)$\Leftrightarrow$(ii) in Theorem 7.15. □

In particular, if f is a convex and Fréchet differentiable function defined on a nonempty open convex subset U of a Banach space X, the (single-valued) mapping $\partial f : U \to X^*$ is $\|\cdot\|$-$\|\cdot\|$-continuous.

Theorem 7.17 *Let f be a convex continuous function defined on a nonempty open convex subset U of a Banach space X and let $x_0 \in U$. Then the following are equivalent:*

(i) *f is Gâteaux differentiable at x_0.*

(ii) $x_n^* \stackrel{w^*}{\to} x_0^*$ *whenever* $x_0^* \in \partial f(x_0)$ *and* $x_n^* \in \partial_{\varepsilon_n} f(x_0)$ *for all* $n \in \mathbb{N}$, *where* $\varepsilon_n \downarrow 0$.
(iii) $\partial f(x_0)$ *is a singleton.*

Proof: (i)$\Longrightarrow$(ii): Assume that f is Gâteaux differentiable at x_0 and let $x_0^* \in X^*$ be its Gâteaux derivative at x_0. Let $\delta_0 > 0$ such that $B(x_0, \delta_0) \subset U$. Fix, for all $n \in \mathbb{N}$, an element $x_n \in \partial_{\varepsilon_n} f(x_0)$, where $\varepsilon_n \downarrow 0$. If (ii) fails for this choice, we can find, by passing to a subsequence if necessary, $h \in S_X$ and $\mu > 0$ such that $\langle x_n^*, h\rangle \geq \langle x_0^*, h\rangle + \mu$ for all $n \in \mathbb{N}$. The definition of Gâteaux differentiability gives $\delta := \delta(h, \mu/2)$ $(< \delta_0)$ such that, for $0 < |t| \leq \delta$,

$$\left| \frac{f(x_0 + th) - f(x_0)}{t} - \langle x_0^*, h\rangle \right| \leq \frac{\mu}{2} \tag{7.4}$$

Since $x_n^* \in \partial_{\varepsilon_n} f(x_0)$, we have $f(x_0 + th) - f(x_0) \geq \langle x_n^*, th\rangle - \varepsilon_n$ for all $n \in \mathbb{N}$ and all t such that $x_0 + th \in U$. Hence, for $t > 0$ with $x_0 + th \in U$ we get

$$\frac{f(x_0 + th) - f(x_0)}{t} \geq \langle x_n^*, h\rangle - \frac{\varepsilon_n}{t}.$$

Fix $t := \delta(h, \mu/2)$ and find $n \in \mathbb{N}$ big enough such that $\varepsilon_n/t < \eta/2$. Then, for this n,

$$\frac{f(x_0 + th) - f(x_0)}{t} \geq \langle x_n^*, h\rangle - \frac{\varepsilon_n}{t} \geq \langle x_0^*, h\rangle + \mu - \frac{\varepsilon_n}{t} > \langle x_0^*, h\rangle + \frac{\mu}{2},$$

and this contradicts (7.4).

(ii)$\Longrightarrow$(iii): If $x^* \in \partial f(x_0)$, then, taking $x_n^* = x^*$ for all $n \in \mathbb{N}$, the sequence $\{x_n^*\}$ satisfies (ii) (for any decreasing null sequence (ε_n)), so $x^* = x_0^*$.

(iii)$\Longrightarrow$(i): Define $p(h) = \lim_{t \downarrow 0} \frac{1}{t}\big(f(x_0 + th) - f(x_0)\big)$, $h \in X$. Then, easily, $p(th) = tp(h)$ for every $h \in X$ and every $t > 0$. Also, p is subadditive, that is, $p(h_1 + h_2) \leq p(h_1) + p(h_2)$ for all $h_1, h_2 \in X$; this follows from the convexity of f. It remains to prove that $p(-h) = -p(h)$ for every $h \in X$. Indeed, then p will be a linear functional, and, by Lemma 7.3, continuous; hence the Gâteaux differentiability of f at x_0 will be proved. So, assume that $p(-h_0) \neq -p(h_0)$ for some $h_0 \in X$. Define

$$\xi(th_0) = tp(h_0) \quad \text{and} \quad \eta(th_0) = -tp(-h_0), \quad t \in R.$$

Then ξ and η are linear functionals on the subspace $\mathbb{R}h_0$, and $\xi(h) \leq p(h)$, $\eta(h) \leq p(h)$ for every $h \in \mathbb{R}h_0$. Hence, Hahn-Banach Theorem 2.1 provides $x^*, y^* \in X^*$ such that $x^*(h) = \xi(h)$ and $y^*(h) = p(h)$ for every $h \in \mathbb{R}h_0$, and moreover $x^* \leq p$, $y^* \leq p$ on X. Therefore, $x^*, y^* \in \partial f(x_0)$. But $x^*(h_0) = \xi(h_0) =$

$p(h_0) \neq -p(-h_0) = \eta(h_0) = y^*(h_0)$. This shows that $\partial f(x_0)$ is not a singleton, which violates (iii). □

Remark: As in the case of Theorem 7.15, the implication (i)$\Longrightarrow$(ii) does not depend on the convexity of g, so it applies to every function f that is Gâteaux differentiable at some point x_0.

We shall be interested, in particular, in continuous convex functions given by the Minkowski functional of a closed convex neighborhood of 0. In order to apply the former results to this situation, we need the two following lemmas.

Lemma 7.18 *Let X be a real Banach space. Let U be a closed convex neighborhood of 0 and let p_U be its Minkowski functional. Then $p_U(x) = \sup\{\langle u^*, x\rangle : u^* \in U^\circ\}$ for every $x \in X$.*

Proof: Fix $x \in X$ and let $\lambda > 0$ such that $x \in \lambda U$. Then we can find $u \in U$ such that $x = \lambda u$. If $u^* \in U^\circ$ we have $\langle u^*, x\rangle = \langle u^*, \lambda u\rangle = \lambda\langle u^*, u\rangle \leq \lambda$. Since this is true for every $u^* \in U^\circ$, we get $\sup\{\langle u^*, x\rangle : u^* \in U^\circ\} \leq \lambda$. This is so for every $\lambda > 0$ such that $x \in \lambda U$, so it is true for the infimum, i.e., $\sup\{\langle u^*, x\rangle : u^* \in U^\circ\} \leq p_U(x)$. Assume now that for some $x \in X$, $(0 \leq) \sup\{\langle u^*, x\rangle : u^* \in U^\circ\} < p_U(x)$. Select t such that $\sup\{\langle u^*, x\rangle : u^* \in U^\circ\} < t < p_U(x)$. In particular, $x \notin tU$. By the separation theorem, we can find $x^* \in X^*$ and $\alpha \in \mathbb{R}$ such that $\langle x^* t, \alpha\rangle < \alpha < \langle x^*, x\rangle$ for every $u \in U$. As $0 \in U$, we may and do assume that $\alpha = 1$. Then $\langle tx^*, u\rangle < 1 < \langle x^* x\rangle$ for all $u \in U$. Hence $tx^* \in U^\circ$ and $\langle tx^*, x\rangle > t$, a contradiction with the fact that $\sup\left\{\langle u^*, x\rangle : u^* \in U^\circ\right\} < t$. □

Lemma 7.19 *Let X be a Banach space and let U be a closed convex neighborhood of 0. Then, for $\varepsilon \geq 0$ and $x_0 \in X$ we have $\partial_\varepsilon p_U(x_0) = \{x^* \in U^\circ : \langle x^*, x_0\rangle \geq s - \varepsilon\}$, where $s := \sup\{\langle u^*, x_0\rangle : u^* \in U^\circ\}$.*

Proof: Assume first that $x^* \in \partial_\varepsilon p_U(x_0)$. Then, since p_U is subadditive,

$$p_U(x_0) + p_U(h) \geq p_U(x_0 + h) \geq p_U(x_0) + \langle x^*, h\rangle - \varepsilon, \text{ for all } h \in X. \quad (7.5)$$

Hence, $p_U(h) \geq \langle x^*, h\rangle - \varepsilon$ for all $h \in X$. Since p_U is positively homogeneous, for all $h \in X$ and $t > 0$, $p_U(th) \geq \langle x^*, th\rangle - \varepsilon$, i.e., $p_U(h) \geq \langle x^*, h\rangle - \varepsilon/t$. This implies $p_U(h) \geq \langle x^*, h\rangle$ for all $h \in X$. In particular, for $u \in U$, $1 \geq p_U(u) \geq \langle x^*, u\rangle$, so $x^* \in U^\circ$.

Letting $h = -x_0$ in (7.5), we get $0 \geq p_U(x_0) - \langle x^*, x_0\rangle - \varepsilon$, i.e., $\langle x^*, x_0\rangle \geq p_U(x_0) - \varepsilon = s - \varepsilon$, since, due to Lemma 7.18, $p_U(x_0) = \sup\{\langle u^*, x_0\rangle : u^* \in U^\circ\}$ $(= s)$. This proves that $\partial_\varepsilon p_U(x_0) \subset \{x^* \in U^\circ : \langle x^*, x_0\rangle \geq s - \varepsilon\}$.

Assume now that $x^* \in U^\circ$ and, moreover, $\langle x^*, x_0\rangle \geq s - \varepsilon$. Recall again that $p_U(x) = \sup\{\langle u^*, x\rangle : u^* \in U^\circ\}$ (Lemma 7.18). Then, for all $x \in X$,

$$p_U(x) \geq \langle x^*, x\rangle = \langle x^*, x - x_0\rangle + \langle x^*, x_0\rangle \geq \langle x^*, x - x_0\rangle + p_U(x_0) - \varepsilon,$$

and this proves that $x^* \in \partial_\varepsilon p_U(x_0)$. □

Remark: In particular, if X is a normed space and $U := B_X$, then $p_U = \|\cdot\|$, and so $\partial\|\cdot\|(x) = \{x^* \in B_{X^*} : \langle x^*, x\rangle = \|x\|\} = \{x^* \in S_{X^*} : \langle x^*, x\rangle = \|x\|\}$ if $x \neq 0$ (if $x = 0$, then $\partial\|\cdot\|(0) = B_{X^*}$). The set-valued mapping $J := \partial\|\cdot\| : X \to 2^{B_{X^*}}$ is called the *duality mapping*.

Corollary 7.20 (Šmulyan lemma [Smul2]) *Let $(X, \|\cdot\|)$ be a Banach space, let U be a closed convex neighborhood of* 0 *and let $x \in X$, $x \neq 0$. Put $s = \sup\{\langle u^*, x\rangle : u^* \in U^\circ\}$.*
(i) *p_U is Fréchet differentiable at x if and only if $\lim_{n\to\infty} \|f_n - g_n\| = 0$ whenever $f_n, g_n \in U^\circ$ satisfy $\lim_{n\to\infty} f_n(x) = \lim_{n\to\infty} g_n(x) = s$, if and only if $\{f_n\} \subset U^\circ$ is convergent whenever $\lim_{n\to\infty} f_n(x) = s$.*
(ii) *p_U is Gâteaux differentiable at x if and only if $(f_n - g_n) \overset{w^*}{\to} 0$ in X^* whenever $f_n, g_n \in U^\circ$ satisfy $\lim_{n\to\infty} f_n(x) = \lim_{n\to\infty} g_n(x) = s$ if and only if there is a unique $f \in U^\circ$ such that $f(x) = s$.*

Proof: It follows directly from Theorem 7.15 (respectively, Theorem 7.17), together with Lemma 7.19. □

Corollary 7.21 *Let X be a Banach space. Let M be a bounded subset of X. Then, the function $f : X^* \to \mathbb{R}$ defined as $f(x^*) = \sup_M x^*$ for $x^* \in X^*$ is Fréchet differentiable at $x_0^* \in X^*$ if and only if x_0^* strongly exposes M.*

Proof: Without loss of generality, we may assume that $0 \in M$. By Lemma 7.18, f is the Minkowski functional p_{M° of M°. Obviously, M° is a closed convex neighborhood of 0 in $(X^*, \|\cdot\|)$. Using Corollary 7.20, we get that f is Fréchet differentiable at x_0^* if and only if x_0^* strongly exposes $M^{\circ\circ}$ $\big(= \overline{\text{conv}}^{w^*}(M)\big)$, where the closure is in (X^{**}, w^*). Now, Exercise 3.148 shows that M has slices of arbitrary small diameter if and only if $M^{\circ\circ}$ has w^*-slices of arbitrary small diameter. This proves the statement. □

Corollary 7.20 applies, in particular, to the case $U := B_X$. In this situation, p_U is, obviously, the original norm of the Banach space X. So we have the following corollary.

Corollary 7.22 (Šmulyan lemma [Smul2]) *Let $(X, \|\cdot\|)$ be a Banach space.*
(i) *The norm $\|\cdot\|$ is Fréchet differentiable at $x \in S_X$ if and only if $\|f_n - g_n\| \to 0$ whenever $\{f_n\}$ and $\{g_n\}$ are sequences in B_{X^*} such that $f_n(x) \to 1$ and $g_n(x) \to 1$, if and only if a sequence $\{f_n\}$ in B_{X^*} converges in norm whenever $f_n(x) \to 1$.*
(ii) *the norm $\|\cdot\|$ is Gâteaux differentiable at $x \in S_X$ if and only if $f_n - g_n \overset{w^*}{\to} 0$ whenever $\{f_n\}$ and $\{g_n\}$ are sequences in B_{X^*} such that $f_n(x) \to 1$ and $g_n(x) \to 1$, if and only if there is a unique $f \in B_{X^*}$ such that $f(x) = 1$.*
(iii) *The dual norm $\|\cdot\|$ in X^* is Fréchet differentiable at $f \in S_{X^*}$ if and only if $\|x_n - y_n\| \to 0$ whenever $\{x_n\}$ and $\{y_n\}$ are sequences in B_X such that $f(x_n) \to 1$*

and $f(y_n) \to 1$, *if and only if a sequence* $\{x_n\}$ *in* B_X *converges in norm whenever* $f(x_n) \to 1$.
(iv) *The dual norm* $\|\cdot\|$ *in* X^* *is Gâteaux differentiable at* $f \in S_{X^*}$ *if and only if* $x_n - y_n \xrightarrow{w} 0$ *whenever* $\{x_n\}$ *and* $\{y_n\}$ *are sequences in* B_X *such that* $f(x_n) \to 1$ *and* $f(y_n) \to 1$, *if and only if* $\{x_n\}$ *is weakly Cauchy whenever* $\{x_n\}$ *is a sequence in* B_X *such that* $f(x_n) \to 1$.

Corollary 7.23 *Let X be a Banach space.*
(i) *If the dual norm of* X^* *is strictly convex, then the norm of X is Gâteaux differentiable.*
(ii) *If the dual norm of* X^* *is Gâteaux differentiable, then the norm of X is strictly convex.*

Proof: (i) Let $x \in S_X$ and $f, g \in S_{X^*}$ be such that $f(x) = g(x) = 1$. Then $\frac{f+g}{2} \in B_{X^*}$ and $\frac{f+g}{2}(x) = 1$, hence $\|\frac{f+g}{2}\| = 1$. By the strict convexity of $\|\cdot\|$ in X^*, we get $f = g$, so $\|\cdot\|$ in X is Gâteaux differentiable according to Corollary 7.22.
(ii) Assume that x, y, and $\frac{x+y}{2} \in S_X$. Let $f \in S_{X^*}$ such that $f\left(\frac{x+y}{2}\right) = 1$. Then $f(x) = f(y) = 1$. Since $\|\cdot\|$ in X^* is Gâteaux differentiable, Corollary 7.22 gives $x = y$, hence $\|\cdot\|$ in X is strictly convex. □

Remark: The converse to (i) in Corollary 7.23 does not hold (see Exercises 7.70 and 8.63). Neither the converse to (ii) is true (consider $X := \ell_1$ and check Exercise 7.64).

Corollary 7.24 *If the norm* $\|\cdot\|$ *of a Banach space X is Fréchet differentiable, then it is* C^1*-smooth on* $X\setminus\{0\}$.

Proof: Let $x, x_n \in X\setminus\{0\}$, $x_n \to x$. Denote by f_n, respectively f the derivative of $\|\cdot\|$ at x_n, respectively x. Then $f, f_n \in S_X$, $f_n(x_n) = \|x_n\|$ and $f(x) = \|x\|$. Set $y_n = \frac{x_n}{\|x_n\|}$, $y = \frac{x}{\|x\|}$, then $f_n(y_n) = 1$, $f(y) = 1$.
We easily check that $f_n(y) \to 1$, so by Corollary 7.22 we get $f_n \to f$. □

Corollary 7.25 *Let* $(X, \|\cdot\|)$ *be a Banach space. If the dual norm* $\|\cdot\|$ *in* X^* *is locally uniformly rotund then* $\|\cdot\|$ *in X is Fréchet differentiable.*

Proof: Let $x \in S_X$, choose $f \in S_{X^*}$ such that $f(x) = 1$. Let $f_n \in S_{X^*}$ satisfy $\lim f_n(x) = 1$. We have

$$2 \geq \|f + f_n\| \geq (f + f_n)(x) \to 2.$$

Therefore $\lim(2\|f_n\|^2 + 2\|f\|^2 - \|f + f_n\|^2) = 0$ and by the LUR property $\lim \|f_n - f\| = 0$. By the Šmulyan Lemma 7.22, $\|\cdot\|$ in X is Fréchet differentiable. □

Corollary 7.26 *Let X be a Banach space. If the dual norm of* X^* *is Fréchet differentiable (respectively Gâteaux differentiable), then X is reflexive (respectively, it does not contain an isomorphic copy of* ℓ_1*).*

Proof: For the Fréchet differentiability statement, by Corollary 3.131 it is enough to show that every $f \in X^*$ attains its norm on B_X. Given $f \in S_{X^*}$, choose $x_n \in S_X$ such that $f(x_n) \to 1$. By Corollary 7.22, $\{x_n\}$ is convergent to some $x \in S_X$. Clearly $f(x) = 1$.

For the Gâteaux differentiability statement, first we show that if Y is a subspace of X, then the dual norm of Y^* is Gâteaux differentiable. Indeed, let $f \in S_{Y^*}$ and x_n and $y_n \in S_Y$, $n = 1, 2, \ldots$ be such that $f(x_n) \to 1$ and $f(y_n) \to 1$. Let $\widetilde{f} \in S_{X^*}$ be a Hahn–Banach extension of f. According to Corollary 7.22, it follows that $\lim\,(x_n - y_n) = 0$ in the weak topology of X and therefore in the weak topology of Y. From the same criterion, it follows that f is a point of Gâteaux differentiability of the dual norm of Y^*. Assume now that there is a subspace $Y \subset X$ such that Y is isomorphic to ℓ_1. Then ℓ_1 admits a norm whose dual is Gâteaux differentiable. Since ℓ_1 is weakly sequentially complete (Theorem 5.36), from Exercise 7.17 it follows that ℓ_1 is reflexive. This contradiction concludes the proof. □

The following is a partial version of the Šmulyan's Lemma 7.20 for the case of uniform Fréchet differentiability and will be enough for our purposes here.

Theorem 7.27 *Let f be a convex continuous function defined on a nonempty open convex subset U of a Banach space X and Fréchet differentiable on U. Then we have*

(i) *If, for some $r > 0$ and a subset S of U such that $S + rB_X \subset U$, the function f is uniformly Fréchet differentiable on S, then*

$$\lim_{\varepsilon \downarrow 0} \sup\{\operatorname{diam} \partial_\varepsilon f(x) : x \in S\} = 0. \tag{7.6}$$

(ii) *If (7.6) holds for $f := \|\cdot\|$ and $S := S_X$, then $\|\cdot\|$ is uniformly Fréchet differentiable on S_X* (see Fig. 7.6).

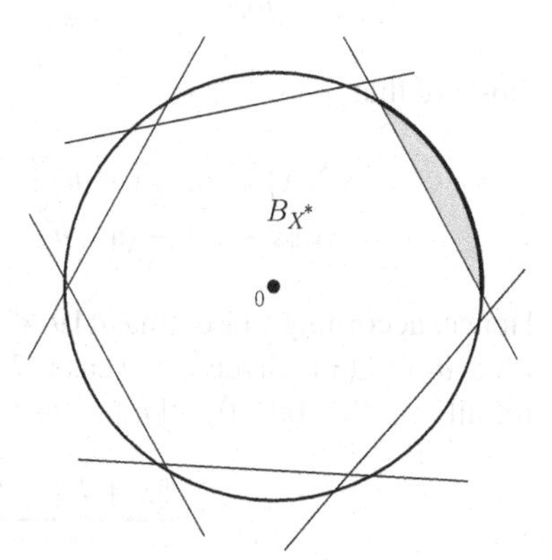

Fig. 7.6 (UF) norm$\Longleftrightarrow$all sections of B_{X^*} with the same depth have the "same" diameter

Proof: (i) Suppose that (7.6) does not hold. We can find $\eta > 0$ such that for every $\varepsilon > 0$ there exists $x_\varepsilon \in V$ with $\operatorname{diam} \partial_\varepsilon f(x_\varepsilon) > \eta$, i.e., two elements $x_\varepsilon^*, y_\varepsilon^*$ in $\partial_\varepsilon f(x_\varepsilon)$ exist such that $\|x_\varepsilon^* - y_\varepsilon^*\| > \eta$. Find $h_\varepsilon \in S_X$ such that $\langle x_\varepsilon^* - y_\varepsilon^*, h_\varepsilon \rangle > \eta$. Due to the fact that $x_\varepsilon^*, y_\varepsilon^*$ are in $\partial_\varepsilon f(x_\varepsilon)$, we have, for every $t \in (-r, r)$,

$$f(x_\varepsilon + th_\varepsilon) - f(x) \geq \langle x_\varepsilon^*, th_\varepsilon \rangle - \varepsilon.$$
$$f(x_\varepsilon - th_\varepsilon) - f(x) \geq \langle y_\varepsilon^*, -th_\varepsilon \rangle - \varepsilon.$$

Fix $t \in (0, r)$. For $\varepsilon := t\eta/4$ find, as above, $x_\varepsilon, x_\varepsilon^*, y_\varepsilon^*$ and h_ε. Then

$$\frac{f(x_\varepsilon + th_\varepsilon) + f(x_\varepsilon - th_\varepsilon) - 2f(x_\varepsilon)}{t}$$
$$\geq \frac{\langle x_\varepsilon^*, th_\varepsilon \rangle - \langle y_\varepsilon^*, th_\varepsilon \rangle - 2\varepsilon}{t} = \langle x_\varepsilon^* - y_\varepsilon^*, h_\varepsilon \rangle - \frac{2\varepsilon}{t} > \eta - \frac{\eta}{2} = \frac{\eta}{2}.$$

Since this is true for every $t \in (0, r)$, it contradicts the fact that f is uniformly Fréchet differentiable on S.

(ii) Take $x \in S_X$ and $0 \neq h \in X$. Let $u^* \in S_{X^*}$ such that $\langle u^*, x + h \rangle = \|x + h\|$ and $v^* \in S_{X^*}$ such that $\langle v^*, x - h \rangle = \|x - h\|$ (i.e., $\{u^*\} = \partial\| \cdot \|(x + h)$ and $\{v^*\} = \partial\| \cdot \|(x - h)$. Then

$$\begin{aligned} \|x + h\| - \|x\| &= \langle u^*, x + h \rangle - \|x\| \\ &= \langle u^*, x \rangle + \langle u^*, h \rangle - \|x\| \leq \|x\| - \langle u^*, h \rangle - \|x\| = \langle u^*, h \rangle, \end{aligned}$$

and, similarly,

$$\begin{aligned} \|x - h\| - \|x\| &= \langle v^*, x - h \rangle - \|x\| \\ &= \langle v^*, x \rangle - \langle v^*, h \rangle - \|x\| \leq \|x\| - \langle v^*, h \rangle - \|x\| = -\langle v^*, h \rangle. \end{aligned}$$

Hence

$$\frac{\|x + h\| + \|x - h\| - 2\|x\|}{\|h\|} \leq \frac{\langle u^*, h \rangle - \langle v^*, h \rangle}{\|h\|} = \left\langle u^* - v^*, \frac{h}{\|h\|} \right\rangle. \tag{7.7}$$

Observe that

$$\begin{aligned} \langle u^*, x \rangle &= \langle u^*, x + h \rangle - \langle u^*, h \rangle \\ &= \|x + h\| - \langle u^*, h \rangle \geq \|x\| - \|h\| - \|h\| = \|x\| - 2\|h\|. \end{aligned}$$

Hence, according to Lemma 7.19, if $\|h\| \leq \varepsilon/2$ then $u^* \in \partial_\varepsilon\| \cdot \|(x)$. Analogously, $v^* \in \partial_\varepsilon\| \cdot \|(x)$. Given $\eta > 0$ use (7.6) to find $\varepsilon > 0$ such that $\operatorname{diam} \partial_\varepsilon\| \cdot \|(x) < \eta$ for all $x \in S_X$. Take $0 < \|h\| < \varepsilon/2$. Then $\|u^* - v^*\| < \eta$, so, from (7.7) we get

$$\frac{\|x + h\| + \|x - h\| - 2\|x\|}{\|h\|} \leq \eta.$$

This shows that $\|\cdot\|$ is uniformly Fréchet differentiable (see the remark after Definition 7.5). □

The following result is useful in several arguments related to Lipschitz functions.

Theorem 7.28 (Fitzpatrick, [Fitz]) *Let $(X, \|\cdot\|)$ be a Banach space, let $r > 0$, and let $f : rB_X \to \mathbb{R}$ be a function. Assume that there exists $u \in S_X$ such that the norm $\|\cdot\|$ is Fréchet differentiable at u and that (the directional derivative) $Df(0)u$ exists, is finite, and is equal to*

$$L_f(0) := \lim_{\delta \downarrow 0} \sup \left\{ \frac{f(x) - f(y)}{\|x - y\|} : x, y \in \delta B_X,\ x \neq y \right\}.$$

Then f is Fréchet differentiable at 0 and $f'(0) = L_f(0)\|\cdot\|'(u)$.

Proof: For $\delta \in [0, r]$ we put

$$\begin{aligned} o_1(\delta) &= \sup\{f(x) - f(y) - L_f(0)\|x - y\| : x, y \in \delta B_X\}, \\ o_2(\delta) &= \sup\{f(0) + Df(0)(tu) - f(tu)) : 0 \le t \le \delta\}, \\ o_3(\delta) &= \sup\{\|u + w\| - \|u\| - \|\cdot\|'(u)w : w \in \delta B_X\}. \end{aligned}$$

From corresponding definitions, it follows that $\lim_{\delta \downarrow 0} \frac{o_i(\delta)}{\delta} = 0,\ i = 1, 2, 3$. Using these functions, for every $t \in (0, r]$ and for every $w \in rB_X$ we can estimate

$$L_f(0)\|tu - w\| + o_1(t \vee \|w\|) \ge f(tu) - f(w) \ge f(0) + Df(0)(tu) - o_2(t) - f(w),$$

and taking into account that $L_f(0) = Df(0)u$ and that $\|u\| = 1$, we can continue

$$\begin{aligned} f(w) - f(0) &\ge -L_f(0)\|tu - w\| + Df(0)(tu) - o_1(t \vee \|w\|) - o_2(t) \\ &= -L_f(0)t\left(\left\|u - \tfrac{1}{t}w\right\| - \|u\|\right) - o_1(t \vee \|w\|) - o_2(t) \\ &\ge -L_f(0)t\Big(\|\cdot\|'(u)\big(-\tfrac{1}{t}w\big) + o_3\big(\tfrac{1}{t}\|w\|\big)\Big) - o_1(t \vee \|w\|) - o_2(t). \end{aligned}$$

Hence, for all $t \in (0, r]$ and all $w \in rB_X$ we have

$$f(w) - f(0) - L_f(0)\|\cdot\|'(u)w \ge -L_f(0)t o_3\big(\tfrac{1}{t}\|w\|\big) - o_1(t \vee \|w\|) - o_2(t).$$

Now fix any $\gamma > 0$. In the latter inequality, consider any $0 \neq w \in \min\{r, \gamma r\}B_X$ and then put $t = \frac{1}{\gamma}\|w\|$; thus $t \in (0, r]$. We get

$$\frac{f(w) - f(0) - L_f(0)\|\cdot\|'(u)w}{\|w\|} \ge -L_f(0)\frac{o_3(\gamma)}{\gamma} - \frac{o_1\big((\gamma^{-1} \vee 1)\|w\|\big)}{\|w\|} - \frac{o_2(\gamma^{-1}\|w\|)}{\|w\|}.$$

Therefore

$$\liminf_{\|w\|\to 0} \frac{f(w) - f(0) - L_f(0)\|\cdot\|'(u)w}{\|w\|} \geq -L_f(0)\frac{o_3(\gamma)}{\gamma}.$$

Finally, letting $\gamma \downarrow 0$ in the latter inequality, we get $\liminf_{\|w\|\to 0} \frac{f(w)-f(0)-L_f(0)\|\cdot\|'(u)w}{\|w\|} \geq 0$. The inequality $\limsup_{\|w\|\to 0} \frac{f(w)-f(0)-L_f(0)\|\cdot\|'(u)w}{\|w\|} \leq 0$ can be deduced similarly. □

We introduce now the concept of Fenchel duality and present some properties of the Fenchel conjugate. This will be used in Chapter 11.

Definition 7.29 *Let f be a proper function from a Banach space X into $(-\infty, +\infty]$. The* Fenchel conjugate *f^* of f is defined by*

$$f^*(x^*) = \sup\{\langle x^*, x\rangle - f(x) : x \in X\}, \textit{ for } x^* \in X^*.$$

Observe that, due to the fact that f is proper, we have $f^*(x^*) \neq -\infty$ for each $x^* \in X^*$, so f^* is a function from X^* into $(-\infty, +\infty]$. Note, too, that we can have $f^* \equiv +\infty$ (consider the function $f : \mathbb{R} \to \mathbb{R}$ given by $f(x) = x$ for $x < 0$, and $f(x) = (1/2)x$ for $x \geq 0$). Fenchel conjugates of some simple functions are listed in Exercise 7.51.

It is obvious that f^* *is proper if f is bounded below*. Another instance of a proper conjugate appears *when f is proper, convex, and lower semicontinuous*: indeed, the separation theorem applied to an element $(x_0, t) \in (X \times \mathbb{R})\setminus \operatorname{epi} f$, where $x_0 \in \operatorname{dom} f$, and the closed convex set $\operatorname{epi} f \subset X \times \mathbb{R}$, gives $x^* \in X^*$, k, α and β in $\mathbb{R}$ such that

$$\langle (x^*, k), (x, f(x))\rangle \leq \alpha < \beta \leq \langle (x^*, k), (x_0, t)\rangle, \quad \text{for all } x \in \operatorname{dom} f,$$

i.e.,

$$\langle x^*, x\rangle + kf(x) \leq \alpha < \beta \leq \langle x^*, x_0\rangle + kt, \quad \text{for all } x \in \operatorname{dom} f.$$

If $k = 0$ we get $\langle x^*, x\rangle \leq \alpha < \beta \leq \langle x^*, x_0\rangle$ for all $x \in \operatorname{dom} f$, a contradiction since $x_0 \in \operatorname{dom} f$. We can choose then, without loss of generality, $k = -1$. In particular we get $\langle x^*, x\rangle - f(x) \leq \alpha$ for all $x \in \operatorname{dom} f$, so $f^*(x^*) \leq \alpha$, and then f^* is proper.

As a supremum of convex w^*-continuous functions on X^*, f^* is convex and w^*-lower semicontinuous.

Assume that f^* is proper. Then we can consider the function $f^{**} : X^{**} \to (-\infty, +\infty]$. It is again a convex and w^*-lower semicontinuous function from X^{**} into $(-\infty, +\infty]$, and it is proper as it follows from the previous argument: note that f^* is convex and also $\|\cdot\|$-lower semicontinuous.

We list in the following statement some easy facts about conjugate functions. The proof is standard.

Proposition 7.30 *Let X be a Banach space. Let f and g be two proper functions from a Banach space X into $(-\infty, +\infty]$. Then, given $x \in X$ and $x^* \in X^*$,*
(i) $f(x) + f^*(x^*) \geq \langle x^*, x\rangle$.
(ii) *If f^* is proper, then $f^{**}|_X \leq f$.*
(iii) *If $f \leq g$, then $f^* \geq g^*$.*

Proposition 7.31 ([BeMo]) *Let X be a Banach space. Let $f : X \to (-\infty, +\infty]$ be a proper function such that f^* is also proper. Then* $\operatorname{epi} f^{**} = \overline{\operatorname{conv}}^{w^*}(\operatorname{epi} f)$.

Proof: First, assume $f \geq 0$. The inclusion $\overline{\operatorname{conv}}^{w^*}(\operatorname{epi} f) \subset \operatorname{epi} f^{**}$ follows from $\operatorname{epi} f \subset \operatorname{epi} f^{**}$ (use (ii) in Proposition 7.30), the convexity, and the w^*-lower semicontinuity of f^{**}. Let $(x_0^{**}, \lambda_0) \in \operatorname{epi} f^{**}$. Suppose that $(x_0^{**}, \lambda_0) \notin \overline{\operatorname{conv}}^{w^*}(\operatorname{epi} f)$. By the separation theorem, there is $x_0^* \in X^*$ and $k, \alpha, \beta \in \mathbb{R}$ such that

$$x_0^{**}(x_0^*) + k\lambda_0 < \alpha < \beta < x^{**}(x_0^*) + k\lambda \text{ for all } (x^{**}, \lambda) \in \overline{\operatorname{conv}}^{w^*}(\operatorname{epi} f).$$

From these inequalities we get $k \geq 0$ (if $k < 0$, it is enough to take $x \in \operatorname{dom} f$ and $\lambda \to +\infty$ in order to obtain a contradiction). In particular we get $x_0^*(x)+kf(x) > \beta$ for all $x \in \operatorname{dom} f$. Take $\varepsilon > 0$. Since $f \geq 0$, we get

$$-\frac{x_0^*(x)}{k+\varepsilon} - f(x) < -\frac{\beta}{k+\varepsilon} \text{ for all } x \in \operatorname{dom} f,$$

hence $f^*\left(-\frac{x_0^*}{k+\varepsilon}\right) \leq -\frac{\beta}{k+\varepsilon}$. Then

$$\begin{aligned} f^{**}(x_0^{**}) &\geq x_0^{**}\Big(-\frac{x_0^*}{k+\varepsilon}\Big) - f^*\Big(-\frac{x_0^*}{k+\varepsilon}\Big) \geq x_0^{**}\Big(-\frac{x_0^*}{k+\varepsilon}\Big) + \frac{\beta}{k+\varepsilon} \\ &= \frac{1}{k+\varepsilon}(\beta - x_0^{**}(x_0^*)) > \frac{\beta - \alpha + k\lambda_0}{k+\varepsilon}. \end{aligned}$$

If $k = 0$, then $f^{**}(x_0^{**}) > (\beta - \alpha)/\varepsilon$. As $\varepsilon > 0$ was arbitrary, we get $x_0^{**} \notin \operatorname{dom} f^{**}$, a contradiction. Therefore, $k > 0$. Since $\varepsilon > 0$ was arbitrary, we get $f^{**}(x_0^{**}) \geq (\beta - \alpha + k\lambda_0)/k > \lambda_0$. This contradicts $(x_0^{**}, \lambda_0) \in \operatorname{epi} f^{**}$. The statement is proved for a non-negative function f.

Now, if $f\colon X \to \mathbb{R} \cup \{+\infty\}$ is an arbitrary proper function such that f^* is also proper, choose $x_0^* \in \operatorname{dom} f^*$. Consider $g\colon X \to \mathbb{R} \cup \{+\infty\}$ given by $g(x) = f(x) + f^*(x_0^*) - x_0^*(x)$. This function is proper, and its Fenchel conjugate g^* is also proper. Moreover, $\operatorname{dom} f = \operatorname{dom} g$, and $g \geq 0$. Observe that $g^{**}(x^{**}) = f^{**}(x^{**}) + f^*(x_0^*) - x^{**}(x_0^*)$ for all $x^{**} \in X^{**}$. By the first part of the proof, the proposition holds for g, and hence for f. □

We omit the simple proof of the two following consequences.

Corollary 7.32 *Let X be a Banach space. Let $f : X \to (-\infty, +\infty]$ be a proper convex function such that f^* is also proper. Then $f^{**}|_X = f$ if and only if f is lower semicontinuous.*

Corollary 7.33 *Let X be a Banach space. Then, every convex and w^*-lower semicontinuous function $g : X^* \to (-\infty, +\infty]$ is the Fenchel conjugate of $g^*\big|_X$.*

Proposition 7.34 *Let X be a Banach space. Let f be a proper function from a Banach space X into $(-\infty, +\infty]$. Let $x \in X$, $x^* \in X^*$ and $\varepsilon > 0$. Then*
(i) $x^* \in \partial_\varepsilon f(x)$ if and only if $f(x) + f^*(x^*) \leq \langle x^*, x\rangle + \varepsilon$.
(ii) $x^* \in \partial f(x)$ if and only if $f(x) + f^*(x^*) = \langle x^*, x\rangle$.

Proof: To prove (i), observe that $x^* \in \partial_\varepsilon f(x)$ if and only if $\langle x^*, y - x\rangle \leq f(y) - f(x) + \varepsilon$ for all $y \in X$, if and only if $\langle x^*, y\rangle - f(y) \leq \langle x^*, x\rangle - f(x) + \varepsilon$ for all $y \in X$, if and only if $f^*(x^*) \leq \langle x^*, x\rangle - f(x) + \varepsilon$, if and only if $f(x) + f^*(x^*) \leq \langle x^*, x\rangle + \varepsilon$. This proves (i), and (ii) follows from (i). □

Observe that if $x^* \in \partial_\varepsilon f(x)$ for some $\varepsilon \geq 0$ and some $x \in X$, the function f^* is, in particular, proper.

Corollary 7.35 *Let X be a Banach space. Let $f : X \to (-\infty, +\infty]$ be a proper function. Let $x \in X$, $x^* \in X^*$ and $\varepsilon \geq 0$.*
(i) If $x^* \in \partial_\varepsilon f(x)$, then $x \in \partial_\varepsilon f^*(x^*)$.
(ii) If f is, moreover, convex and lower semicontinuous, and $x \in \partial_\varepsilon f^*(x^*)$, then $x^* \in \partial_\varepsilon f(x)$.

Proof: Assume that $x^* \in \partial_\varepsilon f(x)$ for some $x \in X$. Then, by (i) in Proposition 7.34 we get $f(x) + f^*(x^*) \leq \langle x^*, x\rangle + \varepsilon$. The function f^* is proper, and, by (ii) in Proposition 7.30, $f^{**}(x) + f^*(x^*) \leq f(x) + f^*(x^*) \leq \langle x^*, x\rangle + \varepsilon$. Again by (i) in Proposition 7.34 we get $x \in \partial_\varepsilon f^*(x^*)$. This proves (i).

To prove (ii), observe that f^* is proper. By (i) in Proposition 7.34, we have $f^{**}(x) + f^*(x^*) \leq \langle x^*, x\rangle + \varepsilon$. By Corollary 7.32, we have $f^{**}(x) = f(x)$, and (ii) follows again by (i) in Proposition 7.34. □

Let A be a subset of a Banach space. Let $x^* \in X^*$ and $\varepsilon > 0$. We denote by $S(A, x^*, \varepsilon)$ the *section* of width ε determined in A by x, i.e., $S(A, x^*, \varepsilon) := \{x \in A : \langle x^*, x\rangle \geq s - \varepsilon\}$, where $s := \sup\{\langle x^*, x\rangle : x \in A\}$.

Proposition 7.36 *Let X be a Banach space. Let $f : X \to (-\infty, +\infty]$ be a proper function. Let $x_0 \in X$, $x_0^* \in X^*$ and $\varepsilon > 0$.*
(i) If $x_0^* \in \partial_\varepsilon f(x_0)$ then $(x_0^*, f^*(x_0^*)) \in S(\text{epi } f^*, (x_0, -1), \varepsilon)$.
(ii) If, moreover, the function f is convex and lower semicontinuous, and $(x_0^*, f^*(x_0^*)) \in S(\text{epi } f^*, (x_0, -1), \varepsilon)$, then $x_0^* \in \partial_\varepsilon f(x_0)$.

Proof: (i) Observe that $s := \sup\{\langle (x^*, t), (x_0, -1)\rangle : (x^*, t) \in \text{epi } f^*\} = \sup\{\langle x^*, x_0\rangle - f^*(x^*) : x^* \in X^*\} = f^{**}(x_0)$. Since $x_0^* \in \partial_\varepsilon f(x_0)$, we have $\langle x_0^*, x - x_0\rangle \leq f(x) - f(x_0) + \varepsilon$, for all $x \in X$. Then $\langle x_0^*, x\rangle - f(x) \leq \langle x_0^*, x_0\rangle - f(x_0) + \varepsilon$. By taking the supremum for $x \in X$ in the left hand side of the previous inequality, we get $f^*(x_0^*) \leq \langle x_0^*, x_0\rangle - f(x_0) + \varepsilon$, so $\langle x_0^*, x_0\rangle - f^*(x_0^*) \geq f(x_0) - \varepsilon \geq f^{**}(x_0) - \varepsilon = s - \varepsilon$. This proves (i).

To prove (ii), let $(x_0^*, f^*(x_0^*)) \in S(\text{epi } f^*, (x_0, -1), \varepsilon)$. Then

$$\langle (x_0^*, f^*(x_0^*)), (x_0, -1) \rangle \geq s - \varepsilon = f^{**}(x_0) = f(x_0) - \varepsilon,$$

as it follows from the description of s above and Corollary 7.32. The result is now a consequence of the definition of $f^*(x_0^*)$. □

Corollary 7.37 *Let X be a Banach space. Let $f : X \to (-\infty, +\infty]$ be a proper function. Assume that f^* is Fréchet differentiable at some $x_0^* \in X^*$. Then $\partial f^*(x_0^*) \in X$.*

Proof: It follows from Proposition 7.36 and Theorem 7.15. □

The following concept plays an important role in differentiability theory and structure of Banach spaces. It will be used frequently in what follows.

Definition 7.38 *A Banach space X is called* Asplund *if for every convex continuous function $f : X \to \mathbb{R}$ there exists a dense G_δ set $D \subset X$ such that f is Fréchet differentiable at every point $x \in D$.*

7.3 Ekeland Principle and Bishop–Phelps Theorem

The classical way to find minima of a real-valued function in the presence of differentiability is to check points where the derivative of the function is 0, which amounts to "support" the epigraph of the function "from below" with a half-space. In the general case, in absence of differentiability, we can replace the maybe non-existent half-space with a cone that "touches" the epigraph at a single point from below. If the cone is "wide enough" (i.e., if ε (> 0) in Theorem 7.39 is small enough), the single common point to the cone and the epigraph behaves *almost as a true minimum* for the function f (see Fig. 7.7). This is the purpose of the following result.

In its proof the following notation will be used: Let X be a Banach space. Given $\varepsilon > 0$, put $K_\varepsilon = \{(x, t) \in X \times \mathbb{R} : t \leq -\varepsilon \|x\|\}$. This set is a downward closed

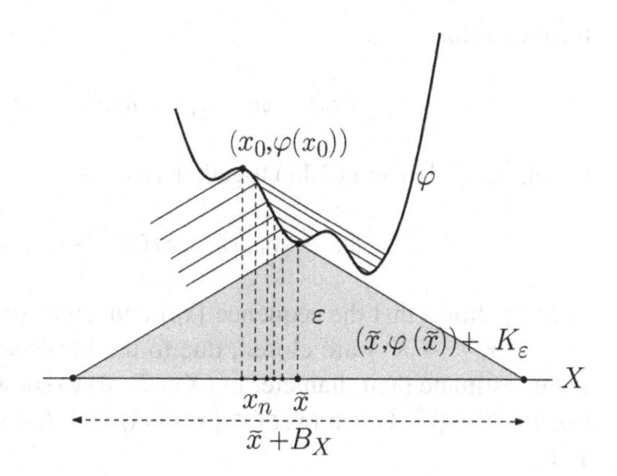

Fig. 7.7 Searching for an "almost" minimum (Ekeland)

cone in $X \times \mathbb{R}$ with vertex at $(0, 0)$ (see Fig. 7.7). We equip $X \times \mathbb{R}$ with the metric ρ defined by $\rho\big((x,t),(y,s)\big) = \|x-y\| + |t-s|$.

Theorem 7.39 (Ekeland variational principle [Ekel], see also [BeLi, p. 87] and [DGZ3, p. 10]) *Let X be a Banach space, and let $\varphi : X \to (0, +\infty]$ be a proper, lower semicontinuous, and bounded below function. Then, given $\varepsilon > 0$ and $\delta > 0$, there exists $\widetilde{x} \in X$ such that $\varphi(\widetilde{x}) < \varphi(x) + \varepsilon\|x - \widetilde{x}\|$ for all $x \in X$ such that $x \neq \widetilde{x}$. Moreover, if $x_0 \in X$ satisfies $\varphi(x_0) < b + \delta$, where $b = \inf\{\varphi(x) : x \in X\}$, then $\widetilde{x}$ can be chosen so that $\|x_0 - \widetilde{x}\| < \delta/\varepsilon$.*

Proof: (See Fig. 7.7) Fix $\delta > 0$. We shall define by induction a sequence $\{x_n\}_{n=0}^{\infty}$ in X that will converge to the sought point $\widetilde{x}$. Let $b = \inf\{\varphi(x) : x \in X\}$. Choose $x_0 \in X$ such that $(b \leq)\ \varphi(x_0) < b + \delta/2$. If x_k has already been defined for $k = 0, 1, 2, \ldots, n$, put

$$b_n = \inf\{t : t \in \mathbb{R}, \text{ there exists } x \in X \text{such that } (x,t) \in \operatorname{epi}\varphi \cap \big((x_n, \varphi(x_n)) + K_\varepsilon\big)\},$$

and choose then $x_{n+1} \in X$ such that

$$(x_{n+1}, \varphi(x_{n+1})) \in (x_n, \varphi(x_n)) + K_\varepsilon, \tag{7.8}$$

$$\varphi(x_{n+1}) < b_n + \delta/2^{n+2}. \tag{7.9}$$

Equation (7.8) is equivalent to

$$(0 \leq)\ \varepsilon\|x_n - x_{n+1}\| \leq \varphi(x_n) - \varphi(x_{n+1}). \tag{7.10}$$

Observe that, in particular, the sequence $\{\varphi(x_n)\}_{n=0}^{\infty}$ is nonincreasing. Due to the fact that $\{(x_n, \varphi(x_n)) + K_\varepsilon\}_{n=0}^{\infty}$ is a nested sequence of sets, the sequence $\{b_n\}_{n=0}^{\infty}$ is nondecreasing. We have $b \leq \varphi(x_1) \leq \varphi(x_0) < b + \delta/2$. This shows that $\varphi(x_0) - \varphi(x_1) < \delta/2$. Moreover, for every $n \in \mathbb{N}$, we have

$$b_{n-1} \leq b_n \leq \varphi(x_{n+1}) \leq \varphi(x_n) < b_{n-1} + \delta/2^{n+1}.$$

It follows that

$$\varphi(x_n) - \varphi(x_{n+1}) < \delta/2^{n+1}, \quad n = 0, 1, 2, \ldots \tag{7.11}$$

Equations (7.10) and (7.11) together give

$$\|x_n - x_{n+1}\| < \delta/(2^{n+1}\varepsilon), \ n = 0, 1, 2, \ldots \tag{7.12}$$

and this shows that the sequence $\{x_n\}$ converges (to some $\widetilde{x} \in X$). The sets $\operatorname{epi}\varphi \cap \big((x_n, \varphi(x_n)) + K_\varepsilon\big)$ are closed, due to the lower semicontinuity of the function φ. Let us estimate their diameter in $(X \times \mathbb{R}, \rho)$ to show that they form a null sequence. Fix $n \in \mathbb{N} \cup \{0\}$. Given $(x,t) \in \big(x_n, \varphi(x_n)\big) + K_\varepsilon$, we have $\varepsilon\|x - x_n\| \leq \varphi(x_n) - t$, and

$$b_{n-1} \leq b_n \leq t \leq \varphi(x_n) < b_{n-1} + \frac{\delta}{2^{n+1}},$$

so $0 \leq \varphi(x_n) - t < \delta/2^{n+1}$, hence $\|x - x_n\| < \delta/(\varepsilon 2^{n+1})$. This proves that $\operatorname{diam}\big(\operatorname{epi}\varphi \cap (x_n, \varphi(x_n)) + K_\varepsilon\big) \to 0$ as $n \to \infty$. It follows that $\operatorname{epi}\varphi \cap \bigcap_{n=0}^{\infty}\big((x_n, \varphi(x_n)) + K_\varepsilon\big)$ reduces to a single point, precisely $(\widetilde{x}, \varphi(\widetilde{x}))$. Obviously, $\operatorname{epi}\varphi \cap \big((\widetilde{x}, \varphi(\widetilde{x})) + K_\varepsilon\big) = (\widetilde{x}, \varphi(\widetilde{x}))$, hence $\varphi(\widetilde{x}) < \varphi(x) + \varepsilon\|x - \widetilde{x}\|$ for all $x \in X$ such that $x \neq \widetilde{x}$.

Moreover, from (7.12) we get $\|x_0 - x_n\| < \frac{\delta}{\varepsilon}(1/2 + 1/2^2 \ldots + 1/2^n)$ for $n \in \mathbb{N}$, so $\|x_0 - \widetilde{x}\| \leq \delta/\varepsilon$. □

Remark: The estimation $\|x_0 - \widetilde{x}\| \leq \delta/\varepsilon$ given in the statement may appear poor at first glance, given that ε appears in the denominator. This, however, can be turned into a sharp estimate just by choosing, for example, $\delta = \varepsilon^2$, and $\varepsilon > 0$ conveniently small.

An application of Ekeland's variational principle is the following corollary.

Corollary 7.40 (Palais–Smale minimizing sequences, see, e.g., [DeGh, p. 423]) *Let X be a Banach space and $\varphi : X \to \mathbb{R}$ be a Gâteaux differentiable function that is bounded below. Then there exists a sequence $\{x_n\}$ in X such that $\varphi(x_n) \to \inf_{x \in X} \varphi(x)$ and $\|\varphi'(x_n)\| \to 0$.*

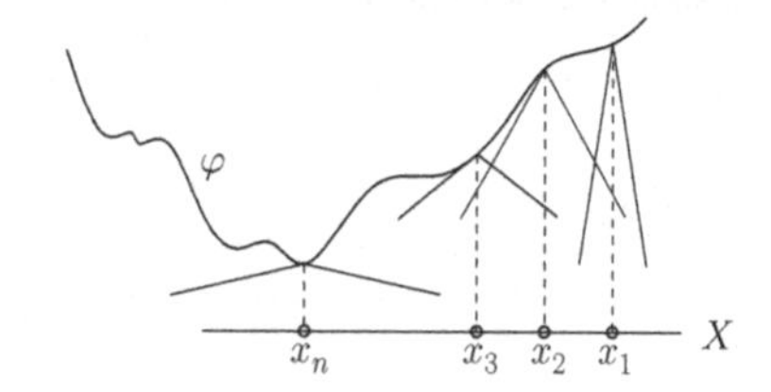

Fig. 7.8 Proof of Corollary 7.40

Proof: Apply Ekeland variational principle for $\varepsilon = \delta := 1/n$ to find $x_n \in X$ such that $\varphi(x_n) < \inf_{x \in X} \varphi(x) + 1/n$ and such that $\varphi(x) \geq \varphi(x_n) - (1/n)\|x - x_n\|$ for every $x \in X$. It is clear that $\|\varphi'(x_n)\| \to 0$ (see Fig. 7.8). □

Consider a functional $f \in X^*$. We say that it *attains its supremum over C* if there is $x \in C$ such that $f(c) = \sup\{f(x) : c \in C\}$.

We say that $f \in X^*$ *attains its norm* if there is $b \in B$ such that $f(b) = \|f\|$.

Theorem 7.41 (Bishop–Phelps [BiPh], see also [BeLi, p. 87] and [DGZ3, p. 14]) *Let C be a nonempty closed convex and bounded subset of a real Banach space X. Then the set of all continuous linear functionals on X that attain their maximum on C is dense in X^*. In particular, the set of all continuous linear functionals on X that attain their norm (i.e., their maximum on B_X) is dense in X^*.*

Proof: Let $f \in X^*$ and $\varepsilon > 0$. Apply Theorem 7.39 to the function $\widetilde{f} : X \to \mathbb{R} \cup \{+\infty\}$ defined by $\widetilde{f}(c) = -f(c)$ for all $c \in C$ and $\widetilde{f}(x) = +\infty$ for all $x \in X \backslash C$ to obtain $x_0 \in C$ such that $\widetilde{f}(x_0) - \varepsilon\|x - x_0\| \leq \widetilde{f}(x)$ for all $x \in X$

(see Fig. 7.9). In particular $f(x) \le f(x_0) + \varepsilon\|x - x_0\|$ for all $x \in C$. Consider the following two convex sets in $X \oplus \mathbb{R}$:

$$K_1 := \{(x, t) : x \in C, t \le f(x)\},$$
$$K_2 := \{(x, t) : x \in X, t \ge f(x_0) + \varepsilon\|x - x_0\|\}.$$

The interior of K_2 is nonempty and disjoint from K_1. Therefore there is a non-zero functional in $X^* \oplus \mathbb{R}$ that separates K_1 and K_2, i.e., there is $x^* \in X^*$ and a real number β such that

$$\langle x^*, x\rangle + \alpha t \le \beta \text{ if } (x, t) \in K_1, \text{ and}$$
$$\langle x^*, x\rangle + \alpha t \ge \beta \text{ if } (x, t) \in K_2.$$

Note that α cannot be negative as otherwise $\langle x^*, x\rangle + \alpha t < \beta$ if $(x, t) \in K_2$ and t is large enough. If $\alpha = 0$, then $\langle x^*, x\rangle \ge \beta$ for any $x \in X$, which gives $x^* = 0$, and this is not the case. Hence $\alpha > 0$, and we can normalize so that $\alpha = 1$. Since $(x_0, f(x_0)) \in K_1 \cap K_2$, we have $\langle x^*, x_0\rangle + f(x_0) = \beta$. Therefore, $\langle x^*, x\rangle + f(x) \le \langle x^*, x_0\rangle + f(x_0)$ for all $x \in C$, so $x^* + f$ attains its maximum on C at x_0.

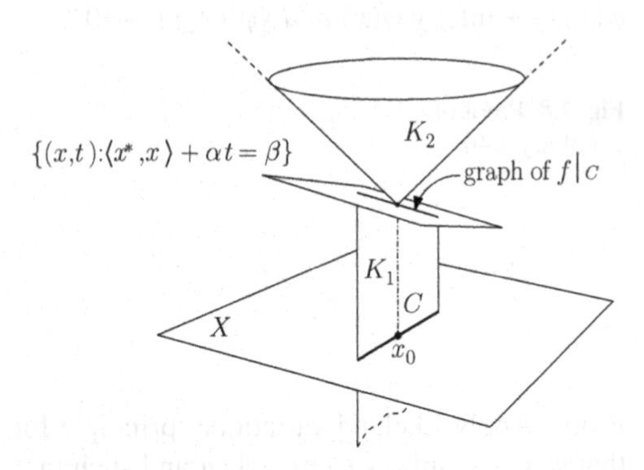

Fig. 7.9 The proof of the Bishop–Phelps theorem

On the other hand, if $x \in X$ and $t := f(x_0) + \varepsilon\|x - x_0\|$, then $(x, t) \in K_2$, hence

$$\langle x^*, x_0\rangle + f(x_0) = \beta \le \langle x^*, x\rangle + f(x_0) + \varepsilon\|x - x_0\|, \text{ for all } x \in X.$$

Therefore for all $x \in X$ we have $\langle x^*, (x_0 - x)\rangle \le \varepsilon\|x - x_0\|$, which gives $|\langle x^*, z\rangle| \le \varepsilon\|z\|$ for all $z \in X$, i.e., $\|x^*\| \le \varepsilon$. Hence $f + x^*$ attains its maximum on C and $\|x^*\| \le \varepsilon$. □

We refer to Exercise 7.53 for a strengthening of Theorem 7.41. We remark that if C is a nonempty closed convex subset of a Banach space, then the support points of C are dense in the boundary of C (see [Phelps, p. 48]). For another extension of Theorem 7.41, see Lemma 11.2.

Katznelson proved that in the space of all polynomials on [0, 1] the set of norm-attaining functionals is not dense in the dual space (see [Megg, p. 271]). Actually, each incomplete separable normed space X contains a closed convex bounded subset C such that no nonzero continuous linear functional on X attains its maximum on C ([Fonf]).

The normed space of finitely supported vectors in ℓ_1 admits a C^∞ norm (see Exercise 10.7). Using this we can give another example that the Bishop–Phelps theorem on density of support functionals does not hold for incomplete spaces. In fact, in other case ℓ_1^* would be separable.

7.4 Smooth Variational Principle

In Theorem 7.43 below we state the Deville–Godefroy–Zizler version of the Borwein–Preiss smooth variational principle ([DGZ3], see [BoPr]).

Definition 7.42 *A function $\varphi\colon X \to \mathbb{R}$ is called a* bump (function) *on X if it has a bounded nonempty support.*

Note that a bump can always be constructed starting from a norm. The argument in Fact 10.4 proves this in such a way that extra conditions of the norm can be carried to the bump function.

Theorem 7.43 (Smooth variational principle [DGZ2], see, e.g., [DGZ3]) *Let X be a Banach space that admits a Fréchet differentiable (respectively Gâteaux differentiable) Lipschitz bump function. Then for every lower semicontinuous and bounded below proper function f on X and for every $\varepsilon > 0$ there exists a function g which is Lipschitz and Fréchet differentiable (respectively Gâteaux differentiable) on X and such that* $\sup\{\|g(x)\| : x \in X\} \leq \varepsilon$, $\sup\{\|g'(x)\| : x \in X\} \leq \varepsilon$ *and $(f - g)$ attains its minimum on X.*

Proof: We will consider the case of Fréchet differentiability, the proof for the Gâteaux differentiability case is similar. Let Y be the normed space of all Lipschitz and Fréchet differentiable functions h on X normed by $|||h||| = \|h\|_\infty + \|h'\|_\infty$. It is standard to show that Y is a Banach space, see Exercise 1.25. For a given lower semicontinuous and bounded below function f defined on X, consider the set

$$U_n = \Big\{g \in Y : \text{ there is } x_0 \in X \text{ such that} \\ (f - g)(x_0) < \inf\big\{(f - g)(x),\ x \in X \backslash B_X^O\big(x_0, \tfrac{1}{n}\big)\big\}\Big\},$$

where $B_X^O\big(x_0, \frac{1}{n}\big)$ is the open ball with radius $\frac{1}{n}$ centered at x_0.

We claim that U_n is an open dense subset of Y. Indeed, U_n is open in Y as $|||\cdot||| \geq \|\cdot\|_\infty$. Fix any $g \in Y$ and $\varepsilon > 0$. We need to find $h \in Y$, $|||h||| < \varepsilon$ and $x_0 \in X$ such that

$$(f - g - h)(x_0) < \inf\big\{(f - g - h)(x) : x \in X \backslash B_X^O\big(x_0, \tfrac{1}{n}\big)\big\}.$$

By using the assumption on the existence of a Lipschitz Fréchet differentiable bump function on X and the shift and multiplication arguments, find $b \in Y$ such that $b(0) > 0$, $\|b\| < \varepsilon$ and $b(x) = 0$ whenever $\|x\| \geq \frac{1}{n}$. Since $f - g$ is bounded below, we can find $x_0 \in X$ such that

$$(f - g)(x_0) < \inf\{(f - g)(x) : x \in X\} + b(0).$$

Let $h(x) = b(x - x_0)$. Then $h \in Y$ and $\|h\| < \varepsilon$. Moreover,

$$(f - g - h)(x_0) = (f - g)(x_0) - b(0) < \inf_X (f - g).$$

If $x \in X \setminus B_X^O(x_0, \frac{1}{n})$, then $(f - g - h)(x) = (f - g)(x) \geq \inf_X (f - g)$. Hence $g + h \in U_n$ and this shows that U_n is dense in Y. Consequently, by the Baire category theorem, $G = \bigcap_{n \geq 1} U_n$ is a dense G_δ subset of the Banach space Y.

We claim that if $g \in G$, then $(f - g)$ attains its minimum on X. Indeed, for $n \geq 1$, using the definition of U_n find $x_n \in X$ such that

$$(f - g)(x_n) < \inf\bigl\{(f - g)(x) : x \in X \setminus B_X^O\bigl(x_n, \tfrac{1}{n}\bigr)\bigr\}.$$

We have $x_p \in B_X^O\bigl(x_n, \frac{1}{n}\bigr)$ for $p \geq n$, as otherwise, by the choice of x_n, we would have $(f - g)(x_p) > (f - g)(x_n)$. Since then $\|x_n - x_p\| \geq \frac{1}{n} \geq \frac{1}{p}$ and considering the choice of x_p, we would then have $(f - g)(x_n) > (f - g)(x_p)$, a contradiction. Hence $\{x_n\}$ is Cauchy in X and thus converging to some $x_\infty \in X$. We will show that $(f - g)$ attains its minimum on X at x_∞.

By the lower semicontinuity of $(f - g)$ we have

$$\begin{aligned}(f - g)(x_\infty) &\leq \liminf_{n \to \infty} (f - g)(x_n) \\ &\leq \liminf_{n \to \infty} \Bigl(\inf\bigl\{(f - g)(x) : x \in X \setminus B_X^O\bigl(x_n, \tfrac{1}{n}\bigr)\bigr\}\Bigr) \\ &\leq \inf\bigl\{(f - g)(x) : x \in X \setminus \{x_\infty\}\bigr\}.\end{aligned}$$

The last inequality follows since if $y \neq x_\infty$, then $\|x_n - y\| > \frac{1}{n}$ for large n, and so for such n, $\inf\bigl\{(f - g)(x) : x \in X \setminus B_X^O\bigl(x_n, \frac{1}{n}\bigr)\bigr\} \leq (f - g)(y)$. Therefore $(f - g)$ attains its infimum on X at x_∞. □

Corollary 7.44 *If a Banach space X admits a Lipschitz Fréchet differentiable (respectively Gâteaux differentiable) bump function, then for every convex continuous function f on X, the set of all points of Fréchet (respectively Gâteaux) differentiability of f is dense in X.*
In particular, if a Banach space X admits a Lipschitz Fréchet differentiable bump function, then X is an Asplund space.

On every nonseparable reflexive space, there are convex continuous functions whose set of points of Gâteaux differentiability is not G_δ ([HSZ]), though it must be residual.

It is not known whether, in general, every Asplund space admits a Lipschitz Fréchet differentiable bump function. In particular, it is not known whether $C(K)$ admits a Fréchet-smooth bump if K is the Kunen compact space (see the remark after Corollary 14.48). However, very recently Todorcevic and López-Abad proved, under an axiom, that there is an Asplund space with the RNP that has no norm with the Mazur Intersection Property; thus, by [DGZ4], this space does not admit any Fréchet bump function.

A Banach space is said to be *weak Asplund* if every convex continuous function defined on it is Gâteaux differentiable at the points of a G_δ dense subset. It is known that X is a weak Asplund space if it admits a Lipschitz Gâteaux differentiable bump function (see [DGZ1]). There exists a non-weak Asplund space that is a *Gâteaux differentiability space*, i.e., every convex continuous function defined on it is Gâteaux differentiable at the points of a dense set [MoSo].

Proof of Corollary 7.44: We will present the proof for the case of Fréchet differentiability. The proof for Gâteaux differentiability is similar.

Let f be a concave continuous function on X. Let b be a Lipschitz and Fréchet differentiable bump function on X with $b(0) \neq 0$ and $b(x) = 0$ for $x \notin B_X$. Let $x_0 \in X$. Choose $\delta > 0$ such that $f(x) > f(x_0) - 1$ for $x \in B_X^O(x_0, \delta)$. Choose $m > \frac{1}{\delta}$. If $\|x - x_0\| \geq \delta$ then $\|m(x - x_0)\| > 1$ and $b(m(x - x_0)) = 0$. Define a function φ on X by

$$\varphi(x) = \begin{cases} b(m(x - x_0))^{-2} & \text{if } b(m(x - x_0)) \neq 0, \\ +\infty & \text{otherwise.} \end{cases}$$

Consider $f + \varphi$. Then $f + \varphi = +\infty$ outside $B_X^O(x_0, \delta)$ and is bounded below by $f(x_0) - 1$ on $B_X^O(x_0, \delta)$. Hence $f + \varphi$ is bounded below on X.

It is easy to check that $f + \varphi$ is lower semicontinuous. By Theorem 7.43, there is a Lipschitz Fréchet differentiable function g on X such that $f + \varphi - g$ attains its minimum at some point y_0 (that necessarily belongs to $B_X^O(x_0, \delta)$). On some neighborhood U of y_0, the function $b(m(x-x_0))$ is not zero. So φ is Fréchet differentiable on U. For $y \in U$ we have $f(y) + \varphi(y) - g(y) \geq f(y_0) + \varphi(y_0) - g(y_0)$ and thus

$$\begin{aligned} 0 &\leq (-f)(y_0 + h) + (-f)(y_0 - h) - 2(-f)(y_0) \\ &\leq \varphi(y_0 + h) + \varphi(y_0 - h) - 2\varphi(y_0) \\ &\quad -g(y_0 + h) - g(y_0 - h) + 2g(y_0) \leq o(\|h\|) + o(\|h\|) \end{aligned}$$

for small h and f is thus Fréchet differentiable at y_0. □

Lemma 7.45 *Let f be a continuous convex function on a Banach space X. Then the set G of all points in X where f is Fréchet differentiable (possibly empty) is a G_δ set in X.*

Proof: For $n \in \mathbb{N}$ define

$$G_n = \Big\{ x \in X : \inf_{\delta > 0} \sup_{y \in S_X} \frac{f(x+\delta y) + f(x-\delta y) - 2f(x)}{\delta} < \frac{1}{n} \Big\}.$$

Since f is convex, $\frac{f(x+\delta y)+f(x-\delta y)-2f(x)}{\delta}$ is decreasing as $\delta \searrow 0^+$ and we get $G = \bigcap G_n$. Hence it suffices to show that each G_n is an open subset of X. To this end, let $x \in G_n$ and f be L-Lipschitz on $B^O(x, \alpha)$. There exists $C < \frac{1}{n}$ and $\delta < \frac{\alpha}{2}$ such that

$$\sup_{y \in S_X} \frac{f(x+\delta y) + f(x-\delta y) - 2f(x)}{\delta} < C.$$

Choose $0 < \varepsilon < \min\{\frac{\delta}{2}, \frac{\delta}{4L}(\frac{1}{n} - C)\}$. We claim that $B^O(x, \varepsilon) \subset G_n$. Indeed, given $z \in B(x, \varepsilon)$ we have for all $y \in S_X$:

$$\begin{aligned} &\frac{f(z+\delta y) + f(z-\delta y) - 2f(x)}{\delta} \\ &\le \frac{f(x+\delta y) + L\|z-x\| + f(x-\delta y) + L\|z-x\| - 2f(x) + 2L\|z-x\|}{\delta} \\ &\le C + \tfrac{4L\varepsilon}{\delta} < C + \left(\tfrac{1}{n} - C\right) = \tfrac{1}{n}. \end{aligned}$$

□

Remark: If the convexity is dropped, Lemma 7.45 may fail, even in the case of Lipschitz functions. See Exercise 7.34.

Theorem 7.46 *Let X be a Banach space. If φ is a continuous Gâteaux differentiable bump function on X, then $\overline{\text{span}}\{\varphi'(x) : x \in X\} = X^*$.*.

Proof: Define $\psi : X \to \mathbb{R} \cup \{+\infty\}$ by $\psi(x) = \varphi^{-2}(x)$ if $\varphi(x) \neq 0$ and $\psi(x) = +\infty$ otherwise. Given $f \in X^*$ and $\varepsilon > 0$, the function $\psi - f$ is lower semicontinuous and bounded below on X and thus by Theorem 7.39, there is $x_0 \in X$ such that for every $h \in X$ and $t > 0$,

$$\psi(x_0 + th) - f(x_0 + th) \ge \psi(x_0) - f(x_0) - \varepsilon t\|h\|.$$

Hence for $h \in X$ and $t > 0$,

$$\frac{\psi(x_0 + th) - \psi(x_0)}{t} \ge \frac{f(x_0 + th) - f(x_0)}{t} - \varepsilon\|h\| = f(h) - \varepsilon\|h\|.$$

As $\psi(x_0) \neq \infty$, we have $\varphi(x_0) \neq 0$ and ψ is thus Gâteaux differentiable at x_0. Hence from the latter inequality it follows that for all $h \in X$,

$$\psi'(x_0)(h) = \lim_{t \to 0^+} \frac{\psi(x_0 + th) - \psi(x_0)}{t} \ge f(h) - \varepsilon\|h\|.$$

Thus $\psi'(x_0)(h) - f(h) \geq -\varepsilon\|h\|$ for all $h \in X$. Considering $\pm h$, it follows that $|\psi'(x_0)(h) - f(h)| \leq \varepsilon\|h\|$ for all $h \in X$ and hence $\|\psi'(x_0) - f\| = \left\|-\frac{1}{\varphi(x_0)^3}\varphi'(x_0) - f\right\| \leq \varepsilon$. $\square$

7.5 Norm-Attaining Operators

We now discuss norm attaining operators. We introduce the following notation: Let X, Y be Banach spaces and $T \in \mathcal{B}(X, Y)$. For a bounded set C in X we put $\|T\|_C = \sup\{\|T(x)\| : x \in C\}$. If there is $c \in C$ such that $\|T(c)\| = \|T\|_C$, we say that T *attains its supremum* over C.

The following version of Lindenstrauss result is in [Zizl1b].

Theorem 7.47 (Lindenstrauss [Lind4]) *Let X, Y be Banach spaces. If C is a w^*-compact subset of X^*, then the set of all dual bounded operators from X^* to Y^* that attain their suprema over C is dense in the Banach space of all dual bounded operators from $\mathcal{B}(X^*, Y^*)$.*

In the proof we will use the following lemma.

Lemma 7.48 *Let $T^*: X^* \to Y^*$ be a dual bounded operator and C be a w^*-compact subset of X^*. The following are equivalent:*
(i) *T^* attains its supremum over C.*
(ii) *There exist $\{f_k\} \subset C$ and $\{y_j\} \subset S_Y$ such that $|T^*(f_k)(y_j)| \geq \|T^*\|_C - \frac{1}{j}$ for all $j, k \in \mathbb{N}$, $k \geq j$.*

Proof: Assume that (ii) holds and f is a w^*-limit point of $\{f_k\}$. Since T^* is w^*-w^*-continuous, we get that $|T^*(f)(y_j)| \geq \|T^*\|_C - \frac{1}{j}$ for every j. Therefore $\|T^*(f)\| = \|T^*\|_C$.

If $\|T^*(f)\| = \|T^*\|_C$ for some $f \in C$, put $f_k = f$ for $k \in \mathbb{N}$ and pick $y_j \in S_Y$, $j \in \mathbb{N}$ such that $|T^*(f)(y_j)| \geq \|T^*(f)\| - \frac{1}{j}$. $\square$

Proof of Theorem 7.47: Let $T^*: X^* \to Y^*$ be a norm-one dual bounded operator, let $C \subset B_{X^*}$ be a w^*-compact set. We put $c = \|T^*\|_C \leq 1$ and assume that $c > 0$.

Take any $\varepsilon \in \left(0, \min\{\frac{1}{3}, \frac{c}{9}\}\right)$ and choose a decreasing sequence $\{\varepsilon_k\}_{k=1}^{\infty}$ of positive numbers such that $2\sum_{i=1}^{\infty}\varepsilon_i < \varepsilon$, $2\sum_{i=k+1}^{\infty}\varepsilon_i < \varepsilon_k^2$ and $2\varepsilon_k^2 c + 14\varepsilon_k < \frac{1}{k+1}$ for every k. We set $T_1^* = T^*$. By induction, having constructed a dual operator $T_k^* \in \mathcal{B}(X^*, Y^*)$, find $f_k \in C$ such that $\|T_k^*(f_k)\| > \|T_k^*\|_C - \varepsilon_k^2 c$ and $y_k \in S_Y$ so that $T_k^*(f_k)(y_k) > \|T_k^*(f_k)\| - \varepsilon_k^2 c$, and finally define a dual operator T_{k+1}^* by

$$T_{k+1}^*(f) = T_k^*(f) + \varepsilon_k T_k^*(f)(y_k) T_k^*(f_k).$$

We claim that $\{T_k^*\}$ satisfy the following properties

(1) $\frac{2}{3} \leq \|T_k^*\| \leq \frac{4}{3}$ and $\frac{2}{3}c \leq \|T_k^*\|_C \leq \frac{4}{3}c$ for all k, $\|T_j^* - T_k^*\|_C \leq 2\sum_{i=j}^{k}\varepsilon_i c$ and $\|T_j^* - T_k^*\| \leq 2\sum_{i=j}^{k-1}\varepsilon_i$ for $j < k$,

(2) $\|T_{k+1}^*\|_C \geq \|T_k^*\|_C + \varepsilon_k \|T_k^*\|_C^2 - 4\varepsilon_k^2 c$ for all k,
(3) $\|T_k^*\|_C \geq \|T_j^*\|_C \geq c$ for $j < k$,
(4) $|T_j^*(f_k)(y_j)| \geq \|T_j^*\|_C - 14\varepsilon_j$ for $j < k$.
Indeed, (1) follows by induction. To see (2), write

$$\begin{aligned}
\|T_{k+1}^*\|_C &\geq \|T_{k+1}^*(f_k)\| = \|(1 + \varepsilon_k T_k^*(f_k)(y_k))T_k^*(f_k)\| \\
&\geq \|T_k^*(f_k)\|(1 + \varepsilon_k(\|T_k^*(f_k)\| - \varepsilon_k^2 c)) \geq (\|T_k^*\|_C - \varepsilon_k^2 c)(1 + \varepsilon_k(\|T_k^*\|_C - 2\varepsilon_k^2 c)) \\
&= \|T_k^*\|_C - \varepsilon_k^2 c + \varepsilon_k \|T_k^*\|_C^2 - 3\varepsilon_k^3 c\|T_k^*\|_C + 2\varepsilon_k^5 c^2 \geq \|T_k^*\|_C + \varepsilon_k \|T_k^*\|_C^2 - 4\varepsilon_k^2 c.
\end{aligned}$$

(3) follows from (1) and (2). To see (4), write

$$\begin{aligned}
\|T_{j+1}^*(f_k)\| &\geq \|T_k^*(f_k)\| - \|T_k^* - T_{j+1}^*\|_C \\
&\geq \|T_k^*\|_C - \varepsilon_k^2 c - 2\sum_{i=j+1}^{k-1} \varepsilon_i c \geq \|T_{j+1}^*\|_C - 3\varepsilon_j^2 c.
\end{aligned}$$

Using this, (2), and the definition of T_{k+1}^* we have for $j < k$:

$$\begin{aligned}
\varepsilon_j |T_j^*(f_k)(y_j)|\, \|T_j^*\|_C + \|T_j^*\|_C &\geq \|T_{j+1}^*(f_k)\| \geq \|T_{j+1}^*\|_C - 3\varepsilon_j^2 c \\
&\geq \|T_j^*\|_C + \varepsilon_j \|T_j^*\|_C^2 - 4\varepsilon_j^2 c - 3\varepsilon_j^2 c = \|T_j^*\|_C + \varepsilon_j \|T_j^*\|_C^2 - 7\varepsilon_j^2 c.
\end{aligned}$$

Therefore

$$|T_j^*(f_k)(y_j)| \geq \|T_j^*\|_C - 7\varepsilon_j c / \|T_j^*\|_C \geq \|T_j^*\|_C - 14\varepsilon_j.$$

It follows from (1) that $\{T_j^*\}$ converges in norm to some operator $\widetilde{T} \in \mathcal{B}(X^*, Y^*)$ satisfying $\|T^* - \widetilde{T}\| < \varepsilon$ and $\|\widetilde{T} - T_j^*\|_C \leq \varepsilon_{j-1}^2 c$ for every j. Since T_j^* are w^*–w^* continuous, we obtain that $\widetilde{T}$ is w^*–w^*-continuous on B_{X^*}. By the Banach–Dieudonné theorem, this means that $\widetilde{T}$ is w^*–w^*-continuous on X^*, so $\widetilde{T}$ is a dual operator.

If $1 < j < k$, then, according to (4),

$$\begin{aligned}
|\widetilde{T}(f_k)(y_j)| &\geq |T_j^*(f_k)(y_j)| - \|T_j^* - \widetilde{T}\|_C \\
&\geq \|T_j^*\|_C - 14\varepsilon_j - \varepsilon_{j-1}^2 c \geq \|\widetilde{T}\|_C - \tfrac{1}{j}.
\end{aligned}$$

By Lemma 7.48 this means that $\|\widetilde{T}\|_C$ is attained at some $f \in C$. □

Let T be a bounded operator from a Banach space X into a Banach space Y. We say that T *attains its norm* if there is $x_0 \in B_X$ such that $\|T\| = \|T(x_0)\|$. From the proof of Theorem 7.47, it follows

Corollary 7.49 *Let X and Y be Banach spaces and C be a weakly compact convex subset of X. Then the set of all operators in $\mathcal{B}(X, Y)$ that attain their suprema over C is norm-dense in $\mathcal{B}(X, Y)$.*

We will now show a result related to variational principles.

Proposition 7.50 ([WhZi]) *Let X be a Banach space that does not contain any isomorphic copy of c_0. For every bounded closed convex set C in X there is a compact set $K \subset C$ such that there is no $h \in X\backslash\{0\}$ with $K \pm h \subset C$.*

This property of K should be compared with the notion of an extreme point.

Proof: Assume that such $K \subset C$ does not exist. Choose $x_1 \in C$. Then there is $x_2' \in X\backslash\{0\}$ such that $\pm x_1 \pm x_2' \in C$. Among all such x_2', choose x_2 such that $\|x_2\| > \frac{1}{2}\sup\{\|x_2'\| : \pm x_1 \pm x_2' \in C\}$. Having chosen x_2, consider the set $\{\pm x_1 \pm x_2\}$. There is $x_3 \in X\backslash\{0\}$ such that $\|x_3\| > \frac{1}{2}\sup\{\|x_3'\| : \pm x_1 \pm x_2 \pm x_3' \in C\}$ and also $\pm x_1 \pm x_2 \pm x_3 \in C$. By induction we obtain a sequence $\{x_i\}$ in $X\backslash\{0\}$ such that $S = \left\{\sum_{i=1}^{n} \varepsilon_i x_i : \varepsilon_i = \pm 1, n \in \mathbb{N}\right\}$ is in C and hence is bounded. Since X contains no isomorphic copy of c_0, $\sum x_i$ is unconditionally convergent by Corollary 4.52 and the set S is relatively compact in X. By our assumption, there is $h \in X\backslash\{0\}$ such that $\overline{S} \pm h \subset C$. Since this h could have been used in all steps in constructing x_i, we have $\|x_i\| \geq \frac{1}{2}\|h\| > 0$ for all i, contradicting the convergence of $\sum x_i$. □

7.6 Michael's Selection Theorem

Let X be a topological space. A family $\mathcal{F}$ of subsets of S is called *locally finite* if every $x \in S$ has a neighborhood that intersects only a finite number of elements in $\mathcal{F}$.

An *open cover* of a topological space X is a family $\mathcal{V}$ of open subsets of X such that $\bigcup_{V \in \mathcal{V}} V = X$. An open cover $\mathcal{W}$ of X is a *refinement* of $\mathcal{V}$ if every $W \in \mathcal{W}$ is contained in some $V \in \mathcal{V}$.

Definition 7.51 A locally finite partition of unity on X *is a family* $\{f_\alpha : \alpha \in A\}$ *of continuous functions from X to* $[0, 1]$ *such that*
(i) *the sets* $\{x \in X : f_\alpha(x) > 0\}$ *form a locally finite open cover of X,*
(ii) $\sum_{\alpha \in A} f_\alpha(x) = 1$ *for all $x \in X$.*
We shall say that the partition of unity is subordinated *to a cover $\mathcal{V}$ if the cover defined in* (i) *is a refinement of $\mathcal{V}$.*

Observe that, for each point $x \in X$ there is a neighborhood $U(x)$ of x where the sum in (ii) is a finite sum.

Definition 7.52 *A topological space X is said to be* paracompact *if every open cover of X admits a locally finite open refinement.*

To the class of paracompact spaces belong all compact topological spaces (this follows right from the definition) and all metrizable topological spaces (this is a consequence of a result of Stone, see, e.g., [Enge, Theorem 4.4.1], or [BePe2, Corollary II.2.2]).

The following theorem applies then, in particular, to metric spaces or to compact spaces.

Theorem 7.53 (Michael selection theorem [Mich]) *Assume that X is a paracompact topological space, E is a Banach space and Φ is a set-valued mapping from X to subsets of E. Assume that Φ is lower semicontinuous. Moreover, assume that $\Phi(x)$ is a closed convex subset of E for every $x \in X$. Then Φ admits a continuous selection, i.e., there is a continuous function $\varphi : X \to E$ such that $\varphi(x) \in \Phi(x)$ for every $x \in X$.*

The key result in the proof is the following lemma.

Lemma 7.54 *Let X, E and Φ be as in the statement of Theorem 7.53, only sets $\Phi(x)$ are not assumed to be closed. Then, for every $\varepsilon > 0$, there is a continuous function $\varphi : X \to E$ such that $\operatorname{dist}(\varphi(x), \Phi(x)) < \varepsilon$ for every $x \in X$.*

Proof of Lemma 7.54: For every $v \in E$, let $B(v, \varepsilon)$ be the open ball in E centered at v and having radius ε.

Put

$$G_v = \{x \in X : \Phi(x) \cap B(v, \varepsilon) \neq \emptyset\}.$$

Then $\{G_v\}_{v \in E}$ is an open cover of X. Let $\{\varphi_i\}_{i \in I}$ be a locally finite partition of unity on X subordinated to $\{G_v\}$ (see Theorem 17.21). For each $i \in I$, choose $v_i \in E$ such that φ_i is supported by G_{v_i}.

Define, for $x \in X$,

$$\varphi(x) = \sum_{i \in I} \varphi_i(x) v_i.$$

The sum is locally finite. Hence φ is a well-defined continuous function from X into E.

If $\varphi_i(x) \neq 0$ for some $x \in X$ and some $i \in I$, then $\operatorname{dist}(v_i, \Phi(x)) < \varepsilon$. Therefore, for every $x \in X$, $\varphi(x)$ is a convex combination of points in the ε-neighborhood of a convex set $\Phi(x)$. Thus $\varphi(x)$ also belongs to this neighborhood. This finishes the proof of the lemma. □

Proof of Theorem 7.53: (See [BeLi, p. 22]) In order to prove it, we use Lemma 7.54 and an iterative procedure. Put $\Phi_0 = \Phi$ and let φ_0 be the continuous mapping obtained from the lemma for $\varepsilon = 1$. Put $\Phi_1(x) = \Phi_0(x) \cap B(\varphi_0(x), 1)$ for $x \in X$. Then the values $\Phi_1(x)$ are convex and nonempty. The mapping Φ_1 is lower semicontinuous. Indeed, let G be an open set in E and choose a point $z \in \{x \in$

$X : \Phi_1(x) \cap G \neq \emptyset\}$. Then there is an $r < 1$ such that the open set $\{x \in X : \Phi_0(x) \cap B(\varphi_0(x), r) \cap G \neq \emptyset\}$ contains z. By the continuity of φ_0, the set $\{x \in X : \Phi_0(x) \cap B(\varphi_0(x), 1) \neq \emptyset\}$ also contains a neighborhood of z. Therefore Φ_1 satisfies the assumptions of the lemma for $\varepsilon = 1/2$.

In the same way, we obtain a sequence of continuous mappings $\varphi_n : X \to E$ for which $\|\varphi_n(x) - \varphi_{n-1}(x)\| \leq 2^{-n+1} + 2^{-n}$ and $\operatorname{dist}(\varphi_n(x), \Phi(x)) < 2^{-n}$ for all $x \in X$. Let φ be the limit of (φ_n). Then φ is a continuous mapping from X into E and $\varphi(x) \in \Phi(x)$ for all $x \in X$ as $\Phi(x)$ is a closed set for each $x \in X$. Thus, φ is a continuous selection of Φ. □

Corollary 7.55 (Tietze) *Let X be a paracompact space, A be a closed subset of X and f be a continuous real-valued function defined on A. Then there is a continuous real-valued function $\widetilde{f}$ on X such that $\widetilde{f} = f$ on A and*

$$(-\infty \leq)\ m := \inf_A f \leq \widetilde{f} \leq M := \sup_A f\ (\leq +\infty).$$

Proof: Define a set-valued mapping $\Phi : X \to \mathcal{P}(\mathbb{R})$ by

$$\Phi(x) = \begin{cases} \{f(x)\} & \text{if } x \in A, \\ [m, M] & \text{if } x \notin A \text{ and } -\infty < m \leq M < +\infty, \\ [m, +\infty) & \text{if } x \notin A \text{ and } -\infty < m < M = +\infty, \\ [-\infty, M] & \text{if } x \notin A \text{ and } -\infty = m < M < +\infty, \\ ]-\infty, +\infty[& \text{if } x \notin A \text{ and } -\infty = m < M = +\infty. \end{cases}$$

It is easy to show that Φ is lower semicontinuous. Then use Theorem 7.53 to find a continuous selection $\widetilde{f}$ for Φ. This function satisfies the requirements. □

Corollary 7.56 (Bartle–Graves selectors, [BarGra], [Mich], see, e.g., [BeLi, p. 23]) *Let Y be a closed subspace of a Banach space X. Then the quotient mapping $\pi : X \to X/Y$ admits a continuous positive homogeneous lifting, i.e., there is a continuous mapping $\varphi : X/Y \to X$ such that $\pi \circ \varphi$ is the identity on X/Y and such that $\varphi(\lambda\hat{x}) = \lambda\varphi(\hat{x})$ for all $\hat{x} \in X/Y$ and $\lambda > 0$. Moreover, given $\varepsilon > 0$, φ can be chosen so that $\varphi(B_{X/Y}) \subset (1 + \varepsilon)B_X$.*

Proof: [BeLi, p. 23] Given $\varepsilon > 0$ put $B : \{x \in X : \|x\| < 1 + \varepsilon\}$. For $\hat{x} \in S_{X/Y}$ put $\Phi(x) = B \cap \pi^{-1}(\hat{x})$. Then Φ is a convex-valued mapping from $S_{X/Y}$ into the subsets of B that is lower semicontinuous. Indeed (see Exercise 7.75), assume that $\hat{x}_i \to \hat{x}$ in $S_{X/Y}$ and let $y \in \Phi(\hat{x})$. By the definition of the quotient norm, there are $z_i \in X$ so that $\pi z_i = \hat{x}_i - \hat{x}$ and $\|z_i\| \leq (1 + \varepsilon)\|\hat{x}_i - \hat{x}\|$, $i \in \mathbb{N}$. In particular, $z_i \to 0$ in X and we can assume that $\|z_i\| < 1 + \varepsilon - \|y\|$ for every i. Then $y_i := y + z_i \to y$ and $y_i \in \Phi(\hat{x}_i)$ because $\pi y_i = \hat{x}_i$ and $\|y_i\| < 1+\varepsilon$ for all i. Put $\Psi(\hat{x}) = \Phi(\hat{x})$ for $\hat{x} \in S_{X/Y}$. By Theorem 7.53, Ψ admits a continuous selection $\psi : S_{X/Y} \to (1+\varepsilon)B_X$. Finally, define φ to be the positive homogeneous extension of ψ to all of X/Y. The continuity of φ at 0 follows from the boundedness of φ on the unit sphere and the continuity elsewhere is obvious. □

Note: An example of a Banach space X and a subspace Y such that there is no uniformly continuous lifting of the quotient mapping π is in [BeLi, p. 24], see also Exercise 12.50. Note also that if φ is a bounded linear selector for π then Y is complemented in X. Indeed, the mapping $Px = x - \varphi(\hat{x})$ is a projection onto Y, since $\widehat{Px} = \hat{x} - \widehat{\varphi(\hat{x})} = 0$, and if $y \in Y$, then $Py = y - \varphi(\hat{y}) = y$. If φ is just a continuous selector, the mapping P defined above gives a (in general, non-linear) continuous projection from X onto Y; this gives an alternative proof of Corollary 5.28.

7.7 Remarks and Open Problems

Remarks

1. We refer, e.g., to [Phelps], [DeGh], [Gile], [BeLi], [Gode4], and [HMVZ] for more on the subject of this chapter.
2. Bourgain proved in [Bou] that if X has the RNP property (see Definition 11.14) and Y is an arbitrary Banach space, then the norm-attaining operators in $\mathcal{B}(X, Y)$ are norm-dense in $\mathcal{B}(X, Y)$.
3. If C is a closed convex balanced bounded set in a complex Banach space X it follows easily from the real case that the set of all continuous linear functionals f such that $|f|$ attains its supremum on C is norm-dense in X^*. This is no longer true for a general closed convex bounded set C, see [Lom].
4. Corson proved that there is a compact convex set C in $\mathbb{R}^3$ such that its set of exposed points of C is of first category in the set of extreme points of C [Cors2].
5. For a localization of the Asplund property, i.e., for considering particular convex functions only, we refer to [Tan0].
6. Borwein and Fabian [BoFa] proved that if X is an infinite-dimensional separable Banach space, then there is a (uniformly) Gâteaux differentiable norm on X that is somewhere not Fréchet differentiable.

Open Problems

1. It is not known whether a Banach space X admits a Lipschitz continuously Fréchet differentiable bump if X admits a Fréchet differentiable bump (see, e.g., [DGZ3, p. 89]).
2. Assume that X is a separable Banach space that does not contain an isomorphic copy of ℓ_1. (a) Is it true that X admits an equivalent norm whose dual norm is Gâteaux differentiable? (b) Is it true that any equivalent norm on X has the property that its second dual norm has a point of Fréchet differentiability? (c) Is it true that the dual space admits an equivalent LUR norm? Question (a) was answered in the positive for the space JT [Haje2]. Question (b) was answered in the positive for JT in [Schc]. For Question (c), see [MOTV2, Question 6.10].

3. For a survey on some results related to Gâteaux differentiability, including a list of open problems, see [HMZ].

Exercises for Chapter 7

7.1 Let f be a finite convex function on a finite-dimensional Banach space X. Show that f is continuous on X.
Hint. f is bounded above on the symmetric convex hull of the basis vectors which contains the origin as an interior point. Then use Lemma 7.3.

7.2 Show that a finite lower semi-continuous convex function f that is defined on a whole Banach space must be continuous.
Hint. Use the Baire category theorem, a cone argument, and Lemma 7.3.

7.3 Can $\sin x$ be the difference of two convex continuous functions on the real line?
Hint. $x^2 + \sin x$ and x^2.

7.4 Let U be a convex subset of a vector space X, $f\colon U \to \mathbb{R}$. Show that f is convex if and only if the function $t \mapsto \frac{f(x+ty)-f(x)}{t}$ is increasing in t for all $x \in U$, $y \in X$ and all t for which $x + ty \in U$ (see Fig. 7.2).
Hint. Assume f convex, take $x \in U$, $y \in X$ and $0 < t < s$ such that $x + sy \in U$. Then $(x + ty) = \lambda x + (1 - \lambda)(x + sy)$ for $\lambda = 1 - t/s$. Applying convexity we obtain $\frac{f(x+ty)-f(x)}{t} \le \frac{f(x+sy)-f(x)}{s}$.

Conversely, given $x, y \in U$ and $\lambda \in (0, 1)$, set $z = y - x$; then $\lambda x + (1 - \lambda)y = x + (1 - \lambda)z$, and $y = x + 1z$. Use the fact that the incremental quotient is increasing in t to deduce the convexity of f.

7.5 Let f, g be convex functions on a Banach space X. The *inf convolution of f and g* is defined by $(f \diamond g)(x) = \inf\{f(y) + g(x - y) : y \in X\}$ (it is defined so that its epigraph is the algebraic sum of the epigraphs of functions involved).

Assume that X is a reflexive Banach space and $f(x) = \|x\|_1^2$, $g(x) = \|x\|_2^2$ for some equivalent norms $\|\cdot\|_1$ and $\|\cdot\|_2$ on X such that $\|\cdot\|_1$ is Fréchet-smooth. Show that then $f \diamond g$ is a Fréchet-smooth convex function on X.
Hint. By the reflexivity of X, given $x_0 \in X$, choose $y_0 \in X$ such that $f \diamond g(x_0) = f(y_0) + g(x_0 - y_0)$. Then for $h \in X$:

$$\begin{aligned}
&f \diamond g(x_0 + h) + f \diamond g(x_0 - h) - 2f \diamond g(x_0)\\
&\le f(y_0) + g(x_0 + h - y_0) + f(y_0) + g(x_0 - h - y_0) - 2\big(f(y_0) + g(x_0 - y_0)\big)\\
&= g(x_0 - y_0 + h) + g(x_0 - y_0 - h) - 2g(x_0 - y_0).
\end{aligned}$$

From the differentiability of g at $(x_0 - y_0)$ the differentiability of $f \diamond g$ at x_0 follows.

7.6 Let f be a continuous convex function defined on an open convex subset C of a Banach space X. Show that for every $x_0 \in C$ then there is a continuous convex function $\tilde{f}$ defined on X and such that $\tilde{f} = f$ on some neighborhood of x_0.

Hint. Define f by $+\infty$ outside C and consider its inf convolution with the function $\phi_n(x) = n\|x\|$. Since f is locally Lipschitz, $\tilde{f} = f \diamond \phi_n$ close to x_0 for n large enough.

7.7 Let X be an infinite-dimensional Banach space. Show that there is a continuous convex function f on X such that f is unbounded on B_X.
Hint. (Vanderwerff) Let τ_n be even continuous convex functions such that τ_n are non-decreasing on $[0, \infty)$, $\tau_n = 0$ on $[0, 1/2]$ and $\tau_n(1) = n$. Let $f_n \in S_{X^*}$ be such that $f_n \xrightarrow{w^*} 0$ (Josefson–Nissenzweig, see Exercise 3.39). Consider $f(x) := \sum \tau_n(f_n(x))$. Note that the sum is locally finite.

7.8 Prove that every convex function f defined on an open interval $I \subset \mathbb{R}$ is differentiable at all but (at most) countably many points of I.
Hint. Observe that $d^+ f(x)(1) = \lim\limits_{t \to 0+} \frac{f(x+t)-f(x)}{t}$, the derivative of f at x from the right, is a non-decreasing function of x. Prove then that at any point where f fails to be differentiable, the monotone function $x \to d^+ f(x)(1)$ has a jump. As there are not more than a countable number of jumps, the conclusion follows.

7.9 Let f, g be functions defined on an open subset U of a Banach space X such that f is convex, $f \leq g$ on X, and $f(x_0) = g(x_0)$ for some $x_0 \in U$. Assume that g is Fréchet (respectively Gâteaux) differentiable at x_0. Show that f is Fréchet (respectively Gâteaux) differentiable at x_0.
Hint. Observe, first, that for every $\varepsilon \geq 0$, $\partial_\varepsilon f(x_0) \subset \partial_\varepsilon g(x_0)$. Apply now Theorem 7.15 and the remark following this same theorem in the Fréchet case, and Theorem 7.17 and the remark after it in the Gâteaux case.

7.10 Let f, g be convex continuous functions on a Banach space X and assume that f is not Fréchet differentiable at $x \in X$. Show that $f+g$ is not Fréchet differentiable at x.
Hint. By the assumption, $f(x+h) + f(x-h) - 2f(x) \geq \varepsilon h$.

7.11 Show that the norm $\|\cdot\|$ on a Banach space X is LUR if and only if $\lim \|x_n - x\| = 0$ whenever $x_n, x \in S_X$ satisfy $\lim \|x_n + x\| = 2$.
Hint. Using the proof of Fact 7.7, (ii)$\Longrightarrow$(iii), show first that for x_n, x as in the definition of LUR we have $\|x_n\| \to \|x\|$.

7.12 Define a norm $|\!|\!|\cdot|\!|\!|$ on $C[0, 1]$ by $|\!|\!|x|\!|\!|^2 = \|x\|_\infty^2 + \|x\|_2^2$, where $\|\cdot\|_\infty$ respectively $\|\cdot\|_2$ denote the norms of $C[0, 1]$ and $L_2[0, 1]$, respectively. Show that $|\!|\!|\cdot|\!|\!|$ is strictly convex (rotund) but not LUR on $C[0, 1]$.
Hint. To see that $|\!|\!|\cdot|\!|\!|$ is not LUR, consider $x = 1$ identically and x_n is a function of the broken line through the points $(0, 0)$, $(\frac{1}{n}, 1)$, $(1, 1)$. $|\!|\!|\cdot|\!|\!|$ is strictly convex by Fact 7.7

7.13 Let X be a Banach space. Show that if E, F are finite-dimensional subspaces of X such that $\dim(F) > \dim(E)$ (no assumption on inclusion of E and F), then there is $x \in F$ with $\|x\| = \mathrm{dist}(x, E) = 1$.
Hint. [LiTz3, p. 77] First assume that the norm of X is strictly convex (rotund). For every $x \in F$ with $\|x\| = 1$, let $p(x)$ be the unique point in E for which $\|x - p(x)\| = \mathrm{dist}(x, E)$. Then p is continuous (see above and recall the compactness of the ball in the finite-dimensional space). We have $p(-x) = -p(x)$. Since $\dim(F) > \dim(E)$, by the Borsuk antipodal theorem (see [Dugu2, p. 349]) we get x with $\|x\| = 1$ in F such that $p(x) = 0$, i.e., $\mathrm{dist}(x, E) = 1$. The general case follows by approximating the norm of X by strictly convex norms and taking the limit of the points obtained.

7.14 Prove that if $f : U \to \mathbb{R}$ is a convex continuous function on an open convex subset U of a Banach space $(X, \|\cdot\|)$, then $\partial f(x) \neq \emptyset$ for every $x \in U$ (use the separation theorem in $X \times \mathbb{R}$). For an analytic approach see Exercise 7.16
Hint. Equip $X \times \mathbb{R}$ with the norm $\|(x, t)\|_1 := \|x\| + |t|$. Since f is continuous, we can easily see that $\mathrm{epi}\, f$ ($\subset X \times \mathbb{R}$) has a non-empty interior (and it is convex, due to the convexity of f). Let $(x^*, r) \in X^* \times \mathbb{R}$ separate $\mathrm{epi}\, f$ and $\big(x_0, f(x_0)\big)$, i.e., $\langle (x^*, r), (x, t)\rangle \geq \langle (x^*, r), \big(x_0, f(x_0)\big)\rangle$ for all $x \in X$, $t \geq f(x)$ (see (i) in Theorem 3.32). Necessarily $r > 0$ (observe that t can be taken arbitrarily big; this shows that $t \geq 0$. If $r = 0$ we reach a contradiction because a linear functional can not be bounded below), and we may assume, without loss of generality, that $r = 1$. Then $f(x) - f(x_0) \geq \langle x^*, x_0 - x\rangle$ for all $x \in X$, so $-x^* \in \partial f(x_0)$.

7.15 Prove that if $f : U \to \mathbb{R}$ is a convex and lower semicontinuous function on an open convex subset U of a Banach space $(X, \|\cdot\|)$, then we have $\partial_\varepsilon f(x) \neq \emptyset$ for every $x \in U$ and for every $\varepsilon > 0$ (use the separation theorem in $X \times \mathbb{R}$).
Hint. Follow the idea in Exercise 7.14. Now, the epigraph has, in general, no longer a nonempty interior; however, it is closed, and it does not contain the point $(x, f(x) - \varepsilon)$. This is enough to apply (ii) in the separation Theorem 3.32 and proceed as in Exercise 7.14.

7.16 Prove that if $f : U \to \mathbb{R}$ is a convex continuous function on an open convex subset U of a Banach space $(X, \|\cdot\|)$, then $\partial f(x) \neq \emptyset$ for every $x \in U$ (use the Hahn–Banach extension theorem). For a geometrical approach see Exercise 7.14
Hint. Let $f : U \to \mathbb{R}$ be a continuous convex function defined on an open convex subset U of a Banach space X. Let $x \in U$ and put,

$$d^+ f(x)(h) = \lim_{t \to 0+} \frac{f(x+t) - f(x)}{t} \quad \text{for } h \in X. \tag{7.13}$$

This limit exists due to the non-decreasing character of the incremental quotient in (7.13) with respect to t (see the paragraph after Lemma 7.3). It is simple to prove that $d^+ f(x)$ is positively homogeneous and subadditive. The positive homogeneity is clear. The subadditivity is a consequence of the convexity of f. Indeed, given h_1, $h_2 \in X$, and $t > 0$,

$$x + \frac{t}{2}(h_1 + h_2) = \frac{(x + th_1) + (x + th_2)}{2},$$

hence

$$f\left(x + \frac{t}{2}(h_1 + h_2)\right) \leq \frac{f(x + th_1) + f(x + th_2)}{2},$$

so

$$\frac{f\left(x + \frac{t}{2}(h_1 + h_2)\right) - f(x)}{t/2} \leq \frac{f(x + th_1) - f(x)}{t} + \frac{f(x + th_2) - f(x)}{t},$$

and the conclusion is obtained by taking limits for $t \downarrow 0$ at both sides of the previous inequality.

Now fix $h_0 \in X \setminus \{0\}$ and define a functional l on $\text{span}\{h_0\}$ as $l(th_0) = td^+ f(x)(h_0)$ for $t \in \mathbb{R}$. This mapping satisfies $l(th_0) \leq d^+ f(x)(th_0)$ for every $t \in \mathbb{R}$. The Hahn-Banach Theorem 2.1 gives an extension (say x^*) of l to X such that $x^*(h) \leq d^+ f(x)(h)$ for all $h \in X$ —in particular, $x^*(h) \leq f(x + h) - f(x)$. Lemma 7.3 guarantees that $d^+ f(x)(h) \leq L\|h\|$ for every $h \in X$ and a suitable fixed $L > 0$; hence $x^* \in X^*$, and so $x^* \in \partial f(x)$ (see Exercise 3.3).

7.17 Let X be a w-sequentially complete Banach space whose dual norm is Gâteaux differentiable. Prove that X is reflexive.
Hint. Corollary 7.22 and the proof of Corollary 7.26.

7.18 Show that the norm $\|\cdot\|$ of a finite-dimensional Banach space is Fréchet differentiable at x if it is Gâteaux differentiable at x.
Hint. Šmulyan's Lemma 7.22 or a direct computation.

7.19 Let a function f be Lipschitz on $\mathbb{R}^n$ and Gâteaux differentiable at x. Show that f is Fréchet differentiable at x. Is this also true for Lipschitz mappings from $\mathbb{R}^n$ to a Banach space Y?
Hint. Proceed by contradiction, using the compactness of the unit ball and the Lipschitz property of f. The second part: Yes.

7.20 Show that $f(x) = x^2 \sin(1/x)$ is a Lipschitz function on $\mathbb{R}$ which is differentiable everywhere, yet its derivative is not a continuous function. So Corollary 7.24 does not work for Lipschitz functions.

7.21 Is it true that a Lipschitz function on $\mathbb{R}^2$ is Gâteaux differentiable at $x_0 \in \mathbb{R}^2$ if it is differentiable at x_0 in all directions?
Hint. No. The directional derivative need not be linear, check

$$f(x,y) = \begin{cases} \frac{xy}{\sqrt{x^2+y^2}} & \text{for } (x,y) \neq (0,0), \\ 0 & \text{for } (x,y) = (0,0). \end{cases}$$

7.22 Let x_i be vectors in a Banach space X such that $\overline{\text{span}}\{x_i\} = X$. Assume that f is a continuous convex function on X such that at all points of X, all directional derivatives in the directions of $\{x_i\}$ exist. Is f Gâteaux differentiable on X?
Hint. Yes, use the uniqueness of the supporting functional which is sufficient in the directions of $\{x_i\}$.

7.23 Let f be a Lipschitz function on a Banach space X and $x_0 \in X$. Assume that f is differentiable at x_0 in a dense set of directions. Show that f is differentiable at x_0 in all directions.
Hint. Show that the differentiation quotients are Cauchy in each direction. Proceed by contradiction, using the Lipschitz property.

7.24 Let B_1 and B_2 be the unit balls of two equivalent norms on a reflexive Banach space and assume that B_2 is the ball of a Gâteaux differentiable norm. Show that then $B_1 + B_2$ is the ball of a Gâteaux differentiable norm.

This fact is behind most first order differentiable renormings.
Hint. $B = B_1 + B_2$ is closed as B_i are w-compact. Let $x \in B$ and $f \in X^*$ have the property that $f(x) = \max\{f(y) : y \in B\}$. Then $x = x_1 + x_2$, where $x_i \in B_i$. Note that $\max\limits_{B_1+B_2}(f) = f(x_1) + f(x_2) = \max\limits_{B_1}(f) + \max\limits_{B_2}(f)$, so $f(x_i) = \max\{f(y) : y \in B_i\}$. Since B_2 is the unit ball of a smooth norm, such f is unique.

7.25 Let X be a Banach space with a Gâteaux differentiable norm. Let $f \in S_{X^*}$ does not attain its norm. Define φ on X by $\varphi(x) = \|x\|^2 - f(x)$ and $M = \{x \in X : \varphi(x) \leq 0\}$. Show that M is a closed bounded convex set, φ is a differentiable convex function such that $\varphi(x) = 0$ on the boundary $\partial M = \{x : \varphi(x) = 0\}$ but $\varphi' \neq 0$ on $\text{Int}(M)$.

Thus Rolle's theorem is not true in infinite-dimensional spaces.
Hint. Assume that $\varphi'(x_0) = 0$ for some $x_0 \in \text{Int}(M)$. Then φ attains its minimum at x_0 by the convexity of φ. Thus $x_0 \neq 0$ and $2\|x_0\|\,\|x_0\|' = f$. Thus f is a multiple of the derivative of the norm and as such it attains its norm, a contradiction.

7.26 Let X be a Banach space and let f be a continuous convex function on X^* that is w^*-lower semicontinuous. Show that if f is Fréchet differentiable at $x^* \in X^*$, then $f'(x^*) \in X$.
Hint. The derivative, as a uniform limit of quotients in B_{X^*}, is also w^*-lower semicontinuous. Then use its linearity to see that $f'(x^*)$ is a functional that is w^*-continuous on B_{X^*} and apply Theorem 3.92.

7.27 (Godefroy) Let X, Y be Banach spaces such that there is an isometry T of X^* onto Y^*. Show that if the dual norm of X^* is Fréchet differentiable on a dense set in X^*, then T is the dual operator to some isometry S of Y onto X.

In short, if the dual norm of X^* is Fréchet differentiable on a dense set in X^*, then X is an isometrically unique predual of X^*.
Hint. Let D be the set of $\|x^*\|'$ for points $x^* \in S_{X^*}$ of Fréchet differentiability of the dual norm of X^*. Note that $D \subset S_X$ (Exercise 7.26). We claim that $\tilde{D} = \overline{\text{conv}}^{w^*}(D) = B_{X^{**}}$. This follows from the separation theorem and from the fact that for every $x^* \in S_{X^*}$ we have $\sup\{f(x^*) : f \in D\} = 1$ (we use here that the set of the points of Fréchet differentiability of the dual norm is dense in X^*). From this and the fact that $D \subset S_X$ it follows that $\overline{\text{conv}}(D) = B_X$. Therefore $\overline{\text{span}}(D) = X$. Since isometry preserves the points of the Fréchet differentiability, we have that an analogous reasoning can be applied to Y^*. Let $y \in Y$ be a Fréchet derivative at some point $y^* \in S_{Y^*}$. Then the functional $T(y)$ is the Fréchet derivative of the dual norm of X^* at the point $T^{-1}(y^*)$ and as such it is from X. Therefore T^* maps Y into X and T is w^*-w^*-continuous. Then $(T^*|_Y)^* = T$.

7.28 Let Y be a closed subspace of a Banach space X and $x \in X$. We say that x is *orthogonal to* Y if $\|x+y\| \geq \|x\|$ for every $y \in Y$. Note that in a Hilbert space, this is equivalent to $x \perp Y$ (Lemma 15.46).

Prove the following: Let $(X, \|\cdot\|)$ be a Banach space such that the dual norm $\|\cdot\|^*$ is Fréchet differentiable. Let Z be a subspace of X of finite codimension. Let M be the set of all elements in S_X that are orthogonal to Z. Then M is a compact set.
Hint. Note that $(X/Z)^* = Z^\perp$ and $M = \{x \in S_X : \|\hat{x}\|_{X/Z} = \|x\|\}$. So by the Hahn–Banach theorem, M is formed exactly by those elements $x \in S_X$ for which there exists $f \in S_{X^*} \cap Z^\perp$ such that $f(x) = 1$.

Given $f \in S_{X^*} \cap Z^\perp$, there is $x \in S_X$ such that $f(x) = 1$ (find appropriate $\hat{x}$ in X/Z which is finite-dimensional) and it is unique, as $\|\cdot\|^*$ is Fréchet differentiable. Call $\varphi(f) := x$. By Corollary 7.16, $\varphi\colon S_{X^*} \cap Z^\perp \to S_X$ is norm–norm continuous and $M = \varphi(S_{X^*} \cap Z^\perp)$, so M is compact.

7.29 Let $(X, \|\cdot\|)$ be a normed space such that its norm is LUR. Prove that $\|\cdot\|$ in X^* is Fréchet differentiable at every point $x_0^* \in X^*\setminus\{0\}$ that attains its norm on B_X.
Hint. If $\|x_0^*\| = 1 = \langle x_0^*, x_0\rangle$ and $x_n \in B_X$ are such that $\langle x_0^*, x_n\rangle \to 1$ then $\langle x_0^*, x_n + x_0\rangle \to 2$ and $\|x_n + x_0\| \leq 2$, so by LUR, $\|x_n - x_0\| \to 0$, and use Corollary 7.20.

7.30 Let $(X, \|\cdot\|)$ be a Banach space. Prove that

(i) The norm $\|\cdot\|$ is UG if and only if the dual norm $\|\cdot\|^*$ is W*UR.

(ii) The norm $\|\cdot\|$ is WUR if and only if the dual norm $\|\cdot\|^*$ is UG.

Hint. Given $h \in S_X$ and $\varepsilon > 0$, there is $\delta > 0$ such that

$$\|x+th\| + \|x-th\| \leq 2 + \varepsilon|t| \quad \text{if } x \in S_X^* \text{ and } |t \leq \delta. \tag{7.14}$$

If $f_n, g_n \in S_{X^*}, n = 1, 2, \ldots$ are such that $\|f_n + g_n\| \to 2$, then for some $x_n \in S_X$, $n = 1, 2, \ldots, x_n(f_n + g_n) \to 2$ and thus $f_n(x_n) \to 1$ and $g_n(x_n) \to 1$. We have from (7.14)

$$f_n(x_n + th) + g_n(x_n - th) \leq 2 + \varepsilon|t|, \text{ for each } n \text{ and } |t| \leq \delta.$$

Thus there is n_0 such that

$$(f_n - g_n)(th) \leq 2 + \varepsilon\delta - f_n(x_n) - g_n(x_n) \leq 2\varepsilon\delta, \text{ for each } n \geq n_0 \text{ and } |t| \leq \delta.$$

Thus, in particular,

$$\delta(f_n - g_n)(h) \leq 2\varepsilon\delta \quad \text{for each } n \geq n_0.$$

This proves that $(f_n - g_n) \xrightarrow{w^*} 0$.

If $\|\cdot\|$ is not UG, then by convexity there are $h \in S_X$, $\varepsilon > 0$, $x_n \in S_X$ and $t_n > 0$, $t_n \to 0$ such that

$$\|x_n + t_n h\| + \|x_n - t_n h\| \geq 2 + \varepsilon t_n. \tag{7.15}$$

Choose $f_n, g_n \in S_{X^*}$, $n = 1, 2, \ldots$, such that

$$f_n(x_n + t_n h) = \|x_n + t_n h\|, \text{ and } g_n(x_n - t_n h) = \|x_n - t_n h\|.$$

Note that

$$f_n(x_n) = f_n(x_n + t_n h) - t_n f_n(h) \to 1 \quad \text{and similarly} \quad g_n(x_n) \to 1.$$

Hence $\lim \|f_n + g_n\| = 2$.

From (7.15) and the choice of f_n and g_n it follows

$$f_n(x_n + t_n h) + g_n(x_n - t_n h) \geq 2 + \varepsilon t_n \quad \text{for each } n.$$

Therefore, for each n,

$$(f_n - g_n)(t_n h) \geq t_n + 2 - f_n(x_n) - g_n(x_n) \geq \varepsilon t_n.$$

Therefore $\|\cdot\|^*$ is not W*UR.

7.31 The point $f_0 := (\frac{1}{2}, \frac{(15)^{1/4}}{2})$ is in $S_{\ell_4^2}$. Find an element x_0 in $S_{\ell_{4/3}}$ at which f_0 attains its norm.

Hint. Differentiate $x^4 + y^4 = 1$ implicitly at f_0. The result: $(\frac{1}{2^3}, \frac{1}{2^3}(15)^{3/4})$.

7.32 Let $\{r_n\}$ be a sequence of all rational numbers in $(0, 1)$. Define a function f on $(0, 1)$ by $f(x) = \sum_{n=1}^{\infty} 2^{-n}|x - r_n|$. Show that f is a convex continuous function on $(0, 1)$ which is differentiable exactly at irrational points of $(0, 1)$.

Hint. Use Lemma 7.4 to see that f is not differentiable at rational points. For irrational points the fraction can be made arbitrarily small, first working with the tail and then with the remaining terms.

7.33 Define a mapping ϕ from $[0, 1]$ into $L_1[0, 1]$ by $\phi(t) = \chi_{[0,t]}$. Show that ϕ is Lipschitz and nowhere differentiable.
Hint. Differential quotients are not Cauchy.

7.34 Let A be a subset of $\mathbb{R}$ with $\lambda(A) = 0$ (Lebesgue measure). Follow the hint to show that there is a Lipschitz function f on $\mathbb{R}$ not differentiable at points of A ([Zaho]).

Since there is a residual set of measure 0 in $\mathbb{R}$, this shows that the set of differentiability points of a Lipschitz function need not be residual.
Hint. ([BeLi, p. 165]) Take open sets $G_1 \supset G_2 \supset \ldots \supset A$ such that $\lambda(G_n) \leq 2^{-n}$ and $\lambda\big((a, b) \cap G_{n+1}\big) \leq (b - a)/3$ for every component (a, b) of G_n and for all $n \in \mathbb{N}$. Put $f_n(x) = \lambda\big((-\infty, x) \cap G_n\big)$, $n \in \mathbb{N}$, and $f = \sum(-1)^{n+1} f$. Then f is Lipschitz.

If $x \in \bigcap G_n$, then f is not differentiable at x. Indeed, let (α_n, β_n) be the components of G_n containing x. Fix n, define $\gamma_{j,n} = \frac{f_j(\beta_n)-f_j(\alpha_n)}{\beta_n-\alpha_n}$. Observe that $\gamma_{j,n} = \beta_n - \alpha_n$ for $j \leq n$, while $\gamma_{j,n} = \beta_j - \alpha_j$ if $j > n$. So, for a fixed n, the sequence $\{\gamma_{j,n}\}_{j=1}^{\infty}$ is decreasing, $\gamma_{1,n} = \cdots = \gamma_{j,n} = 1$ and $\gamma_{n+1,n} \leq \frac{1}{3}$. Thus if n is even, $\frac{f(\beta_n)-f(\alpha_n)}{\beta_n-\alpha_n} = 1 - 1 + \cdots - 1 + \gamma_{n+1,n} - \gamma_{n+2,n} + \cdots \leq \frac{1}{3}$, while if n is odd, $\frac{f(\beta_n)-f(\alpha_n)}{\beta_n-\alpha_n} = 1 - 1 + \cdots + 1 - \gamma_{n+1,n} + \gamma_{n+2,n} - \cdots \geq \frac{2}{3}$.

7.35 Let $(X, \|\cdot\|)$ be a Banach space. Show that $\|\cdot\|^2$ is Fréchet differentiable at 0 and $\big(\|\cdot\|^2\big)'(0) = 0$.

Let $(H, \|\cdot\|)$ be a Hilbert space H. Show that $\|\cdot\|^2$ is Fréchet differentiable at every point of H. Show that $\|\cdot\|$ is Fréchet differentiable at every point $x \neq 0$.
Hint. The first part follows by direct calculation. For the second, check $F(h) = 2(x, h)$ in the notation of Definition 7.1. For the last part, use the Chain Rule.

7.36 Show that the supremum norm of c_0 is Fréchet differentiable at $x = (x_i) \in S_{c_0}$ if and only if $\|x\| = \max |x_i|$ is attained at exactly one i.
Hint. If the condition is satisfied, $\frac{\|x+th\|+\|x-th\|-2\|x\|}{t} = 0$ for t small enough and every $h \in S_{c_0}$. If this condition fails and say $x_1 = x_2 = 1$, then e_1 and e_2 are two support functionals in ℓ_1 to B_{c_0} at x. Note that this means that the norm of c_0 is Gâteaux differentiable at x if and only if it is Fréchet differentiable at x.

7.37 Show that the canonical norm of ℓ_1 is nowhere Fréchet differentiable and is Gâteaux differentiable at $x = (x_i)$ if and only if $x_i \neq 0$ for every i.

If Γ is uncountable, show that the canonical norm of $\ell_1(\Gamma)$ is not Gâteaux differentiable at any point.

Hint. Let $x \in S_X$. Given $\varepsilon > 0$, find i such that $|x_i| < \varepsilon/2$ and consider $h = \varepsilon e_i$. Show that $\|x \pm h\| \geq 1 + \varepsilon/2$ and use Lemma 7.4.

Note that every vector in $\ell_1(\Gamma)$ has a countable support, choose a standard unit vector outside this support and use Lemma 7.4.

7.38 Show that the norm of $C[0, 1]$ is nowhere Fréchet differentiable.

Show that the norm of $C[0, 1]$ is Gâteaux differentiable at $x \in S_{C[0,1]}$ if and only if $|x|$ attains its maximum at exactly one point of $[0, 1]$.

Hint. Note that the distance between two different Dirac measures in $C[0, 1]^*$ is two. Given $x \in S_{C[0,1]}$, choose $t_0 \in [0, 1]$ such that $x(t_0) = 1$. Then choose $t_n \neq t_0$ such that $x(t_n) \to 1$. By the Šmulyan Lemma 7.22, x is not a point of Fréchet differentiability of the supremum norm on $C[0, 1]$.

For the second part, assume that $x \in S_{C[0,1]}$ is such that $x(t_0) = 1$ and $|x(t)| < 1$ for every $t \neq t_0$. Put $H = \{f \in C[0, 1]^* : \|f\| \leq 1, f(x) = 1\}$. If $H \cap B_{C[0,1]^*} \neq \{\delta_{t_0}\}$, then this intersection would have at least two extreme points that would be extreme points of $B_{C[0,1]^*}$. All the extreme points of $B_{C[0,1]^*}$ are $\pm$ Dirac measures (Lemma 3.116).

7.39 Let $p \in (1, \infty)$. Show that the norm of $L_p[0, 1]$ is Fréchet differentiable and calculate its Fréchet derivative.

Hint. By the standard rules (use the monotonicity in the differential quotient), we get $\|\cdot\|'_x(h) = \|x\|^{1-p} \int |x(t)|^{p-1} \operatorname{sign}\big(x(t)\big)h(t)\,dt$. The convergence of the integral follows from Hölder's inequality (1.1).

7.40 Show that the norm of ℓ_∞ is Fréchet differentiable on a dense set in ℓ_∞.

Hint. Show that the norm is Fréchet differentiable at all points $x \in S_{\ell_\infty}$ such that $|x_i| = 1$ for some i and $\sup\{|x_j| : j \neq i\} < 1$. Such points are dense in S_{ℓ_∞}.

7.41 Show that the norm of $L_\infty[0, 1]$ is nowhere Gâteaux differentiable. Thus $L_\infty[0, 1]$ is not isometric to ℓ_∞ as the norm of ℓ_∞ is Fréchet differentiable on a dense set (see the Exercise 7.40). However, these two spaces are isomorphic (Exercise 4.42).

Hint. Let $\|f\| = 1$. Assume that there is a sequence of pairwise disjoint sets I_n of positive measure such that $|f(t) - 1| \leq \frac{1}{n}$ for all $t \in I_n$. Consider functionals $F_n(g) = \mu(I_n)^{-1} \int_{I_n} f\,d\mu$. Let H and G be w^*-cluster points of $\{F_{2n}\}$ and $\{F_{2n+1}\}$. They give two supporting hyperplanes to the unit ball at f. As checked on the characteristic function of the union of $\{I_{2n}\}$, H and G are distinct.

7.42 Let $\|\cdot\|_\infty$ denote the canonical of ℓ_∞ and set $p(x) = \limsup |x_i|$. Define $|||x||| = \|x\|_\infty + p(x)$ for $x \in \ell_\infty$. Show that $|||\cdot|||$ is nowhere Gâteaux differentiable.

Hint. It is enough to show that p is nowhere differentiable. If $x = (x_i) \in \ell_\infty$ and $x_{n_k} \to 1 = p(x)$, consider the direction $h = \sum (-1)^k e_{n_k}$.

7.43 Show that the restriction of the canonical norm of ℓ_1 to any two-dimensional subspace of ℓ_1 is not Gâteaux differentiable .
Hint. Let $a = (a_i) \in \ell_1$ and $b = (b_i) \in \ell_1, b \neq 0$. Define a function for $\lambda \in \mathbb{R}$ by $f(\lambda) = \sum |a_i + \lambda b_i|$. Show that f is not differentiable for some λ.

7.44 Show that ℓ_1 has two-dimensional subspaces that are strictly convex in the canonical norm of ℓ_1. This Lindenstrauss's idea ([Lind5]) was extended by Fonf and Kadec who found infinite-dimensional w^*-closed subspaces of ℓ_1 with the same property ([FoKa]).
Hint. Let $\{a_i\}$ and $\{b_i\}$ be such that $\{\frac{a_i}{b_i}\}$ is dense in $\mathbb{R}$. Show that the function $f(\lambda) = \sum |a_i + \lambda b_i|$ is then not affine on any non-degenerate interval in $\mathbb{R}$.

7.45 Show that c_0 contains no two-dimensional subspace on which the standard norm is Gâteaux differentiable.
Hint. Then some quotient Q of ℓ_1 would have uncountably many extreme points as the dual to a two-dimensional smooth space is strictly convex. Every point of the sphere of Q which is identified with the restriction to the two-dimensional subspace in question extends to an extreme point of the sphere in ℓ_1 by the Krein–Milman theorem, considering the face of all the extensions. Thus there is uncountably many such extreme points of the ball of the standard norm in ℓ_1, which is a contradiction as the extreme points of the ball of ℓ_1 are exactly $\pm e_i$.

7.46 Let X be a reflexive strictly convex Banach space, C a closed convex set in X. Show that there is a unique nearest point to x in C.
Hint. x has a nearest point as X is reflexive (Exercise 3.165). Assume that $x = 0$ and $\operatorname{dist}(x, C) = 1$. Let $c_1 \neq c_2 \in C$ satisfy $\|c_1\| = \|c_2\| = 1$. Then $\frac{1}{2}(c_1+c_2) \in C$, yet $\|\frac{1}{2}(c_1 + c_2)\| < 1$ by the strict convexity, a contradiction.

7.47 Let C be a convex closed set in a reflexive Banach space whose norm is locally uniformly convex. To every $x \in X$ assign $p(x)$, the closest point of C to x. Show that p is continuous.
Hint. Let $\operatorname{dist}(0, C) = 1$ and $x_n \to 0$. Let $y = p(0)$ and $y_n = p(x_n)$. Then since $\|x - p(x)\| = \operatorname{dist}(x, C)$ is Lipschitz function, we have that $\|y\| = 1$, $\|y_n\| \to 1$. Because of the convexity of C we have $\frac{1}{2}(y + y_n) \in C$ and thus $\|\frac{1}{2}(y + y_n)\| \geq \operatorname{dist}(0, C) = 1$. From the triangle inequality we have $\|y_n\| + \|y\| \geq \|y + y_n\|$, so $\|\frac{1}{2}(y + y_n)\| \to 2$. From the local uniform rotundity, we have $\|y_n - y\| \to 0$.

7.48 Let C be a closed convex subset of a real Hilbert space H. Denote by P the nearest point mapping of H onto C. Show that P is 1-Lipschitz, and if we define $f(x) = \frac{1}{2}(\|x\|^2 - \|x - P(x)\|^2)$ then f is convex and $f'(x) = P(x)$ in the Fréchet sense.
Hint. $\big(x - P(x), z - P(x)\big) \leq 0$ for all $z \in C$. Indeed, if $z \in C$ and $0 < t < 1$, then $z_t = tz + (1-t)P(x) \in C$ and thus $\|x - P(x)\| \leq \|x - z_t\| = \|(x - P(x)) - t(z - P(x))\|$. Taking the square and expanding gives $0 \leq -2t\big(x - P(x), z - P(x)\big) +$

$t^2\|z - P(x)\|^2$. Dividing by t and taking the limit as $t \to 0$ establishes the claim. Thus $(x - P(x), P(x) - P(y)) \geq 0$ and $(P(y) - y, P(x) - P(y)) \geq 0$. Adding them we get $(x - y, P(x) - P(y)) \geq \|P(x) - P(y)\|^2$, which implies that P is 1-Lipschitz.

We have $2f(x) = \|x\|^2 - \inf\{\|x-y\|^2 : y \in C\} = \sup\{2(x, y) - \|y\|^2 : y \in C\}$, and thus f is convex as a supremum of affine functions. Fix $x \in H$. For every $y \in H$ we have $\|(x + y) - P(x + y)\| \leq \|(x + y) - P(x)\|$, so $\|(x + y) - P(x + y)\|^2 \leq \|x + y\|^2 - 2(x + y, P(x)) + \|P(x)\|^2 = \|x + y\|^2 + \|x - P(x)\|^2 - \|x\|^2 - 2(y, P(x))$, hence $f(x + y) - f(x) - (P(x), y) \geq 0$. On the other hand, since $\|x - P(x)\| \leq \|x - P(x + y)\|$, we get $f(x + y) - f(x) - (P(x), y) \leq (y, P(x + y) - P(x)) \leq \|y\| \cdot \|P(x + y) - P(x)\| \leq \|y\|^2$, which gives the differentiability assertion.

7.49 Let $C = \{x = (x_\gamma) \in \ell_2(\Gamma) : x_\gamma \geq 0 \text{ for all } \gamma \in \Gamma\}$, i.e., C is the positive cone in $\ell_2(\Gamma)$. Let P assign to $x \in \ell_2(\Gamma)$ its nearest point in C. Then P is a Lipschitz mapping (the previous exercise). Show that if Γ is uncountable then P is nowhere Gâteaux differentiable. If Γ is infinite, then P is nowhere Fréchet differentiable. Note that Preiss [Prei2] proved that every real-valued Lipschitz function on $\ell_2(\Gamma)$ is Fréchet differentiable on a dense set.
Hint. $P(x)_\gamma = x_\gamma^+$ and $(P(x))' = \operatorname{sign}(x^+)$. To see the Fréchet case, if for example $x = (x_\gamma)$ and $x_\gamma > 0$ for all γ, let $\alpha_\gamma = -2x_\gamma$ for all γ and it is not true that $\alpha^{-1}[P(x + \alpha_\gamma e_\gamma) - P(x) - \alpha_\gamma (P(x))'(e_\gamma)] \to 0$. The Gâteaux case follows, as every $x \in \ell_2(\Gamma)$ is countably supported and we use the formula for $P(x)$.

7.50 Let the norm $\|\cdot\|$ of a Banach space X be Gâteaux differentiable. Let P be a norm-one projection of X onto $P(X) \subset X$. Show that if $x \in S_{P(X)}$ then $\|x\|' \in P^*(X^*) = P(X)^*$.
Hint. $\|x\|'$ and $P^*(\|x\|')$ are two supporting functionals of B_X at x.

7.51 Let $(X, \|\cdot\|)$ be a Banach space. Check that the formulas for the Fenchel conjugates below are correct. (i) If $f := \|\cdot\|$ then $f^*(x^*) = 0$ for all $x^* \in B_{X^*}$, and $f^*(x^*) = +\infty$ for all $x^* \in X^* \setminus B_{X^*}$. (ii) If $f(x) = 0$ for all $x \in B_X$ and $f(x) = +\infty$ for all $x \in X \setminus B_X$, then $f^* = \|\cdot\|^*$, where $\|\cdot\|^*$ denotes here the canonical dual norm in X^*. (iii) If $f := x_0^*$ for some $x_0^* \in X^*$, then $f^*(x_0^*) = 0$ and $f^*(x^*) = +\infty$ for all $x^* \in X^* \setminus \{x_0^*\}$. (iv) For $x_0 \in X$ let $f(x_0) = 0$ and $f(x) = +\infty$ for all $x \in X \setminus \{x_0\}$. Then $f^* = x_0$. (v) If $f : \mathbb{R} \to \mathbb{R}$ is given by $f(x) = |x|^p/p$, $p > 1$, then $f^*(y) = |y|^q/q$ for $y \in \mathbb{R}$, where $(1/p) + (1/q) = 1$.
Hint. Standard computation.

7.52 Let $(X, \|\cdot\|)$ be a Banach space. Define a function f on X by $f(x) = \frac{1}{2}\|x\|^2$. Calculate its conjugate f^* on X^*.
Hint. Note that $f^*(x^*) = \sup_{r>0}\{\sup\{x^*(x) - f(x) : \|x\| = r\}\}$. Use the definition of $\|x^*\|$ to calculate the "inside" supremum and then use elementary calculus to obtain the final result.

7.53 (Bishop, Phelps, Bollobás) Prove the following strengthening of Theorem 7.41: *Let X be a Banach space, and $x_n \in S_X$ and $f_n \in S_{X^*}$, $n \in \mathbb{N}$, are such that $f_n(x_n) \to 1$. Then there exist $g_n \in S_{X^*}$ and $y_n \in S_X$, $n \in \mathbb{N}$, such that $\|f_n - g_n\| \to 0$, $\|x_n - y_n\| \to 0$, and $g_n(y_n) = 1$, $n \in \mathbb{N}$.* Note that given $f_0 \in S_{X^*}$, letting $f_n := f_0$ for all $n \in \mathbb{N}$ and taking $x_n \in S_X$ such that $f_0(x_n) \to 1$, we can see that the original Bishop–Phelps theorem follows from this result.

Hint. We follow [Boll2, p. 122]. It is enough to prove the following lemma: *Let X be a Banach space and let $x_0 \in S_X$ and $f_0 \in S_{X^*}$ be such that $f_0(x_0) > \frac{3}{4}$. Then there exist $x_1 \in S_X$ and $f_1 \in S_{X^*}$ such that $f_1(x_1) = 1$, $\|f_0 - f_1\| \le 2\sqrt{1 - f_0(x_0)}$, and $\|x_0 - x_1\| \le 2\sqrt{1 - f_0(x_0)}$.* To this end, put $\varepsilon = 2\sqrt{1 - f_0(x_0)}$ (< 1). If $\varepsilon = 0$ there is nothing to prove. Otherwise, put $\delta = \frac{1}{4}\varepsilon^2$, $C := \{x \in X : f_0(x) = 0, \|x\| \le 2\varepsilon^{-1}\}$, and $K := (f_0(x_0))^{-1}(1 + 2\varepsilon^{-1})$. Let D be the convex hull of the union of B_X and C. (See Figure 7.10.)

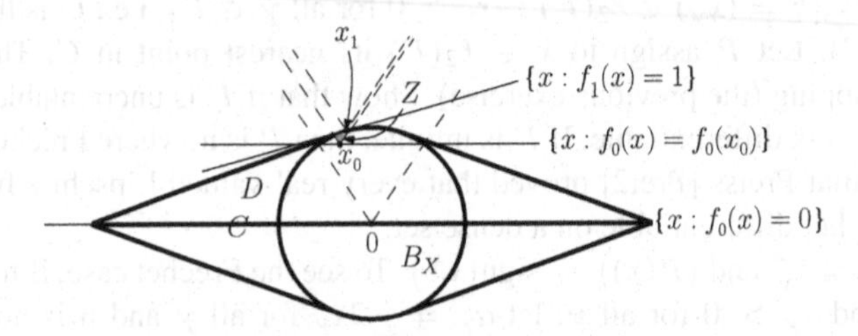

Fig. 7.10 The proof of the lemma in Exercise 7.53. The order is defined by the dashed cone

We need to show that there is $x_1 \in S_X$ and $f_1 \in S_{X^*}$ such that

$$f_1(x_1) = 1, \ \|f_0 - f_1\| \le \varepsilon, \ \text{and} \ \|x_0 - x_1\| \le \varepsilon.$$

Define a partial ordering $\le$ on B_X as follows: We say that $y \le x$ (alternatively, $x \ge y$) if

$$\|x - y\| \le K\big(f(x) - f(y)\big). \tag{7.16}$$

Let $Z := \{x \in B_X : x \ge x_0\}$. We shall prove that $(Z, \le)$ has a maximal element. For this, assume that W is a totally ordered subset of $(Z, \le)$. Obviously, the net of real numbers $\{f(x) : x \in W, \le\}$ is bounded. It is also non-decreasing, due to (7.16), so it converges to its supremum. Again from (7.16) it follows that W is a Cauchy net. By the completeness of X, W converges to a point $y \in B_X$, so y is an upper bound for W. Now, by Zorn's lemma there exists a maximal element x_1 of $(Z, \le)$. In particular, $x_0 \le x_1$, and $x_1 \in B_X$.

We shall show that $x_1 \in \partial D$ (i.e., the topological boundary of D). If we assume, on the contrary, that x_1 is an interior point of D, then $p_D(x) < 1$, where p_D denotes the Minkowski functional of D. It follows that

$$x_1 = sb + (t - s)c, \ \text{where } b \in B_X, \ c \in C, \ \text{and } 0 < s < t < 1. \tag{7.17}$$

Note that

$$f_0(x_0) \le f_0(x_1) = s f_0(b) < f_0(b). \tag{7.18}$$

From (7.18) we get

$$f_0(b - x_1) = f_0(b) - s f_0(b) = (1 - s) f_0(b) > (1 - s) f_0(x_0) \ (> 0). \tag{7.19}$$

On the other hand,

$$\begin{aligned} \|b - x_1\| &= \|b - sb - (t - s)c\| < (1 - s) + (t - s)\|c\| \\ &\le (1 - s) + (t - s)\frac{2}{\varepsilon} < (1 - s)\left(1 + \frac{2}{\varepsilon}\right) \\ &= (1 - s)\frac{1 + 2/\varepsilon}{f_0(x_0)} f_0(x_0) = (1 - s) K f_0(x_0) < K f_0(b - x_1), \end{aligned} \tag{7.20}$$

where the last inequality follows from (7.19). This shows that $(x_0 \le)\ x_1 \le b$ $(\in B_X)$, hence $x_1 = b$, and this is a contradiction with (7.17), proving that, indeed, $x_1 \in \partial D$.

Since D has nonempty interior, by the separation theorem there is $f_1 \in S_{X^*}$ such that $\sup\{f_1(x) : x \in D\} = f_1(x_1)$. It follows that

$$1 = \sup_{B_X} f_1 \le \sup_{D} f_1 = f_1(x_1) \le 1,$$

since $x_1 \in B_X$, and thus $f_1(x_1) = 1 = \|x_1\|$.

We now show that $\|x_1 - x_0\| \le \varepsilon$.

This can be seen as follows. Since $x_1 \ge x_0$, we have

$$\begin{aligned} \|x_1 - x_0\| &\le K f_0(x_1 - x_0) \le K\big(1 - f_0(x_0)\big) \\ &= K\delta = \frac{1 + 2/\varepsilon}{f_0(x_0)}\delta = \frac{\varepsilon}{2 - \varepsilon} < \varepsilon. \end{aligned}$$

It remains to show that $\|f_0 - f_1\| \le \varepsilon$. Since $f_1(x) \le 1$ for all $x \in C$, by Exercise 2.13 we have that either $\|f_0 - f_1\| \le \varepsilon$ or $\|f_0 + f_1\| \le \varepsilon$.

But

$$\begin{aligned} \|f_0 + f_1\| &\ge (f_0 + f_1)(x_1) = f_0(x_1) + 1 = f_1(x_1) + 1 + f_0(x_1) - f_1(x_1) \\ &= 2 + (f_0 - f_1)(x_1) \ge 2 - \|f_0 - f_1\| \ge 2 - \varepsilon > \varepsilon, \end{aligned}$$

since $\varepsilon < 1$. This completes the proof of the lemma.

7.54 Assume X a separable Banach space. Assume that there is no 1-norming proper closed subspace in X^*. Show that X^* is separable.

Hint. Let $\{x_n\}$ be dense in S_X and for each n, let $f_n \in S_{X^*}$ be such that $f_n(x_n) = 1$. Consider the norm-closed linear hull of $\{f_n\}$.

7.55 Assume that the norm of a separable X is Fréchet differentiable. Does there exist a 1-norming proper closed subspace M of X^*?
Hint. No. Let $f \in S_{X^*}$ attain its norm at $x \in S_X$. If $f_n \in M \cap S_{X^*}$ are such that $f_n \to f$ in the w^*-topology, then $f_n(x) \to 1$ and, by the Šmulyan lemma 7.22, $\|f_n - f\| \to 0$. Then use the Bishop–Phelps theorem.

7.56 Find an example of a space with separable dual and a proper 1-norming closed subspace of its dual.
Hint. Any nonreflexive space X whose second dual is separable. Then X is 1-norming in X^{**} by Goldstine's theorem.

7.57 (Lindenstrauss, [Lind9b]) Let X be a Banach space. Show that the following two statements are equivalent:

(i) Every bounded closed and convex subset of X has an extreme point.

(ii) Every bounded closed and convex subset of X is the closed convex hull of its extreme points.

Hint. Clearly (ii) implies (i). Assume that (i) holds and that K is a bounded closed and convex subset of X. Let K_0 be the closed convex hull of the extreme points of K. If $K_0 \neq K$, then, by the separation theorem, we can find f_0 that separates K_0 and some point in $K \setminus K_0$, and then, by the Bishop–Phelps theorem, we can find $f \in X^*$ and $y \in K$ such that $f(y) = \sup_{x \in K} f(x) > \sup_{x \in K_0}$. The set $H := \{x \in X;\ f(x) = f(y)\}$ is a supporting manifold of K (see Lemma 3.61), and $H \cap K_0 = \emptyset$. The set $K_1 := H \cap K$ satisfies $K_0 \supset \text{Ext}(K) \supset \text{Ext}(K_1) \neq \emptyset$ (by (i) and Lemma 3.62), so we get a contradiction with the fact that K_0 is disjoint from K_1.

7.58 Assume that C is a bounded closed convex set in a Banach space X. Show that the closed convex hull of points y of C for which there is a nonzero $f \in X^*$ that attains its supremum on C at x is the whole set C (such point x is called a *support point of* C)
Hint. Denote the closed convex hull in question by C_1. If $x_0 \in C \setminus C_1$, then there is, by the Bishop–Phelps theorem, a $g \in X^*$ that separates C_1 from x_0 (in the sense that $g(x_0) > \sup_{x \in C_1} g(x)$) and there is $y_0 \in C$ such that g attains its supremum on C at y_0, i.e., that $y_0 \in C_1$. Then $g(y_0) \geq g(x_0) > \sup_{x \in C_1} g(x) \geq g(y_0)$, a contradiction.

Note that the existence of support points in general bounded closed convex sets was an open problem (posed by V. Klee) before the Bishop–Phelps theorem was proved.

7.59 (Lindenstrauss, [Lind9b]) Let K be a bounded closed and convex subset of ℓ_1 and $\varepsilon > 0$ be given. Show that there is a closed face F (i.e., the intersection

of a closed supporting manifold with K) of K and an integer n such that if $x = (x_1, x_2, \cdots) \in F$, then $\sum_{i>n} |x_i| \le \varepsilon$.
Hint. Let $M = \sup_{x \in K} \|x\|$ and choose a $y = (y_1, y_2, \cdots)$ in K such that $\|y\| \ge M - \varepsilon/4$. Let n be such that $\sum_{i=1}^{n} |y_i| \ge M - \varepsilon/2$ and let $f \in \ell_1^*$ be defined by

$$f(x_1, x_2, \cdots) = \sum_{i=1}^{n} (\text{sign } y_i) x_i,$$

where sign $t = 1$ if $t \ge 0$ and sign $t = -1$ if $t < 0$. By the Bishop–Phelps theorem, there is a $g \in \ell_1^*$ such that $\|g - f\| < \varepsilon/4M$ and $F := \{x \in K : g(x) = \sup_{u \in K} g(u)\}$ is not empty. Then F has the required properties. Indeed, let $x \in F$. Then

$$\begin{aligned}
\sum_{i=1}^{n} |x_i| &\ge f(x) \ge g(x) - \|f - g\|\|x\| \ge g(y) - \|f - g\|\|x\| \\
&\ge f(y) - \|f - g\|(\|x\| + \|y\|) = \sum_{i=1}^{n} |y_i| - \|f - g\|(\|x\| + \|y\|) \\
&\ge M - \varepsilon/2 - (\varepsilon/4M)(2M) = M - \varepsilon.
\end{aligned}$$

Since $\|x\| \le M$, we have that $\sum_{i>n} |x_i| = \sum_{i=1}^{\infty} - \sum_{i \le n} |x_i| \le M - (M - \varepsilon) = \varepsilon$.

7.60 (Lindenstrauss, [Lind9b]) Show that in ℓ_1, every closed bounded and convex set is the closed convex hull of its extreme points.
Hint. Let K be a closed convex and bounded set in ℓ_1. By Exercise 7.57, we need only to show that $\text{Ext}(K) \ne \emptyset$. By Exercise 7.59, there is a sequence $\{F_i\}_{i=1}^{\infty}$ of closed faces of K and a sequence of integers $\{n_i\}_{i=1}^{\infty}$ such that F_{i+1} is a face of F_i and $x = (x_1, x_2, \cdots) \in F_i$ implies that $\sum_{j>n_i} |x_j| < \frac{1}{i}$. Let $\{y^i\}$ be a sequence of points in ℓ_1 such that $y^i \in F_i$. From the properties of F_i, it follows that the set $\{y^i\}_{i=1}^{\infty}$ is totally bounded. Hence $F := \bigcap_{i=1}^{\infty} F_i$ is a nonempty compact face of K. By the Krein–Milman theorem, $\text{Ext}(F) \ne \emptyset$ and hence $\text{Ext}(K) \ne \emptyset$. This concludes the proof.

Lindenstrauss' result in this exercise became a cornerstone in the development of the Radon–Nikodým property (see Definition 11.14). It was extended to separable dual spaces by Bessaga and Pełczyński [BePe1a] by using *nonlinear homeomorphism* [Kade4] and [Klee2]. These results became starting points in the development of the Krein–Milman and Radon–Nikodým properties.

7.61 Let X, Y be Banach space, let h be a Gâteaux (Fréchet) differentiable homeomorphism of X into Y. Show that if Y admits a Gâteaux (Fréchet) smooth bump then so does X.
Hint. Use the composition of mappings and the chain rule for differentiation.

7.62 Assume that a Banach space X admits a Gâteaux (Fréchet) differentiable bump b. Show that X admits a bounded Gâteaux (Fréchet) differentiable bump.

It is not known whether X admits a Lipschitz Fréchet differentiable bump if X admits a Fréchet differentiable bump. See the open problem 7.1

Hint. Assume that $b(0) = 1$ and $b(x) = 0$ for $\|x\| \geq 1$. Put $\psi(x) = 1 - e^{-b^2(x)}$ for $x \in X$.

7.63 Use Theorem 7.46 to show that ℓ_∞ does not admit any continuous Gâteaux differentiable bump function.

Hint. $\operatorname{dens}(\ell_\infty^*) \geq \operatorname{card}(\beta\mathbb{N}) = 2^c > \operatorname{card}(\ell_\infty) = c$.

7.64 Show that ℓ_∞ admits no equivalent Gâteaux differentiable norm. (See also Exercise 7.63.)

Hint. The norm $|||\cdot|||$ defined on ℓ_∞ by $|||x||| = \|x\|_\infty + \limsup_{i\to\infty} |x_i|$ is nowhere Gâteaux differentiable (Exercise 7.42). If ℓ_∞ admitted a Lipschitz Gâteaux differentiable bump function, then by Corollary 7.44 we would have that $|||\cdot|||$ must be Gâteaux differentiable on a dense set of points.

7.65 Show that $\ell_1(\Gamma)$ admits no Lipschitz Gâteaux differentiable bump function if Γ is uncountable.

Hint. The standard norm of $\ell_1(\Gamma)$ is nowhere Gâteaux differentiable (Exercise 7.37). Then use the same argument as in the previous exercise.

7.66 Use Theorem 11.6 to prove Pitt's theorem: Given $1 \leq p < q < \infty$, then every $T \in \mathcal{B}(\ell_q, \ell_p)$ is a compact operator.

Hint. Let $\{x_i\}$ be a bounded sequence in ℓ_q. We may assume that $x_i \overset{w}{\to} y$, then $T(y_i) \overset{w}{\to} T(y)$. Apply Theorem 11.6 to $x \mapsto \|x\|_q^q - \|T(x)\|_p^p$. Then for $h = t(x_i - y)$, $t > 0$, $i \in \mathbb{N}$ we get

$$\begin{aligned}&\|x + t(x_i - y)\|_q^q + \|x - t(x_i - y)\|_q^q - 2\|x\|_q^q \\ &\geq \|T(x) + tT(x_i - y)\|_p^p + \|T(x) - tT(x_i - y)\|_p^p - 2\|T(x)\|_p^p,\end{aligned}$$

and $2t^q \limsup_{i\to\infty} \|x_i - y\|_q^q \geq 2t^p \limsup_{i\to\infty} \|T(x_i - y)\|_p^p$ for $t > 0$. Hence $\|T(x_i - y)\|_p \to 0$ as $i \to \infty$.

7.67 Assume that the norm $\|\cdot\|$ of a separable Banach space X can be approximated by a Lipschitz C^1-smooth function φ on X such that $|\|x\| - \varphi(x)| < 1$ for every $x \in X$. Is X^* necessarily separable?

Hint. Yes. By considering a C^1-smooth function τ on $\mathbb{R}$ such that $\tau(t) = 2$ whenever $t \in [-1, 1]$ and $\tau(t) = 0$ for $|t| \geq 4$, we have $\tau(\varphi(0)) = 2$, $\tau(\varphi(x)) = 0$ for $\|x\| \geq 5$ and $\tau \circ \varphi$ is C^1-smooth. Use Corollary 7.44.

7.68 Assume that φ is a non-negative C^1-smooth bump on $\mathbb{R}^n$ such that $\int_{\mathbb{R}^n} \varphi(t)\,dt = 1$ and $\operatorname{supp}(\phi) \subset \delta B_{\mathbb{R}^n}$.

Show that f is a uniformly continuous convex function on $\mathbb{R}^n$, then $f_\varphi(x) = \int_{\mathbb{R}^n} f(x-t)\varphi(t)\,dt$ provides a good convex smooth approximation of f in the sense that $\|f_\varphi - f\|_{C(\mathbb{R}^n)} \to 0$ as $\delta \to 0$.

This method cannot be used in infinite dimensions.

Hint. Estimate $|f(x) - f_\delta(x)| \le \int_{\mathbb{R}^n} |f(x) - f(x-t)|\varphi(t)\,dt$.

7.69 (Phelps, [Phel0]) Prove that, if X is a Banach space, then the dual norm of X^* is strictly convex if and only if for every subspace Y of X and every $f \in S_{Y^*}$, there is a unique $\widetilde{f} \in S_{X^*}$ that extends f.

Hint. Let Y be a subspace of X and $f \in S_{Y^*}$. Assume that $\widetilde{f_1}$ and $\widetilde{f_2}$ are two distinct elements in S_{X^*} that both extend f. Then $1 \ge \|(1/2)(\widetilde{f_1} + \widetilde{f_2})\| \ge \|(1/2)(f + f)\| = 1$, hence the dual norm of X^* is not strictly convex.

Assume now that the dual norm is not strictly convex. Then S_{X^*} contains a line segment between two distinct elements f_0 and g_0 in S_{X^*}. Let $M = \operatorname{span}\{f_0 - g_0\}$. Then $M_\perp$ is a hyperplane in X. Put $h = (1/2)(f_0 + g_0)$. Then $\operatorname{dist}(h, M) = 1$, since the distance of 0 to the line through f_0 and g_0 is 1. Since $(M_\perp)^*$ is isometric to $X^*/(M_\perp^\perp)$ $(= X^*/M)$ (Proposition 2.6), we have that $\|h|_{M_\perp}\|_{(M_\perp)^*} = \operatorname{dist}(h, M) = 1$. Note, too, that f_0 and g_0 are norm-1 extensions of h. Indeed, $\|f_0\| = \|g_0\| = 1$ and, say $f_0 = h + (1/2)(f_0 - g_0)$, with $f_0 - g_0 \in M$ $(= (M_\perp)^\perp)$. So $f_0 = h$ on $M_\perp$. Similarly, $g_0 = h$ on $M_\perp$. This finishes the proof.

7.70 Follow the hint to prove that there is no equivalent norm on $C[0, \omega_1]$ such that its dual norm is rotund (however, it is proved in [Hayd2] that for every ordinal μ, the space $C[0, \mu]$ admits a C^∞-smooth norm).

Hint. Assume that $|||\cdot|||$ is a dual rotund norm on $C[0, \omega_1]^*$. For $\alpha \in [0, \omega_1]$, let δ_α be the Dirac measure corresponding to α. Then the function $\alpha \mapsto |||\delta_\alpha|||$ is lower semicontinuous on $[0, \omega_1]$. Thus it is constant, equal to, say, a on a closed cofinal subset A of $[0, \omega_1)$. For $\alpha \in A$, let α' be the immediate successor of α in A. The map $\alpha \mapsto |||(\delta_\alpha + \delta_{\alpha'})/2|||$ is lower semicontinuous on A and thus equal to, say, b on a closed cofinal subset B of A. Let $\{a_n\}$ be a strictly increasing sequence in B such that $\alpha = \lim \alpha_n$. Using the facts that δ_α is the w^*-limit of $(\delta_{\alpha_n} + \delta_{\alpha_n'})/2$ and that $|||\cdot|||$ is w^*-lower semicontinuous, we get that $\lim |||(\delta_{\alpha_n} + \delta_{\alpha_n'})/2||| \ge |||\delta_\alpha|||$. Hence $b \ge a$. On the other hand, $|||(\delta_\alpha + \delta_{\alpha'})/2||| \le |||\delta_\alpha|||/2 + |||\delta_{\alpha'}|||/2 = |||\delta_\alpha|||$. Hence $a = b$ and, for every $\alpha \in B$, $|||(\delta_\alpha + \delta_{\alpha'})/2||| = (|||\delta_\alpha||| + |||\delta_{\alpha'}|||)/2 = a$. In particular, $|||\cdot|||$ is not rotund.

7.71 Let X be a reflexive Banach space with Gâteaux differentiable norm. If Y is a subspace of X and $h \in S_{Y^*}$, prove that there is a unique norm-1 extension of h to X.

Hint. The dual norm on X^* is strictly convex by Corollary 7.23. Apply now Exercise 7.69.

7.72 Let S be a nonempty subset of a Banach space X. Prove that an exposed point of S is an extreme point of S, and find a point that is extreme but not exposed.

Hint. We may assume without loss of generality that X is a real space. Let e_0 be an exposed point (exposed by $f \in X^*$) of S. Assume that $e_0 \in (x, y) \subset S$, where (x, y) denotes an open interval. Then f must be constant on (x, y). This is in contradiction with the fact that e_0 is exposed by f. Consider the convex hull of $B_{\ell_\infty^2}$ and the circle centered at $(1, 0)$ and having radius 1. Check the point $(1, 1)$.

7.73 (Lindenstrauss, [Lind4]) In ℓ_2, let $x_n = (1 - \frac{1}{n}, 0, \ldots, 1, 0 \cdots)$ for $n = 2, 3, \ldots$ (where the number 1 stands in the nth place). Let C be the closed convex hull of $\bigcup_{n=2}^{\infty} \{x_n\} \cup B_{\ell_2}$. Show that $e_1 := (1, 0, 0, \ldots)$ is an exposed point of C that is not strongly exposed.
Hint. The point e_1 is the only point in C whose first coordinate is 1. It is not strongly exposed, since the first coordinate of x_n tends to 1 and $\|e_1 - x_n\| \geq 1$.

7.74 Show that if x is a strongly exposed point of B_X, then it is a strongly exposed point of $B_{X^{**}}$. The same for a Fréchet smooth point.
Hint. The definition and Goldstine's Theorem 3.96.

7.75 (Kuratowski) If X is a metric space, E is a topological space and $\Phi : X \to E$ is a set-valued mapping, show that Φ is lower semicontinuous if and only if given any sequence $\{x_n\}$ in X such that $x_n \to x_0$ in X, and given any $y_0 \in \Phi(x_0)$, then there are $y_n \in \Phi(x_n)$ such that $y_n \to y_0$.
Hint. This is standard.

7.76 Let X be a separable Banach space and $f_1 \leq f_2$ be two real-valued functions on X such that f_1 is upper semicontinuous and f_2 is lower semicontinuous. Then there is a continuous real-valued function f on X such that $f_1 \leq f \leq f_2$. Note that this is the separable Banach space version of a theorem of Katetov that says that the former statement holds in any normal topological space X (see, e.g., [Jmsn, Theorem 12.16]).
Hint. Consider the mapping Φ from X into subsets of the real line defined for $x \in X$ by $\Phi(x) = [f_1(x), f_2(x)]$ and use Michael selection theorem.

7.77 Consider a Banach space $(X, \|\cdot\|)$ with Fréchet smooth norm. Let M be a closed subset of X such that $M \neq \emptyset$ and $M \neq X$. Define the *distance function* as follows

$$X \ni x \longmapsto f(x) = \inf\{\|x - y\| : y \in M\}.$$

Assume that the set $\{x \in X : f(x) = \|x - y\| \text{ for some } y \in M\}$ is dense in X. Prove that, then, the set of all $x \in X$ where f is Fréchet differentiable is dense in X.
Hint. Use Theorem 7.28. If $x \in X \backslash M$ is such that $f(x) = \|x - y\|$ for some $y \in M$, then $Df\big((\lambda x + (a - \lambda)y\big)(x - y) = 1 = L_f(\lambda x + (a - \lambda)y\big)$ whenever $0 < \lambda < 1$, where L_f is defined in the statement of Theorem 7.28.

Chapter 8
C^1-Smoothness in Separable Spaces

In this chapter we study separable Asplund spaces, i.e., Banach spaces with a separable dual space. These spaces admit many equivalent characterizations, in particular by means of C^1-smooth renormings and differentiability properties of convex functions. Asplund spaces also play an important role in applications. We study basic results in smooth approximation and ranges of smooth nonlinear operators.

8.1 Smoothness and Renormings in Separable Spaces

The concept of a locally uniformly rotund norm (LUR, in short), see Definition 7.9, was introduced by Lovaglia [Lov].

Theorem 8.1 (Kadec [Kade3], [Kade4], [DaJo], [Klee2], see also [LiTz3, p. 12]) *Every separable Banach space admits an equivalent locally uniformly rotund norm.*

Proof: ([DaJo]) Let $(X, \|\cdot\|)$ be a separable Banach space. Let $\{y_n : n \in \mathbb{N}\} \subset S_X$ be dense in S_X and let $\{f_n : n \in \mathbb{N}\} \subset S_{X^*}$ be a separating family for X. For $n \in \mathbb{N}$, put $F_n = \operatorname{span}\{y_1, \ldots, y_n\}$ and note that $\operatorname{dist}(x, F_n) \to 0$ for all $x \in X$. Define a norm $|\!|\!|\cdot|\!|\!|$ on X by

$$|\!|\!|x|\!|\!|^2 = \|x\|^2 + \sum_{n=1}^{\infty} 2^{-n} \operatorname{dist}(x, F_n)^2 + \sum_{n=1}^{\infty} 2^{-n} f_n^2(x),$$

where the distance is in the norm $\|\cdot\|$. As $\operatorname{dist}(x, F_i)$ is a positive homogeneous subadditive function, $|\!|\!|\cdot|\!|\!|$ is an equivalent norm on X.

We will show that $|\!|\!|\cdot|\!|\!|$ is LUR. To this end, assume that $x_k, x \in X$ are such that $\lim_{k\to\infty} (2|\!|\!|x|\!|\!|^2 + 2|\!|\!|x_k|\!|\!|^2 - |\!|\!|x + x_k|\!|\!|^2) = 0$.

Since the expression in this limit is non-negative for all the functions involved in the definition of $|\!|\!|\cdot|\!|\!|$, it implies the existence of the following limits:

M. Fabian et al., *Banach Space Theory*, CMS Books in Mathematics,
DOI 10.1007/978-1-4419-7515-7_8, © Springer Science+Business Media, LLC 2011

$$\lim_{k\to\infty}\left(2\|x\|^2+2\|x_k\|^2-\|x+x_k\|^2\right)=0,$$
$$\lim_{k\to\infty}\left(2\operatorname{dist}(x,F_n)^2+2\operatorname{dist}(x_k,F_n)^2-\operatorname{dist}(x+x_k,F_n)^2\right)=0 \text{ for every } n,$$
$$\lim_{k\to\infty}\left(2f_n^2(x)+2f^2(x_k)-f_n^2(x+x_k)\right)=0 \text{ for every } n.$$

As in the proof of (ii)$\Longrightarrow$(iii) in Fact 7.7, we conclude that then

$$\lim_{k\to\infty}\|x_k\|=\|x\|, \tag{8.1}$$
$$\lim_{k\to\infty}\left(\operatorname{dist}(x_k,F_n)\right)=\operatorname{dist}(x,F_n) \text{ for every } n, \tag{8.2}$$
$$\lim_{k\to\infty}\left(f_n(x_k)\right)=f_n(x) \text{ for every } n. \tag{8.3}$$

Since $\{f_n\}$ is separating, the topology of pointwise convergence on $\{f_n\}$ is a Hausdorff topology in X. We will show below that $\overline{\{x_k\}\cup\{x\}}$ is norm compact. Therefore on $\overline{\{x_k\}\cup\{x\}}$ the topology of pointwise convergence on $\{f_n\}$ is equivalent to the norm topology. Thus (8.3) implies that $\lim\|x_k-x\|=0$ and the proof is complete. It remains to show that $\overline{\{x_k\}\cup\{x\}}$ is norm compact.

Using (8.1), choose $K>0$ such that $\|x_k\|\le K$ for every k. Let $\varepsilon\in(0,1)$ be given. Choose $n\in\mathbb{N}$ such that $\operatorname{dist}(x,F_n)<\varepsilon$ and choose a finite ε-net F in $(K+1)B_{F_n}$. Using (8.2), choose k_0 such that $\operatorname{dist}(x_k,F_n)<\varepsilon$ for $k>k_0$. We claim that $\{x_1,x_2,\dots,x_{k_0}\}\cup F$ is a 2ε-net for $\{x_k\}$. Indeed, for every $k>k_0$ there is $x_k'\in F_n$ such that $\|x_k-x_k'\|<\varepsilon$. Since $\|x_k\|\le K$ for every k and $\varepsilon<1$, we have that $\|x_k'\|<K+1$. As F is an ε-net for $(K+1)B_{F_n}$, there is $x_k''\in F$ such that $\|x_k'-x_k''\|<\varepsilon$. This completes the proof. □

The second statement in the following theorem is due Mazur [Mazu2].

Theorem 8.2 *Every separable Banach space X admits an equivalent Gâteaux differentiable locally uniformly rotund norm and every continuous convex function on X is Gâteaux differentiable at the points of a G_δ-dense subset of X.*

Proof: ([JoZi3]) Let $\|\cdot\|$ be an equivalent LUR norm on X (Theorem 8.1). Let $\{y_i : i\in\mathbb{N}\}$ be a dense subset of S_X. We define equivalent dual norms $\|\cdot\|_n^*$ on X^* by

$$\|f\|_n^{*2}=(\|f\|^*)^2+\frac{1}{n}\Big(\sum_{i=1}^{\infty}2^{-i}f^2(y_i)\Big).$$

Then $\|\cdot\|_n^*$ are R and their predual norms $\|\cdot\|_n$ converge to $\|\cdot\|$ uniformly on bounded sets. Define a norm $\|\cdot\|_0$ on X by $\|x\|_0^2=\sum_{n=1}^{\infty}2^{-n}\|x\|_n^2$ for $x\in X$. Then $\|\cdot\|_0$ is an equivalent Gâteaux differentiable norm on X. The differentiability follows by a standard argument as the derivatives of the norms $\|\cdot\|_n$ on X are uniformly bounded. We will check that $\|\cdot\|_0$ is LUR. To this end, let $\lim\limits_{k\to\infty}(2\|x_k\|_0^2+2\|x\|_0^2-\|x+x_n\|_0^2)=0$. Then $\{x_k\}$ is bounded and a similar limit relation is true for each norm $\|\cdot\|_n$ as shown in Theorem 8.1. Since the norms

$\|\cdot\|_n$ converge uniformly on bounded sets to $\|\cdot\|$, we get by the standard limit interchanging rule that $\lim_{k\to\infty}(2\|x\|^2+2\|x_k\|^2-\|x+x_k\|^2)=0$. As $\|\cdot\|$ is LUR, we get $\|x-x_k\|\to 0$.

As in the Fréchet case, the smooth variational principle implies that every convex continuous function f is Gâteaux differentiable at a dense set of points. From the convexity of f, we get that the set of all points where f is Gâteaux differentiable is equal to $\bigcap_{n.m} G_{n,m}$, where

$$G_{n,m} := \left\{x \in X : \exists \delta > 0 \text{ so that } f(x+\delta x_m) - f(x-\delta x_m) - 2f(x) < \frac{\delta}{n}\right\}.$$

Each $G_{n,m}$ is open in X and thus the set of all points where f is Gâteaux differentiable is a G_δ-set. □

8.2 Equivalence of Separable Asplund Spaces

In order to formulate Theorem 8.6 below, we need two new concepts: the notion of a *rough* norm, and the notion of *Szlenk index.*

Definition 8.3 *Let $\delta > 0$. A norm $\|\cdot\|$ on a normed space X is called δ-*rough *if for all $x \in X$,*

$$\limsup_{\|h\|\to 0,\ h\neq 0} \frac{\|x+h\|+\|x-h\|-2\|x\|}{\|h\|} \geq \delta.$$

The norm is said to be rough *if it is δ-rough for some $\delta > 0$.*

Note that the canonical norm of ℓ_1 is 2-rough, and that a rough norm is nowhere Fréchet differentiable.

We characterize now δ-roughness by the diameter of the w^*-sections of the dual unit ball.

Proposition 8.4 *Let $(X, \|\cdot\|)$ be a normed space and let $\delta > 0$. The norm $\|\cdot\|$ is δ-rough if and only if for every $x \in S_X$,* $\operatorname{diam}\{x^* \in B_{X^*} : \langle x^*, x\rangle \geq 1-\alpha\} \geq \delta$ *for every $\alpha > 0$.*

Proof: Recall that Lemma 7.19 ensures that

$$\{x^* \in B_{X^*} : \langle x^*, x\rangle \geq 1-\alpha\} = \partial_\alpha\|\cdot\|(x), \text{ for every } \alpha > 0.$$

Assume that $\|\cdot\|$ is δ-rough. Fix $x \in S_X$. Then there exists a sequence $\{h_n\}_{n=1}^\infty$ in $X\backslash\{0\}$ such that $h_n \to 0$ and

$$\|x+h_n\|+\|x-h_n\|-2 \geq (\delta-1/n)\|h_n\|, \text{ for all } n \in \mathbb{N}. \tag{8.4}$$

Take, for each $n \in \mathbb{N}$, $x_n^* \in \partial\|\cdot\|(x+h_n)$ $\big(\subset \partial_{2\|h_n\|}\|\cdot\|(x)\big)$ and $y_n^* \in \partial\|\cdot\|(x-h_n)$ $\big(\subset \partial_{2\|h_n\|}\|\cdot\|(x)\big)$ (the inclusions come from Lemma 7.13).

Then

$$\langle x_n^*, x + h_n \rangle + \langle y_n^*, x - h_n \rangle = \|x + h_n\| + \|x - h_n\| \geq (\delta - 1/n)\|h_n\| + 2.$$

This implies

$$\langle x_n^* - y_n^*, h_n \rangle \geq \left(\delta - \frac{1}{n}\right)\|h_n\| + 2 - \langle x_n^*, x \rangle - \langle y_n^*, x \rangle$$

$$\geq \left(\delta - \frac{1}{n}\right)\|h_n\| + 2 - 1 - 1 = \left(\delta - \frac{1}{n}\right)\|h_n\|,$$

hence $\|x_n^* - y_n^*\| \geq (\delta - \frac{1}{n})$, so $\operatorname{diam} \partial_{2\|h_n\|}\|\cdot\|(x) \geq (\delta - \frac{1}{n})$. Since $\delta_\alpha\|\cdot\|(x) \subset \delta_\beta\|\cdot\|(x)$ for all $0 \leq \alpha \leq \beta$, we get the conclusion.

Assume now that all w^*-sections of B_{X^*} have diameter greater than or equal to δ. Fix $x \in S_X$ and $n \in \mathbb{N}$. Then there exist x_n^* and y_n^* in $\partial_{1/n^2}\|\cdot\|(x)$ such that $\|x_n^* - y_n^*\| > \delta - 1/n$. Therefore we can find $h_n \in S_X$ such that $\langle x_n^* - y_n^*, h_n \rangle > \delta - 1/n$. We have

$$\left\|x + \frac{h_n}{n}\right\| - \|x\| \geq \left\langle x_n^*, \frac{h_n}{n} \right\rangle - \frac{1}{n^2}, \tag{8.5}$$

$$\left\|x - \frac{h_n}{n}\right\| - \|x\| \geq \left\langle x_n^*, -\frac{h_n}{n} \right\rangle - \frac{1}{n^2}. \tag{8.6}$$

Adding Equations (8.5) and (8.6) we get

$$\left\|x + \frac{h_n}{n}\right\| - \left\|x - \frac{h_n}{n}\right\| - 2\|x\| \geq \frac{1}{n}\langle x_n^* - y_n^*, h_n \rangle - \frac{2}{n^2},$$

so

$$\frac{\left\|x + \frac{h_n}{n}\right\| - \left\|x - \frac{h_n}{n}\right\| - 2\|x\|}{\frac{1}{n}} \geq \langle x_n^* - y_n^*, h_n \rangle - \frac{2}{n} \geq \delta - \frac{3}{n}.$$

Since this is true for every $n \in \mathbb{N}$ and for every $x \in S_X$, we get that $\|\cdot\|$ is δ-rough.

$\square$

Definition 8.5 *Assume that X is an infinite-dimensional Banach space and K be a w^*-compact subset of X^*. For any $\varepsilon > 0$ let Γ be the set of all w^*-open subsets V of K such that the norm-diameter of V is less than ε, and put $S_\varepsilon^0(K) = K$, $S_\varepsilon(K) = K \setminus \bigcup\{V : V \in \Gamma\}$. Then define inductively $S_\varepsilon^\alpha(K)$ for any ordinal α by $S_\varepsilon^{\alpha+1}(K) = S_\varepsilon(S_\varepsilon^\alpha(K))$ and $S_\varepsilon^\alpha(K) = \bigcap_{\beta<\alpha} S_\varepsilon^\beta(K)$ if α is a limit ordinal. The symbol $Sz(X, \varepsilon)$ denotes the least ordinal α such that $S_\varepsilon^\alpha(B_{X^*}) = \emptyset$ if such an ordinal exists. Otherwise we write $Sz(X, \varepsilon) = \infty$. The* Szlenk index *is defined by $Sz(X) = \sup_{\varepsilon>0} Sz(X, \varepsilon)$. If K is a w^*-compact convex set in X^*, we call* *a* w^*-slice *of* K *any nonempty set of the form* $S = \{x^* \in K : x^*(x) > t\}$, *where* $x \in X$ *and* $t \in \mathbb{R}$. *We denote by* Φ *the set of all* w^**-slices of* K *of norm-diameter less than* ε *and put* $D_\varepsilon^0(K) = K$, $D_\varepsilon(K) = K \setminus \bigcup\{S : S \in \Phi\}$. *Then, similarly as in the Szlenk index case, we define* $D_\varepsilon^\alpha(K)$, $D_Z(X, \varepsilon)$ *and put* $D_Z(X) := \sup_{\varepsilon>0} D_Z(X, \varepsilon)$ *and call it the* dual w^*-dentability index *of* X.

Note that, because of the w^*-compactness of each $S_\varepsilon^\alpha(K)$, the ordinal $S_Z(X, \varepsilon)$ cannot be a limit ordinal. This, in particular, implies that if $S_Z(X) = \omega_0$ then $S_Z(X, \varepsilon)$ is finite for each $\varepsilon > 0$.

The following result, due to many mathematicians, presents a set of equivalent conditions for the separability of the dual space of a Banach space that are related to smoothness.

From the pioneering papers in this direction, let us mention: [Kurz], [Day1], [Lind4] [Aspl], and [LeWh]. In particular, it was proved in [Kurz] that $C[0, 1]$ does not admit any Fréchet differentiable bump function. Similar ideas were independently found in [Day1].

Recall that (v) below means that the space X is Asplund (see Definition 7.38). Some of the conditions will also be discussed later in Chapters 11, 13, and 14. Note that not all conditions are equivalent in the setting of non-separable Banach spaces. For example, the space in Exercise 14.63 satisfies (v) and not (iii), as it is shown there.

Theorem 8.6 *Let* X *be a separable Banach space. Then the following are equivalent.*

(i) X^* *is separable.*
(ii) X *admits an equivalent norm whose dual norm is LUR.*
(iii) X *admits an equivalent Fréchet differentiable norm.*
(iv) X *admits a Lipschitz continuously Fréchet differentiable bump function.*
(iv') X *admits a Fréchet differentiable bump function.*
(v) *Every continuous convex function on* X *is Fréchet differentiable at every point of a dense* G_δ *set in* X.
(vi) *There is no equivalent rough norm on* X.
(vii) *The Szlenk index of* X *is well defined (and it is less than* ω_1 *in this case).*
(viii) *The dual dentability index of* X *is well defined (and it is less than* ω_1 *in this case).*
(ix) X *admits an equivalent weakly uniformly rotund norm.*
(x) X *admits an equivalent norm that is locally uniformly rotund, weakly uniformly rotund and Fréchet differentiable.*
(xi) X *admits an equivalent norm so that the identity mapping from* (B_{X^*}, w^*) *to* $(B_{X^*}, \|\cdot\|)$ *is the pointwise limit of a sequence of* w^*-$\|\cdot\|$*-continuous mappings. (Then all equivalent norms on* X *have this property.)*

We shall proceed with the following chain of implications (see Fig. 8.1).

Proof: (i)$\Longrightarrow$(ii): We shall need the following result, the version of Theorem 8.1 for the dual space (under the extra assumption of separability of the dual). The proof is

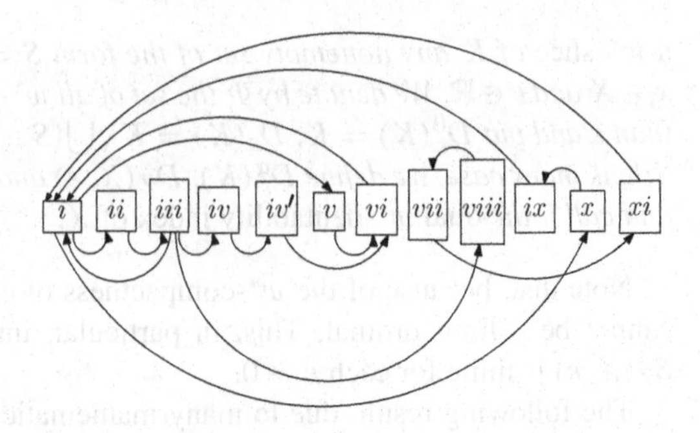

Fig. 8.1 Paths to prove Theorem 8.6

very similar to the one provided for that theorem. We include it here for the sake of completeness.

Theorem 8.7 (Klee, Kadec, and Asplund [Klee2], [Kade4], [Aspl]) *If X is a Banach space and X^* is separable, then X admits an equivalent norm whose dual norm is* LUR.

Proof: Let $\{x_n : n \in \mathbb{N}\}$ be a dense subset of S_X, and let $\{f_n : n \in \mathbb{N}\}$ be a norm-dense subset of S_{X^*}. For $n \in \mathbb{N}$, put $F_n = \operatorname{span}\{f_1, \dots, f_n\}$ and note that $\operatorname{dist}(f, F_n) \to 0$ for all $f \in X^*$. Define a norm $|||\cdot|||$ on X^* by

$$|||f|||^2 = \|f\|^2 + \sum_{n=1}^{\infty} 2^{-n} \operatorname{dist}(f, F_n)^2 + \sum_{n=1}^{\infty} 2^{-n} f^2(x_n), \quad f \in X^*,$$

where $\|\cdot\|$ is the original dual norm of X^* and the distance is in the norm $\|\cdot\|$. As $\operatorname{dist}(f, F_i)$ is a positive homogeneous subadditive function, $|||\cdot|||$ is an equivalent norm on X^*.

We claim that $|||\cdot|||$ is w^*-lower semicontinuous. Indeed, note that $F_n + B_{X^*}$ is w^*-closed as F_n is w^*-closed and B_{X^*} is w^*-compact by Alaoglu's theorem. Since $\{f \in X^* : \operatorname{dist}(f, F_n) \le 1\} = F_n + B_{X^*}$, the functions $\operatorname{dist}(f, F_n)$ are w^*-lower semicontinuous and so is the supremum of their weighted partial sums. Consequently $|||\cdot|||$ is w^*-lower semicontinuous and thus it is the dual norm to an equivalent norm (denoted again by $|||\cdot|||$) on X by Lemma 3.97.

We will show that $|||\cdot|||$ in X^* is LUR. To this end, assume that $f_k, f \in X^*$ are such that $\lim_{k\to\infty} (2|||f|||^2 + 2|||f_k|||^2 - |||f + f_k|||^2) = 0$.

Since the expression in this limit is non-negative for all the functions involved in the definition of $|||\cdot|||$, it implies the existence of the following limits:

$$\lim_{k\to\infty} \left(2\|f\|^2 + 2\|f_k\|^2 - \|f + f_k\|^2\right) = 0,$$

$$\lim_{k\to\infty} \left(2\operatorname{dist}(f, F_n)^2 + 2\operatorname{dist}(f_k, F_n)^2 - \operatorname{dist}(f + f_k, F_n)^2\right) = 0 \text{ for every } n,$$

$$\lim_{k\to\infty} \left(2f^2(x_n) + 2f_k^2(x_n) - (f + f_k)^2(x_n)\right) = 0 \text{ for every } n.$$

As in the proof of (ii)$\Longrightarrow$(iii) in Fact 7.7, we conclude that then

$$\lim_{k\to\infty} \|f_k\| = \|f\|, \tag{8.7}$$

$$\lim_{k\to\infty} \operatorname{dist}(f_k, F_n) = \operatorname{dist}(f, F_n) \text{ for every } n, \tag{8.8}$$

$$\lim_{k\to\infty} f_k(x_n) = f(x_n) \text{ for every } n. \tag{8.9}$$

Since $\{x_n : n \in \mathbb{N}\}$ is dense in S_X, the topology of pointwise convergence on $\{x_n : n \in \mathbb{N}\}$ is a Hausdorff topology in X^*. We will show below that $\overline{\{f_k\} \cup \{f\}}$ is norm-compact. Therefore on $\overline{\{f_k\} \cup \{f\}}$ the topology of pointwise convergence on $\{x_n\}$ is equivalent to the norm topology. Thus (8.9) implies that $\lim \|f_k - f\| = 0$ and the proof is complete. It remains to show that $\overline{\{f_k\} \cup \{f\}}$ is norm-compact.

Using (8.7), choose $K > 0$ such that $\|f_k\| \le K$ for every k. Let $\varepsilon \in (0, 1)$ be given. Choose $n \in \mathbb{N}$ such that $\operatorname{dist}(f, F_n) < \varepsilon$ and choose a finite ε-net F in $(K+1)B_{F_n}$. Using (8.8), choose k_0 such that $\operatorname{dist}(f_k, F_n) < \varepsilon$ for $k > k_0$. We claim that $\{f_1, f_2, \dots, f_{k_0}\} \cup F$ is a 2ε-net for $\{f_k\}$. Indeed, for every $k > k_0$ there is $f_k' \in F_n$ such that $\|f_k - f_k'\| < \varepsilon$. Since $\|f_k\| \le K$ for every k and $\varepsilon < 1$, we have that $\|f_k'\| < K + 1$. As F is an ε-net for $(K+1)B_{F_n}$, there is $f_k'' \in F$ such that $\|f_k' - f_k''\| < \varepsilon$. This completes the proof. □

This concludes the proof of (i)⟹(ii).

(ii)⟹(iii) is contained in Corollary 7.25.

(iii)⟹(i) is the following theorem.

Theorem 8.8 (Kadec [Kade4], Restrepo [Rest]) *Let X be a separable Banach space. If X admits an equivalent Fréchet differentiable norm then X^* is separable .*

Proof: Observe that the set $B = \{\|x\|' : x \in X,\ x \neq 0\}$ is norm-separable, where $\|x\|'$ denotes the derivative of $\|\cdot\|$ at x (see Corollary 7.24). The set B contains all norm-attaining functionals, and is thus norm-dense in X^* by the Bishop–Phelps theorem. □

(iii)⟹(iv): See Corollary 7.24 and Fact 10.4.

(iv)⟹(iv') is trivial.

(iv')⟹(vi): This is a consequence of the following result.

Lemma 8.9 *Let X be a Banach space and $\varepsilon > 0$. Suppose that there is on X an equivalent ε-rough norm $\|\cdot\|$. Then there is no Fréchet differentiable bump function on X.*

Proof: Assume that there is a Fréchet differentiable bump function g on X. Let $f : X \to \mathbb{R} \cup \{+\infty\}$ be defined by:

$$f(x) = \begin{cases} g(x)^{-2} - \|x\| & \text{if } g(x) \neq 0 \\ +\infty & \text{otherwise.} \end{cases}$$

The function f satisfies the hypothesis of Ekeland's Theorem 7.39 and thus there exists $x_0 \in D(f)$ such that

$$f(x_0+h) \geq f(x_0) - \frac{\varepsilon}{4}\|h\|, \quad \text{for every } h \in X.$$

For $\|h\|$ small enough, $x_0 + h$ and $x_0 - h$ are in $D(f)$, and if we write down the above inequality for h and for$-h$ and add them, we obtain

$$g(x_0+h)^{-2} + g(x_0-h)^{-2} - \|x_0+h\| - \|x_0-h\| \geq 2g(x_0)^{-2} - 2\|x_0\| - \frac{\varepsilon}{2}\|h\|.$$

Hence

$$\limsup_{\|h\|\to 0} \frac{g(x_0+h)^{-2} + g(x_0-h)^{-2} - 2g(x_0)^{-2}}{\|h\|} \geq$$
$$\limsup_{\|h\|\to 0} \frac{\|x_0+h\| + \|x_0-h\| - 2\|x_0\|}{\|h\|} - \frac{\varepsilon}{2} > 0.$$

This contradicts the Fréchet differentiability of g at x_0 (see Lemma 7.4). □

(i)⟹(v) (Lindenstrauss [Lind4], Asplund, [Aspl]): Assume that f is a convex continuous function on X, $x_0 \in X$ and $\delta > 0$ be such that f is Lipschitz on $B(x_0, \delta)$ (use Lemma 7.3). For $x \in X$ define a function ψ by

$$\psi(x) = \begin{cases} -f(x), & \text{if } x \in B(x_0, \delta), \\ +\infty & \text{otherwise.} \end{cases}$$

Let b be a nonnegative C^1-smooth Lipschitz function on X such that $b(x_0) - f(x_0) > 0$ and $b(x) = 0$ if $x \notin B(x_0, \delta/2)$ (use the equivalence of (i) and (iv) in this theorem). Then $b + \psi$ is lower semicontinuous and bounded below. By the smooth variational principle (Theorem 7.43), there is a Lipschitz and C^1-smooth function g on X such that $b + \psi + g$ attains its minimum at some point z_0. Clearly, $z_0 \in B(x_0, \delta/2)$.

So, for every $x \in X$,

$$b(x) + \psi(x) + g(x) \geq b(z_0) + \psi(z_0) + g(z_0).$$

On a neighborhood U of z_0 we thus have

$$b(x) - f(x) + g(x) \geq b(z_0) - f(z_0) + g(z_0),$$

so,

$$f(x) \leq b(x) + g(x) - b(z_0) + f(z_0) - g(z_0).$$

Put

$$h(x) = b(x) + g(x) - b(z_0) + f(z_0) - g(z_0)$$

for $x \in U$.

Then $f(x) \le h(x)$ on U and $f(z_0) = h(z_0)$ and f is convex. Therefore, by Exercise 7.9 we get that f is Fréchet differentiable at z_0. The last statement in (v) follows by using Lemma 7.45.

For an alternative proof of (i)$\Longrightarrow$(v) see Exercise 8.17.

(v)$\Longrightarrow$(vi) is trivial.

(vi)$\Longrightarrow$(i): Assume that X is separable and X^* is not separable. An equivalent rough norm on X is constructed in the proof of the following result.

Theorem 8.10 (Leach, Whitfield [LeWh]) *Let X be a separable Banach space. If X^* is not separable, then X admits an equivalent norm that is rough.*

Proof: Since B_{X^*} is not norm-separable, there is n such that B_{X^*} contains no countable $\frac{1}{n}$-net. Thus a maximal $\frac{1}{n}$-separated set S obtained by Zorn's lemma is uncountable. On the other hand, B_{X^*} in the w^*-topology is a metrizable compact space, and thus a separable space. Therefore (B_{X^*}, w^*) is second countable.

Let $\mathcal{A} = \{A_k\}$ be a countable base for (B_{X^*}, w^*) (i.e., every w^*-open set is a union of a subcollection of $\mathcal{A}$). Collect all A_k such that $S \cap A_k$ is countable. Denote the collection of these A_k by $\mathcal{B}$. Let $\tilde{S}$ be the set of all elements of S that lie in some of A_k, $A_k \in \mathcal{B}$. Clearly, $\tilde{S}$ is countable, and if $s \in S \backslash \tilde{S}$, then every w^*-neighborhood of s contains an uncountable number of elements of S.

Let $C = \overline{\operatorname{conv}}^{w^*}\big[(S\backslash\tilde{S}) \cup \big(-(S\backslash\tilde{S})\big)\big]$ and $U^* = B_{X^*} + C$. Both B_{X^*} and C are w^*-compact, so U^* is a w^*-compact convex and symmetric set. Its Minkowski functional is thus w^*-lower semicontinuous, hence U^* is the dual ball of some equivalent norm $\|\cdot\|$ on X by Lemma 3.97. We will show that $\|\cdot\|$ is nowhere Fréchet differentiable. It is enough to show that every set of the form $U^* \cap \{f : f(x) > \|x\| - \delta\}$, where $x \in X$ and $\delta > 0$ are arbitrary, has diameter greater than or equal to $\frac{1}{n}$.

Let D be such a set. Then D contains an element of the form $b + c$, where $b \in B_{X^*}$ and $c = \sum \lambda_i c_i$ with $\lambda_i \ge 0$, $\sum \lambda_i = 1$ and $c_i \in (S\backslash\tilde{S}) \cup \big(-(S\backslash\tilde{S})\big)$. Thus $b + c = \sum \lambda_i (b + c_i)$. Since $\{f : f(x) > \|x\| - \delta\} \subset X^*$ contains a convex combination of points $b + c_i$, it must contain at least one of them. Therefore, there is i such that $b + c_i$ is in D. Since $\{f : f(x) > \|x\| - \delta - b(x)\}$ is a w^*-neighborhood of c_i, it also contains $d \in B_{X^*}$ with $\|d - c_i\| > \frac{1}{n}$. Then $\|(b + c_i) - (b + d)\| > \frac{1}{n}$ and $(b + d) \in D$, which completes the proof. $\square$

(iii)$\Longrightarrow$(viii): We need first the following result. Although it is a particular case of a later one (note that we already proved (iii)$\Longrightarrow$(i)$\Longrightarrow$(v) here, so every Banach space with a Fréchet differentiable norm is Asplund and we can use now (iv)$\Longrightarrow$(i) in Proposition 11.8), we shall present it here for keeping the proof of Theorem 8.6 self-contained. The definition of a weak*-dentable space will be given at the beginning of Section 11.2.

Proposition 8.11 *If X is a Banach space with a Fréchet differentiable norm, then X^* is w^*-dentable.*

Proof: Let M be a nonempty bounded subset of X^* (that we may, without loss of generality, assume inside B_{X^*}). Fix $\varepsilon > 0$. Put $r = \sup\{\|x^*\| : x^* \in M\}$. Fix

$n \in \mathbb{N}$. We can find $x^* \in M$ such that $\|x^*\| > r - 1/n$. By the Bishop–Phelps theorem, there exists $x_0^* \in X^*$ so that $\|x^* - x_0^*\| < 1/n$ and $x_0 \in S_X$ such that $\langle x_0^*, x_0\rangle = \|x_0^*\|$. We shall prove that, by choosing $n \in \mathbb{N}$ small enough, a certain section of M defined by x_0 is nonempty and has diameter less than ε. Let $d^* \in M$ such that $\langle d^*, x_0\rangle > \sup_M x_0 - 1/n$. Then

$$\begin{aligned}\langle d^*, x_0\rangle &> \sup_M x_0 - \frac{1}{n} \geq \langle x^*, x_0\rangle - \frac{1}{n}\\ &\geq \langle x_0^*, x_0\rangle - \frac{2}{n} = \|x_0^*\| - \frac{2}{n} \geq \|x^*\| - \frac{3}{n} \geq r - \frac{4}{n}.\end{aligned} \tag{8.10}$$

Certainly, $\sup_{rB_{X^*}} x_0 = r$, and $M \subset rB_{X^*}$. Then, Equation (8.10) shows that $d^* \in \{z^* \in rB_{X^*} : \langle z^*, x_0\rangle > \sup_{rB_{X^*}} x_0 - 4/n\}$. Since $\|\cdot\|$ is Fréchet differentiable at x_0 it follows that, for some $n \in \mathbb{N}$, the diameter of the section $\{z^* \in rB_{X^*} : \langle z^*, x_0\rangle > \sup_{rB_{X^*}} x_0 - 4/n\}$ of the ball rB_{X^*} is less than ε, hence the diameter of the section $\{d^* \in M : \langle d^*, x_0\rangle > \sup_M x_0 - 1/n\}$ is also less than ε. □

In the next few steps we follow [Lanc2].

To finalize the proof of the implication (iii)$\Longrightarrow$(viii) observe that, due to Proposition 8.11, for any $\varepsilon > 0$, $\{D_\varepsilon^\alpha(B_{X^*})\}_\alpha$ is a strictly decreasing family of w^*-closed subsets of B_{X^*} (see Definition 8.5). Since (B_{X^*}, w^*) is separable, we get that for any $\varepsilon > 0$, $D_Z(X, \varepsilon) < \omega_1$. Thus, $D_Z(X) = \sup_n D_Z(X, 1/n) < \omega_1$.

(viii)$\Longrightarrow$(vii) is trivial.

(vii)$\Longrightarrow$(xi): Let $\Phi = \mathrm{Id} : (B_{X^*}, w^*) \to (B_{X^*}, \|\cdot\|)$. If $S_Z(X) < \omega_1$, then it follows that for any nonempty w^*-closed subset F of B_{X^*}, the restriction $\Phi\big|_F$ has a point of continuity. Indeed, otherwise, by Baire's category theorem, there would exist a w^*-closed subset F of B_{X^*} and $n \in \mathbb{N}$ such that the set F_n of all x^* in F such that the oscillation of $\Phi\big|_F$ at x^* is at least $1/n$, has a nonempty $\|\cdot\|$-interior in F (the oscillation at x^* is defined by $\inf_{U\in\mathcal{U}}\{\operatorname{diam} \Phi\big|_F U\}$, where the infimum is taken on all neighborhoods of x^* in the w^*-topology of F).

Thus, by induction, we get that for any ordinal α and any $\varepsilon < 1/n$, the interior of F_n is included in $S_\varepsilon^\alpha(F)$ and therefore in $S_\varepsilon^\alpha(B_{X^*})$, which is a contradiction with $S_Z(X) < \omega_1$. Therefore, by Baire's Great Theorem 17.12, Φ is the pointwise limit of a sequence of w^*-to-norm continuous mappings.

(xi)$\Longrightarrow$(i): Let (f_n) be a sequence of continuous functions from (B_{X^*}, w^*) into $(X^*, \|\cdot\|)$ such that for any $x^* \in B_{X^*}$, $\|f_n(x^*) - x^*\| \to 0$. Since B_{X^*} is w^*-separable, we have that for every $n \in \mathbb{N}$, $f_n(B_{X^*})$ is norm-separable. Thus B_{X^*}, which is included in the norm-closure of $\bigcup_n f_n(B_{X^*})$, is norm-separable.

(ix)$\Longrightarrow$(i): This is contained in the following result.

Theorem 8.12 ([Haje2]) *If the norm of a separable Banach space X is WUR, then X^* is separable.*

Proof: (Godefroy) For $n \in \mathbb{N}$ put

$$V_n = \left\{f \in B_{X^*} : |f(x - y)| \leq \tfrac{1}{3} \text{ for } x, y \in B_X \text{ such that } \|x + y\| \geq 2 - \tfrac{1}{n}\right\}.$$

Note that $B_{X^*} = \bigcup_{n\in\mathbb{N}} V_n$.

Since (B_{X^*}, w^*) is a metric compact space, for every $n \in \mathbb{N}$ there is a countable w^*-dense set S_n in V_n. We claim that $\overline{\operatorname{span}}\{\bigcup_{n\in\mathbb{N}} S_n\} = X^*$.

Assume that this is not the case and find $F \in S_{X^{**}}$ such that $F(f) = 0$ for all $f \in \bigcup_{n\in\mathbb{N}} S_n$. Choose $f_0 \in S_{X^*}$ such that $F(f_0) > \frac{8}{9}$, and let $n_0 \in \mathbb{N}$ such that $f_0 \in V_{n_0}$. Choose a net $\{x_\alpha\}$ in B_X such that $x_\alpha \overset{w^*}{\rightarrow} F$ (Goldstine). From the w^*-lower semicontinuity of the second dual norm (see Exercise 3.31), we can find α_0 so that $\|x_\alpha + x_\beta\| \geq 2 - \frac{1}{n}$ for every $\alpha, \beta \geq \alpha_0$, hence $|f(x_\alpha) - f(x_\beta)| \leq \frac{1}{3}$ for all $\alpha, \beta \geq \alpha_0$ and $f \in V_{n_0}$.

As $x_\beta \overset{w^*}{\rightarrow} F$, we have $|f(x_{\alpha_0}) - F(f)| \leq \frac{1}{3}$ for all $f \in V_{n_0}$. Thus for $f \in S_{n_0}$, we have

$$\begin{aligned}|(f - f_0)(x_{\alpha_0})| &= |F(f) - F(f_0) + f(x_{\alpha_0}) - F(f) + F(f_0) - f_0(x_{\alpha_0})|\\ &\geq |F(f) - F(f_0)| - |f(x_{\alpha_0}) - F(f)| - |F(f_0) - f_0(x_{\alpha_0})| \geq \tfrac{8}{9} - \tfrac{6}{9} = \tfrac{2}{9},\end{aligned}$$

which contradicts the fact that $f_0 \in \overline{S_{n_0}}^{w^*}$. □

(i)⟹(x): This implication will be proved in several steps. We start with a Banach space $(X, \|\cdot\|)$ with a separable dual.

First, thanks to the separability of X there exists, by Theorem 8.1, an *equivalent LUR norm* $\|\cdot\|_1$ *on* X.

Second, we shall construct *an equivalent WUR norm* $\|\cdot\|_2$ *on* X. To this end, let $\{f_n : n \in \mathbb{N}\}$ be a $\|\cdot\|$-dense subset of S_{X^*}. Define $\|\cdot\|_2$ on X by

$$\|x\|_2^2 = \|x\|^2 + \sum_{n=1}^{\infty} \frac{f_n^2(x)}{2^n}, \quad x \in X.$$

Assume that $\{x_n\}_{n\in\mathbb{N}}$ and $\{y_n\}_{n\in\mathbb{N}}$ are bounded sequences in X such that $2\|x_n\|_2^2 + 2\|y_n\|_2^2 - \|x_n + y_n\|_2^2 \to 0$ as $n \to \infty$. Then, for each $i \in \mathbb{N}$, $f_i(x_n) - f_i(y_n) \to 0$ as $n \to \infty$. This gives $x_n - y_n \overset{w}{\rightarrow} 0$ and proves that $\|\cdot\|_2$ on X is WUR.

Third, let $\|\cdot\|_3$ on X be defined by

$$\|x\|_3^2 = \|x\|_1^2 + \|x\|_2^2, \quad \text{for every } x \in X.$$

Again by a convexity argument, it is elementary to check that $\|\cdot\|_3$ is an equivalent norm on X that is, simultaneously, LUR and WUR.

Fourth, Theorem 8.7 gives an equivalent norm $\|\cdot\|_4$ on X such that its dual norm is LUR.

Fifth, for $n = 5, 6, , \ldots$ put

$$\|x^*\|_n^2 = \|x^*\|_3^2 + \frac{1}{n}\|x^*\|_4^2, \quad \text{for every } x^* \in X^*.$$

This defines equivalent dual LUR norms $\|\cdot\|_5, \|\cdot\|_6, \ldots$ on X^* (so, by Corollary 7.25, the predual norms $\|\cdot\|_5, \|\cdot\|_6, \ldots$ on X are all Fréchet differentiable). Moreover, the sequence $\{\|\cdot\|_n\}_{n=5}^{\infty}$ converges uniformly on bounded sets to $\|\cdot\|_3$.

Finally, put

$$\|x\|_0^2 = \sum_{n=5}^{\infty} 2^{-n} \|x\|_n^2, \quad \text{for every } x \in X. \tag{8.11}$$

We shall prove that this norm satisfies all the requirements in (x). Indeed, $\|\cdot\|_0$ is Fréchet differentiable because the derivatives of the summands in (8.11) are uniformly bounded on bounded sets and all summands are Fréchet differentiable. The norm $\|\cdot\|_0$ is, moreover, LUR and WUR. To show this, let $x \in X$ and let $\{x_n\}$ be a sequence in X such that $2\|x\|_0^2 + 2\|x_n\|_0^2 - \|x + x_n\|_0^2 \to 0$ as $n \to \infty$. Then the sequence $\{x_n\}$ is bounded, and a similar limit relation is true for each $\|\cdot\|_n$, $n = 5, 6, \ldots$ by a convexity argument (see the proof of Theorem 8.1). Since $\{\|\cdot\|_n\}_{n=5}^{\infty}$ converges uniformly on bounded sets to $\|\cdot\|_3$ we get, by the standard limit-interchanging rule, that $2\|x\|_3^2 + 2\|x_n\|_3^2 - \|x + x_n\|_3^2 \to 0$ as $n \to \infty$, and thus $\|x_n - x\|_3 \to 0$ as $n \to \infty$ due to the fact that $\|\cdot\|_3$ is LUR. This proves that $\|\cdot\|_0$ is LUR. A similar argument proves that $\|\cdot\|_0$ is WUR, too.

(x)$\Longrightarrow$(ix) is trivial.

This concludes the proof of Theorem 8.6. □

Remark: Note that norms constructed in Theorem 8.6 can be chosen so that the equivalence constant (to the original norm) is arbitrarily close to 1. In the non-separable setting, the situation is different, see, e.g., Remark 11.5.

8.3 Applications in Convexity

The following result is a strengthening of the Krein–Milman Theorem 3.65 in the context of Banach spaces, due to the fact that every strongly exposed point (even every exposed point) of a closed convex set is an extreme point (Exercise 7.72). The proof given here relies on the theory of operators that attain their norm, more precisely on Corollary 7.49. Theorem 8.13 will be proved by a different method in Theorem 11.11. Below, in Corollary 8.21, we provide a proof for the case of Banach spaces with a separable dual that depends on the concept of farthest point and the so-called Mazur intersection property. Pioneering papers in this area were [Stra], [Lind4], [BePe1a], and [Aspl2].

Theorem 8.13 (Lindenstrauss [Lind4], Troyanski [Troy1]) *Let W be a weakly compact convex set in a Banach space $(X, \|\cdot\|)$. Then W is the closed convex hull of its strongly exposed points.*

Proof: (Lindenstrauss [Lind4]) (see Fig. 8.2). Assume without loss of generality that $W \subset B_X$, that $X = \overline{\operatorname{span}} W$ and that the norm $\|\cdot\|$ of X is LUR (indeed, the space X is then weakly compactly generated, see Definition 13.1, and we can use Theorem 13.25; if we assume just that X is separable, use Theorem 8.1). Let C be the closed convex hull of the set of all strongly exposed points of W. Suppose that $C \neq W$. Then there is $f \in X^*$ with $\sup_{x\in W} f(x) = 1$ and $0 < \delta < 1$ such that $f(x) < 1 - \delta$ for $x \in C$. Let Y be the space $X \oplus \mathbb{R}$ equipped with the norm $\|(x, r)\| = (\|x\|^2 + r^2)^{\frac{1}{2}}$. Obviously, Y is also LUR.

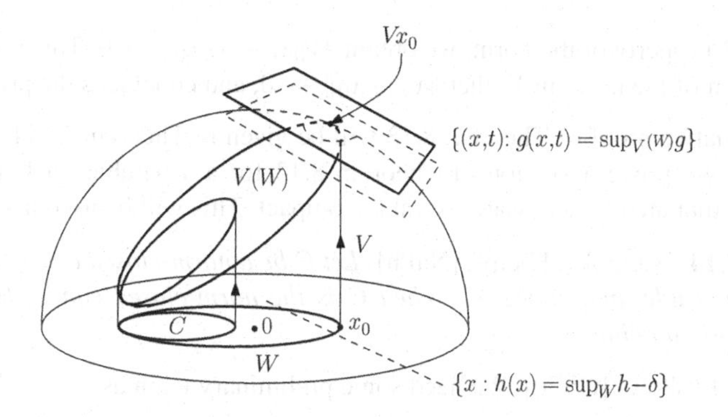

Fig. 8.2 The construction in Theorem 8.13

Let V be an operator from X into Y defined by $V(x) = \big(x, Mf(x)\big)$, where $M > \frac{2}{\delta}$. Then V is an isomorphism into, and so is every operator sufficiently close to V in the operator norm. It is easy to check that

$$\sup_{x\in W} \|Vx\| \geq M, \qquad \sup_{x\in C} \|Vx\| \leq \big(1 + (M-2)^2\big)^{\frac{1}{2}} \leq M - 1.$$

It follows that, for operators V_0 sufficiently close to V in the operator norm, it cannot exist $x_0 \in C$ such that $\sup_{x\in W} \|V_0 x\| = \|V_0 x_0\|$. By Corollary 7.49, the set of operators that attain their suprema on W is dense in $\mathcal{B}(X, Y)$ so, by the previous observation, we may find $V_0 \in \mathcal{B}(X, Y)$ sufficiently close to V that attains its supremum on W at some $x_0 \in W$ (and $x_0 \notin C$).

To finish the proof, it thus suffices to show that such x_0 is necessarily a strongly exposed point of W. This will contradict that all strongly exposed points of W are in C.

For it, let $g \in S_{Y^*}$ satisfy $g(V_0 x_0) = \|V_0 x_0\|$ ($\geq g(V_0 x)$ for all $x \in W$). Put $h = g \circ V_0 \in X^*$. Then $h(x_0) \geq h(x)$ for all $x \in W$. We will show that h strongly exposes W at x_0 (a contradiction, since then $x_0 \in C$).

Indeed, let $\{x_n\}$ be a sequence in W such that $h(x_n) \to h(x_0)$. Then

$$g\left(\frac{V_0 x_n + V_0 x_0}{2}\right) = \frac{h(x_n) + h(x_0)}{2} \to h(x_0) = g(V_0 x_0) = \|V_0 x_0\|.$$

Moreover,

$$g\left(\frac{V_0x_n + V_0x_0}{2}\right) \le \left\|\frac{V_0x_n + V_0x_0}{2}\right\| \le \|V_0x_0\|,$$

so

$$\left\|\frac{V_0x_n + V_0x_0}{2}\right\| \to \|V_0x_0\|.$$

By the LUR property of the norm, we obtain $\|V_0x_n - V_0x_0\| \to 0$. This implies, by the definition of the norm in Y, that $\|x_n - x_0\| \to 0$, and concludes the proof. □

An alternative proof of Theorem 8.13 will be given in Theorem 11.11. Now we prove Now we prove a version of Theorem 8.13 for a separable dual space and convex sets that are not necessarily weakly compact. This will be used later.

Theorem 8.14 (Namioka, Phelps, [NaPh]) *Let C be a norm-closed convex bounded set in a separable dual space X^*. Then C is the norm-closed convex hull of its strongly exposed points.*

The proof follows [CoEd]. We need some preliminary lemmas.

Lemma 8.15 *Let X be a Banach space. For every $x \in X\backslash\{0\}$ and $a \in X^{**}\backslash\{0\}$ there is an isomorphism T of X^* onto X^* such that $T^*(x) = a$.*

Proof: Consider the subspace $Z := \text{span}\{x, a\}$ of X^{**}. Working in the w^*-topology, we get that Z is complemented in X^{**} by a w^*-continuous projection P. Let A be a one-to-one operator on Z such that $A(x) = a$ and $A(a) = x$. Then $A \circ A = I_Z$. Let G be an operator on X^{**} defined by $G = I_{X^{**}} - P + A \circ P$. Observe that $G \circ G = I_{X^{**}}$, hence G is an isomorphism of X^{**} onto X^{**}, and $G(x) = a$. It is clearly w^*-w^*-continuous, so $G = T^*$ for some isomorphism $T : X^* \to X^*$. □

In the following lemma, p_S denotes the Minkowski functional of a set S.

Lemma 8.16 *Let X be a Banach space. Let C be a bounded closed convex subset of X^* such that $0 \in C$. Let T be an isomorphism of X^* onto X^*. Define a function F on X^{**} by $F(x^{**}) = \sup_C x^{**}$ and a function H on X by $H(x) = \sup_{T(C)} x$ (x is considered a function on X^*). If H is Fréchet differentiable at x_0, then F is Fréchet differentiable at $T^*(x_0)$.*

Proof: By Lemma 7.18, $F = p_{C^\circ}$, i.e., the Minkowski functional of the set C° (the polar set of C in X^{**}), and $H = p_{T(C)_\circ}$, where $T(C)_\circ$ is the polar set of $T(C)$ in X (indeed, $\sup_{T(C)} x = \sup_{\overline{T(C)}^{w^*}} x$ for every $x \in X$; use then the bipolar theorem). Since H is Fréchet differentiable, the Šmulyan lemma (Theorem 7.20) gives a single $f_0 \in \overline{T(C)}^{w^*}$ that is strongly supported by x_0 (see Fig. 8.3). In particular, $f_0 \in T(C)$ (if $\{f_i\}$ is a net in $T(C)$ such that $\langle f_i - f_0, x_0\rangle \to_i 0$, then $\|f_i - f_0\| \to_i 0$).

Find $g_0 \in C$ such that $T(g_0) = f_0$. Obviously, $T^*(x_0)$ supports C at g_0. In fact, $T^*(x_0)$ *strongly supports* C at g_0. Indeed, if $\{g_n\}$ is a sequence in C such that $\langle g_n - g_0, T^*(x_0)\rangle \to_n 0$, we have $\langle T(g_n) - f_0, x_0\rangle \to_n 0$, hence $\|T(g_n) - f_0\| \to 0$

Fig. 8.3 The construction in Lemma 8.16

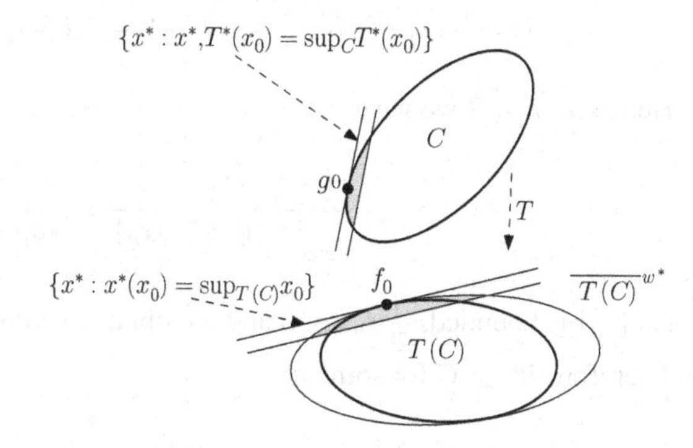

and so $\|g_n - g_0\| \to 0$. This proves, again using Šmulyan lemma (Theorem 7.20), that F is Fréchet differentiable at $T^*(x_0)$. □

Proof of Theorem 8.14 [CoEd]: Let S be the set of all strongly exposed points of C. We will show that $\overline{\text{conv}}(S) = C$. Assume the contrary. By the separation theorem, there is $x_0^{**} \in X^{**}$ that strongly separates $\overline{\text{conv}}(S)$ and a certain point $c^* \in C$. Fix an arbitrary $x_0 \in X \setminus \{0\}$ and find, according to Lemma 8.15, an isomorphism $T : X^* \to X^*$ such that $T^*(x_0^*) = x_0^{**}$. The space X is Asplund. Due to Theorem 8.6, the convex continuous function $H : X \to \mathbb{R}$ given by $H(x) = \sup_{T(C)} x$ is Fréchet differentiable on a G_δ dense subset of X. In particular, we can find in X points x arbitrarily close to x_0 where H is Fréchet differentiable. According to Lemma 8.16, the function $F : X^{**} \to \mathbb{R}$ given by $F(x^{**}) = \sup_C x^{**}$ has points of Fréchet differentiability $T^*(x)$ arbitrarily close to x_0^{**} (so strongly separating $\overline{\text{conv}}(S)$ and c^*). Elements in C where such an element $T^*(x)$ attain its supremum are, due to the Šmulyan lemma (Theorem 7.20), strongly exposed points of C, arbitrarily close to c^* and so out of S, a contradiction. □

Definition 8.17 *A Banach space is said to have the* Mazur intersection property *(MIP, in short) if every bounded closed convex subset is an intersection of balls.*

Theorem 8.18 (Mazur, Phelps, see, e.g., [Phelps, p. 112]) *Every Banach space X whose norm is Fréchet differentiable has the Mazur intersection property.*

Proof: (see Fig. 8.4). Let C be a bounded closed convex subset of X. Assume that $0 \notin C$. We will find a ball $B(x, \rho)$ such that $C \subset B(x, \rho)$ and $0 \notin B(x, \rho)$.

Using the separation theorem, we find $f_0 \in S_{X^*}$ such that $\inf_C f_0 > 0$. By the Bishop–Phelps theorem we may assume that $f_0 = \|x_0\|'$ for some $x_0 \in S_X$. Put $\varepsilon = \frac{1}{2} \inf_C f_0$ and $B^n = B(n\varepsilon x_0, (n-1)\varepsilon)$.

Note that $0 \notin B^n$ for every $n \geq 2$. Hence it suffices to show that $C \subset B^n$ for some $n \geq 2$.

Assume that for every $n \geq 2$ there is $x_n \in C \setminus B^n$. Then $\|x_n - n\varepsilon x_0\| > (n-1)\varepsilon$ and thus $\|x_0 - \frac{1}{n\varepsilon} x_n\| > 1 - \frac{1}{n}$ for every $n \geq 2$. As the norm is Fréchet differentiable and $f_0 = \|x_0\|'$, for $h \in X$ we have $\|x_0 + h\| - \|x_0\| - f_0(h) = r(h)$, where $\lim\limits_{h \to 0} \frac{r(h)}{\|h\|} = 0$. Using this for $h = -\frac{1}{n\varepsilon} x_n$, we obtain

$$r\left(-\tfrac{1}{n\varepsilon}x_n\right) = \|x_0 - \tfrac{1}{n\varepsilon}x_n\| - 1 + f_0\left(\tfrac{1}{n\varepsilon}x_n\right) > -\tfrac{1}{n} + \tfrac{2\varepsilon}{\varepsilon n} = \tfrac{1}{n}.$$

Hence for $n \geq 2$ we have

$$\frac{r\left(-\frac{1}{n\varepsilon}x_n\right)}{\|\frac{1}{n\varepsilon}x_n\|} \geq \frac{\varepsilon}{\|x_n\|} \geq \frac{\varepsilon}{\sup \|x_n\|}.$$

As $\{x_n\}$ is bounded, $\frac{1}{n\varepsilon}x_n \to 0$ and we obtained a contradiction with $\lim_{h\to 0} \frac{r(h)}{\|h\|} = 0$. Therefore $B^n \supset C$ for some n. □

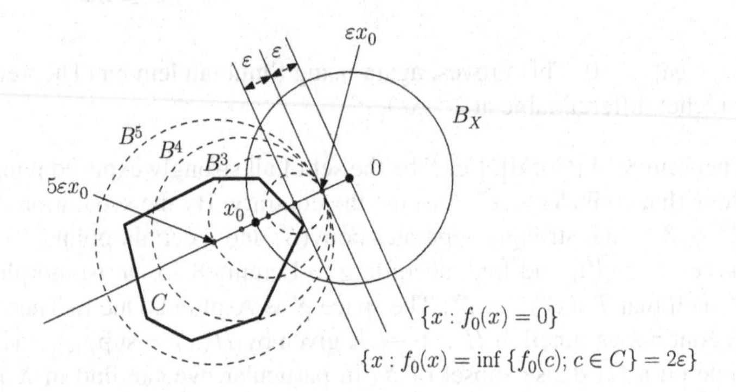

Fig. 8.4 The proof of the Mazur–Phelps theorem

If C is a weakly compact subset of a Banach space X, define the "farthest distance" function r for $x \in X$ by the expression

$$r(x) = \sup\{\|x - y\| : y \in C\}.$$

The function r is convex and 1-Lipschitz, as a supremum of such functions.

For $x \in X$, let $\partial r(x)$ denote the subdifferential of r at x. By Fact 7.12, we get that $\partial r(x) \subset B_{X^*}$ for every $x \in X$. We have that $c \in C$ is a *farthest point in C to $x \in X$* (in short, a *farthest point*) if $r(x) = \|x - c\|$.

In this notation, we have the following result.

Theorem 8.19 (Lau [Lau]) *Let C be a weakly compact set in a Banach space X. Then the set F of points in X that have farthest points in C contains a G_δ dense set in X.*

Proof: We shall prove the following

Claim: *The set*

$$G := \{x \in X : \sup\{x^*(x - z) : z \in C\} = r(x) \quad \textit{for every } x^* \in \partial r(x)\}$$

is G_δ dense in X. Furthermore, if $x \in G$, then x has a farthest point in C.

To this end, for $n \in \mathbb{N}$ put

$$F_n = \{x \in X : \inf_{y \in C} x^*(y - x) \geq -r(x) + \frac{1}{n}, \quad \text{for some } x^* \in \partial r(x)\}.$$

Note that F_n is closed for each n. Indeed, let $x_j \in F_n$, $j = 1, 2, \ldots$ with limit x and $x_j^* \in \partial r(x_j)$ be chosen according to the definition of F_n. Let x^* be a w^*-limit point of $\{x_j^*\}$. By passing to the limit, we have $x^*(y - x) \geq -r(x) + \frac{1}{n}$ for all $y \in C$ and that $x^* \in \partial r(x)$. Therefore $x \in F_n$ and F_n is thus closed. Because $X \backslash G = \bigcup_{n=1}^{\infty} F_n$, to prove the first statement in the Claim we need only to show that $\operatorname{Int} F_n = \emptyset$ for every $n \in \mathbb{N}$. Suppose that for some n there is a ball U centered at $y_0 \in F_n$ of radius $2\lambda r(y_0)$ for some $\lambda > 0$ such that $U \subset F_n$. Put $\varepsilon = \frac{\lambda}{4(1+\lambda)n} \min\{r(y_0), 1\}$. Choose $z_0 \in C$ such that $\|y_0 - z_0\| > r(y_0) - \varepsilon > r(y_0)/2$ and put $x_0 = y_0 + \lambda(y_0 - z_0)$. Note that $\|x_0 - y_0\| = \lambda\|y_0 - z_0\| > \lambda r(y_0)/2 > \varepsilon$. Pick x_1 in the line segment $[x_0, y_0]$ such that $\|x_0 - x_1\| = \varepsilon$. Because $\|x_0 - y_0\| = \lambda\|y_0 - z_0\| \leq \lambda r(y_0)$, we have that x_0 and thus $x_1 \in U \subset F_n$. Hence, by the definition of F_n, there exists $x_1^* \in \partial r(x_1)$ such that

$$\inf\{x_1^*(y - x_1) : y \in C\} \geq -r(x_1) + \frac{1}{n}.$$

By using this inequality, it follows that

$$\begin{aligned}
r(y_0) - r(x_1) &< \|y_0 - z_0\| + \varepsilon - r(x_1) = \frac{1}{1+\lambda}\|x_0 - z_0\| + \varepsilon - r(x_1) \\
&\leq \frac{1}{1+\lambda} r(x_0) + \varepsilon - r(x_1) \leq \frac{1}{1+\lambda} r(x_1) + 2\varepsilon - r(x_1) \leq \frac{\lambda}{1+\lambda}\Big(x_1^*(z_0 - x_1) - \frac{1}{n}\Big) + 2\varepsilon \\
&\leq \frac{\lambda}{1+\lambda}\Big(x_1^*(z_0 - x_0) - \frac{1}{n}\Big) + 3\varepsilon = \frac{1}{1+\lambda}\Big(x_1^*(\lambda z_0 - \lambda x_0) - \frac{\lambda}{n}\Big) + 3\varepsilon \\
&= \frac{1}{1+\lambda} x_1^*\big((1+\lambda)y_0 - (1+\lambda)x_0\big) - \frac{\lambda}{(1+\lambda)n} + 3\varepsilon = x_1^*(y_0 - x_0) - \frac{\lambda}{(1+\lambda)n} + 3\varepsilon \\
&\leq x_1^*(y_0 - x_1) - \frac{\lambda}{(1+\lambda)n} + 4\varepsilon \leq x_1^*(y_0 - x_1).
\end{aligned}$$

Hence $r(y_0) < r(x_1) + x_1^*(y_0 - x_1)$, which contradicts $x_1^* \in \partial r(x_1)$ and proves the first part of the Claim.

If $x \in G$, pick $x^* \in \partial r(x)$. Using the weak compactness of C, pick $z_0 \in C$ such that $\sup\{x^*(x - z) : z \in C\} = x^*(x - z_0)$. Then

$$r(x) = x^*(x - z_0) \leq \|x^*\| \cdot \|x - z_0\| \leq \|x - z_0\| \leq r(x).$$

Therefore $\|x - z_0\| = r(x)$, which completes the proof of the Claim and so of the theorem. □

Proposition 8.20 (Edelstein, see [Lau]) *Let X be a Banach space with the Mazur intersection property. Let C be a nonempty convex and weakly compact subset of X. Then C is the closed convex hull of the subset of C consisting of points that are farthest points.*

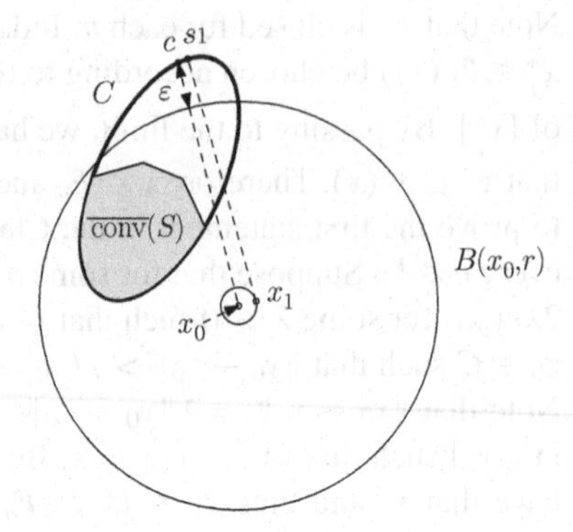

Fig. 8.5 The proof of Proposition 8.20

Proof: (see Fig. 8.5) Let S be the set of all points in C that are farthest points. Assume that $\overline{\text{conv}}(S) \neq C$. Since X has the Mazur intersection property, there exists a closed ball $B(x_0, r)$ in X and some $c \in C$ such that $\overline{\text{conv}}(S) \subset B(x_0, r)$ and $c \notin B(x_0, r)$. Put $\varepsilon = \text{dist}(c, B(x_0, r))$. By Theorem 8.19, there exists a point $x_1 \in B(x_0, \varepsilon/4)$ that has a farthest point s_1 in C. We get

$$\begin{aligned}\|s_1 - x_0\| &\geq \|s_1 - x_1\| - \|x_1 - x_0\| \geq \|c - x_1\| - \|x_1 - x_0\| \\ &\geq \|c - x_0\| - \|x_0 - x_1\| - \|x_0 - x_1\| \geq r + \varepsilon - \varepsilon/2 = r + \varepsilon/2.\end{aligned}$$

Therefore $s_1 \notin \overline{\text{conv}}(S)$, a contradiction. □

Now we are ready to prove a particular case of Theorem 8.13, namely for Banach spaces with a separable dual. It is proved only by using the technique of farthest points and the renorming result contained in Theorem 8.6.

Corollary 8.21 *Let X be a Banach space with a separable dual. Then every weakly compact convex subset of X is the closed convex hull of its strongly exposed points.*

Proof: By (x) in Theorem 8.6, there exists an equivalent norm $\|\cdot\|$ in X such that $(X, \|\cdot\|)$ is LUR and Fréchet differentiable. Observe that, if the norm of a Banach space is LUR and C is a weakly compact convex subset of X, then every farthest point in C is a strongly exposed point. Indeed, let $c \in C$ be a farthest point to some $x \in X$, and put $r = \|x - c\|$. Let $f \in S_{X^*}$ be a support functional of $B(x, r)$ at c. If $\{c_n\}$ is a sequence in C such that $f(c_n) \to f(c)$, then $f(c + c_n) \to 2f(c)$, and by the LUR property of $\|\cdot\|$, we get $c_n \to c$. This proves that c is a strongly exposed point of C. It is enough now to apply Proposition 8.20, since, by Theorem 8.18, $(X, \|\cdot\|)$ has the Mazur intersection property. □

The following proposition shows another application of smoothness.

Proposition 8.22 *Let X be a Banach space with a Fréchet differentiable norm $\|\cdot\|$. Let $\{e_i\}_{i=1}^{\infty}$ be a Schauder basis of X with associated projections P_n. If $\lim \|P_n\| = 1$, then $\{e_i\}$ is shrinking.*

Proof: Let $x^* \in S_{X^*}$. Given $\varepsilon > 0$, by the Bishop–Phelps theorem find a support point $u^* \in S_{X^*}$ (supported by some $u \in S_X$) such that $\|u^* - x^*\| < \varepsilon$. As $\|\cdot\|$ is Fréchet differentiable, there exists $0 < \delta < \varepsilon$ such that $\operatorname{diam} S(u, \delta) < \varepsilon/2$, where $S(u, \delta) = \{y^* \in B_{X^*} : u(y^*) > 1 - \delta\}$. Observe, that $\|P_n^*\| \to 1$ and that $P_n^*(u^*) \overset{w^*}{\to} u^*$. Choose $n_0 \in \mathbb{N}$ such that for every $n \geq n_0$, $|P_n^*(u^*)(u) - 1| < \frac{\delta}{2}$ and $\|P_n^*(u^*)\| < 1 + \frac{\delta}{2}$. Let $b^* \in B_{X^*}$, $\|b^* - P_n^*(u^*)\| < \frac{\delta}{2}$. Then $|b^*(u) - 1| < \delta$, so $\|b^* - u^*\| < \frac{\varepsilon}{2}$ and $\|P_n^*(u^*) - u^*\| < \varepsilon$ for every $n \geq n_0$. This shows that $\{e_i\}$ is shrinking. □

An application of Theorem 8.7 gives another proof of the fact that every Banach space with a separable dual has a shrinking basic sequence (see Theorem 4.19). Indeed, if X^* is separable, we may assume without loss of generality that the norm of X is Fréchet differentiable (Theorem 8.7 and Corollary 7.25). By Proposition 8.22, the basic sequence that is constructed in Theorem 4.19 is shrinking.

The concept of the Cantor derivative K' of a compact topological space K is introduced right after Definition 14.19. There, the notion of height of K is also defined.

The property of being c-Lipschitz Kadec–Klee smooth is introduced in Definition 12.60.

Definition 8.23 *We say that a Banach space X is* uniformly Kadec–Klee smooth *(or that its norm is* uniform Kadec–Klee smooth*) if for every $\varepsilon > 0$ there exists $\delta > 0$ such that, given $x^* \in X^*$ with the property that* $\operatorname{diam}(V \cap B_{X^*}) > \varepsilon$ *for each w^*-neighborhood V of x^*, then $\|x^*\| < 1 - \delta$.*

Theorem 8.24 *Let K be an infinite compact metric space. Then the following are equivalent.*
(i) *The space K is of finite height.*
(ii) *$C(K)$ is isomorphic to c_0.*
(iii) *$C(K)$ admits an equivalent uniformly Kadec–Klee smooth norm.*
(iv) *$C(K)$ admits an equivalent 1-Lipschitz Kadec–Klee smooth norm.*
(v) *$Sz(C(K)) = \omega_0$.*

Proof: (i)$\Longrightarrow$(ii): By induction on the height n of K. If $n = 1$, K is finite and (ii) holds true. Assume that the implication holds true if the height of K in n and pick K such that $K^{(n+1)} = \emptyset$. Put $L = K'$ and put $X = \{f \in C(K) : f|_L \equiv 0\}$. The space X is clearly isometric to $c_0(K \setminus L)$. Observe that, since K is an infinite compact metric space, then $K \setminus L$ is finite or countable. Moreover, $C(K)/X$ is, by Tietze's theorem, isometric to $C(L)$, which in turn is isomorphic to some $c_0(\Gamma)$, Γ being finite or $\mathbb{N}$ due to the induction hypothesis. The space X is complemented in $C(K)$ by Sobczyk's theorem as $C(K)$ is separable. Thus $C(K)$ is isomorphic to $X \oplus C(L)$, which is isomorphic to $c_0(K \setminus L) \oplus c_0(\Gamma)$.

(ii)$\Longrightarrow$(iv): This follows from Proposition 12.64.

(iv)$\Longrightarrow$(iii) is trivial.

(iii)$\Longrightarrow$(v): Let $\|\cdot\|$ be a uniformly Kadec–Klee smooth norm on $X := C(K)$. Given $\varepsilon > 0$, find $\delta > 0$ accordingly (see Definition 8.5). Then $S_\varepsilon(B_{X^*}) \subset (1-\delta)B_{X^*}$ and so, by an homogeneity argument, $S_Z(X, \varepsilon)$ must be finite. Since this is true for every $\varepsilon > 0$, the conclusion follows.

(v)$\Longrightarrow$(i): Since K is a w^*-compact set in $B_{C(K)^*}$, this implication is trivial: we mentioned after Definition 8.5 that, in this case, $S_Z(X, \varepsilon) \in \mathbb{N}$ for every $\varepsilon > 0$. The finiteness of $S_Z(\cdot, \varepsilon)$ is hereditary to w^*-compact sets in the dual space and $\|s - t\|_{C(K)^*} = 1$ whenever s, t are distinct elements in K, so nonempty w^*-open subsets of K must consists of a single point. □

8.4 Smooth Approximation

Theorem 8.25 *Let X be a separable Banach space. Then, for every open cover $\mathcal{V}$ of X there is a locally finite partition of unity, consisting of continuous, Gâteaux differentiable functions, and subordinated to $\mathcal{V}$. If X^* is separable, the word "Gâteaux" can be replaced by the word "Fréchet."*

Proof: Assume without loss of generality that the norm $\|\cdot\|$ of X is Gâteaux differentiable (Theorem 8.2). Let τ be a C^1-smooth function from the real line $\mathbb{R}$ into $[0, 1]$ such that $\tau = 1$ on $[-1, 1]$ and $\tau = 0$ outside $[-2, 2]$.

If B is an open ball $B(x_0, r)$ in X, we will denote by $2B$ the open ball $B(x_0, 2r)$. For each $x \in X$, choose an open ball B containing x such that there is $V \in \mathcal{V}$ with $2B \subset V$. Since X is separable, there is a countable family $(B_i)_{i=1}^\infty$ of these balls such that $\bigcup_i B_i = X$.

For each $i \in \mathbb{N}$, define a bump function h_i on X as follows: If $B_i = B(x_i, r_i)$ put

$$h_i(x) = \tau(\|x - x_i\|).$$

Then h_i is a continuous function on X such that is Gâteaux differentiable.

Put $g_1 = h_1$ and

$$g_n = h_n \prod_{i<n} (1 - h_i), \text{ for } n > 1.$$

Then all g_n are continuous, Gâteaux differentiable functions on X.

Given $i \in \mathbb{N}$, if $x \in B_i$ and $n > i$, then $h_i(x) = 1$ and thus $g_n(x) = 0$. This implies that $\{x \in X : g_n(x) > 0\}_{n=1}^\infty$ is a locally finite family. Therefore, $g := \sum_{n=1}^\infty g_n$ is a continuous, Gâteaux differentiable function on X. If $g_n(x) > 0$ for some $x \in X$ and some $n \in \mathbb{N}$, then $h_n(x) > 0$ and thus $x \in 2B_n \subset V$ for some $V \in \mathcal{V}$. Therefore the partition $\{g_n\}$ is subordinated to $\mathcal{V}$.

Assume that $\sum_n g_n(x) = 0$ for some $x \in X$. Then $g_n(x) = 0$ for all $n \in \mathbb{N}$. We will show by induction that then $h_n(x) = 0$ for all n. This would mean that

$x \notin B_n$ for all n, as $h_n = 1$ on B_n for all n. To perform the induction, if $n = 1$ then $h_1(x) = g_1(x) = 0$. Assume that $h_1(x) = h_2(x) = \ldots = h_n(x) = 0$. Then

$$0 = g_{n+1}(x) = h_{n+1}(x) \prod_{i<n+1} (1 - h_i(x))) = h_{n+1}(x).$$

If X^* is separable, we start with an equivalent Fréchet differentiable norm on X by using Theorem 8.6. □

Theorem 8.26 (Bonic and Frampton [BoFr]) *Let X be a Banach space with separable dual, and Y be a Banach space. Then any continuous mapping f from X into Y can be uniformly approximated by C^1-smooth mappings.*

Proof: Fix $\varepsilon > 0$ and for every $x_0 \in X$, choose a neighborhood U_{x_0} such that the oscillation of f on U_{x_0} is smaller than ε. Let $\{\varphi_\alpha\}$ be a C^1-partition of unity subordinated to this cover (Theorem 8.25). Fix points x_α with $\varphi_\alpha(x_\alpha) \neq 0$. Put for $x \in X$,

$$g(x) = \sum \varphi_\alpha(x) f(x_\alpha).$$

Then

$$\|f(x) - g(x)\| \leq \sum \varphi_\alpha(x) \|f(x) - f(x_\alpha)\| < \varepsilon$$

because $\sum \varphi_\alpha(x) = 1$ and a term in the sum is nonzero only if x and x_α belong to the same U_{x_0} and then $\|f(x) - f(x_\alpha)\| < \varepsilon$. □

Corollary 8.27 *Let X be a Banach space such that X^* is separable. If A, B are closed disjoint subsets of X, then there is a C^1-smooth function φ on X such that $\varphi = 0$ on A and $\varphi = 1$ on B.*

Proof: Let $f(x) = \frac{\text{dist}(x,A)}{\text{dist}(x,A)+\text{dist}(x,B)}$. By Theorem 8.26 there is a C^1-smooth function g on X such that $|f(x) - g(x)| < \frac{1}{4}$ for all $x \in X$. Let τ be a C^1-smooth function on $\mathbb{R}$ such that $\tau(x) = 0$ whenever $|x| < \frac{1}{4}$ an $\tau(x) = 1$ whenever $\frac{3}{4} < x < \frac{5}{4}$. Then $\varphi = \tau \circ g$ satisfies the requirements. □

Theorem 8.28 ([BoFr], see, e.g., [DGZ3, p. 101]) *Let X and Y be separable Banach spaces such that X^* is nonseparable and Y^* is separable. Let $\varphi : X \to Y$ be a Lipschitz Fréchet continuously differentiable mapping. If B is an open ball in X, then $\varphi(\partial B)$ is norm-dense in $\varphi(\overline{B})$, where ∂B is the boundary of B.*

Proof: Without loss of generality, we assume that $0 \in B$. We will show that $\varphi(0) \in \overline{\varphi(\partial B)}$. Assume that $\varphi(0) \notin \overline{\varphi(\partial B)}$. Let $b : Y \to \mathbb{R}$ be a Lipschitz,

Fréchet continuously differentiable function such that $b(\varphi(0)) = 1$ and $b = 0$ on a neighborhood of $\varphi(\partial B)$ (Theorem 8.6).

Put

$$\Phi(x) = \begin{cases} b(\varphi(x)), & \text{if } x \in \overline{B}, \\ 0, & \text{if } x \in X \setminus \overline{B}. \end{cases}$$

Then Φ is a Lipschitz, Fréchet continuously differentiable bump function on X, a contradiction with Theorem 8.6. □

Theorem 8.29 (Azagra and Ferrera [AzFe]) *Let X be a separable Banach space and let C be a closed convex subset of X. Then there is a C^∞-smooth convex function $f : X \to [0, \infty)$ such that $C = f^{-1}(0)$.*

Proof: Let C° be the polar set of C in X^*. Since for each $n \in \mathbb{N}$ the set $C^{circ} \cap nB_{X^*}$ is w^*-separable, so it is C°. Let $\{\varphi_n : n \in \mathbb{N}\}$ be a w^*-dense subset of C°. Obviously, $C \subset \bigcap_{n=1}^\infty \varphi_n^{-1}(-\infty, 1]$. If $x_0 \notin C$, then there is $\varphi \in X^*$ such that $\varphi(x_0) > 1 > \sup_C \varphi$. Then $\varphi \in C^\circ$. From the w^*-density of $\{\varphi_n : n \in \mathbb{N}\}$ in C°, we get that there is φ_n such that $\varphi_n(x_0) > 1$, thus $C = \bigcap_{n=1}^\infty \varphi_n^{-1}(-\infty, 1]$. By scaling, we can represent $C = \bigcap_{n=1}^\infty \psi(-\infty, \alpha_n]$, where $\|\psi_n\| = 1$ for all n. Choose $\theta : \mathbb{R} \to [0, \infty)$ a C^∞-smooth convex function such that $\theta(t) = 0$ for all $t \le 0$ and $\theta(t) = t + b$ for all $t > 1$, where $b \in (-1, 0)$. The desired C^∞-smooth convex function f is defined by

$$f(x) = \sum_{n=1}^\infty \frac{\theta\big(\psi_n(x) - \alpha_n\big)}{(1 + |\alpha_n|)2^n}, \quad \text{for } x \in X.$$

□

Note that it was proved in [Haje3] that, *if Γ is an uncountable set, then $c_0(\Gamma)$ admits no C^2-smooth function that would attain its minimum at exactly one point.*

Example: A construction of a coordinatewise Fréchet smooth homeomorphism φ of a separable Asplund space into $\ell_2(\mathbb{N})$.

Let $\|\cdot\|$ be a norm on X which is locally uniformly rotund and Fréchet differentiable (Theorem 8.6). Let $\{f_i\}_{i=1}^\infty \subset S_{X^*}$ be dense in S_{X^*}. Finally, let a mapping φ of X into $\ell_2(\mathbb{N})$ be defined for $x \in X$ and $i \in \mathbb{N}$ by

$$\varphi(x)(i) = \begin{cases} \|x\|^2 & \text{if } i = 1 \\ 2^{-i+1} f_{i-1}(x) & \text{if } i > 1. \end{cases}$$

Then φ is a continuous and one-to-one mapping of X onto $\varphi(X) \subset \ell_2(\mathbb{N})$, such that for each i, the mapping $x \to \varphi(x)(i)$ is a continuously Fréchet differentiable function on X. If $x_n, x \in X$, $n = 1, 2, \ldots$ are such that $\lim \varphi(x_n) = \varphi(x)$ in $\ell_2(\mathbb{N})$, then $\lim \|x_n\| = \|x\|$ and $\lim f_i(x_n) = f_i(x)$ for each i. Since $\{f_i\}_{i=1}^\infty$ is dense in S_{X^*}, it follows that $x_n \to x$ in the weak topology of X. This, with the fact that the norm $\|\cdot\|$ has the Kadec–Klee property (Exercise 8.45), implies that $\lim \|x_n - x\| = 0$. This completes the proof that φ is a homeomorphism of X into $\ell_2(\mathbb{N})$.

8.5 Ranges of Smooth Maps

In this section we are concerned with the classical problem on existence of (non-linear) mappings—having a certain degree of differentiability—from a Banach space onto another. A variant of this problem consists of building a bump function on a given Banach space having a derivative whose range contains the entire dual unit ball. We provide here a unified approach of those two questions.

Let X be a separable infinite-dimensional Asplund space. By Theorem 8.6 and Corollary 7.24, X has a C^1-Fréchet smooth equivalent norm. Composing this norm with a suitable smooth real function, we see that there exists a C^1-Fréchet smooth and K-Lipschitz bump function $b : X \to [0, 1]$ such that $b(x) = 1$ for every $x \in (2/3)B_X) = 1$, and $b(x) = 0$ for every $x \in X\backslash B_X$. Let $\{e_n\}_{n=1}^{\infty}$ be a normalized monotone Schauder basic sequence in X, and $\tau \in (0, 1)$.

For $n \in \mathbb{N}$ and $\tau \in (0, 1)$, consider the set $\{e_\sigma : \sigma \in \mathbb{N}^n\}$ $(\subset X)$, where

$$e_\sigma := \sum_{i=1}^{n} \frac{\tau^i}{3^{i-1}} e_{\sigma(i)}. \tag{8.12}$$

Note that $\|e_\sigma\| \le 3/2$ for every $\sigma \in \mathbb{N}^n$.

Consider, too, the set $\{b_\sigma : \sigma \in \mathbb{N}^n\}$, where $b_\sigma : X \to \mathbb{R}$ is the function defined by

$$b_\sigma(x) = \frac{\tau^{2n}}{3^n} b\left(\frac{2C3^n}{\tau^n}(x - e_\sigma)\right). \tag{8.13}$$

Obviously, for any $\sigma \in \mathbb{N}^n$ we have

$$\operatorname{supp}(b_\sigma) \subset B\left(e_\sigma, \frac{\tau^n}{2C3^n}\right), \quad |b_\sigma(x)| \le 3^{-n}\tau^{2n}, \text{ and } |b'_\sigma(x)| \le 2CK\tau^n.$$

Observe too that, for $\sigma \in \mathbb{N}^n$,

$$b_\sigma(x) = \begin{cases} \frac{\tau^{2n}}{3^n}, & \text{if } \|x - e_\sigma\| \le \frac{\tau^n}{C3^{n+1}}, \\ 0, & \text{if } \|x - e_\sigma\| \ge \frac{\tau^n}{2C3^n}. \end{cases} \tag{8.14}$$

Lemma 8.30 *The supports of b_σ and $b_{\sigma'}$ are disjoint if $\sigma \neq \sigma'$ in $\mathbb{N}^n$.*

Proof: Let j be the first index in $\{1, 2, \ldots, n\}$ such that $\sigma(j) \neq \sigma'(j)$. Put $f = e^*_{\sigma(j)}$. If $j < n$,

$$\|e_\sigma - e_{\sigma'}\| \geq \frac{1}{\|f\|}\langle f, e_\sigma - e_{\sigma'}\rangle \geq \frac{1}{C}\left(\frac{\tau^j}{3^{j-1}}\right)\left(1 - \sum_{i=1}^{\infty}\left(\frac{\tau}{3}\right)^i\right) \geq \frac{1}{2C}\frac{\tau^j}{3^{j-1}}.$$

In case $j = n$ we get

$$\|e_\sigma - e_{\sigma'}\| = \frac{\tau^n}{3^{n-1}}\|e_{\sigma(n)} - e_{\sigma'(n)}\| \geq \frac{\tau^n}{3^{n-1}}\frac{1}{C}.$$

Thus, in both cases,

$$\|e_\sigma - e_{\sigma'}\| \geq \frac{1}{2C}\frac{\tau^n}{3^{n-1}}. \tag{8.15}$$

In particular, given σ and σ' in $\mathbb{N}^n$ such that $\sigma \neq \sigma'$,

$$B\left(e_\sigma, \frac{\tau^n}{2C3^n}\right) \cap B\left(e_{\sigma'}, \frac{\tau^n}{2C3^n}\right) = \emptyset.$$

□

Given $\sigma \in \mathbb{N}^{\mathbb{N}}$ we set

$$e_\sigma = \sum_{j=1}^{\infty}\frac{\tau^i}{3^{i-1}}e_{\sigma(i)}.$$

Observe that $\|e_\sigma\| \leq 3/2$. Denote by $\sigma^n \in \mathbb{N}^n$ the initial segment of length n of σ. Then,

$$\|e_\sigma - e_{\sigma^n}\| \leq \tau \sum_{i=n+1}^{\infty}\left(\frac{\tau}{3}\right)^{i-1} = \tau\frac{(\tau/3)^n}{1-(\tau/3)}.$$

Choose $0 < \tau < \frac{3}{1+9C}$. Then $\frac{3\tau}{3-\tau} \leq \frac{1}{3C}$, and this implies

$$(\|e_\sigma - e_{\sigma^n}\| \leq)\ \tau\frac{(\tau/3)^n}{(1-\tau/3)} \leq \frac{\tau^n}{C3^{n+1}}. \tag{8.16}$$

In view of (8.14) we get

$$b_{\sigma^n}(e_\sigma) = \frac{\tau^{2n}}{3^n}. \tag{8.17}$$

Denote by T_τ the set $\{e_\sigma : \sigma \in \mathbb{N}^{\mathbb{N}}\}$ $(\subset (3/2)B_X)$. The estimate (8.16) implies that for $n \in \mathbb{N}$,

$$T_\tau \subset \bigcup_{\sigma \in \mathbb{N}^n} B\left(e_\sigma, \frac{\tau^n}{C3^{n+1}}\right). \tag{8.18}$$

Let $\{f_n\}_{n=1}^\infty$ be a sequence of continuous affine mappings from X into some Banach space Y, such that

$$\sup_{n\in\mathbb{N}} \sup_{x\in B_X} \{\|f_n(x)\|, \|f_n'(x)\|\} \leq 1. \tag{8.19}$$

Take $n \in \mathbb{N}$. Lemma 8.30 shows that in the following expression, the summands have mutually disjoint supports.

$$b_n(x) := \sum_{\sigma\in\mathbb{N}^n} b_\sigma(x) f_{\sigma(n)}(x), \quad x \in X. \tag{8.20}$$

In particular, for $x \in X$,

$$\|b_n(x)\| \leq \frac{\tau^{2n}}{3^n}\|x\|, \quad \|b_n'(x)\| \leq \left(2CK\tau^n + \frac{\tau^{2n}}{3^n}\right)\|x\|. \tag{8.21}$$

Let

$$\widetilde{b} := \sum_{n=1}^\infty b_n. \tag{8.22}$$

Due to the estimations in (8.21), the series in (8.22) defines a Lipschitz and C^1-Fréchet smooth mapping $\widetilde{b}$ from X to Y, and $\text{supp}(\widetilde{b}) \subset 2B_X$.

Take $\sigma \in \mathbb{N}^{\mathbb{N}}$. Due to (8.16) and (8.14) we get, for all $n \in \mathbb{N}$, $b_n(e_\sigma) = \frac{\tau^{2n}}{3^n} f_{\sigma(n)}(e_\sigma)$, hence

$$\widetilde{b}(e_\sigma) = \sum_{n=1}^\infty \frac{\tau^{2n}}{3^n} f_{\sigma(n)}(e_\sigma),$$

and

$$\widetilde{b}'(e_\sigma) = \sum_{n=1}^\infty \frac{\tau^{2n}}{3^n} f_{\sigma(n)}'(e_\sigma).$$

Lemma 8.31 *Let X be a Banach space endowed with a Fréchet smooth norm $\|\cdot\|$, and assume that X admits a Schauder basis $\{e_n\}_{n=1}^\infty$. Let Y be Banach space.*

(i) *If $\{f_i\}_{i=1}^\infty$ is a sequence of constant functions on X, say $f_i(x) = y_i$ for all $x \in X$ and $i \in \mathbb{N}$, where $\{y_i : i \in \mathbb{N}\}$ is dense in B_Y, then $\frac{\tau^2}{3} B_Y \subset \widetilde{b}(T_\tau)$.*

(ii) *If $Y = \mathbb{R}$, and $\{f_i\}_{i=1}^\infty$ is a dense subset of B_{X^*}, then $\frac{\tau^2}{3} B_{X^*} \subset \widetilde{b}'(T_\tau)$.*

Proof: The proofs of (i) and (ii) are similar, hence we shall show (i). Let $y \in \frac{\tau^2}{3} B_Y$. Choose $\gamma_1 \in \mathbb{N}$ so that $\|y - \frac{\tau^2}{3} y_{\gamma_1}\| < \frac{\tau^4}{3^2}$. Then we can find $\gamma_2 \in \mathbb{N}$ such that $\|y - \frac{\tau^2}{3} y_{\gamma_1} - \frac{\tau^4}{3^2} y_{\gamma_2}\| < \frac{\tau^6}{3^3}$. Proceed inductively, at each step choosing $\gamma_n \in \mathbb{N}$ such that $\|y - \sum_{j=1}^n \frac{\tau^{2j}}{3^j} y_{\gamma_j}\| < \frac{\tau^{2(n+1)}}{3^{n+1}}$. Clearly, $y = \sum_{j=1}^\infty \frac{\tau^{2j}}{3^j} y_{\gamma_j}$ $(= \widetilde{b}(e_\sigma)$, where $\sigma := (\gamma_1, \gamma_2, \ldots))$, and this shows (i). □

Theorem 8.32 (Bates, [Bate], see, e.g., [BeLi, p. 261]) *Let X, Y be separable infinite-dimensional Banach spaces. Then there exists a C^1-Fréchet smooth and Lipschitz mapping from X onto Y.*

Proof: By Theorem 4.28, there exists a linear quotient mapping $Q : X \to Z$, where Z is a Banach space with a Schauder basis $\{e_n\}_{n=1}^\infty$. Assume, without loss of generality, that the basis is seminormalized and contained in the image $Q(B_X)$. Denote by $I : Z \to c_0$ the bounded operator $I(\sum_{i=1}^\infty a_i e_i) := (a_i)_{i=1}^\infty$. Fix $\tau = 3/7$ and the corresponding $T_\tau \subset c_0$. Note that $T_\tau \subset I \circ Q(B_X)$. By Lemma 8.31 there exist a C^1-Fréchet smooth and Lipschitz mapping $\widetilde{b}$ from c_0 to Y, with $\operatorname{supp}(\widetilde{b}) \subset 2B_{c_0}$ and a positive constant $c > 0$ such that $cB_Y \subset \widetilde{b}(T_\tau)$. It is now easy to verify that

$$\widetilde{f}(x) = \sum_{n=1}^{\infty} n\widetilde{b}\left(\frac{4^n e_1 + I \circ Q(x)}{n}\right)$$

is the sought surjective mapping. □

Theorem 8.33 (Azagra, Deville, [AzDe2]) *Let X be a separable infinite-dimensional Asplund space. Then there exists a C^1-Fréchet smooth and Lipschitz bump function on X whose range of the derivative contains the whole B_{X^*}.*

Proof: It follows from Lemma 8.31, by using the fact that the X contains a Schauder basic sequence (Theorem 4.19), and it admits a C^1-Fréchet smooth and Lipschitz bump function since X^* is separable (Theorem 8.6). □

8.6 Remarks and Open Problems

Remarks

1. The proof of Theorem 8.32 should be compared with the classical Morse–Sard theorem, see, e.g., [Stern]: *If $n \geq m$ and if $\varphi : \mathbb{R}^n \to \mathbb{R}^m$ is C^{n-m-1}, then the set of critical values of φ has measure zero in $\mathbb{R}^m$*. (A point $y \in \mathbb{R}^m$ is a *critical value* of φ if $y = \varphi(x)$, where the rank of $\varphi'(x)$ is smaller than m.) In particular, if φ is surjective, it follows that almost every $y \in \mathbb{R}^m$ is a regular (i.e., non-critical) value. The proof of Theorem 8.32 shows that the Morse–Sard theorem has no chance to hold true in infinite-dimensional situations.

2. It is shown in [Haje4] that if a Banach space E has an unconditional basis and if there is a mapping from c_0 onto E that has a locally uniformly continuous derivative, then E necessarily contains an isomorphic copy of c_0.
3. It is shown in [Bate], see, e.g., [BeLi, p. 262], that every separable Banach space is the image, under a surjective C^{∞}-mapping, of a Hilbert space.
4. The space ℓ_∞ does not admit any equivalent LUR norm ([Troy1] and [Lind13], see, e.g., [DGZ3, Theorem II.7.10]). If Γ is uncountable, then $\ell_\infty(\Gamma)$ does not admit any equivalent rotund norm (Day, see, e.g., [DGZ3, Cor. II.7.13]). Bourgain proved in [Bou2] that ℓ_∞/c_0 has no equivalent strictly convex norm.
5. Hagler proved in [Hag] that if X is a separable Banach space such that X^* is non-separable, then X contains a subspace Y with a non-shrinking Schauder basis.
6. It is shown, e.g., in [DGZ3, p. 26] that X is an Asplund space if and only if for every convex continuous function f defined on an open convex set $U \subset X$, there is a selector φ for the subdifferential mapping ∂f that is the pointwise limit of a sequence of norm-to-norm continuous mappings from X into X^*. Such selectors are called Jayne–Rogers selectors, see [JaRo].
7. We remark that if X is a separable nonreflexive Banach space, then there is an equivalent Gâteaux differentiable norm $\|\cdot\|$ on X such that some points of $S_{(X,\|\cdot\|)}$ are no longer points of Gâteaux differentiability of the second dual norm $\|\cdot\|^{**}$ on X^{**} ([Godu]). If X is a separable nonreflexive Banach space, then X admits an equivalent strictly convex norm $\|\cdot\|$ such that some points of $S_{(X,\|\cdot\|)}$ are no longer extreme points of $B_{(X^{**},\|\cdot\|^{**})}$ ([Godu2]). If a separable Banach space has a subspace isomorphic to c_0 then it admits an equivalent strictly convex norm $\|\cdot\|$ in which no element of $S_{(X,\|\cdot\|)}$ is an extreme point of $B_{(X^{**},\|\cdot\|^{**})}$ ([Morr]). In [SchSerWer] it is shown that if a separable Banach space X fails the RNP property (see Definition 11.14), then for every $\varepsilon > 0$, there is a symmetric closed convex set C in the unit ball of X so that $\operatorname{dist}(x^{**}, X) \geq 1 - \varepsilon$ for every extreme point x^{**} of the w^*-closure of C in X^{**}. If a separable Banach space X contains an isomorphic copy of ℓ_1, then X admits an equivalent Gâteaux differentiable norm $\|\cdot\|$ such that no point of $S_{(X,\|\cdot\|)}$ is a point of Gâteaux differentiability of $\|\cdot\|^{**}$ ([Tan1]).
8. The renorming by an LUR norm is a three-space property [GTWZ2] see, e.g., [DGZ3, p. 299] or [CasGon].
9. We refer to, e.g., [HMVZ, Ch. 8] for more on the Mazur intersection property.

Open Problems

1. Is it true that an equivalent Fréchet differentiable norm in a subspace of a separable and reflexive Banach space can be extended to an equivalent Fréchet differentiable norm in the whole space?
2. Is every norm on $\ell_2(\omega_1)$ approximable by C^{∞}-smooth norms?

3. Is it true that given density α, all spaces of this density are C^1-smooth images of each other?
4. Is it true that for density c, the previous question holds even for C^∞-smooth functions?
5. We refer, e.g., to [HaZi] and [FMZ5] for a list of some open problems in this area.

Exercises for Chapter 8

8.1 Let $x_n \in S_X$ and $y_n \in S_X$ be such that $\lim_{n\to\infty} \|x_n + y_n\| = 2$. Let z_n be a point on the line segment between x_n and y_n for every n. Show that $\lim_{n\to\infty} \|z_n\| = 1$.
Hint. Let $f_n \in S_{X^*}$ be such that $f_n(\frac{1}{2}(x_n + y_n)) = \frac{1}{2}\|x_n + y_n\| \to 1$. Observe that $f(x_n) \to 1$ and $f(y_n) \to 1$. Indeed, if $f(x_n) \le 1-\delta$ for all n, then $f(y_m) > 1+\delta/2$ for m large enough, contradicting $\|f\| = 1$, $\|y_m\| = 1$. Then also $f(z_n) \to 1$ and hence $1 \ge \|z_n\| \ge f_n(z_n) \to 1$.

8.2 [MoTo] Let X be a Banach space. Given $x \notin B_X$ define $R(x, B_X) = \operatorname{conv}(\{x\} \cup B_X)\backslash B_X$ (see Exercise 8.7). Prove that X is LUR if and only if $\lim_{t\to 1^+} \operatorname{diam} R(tx, B_X) = 0$ for all $x \in S_X$.
Hint. Given $x \in S_X$ and $\varepsilon \in (0, 2]$, define $\delta(x, \varepsilon) = \inf\{1-\|\frac{x+y}{2}\| : y \in S_X, \|x - y\| \ge \varepsilon\}$ (see Definition 9.1). Define $\Delta(x, \varepsilon) = \frac{\varepsilon}{2}\delta(x, \varepsilon)$; it is a strictly increasing function of ε. Then prove that $\operatorname{diam} R(tx, B_X) \le 2\big[t\big(\Delta(x, \cdot)\big)^{-1}\big(2\frac{t-1}{t}\big) + t - 1\big]$. This proves the only if part.

Now assume that X is not LUR. Find $x \in S_X$, $\varepsilon > 0$ and a sequence $\{x_n\}$ in S_X such that $\|x_n - x\| \ge \varepsilon$ and $\left\|\frac{x_n+x}{2}\right\| > 1 - \frac{1}{n}$, $n \in \mathbb{N}$. For a given $n = 3, 4, 5, \ldots$ take $t \in \left(\frac{n}{n-1}, 2\right)$. Then $t\frac{x_n+x}{2}$ is a convex combination of x_n and $\frac{t}{2-t}x$. Evaluate the distance between $t\frac{x_n+x}{2}$ and $\frac{t}{2-t}x$.

8.3 Assume that X is a reflexive Banach space whose norm is LUR (respectively Fréchet differentiable). Let Y be a closed subspace of X. Show that the quotient norm of X/Y is LUR (respectively Fréchet differentiable).
Hint. The LUR case: Let $\hat{x}, \hat{x}_n \in S_{X/Y}$ and $\|\hat{x} + \hat{x}_n\| \to 2$. Choose $x_n \in (1 + \varepsilon)S_X \cap \hat{x}_n$ and $x \in S_X \cap \hat{x}$. Then $2 + \frac{2}{\varepsilon} \ge \|x + x_n\| \ge \|\hat{x} + \hat{x}_n\| \to 2$. Use LUR to get $\|x_n - x\| \to 0$. This implies $\|\hat{x} - \hat{x}_n\| \to 0$. Similarly in the Fréchet case, use Lemma 7.4.

8.4 Let $(X_n, \|\cdot\|_n)$, $n \in \mathbb{N}$, be Banach spaces. If their norms are LUR, show that the canonical norm of $\big(\sum(X_n, \|\cdot\|_n)\big)_2$ is LUR.
Hint. Let $\big(2\|x^n\|^2 + 2\|x\|^2 - \|x + x^n\|^2\big) \to 0$. Given $\varepsilon > 0$, fix j_0 such that $\sum_{j=j_0}^{\infty} \|x_j\|^2 < \varepsilon$. From the convexity it follows that also $\sum_{i=j_0}^{\infty} \|x_j^n\|^2 < \varepsilon$ for large n. Then use LUR on the first j_0 coordinates.

8.5 Assume that the norm $\|\cdot\|$ of a Banach space X and its dual norm on X^* are both Fréchet differentiable. Then the norm $\|\cdot\|$ and its dual norm are both locally uniformly rotund.
Hint. Let $x, x_1, x_2, \dots \in S_X$ be such that $\|x + x_n\| \to 2$. Let f_n be the derivative of $\|\cdot\|$ at $\frac{x+x_n}{2}$ and f be the derivative of $\|\cdot\|$ at x. Then $\|f_n\| = 1$, $f_n(x+x_n) = \|x+x_n\|$ and $\|f\| = 1 = f(x)$. If $\liminf f_n(x) < 1$, then for some $0 < q < 1$ and some subsequence n_k, $f_{n_k}(x) < q$ for all k. Then $f_{n_k}(x+x_{n_k}) < q + f_{n_k}(x_{n_k}) < 1+q < 2$, a contradiction with $f_{n_k}(x + x_{n_k}) = \|x + x_{n_k}\| \to 2$. Similarly, $f_n(x_n) \to 1$. Thus $\liminf f_n(x) = 1$. As $f_n(x) \leq \|f_n\|\,\|x\| = 1$ for all n, $\lim f_n(x) = 1$.

As $\|\cdot\|$ is Fréchet differentiable at x, from the Šmulyan Lemma 7.22 we have $\|f_n - f\| \to 0$. Thus $|(f_n - f)(x_n)| \leq \|f - f_n\|\|x_n\| \to 0$. Hence $f(x_n) = f_n(x_n) + (f - f_n)(x_n) \to 1$. As the dual norm of $\|\cdot\|$ is Fréchet differentiable at f and $f(x) = 1$, from the Šmulyan lemma 7.22 we have $\|x - x_n\| \to 0$. This shows that $\|\cdot\|$ is locally uniformly rotund. Since the dual norm is Fréchet differentiable, X is reflexive and we can follow the first part of this proof to show that the dual norm of $\|\cdot\|$ is locally uniformly rotund.

8.6 A norm $\|\cdot\|$ of a Banach space X is said to have the $(2R)$-*property* if $\{x_n\}$ is a convergent sequence whenever $\|x_n + x_m\| \to 2$.

Show that every space whose norm has the $(2R)$-property is reflexive.

We remark that every separable reflexive space has an equivalent norm with the $(2R)$-property ([OdSc3]).
Hint. If $f \in S_{X^*}$ and $x_n \in S_X$ satisfy $f(x_n) \to 1$, then $\|x_n + x_m\| \to 2$. Use Corollary 3.131.

8.7 [Mont2], [Mont3] Let X be a Banach space. The *drop* defined by $x \in X \setminus B_X$ is the set $D(x, B_X) = \operatorname{conv}(\{x\} \cup B_X)$. The Banach space X is said to have the *drop property* if given any closed set $S \subset X$ such that $S \cap B_X = \emptyset$, there exists $s \in S$ such that $D(s, B_X) \cap S = \{x\}$.

(i) Let X be a Banach space. Show that given a closed set $A \subset X$ such that $\operatorname{dist}(A, B_X) > 0$, there exists $a \in A$ such that $D(a, B_X) \cap A = \{a\}$ ([Dane]).

(ii) Prove that X has the drop property if and only if X is reflexive and has the Kadec property (that is, norm and weak convergent sequences in S_X are the same).
Hint. (i) Use Theorem 7.39.

(ii) Assume that X has the drop property. If X is not reflexive, by Corollary 3.131, there exists $f \in S_{X^*}$ that does not attain its norm. Let $\varepsilon_n > 0$, $\varepsilon_n \to 0$. Choose $x_1 \in X$ such that $f(x_1) > 1$, then $b_1 \in B_X$ such that $f(b_1) = 1 - \varepsilon_1$. Choose $x_2 \in [b_1, x_1]$ such that $f(x_2) = 1 + \varepsilon_1$, then $b_2 \in B_X$ such that $f(b_2) = 1 - \varepsilon_2$ and so on. The set $\{x_n\}$ is closed and does not intersect B_X; however, $x_m \in D(x_n, B_X)$ for $m \geq n$, a contradiction.

If X does not have the Kadec property, we can find a sequence $\{x_n\}$ in S_X and $x \in S_X$ such that $x_n \overset{w}{\to} x$ but nor $x_n \to x$. Let $\varepsilon_n > 0$, $\varepsilon_n \to 0$. Define $z_1 = \varepsilon_0 x_1$, choose $z_2 \in [x_1, z_1]$, $\|x_1 - z_2\| < \varepsilon_2$, then choose $z_3 \in [x_2, z_2]$, $\|x_2 - z_3\| < \varepsilon_2$, and so on. Apply the same argument as before.

Assume that X is reflexive and has the Kadec property. Let $\{x_n\}$ be a sequence in X that is not eventually constant such that $x_{n+1} \in D(x_n, B_X)$ for every n. Suppose first that $\|x_n\| \not\to 1$. By (i), it must have a convergent subsequence. If $\|x_n\| \to 1$, by the Eberlein–Šmulyan theorem there exists a subsequence which w-converges to some x_0. As $x_0 \in \overline{\text{conv}}\{x_n\}$, it follows that $\|x_0\| = 1$. By the Kadec property, $x_n \to x_0$ and this implies that X has the drop property.

8.8 Let Y be a closed subspace of a Banach space X. Assume that the dual norm of X^* is LUR. To every $f \in S_{Y^*}$ assign as $\Phi(f)$ the unique extension of f on X (see Exercise 7.69). Show that Φ is a continuous mapping from S_{Y^*} into S_{X^*}.
Hint. If $f_n \in S_{Y^*} \to f \in S_{Y^*}$, then $\|\Phi(f_n) + \Phi(f)\| \geq \|f_n + f\| \to 2$.

8.9 Let $\{e_i\}$ be a Schauder basis of a Banach space X.
(i) Show that there is an equivalent locally uniformly rotund norm $|||\cdot|||$ on X such that $\{e_i\}$ is monotone in $|||\cdot|||$.
(ii) Assume that X^* is separable. Does there exist an equivalent Fréchet differentiable norm $|||\cdot|||$ on X such that $\{e_i\}$ is monotone in $|||\cdot|||$?
Hint. (i) First renorm the space X by $\|x\|_1 = \sup_{n\leq m} \left\|\sum_{i=n}^{m} a_i e_i\right\|$ for $x = \sum a_i e_i$. Then use the same method of proof as in Kadec's renorming theorem, instead of the distances we use the functions $\rho_n(x) = \|x - P_n(x)\|$, where P_n are the canonical projections for $\{e_i\}$.
(ii) Not in general. Indeed, then $\{e_i\}$ would be shrinking by Proposition 8.22. However, James' space J has a separable dual and a non-shrinking basis.

8.10 Does there exist a bounded operator from $C[0, 1]$ onto ℓ_1? Does there exist a bounded operator from $C[0, 1]$ onto c_0?
Hint. No for the first question. Otherwise ℓ_∞ is isomorphic to a subspace of $C[0, 1]^*$, which is not the case. One of the reasons for this is that $C[0, 1]^*$ admits an LUR norm and ℓ_∞ does not (see Remark 4 to this chapter). Therefore $C[0, 1]$ does not have a complemented subspace isomorphic to ℓ_1.
Yes for the second question (Sobczyk's theorem).

8.11 Define the following equivalent norm on ℓ_1: $\|x\| = \|x\|_1 + \|x\|_2$. Then $\|\cdot\|$ is the dual norm to some norm (denoted again by $\|\cdot\|$) on c_0. Show that $\|\cdot\|$ on ℓ_1 is strictly convex and on its unit sphere the norm- and w^*-topology coincide. Consequently, $\|\cdot\|$ on c_0 is Fréchet differentiable, yet its dual norm $\|\cdot\|$ on ℓ_1 is not LUR.
Hint. To see that $\|\cdot\|$ on ℓ_1 is not LUR, use $(\frac{1}{2}, 0, 0 \dots)$ and $(0, \frac{1}{n}, \dots, \frac{1}{n}, 0, \dots)$, where there are n coordinates equal to $\frac{1}{n}$. The rest is standard.

8.12 (Yost, [Yost0]) Let the norm on ℓ_2 be defined by $|||x|||^2 = (|x_1| + \|(0, x_2, x_3 \cdots)\|_2)^2 + \sum_{n=2}^{\infty} 2^{-n} x_n^2$, for $x = (x_1, x_2, \cdots) \in \ell_2$ where $\|\cdot\|_2$ is the norm in ℓ_2.

Show that $|||\cdot|||$ is an equivalent strictly convex norm on ℓ_2 that is not LUR and all the points on its unit sphere are strongly exposed points.
Hint. To show that $|||\cdot|||$ is rotund, use the last term in the definition of $|||\cdot|||$, the fact that $\|\cdot\|_2$ is rotund and thus you deal then only in the linear hull of the first standard unit vector in ℓ_2. Then show the standard fact that the sum of two seminorms that share Kadec-Klee property of the coincidence of the weak and norm topologies on the sphere also shares this property. The predual of the norm $|||\cdot|||$ on ℓ_2 is thus Gâteaux differentiable and because of the Kadec-Klee property of $|||\cdot|||$, the predual norm is thus Fréchet differentiable by the Šmulyan Lemma 7.22. Thus, by the Šmulyan lemma applied again, all points on the unit sphere of $|||\cdot|||$ are strongly exposed. To see that $|||\cdot|||$ is not LUR, use the standard unit vectors. Note the connection with Exercise 7.72.

8.13 Find an example of a norm on a separable Banach space X that is Fréchet differentiable at the points of a dense set and yet X^* is non-separable.
Hint. Let $\|\cdot\|$ be an equivalent LUR norm on c_0 whose dual norm is LUR. Consider its dual norm $\|\cdot\|^*$ on $X = \ell_1$.

If $f \in S_{(\ell_1,\|\cdot\|^*)}$ attains its norm at $x \in S_{(c_0,\|\cdot\|)}$ and $x_n \in S_{(c_0,\|\cdot\|)}$ are such that $f(x_n) \to 1$, then $2 \geq \|x+x_n\| \geq f(x+x_n) \to 2$ and thus by LUR of $\|\cdot\|$ we have $\|x - x_n\| = 0$. By Šmulyan's Lemma 7.22, f is a point of Fréchet smoothness. By the Bishop–Phelps theorem, $\|\cdot\|^*$ is Fréchet differentiable on a dense set in ℓ_1 and ℓ_1^* is not separable.

8.14 ([Sing2]) Show that if X is a separable Banach space such that its second dual norm is Gâteaux differentiable, then X^* is separable.
Hint. We need to show that X^* is w-separable (Proposition 3.105). It suffices to show that the duality mapping (see the definition right after Lemma 7.19) $x \in S_X \mapsto \|\cdot\|'(x) \in S_{X^*}$ is norm to weak continuous.

Let $x_n, x \in S_X$ and $x_n \to x$. Let $f_n, f \in S_{X^*}$ be such that $f_n(x_n) = 1$ and $f(x) = 1$. Then $f_n(x) = f_n(x_n) + f_n(x - x_n) \to 1$. Since $x \in S_{X^{**}}$ is a point of Gâteaux differentiability of the second dual norm, by the Šmulyan's Lemma 7.22 we have $f_n \overset{w^*}{\to} f$ in X^{***}. Since $f_n, f \in X^*$, we get $f_n \overset{w}{\to} f$ in X^*.

8.15 (Giles, Kadec, Phelps) Show that if X is a Banach space the third dual norm of which is Gâteaux differentiable, then X is reflexive.
Hint. By the Bishop–Phelps theorem, it is enough to show that every $F \in S_{X^{**}}$ that attains its norm on S_{X^*} is from X. Let $F \in S_{X^{**}}$ and $f \in S_{X^*}$ be such that $F(f) = 1$. Let $x_n \in S_X$ be such that $f(x_n) \to 1$. Consider f as an element of $S_{X^{***}}$ and $F \in S_{X^{****}}$. Since the third dual norm is Gâteaux differentiable at f, we get from the Šmulyan's Lemma 7.22 that $x_n \overset{w^*}{\to} F$ in X^{****}. Since $x_n, F \in X^{**}$, we get that $x_n \overset{w}{\to} F$ in X^{**}. Since $x_n \in X$ and X is w-closed in X^{**}, we have that $F \in X$.

8.16 Let X be a separable Banach space. Show that X^* is separable if every continuous real-valued function can be uniformly approximated by C^1-functions.

Hint. See Theorem 8.6, (i)$\Longleftrightarrow$(iv').

8.17 Follow the hint for an alternative proof of (i)$\Longrightarrow$(v) in Theorem 8.6.
Hint. (Preiss-Zajíček, [PrZa]) Let A be the set of all points in X where f is not Fréchet differentiable. We will show that A is of first category in X. Consider the epigraph of f, i.e., $\{(x, y) \in X \oplus \mathbb{R};\ y \geq f(x)\}$. Using the separation theorem in $X \oplus \mathbb{R}$ (note that its dual is isomorphic to $X^* \oplus \mathbb{R}$), for every $x \in X$ we find a functional $p^x \in X^*$ such that $f(x+h) - f(x) \geq p^x(h)$ for every $h \in X$. Since f is not Fréchet differentiable at points of A, for every $x \in A$ we find $m_x \in \mathbb{N}$ such that

$$\limsup_{h \to 0} \frac{f(x+h) - f(x) - p^x(h)}{\|h\|} > \frac{1}{m_x}.$$

For $m \in \mathbb{N}$ put $A_m = \{x \in A;\ m_x = m\}$. Given $m \in \mathbb{N}$, consider the cover of X^* by all open balls in X^* of radius $\frac{1}{24m}$. Since X^* is separable, by the Lindelöf property, let $\{B_k^m\}_k$ be a countable subfamily of these balls that covers X^*. For $k \in \mathbb{N}$ define $A_{m,k} = \{x \in A_m;\ p^x \in B_k^m\}$.

We have $A = \bigcup_{m,k} A_{m,k}$. Hence it is enough to show that $A_{m,k}$ is nowhere dense for each m, k. Fix m and k, choose any $x \in A_{m,k}$ and a neighborhood U of x. We will show that there is a point $y \in U$ that has a neighborhood V such that $V \cap A_{m,k} = \emptyset$.

By Lemma 7.3, assume that U is of the form $U = B_X^O(x, r)$, the open r-ball centered at x, where r is chosen so that f is Lipschitz with constant $K > 1/m$ on $B_X^O(x, r)$. Since $x \in A_m$, there is $h \in X$, $\|h\| < r$ such that

$$f(x+h) - f(x) > \frac{\|h\|}{m} + p^x(h). \tag{8.23}$$

We will show that $B_X^O(x+h, \|h\|/12Km) \cap A_{m,k} = \emptyset$. Assume that there is $z \in B_X^O(x+h, \|h\|/12Km) \cap A_{m,k}$. As $z \in A_{m,k}$ and $x \in A_{m,k}$, by the definition of $A_{m,k}$ we obtain $\|p^x - p^z\| < \frac{1}{12m}$. By the choice of p^z we have

$$f(x) - f(z) \geq p^z(x - z). \tag{8.24}$$

Adding (8.23) and (8.24) we have

$$\begin{aligned} f(x+h) - f(z) &> p^z(x-z) + \frac{\|h\|}{m} + p^x(h) \\ &= p^x(x+h-z) + (p^z - p^x)(x-z) + \frac{\|h\|}{m}. \end{aligned} \tag{8.25}$$

Since $\|x + h - z\| \leq \frac{\|h\|}{12Km}$ and $\|p^x\| \leq K$, we have $|p^x(x+h-z)| \leq \frac{\|h\|}{12m}$. Furthermore, $\|z - x\| \leq \|z - (x+h) + h\| \leq \frac{\|h\|}{12Km} + \|h\| \leq 2\|h\|$ and $\|p^x - p^z\| < \frac{1}{12m}$. Therefore $|(p^z - p^x)(x - y)| \leq \frac{1}{12m} \cdot 2\|h\| = \frac{\|h\|}{6m}$. Hence from (8.25) we obtain

$$f(x+h) - f(z) > \frac{\|h\|}{m} - \frac{\|h\|}{6m} - \frac{\|h\|}{3m} = \frac{\|h\|}{2m}.$$

This contradicts the fact that $\big|\|x+h\| - \|z\|\big| \leq \|x+h-z\| \leq \frac{\|h\|}{12mK}$. Therefore f is Fréchet differentiable on a residual set in X. The fact that it is actually Fréchet differentiable on a dense G_δ set in X follows from Lemma 7.45.

8.18 Find an example of a Gâteaux differentiable norm on a Banach space that is not Fréchet differentiable at some points.
Hint. Any equivalent renorming by a Gâteaux differentiable norm of ℓ_1 (Theorem 8.2) satisfies the requirement (Theorem 8.6).

8.19 Let X be a separable Banach space whose norm is Fréchet differentiable. Show that if $Y \subset X^*$ is a closed 1-norming subspace of X^*, then $Y = X^*$.
Hint. Given $x \in S_X$, put $z = \|x\|'$. Let $\{f_n\} \in B_Y$ be such that $f_n(x) \to 1$. By the Šmulyan's Lemma 7.22, $f_n \to z$, hence $z \in Y$. Thus Y contains a James boundary $\{\|x\|' : x \in S_X\}$ of X. By Theorem 3.122, $Y = X^*$.

8.20 Show that the canonical norm of $C[0, 1]$ is nowhere weak Hadamard differentiable, i.e., differentiable uniformly in directions in weakly compact sets in $C[0, 1]$.
Hint. Show that the Dirac measures $\delta_{1/n}$ do not converge to δ_0 in the topology of the uniform convergence on weakly compact sets in $C[0, 1]$; to see this, let $f_n \in C[0, 1]$ be such that $\|f_n\| = 1$, $f(1/n) = 1$, $f(0) = 0$ and $f_n \to 0$ pointwise. Then $f_n \to 0$ weakly and $(\delta_{1/n})$ does not converge uniformly on $\{f_n : n \in \mathbb{N}\}$. Then use the Šmulyan lemma (Corollary 7.20) for the weak Hadamard differentiability.

8.21 Show that every exposed point of a convex set is extreme and give an example of an extreme point that is not exposed.
Hint. Consider the function f that is identically zero on $[-\infty, 0]$ and $f(x) = x^2$ for $x > 0$. Check the origin and the epigraph of f.

A modification of this example also shows that an extreme point cannot in general be replaced by an exposed point in Proposition 3.64.

8.22 Let X be a Banach space. Show that, if every point of S_X is an extreme point, then every such point is an exposed point.
Hint. Standard.

8.23 Show that none of the spaces $C[0, 1]$, c_0 or $L_1[0, 1]$ is isomorphic to a dual space.
Hint. The unit balls of these spaces are not the closed convex hulls of their extreme points. Then use Theorem 8.14.

8.24 Let X be a Banach space with a strictly convex (rotund) norm. Show that all points of S_X are exposed points of B_X.

Hint. If $f(x) = f(y) = 1$ for $x \neq y \in S_X$ then $f\big(\frac{1}{2}(x+y)\big) = 1$ and thus $\|\frac{x+y}{2}\| = 1$.

8.25 Let $C \neq \emptyset$ be a convex compact set in a Hilbert space H. Show that C has at least one exposed point.
Hint. Assume that $\|x\| = \sup\{\|y\| : y \in C\}$. Then x is an exposed point of $\|x\| B_H$ (previous exercise) and thus an exposed point of C.

8.26 Show that exposed points need not form a James boundary.
Hint. For a two-dimensional example, consider the ball in $\mathbb{R}^2$ that is the intersection of two discs of radius 2 centered at the points $(1, 0)$ and $(-1, 0)$ and draw a picture of the dual ball.

8.27 Let X be a Banach space. Show that if the norm of X is LUR (in particular if X is a Hilbert space), then every $x \in S_X$ is a strongly exposed point of B_X.
Hint. If $x \in B_X$, $f \in B_{X^*}$ satisfy $f(x) = 1$, and $x_n \in B_X$ are such that $f(x_n) \to 1$, then $2 \geq \|x + x_n\| \geq f(x + x_n) \to 2$, hence $\|x + x_n\| \to 2$ and thus $\|x - x_n\| \to 0$ by the local uniform rotundity or the parallelogram law.

8.28 Let $K = \overline{\text{conv}}\{\{0\} \cup \{e_n\}_{n=1}^{\infty}\} \subset \ell_2$, where e_n are the standard unit vectors. Show that there does not exist a probability measure μ on K supported by $\{e_n\}$ that represents 0. This shows that Choquet's representation theorem cannot give a measure supported by the strongly exposed points $\{e_n\}$ of K.
Hint. Check that $\{e_n\}$ is the set of all strongly exposed points of K. Choose $f = \big(\frac{1}{n}\big)$.

8.29 Show that every standard unit vector e_i is a strongly exposed point of B_{ℓ_1}.
Hint. The unit vectors e_i are points of Fréchet differentiability of the norm of c_0. Then use the Šmulyan's Lemma 7.22.

8.30 Let $C = \{x \in B_{\ell_2} : x_i \geq 0 \text{ for all } i\}$. Show that 0 is an exposed point of C but a not strongly exposed point of C.

If C is defined in an analogous way for a nonseparable $\ell_2(\Gamma)$, then 0 is not even exposed. Is it extreme?

Hint. Exposed: Consider the functional (2^{-i}). Strongly exposed: $e_n \xrightarrow{w} 0$, so $f(e_n) \to f(0)$, yet $e_n \not\to 0$. In the nonseparable case, every continuous linear functional has a countable support and thus must vanish at some unit vectors. Hence 0 cannot be exposed.

However, 0 is extreme as we check by coordinates. Note that 0 is then not even a weak G_δ-point in C (i.e., an intersection of a countable family of weakly open sets) for the similar reason as for non-exposedness.

8.31 Let C be a compact convex set in a Banach space X and x be an exposed point of C. Show that x is a strongly exposed point of C.
Hint. Use the definition and compactness.

8.32 Define $C = \{x \in B_{\ell_1} : x_i \geq 0\}$. Show that C is not the closed convex hull of its w^*-strongly exposed points. A strongly exposed point of $C \subset X^*$ is called *w^*-strongly exposed* if it is strongly exposed by a functional from X.

Hint. The only w^*-strongly exposed points are the standard unit vectors. The point 0 is not in their norm closed convex hull as it is separated from them by the functional $(1, 1, \dots)$. This functional strongly exposes 0 in C.

8.33 Prove that every extreme point of B_{ℓ_∞} is a w^*-exposed point of B_{ℓ_∞}. An exposed point of $C \subset X^*$ is called *w^*-exposed* if it is exposed by a functional from X.

Hint. $\text{Ext}(B_{\ell_\infty}) = \{(x_i) : x_i = \pm 1\}$, use $y \in B_{\ell_1}$ with $\sum y_i \operatorname{sign}(x_i) = 1$.

8.34 (Lindenstrauss, Phelps [LiPh]) Show that if X is a separable reflexive Banach space, then there is an equivalent norm $\|\cdot\|$ on X such that the unit ball in $\|\cdot\|$ has only countably many strongly exposed points.

Hint. Let N be a maximal $\frac{1}{2}$-separated set in S_X. By contradiction prove that $C = \overline{\text{conv}}(N)$ contains $\frac{1}{2}B_X$: Let $x \in \frac{1}{2}B_X \setminus C$. By the separation theorem, there is $f \in S_{X^*}$ such that $f(x) > \sup_N(f)$. For any $\delta > 0$, there is $y \in S_X$ such that $f(y) > 1 - \delta$. By the maximality of N, there is $z \in N$ with $\frac{1}{2} > \|y - z\| \geq f(y) - f(z)$. Thus $\sup_N(f) \geq f(z) > f(y) - \frac{1}{2} > 1 - \delta - \frac{1}{2}$. Hence $1 - \frac{1}{2} - \delta \leq \sup_N(f) < f(x) \leq \|x\| \leq 1 - \frac{1}{2}$, i.e., $1 - \frac{1}{2} \leq \sup_N(f) < 1 - \frac{1}{2}$.

Put $\|x\| = \mu_C(x)$. Since X is reflexive, C is the closed convex hull of its strongly exposed points, so the set of the strongly exposed points is infinite. As N is in a separable space, we have $\text{card}(N) \leq \aleph_0$, so we just show that every strongly exposed point of C lies in N.

Let $x \in C$ be a strongly exposed point of C. Then there is $f \in X^*$ such that $f(x) = \sup_C(f)$ and whenever $f(x_n) \to f(x)$ for some $x_n \in C$ then $x_n \to x$. Note that $\sup_C(f) = \sup_N(f)$ and thus there are $y_n \in N$ such that $f(y_n) \to f(x)$. Thus $y_n \to x$. As N is closed (it is $\frac{1}{2}$-separated), $x \in N$.

8.35 (Lindenstrauss, Phelps [LiPh]) Show that if X is an infinite-dimensional reflexive Banach space, then $\text{Ext}(B_X)$ is uncountable.

Hint. Assume $\text{Ext}(B_X) = \{x_n\}$. Put $F_n = \{f \in B_{X^*} : f(x_n) = \|f\|\}$. Show that each F_n is weakly closed. From Proposition 3.64 we get that B_{X^*} is the union of $\{F_n\}$. By the Baire category theorem, one of F_n, say F_1, has a relative interior point f_0 in B_{X^*} in its w-topology. Assume $\|f_0\| < 1$. Thus there are $y_1, \dots, y_n \in X$ and $\varepsilon > 0$ such that $f \in F_1$ whenever $\|f\| \leq 1$ and $|(f - f_0)(y_i)| < \varepsilon$ for all $i = 1, \dots, n$. Let

$$N = \{f \in X^* : f(y_i) = f_0(y_i) \text{ for } i = 1, \dots, n \text{ and } f(x_1) = f_0(x_1)\}.$$

As X is infinite-dimensional, there is $g \in N \cap S_{X^*}$. As $g \in N$, we have $g \in F_1$ and thus $1 = \|g\| = g(x_1) = f_0(x_1) = \|f_0\|$, a contradiction.

8.36 Let X be a separable reflexive Banach space. Show that extreme points of B_X are not all isolated in the norm topology.
Hint. Every set of isolated points in a separable metric space is countable. Then use the previous exercise.

8.37 Let C be a closed convex bounded set in a Banach space X. Show that C is weakly compact if and only if C and $\overline{C}^{w^*} \subset X^{**}$ have the same extreme points.
Hint. If C is not weakly compact, by Theorem 3.130, let $f \in X^*$ does not attain its supremum on C. By the Krein–Milman theorem, there is an extreme point $\tilde{x}$ of $\overline{C}^{w^*}$ such that $f(\tilde{x}) = \sup_C(f)$ and so $\tilde{x} \notin C$.

8.38 ([Mont1]) We say that a subset A of a Banach space X has *property* $(*)$ if A is a nonempty closed convex and bounded subset of X and every point $a \in A$ is a *proper support point*, that is, given $a \in A$ there exists a^* in X^* such that $a^*(a) = \sup_A(a^*)$ and there is $x \in A$ such that $a^*(x) < \sup_A(a^*)$. Show that:
(i) $C[0, 1]^*$ contains a subset with property $(*)$.
(ii) Given an uncountable compact set K, $C(K)^*$ contains a subset with property $(*)$.
(iii) If Γ is an infinite set, $\ell_\infty(\Gamma)$ contains a subset with property $(*)$.
(iv) If X contains an isomorphic copy of $\ell_1(\Gamma)$, then X^* contains a subset with property $(*)$.
Hint. (i) Define $A = \{\mu \in C[0, 1]^* : \|\mu\| = 1, \mu \geq 0, \mu \text{ atomic }\}$. A is a convex and bounded set. Prove that A is also closed. To check that every $\mu_0 \in A$ is a proper support point, find a countable set $D \subset [0, 1]$ such that $\mu_0([0, 1]\setminus D) = 0$. Define a continuous linear functional L on $C[0, 1]^*$ by $L(\mu) = \mu(D)$, $\mu \in C[0, 1]^*$. Then L supports A properly at μ_0.
(ii) Use Milyutin's theorem: All $C(K)$ spaces are isomorphic for K an uncountable compact metric space.
(iii) Assume first that Γ is countable. $C[0, 1]$ is then isometric to a quotient of $\ell_1(\Gamma)$. Then $\ell_\infty(\Gamma)$ has a closed subspace isometric to $C[0, 1]^*$ and the result follows from (i). If Γ is uncountable, ℓ_∞ is isometric to a closed subspace of $\ell_\infty(\Gamma)$ and the result follows from the first part.
(iv) X^* then has a quotient which is isomorphic to $\ell_\infty(\Gamma)$. Let q be the quotient mapping. Using the previous results, find a subset A_q with property $(*)$ in this quotient and define $A = q^{-1}(A_q) \cap (M + \varepsilon)B_{X^*}$, where $\varepsilon > 0$ is arbitrary and M is a bound for A_q in the norm.

8.39 Let X be a reflexive Banach space whose norm is Fréchet differentiable. Show that if A_1, A_2 are bounded closed convex subsets of X such that $A_1 \cap A_2 = \emptyset$, then there are balls B_1, B_2 such that $A_1 \subset B_1$, $A_2 \subset B_2$, and $B_1 \cap B_2 = \emptyset$.
Hint. Proof of Theorems 8.18 and 3.35.

8.40 ([CoLi], [GoKa1]) Let K be a weakly compact set in a Banach space X. Prove that K is the intersection of a family of finite unions of balls.

Hint. We will present the Corson–Lindenstrauss proof of their version of this result, when X is separable and reflexive. Let $\mathcal{T}$ be the topology on B_X in which the closed sets are exactly the intersections of finite unions of balls in X. Note that $\mathcal{T}$ is weaker than the weak topology on B_X. Since B_X is weakly compact, to prove the result it is enough to show that $\mathcal{T}$ is Hausdorff as $\mathcal{T}$ and the weak topology then have to coincide on B_X.

So let $y_1, y_2 \in B_X$, $y_1 \neq y_2$. It is enough to find two balls $B(x_1, r_1)$ and $B(x_2, r_2)$ in X such that $y_i \notin B(x_i, r_i)$, $i = 1, 2$ and $B(x_1, r_1) \cup B(x_2, r_2) \supset B_X$. Indeed, then $B_X \setminus B(x_i, r_i)$, $i = 1, 2$, are two disjoint open sets in $\mathcal{T}$ and $y_i \in B_X \setminus B(x_i, r_i)$ for $i = 1, 2$, showing that $\mathcal{T}$ is a Hausdorff topology.

Put $z = y_1 - y_2$ and find $u \in S_X$ be a point of the Fréchet differentiability of the norm of X such that $\|u - \frac{z}{\|z\|}\| < \frac{1}{6}$. Indeed, X^* is separable, so we can use Theorem 8.6. For $n \in \mathbb{N}$ we define $x_1^n = y_2 - \big(n - \frac{2}{3}\|z\|\big)u$ and $x_2^n = y_1 + \big(n - \frac{2}{3}\|z\|\big)u$. We then have

$$\begin{aligned}\|x_1^n - y_1\| &= \Big\|z + \big(n - \tfrac{2}{3}\|z\|\big)u\Big\| = \Big\|z + nu - \tfrac{2}{3}\|z\|u\Big\| \\ &= \Big\|nu + \tfrac{\|z\|}{3}u - \|z\|u + z\Big\| = \Big\|nu + \tfrac{\|z\|}{3}u - \big(\|z\|u - z\big)\Big\| \\ &\geq \Big\|\big(n + \tfrac{\|z\|}{3}\big)u\Big\| - \|z\|\Big\|u - \tfrac{z}{\|z\|}\Big\| \geq n + \tfrac{\|z\|}{3} - \tfrac{\|z\|}{6} > n.\end{aligned}$$

Similarly we show that $\|x_2^n - y_2\| > n$ for all $n \in \mathbb{N}$, so $y_i \notin B(x_i^n, n)$ for $i = 1, 2$.

The proof will be complete when we show that $B\big(x_1^n, n\big) \cup B\big(x_2^n, n\big) \supset B_X$ for some $n \in \mathbb{N}$.

Suppose this is not the case. Then for every $n \in \mathbb{N}$ there is $z^n \in B_X$ such that $\|z^n - y_2 + \big(n - \frac{2}{3}\|z\|\big)u\| \geq n$ and $\|z^n - y_1 - \big(n - \frac{2}{3}\|z\|\big)u\| \geq n$. Thus for all $n \in \mathbb{N}$,

$$\Big\|u + \frac{1}{n}\big(z^n - y_2 - \frac{2}{3}\|z\|u\big)\Big\| \geq 1, \tag{8.26}$$

$$\Big\|u - \frac{1}{n}\big(y_1 - z^n - \frac{2}{3}\|z\|u\big)\Big\| \geq 1. \tag{8.27}$$

Let u^* be the Fréchet derivative of the norm of X at u. Then from the Fréchet differentiability we have $\|u + y\| = 1 + u^*(y) + o(\|y\|)$ and thus by (8.26) and (8.27), and using the fact that $\{z^n\} \subset B_X$ for $n \in \mathbb{N}$, we have

$$\tfrac{1}{n}u^*\big(z^n - y_2 - \tfrac{2}{3}\|z\|u\big) = \Big\|u + \tfrac{1}{n}\big(z^n - y_2 - \tfrac{2}{3}\|z\|u\big)\Big\| - 1 - o\big(\tfrac{1}{n}\big) \geq -o\big(\tfrac{1}{n}\big)$$

and similarly $\frac{1}{n}u^*\big(y_1 - z^n - \frac{2}{3}\|z\|u\big) \geq -o\big(\frac{1}{n}\big)$.

By adding the latter two inequalities, we obtain

$$\frac{1}{n}u^*\big(y_1 - y_2 - \tfrac{4}{3}\|z\|u\big) \geq -o\big(\tfrac{1}{n}\big).$$

Given $\varepsilon > 0$, there is n_0 such that $-o(\frac{1}{n}) > -\frac{\varepsilon}{n}$ for $n \geq n_0$. Thus for $n \geq n_0$ we have $\frac{1}{n}u^*(y_1 - y_2 - \frac{4}{3}\|z\|u) \geq -\frac{\varepsilon}{n}$ and hence $u^*(y_1 - y_2 - \frac{4}{3}\|z\|u) \geq -\varepsilon$.

As this holds for all $\varepsilon > 0$, we have $u^*(y_1 - y_2 - \frac{4}{3}\|z\|u) \geq 0$. Hence

$$u^*(z) - \tfrac{4}{3}\|z\| = u^*(z) - \tfrac{4}{3}\|z\|u^*(u) = u^*(y_1 - y_2 - \tfrac{4}{3}\|z\|u) \geq 0.$$

Thus $u^*(z) \geq \frac{4}{3}\|z\|$, which contradicts the fact that $\|u^*\| = 1$.

For more in this direction see [GoKa1].

8.41 Show that the set $A := \{x \in \ell_2 : \sum(1 + \frac{1}{i})x_i^2 \leq 1\}$ does not contain an element with norm equal to $\sup\{\|x\| : x \in A\}$.

Hint. If $x \in A$, then $\|x\|^2 = \sum x_i^2 < \sum(1 + \frac{1}{i})x_i^2 \leq 1 = \sup_A \|x\|$.

8.42 Let $\{c_n\}$ be a strictly increasing sequence of positive numbers tending to 1 and $\{e_n\}$ be the sequence of standard unit vectors in ℓ_2. Put $C := \{c_n e_n : n \in \mathbb{N}\}$. Show that $D := \overline{\text{conv}}\,(C)$ is a subset of the *open* unit ball of ℓ_2. If $c_1 > \sqrt{2} - 1$, find a point y in ℓ_2 that has a farthest point in C.

Hint. If $x \in D$ and $\|x\| = 1$, the Hahn–Banach functional supporting B_{ℓ_2} at x is again x. Then $\sup_D x = \sup_C x = 1$. This is impossible due to the fact that, if $x = (x_n)$, then $\langle x, c_n e_n\rangle = c_n x_n \to 0$. For the second question, take $y = -e_1$.

8.43 Find a bounded closed convex set and an equivalent norm on c_0 such that no point in c_0 has a farthest point in the set in the new norm.

Hint. Let $\|\cdot\|$ be an equivalent strictly convex norm on c_0 (see Theorem 13.27) Let B be the unit ball in the sup norm on c_0. If x is a farthest point in the norm $\|\cdot\|$ to some point in c_0, then it is easy to observe that x is an extreme point of B. However, B has no extreme point (see Exercise 3.131).

8.44 (Deville) Show that that there is a weak*-compact convex set in ℓ_1 such that no point in ℓ_1 has a farthest point in it.

Hint. Let $C = \{x \in \ell_1 : \sum(|x_i| + |x_i|^2) \leq 1\}$. If we denote by $r(x) = \sup\{\|x - y\| : y \in C\}$, where $\|\cdot\|$ denotes the standard norm on ℓ_1, then $r(x) = 1 + \|x\|$ for every $x \in \ell_1$. Indeed, if $y \in C$, then $\|y\| < 1$. This shows that if $x \in \ell_1$ and $y \in C$, then $\|x - y\| \leq \|x\| + \|y\| < 1 + \|x\|$, so, $r(x) \leq 1 + \|x\|$ for every $x \in \ell_1$. So it remains to show that given $x \in \ell_1$ and $\varepsilon > 0$, there is apoint $y \in C$ such that $\|x - y\| \geq 1 + \|x\| - \varepsilon$. For it, choose $p \in I\!N$ so that $\sum_{i=p+1}^{\infty} |x_i| < \varepsilon$. For each $n \in I\!N$ choose $\delta_n > 0$ so that $n(\delta_n + \delta_n^2) = 1$. Note that $\delta_n < \frac{1}{n}$ and $n\delta_n \to 1$. For each n choose $y_n = (\delta_n, \cdots \delta_n, 0, \cdots)$ where δ_n is repeated n times. Note that $y_n \in C$ by the choice of δ_n. Then for $n > p$,

$$\|x - y_n\| = \sum_{i=1}^{p} |x_i - \delta_n| + \sum_{i=p+1}^{n} |x_i - \delta_n| + \sum_{n+1}^{\infty} |x_i|$$

$$\geq \sum_{1} x_i| - p\delta_n + (n-p)\delta_n - \sum_{i=p+1}^{n} |x_i| \geq$$

$$\|x\| - \frac{\varepsilon}{3} - p\delta_n + (n-p)\delta_n - \frac{\varepsilon}{3} \geq \|x\| + 1 - \varepsilon$$

if n is chosen so that $(n - 2p)\delta_n > 1 - \frac{\varepsilon}{3}$.

8.45 Let X be a Banach space. We say that X has the *Kadec–Klee property* if the weak and norm topologies coincide on S_X.

We say that X^* has the *w^*-Kadec–Klee property* if the w^*- and norm topologies coincide on S_{X^*}.

(i) Let X be a locally uniformly rotund space. Show that X has the Kadec–Klee property.

(ii) Let X be a Banach space such that the dual norm of X^* is LUR. Show that X^* has the w^*-Kadec–Klee property.

Hint. (i) If $x_n \overset{w}{\to} x$ in S_X, take $f \in S_{X^*}$ such that $f(x) = 1$. Then $f(x_n) \to 1$ and $\|x + x_n\| \geq f(x + x_n) \to 2$. By LUR, $x_n \to x$.

(ii) Assume that $f_n, f \in S_{X^*}$ satisfy $f_n \overset{w^*}{\to} f$. Given $\varepsilon > 0$, get $x \in S_X$ such that $f(x) > 1 - \varepsilon$. Then $\|f_n + f\| \geq (f_n + f)(x) > 2 - \varepsilon$ for n large enough. Thus $\|f_n + f\| \to 2$ and use LUR.

8.46 Let X be an infinite-dimensional Banach space. Show that the weak and norm topologies do not coincide on B_X.

Hint. 0 is in the weak closure of S_X (Exercise 3.46).

8.47 Show that no point $x \in B_{c_0}$ has the property that the norm and the weak topology of B_{c_0} coincide at x.

Hint. Given $x \in B_{c_0}$ and $f_1, \ldots, f_n \in c_0^*$, use the fact that each f_i is given by a summable sequence to find a point $x' \in B_{c_0}$ such that $\max |f_i(x - x')| < \varepsilon$ and $\|x - x'\|_\infty$ is larger than $\frac{1}{2}$.

8.48 Let X be a Banach space. Let $\|\cdot\|$ be an equivalent norm on X^* such that the w^*- and norm topologies coincide on S_{X^*}. Show that then $\|\cdot\|$ is a dual norm on X^*.

Hint. Assume, by contradiction, that a net $\{f_\lambda\}$ in S_{X^*} converges to f in the w^*-topology and $\|f\| = 1 + \varepsilon$ for some $\varepsilon > 0$. Let $\tilde{f}_\lambda \in (1+\varepsilon)S_{X^*} \cap \{f + t(f_\lambda - f) : t > 0\}$. Since $\|f_\lambda - f\| \geq \varepsilon$, the numbers t_λ that define $\tilde{f}_\lambda$ are bounded and thus we have $\tilde{f}_\lambda \overset{w^*}{\to} f$ as $\tilde{f}_\lambda, f \in (1+\varepsilon)S_{X^*}$. Thus $\tilde{f}_\lambda \to f$, which is a contradiction as $\|\tilde{f}_\lambda - f\| \geq \|f_\lambda - f\| \geq \varepsilon$ (draw a picture in two dimensions).

8.49 Let X be a separable Banach space. Show that if X^* has the w^*-Kadec–Klee property, then X^* is separable. Show that, on the other hand, if X^* is separable then X has an equivalent norm such that X^* has the w^*-Kadec–Klee property.
Hint. Let x_n be dense in S_X and $f_n \in S_X^*$ be such that $f_n(x_n) = 1$. Since $\{f_n\}$ is separating, $\overline{\text{conv}}^{w^*}\{f_n\} = B_{X^*}$, so given $g \in S_{X^*}$, there are $g_n \in \overline{\text{conv}}\{f_n\}$ such that $g_n \overset{w^*}{\to} g$. Then $g_n \to g$ by the w^*-Kadec–Klee property. Since $\overline{\text{conv}}\{f_n\}$ is norm-separable, the first part of the statement follows. The second part follows from Exercise 8.45, (ii).

8.50 Let X be a separable Banach space and assume that X^* has the w^*-Kadec–Klee property. Show that if C is a w^*-compact convex set in X^*, then C has a point where the identity mapping of C into C is w^*-to-norm continuous.
Hint. Consider the dual norm $\|\cdot\|^*$ on C. As it is lower semicontinuous on (C, w^*), which is a complete metrizable space, by the Baire category theorem it is continuous at some point of C. At this point, due to the w^*-Kadec–Klee property, the w^*- and norm topology coincide.

8.51 Let X be a Banach space. Show that if the dual norm of X^* is LUR, then every point of S_{X^*} has a neighborhood base of the relative norm topology of B_{X^*} formed by slices given by functionals from X.
Hint. LUR implies w^*-Kadec–Klee. The w^*-topology has a neighborhood base formed by slices given by elements of X by Choquet's lemma.

8.52 (Troyanski) Show that if a Banach space X admits a strictly convex (i.e., rotund) norm and also a norm with the Kadec–Klee property, then X admits an LUR norm.

Note that there are spaces that admit Kadec–Klee norms and no norm that is LUR (Haydon, see, e.g., [DGZ3]). The space ℓ_∞ admits a strictly convex norm but no LUR norm (see Remarks to this chapter).
Hint. (Raja [Raja3]) First, let $\|\cdot\|$ be a norm of X that is both strictly convex and Kadec–Klee (Asplund averaging, see, e.g., [DGZ3]). Then one can use Exercise 3.146 to see that each point of the new unit sphere is *denting*, that is, for every $x \in S_X$ and $\varepsilon > 0$ there is a half-space H such that $x \in H$ and $\text{diam}(B_X \cap H) < \varepsilon$. For $m \in \mathbb{N}$ put

$$A_m = \left\{x \in B_X : \text{diam}(B_X \cap H) > \tfrac{1}{m} \text{ for all half-spaces } H \text{ containing } x\right\}.$$

Check that A_m is closed, convex, symmetric, and that $0 \in \text{Int}(A_m)$.

Let f_m be the Minkowski functional of the set A_m, which is an equivalent norm on X. Let $a_m > 0$ be such that $a_m f_m^2(x) \le 2^{-m}\|x\|^2$ for all $x \in X$. Define $|x|^2 = \sum a_m f_m^2(x) + \|x\|^2$. Then $|\cdot|$ is an equivalent norm on X.

In fact, $|\cdot|$ is an LUR: Let $x, x_k \in X$ be such that

$$2|x|^2 + 2|x_k|^2 - |x + x_k|^2 \to 0. \tag{8.28}$$

We have to show that $|x - x_k| \to 0$.

For $x = 0$ it is clear, so assume $x \neq 0$. Multiplying x and x_k by $\frac{1}{\|x\|}$ we assume that $\|x\| = 1$. From $(*)$ we have that $\|x_k\| \to \|x\| = 1$, $\|x + x_k\|/2 \to 1$. Replacing x_k by $x_k' = x_k/\|x_k\|$ we get $\|x_k'\| = 1$ and $1 \geq \|x + x_k'\|/2 \to 1$. Fix any $\varepsilon > 0$. Find $m \in \mathbb{N}$ with $m > \frac{2}{\varepsilon}$. Since x, x_k' are denting, $x, x_k' \notin A_m$. As A_m is closed, $f_m(x) > 1$ and $f_m(x_k') > 1$. From (8.28) we then have $f_m\big((x+x_k')/2\big) > 1$ for large $k \in \mathbb{N}$. Hence for these k we have $(x + x_k')/2 \notin A_m$ and thus there is a half-space H such that $(x + x_k')/2 \in H$ and $\operatorname{diam}(B_X \cap H) \leq \frac{1}{m} < \varepsilon/2$. As $x, x_k' \in B_X$, by a convexity argument we have $\|x - x_k'\|/2 < \varepsilon/2$, i.e., $\|x - x_k'\| < \varepsilon$. This means that $x_k' \to x$ and thus $x_k \to x$.

8.53 Assume that a Banach space X has the Kadec–Klee property. Show that the norm and weak Borel structures on X coincide and X is a Borel set in (X^{**}, w^*).

Note that the weak and norm Borel structures do not coincide on ℓ_∞ (Talagrand).

Hint. (Schachermayer) The mapping $(t, x) \mapsto tx$ is a Borel homeomorphism $(0, \infty) \times (S_X, w) \to (X\backslash\{0\}, w)$ (that is, the mapping and its inverse map Borel sets onto Borel sets). This follows from the fact that the first coordinate of the inverse $y \mapsto (\|y\|, y/\|y\|)$ is w-lower semicontinuous and hence Borel, while the second coordinate is the product of the w-upper semicontinuous function $y \mapsto \frac{1}{\|y\|}$ and the continuous identity mapping.

Moreover, as (S_X, w) is completely metrizable by Kadec–Klee, it is a G_δ set in its w^*-compactification. By Goldstine, this compactification is $B_{X^{**}}$. Thus X is a Borel set in (X^{**}, w^*).

8.54 (Szlenk [Szle]) There is no separable reflexive Banach space X so that every separable reflexive Banach space is isomorphic to a subspace of X.

Hint. Since every separable reflexive space has a countable Szlenk index and this is inherited by subspaces, it suffices to construct a family $(X_\alpha)_{\alpha<\omega_1}$ of separable spaces such that for every countable ordinal α, $S_Z(X_\alpha, 1) > \alpha$. For it, put inductively: $X_1 = \ell_2$, $X_{\alpha+1} = X_\alpha \oplus_1 \ell_2$ and $X_\alpha = (\sum_{\beta<\alpha} X_\beta)_{\ell_2}$ if α is a limit ordinal. Then X_α are all separable and reflexive. By induction we can show that: $0 \in s_1^\alpha B_{X^*}$. Indeed, If $x^* \in s_1^\alpha B_{X^*}$ then, for any $n \in \mathbb{N}$, $(x^*, e_n) \in s_1^\alpha(B_{X^* \oplus_\infty \ell_2})$, where (e_n) is the canonical basis of ℓ_2. Since $(0, e_n)$ is w^*-null in $X^* \oplus \ell_2 = (X \oplus_1 \ell_2)^*$, we get that $(x^*, 0) \in s_1^\alpha(B_{(X \oplus_1 \ell_2)^*})$.

8.55 Show that the Szlenk index of $c_0(\Gamma)$ is ω_0 for all Γ.

Hint. The supremum norm of the space $X = c_0(\Gamma)$ is uniformly Kadec–Klee. Therefore for every $\varepsilon > 0$, there is $0 < \theta(\varepsilon) < 1$ such that if we assume that every weak*-neighborhood of a given point $f \in B_{X^*}$ has diameter $> \varepsilon$, then $\|f\| \leq \theta(\varepsilon)$. Then from an argument of homogeneity it follows that the ε-Szlenk index is finite.

8.56 Show that the dentability index of $c_0(\Gamma)$ is countable for all Γ.

Hint. We use the fact that if j is a bijective isometry on a Banach space X, and if K is a weak*-compact convex set in X^*, then $j^*(K_\varepsilon^{(\alpha)}) = K_\varepsilon^{(\alpha)}$.

Put now $K = B_{\ell_1(\Gamma)}$. Take a countable subset D of Γ. For any $y \in \ell_1(\Gamma)$, there is a bijective isometry j_y on $c_0(\Gamma)$ such that the support of of j_y^* is included in D.

This implies that for every $\varepsilon > 0$ and any ordinal α,

$$K_\varepsilon^{(\alpha)} = \cup\{(j^*)^{-1}(K_\varepsilon^{(\alpha)} \cap B_{\ell_1(D)}) : j \text{ bijective isometry on } c_0(\Gamma)\}.$$

Since $c_0(\Gamma)$ is an Asplund space, $(K_\varepsilon^{(\alpha)})_\alpha$ is strictly decreasing as long as $K_\varepsilon^{(\alpha)} \neq \emptyset$ and $K_\varepsilon^{(\alpha)} = \emptyset$ if and only if $K_\varepsilon^{(\alpha)} \cap B_{\ell_1(D)} = \emptyset$. Since $\ell_1(D)$ is separable, there exists an ordinal $\alpha < \omega_1$ such that $K_\varepsilon \cap B_{\ell_1(D)} = \emptyset$. Thus the dentability index of $c_0(\Gamma)$ is less than ω_1.

Note that *the dentability index of* $c_0(\Gamma)$ *is* ω_0^2 *for any* Γ [HaLa].

8.57 Let $\|\cdot\|$ be an LUR norm on a separable Banach space X. Show that the set S of all elements in X^* that attain their norm is dense and G_δ in X^*.
Hint. By the Bishop–Phelps theorem, S is dense in X^*.

If f attains its norm at x and $x_n \in S_X$ satisfy $\|f\| = 1 = \lim_{n\to\infty} f(x_n)$, then $2 \geq \|x + x_n\| \geq f(x + x_n) \to 2$ and thus by LUR, $\|x_n - x\| \to 0$. Thus the dual norm is Fréchet differentiable at f. Hence S coincides with the set D of all points of Fréchet differentiability (the other direction follows from Exercise 7.26). The set D is a G_δ set in X^*.

8.58 Let $T: \ell_2 \to \ell_2$ be a diagonal operator defined for $x = (x_i) \in \ell_2$ by $T(x) = \big((1 - \frac{1}{i})x_i\big)$. Show that T does not attain its norm.
Hint. Clearly $\|T\| = 1$. If $x \in B_{\ell_2}$, then $\sum_{i=1}^{\infty}\big(1 - \frac{1}{i}\big)^2 x_i^2 < \sum x_i^2 \leq 1$.

8.59 (Lindenstrauss, [Lind4]). Let Y be the space c_0 in the standard sup-norm $\|\cdot\|$. Consider an equivalent strictly convex norm $|\!|\!|\cdot|\!|\!|$ on c_0 (it is a separable space), let $Z = (c_0, |\!|\!|\cdot|\!|\!|)$. Set $X = Y \oplus Z$ with the norm $\|(y, z)\| = \max\{\|y\|, |\!|\!|z|\!|\!|\}$. Show that the set of all bounded operators from Z into Z that attain their norm is not dense in $\mathcal{B}(X)$.
Hint. Let T_0 be an isomorphism of Y onto Z with $\|T_0\| = 1$ and define $T \in \mathcal{B}(X)$ by $T(y, z) = (0, T_0(y))$. Denote $\varepsilon \frac{1}{2\|T_0^{-1}\|} < \frac{1}{2}$ and assume that there is $\widetilde{T} \in \mathcal{B}(X)$ with $\|\widetilde{T} - T\| < \varepsilon$ and $\|\widetilde{T}\| = \|\widetilde{T}(y_0, z_0)\|$ for some (y_0, z_0) in B_X. Put $(u, v) = \widetilde{T}(y_0, z_0)$. Then $\|u\| < \varepsilon$ and since $\|\widetilde{T}\| > \varepsilon$, we have $\|u\| < \|\widetilde{T}\| = \|v\|$. Since S_Y has no extreme points, there is $y_1 \in Y\backslash\{0\}$ such that $\|y_1 + y_0\| = \|-y_1 + y_0\| \leq 1$. Hence $\|\widetilde{T}(y_0, z_0) \pm \widetilde{T}(y_1, 0)\| \leq \|\widetilde{T}\|$. Since Z is strictly convex and $\|v\| = \|\widetilde{T}\|$, we have $\widetilde{T}(y_1, 0) = (y_2, 0)$ for some $y_2 \in Y$. Then $\varepsilon\|y_1\| \geq \|T(y_1, 0) - \widetilde{T}(y_1, 0)\| \geq \|T_0(y_1)\| \geq 2\varepsilon\|y_1\|$, a contradiction.

8.60 Let the norm of a Banach space X be Gâteaux differentiable, P be a norm-one projection from X onto PX and $x_0 \in S_{PX}$. Show that $\|\cdot\|'(x_0) \in P^*X^*$, where $\|\cdot\|'(x_0)$ denotes the Gâteaux derivative of the norm $\|\cdot\|$ at x_0.
Hint. Put $f_0 = \|\cdot\|'(x_0)$. Then $\|f_0\| = 1$ and $f_0(x_0) = 1$. Moreover, $(P^* f_0)(x_0) = f_0(Px_0) = f_0(x_0) = 1$, $\|P^* f_0\| \leq \|f_0\| = 1$, and then $\|P^* f_0\| = 1$. By the Gâteaux differentiability of the norm (see Šmulyan's Lemma 7.22), $P^* f_0 = f_0$.

8.61 Let P be a bounded projection from a Banach space X onto PX. Show that PX is closed. Is PB_X closed in general? What about if $\|P\| = 1$?
Hint. $PX = (I - P)^{-1}(0)$. For the next questions: Let $f_0 \in 2S_{X^*}$ be so that f_0 does not attain its norm and choose $x_0 \in X$ such that $f_0(x_0) = 1$. Put $Px = f_0(x)x_0$ for $x \in X$. Then P is a projection from X onto $\operatorname{span}\{x_0\}$, and the norm of P is $2\|x_0\|$. Moreover, $PB_X = (-2x_0, 2x_0)$. If $\|P\| = 1$, then $PB_X = B_{PX}$. Indeed, clearly $PB_X \subset B_{PX}$ and if $y \in B_{PX}$ then $Py = y$, and $\|y\| \leq 1$, so $y \in PB_X$.

8.62 (Phelps [Phel0]) Show that ℓ_1 can be renormed so that the new norm is everywhere (outside the origin) Gâteaux differentiable but nowhere Fréchet differentiable.
Hint. Let $\|\cdot\|$ be an equivalent norm on ℓ_∞ defined by $\|x\| = \|x\|_\infty + (\sum_{i=1}^{\infty} \frac{1}{2^i}|x_i|^2)^{\frac{1}{2}}$. Then $\|\cdot\|$ is a dual strictly convex norm on ℓ_∞. We will show that the predual norm (denoted also by $\|\cdot\|$) in ℓ_1 is nowhere Fréchet differentiable by using the Šmulyan Lemma 7.22. For this, let $x \in S_{\ell_1}$ and $f_n \in S_{\ell_\infty}$ in the new norms be such that $f_n(x) \to 1$. Then change f_n on large coordinates only so that both ℓ_∞ norm and the "summation" norm do not change too much and yet new $f_n'(x) \to 1$. In the last statement we use the fact that $x_i \to 0$. Then the new f_n' violate the conclusion in Šmulyan's lemma.

8.63 (Klee, [Klee1]) Show that every separable nonreflexive Banach space $(X, \|\cdot\|)$ can be equivalently renormed so that the new norm is Gâteaux differentiable but its dual norm is not strictly convex.
Hint. See Figures 8.6 and 8.7. Let L, J, and H be closed linear subspaces of X such that $L \subset J \subset H$, J a hyperplane of H, and H a hyperplane of X. Let $p \in H$ such that $\operatorname{dist}(p, J) \geq 2$ and $q \in X$ such that $\operatorname{dist}(q, H) \geq 1$. Let Q_1 and Q_2 be the closed half-spaces of X containing 0 and bounded by the translated hyperplanes determined by $\operatorname{span}(J + q) \cup \{-p\}$ and $\operatorname{span}(J + q) \cup \{p\}$, respectively. We shall produce a smooth absolutely convex body B in X such that $B \subset Q_1 \cap Q_2$, $B \cap (L + q) = \emptyset$, but $\operatorname{dist}(B, L + q) = 0$. Then, if $|\cdot|_B$ denotes the Minkowski functional of B in X (an equivalent norm in X), as well as the corresponding norm in X/L and the dual norm in X^*, the closed unit ball of the space $(X/L, |\cdot|_B)$ admits two

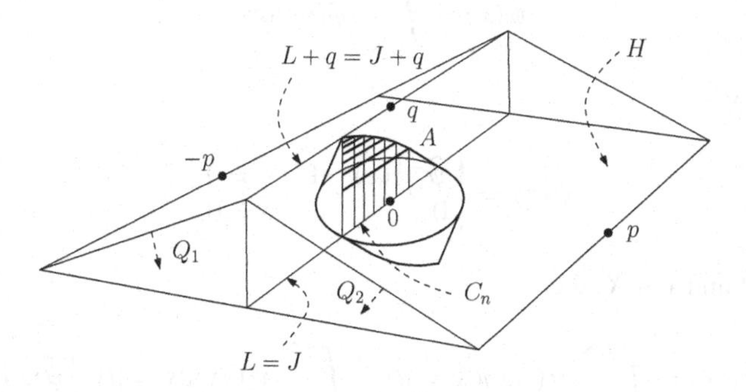

Fig. 8.6 Construction of the set A (for simplicity, we assumed $L = J$)

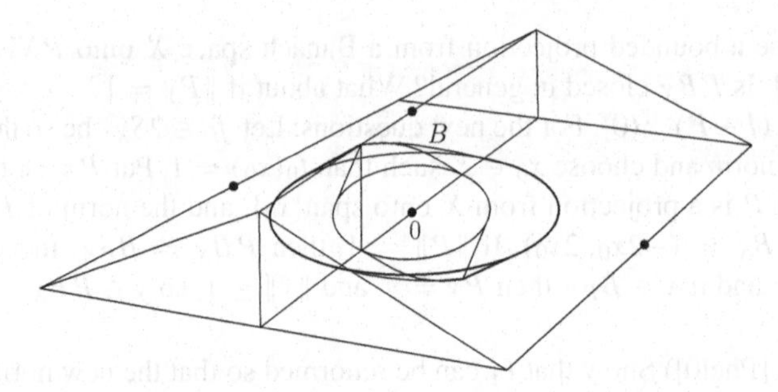

Fig. 8.7 Construction of the set B

distinct supporting hyperplanes at the point $q + L$. This implies, in particular, that $|\cdot|_B$ in X^* is not strictly convex.

To construct such B, let C_0 be the closed unit ball of $(H, \|\cdot\|)$. There exists a decreasing sequence $\{C_n\}$ of bounded closed convex in L whose intersection is empty (note that L is not reflexive). Consider $A := \overline{\Gamma}\big(\bigcup_{n=0}^{\infty}(C_n + (1 - 2^{-n})q)\big)$, where $\Gamma(S)$ denotes the convex and balanced hull of a set S. Then $A \subset X \setminus (L + q)$ and $\operatorname{dist}(A, L + q) = 0$.

There exists a compact absolutely convex smooth subset K that is contained in the open unit ball in $(X, \|\cdot\|)$ (for example $T^* B_{\ell_2}$ where T is a compact one-to-one operator from X into ℓ_2). For each $t \in (-1, 1)$, let $A_t = A \cap (H + tq)$. Finally, let $B = \bigcup_{t\in(-1,1)} \big(A_t + (1 - t)K\big)$.

8.64 Assume that a Banach space X admits a C^1-smooth bump function. Prove then that there is a function ψ on X such that

1. ψ is C^1-smooth away from the origin.
2. $\psi(tx) = |t|\psi(x)$ for $x \in X$ and $t \in \mathbb{R}$.
3. There are constants $a > 0$ and $b > 0$ such that $a\|x\| \le \psi(x) \le b\|x\|$ for $x \in X$.

Hint. Let φ be a continuously Fréchet differentiable bump function on X such that $\varphi(0) \neq 0$. Put $\widetilde{\varphi} = 1 - \exp(-\varphi^2)$ on X and define a function $\widetilde{\psi}$ for $x \in X$ by

$$\widetilde{\psi}(x) = \int_{-\infty}^{+\infty} \widetilde{\varphi}(sx)\mathrm{d}s.$$

Set for $x \in X$,

$$\psi(x) = \begin{cases} \widetilde{\psi}^{-1}(x) & \text{if} \quad x \neq 0 \\ 0 & \text{if} \quad x = 0. \end{cases}$$

If $t \in \mathbb{R}\setminus\{0\}$ and $x \in X$, then

$$\widetilde{\psi}(tx) = \int_{-\infty}^{+\infty} \widetilde{\psi}(stx)\mathrm{d}s = |t|^{-1} \int_{-\infty}^{+\infty} \widetilde{\psi}(sx)\mathrm{d}s = |t|^{-1}\widetilde{\psi}(x)$$

and therefore $\psi(tx) = |t|\psi(x)$ for $t \in \mathbb{R}$ and $x \in X$.

Note that $\widetilde{\varphi}(0) > 0$, $0 \leq \widetilde{\varphi} \leq 1$ on X and $\widetilde{\varphi}$ is a bump function on X. There exist an $a > 0$ and a $b > 0$ such that $\widetilde{\varphi}(x) > \widetilde{\varphi}(0)/2$ for $\|x\| \leq a$ and $\widetilde{\varphi}(x) = 0$ for $\|x\| \geq b$.

Given $x \neq 0$, the function $s \to \widetilde{\varphi}(sx)$ is continuous on $\mathbb{R}$, vanishes for $|s| \geq b/\|x\|$ and hence

$$\widetilde{\psi}(x) = \int_{-b/\|x\|}^{b/\|x\|} \widetilde{\varphi}(sx)\, \mathrm{d}s. \tag{8.29}$$

Since $0 \leq \widetilde{\varphi} \leq 1$ on X, it follows that for $x \neq 0$, $\widetilde{\psi}(x) \leq 2b/\|x\|$. Because $\widetilde{\varphi} \geq 0$ on X, $\widetilde{\psi}(x) \geq \int_{-a/\|x\|}^{a/\|x\|} \widetilde{\varphi}(sx)\, \mathrm{d}s \geq a\widetilde{\varphi}(0)/\|x\|$ for $x \neq 0$. Hence for $x \in X$,

$$\|x\|/2b \leq \psi(x) \leq \|x\|/a\widetilde{\varphi}(0). \tag{8.30}$$

Note that (8.29) implies that given $x \neq 0$, there is $\varepsilon > 0$ and $K > 0$ such that for $\|x' - x\| < \varepsilon$, $\widetilde{\psi}(x') = \int_{-K}^{K} \widetilde{\varphi}(sx')\, \mathrm{d}s$. Since $\widetilde{\varphi}$ is continuously Fréchet differentiable on X, from the rules of differentiation it follows that if $x, h \in X$, $x \neq 0$, then $\widetilde{\psi}'(x)(h) = \int_{-K}^{K} s\widetilde{\varphi}'(sx)(h)\, \mathrm{d}s$. Therefore, $\widetilde{\psi}$ is continuously Fréchet differentiable at any $x \neq 0$ and so is ψ.

Chapter 9
Superreflexive Spaces

In this chapter we study superreflexive Banach spaces. These spaces admit many characterizations by means of equivalent renormings, local properties, uniform smoothness, and dentability properties. We also discuss the structure of these spaces and basic sequences in them.

9.1 Uniform Convexity and Uniform Smoothness, ℓ_p and L_p Spaces

Definition 9.1 *Let* $(X, \|\cdot\|)$ *be a Banach space. For every* $\varepsilon \in (0, 2]$, *we define the* modulus of convexity *(or* rotundity*) of* $\|\cdot\|$ *by*

$$\delta_X(\varepsilon) = \inf\Big\{1 - \Big\|\frac{x+y}{2}\Big\| : x, y \in B_X, \|x - y\| \geq \varepsilon\Big\}.$$

The norm $\|\cdot\|$ *is called* uniformly convex (UC) *(or* uniformly rotund (UR)*) if* $\delta_X(\varepsilon) > 0$ *for all* $\varepsilon \in (0, 2]$. *The space* $(X, \|\cdot\|)$ *is then called a* uniformly convex space.

Note that $\delta_X(\varepsilon) = \inf\{\delta_Y(\varepsilon) : Y$ is a 2-dimensional subspace of $X\}$. The modulus of convexity in Definition 9.1 was introduced by Clarkson [Cla].

Lemma 9.2 *Let* $(X, \|\cdot\|)$ *be a Banach space and let* $\delta(\varepsilon)$ *be the modulus of convexity of* $\|\cdot\|$. *Then* $\delta(\varepsilon) = \inf\big\{1 - \|\frac{x+y}{2}\| : x, y \in S_X, \|x - y\| = \varepsilon\big\}$.

Proof: (see, e.g., [Figi1]) First note that if $x, y \in B_X$ and $\|x - y\| \geq \varepsilon$, then there are x', y' on the segment $[x, y]$ such that $\frac{x'+y'}{2} = \frac{x+y}{2}$ and $\|x' - y'\| = \varepsilon$. So it is enough to consider $x, y \in B_X$ with $\|x - y\| = \varepsilon$. It remains to prove that

$$\sup\{\|x+y\| : x, y \in B_X, \|x-y\| = \varepsilon\} = \sup\{\|x+y\| : x, y \in S_X, \|x-y\| = \varepsilon\}.$$

We may assume that X is 2-dimensional, so that the suprema are attained. Assume that $u_0, v_0 \in B_X$ maximize the left-hand-side supremum. We will show that $u_0, v_0 \in S_X$. By contradiction, assume that $\|v_0\| < 1$.

M. Fabian et al., *Banach Space Theory*, CMS Books in Mathematics,
DOI 10.1007/978-1-4419-7515-7_9,

Denote $A = \{w \in B_X : \|w - u_0\| = \varepsilon\}$. Let $x^* \in S_{X^*}$ be a functional such that $x^*(u_0 + v_0) = \|u_0 + v_0\|$. Then for $w \in A$ we have $x^*(u_0 + w) \le \|u_0 + w\| \le \|u_0 + v_0\| = x^*(u_0 + v_0)$. Since $\|v_0\| < 1$, it follows that x^* attains a local maximum with respect to A at v_0. This implies that x^* norms $(v_0 - u_0)$, so $x^*(v_0 - u_0) = \|v_0 - u_0\| = \varepsilon$. Then we get $\|u_0\| \le \frac{1}{2}\big(\|u_0 + v_0\| + \|v_0 - u_0\|\big) = \frac{1}{2}\big(x^*(u_0 + v_0) + x^*(v_0 - u_0)\big) = x^*(v_0) < 1$. This is not possible. Indeed, take $\delta = \frac{1}{2}\min(1 - \|u_0\|, 1 - \|v_0\|) > 0$, then $u' = u_0 + \delta(u_0 + v_0) \in B_X$, $v' = v_0 + \delta(u_0 + v_0) \in B_X$, $\|u' - v'\| = \varepsilon$ and $\|u' + v'\| = (1 + 2\delta)\|u_0 + v_0\| > \|u_0 + v_0\|$, contradicting the maximality. $\square$

Note that if $\|x\| = \|y\| = 1$ and $\|x - y\| = \varepsilon$, then

$$\Big\|\frac{x + y}{2}\Big\| = \Big\|x + \frac{y - x}{2}\Big\| \ge \|x\| - \Big\|\frac{y - x}{2}\Big\| \ge 1 - \frac{\varepsilon}{2}$$

and thus we have $\delta(\varepsilon) \le \frac{\varepsilon}{2}$ for every $\varepsilon \in [0, 2]$.

It is easy to see that ℓ_1^k and ℓ_∞^k are not uniformly convex. Consequently, c_0, ℓ_1, and ℓ_∞ are not uniformly convex. On the other hand, if H is a Hilbert space and $\|\cdot\|$ is the Hilbertian norm of H, then $\|\cdot\|$ is uniformly convex.

Indeed, using the parallelogram equality we have for $\varepsilon \in (0, 2]$,

$$\begin{aligned}\delta(\varepsilon) &= \inf\Big\{1 - \Big\|\frac{x + y}{2}\Big\| : x, y \in S_X, \|x - y\| = \varepsilon\Big\}\\ &= \inf\Big\{1 - \sqrt{\frac{1}{2}\|x\|^2 + \frac{1}{2}\|y\|^2 - \Big\|\frac{x - y}{2}\Big\|^2} : x, y \in S_X, \|x - y\| = \varepsilon\Big\}\\ &= 1 - \sqrt{1 - \tfrac{\varepsilon^2}{4}} > 0.\end{aligned}$$

Theorem 9.3 (Clarkson, see, e.g., [Dies1]) *Let (Ω, μ) be a measure space. If $p \in (1, \infty)$, then $L_p(\mu)$ is uniformly convex.*

In the proof, we will use the following one-dimensional fact.

Fact 9.4 *Let $p \in (1, \infty)$ and $\varepsilon > 0$. There is $\tilde{\delta}(\varepsilon) > 0$ such that if numbers $x, y \in \mathbb{R}$ satisfy $|x - y| \ge \varepsilon \max\{|x|, |y|\}$, then $\big|\frac{x+y}{2}\big|^p < (1 - \tilde{\delta}(\varepsilon))\big(\frac{|x|^p + |y|^p}{2}\big)$.*

Proof: By homogeneity, we may assume that $x = 1$ and $1 - \varepsilon \ge y \ge 0$. We have $\big[\big(\frac{1+y}{2}\big)^p\big]' = \frac{p}{2}\big(\frac{1+y}{2}\big)^{p-1} > \frac{p}{2}y^{p-1} = \big[\frac{1+y^p}{2}\big]'$ for $y \in (0, 1)$. Consequently, $f(y) = \frac{1+y^p}{2} - \big(\frac{1+y}{2}\big)^p$ is a decreasing function on $[0, 1]$. Thus $f(y) \ge f(1 - \varepsilon) > f(1) = 0$ for $y \in (0, 1 - \varepsilon)$ and the existence of $\tilde{\delta}(\varepsilon)$ follows. $\square$

Proof of Theorem 9.3: Let $\varepsilon \in (0, 2]$ be given and let $\delta = \tilde{\delta}\big(\varepsilon \cdot 4^{-\frac{1}{p}}\big)$ be from Fact 9.4. Let $x, y \in L_p(\mu)$, $\|x\|, \|y\| \le 1$ and $\|x - y\| \ge \varepsilon$, where $\|\cdot\|$ is the canonical norm of $L_p(\mu)$. Put

$$M = \{\omega : \varepsilon^p(|x(\omega)|^p + |y(\omega)|^p) \le 4|x(\omega) - y(\omega)|^p\}.$$

We claim that $\max\{\int_M |x|^p \, d\mu, \int_M |y|^p \, d\mu\} \ge \dfrac{\varepsilon^p}{2^{\frac{1}{p}+1}}$.

Assuming this claim is true, we can finish the proof as follows: Using the convexity of the function $|x|^p$, Fact 9.4 and the claim we have

$$\begin{aligned}
&\int \Big(\frac{|x(\omega)|^p + |y(\omega)|^p}{2} - \Big| \frac{x(\omega) + y(\omega)}{2} \Big|^p \Big) \, d\mu \\
&\ge \int_M \Big(\frac{|x(\omega)|^p + |y(\omega)|^p}{2} - \Big| \frac{x(\omega) + y(\omega)}{2} \Big|^p \Big) \, d\mu \\
&\ge \int_M \delta \Big(\frac{|x(\omega)|^p + |y(\omega)|^p}{2} \Big) \, d\mu \ge \frac{\delta \varepsilon^p}{2^{\frac{1}{p}+1}}.
\end{aligned}$$

Therefore

$$\int \Big| \frac{x(\omega) + y(\omega)}{2} \Big|^p \, d\mu \le \int \frac{|x(\omega)|^p + |y(\omega)|^p}{2} \, d\mu - \delta \frac{\varepsilon^p}{2^{\frac{1}{p}+2}} \le 1 - \delta \frac{\varepsilon^p}{2^{\frac{1}{p}+2}}.$$

Hence $\frac{1}{2}\|x + y\| \le \Big(1 - \delta \frac{\varepsilon^p}{2^{\frac{1}{p}+2}}\Big)^{\frac{1}{p}}$, so $\delta(\varepsilon) \ge 1 - \Big(1 - \delta \frac{\varepsilon^p}{2^{\frac{1}{p}+2}}\Big)^{\frac{1}{p}} > 0$.

To prove the claim, consider the complement M^c of the set M. We have

$$\begin{aligned}
\int_{M^c} |x(\omega) - y(\omega)|^p \, d\mu &\le \frac{\varepsilon^p}{4} \int_{M^c} \big(|x(\omega)|^p + |y(\omega)|^p\big) \, d\mu \\
&\le \frac{\varepsilon^p}{4} \int \big(|x(\omega)|^p + |y(\omega)|^p\big) \, d\mu \le \frac{\varepsilon^p}{2}.
\end{aligned}$$

Therefore, using $\|x - y\| \ge \varepsilon$ we have $\int_M |x(\omega) - y(\omega)|^p \, d\mu \ge \frac{\varepsilon^p}{2}$. Hence $\|x - y\|_M \ge \varepsilon 2^{-\frac{1}{p}}$, where $\|\cdot\|_M$ denotes the norm of $L_p(M, \mu)$. Thus

$$\max\{\|x\|_M, \|y\|_M\} \ge \frac{1}{2} \cdot \frac{\varepsilon}{2^{\frac{1}{p}}}.$$

□

Fact 9.5 *Let $(X, \|\cdot\|)$ be a Banach space. The following are equivalent:*
(i) *X is uniformly convex.*
(ii) *If $x_n, y_n \in X$, $n \in \mathbb{N}$, $\lim_{n\to\infty} (2\|x_n\|^2 + 2\|y_n\|^2 - \|x_n + y_n\|^2) = 0$, and $\{x_n\}$ is bounded, then $\lim_{n\to\infty} \|x_n - y_n\| = 0$.*
(iii) *If $x_n, y_n \in B_X$, $n \in \mathbb{N}$, and $\lim_{n\to\infty} \|x_n + y_n\| = 2$, then $\|x_n - y_n\| = 0$.*

Proof: Only (iii)$\Longrightarrow$(ii) needs more work. Let $\{x_n\}, \{y_n\} \subset X$, $\{x_n\}$ bounded and $\lim\limits_{n\to\infty} (2\|x_n\|^2 + 2\|y_n\|^2 - \|x_n + y_n\|^2) = 0$. Using the estimate

$$2\|x_n\|^2+2\|y_n\|^2-\|x_n+y_n\|^2 \geq 2\|x_n\|^2+2\|y_n\|^2-(\|x_n\|+\|y_n\|)^2 = (\|x_n\|-\|y_n\|)^2 \geq 0$$

it follows that $\lim\limits_{n\to\infty} (\|x_n\| - \|y_n\|) = 0$ and thus $\{y_n\}$ is bounded. By passing to a subsequence, we may assume that $\lim\limits_{n\to\infty} \|x_n\| = \lim\limits_{n\to\infty} \|y_n\| = a$. If $a = 0$, we are done.

Assume $a > 0$, then $\|x_n + y_n\| \to 2a$. We have $\frac{x_n}{\|x_n\|}, \frac{y_n}{\|y_n\|} \in B_X$ and $\left\| \frac{x_n}{\|x_n\|} + \frac{y_n}{\|y_n\|} \right\| \to 2$. By (iii), $\left\| \frac{x_n}{\|x_n\|} - \frac{y_n}{\|y_n\|} \right\| \to 0$, hence $\|x_n - y_n\| \to 0$. □

We remark that the assumption of boundedness in Fact 9.5(ii) is needed, see [GuiHa1].

Example: Define an equivalent norm $|||f|||$ on $C[0, 1]$ by

$$|||f|||^2 = \|f\|_\infty^2 + \|f\|_2^2,$$

where $\|\cdot\|_\infty$ denotes the standard supremum norm of $C[0, 1]$ and $\|\cdot\|_2$ denotes the canonical norm of $L_2[0, 1]$. It was shown in Exercise 7.12 that $|||\cdot|||$ is a strictly convex norm on $C[0, 1]$. Consider functions $f_n \equiv 1$ and g_n for every n, where the graph of g_n is the broken line determined by the points $(0, 0)$, $\left(\frac{1}{n}, 1\right)$, $(1, 1)$. It is easy to verify that f_n, g_n fail the property of uniform convexity. Thus $|||\cdot|||$ is not uniformly convex.

Definition 9.6 *Let $(X, \|\cdot\|)$ be a Banach space. For $\tau > 0$ we define the* modulus of smoothness *of $\|\cdot\|$,*

$$\rho(\tau) = \sup\left\{\frac{\|x + \tau h\| + \|x - \tau h\| - 2}{2} : \|x\| = \|h\| = 1\right\}.$$

We say that $\|\cdot\|$ is uniformly smooth *if* $\lim\limits_{\tau\downarrow 0} \frac{\rho(\tau)}{\tau} = 0$ (see Fig. 9.1). *We then say that $(X, \|\cdot\|)$ is* uniformly smooth.

Note that $2\|x\| = \|(x + \tau h) + (x - \tau h)\| \leq \|x + \tau h\| + \|x - \tau h\|$, so ρ is a non-negative function.

A norm $\|\cdot\|$ is uniformly smooth if for every $\varepsilon > 0$ there is $\delta > 0$ such that for all $x \in S_X$ and $y \in \delta B_X$ we have $\|x + y\| + \|x - y\| \leq 2 + \varepsilon\|y\|$.

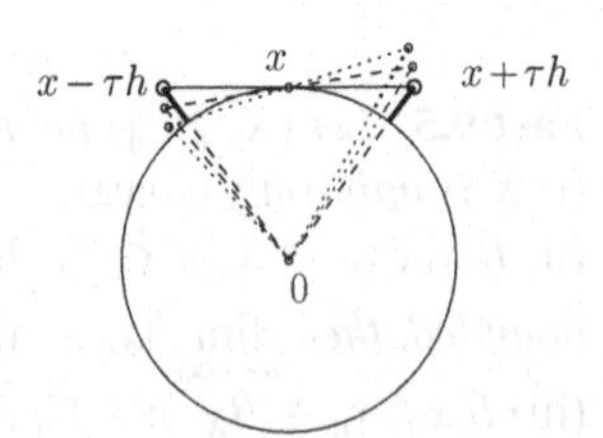

Fig. 9.1 The modulus of smoothness is the supremum of the average of the excess

Clearly a subspace of a uniformly smooth space is uniformly smooth.

The modulus of smoothness in Definition 9.6 was introduced by Day [Day0].

Fact 9.7 *Let $(X, \|\cdot\|)$ be a Banach space. The following are equivalent:*
(i) *X is uniformly smooth.*
(ii) *The norm is uniformly Fréchet differentiable on S_X.*
(iii) *The norm is Fréchet differentiable on S_X and the mapping $x \mapsto \|x\|'$ from S_X into S_X^* is uniformly continuous.*

We will omit the proof of Fact 9.7 as (i)$\Longleftrightarrow$(ii) follows a pattern similar to that of Lemma 7.4, and (i)$\Longleftrightarrow$(iii) follows from Theorem 7.27.

Lemma 9.8 (Lindenstrauss, see, e.g., [LiTz4]) *Let $(X, \|\cdot\|)$ be a Banach space, let $\delta(\varepsilon)$ be the modulus of convexity of $\|\cdot\|$ and $\rho^*(\tau)$ be the modulus of smoothness of the dual norm $\|\cdot\|^*$. Then for every $\tau > 0$,*

$$\rho^*(\tau) = \sup\Big\{\tau\frac{\varepsilon}{2} - \delta(\varepsilon) : 0 < \varepsilon \le 2\Big\}.$$

Similarly, let $\rho(\tau)$ be the modulus of smoothness of $\|\cdot\|$ and $\delta^*(\varepsilon)$ be the modulus of convexity of the dual norm $\|\cdot\|^*$. Then for every $\tau > 0$,

$$\rho(\tau) = \sup\Big\{\tau\frac{\varepsilon}{2} - \delta^*(\varepsilon) : 0 < \varepsilon \le 2\Big\}.$$

Proof: We claim that for $\varepsilon \in (0, 2]$ and $\tau > 0$ we have $\delta(\varepsilon) + \rho^*(\tau) \ge \tau\frac{\varepsilon}{2}$.

Indeed, let $x, y \in S_X$ be such that $\|x - y\| \ge \varepsilon$. Choose $f, g \in S_{X*}$ such that $f(x+y) = \|x+y\|$ and $g(x-y) = \|x-y\|$. From the definition of $\rho^*(\tau)$ we have

$$\begin{aligned}2\rho^*(\tau) &\ge \|f + \tau g\|^* + \|f - \tau g\|^* - 2 \ge (f + \tau g)(x) + (f - \tau g)(y) - 2\\ &= f(x + y) + \tau g(x - y) - 2 = \|x + y\| + \tau\|x - y\| - 2 \ge \|x + y\| + \tau\varepsilon - 2.\end{aligned}$$

Hence $2 - \|x + y\| \ge \tau\varepsilon - 2\rho^*(\tau)$. Thus from the definition of $\delta(\varepsilon)$ we have $\delta(\varepsilon) + \rho^*(\tau) \ge \tau\frac{\varepsilon}{2}$. Consequently, $\rho^*(\tau) \ge \sup\{\tau\frac{\varepsilon}{2} - \delta(\varepsilon) : 0 < \varepsilon \le 2\}$.

To prove the converse inequality, let $\tau > 0$ and $f, g \in S_{X^*}$. For $\eta > 0$ there exist $x, y \in S_X$ such that $(f+\tau g)(x) \ge \|f+\tau g\|^* - \eta$ and $(f-\tau g)(y) \ge \|f-\tau g\|^* - \eta$. Hence

$$\begin{aligned}&\tfrac{1}{2}\big(\|f + \tau g\|^* + \|f - \tau g\|^* - 2\big)\\ &\le \tfrac{1}{2}\big(f(x + y) - 2\big) + \tfrac{\tau}{2}g(x - y) + \eta \le \Big(\Big\|\frac{x + y}{2}\Big\| - 1\Big) + \tfrac{\tau}{2}\|x - y\| + \eta\\ &\le -\delta(\|x - y\|) + \tfrac{\tau}{2}\|x - y\| + \eta \le \sup\big\{t\tfrac{\varepsilon}{2} - \delta(\varepsilon) : 0 < \varepsilon \le 2\big\} + \eta.\end{aligned}$$

Thus $\rho^*(\tau) \le \sup\big\{\tau\frac{\varepsilon}{2} - \delta(\varepsilon) : 0 < \varepsilon \le 2\big\}$ as $\eta > 0$ was arbitrary.

The dual statement is obtained similarly. □

For a Hilbert space H, we easily calculate $\rho_H(\tau) = \sqrt{1+\tau^2} - 1$, in particular H is uniformly smooth. In fact, Hilbert spaces are the "most uniformly convex" and "most uniformly smooth" spaces; precisely, for every Banach space X we have $\delta_X(\varepsilon) \le 1 - \sqrt{1 - \frac{\varepsilon^2}{4}}$ and $\rho_X(\tau) \ge \sqrt{1+\tau^2} - 1$ (Nördlander, see, e.g., [Dies1]).

Theorem 9.9 (Lindenstrauss [Lind3]) *Let $(X, \|\cdot\|)$ be a Banach space and $(X^*, \|\cdot\|^*)$ its dual.*
(i) *The norm $\|\cdot\|$ is uniformly convex if and only if $\|\cdot\|^*$ is uniformly Fréchet differentiable.*
(ii) *The norm $\|\cdot\|$ is uniformly Fréchet differentiable if and only if $\|\cdot\|^*$ is uniformly convex.*

Proof: (i) Let $\|\cdot\|$ be uniformly convex with the modulus of convexity $\delta(\varepsilon)$ and let $\rho^*(\tau)$ be the modulus of smoothness of $\|\cdot\|^*$. Let $\varepsilon_0 > 0$ be given. From the definition of $\delta(\varepsilon)$, we have $\delta(\varepsilon) \ge \delta(\varepsilon_0) > 0$ for every $\varepsilon \in [\varepsilon_0, 2]$.

Let $\tau \in \big(0, \delta(\varepsilon_0)\big)$. For $\varepsilon \in [\varepsilon_0, 2]$ we have $\frac{\varepsilon}{2} - \delta(\varepsilon)/\tau \le \frac{\varepsilon}{2} - \delta(\varepsilon_0)/\tau \le \frac{\varepsilon}{2} - 1 \le 0$, so by Lemma 9.8,

$$\frac{\rho^*(\tau)}{\tau} = \sup_{0<\varepsilon\le\varepsilon_0}\left(\frac{\varepsilon}{2} - \frac{\delta(\varepsilon)}{\tau}\right) \le \sup_{0\le\varepsilon\le\varepsilon_0}\left(\frac{\varepsilon}{2}\right) = \frac{\varepsilon_0}{2}.$$

Hence $\lim\limits_{\tau\to 0} \frac{\rho^*(\tau)}{\tau} = 0$.

Conversely, if $\|\cdot\|$ is not uniformly convex, then there is $\varepsilon_0 > 0$ such that $\delta(\varepsilon_0) = 0$. Then from Lemma 9.8, we have for every $\tau > 0$

$$\rho^*(\tau) = \sup_{0<\varepsilon\le 2}\left\{\frac{\varepsilon\tau}{2} - \delta(\varepsilon)\right\} \ge \frac{\varepsilon_0\tau}{2}.$$

Therefore $\limsup\limits_{\tau\to 0} \frac{\rho^*(\tau)}{\tau} \ge \varepsilon_0$ and $\|\cdot\|^*$ is not uniformly Fréchet differentiable.

(ii) It is proved similarly using the second statement in Lemma 9.8. □

From Theorems 9.3 and 9.9 we get the following.

Theorem 9.10 *Let (Ω, μ) be a measure space with a σ-finite measure μ. If $p \in (1, \infty)$, then $L_p(\mu)$ is uniformly smooth.*

As duals of spaces whose canonical norms are not uniformly convex, ℓ_1^k and ℓ_∞^k are not uniformly smooth. Consequently, c_0, ℓ_1, and ℓ_∞ are not uniformly smooth.

Theorem 9.11 (Milman, Pettis) *Let $(X, \|\cdot\|)$ be a Banach space. If $\|\cdot\|$ is uniformly convex or uniformly Fréchet differentiable, then X is reflexive.*

Proof: Let $(X, \|\cdot\|)$ be a uniformly convex space and $f \in S_{X^*}$ be given. Choose $x_n \in S_X$ such that $\lim\limits_{n\to\infty} f(x_n) = 1$. Then given $\varepsilon > 0$, for n, m greater than some

n_0 we have $2 \geq \|x_n + x_m\| \geq f(x_n + x_m) \geq 2 - \varepsilon$. Hence $\{x_n\}$ is Cauchy by the uniform convexity of $\|\cdot\|$ and $\lim_{n\to\infty} x_n = x$ for some $x \in S_X$. Clearly $f(x) = 1$, so f attains its norm. By Corollary 3.131, X is reflexive.

Assume now that the norm is uniformly Fréchet differentiable. Then the norm of X^* is uniformly convex by Theorem 9.9. Therefore X^* is reflexive by the first part of this proof and thus X is reflexive. □

For an argument proving reflexivity that does not use James' theorem, see Exercise 9.3.

9.2 Finite Representability, Superreflexivity

Proposition 9.12 *Let the norm $\|\cdot\|_X$ of a Banach space X be uniformly convex (respectively uniformly Fréchet differentiable). If a Banach space Y is crudely finitely representable in X, then Y admits an equivalent norm that is uniformly convex (respectively uniformly Fréchet differentiable).*

We will prove separable versions; the general case requires only a minor adjustment.

Proof: Assume that $\|\cdot\|_X$ is uniformly convex and a separable Banach space Y is crudely finitely representable in X with constant $K > 1$. Let $\{x_n\}$ be dense in Y. For every $n \in \mathbb{N}$, put $F_n = \operatorname{span}\{x_i\}_{i=1}^n$ and let $T_n \colon F_n \to X$ be an operator such that $\|T_n\| \leq K$ and $\|T_n^{-1}\| = 1$. Define a norm $\|\cdot\|_n$ on F_n by $\|x\|_n = \|T_n(x)\|_X$. As $\|x\|_Y = \|T_n^{-1}T_n(x)\|_X \leq \|T_n(x)\|_X$ and $\|T_n(x)\|_X \leq \|T_n\| \cdot \|x\|_Y$, we have $\|x\|_Y \leq \|x\|_n \leq K\|x\|_Y$ for every $x \in F_n$. Extend $\|\cdot\|_n$ by 0 to Y, and by the Cantor diagonal argument assume that a subsequence $\{\|\cdot\|_{n_k}\}$ is convergent at each point of $\{x_n\}$. Because of the uniform equicontinuity of all $\|\cdot\|_{n_k}$ on F_ns, we have that the sequence $\|\cdot\|_{n_k}$ is convergent at every point of Y and its limit is an equivalent norm $\|\cdot\|_0$ that satisfies $\|x\|_Y \leq \|x\|_0 \leq K\|x\|_Y$ for every $x \in Y$. We claim that $\|\cdot\|_0$ is uniformly convex.

Indeed, from the uniform convexity of the norm $\|\cdot\|_X$ on X we have that given $\varepsilon > 0$, there is $\delta > 0$ such that if $u, v \in X$ satisfy $\|u\| \in (1-\delta, 1+\delta)$, $\|v\| \in (1-\delta, 1+\delta)$ and $\|\frac{u+v}{2}\| \in (1-\delta, 1+\delta)$, then $\|u - v\| < \varepsilon$. Given x, y in Y such that $\|x\|_0 = \|y\|_0 = 1$ and $\|\frac{x+y}{2}\|_0 > 1 - \frac{\delta}{2}$, then for k large enough we have that $\|x\|_{n_k} = \|T_{n_k}(x)\|_X \in (1-\delta, 1+\delta)$, $\|y\|_{n_k} = \|T_{n_k}(y)\|_X \in (1-\delta, 1+\delta)$, $\big\|\frac{x+y}{2}\big\|_{n_k} = \big\|T_{n_k}\big(\frac{x+y}{2}\big)\big\|_X > 1 - \delta$, and $\big|\|x-y\|_0 - \|x-y\|_{n_k}\big| = \big|\|x-y\|_0 - \|T_{n_k}(x-y)\|_X\big| < \frac{\varepsilon}{2}$. Since from the above we get $\|T_{n_k}(x-y)\|_X < \varepsilon$, we have $\|x - y\|_0 < \varepsilon$.

The proof for the other cases is similar. □

Note that it follows from Proposition 9.12 and Theorem 9.11 that if the norm of a Banach space X is uniformly convex or uniformly Fréchet differentiable and Y is crudely finitely representable in X, then Y is reflexive. Since, for instance, c_0 is not reflexive and is finitely representable in the reflexive space $\left(\sum \ell_\infty^n\right)_2$ by

Theorem 6.2, $\left(\sum \ell_\infty^n\right)_2$ does not admit any norm that is uniformly convex or uniformly Fréchet differentiable.

Definition 9.13 *A Banach space X is said to be* superreflexive *if every Banach space finitely representable in X is reflexive.*

Note that $X = \left(\sum \ell_\infty^n\right)_2$ is a reflexive space which is not superreflexive, as c_0 is finitely representable in X. There is a reflexive separable Banach space with unconditional basis, constructed by Tsirelson [Tsir], and denoted by T (see Exercises 9.29, 9.30, and 9.31), such that c_0 is finitely representable in all its infinite-dimensional subspaces. This example will be discussed in Exercises 9.29, 9.30, and 9.31.

The following result is due to the contribution of several mathematicians.

Theorem 9.14 *Let X be a Banach space. The following are equivalent:*
(i) *X is superreflexive.*
(ii) *X admits an equivalent uniformly convex norm.*
(iii) *X admits an equivalent uniformly Fréchet differentiable norm.*
(iv) *X admits an equivalent norm which is uniformly convex and uniformly Fréchet differentiable.*
(v) *X admits a uniformly Fréchet differentiable bump function.*
(vi) *X does not contain an isomorphic copy of c_0 and it admits a bump with locally uniformly continuous derivative.*
(vii) *The dual dentability index of X is ω_0.*

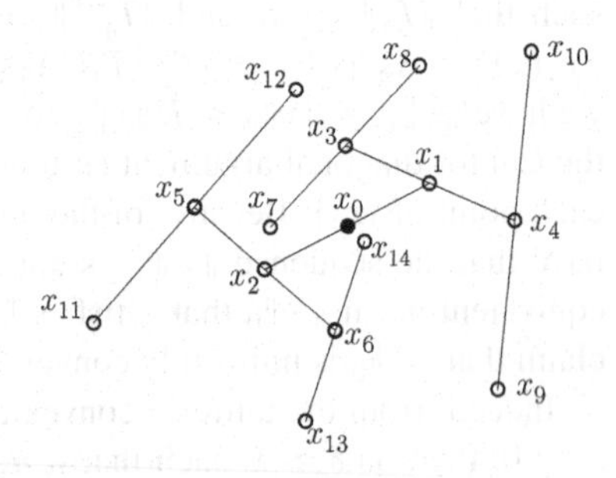

Fig. 9.2 The elements of a $(3, \varepsilon)$-tree with vertex x_0

We remark that if a superreflexive Banach space X has a Schauder basis $\{e_i\}_{i=1}^{\infty}$, then X admits an equivalent uniformly convex (respectively uniformly Fréchet differentiable) norm in which the basis is monotone (see [GuiHa2]).

To prove Theorem 9.14, we need to do some preparatory work. First, we need to introduce *trees* in a Banach space. For $n = 0, 1, \ldots$ and $\varepsilon > 0$, an *(n, ε)-tree* in a Banach space X is defined inductively as follows (see Fig. 9.2): For $n = 0$, any point $x_0 \in X$ is a $(0, \varepsilon)$-tree. Assume that an (n, ε)-tree was formed by adding the points $x_1, \ldots, x_{2^n}$ to an $(n-1, \varepsilon)$-tree, then an $(n+1, \varepsilon)$-tree results from adding 2^{n+1} points $y_i, z_i, i = 1, \ldots, 2^n$, to the (n, ε)-tree in such a way that $x_i = \frac{1}{2}(y_i + z_i)$ and $\|y_i - z_i\| \geq \varepsilon$ for each i. Therefore an (n, ε)-tree has $2^0 + \cdots + 2^n = 2^{n+1} - 1$ elements.

We arrange the elements of an (n, ε)-tree into a sequence as follows. x_0 is the vertex. By x_1, x_2 we denote the elements added at the second step, so we have $x_0 = (x_1 + x_2)/2$, $\|x_1 - x_2\| \geq \varepsilon$. By x_3, x_4, x_5, x_6 we denote the elements added at the third step, so we have $x_1 = (x_3 + x_4)/2$, $\|x_3 - x_4\| \geq \varepsilon$, and $x_2 = (x_5 + x_6)/2$, $\|x_5 - x_6\| \geq \varepsilon$. In general, $x_i = (x_{2i+1} + x_{2i+2})/2$ and $\|x_{2i+1} - x_{2i+2}\| \geq \varepsilon$ for every $i \leq 2^n - 2$.

An (∞, ε)*-tree* is defined by repeating the above construction countably many times.

Note that if $n \in \mathbb{N}$, we can build an $(n, 2)$-tree in $B_{\ell_\infty^n}$ as follows: set $x_0 = (0, \ldots, 0)$, $x_1 = (1, 0, \ldots, 0)$, $x_2 = (-1, 0, \ldots, 0)$, $x_3 = (1, 1, 0, \ldots)$, $x_4 = (1, -1, 0, 0, \ldots, 0)$, etc. Therefore $X = \left(\sum \ell_\infty^n\right)_2$ has the property that for every n, B_X contains an $(n, 2)$-tree. However, X cannot contain a bounded (∞, ε)-tree for any $\varepsilon > 0$. Indeed, we have the following result.

Theorem 9.15 *A reflexive Banach space does not contain a bounded* (∞, ε)*-tree for any* $\varepsilon > 0$.

Proof: By contradiction, assume that a reflexive space X contains a bounded (∞, ε)-tree T. As $\overline{T}$ is separable, we may assume that X is separable. Put $C = \overline{\text{conv}}(T)$. Then C is weakly compact and convex and thus contains a strongly exposed point $x \in C$ (Theorem 8.14). Thus there is $f \in X^*$ such that $f(x) = \sup_C(f) = \sup_T(f)$ and there is $\delta > 0$ such that $\|z - x\| < \frac{\varepsilon}{8}$ whenever $z \in T$ and $f(z) > f(x) - \delta$. Since $\sup_T(f) = f(x)$, there is $t \in T$ such that $f(t) > f(x) - \delta$. Let $t = (t_1 + t_2)/2$, $\|t_1 - t_2\| \geq \varepsilon$, $t_1, t_2 \in T$. Then for one element of $\{t_1, t_2\}$, say t_1, we have $f(t_1) > 1 - \delta$. Thus $\|t - x\| < \frac{\varepsilon}{8}$ and $\|t_1 - x\| < \frac{\varepsilon}{8}$, hence $\|t_1 - t\| < \frac{\varepsilon}{4}$, a contradiction with $\|t - t_1\| = \frac{1}{2}\|t_1 - t_2\| > \frac{\varepsilon}{2}$. □

Theorem 9.16 (James [Jame6]) *Let* X *be a Banach space. If there is* $\varepsilon > 0$ *such that* B_X *contains an* (n, ε)*-tree for every* $n \in \mathbb{N}$*, then there is a Banach space* Y *finitely representable in* X *such that* B_Y *contains an* (∞, ε)*-tree.*

Proof: For $n \in \mathbb{N}$, let $\{x_i^n : i = 0, 1, \ldots, 2^{n+1} - 2\}$ be an (n, ε)-tree in B_X. For $(\alpha_i) \in c_{00}$ put $\|(\alpha_i)\|_n = \left\| \sum_{i=0}^{2^{n+1}-2} \alpha_i x_i^n \right\|$. By the separability of c_{00} and the boundedness of $\{x_i^n\}$, we can find a sequence $\{n_k\}$ in $\mathbb{N}$ such that for every $(\alpha_i) \in c_{00}$, the limit $\|(\alpha_i)\| = \lim \|(\alpha_i)\|_{n_k}$ exists for every $(\alpha_i) \in c_{00}$.

Denote by $\widetilde{Y}$ the space $(c_{00}, \|\cdot\|)$. Note that $\|\cdot\|$ is a seminorm. Consider the closed subspace $N = \{v \in \widetilde{Y} : \|v\| = 0\}$ and put $Y_1 = \widetilde{Y}/N$. Clearly, Y_1 is a normed space that is finitely representable in X. By $\{e_i\}$ we denote the images of the canonical basis of c_{00} in Y_1 under the quotient mapping. Let Y be the completion of Y_1.

Note that if W is a Banach space and Z is a dense subspace of W, then W is finitely representable in Z. Indeed, let $\varepsilon > 0$ and let F be a finite-dimensional subspace of W with a normalized basis $\{w_1, \ldots, w_n\}$, let $K > 0$ be such that for every $w = \sum \alpha_i w_i$ we have $\frac{1}{K} \sum |\alpha_i| \leq \|w\| \leq K \sum |\alpha_i|$. Choose $z_i \in Z$ such that $|z_i - w_i| \le \varepsilon/K$ and define an operator $T\colon F \to \operatorname{span}\{z_i\}$ by $T(w) = \sum \alpha_i z_i$ for $w = \sum \alpha_i w_i$. Then $\big|\|w\| - \|T(w)\|\big| \le \sum |\alpha_i| \cdot \|w_i - z_i\| \le K\|w\| \cdot \varepsilon/K$. Therefore $\|T\| \cdot \|T^{-1}\|$ can be made arbitrarily close to 1.

This means that Y is finitely representable in Y_1, hence in X. We claim that $\{e_i\}_{i\in\mathbb{N}}$ forms a bounded (∞, ε)-tree in Y. Indeed, given $i \in \mathbb{N}$, we have $\|e_i\|_{n_k} = \|x_i^{n_k}\| \le 1$, hence $\{e_i\}$ is bounded.

Next, given $i \in \mathbb{N}$, for every n such that $2^n - 2 \ge i$ (that is, x_i^n is not in the last row of the (n, ε)-tree) we have $x_i^n - (x_{2i+1}^n + x_{2i+2}^n)/2 = 0$. Therefore for k large enough we have $\|e_i - (e_{2i+1} + e_{2i+2})/2\|_{n_k} = 0$, so passing to the limit with k we get $\|e_i - (e_{2i+1} + e_{2i+2})/2\| = 0$. This means that $e_i = (e_{2i+1} + e_{2i+2})/2 \in Y_1$, hence in Y.

Similarly we show that $\|e_{2i+1} - e_{2i+2}\| \ge \varepsilon$, which completes the proof. □

Remark: Note that from Theorems 9.16 and 9.15 it follows that given a superreflexive space X and $\varepsilon > 0$, there is n_0 such that B_X contains no (n, ε)-tree for $n \ge n_0$.

For a Banach space X and $x_0 \in X$, consider a sequence $\{x_0, \ldots, x_{2^{n+1}-2}\}$ of vectors such that $x_i = x_{2i+1} + x_{2i+2}$, $\|x_{2i+1}\| = \|x_{2i+2}\|$, and $\left\| \frac{x_{2i+1}}{\|x_{2i+1}\|} - \frac{x_{2i+2}}{\|x_{2i+2}\|} \right\| \ge \varepsilon$ for every $0 \le i \le 2^n - 2$.

Clearly, $x_0 = \sum_{i=2^k-1}^{2^{k+1}-2} x_i$ for $0 \le k \le n$. The 2^n elements $\{x_{2^n-1}, \ldots, x_{2^{n+1}-2}\}$ are then called the *(n, ε)-decomposition* of x_0. As $\|x_{2i+1}\| = \|x_{2i+2}\| \ge \frac{\|x_i\|}{2}$ for $0 \le i \le 2^n - 2$, we have by induction that $\|x_j\| \ge \|x_0\|/2^k$ for every $2^k - 1 \le j \le 2^{k+1} - 2$. It is easy to show that $\{x_0, 2x_1, 2x_2, \ldots, 2^n x_{2^{n+1}-2}\}$, where the coefficient of x_i is 2^k for $2^k - 1 \le i \le 2^{k+1} - 2$, forms an $(n, \varepsilon\|x_0\|)$-tree in X.

If X is superreflexive and $\varepsilon > 0$, by the Remark after Theorem 9.16 there exists $n \in \mathbb{N}$ such that for every (n, ε)-decomposition $\{x_i\}$ of x_0 we have $\|2^n x_j\| > 2\|x_0\|$ for some $2^n - 1 \le j \le 2^{n+1} - 2$. Therefore

$$\sum_{i=2^n-1}^{2^{n+1}-2} \|x_i\| = \|x_j\| + \sum_{i=2^n-1, i\neq j}^{2^{n+1}-2} \|x_i\| \ge \frac{\|x_0\|}{2^{n-1}} + (2^n - 1)\frac{\|x_0\|}{2^n}$$

$$\ge \|x_0\| + \frac{\|x_0\|}{2^{n-1}} - \frac{\|x_0\|}{2^n} \ge \left(1 + \frac{1}{2^{n-1}}\right)\|x_0\|.$$

By a telescopic argument, we obtain

$$\sum_{i=2^{kn}-1}^{2^{kn+1}-2} \|x_i\| \ge \left(1 + \frac{1}{2^{n-1}}\right)^k \|x_0\|$$

for every (kn, ε)-decomposition of x_0. Thus if X is superreflexive, $\varepsilon > 0$ is given and $K > 0$ is a constant, there is $M = M(\varepsilon, K)$ such that for all (M, ε)-decompositions $z = \sum_{i=1}^{2^M} x_i$ of $z \in X\setminus\{0\}$ we have $\sum_{i=1}^{2^M} \|x_j\| \ge K\|z\|$.

Lemma 9.17 (Enflo [Enfl1]) *Let $(X, \|\cdot\|)$ be a superreflexive space, $\varepsilon \in (0, \frac{1}{8})$. Then there is a non-negative function $|\cdot|$ on X such that:*
(a) $|\alpha x| = |\alpha||x|$ *for all $x \in X$ and scalars α,*

(b) $\left(1-\frac{\varepsilon}{4}\right)\|x\| \leq |x| \leq \left(1-\frac{\varepsilon}{12}\right)\|x\|$ *for every* $x \in X$,
(c) *There exists* $\eta > 0$ *such that if* $\|x\| = \|y\| = 1$ *and* $\|x-y\| \geq \varepsilon$, *then* $|x+y| \leq |x|+|y|-\eta$.

Proof: We define $|\cdot|$ for $x \in X$ by

$$|x| = \inf\left\{\frac{\sum_{j=1}^{2^n}\|u_j\|}{1+\frac{\varepsilon}{6}\left(1-\frac{1}{4^{n+1}}\right)}\right\},$$

where the infimum runs through all $n \in \mathbb{N}_0$ and all (n,ε)-decompositions $x = \sum_{j=1}^{2^n} u_j$ of x.

It is easy to check (a). To get (b), note that $|x| \geq \dfrac{\|x\|}{1+\frac{\varepsilon}{6}} \geq \|x\|\left(1-\frac{\varepsilon}{4}\right)$. The $(0,\varepsilon)$-decomposition $\{x\}$ yields $|x| \leq \dfrac{\|x\|}{1+\frac{\varepsilon}{8}} \leq \|x\|\left(1-\frac{\varepsilon}{12}\right)$.

We now prove (c). Let $M = M\left(\varepsilon, 1+\frac{\varepsilon}{8}\right)$. By (b), we may assume in the definition of $|x|$ and $|y|$ that $n < M$. Fix $\delta > 0$ and let $x = u_1+u_2+\cdots+u_{2^k}$, $y = v_1+\cdots+v_{2^l}$ give approximations of $|x|$ respectively $|y|$ up to δ, $k, l < M$. Assume $l \geq k$. By summing up the appropriate vectors from the (l,ε)-decomposition $\{v_i\}_{i=1}^{2^l}$ of y, we obtain a (k,ε)-decomposition $\{w_i\}_{i=1}^{2^k}$ of y satisfying:

$$\frac{\sum_{j=1}^{2^k}\|w_j\|}{1+\frac{\varepsilon}{6}} \leq \frac{\sum_{j=1}^{2^l}\|v_j\|}{1+\frac{\varepsilon}{6}\left(1-\frac{1}{4^{l+1}}\right)} \leq |y|+\delta.$$

Since $\|x\| = \|y\| = 1$ and $\|x-y\| \geq \varepsilon$, we have that $x+y = u_1+\cdots+u_{2^k}+w_1+\cdots+w_{2^k}$ is a $(k+1,\varepsilon)$-decomposition of $x+y$ and thus

$$\frac{\sum\|u_j\|+\sum\|w_j\|}{1+\frac{\varepsilon}{6}\left(1-\frac{1}{4^{k+2}}\right)} \geq |x+y|.$$

Therefore we have

$$|x|+|y|-|x+y| \geq \frac{\sum\|u_j\|}{1+\frac{\varepsilon}{6}\left(1-\frac{1}{4^{k+1}}\right)} + \frac{\sum\|w_j\|}{1+\frac{\varepsilon}{6}} - \frac{\sum\|u_j\|+\sum\|w_j\|}{1+\frac{\varepsilon}{6}\left(1-\frac{1}{4^{k+2}}\right)} - 2\delta$$

$$= \frac{\frac{\varepsilon}{2}\frac{1}{4^{k+2}}\sum\|u_j\|}{\left(1+\frac{\varepsilon}{6}\left(1-\frac{1}{4^{k+1}}\right)\right)\left(1+\frac{\varepsilon}{6}\left(1-\frac{1}{4^{k+2}}\right)\right)} - \frac{\frac{\varepsilon}{6}\frac{1}{4^{k+2}}\sum\|w_j\|}{\left(1+\frac{\varepsilon}{6}\right)\left(1+\frac{\varepsilon}{6}\left(1-\frac{1}{4^{k+2}}\right)\right)} - 2\delta$$

$$\geq \frac{\varepsilon}{2}\cdot\frac{1}{4^{k+2}}\frac{|x|-\frac{1}{3}(|y|+\delta)}{1+\frac{\varepsilon}{6}\left(1-\frac{1}{4^{k+2}}\right)} - 2\delta \geq \frac{\varepsilon}{2\cdot 4^{k+2}}\left(\frac{1-\frac{\varepsilon}{4}}{2}-\frac{1}{3}\right) - 2\delta$$

$$\geq \frac{\varepsilon}{32\cdot 4^M}\cdot\left(\frac{1}{6}-\frac{\varepsilon}{8}\right) - 2\delta \left(\text{as } 0 < \varepsilon \leq \frac{1}{8}\right) \geq \frac{\varepsilon}{32\cdot 4^M}\cdot\left(\frac{1}{6}-\frac{1}{64}\right) - 2\delta = \eta > 0$$

for δ sufficiently small. □

Lemma 9.18 (Enflo [Enfl1]) *Let $(X, \|\cdot\|)$ be a superreflexive space, $\varepsilon \in (0, \frac{1}{8})$. Let $|\cdot|$ be a non-negative function on X with properties* (a)–(c) *in Lemma* 9.17. *Then X admits an equivalent norm $|||\cdot|||$ which has the following properties:*
(A) $\left(1-\frac{\varepsilon}{4}\right)\|x\| \leq |||x||| \leq \left(1-\frac{\varepsilon}{12}\right)\|x\|$ for every $x \in X$,
(B) If $\|x\| = \|y\| = 1$ and $\|x-y\| \geq 5\varepsilon$ then $|||x+y||| \leq |||x||| + |||y||| - \eta\varepsilon$, where η is as in Lemma 9.17 (c).

Proof: Define for $x \in X$,

$$|||x||| = \inf\Big\{\sum_{j=1}^{n} |x_j - x_{j-1}| : n \in \mathbb{N}, \{x_j\}_{j=0}^{n} \subset X \text{ with } x_0 = 0, x_n = x\Big\}.$$

We clearly have $|||\alpha x||| = |\alpha|\,|||x|||$ for every $x \in X$ and every scalar α. The function $|||\cdot|||$ satisfies the triangle inequality: Choose any $\delta > 0$, let $0 = x_0, x_1, \ldots, x_n = x$ and $0 = y_0, y_1, \ldots, y_m = y$ approximate $|||x|||$ and $|||y|||$ up to δ. Then by considering the sequence $x_0, x_1, \ldots, x_n = x_n + y_0, x_n + y_1, x_n + y_2, \ldots, x_n + y_m$ for $x + y$ we get

$$\begin{aligned}|||x+y||| &\leq |x_1 - x_0| + |x_2 - x_1| + \cdots |x_n - x_{n-1}| + \\ &+ |y_1 - y_0| + |y_2 - y_1| + \cdots + |y_m - y_{m-1}| \leq |||x||| + |||y||| + 2\delta.\end{aligned}$$

Since δ was arbitrary, the triangle inequality follows. Thus $|||\cdot|||$ is an equivalent norm.

The sequence $\{0, x\}$ gives $|||x||| \leq |x| \leq \left(1-\frac{\varepsilon}{12}\right)\|x\|$. On the other hand, for every sequence we have

$$\sum_{j=1}^{n} |x_j - x_{j-1}| \geq \sum_{j=1}^{n}\Big(1-\frac{\varepsilon}{4}\Big)\|x_j - x_{j-1}\| \geq \Big(1-\frac{\varepsilon}{4}\Big)\|x\|$$

and (A) follows.

To prove (B), let $\|x\| = \|y\| = 1$ and $\|x-y\| \geq 5\varepsilon$, and let $0 = x_0, x_1, \ldots, x_n = x$ and $0 = y_0, y_1, \ldots, y_m = y$ give an approximation of $|||x|||$ and $|||y|||$ up to $\delta \in \left(0, \frac{\varepsilon}{2}\right)$. Note that if x lies on the segment $[x_i, x_{i+1}]$, then the sequence $\tilde{x}_0 = x_0, \ldots, \tilde{x}_i = x_i, \tilde{x}_{i+1} = x, \tilde{x}_{i+2} = x_{i+1}, \ldots, \tilde{x}_{n+1} = x_n$ satisfies $\sum_{j=1}^{n} |x_j - x_{j-1}| = \sum_{j=1}^{n+1} |\tilde{x}_j - \tilde{x}_{j-1}|$. By inserting division points in this manner, we may assume $\|y_j - y_{j-1}\| = \|x_j - x_{j-1}\|$ for $j \leq \min(m, n)$. Assume that $n \leq m$. Then we have

$$1 = \|x\| \leq \sum_{j=1}^{n} \|x_j - x_{j-1}\| \leq \frac{1}{1-\frac{\varepsilon}{4}} \sum_{j=1}^{n} |x_j - x_{j-1}| \leq \frac{|||x||| + \frac{\varepsilon}{2}}{1-\frac{\varepsilon}{4}} \leq \frac{\left(1-\frac{\varepsilon}{12}\right) + \frac{\varepsilon}{2}}{1-\frac{\varepsilon}{4}} < 1+\varepsilon.$$

Similarly we have $1 \leq \sum_{j=1}^{m} \|y_j - y_{j-1}\| < 1 + \varepsilon$.

Since $\sum_{j=1}^{n} \|x_j - x_{j-1}\| = \sum_{j=1}^{n} \|y_1 - y_{j-1}\|$, we have $\sum_{j=n+1}^{m} \|y_j - y_{j-1}\| < \varepsilon$. Hence

$$\sum_{i=1}^{n} \|(x_i - x_{i-1}) - (y_i - y_{i-1})\| \geq \Big\| \sum_{i=1}^{n} \big((x_i - x_{i-1}) - (y_i - y_{i-1})\big) \Big\|$$
$$= \|x_n - y_n\| = \|x - y_n\| \geq \|x - y\| - \|y - y_n\| \geq 5\varepsilon - \varepsilon = 4\varepsilon.$$

Denote $J = \big\{i \in \{1, \ldots, n\} : \|(x_i - x_{i-1}) - (y_i - y_{i-1})\| \geq \varepsilon \|x_i - x_{i-1}\|\big\}$. Then

$$2\sum_{i\in J} \|x_i - x_{i-1}\| \geq \sum_{i\in J} \|(x_i - x_{i-1}) - (y_i - y_{i-1})\|$$
$$= \sum_{i=1}^{n} \|(x_i - x_{i-1}) - (y_i - y_{i-1})\| - \sum_{i\notin J} \|(x_i - x_{i-1}) - (y_i - y_{i-1})\|$$
$$\geq 4\varepsilon - \sum_{i\notin J} \|(x_i - x_{i-1}) - (y_i - y_{i-1})\| \geq 4\varepsilon - \sum_{i\notin J} \varepsilon \|x_i - x_{i-1}\|$$
$$\geq 4\varepsilon - \sum_{i=1}^{n} \varepsilon \|x_i - x_{i-1}\| \geq 4\varepsilon - \varepsilon(1+\varepsilon) \geq 2\varepsilon.$$

Also, if $i \in J$ then

$$|(x_i - x_{i-1}) + (y_i - y_{i-1})| < |x_i - x_{i-1}| + |y_i - y_{i-1}| - \eta\|x_i - x_{i-1}\|.$$

Consider a sequence $\{z_k\} \subset X$ of $N = n + m - |J|$ points such that $z_0 = 0$, $z_N = x+y$, and the differences of two consecutive points are $(x_{i-1}+y_{i-1})-(x_i+y_i)$ for $i \in J$, $x_{i-1} - x_i$ and $y_{i-1} - y_i$ for $i \notin J$, and $y_{i-1} - y_i$ for $i > n$. Note that the order in which these differences occur does not matter in the following argument. Then

$$|\!|\!| x + y |\!|\!| \leq \sum_{i\in J} |(x_i - x_{i-1}) + (y_i - y_{i-1})| + \sum_{i\leq n, i\notin J} |x_i - x_{i-1}|$$
$$+ \sum_{i\leq n, i\notin J} |y_i - y_{i-1}| + \sum_{i>n} |y_i - y_{i-1}|$$
$$\leq \sum_{i\in J} |x_i - x_{i-1}| + \sum_{i\in J} |y_i - y_{i-1}| - \eta \sum_{i\in J} \|x_i - x_{i-1}\|$$
$$+ \sum_{i\leq n, i\notin J} |x_i - x_{i-1}| + \sum_{i\leq n, i\notin J} |y_i - y_{i-1}| + \sum_{i>n} |y_i - y_{i-1}|$$
$$\leq |\!|\!| x |\!|\!| + \delta + |\!|\!| y |\!|\!| + \delta - \eta\varepsilon.$$

Since η and ε do not depend on δ and δ was arbitrary, we have

$$|\!|\!| x + y |\!|\!| \leq |\!|\!| x |\!|\!| + |\!|\!| y |\!|\!| - \eta\varepsilon.$$

This proves Lemma 9.18. □

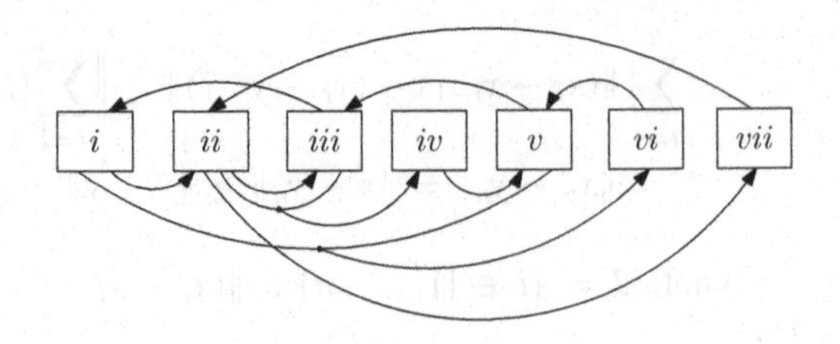

Fig. 9.3 The paths to follow for the proof of Theorem 9.14

Proof of Theorem 9.14 (see Fig. 9.3):

(i)$\Longrightarrow$(ii): (Enflo, [Enfl1]) For $\varepsilon_n = 2^{-n-3}$, $n \in \mathbb{N}$, we find by Lemma 9.18 an equivalent norm $\|\!|\cdot|\!\|_n$ on X satisfying $\left(1-\frac{\varepsilon_n}{4}\right)\|x\| \le \|\!|x|\!\|_n \le \left(1-\frac{\varepsilon_n}{12}\right)\|x\|$ and $\eta_n > 0$ such that whenever $\|x\| = \|y\| = 1$ and $\|x-y\| \ge 5\varepsilon_n$, we have $\|\!|x+y|\!\|_n \le \|\!|x|\!\|_n + \|\!|y|\!\|_n - \eta_n\varepsilon_n$. Define a norm $\|\!|\cdot|\!\|_0$ on X by $\|\!|x|\!\|_0 = \sum_{n=1}^{\infty}\frac{\|\!|x|\!\|_n}{2^n}$.

Finally, define an equivalent norm $\|\cdot\|_0$ by $\|x\|_0 = \left(\|\!|x|\!\|_0^2 + \|x\|^2\right)^{\frac{1}{2}}$. We claim that $\|\cdot\|_0$ is uniformly convex. Let $x_m, y_m \in X$ be such that $\lim_{m\to\infty}(2\|x_m\|_0^2 + 2\|y_m\|_0^2 - \|x_m+y_m\|_0^2) = 0$ and $\{x_m\}$ is bounded. Then $\lim_{m\to\infty}\left(\|\!|x_m|\!\|_0 - \|\!|y_m|\!\|_0\right) = 0$ and $\lim_{m\to\infty}\left(\|x_m\| - \|y_m\|\right) = 0$. Assume that $\lim_{m\to\infty}\|\!|x_m|\!\|_0 = \lim_{m\to\infty}\|\!|y_m|\!\|_0 = 1$ and $\lim_{m\to\infty}\|x_m\| = \lim_{m\to\infty}\|y_m\| = a > 0$. Consider the vectors $u_m = \dfrac{x_m}{\|x_m\|}$ and $v_m = \dfrac{y_m}{\|y_m\|}$. It is enough to show that $\lim_{m\to\infty}\|u_m - v_m\|_0 = 0$. Indeed, having this,

$$\|x_m - au_m\|_0 = \big|\|x_m\| - a\big|\|u_m\|_0 \to 0$$

as $\{u_m\}$ is bounded. Similarly $\|y_m - av_m\|_0 \to 0$, and so $\|x_m - y_m\|_0 \to 0$.

By contradiction, assume that $\|u_m - v_m\|_0 \ge \delta > 0$ for some $\delta > 0$ and all m. Fix n_0 such that $\delta \in \left[\frac{5}{2^{n_0+3}}, \frac{5}{2^{n_0+2}}\right]$. From Lemma 9.18 and a convexity argument, we have

$$\|\!|u_m|\!\|_0 + \|\!|v_m|\!\|_0 - \|\!|u_m+v_m|\!\|_0 \ge \frac{1}{2^{n_0}}\left(\|\!|u_m|\!\|_{n_0} + \|\!|v_m|\!\|_{n_0} - \|\!|u_n+v_m|\!\|_{n_0}\right) \ge \frac{\eta_{n_0}\varepsilon_{n_0}}{2^{n_0}}.$$

This is a contradiction with the fact that $\|\!|u_m|\!\|_0 \to 1/a$, $\|\!|v_m|\!\|_0 \to 1/a$ and $\|\!|u_m + v_m|\!\|_0 = \left\|\!\left|\dfrac{x_m}{\|x_m\|} + \dfrac{y_m}{\|y_m\|}\right|\!\right\|_0 \to \dfrac{2}{a}$. Thus (i) implies (ii) in Theorem 9.14.

(ii)$\Longrightarrow$(iii): If X admits an equivalent uniformly convex norm, then X^* admits an equivalent uniformly Fréchet differentiable norm by Theorem 9.9. By Proposition 9.12, every space Y that is finitely representable in X^* admits an equivalent uniformly Fréchet differentiable norm, and thus by Theorem 9.11, every such space Y is reflexive. Therefore X^* is superreflexive and by (i) implies (ii) we have that X^* admits an equivalent uniformly convex norm $\|\cdot\|$. The predual norm to $\|\cdot\|$ on X is then uniformly Fréchet differentiable by Theorem 9.9.

(iii)$\Longrightarrow$(i): Let $|||\cdot|||$ be an equivalent uniformly Fréchet differentiable norm on $(X, \|\cdot\|)$. Let Y be finitely representable in $(X, \|\cdot\|)$. Then Y is crudely finitely representable in $(X, |||\cdot|||)$ and by Proposition 9.12, Y admits an equivalent norm that is uniformly Fréchet differentiable. Therefore Y is reflexive by Theorem 9.11. Hence X is superreflexive.

(ii)–(iii)$\Longrightarrow$(iv): Let $\|\cdot\|$ be an equivalent uniformly convex norm on X and let $\|\cdot\|_0$ be an equivalent uniformly Fréchet differentiable norm on X. Define $\|\cdot\|_n^*$ on X^* by $\|f\|_n^{*2} = \|f\|^{*2} + \frac{1}{n}\left(\|f\|_0^*\right)^2$, where $\|\cdot\|_0^*$ is the dual norm to $\|\cdot\|_0$. As $\|\cdot\|_0^*$ is uniformly convex by Theorem 9.9, we get that $\|\cdot\|_n^*$ are uniformly convex and their predual norms $\|\cdot\|_n$ on X are thus uniformly Fréchet differentiable by Theorem 9.9. Define a norm on X by $|||x|||^2 = \sum 2^{-n}\|x\|_n^2$.

Note that the derivatives of $\|\cdot\|_n^2$ are bounded on bounded sets and thus $\|\cdot\|$ is uniformly Fréchet differentiable as all $\|\cdot\|_n$ are.

If $x_m, y_m \in X$ satisfy $(2|||x_m|||^2 + 2|||y_m|||^2 - |||x_m + y_m|||^2) \to 0$ and $\{x_n\}$ is bounded, then the same is true for all the norms $\|\cdot\|_n$, and since $\|\cdot\|_n$ converge uniformly on bounded sets to $\|\cdot\|$, we have that $(2\|x_m\|^2+2\|y_m\|^2-\|x_m+y_m\|^2) \to 0$. Since $\|\cdot\|$ is uniformly convex, we have $\lim\limits_{m\to\infty} \|x_m - y_m\| = 0$. Hence $|||\cdot|||$ is uniformly convex.

(iv)$\Longrightarrow$(v) is trivial (see Fact 10.4).

(v)$\Longrightarrow$(iii): This is contained in the following result.

Theorem 9.19 ([FWZ]) *Let X be a Banach space. If X admits a uniformly Fréchet differentiable (UF, in short) bump function, then X admits an equivalent uniformly Fréchet differentiable norm.*

Proof: Let B_X^O denote the open unit ball of X. Let τ be a Lipschitz C^1-smooth function on the real line $\mathbb{R}$ such that $\tau(0) = 0$, $\tau(1) = -1$, $\tau(\mathbb{R}) = [0, -1]$, and $\tilde{\phi}$ be a symmetric UF-smooth bump function on X such that $\tilde{\phi}'$ is uniformly continuous on X, $\tilde{\phi}(0) = 1$ and $\tilde{\phi}(x) = 0$ whenever $\|x\| \geq \frac{1}{3}$. Set $\phi = \tau \circ \tilde{\phi}$. Then $\phi \leq 0$, $\inf(\phi) = -1 = \phi(0)$ and $\operatorname{supp}(\phi) \subset \frac{1}{2}B_X^O$. Note that $2\|x\| - 2 \leq \phi(x)$ for all $x \in B_X$. For $t > 0$ define

$$\omega(t) = \sup\left\{\frac{\phi(x+h) + \phi(x-h) - 2\phi(x)}{\|h\|} : x \in X, \|h\| < t\right\}.$$

Define $\psi\colon B_X^O \to \mathbb{R}$ by

$$\psi(x) = \inf\left\{\sum_{i=1}^{n} \alpha_i\phi(x_i) : \sum \alpha_i x_i = x, \alpha_i \geq 0, \sum \alpha_i = 1, x_i \in B, n \in \mathbb{N}\right\}.$$

Note that ψ is convex. Let $G = \{x \in B_X^O : \psi(x) < -\frac{1}{2}\}$. We claim that ψ' is uniformly continuous on G.

To see this, let $x \in G, h \in X, \|h\| < \frac{1}{4}, h \neq 0$ and $0 < \varepsilon < -\frac{1}{2} - \psi(x)$ be given. Let $x_i \in B_X^O, \alpha_i \geq 0$ with $\sum \alpha_i = 1$ be such that $\sum \alpha_i\phi(x_i) < \psi(x)+\varepsilon\|h\| < -\frac{1}{2}$, assume that $\phi(x_i) < 0$ for $i = 1, \dots, k$ and $\phi(x_{k+1}) = \dots = \phi(x_n) = 0$. Then $-\frac{1}{2} > \sum_{i=1}^{k} \alpha_i \phi(x_i) \geq -\sum_{i=1}^{k} \alpha_i$. Thus $1 \geq \alpha = \sum_{i=1}^{k} \alpha_i \geq \frac{1}{2}$ and $\|x_i + \frac{1}{\alpha} h\| \leq \|x_i\| + 2\|h\| < 1$ for $i = 1, \dots, k$. Consequently,

$$\begin{aligned}
&\psi(x+h) + \psi(x-h) - 2\psi(x) \\
&\leq \sum_{i=1}^{k} \alpha_i \phi(x_i + \tfrac{1}{\alpha} h) + \sum_{i=1}^{k} \alpha_i \phi(x_i - \tfrac{1}{\alpha} h) - 2\sum_{i=1}^{k} \alpha_i \phi(x_i) + 2\varepsilon\|h\| \\
&= \sum_{i=1}^{k} \alpha_i \big(\phi(x_i + \tfrac{1}{\alpha} h) + \phi(x_i - \tfrac{1}{\alpha} h) - 2\phi(x_i)\big) + 2\varepsilon\|h\| \\
&\leq \alpha\omega\big(\tfrac{\|h\|}{\alpha}\big)\tfrac{\|h\|}{\alpha} + 2\varepsilon\|h\|.
\end{aligned}$$

Since $\varepsilon > 0$ was arbitrary, $\psi(x+h) + \psi(x-h) - 2\psi(x) < 2\omega(2\|h\|) \cdot \|h\|$.

This shows the existence of $\psi'(x)$. We will now show that ψ' is uniformly continuous on G. To this end, let $x, y \in G, 0 < \|x - y\| < \frac{1}{8}$ be given. Then for $h \in X$ with $\|h\| = \|x - y\|$ we have $x + h \in B_X^O$, $y - (x + h - y) \in B_X^O$ and by convexity,

$$\begin{aligned}
&\big(\psi'(x) - \psi'(y)\big)(h) \leq \psi(x+h) - \psi(x) - \psi'(y)(h) \\
&= \psi(x+h) - \psi(y) - \psi'(y)(x+h-y) + \psi(y) - \psi(x) + \psi'(y)(x-y) \\
&\leq \psi(x+h) - \psi(y) - \psi'(y)(x+h-y) \\
&\leq \psi(y + (x+h-y)) + \psi(y - (x+h-y)) - 2\psi(y) \\
&\leq 2\omega(2\|x+h-y\|)\|x+h-y\| \leq 4\omega(4\|x-y\|)\|h\|.
\end{aligned}$$

By taking the supremum over all $h \in X$, $\|h\| = \|x - y\|$, we obtain that $\|\psi'(x) - \psi'(y)\| \leq 4\omega(4\|x - y\|)$ for $x, y \in G$ and $0 < \|x\| < \frac{1}{8}$.

Let $Q = \{x \in B_X^O : \psi(x) \leq -\frac{3}{4}\}$ and let q be the Minkowski functional of Q. If $q(x) = 1$, then $\psi'(x)(x) \geq \psi(x) - \psi(0) = \frac{1}{4}$. From the implicit function theorem, it follows that the derivative of the norm $|\!|\!|x|\!|\!| = q(x) + q(-x)$ is uniformly continuous on the sphere. The statement now follows from Theorem 9.14. □

(i)–(v)$\Longrightarrow$(vi): This is trivial.

(vi)$\Longrightarrow$(v): We shall prove first the following *compact variational principle*.

Theorem 9.20 (see, e.g., [Gode4]) *Let X be a Banach space that does not contain any isomorphic copy of c_0. Let U be a bounded symmetric open subset of X containing the origin. Let f be a continuous function on $\overline{U}$ such that $f(0) \leq 0$, $f(x) = f(-x)$ for all $x \in \overline{U}$, and $m = \inf\{f(x) : x \in \partial U\} > 0$.*
Then there are a compact symmetric set $K \subset U$ and $\Delta > 0$ such that $K + \Delta B_X \subset U$ and for every $\delta \in (0, \Delta)$ there is a finite subset K_δ of K such that

$$\inf\{f_{K_\delta}(x) : x \in V, \delta \leq \|x\| \leq \Delta\} > f_{K_\delta}(0),$$

where $f_{K_\delta}(x) = \sup\{f(x+k) : k \in K_\delta\}$.

Proof: Note that since K_δ is finite, f_{K_δ} is a continuous function.

Put $x_0 = 0$ and if $x_0, \ldots, x_n$ were chosen, set $K_n = \left\{\sum_{i=0}^n \varepsilon_i x_i \,:\, \varepsilon_i = \pm 1\right\}$ and

$$E_n = \left\{x \in X : x + k \in U \text{ and } f(x+k) \le \tfrac{m}{2}\left(1 - \tfrac{1}{2^n}\right) \text{ for all } k \in K_n\right\}.$$

Note that E_n is a symmetric set. Set $\alpha_n = \sup\{\|x\| : x \in E_n\}$ and choose $x_{n+1} \in E_n$ such that $\|x_{n+1}\| \ge \frac{\alpha_n}{2}$.

Once the sequence $\{x_n\}$ is constructed, we define $K = \overline{\bigcup_{n \ge 0} K_n}$. Then $K \subset \overline{U}$ and thus K is bounded. As X does not contain any isomorphic copy of c_0, it follows from Corollary 4.52 that K is compact and $x_n \to 0$. By the choice of $\{x_n\}$ also $\alpha_n \to 0$. From the continuity of f on $\overline{U}$ we get $f(x) \le \frac{m}{2}$ for all $x \in K$, hence $\partial U \cap K = \emptyset$. Therefore, $K \subset U$ and from the compactness of K we get Δ such that $K + \Delta B_X \subset U$.

Now fix any $\delta \in (0, \Delta)$. Choose $n \in \mathbb{N}$ such that $\alpha_n < \delta$. If x is such that $\delta \le \|x\| \le \Delta$, then by the definition of α_n and using $\alpha_n < \delta$ we have $f_{K_n}(x) > \frac{m}{2}(1 - \frac{1}{2^n})$. On the other hand,

$$f_{K_n}(0) = \sup_{K_n}(f) = \max\{f_{K_{n-1}}(x_n), f_{K_{n-1}}(-x_n)\} = f_{K_{n-1}}(x_n) \le \tfrac{m}{2}(1 - \tfrac{1}{2^{n-1}}).$$

By combining these inequalities, we get $f_{K_n}(x) > f_{K_n}(0) + \frac{m}{2^{n+1}}$, for all x with $\delta \le \|x\| \le \Delta$ and the statement follows. □

As a consequence, we have the following result; its proof shows the sought implication (vi)$\Longrightarrow$(v).

Theorem 9.21 ([FWZ]) *Let X be a Banach space that admits a bump function f with locally uniformly continuous derivative. If X does not contain any isomorphic copy of c_0, then X is superreflexive.*

Proof: (Sketch) Applying Theorem 9.20 to f, $U = B_X^O$ and $\delta > 0$ we obtain K_δ. Consider the function

$$\psi(x) = \sum_{y \in K_\delta} \big(f(x+y) - f(y)\big)^2.$$

Then $\psi(0) = 0$ and $\psi(x) \ge \varepsilon^2$ on some $\{x \in X : \delta \le \|x\| < \beta\}$. Let τ be a C^∞-smooth function on the real line such that $\tau(0) = 1$ and $\tau(t) = 0$ for $t > \frac{\varepsilon^2}{2}$. Then put

$$g(x) = \begin{cases} \tau(\psi(x)) & \text{if } \|x\| < \delta, \\ 0 & \text{if } \|x\| \ge \delta, \end{cases}$$

and use Theorem 9.19. □

(ii)$\Longleftrightarrow$(vii) is what the following result shows.

Theorem 9.22 (Lancien [Lanc]) *A Banach space X is superreflexive if and only if the dual dentability index $D_Z(X)$ is ω_0.*

Proof: We will first prove the "if" implication. Let $\|\cdot\|$ be the norm of X and $\varepsilon > 0$ and $\Delta > 0$ be given. For a convex set $D \subset B_{X^*}$ put

$$D'_\varepsilon = \{f \in D : \operatorname{diam} S \geq \varepsilon \text{ for each } w^*\text{-slice } S \text{ of } D \text{ containing } f\},$$

where a w^*-slice of a set D is the intersection of D with a w^*-open half-space in X^*.

Put $D_0 = B_{X^*}$ and $r = D_Z(X, \varepsilon)$. For $j \in \{1, 2, \dots, r\}$ put $D_j = (D_0)_\varepsilon^{(j)}$. Define $F : X^* \to [0, +\infty)$ by

$$F(f) = \|f\| + \Delta \sum_{j=0}^{r-1} \frac{1}{2^{j+1}} \operatorname{dist}(f, D_j)$$

for $f \in X^*$.

We will show the following

Claim 1: *Let (f_n) and (g_n) be sequences in B_{X^*} such that*

$$\frac{1}{2}F(f_n) + \frac{1}{2}F(g_n) - F\left(\frac{1}{2}(f_n + g_n)\right) \to 0$$

as $n \to \infty$. Then $\limsup_n \|f_n - g_n\| \leq 2\varepsilon$.

Assuming Claim 1 shown, we will prove the following:

Claim 2: *The space X^* admits a dual norm $|||\cdot|||$ such that $\|\cdot\| \leq |||\cdot||| \leq (1 + \Delta)\|\cdot\|$ and* $\limsup_n \|f_n - g_n\| \leq 2\varepsilon$ *whenever $f_n, g_n \in B_{(X^*, |||\cdot|||)}$, $n \in \mathbb{N}$, satisfy* $\lim_n |||f_n + g_n||| = 2$.

Proof of Claim 2: Let $|||\cdot|||$ be the Minkowski functional of the set $\{f \in X^* : F(f) \leq 1\}$. Then $|||\cdot|||$ is a dual norm on X^* and $\|f\| \leq |||f||| \leq (1 + \Delta)\|f\|$ for every $f \in X^*$. Let $f_n, g_n \in B_{(X^*, |||\cdot|||)}$, $n \in \mathbb{N}$, be such that $\lim_n |||f_n + g_n||| = 2$. Then $F(f_n) \leq 1$, $F(g_n) \leq 1$ for $n \in \mathbb{N}$, and from the uniform continuity of F on bounded sets we get

$$F\left(\frac{f_n + g_n}{2}\right) - F\left(\frac{f_n + g_n}{|||f_n + g_n|||}\right) \to 0, \quad \text{as } n \to \infty.$$

Therefore, by Claim 1,

$$\limsup_n \|f_n - g_n\| \leq 2\varepsilon.$$

This finishes the proof of Claim 2. $\triangle$

Still assuming Claim 1 shown, we will now finish the proof of the "if" implication as follows:

For $n \in \mathbb{N}$, let $|||\cdot|||_m$ be a dual norm on X^* such that

$$\|f\| \leq |||f|||_m \leq 2\|f\|, \text{ for all } f \in X^*$$

and that

$$\limsup_n \|f_n - g_n\| \leq \frac{2}{m}$$

whenever $f_n, g_n \in B_{(X^*, |||\cdot|||_m)}$ satisfy $\lim \|f_n + g_n\| = 2$.

Define

$$|||f|||^2 = \sum_{m=1}^{\infty} 2^{-m} |||f|||_m^2, \text{ for } f \in X^*.$$

We have

$$\frac{1}{2}\|f\|^2 \leq |||f|||^2 \leq 4\|f\|^2, \text{ for all } f \in X^*.$$

We will show that $|||\cdot|||$ is uniformly convex on X^*.

For it, let $f_n, g_n \in B_{(X^*, |||\cdot|||)}$, $n \in \mathbb{N}$, satisfy $\lim |||f_n + g_n||| = 2$.

Assume that for some $\varepsilon > 0$ and for an increasing sequence (n_i) in $\mathbb{N}$ we have $\|f_{n_i} - g_{n_i}\| > \varepsilon$ for all $i \in \mathbb{N}$. Pick $m < 5/\varepsilon$ and fix it. Since $f_n, g_n \in B_{(X^*, |||\cdot|||)}$ are such that $|||f_n + g_n||| \to 2$, a standard convexity argument gives

$$2|||f_{n_i}|||^2 + 2|||g_{n_i}|||_m^2 - 2|||f_{n_i} + g_{n_i}|||_m^2 \to 0$$

as $i \to \infty$.

Hence

$$|||f_{n_i}|||_m - |||g_{n_i}|||_m \to 0$$

and

$$|||f_{n_i} + g_{n_i}|||_m - 2|||f_{n_i}||| \to 0$$

as $i \to \infty$.

Note that $|||f_{n_i}||| \leq 2\|f_{n_i}\| \leq |||f_{n_i}|||_m$ and that $|||f_{n_i}||| \to 1$ as $i \to \infty$. Put $f_i' = f_{n_i}/|||f_{n_i}|||_m$ and $g_i' = g_{n_i}/|||g_{n_i}|||_m$ for $i \in \mathbb{N}$. The sequence (f_i') and (g_i') lie in $S_{(X^*, |||\cdot|||_m)}$ and $|||f_i' + g_i'|||_m \to 2$ as $i \to \infty$.

Therefore

$$\limsup_i \|f_i' - g_i'\| \leq \limsup_i \|f_i' - g_i'\|_m \leq \frac{2}{m}.$$

Since $|||f_{n_i}||| - |||g_{n_i}|||_m \to 0$ and $|||\cdot||| \leq 2\|\cdot\| \leq 2|||\cdot|||$, we get that $\limsup_i \|f_{n_i} - g_{n_i}\| \leq 5/m < \varepsilon$, a contradiction that shows the "if" implication provided Claim 1 is proved. ∎

We will now show Claim 1.

Assume that the statement in Claim 1 fails. Then, without loss of generality, assume that for some $\delta > 0$ and any n, $\|f_n - g_n\| > 2\varepsilon + \delta$. Then we will show the following

Subclaim: *For every* $j \in \{0, 1, \ldots, r-1\}$ *we have* $\text{dist}(f_n, D_j) \to 0$ *and* $\text{dist}(g_n, D_j) \to 0$ *as* $n \to \infty$.

Proof of the Subclaim: For $j = 0$ the statement is trivial. So assume $r > 1$, fix $k \in \{0, 1, \ldots, r-2\}$ and assume that the statement holds for $j = k$. Fix $n \in \mathbb{N}$. Find $f_n', g_n' \in D_k$ such that $\|f_n' - f_n\| \leq 2\,\text{dist}(f_n, D_k)$ and $\|g_n' - g_n\| \leq 2\,\text{dist}(g_n, D_k)$. Then

$$\begin{aligned}\|f_n - g_n\| &\geq \|f_n - g_n\| - \|f_n' - f_n\| - \|g_n' - g_n\|\\ &> 2\varepsilon + \delta - 2\,\text{dist}(f_n, D_k) - 2\,\text{dist}(g_n, D_k).\end{aligned}$$

Hence

$$\|f_n' - g_n'\| > 2\varepsilon, \text{ for all large } n \in \mathbb{N}.$$

Assume without loss of generality that it holds for all n. Now, since any w^*-slice S of D_k containing $(1/2)(f_n' + g_n')$ contains necessarily either f_n' or g_n' by a standard convexity argument, we have

$$\text{diam}\, S \geq \left\| f_n' - \frac{f_n' + g_n'}{2} \right\| = \left\| g_n' - \frac{f_n' + g_n'}{2} \right\| = \frac{1}{2}\|f_n' - g_n'\| > \varepsilon$$

for all n.

Therefore

$$\frac{1}{2}(f_n' + g_n') \in D_{k+1}, \text{ for all } n \in \mathbb{N}.$$

From the definition if F and from a standard convexity argument, we get

$$\frac{1}{2}\,\text{dist}(f_n, D_{k+1}) + \frac{1}{2}\,\text{dist}(g_n, D_{k+1}) - \text{dist}(\frac{1}{2}(f_n + g_n), D_{k+1}) \to 0$$

as $n \to \infty$.

Since $f_n - f_n' \to 0$, $g_n - g_n' \to 0$ and $(1/2)(f_n' + g_n') \in D_{k+1}$, we get that $\text{dist}(f_n, D_{k+1}) \to 0$ as $n \to \infty$. Thus the statement in the Subclaim holds. △

We will now prove Claim 1.

For $j = r-1$, we have by the Subclaim that $\text{dist}(f_n, D_{r-1}) \to 0$ and $\text{dist}(g_n, D_{r-1}) \to 0$ as $n \to \infty$. For $n \in \mathbb{N}$, find $f_n', g_n' \in D_{r-1}$ such that

$\|f_n' - f_n\| \leq 2\,\mathrm{dist}(f_n, D_{r-1})$ and $\|g_n' - g_n\| \leq 2\,\mathrm{dist}(g_n, D_{r-1})$. Fix any $n \in \mathbb{N}$. Since $D_r = \emptyset$, there must exist a w^*-slice S of D_{r-1} containing $(1/2)(f_n' + g_n')$ such that $\mathrm{diam}\, S < \varepsilon$. Then either f_n' or g_n' is in S. Then

$$\left\| f_n' - \frac{1}{2}(f_n' + g_n') \right\| = \left\| g_n' - \frac{1}{2}(f_n' + g_n') \right\| < \varepsilon.$$

Thus $\|f_n' - g_n'\| < 2\varepsilon$.

Therefore

$$\limsup_n \|f_n - g_n\| = \limsup_n \|f_n' - g_n\| \leq 2\varepsilon,$$

a contradiction. This proves Claim 1 and finishes the proof of the "only if" implication in the statement of the theorem. $\triangle$

Assume now that the dual norm of X^* is uniformly convex (as the notion of dual dentability index is isomorphically invariant). We need to show that for every $\varepsilon > 0$, $D_Z(X, \varepsilon)$ is finite.

From the uniform convexity of the dual norm of X^*, given $\varepsilon > 0$, find $\delta > 0$ such that $\|f - g\| < \varepsilon$ whenever $f, g \in B_{X^*}$ are such that $\|f + g\| > 2 - 2\delta$. Then $(B_{X^*})_\varepsilon' \subset (1 - \delta)B_{X^*}$, where $(B_{X^*})_\varepsilon'$ denotes as above the set of all points in B_{X^*}, each w^*-slice of which has norm-diameter greater than or equal to ε.

Indeed, assume that there is $f_0 \in (B_{X^*})_\varepsilon' \backslash (1 - \delta)B_{X^*}$. Find $x_0 \in S_X$ such that $f(x_0) > 1 - \delta$. Put $S = \{f \in B_{X^*} : f(x_0) > 1 - \delta\}$. This is a w^*-slice of B_{X^*} and $f_0 \in S$. If $f, g \in S$, then $\|f + g\| \geq f(x_0) + g(x_0) > 2 - 2\delta$. Therefore $\|f - g\| < \varepsilon - \delta$. Thus $f_0 \notin (B_{X^*})_\varepsilon'$. This proves $(B_{X^*})_\varepsilon' \subset (1 - \delta)B_{X^*}$. Using this and a homogeneity argument, we get that

$$(B_{X^*})_\varepsilon^{(2)} \subset (1 - \delta)^2 B_{X^*}, \quad (B_{X^*})_\varepsilon^{(3)} \subset (1 - \delta)^3 B_{X^*}, \ldots$$

If $k \in \mathbb{N}$ is big enough, then $\mathrm{diam}(1 - \delta)^k B_{X^*} < \varepsilon$, and thus $(B_{X^*})_\varepsilon^{k+1} = \emptyset$. Therefore $D_Z(X, \varepsilon)$ is finite and $D_Z(X) = \omega_0$. This finishes the proof of Theorem 9.22. $\square$

This completes the proof of Theorem 9.14 $\square$

9.3 Applications

Theorem 9.23 (Kadec [Kade1]) *Let $(X, \|\cdot\|)$ be a uniformly convex Banach space with modulus of convexity $\delta(\varepsilon)$. If $\sum x_i$ converges unconditionally in X, then $\sum_{j=1}^{\infty} \delta(\|x_j\|) < \infty$.*

The proof is based on the following lemma.

Lemma 9.24 *In the same notation, let* $z_1, \dots, z_n \in X$ *be such that* $\max_{\varepsilon_i=\pm1} \left\| \sum_{i=1}^n \varepsilon_i z_i \right\| \le 2$. *Then* $\sum_{j=1}^n \delta(\|z_j\|) \le 1$.

Proof: Assume without loss of generality that for $s_n = \sum_{j=1}^n z_j$ we have $\|s_n\| \ge \left\| \sum_{j=1}^n \varepsilon_j z_j \right\|$ for every choice of $\varepsilon_j = \pm1$. Then for $j \in \{1, \dots, n\}$ we have by our assumption

$$\|z_j\| \le \frac{2\|z_j\|}{\|s_n\|} = \left\| \frac{s_n}{\|s_n\|} - \frac{(s_n - 2z_j)}{\|s_n\|} \right\|.$$

Since $\delta(\varepsilon)$ is a non-decreasing function and since $\frac{s_n}{\|s_n\|}$ and $\frac{(s_n - 2z_j)}{\|s_n\|}$ are both in B_X by the assumption, we have

$$\begin{aligned}\delta(\|z_j\|) &\le \delta\left(\left\| \frac{s_n}{\|s_n\|} - \frac{(s_n - 2z_j)}{\|s_n\|} \right\| \right) \\ &\le 1 - \left\| \frac{1}{2}\left(\frac{s_n}{\|s_n\|} + \frac{(s_n - 2z_j)}{\|s_n\|} \right) \right\| = 1 - \frac{\|s_n - z_j\|}{\|s_n\|}.\end{aligned}$$

Therefore

$$\begin{aligned}\sum_{j=1}^n \delta(\|z_j\|) &\le \sum_{j=1}^n \left(1 - \frac{\|s_n - z_j\|}{\|s_n\|} \right) = n - \sum_{j=1}^n \frac{\|s_n - z_j\|}{\|s_n\|} \\ &\le n - \frac{1}{\|s_n\|} \left\| \sum_{j=1}^n (s_n - z_j) \right\| = n - \frac{1}{\|s_n\|} \|(n-1)s_n\| = 1.\end{aligned}$$

□

Proof of Theorem 9.23: There is n_0 such that $\left\| \sum_{j=n_0}^n \varepsilon_j x_j \right\| < 2$ for every $n > n_0$ and every $\varepsilon_j = \pm1$. Therefore $\sum_{j=n_0}^n \delta(\|x_j\|) < 1$ for every $n > n_0$, hence $\sum_{j=1}^\infty \delta(\|x_j\|) < \infty$. □

There is also a converse result [Lind3] to Theorem 9.23: *Let X be a superreflexive Banach space with a modulus of smoothness* ρ. *If for some sequence* $\{x_n\}$ *in* X, $\sum \rho(\|x_n\|) < \infty$, *then* $\sum \varepsilon_i x_i$ *converges for some choice of* $\varepsilon_i = \pm1$.

Theorem 9.25 (Gurarii, Gurarii [GuGu], James [Jame8]) *Let* $(X, \|\cdot\|)$ *be a superreflexive Banach space with a seminormalized Schauder basis* $\{x_i\}$. *Then there exist* $p, q \in (1, \infty)$ *and* $K > 0$ *such that for every* $x = \sum_{i=1}^\infty \alpha_i x_i \in X$ *we have*

$$\frac{1}{K} \left(\sum_{i=1}^\infty |\alpha_i|^q \right)^{\frac{1}{q}} \le \|x\| \le K \left(\sum_{i=1}^\infty |\alpha_i|^p \right)^{\frac{1}{p}}.$$

Theorem 9.25 will follow from Theorem 9.14 and Lemmas 9.26 and 9.27.

Lemma 9.26 *Let $(X, \|\cdot\|)$ be a Banach space with a seminormalized Schauder basis $\{x_i\}$. If X is uniformly convex, then there is $p \in (1, \infty)$ and $K > 0$ such that for every $x = \sum_{i=1}^{\infty} \alpha_i x_i$ in X we have $\|x\| \le K\Big(\sum_{i=1}^{\infty} |\alpha_i|^p\Big)^{\frac{1}{p}}$.*

Proof: We may assume without loss of generality that the basis is normalized. Fix some $\varepsilon \in \big(0, \frac{1}{\mathrm{bc}\{x_i\}}\big)$ and set $\lambda = 2(1-\delta(\varepsilon))$, where $\delta(\varepsilon)$ is the modulus of convexity of $\|\cdot\|$. Fix $p \in (1, \log_\lambda 2)$. First we will show that there is $\alpha \in (0, 1)$ such that $\|x+ty\|^p < 1+t^p$ whenever $|t-1| \le \alpha$ and $x, y \in S_X$ are finitely and consecutively supported vectors.

Assume without loss of generality that $l = \max\big(\mathrm{supp}(x)\big) < \min\big(\mathrm{supp}(y)\big)$. Let P_l be the canonical projection associated with $\{x_i\}$. Then $\mathrm{bc}\{x_i\}\|x - y\| \ge \|P_l(x - y)\| = \|x\| = 1$. Thus $\|x - y\| \ge \frac{1}{\mathrm{bc}\{x_i\}} > \varepsilon$ and we have $\|x + y\| \le \lambda$. Since $\lambda^p < 2$, we have $\|x + 1 \cdot y\|^p < 1 + 1^p$.

Since the function $\|x + ty\|^p$ is uniformly continuous at $t = 1$ with respect to $x, y \in S_X$, we have that there is $\alpha \in (0, 1)$ such that $\|x + ty\|^p < 1 + t^p$ for $|t - 1| < \alpha$ and x, y as stated.

Now set $K = \frac{2}{\alpha}$. It is enough to show that $\|z_n\| \le K \cdot \Big(\sum_{i=1}^{n} |\alpha_i|^p\Big)^{\frac{1}{p}}$ for every finitely supported vector $z_n = \sum_{i=1}^{n} \alpha_i x_i$. We proceed by induction on $n \in \mathbb{N}$. For $n = 1$ the result clearly holds. Suppose that it is valid for n. Consider a non-zero vector $z_{n+1} = \sum_{i=1}^{n+1} \alpha_i x_i$. We distinguish two cases:

(1) $|\alpha_i| \le \frac{1}{K}\|z_{n+1}\|$ for all $i = 1, 2, \ldots, n + 1$,

(2) $|\alpha_{i_0}| > \frac{1}{K}\|z_{n+1}\|$ for some $i_0 \in \{1, \ldots, n + 1\}$.

In the first case, put $z_0 = 0$, $z_s = \sum_{i=1}^{s} \alpha_i x_i$ for $s = 1, 2, \ldots, n + 1$, $y_s = \sum_{i=s+1}^{n+1} \alpha_i x_i$ for $s = 0, 1, \ldots, n$, $y_{n+1} = 0$. Then $\|z_0\| < \|y_0\|$, $\|z_{n+1}\| > \|y_{n+1}\|$, $\big|\|z_{i+1}\| - \|z_i\|\big| \le \frac{1}{K}\|z_{n+1}\|$, and $\big|\|y_{i+1}\| - \|y_i\|\big| \le \frac{1}{K}\|z_{n+1}\|$ for $i = 1, \ldots, n$.

Claim:
Let $\{\xi_i\}_{i=1}^n$ and $\{\eta_i\}_{i=1}^n$ be two sets of real numbers and $\varepsilon > 0$. Assume that $\xi_1 < \eta_1$, $\xi_n > \eta_n$, and $|\xi_{i+1} - \xi_i| < \varepsilon$, $|\eta_{i+1} - \eta_i| < \varepsilon$ for $i = 1, 2, \ldots, n - 1$. Then there is an index $i_0 \in \{1, \ldots, n\}$ such that $|\xi_{i_0} - \eta_{i_0}| < \varepsilon$.

This follows by considering the first index j for which $\xi_j > \eta_j$. △

By this claim there is $r \in \{1, \ldots, n\}$ such that $\big|\|z_r\| - \|y_r\|\big| \le \frac{1}{K}\|z_{n+1}\|$. By interchanging z_r and y_r if needed, we can assume that $\|z_r\| \ge \|y_r\|$. By the homogeneity, assume without loss of generality that $\|z_r\| = 1$. Since $\|y_r\| \le \|z_r\|$ and $z_{n+1} = z_r + y_r$, we have $\|z_{n+1}\| \le 2$ and therefore $\big|1 - \|y_r\|\big| \le \frac{2}{K} = \alpha$.

Put $t = \|y_r\|$ and $\tilde{y}_r = \frac{y_r}{\|y_r\|}$. Applying the first part of this proof to $x = z_r$ and $y = \tilde{y}_r$ (note that $|1 - t| \le \alpha$) yields $\|z_r + t\tilde{y}_r\|^p < 1 + t^p$. Therefore

$$\|z_{n+1}\|^p = \|z_r + y_r\|^p = \|z_r + t\tilde{y}_r\|^p < 1 + t^p = \|z_r\|^p + \|y_r\|^p.$$

By the induction hypothesis,

$$\|z_r\|^p + \|y_r\|^p \le K^p \sum_{i=1}^{r} |\alpha_i|^p + K^p \sum_{i=r+1}^{n+1} |\alpha_i|^p.$$

Therefore we have $\|z_{k+1}\|^p \le K^p \sum_{i=1}^{n+1} |\alpha_i|^p$, which concludes the first case.

In the second case, assuming $|\alpha_{i_0}| > \frac{1}{K}\|z_{n+1}\|$ we get

$$\|z_{n+1}\| \le K \cdot |\alpha_{i_0}| \le K \cdot \Big(\sum_{i=1}^{n+1} |\alpha_i|^p\Big)^{\frac{1}{p}}.$$

Therefore the induction step to $n+1$ is justified and Lemma 9.26 is proved. □

Lemma 9.27 *Let $(X, \|\cdot\|)$ be a Banach space with a seminormalized Schauder basis $\{x_i\}$. If $\|\cdot\|$ is uniformly smooth, then there is $q \in (1, \infty)$ and $L > 0$ such that for every $x = \sum_{i=1}^{\infty} \alpha_i x_i$ in X we have $\|x\| \ge L\Big(\sum_{i=1}^{\infty} |\alpha_i|^q\Big)^{\frac{1}{q}}$.*

Proof: Let f_i be the biorthogonal functionals to x_i. Then $\{f_i\}$ is a Schauder basis of X^* as X is reflexive by Theorem 9.11. We have $\|f_i\| \ge \frac{f_i(x_i)}{\|x_i\|} \ge \frac{1}{\sup \|x_i\|}$, $\|f_i\| \le 2\,\mathrm{bc}\{x_i\}$ so $\{f_i\}$ is a seminormalized basis.

Since the dual norm $\|\cdot\|^*$ of X^* is uniformly convex by Theorem 9.9, we can use Lemma 9.26 for the Schauder basis $\{f_i\}$ of X^*. Therefore, there is $p \in (1, \infty)$ and $K > 0$ such that for every $f = \sum_{i=1}^{\infty} \beta_i f_i \in X^*$ we have

$$\|f\|^* \le K \cdot \Big(\sum_{i=1}^{\infty} |\beta_i|^p\Big)^{\frac{1}{p}}.$$

We claim that $L = \frac{1}{K}$ and the dual index $q = \frac{p}{p-1}$, satisfy the conclusion of Lemma 9.27. Indeed, let $z_n = \sum_{i=1}^{n} \alpha_i x_i$ and let $g_n = \sum_{i=1}^{n} \beta_i f_i \in X^*$. Then $\frac{|g_n(z_n)|}{\|z_n\|} \le \|g_n\| \le K \cdot \Big(\sum_{i=1}^{n} |\beta_i|^p\Big)^{\frac{1}{p}}$. Therefore

$$\|z_n\| \ge \frac{1}{K} \cdot \frac{|g_n(z_n)|}{\Big(\sum_{i=1}^{n} |\beta_i|^p\Big)^{\frac{1}{p}}} = \frac{1}{K} \cdot \frac{\Big|\sum_{i=1}^{n} \alpha_i \beta_i\Big|}{\Big(\sum_{i=1}^{n} |\beta_i|^p\Big)^{\frac{1}{p}}}.$$

Choose $\beta_i = |\alpha_i|^{\frac{1}{p-1}} \operatorname{sign}(\alpha_i)$ for $i = 1, \ldots, n$. Then we have

$$\|z_n\| \ge \frac{1}{K} \frac{\sum_{i=1}^{n} |\alpha_i|^{\frac{p}{p-1}}}{\Big(\sum_{i=1}^{n} |\alpha_i|^{\frac{p}{p-1}}\Big)^{\frac{1}{p}}} = \frac{1}{K}\Big(\sum_{i=1}^{n} |\alpha_i|^q\Big)^{\frac{1}{q}}.$$

□

This completes the proof of Theorem 9.25. □

9.4 Remarks

1. The property of superreflexivity is a quite stable property. For example, it is a three space property [ELP]. In Theorem 12.67 we will show that it is preserved by uniform homeomorphisms (nonlinear, in general). We refer to, e.g., [AlKa] and [Gode4] for more results concerning finite representability and superreflexivity.
2. The calculated moduli of convexity and smoothness of L_p spaces are the following (see [Meir1], [Meir2], [Kade1], and [Hann]):

$$\delta_{L_p}(\varepsilon) = \begin{cases} (p-1)\frac{\varepsilon^2}{8} + o(\varepsilon^2), & \text{if } 1 < p \leq 2, \\ \frac{\varepsilon^p}{p 2^p} + o(\varepsilon^p), & \text{if } 2 \leq p < \infty. \end{cases}$$

$$\rho_{L_p}(\tau) = \begin{cases} \frac{\tau^p}{p} + o(\tau^p), & \text{if } 1 < p \leq 2, \\ (p-1)\frac{\tau^2}{2} + o(\tau^2), & \text{if } 2 \leq p < \infty. \end{cases}$$

3. Kakutani proved in [Kaku1], see also [Dies2, p. 125], that every uniformly convex space X has the Banach–Saks property, i.e., each bounded sequence (x_n) in X has a subsequence (x_{n_k}) such that $\sum_l \frac{1}{l}(x_{n_1} + x_{n_2} + \ldots x_{n_l})$ is norm-convergent in X.
4. Pisier proved that every superreflexive space can be renormed so that its modulus of convexity satisfies $\delta(\varepsilon) \geq k\varepsilon^p$ for some $p \geq 2$ ([Pisi1]).
   If the modulus of convexity of a Banach space X satisfies $\delta(\varepsilon) \geq k\varepsilon^p$ for $p \geq 2$, then X is of cotype p, see, e.g., [DGZ3]. The canonical basis of ℓ_p shows that ℓ_p is not of any cotype $q < p$. The space $L_1(\mu)$ is of cotype 2 (see, e.g., [LiTz4], see also [AlKa, p. 140]).
   As a corollary to these results and Theorem 9.23, we obtain that if $1 \leq p \leq 2$ and $\sum x_i$ is an unconditionally convergent series in $L_p([0,1])$, then $\sum \|x_i\|^2 < \infty$. This is a result of Orlicz.
5. We refer to [Figi1] for more on moduli of convexity.
6. James showed in [James10] that there are nonreflexive spaces of type 2.
7. James showed in [Jame9b] that there is a reflexive, not superreflexive, Banach space X such that ℓ_1 is not finitely represented in X.
8. For characterizations of superreflexivity in terms of "geodesics" on the unit sphere see [JamSch].
9. We refer to, e.g., [HMVZ, Ch. 2] for more on the Szlenk index and its applications.

Exercises for Chapter 9

9.1 Show that a norm of a finite-dimensional space is uniformly convex if it is strictly convex.
Hint. Use a compactness argument.

9.2 Show that a norm $\|\cdot\|$ of a Banach space X is uniformly smooth if and only if for every $\varepsilon > 0$ there is $\delta > 0$, such that $\|x+y\| + \|x-y\| \leq 2 + \varepsilon\|y\|$ whenever $\|x\| = 1, \|y\| < \delta$.
Hint. Directly from the definition.

9.3 Prove the reflexivity of a uniformly convex Banach space (Theorem 9.11) without using James' theorem.
Hint. Let $x^{**} \in S_{X^{**}}$. There exists a net $\{x_i\}_{i\in I}$ in S_X that w^*-converges to x^{**}. Assume that $\{x_i\}_{i\in I}$ is not $\|\cdot\|$-Cauchy. Then, there exists $\varepsilon > 0$ so that for every $i \in I$ we can find $\alpha(i), \beta(i) \in I$ with $\alpha(i) \geq i$, $\beta(i) \geq i$, and $\|x_{\alpha(i)} - x_{\beta(i)}\| \geq \varepsilon$. For $i \in I$ put $z_i = (1/2)(x_{\alpha(i)} + x_{\beta(i)})$; then $z_i \in (1-\delta)B_X$, where δ is given by the definition of uniform convexity of X. Obviously, the net $\{z_i\}_{i\in I}$ is w^*-convergent to x^{**}, hence $\|x^{**}\| \leq 1 - \delta$, a contradiction. Thus $\{x_i\}_{i\in I}$ is $\|\cdot\|$-Cauchy, hence it converges to some $x \in X$. Necessarily, $x^{**} = x\ (\in X)$.

9.4 Let norms $\|\cdot\|_n$ of Banach spaces X_n be uniformly convex with moduli of convexity $\delta_n(\varepsilon)$, $n \in \mathbb{N}$. Assume that $\delta(\varepsilon) = \inf_n\big(\delta_n(\varepsilon)\big) > 0$ for every $\varepsilon > 0$. Show that then the canonical norm of $\big(\sum(X_n, \|\cdot\|_n)\big)_2$ is uniformly convex.
Hint. Direct calculation.

9.5 For $n \in \mathbb{N}$, let $\|\cdot\|_n$ be a strictly convex norm on ℓ_∞^n such that $\|\cdot\|_\infty \leq \|\cdot\|_n \leq 2\|\cdot\|_\infty$. Let $X = \big(\sum(\ell_\infty^n, \|\cdot\|_n)\big)_2$. Show that X is locally uniformly rotund.
Hint. Direct calculation.

9.6 Let X be a uniformly convex infinite-dimensional Banach space whose modulus of convexity satisfies $\delta(\varepsilon) \geq k\varepsilon^p$ for some $k, p > 0$ and all $\varepsilon(0, 2]$. Show that $p \geq 2$.
Hint. Use Theorem 6.2 (iii).

9.7 Let the norm $\|\cdot\|$ of a Banach space X be uniformly convex (respectively uniformly Fréchet differentiable). Assume that Y is a closed subspace of X. Show that the canonical norm of X/Y is uniformly convex (respectively uniformly Fréchet differentiable).
Hint. $(X/Y)^*$ is isometric to the subspace $Y^\perp$ of X^*. Use Theorem 9.9.

9.8 Let X be a uniformly convex space with modulus of convexity $\delta(\varepsilon)$. Take $f \in S_{X^*}$ and consider the affine hyperplane $H = (f^{-1}(0) + z)$ for some $z \in X$. Show that if $\operatorname{dist}(0, H) \geq 1 - \delta(\varepsilon)/2$ then $\operatorname{diam}(B_X \cap H) \leq \varepsilon$.
Hint. If $x, y \in B_X \cap H$, then $(x+y)/2 \in H \cap B_X$.

9.9 Show that the reflexive separable space $X := \big(\sum_n \ell_{(1+1/n)}\big)_2$ does not have an equivalent uniformly Kadec–Klee smooth norm.
Hint. First we show that the canonical norm on X is not uniformly Kadec–Klee smooth. If $\varepsilon > 0$ is given, denote by $\eta_\varepsilon(B_{X^*})$ the collection of such points f in B_{X^*} so that there is a sequence of points (f_n) in B_{X^*} such that $\|f_n - f\| \geq \varepsilon$ and $f_n \to$

f in the w^* topology. From the definition of uniform Kadec–Klee smoothness, it follows that there is $\delta > 0$ so that $\eta_\varepsilon(B_{X^*}) \subset \delta B_{X^*}$. By iterating this procedure and using the homogeneity, we get that, for large n, the iterated $\eta_\varepsilon^{(n)}(B_{X^*}) = \emptyset$. Thus, for proving that the canonical norm of X is not uniformly Kadec–Klee smooth it suffices to show the following Claim. *If* $Y = \ell_p$, *where* $\frac{1}{p} + \frac{1}{q} = 1$ *and* $m \le 2^p$, *then* $\eta_{\frac{1}{2}}^{(n)}(B_{Y^*}) \ne \emptyset$. In order to see the claim, first note that if e_k are the unit vectors in ℓ_p and $n_1 < n_2 < ...n_m$, then $\| \sum_{k=1}^m \frac{1}{2} e_k \| = (\frac{m}{2^p})^{\frac{1}{p}} \le 1$. Thus $\sum_{k=1}^{m-1} \frac{1}{2} e_k$ is in $\eta_{\frac{1}{2}}(B_{Y^*})$, as it is a w^* limit of $\sum_{k=1}^{m-1} \frac{1}{2} e + \frac{1}{2} e_i$ when $i \to \infty$ since the latter points are in B_{Y^*} and have distance $\frac{1}{2}$ to $\sum_{k=1}^{m-1} \frac{1}{2} e_k$. By iterating, we get that any $\sum_{k=1}^{m-n} \frac{1}{2} e_k \in \eta_{\frac{1}{2}}^n(B_{Y^*})$ for each $n < m$. Since 0 is in the weak* closure of the collection of such elements, we get that 0 is in the w^*-closure of $\eta_{\frac{1}{2}}^n(B_{Y^*})$. This by the above means that the canonical norm of X is not uniformly Kadec–Klee smooth. It cannot have such an equivalent norm either since if $B_1 \subset B_2$, then obviously $\eta_\varepsilon(B_1) \subset \eta_\varepsilon(B_2)$.

9.10 Show that the norm on a Banach space X is uniformly Kadec–Klee smooth if it is uniformly Fréchet smooth.
Hint. Use Theorem 7.27: given $\varepsilon > 0$ find $\delta > 0$ according to (7.6) in this lemma. Let $x^* \in X^*$ such that $\operatorname{diam}(V \cap B_{X^*}) > \varepsilon$ for every w^*-neighborhood V of x^*. In particular, $x^* \in B_{X^*}$ and x^* cannot lie in $\{y^* \in B_{X^*} : \langle y^*, x\rangle > 1 - \delta\}$ for any $x \in S_X$. This implies $\|x^*\| \le 1 - \delta$.

9.11 Show that if a Banach space Y is finitely representable in a uniformly convex Banach space X, then the norm of Y is uniformly convex.
Hint. See [DGZ3, p. 133].

9.12 Let X be a superreflexive space. Show that if Y is isomorphic to X then Y is superreflexive.
Hint. If $|||\cdot|||$ is uniformly convex norm on X (Theorem 9.14) and T is an isomorphism of X onto Y, then the norm $|||\cdot|||_1$ defined for $y \in Y$ by $|||y|||_1 = |||T^{-1}(y)|||$ is an equivalent uniformly convex norm on Y.

9.13 Show that a Banach space is superreflexive if and only if every space Z that is crudely finitely representable in X is reflexive.
Hint. Proposition 9.12 and Theorem 9.14.

9.14 Show that if $\|\cdot\|$ is a uniformly smooth norm of a Banach space X, then $(\|\cdot\|^2)'$ is uniformly continuous on bounded sets.
Hint. Direct calculation, the chain rule, use the fact that the first derivative of the norm is bounded.

9.15 Let X be a Banach space. Show that X is superreflexive if and only if X admits an equivalent norm $\|\cdot\|$ such that the derivative $(\|\cdot\|^2)'$ is uniformly continuous on a neighborhood of the origin.
Hint. One direction: see the previous exercise. If $(\|\cdot\|^2)'$ is uniformly continuous on a neighborhood of the origin, then the Minkowski functional of the appropriate level set of $\|\cdot\|^2$ gives an equivalent uniformly Fréchet differentiable norm on X.

9.16 Show that X is superreflexive if it admits a norm that is at the same time locally uniformly rotund and Fréchet differentiable with locally uniformly continuous derivative.
Hint. Take a point x_0 on the unit sphere and f_0 be a support functional to that point. Consider $\|\cdot\| - f(\cdot)$ around the point x_0. For small $\delta > 0$, this function is bounded from zero on $\{x : \|x - x_0\| = \delta\}$ by the LUR property of the norm. From this, construct a bump on X with uniformly continuous derivative and use Theorem 9.14, (v).

9.17 Let X be a superreflexive Banach space with a normalized Schauder basis $\{x_i\}$. Show that the series $\sum_{n=1}^{\infty} \frac{x_n}{n}$ converges in X but $\sum_{n=1}^{\infty} \frac{x_n}{\ln(n+1)}$ does not converge.
Hint. $\sum n^{-p} < \infty$ for every $p > 1$. Use Theorem 9.25.

If the second series did converge, then $\sum \frac{1}{\ln^q(1+n)} < \infty$ for some $q \in (1, \infty)$ by Theorem 9.25, which is not the case.

9.18 A norm $\|\cdot\|$ of a Banach space X is called *uniformly rotund in every direction (URED)* if for every $z \in S_X$ and all bounded sequences $\{x_n\}, \{y_n\} \subset X$ such that $2\|x_n\|^2 + 2\|y_n\|^2 - \|x_n + y_n\|^2 \to 0$ and $x_n - y_n = \lambda_n z$ for some λ_n, we have $\lambda_n \to 0$.

Let Γ be an uncountable set. Show that $c_0(\Gamma)$ has no equivalent URED norm [DJS].
Hint. Let $\|\cdot\|_\infty$ be the supremum norm of $c_0(\Gamma)$ and assume that $\|\cdot\|$ is an equivalent norm on $c_0(\Gamma)$. Set $M = \sup_{\|x\|_\infty \le 1} \|x\|$. Let $x_n \in c_0(\Gamma)$ satisfy $\|x_n\|_\infty = 1$ and $\|x_n\| \to M$. Let z be such that $\|z\|_\infty = 1$ and its support is disjoint from all the supports of x_n. Then $\|x_n \pm z\|_\infty = 1$ for every n.

Since $\|(x_n + z + x_n - z)/2\| \to M$, we have $\|x_n \pm z\| \to M$ and also $2\|x_n + z\|^2 + 2\|x_n - z\|^2 - \|x_n + z + x_n - z\|^2 \to 0$. Thus $\|\cdot\|$ is not URED.

9.19 Let Γ be uncountable. Use the preceding exercise to show that there is no bounded one-to-one operator from $c_0(\Gamma)$ into any URED Banach space X.
Hint. Let $T\colon c_0(\Gamma) \to X$ be such an operator. Let $\|\cdot\|$ be an equivalent URED norm on X and define an equivalent norm $|||\cdot|||$ on $c_0(\Gamma)$ by $|||x|||^2 = \|x\|_\infty^2 + \|T(x)\|^2$. Let $\{x_n\}, \{y_n\}$ be bounded sequences in $c_0(\Gamma)$, $z \in c_0(\Gamma)$ be such that $\|z\|_\infty = 1$, $x_n - y_n = \lambda_n z$ for some λ_n and $2|||x_n|||^2 + 2|||y_n|||^2 - |||x_n + y_n|||^2 \to 0$. Then as in Chapter 8 we have $2\|T(x_n)\|^2 + 2\|T(y_n)\|^2 - \|T(x_n + y_n)\|^2 \to 0$. Since $\|\cdot\|$ is

uniformly convex, we have $T(x_n - y_n) = \lambda_n T(z) \to 0$. Since $T(z) \neq 0$, we have $\lambda_n \to 0$. Thus $|||\cdot|||$ is URED, a contradiction with the previous exercise.

9.20 ([Zizl1]) Show that every separable Banach space has an equivalent URED norm.
Hint. Let $(X, \|\cdot\|)$ be a separable Banach space. Assume that $\{f_n\}_{n=1}^{\infty}$ is a sequence in S_{X^*} that is separating for X. Consider the norm $|||\cdot|||$ on X given by $|||x|||^2 := \|x\|^2 + \sum_{i=1}^{\infty} f_i^2(x)/2^i$ for $x \in X$. If $2|||x_n|||^2 + 2|||y_n|||^2 - |||x_n + y_n|||^2 \to 0$, where $\{x_n\}$ and $\{y_n\}$ are bounded sequences, then

$$f_i^2(x_n - y_n) = 2f_i^2(x_n) + 2f_i^2(y_n) - f_i^2(x_n + y_n) \to 0$$

for each i, so $\lambda_n \to 0$ if $x_n - y_n = \lambda_n z$ for all n and for some $z \neq 0$.

9.21 Show that the space ℓ_1 does not contain any bounded (∞, ε)-tree for any $\varepsilon > 0$.
Hint. Norm-closed bounded sets in separable duals have strongly exposed points (see Theorem 8.14).

9.22 Since ℓ_1 is not superreflexive as it is not reflexive, there is $\varepsilon > 0$ such that B_{ℓ_1} has arbitrary large (n, ε)-trees. Show this fact by a concrete example.
Hint. Build a $(4, 1)$-tree as follows: Split $\left(\frac{1}{4}, \frac{1}{4}, \frac{1}{4}, \frac{1}{4}, 0, \ldots\right)$ into the pair $\left(\frac{1}{2}, 0, \frac{1}{4}, \frac{1}{4}, 0, \ldots\right)$ and $\left(0, \frac{1}{2}, \frac{1}{4}, \frac{1}{4}, 0, \ldots\right)$.
Then split $\left(\frac{1}{2}, 0, \frac{1}{4}, \frac{1}{4}, 0, \ldots\right)$ into the pair $\left(\frac{1}{2}, 0, \frac{1}{2}, 0, \ldots\right)$ and $\left(\frac{1}{2}, 0, 0, \frac{1}{2}, 0, \ldots\right)$, etc.

9.23 Let X be a Banach space. Assume that every separable closed subspace of X admits an equivalent uniformly convex norm. Show that X has an equivalent uniformly convex norm.
Hint. Assume the contrary. Then there is $\varepsilon > 0$ such that for every n there is an (n, ε)-tree in B_X. The union of these trees lies in a separable closed subspace Z of X. Then B_Z contains (n, ε)-trees for all n, hence Z is not superreflexive.

Exercises 9.24, 9.25, and 9.26 below provide proofs to the implications (v)$\Longrightarrow$(iii)$\Longrightarrow$(vi) in Theorem 11.8. The techniques here, which have obvious similarities with those there, use the intermediate concept of a *bounded ε-tree*.

9.24 Let $\{K_n\}$ be a sequence of nonempty compact convex subsets of a topological vector space such that $K_{2n} \cup K_{2n+1} \subset K_n$ for all n. Show that then there exists an infinite tree $\{x_n\}$ such that $x_n \in K_n$ for all n.
Hint. Consider $K = \prod K_n$. Define for $n \in \mathbb{N}$,

$$A_n = \{x = \{x_k\} \in K : x_n = (x_{2n} + x_{2n+1})/2\}.$$

Note that each A_n is closed and hence compact in the compact set K and that $x = \{x_n\} \in K$ is an infinite tree if $x \in \bigcap A_n$. Thus we only need to prove that $\bigcap A_n \neq \emptyset$. By compactness, it suffices to show that $A_1 \cap A_2 \cap \cdots \cap A_k \neq \emptyset$ for every k. To this end, fix k and define $x \in K$ as follows: If $n > k$, let x_n be an arbitrary element in K_n. From $n = k$ to $n = 1$ we proceed by induction: If for $m = n+1, n+2, \ldots$ points $x_m \in K_m$ were chosen, define $x_n = \frac{1}{2}(x_{2n} + x_{2n+1})$. The resulting x is in $A_1 \cap A_2 \cap \cdots \cap A_k$.

9.25 Let X be a Banach space. Show that if X contains a separable closed subspace Y such that Y^* is non-separable, then there exist $\varepsilon > 0$ and a bounded set A in X^* such that every nonempty relatively w^*-open subset of A has diameter greater than ε.

Hint. By the proof of Theorem 8.10, there is such a set C in Y^* which is w^*-closed and contained in B_{Y^*}. Let R be the restriction mapping of X^* onto Y^*. Note that R is w^*-w^*-continuous and maps B_{X^*} onto B_{Y^*}. Let A be a minimal w^*-compact subset of B_{X^*} such that $R(A) = C$. Let U be a relatively w^*-open subset of A. Then $C_1 = R(A \setminus U)$ is a w^*-compact set in C which is a proper subset of C due to the minimality of A. Then $C \setminus C_1$ is a relatively w^*-open set in C and thus it has diameter greater than ε. Since R is a 1-Lipschitz mapping, it follows that $\operatorname{diam} U > \varepsilon$.

9.26 (Stegall) Assume that a Banach space X has a separable closed subspace Y of X such that Y^* is non-separable. Show that X^* contains a bounded infinite ε-tree for some $\varepsilon > 0$.

Hint. (van Dulst–Namioka) Let A be the set from Exercise 9.25. Construct a sequence $\{U_n\}$ of relatively w^*-open subsets of A and a sequence $\{x_n\} \in S_X$ such that $U_{2n} \cup U_{2n+1} \subset U_n$ for every n and $(x^* - y^*)(x_n) \geq \varepsilon$ for every $x^* \in U_{2n}$, $y^* \in U_{2n+1}$ and n. We construct such a sequence by induction. To see the few steps, put $U_1 = A$. Since $\operatorname{diam} A_1 > \varepsilon$, there are z_0^* and z_1^* in U_1 such that $\|z_0^* - z_1^*\| > \varepsilon$. Choose a point $x_1 \in S_X$ such that $(z_0^* - z_1^*)(x_1) = \varepsilon + \delta$ for some $\delta > 0$. Let

$$U_2 = \{x^* \in U_1 : x^*(x_1) > z_0^*(x_1) - \delta/2\},$$
$$U_3 = \{y^* \in U_1 : y^*(x_1) < z_1^*(x_1) + \delta/2\}.$$

We use the following observation: *If J_1 and J_2 are sets in a topological vector space, then* $\overline{\operatorname{conv}}(J_1) - \overline{\operatorname{conv}}(J_2) \subset \overline{\operatorname{conv}}(J_1 - J_2)$. Indeed, $a - \overline{\operatorname{conv}}(J_2) = \overline{\operatorname{conv}}(a - J_2) \subset \overline{\operatorname{conv}}(J_1 - J_2)$ for every $a \in J_1$. Having the sets U_n constructed as above, define $K_n = \overline{\operatorname{conv}}^{w^*}(U_n)$ for every n. We have $K_{2n} \cup K_{2n+1} \subset K_n$ for every n. By Exercise 9.24, there is a tree $\{x_n^*\}$ in X^* such that $x_n^* \in K_n$ for each n. By the observation above, we have

$$(x_{2n}^* - x_{2n+1}^*) \in (K_{2n} - K_{2n+1}) \subset \overline{\operatorname{conv}}^{w^*}(U_{2n} - U_{2n+1}),$$

so we have $(x_{2n}^* - x_{2n+1}^*)(x_n) \geq \varepsilon$. Hence $\|x_{2n}^* - x_{2n+1}^*\| \geq \varepsilon$ and $\{x_n^*\}$ is a bounded infinite ε-tree in X^*.

Note that it follows from this result that if X is a separable space with non-separable dual, then there is a bounded convex norm-closed set C in X^* that contains no strongly exposed point. Indeed, the norm-closed convex hull of the tree in Stegall's result is such a set. We remark that there exists a Banach space X that does not have RNP (see Definition 11.14) but whose closed unit ball does not contain any (∞, ε)-tree [BoRo]. In this direction, see also [Tala3a].

9.27 (Daugavet) Let A be a compact operator in $C[0, 1]$. Show that $\|I + A\| = 1 + \|A\|$ (we say that A *satisfies the Daugavet equation*).
Hint. Assume that A is a one-dimensional norm-one operator $x \mapsto f(x)y$, where $y \in C[0, 1]$ and $f \in C[0, 1]^*$ correspond via the Riesz representation to a function φ with bounded variation. Let $x_0 \in S_{C[0,1]}$ satisfy $\|A(x_0)\| > 1-\varepsilon$. Let $y_0 = A(x_0)$. Let J be an interval where y_0 is greater than $1-\varepsilon$ and the variation of φ on J is small enough. Change x_0 to x_1 to have $x_0 = x_1$ outside J, $x_1(t_0) = 1$ for some $t_0 \in J$ and x_1 is a broken line on J. Put $y_1 = A(x_1)$. Due to the smallness of the variation of φ on J, we have $\|y_0 - y_1\| \leq \varepsilon$. Then $\|I + A\| \geq x_1(t_0) + y_1(t_0) > 1 + 1 - 2\varepsilon$.

9.28 ([AAB]) Let X be a uniformly convex Banach space and $T \in \mathcal{B}(X)$. Show that T satisfies the Daugavet equation if and only if $\|T\|$ lies in the approximate point spectrum $\sigma_{\text{ap}}(T)$ of T (this concept is introduced after Definition 15.13).
Hint. If $\|T\| \in \sigma_{\text{ap}}(T)$, choose $\{x_n\} \subset S_X$ such that $\big\| \|T\|x_n - T(x_n)\big\| \to 0$. Then

$$\|I+T\| \geq \|(I+T)(x_n)\| \geq \big\|x_n+\|T\|x_n\big\|-\big\|\|T\|x_n-T(x_n)\big\| = 1+\|T\|-\big\|\|T\|x_n-T(x_n)\big\|.$$

Let now T satisfy the Daugavet equation. Then $S = T/\|T\|$ satisfies the Daugavet equation and $\|I + S\| = 1 + \|S\| = 2$.

Hence there is a sequence $\{x_n\} \subset S_X$ such that $\|x_n + S(x_n)\| = 2$. As $\|x_n\| = 1$ and $\|S(x_n)\|\|x_n\| \leq 1$, we have from the uniform convexity of X that $\|S(x_n) - x_n\| \to 0$. By multiplying through by $\|T\|$, $\big\|T(x_n) - \|T\|x_n\big\| \to 0$. Thus $\|T\| \in \sigma_{\text{ap}}(T)$.

In the following exercises, we will construct and investigate the Tsirelson space T, see [Tsir]. All notation and definitions made in one exercise apply to the following exercises as well.

Let $\{e_i\}$ be the canonical basis of c_0, we will use $x(i)$ for the ith coefficient of $x \in c_0$, that is, $x = \sum x(i)e_i$. For $n \in \mathbb{N}$ we define a tail projection on c_0 by $P_n\Big(\sum_{i=1}^{\infty} x(i)e_i\Big) = \sum_{i=n}^{\infty} x(i)e_i$. Let $\{v_i\}$ be a finite sequence of vectors. If they are consecutively supported (which will be denoted by $v_1 < \cdots < v_n$), we write $(v_1, \ldots, v_n)$ for $\sum_{i=1}^{n} v_i$. When we use $(v_1, \ldots, v_n)$, it is automatically assumed that v_i are successively supported.

Let A be a subset of c_0. We consider the following set of conditions:

(1) $A \subset B_{c_0}$ and $\{e_i\} \subset A$.
(2) If $x = \sum x(i)e_i \in A$ and $|y(i)| \leq |x(i)|$ for all i, then $\sum y(i)e_i \in A$.
(3) If $v_1 < \cdots < v_n \in A$, then $\frac{1}{2}P_n\big((v_1, \ldots, v_n)\big) \in A$.
(4) For every $x \in A$ there is $n \in \mathbb{N}$ such that $2P_n(x) \in A$.

9.29 Show that there is a *weakly compact* set $K \subset c_0$ satisfying (1)–(4) above.
Hint. Put $A_1 = \{\alpha e_i : |\alpha| \leq 1, i \in \mathbb{N}\}$ and

$$A_{n+1} = A_n \cup \big\{\tfrac{1}{2}P_N\big((v_1, \ldots, v_N)\big) : N \in \mathbb{N}, v_1 < \cdots < v_N \in A_n\big\}.$$

Let K be the pointwise closure of $\bigcup_{n=1}^{\infty} A_n$. Only (4) needs an explanation. Assume that $v \in K \setminus A_1$, $v(i_0) \neq 0$ and $v(i) = 0$ for $i < i_0$. Choose a sequence $\{v^i\}$ in $\bigcup_{n=1}^{\infty} A_n$ pointwise convergent to v. We may assume that $v^i(j) = 0$ for all i and $j < i_0$ and that $v^i \notin A_1$. Thus $v^i = \frac{1}{2}P_{N_i}\big((v_1^i, \ldots, v_{N_i}^i)\big)$, hence $N_i \leq i_0$. By passing to a subsequence, we may assume that $v^i = \frac{1}{2}P_N\big((v_1^i, \ldots, v_N^i)\big)$ for some $N \in \mathbb{N}$. Thus $v = \frac{1}{2}P_N\big((v_1, \ldots, v_N)\big)$, where every v_j is a pointwise limit of $\{v_j^i\}_i$. Put $M = \max\{i : v_i \neq 0\}$, $m = \min\big(\mathrm{supp}(v_M)\big)$. Then by (2), $P_N\big((v_1, \ldots, v_N)\big) \in K$, so $2P_N(v) \in K$.

The weak compactness of K follows as K is pointwise sequentially compact. This can be seen by estimating the distribution of those coordinates of $x \in K$ that are larger than a given $\varepsilon > 0$.

9.30 Show that $V = \overline{\mathrm{conv}}(K)$ also satisfies (1)–(4) above.
Hint. Only (3) and (4) need an explanation. To see (3), consider $x_i = \alpha_i^1 x_i^1 + \cdots + \alpha_i^n x_i^n$, where $x_i^j \in K$, for $m > i$ the supports of all x_i^j precede the supports of all x_m^l, and $x = (x_1, \ldots, x_n)$. Then $\frac{1}{2}P_N(x)$ is a convex combination of $\frac{1}{2}P_N\big((\alpha_1^{j_1} x_1^{j_1}, \ldots, \alpha_n^{j_n} x_n^{j_n})\big)$ and so $\frac{1}{2}P_N(x) \in V$.

To see (4), set $D_n = \{x \in K : 4P_n(x) \in K\}$. Clearly, $D_n \subset D_{n+1}$ and $K = \bigcup_{n \in \mathbb{N}} D_n$. For every $x_0 \in V$ there exists a Borel measure μ in the weak topology of V such that $f(x_0) = \int_K f \,\mathrm{d}\mu$ for $f \in c_0^*$. Thus for n_0 large enough we have $\mu(D_{n_0}) \geq \frac{3}{4}$. We claim that $2P_{n_0}(x_0) \in V$. Indeed, otherwise, consider the set $W = \{x \in V : 2P_{n_0}(x) \in V\}$. Choose $f \in c_0^*$ such that $f(x_0) > 1$ and $|f(x)| \leq 1$ whenever $x \in W$. Then

$$1 < f(x_0) = \int_{D_{n_0}} f \,\mathrm{d}\mu + \int_{K \setminus D_{n_0}} f \,\mathrm{d}\mu \leq \frac{1}{2}\mu(D_{n_0}) + 2\mu(K \setminus D_{n_0}) \leq \frac{1}{2} + \frac{1}{2} = 1,$$

a contradiction.

9.31 Let $T = \mathrm{span}(V)$ (span taken in c_0) and consider the norm given on X by the Minkowski functional of V. Show that T is a reflexive Banach space with an unconditional basis $\{e_i\}$. Show that c_0 is finitely representable in every infinite-dimensional closed subspace of T. In particular, T contains no isomorphic copy of c_0, ℓ_1, or a superreflexive space.
Hint. T is a Banach space as V is weakly compact in c_0 (Theorem 3.133). From (2) and (4) above it follows that $\{e_i\}$ is an unconditional basis of T. The basis $\{e_i\}$ is shrinking. Indeed, suppose that $\|P_n^*(f)\| \geq 2\varepsilon$ for some $f \in T^*$, $\varepsilon > 0$ and $n \in \mathbb{N}$. We choose $x_1 < x_2 < \cdots$ such that $f(x_i) > \varepsilon$. Since P_N does not effect

$x_{N+1} + \cdots + x_{2N}$ and thus $\|x_{N+1} + \cdots + x_{2N}\| \le 2$, we get a contradiction with the fact that $f(x_{N+1} + \cdots + x_{2N}) > N\varepsilon$.

Having shown that $\{e_i\}$ is shrinking, we show that V is weakly compact in T. Indeed, $\operatorname{span}\{e_i^*\}$ is dense in T^*, so the w-topology of T restricted to V coincides with the w-topology of c_0 restricted to V, which is compact by Krein's theorem.

If $\{x_i\}$ is a normalized block basic sequence of $\{e_i\}$, then $\frac{1}{2}P_N(x_1+\cdots+x_N) \in V$ for every N by (3). Thus by (2), $\|P_N(\lambda_1 x_1 + \cdots + \lambda_N x_N)\| \le 2\max|\lambda_i|$. Consequently,

$$\max|\lambda_i| \le \|\lambda_1 x_{N+1} + \cdots + \lambda_N x_{2N}\| \le 2\max|\lambda_i|.$$

Using the technique from Theorem 4.26, we find that c_0 is crudely finitely representable in every infinite-dimensional closed subspace of T. Thus none of infinite-dimensional closed subspaces of T can be superreflexive, hence it does not contain subspaces isomorphic to ℓ_p for $p \in (1,\infty)$. Since T is reflexive, it does not contain an isomorphic copy of c_0 or ℓ_1.

It turned out that the dual of Tsirelson space T is a reflexive space that does not contain ℓ_p spaces. This dual space is actually easier to handle analytically than T, [FiJo2].

9.32 ([Cepe]) Let f be a Lipschitz function on a uniformly convex Banach space X. Then there is a sequence $\{f_n\}$ of Δ-convex functions (i.e., differences of convex functions) that are bounded on bounded sets of X and converge to f uniformly on bounded sets.

Hint. Assume that the Lipschitz constant of f is 1. For $n \in \mathbb{N}$ and $x \in X$ define $f_n(x) = \inf_{y\in X}\{f(y) + 2n\|x\|^2 + 2n\|y\|^2 - n\|x+y\|^2\}$.

Then $f_n = c_n - d_n$, where $c_n = 2n\|x\|^2$ and $d_n(x) = \sup_{y\in X}\{n\|x+y\|^2 - 2n\|y\|^2 - f(y)\}$. The functions c_n and d_n are convex and by taking $y = x$ in the infimum, we can see that $f_n \le f$ for all n. Since $2\|x\|^2 + 2\|y\|^2 - \|x+y\|^2 \ge 2\|x\|^2 + 2\|y\|^2 - (\|x\| + \|y\|)^2 = (\|x\| - \|y\|)^2 \ge 0$, f_n is an increasing sequence of functions.

Let y be such that $f(y) + 2n\|x\|^2 + 2n\|y\|^2 - n\|x+y\|^2 \le f(x)$. Then since f is 1-Lipschitz,

$$n(\|x\| - \|y\|)^2 \le 2n\|x\|^2 + 2n\|y\|^2 - n\|x+y\|^2 \le f(x) - f(y) \le \|x-y\|.$$

Suppose that $\|y\| \ge 1 + \|x\|$. Then

$$1 \le \big|\|x\| - \|y\|\big| \le (\|x\| - \|y\|)^2 \le \tfrac{1}{n}\|x-y\| \le \tfrac{1}{n}\|x\| + \tfrac{1}{n}\|y\|.$$

Hence for $n \ge 3$ we have $\|y\| \le \frac{n+1}{n-1}\|x\| \le 2\|x\|$. Thus in any case we have $\|y\| \le 2(1+\|x\|)$. This implies that

$$f_n(x) = \inf_{\|y\|\le 2(1+\|x\|)}\{f(y) + 2n\|x\|^2 + 2n\|y\|^2 - n\|x+y\|^2\}.$$

Hence in particular f_n is bounded on bounded sets and so is $d_n = c_n - f_n$. Moreover, for such y we have

$$0 \leq 2\|x\|^2 + 2\|y\|^2 - \|x+y\|^2 \leq \tfrac{1}{n}\|x-y\| \leq \tfrac{1}{n}\|x\| + \tfrac{1}{n}\|y\| < \tfrac{3}{n}(1+\|x\|).$$

Assume that it is not true that $f_n \to f$ uniformly on bounded sets. Then there is $\varepsilon > 0$ and a bounded sequence $\{x_n\}$ such that $f_n(x_n) + \varepsilon < f(x_n)$ for all $n \in \mathbb{N}$. For each n choose y_n such that

$$f_n(y_n) + \varepsilon \leq f(y_n) + 2n\|x_n\|^2 + 2n\|y_n\|^2 - n\|x_n + y_n\|^2 + \varepsilon < f(x_n).$$

Then we have $\|x_n - y_n\| \geq f(x_n) - f(y_n) > \varepsilon$ and

$$0 \leq 2\|x_n\|^2 + 2\|y_n\|^2 - \|x_n + y_n\|^2 < \tfrac{3}{n}(1 + \|x_n\|) \to 0$$

as $n \to \infty$. Thus X is not uniformly convex, a contradiction.

9.33 Suppose X is a Banach space and let f be a convex function on X which is bounded on bounded sets. If $\{g_k\}$ is a sequence of convex functions such that $g_k(0) \leq 1/k$ and $g_k(x) > k\|x\| - (1/k)$ for all $x \in X$, then $f \diamond g_k \to f$ uniformly on bounded subsets of X. For the definition of $\diamond$ see Exercise 7.5.
Hint. ([MPVZ]) Let $r > 0$ and suppose that f has Lipschitz constant K on B_{r+1}. For $x_0 \in B_r$ fixed and for each k we can choose y_k so that $f \diamond g_k(x_0) \geq f(y_k) + g_k(x_0 - y_k) - 1/k$. For any $k \geq K+1$ with $k \geq 3$ we have

$$\begin{aligned} f(x_0) + \frac{1}{k} &\geq f(x_0) + g_k(0) \geq f \diamond g_k(x_0) \\ &\geq f(y_k) + g_k(x_0 - y_k) - \frac{1}{k} \geq f(y_k) + k\|x_0 - y_k\| - \frac{2}{k}. \end{aligned} \tag{9.1}$$

Let $\Lambda_0 \in \partial f(x_0)$, then $\|\Lambda_0\|^* \leq K$ since f has Lipschitz constant K on B_{r+1}. Because $f(y_k) - f(x_0) \geq \Lambda_0(y_k) - \Lambda_0(x_0)$, we have

$$f(x_0) - f(y_k) \leq \|\Lambda_0\|^*\|y_k - x_0\| \leq K\|y_k - x_0\|.$$

Thus it follows from (9.1) that

$$K\|y_k - x_0\| + \frac{3}{k} \geq k\|x_0 - y_k\|.$$

In other words,

$$\|x_0 - y_k\| \leq \frac{3}{k(k-K)}.$$

In particular, $y_k \in B_{r+1}$ and so $|f(y_k) - f(x_0)| \leq K\|y_k - x_0\|$. From this we obtain

$$
\begin{aligned}
& f(y_k) + k\|x_0 - y_k\| - \frac{2}{k} \\
& \geq f(x_0) - K\|x_0 - y_k\| + k\|x_0 - y_k\| - \frac{2}{k} \geq f(x_0) - \frac{2}{k}.
\end{aligned} \tag{9.2}
$$

Clearly the lemma follows from (9.1) and (9.2).

9.34 Let $(X, \|\cdot\|)$ be a Hilbert space. Then any convex function f which is bounded on bounded subset of X can be approximated uniformly on bounded sets by convex functions with Lipschitz derivatives.
Hint. ([MPVZ]) Observe that $\|\cdot\|^2$ has Lipschitz derivative on all of X; hence so does g_k, where $g_k(x) := k^4\|x\|^2$. Easily $g_k(x) \geq k\|x\| - (1/k)$ for all k and $g_k(0) = 0$, therefore $f \diamond g_k \to f$ uniformly on bounded sets by Exercise 9.33.

To see that $f_k := f \diamond g_k$ has Lipschitz derivative for each k we use the mean value theorem to choose $C_k > 0$ such that

$$
g_k(x+h) + g_k(x-h) - 2g_k(x) \leq C_k\|h\|^2 \tag{9.3}
$$

for all $x, h \in X$. Fix an arbitrary $x_0 \in X$. Since X is reflexive, we choose y_k so that $f_k(x_0) = f(y_k) + g_k(x_0 - y_k)$. Then, using (9.3), for any $h \in X$ we have

$$
\begin{aligned}
& f_k(x_0 + h) + f_k(x_0 - h) - 2f_k(x_0) \\
& \leq f(y_k) + g_k(x_0 + h - y_k) + f(y_k) + \\
& + g_k(x_0 - h - y_k) - 2(f(y_k) + g_k(x_0 - y_k)) \\
& = g_k(x_0 - y_k + h) + g_k(x_0 - y_k - h) - 2g_k(x_0 - y_k) \leq C_k\|h\|^2.
\end{aligned}
$$

Since C_k does not depend on x_0, it follows from [DGZ3, Ch. V] that f'_k is Lipschitz.

9.35 Show that every real-valued Lipschitz function on a Hilbert space can be approximated uniformly on bounded sets by functions with Lipschitz derivatives.
Hint. Use Exercises 9.32, 9.33, and 9.34.

Chapter 10
Higher Order Smoothness

In this chapter, we will first discuss the properties of smoothness in ℓ_p spaces and in Hilbert spaces. Then we study spaces that have countable James boundary in connection with their higher order smoothness, and its applications. In particular, we study spaces of continuous functions on countable compact spaces.

10.1 Introduction

Definition 10.1 *Let X be a Banach space, $n \in \mathbb{N}$. A function $M\colon X^n \to \mathbb{R}$ is called an* n-linear form *on X if it satisfies*

$$M(x_1, \dots, x_{k-1}, \alpha x + \beta y, x_{k+1}, \dots, x_n)$$
$$= \alpha M(x_1, \dots, x_{k-1}, x, x_{k+1}, \dots, x_n) + \beta M(x_1, \dots, x_{k-1}, y, x_{k+1}, \dots, x_n)$$

for every $k \in \{1, \dots, n\}$, $x, y, x_i \in X$ and $\alpha, \beta \in \mathbb{R}$.
If $M\big((B_X)^n\big)$ is bounded, then M is called bounded.
The form is called symmetric *if $M(x_1, \dots, x_n) = M(x_{\pi(1)}, \dots, x_{\pi(n)})$ for every permutation π of $\{1, \dots, n\}$.*

It is routine to check that M is bounded if and only if it is continuous on X^n, and the linear space $\mathcal{L}(^nX)$ of all bounded n-linear forms on X, endowed with the norm $\|M\| = \sup\{|M(x_1, \dots, x_n)| : x_i \in B_X\}$, is a Banach space.

By $\mathcal{L}_s(^nX)$ we denote the closed subspace of $\mathcal{L}(^nX)$ that consists of all symmetric forms. Note that $\mathcal{L}_s(^1X) = \mathcal{L}(^1X) = X^*$.

Multilinear forms allow us to define polynomials on Banach spaces:

Definition 10.2 *We say that $P\colon X \to \mathbb{R}$ is a* polynomial of degree $n \in \mathbb{N}$ *if for $i = 1, \dots, n$ there are bounded i-linear forms B_i such that for all $h \in X$ we have*

$$P(h) = B_1(h) + B_2(h, h) + \cdots + B_n(h, \dots, h).$$

Basic facts about multilinear forms and polynomials can be found, e.g., in [Barr].

We define higher order derivatives by induction, using Fréchet derivative in case $n = 1$.

M. Fabian et al., *Banach Space Theory*, CMS Books in Mathematics,
DOI 10.1007/978-1-4419-7515-7_10,

Definition 10.3 *Let U be an open set in a Banach space X. Let $n \geq 1$ and assume that $f\colon U \to \mathbb{R}$ is n-times differentiable in U. Suppose that the nth derivative $f^{(n)}\colon U \to \mathcal{L}_s(^nX)$ is continuous at $x_0 \in U$.*
If there exists $M \in \mathcal{L}_s(^{n+1}X)$ such that

$$\lim_{t\to 0}\Big(\tfrac{f^{(n)}(x_0+th_0)(h_1,\ldots,h_n)-f^{(n)}(x_0)(h_1,\ldots,h_n)}{t} - M(h_0,\ldots,h_n)\Big) = 0$$

and the limit is uniform for $h_0,\ldots,h_n \in B_X$, we say that $M = f^{(n+1)}(x_0)$ is the $(n+1)$th derivative *of f at x_0.*
We say that f is F^n-smooth *(respectively C^n*-smooth*) on U if $f^{(n)}$ exists (respectively $f^{(n)}$ is continuous) at all points of U.*
We say that f is C^∞-smooth *on U if it is F^n-smooth on U for all $n \in \mathbb{N}$.*

Note that if a function f is F^{n+1}-smooth on some open set U, then it is C^n-smooth on U.

Fact 10.4 *Let X be a Banach space. If X has a C^n-smooth norm, then X admits a C^n-smooth bump.*
If X has a C^∞-smooth norm, then X admits a C^∞-smooth bump.

Proof: Consider any C^∞-smooth real-valued function τ on $\mathbb{R}$ such that $\tau([-\frac{1}{2},\frac{1}{2}]) = 1$ and $\tau(t) = 0$ for $\tau \geq 1$. The composition function $\varphi(x) = \tau(\|x\|)$ is then a C^n-smooth (respectively C^∞-smooth) function on X such that $\varphi(0) = 1$ and $\varphi(x) = 0$ whenever $x \notin B_X$ (see, e.g., [Died]). $\square$

Haydon constructed a nonseparable Banach space X that has C^∞-smooth Lipschitz bump function but does not admit an equivalent Gâteaux differentiable norm ([Hayd3]).

The following open question is related to variational principles. If f is a convex continuous function on c_0, it is not known whether there are $x_0 \in c_0$ and $K, \delta > 0$ such that $f(x_0+h) + f(x_0-h) - 2f(x_0) \leq K\|h\|^2$ for all $h \in \delta B_{c_0}$.

10.2 Smoothness in ℓ_p

Theorem 10.5 (Meshkov [Mesh], see, e.g., [DGZ3, p. 209]) *Let X be a Banach space. If both X and X^* admit a C^2-smooth bump, then X is isomorphic to a Hilbert space.*

We postpone the proof of Theorem 10.5 to Chapter 11 (Theorem 11.7).

Lemma 10.6 *Let $p \in (1,\infty)$, let $\{e_i\}_{i=1}^\infty$ be the canonical basis of ℓ_p.*
(i) *If $x \in \ell_p$, $x \neq 0$, then for $t \in (0, \|x\|_p)$ we have*

$$\lim_{i\to\infty}\|x+te_i\|_p = \big(\|x\|_p^p + t^p\big)^{1/p} > \|x\|_p + \tfrac{1}{2p}\|x\|_p^{1-p}t^p.$$

(ii) *If* $P: X \to \mathbb{R}$ *is a polynomial of degree less than p, then* $\lim_{i\to\infty} P(e_i) = 0$.

Proof: (i) The equality is proven in a standard way, first considering x with a finite support. The inequality follows from the strict concavity of the function $t \mapsto t^{1/p}$, $t > 0$.

(ii) Since $e_i \xrightarrow{w} 0$, $P(e_i) \to 0$ for every polynomial of degree 1. Take any integer $1 < n < p$ and assume we have already verified (ii) for all polynomials of degree less than n. Let P be a polynomial of degree n. By contradiction, assume that $\limsup_{i\to\infty} |P(e_i)| = 3a > 0$. Without loss of generality, we will assume that $P(e_i) > 2a$ for all $i \in \mathbb{N}$. We observe that $P(x+h) = P(x) + Q(x,h) + P(h)$ for $x, h \in \ell_p$, where $Q(x,\cdot)$ is a polynomial of degree less than n. Put $x_1 = e_1$. If x_i has been constructed, using the induction assumption find $j \in \mathbb{N}$ so that $Q(x_i, e_j) > -a$. Put $x_{i+1} = x_i + e_j$. Then

$$P(x_{i+1}) = P(x_i) + Q(x_i, e_j) + P(e_j) > P(x_i) + a > \cdots > (i+2)a.$$

Note that $\|x_i\|_p = i^{1/p}$ for all $i \in \mathbb{N}$. Therefore

$$\frac{P(x_i)}{\|x_i\|_p^n} > \frac{ia}{i^{n/p}} = i^{1-n/p}a \to \infty \text{ as } i \to \infty,$$

contradicting the the fact that $\{\frac{\|P(h)\|}{\|h\|^n} : \|h\| \geq 1\}$ is bounded. $\square$

We note that there is a reflexive separable infinite-dimensional Banach space X such that all polynomials on X are weakly sequentially continuous (the *Tsirelson space*, see [AlArDi]).

Theorem 10.7 (Kurzweil [Kurz]) *Let* $p \in (1,\infty)$. *If p is not an even integer, then the space* ℓ_p *does not admit any continuous* C^p*-smooth bump.*

Proof: ([FaZi2]) By contradiction, assume that $b: \ell_p \to \mathbb{R}$ is a C^p-smooth continuous bump. Applying Theorem 11.6 to $\varphi = b^{-2} - \|\cdot\|_p$ (and to $Z^* = {\ell_p}^* = \ell_q$, $q = \frac{p}{p-1}$) we get $x \in \ell_p$ such that $b^{-2}(x+h) + b^{-2}(x-h) - 2b^{-2}(x) \geq \|x+h\|_p + \|x-h\|_p - 2\|x\|_p$ for all $h \in \ell_p$. Then $b(x) \neq 0$ and we check that b^{-2} is C^p-smooth at x. By the above inequality and definition of C^p-smoothness, for all $h \in \ell_p$ we have

$$P(h) + o(\|h\|^p) \geq \|x+h\|_p + \|x-h\|_p - 2\|x\|_p.$$

where P is a polynomial of even degree, say n, with $n \leq p$. (If P is a polynomial of odd degree, then $h \mapsto P(h) + P(-h)$ is a polynomial of smaller and even degree satisfying the above property.) Fix any $t > 0$ and consider the last inequality with $h = te_i$, $i \in \mathbb{N}$. Since p is not and even integer, then $n < p$ and Lemma 10.6 (ii) yields

$$o(t^p) = \liminf_{i\to\infty}\big(P(te_i) + o(\|te_i\|^p)\big) \geq \lim_{i\to\infty}\big(\|x + te_i\|_p + \|x - te_i\|_p - 2\|x\|_p\big).$$

Now, if $x = 0$ then $o(t^p) \geq 2t$, a contradiction. If $x \neq 0$, then for $0 < t < \|x\|_p$ we have by Lemma 10.6 (i) that $o(t^p) \geq \frac{1}{p}\|x\|_p^{1-p}t^p$, which is again a contradiction. □

Theorem 10.8 *Let $p \in (1, \infty)$.*
If p is even then the canonical norm of ℓ_p is C^∞-smooth.
If p is odd then the canonical norm of ℓ_p is C^{p-1}-smooth.
If p is not an integer then the canonical norm of ℓ_p is $C^{[p]}$-smooth, where $[p]$ is the integer part of p.

We refer to [DGZ3] for a standard proof.

The best order of Gâteaux smoothness for an equivalent renorming of ℓ_p is not known.

10.3 Countable James Boundary

Typical example of a Banach space with a countable James boundary is c_0 or $C(K)$ for some countable compact space K.

Theorem 10.9 (see, e.g., [FLP] or [PWZ]) *Let X be an infinite-dimensional Banach space. If X has a countable James boundary, then X is saturated with subspaces isomorphic to c_0, that is, every infinite-dimensional closed subspace of X contains a subspace isomorphic to c_0.*

Corollary 10.10 (Bessaga, Pełczyński, [BePe1]) *If K is a countable compact space, then $C(K)$ is saturated with subspaces isomorphic to c_0.*

In the proof of Theorem 10.9, we will use the following notion:

Definition 10.11 *We say that a real-valued function φ on a Banach space X* locally depends on finitely many functionals *(or* coordinates*) if for every $x \in X$ there is a neighborhood U of x, $f_1, \ldots, f_n \in X^*$ and a continuous real-valued function ψ on $\mathbb{R}^n$ such that $\varphi(z) = \psi\big(f_1(z), f_2(z), \ldots, f_n(z)\big)$ for all $z \in U$.*

The supremum norm in c_0 is an example of a function that locally depends on finitely many coordinates away from the origin.

Proof of Theorem 10.9: Observe that the restrictions of the elements of a James boundary to a closed subspace form a James boundary of that subspace. Therefore it suffices to show that X contains a subspace isomorphic to c_0.

Assume that this is not true. Let $B = \{y_n^*\}$ be a countable James boundary of X. Define a norm $|||\cdot|||$ on X by

$$|||x||| = \sup\big\{\big(1 + \tfrac{1}{n}\big)y_n(x) : n \in \mathbb{N}\big\}.$$

If $x \in X$ is such that $\|x\| = 1$ ($\|\cdot\|$ denotes the original norm of X), then there is n such that $y_n^*(x) = 1$ and thus $|||x||| \geq \big(1 + \frac{1}{n}\big)y_n^*(x) > 1$. On the other hand, if

$\|x\| = 1$, then $\left(1 + \frac{1}{n}\right)y_n^*(x) \leq \left(1 + \frac{1}{n}\right)\|y_n^*\|\|x\| \leq 2$ for every n. Hence $|||\cdot|||$ is an equivalent norm on X. We will show that away from the origin, the norm $|||\cdot|||$ locally depends on finitely many $\{y_n^*\}$. Let $n_0 \in \mathbb{N}$ be such that $y_{n_0}^*(x) = 1$. For $n > n_0$ we have

$$\left(1 + \tfrac{1}{n}\right)y_n^*(x) \leq 1 + \tfrac{1}{n} \leq 1 + \tfrac{1}{n_0+1} < 1 + \tfrac{1}{n_0}.$$

Hence

$$\sup\left\{\left(1 + \tfrac{1}{n}\right)y_n^*(x) : n > n_0\right\} \leq 1 + \tfrac{1}{n_0+1} < 1 + \tfrac{1}{n_0}.$$

On the other hand,

$$\sup\left\{\left(1 + \tfrac{1}{n}\right)y_n^*(x) : n \leq n_0\right\} \geq \left(1 + \tfrac{1}{n_0}\right)y_{n_0}^*(x) = 1 + \tfrac{1}{n_0}.$$

Since the supremum of a family of 2-Lipschitz functions is a 2-Lipschitz function, from the last two inequalities we deduce that there is a neighborhood U of x such that for all $z \in U$,

$$\sup\left\{\left(1 + \tfrac{1}{n}\right)y_n^*(z) : n \leq n_0\right\} > \sup\left\{\left(1 + \tfrac{1}{n}\right)y_n^*(z) : n > n_0\right\}.$$

Therefore $|||z||| = \sup\left\{\left(1 + \frac{1}{n}\right)y_n^*(z) : n \leq n_0\right\}$ for all $z \in U$, showing that away from the origin, the norm $|||\cdot|||$ locally depends on finitely many $\{y_n^*\}$.

Let τ be a continuous real-valued function on the real line such that $\tau(x) = 1$ for $x \in \left(-\frac{1}{2}, \frac{1}{2}\right)$ and $\tau(x) = 0$ if $|x| \geq 1$. Define a function on X by $g(x) = 1 - \tau(|||x|||)$. Then g is continuous and locally depends on finitely many $\{y_n^*\}$. Note that $g(0) = 0$ and $g(x) = 1$ for $|||x||| > 1$.

We will construct inductively a sequence $\{x_n\}_{n=0}^{\infty} \subset X$ as follows: set $x_0 = 0$ and if $x_0, x_1, \ldots, x_n$ have been chosen, choose x_{n+1} so that

(1) $g\left(\sum_{i=0}^{n} \varepsilon_i x_i + \varepsilon_{n+1}x_{n+1}\right) = 0$ for $\varepsilon_i = \pm 1, i = 1, \ldots, n+1$,

(2) $\|x_{n+1}\| \geq \frac{1}{2}M_n = \frac{1}{2}\sup\|y\|$,

where the supremum is taken over all $y = x_{n+1}$ satisfying (1). Since $g\left(\sum_{i=1}^{k} \varepsilon_i x_i\right) = 0$ for every $k \in \mathbb{N}$, we get $\left\|\sum_{i=1}^{k} \varepsilon_i x_i\right\| \leq 1$ for all $k \in \mathbb{N}$. Since we assumed that X does not contain an isomorphic copy of c_0, by Corollary 4.52 we have that $\sum x_i$ is unconditionally convergent and $S = \left\{\sum_{i=1}^{n} \varepsilon_i x_i : \varepsilon_i = \pm 1, n \in \mathbb{N}\right\}$ is relatively compact in X (Exercise 1.42).

From the local dependence of g on finitely many $\{y_n^*\}$ we have that for each $z \in \overline{S}$ there is $\delta_z > 0$ and a finite set K_z of natural numbers such that $g(w \pm \delta x) = g(w)$ for all $w \in B(z, \delta_z)$, $\delta \leq \delta_z$ and $x \in \bigcap_{i \in K_z} (y^*)_i^{-1}(0)$. By compactness, $\overline{S} \subset \bigcup_{i=1}^{k} B(z_i, \delta_{z_i})$

for some $z_i \in X$. Let $\delta = \min\{\delta_{z_i} : i = 1, \dots, k\}$ and $K = \bigcup_{i=1}^{k} K_{z_i}$. Let $x \in \bigcap_{i \in K} (y^*)_i^{-1}(0)$, $0 < \|x\| < \delta$. If $w \in S$ then $w \in B(z_i, \delta_{z_i})$ for some i and hence $g(w \pm \delta x) = g(w)$. Thus $\inf(M_n) \geq \delta > 0$ for all $n \in \mathbb{N}$, which contradicts the convergence of $\sum x_n$. □

The equivalence (i)$\Longleftrightarrow$(iii) in the following result was shown in [Fonf0].

Theorem 10.12 ([Haje1]) *Let $(X, \|\cdot\|)$ be a separable normed space. Then the following are equivalent.*
(i) *X admits an equivalent norm having a countable James boundary.*
(ii) *X admits an equivalent norm with a James boundary B, such that there is a sequence $\{K_n\}_{n\in\mathbb{N}}$ of norm compact sets in X^* satisfying $B \subset \bigcup_{n\in\mathbb{N}} K_n$.*
(iii) *X admits an equivalent norm that depends locally on finitely many coordinates (away from the origin).*
(iv) *X admits an equivalent norm that is C^∞-smooth away from the origin and depends locally on finitely many coordinates.*

Proof: (iv)$\Longrightarrow$(iii) is trivial.

(iii)$\Longrightarrow$(ii): Let $\|\cdot\|$ be a norm on X that depends locally on finitely many coordinates. Since S_X is Lindelöf, every open cover of S_X has a countable subcover. Therefore, there exist a system $\{S_n\}_{n\in\mathbb{N}}$ of open sets in S_X, a system $\{\Phi_n\}_{n\in\mathbb{N}}$ of finite subsets of S_{X^*}, $\Phi_n = \{\phi_1^n, \dots, \phi_{k_n}^n\}$ and a system of functions $\{f_n\}_{n\in\mathbb{N}}$, $f_n : \mathbb{R}^{k_n} \to \mathbb{R}$ such that the following holds.

(a) $S_X \subset \bigcup_{n\in\mathbb{N}} S_n$.
(b) $\|y\| = f_n(\phi_1^n(y), \dots, \phi_{k_n}^n(y))$ for $y \in S_n$.

Consider the duality mapping $J : S_X \to \exp(S_{X^*})$, where $\exp(M)$ denotes the set of all subsets of M. Denote by

$$K_n = \operatorname{span}(\phi_1^n, \dots, \phi_{k_n}^n) \cap S_{X^*} \text{ for } n \in \mathbb{N}.$$

Then K_n are norm compact sets. For arbitrary $x \in S_n$, $x^* \in J(x)$ we have $x^*(y) \leq \|y\| \leq 1$ for $y \in S_n$. So $x^*(y) \leq f(\phi_1^n(y), \dots, \phi_{k_n}^n(y)) \leq 1$ for $y \in S_n$. If $h \in \bigcap_{i=1}^{k_n} \operatorname{Ker}(\phi_i^n)$ and $\|h\|$ is small enough, then

$$\|x \pm h\| = f(\phi_1^n(x \pm h), \dots, \phi_{k_n}^n(x \pm h)) = f(\phi_1^n(x), \dots, \phi_{k_n}^n(x)) = 1.$$

Thus $x^*(x \pm h) = 1 \pm x^*(h) \leq 1$ and we have $h \in \operatorname{Ker}(x^*)$. Altogether, $\bigcap_{i=1}^{k_n} \operatorname{Ker}(\phi_i^n) \subset \operatorname{Ker}(x^*)$, therefore $x^* \in K_n$. Hence $J(y) \subset K_n$ for $y \in S_n$. In combination with (a) we obtain $J(S_X) \subset \bigcup_{n\in\mathbb{N}} K_n$ and the implication follows.

(ii)$\Longrightarrow$(i): Let $\|\cdot\|$ be a norm as in (ii), and without loss of generality assume that $K_n \subset B_{X^*}$ for each $n \in \mathbb{N}$. Take a decreasing sequence $\{\varepsilon_n\}_{n\in\mathbb{N}}$ such that

$\varepsilon_1 < 1, \varepsilon_n \downarrow 0$. Then there exists an increasing sequence $\{i_n\}_{n=1}^\infty$ of integers with $i_1 = 1$, and a mapping $I : \mathbb{N} \to B_{X^*}$ such that $I([i_n, i_{n+1}))$ forms an $\frac{\varepsilon_n}{4}$-net in K_n. This mapping gives rise to a linear mapping $M : X \to \ell_\infty(\mathbb{N}) = \ell_\infty$ defined by the formula

$$M(x)(n) = (1 + \varepsilon_k)I(n)(x), \text{ where } n \in [i_k, i_{k+1}). \tag{10.1}$$

Take an arbitrary $x \in B_X$. Then $|I(n)(x)| \leq 1$ for each $n \in \mathbb{N}$ and so $\|M(x)\|_\infty \leq 1 + \varepsilon_1$. On the other hand, there exist $n_0 \in \mathbb{N}$ and an element $b \in B$ such that $b \in K_{n_0}$, $b(x) = 1$. Hence, for some $m \in [i_{n_0}, i_{n_0+1})$ we have

$$\|I(m) - b\|^* < \frac{\varepsilon_{n_0}}{4},$$

and so $I(m)(x) \geq 1 - \frac{\varepsilon_{n_0}}{4}$. Consequently

$$M(x)(m) \geq (1 + \varepsilon_{n_0})(1 - \frac{\varepsilon_{n_0}}{4}) \geq 1 + \frac{\varepsilon_{n_0}}{2}.$$

This proves that M is an isomorphism from X onto some closed subspace of ℓ_∞. Since $\limsup_{n\to\infty} M(x)(n) \leq 1$, we have that $M(x)$ attains its norm on $\mathbb{N}$. If we define

$$\|x\|_1 = \|M(x)\|_\infty,$$

then the set $\{M^*(e_n^*)\}_{n\in\mathbb{N}}$, where e_n^* are the dual functionals in ℓ_∞^*, is a countable James boundary of $(X, \|\cdot\|_1)$.

(i)$\Longrightarrow$(iv): By assumption, X is isomorphic to some $Y \subset \ell_\infty$ such that every $y \in Y$ attains its norm, i.e., $|y(n)| = \|y\|_\infty$ for some $n \in \mathbb{N}$. Consider a decreasing sequence $\delta_n \downarrow 0$ and an isomorphism $S : Y \to Z$, $Z \subset \ell_\infty$ defined by:

$$S(y)(n) = (1 + \delta_n)y(n).$$

We renorm the space $(Z, \|\cdot\|_\infty)$ by a C^∞-smooth norm. Note that for each $y \in Y$ with $\|y\|_\infty = 1$ there is $n \in \mathbb{N}$ satisfying $|y(n)| = 1$. Therefore, for each $z \in Z$ there exists $n_z \in \mathbb{N}$ satisfying $|S^{-1}(z)(n_z)| = \|S^{-1}(z)\|_\infty$. Thus

$$|z(n_z)| = (1 + \delta_{n_z})\|S^{-1}(z)\|_\infty$$

and

$$|z(k)| \leq (1 + \delta_{n_z+1})\|S^{-1}(z)\|_\infty \text{ for } k > n_z.$$

Take a sequence $\{b_n\}_{n=1}^\infty$ of C^∞-smooth bump functions $b_n : \mathbb{R} \to \mathbb{R}$, $b_n \geq 0$, $\int_{-\infty}^{\infty} b_n(t)\,\mathrm{d}t = 1$, $\operatorname{supp}(b_n) \subset [\frac{\delta_{n+1}-\delta_n}{4}, \frac{\delta_n-\delta_{n+1}}{4}]$. Define a nondecreasing sequence

$\{F_n\}_{n=0}^{\infty}$ of convex functions on ℓ_∞ by the inductive formula

$$F_0 = \|\cdot\|_\infty,$$

$$F_n(z) = \int_{\frac{\delta_{n+1}-\delta_n}{4}}^{\frac{\delta_n-\delta_{n+1}}{4}} F_{n-1}(z + te_n)b_n(t)\,\mathrm{d}t.$$

Suppose $\|S^{-1}(z)\|_\infty \geq 1$ and $\rho = \frac{\delta_{n_z}-\delta_{n_z+1}}{4}$, $\|y - z\|_\infty < \rho$. Then for $k > n_z$ we have

$$F_k(y) = \int_{\frac{\delta_{k+1}-\delta_k}{4}}^{\frac{\delta_k-\delta_{k+1}}{4}} \cdots \int_{\frac{\delta_2-\delta_1}{4}}^{\frac{\delta_1-\delta_2}{4}} \|y + \sum_{i=1}^{k} t_i e_i\|_\infty b_1(t_1)\ldots b_k(t_k)\,\mathrm{d}t_1 \ldots \mathrm{d}t_k. \tag{10.2}$$

If $n_z \leq l \leq k$ and $b_l(t_l) \neq 0$, then $-\rho \leq t_l \leq \rho$. Since $y(n_z) - y(l) \geq 4\rho$, we have that

$$\|y + t_1e_1 + \cdots + t_ke_k\|_\infty = \|y + t_1e_1 \cdots + t_{n_z}e_{n_z}\|_\infty. \tag{10.3}$$

whenever the integrated function in (10.2) is nonzero. Consequently, if $\|z - y\|_\infty < \rho$ and $k \geq n_z$ then $F_k(y) = F_{n_z}(y)$, i.e.,

$$F_k(y) = \int_{\frac{\delta_{n_z+1}-\delta_{n_z}}{4}}^{\frac{\delta_{n_z}-\delta_{n_z+1}}{4}} \cdots \int_{\frac{\delta_2-\delta_1}{4}}^{\frac{\delta_1-\delta_2}{4}} \max_{i\leq n_z} |y(i) + t_i| \Pi_{i=1}^{n_z} b_i(t_i)\,\mathrm{d}t_1 \ldots \mathrm{d}t_{n_z}. \tag{10.4}$$

It follows that the convex function F on Z defined by $F = \sup_{n\in\mathbb{N}}(F_n)$ is locally dependent on finitely many coordinates (namely the functionals $e_1^*, \ldots, e_{n_z}^*$), and C^∞-smooth on $\{z \in Z : \|z\|_\infty > 1 + \delta_1\}$. Notice that $F_k(z) \leq \|z\|_\infty + \frac{\delta_1-\delta_2}{4}$ for arbitrary $z \in \ell_\infty$ and $k \in \mathbb{N}$. Applying the implicit function theorem, we obtain that the Minkowski functional of the set $\{z \in Z, F(z) \leq 1 + 2\delta_1\}$ introduces a C^∞-Fréchet smooth norm on Z that locally depends on finitely many coordinates. This completes the proof. □

Theorem 10.13 ([PWZ], [FaZi1]) *Let X be a separable Banach space. If X admits a continuous bump φ that locally depends on finitely many coordinates, then X^* is separable and X is saturated with isomorphic copies of c_0.*

Proof: The statement on the saturation follows from the proof of Theorem 10.9. We will prove that X^* is separable. For every $x \in X$ we choose a neighborhood U_x in

X, functionals $f_1^x, \dots, f_{n_x}^x \in X^*$ and a continuous function ψ^x on $\mathbb{R}^{n_x}$ such that for every $z \in U_x$,

$$\varphi(z) = \psi^x\big(f_1^x(z), \dots, f_{n_x}^x(z)\big).$$

By the Lindelöf property of X, let $\{x_n\}_{n=1}^\infty \subset X$ be such that $\{U_{x_n}\}_{n=1}^\infty$ covers X. For $n \in \mathbb{N}$ denote $F_n = \{f_1^{x_n}, \dots, f_{n_{x_n}}^{x_n}\}$ and $F = \bigcup_{n=1}^\infty F_n$. Then φ is a continuous function on X with bounded nonempty support, φ locally depends on finitely many elements of F and F is countable.

Define a function Φ on X by

$$\Phi(x) = \begin{cases} \varphi^{-2}(x) & \text{if } \varphi(x) \neq 0, \\ +\infty & \text{if } \varphi(x) = 0. \end{cases}$$

Then Φ is a bounded below lower semicontinuous function on X such that $S = \{x \in X : \Phi(x) < \infty\}$ is open, and on its domain Φ locally depends on finitely many elements of F.

As F is countable, in order to show that X^* is separable, it suffices to prove that $\overline{\operatorname{span}}(F) = X^*$. So fix any $f \in X^*$ and $\varepsilon > 0$. From Theorem 7.39, it follows that there is $x_0 \in S$ such that

$$(\Phi - f)(x) \geq (\Phi - f)(x_0) - \varepsilon\|x - x_0\| \tag{10.5}$$

for all $x \in X$. Let U, $f_1, \dots, f_n \in F$ and ψ be as in the definition of the local dependence for $\Phi - f$ at x_0. Set $W = \{x \in X : f_1(x) = \dots = f_n(x) = 0\}$. Finally, let $\delta > 0$ be such that $x_0 + h \in U$ whenever $h \in X$, $\|h\| < \delta$. Then if $h \in W$, $\|h\| < \delta$, we have

$$\begin{aligned}\Phi(x_0 + h) - \Phi(x_0) &= \psi\big(f_1(x_0 + h), \dots, f_n(x_0 + h)\big) - \psi\big(f_1(x_0), \dots, f_n(x_0)\big) \\ &= \psi\big(f_1(x_0), \dots, f_n(x_0)\big) - \psi\big(f_1(x_0), \dots, f_n(x_0)\big) = 0.\end{aligned}$$

Hence from (10.5), for $h \in W$, $\|h\| < \delta$, we have

$$f(h) = f(x_0 + h) - f(x_0) \leq \Phi(x_0 + h) - \Phi(x_0) + \varepsilon\|h\| = \varepsilon\|h\|.$$

Let $\tilde{f} = f|_W$. Then by the last inequality, for $\tilde{f}$ as an element of W^* we have $\|\tilde{f}\| \leq \varepsilon$. Let g be a norm-preserving Hahn–Banach extension of $\tilde{f}$ to X. Note that $f - g \in W^\perp$. Since $\operatorname{span}\{f_1, \dots, f_n\}$ is w^*-closed in X^* as a finite-dimensional subspace, by the bipolar theorem $W^\perp = \operatorname{span}\{f_1, \dots, f_n\}$. Hence

$$\operatorname{dist}(f, \operatorname{span}(F)) \leq \operatorname{dist}(f, \operatorname{span}\{f_1, \dots, f_n\}) = \operatorname{dist}(f, W^\perp) \leq \|f - (f - g)\| = \|g\| \leq \varepsilon.$$

Thus $\overline{\operatorname{span}}(F) = X^*$. □

Corollary 10.14 (Haydon [Hayd2]) *Let K be a countable compact space. Then $C(K)$ admits an equivalent C^∞-smooth norm.*
Proof: $C(K)$ has a countable James boundary $B = \{\pm\delta_k : k \in K\}$, where δ_k are Dirac functionals, so we can apply Theorems 10.12 and 10.13. □

10.4 Remarks and Open Problems

Remarks

1. It is usually hard to construct higher order equivalent norms on spaces. This is because the convexity arguments no longer works. On the other hand, we have seen that the existence of such norms provide a powerful information on the structure of spaces. The existence of such norms in some sense often strengthen the power of other properties. For example:
   - Assume that an infinite-dimensional Banach space X admits a C^∞-smooth bump function and that X does not contain an isomorphic copy of c_0. Then the infimum of cotypes of X is an even integer, say $2p$, and X is of cotype $2p$. Moreover, there is an even integer $2q$ such that X contains an isomorphic copy of ℓ_{2q}. These are Deville's results [Devi], see, e.g., [DGZ3, p. 209]. So, there are no Tsirelson-like very smooth spaces. It would be interesting to have more results in this direction (distortions, indecomposability, etc.)
   - If K is a compact space such that $K^{(\omega_1)} = \emptyset$, then $C(K)$ admits an equivalent C^∞-smooth norm [Haje5]. However, there is a compact space K such that $K^{(\omega_1+1)} = \emptyset$ and $C(K)$ does not admit an equivalent Gâteaux smooth norm [Hayd3].
   - If X admits an equivalent C^2-smooth bump, and is saturated by Hilbert spaces, then it is isomorphic to a Hilbert space [Maka], see, e.g., [DGZ3, p. 226].
2. Smooth norms in Orlicz spaces are studied for example in [MaTr].
3. For higher order Gâteaux differentiability of norms, we refer the reader for example to [Troy7] and [Vand3].
4. Banach spaces with countable boundaries relate to polyhedral spaces. Recall that a Banach space X is *polyhedral* if the unit ball of every finite-dimensional subspace of X is an intersection of finitely many half-spaces [FLP].
5. It is shown in [Haje2b] that if K is a compact space, then $C(K)$ admits an equivalent *real analytic norm* (i.e., a norm that can be expressed, away from the origin, by Taylor's series) if and only if K is countable.
6. It was proved in [HaTr] that there is a separable Banach space that admits an equivalent C^∞-norm but admits no equivalent real analytic norm.

Open Problems

1. The best order of Gâteaux smoothness for an equivalent renorming of ℓ_p is not known.

2. Assume that a separable Banach space X admits a twice differentiable bump. Does it admit a twice differentiable norm?
3. Assume that a separable Banach space X admits a separating polynomial (we say that a polynomial P on X is *separating* if $\inf_{x \in S_X} P(x) > 0$). Does it admit a C^∞ norm?
4. Assume that a Banach space X admits a C^k-smooth bump. Does it admit C^k-smooth partitions of unity?
5. Assume that a separable Banach space X admits a C^k norm for all k. Does X admit a C^∞ norm?
6. The following open question is related to variational principles. If f is a convex continuous function on c_0, it is not known whether there are $x_0 \in c_0$ and $K, \delta > 0$ such that $f(x_0 + h) + f(x_0 - h) - 2f(x_0) \leq K\|h\|^2$ for all $h \in \delta B_{c_0}$.
7. Assume that K is a countable compact space and X is a quotient of $C(K)$. Is X c_0 saturated? (See [Rose10, p. 1571].)
8. Assume that a separable space admits a continuous bump that locally depends on finitely many coordinates. Does X admit such a norm?
9. (Troyanski) Does ℓ_3 admit an equivalent four times Gâteaux smooth norm? Does $\ell_{5/2}$ admit an equivalent three times Gâteaux smooth norm?
10. Let K be a scattered compact space. Does $C(K)$ admit a C^∞-bump?
11. We refer to [HaZi] and [FMZ5] for a list of open problems in this area.

Exercises for Chapter 10

10.1 Let X be a Banach space and $n \in \mathbb{N}$. For an n-form Q on X, the following are equivalent. (i) Q is continuous, (ii) Q is bounded, and (iii) Q is continuous at 0.
Hint. For a 2-form Q, $Q(x + x_0, y + y_0) - Q(x_0, y_0) = Q(x, y) + Q(x, y_0) + Q(x_0, y)$, and $Q(x, y_0) = Q(mx, (1/m)y_0)$ for all $m \in \mathbb{N}$.

10.2 Follow the hint to show that every norm on a Hilbert space X is Fréchet differentiable at some points.
Hint. Let $\|\cdot\|$ be a norm on X and let $\|\cdot\|_1$ be the Hilbertian norm of X. By Theorem 11.6 there is $f \in X^*$ such that $\inf\{\|x\|_1^2 - \|x\| + f(x) : x \in X\} = -\delta < 0$ is attained at some point $x_0 \in X$. Therefore $\|x\|_1^2 + f(x) + \delta \geq \|x\|$ for every $x \in X$ and $\|x_0\|_1^2 + f(x_0) + \delta = \|x_0\|$. Since $\|x\|_1^2 + f(x_0)$ is differentiable at x_0 (f is linear), we have that $\|\cdot\|$ is differentiable at x_0 by Exercise 7.9.

10.3 Find an example of a C^2-smooth norm on a Banach space X such that its dual norm is also C^2-smooth but X is not isometric to a Hilbert space.
Hint. Work in $\mathbb{R}^2$.

10.4 Let $(X, \|\cdot\|)$ be a Banach space. Show that if $\|\cdot\|^2$ is twice Fréchet differentiable at 0, then X is isomorphic to a Hilbert space.

Hint. If $\|x\|^2 = t(x) + p(x)$ is the Taylor expansion at 0, note that $t(x) = 0$ and $\|x\|^2 - p(x) = o(\|x\|^2)$. Thus there is $\varepsilon > 0$ such that $p(x) \geq \frac{1}{2}\|x\|^2$ for $\|x\| > \varepsilon$. Therefore $\sqrt{p(x)}$ defines a Hilbertian norm on X.

10.5 Show that all polynomials on c_0 are weakly sequentially continuous.
Hint. Similar to the proof of Lemma 10.6.

10.6 Assume that a Banach space X has a σ-compact James boundary. Show that X can be equivalently renormed to have a countable James boundary.
Hint. Let $\bigcup K_n$ be a James boundary of X, where K_n are compact sets in S_X. From every $n \in \mathbb{N}$ fix $\varepsilon_n > 0$ so that $(1 + \frac{1}{n})(1 - \varepsilon_n) > 1 + \frac{1}{n+1}$. Let F_n be an $\frac{\varepsilon_n}{4}$-finite net for K_n and $F = \bigcup F_n$. Define an equivalent norm $|||\cdot|||$ on X by $|||x||| = \sup_n\{(1 + \frac{1}{n}) \sup_{F_n} x\}$.

For an $x \in X$ there is $n_0 \in \mathbb{N}$ and a neighborhood $U(x)$ of x such that for all $z \in U(x)$,

$$\sup\Big\{(1 + \tfrac{1}{n}) \sup_{F_n}(z) \,:\, n > n_0\Big\} < \sup\Big\{(1 + \tfrac{1}{n}) \sup_{F_n}(z) \,:\, n \leq n_0\Big\}.$$

From this and the fact that F_n is an ε-net for K_n for each n, we get that there is $f \in \bigcup_{n \leq n_0} F_n$ such that $|||x||| = f(x)$. Hence F is a countable James boundary of X in the norm $|||\cdot|||$.

10.7 Show that there is an equivalent norm on ℓ_1 such that its restriction to the subspace of all finitely supported vectors in ℓ_1 is C^∞-smooth away from the origin.
Hint. Let Y be the subspace of ℓ_1 formed by all finitely supported vectors. For $n \in \mathbb{N}$, let $B_n = \{(x_i) \in S_{\ell_\infty} : x_j = 0 \text{ for } j \geq n\}$. Then $\bigcup B_n$ is a James boundary of Y and each B_n is compact. Thus Y has an equivalent norm with a countable James boundary and the result follows from Theorem 10.12, which is also true for normed spaces.

10.8 Show that ℓ_∞/c_0 does not admit a Lipschitz Gâteaux differentiable bump function.
Hint. Use Exercise 14.3 and the fact that $\beta\mathbb{N} \setminus \mathbb{N}$ does not contain any non-trivial convergent sequence.

10.9 No equivalent LUR and Fréchet differentiable norm on an infinite-dimensional Banach space X with separable dual has a James boundary covered by countably many compact sets. Consequently, the set of norms whose James boundary cannot be covered by countably many compact sets is residual in the space of all equivalent norms.
Hint. Let $\|\cdot\|$ be LUR and Fréchet differentiable. Denote by $J : S_X \to S_{X^*}$ the duality mapping. Suppose by contradiction that there exists a sequence $\{K_n\}_{n\in\mathbb{N}}$ of

compact sets forming a James boundary of $\|\cdot\|$. As $\|\cdot\|$ is Fréchet differentiable and LUR, J is a continuous one-to-one mapping and $J(S_X) \subset \bigcup_{n\in\mathbb{N}} K_n$. Denote by $L_n = J^{-1}(K_n)$. By the Baire category theorem for some $n \in \mathbb{N}$, L_n has nonempty interior, i.e., there exists an $x \in X$ and $\varepsilon > 0$ such that $B(x, \varepsilon) \subset L_n$. Show that there exists a $\delta > 0$ such that $x^* \in S_{X^*}$ and $\|x^* - J(x)\| < \delta$ implies $\|J^{-1}(x^*) - x\| < \varepsilon$. Let $x_n^* \to J(x)$, $y_n = J^{-1}(x_n^*)$. Then $\lim_{n\to\infty} x_n^*(x + y_n) = 2$. Therefore,

$$4 - \left(x_n^*(x + y_n)\right)^2 \geq 2\|x\|^2 + 2\|y_n\|^2 - \|x + y_n\|^2 \geq 0.$$

Since $\|\cdot\|$ is LUR, we have $\lim_{n\to\infty} \|x - y_n\| = 0$. By the Bishop–Phelps theorem, $J(S_X)$ is dense in S_{X^*}. It follows that $J(B(x, \varepsilon) \cap S_X)$ is dense in $B(J(x), \delta) \cap S_{X^*}$ in the dual. This is a contradiction with the fact that the former is relatively compact.

10.10 Can ℓ_3^2 be isometric to a subspace of L_4?
Hint. No: the norm of ℓ_3^2 is not 3 times differentiable.

10.11 Assume that the norm $\|\cdot\|$ together with its dual norm are both C^2-smooth. Follow the following argument to get that X is isomorphic to a Hilbert space.
Hint. Put $g(x) = \|x\|^2/2$ for $x \in X$. Then the Fenchel dual function $g^*(x^*) = \|x^*\|^2/2$. Pick any $x_0 \neq 0$ in X. Then if $g'(x_0) = x^*$ then $(g^*)'(x_0^*) = x_0$. Since g^* is C^2, there is $\delta > 0$ such that for some $c > 0$ and all $h^* \in X^*$ with $\|x^*\| < \delta$ one has

$$g^*(x_0^* + h^*) - g^*(x_0^*) - h^*(x_0) \leq c\|h^*\|^2.$$

Then from the Fenchel duality,

$$g(x_0 + h) - g(x_0) - x_0^*(h) \geq \|h\|^2/4c$$

for all $h \in X$ such that $\|h\| < 2c\delta$. Then the Taylor expansion of order 2 shows that

$$g''(x_0)(h, h) \geq \|h\|^2/2c$$

for all $h \in X$ such that $x_0^*(h) = 0$. Therefore $g''(x_0)$ generates the Hilbertian norm on this hyperplane and thus on the whole X.

Chapter 11
Dentability and Differentiability

The main topic of the present chapter is the dentability of bounded sets and the closely related Radon–Nikodým property (RNP) of Banach spaces. This property has several equivalent characterizations and applications. In particular, Asplund spaces are characterized by the Radon–Nikodým property of their dual spaces. As another application, we show that Lipschitz mappings from separable Banach spaces into Banach spaces with RNP are at some points Gâteaux differentiable.

11.1 Dentability in X

Throughout this chapter, $(X, \|\cdot\|)$ will always be a Banach space over the real field $\mathbb{R}$. Moreover, the word "subspace" will always mean "closed subspace". If $x^* \in X^*$ and $a \in \mathbb{R}$ we put $\{x^* > a\} = \{x \in X : \langle x^*, x\rangle > a\}$. Let $\emptyset \neq M \subset X$ be given. By a *slice* of M we understand any intersection $M \cap \{x^* > a\}$ where $x^* \in X^*$ and $a \in \mathbb{R}$. The space X will be always considered as a subspaces of X^{**}.

Definition 11.1 *Let X be a Banach space. We say that a subset M of X is* dentable *if for every $\varepsilon > 0$, there are $x^* \in X^*$ and $a \in \mathbb{R}$ such that the slice $M \cap \{x^* > a\}$ is nonempty and has diameter less than ε. The Banach space X is called* dentable *if every nonempty bounded subset of it is dentable.*

If M is a nonempty subset of X, $x \in M$ and $x^* \in X^*$ are such that $\langle x^*, x\rangle = \sup\langle x^*, M\rangle$ and the diameters of the slices $\{y \in M : \langle x^*, y\rangle > \langle x^*, x\rangle - \delta\}$ go to 0 as $\delta \downarrow 0$, then x is called a *strongly exposed point* of M and we say that x^* *strongly exposes* M at x. Another equivalent definition of a strongly exposed point was given in Definition 7.10.

Consider a function $f : X \to (-\infty + \infty]$. Recall that f is said to be *proper* if its *domain* $\operatorname{dom} f := \{x \in X : f(x) < +\infty\}$ is nonempty. For $x \in \operatorname{dom} f$ the *subdifferential* $\partial f(x)$ of f at x was introduced in Definition 7.11. Assume that the function f is convex and that x lies in the interior of $\operatorname{dom} f$. Then it makes a sense to speak about Fréchet differentiability of f at x. Below, we shall frequently use the fact, see Lemma 7.4, that such an f is Fréchet differentiable at x if and only if for every $\varepsilon > 0$ there is a $\delta > 0$ such that

M. Fabian et al., *Banach Space Theory*, CMS Books in Mathematics,
DOI 10.1007/978-1-4419-7515-7_11,

$$f(x+h)+f(x-h)-2f(x) \le \varepsilon\|h\| \quad \text{whenever} \quad h \in X \quad \text{and} \quad \|h\| < \delta.$$

Lemma 11.2 *Let U be an open convex subset of a dual Banach space X^* and let $f : U \to \mathbb{R}$ be a convex weak*-lower semicontinuous function. Then the set $\{x^* \in U : \partial f(x^*) \cap X \neq \emptyset\}$ is dense in U. Moreover, if f is Fréchet differentiable at some $x^* \in U$, then the derivative $f'(x^*)$ belongs to X.*

Proof: Fix any $x_0^* \in U$ and any $\varepsilon > 0$ so small that $B(x_0^*, \varepsilon) \subset U$. Define $g : X^* \to (-\infty, +\infty]$ by $g(u^*) = f(x_0^* + u^*)$ if $u^* \in \varepsilon B_{X^*}$ and $g(u^*) = +\infty$ if $u^* \in X^* \setminus \varepsilon B_{X^*}$. Note that g is weak*-lower semicontinuous and its domain is weak*-compact, so g is bounded below. Let $g_* = g^*|_X$, where $g^* : X^{**} \to (-\infty, +\infty]$ is the Fenchel conjugate of g, i.e.,

$$g_*(u) = \sup\{\langle u^*, u\rangle - g(u^*) : u^* \in X^*\}, \quad u \in X.$$

This is a convex, proper, and lower semicontinuous function on X, see Definition 7.29 and comments after it. Clearly,

$$-f(x_0^*) \le g_*(u) \le \varepsilon\|u\| - \inf g \quad \text{for every} \quad u \in X.$$

Hence g_* is in fact continuous and $\operatorname{dom} g_* = X$, see Lemma 7.3. Ekeland's variational principle (Theorem 7.39) yields $x \in X$ such that $g_*(u) - g_*(x) \ge -\varepsilon\|u - x\|$ for every $u \in X$. Proposition 2.13 (ii) applied to the (disjoint convex) sets $C_1 := \{(u,t) \in X \times \mathbb{R} : t \ge g_*(u) - g_*(x)\}$ and $C_2 := \{(u,t) \in X \times \mathbb{R} : t < -\varepsilon\|u - x\|\}$ then yields $(\xi, s) \in X^* \times \mathbb{R}$ such that

$$\inf\{\langle \xi, u\rangle + st : (u,t) \in C_1\} \ge \sup\{\langle \xi, u\rangle + st : (u,t) \in C_2\}.$$

It is easy to check that $s \neq 0$. The possibility that $s < 0$ also leads to a contradiction. Hence $s > 0$. Then, we may and do assume that $s = 1$. Thus, for every $u, v \in X$ we have

$$\langle \xi, u\rangle + g_*(u) - g_*(x) \ge \langle \xi, v\rangle - \varepsilon\|v - x\|. \tag{11.1}$$

Considering here any $u \in X$ and $v := x$, we get that $\langle \xi, u\rangle + g_*(u) - g_*(x) \ge \langle \xi, x\rangle$, which means that $-\xi \in \partial g_*(x)$. For $u := x$ and any $v \in X$ in (11.1), we get that $\langle \xi, x\rangle \ge \langle \xi, v\rangle - \varepsilon\|v - x\|$, and hence $\|\xi\| \le \varepsilon$.

Now, once having that $-\xi \in \partial g_*(x)$, Corollary 7.35 (ii) says that $x \in \partial(g_*)^*(-\xi)$. By Corollary 7.33, $(g_*)^* = g$, so $x \in \partial g(-\xi) \ \big(= \partial f(x_0^* - \xi)\big)$; thus $\partial f(x_0^* - \xi) \cap X \neq \emptyset$. Moreover, $x_0^* - \xi \in B(x_0^*, \varepsilon)$ $(\subset U)$. We thus proved the first part of our lemma.

The second statement follows from Corollary 7.37. □

Theorem 11.3 (Phelps, [Phel1b], Bourgain [Bou]) *For a Banach space $(X, \|\cdot\|)$ and for a nonempty bounded closed convex subset W of X the following assertions are equivalent:*

(i) *Every nonempty subset of W is dentable.*
(ii) *For every nonempty open convex set $U \subset X^*$ and for every convex weak*-lower semicontinuous function $f : U \to \mathbb{R}$, with $\partial f(U) \cap X \subset W$, the set D of all points at which f is Fréchet differentiable is dense G_δ in U.*
(iii) *For every nonempty closed convex subset M of W, the set E^* of all $x^* \in X^*$ that strongly expose M is dense G_δ in X^*.*
(iv) *Every nonempty closed convex subset M of W is equal to the closed convex hull of the set E of all its strongly exposed points.*

Proof: (i)$\Longrightarrow$(ii). Let U and f be as in (ii). Put $L = \sup\{\|w\| : w \in W\}$. For $n \in \mathbb{N}$ define

$$G_n = \Big\{x^* \in X^* : \text{ there exists an open subset } V \subset U \text{ such that}\} \\ x^* \in V \text{ and } \operatorname{diam}\big(\partial f(V) \cap X\big) < \tfrac{1}{n}\Big\}.$$

The sets G_n are clearly open. We shall show they are also dense in U. So, fix any $n \in \mathbb{N}$ and let $\Omega \subset U$ be any nonempty open convex set. Lemma 11.2 guarantees that the set $\partial f(\Omega) \cap X$ is nonempty. And, as $\partial f(\Omega) \cap X \subset W$, (i) provides $x^* \in X^*$ and $a > 0$ such that the slice $\partial f(\Omega) \cap X \cap \{x^* > a\}$ is nonempty and has diameter less than $\frac{1}{n}$. Pick an x in this slice. Then $x \in \partial f(y^*)$ for some $y^* \in \Omega$. Find $t > 0$ so small that $y^* + tx^* \in \Omega$. Since $x \in \{x^* > a\}$, there is $b > a$ such that $\langle x^*, x\rangle > b$. Find $0 < \delta < \frac{(b-a)t}{2L}$ so small that $B(y^* + tx^*, \delta) \subset \Omega$. We shall prove that

$$\partial f\big(B(y^* + tx^*, \delta)\big) \cap X \subset \partial f(\Omega) \cap X \cap \{x^* > a\}; \tag{11.2}$$

then $y^* + tx^* \subset \Omega \cap G_n$ and the density of G_n will be proved. In order to prove (11.2), consider any $y \in \partial f\big(B(y^* + tx^*, \delta)\big) \cap X$. Find $z^* \in B(y^* + tx^*, \delta)$ such that $\partial f(z^*) \ni y$. Then, using the convexity of f, we have

$$0 \le \langle z^* - y^*, y - x\rangle \le \langle tx^*, y - x\rangle + \delta\|y - x\| < t\langle x^*, y\rangle - tb + 2\delta L,$$

and hence $\langle x^*, y\rangle > b - \frac{2\delta L}{t}$ $(> a)$. Therefore, $y \in \partial f(\Omega) \cap X \cap \{x^* > a\}$ and (11.2) is proved.

Put $D = \bigcap_{n=1}^\infty G_n$; this is a dense (and G_δ) subset of U by the Baire's category theorem. Fix any $x^* \in D$. We shall show that f is Fréchet differentiable at x^*. Let any $\varepsilon > 0$ be given. Find $n \in \mathbb{N}$ such that $n > \frac{1}{\varepsilon}$. As $x^* \in G_n$, there is $\delta > 0$ so small that $B(x^*, \delta) \subset U$ and that $\operatorname{diam}\big(\partial f(B(x^*, \delta)) \cap X\big) < \frac{1}{n}$ $(< \varepsilon)$. Pick any $h^* \in \delta B_{X^*}$. Using Lemma 11.2, for $i = 1, 2, \ldots$ find $u_i^*,\ v_i^* \in B(x^*, \delta)$ such that

$$\|x^* + h^* - u_i^*\| < \tfrac{1}{i}, \quad \|x^* - h^* - v_i^*\| < \tfrac{1}{i}$$

and that $\partial f(u_i^*) \cap X \neq \emptyset,\ \partial f(v_i^*) \cap X \neq \emptyset$; pick further $u_i \in \partial f(u_i^*) \cap X,\ v_i \in \partial f(v_i^*) \cap X$. Then, using the convexity and the lower semicontinuity of f, we can estimate

$$
\begin{aligned}
&f(x^*+h^*)+f(x^*-h^*)-2f(x^*)\\
&\le \liminf_{i\to\infty} f(u_i^*)+\liminf_{i\to\infty} f(v_i^*)-2f(x^*)\le \liminf_{i\to\infty}\big(\langle u_i^*-x^*,u_i\rangle+\langle v_i^*-x^*,v_i\rangle\big)\\
&\le \liminf_{i\to\infty}\big(\langle h^*,u_i\rangle+\langle -h^*,v_i\rangle+\|u_i^*-x^*-h^*\|\|u_i\|+\|v_i^*-x^*+h^*\|\|v_i\|\big)\\
&\le \liminf_{i\to\infty}\big(\|h^*\|\|u_i-v_i\|+\tfrac{2L}{i}\big)\le \tfrac{1}{n}\|h^*\|\ \big(<\varepsilon\|h^*\|\big).
\end{aligned}
$$

Therefore, since $h^* \in \delta B_{X^*}$ was arbitrary, f is Fréchet differentiable at x^*.

Now, assume that f is Fréchet differentiable at some $x^* \in U$. We shall show that $x^* \in D$. To this end, fix any $n \in \mathbb{N}$. Find $\delta > 0$ so small that $B(x^*, 2\delta) \subset U$ and that

$$f(x^*+h^*)-f(x^*)-f'(x^*)h^* \le \tfrac{1}{5n}\|h^*\| \quad \text{whenever} \quad h^* \in 2\delta B_{X^*}.$$

Then $\operatorname{diam}\big(\partial f(B(x^*,\delta))\cap X\big) < \frac{1}{n}$, and hence $x^* \in G_n$. Indeed, pick any $y \in \partial f\big(B(x^*,\delta)\big)\cap X$. Find $y^* \in B(x^*,\delta)$ such that $\partial f(y^*) \ni y$. Take any $h^* \in B_{X^*}$. We can estimate

$$
\begin{aligned}
\delta\langle h^*, y-f'(x^*)\rangle &\le f(y^*+\delta h^*)-f(y^*)-\langle \delta h^*, f'(x^*)\rangle\\
&=\big(f(y^*+\delta h^*)-f(x^*)-\langle y^*+\delta h^*-x^*, f'(x^*)\rangle\big)\\
&\quad+\big(f(x^*)-f(y^*)+\langle y^*-x^*, f'(x^*)\rangle\big)\\
&\le \tfrac{1}{5n}\|y^*+\delta h^*-x^*\| \le \tfrac{2\delta}{5n}.
\end{aligned}
$$

Taking here supremum throughout all $h^* \in B_{X^*}$ and then dividing by δ, we get $\|y-f'(x^*)\| \le \frac{2}{5n}$. Therefore, $\operatorname{diam}\big(\partial f(B(x^*,\delta))\cap X\big) \le \frac{4}{5n} < \frac{1}{n}$. We thus proved that $x^* \in G_n$ for every $n \in \mathbb{N}$, and so $x^* \in D$. Therefore, D coincides with the set of all points of Fréchet differentiability of f.

(ii)$\Longrightarrow$(iii). Let M be as in (iii). Define $f : X^* \to \mathbb{R}$ by $f(x^*) = \sup\langle x^*, M\rangle$, $x^* \in X^*$. Clearly, f is a convex and weak*-lower semicontinuous function on X^*. Use Corollary 7.21 to conclude that f is Fréchet differentiable at $x_0^* \in X^*$ if and only if x_0^* strongly exposes M. Hence, (ii) gives that E^* is a dense G_δ subset of X^*.

(iii)$\Longrightarrow$(iv). As we already know that $E^* \neq \emptyset$, we can deduce that $E \neq \emptyset$. Assume that the equality $\overline{\operatorname{conv}}\, E = M$ is not true. Then the separation theorem provides $y^* \in X^*$ such that $\sup\langle y^*, M\rangle > \sup\langle y^*, E\rangle$. (iii) guarantees that the set E^* $(= D)$ is dense in X^*, Hence, there is $x^* \in E^*$ so close to y^* that we still have $\sup\langle x^*, M\rangle > \sup\langle x^*, E\rangle$. However, this contradicts with the fact proved above that $f'(x^*) \in E$. The proof of (iv) is thus completed.

(iv)$\Longrightarrow$(i). Consider any $\emptyset \neq M \subset W$ and any $\varepsilon > 0$. (iv) applied to $\overline{\operatorname{conv}}\, M$ yields $x^* \in X^*$ and $a \in \mathbb{R}$ such that the slice $\overline{\operatorname{conv}}\, M \cap \{x^* > a\}$ is nonempty and has diameter less than ε. Then, of course, the slice $M \cap \{x^* > a\}$ is also nonempty (and has diameter less than ε). □

Remark: In the situation described in (ii) in Theorem 11.3, $f'(x^*) \in W$ for every $x^* \in D$. This is a consequence of Corollary 7.37.

Corollary 11.4 *Let X be a dentable Banach space. Then every bounded closed convex set in X is equal to the closed convex hull of the set of its strongly exposed points.*

Let $f : X \to (-\infty, +\infty]$ be a proper function. We recall that its conjugate $f^* : X \to (-\infty, +\infty]$ is defined by $X^* \ni x^* \longmapsto f^*(x^*) = \sup_{u \in X} \big(\langle x^*, u\rangle - f(u)\big)$, see Definition 7.29. The function $f^*{}_* : X \to (-\infty, +\infty]$ is defined by $f^*{}_*(x) = \sup_{u^* \in X^*} \big(\langle u^*, x\rangle - f^*(u^*)\big)$. If f^* is also proper, we can compute f^{**}, a function from X^{**} into $(-\infty, +\infty]$. Obviously, $f^*{}_* = f^{**}|_X$.

Lemma 11.5 *Let $f : X \to (-\infty, +\infty]$ be a proper lower semicontinuous (not necessarily convex) function on a Banach space $(X, \|\cdot\|)$ such that $\inf f > -\infty$. Assume that f^* is Fréchet differentiable at $x_0^* \in X^*$. Then the derivative $(f^*)'(x_0^*) =: x_0$ belongs to X, we have $f^*{}_*(x_0) = f(x_0) \in \mathbb{R}$, and for every $\Delta > 0$ there exists $\delta > 0$ such that $\|x - x_0\| < \Delta$ whenever $x \in X$ and $f(x) - f(x_0) - \langle x_0^*, x - x_0\rangle < \delta$.*

Proof: Since the function f^* is convex and weak*-lower semicontinuous (see the paragraphs after Definition 7.29), Lemma 11.2 guarantees that $x_0 \in X$. In order to prove that $f^*{}_*(x_0) = f(x_0)$, note first that $\operatorname{epi} f^*{}_* = \operatorname{epi} f^{**} \cap (X \times \mathbb{R})$. Since $\operatorname{epi} f^{**} = \overline{\operatorname{conv}}^{w^*}(\operatorname{epi} f)$ (see Proposition 7.31), we get $\operatorname{epi} f^*{}_* = \overline{\operatorname{conv}}\,(\operatorname{epi} f)$.

Now, we are ready to prove that $f^*{}_*(x_0) = f(x_0)$. By Proposition 7.30, we have $f^*{}_*(x_0) \le f(x_0)$. In order to prove the reverse inequality, we fix any $\varepsilon > 0$. The lower semi-continuity of f yields $0 < \Delta < \varepsilon$ such that $f(x) > f(x_0) - \varepsilon$ whenever $x \in X$ and $\|x - x_0\| < \Delta$. The Fréchet differentiability of f^* at x_0^* yields $0 < \beta < 1$ so small that

$$f^*(x^*) - f^*(x_0^*) - \langle x^* - x_0^*, x_0\rangle < \Delta\beta \text{ whenever } x^* \in X^* \text{ and } \|x^* - x_0^*\| \le 2\beta.$$

Since $\big(x_0, f^*{}_*(x_0)\big) \in \overline{\operatorname{conv}}\,(\operatorname{epi} f)$, there is $(x, t) \in \operatorname{conv}(\operatorname{epi} f)$ such that $t - f^*{}_*(x_0) - \langle x_0^*, x - x_0\rangle < \Delta\beta$. Find $m \in \mathbb{N}$, $\alpha_i \ge 0$, and $x_i \in X$, $i = 1, \ldots, m$, such that $(x, t) = \sum_{i=1}^m \alpha_i (x_i, t_i)$, where $(x_i, t_i) \in \operatorname{epi} f$, i.e., $t_i \ge f(x_i)$, for all $i = 1, 2, \ldots, m$. Then

$$\sum_{i=1}^m \alpha_i f(x_i) - f^*{}_*(x_0) - \Big\langle x_0^*, \sum_{i=1}^m \alpha_i x_i - x_0\Big\rangle < \Delta\beta,$$

and hence, $f(x_i) - f^*{}_*(x_0) - \langle x_0^*, x_i - x_0\rangle < \Delta\beta$ for some $i \in \{1, \ldots, m\}$. We claim that $\|x_i - x_0\| < \Delta$. Indeed, since $x_0 \in \partial f^*(x_0^*)$, we get from (ii) in Proposition 7.34 that $f^*{}_*(x_0) + f^*(x_0^*) = \langle x_0^*, x_0\rangle$. We can estimate

$$\begin{aligned}
\Delta\beta &> f(x_i) - f^*{}_*(x_0) - \langle x_0^*, x_i - x_0\rangle \ge f^*{}_*(x_i) + f^*(x_0^*) - \langle x_0^*, x_i\rangle \\
&\ge \sup\big\{\langle x^*, x_i\rangle - f^*(x^*) : x^* \in X^*,\ \|x^* - x_0^*\| = 2\beta\big\} + f^*(x_0^*) - \langle x_0^*, x_i\rangle \\
&= -\inf\big\{\big(f^*(x^*) - f^*(x_0^*) - \langle x^* - x_0^*, x_0\rangle\big) - \langle x^* - x_0^*, x_i - x_0\rangle : \\
&\qquad\qquad x^* \in X^*,\ \|x^* - x_0^*\| = 2\beta\big\} \\
&\ge -\big(\Delta\beta - 2\beta\|x_i - x_0\|\big).
\end{aligned}$$

Thus $\|x_i - x_0\| < \Delta$ ($< \varepsilon$), and so

$$f(x_0) < f(x_i) + \varepsilon < \Delta\beta + f^*{}_*(x_0) + \langle x_0^*, x_i - x_0\rangle + \varepsilon$$
$$< \varepsilon + f^*{}_*(x_0) + \|x_0^*\|\varepsilon + \varepsilon.$$

Hence, letting $\varepsilon \downarrow 0$, we conclude that $f(x_0) \le f^*{}_*(x_0)$.

The last statement of our lemma, with $\delta := \Delta\beta$, follows from the proof of the claim above. □

Theorem 11.6 (Stegall's variational principle, [Steg2]) *A Banach space* $(X, \|\cdot\|)$ *is dentable (if and) only if the following holds:*
For every $\varepsilon > 0$ *and for every proper lower semi-continuous function* $f : X \to (-\infty, +\infty]$ *such that* $\inf f > -\infty$ *and* $\lim_{\|x\|\to\infty} \frac{f(x)}{\|x\|} > 0$, *there exist* $x_0 \in X$ *and* $x_0^* \in X^*$ *such that* $\|x_0^*\| < \varepsilon$ *and*

$$f(x) \ge f(x_0) + \langle x_0^*, x - x_0\rangle \quad \textit{for every} \quad x \in X, \tag{11.3}$$

and moreover, for every $\Delta > 0$ *there is* $\delta > 0$ *so that* $\|x - x_0\| < \Delta$ *whenever* $x \in X$ *and* $f(x) - f(x_0) - \langle x_0^*, x - x_0\rangle < \delta$.

Proof: `Necessity`. Assume that X is dentable, fix any $\varepsilon > 0$, and let f be a function as in Stegall's variational principle. Find $a, b > 0$ so that $f(x) > a\|x\|$ whenever $x \in X$ and $\|x\| > b$. Then for every $x^* \in X^*$, with $\|x^*\| \le a$, we have that $f^*(x^*) \le \max\{0, b\|x^*\| - \inf f\}$. Hence, aB_{X^*} lies in dom f^*. Theorem 11.3 provides $x_0^* \in X^*$, with $\|x_0^*\| < a$, where f^* is Fréchet differentiable; put $x_0 = (f^*)'(x_0^*)$. By Lemma 11.5, x_0 is an element of X and $f^*{}_*(x_0) = f(x_0)$. As $x_0 \in \partial f^*(x_0^*)$, we have that $x_0^* \in \partial f^*{}_*(x_0)$. Putting all this together yields

$$f(x) - f(x_0) - \langle x_0^*, x - x_0\rangle \ge f^*{}_*(x) - f^*{}_*(x_0) - \langle x_0^*, x - x_0\rangle \quad (\ge 0)$$

for every $x \in X$. The last statement of the principle follows from Lemma 11.5.

`Sufficiency`. Assume that Stegall's principle holds in the space X. Consider any nonempty set $M \subset B_X$. Define $f : X \to \{0, +\infty\}$ by

$$f(x) = \begin{cases} 0 & \text{if } x \in \overline{M}; \\ +\infty & \text{if } x \in X \setminus \overline{M}. \end{cases}$$

It is easy to check that this f satisfies all the conditions required in Stegall's principle. Find $x_0 \in X$ and $x_0^* \in X^*$ so that $f(x) \ge f(x_0) + \langle x_0^*, x - x_0\rangle$ for every $x \in X$. Then, necessarily, $x_0 \in \overline{M}$ and $\langle x_0^*, x - x_0\rangle \le 0$ for every $x \in M$. Let $\varepsilon > 0$ be arbitrary. Find $\delta > 0$ corresponding to $\Delta := \frac{\varepsilon}{2}$. Then we can immediately see that the diameter of the (nonempty) slice $M \cap \{x_0^* > \langle x_0^*, x_0\rangle - \delta\}$ is at most 2Δ $(= \varepsilon)$. □

The utility of Stegall's principle is illustrated in the proof of the following result.

Theorem 11.7 (Meshkov [Mesh], see [DGZ3, p. 209]) *Let $(X, \|\cdot\|)$ be a Banach space such that both X and X^* admit C^2-smooth bumps. Then X is isomorphic to a Hilbert space.*

Proof: ([FaZi2]) Let $f: X \to \mathbb{R}$ and $g: X^* \to \mathbb{R}$ be C^2-smooth bumps. By Corollary 7.44 and Exercise 11.20, X is an Asplund space. Hence the dual X^* is dentable according to Theorem 11.8. We shall use the convention that $\frac{1}{0} = +\infty$. Put $\varphi(x) = f^{-2}(x)$, $x \in X$; this is a proper function. Applying Theorem 11.6 to the (proper and lower semicontinuous) function $-\varphi^* + g^{-2}$, we find $x^* \in X^*$ and $x^{**} \in X^{**}$ such that

$$-\varphi^*(x^* + h^*) + g^{-2}(x^* + h^*) - \big(-\varphi^*(x^*) + g^{-2}(x^*)\big) \geq \langle x^{**}, h^* \rangle$$

for all $h^* \in X^*$. Then

$$-\varphi^*(x^*+h^*)+g^{-2}(x^*+h^*)-\varphi^*(x^*-h^*)+g^{-2}(x^*-h^*)+2\varphi^*(x^*)-2g^{-2}(x^*) \geq 0,$$

and hence

$$\varphi^*(x^*+h^*)+\varphi^*(x^*-h^*)-2\varphi^*(x^*) \leq g^{-2}(x^*+h^*)+g^{-2}(x^*-h^*)-2g^{-2}(x^*)$$

for all $h^* \in X^*$. Now, g^{-2} is C^2-smooth at x^* as $-\varphi^*(x^*) + g^{-2}(x^*) \in \mathbb{R}$. Therefore, there are $c > 0$ and $\delta > 0$ such that

$$g^{-2}(x^* + h^*) + g^{-2}(x^* - h^*) - 2g^{-2}(x^*) \leq c\|h^*\|^2$$

for all $h^* \in \delta B_{X^*}$. Thus

$$\varphi^*(x^* + h^*)+\varphi^*(x^* - h^*) - 2\varphi^*(x^*) \leq c\|h^*\|^2 \text{ for every } h^* \in \delta B_{X^*}. \quad (11.4)$$

This inequality and the convexity of φ^* immediately imply that φ^* is Fréchet differentiable at x^*; put $x = (\varphi^*)'(x^*)$. By Lemma 11.5, we know that $x \in X$ and that ${\varphi^*}_*(x) = \varphi(x) \in \mathbb{R}$. Using the already proved identity ${\varphi^*}_*(x) + \varphi^*(x^*) = \langle x^*, x \rangle$ and (11.4), we then have

$$\begin{aligned}
&{\varphi^*}_*(x + h) + {\varphi^*}_*(x - h) - 2{\varphi^*}_*(x) = \sup\big\{\langle x^* + h^*, x + h\rangle - \varphi^*(x^* + h^*) : h^* \in X^*\big\} \\
&\quad + \sup\big\{\langle x^* + k^*, x - h\rangle - \varphi^*(x^* + k^*) : k^* \in X^*\big\} - 2\varphi(x) \\
&\geq \sup\big\{\langle x^* + h^*, x + h\rangle - \varphi^*(x^* + h^*) \\
&\quad + \langle x^* - h^*, x - h\rangle - \varphi^*(x^* - h^*) : h^* \in \delta B_{X^*}\big\} - 2\varphi(x) \\
&\geq \sup\big\{2\langle x^*, x\rangle + 2\langle h^*, h\rangle - 2\varphi^*(x^*) - c\|h^*\|^2 : h^* \in \delta B_{X^*}\big\} - 2\varphi(x) \\
&= \sup\{2\langle h^*, h\rangle - c\|h^*\|^2 : h^* \in \delta B_{X^*}\} = \sup\{2\|h\|s - cs^2 : 0 \leq s \leq \delta\} = \tfrac{1}{c}\|h\|^2
\end{aligned}$$

for all $h \in X$ such that $\|h\| \leq \delta c$. And as $\varphi \geq {\varphi^*}_*$, we have

$$\varphi(x+h) + \varphi(x-h) - 2\varphi(x) \geq \frac{1}{c}\|h\|^2 \quad \text{whenever} \quad h \in \delta c B_X.$$

Now, φ $(= f^{-2})$ is C^2-smooth at x, as $\varphi(x) \in \mathbb{R}$. Hence, the latter inequality implies that $\varphi''(x)(h,h) \geq \frac{1}{c}\|h\|^2$ for every $h \in X$. And, because $\varphi''(x)$ is a bounded bilinear form on X, the assignment $X \ni h \longmapsto \big(\varphi''(x)(h,h)\big)^{1/2}$ is an equivalent Hilbertian norm on X. □

Remark: Some extra effort reveals that the assumptions of the just proved theorem can be weakened, see [FaZi2].

11.2 Dentability in X^*

The dentability of dual Banach spaces has several additional interesting features. We shall need some more concepts. Let $(X, \|\cdot\|)$ be a Banach space. We say that the dual X^* is *weak**-*dentable* if for every nonempty bounded subset $M \subset X^*$ and for every $\varepsilon > 0$ there are $u \in X$ and $a \in \mathbb{R}$ such that the slice $\{x^* \in M : \langle x^*, u\rangle > a\}$ is nonempty and has diameter less than ε. We say that X^* is *weak**-*fragmentable* if for every nonempty bounded subset $M \subset X^*$ and for every $\varepsilon > 0$ there is a weak*-open set $V \subset X^*$ such that the intersection $M \cap V$ is nonempty and has diameter less than ε. Let $\emptyset \neq M \subset X^*$ be given. We recall that, if $x^* \in M$ and $x \in X$ are such that $\langle x^*, x\rangle = \sup\langle M, x\rangle$ and the diameters of the slices $\{y^* \in M : \langle y^*, x\rangle > \langle x^*, x\rangle - \delta\}$ go to 0 as $\delta \downarrow 0$, then x^* is called a *weak**-*strongly exposed point* of M (and we say that x *strongly exposes* M at x^*), see Exercise 8.32. The class of Asplund spaces was introduced in Definition 7.38.

Theorem 11.8 (Namioka, Phelps, [NaPh], see, e.g., [Phelps]) *Let* $(X, \|\cdot\|)$ *be a Banach space. Then the following assertions are equivalent:*
(i) X^* *is weak**-*dentable.*
(ii) X^* *is dentable.*
(iii) X^* *is weak**-*fragmentable.*
(iv) X *is an Asplund space.*
(v) *Every nonempty convex weak**-*compact subset* $M \subset X^*$ *is equal to the weak**-*closed convex hull of the set* E *of all its weak**-*strongly exposed points.*
(vi) *Every separable subspace of* X *has a separable dual.*

Proof: (i)$\Longrightarrow$(ii) is trivial.

(ii)$\Longrightarrow$(iii). (van Dulst and Namioka [DuNa]) Assume that (ii) holds and that X^* is not weak*-fragmentable. Find then an $\varepsilon > 0$ and a bounded set $M \subset X^*$ whose each nonempty weak*-relatively open subset has diameter greater than ε (see Fig. 11.1). Denote $\mathcal{D} = \{\emptyset\} \cup \{0,1\} \cup \{0,1\}^2 \cup \ldots$ For $d \in \mathcal{D}$ we shall construct weak*-relatively open sets $U_d \subset M$, and norm-one vectors $h_d \in X$ such that $U_{d0} \cup U_{d1} \subset U_d$ and $\inf\langle U_{d0} - U_{d1}, h_d\rangle > \varepsilon$; here and further we put $di = (d_1, d_2, \ldots, d_n, i)$ if $d = (d_1, \ldots, d_n) \in \mathcal{D}$ and $i \in \{0,1\}$. Let $|d|$ be the "length" of d, that is, $|d| = n$ whenever $d \in \{0,1\}^n$. Put $U_\emptyset = M$. Consider any $d \in \mathcal{D}$ and assume that U_d

has already been constructed. We know that $\operatorname{diam} U_d > \varepsilon$. Find a norm-one vector $h_d \in X$ and $\xi_0, \xi_1 \in U_d$ such that $\langle \xi_0 - \xi_1, h_d \rangle > \varepsilon$. Then find weak*-relatively open sets $U_{di} \subset U_d$ such that $\xi_i \in U_{di}$, $i = 0, 1$, and $\inf \langle U_{d0} - U_{d1}, h_d \rangle > \varepsilon$. This finishes the induction step.

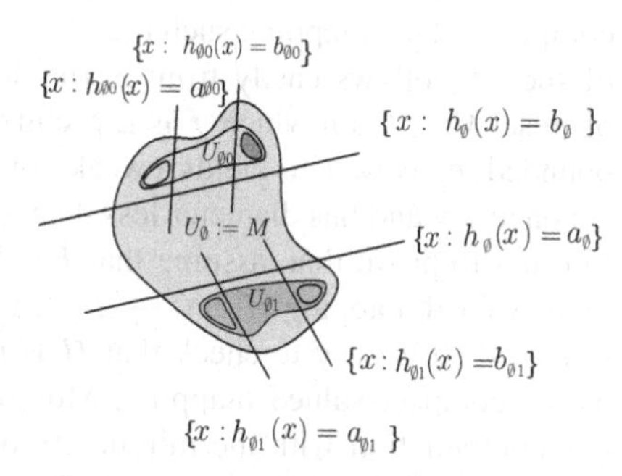

Fig. 11.1 The construction in the proof of (ii)$\Longrightarrow$(iii) in Theorem 11.8

For $d \in \mathcal{D}$ let K_d denote the weak*-closed convex hull of U_d. We note that $K_{d0} \cup K_{d1} \subset K_d$, and hence $\frac{1}{2}(K_{d0} + K_{d1}) \subset K_d$ for every $d \in \mathcal{D}$. We claim that there exists $t = (t_d)_{d \in \mathcal{D}} \in \prod_{d \in \mathcal{D}} K_d$ such that $t_d = \frac{1}{2}(t_{d0} + t_{d1})$ for every $d \in \mathcal{D}$. Clearly, in order to prove the claim, it is enough to show that $\bigcap_{d \in \mathcal{D}} A_d \neq \emptyset$, where

$$A_d = \Big\{ (t_d)_{d \in \mathcal{D}} \in \prod_{d \in \mathcal{D}} K_d : \ t_d = \tfrac{1}{2}(t_{d0} + t_{d1}) \Big\}, \quad d \in \mathcal{D}.$$

Using an argument of compactness, it is enough to prove that $\bigcap \big\{ A_d : \ d \in \mathcal{D},\ |d| \leq n \big\}$ is nonempty for every $n \in \mathbb{N}$. So fix one $n \in \mathbb{N}$. For $d \in \mathcal{D}$, with $|d| > n$ let t_d be any element of K_d. For the definition of t_d, $d \in \mathcal{D}$, with $|d| \leq n$, we shall use a downward induction. Fix any $d \in \mathcal{D}$ with $|d| \leq n$, and assume that we have already defined $t_{d0} \in K_{d0}$ and $t_{d1} \in K_{d1}$. Put then $t_d = \frac{1}{2}(t_{d0} + t_{d1})$; hence $t_d \in K_d$. Thus, we subsequently construct t_d for every $d \in \mathcal{D}$. It is clear that every $(t_d)_{d \in \mathcal{D}}$ from the nonempty set $\bigcap_{n=1}^{\infty} \big\{ A_d : \ d \in \mathcal{D},\ |d| \leq n \big\}$ satisfies the claim.

Pick some $(t_d)_{d \in \mathcal{D}} \in \prod_{d \in \mathcal{D}} K_d$. By (ii), there are $x^{**} \in X^{**}$ and $a \in \mathbb{R}$ such that the slice $\{t_d : \ d \in \mathcal{D}$ and $\langle x^{**}, t_d \rangle > a\}$ is nonempty and has diameter less than $\frac{\varepsilon}{2}$. Take $d \in \mathcal{D}$ so that $\langle x^{**}, t_d \rangle > a$. Then also $\langle x^{**}, t_{di} \rangle > a$ for a suitable $i \in \{0, 1\}$. Hence $\|t_d - t_{di}\| < \frac{\varepsilon}{2}$. However, from the construction of the sets U_{d0}, U_{d1} we have

$$\varepsilon < \langle t_{d0} - t_{d1}, h_d \rangle = 2 |\langle t_d - t_{di}, h_d \rangle| \leq 2 \|t_d - t_{di}\| < 2 \cdot \tfrac{\varepsilon}{2} = \varepsilon,$$

a contradiction. We thus proved (iii).

(iii)$\Longrightarrow$(iv). Let $f : X \to \mathbb{R}$ be a convex continuous function. For $n \in \mathbb{N}$, let G_n be the set of all points $x \in X$ such that

$$\tfrac{1}{t} \sup \big\{ \big(f(x + th) + f(x - th) - 2f(x) : \ h \in B_X \big\} < \tfrac{1}{n}$$

for some $t > 0$ (see Lemma 7.4 and the remark after its proof). We shall show that each G_n is dense and open. In order to prove the density of G_n, let $U \subset X$ be any nonempty open set. Let $\partial f : X \to 2^{X^*}$ be the subdifferential of f defined in (7.2). The mapping ∂f is norm-to-weak*-upper semicontinuous, see Proposition 7.14. Let $F : X \to 2^{X^*}$ be a "minimal" norm-to-weak*-upper semicontinuous and weak*-compact-valued mapping such that $\emptyset \neq Fx \subset \partial f(x)$ for every $x \in X$; the existence of such F follows easily from Zorn's lemma. Since f is continuous, there is an open set $V \subset U$ on which f is Lipschitz, see Lemma 7.3. By Fact 7.12, $F(V)$ is a bounded set. Now, (iii) yields a weak*-open set $W \subset X^*$ such that the set $F(V) \cap W$ is nonempty and has diameter less than $\frac{1}{n}$. We claim that $Fx \subset W$ for some $x \in V$. In order to prove that, assume that $Fx \setminus W \neq \emptyset$ for every $x \in V$. Define then the multi-valued mapping $H : X \to 2^{X^*}$ as $Hx = Fx$ if $x \in X \setminus V$ and $Hx = Fx \setminus W$ if $x \in V$. It is easy to check that H is norm-to-weak*-upper semicontinuous and weak*-compact-valued mapping. Moreover, H is strictly "smaller" than F. This is a contradiction with the minimality of F. Therefore, there is $x \in V$ such that $Fx \subset W$ and the claim is proved. The norm-to-weak*-upper semicontinuity of F yields an open set Ω such that $x \in \Omega \subset V$ and $F(\Omega) \subset W$. Then $\operatorname{diam} F(\Omega) < \frac{1}{n}$. Find $t > 0$ so small that $x + tB_X \subset \Omega$. Then for every $h \in B_X$, we have

$$\tfrac{1}{t}\big(f(x+th) + f(x-th) - 2f(x)\big) \leq \langle h, \xi - \eta \rangle \leq \operatorname{diam} F(\Omega) < \tfrac{1}{n},$$

where $\xi \in F(x+th)$ and $\eta \in F(x-th)$. We thus proved that $x \in \Omega \cap G_n$ ($\subset U \cap G_n$), which implies the density of G_n. That the set G_n is open is an easy exercise using the local Lipschitz property of f. Now, put $D = \bigcap_{n=1}^{\infty} G_n$. This is a G_δ dense set, by Baire's category theorem. Clearly, f is then Fréchet differentiable at every point of D. (iv) is thus proved.

(iv)$\Longrightarrow$(v). Let M and E be as in (v). Without loss of generality we may assume that $0 \in M$. Define $f(\cdot) = \sup \langle M, \cdot \rangle$; this is a convex continuous function on X. Put $U = \{x \in X : \sup \langle M, x \rangle \leq 1\}$; then U is a neighborhood of 0. The bipolar Theorem 3.38 yields that $U^\circ = M$. By Lemma 7.18, f is the Minkowski functional of U. Consider any point $x \in X$ where f is Fréchet differentiable; it exists by (iv). Put $\xi = f'(x)$. Therefore, by Lemma 7.19 and Corollary 7.20 we get $\xi \in E$.

It remains to prove that the weak*-closed convex hull of E is equal to M. Assume that there is $x^* \in M \setminus \overline{\operatorname{conv}}^{w^*}(E)$. Corollary 3.34 yields $h \in X$ such that $\langle x^*, h \rangle > \sup \langle E, h \rangle$. From (iv) find $u \in X$ such that f is Fréchet differentiable at u and that still $\langle x^*, u \rangle > \sup \langle E, u \rangle$. Denote $\eta = f'(u)$. We already know that $\eta \in E$ and that $\langle \eta, u \rangle = f(u)$. Therefore, $f(u) = \langle \eta, u \rangle < \langle x^*, u \rangle \ \big(\leq f(u)\big)$; a contradiction.

(v)$\Longrightarrow$(i). It is enough to realize that every weak*-strongly exposed point of a set $M \subset X^*$ lies in "weak*-slices" of M with an arbitrarily small diameter.

(iii)$\Longrightarrow$(vi). We shall follow the argument from [NaPh, p. 742]. Fix any separable subspace Y of X and let $q : X^* \to Y^*$ be the canonical quotient mapping. Let $\varepsilon > 0$ be arbitrary. By Zorn's lemma, we find a maximal ε-separated subset S of B_{Y^*} (that is, $\|y_1^* - y_2^*\| \geq \varepsilon$ for all distinct $y_1^*, y_2^* \in S$). The maximality of S yields that for every $y^* \in B_{Y^*}$ there is $s^* \in S$ so that $\|y^* - s^*\| < \varepsilon$. Assume that the set S is uncountable. We observe that the weak* topology on B_{Y^*} has a countable basis. Let S_0 denote the set of all $y^* \in S$ such that for every weak*-open subset V of Y^* such that $y^* \in V$, the intersection $S \cap V$ is uncountable. It is simple to prove that the set S_0 is also uncountable. Since X^* is weak*-fragmentable, there is a weak*-open set $W \subset X^*$ such that $q^{-1}(S_0) \cap B_{X^*} \cap W$ is nonempty and has diameter less than ε. But this leads to a contradiction because $S_0 \cap q(W)$ is not a singleton (see Corollary 3.51). Therefore, S must be at most countable. This holds for every $\varepsilon > 0$. Therefore, Y^* is separable.

(vi)$\Longrightarrow$(iii). This argument goes back to I. Namioka. Let $M \subset B_{X^*}$ be a nonempty set. Let $\{0\} \neq Y_0$ be a fixed separable subspace of X. Let $W_i^0 \subset B_{X^*}$, $i \in \mathbb{N}$, be a countable basis for the (possibly non-Hausdorff) topology on B_{X^*} of pointwise convergence on Y_0. For every $i \in \mathbb{N}$ we find a countable set $A_i^0 \subset B_X$ such that A_i^0-diam$(M \cap W_i^0) = \operatorname{diam}(M \cap W_i^0)$, where A-diam means the diameter in the seminorm $X^* \ni x^* \longmapsto \sup|\langle x^*, A\rangle|$. Let then Y_1 be the closed linear span of the set $Y_0 \cup \bigcup_{i\in\mathbb{N}} A_i^0$. Generally, consider any fixed $n \in \mathbb{N}$ and assume we have already found separable subspaces $Y_0 \subset Y_1 \subset \cdots \subset Y_n \subset X$, sets $A_i^0, A_i^1, \ldots, A_i^{n-1} \subset B_X$, $i \in \mathbb{N}$, and relatively weak*-open sets $W_i^0, W_i^1, \ldots, W_i^{n-1} \subset B_{X^*}$, $i \in \mathbb{N}$. Let $W_i^n \subset B_{X^*}$, $i \in \mathbb{N}$, be a (countable) basis for the topology on B_{X^*} of the pointwise convergence on Y_n. For every $i \in \mathbb{N}$ we find a countable set $A_i^n \subset B_X$ such that A_i^n-diam$(M \cap W_i^n) = \operatorname{diam}(M \cap W_i^n)$. Let then Y_{n+1} be the closed linear span of the set $Y_n \cup \bigcup_{i\in\mathbb{N}} A_i^n$. This completes the induction step. Finally, let Y be the closure of $\bigcup_{i\in\mathbb{N}} Y_n$, and put $A = \bigcup_{i,n\in\mathbb{N}} A_i^n$. We observe that Y is a separable subspace of X and that A is a countable set.

Now, fix any $\varepsilon > 0$. We shall show that $\operatorname{diam}(M \cap W) < \varepsilon$ for a suitable weak*-open set $W \subset X^*$. From the hypothesis we find a countable set $C \subset B_{X^*}$ such that for every $x^* \in B_{X^*}$ there is $c \in C$ satisfying $\sup\langle x^* - c, B_Y\rangle < \frac{\varepsilon}{2}$. Let $\mathcal{T}$ denote the (possibly non-Hausdorff) topology on B_{X^*} of pointwise convergence on Y, and let $\overline{M}^{\mathcal{T}}$ mean the $\mathcal{T}$-closure of M. We note that $(B_{X^*}, \mathcal{T})$ is a compact space. Hence, the set $\overline{M}^{\mathcal{T}} \cap B_{X^*}$ is $\mathcal{T}$-compact. We can write

$$\overline{M}^{\mathcal{T}} \cap B_{X^*} = \bigcup_{c\in C} \left\{x^* \in \overline{M}^{\mathcal{T}} \cap B_{X^*} : \ \sup|\langle x^* - c, A\rangle| \leq \tfrac{\varepsilon}{2}\right\}.$$

Baire's category theorem yields $c \in C$, and a $\mathcal{T}$-open set $V \subset B_{X^*}$ so that $\emptyset \neq \overline{M}^{\mathcal{T}} \cap V \subset \left\{x^* \in \overline{M}^{\mathcal{T}} \cap B_{X^*} : \ \sup\langle x^* - c, B_Y\rangle \leq \frac{\varepsilon}{2}\right\}$. Therefore, the (nonempty) set $M \cap V$ has A-diameter $\leq 2\cdot\frac{\varepsilon}{2} = \varepsilon$. We may and do assume that V is of the form $V = \left\{x^* \in B_{X^*} : \ \langle y_i, x^*\rangle < \alpha_i, \ i = 1, \ldots, l\right\}$ for suitable $n, l \in \mathbb{N}$, $y_1, \ldots, y_l \in Y_n$, and $\alpha_1, \ldots, \alpha_l \in \mathbb{R}$. Thus, V is an open set in the topology of pointwise convergence on Y_n. Hence, there must exist $i \in \mathbb{N}$ such that $\emptyset \neq M \cap W_i^n \subset M \cap V$. Find a weak*-open set $W \subset X^*$ such that $W_i^n = W \cap B_{X^*}$. Then $M \cap W = M \cap W_i^n$ and hence

$$\begin{aligned}\operatorname{diam}(M \cap W) &= \operatorname{diam}(M \cap W_i^n) = A_i^n\text{-diam}(M \cap W_i^n)\\ &\leq A\text{-diam}(M \cap W_i^n) \leq A\text{-diam}(M \cap V) < \varepsilon.\end{aligned}$$

This proves that M is weak*-fragmentable. □

Remark: In Exercises 9.24, 9.25, and 9.26, we proved (v)$\Longrightarrow$(iii)$\Longrightarrow$(vi) in Theorem 11.8 by using the intermediate notion of a *bounded ε-tree*. Some of the arguments used there are similar to the ones used here.

Corollary 11.9 *Let X be a Banach space. Then X is an Asplund space if and only if each norm-closed convex bounded set in X^* is the norm-closed convex hull of the set of its strongly exposed points.*

Corollary 11.10 *Reflexive Banach spaces are dentable and Asplund.*

Proof: By Proposition 3.114, every separable subspace of a reflexive space is reflexive, and hence, it has a separable dual. Thus, Theorem 11.8 concludes the proof. □

Theorem 11.11 (Lindenstrauss [Lind4], Troyanski [Troy1]) *Every weakly compact convex subset W of a Banach space X is dentable; it is actually equal to the closed convex hull of its strongly exposed points.*

Proof: Theorem 13.22, of Davis–Figiel–Johnson–Pełczyński, yields a reflexive Banach space Y and an injective and bounded operator $T : Y \to X$ such that $TB_Y \supset W$ and that T^*X^* is dense in Y^*. Consider any $\emptyset \neq M \subset W$. The space Y is dentable by Corollary 11.10. Moreover, $T^{-1}(M) \subset B_Y$. Hence, there is a nonempty slice $T^{-1}(M) \cap \{y^* > a\}$, with suitable $y^* \in Y^*$ and $a \in \mathbb{R}$, whose diameter is less than any a priori given $\varepsilon > 0$. From the density of T^*X^* in Y^* find $x^* \in X^*$ and $b > a$ so that

$$\emptyset \neq T^{-1}(M) \cap \{T^*x^* > b\} \subset T^{-1}(M) \cap \{y^* > a\}.$$

Then the slice $M \cap \{x^* > b\}$ $\big(= T\big(T^{-1}(M) \cap \{T^*x^* > b\}\big)\big)$ is nonempty and has diameter less than $\varepsilon\|T\|$. We thus proved the dentability of the set M. The conclusion now follows from Theorem 11.3. □

11.3 The Radon–Nikodým Property

Let λ mean the Lebesgue measure on $[0, 1)$, and let $\mathcal{L}$ denote the σ-algebra of Lebesgue measurable subsets of $[0, 1)$. Saying "almost everywhere" and "almost all" will be always related to the measure λ. Denote $\mathcal{L}^+ = \{E \in \mathcal{L} : \lambda(E) > 0\}$. For $A \in \mathcal{L}^+$ we put $\mathcal{L}^+(A) = \{E \in \mathcal{L}^+ : E \subset A\}$. All integrals of real-valued functions occurred in the rest of this chapter are considered in the Lebesgue sense. We assume that the reader is familiar with the theory of Lebesgue measure, Lebesgue measurable functions, and Lebesgue integral as they are explained, for instance, in [Rudi2, Chapters 1 and 2] and [LuMa]. See also Section 17.13.1.

We shall need the following concept. Let $\mathcal{P}$ and $\mathcal{Q}$ be two families of subsets of certain nonempty set. We say that $\mathcal{P}$ *is cofinal* in $\mathcal{Q}$ if $\mathcal{P} \subset \mathcal{Q}$ and for every $Q \in \mathcal{Q}$

there is $P \in \mathcal{P}$ such that $P \subset Q$. The next statement will be of use in arguments below.

Lemma 11.12 *(Principle of exhaustion) Let $J \in \mathcal{L}^+$ be given and assume that a subfamily $\mathcal{P}$ is cofinal in $\mathcal{L}^+(J)$. Then there exists an (at most countable) $\mathcal{B} \subset \mathcal{P}$, consisting of pairwise disjoint sets, such that $\sum_{B\in\mathcal{B}} \lambda(B) = \lambda(J)$.*

Proof: If there exists a finite family $\mathcal{B} \subset \mathcal{P}$, consisting of pairwise disjoint sets, such that $\sum_{B\in\mathcal{B}} \lambda(B) = \lambda(J)$, we are done. Further assume that this is not the case. By induction, we shall construct an infinite sequence $B_1, B_2, \dots$ of pairwise disjoint elements of $\mathcal{P}$ as follows. Pick some $B_1 \in \mathcal{P}$. Consider any $i \in \mathbb{N}$, and assume that we have already found pairwise disjoint sets $B_1, \dots, B_i \in \mathcal{P}$, and positive numbers $a_1, \dots, a_{i-1}$. Put $a_i = \sup\big\{\lambda(B) : B \in \mathcal{L}^+\big(J\backslash(B_1 \cup \dots \cup B_i)\big) \cap \mathcal{P}\big\}$. Pick then $B_{i+1} \in \mathcal{L}^+\big(J\backslash(B_1 \cup \dots \cup B_i)\big) \cap \mathcal{P}$ such that $\lambda(B_{i+1}) > \frac{1}{2}a_i$. This finishes the induction step. We thus constructed an infinite sequence $B_1, B_2, \dots$ of pairwise disjoint elements of $\mathcal{P}$. If $\sum_{i=1}^{\infty} \lambda(B_i) < \lambda(J)$, there is, by our assumption, $B \in \mathcal{L}^+\big(J\backslash(B_1 \cup B_2 \cup \cdots)\big) \cap \mathcal{P}$; hence $0 < \lambda(B) \le a_i$ for every $i \in \mathbb{N}$, and so

$$\lambda(J) > \sum_{i=2}^{\infty} \lambda(B_i) > \frac{1}{2}\sum_{i=1}^{\infty} a_i = +\infty,$$

a contradiction. Therefore $\sum_{i=1}^{\infty} \lambda(B_i) = \lambda(J)$, and it remains to put $\mathcal{B} = \{B_1, B_2, \dots\}$. □

Let $(X, \|\cdot\|)$ be a Banach space. Consider a function $f : [0, 1) \to X$. We put $v_f(0) = 0$ and for $t \in (0, 1]$ we define $v_f(t)$ as the supremum of all possible sums $\sum_{i=1}^{n} \|f(a_i) - f(a_{i-1})\|$, where $n \in \mathbb{N}$ and $0 = a_0 < a_1 < a_2 < \cdots < a_n < t$. The function $v_f : [0, 1] \to [0, +\infty]$ is then called the *variation* of f. We say that the function f is *absolutely continuous* if for every $\varepsilon > 0$ there is $\delta > 0$ such that $\sum_{i=1}^{n} \|f(b_i) - f(a_i)\| < \varepsilon$ whenever $n \in \mathbb{N}$, $0 \le a_1 < b_1 < a_2 < b_2 < \cdots < a_n < b_n < 1$, and $\sum_{i=1}^{n} (b_i - a_i) < \delta$. Note that, in particular, Lipschitz functions have this property.

For the definition and some simple fact about vector measures, see Section 17.13.1 and Exercise 11.23.

Proposition 11.13 *Let $(X, \|\cdot\|)$ be an arbitrary Banach space and let $f : [0, 1) \to X$ be an absolutely continuous function. Then there exists a unique λ-absolutely continuous vector measure $\tau : \mathcal{L} \to X$ such that for every $t \in (0, 1)$ we have*

$$f(t) = f(0) + \tau([0, t)) \quad \textit{and} \quad |\tau|([0, t)) = v_f(t) \le v_f(1) < +\infty. \tag{11.5}$$

Proof: We shall define a mapping $\tau : \mathcal{L} \to X$, which will be subsequently shown to be a *vector measure, λ-absolutely continuous, and of bounded variation.* We put $\tau(\emptyset) = 0$ and $\tau\big((a, b)\big) = f(b) - f(a)$ whenever $0 \le a < b \le 1$. Let $G \subset (0, 1)$ be any nonempty open set. Then there is a unique at most countable family $\mathcal{J}$ of open pairwise disjoint intervals in $(0, 1)$ such that $G = \bigcup \mathcal{J}$; see Exercise 11.25. For $J \in \mathcal{J}$ write $J = (a_J, b_J)$. We then define

$$\tau(G) = \sum_{J \in \mathcal{J}} \tau\big((a_J, b_J)\big). \tag{11.6}$$

We note that the absolute continuity of f guarantees that, for every $\varepsilon > 0$ there is a finite $\mathcal{F} \subset \mathcal{J}$ such that $\sum_{J \in \mathcal{J} \setminus \mathcal{F}} \|f(b_J) - f(a_J)\| < \varepsilon$. Thus, the sum in (11.6) converges even absolutely.

Now, let E be any element in $\mathcal{L}$. We shall define $\tau(E)$. Using the regularity of λ, we find a sequence of open sets $(0, 1) \supset G_1 \supset G_2 \supset \cdots \supset E \setminus \{0\}$ such that $\lambda(G_n) \downarrow \lambda(E)$ as $n \to \infty$. Put then

$$\tau(E) = \lim_{n \to \infty} \tau(G_n). \tag{11.7}$$

In what follows we must show that this limit does exist, that it does not depend on the specific sequence $G_1, G_2, \ldots$, and that, for open sets E, the vectors $\tau(E)$ defined in the formulas (11.6) and (11.7) are the same. We shall need the following

Claim 1: *For every $\Delta > 0$ there is $\delta > 0$ such that $\|\tau(G) - \tau(G')\| < \Delta$ whenever $G, G' \subset (0, 1)$ are open sets such that $G \supset G'$ and $\lambda(G \setminus G') < \delta$.*

Proof: Fix any $\Delta > 0$. From the absolute continuity of f find $\delta > 0$ such that

$$\sum_{i=1}^{n} \big\| f(b_i) - f(a_i) \big\| < \tfrac{\Delta}{3} \quad \text{whenever } n \in \mathbb{N},$$

$$0 \le a_1 < b_1 < \cdots < a_n < b_n < 1, \text{ and } \sum_{i=1}^{n} \big(b_i - a_i\big) < 2\delta. \tag{11.8}$$

Fix any G, G' as in the premise. Let $G = \bigcup \mathcal{J}$ and $G' = \bigcup \mathcal{J}'$ be the unique representations of G and G' by at most countable families $\mathcal{J}$ and $\mathcal{J}'$ of pairwise disjoint open intervals, respectively. Find a finite family $\mathcal{F}' \subset \mathcal{J}'$ such that $\sum_{J' \in \mathcal{J}' \setminus \mathcal{F}'} \lambda(J') < \delta$. Thus, by (11.6) and (11.8),

$$\Big\| \tau(G') - \sum_{J' \in \mathcal{F}'} \tau(J') \Big\| = \Big\| \sum_{J' \in \mathcal{J}' \setminus \mathcal{F}'} \tau(J') \Big\| \le \sum_{J' \in \mathcal{J}' \setminus \mathcal{F}'} \|\tau(J')\| \le \tfrac{\Delta}{3}.$$

Now for every $J' \in \mathcal{F}'$ find $J \in \mathcal{F}$ such that $J \supset J'$ and denote the family of all such J by $\mathcal{F}$. Add to $\mathcal{F}$ some elements of $\mathcal{J}$ so that the new family, still denoted by $\mathcal{F}$, will satisfy the inequality $\sum_{J \in \mathcal{J} \setminus \mathcal{F}} \lambda(J) < \delta$. Then $\big\|\tau(G) - \sum_{J \in \mathcal{F}} \tau(J)\big\| \le \frac{\Delta}{3}$, and so

$$\|\tau(G) - \tau(G')\| \le \tfrac{2}{3}\Delta + \Big\| \sum_{J \in \mathcal{F}} \tau(J) - \sum_{J' \in \mathcal{F}'} \tau(J') \Big\|. \tag{11.9}$$

It remains to estimate the last term in (11.9). Write every $J \in \mathcal{F}$ as (a_J, b_J) and put $\mathcal{F}'_J = \{J' \in \mathcal{F}' : J' \subset J\}$. Clearly, the families $\mathcal{F}'_J$ are pairwise disjoint and $\bigcup_{J\in\mathcal{F}} \mathcal{F}'_J = \mathcal{F}'$. Further, for $J \in \mathcal{F}$ write $\mathcal{F}'_J$ as $\{(a^J_1, b^J_1), \dots, (a^J_{k_J}, b^J_{k_J})\}$ where

$$a_J \le a^J_1 < b^J_1 < a^J_2 < b^J_2 < \cdots < a^J_{k_J} < b^J_{k_J} \le b_J.$$

We observe that

$$\begin{aligned}
&\sum_{J\in\mathcal{F}} \left((a^J_1 - a_J) + (a^J_2 - b^J_1) + \cdots + \left(a^J_{k_J} - b^J_{k_J-1}\right) + (b_J - b^J_{k_J})\right)\\
&= \sum_{J\in\mathcal{F}} \Big(\lambda(J) - \sum_{J'\in\mathcal{F}'_J} \lambda(J')\Big) = \sum_{J\in\mathcal{F}} \lambda(J) - \sum_{J'\in\mathcal{F}'} \lambda(J')\\
&= \lambda(G) - \lambda(G') - \sum_{J\in\mathcal{J}\setminus\mathcal{F}} \lambda(J) + \sum_{J'\in\mathcal{J}'\setminus\mathcal{F}'} \lambda(J') < \delta + \delta = 2\delta.
\end{aligned}$$

Hence, we can estimate

$$\begin{aligned}
&\Big\|\sum_{J\in\mathcal{F}} \tau(J) - \sum_{J'\in\mathcal{F}'} \tau(J')\Big\| \le \sum_{J\in\mathcal{F}} \Big\|\tau(J) - \sum_{J'\in\mathcal{F}'_J} \tau(J')\Big\|\\
&\le \sum_{J\in\mathcal{F}} \Big\| f(b_J) - f(a_J) - \sum_{i=1}^{k_J} \big(f(b^J_i) - f(a^J_i)\big)\Big\|\\
&\le \sum_{J\in\mathcal{F}} \Big(\|f(a^J_1) - f(a_J)\| + \|f(a^J_2) - f(b^J_1)\|\\
&\quad + \cdots + \big\|f(a^J_{k_J}) - f\big(b^J_{k_J-1}\big)\big\| + \big\|f(b_J) - f\big(b^J_{k_J}\big)\big\|\Big) < \tfrac{\Delta}{3}
\end{aligned}$$

by (11.8). This, together with (11.9), yields that $\|\tau(G) - \tau(G')\| < \Delta$. △

Claim 2: *The limit* $\lim_{n\to\infty} \tau(G_n)$ *in (11.7) does exist. If* $(0, 1) \supset G'_1 \supset G'_2 \supset \cdots \supset E\setminus\{0\}$ *is another sequence of open sets such that* $\lambda(G'_n) \downarrow \lambda(E)$ *as* $n \to \infty$*, then* $\lim_{n\to\infty} \tau(G'_n) = \lim_{n\to\infty} \tau(G_n)$*. In particular, if* $E \subset (0, 1)$ *is open, then* $\tau(E)$ *defined in (11.7) coincides with that defined in (11.6).*

Proof: We have $\lambda(G_n \setminus G_{n+m}) \to 0$ as $n, m \to \infty$. Claim 1 yields that $\|\tau(G_n) - \tau(G_{n+m})\| \to 0$ as $n, m \to \infty$. Hence $\lim_{n\to\infty} \tau(G_n)$ does exist. Having another sequence $G'_1, G'_2, \dots$, we have $G_1 \cap G'_1 \supset G_2 \cap G'_2 \supset \cdots \supset E$, and for $n \to \infty$ we subsequently get

$$\lambda\big(G_n\setminus(G_n \cap G'_n)\big) \le \lambda(G_n\setminus E) \to 0, \quad \lambda\big(G'_n\setminus(G_n \cap G'_n)\big) \to 0,$$

and hence

$$\|\tau(G_n) - \tau(G'_n)\| \le \|\tau(G_n) - \tau(G_n \cap G'_n)\| + \|\tau(G'_n) - \tau(G_n \cap G'_n)\| \to 0.$$

$\triangle$

Claim 3: *For every $\varepsilon > 0$ there is $\delta > 0$ such that $\|\tau(E) - \tau(F)\| < \varepsilon$ whenever $E, F \in \mathcal{L}$, $E \supset F$, and $\lambda(E \setminus F) < \delta$; in particular, $\|\tau(E)\| < \varepsilon$ whenever $E \in \mathcal{L}$ and $\lambda(E) < \delta$.*

Proof: By Claim 1, find $\delta > 0$ corresponding to $\Delta := \frac{\varepsilon}{2}$. Find open sets $(0, 1) \supset G_1 \supset G_2 \supset \cdots \supset E\setminus\{0\}$ and $(0, 1) \supset H_1 \supset H_2 \supset \cdots \supset F\setminus\{0\}$ such that $\lambda(G_n) \downarrow \lambda(E)$ and $\lambda(H_n) \downarrow \lambda(F)$ as $n \to \infty$. Since $\lambda\big(G_n\setminus(G_n \cap H_n)\big) \le \lambda(G_n) - \lambda(F)$, Claim 1 yields that $\big\|\tau(G_n) - \tau(G_n \cap H_n)\big\| < \frac{\varepsilon}{2}$ for all large $n \in \mathbb{N}$. Therefore, since $\lambda(G_n \cap H_n) \downarrow \lambda(F)$ as $n \to \infty$, using Claim 2, we have

$$\|\tau(E) - \tau(F)\| = \lim_{n\to\infty} \big\|\tau(G_n) - \tau(G_n \cap H_n)\big\| \le \tfrac{\varepsilon}{2} < \varepsilon.$$

$\triangle$

Claim 4: *τ is additive, that is, $\tau(E_1 \cup E_2) = \tau(E_1) + \tau(E_2)$ for every disjoint $E_1, E_2 \in \mathcal{L}$; in particular, $\tau([0, 1)\setminus E) = f(1) - f(0) - \tau(E)$ for every $E \in \mathcal{L}$.*

Proof: Fix any $\varepsilon > 0$. Let $\delta > 0$ be found in Claim 3 for our ε. The regularity of λ yields closed sets $C_1 \subset E_1$, $C_2 \subset E_2$ such that $0 \notin C_i$ and $\lambda(E_i\setminus C_i) < \frac{\delta}{2}$, $i = 1, 2$. Then $\lambda(E_1 \cup E_2\setminus(C_1 \cup C_2)) < \delta$, and, by Claim 3, we have that

$$\|\tau(E_1 \cup E_2) - \tau(C_1 \cup C_2)\| < \varepsilon, \text{ and } \|\tau(E_i) - \tau(C_i)\| < \varepsilon, \quad i = 1, 2. \tag{11.10}$$

Now, C_1, C_2 being two disjoint compact sets, we can find open disjoint sets G_1, $G_2 \subset (0, 1)$ such that $G_i \supset C_i$, $i = 1, 2$. By diminishing them, if necessary, we may and do assume that $\lambda(G_i\setminus C_i) < \frac{\delta}{2}$, $i = 1, 2$. Then $\lambda(G_1 \cup G_2\setminus(C_1 \cup C_2)) < \delta$, and Claim 3 yields

$$\|\tau(G_1 \cup G_2) - \tau(C_1 \cup C_2)\| < \varepsilon \quad \text{and} \quad \|\tau(G_i) - \tau(C_i)\| < \varepsilon, \; i = 1, 2. \tag{11.11}$$

Combining (11.10) and (11.11) we get

$$\|\tau(E_1 \cup E_2) - \tau(E_1) - \tau(E_2)\| < 6\varepsilon + \|\tau(G_1 \cup G_2) - \tau(G_1) - \tau(G_2)\| = 6\varepsilon,$$

where the last equality followed easily from the very definition (11.6). As $\varepsilon > 0$ was arbitrary, we are done. $\triangle$

Claim 5: *τ is a vector measure.*

Proof: It remains to verify that τ is σ-additive. So, consider any sequence $E_1, E_2, \ldots$ consisting of pairwise disjoint elements of $\mathcal{L}$. Using Claim 4 repeatedly, we get

$$\tau\Big(\bigcup_{i=1}^{\infty} E_i\Big) = \sum_{i=1}^{n} \tau(E_i) + \tau\Big(\bigcup_{i=n+1}^{\infty} E_i\Big)$$

for every $n \in \mathbb{N}$. Now, Claim 3 yields $\big\|\tau\big(\bigcup_{i=n+1}^{\infty} E_i\big)\big\| \to 0$ as $n \to \infty$. $\triangle$

Claim 6: τ *is* λ*-absolutely continuous.*

Proof: Let $E \in \mathcal{L}$ be such that $\lambda(E) = 0$. By Claim 3, $\tau(F) = 0$ for every $F \in \mathcal{L}$ such that $F \subset E$. It follows that $|\tau|(E) = 0$. $\triangle$

It remains to prove (11.5). So fix any $t \in (0, 1)$. Of course, the first equality follows trivially from (11.6). As regards the second equality, for any $n \in \mathbb{N}$ and any $0 \le a_1 < b_1 < \cdots < a_n < b_n < t$ we have $\sum_{i=1}^n \|f(b_i) - f(a_i)\| = \sum_{i=1}^n \|\tau((a_i, b_i))\| \le |\tau|((0, t))$ by the very definition of $|\tau|$. Hence $v_f(t) \le |\tau|((0, t))$. To prove the reverse inequality, fix any $\varepsilon > 0$. Consider any finite family $\mathcal{F} \subset \mathcal{L}$ of pairwise disjoint subsets of $(0, t)$. Using the regularity of λ and Claim 3, find closed sets $C_E \subset E$, $E \in \mathcal{F}$, such that $\sum_{E\in\mathcal{F}} \|\tau(E)\| < \sum_{E\in\mathcal{F}} \|\tau(C_E)\| + \varepsilon$. Then find open pairwise disjoint sets $C_E \subset G_E \subset (0, t)$, $E \in \mathcal{F}$; this is easy. Now, since each G_E is the union of at most countably many open intervals (Exercise 11.25), the compactness of C_E guarantees that we may and do assume that G_E is the union of a finite family $\mathcal{J}_E$ of open pairwise disjoint intervals for every $E \in \mathcal{F}$. Then, clearly, $\bigcup_{E\in\mathcal{F}} \mathcal{J}_E$ is a finite family of pairwise disjoint intervals in $(0, t)$, and

$$\sum_{E\in\mathcal{F}} \|\tau(E)\| < \sum_{E\in\mathcal{F}} \|\tau(G_E)\| + \varepsilon \le \sum_{E\in\mathcal{F}} \sum_{J\in\mathcal{J}_E} \|\tau(J)\| + \varepsilon \le v_f(t) + \varepsilon.$$

Since $\varepsilon > 0$ and $\mathcal{F}$ were arbitrary, we get that $|\tau|((0, t)) \le v_f(t)$. This proves the second equality in (11.5).

In order to prove that $v_f(1) < +\infty$, find $\delta > 0$ corresponding to $\varepsilon := 1$ in the definition of the absolute continuity of our f. Consider any $n \in \mathbb{N}$ and any $0 = a_0 < a_1 < a_2 < \cdots < a_n < 1$. Pick $k \in \mathbb{N}$ so big that $\frac{1}{k} < \delta$. Find $m \in \mathbb{N}$ and $0 = b_0 < b_1 < \cdots < b_m < 1$ such that $\{b_0, b_1, \ldots, b_m\} = \{a_0, a_1, \ldots, a_n\} \cup \{\frac{1}{k}, \frac{2}{k}, \ldots, \frac{k-1}{k}\}$. Then $\sum_{i=1}^n \|f(a_i) - f(a_{i-1})\| \le \sum_{i=1}^m \|f(b_i) - f(b_{i-1})\| < k\cdot 1$. Therefore $v_f(1) \le k$ $(< +\infty)$.

Finally, let $\nu : \mathcal{L} \to X$ be another λ-absolutely continuous vector measure satisfying (11.5), where τ is replaced by ν. Then (11.6) implies that $\nu(G) = \tau(G)$ for every open set $G \subset (0, 1)$, and (11.7), together with the λ-absolute continuity of ν, yield that $\nu(E) = \lim_{n\to\infty} \nu(G_n) = \lim_{n\to\infty} \tau(G_n) = \tau(E)$ for every $E \in \mathcal{L}$. $\square$

Let $(X, \|\cdot\|)$ be a Banach space and let $f : [0, 1) \to X$ be a Lebesgue measurable function. We observe that the real-valued function $\|f(\cdot)\|$ is Lebesgue measurable. For more information about measurable functions, see Section 17.13.1 and Exercise 11.28.

Definition 11.14 *We say that a Banach space* $(X, \|\cdot\|)$ *has the* Radon–Nikodým property *(RNP, in short) if for every* λ*-absolutely continuous vector measure* $\tau : \mathcal{L} \to X$ *of bounded variation, there exists a measurable function* $g : [0, 1) \to X$ *such that for every* $x^* \in X^*$ *the composition* $x^* \circ g$ *is Lebesgue integrable and for every* $E \in \mathcal{L}$

$$\langle x^*, \tau(E)\rangle = \int_E \langle x^*, g(t)\rangle \mathrm{d}\lambda(t). \tag{11.12}$$

In this case, g is called the Radon–Nikodým derivative *of* τ.

Remark: It follows from Exercise 11.32 that, if $g : [0, 1) \to X$ is measurable and (11.12) holds for every $x^* \in X^*$ and every $E \in \mathcal{L}$, then

$$|\tau|(E) = \int_E \|g(t)\| \,\mathrm{d}\lambda(t) \text{ for every } E \in \mathcal{L}. \tag{11.13}$$

Let $f : [0, 1) \to X$ be a function with values in a Banach space X. We say that f is *differentiable* at $t \in (0, 1)$ if the limit $\lim_{h\to 0} \frac{1}{h}\big(f(t+h) - f(t)\big)$, in the norm topology of X, exists; this limit is then denoted by $f'(t)$ and called the *derivative* of f at t. The concept of dentable Banach space was introduced in Definition 11.1.

In the following result, (i)$\Longrightarrow$(ii) is in [Rieff], (ii)$\Longrightarrow$(i) in [May], [DaPh], and [Hu]. For (ii)$\Longleftrightarrow$(iii)$\Longleftrightarrow$(iv) see [Qu].

Theorem 11.15 *For a Banach space* $(X, \|\cdot\|)$ *the following assertions are equivalent:*

(i) *X is dentable.*

(ii) *X has the Radon–Nikodým property.*

(iii) *Every absolutely continuous function* $f : [0, 1) \to X$ *is differentiable almost everywhere, the derivative* f' *is a measurable function, and for every* $t \in [0, 1)$ *and every* $x^* \in X^*$ *we have*

$$\left.\begin{array}{l} \int_0^t \|f'(u)\|\mathrm{d}\lambda(u) = v_f(t) < +\infty \quad \text{and} \\ \langle x^*, f(t)\rangle = \langle x^*, f(0)\rangle + \int_0^t \langle x^*, f'(u)\rangle \mathrm{d}\lambda(u). \end{array}\right\} \tag{11.14}$$

(iv) *For every Lipschitz function* $f : [0, 1) \to X$ *and for every* $\varepsilon > 0$ *there are* $t \in (0, 1)$, $y \in X$, *and* $\delta > 0$ *such that* $\big\|\frac{1}{s}(f(t+s) - f(t)) - y\big\| < \varepsilon$ *whenever* $0 \neq s \in \mathbb{R}$ *and* $|s| < \delta$.

Proof: (i)$\Longrightarrow$(ii). Assume that X is dentable, and let $\tau : \mathcal{L} \to X$ be a λ-absolutely continuous vector measure of bounded variation. The proof will be divided into several steps.

Claim 1: *For every* $J \in \mathcal{L}^+$ *there is* $A \in \mathcal{L}^+(J)$ *such that* $\sup\{|\tau|(E)/\lambda(E) : E \in \mathcal{L}^+(A)\} < +\infty$.

Proof: Fix any $J \in \mathcal{L}^+$ and assume that $\sup\{|\tau|(E)/\lambda(E) : E \in \mathcal{L}^+(A)\} = +\infty$ for every $A \in \mathcal{L}^+(J)$. Denote $\mathcal{P} = \big\{E \in \mathcal{L}^+(J) : |\tau|(E)/\lambda(E) > 2|\tau|(J)/\lambda(J)\big\}$. Then $\mathcal{P}$ is cofinal in $\mathcal{L}^+(J)$. Indeed, given any $A \in \mathcal{L}^+(J)$, there is $E \in \mathcal{L}^+(A)$ $(\subset \mathcal{L}^+(J))$ so that $|\tau|(E)/\lambda(E) > 2|\tau|(J)/\lambda(J)$, and hence $E \in \mathcal{P}$. Now, the principle of exhaustion formulated in Lemma 11.12 yields an at most countable subfamily $\mathcal{B} \subset \mathcal{P}$, consisting of pairwise disjoint sets, such that $\sum_{E\in\mathcal{B}} \lambda(E) = \lambda(J)$. But then, using the λ-absolute continuity of τ, we have

$$|\tau|(J) = \sum_{E\in\mathcal{B}} |\tau|(E) > \frac{2|\tau|(J)}{\lambda(J)} \sum_{E\in\mathcal{B}} \lambda(E) = 2|\tau|(J),$$

a contradiction.

△

Claim 2: (crucial) *Let* $\varepsilon > 0$ *and let* $A \in \mathcal{L}^+$ *be such that the set* $\{\tau(E)/\lambda(E) : E \in \mathcal{L}^+(A)\}$ *is bounded. Then there exists* $B \in \mathcal{L}^+(A)$ *such that* $\operatorname{diam}\{\tau(E)/\lambda(E) : E \in \mathcal{L}^+(B)\} \le 2\varepsilon$.

Proof: For $E \in \mathcal{L}^+$ we put $\widetilde{\tau}(E) = \tau(E)/\lambda(E)$. Assume that the conclusion is false, that is, for every $B \in \mathcal{L}^+(A)$ the diameter of the set $\{\widetilde{\tau}(E) : E \in \mathcal{L}^+(B)\}$ is greater than 2ε. Put $M = \{\widetilde{\tau}(E) : E \in \mathcal{L}^+(A)\}$. This set is bounded. Hence, the dentability of X provides $x^* \in X^*$ and $a \in \mathbb{R}$ such that the slice $M \cap \{x^* > a\}$ is nonempty and has diameter less than ε. Pick $x \in M \cap \{x^* > a\}$. If $y \in M \backslash B(x, \varepsilon)$, then $y \notin \{x^* > a\}$. It follows that $M \backslash B(x, \varepsilon) \subset \{x^* \le a\}$, hence $\overline{\operatorname{conv}}(M \backslash B(x, \varepsilon)) \subset \{x^* \le a\}$, and this implies that $x \notin \overline{\operatorname{conv}}(M \backslash B(x, \varepsilon))$. Now, find $J \in \mathcal{L}^+(A)$ so that $\widetilde{\tau}(J) = x$; thus $\widetilde{\tau}(J) \notin \overline{\operatorname{conv}}(M \setminus B(\widetilde{\tau}(J), \varepsilon))$. We observe that the family $\mathcal{P} := \{E \in \mathcal{L}^+(J) : \|\widetilde{\tau}(E) - \widetilde{\tau}(J)\| > \varepsilon\}$ is cofinal in $\mathcal{L}^+(J)$. Indeed, given any $B \in \mathcal{L}^+(J)$, we know that $\operatorname{diam}\{\widetilde{\tau}(E) : E \in \mathcal{L}^+(B)\} > 2\varepsilon$, and hence, necessarily, there exists $E \in \mathcal{L}^+(B)$ $(\subset \mathcal{L}^+(J))$ such that $\|\widetilde{\tau}(E) - \widetilde{\tau}(J)\| > \varepsilon$, and so, $E \in \mathcal{P}$. Now, the principle of exhaustion, Lemma 11.12, yields an at most countable family $\mathcal{Q} \subset \mathcal{P}$, consisting of pairwise disjoint sets, such that $\sum_{E\in\mathcal{Q}} \lambda(E) = \lambda(J)$. Having this, the λ-absolute continuity of τ guarantees

$$\widetilde{\tau}(J) = \frac{1}{\lambda(J)} \sum_{E\in\mathcal{Q}} \tau(E) = \sum_{E\in\mathcal{Q}} \frac{\lambda(E)}{\lambda(J)} \widetilde{\tau}(E). \tag{11.15}$$

If $\mathcal{Q}$ is finite, then (11.15) says that $\widetilde{\tau}(J)$ belongs to the convex hull of $M \backslash B(\widetilde{\tau}(J), \varepsilon)$, a contradiction. If $\mathcal{Q}$ is infinite, then (11.15), together with (ii) in Exercise 1.67, guarantees that $(x =)$ $\widetilde{\tau}(J) \in \overline{\operatorname{conv}}(M \backslash B(\widetilde{\tau}(J), \varepsilon))$, a contradiction again.

△

Claim 3: *For every* $\varepsilon > 0$ *and for every* $J \in \mathcal{L}^+$ *there exists* $B \in \mathcal{L}^+(J)$ *such that* $\operatorname{diam}\{\tau(E)/\lambda(E) : E \in \mathcal{L}^+(B)\} < \varepsilon$.

Proof: Just put together Claims 1 and 2.

△

Claim 4: *For every* $\varepsilon > 0$, *there exists an at most countable family* $\mathcal{B} \subset \mathcal{L}^+$ *consisting of pairwise disjoint sets, such that* $\sum_{B\in\mathcal{B}} \lambda(B) = 1$ *and that* $\operatorname{diam}\{\tau(E)/\lambda(E) : E \in \mathcal{L}^+(B)\} < \varepsilon$ *for every* $B \in \mathcal{B}$.

Proof: The family $\mathcal{P} := \{B \in \mathcal{L}^+ : \operatorname{diam}\{\tau(E)/\lambda(E) : E \in \mathcal{L}^+(B)\} < \varepsilon\}$ is cofinal in $\mathcal{L}^+$. Indeed, take any $J \in \mathcal{L}^+$. Claim 3 yields $B \in \mathcal{L}^+(J)$ such that $B \in \mathcal{P}$. Now, the principle of exhaustion applied to $J := [0, 1)$ concludes the proof.

△

Claim 5: *There are families* $\mathcal{C}_n \subset \mathcal{L}^+$, $n \in \mathbb{N}$, *each one consisting of at most countably many pairwise disjoint sets, such that* $\sum_{C\in\mathcal{C}_n} \lambda(C) = 1$ *for every* $n \in \mathbb{N}$, *that* diam $\{\tau(E)/\lambda(E):\ E \in \mathcal{L}^+(C)\} < \frac{1}{n}$ *for every* $n \in \mathbb{N}$ *and every* $C \in \mathcal{C}_n$, *and that, for every* $n, m \in \mathbb{N}$, *if* $n < m$, *and* $C \in \mathcal{C}_m$, *then there is* $C' \in \mathcal{C}_n$ *such that* $C' \supset C$.

Proof: For $n \in \mathbb{N}$ let $\mathcal{B}_n$ be the family $\mathcal{B}$ found in Claim 5 for $\varepsilon := \frac{1}{n}$. Now, for every $n \in \mathbb{N}$, consider all possible intersections $B_1 \cap \cdots \cap B_n$, where $B_1 \in \mathcal{B}_1, \ldots, B_n \in \mathcal{B}_n$, which have positive Lebesgue measure, and denote the family so obtained as $\mathcal{C}_n$. It is clear that these $\mathcal{C}_n$ work. $\triangle$

Claim 6: *For* $n \in \mathbb{N}$ *define* $g_n = \sum_{C\in\mathcal{C}_n} \frac{\tau(C)}{\lambda(C)}\chi_C$, *where* $\mathcal{C}_n$, $n \in \mathbb{N}$, *are the families found in Claim 5. Then there exists a set* $N \subset [0, 1)$, *with* $\lambda(N) = 0$, *such that*

$$\|g_n(t) - g_m(t)\| < \tfrac{1}{n} \quad \textit{whenever} \quad n, m \in \mathbb{N},\ n < m, \ \textit{ and } \ t \in [0, 1)\backslash N.$$

Proof: Using Claim 5, put $N = \bigcup_{n=1}^{\infty}\big([0, 1)\backslash\big(\bigcup \mathcal{C}_n\big)\big)$. $\triangle$

Claim 7: *Defining* $g(t) = \lim_{n\to\infty} g_n(t)$ *if* $t \in [0, 1)\backslash N$ *and* $g(t) = 0$ *if* $t \in N$, *then the function* $g : [0, 1) \to X$ *is measurable.*

Proof: From Claim 5, for every $n \in \mathbb{N}$ find $E_n \in \mathcal{L}$ so that the function $h_n := g_n - g_n \cdot \chi_{E_n}$ is simple and that $\lambda(E_n) < 2^{-n}$. Put $M = \bigcap_{n=1}^{\infty} (E_n \cup E_{n+1} \cup \ldots)$; then $\lambda(M) = 0$. We shall show that $\lim_{n\to\infty} h_n(t) = g(t)$ for every $t \in [0, 1)\backslash(N \cup M)$, which will imply that g is measurable. So, consider any such t. Fix any $\varepsilon > 0$. Find $m \in \mathbb{N}$ so big that $m > \frac{1}{\varepsilon}$ and that $t \notin E_m \cup E_{m+1} \cup \cdots$. Then for every $n \geq m$ we have $t \notin E_n$; hence $h_n(t) = g_n(t)$ and Claim 6 yields

$$\|h_n(t) - g(t)\| = \|g_n(t) - g(t)\| = \lim_{i\to\infty} \|g_n(t) - g_i(t)\| \leq \tfrac{1}{n} \leq \tfrac{1}{m} < \varepsilon.$$

We thus proved that g is the pointwise limit, almost everywhere, of the sequence of simple functions, that is, it is measurable. $\triangle$

Claim 8: *For every* $x^* \in X^*$ *and every* $E \in \mathcal{L}$ *Equation (11.12) holds.*

Proof: Fix any such x^* and E. For every $n \in \mathbb{N}$ we have

$$\Big|\langle x^*, \tau(E)\rangle - \int_E \langle x^*, g_n(t)\rangle \mathrm{d}\lambda(t)\Big| = \Big|\langle x^*, \tau(E)\rangle - \sum_{C\in\mathcal{C}_n} \langle x^*, \widetilde{\tau}(C)\rangle\lambda(C \cap E)\Big|$$

$$\leq \|x^*\| \Big\|\tau(E) - \sum_{C\in\mathcal{C}_n} \widetilde{\tau}(C)\lambda(C \cap E)\Big\| \leq \|x^*\| \sum_{C\in\mathcal{C}_n} \big\|\widetilde{\tau}(C \cap E) - \widetilde{\tau}(C)\big\|\lambda(C \cap E)$$

$$\leq \frac{1}{n}\|x^*\| \sum_{C\in\mathcal{C}_n} \lambda\big(C \cap E\big) = \frac{1}{n}\|x^*\|\lambda(E).$$

and also, by Claims 6 and 7,

$$\Big|\int_E \langle x^*, g_n(t)\rangle \mathrm{d}\lambda(t) - \int_E \langle x^*, g(t)\rangle \mathrm{d}\lambda(t)\Big| \leq \|x^*\| \int_E \big\|g_n(t) - g(t)\big\| \mathrm{d}\lambda(t) \leq \frac{1}{n}\|x^*\|\lambda(E).$$

Hence $\left|\langle x^*, \tau(E)\rangle - \int_E \langle x^*, g(t)\rangle \mathrm{d}\lambda(t)\right| \leq \frac{2}{n}\|x^*\|\lambda(E)$ for every $n \in \mathbb{N}$. And letting $n \to \infty$, we get (11.12). △

We proved (ii).

(ii)⟹(iii). Let f be as in (iii). Let $\tau : \mathcal{L} \to X$ be the vector measure found in Proposition 11.13 for this f. Note that τ is λ-absolutely continuous and of bounded variation. (ii) applied to this τ yields a measurable function $g : [0, 1) \to X$ such that (11.12) holds for every $E \in \mathcal{L}$ and every $x^* \in X^*$. According to the Remark after Definition 11.14, the real-valued function $\|g(\cdot)\|$ is integrable. Then, in particular, we have

$$\langle x^*, f(t)\rangle - \langle x^*, f(0)\rangle = \langle x^*, \tau((0, t))\rangle = \int_0^t \langle x^*, g(u)\rangle \mathrm{d}\lambda(u) \tag{11.16}$$

and

$$|\tau|([0, t)) = \int_0^t \|g(u)\|\mathrm{d}\lambda(u), \tag{11.17}$$

for every $x^* \in X^*$ and every $t \in [0, 1)$. Since g is measurable, it is easy to find a Lebesgue negligible set $N \subset [0, 1)$ such that the set $g([0, 1)\setminus N)$ is separable; let D be a countable dense subset of it. Fix any $y \in D$. Since the function $[0, 1) \ni t \longmapsto g(t) - y$ is the limit of a sequence of simple functions almost everywhere, the function $[0, 1) \ni t \longmapsto \|g(t) - y\|$ is Lebesgue measurable. It is actually Lebesgue integrable by (11.17). Now, we shall use the following theorem (see, e.g., [Rudi2, Theorem 8.17]):

Let $\varphi : [0, 1) \to \mathbb{R}$ be a Lebesgue integrable function and put $\psi(t) = \int_0^t \varphi \,\mathrm{d}\lambda,\ t \in [0, 1)$. Then ψ is differentiable almost everywhere in $(0, 1)$, and moreover, $\psi'(t) = \varphi(t)$ for almost every $t \in (0, 1)$.

Thus, we get that

$$\frac{1}{h}\int_t^{t+h} \|g(u) - y\|\mathrm{d}\lambda(u) \to \|g(t) - y\| \quad \text{as} \quad 0 \neq h \to 0$$

for almost all $t \in (0, 1)$. Hence, there exists a Lebesgue negligible set $M \subset (0, 1)$ such that the convergence above holds for every $y \in D$ and for every $t \in (0, 1)\setminus M$. From this, we can easily deduce that

$$\frac{1}{h}\int_t^{t+h} \|g(u) - g(t)\|\mathrm{d}\lambda(u) \to 0 \quad \text{as} \quad 0 \neq h \to 0 \tag{11.18}$$

for every $t \in (0, 1)\setminus(N \cup M)$. Fix now any $t \in (0, 1)\setminus(N \cup M)$. Then for $0 \neq h \in \mathbb{R}$ such that $t \pm h \in (0, 1)$ we have

$$\left\|\frac{1}{h}\big(f(t+h)-f(t)\big)-g(t)\right\| = \sup_{x^*\in B_{X^*}} \left\langle x^*, \frac{1}{h}\big(f(t+h)-f(t)\big)-g(t)\right\rangle$$

$$= \sup_{x^*\in B_{X^*}} \frac{1}{h}\int_t^{t+h} \langle x^*, g(u)-g(t)\rangle \mathrm{d}\lambda(u) \le \frac{1}{h}\int_t^{t+h} \|g(u)-g(t)\|\mathrm{d}\lambda(u) \to 0$$

as $0 \neq h \to 0$, by (11.18). This means that f is differentiable at our t and that $f'(t) = g(t)$. Thus, $f'(t) = g(t)$ for almost all $t \in [0, 1)$. Hence, f' is measurable and (11.16) means the second equality in (11.14). Finally, (11.17) now reads as $|\tau|((0,t)) = \int_0^t \|f'(u)\|\mathrm{d}\lambda(u)$. Proposition 11.13 then gives the first equality in (11.14). The proof of (iii) is complete.

(iii)⟹(iv) is trivial.

(iv)⟹(i). Assume that (iv) holds and that X is not dentable. Then there exists a nonempty subset $M \subset B_X$ and $\varepsilon > 0$ such that every nonempty slice of it has diameter greater than 2ε. We observe that then *every* $x \in M$ *belongs to* $\overline{\mathrm{conv}}\,\big(M\backslash B(x,\varepsilon)\big)$. Indeed, assume there is $x \in M$ such that $x \notin \overline{\mathrm{conv}}\,\big(M\backslash B(x,\varepsilon)\big)$. The separation Theorem 2.12 provides $x^* \in X^*$ and $a \in \mathbb{R}$ such that $\langle x^*, x\rangle > a > \sup\langle x^*, \overline{\mathrm{conv}}\,\big(M\backslash B(x,\varepsilon)\big)\rangle$. Thus, if $y \in M \cap \{x^* > a\}$, we have $y \notin M\backslash B(x,\varepsilon)$, that is, $y \in B(x,\varepsilon)$, and so, $\mathrm{diam}\,\big(M \cap \{x^* > a\}\big) \le 2\varepsilon$, a contradiction.

Next, we shall construct simple functions $g_1, g_2, \ldots : [0,1) \to X$, and then Lispchitz functions $f_1, f_2, \ldots : [0,1) \to X$ as follows. By "interval" we shall always mean $[a,b)$ where $0 \le a < b \le 1$. Pick some $x_1^1 \in M$ and put $g_1 = x_1^1\chi_{I_1^1}$ where $I_1^1 = [0,1)$. Assume that $n \in \mathbb{N}$ is given and that we already found a simple function $g_n : [0,1) \to X$ of form $g_n = \sum_{i=1}^{k_n} x_i^n \chi_{I_i^n}$ where $k_n \in \mathbb{N}$, $x_1^n, \ldots, x_{k_n}^n \in M$, and $I_1^n, \ldots, I_{k_n}^n$ are pairwise disjoint intervals such that $I_1^n \cup \cdots \cup I_{k_n}^n = [0,1)$ and $\lambda(I_i^n) \le 2^{-n+1}$ for every $i = 1, \ldots, k_n$. Fix any $i \in \{1, \ldots, k_n\}$ for a while. As $x_i^n \in \overline{\mathrm{conv}}\big(M\backslash B(x_i^n, \varepsilon)\big)$, there are a finite set $A_i^n \subset (0,1]$, with $\sum_{a\in A_i^n} a = 1$, and points $x_a \in M\backslash B(x_i^n, \varepsilon)$, $a \in A_i^n$, such that

$$\left\|x_i^n - \sum\left\{a x_a : a \in A_i^n\right\}\right\| < 2^{-n}. \tag{11.19}$$

Find then pairwise disjoint intervals $I_a^{n,i} \subset I_i^n$, $a \in A_i^n$, such that $\bigcup_{a\in A_i^n} I_a^{n,i} = I_i^n$, and that $\lambda(I_a^{n,i}) = a\lambda(I_i^n)$ for every $a \in A_i^n$. We may easily tune the construction in such a way that $\lambda(I_a^{n,i}) \le 2^{-n}$ for every $a \in A_i^n$. Doing this for every i, we get intervals $I_a^{n,i}$, $a \in A_i^n$, $i \in \{1, \ldots, k_n\}$. Clearly, they are pairwise disjoint and their union is equal to $[0,1)$. Let us enumerate them as $I_1^{n+1}, I_2^{n+1}, \ldots, I_{k_{n+1}}^{n+1}$, and then we enumerate the set $A_1^n \cup \cdots \cup A_{k_n}^n$ as $x_1^{n+1}, x_2^{n+1}, \ldots, x_{k_{n+1}}^{n+1}$, in such a way that, if $a \in A_i^n$ is named as x_j^{n+1}, then I_j^{n+1} means $I_a^{n,i}$. Define then $g_{n+1} = \sum_{i=1}^{k_{n+1}} x_i^{n+1}\chi_{I_i^{n+1}}$. This finishes the induction step. By permuting the sets $\{1, \ldots, k_n\}$, we may and do assume that for every $n \in \mathbb{N}$ there exist $0 = b_0^n < b_1^n < b_2^n < \cdots < b_{k_n}^n = 1$ such that $I_1^n = [b_0^n, b_1^n)$, $I_2^n =$

$[b_1^n, b_2^n), \ldots, I_{k_n}^n = [b_{k_n-1}^n, b_{k_n}^n)$. We observe that $\|g_{n+1}(t) - g_n(t)\| \geq \varepsilon$ for every $n \in \mathbb{N}$ and every $t \in [0, 1)$.

Given any simple function $h : [0, 1) \to X$, of form $h = \sum_{i=1}^m x_i \chi_{E_i}$, with $E_1, \ldots, E_m \in \mathcal{L}$ pairwise disjoint, we define

$$\int_E h = \sum_{i=1}^m x_i \lambda(E \cap E_i) \quad \text{and} \quad \int_s^t h = \int_{[s,t)} h$$

for all $E \in \mathcal{L}$ and all $0 \leq s < t \leq 1$. For $n \in \mathbb{N}$ then define $f_n(t) = \int_0^t g_n$, $t \in [0, 1)$. Since the functions g_n have values in B_X, the functions f_n are 1-Lipschitz.

Fix any $n \in \mathbb{N}$ and any $i \in \{1, \ldots, k_n\}$ for a while. Find $r \in \{1, \ldots k_{n+1}\}$ so that $b_i^n = b_r^{n+1}$. We have $f_n(b_i^n) = \sum_{j=1}^i \lambda(I_j^n) x_j^n$ and

$$f_{n+1}(b_i^n) = f_{n+1}(b_r^{n+1}) = \sum_{j=1}^{i} \sum_{l \in N_j} \lambda(I_l^{n+1}) x_l^{n+1}$$

where $N_1, \ldots, N_i$ are suitable "segments" of $\{1, \ldots, k_{n+1}\}$ such that $\max N_1 = \min N_2 - 1, \ldots, \max N_{i-1} = \min N_i - 1$ and $N_1 \cup \cdots \cup N_i = \{1, \ldots, r\}$. Thus

$$\|f_n(b_i^n) - f_{n+1}(b_i^n)\| \leq \sum_{j=1}^{i} \Big\|\lambda(I_j^n) x_j^n - \sum_{l \in N_j} \lambda(I_l^{n+1}) x_l^{n+1}\Big\| < \sum_{j=1}^{i} 2^{-n} \lambda(I_j^n) = 2^{-n} b_i^n \leq 2^{-n}$$

by (11.19). Therefore, since $b_i^n - b_{i-1}^n = \lambda(I_i^n) \leq 2^{-n}$ for every $i = 1, \ldots, k_n$, and f_n, f_{n+1} are 1-Lipschitz, we conclude that $\|f_n(t) - f_{n+1}(t)\| < 2^{-n+2}$ for every $t \in [0, 1)$. This implies that the functions f_n converge to a Lipschitz function, f, say, uniformly on $[0, 1)$.

Fix again any $n \in \mathbb{N}$ and any $i \in \{1, \ldots, k_n\}$ for a while. Find a "segment" $N_i \subset \{1, \ldots, k_{n+1}\}$ such that $I_i^n = \bigcup_{l \in N_i} I_l^{n+1}$; then $\lambda(I_i^n) = \sum_{l \in N_i} \lambda(I_l^{n+1})$. We have $\int_{I_i^n} g_n = \lambda(I_i^n) x_i^n$, and

$$\int_{I_i^n} g_{n+1} = \sum_{l \in N_i} \int_{I_l^{n+1}} g_{n+1} = \sum_{l \in N_i} \lambda(I_l^{n+1}) x_l^{n+1}$$

Hence, by (11.19),

$$\Big\|\int_{I_i^n} g_n - \int_{I_i^n} g_{n+1}\Big\| = \Big\|\lambda(I_i^n) x_i^n - \sum_{l \in N_i} \lambda(I_l^{n+1}) x_l^{n+1}\Big\| < 2^{-n} \lambda(I_i^n).$$

Replacing here n by $n + 1$, we get that $\| \int_{I_l^{n+1}} g_{n+1} - \int_{I_l^{n+1}} g_{n+2}\| < 2^{-n-1} \lambda\big(I_l^{n+1}\big)$ for every $l = 1, \ldots, k_{n+1}$. Hence

$$\left\| \int_{I_i^n} g_n - \int_{I_i^n} g_{n+2} \right\| \leq \left\| \int_{I_i^n} g_n - \int_{I_i^n} g_{n+1} \right\| + \sum_{l \in N_i} \left\| \int_{I_l^{n+1}} g_{n+1} - \int_{I_l^{n+1}} g_{n+2} \right\|$$
$$< 2^{-n}\lambda(I_i^n) + 2^{-n-1} \sum_{l \in N_i} \lambda(I_l^{n+1}) = (2^{-n} + 2^{-n-1})\lambda(I_i^n).$$

Therefore, by induction,

$$\left\| \int_{I_i^n} g_n - \int_{I_i^n} g_{n+m} \right\| < (2^{-n} + \cdots + 2^{-n-m+1})\lambda(I_i^n) < 2^{-n+1}\lambda(I_i^n).$$

This, translated into terms of f_n and b_i^n, reads as

$$\left\| f_n(b_i^n) - f_n(b_{i-1}^n) - \left(f_{n+m}(b_i^n) - f_{n+m}(b_{i-1}^n)\right) \right\| < 2^{-n+1}\lambda(I_i^n).$$

Letting $m \to \infty$ here, we get

$$\left\| f_n(b_i^n) - f_n(b_{i-1}^n) - \left(f(b_i^n) - f(b_{i-1}^n)\right) \right\| \leq 2^{-n+1}\lambda(I_i^n). \tag{11.20}$$

This holds for every $n \in \mathbb{N}$ and every $i = 1, \ldots, k_n$.

Now, (iv) applied to our f yields $t \in (0, 1)$, $y \in X$, and $\delta > 0$ such that

$$\left\| \frac{f(s) - f(t)}{s - t} - y \right\| < \frac{\varepsilon}{4} \quad \text{whenever} \quad s \in [0, 1) \quad \text{and} \quad 0 < |s - t| < \delta.$$

Then, for every $s_1 \in (t - \delta, t] \cap [0, 1)$ and every $s_2 \in (t, t + \delta) \cap [0, 1)$ we have

$$\left\| f(s_2) - f(s_1) - (s_2 - s_1)y \right\| \leq \left\| f(s_2) - f(t) - (s_2 - t)y \right\|$$
$$+ \left\| f(s_1) - f(t) - (s_1 - t)y \right\| < \tfrac{\varepsilon}{4}(s_2 - t) + \tfrac{\varepsilon}{4}(t - s_1) = \tfrac{\varepsilon}{4}(s_2 - s_1),$$

and hence

$$\left\| \frac{f(s_2) - f(s_1)}{s_2 - s_1} - y \right\| < \frac{\varepsilon}{4}. \tag{11.21}$$

Find $n \in \mathbb{N}$ so big that $2^{-n+3} < \varepsilon$ and that $2^{-n} < \delta$. Find then $i \in \{1, \ldots, k_n\}$ so that $t \in I_i^n$ $\left(= [b_{i-1}^n, b_i^n)\right)$, and $j \in \{1, \ldots, k_{n+1}\}$ so that $t \in I_j^{n+1}$ $\left(= [b_{j-1}^{n+1}, b_j^{n+1})\right)$. Realizing that

$$f_n(b_i^n) - f_n(b_{i-1}^n) = \int_{b_{i-1}^n}^{b_i^n} g_n = \lambda(I_i^n)x_i^n$$

and

$$f_{n+1}(b_j^{n+1}) - f_{n+1}(b_{j-1}^{n+1}) = \int_{b_{j-1}^{n+1}}^{b_j^{n+1}} g_{n+1} = \lambda(I_j^{n+1})x_j^{n+1},$$

(11.20) yields

$$\left\|\lambda(I_i^n)x_i^n - \left(f(b_i^n) - f(b_{i-1}^n)\right)\right\| < 2^{-n+1}\lambda(I_i^n)$$

and

$$\left\|\lambda(I_j^{n+1})x_j^{n+1} - \left(f(b_j^{n+1}) - f(b_{j-1}^{n+1})\right)\right\| < 2^{-n}\lambda(I_j^{n+1}),$$

that is,

$$\left\|x_i^n - \frac{f(b_i^n) - f(b_{i-1}^n)}{b_i^n - b_{i-1}^n}\right\| < 2^{-n+1} \quad \text{and} \quad \left\|x_j^{n+1} - \frac{f(b_j^{n+1}) - f(b_{j-1}^{n+1})}{b_j^{n+1} - b_{j-1}^{n+1}}\right\| < 2^{-n}.$$

Therefore, using (11.21),

$$\left\|x_i^n - y\right\| < \tfrac{\varepsilon}{4} + 2^{-n+1} \quad \text{and} \quad \left\|x_j^{n+1} - y\right\| < \tfrac{\varepsilon}{4} + 2^{-n},$$

and so $\|x_i^n - x_j^{n+1}\| < \frac{\varepsilon}{2} + 2^{-n+2} < \varepsilon$. But we know that $x_j^{n+1} \in M\backslash B(x_i^n, \varepsilon)$, a contradiction. This finishes the proof of the implication (iv)$\Longrightarrow$(i). □

The next statement will be needed in Chapter 16.

Proposition 11.16 *Let $(X, \|\cdot\|)$ be a Banach space and let $K \subset X^*$ be a weak*-compact set such that every subset of it has weak* slices of arbitrarily small diameter. Let μ be a non-negative regular Borel measure on K. Then for every $\varepsilon > 0$ there is a weak*-closed subset $K_\varepsilon \subset K$ such that $\mu(K\backslash K_\varepsilon) < \varepsilon$ and the inclusion mapping $K_\varepsilon \hookrightarrow K$ is weak*-$\|\cdot\|$-continuous.*

Proof: Fix any $\Delta > 0$. Let $\mathcal{B}$ denote the σ-algebra of Borel subsets of (K, w^*). Let $\mathcal{P}$ denote the collection of all $E \in \mathcal{B}$ whose diameter is less than Δ. Then $\mathcal{P}$ is cofinal in $\mathcal{B}$ since, given any $E \in \mathcal{B}$, there is a weak*-open halfspace $H \subset X^*$ such that $H \cap E$ is nonempty and has diameter less than Δ, and thus $H \cap E \in \mathcal{P}$. Now the principle of exhaustion like in Lemma 11.12 where $\mathcal{L}^+(J)$ is replaced by $\mathcal{B}$ and λ by μ yields an at most countable family $\mathcal{R} \subset \mathcal{P}$ consisting of pairwise disjoint sets, such that $\sum_{E\in\mathcal{R}} \mu(E) = \mu(K)$. Find then a finite $\mathcal{F} \subset \mathcal{R}$ such that $\sum_{E\in\mathcal{F}} \mu(E) > \mu(K) - \Delta$. Now the regularity of μ yields weak*-closed sets $C_E \subset E$, $E \in \mathcal{F}$, such that we still have $\sum_{E\in\mathcal{F}} \mu(C_E) > \mu(K) - \Delta$. Put then $D_\Delta = \bigcup_{E\in\mathcal{F}} C_E$; this is a weak*-closed set, $\mu(D_\Delta) = \sum_{E\in\mathcal{F}} \mu(C_E) > \mu(K) - \Delta$, and for every $x^* \in C_E$ there is a weak*-open set $W \subset X^*$ so that $W \ni x^*$ and $W \cap C_{E'} = \emptyset$ for every $E' \in \mathcal{F}\backslash\{E\}$; hence $\operatorname{diam}(D_\Delta \cap W) < \Delta$.

Now, let $\varepsilon > 0$ be fixed. Put $K_\varepsilon = \bigcap_{n=1}^\infty D_{\varepsilon 2^{-n}}$. Then $\mu(K\backslash K_\varepsilon) = \mu\left(\bigcup_{n=1}^\infty (K \setminus D_{\varepsilon 2^{-n}})\right) < \sum_{n=1}^\infty \varepsilon 2^{-n} = \varepsilon$. Finally, fix any $x^* \in K_\varepsilon$ and any $\gamma > 0$.

Find $n \in \mathbb{N}$ so big that $\varepsilon 2^{-n} < \gamma$. As $x^* \in D_{\varepsilon 2^{-n}}$, there is a weak*-open set W, containing x^*, such that $\operatorname{diam}\left(D_{\varepsilon 2^{-n}} \cap W\right) < \varepsilon 2^{-n}$, and hence $\operatorname{diam}(K_\varepsilon \cap W) < \gamma$. □

11.4 Extension of Rademacher's Theorem

In the previous section, we proved that a Lipschitz function from an interval into a dentable Banach space is almost everywhere differentiable (Theorem 11.15, (iii)). Here, we shall extend this statement for Lipschitz mappings defined first on the *Hilbert cube* $[0, 1]^{\mathbb{N}}$, and then on an arbitrary separable Banach space. We recall here the classical result of Rademacher (see, e.g., [Fede, 3.1.6] or [BoVa, Theorem 2.5.4]) stating that *every Lipschitz function from an open subset of* $\mathbb{R}^n$ *into* $\mathbb{R}$ *is almost everywhere differentiable*. For a simple proof of this, based on Fubini's theorem, we refer to [NeZa].

Below, we shall follow an approach due to Mankiewicz [Mank]. Let $n \in \mathbb{N}$. The (σ-algebra) of Lebesgue measurable sets in $\mathbb{R}^n$ is denoted by $\mathcal{L}^n$ and the corresponding n-dimensional Lebesgue measure by λ_n; see, e.g., [Fede, 2.6], [LuMa, Section 26], or [Rudi2, Paragraphs 2.19–2.21]. We recall that $\mathcal{L}^n$ contains the σ-algebra of Borel sets in $\mathbb{R}^n$. Note that λ_1 was formerly denoted by λ.

Lemma 11.17 *Let* $n \in \mathbb{N}\backslash\{1\}$, *let* $N \in \mathcal{L}^n$, *and assume that there exists a vector* $h = (h_1, \ldots, h_n) \in \mathbb{R}^n$, *with* $\|h\|_{\ell_2} = 1$, *such that* $\lambda_1\left(\{s \in \mathbb{R} : x + sh \in N\}\right) = 0$ *for every* $x \in \mathbb{R}^n$. *Then* $\lambda_n(N) = 0$.

Proof: The Gram–Schmidt orthogonalization process yields vectors $w_2, \ldots, w_n$ in $\mathbb{R}^n$ such that $h, w_2, \ldots, w_n$ are mutually orthogonal. The matrix A with rows $h, w_2, \ldots, w_n$ defines a linear isometry (denoted again by A) from $\mathbb{R}^n$ onto $\mathbb{R}^n$ such that $Ah = (1, 0, \ldots, 0)$. (If $n = 2$, we can define $Au = (h_1u_1 + h_2u_2, -h_2u_1 + h_1u_2)$ for $u = (u_1, u_2) \in \mathbb{R}^2$.) Then for every $v = (v_2, \ldots, v_n) \in \mathbb{R}^{n-1}$ we have

$$\begin{aligned}\{s \in \mathbb{R} : (s, v) \in A(N)\} &= \{s \in \mathbb{R} : (0, v) + sAh \in A(N)\} \\ &= \left\{s \in \mathbb{R} : A^{-1}(0, v) + sh \in N\right\}.\end{aligned}$$

By the assumption, the latter set is Lebesgue-negligible. Then Fubini's theorem yields

$$\lambda_n\big(A(N)\big) = \int \lambda_1\big(\{s \in \mathbb{R} : (s, v) \in A(N)\}\big) \mathrm{d}\lambda_{n-1}(v) = 0$$

and [Rudi2, Theorem 8.26 (d)] implies that $\lambda_n(N) = \left|\det A^{-1}\right| \lambda_n\big(A(N)\big)$ $(= 0)$. □

Given a Banach space $(X, \|\cdot\|)$, a set $M \subset X$ is called *porous* if there exists $\Delta \in (0, 1)$ such that for every $x \in M$ and every $r > 0$ there is $\widetilde{x} \in X$ such that $\|\widetilde{x} - x\| \leq r$ and $B(\widetilde{x}, \Delta\|\widetilde{x} - x\|) \cap M = \emptyset$. The set M is called *σ-porous* if it can be written as the union of countably many porous sets. It is easy to check that porous sets are nowhere dense, hence σ-porous sets are of the first Baire category. Also, if $n \in \mathbb{N}$ and a set $M \in \mathcal{L}^n$ is porous, then $\lambda_n(M) = 0$. Indeed, assume that $\lambda_n(M) > 0$. Lebesgue's density theorem ([Rudi2, p. 189, exercise], [Fede, 2.9.13], or [LuMa, Theorem 29.2]) then yields an $x \in M$ such that

$$\lim_{r \downarrow 0} \frac{\lambda_n\big(M \cap B(x,r)\big)}{\lambda_n\big(B(x,r)\big)} = 1.$$

Let $\Delta > 0$ witness for the porosity of M. Find $r > 0$ so small that

$$\lambda_n\big(M \cap B(x,s)\big) > \Big(1 - \frac{\Delta^n}{2^n}\Big)\lambda_n(B(x,s)) \quad \text{whenever} \quad 0 < s < r.$$

Find $\widetilde{x} \in X$ such that $\|\widetilde{x} - x\| < \frac{r}{2}$ and $B(\widetilde{x}, \Delta\|\widetilde{x} - x\|) \cap M = \emptyset$. Then

$$\begin{aligned}
\Big(1 - \frac{\Delta^n}{2^n}\Big)\lambda_n\big(B(x, 2\|\widetilde{x} - x\|)\big) &< \lambda_n\big(M \cap B(x, 2\|\widetilde{x} - x\|)\big) \\
&\le \lambda_n\big(B(x, 2\|\widetilde{x} - x\|)\backslash B(\widetilde{x}, \Delta\|\widetilde{x} - x\|)\big) \\
&= \Big(1 - \frac{\Delta^n}{2^n}\Big)\lambda_n\big(B(x, 2\|\widetilde{x} - x\|)\big),
\end{aligned}$$

a contradiction. Therefore $\lambda_n(M) = 0$.

Given Banach spaces X, Y and a mapping g from an open subset U of X into Y, the *directional derivative* of g at $x \in U$ in a direction $h \in X$ is defined as

$$Dg(x)h = \lim_{s \to 0} \frac{g(x + sh) - g(x)}{s}$$

if the latter limit, in the norm topology of Y, exists; see Definition 7.2.

Proposition 11.18 (Preiss-Zajíček [PrZa]) *Let U be a nonempty open subset of a separable Banach space X, let Y be any Banach space, let $g : U \to Y$ be a Lipschitz mapping, and let $h_1, h_2 \in X$ be two non-zero vectors. Then the set M of all $x \in U$ such that the directional derivatives $Dg(x)h_1$, $Dg(x)h_2$, $Dg(x)(h_1+h_2)$ exist and yet $Dg(x)(h_1 + h_2) \neq Dg(x)h_1 + Dg(x)h_2$ is σ-porous.*

Proof: Let $L > 0$ be a Lipschitz constant of g. Let $Z \subset Y$ be a countable dense subset of the linear span of $g(U)$; this is possible since the latter set is separable. For $m, j \in \mathbb{N}$ and for $z_1, z_2 \in Z$ let M_{m,j,z_1,z_2} denote the set of all $x \in U$ such that

$$\Big\| \frac{g\big(x + s(h_1 + h_2)\big) - g(x)}{s} - z_1 - z_2 \Big\| > \frac{3}{m} \quad \text{and} \tag{11.22}$$

$$\Big\| \frac{g(x + sh_i) - g(x)}{s} - z_i \Big\| < \frac{1}{m}, \quad i = 1, 2, \tag{11.23}$$

whenever $0 < s \le \frac{1}{j}$.

We shall prove that all these sets are porous. To this end, fix any quadruple $(m, j, z_1, z_2) \in \mathbb{N}^2 \times Z^2$. Put $\Delta = \frac{1}{2Lm\|h_1\|}$. Fix any $x \in M_{m,j,z_1,z_2}$ and any $r > 0$. Put $s = \min\{\frac{1}{j}, \frac{r}{\|h_1\|}\}$. We shall first show that $B(x + sh_1, \frac{s}{2Lm}) \cap M_{m,j,z_1,z_2} = \emptyset$. In order to prove this, take any $w \in X$ with $\|w\| \le \frac{s}{2Lm}$. We are ready to estimate

$$
\begin{aligned}
&\Big\| \frac{g\big((x + sh_1 + w) + sh_2\big) - g(x + sh_1 + w)}{s} - z_2 \Big\| \\
&\ge \Big\| \frac{g\big(x + s(h_1 + h_2)\big) - g(x + sh_1)}{s} - z_2 \Big\| - 2L\frac{\|w\|}{s} \\
&= \Big\| \Big(\frac{g\big(x + s(h_1 + h_2)\big) - g(x)}{s} - z_1 - z_2 \Big) - \Big(\frac{g(x + sh_1) - g(x)}{s} - z_1 \Big) \Big\| - 2L\frac{\|w\|}{s} \\
&> \frac{3}{m} - \frac{1}{m} - 2L\frac{\|w\|}{s} \ge \frac{2}{m} - 2L\frac{1}{2Lm} = \frac{1}{m},
\end{aligned}
$$

and hence $x + sh_1 + w \notin M_{m,j,z_1,z_2}$. Therefore $B\big(x + sh_1, \frac{s}{2Lm}\big) \cap M_{m,j,z_1,z_2} = \emptyset$. Now, put $\widetilde{x} = x + sh_1$; then $\|\widetilde{x} - x\| \le r$. Also $\Delta\|\widetilde{x} - x\| = \frac{1}{2Lm\|h_1\|} \cdot s\|h_1\| = \frac{s}{2Lm}$. Therefore, $B(\widetilde{x}, \Delta\|\widetilde{x} - x\|) \cap M_{m,j,z_1,z_2} = \emptyset$, and the porosity of M_{m,j,z_1,z_2} was proved.

It remains to prove that $M \subset \bigcup\{M_{m,j,z_1,z_2} : (m, j, z_1, z_2) \in \mathbb{N}^2 \times Z^2\}$. So, fix any $x \in M$. Find $m \in \mathbb{N}$ such that $\|Dg(x)(h_1+h_2) - Dg(x)h_1 - Dg(x)h_2\| > \frac{5}{m}$. Find $z_1, z_2 \in Z$ such that

$$
\|Dg(x)h_1 - z_1\| < \tfrac{1}{m} \quad \text{and} \quad \|Dg(x)h_2 - z_2\| < \tfrac{1}{m}\,.
$$

Then $\|Dg(x)(h_1 + h_2) - z_1 - z_2\| > \frac{3}{m}$. From this, we can find $j \in \mathbb{N}$ so big that (11.22) and (11.23) hold whenever $0 < s \le \frac{1}{j}$. Then, obviously, $x \in M_{m,j,z_1,z_2}$. □

For studying the Gâteaux differentiability of Lipschitz mappings with infinite-dimensional domain, we shall need some tools from measure theory and integration on the Hilbert cube. Let $n \in \mathbb{N}$. From now on the symbol $\mathcal{L}^n$ will denote the σ-algebra of Lebesgue measurable subsets of $[0, 1]^n$. Consider the Hilbert cube $[0, 1]^{\mathbb{N}}$ and endow it with the product topology; it becomes a compact space. Let $\mathcal{C}$ denote the family of all "cylinders" $E \times [0, 1]^{\mathbb{N}}$, where $E \in \mathcal{L}^n$ and $n \in \mathbb{N}$. We shall always assume that these cylinders are identified with subsets of $[0, 1]^{\mathbb{N}}$ via the injections $E \times [0, 1]^{\mathbb{N}} \ni (e, t) \longmapsto (e_1, \ldots, e_n, t_1, t_2, \ldots) \in [0, 1]^{\mathbb{N}}$. Clearly, $[0, 1]^{\mathbb{N}} \in \mathcal{C}$, $[0, 1]^{\mathbb{N}} \setminus C \in \mathcal{C}$ if $C \in \mathcal{C}$, and also $C_1 \cup C_2 \in \mathcal{C}$ if $C_1, C_2 \in \mathcal{C}$. This means that $\mathcal{C}$ is an algebra of subsets of $[0, 1]^{\mathbb{N}}$. Define $\nu : \mathcal{C} \to [0, 1]$ as $\nu(C) = \lambda_n(E)$ if C is of the form $E \times [0, 1]^{\mathbb{N}}$ and $E \in \mathcal{L}^n$. Fubini's theorem ([Rudi2, Theorem 7.8], or [LuMa, Theorem 26.9]) immediately guarantees that ν is well defined. Clearly, $\nu(\emptyset) = 0$, $\nu([0, 1]^{\mathbb{N}}) = 1$, and $\nu(C_1 \cup C_2) = \nu(C_1) + \nu(C_2)$ whenever $C_1, C_2 \in \mathcal{C}$ are disjoint. Let Σ be the (unique) smallest σ-algebra of subsets of $[0, 1]^{\mathbb{N}}$ containing $\mathcal{C}$; it is simple to verify that such a Σ exists. The very definition of the topology on $[0, 1]^{\mathbb{N}}$ shows that every open subset of it can be obtained as the union of an at most countable family of elements of $\mathcal{C}$. Thus

Σ contains all open subsets, and hence also all Borel subsets of $[0,1]^{\mathbb{N}}$. We shall check that ν is regular. To this end, consider any $\varepsilon > 0$ and any $C \in \mathcal{C}$. Find $n \in \mathbb{N}$ and $E \in \mathcal{L}^n$ so that $C = E \times [0,1]^{\mathbb{N}}$. The measure λ_n is regular, see [Rudi2, Theorem 2.20] or [LuMa, Theorem 26.1]. Hence, there is a closed set $F \subset E$ such that $\lambda_n(E\setminus F) < \varepsilon$. Then $F \times [0,1]^{\mathbb{N}}$ is a closed subset of $[0,1]^{\mathbb{N}}$ and

$$\nu\big(E \times [0,1]^{\mathbb{N}} \setminus (F \times [0,1]^{\mathbb{N}})\big) = \nu\big((E\setminus F) \times [0,1]^{\mathbb{N}}\big) = \lambda_n(E\setminus F) < \varepsilon.$$

Once ν is regular, a theorem of A.D. Alexandrov [DuSc, Theorem III.5.13] says that ν is σ-additive on the algebra $\mathcal{C}$, that is, for every infinite sequence $C_1, C_2, \ldots$ of pairwise disjoint elements of $\mathcal{C}$, with $\bigcup_{i=1}^{\infty} C_i \in \mathcal{C}$, we have $\nu\big(\bigcup_{i=1}^{\infty} C_i\big) = \sum_{i=1}^{\infty} \nu(C_i)$. Having at hand the σ-additive function $\nu : \mathcal{C} \to [0,1]$, Hahn's extension theorem [DuSc, Theorem III.5.5] provides a (unique) measure $\mu : \Sigma \to [0,1]$ such that $\mu(C) = \nu(C)$ for every $C \in \mathcal{C}$. A function $f : [0,1]^{\mathbb{N}} \to \mathbb{R}$ is called *μ-measurable* if there is $S \in \Sigma$ such that $\mu(S) = 1$ and $f^{-1}([a,+\infty)) \cap S \in \Sigma$ for every $a \in \mathbb{R}$. Having the measure μ on $([0,1]^{\mathbb{N}}, \Sigma)$, we can build the Lebesgue integral, denoted by $\int f(t)\mathrm{d}\mu(t)$, or just by $\int f\mathrm{d}\mu$, see [Rudi2, Chapter 1] and Section 17.13.1. In particular, $\int f\mathrm{d}\mu$ makes sense and is finite provided that $f : [0,1]^{\mathbb{N}} \to \mathbb{R}$ is any μ-measurable and bounded function.

For $S \in \Sigma$, $t \in [0,1]^{\mathbb{N}}$, and $n \in \mathbb{N}$, we put $S^t = \{x \in [0,1]^n : (x,t) \in S\}$. Now, we are ready to present a Fubini-like theorem.

Proposition 11.19 *Let $n \in \mathbb{N}$ and $S \in \Sigma$. Then:*

(i) *there exists $N \in \Sigma$, with $\mu(N) = 0$, such that $S^t \in \mathcal{L}^n$ for every $t \in [0,1]^{\mathbb{N}}\setminus N$;*

(ii) *the function $[0,1]^{\mathbb{N}} \ni t \longmapsto \lambda_n(S^t)$ is μ-measurable; and*

(iii) *$\mu(S) = \int \lambda_n(S^t)\mathrm{d}\mu(t)$.*

Proof: Denote by $\mathcal{A}$ the set of all $S \in \Sigma$ that satisfy (i), (ii), and (iii). We shall first show that $\mathcal{C} \subset \mathcal{A}$. So fix any $C \in \mathcal{C}$. Find $k \in \mathbb{N}$ and $E \in \mathcal{L}^{n+k}$ so that $C = E \times [0,1]^{\mathbb{N}}$. (Yes, such a k always exists.) According to Fubini's theorem applied for the product $([0,1]^n, \mathcal{L}^n, \lambda_n) \times ([0,1]^k, \mathcal{L}^k, \lambda_k)$, we know that the set $E^{(t_1,\ldots,t_k)} := \{x \in [0,1]^n : (x, t_1, \ldots, t_k) \in E\}$ belongs to $\mathcal{L}^n$ for λ_k-almost all $(t_1, \ldots, t_k) \in [0,1]^k$, that the mapping $[0,1]^k \ni (t_1, \ldots, t_k) \longmapsto \lambda_n\big(E^{(t_1,\ldots,t_k)}\big)$ is $\mathcal{L}^k$-measurable, and that $\lambda_{n+k}(E) = \int \lambda_n\big(E^{(t_1,\ldots,t_k)}\big)\mathrm{d}\lambda_k(t_1, \ldots, t_k)$. Put $M = \{(t_1, \ldots, t_k) \in [0,1]^k : E^{(t_1,\ldots,t_k)} \notin \mathcal{L}^k\}$; then $\lambda_k(M) = 0$. Hence, denoting $N = M \times [0,1]^{\mathbb{N}}$, we have $N \in \Sigma$, $\mu(N) = 0$, and so $C^t = E^{(t_1,\ldots,t_k)} \in \mathcal{L}^n$ for every $t \in [0,1]^{\mathbb{N}}\setminus N$. Further, for every $a \geq 0$ we have $\{t \in [0,1]^{\mathbb{N}} : \lambda_n(C^t) \geq a\} = \{(t_1, \ldots, t_k) \in [0,1]^k : \lambda_n\big(E^{(t_1,\ldots,t_k)}\big) \geq a\} \times [0,1]^{\mathbb{N}} \in \mathcal{C} \subset \Sigma$, and so (ii) is also satisfied. Finally,

$$\begin{aligned}\mu(C) = \lambda_{n+k}(E) &= \int \lambda_n\big(E^{(t_1,\ldots,t_k)}\big)\mathrm{d}\lambda_k(t_1, \ldots, t_k) \\ &= \int \lambda_n\big(E^{(t_1,\ldots,t_k)}\big)\mathrm{d}\mu(t_1, t_2, \ldots) = \int \lambda_n(C^t)\mathrm{d}\mu(t).\end{aligned}$$

Here, we used the equality

$$\int \varphi(t_1, \ldots, t_k) \mathrm{d}\lambda_k(t_1, \ldots, t_k) = \int \varphi(t_1, \ldots, t_k) \mathrm{d}\mu(t_1, t_2 \ldots)$$

valid for all $\mathcal{L}^k$-measurable functions $\varphi : [0, 1]^k \to [0, 1]$. We observe that this equality is surely true for simple functions of form $a_1\chi_{E_i} + \cdots + a_m\chi_{E_m}$, $a_i \geq 0$, $E_i \in \mathcal{L}^k$, $i = 1, \ldots, m$. For a general φ we find a sequence $0 \leq s_1 \leq s_2 \leq \cdots \leq \varphi$ of simple functions, converging to φ and then use [Rudi2, Theorem 1.26].) This way we verified (i)–(iii) for our C and thus we proved that $\mathcal{C} \subset \mathcal{A}$.

It remains to show that $\mathcal{A}$ is a σ-algebra. If $S \in \mathcal{A}$, then $([0, 1]^{\mathbb{N}} \setminus S)^t = [0, 1]^n \setminus S^t$ for all $t \in [0, 1]^{\mathbb{N}}$, and thus we can easily verify that $S^t \in \mathcal{A}$. Further, consider an infinite sequence $S_1 \subset S_2 \subset \cdots$ of elements of $\mathcal{A}$ and put $S = \bigcup_{i=1}^{\infty} S_i$. Then $S^t = \bigcup_{i=1}^{\infty} S_i{}^t$, and hence (i) holds. (ii) is valid since the limit of a sequence of μ-measurable functions is also μ-measurable. Now, for $i \to \infty$ we have $\mu(S_i) \uparrow \mu(S)$, $\lambda_n(S_i{}^t) \uparrow \lambda_n(S^t)$ for μ-almost all $t \in [0, 1]^{\mathbb{N}}$, and, by [Rudi2, Theorem 1.26], $\int \lambda_n(S_i{}^t) \mathrm{d}\mu(t) \uparrow \int \lambda_n(S^t) \mathrm{d}\mu(t)$. (iii) is thus verified, and hence $S \in \mathcal{A}$. Now, knowing that $\mathcal{A}$ is a σ-algebra, it must contain Σ. □

Let H denote the (countable) set of all vectors $h = (h_i) \in \mathbb{R}^{\mathbb{N}}$, with rational entries, and such that $h_n = h_{n+1} = \cdots = 0$ for some $n \in \mathbb{N}$.

Lemma 11.20 *Let Y be a dentable Banach space and let $f : [-1, 2]^{\mathbb{N}} \to Y$ be a mapping such that $\|f(t) - f(s)\| \leq \sum_{i=1}^{\infty} 2^{-i}|t_i - s_i|$ for every $t, s \in [-1, 2]^{\mathbb{N}}$. Then there exists a Borel set $\Omega \subset [0, 1]^{\mathbb{N}}$ such that $\mu([0, 1]^{\mathbb{N}} \setminus \Omega) = 0$ and for every $t \in \Omega$ the mapping $H \ni h \longmapsto Df(t)h \in Y$ is well defined and additive.*

Proof: For $h \in H$ let M_h denote the set of all $t \in [0, 1]^{\mathbb{N}}$ where $Df(t)h$ exists. Fix for a while any $0 \neq h \in H$. For $i \in \mathbb{N}$ and for $r, s \in (-1/\|h\|_\infty, 0) \cup (0, 1/\|h\|_\infty)$ put

$$A_{i,r,s} = \Big\{t \in [0, 1]^{\mathbb{N}} : \tfrac{1}{r}\big(f(t + rh) - f(t)\big) - \tfrac{1}{s}\big(f(t + sh) - f(t)\big) < \tfrac{1}{i}\Big\};$$

this is a Borel subset of $[0, 1]^{\mathbb{N}}$ since the mapping f is continuous on the space $[-1, 2]^{\mathbb{N}}$. We observe that

$$M_h = \bigcap_{i=1}^{\infty} \bigcup_{j=1}^{\infty} \bigcap \Big\{A_{i,r,s} : r, s \text{ are rational}, \ rs > 0, \text{ and } |r|, |s| < \frac{1}{j\|h\|_\infty}\Big\}.$$

Therefore, M_h is also a Borel set; thus $M_h \in \Sigma$. Find $n \in \mathbb{N}$ so big that $h_{n+1} = h_{n+2} = \cdots = 0$ and put then $k = (h_1, \ldots, h_n)$. Fix for a while any $t \in [0, 1]^{\mathbb{N}}$ and define $g : [-1, 2]^n \to Y$ by $g(x) = f(x, t)$, $x \in [-1, 2]^n$. Fix for a while any $x \in [0, 1]^n$ and put $J = \{s \in \mathbb{R} : x + sk \in [0, 2]^n\}$; this is a nonempty interval. Define $\varphi : J \to Y$ by $\varphi(s) = g(x + sk)$, $s \in J$. It is easy to check that φ is a Lipschitz function. We observe that, given any $s \in J$, then the derivative $\varphi'(s)$ exists if and only if the directional derivative $Dg(x+sk)k$ exists, if and only if $Df(x+sk,t)h$ exists, if and only if $(x+sk,t) \in M_h$, if and only if $x+sk \in \left(M_h\right)^t$. Now, Theorem 11.15 guarantees that $\lambda_1\left(\{s \in J : \varphi'(s) \text{ does not exist}\}\right) = 0$. Hence $\lambda_1\left(\{s \in J : x+sk \in [0,1]^n \setminus \left(M_h\right)^t\}\right) = 0$. This holds for every $x \in [0,1]^n$. Lemma 11.17 then yields that $\lambda_n\left([0,1]^n \setminus (M_h)^t\right) = 0$. Therefore $\lambda_n\left(([0,1]^{\mathbb{N}} \setminus M_h)^t\right) = 0$, and finally, Proposition 11.19 gives that $\mu\left([0,1]^{\mathbb{N}} \setminus M_h\right) = 0$.

Put $M = \bigcap_{h \in H} M_h$; this is a Borel set and $\mu\left([0,1]^{\mathbb{N}} \setminus M\right) = 0$ as well. For a fixed $h \in H$ we have that $Df(t)h = \lim\limits_{m \to \infty} m\left(f(t + \frac{1}{m}h) - f(t)\right)$ for every $t \in M$; hence the mapping $M \ni t \longmapsto Df(t)h \in Y$ is Borel. Therefore, if for $h, k \in H$ we put $\Omega_{h,k} = \{t \in M : Df(t)h + Df(t)k = Df(t)(h+k)\}$ (recall that $h + k \in H$), this will be a Borel set. Fix any $h, k \in H$. Find $n \in \mathbb{N}$ so big that $h_{n+1} = k_{n+1} = h_{n+2} = k_{n+2} = \cdots = 0$ and put $u = (h_1, \ldots, h_n)$, $v = (k_1, \ldots, k_n)$. Fix any $t \in [0,1]^{\mathbb{N}}$ and define $g : [-1,2]^n \to Y$ by $g(x) = f(x,t)$, $x \in [-1,2]^n$; this is a Lipschitz mapping. Then $x \in \left(M \setminus \Omega_{h,k}\right)^t$, if and only if $Df(x,t)h$, $Df(x,t)k$, $Df(x,t)(h+k)$ exist and $Df(x,t)h + Df(x,t)k \neq Df(x,t)(h+k)$, if and only if $Dg(x)u$, $Dg(x)v$, $Dg(x)(u+v)$ exist and $Dg(x)u + Dg(x)v \neq Dg(x)(u+v)$. Then, by Proposition 11.18 $\left(M \setminus \Omega_{h,k}\right)^t$ is a σ-porous set in $\mathbb{R}^n$, and hence $\lambda_n\left(\left(M \setminus \Omega_{h,k}\right)^t\right) = 0$. Here we profited from the fact that $\left(M \setminus \Omega_{h,k}\right)^t$ is a Borel set and hence belongs to $\mathcal{L}^n$. Now, Proposition 11.19 yields that $\mu\left(M \setminus \Omega_{h,k}\right) = 0$. Put $\Omega = \bigcap_{h,k \in H} \Omega_{h,k}$; this is still a Borel set. Then $\mu(M \setminus \Omega) = 0$ and finally $\mu\left([0,1]^{\mathbb{N}} \setminus \Omega\right) = \mu\left([0,1]^{\mathbb{N}} \setminus M\right) + \mu(M \setminus \Omega) = 0 + 0 = 0$. We have proved that the mapping $H \ni h \longmapsto Df(t)h \in Y$ is well defined and additive for every $t \in \Omega$. □

Theorem 11.21 (Aronszajn [Aro], Christensen [Chri], Mankiewicz [Mank]) *Let U be a nonempty open subset of a separable Banach space X, let Y be a dentable Banach space, and let $F : U \to Y$ be a Lipschitz mapping. Then F is Gâteaux differentiable at densely many points of U.*

Proof: Clearly, it will be enough to find at least one point in U where F is Gâteaux differentiable. We may and do assume that $2B_X \subset U$. Pick $e_1, e_2, \ldots \in B_X$ such that $\overline{\{e_1, e_2, \ldots\}} = B_X$. Define $\varphi : \ell_\infty \to X$ by $\varphi(t) = \sum_{i=1}^{\infty} 2^{-i} t_i e_i$, $t = (t_1, t_2, \ldots) \in \ell_\infty$. Clearly, this mapping is well defined, linear, and $\|\varphi(t)\| \leq \|t\|_\infty$ for every $t \in \mathrm{l}_\infty$. Define $f : [-1,2]^{\mathbb{N}} \to Y$ by $f(t) = F\left(\varphi(t)\right)$, $t \in [-1,2]^{\mathbb{N}}$. If L denotes a Lipschitz constant of F, then $\|f(t) - f(s)\| \leq L \sum_{i=1}^{\infty} 2^{-i} |t_i - s_i|$ for every $t, s \in [-1,2]^{\mathbb{N}}$.

Let Ω be the (nonempty) set found in Lemma 11.20 for our f. Pick one $t \in \Omega$ and denote $x = \varphi(t)$. Then $\|x\| \leq 1$ and so $x \in U$. We shall show that the mapping F is Gâteaux differentiable at x. Indeed, we can immediately see that the derivative $DF(x)\left(\varphi(h)\right) = Df(t)h$ for every $h \in H$. Moreover

$$
\begin{aligned}
DF(x)\left(\varphi(h) + \varphi(k)\right) &= DF(x)\left(\varphi(h+k)\right) = Df(t)(h+k) \qquad (11.24)\\
&= Df(t)h + Df(t)k = DF(x)\left(\varphi(h)\right) + DF(x)\left(\varphi(k)\right)
\end{aligned}
$$

for every $h, k \in H$. Now, realizing that $\varphi(H)$ is dense in X, and that F is Lipschitz, we get that for every $u \in X$ the directional derivative $DF(x)u$ exists and can be obtained as $\lim_{n\to\infty} DF(x)\big(\varphi(h_n)\big)$ for any sequence $h_1, h_2, \ldots$ from H such that $\lim_{n\to\infty} \|\varphi(h_n) - u\| = 0$. Finally, using this and (11.24), we can conclude that $DF(x)(u + v) = DF(x)u + DF(x)v$ for every $u, v \in X$. □

11.5 Remarks and Open Problems

Remarks

1. For a more detailed account on the dentability, Asplund spaces, and the Radon–Nikodým property, together with corresponding references, we recommend to read the books of Phelps [Phelps], Diestel and Uhl, Jr. [DiUh], Bourgain [Bou], and of Bourgin [Bour].
2. While the proof that every separable dual Banach space has the Radon–Nikodým property goes back to Dunford and Pettis, the converse was proved for the first time by Stegall in [Stegl]. (We get here Stegall's result by combining Theorems 11.8 and 11.15). In [GhMa1] and [GhMa2] it is shown that every infinite-dimensional Banach space with the Radon–Nikodým property contains a subspace isometric to a separable infinite-dimensional dual Banach space. In this direction we refer to [McOb], where it is constructed a separable Banach space with the Radon–Nikodým property that is not isomorphic to a subspace of a separable dual space.
3. If X does not have the Radon–Nikodým property, it does not mean that it necessarily contains a bounded infinite ε-dyadic tree [BoRo], but it does contain a so-called infinite bounded ε-bush for some $\varepsilon > 0$, see [BeLi, p. 111].
4. ([SchSerWer]) If a Banach space X fails the Radon–Nikodým property, then for every $\varepsilon > 0$, there is a closed convex subset C of X with $\operatorname{diam} C = 1$ and $\operatorname{diam} S > 1 - \varepsilon$ for any slice S of C.
5. A seminal paper for the development of the RNP property was [Rieff].
6. The Asplund property as well as the RNP property are three-space properties, see, e.g., [CasGon].
7. If a separable space contains an isomorphic copy of ℓ_1 then its dual space contains a w^*-compact convex nondentable subset, see [SchSerWer].

Open Problems

1. A Banach space X is said to have the *Krein–Milman property* if each nonempty bounded closed convex subset of it has an extreme point. This is the same as to say that every closed convex and bounded subset of X is the closed convex hull of the set of its extreme points, see Exercise 7.57. According to Theorem 11.3, dentable spaces have the Krein–Milman property. The converse implication is an

open problem. Huff and Morris [HuMo1], using Stegall's construction in [Stegl], proved that if X is separable and X^* is non-separable, then X^* lacks even the Krein–Milman property.

2. Assume that the norm of a separable Banach space X is such that the restriction of it to every subspace of X is Fréchet differentiable at a point. Must X^* be separable?
3. Does every Asplund space admit a Fréchet smooth bump? Recall that that an Asplund space may not even admit any equivalent Gâteaux smooth norm [Hayd1], [Hayd3].
4. [ArMe1] Assume that, for a Banach space X, the closed dual unit ball B_{X^*} in the weak* topology is Corson compact (see Definition 14.40). Is every continuous convex function on X Gâteaux differentiable at some point?
5. Is the set of all Fréchet smooth equivalent norms on $C[0, \omega_1]$ dense in the set of all equivalent norms on this space?
6. Assume that a Banach space X is such that every Gâteaux smooth continuous convex function on X is Fréchet differentiable at some point. Is the same true for all Gâteaux differentiable Lipschitz functions on X?
7. [DGZ3, p. 176] Suppose that f is a convex continuous function defined on a separable Hilbert space X. Do there exist $x, y \in X$ and a continuous bilinear form B on X such that for all $h \in X$ we have

$$f(x + th) - f(x) - \langle y, th\rangle - t^2 B(h, h) = o(t^2) \quad \text{as} \quad t \to 0\,? \qquad (11.25)$$

Alexandrov's theorem gives a positive answer to this question in finite-dimensional spaces ([Alex], see also [Rock], or [BoVa, Theorem 2.6.4]). Recall that Alexandrov's theorem states that *a convex continuous function in $\mathbb{R}^n$ has second order expansions almost everywhere*, that is, for almost every $x \in \mathbb{R}^n$, there exists $p \in \mathbb{R}^n$ and a bilinear form B or $\mathbb{R}^n \times \mathbb{R}^n$ such that $f(x + h) = f(x) + \langle p, h\rangle + B(h, h) + o(\|h\|^2)$ as $h \to 0$. We refer to [BoNo]. Here, among other things, examples are discussed to show that, in general, no point can be found where the limit involved in (11.25) is uniform for $h \in S_X$ if $X = \ell_2(\mathbb{N})$. It is shown here that our question has a negative answer if $X = \ell_2(\Gamma)$ and Γ is uncountable.

8. Is the renorming by an equivalent Fréchet smooth norm a three space property?
9. It is an open problem whether a separable space must contain ℓ_1 if X^* has a w^*-compact convex non- dentable subset [SchSerWer].

Exercises for Chapter 11

11.1 The spaces c_0 and L_1 are not dentable.

Hint. Let B denote the unit ball of either of the spaces above with respect to their canonical norms. Assume that there exists a slice S of B with diameter less than

$\frac{1}{2}$. Pick $x \in S$, with $\|x\| = 1$. Find $x_1, x_2 \in B$ such that $x = \frac{1}{2}(x_1 + x_2)$, and $\|x_1 - x\| > \frac{1}{2}$. Then either $x_1 \in S$ or $x_2 \in S$, a contradiction.

11.2 Find a nonreflexive Banach space which is simultaneously dentable and Asplund.
Hint. James' space, Definition 4.43.

11.3 Deduce the Bishop–Phelps Theorem 7.41 from Lemma 11.2.
Hint. Given a closed, convex and bounded subset C of X, put $f := p_{C^\circ}$, the Minkowski functional of C°, and use Lemma 7.19 together with Lemma 11.2.

11.4 Using Theorem 3.92, reprove the last statement in Lemma 11.2.
Hint. Assume that f is Fréchet differentiable at $x^* \in U$. Put $x = f'(x^*)$ and find $\delta > 0$ so small that $x^* + \delta B_{X^*} \subset U$. Observe that for $n \in \mathbb{N}$ the functions $\delta B_{X^*} \ni h^* \longmapsto n\big(f(x^* + \frac{1}{n}h^*) - f(x^*)\big)$ are weak*-lower semicontinuous and for $n \to \infty$ they converge uniformly to $x\big|_{\delta B_{X^*}}$. Thus $x\big|_{\delta B_{X^*}}$ is also weak*-lower semicontinuous. And since x is linear, $x\big|_{\delta B_{X^*}}$ is weak*-continuous. Thus, by the aforementioned theorem, the whole x is weak*-continuous, and hence must belong to X.

11.5 In Theorem 11.3, the condition (ii) can be weakened to:
(ii') *Every convex weak*-lower semicontinuous function $f : U \to \mathbb{R}$, with $\partial f(U) \cap X \subset W$, is Fréchet differentiable at least at one point;*
and (iii) can be weakened to:
(iii') *Every nonempty closed convex subset of W has at least one strongly exposed point.*
Hint. Check the proof of Theorem 11.3.

11.6 (Phelps) There exists a closed subset A of the open unit ball of c_0 (in its maximum norm) so that $\overline{\text{conv}}\, A$ equals to the closed unit ball of c_0.
Hint. Put

$$A = \left\{\frac{n}{n+1}(\varepsilon_1, \ldots, \varepsilon_n, 0, 0, \ldots) : n \in \mathbb{N}, (\varepsilon_1, \ldots, \varepsilon_n) \in \{-1, 1\}^n\right\}.$$

We remark that Huff and Morris constructed, in every non-dentable Banach space, an equivalent norm and a set A having the same properties, see [HuMo2] and [Bour, Theorem 3.7.8].

11.7 Show that a Banach space X is not dentable if and only if there is a bounded closed set A in X such that no $f \in S_{X^*}$ attains its supremum on A.
Hint. If X is not dentable, then use the remark in Exercise 11.6. If X is dentable, apply Theorem 11.3 (iii) to the set $M := \overline{\text{conv}}\, A$.

11.8 If X is not dentable, show that there is an equivalent norm on X whose closed unit ball is not dentable.
Hint. If $M \subset X$ is a bounded closed set whose each nonempty slice has diameter greater than some fixed $\varepsilon > 0$, then the closure of the set $\operatorname{conv}(M \cup -M) + B_X$ is the closed unit ball of the desired equivalent norm.

11.9 Concerning Theorem 11.6, show that *there exists $\varepsilon > 0$ such that the set of points $x_0^* \in \varepsilon B_{X^*}$ for which there is $x_0 \in X$ such that* (11.3) *holds is actually G_δ dense in εB_{X^*}.*
Hint. Observe that the points x_0^* are the points of Fréchet differentiability of a convex weak*-lower semicontinuous function; use then Theorem 11.3.

11.10 For the proof of the sufficiency in Theorem 11.6, we do not need the strong attainment of $\inf(f - x_0^*)$ at x_0; actually, the (not necessarily strong) attainment is enough.
Hint. This fact follows from a result of Huff and Morris [HuMo2]; for more details see [Steg4b, Theorem 16] or [FabFin].

11.11 Prove the following statement, close to the original formulation of Stegall's variational principle: [Steg2] *X is dentable (if and) only if for every closed bounded set $D \subset X$ and for every proper lower semicontinuous function $f : D \longrightarrow (-\infty, +\infty]$ the set of all $x_0^* \in X^*$ such that the function $f - x_0^*$ attains its infimum strongly at some point of D is dense G_δ in X^*.*
Hint. Define $\tilde{f} : X \to (-\infty, +\infty]$ by $\tilde{f}(x) = f(x)$ if $x \in D$ and $\tilde{f}(x) = +\infty$ if $x \in X \setminus D$. Use then Theorem 11.6.

11.12 If X is a dual dentable Banach space, then in Theorem 11.6, the functional x_0^* can be found in the predual of X.
Hint. Read the proof of Theorem 11.6, see also the proof of [FHHMPZ, Theorem 10.20].

11.13 *Given distinct $p, q \in [1, +\infty)$, then ℓ_q does not contain any isomorphic copy of ℓ_p.*
Hint. Assume there exists a linear isomorphism T from ℓ_p into ℓ_q. If $p < q$, apply Theorem 11.6 to the function $\ell_p \ni x \longmapsto \|Tx\|_q{}^q - \|x\|_p{}^p$. If $p > q$, apply Theorem 11.6 to the function $\ell_p \ni x \longmapsto \|x\|_p{}^p - \|Tx\|_q{}^q$. See also Proposition 4.49.

11.14 Prove directly that X^* *is weak*-dentable if it is weak*-fragmentable.*
Hint. Use Milman's theorem applied to the set of extremal points; see [NaPh, p. 738].

11.15 Given a Banach space X, assume that B_X is "ε-separable" for some $0 < \varepsilon < 1$. Prove that then X is separable. This observation simplifies a little the proof of the implication (iii)$\Longrightarrow$(vi) in Theorem 11.8.
Hint. Use, for instance, Riesz' Lemma 1.37.

11.16 A multi-valued mapping $T : X \to 2^{X^*}$ is called *monotone* if $\langle x^* - y^*, x - y\rangle \geq 0$ whenever $x, y \in X$, $x^* \in Tx$, and $y^* \in Ty$. Prove that once $f : X \to \mathbb{R}$ *is convex and continuous, then the subdifferential mapping* $\partial f : X \to 2^{X^*}$ *is monotone.* Prove that, *if* X^* *is weak*-dentable, then every monotone mapping* $T : X \to 2^{X^*}$ *is single-valued and norm-to-norm upper semicontinuous at the points of a dense* G_δ *subset of* X. *The same is true if* $T : X \to 2^{X^*}$ *is any minimal norm-to-weak* usco (i.e., upper semicontinuous and compact valued) mapping.*
Hint. See the proof of [Phelps, Theorem 2.30].

11.17 Prove directly that *a Banach space is Asplund if its dual is weak*-dentable* [Ken].
Hint. Use the monotonicity of the multi-valued mapping ∂f; see Exercise 11.16.

11.18 Let U be an open convex set in a Banach space $(X, \|\cdot\|)$ and let $f : U \to \mathbb{R}$ be a convex continuous function. Then for every $x_0 \in U$ there are an open set $x_0 \in V \subset U$ and a convex (globally) Lipschitz function $g : X \to \mathbb{R}$ such that $f\big|_V = g\big|_V$.
Hint. By Lemma 7.3, find a convex open set $x_0 \in V \subset U$ such that $f\big|_V$ is Lipschitz, with Lipschitz constant, $L > 0$, say. For $x \in X$ put $g(x) = \inf\{f(y) + L\|x - y\| : y \in V\}$. A consequence: If a Banach space X is Asplund (weak Asplund), $U \subset X$ is an open convex set, and $f : U \to X$ is convex continuous, then f if Fréchet (Gâteaux) differentiable at the points of a dense G_δ subset of U.

11.19 The condition (v) in Theorem 11.8 can be enhanced in the spirit of the condition (iii) in Theorem 11.3.
Hint. See the proof of Theorem 11.3.

11.20 In Theorem 11.8, the conditions (iv) and (v) can be weakened in the spirit of Exercise 11.5.

11.21 There is an alternate, a bit longer, proof of Lemma 11.12: If we do not profit from the $\frac{1}{2}$-business, then we construct a "longer" sequence (B_α), indexed by ordinal numbers, of pairwise disjoint elements of $\mathcal{P}$. The process must stop at some `countable` ordinal, since each B_α has positive Lebesgue measure.

11.22 Let $\tau : \mathcal{L} \to X$ be a vector measure and consider a pairwise disjoint sequence $E_1, E_2, \ldots$ in $\mathcal{L}$. Prove that the series $\sum_{i=1}^{\infty} \tau(E_i)$ *converges unconditionally to* $\tau\big(\bigcup_{i=1}^{\infty} E_i\big) =: x$, which means that for every permutation $\pi : \mathbb{N} \to \mathbb{N}$ we have $\big\| \sum_{i=1}^{n} \tau(E_{\pi(i)} - x \big\| \to 0$ as $n \to \infty$. If, moreover, τ is of bounded variation, then this series converges even absolutely, that is, $\sum_{i=1}^{\infty} \|\tau(E_i)\| < +\infty$.

11.23 Let $(X, \|\cdot\|)$ be a Banach space and $\tau : \mathcal{L} \to X$ a vector measure. Prove that:
(i) *If* $A, B \in \mathcal{L}$ *and* $A \subset B$, *then* $|\tau|(A) \leq |\tau|(B)$.
(ii) $|\tau|$ *is additive, that is,* $|\tau|(A \cup B) = |\tau|(A) + |\tau|(B)$ *whenever* $A, B \in \mathcal{L}$ *and* $A \cap B = \emptyset$.

(iii) $|\tau|$ *is countably additive.*
(iv) *If* τ *is of bounded variation, and* $E_1, E_2, \ldots \in \mathcal{L}$ *is a decreasing sequence, then* $|\tau|(\bigcap_{n=1}^{\infty} E_n) = \lim\limits_n |\tau|(E_n)$.
(v) *If* τ *is of bounded variation, then* τ *is* λ*-absolutely continuous if and only if for every* $\varepsilon > 0$ *there exists* $\delta > 0$ *such that* $E \in \mathcal{L}$ *and* $\lambda(E) < \delta$ *imply* $|\tau|(E) < \varepsilon$.
Hint. (i) For any finite partition $\mathcal{F} \subset \mathcal{L}$ of A we have $\sum_{E\in\mathcal{F}} \|\tau(E)\| \leq \sum_{E\in\mathcal{F}} \|\tau(E)\| + \|\tau(B\backslash A)\| \leq |\tau|(B)$. Hence $|\tau|(A) \leq |\tau|(B)$.

(ii) Let $E_1, E_2 \in \mathcal{L}$ be such that $E_1 \cap E_2 = \emptyset$. If $\mathcal{F}_1 \subset \mathcal{L}$, $\mathcal{F}_2 \subset \mathcal{L}$ are finite partitions of E_1, E_2 respectively, then $\sum_{E\in\mathcal{F}_1} \|\tau(E)\| + \sum_{E\in\mathcal{F}_2} \|\tau(E)\| \leq |\tau|(E_1 \cup E_2)$, and hence $|\tau|(E_1) + |\tau|(E_2) \leq |\tau|(E_1 \cup E_2)$. Now, let $\mathcal{F} \subset \mathcal{L}$ be any finite partition of $E_1 \cup E_2$. Then

$$\sum_{E\in\mathcal{F}} \|\tau(E)\| \leq \sum_{E\in\mathcal{F}} \|\tau(E \cap E_1)\| + \sum_{E\in\mathcal{F}} \|\tau(E \cap E_2)\| \leq |\tau|(E_1) + |\tau|(E_2).$$

Hence $|\tau|(E_1 \cup E_2) \leq |\tau|(E_1) + |\tau|(E_2)$.

(iii) Let $E_1, E_2, \ldots \in \mathcal{L}$ be any sequence of pairwise disjoint sets. For every $n \in \mathbb{N}$, (i) and (ii) yield that $\sum_{i=1}^{n} |\tau|(E_i) = |\tau|\big(\bigcup_{i=1}^{n} E_i\big) \leq |\tau|\big(\bigcup_{i=1}^{\infty} E_i\big)$. Hence $\sum_{i=1}^{\infty} |\tau|(E_i) \leq |\tau|\big(\bigcup_{i=1}^{\infty} E_i\big)$. In order to prove the reverse inequality, consider any finite partition $\mathcal{F} \subset \mathcal{L}$ of the set $\bigcup_{i=1}^{\infty} E_i$. We observe that for every $i \in \mathbb{N}$ the family $\{E \cap E_i : E \in \mathcal{F}\}$ is a finite partition of E_i. Now, we are ready to estimate

$$\begin{aligned}\sum_{i=1}^{\infty} |\tau|(E_i) &\geq \sum_{i=1}^{\infty} \sum_{E\in\mathcal{F}} \|\tau(E \cap E_i)\| = \sum_{E\in\mathcal{F}} \sum_{i=1}^{\infty} \|\tau(E \cap E_i)\| \\ &\geq \sum_{E\in\mathcal{F}} \left\| \tau\Big(\bigcup_{i=1}^{\infty} (E \cap E_i)\Big) \right\| = \sum_{E\in\mathcal{F}} \|\tau(E)\|.\end{aligned}$$

Therefore, $\sum_{i=1}^{\infty} |\tau|(E_i) \geq |\tau|\big(\bigcup_{i=1}^{\infty} E_i\big)$.

(iv) For every $n \in \mathbb{N}$ we have from (iii) that $|\tau|(E_n) = |\tau|\Big(\bigcap_{i=1}^{\infty} E_i\Big) + \sum_{i=n}^{\infty} |\tau|\big(E_i \backslash E_{i-1}\big)$. Now, it remains to let n go to ∞.

(v) The sufficiency is obvious. To prove the necessity, proceed by contradiction: if it fails, there exists $\varepsilon > 0$ such that no $\delta > 0$ works. We can find then a sequence $E_1, E_2, \ldots$ in $\mathcal{L}$ such that $\lambda(E_n) < 2^{-n}$ and $|\tau|(E_n) \geq \varepsilon$ for every $n \in \mathbb{N}$. The set $E := \bigcap_{n=1}^{\infty} \bigcup_{i=n}^{\infty} E_i$ satisfies $\lambda(E) = 0$ and $|\tau|(E) \geq \varepsilon$ by (iv). □

11.24 Under the assumptions of Proposition 11.16, there exists a closed convex `separable` set $S \subset K$ such that $\mu(S) = 1$.
Hint. Observe that K_ε is separable for every $\varepsilon > 0$. Put then $S = \overline{\text{conv}} \bigcup_{n=1}^{\infty} K_{1/n}$.

11.25 Let $G \subset (0, 1)$ be an open set. Show that there exists a unique at most countable family $\mathcal{J}$ of open pairwise disjoint intervals such that $G = \bigcup \mathcal{J}$. Show also that it may happen that the complement $(0, 1)\backslash G$ is an uncountable set.

Hint. Place each point of G in a maximal open interval lying in G. For countability, look at the countable dense subset of $(0, 1)$ consisting of rational numbers. For and example of the situation described, consider the complement of the Cantor set, see Section 17.7.

11.26 Consider $(0, 1)$ with the Lebesgue measure λ. Prove that if $E \in \mathcal{L}$ has full measure, that is $\lambda(E) = 1$, then E is dense in $(0, 1)$. However, if E is dense in $(0, 1)$, and even open, then it may still happen that $\lambda(E)$ is small.
Hint. For proving the second statement, "envelop" rational numbers by sufficiently small neighborhoods.

11.27 Let X be a Banach space, let $\tau : \mathcal{L} \to X$ be a λ-absolutely continuous vector measure of bounded variation, and define $f : [0, 1) \to X$ by $f(t) = \tau\big([0, t)\big)$, $t \in [0, 1)$. Then f is absolutely continuous and $v_f(t) = |\tau|\big([0, t)\big)$, $t \in [0, 1)$. Using this, we get that (iii)$\Longrightarrow$(ii) in Theorem 11.15.
Hint. for the inequality "$\geq$". Fix any $\varepsilon > 0$. Find a finite partition $\mathcal{F} \subset \mathcal{L}$ of $(0, t)$ such that $|\tau|([0, t)) - \varepsilon < \sum_{E\in\mathcal{F}} \|\tau(E)\|$. From the λ-absolute continuity of τ and regularity of λ, using Exercise 11.23 (v), find closed sets $C_E \subset E,\ E \in \mathcal{F}$, such that still $|\tau|([0, t)) - \varepsilon < \sum_{E\in\mathcal{F}} \|\tau(C_E)\|$. Find then pairwise disjoint open sets $C_E \subset G_E \subset (0, t),\ E \in \mathcal{F}$. Using Exercise 11.25 and compactness, we may and do assume that each G_E is equal to the union of a finite family $\mathcal{J}_E$ of open pairwise disjoint intervals. Then $|\tau|([0, t)) - \varepsilon < \sum_{E\in\mathcal{F}} \sum_{J\in\mathcal{J}_E} \|\tau(J)\| \leq v_f(t)$.

Exercises 11.28 and 11.29 shed more light on the concept of a measurable function.

11.28 Prove the following result: *Given a Banach space* $(X, \|\cdot\|)$ *and a function* $f : [0, 1) \to X$*, then the following statements are equivalent:*
(i) f is measurable.
(ii) $f^{-1}(B) \in \mathcal{L}$ for every closed ball B in X, and $f([0, 1)\backslash N)$ is a separable set for a suitable Lebesgue-zero set $N \subset [0, 1)$.
(iii) (Pettis) The composition $x^* \circ f$ is Lebesgue measurable for every $x^* \in X^*$ and $f([0, 1)\backslash N)$ is a separable set for a suitable Lebesgue-zero set $N \subset [0, 1)$.
Hint. (i)$\Longrightarrow$(ii). The verification of the "almost separability" of the range of f is simple. The rest follows from the easily verifiable formula below

$$f^{-1}(B)\backslash N = \bigcap_{k=1}^{\infty} \bigcup_{m=1}^{\infty} \bigcap_{n=m}^{\infty} f_n^{-1}\big(B + \tfrac{1}{k} B_X\big)\backslash N. \tag{11.26}$$

valid for every closed ball $B \subset X$ and every sequence $f_1, f_2, \ldots : [0, 1) \to X$ converging almost everywhere to f. The proof of (i)$\Longrightarrow$(iii) is the same.

(iii)$\Longrightarrow$(ii). Find $N \subset [0, 1)$ such that $\lambda(N) = 0$ and that the set $f\big([0, 1)\backslash N\big)$ is separable. Find norm-one vectors $x_1^*, x_2^*, \ldots \in X^*$ such that $\|x\| = \sup_{i\in\mathbb{N}} \langle x_i^*, x\rangle$ for every $x \in f\big([0, 1)\backslash N\big)$. Take any $r > 0$. Then $f^{-1}(rB_X)\backslash N = \bigcap_{i=1}^{\infty} (x_i^* \circ$

$f)^{-1}[-r,r]\backslash N$, and so $f^{-1}(rB_X) \in \mathcal{L}$. And, if $B \subset X$ is a closed ball with center at $x_0 \in X$, we apply what we have just proved and (iii) where f is replaced by the function $[0,1) \ni t \longmapsto f(t) - x_0$.

(ii)$\Longrightarrow$(i). Find $N \subset [0,1)$ such that $\lambda(N) = 0$ and that the set $f\big([0,1)\backslash N\big)$ is separable. Fix any $n \in \mathbb{N}$. Let $\mathcal{B}$ be an at most countable family of balls in X, of radius less than $\frac{1}{2n}$, that covers $f\big([0,1)\backslash N\big)$. Find a finite subfamily $\mathcal{F} \subset \mathcal{B}$ such that $\lambda\big(f^{-1}(\bigcup \mathcal{F})\big) > 1 - 2^{-n}$. Enumerate $\mathcal{F}$ as $B_1, \ldots, B_{k_n}$ and put $E_1 = f^{-1}(B_1)\backslash N$, $E_2 = f^{-1}(B_2)\backslash(E_1 \cup N), \ldots, E_{k_n} = f^{-1}(B_{k_n})\backslash\big(E_1 \cup \cdots \cup E_{k_n-1} \cup N\big)$. For $i = 1, \ldots, k_n$ pick $t_i \in E_i$ and define then the (simple) function $s_n = f(t_1)\chi_{E_1} + \cdots + f(t_k)\chi_{E_{k_n}}$. We observe that $\big\|s_n(t) - f(t)\big\| < \frac{1}{n}$ for every $t \in E_1 \cup \cdots \cup E_{k_n}$. Put $M_n = [0,1)\backslash\big(E_1 \cup \cdots \cup E_{k_n}\big)$. Then $\lambda\big([0,1)\backslash M_n\big) = \lambda\big(f^{-1}(\bigcup \mathcal{F})\big) > 1 - 2^{-n}$. Performing the construction above for every $n \in \mathbb{N}$, put $M = \bigcap_{n=1}^{\infty} M_n$; note that $\lambda(M) = 0$. Then $\|s_n(t) - f(t)\| \to 0$ as $n \to \infty$ for every $t \in [0,1)\backslash(N \cup M)$, and hence f is measurable. Indeed, fix any such t and any $\varepsilon > 0$. Find $m \in \mathbb{N}$ so big that $t \notin M_m \cup M_{m+1} \cup \cdots$ and that $\frac{1}{m} < \varepsilon$. Then, for every $n \in \mathbb{N}$, with $n \geq m$, we have $t \notin M_n$ and so $\|s_n(t) - f(t)\| < \frac{1}{m} \leq \frac{1}{n} < \varepsilon$.

11.29 Prove the following corollary to the result in Exercise 11.28: (i) *Every continuous function $f : [0,1) \to X$ is measurable.* (ii) *The pointwise limit, almost everywhere, of a sequence of measurable functions is measurable.* (iii) *If a measurable function $f : [0,1) \to X$ is almost everywhere differentiable, then f' is a measurable function.*

Hint. (i) $f^{-1}(B)$ is closed for every closed ball $B \subset X$. (ii) See (11.26). (iii) The sequence $n\big(f(\cdot + \frac{1}{n}) - f(\cdot)\big)$, $n \in \mathbb{N}$, of measurable functions converges almost everywhere to f'; hence (ii) applies.

11.30 Prove that every λ*-continuous vector measure of bounded variation has a separable range* $\{\tau(E) : E \in \mathcal{L}\} =: R$. Therefore, *the Radon–Nikodým property is separably determined*, that is, *a Banach space has the Radon–Nikodým property if (and only if) every separable subspace of it has this property.*

Hint. Exercise 11.23 (v) yields that the (countable) set of all vectors $\tau(J)$ where J is an open interval in [0,1) with rational endpoints, is dense in R.

11.31 For the definition of Bochner integral, see, for instance, [BeLi, pp. 99–101], [DiUh, Section II.2], or Section 17.13.1. Once we have this concept at hand, we can easily show that *a Banach space X has the Radon–Nikodým property if and only if every λ-absolutely continuous vector measure $\tau : \mathcal{L} \to X$, of bounded variation, has a Bochner integrable derivative, i.e., there is a Bochner integrable function $g : [0,1) \to X$ such that $\tau(E) = \int_E g \,\mathrm{d}\lambda$ for every $E \in \mathcal{L}$.*

11.32 Let X be any (possibly non-dentable) Banach space, let $\tau : \mathcal{L} \to X$ be a vector measure, and let $g : [0,1) \to X$ be a measurable function such that $\big\langle x^*, \tau(E)\big\rangle = \int_E \langle x^*, g(t)\rangle \mathrm{d}\lambda(t)$ for every $x^* \in X^*$ and for every $E \in \mathcal{L}$. Then

$|\tau|(E) = \int_E \|g(t)\| \mathrm{d}\lambda(t)$ for every $E \in \mathcal{L}$, that is, (11.13) holds. Hence, τ is of bounded variation if and only if $\|g(\cdot)\|$ is Lebesgue integrable.
Hint. Fix any $E \in \mathcal{L}$. For every $x^* \in B_{X^*}$, we have $\langle x^*, \tau(E)\rangle = \int_E \langle x^*, g(t)\rangle \mathrm{d}\lambda(t) \leq \int_E \|g(t)\| \mathrm{d}\lambda(t)$. And, taking here supremum for $x^* \in B_{X^*}$ yields the inequality "$\leq$". As regards the reverse inequality, Exercise 11.28 yields a Lebesgue negligible set $N \subset [0, 1)$ such that $g(E \backslash N)$ is separable. Fix $\varepsilon > 0$. We can then cover $g(E \backslash N)$ by an at most countable family of closed balls of radius ε. Having this cover, Exercise 11.28 yields pairwise disjoint sets $E_1, E_2, \ldots \in \mathcal{L}$, with union equal to $E \backslash N$, and such that diam $g(E_i)$ is at most 2ε for every $i \in \mathbb{N}$. For every $i \in \mathbb{N}$ pick $t_i \in E_i$ and find a unit vector $x_i^* \in X^*$ so that $\langle x_i^*, g(t_i)\rangle = \|g(t_i)\|$. Then

$$\|g(t)\| \leq \|g(t_i)\| + 2\varepsilon = \langle x_i^*, g(t_i)\rangle + 2\varepsilon \leq \langle x_i^*, g(t)\rangle + 4\varepsilon$$

for every $t \in E_i$, and hence, using the premise,

$$\begin{aligned}\int_E \|g(t)\| \mathrm{d}\lambda(t) &\leq \sum_{i=1}^{\infty} \int_{E_i} \big(\langle x_i^*, g(t)\rangle + 4\varepsilon\big) \mathrm{d}\lambda(t) \\ &= \sum_{i=1}^{\infty} \langle x_i^*, \tau(E_i)\rangle + 4\varepsilon\lambda(E) \leq |\tau|(E) + 4\varepsilon\lambda(E).\end{aligned}$$

11.33 Let X be any (possibly non-dentable) Banach space, let $f : [0, 1) \to X$ be an absolutely continuous function, and assume that it is almost everywhere differentiable (this may not be the case, see the proof of (iv)$\Longrightarrow$(i) in Theorem 11.15). Corollary in Exercise 11.28 guarantees that f' is measurable, and so is $\|f'(\cdot)\|$. Show that $\int_0^t \|f'(u)\| \mathrm{d}\lambda(u) = v_f(t)$ for every $t \in [0, 1)$.
Hint. Fix $x^* \in X^*$. Since the composition $x^* \circ f$ is absolutely continuous, [Rudi2, Theorems 8.18 and 8.19] or [LuMa, Theorem 23.2] guarantee that $(x^* \circ f)'$ is a well-defined Lebesgue integrable function and that $\langle x^*, f(t)\rangle - \langle x^*, f(0)\rangle = \int_0^t \big(\langle x^*, f(u)\rangle\big)' \mathrm{d}\lambda(u)$ for every $t \in [0, 1)$. On the other hand, Proposition 11.13 provides a λ-continuous vector measure $\tau : \mathcal{L} \to X$ such that $f(t) - f(0) = \tau([0, t))$ for every $t \in [0, 1)$, and hence

$$\big\langle x^*, \tau([0, t))\big\rangle = \int_0^t \big(\langle x^*, f(u)\rangle\big)' \mathrm{d}\lambda(u) = \int_0^t \langle x^*, f'(u)\rangle \mathrm{d}\lambda(u)$$

by the assumption. The latter equation, the λ-absolute continuity of τ guaranteed by Proposition 11.13, and the Lebesgue integrability of the function $x^* \circ f'$ yield that $\langle x^*, \tau(E)\rangle = \int_E \langle x^*, f'(u)\rangle \mathrm{d}\lambda(u)$ for every $E \in \mathcal{L}$. Exercise 11.32 provides that $|\tau|(E) = \int_E \|f'(t)\| \mathrm{d}\lambda(t)$, and in particular, $\int_0^t \|f'(u)\| \mathrm{d}\lambda(u) = |\tau|([0, t))$ for every $t \in [0, t)$. Now, Proposition 11.13 concludes the proof.

11.34 Let X be any (possibly non-dentable) Banach space and let $g : [0, 1) \to X$ be a measurable function such that the function $\|g(\cdot)\|$ is Lebesgue integrable. Prove that there exists a λ-absolutely continuous vector measure $\tau : \mathcal{L} \to X$ of bounded

variation such that for every $x^* \in X^*$ and every $E \in \mathcal{L}$ we have $\langle x^*, \tau(E)\rangle = \int_E \langle x^*, g(t)\rangle d\lambda(t)$, and hence, by Exercise 11.32, $|\tau|(E) = \int_E \|g(t)\| d\lambda(t)$.
Hint. g has "almost" separable range. Hence X can be assumed to be separable. Then (B_{X^*}, w^*) is metrizable. For any fixed $E \in \mathcal{L}$ let φ denote the assignment $x^* \longmapsto \int_E \langle x^*, g(t)\rangle d\lambda(t)$, $x^* \in X^*$; this is an element of X^{**}. Using Lebesgue's dominated convergence theorem and Banach–Dieudonné Theorem 3.92, show that φ is weak*-continuous. Hence, φ can be represented by an element of X; call it $\tau(E)$. This way we got the formula $\langle x^*, \tau(E)\rangle = \int_E \langle x^*, g(t)\rangle d\lambda(t)$. Using it, we can easily verify that τ is a vector measure. The remaining properties of τ can be deduced from the formula $|\tau|(E) = \int_E \|g(t)\| d\lambda(t)$.

11.35 From Theorem 11.15 and Exercise 11.32, deduce the (scalar) Radon–Nikodým theorem: *If $\mu : \mathcal{L} \to \mathbb{R}$ is a λ-absolutely continuous measure of bounded variation, then there exists a Lebesgue integrable function $g : [0, 1) \to \mathbb{R}$ such that for every $E \in \mathcal{L}$*

$$\mu(E) = \int_E g(t) d\lambda(t) \quad \text{and} \quad |\mu|(E) = \int_E |g(t)| d\lambda(t).$$

Hint. The space $\mathbb{R}$ is dentable.

11.36 A Banach space X is called *Gelfand* provided that every absolutely continuous function $f : [0, 1) \to X$ is almost everywhere differentiable. By Theorem 11.15, *X is a Gelfand space, if and only if, it is dentable, if and only if, it has the Radon–Nikodým property*. Show that *Gelfand spaces are separably determined*, that is, *X is Gelfand if (and only if) every separable subspace of X is Gelfand*. Thus, *the Radon–Nikodým property is separably determined, and so is the dentability*.

11.37 Assume that we have at hand the (scalar) Radon–Nikodým theorem, see Exercise 11.35. Let X be a separable Banach space with separable dual X^*. Prove that X^* has the Radon–Nikodým property. (Dunford–Pettis)
Hint. See [DiUh, pp. 79–81].

11.38 (Lindenstrauss) Define the mapping $f : L_2[0, 1] \to L_2[0, 1]$ by $\big(f(x)\big)(t) = \sin\big(x(t)\big)$, $x \in L_2[0, 1]$, $t \in [0, 1]$. Show that f is everywhere Gâteaux differentiable with derivative $\big(f'(x)h\big)(t) = \cos\big(x(t)\big) \cdot h(t)$, $x, h \in L_2[0, 1]$, $t \in [0, 1]$, and that f is nowhere Fréchet differentiable.
Hint. The Gâteaux differentiability follows from Lebesgue's dominated convergence theorem. For the functions $u_\tau := \chi_{[0,\tau]}$, $\tau \in [0, 1]$, we have $\big\| f(u_\tau) - f(0) - f'(0)h \big\|_{L_2} = (1 - \sin 1)\|u_\tau\|_{L_2}$ while $\|u_\tau\|_{L_2} \to 0$ as $\tau \downarrow 0$. Hence, F is not Fréchet differentiable at 0.

11.39 Let $n \in \mathbb{N}$, let $U \subset \mathbb{R}^n$ be an open set, let $f : U \to \mathbb{R}$ be Lipschitz, and let $x \in U$. Then f is Fréchet differentiable at x if (and only if) f is Gâteaux differentiable at x.

Chapter 12
Basics in Nonlinear Geometric Analysis

In this chapter, we begin by proving the Brouwer and the Schauder fixed-point theorems. Then we turn to results on homeomorphisms of convex sets and spaces. We prove Keller's theorem on homeomorphism of infinite-dimensional compact convex sets in Banach spaces to $\mathbb{I}^{\mathbb{N}}$. We also prove the Kadec theorem on the homeomorphism of every separable reflexive space to a Hilbert space. Then we prove some results on uniform, in particular Lipschitz, homeomorphisms.

12.1 Contractions and Nonexpansive Mappings

Definition 12.1 *Let f be a mapping from a set K into K. A point $x_0 \in K$ is called a* fixed point *of f if $f(x_0) = x_0$.*

Definition 12.2 *A topological space X is said to have the* fixed point property *(FPP, in short) if every continuous mapping $f : X \to X$ has a fixed point.*

Note that the fixed point property is invariant by homeomorphisms.

Let (P, ρ) be a metric space and $f : P \to P$. The mapping f is called a *contraction* if there exists $q < 1$ such that $\rho\big(f(x), f(y)\big) \le q\rho(x, y)$ for all $x, y \in P$. f is called *nonexpansive* if $\rho\big(f(x), f(y)\big) \le \rho(x, y)$ for all $x, y \in P$.

Theorem 12.3 (Banach contraction principle) *Let (P, ρ) be a complete metric space. If f is a contraction from P into P, then there exists a unique fixed point of f.*

Proof: Let $q < 1$ be such that $\rho\big(f(x), f(y)\big) \le q\rho(x, y)$ for $x, y \in P$. First we show the uniqueness of a possible fixed point. If $f(x_1) = x_1$ and $f(x_2) = x_2$ for $x_1, x_2 \in P$, then $\rho(x_1, x_2) = \rho\big(f(x_1), f(x_2)\big) \le q\rho(x_1, x_2)$, which implies $\rho(x_1, x_2) = 0$ as $q < 1$. Therefore $x_1 = x_2$.

To show the existence of a fixed point $x_0 \in P$, we choose an arbitrary point $x_1 \in P$ and define inductively $x_{n+1} := f(x_n) =: f^n(x_1)$, $n \in \mathbb{N}$. For $n \ge 3$ we have

M. Fabian et al., *Banach Space Theory*, CMS Books in Mathematics,
DOI 10.1007/978-1-4419-7515-7_12, © Springer Science+Business Media, LLC 2011

$$\rho(x_{n+1},x_n)=\rho\big(f(x_n),f(x_{n-1})\big)\le q\rho(x_n,x_{n-1})$$
$$\le q^2\rho(x_{n-1},x_{n-2})\le\cdots\le q^{n-1}\rho(x_2,x_1).$$

Therefore, for $n > m \in \mathbb{N}$ we have

$$\rho(x_n,x_m)\le\rho(x_n,x_{n-1})+\rho(x_{n-1},x_{n-2})+\cdots+\rho(x_{m+1},x_m)$$
$$\le\big(q^{n-2}+q^{n-1}+\cdots+q^{m-1}\big)\rho(x_2,x_1)\le\tfrac{q^{m-1}}{1-q}\rho(x_2,x_1). \qquad (12.1)$$

Consequently, $\{x_n\}$ is a Cauchy sequence, so there is $x_0 \in X$ such that $x_n \to x_0$. Since f is a continuous mapping, we have $f(x_0) = \lim f(x_n)$. As $f(x_n) = x_{n+1}$, we have $x_0 = \lim x_{n+1} = \lim f(x_n) = f(x_0)$. □

It is worth to mention that, in the situation of Theorem 12.3, the sequence of iterates $\{x_{n+1} := f^n(x_1)\}$ converges to the unique fixed point x_0 *regardless of the choice of the initial point* x_1. Moreover, letting $n \to \infty$, formula (12.1) gives the following estimate for the error between any of the iterates and the fixed point.

$$\rho(x_0,x_m)\le\tfrac{q^{m-1}}{1-q}\rho(x_2,x_1),\ m\in\mathbb{N}, \qquad (12.2)$$

where q is given by the contractive property of f.

Theorem 12.3 fails if instead of being contractive, f is supposed to be merely nonexpansive, even for P a convex and weakly compact subset of a Banach space endowed with the metric induced by the norm (see Exercise 12.5). It is an open problem whether a nonexpansive mapping on the closed unit ball of a (super)reflexive Banach space has a fixed point.

Let S be a nonempty set. A nonempty subset S_0 of S is called f-*invariant* for a mapping $f : S \to S$ if $f(S_0) \subset S_0$. Let $\mathcal{S}$ be a class of subsets of S. We say that an element $S_0 \in \mathcal{S}$ is $\mathcal{S}$-*minimal for* f if there exists no proper f-invariant subset of S_0 in the class $\mathcal{S}$. We are interested mainly in the case that S is a subset of a Banach space X and $\mathcal{S}$ is the class of weakly compact subsets of X or the class of closed convex subsets of X.

If K is a nonempty weakly compact subset of a Banach space X and $f : K \to K$ is a mapping, it is an easy consequence of Zorn's lemma that K contains an f-invariant subset K_0 that is w-compact-minimal for $f|K_0$ (Exercise 12.6).

Lemma 12.4 *Let X be a Banach space.*
(i) *If S is a w-closed subset of X that is w-closed-minimal for a w-continuous mapping* $f : S \to S$*, then S is w-separable.*
(ii) *If C is a closed convex subset of X that is closed-convex-minimal for a continuous mapping* $f : C \to C$*, then C is separable.*

Proof: (i) Take $s \in S$ and put $S_0 = \{f^{(n)}(s) :\ n = 0, 1, 2, \ldots\}$, where $f^{(0)}$ is the identity mapping and, if $f(m)$ has already been defined for $m = 0, 1, 2, \ldots, n$, then $f^{(n+1)} := f \circ f^{(n)}$. Obviously, S_0 is f-invariant. Given $x \in \overline{S_0}^{w}$, let $\{x_i\}$ be a net in S_0 that w-converges to x. Then $f(x_i) \xrightarrow{w} f(x_0)$ and we get $f(x_0) \in \overline{S_0}^w$. The set $\overline{S_0}^w$ is a w-closed subset of S. By minimality, $S = \overline{S_0}^w$.

(ii) Take $x_0 \in C$. Put $C_0 = \{x_0\}$. If C_m has been already defined for $m = 0, 1, 2, \ldots, n$, let $C_{n+1} = \overline{\text{conv}}\,(C_n \cup f(C_n))$. This defines an increasing sequence $\{C_m\}$ of subsets of C. Put $C_\infty = \bigcup_{m=1}^{\infty} C_m$. This set if obviously f-invariant. The set $\overline{C_\infty}$ is f-invariant, too; indeed, if $x \in \overline{C_\infty}$ and $\{x_n\}$ is a sequence in C_∞ that converges to x, then $f(x_n) \to f(x)$, so $f(x) \in \overline{C_\infty}$. The set $\overline{C_\infty}$ is a separable subset of C; moreover, $\overline{C_\infty}$ is closed and convex. By minimality, $\overline{C_\infty} = C$. □

Lemma 12.5 *Let K be a nonempty closed and convex subset of a Banach space X. If K is closed-convex-minimal for a mapping $f : K \to K$ and $\alpha : K \to \mathbb{R} \cup \{+\infty\}$ is a lower semicontinuous and convex function such that $\alpha\big(f(x)\big) \le \alpha(x)$ for all $x \in K$, then α is a constant function.*

Proof: For every $c \in \mathbb{R}$, the set $\{x \in K : \alpha(x) \le c\}$ is closed and convex. Assume that α is not constant on K. We can find then some $c \in \mathbb{R}$ such that $\emptyset \ne K_0 := \{x \in K : \alpha(x) \le c\} \ne K$. Given $x \in K_0$ we have $\alpha\big(f(x)\big) \le \alpha(x) \le c$, hence $f(x) \in K_0$, so K_0 is f-invariant. This violates the minimality of K. □

We shall investigate now some features of nonexpansive mappings.

As it was mentioned after the proof of Theorem 12.3, a nonexpansive mapping $f : C \to C$, where C is a nonempty closed convex and bounded subset of a Banach space X, does not have, in general, a fixed point (even if C is weakly compact). However (Proposition 12.7), it always has a weaker form of a fixed point, given in the next definition.

Definition 12.6 *Let S be a nonempty subset of a Banach space X. A sequence $\{x_n\}$ in S is called an* approximate fixed point sequence *of a mapping $f : S \to S$ if* $\lim_{n\to\infty} \|x_n - f(x_n)\| = 0$.

Proposition 12.7 *Let C be a nonempty closed convex and bounded subset of a Banach space X. Let $f : C \to C$ be a nonexpansive mapping. Then there exists in C an approximate fixed point sequence of f.*

Proof: By translating we may assume that $0 \in C$. Let $M = \sup\{\|x\| : x \in C\}$. Given $0 < \varepsilon < 1$, let $C_\varepsilon = \{(1-\varepsilon)x : x \in C\}$ (a closed convex subset of C), and let $f_\varepsilon = (1-\varepsilon)f$. The mapping f_ε is a contraction from C_ε into itself so, by Theorem 12.3, it has a fixed point $x_\varepsilon \in C_\varepsilon$. We get

$$x_\varepsilon = f_\varepsilon(x_\varepsilon) = (1-\varepsilon)f(x_\varepsilon) = f(x_\varepsilon) - \varepsilon f(x_\varepsilon),$$

hence

$$\|f(x_\varepsilon) - x_\varepsilon\| = \varepsilon\|f(x_\varepsilon)\| \le \varepsilon M.$$

Since $\varepsilon \in (0, 1)$ was arbitrary, this gives readily the conclusion. □

If C is a nonempty bounded subset of a Banach space X, a point $x \in C$ is called *diametral* if $\sup\{\|x - y\| : y \in C\} = \operatorname{diam} C$.

Lemma 12.8 (Goebel–Karlovitz [Goeb], [Karv]) *Let K be a nonempty weakly compact and convex subset of a Banach space X. Assume that K is closed-convex-minimal for a nonexpansive mapping $f\colon K \to K$. Then,*
(i) *Every point $x \in K$ is diametral.*
(ii) *For every $x \in K$ and for every approximate fixed point sequence $\{x_n\}$ of f in K, the limit $\lim_{n\to\infty} \|x - x_n\|$ exists and*

$$\lim_{n\to\infty} \|x - x_n\| = \operatorname{diam} K. \tag{12.3}$$

Proof: (i) The function $\phi : K \to \mathbb{R}$ given by $\phi(x) = \sup\{\|x - y\| : y \in K\}$ is lower semicontinuous and convex. We claim that $\phi(x) = \sup\{\|x - f(y)\| : y \in K\}$ for every $x \in K$. Indeed, if for some $x \in K$ we have $d_0 := \sup\{\|x - f(y)\| : y \in K\} < \sup\{\|x - y\| : y \in K\}$, the set $K \cap B(x, d_0)$ is a proper weakly compact and convex f-invariant subset of K, a contradiction, so the claim is proved. Therefore, for $x \in K$,

$$\phi\big(f(x)\big) = \sup\{\|f(x) - f(y)\| : y \in K\} \le \sup\{\|x - y\| : y \in K\} = \phi(x).$$

It follows from Lemma 12.5 that ϕ is a constant function. Obviously, its constant value is $\operatorname{diam} K$.

(ii) Fix an approximate fixed point sequence $\{x_n\}$ of f in K. According to Lemma 12.4, the set K is separable. By a diagonal procedure, we can select a subsequence of $\{x_n\}$ (denoted again $\{x_n\}$) such that $\varphi(x) := \lim \|x - x_n\|$ exists for every $x \in K$. We shall prove that this value is $\operatorname{diam} K$ for every $x \in K$ (independently of the original sequence $\{x_n\}$ and the chosen subsequence, so passing to a subsequence will not necessary after all). Indeed, for all $x \in K$,

$$\begin{aligned}\varphi\big(f(x)\big) &= \lim \|f(x) - x_n\| \\ &= \lim \|f(x) - f(x_n)\| \le \lim \|x - x_n\| = \varphi(x).\end{aligned} \tag{12.4}$$

The mapping φ is certainly lower semicontinuous and convex. Since (12.4) holds, Lemma 12.5 ensures that φ is a constant function, and certainly $d := \varphi(x) \le \operatorname{diam} K$ for $x \in K$. By passing to a further subsequence, we may assume that $\{x_n\}$ is weakly convergent, say to $y \in K$. Then, $d := \lim \|x - x_n\| \ge \|x - y\|$ for every $x \in K$. From (i) we know that y is diametral, so $d \ge \operatorname{diam} K$. This proves (ii). □

Definition 12.9 *Let X be a Banach space. We say that X has the* fixed point property for nonexpansive mappings *((FPPNE), in short) if for every nonempty closed* *convex and bounded subset C of X, every nonexpansive mapping $f : C \to C$ has a fixed point. We say that X has the* weak fixed point property for nonexpansive mappings *((w-FPPNE), in short), if this happens for every nonempty weakly compact and convex subset of X.*

We mention without proof the following result.

Theorem 12.10 (Domínguez-Benavides [D-B]) *Every Banach space X such that, for some nonempty set Γ, there exists a one-to-one bounded operator into $c_0(\Gamma)$, can be renormed to have property (FPPNE).*

A nonempty bounded convex subset C of a Banach space is said to have *normal structure* if every nonempty convex subset of C which is not a singleton contains a non-diametral point. A Banach space is said to have *normal structure* if every nonempty convex bounded subset of X has normal structure. Examples of Banach spaces with normal structure are Hilbert spaces and more generally spaces with a uniformly convex norm. Indeed, the norm of such a space is *uniformly convex in every direction* (URED, in short) (for a definition see Exercise 9.18), and we have the following result.

Proposition 12.11 ([Zizl1]) *Every URED Banach space has normal structure.*

Proof: Let C be a nonempty bounded convex subset of a URED Banach space X, not reduced to a point. Take $x \neq y$ in C, and put $z = (1/2)(x + y)$. We claim that z is not diametral. Otherwise, there would be a sequence $\{x_n\}$ in C such that $\|z - x_n\| \to \operatorname{diam} C$. Since $\|x_n - x\| \leq \operatorname{diam} C$, $\|x_n - y\| \leq \operatorname{diam} C$, and

$$\|x_n - z\| = \left\| x_n - \frac{x+y}{2} \right\|$$
$$= \frac{1}{2}\|(x_n - x) + (x_n - y)\| \leq \frac{1}{2}\big(\|x_n - x\| + \|x_n - y\|\big) \leq \operatorname{diam} C,$$

we conclude that $\|x - x_n\| \to \operatorname{diam} C$ and $\|y - x_n\| \to \operatorname{diam} C$. Thus

$$2\|x_n - x\|^2 + 2\|x_n - y\|^2 - \|(x_n - x) + (x_n - y)\|^2 \to 0,$$

and $(x_n - x) - (x_n - y) = (x - y)$, for every $n \in \mathbb{N}$. This contradicts the definition of URED norm. □

Theorem 12.12 *Let X be a Banach space and K a weakly compact convex subset of X. Assume that K has normal structure. Then, every nonexpansive mapping $f : K \to K$ has a fixed point.*

Proof: Let K_0 be a subset of K that is closed-convex-minimal for f. Then (i) in Lemma 12.8 ensures that every point of K_0 is diametral. The fact that K has normal structure forces K_0 to be a singleton, say $\{k_0\}$. Then k_0 is a fixed point for f. □

Corollary 12.13 (Browder, Göhde, Kirk, see [Kirk]) *Let K be a bounded closed convex subset of a uniformly convex space H. If f is a nonexpansive mapping from K into K, then there exists a fixed point for f in K.*

Proof: Combine Proposition 12.11 and Theorem 12.12. □

Remark: The method of using normal structure for fixed points cannot be used for UG [JohaRy]. However, it can be used for UF [Tur].

12.2 Brouwer and Schauder Theorems

The following result is basic for the Fixed Point Theory.

Theorem 12.14 (Brouwer) *Let K be a nonempty compact convex subset of a finite-dimensional Banach space. If $f: K \to K$ is continuous, then f has a fixed point in K.*

In view of Exercises 1.31 and 12.27, and the remark after Definition 12.2, it is enough to prove the result for the closed unit ball of $(\mathbb{R}^n, \|\cdot\|_2)$, where n is an arbitrary positive integer. In this section, the Euclidean norm $\|\cdot\|_2$ on $\mathbb{R}^n$ will be denoted by $\|\cdot\|$, and the Euclidean inner product on $\mathbb{R}^n$ by $\langle\langle\cdot,\cdot\rangle\rangle$. The approach follows [Miln], and it is based in the so-called "Hairy ball theorem" (Theorem 12.15). If G is a nonempty subset of $\mathbb{R}^n$, a mapping $v : G \to \mathbb{R}^n$ is called a *vector field* (or just a *field*) *defined on G*. If G is open, the mapping v is differentiable and the differential dv is a continuous mapping from G into $\mathcal{L}(\mathbb{R}^n)$ (the space of operators from $\mathbb{R}^n$ into $\mathbb{R}^n$), the field is called *continuously differentiable.*

Theorem 12.15 (Hairy Ball Theorem) *For n odd, there is no continuous field of non-zero tangent vectors defined on $S_{\mathbb{R}^n}$.*

The requirement that n should be odd is essential. The vector field $v(z) = iz$, where $z \in \mathbb{C}$ (here $\mathbb{C}$, the set of complex numbers, is identified with $\mathbb{R}^2$), is a continuous (even continuously differentiable) field of unit tangent vectors defined on $S_{\mathbb{R}^2}$ (see Fig. 12.1).

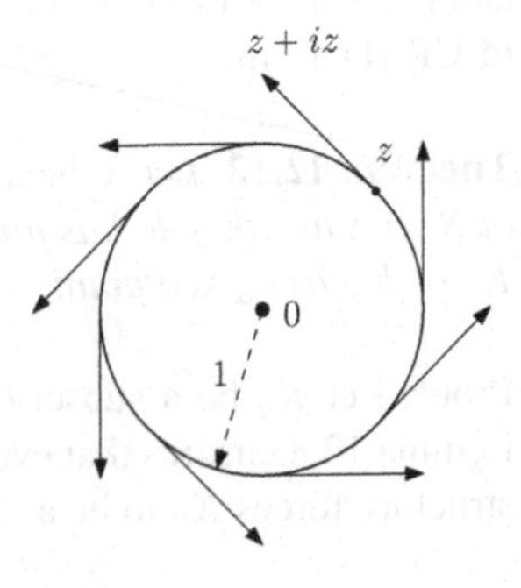

Fig. 12.1 A continuously differentiable field of unit tangent vectors on $S_{\mathbb{R}^2}$

Lemma 12.16 *If there exists a continuous field of non-zero tangent vectors on $S_{\mathbb{R}^n}$, then there exists a continuously differentiable field v on $\mathbb{R}^n\setminus\{0\}$ such that $\|v(u)\| = 1$, $v(ru) = rv(u)$ and $\langle\langle u, v(u)\rangle\rangle = 0$ for all $u \in S_{\mathbb{R}^n}$ and $r > 0$.*

Proof: Assume that $S_{\mathbb{R}^n}$ possesses a continuous field ϕ of non-zero tangent vectors. Let $m = \min\{\|\phi(u)\| : u \in S_{\mathbb{R}^n}\}$. By the Weierstrass approximation theorem, there exists a polynomial mapping $p : S_{\mathbb{R}^n} \to \mathbb{R}^n$ such that

$$\|p(u) - \phi(u)\| < m/2, \text{ for all } u \in S_{\mathbb{R}^n}. \tag{12.5}$$

Let us define a field $w : S_{\mathbb{R}^n} \to \mathbb{R}^n$ by $w(u) = p(u) - \langle\langle p(u), u\rangle\rangle u$, $u \in S_{\mathbb{R}^n}$. Take any $u \in S_{\mathbb{R}^n}$. We have $\langle\langle w(u), u\rangle\rangle = 0$; moreover, $|\langle\langle p(u), u\rangle\rangle| = |\langle\langle p(u) - \phi(u), u\rangle\rangle| \le \|p(u) - \phi(u)\| < m/2$, so $\|w(u) - p(u)\| = |\langle\langle p(u), u\rangle\rangle| < m/2$. From this, Equation (12.5), and the fact that $\|\phi(u)\| \ge m$, we get $w(u) \ne 0$. Thus, the formula

$$v(x) = \frac{\|x\| w\left(\frac{x}{\|x\|}\right)}{\left\| w\left(\frac{x}{\|x\|}\right)\right\|}, \; x \ne 0, \tag{12.6}$$

defines a continuously differentiable field on $\mathbb{R}^n\setminus\{0\}$ that satisfies the requirements. □

Lemma 12.17 *Let K be a nonempty compact subset of $\mathbb{R}^n$, and let $v = (v_1, v_2, \ldots, v_n) : G \to \mathbb{R}^n$ be a continuously differentiable vector field defined on an open subset G of $\mathbb{R}^n$ containing K. For $t \in \mathbb{R}$, we consider the function $f_t : G \to \mathbb{R}^n$ defined by*

$$f_t(x) = x + tv(x), \quad x \in G. \tag{12.7}$$

Then, if $|t|$ is sufficiently small, f_t is one-to-one on K and the compact set $f_t(K)$ has a volume that can be expressed as a polynomial function of t.

Proof: Since K is compact and the field v is continuously differentiable, v satisfies a Lipschitz condition on K, i.e., there exists a positive constant c such that

$$\|v(x) - v(y)\| \le c\|x - y\|, \text{ for all } x, y \in K. \tag{12.8}$$

This is proved as follows. First consider the particular case where K is a cube with edges parallel to the coordinate axes. Passing from x to y in n steps by changing one coordinate at a time, and applying the mean value theorem of differential calculus, it is easy to see that (12.8) holds for some c. Assume now that K is an arbitrary nonempty compact subset of $\mathbb{R}^n$. Then K can be covered by a finite number of open cubes with edges parallel to the coordinate axes. On each of them a Lipschitz condition holds for v; moreover, a positive lower bound exists for the mutual distance between any two points x and y in K sitting in different open cubes. It follows that (12.8) holds (with some other constant c).

Choose any t with $|t| < c^{-1}$. Then f_t is one-to-one on K. Indeed, if for some $x, y \in K$ we have $f_t(x) = f_t(y)$, then $x - y = t(v(y) - v(x))$, and the inequality $\|x - y\| \leq |t|c\|x - y\|$ implies that $x = y$.

The matrix of the first partial derivatives of f_t at $x \in G$ can be written as $I + t[(\partial v_i/\partial x_j)(x)]$, where I is the identity matrix. Hence its determinant is a polynomial function of t of the form $1 + t\sigma_1(x) + \ldots + t^n\sigma_n(x)$, the coefficients being continuous functions of x. The determinant is strictly positive for $|t|$ small enough. Integrating over K we see that the volume of the image region can be expressed as a polynomial function of t,

$$\operatorname{vol} f_t(K) = a_0 + a_1 t + \ldots + a_n t^n, \tag{12.9}$$

with coefficients $a_0 = \operatorname{vol} K$, $a_k = \int \ldots \int_K \sigma_k(x)\, \mathrm{d}x_1 \ldots \mathrm{d}x_n$, $k = 1, 2, \ldots, n$, and $x = (x_1, \ldots, x_n)$. □

Lemma 12.18 *Let $v : G \to \mathbb{R}^n$ be a continuously differentiable field defined on an open neighborhood G of $S_{\mathbb{R}^n}$ that does not contain 0. Assume that v satisfies $\|v(u)\| = 1$ and $\langle\langle u, v(u)\rangle\rangle = 0$ for all $u \in S_{\mathbb{R}^n}$. Then, if $|t|$ is sufficiently small, f_t defined in (12.7) maps $S_{\mathbb{R}^n}$ onto $\sqrt{1+t^2}\,S_{\mathbb{R}_n}$* (see Fig. 12.2).

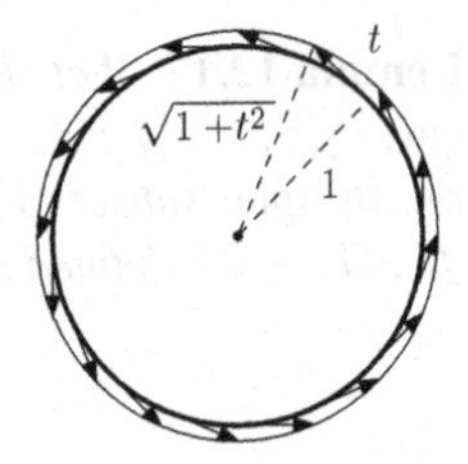

Fig. 12.2 The image of $S_{\mathbb{R}^n}$ under f_t

Proof: We can find $0 < a < 1 < b$ such that $K := \{x \in \mathbb{R}^n : a \leq \|x\| \leq b\} \subset G$. The field w defined on G as $w(x) = \|x\| v(x/\|x\|)$ is again continuously differentiable and coincides with v on $S_{\mathbb{R}_n}$. Moreover, $\|w(x)\| = \|x\|$ and $\langle\langle w(x), x\rangle\rangle = 0$ for every $x \in G$. Put $M = \max\{\|w(x)\| : x \in K\}$. Fix $u_0 \in S_{\mathbb{R}^n}$ and $t \in \mathbb{R}$. Let us define the mapping $\phi_{u_0,t} : K \to \mathbb{R}^n$ by $\phi_{u_0,t}(x) = u_0 - tw(x)$. Then, for $x \in K$ we have $\|\phi_{u_0,t}(x)\| \leq \|u_0\| + |t|.\|w(x)\| \leq 1 + |t|M$. Analogously, $\|\phi_{u_0,t}(x)\| \geq \|u_0\| - |t|.\|w(x)\| \geq 1 - |t|M$. The field w satisfies a Lipschitz condition on K with some constant c (see the proof of Lemma 12.17 above). If $|t| < \min\{(1/M)(1-a), (1/M)(b-1), c^{-1}\}$, then $\phi_{u_0,t}$ maps K into itself and satisfies a Lipschitz condition with constant less than 1. Since K is a complete metrizable space, $\phi_{u_0,t}$ has, by Theorem 12.3, a (unique) fixed point, say x_0, so the equation $x + tw(x) = u_0$ has x_0 as the (unique) solution. Since $x_0 + tw(x_0) = u_0$ and $\langle\langle x_0, w(x_0)\rangle\rangle = 0$, we get $1 = \|u_0\| = \sqrt{1+t^2}\|x_0\|$. This proves that, given $|t|$ sufficiently small, for any $u_0 \in S_{\mathbb{R}^n}$ there exists a (unique) $x_0 \in (\sqrt{1+t^2})^{-1} S_{\mathbb{R}^n}$

such that $x_0 + tw(x_0) = u_0$, i.e., $x_0 + t\|x_0\|v(x_0/\|x_0\|) = u_0$ or, in other words, $x_0/\|x_0\| + tv(x_0/\|x_0\|) = \sqrt{1+t^2}u_0$. This proves the lemma. □

Proof of Theorem 12.15: Assume that there exists a continuous field $\phi : S_{\mathbb{R}^n} \to \mathbb{R}^n$ such that $\langle\langle u, \phi(u)\rangle\rangle = 0$ and $\phi(u) \neq 0$ for every $u \in S_{\mathbb{R}^n}$. By Lemma 12.16, there exists a continuously differentiable field v on $\mathbb{R}^n\backslash\{0\}$ such that $\langle\langle u, v(u)\rangle\rangle = 0$, $v(ru) = rv(u)$ and $\|v(u)\| = 1$ for every $u \in S_{\mathbb{R}^n}$ and $r > 0$. The field v defines, for $t \in \mathbb{R}$, a mapping $f_t : \mathbb{R}^n\backslash\{0\} \to \mathbb{R}^n$ given by (12.7). Observe that $f_t(ru) = rf_t(u)$ for every $u \in S_{\mathbb{R}^n}$ and $r \in \mathbb{R}\backslash\{0\}$. Fixing an annulus $K := \{x \in \mathbb{R}^n : a \leq \|x\| \leq b\}$, where $0 < a < 1 < b$, we get, by Lemma 12.18 and the previous observation, that for $0 < |t|$ sufficiently small, the mapping f_t defined for v as in (12.7) maps K in a one-to-one way onto $\{x \in \mathbb{R}^n : a\sqrt{1+t^2} \leq \|x\| \leq b\sqrt{1+t^2}\}$, so

$$\operatorname{vol} f_t(K) = \left(\sqrt{1+t^2}\right)^n \operatorname{vol} K. \tag{12.10}$$

If n is odd, this contradicts Lemma 12.17. □

Proof of Theorem 12.14: Assume that $f(x) \neq x$ for all $x \in B_{\mathbb{R}^n}$. The mapping

$$w(x) := x - \frac{f(x)(1 - \langle\langle x, x\rangle\rangle)}{1 - \langle\langle x, f(x)\rangle\rangle}, \quad x \in B_{\mathbb{R}^n} \tag{12.11}$$

is well defined. Indeed, the denominator in the previous fraction is a continuous function of x that does not vanish on $B_{\mathbb{R}^n}$, since $1 = \langle\langle x, f(x)\rangle\rangle$ $(\leq \|x\|.\|f(x)\|)$ for some $x \in B_{\mathbb{R}^n}$ would imply $\|x\| = \|f(x)\| = 1$ and the two vectors x $(\in B_{\mathbb{R}^n})$ and $f(x)$ $(\in B_{\mathbb{R}^n})$ would be linearly dependent; necessarily $x = f(x)$, a contradiction.

The mapping w is a continuous vector field on $B_{\mathbb{R}^n}$ such that $w(u) = u$ for all $u \in S_{\mathbb{R}^n}$. Using this field, we shall define a continuous vector field W of non-zero tangent vectors on $S_{\mathbb{R}^{n+1}}$ (a contradiction with Theorem 12.15 if n is even). In order to do that (see Fig. 12.3), we consider $\mathbb{R}^n \subset \mathbb{R}^{n+1}$, where $(x_1, \ldots, x_n)$ is identified with $(x_1, \ldots, x_n, 0)$. Let first $s : \mathbb{R}^n \to S_{\mathbb{R}^{n+1}}$ be the stereographic projection from the north pole $N := (0, \ldots, 0, 1) \in S_{\mathbb{R}^{n+1}}$ to map each point $x \in \mathbb{R}^n$ to the point of $S_{\mathbb{R}^{n+1}}$ where the straight line joining N and x meets $S_{\mathbb{R}^{n+1}}$. For $x \in \mathbb{R}^n$, put $W(s(x)) = ds_x(w(x))$. Since this last expression is the directional derivative of s at x in the direction $w(x)$, it is clear that, on the southern hemisphere $x_{n+1} \leq 0$ of $S_{\mathbb{R}^{n+1}}$, W is a continuous field of non-zero tangent vectors such that $W(u) = (0, \ldots, 0, 1)$ for every $u \in S_{\mathbb{R}^n}$, due to the fact that, for those u, we have $w(u) = u$ (see Fig. 12.3).

Similarly, using stereographic projection from the south pole, the vector field $-w$ gives raise to a vector field on the northern hemisphere that, at points $u \in S_{\mathbb{R}^n}$, takes again the value $(0, \ldots, 0, 1)$. Piecing these two vector fields together, we obtain a continuous vector field of non-zero tangent vectors on $S_{\mathbb{R}^{n+1}}$. This violates Theorem 12.15 if n is even and proves Theorem 12.14 in this case. But this suffices to prove Theorem 12.14 for n odd. Indeed, if $f : B_{\mathbb{R}^n} \to B_{\mathbb{R}^n}$ is continuous and

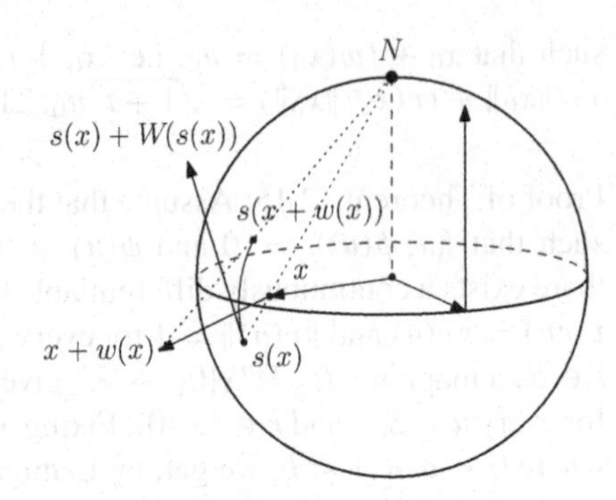

Fig. 12.3 The field W on the southern hemisphere

has no fixed point, the mapping $F : B_{\mathbb{R}^{n+1}} \to B_{\mathbb{R}^{n+1}}$ given by $F(x_1, \ldots, x_{n+1}) = (f(x_1, \ldots, x_n), 0)$ is also continuous and has no fixed point. □

One of the most useful results in Fixed Point Theory is the following theorem. As an application of some results about Keller spaces (see Section 12.3), we shall provide later a proof, without using partitions of unity, of a quite wide-applicable particular case (see Proposition 12.39) that includes the classical Schauder fixed point theorem as a consequence (Corollary 12.40).

Theorem 12.19 (Schauder, [Schau], Tychonoff, [Tych]) *Let $(X, \mathcal{T})$ be a locally convex space. Then, every compact convex subset K of X has the fixed point property, i.e., every continuous mapping $f : K \to K$ has a fixed point.*

Proof: Let $\mathcal{U}(0)$ be the family of all convex balanced open neighborhoods of 0 in X. Since K is compact, for each $U \in \mathcal{U}(0)$ there exists another $V \in \mathcal{U}(0)$, $V \subset U$, such that $f(x) - f(y) \in U$ whenever $x, y \in K$ satisfy $x - y \in V$. Again by compactness, we can find a finite set $A \subset K$ such that $\{a + V : a \in A\}$ is an open cover of K. Let $\{\phi_a : a \in A\}$ be a partition of unity subordinated to this cover (see Theorem 17.21; we need only the trivial fact that every compact topological space is paracompact).

The mapping $f_U(x) := \sum_{a \in A} \phi_a(x) f(a)$ is a continuous function from K into the convex hull of a finite number of points in K. By Brouwer's Theorem 12.14, it has a fixed point x_U. Note that $f_U(x)$ is, for all $x \in K$, a convex combination of points $f(a)$ with $a - x \in V$. Hence $f(a) - f(x) \in U$ for each of them, so $f_U(x) - f(x) \in U$ for every $x \in K$. In particular, $x_U - f(x_U) = f_U(x_U) - f(x_U) \in U$. The net $\{x_U\}_{U \in \mathcal{U}(0)}$ has a convergent subnet, say to x. Fix $U_0 \in \mathcal{U}(0)$ and consider only elements U in $\mathcal{U}(0)$ with $U \subset U_0$. Then $x_U - f(x_U) \in U_0$, so $x - f(x) \in \overline{U_0}$. Since U_0 is an arbitrary element of $\mathcal{U}(0)$, it follows that $f(x) = x$. □

In Exercise 12.22 we shall prove the existence of fixed points for continuous mapping $f : Q \to Q$, where Q is the Hilbert cube, without using partitions of unity. As a consequence of this and of Keller's Theorem 12.37, an alternative proof of the existence of fixed points for continuous functions on a wide class of compact sets in locally convex spaces is given in Proposition 12.39.

Theorem 12.20 (V. Klee) *Let K be a noncompact bounded closed convex subset of a Banach space. Then there is a continuous mapping from K into itself with no fixed points.*

Proof: [BeLi, p. 62] First, observe that K contains a closed subset R which is homeomorphic to the real line. Indeed, since K is not compact, there are $\delta > 0$ and an infinite sequence $(x_n)_{-\infty}^{\infty}$ in K such that $\operatorname{dist}(x_n, \operatorname{span}\{x_i : |i| \leq n, i \neq n\}) > \delta$ for all n. Let R be the polygonal line $\bigcup [x_n, x_{n+1}]$. Let h be the homeomorphism from R onto $\mathbb{R}$ defined in a standard way.

From the Tietze extension theorem we get that there is a continuous retraction τ from K onto R, i.e., a continuous mapping from K into R so that $\tau(r) = r$ for all $r \in R$.

To finalize the proof, let $\psi : R \to R$ be defined by $\psi(r) = h^{-1}(h(r) + 1)$. If $\psi(r) = r$ for some $r \in R$, then $h(\psi(r)) = h(r) + 1 = h(r)$, a contradiction. The sought mapping from K into K is $\psi \circ \tau$. □

We will now show two results concerning common fixed point of a family of mappings. Let K be a set and let $\mathcal{G}$ be a family of mappings $K \to K$. $\mathcal{G}$ is called *commuting* if $T_1 T_2 = T_2 T_1$ for every $T_1, T_2 \in \mathcal{G}$.

Theorem 12.21 (Markov, Kakutani) *Let K be a convex compact subset of a Banach space X. If $\mathcal{G}$ is a commuting family of continuous affine mappings from K to K, then there is $x_0 \in K$ such that $T(x_0) = x_0$ for every $T \in \mathcal{G}$.*

For Werner's proof [Wern] of this result based on the Hahn–Banach theorem, we refer to Exercise 12.19.

By a *semigroup of mappings on K*, we mean a family $\mathcal{S}$ of mappings from K into K such that I_K is in $\mathcal{S}$ and $S_1 S_2 \in \mathcal{S}$ whenever $S_1, S_2 \in \mathcal{S}$.

Theorem 12.22 (Ryll-Nardzewski, see, e.g., [Nami3]) *Let K be a weakly compact convex set in a Banach space X. Let $\mathcal{S}$ be a semigroup of affine weakly continuous mappings from K into K. If for all $x, y \in K$, $x \neq y$, we have $0 \notin \overline{\{s(x) - s(y) : s \in \mathcal{S}\}}$, then $\mathcal{S}$ has a common fixed point.*

Proof: Without loss of generality, we may assume that K is a closed-convex-minimal convex weakly compact set invariant for $\mathcal{S}$ (i.e., invariant for each $s \in \mathcal{S}$). Let M be a weakly compact-minimal weakly compact subset of K invariant for $\mathcal{S}$ (Zorn's lemma). Then $\overline{\operatorname{conv}}(M) = K$, since each $s \in \mathcal{S}$ is affine and then it is easily proved that $\overline{\operatorname{conv}}(M)$ is invariant for $\mathcal{S}$. Due to Lemma 12.4, the set K is w-separable, so K $(= \overline{\operatorname{conv}}(M))$ is w-separable, i.e., $\|\cdot\|$-separable, hence we may also assume that X is a separable Banach space.

We need to show that M is a singleton. If not, then there are $x, y \in M$, $x \neq y$. By our assumption, there is $\varepsilon > 0$ such that

$$\|s(x) - s(y)\| \geq \varepsilon \tag{12.12}$$

for all $s \in \mathcal{S}$. Since K is a convex weakly compact set in a separable Banach space, by Theorem 8.13 there is a strongly exposed point $u \in K$ contained in a relatively weakly open subset V of K with diam $V < \varepsilon$. Observe that u is an extreme point of K (see Exercise 7.72). By Theorem 3.66, $u \in \overline{M}^w$ ($= M$), and it is contained in the set $V \cap M$, a relatively open subset of M with $\operatorname{diam}(V \cap M) < \varepsilon$. Due to the w-compact-minimality of M and the fact that $\mathcal{S}$ is a semigroup, we have $M = \overline{\mathcal{S}(u)}^w$, where $\mathcal{S}(u) := \{s(u) : s \in \mathcal{S}\}$. Hence $s_0(u) \in V$ for some $s_0 \in \mathcal{S}$.

Set $z = \frac{1}{2}(x + y) \in K$. By the minimality of K, $K = \overline{\operatorname{conv}}\big(\overline{\mathcal{S}(z)}^w\big)$. By Theorem 3.66, $u \in \overline{\mathcal{S}(z)}^w$. Hence there is a net $\{s_\alpha\}$ in $\mathcal{S}$ such that $s_\alpha(z) \xrightarrow{w} u$. Since K is w-compact, by taking subnets if needed, we may assume that $s_\alpha(x) \xrightarrow{w} a$ and $s_\alpha(y) \xrightarrow{w} b$. From $s_\alpha(z) = \frac{1}{2}\big(s_\alpha(x) + s_\alpha(y)\big)$ we get $u = \frac{1}{2}(a + b)$ and $a = b = u$ as u is an extreme point of K. It follows that $s_0 s_\alpha(x) \xrightarrow{w} s_0(a) = s_0(u) \in V$. Since $\{s_0 s_\alpha(x)\}$ is a net in M and V is relatively open in M, $s_0 s_\alpha(x) \in V$ eventually. Similarly, $\{s_0 s_\alpha(y)\} \in V$ eventually. So, for some α, $s_0 s_\alpha(x)$ and $s_0 s_\alpha(y)$ are both in V. Then $\|s_0 s_\alpha(x) - s_0 s_\alpha(y)\| \leq \operatorname{diam} V < \varepsilon$, contradicting (12.12). □

Theorem 12.23 (Aronszajn, Smith, see [ArSm]) *Let X be an infinite-dimensional complex Banach space. If T is a compact operator on X, then T has a non-trivial invariant subspace.*

Proof: Assume that $\|T\| = 1$ and put $\mathcal{T} = \{S \in \mathcal{B}(X) : ST = TS\}$. Note that $\mathcal{T}$ is nonempty as $I_X \in \mathcal{T}$. Since $\mathcal{T}$ is a subspace of $\mathcal{B}(H)$, we get $\operatorname{span}\{S(y) : S \in \mathcal{T}\} = \{S(y) : S \in \mathcal{T}\}$ for any $y \in X$. Note that if $S \in \mathcal{T}$ then $ST, TS \in \mathcal{T}$. Thus for any $y \in X \setminus \{0\}$, $Y = \overline{\{S(y) : S \in \mathcal{T}\}}$ is an invariant subspace of T and $Y \neq \{0\}$.

If there is $y \in X \setminus \{0\}$ such that $Y \neq X$, we are done. So assume that no such y exists. Choose $x_0 \in X$ with $\|T(x_0)\| > 1$. Since for any $y \in X \setminus \{0\}$ we have $Y = X$, there is an operator $S \in \mathcal{T}$ such that

$$\|S(y) - x_0\| < 1. \tag{12.13}$$

Denote $B_0 = \{x \in X : \|x - x_0\| \leq 1\}$ and note that $0 \notin B_0 \cup \overline{T(B_0)}$. As $\overline{T(B_0)}$ is a compact set not containing the origin, by (12.13) there are operators $T_1, \ldots, T_n \in \mathcal{T}$ such that for every $y \in \overline{T(B_0)}$ there is i such that $\|T_i(y) - x_0\| < 1$. For every $y \in \overline{T(B_0)}$ and $i \leq n$ put $\lambda_i = \max\{0, 1 - \|T_i(y) - x_0\|\}$ and $\lambda(y) = \sum \lambda_i(y)$. Clearly $\lambda(y) > 0$ for every $y \in \overline{T(B_0)}$. For $y \in \overline{T(B_0)}$ put

$$\widetilde{\psi}(y) = \sum \frac{\lambda_i(y)}{\lambda(y)} T_i(y).$$

The mapping $\psi = \widetilde{\psi} \circ T : B_0 \to B_0$ is continuous and $\psi(B_0)$ is compact, so by Theorem 12.19 (or just use Corollary 12.41) it has a fixed point on $z_0 \neq 0$. Define an operator by $S_0(z) = \sum \frac{\lambda_i\big(T(z_0)\big)}{\lambda\big(T(z_0)\big)} T_i(z)$. Then $S_0 \in \mathcal{T}$ and $Z = \operatorname{Ker}(I_X - S_0 \circ T)$ contains z_0. Since S_0 and T commute, Z is easily seen to be a nontrivial invariant subspace of T. Indeed, if $z \in Z$, then $z = S_0 T(z)$. Thus $T(z) = T S_0 T(z) = (S_0 T)\big(T(z)\big)$ and $T(z) \in Z$. As $T\big|_Z$ has an inverse $S_0\big|_Z$, it is a compact isomorphism and thus $\dim(Z) < \infty$, in particular $Z \neq X$. Moreover, $T|_Z$ has an eigenvalue. □

12.3 The Homeomorphisms of Convex Compact Sets: Keller's Theorem

12.3.1 Introduction

In this section, all topological concepts in ℓ_2 refer to the topology of the norm $\|\cdot\|_2$.

Definition 12.24 A Keller space *is a compact convex subset of a topological vector space affinely homeomorphic to an infinite-dimensional compact convex subset of* ℓ_2.

The main result of this section is Theorem 12.37, which ensures that *all Keller spaces are mutually homeomorphic.*

Observe that *every compact convex set C in a topological vector space X such that there exists a countable subset $\{f_n : n \in \mathbb{N}\}$ of X^* that* separates points of C *(i.e., $x = y$ whenever $x, y \in C$ and $f_n(x) = f_n(y)$ for all $n \in \mathbb{N}$) is a Keller space.* Indeed, we may assume that $\sup_{x \in C} |f_n(x)| < 1/n$ for all $n \in \mathbb{N}$. Then, $\phi(x) = (f_n(x))_{n=1}^{\infty}$, $x \in C$, defines an affine homeomorphism from C into a compact convex subset of ℓ_2. In particular, *every compact convex subset of a metrizable locally convex space —or, more generally, a convex compact and metrizable subset of a locally convex space— is a Keller space* (see Exercise 12.20).

Let $\mathbb{I} = [-1, 1]$. The *Hilbert cube* $Q := \mathbb{I}^{\mathbb{N}}$ is a compact convex subset of the Fréchet space (see Definition 3.28) $(\mathbb{R}^{\mathbb{N}}, \mathcal{T}_p)$, where $\mathcal{T}_p$ is the product topology, so, by the previous observation, Q is a Keller space. It will follow from Theorem 12.37 that every Keller space is homeomorphic to Q. Note that $\mathbb{I}^{\mathbb{N}}$ is affinely homeomorphic to the subset $\{(x_n) : |x_n| \leq 2^{-n}, n \in \mathbb{N}\}$ of $(\ell_2, \|\cdot\|_2)$ defined in Exercise 1.51.

In order to prove Theorem 12.37 it suffices to ensure that all infinite-dimensional compact convex subsets of ℓ_2 are mutually homeomorphic. A natural way to do this is to find a single compact space T (typically, a subset of $\mathbb{R}^{\mathbb{N}}$) such that every infinite-dimensional compact convex subset K of ℓ_2 would be homeomorphic to T. Fix such a K. The set $K - K$ is compact and convex balanced, so there exists $\{v_n : n \in \mathbb{N}\} \subset (K - K)$ such that the set $\{g_n : n \in \mathbb{N}\} \subset \ell_2$ given by the formula $g_n(x) = \langle x, v_n \rangle$, $x \in \ell_2$, $n \in \mathbb{N}$, is minimal for the property of separating points of K (see Exercise 12.21). For $x_0 \in K$, we construct (see Fig. 12.4) a decreasing sequence $(K_n(x_0))_{n=0}^{\infty}$ of compact convex subsets of K such that $\bigcap_{n=0}^{\infty} K_n(x_0) = \{x_0\}$. This is done by induction. Start by putting $K_0(x_0) = K$. Assume that $K_n(x_0)$ has been defined for $k = 0, 1, 2, \ldots, n$. Then put $K_{n+1}(x_0) = \{x \in K_n(x_0) : g_{n+1}(x) = g_{n+1}(x_0)\}$. Let $a_{n+1}(x_0) = \inf\{g_{n+1}(x) : x \in K_n(x_0)\}$, $b_{n+1}(x_0) = \sup\{g_{n+1}(x) : x \in K_n(x_0)\}$ (in particular, $a_1(x_0) = a_1 := \inf\{g_1(x) : x \in K\}$ and, analogously, $b_1(x_0) = b_1 := \sup\{g_1(x) : x \in K\}$ for all $x_0 \in K$; notice that $a_1 < b_1$, otherwise $\{g_n : n \geq 2\}$ would separate points of K).

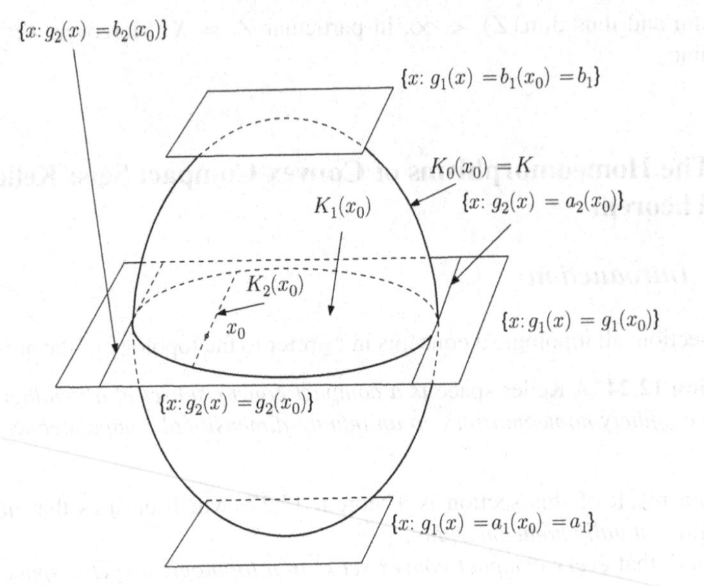

Fig. 12.4 Defining the sequence $\{K_n(x_0)\}$

It is natural now to define a mapping $F : K \to \mathbb{I}^{\mathbb{N}}$ by the formula

$$F(x) = (f_n(x)),\ x \in K, \tag{12.14}$$

where, for $n \in \mathbb{N}$,

$$f_n(x) := \begin{cases} \frac{2g_n(x)-(b_n(x)+a_n(x))}{b_n(x)-a_n(x)}, & \text{if } b_n(x) - a_n(x) \neq 0, \\ 0 & \text{otherwise.} \end{cases} \tag{12.15}$$

The mapping F is one-to-one. Indeed, if $F(x) = F(y)$ for two points x and y in K, then, in particular, $f_1(x) = f_1(y)$. It follows from (12.15) that $g_1(x) = g_1(y)$. This implies that $K_1(x) = K_1(y)$. If g_2 is constant on $K_1(x)$, then $g_2(x) = g_2(y)$. Otherwise, $(a_2(y) =)\ a_2(x) < b_2(x)\ (= b_2(y))$, and from (12.15) we get again $g_2(x) = g_2(y)$. Proceed recursively to get $g_n(x) = g_n(y)$ for all $n \in \mathbb{N}$. The fact that $\{g_n : n \in \mathbb{N}\}$ separates points of K concludes that $x = y$.

It is unlikely that F would map K onto $\mathbb{I}^{\mathbb{N}}$. For example, assume that K has the property that every *proper support functional* (see Definition 12.25) attains its supremum on K at a single point (i.e., K is *elliptically convex*, see Definition 12.27 and Exercise 12.23). Then the sequence $(1, 1, c_2, c_3, \dots)$, for c_n arbitrary in $\mathbb{I}$, $n > 2$, is *not* in the range of F. Indeed, $f_1(x) = 1$ means that x is a proper support point of K (supported by g_1), so $K_1(x) = \{x\}$ and then $a_2(x) = g_2(x) = b_2(x)$. It follows that $f_n(x) = 0$ for $n \geq 2$.

In order to deal with this situation, we will modify homeomorphically our set K to become elliptically convex (this is relatively easy, see Proposition 12.29) and will consider as the range of F an adequate subset T of $\mathbb{I}^{\mathbb{N}}$ reflecting precisely the behavior above. Moreover, T will be compact endowed with a topology $\mathcal{T}$ coarser than (and different from) the product topology $\mathcal{T}_p$, see the description of the space T below.

12.3.2 Elliptically Convex Sets

Definition 12.25 *Let A be a nonempty subset of a normed space $(X, \|\cdot\|)$. An element $f \in X^*$ is a* proper support functional *of A if f is bounded above on A, not constant on A, and there exists $x_0 \in A$ such that $f(x_0) = \sup\{f(x) : x \in A\}$. We say then that x_0 is a* proper support point *of A (*properly supported by f*). The set of properly supported points of A is denoted* psupp A.

Lemma 12.26 *Let C be a nonempty separable and complete convex subset of a normed space $(X, \|\cdot\|)$. Then $C \backslash \text{psupp}\, C \neq \emptyset$.*

Proof: Without loss of generality, we may assume that $0 \in C$. Let (x_n) be a dense sequence in C. We can find a sequence (λ_n) in $\mathbb{R}$ such that $\lambda_n > 0$ for all $n \in \mathbb{N}$, $\sum_{n=1}^{\infty} \lambda_n \leq 1$ and, moreover, $\sum_{n=1}^{\infty} \lambda_n x_n$ converges (to some element x_0 $(\in C)$). We claim that $x_0 \notin \text{psupp}\, C$. Indeed, assume that $f \in X^*$ properly supports C at x_0. Then

$$f(x_0) = \sum_{n=1}^{\infty} \lambda_n f(x_n) \leq \left(\sum_{n=1}^{\infty} \lambda_n\right) . \sup_n f(x_n) \leq f(x_0),$$

hence $f(x_n) = f(x_0)$ for all $n \in \mathbb{N}$, so f is constant on C, a contradiction. □

Recall that, for x, y in a linear space, $[x, y]$ denotes the closed line segment with endpoints x and y, and that, if $x \neq y$, (x, y) denotes the corresponding open line segment.

The following definition describes a class of convex sets which behave in a "strictly convex" way. Certainly, the closed unit ball of a normed space is in this class if and only if it is strictly convex (see Definition 7.6).

Definition 12.27 *A nonempty subset C of a normed space X is called* elliptically convex *if, for every $x \neq y$ in C, we have $(x, y) \cap \text{psupp}\, C = \emptyset$.*

It is easy to prove (see Exercise 12.23) that *a nonempty subset C of a normed space X is elliptically convex if it is convex and every proper support functional $f \in X^*$ attains its supremum on C precisely at a single point.*

Proposition 12.28 *Let $(X, \|\cdot\|)$ be a normed space, and let $C \subset X$ be an elliptically convex set. Then, given $f \in X^*$ and $\alpha \in \mathbb{R}$ such that $\inf_{x \in C} f(x) < \alpha < \sup_{x \in C} f(x)$, the set $C_\alpha := C \cap \{x \in X : f(x) = \alpha\}$ is again elliptically convex.*

Proof: It will be enough to prove that $\text{psupp}(C_\alpha) = \text{psupp}(C) \cap C_\alpha$. The inclusion $\text{psupp}(C_\alpha) \supset \text{psupp}(C) \cap C_\alpha$ is easy, since C_α separates C into two parts. In the other direction, take $x_0 \in \operatorname{psupp}(C_\alpha)$ and let $g \in X^*$ be a proper support functional of C_α at x_0. By translating, we may assume that $x_0 = 0$ (so $\alpha = 0$). Let $C^+ = \{x \in C:\ f(x) > 0\}$, $C^- = \{x \in C:\ f(x) < 0\}$, two nonempty subsets of C. For $t \in \mathbb{R}$ put $g_t = g + tf$, and let

$$A = \{t \in \mathbb{R}:\ \sup_{x \in C^-} g_t(x) > 0\}, \qquad B = \{t \in \mathbb{R}:\ \sup_{x \in C^+} g_t(x) > 0\}.$$

Letting $t \to +\infty$ we get that $B \neq \emptyset$. Similarly, $A \neq \emptyset$, by letting $t \to -\infty$. If $t \in A \cap B$, then there exists $c^+ \in C^+$ and $c^- \in C^-$ such that $g_t(c^+) > 0$ and $g_t(c^-) > 0$, so $g_t(c_0) > 0$, where $\{c_0\} := [c^-, c^+] \cap C_0$, i.e., $g(c_0) > 0$. This contradicts the fact that $g(x) \le 0$ for $x \in C_0$. It follows that $A \cap B = \emptyset$. Since obviously A and B are open subsets of $\mathbb{R}$, there exists $t \in \mathbb{R} \backslash (A \cup B)$. Then $g_t(c^+) \le 0$ for all $c^+ \in C^+$ and $g_t(c^-) \le 0$ for all $c^- \in C^-$. Since g is not constant on C_0, we get that $0 \in \operatorname{psupp} C$ (properly supported by g_t, see Fig. 12.5). □

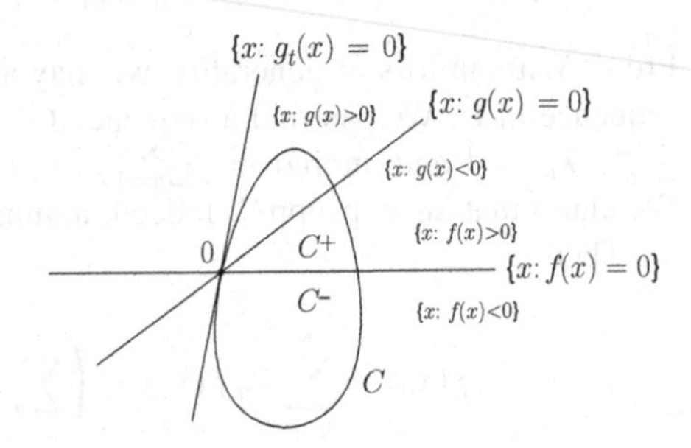

Fig. 12.5 Searching for a "tangent line"

Given a normed space $(X, \|\cdot\|)$, let us define a mapping $R : X \to X$ by $R(x) = (1+\|x\|)^{-1}x$, $x \in X$. Obviously, R is a homeomorphism from X onto B_X^O, the open unit ball of X.

Proposition 12.29 *Let $(X, \|\cdot\|)$ be a normed space. Let C be a convex subset of X such that $0 \in C$. Then $R(C)$ is convex. Moreover, if $0 \in C \backslash \operatorname{psupp} C$ and $\|\cdot\|$ is strictly convex, the set $R(C)$ is elliptically convex* (see Fig. 12.6).

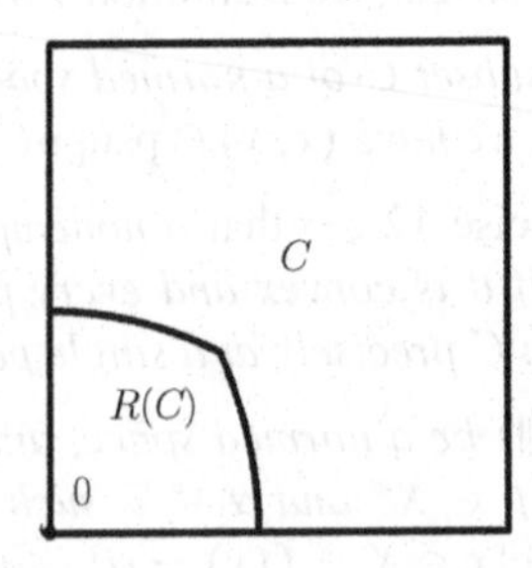

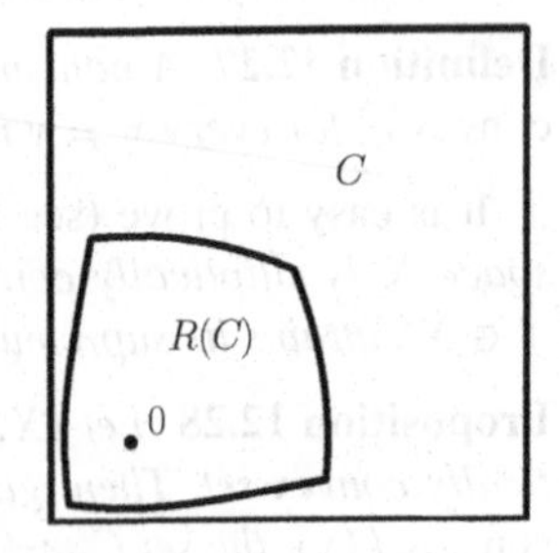

Fig. 12.6 In the second case, the object becomes elliptically convex

Proof: Given two points $x \neq y$ in C and $t, t' > 0$ such that $t + t' = 1$, a simple computation gives that $tR(x) + t'R(y) = R(ax + by)$, where

$$a := t(1 + \|x\|)^{-1}\left(1 - \left\|\frac{tx}{1+\|x\|} + \frac{t'y}{1+\|y\|}\right\|\right)^{-1}$$

and

$$b := t'(1 + \|y\|)^{-1}\left(1 - \left\|\frac{tx}{1+\|x\|} + \frac{t'y}{1+\|y\|}\right\|\right)^{-1}.$$

Obviously, $a > 0$ and $b > 0$. Moreover,

$$\begin{aligned} 1 - \left\|\frac{tx}{1+\|x\|} + \frac{t'y}{1+\|y\|}\right\| &\geq 1 - t\frac{\|x\|}{1+\|x\|} - t'\frac{\|y\|}{1+\|y\|} \\ = t + t' - t\frac{\|x\|}{1+\|x\|} - t'\frac{\|y\|}{1+\|y\|} &= \frac{t}{1+\|x\|} + \frac{t'}{1+\|y\|}. \end{aligned} \tag{12.16}$$

Hence, $a + b \leq 1$. Since C is convex and contains 0, we get $ax + by \in C$, so $tR(x) + t'R(y) \in R(C)$. This proves that $R(C)$ is convex.

Assume now that, additionally, $\|\cdot\|$ is strictly convex and $0 \notin \text{psupp}(C)$. If x and y are linearly independent, the inequality in (12.16) is strict, so $a + b < 1$. Obviously, $\frac{ax+by}{a+b} \in C$ and, since the mapping $t \to t(1+t)$ defined in $[0, +\infty)$ is strictly increasing, $tR(x) + t'R(y) = R(ax + by) \in (0, R(\frac{ax+by}{a+b}))$. If x, y are linearly dependent, we encounter again that $tR(x)+t'R(y) \in (0, z)$ for some $z \neq 0$ in $R(C)$. If $tR(x)+t'R(y)$ is properly supported in $R(C)$ by some $f \in X^*$, we have two possibilities: either $f(tR(x) + t'R(y)) > 0$, so $f(z) > f(tR(x) + t'R(y))$, a contradiction, or $f(tR(x) + t'R(y)) = 0$ and $0 \in \text{psupp}\, R(C)$. This easily implies that $0 \in \text{psupp}(C)$, again a contradiction. It follows that $R(C)$ is elliptically convex. □

Corollary 12.30 *Every closed convex subset of ℓ_2 is homeomorphic to a closed elliptically convex subset of ℓ_2.*

Proof: Let C be a nonempty closed convex subset of ℓ_2. From Lemma 12.26, we can find $c_0 \in C \setminus \text{psupp}(C)$. By translating C we may assume that $c_0 = 0$. It is enough now to apply Proposition 12.29, since certainly $\|\cdot\|_2$ in ℓ_2 is strictly convex (see the paragraph preceding Theorem 9.3). □

12.3.3 The Space T

Given a sequence $0 \neq x = (x(n))$ of real numbers such that for some $n \in \mathbb{N}$ we have $x(m) = 0$ for all $m > n$, let $\text{ord}\, x$ be the smallest n with this property. If no such n exists or if $x = 0$, put $\text{ord}\, x = \infty$.

The set T consists of all sequences $x = (x(n))$ in one or the two following classes.

(i) Either $\operatorname{ord} x = \infty$ and $|x(n)| < 1$ for all $n \in \mathbb{N}$, or

(ii) $\operatorname{ord} x < \infty$, $|x(\operatorname{ord} x)| = 1$ and $|x(n)| < 1$ for every $n \in \mathbb{N}$ such that $n < \operatorname{ord} x$.

Let us define mappings $\varphi_n : T \to \mathbb{I}$, $n \in \mathbb{N}$. For $x \in T$, put

$$\begin{cases} \varphi_1(x) = x(1), \\ \varphi_n(x) = x(n) \prod_{j=1}^{n-1}(1 - |x(j)|), \; n = 2, 3, \dots \end{cases}$$

It is obvious that the family $\{\varphi_n : n \in \mathbb{N}\}$ separates points of T. Therefore, the formula

$$d(x, y) = \sum_{n=1}^{\infty} 2^{-n} |\varphi_n(x) - \varphi_n(y)|, \; x, y \in T$$

defines a metric d on T. An alternative description of the topology $\mathcal{T}$ induced by d is the following. $x_k \xrightarrow{\mathcal{T}} x_0$ *in* T *if and only* $\lim_k x_k(n) = x_0(n)$ *for every* $n \in \mathbb{N}$ *such that* $n \leq \operatorname{ord} x_0$ (see Exercise 12.24).

It is easy to prove that the metric space $(T, \mathcal{T})$ is compact (see Exercise 12.25).

12.3.4 Compact Elliptically Convex Subsets of ℓ_2

Proposition 12.31 *Let K be an infinite-dimensional compact and elliptically convex subset of ℓ_2. Then K is homeomorphic to T.*

We follow the notation introduced in the Introduction to this section. In order to prove Proposition 12.31, we need some simple lemmas.

Lemma 12.32 *Let K be as in Proposition* 12.31. *Let $x \in K$ such that for some $n \in \mathbb{N}$ we have $|f_i(x)| < 1$, $i = 1, 2, \dots, n$. Then $K_i(x)$ is elliptically convex for $i = 1, 2, \dots, n$.*

Proof: The properties of the system $\{v_n : n \in \mathbb{N}\}$ imply that $a_1 \neq b_1$. Since $|f_1(x)| < 1$, we get $a_1 < g_1(x) < b_1$. By Proposition 12.28, $K_1(x)$ is elliptically convex. Assume then that $|f_2(x)| < 1$. If $a_2(x) = b_2(x)$ then $K_2(x) = K_1(x)$ and so $K_2(x)$ is certainly elliptically convex. Otherwise, $a_2(x) < g_2(x) < b_2(x)$, so, again by Proposition 12.28, $K_2(x)$ is also elliptically convex. Proceed recursively while $|f_i(x)| < 1$. □

Lemma 12.33 *Let K be as in Proposition* 12.31. *Then $F(K) \subset T$, where F is the mapping defined in* (12.14).

Proof: Certainly, $F(x) := (f_n(x)) \in \mathbb{I}^{\mathbb{N}}$ for all $x \in K$. If $|f_n(x)| < 1$ for all $n \in \mathbb{N}$, then $F(x) \in T$. Otherwise, let n be the first natural number with $|f_n(x)| = 1$. If

$n = 1$, then $g_1(x) \in \{a_1, b_1\}$. Since $K_0(x)$ $(= K)$ is elliptically convex, it follows that $K_1(x) = \{x\}$ (see Exercise 12.23) and so $f_m(x) = 0$ for all $m > 1$. If $n > 1$, then $|f_i(x)| < 1$ for $i = 1, 2, \dots, n-1$. From Lemma 12.32 we know that $K_{n-1}(x)$ is elliptically convex. Moreover, $a_n(x) < b_n(x)$ (otherwise, $f_n(x) = 0$). Then g_n is not constant on $K_{n-1}(x)$, so $K_n(x) = \{x\}$ (see Exercise 12.23) and $f_m(x) = 0$ for all $m > n$, so again $F(x) \in T$. □

Our next task is to prove surjectivity. Let $K_0 = K$ be a set as in Proposition 12.31. Assume, without loss of generality, that $0 \in K_0$. Fix $n \in \mathbb{N}$ and let $\{t_i : 1 \le i \le n\}$ be a finite subset of $(-1, 1)$. We shall inductively construct a strictly decreasing finite sequence $K_0 \supset K_1 \supset \dots \supset K_n$ of nonempty compact and elliptically convex sets in the following way. Let $a_i = \inf_{x \in K_{i-1}} g_i(x)$, $b_i = \sup_{x \in K_{i-1}} g_i(x)$, for $i = 1, 2, \dots, n$. Then $a_1 < b_1$ (otherwise, $\{g_n : n \ge 2\}$ will separate points of K_0). Let $\gamma_1 \in (a_1, b_1)$ such that $t_1 = \frac{2\gamma_1 - (a_1 + b_1)}{b_1 - a_1}$. Put $K_1 = \{x \in K_0 : g_1(x) = \gamma_1\}$, a nonempty compact and elliptically convex proper subset of K_0 (Proposition 12.28).

We Claim that $a_2 < b_2$. If not, g_2 will be constant on K_1. Assume that $k, k' \in K$ satisfy $g_i(k) = g_i(k')$ for all $i \neq 2$. If $g_1(k)$ $(= g_1(k'))$ $\ge \gamma_1$, take $w_1 \in K_0$ such that $g_1(w_1) = a_1$. Otherwise, take $w_1 \in K_0$ such that $g_1(w_1) = b_1$. Find p, p' in K_1 such that $p \in (w_1, k]$ and $p' \in (w_1, k']$, say $p = \lambda w_1 + (1 - \lambda)k$ and $p' = \lambda' w_1 + (1 - \lambda')k'$, where $\lambda, \lambda' \in [0, 1)$. Since $g_1(p) = g_1(p')$ and $g_1(k) = g_1(k')$ we get $\lambda = \lambda'$. By the assumption, $g_2(p) = g_2(p')$, hence $g_2(k) = g_2(k')$. It follows that $\{g_i : i \neq 2\}$ separates points of K_0, a contradiction. This proves the claim. △

Let $\gamma_2 \in (a_2, b_2)$ such that $t_2 = \frac{2\gamma_2 - (a_2 + b_2)}{b_2 - a_2}$. Put $K_2 = \{x \in K_1 : g_2(x) = \gamma_2\}$, a nonempty compact and elliptically convex proper subset of K_1 (Proposition 12.28).

We Claim that $a_3 < b_3$. If not, g_3 will be constant on K_2. Assume that $k, k' \in K$ satisfy $g_i(k) = g_i(k')$ for all $i \neq 3$. Define w_1 as before, according to the value $g_1(k)$. Find p, p' in K_1 such that $p \in (w_1, k]$ and $p' \in (w_1, k']$. Using that $g_1(p) = g_1(p')$ and $g_i(k) = g_i(k')$ for $i = 1, 2$ we get $g_2(p) = g_2(p')$. Proceed similarly, now with $p, p' \in K_1$, finding $w_2 \in K_1$ where g_2 attains the maximum or minimum on K_1, $q, q' \in K_2$, $q \in (w_2, p]$, $q' \in (w_2, p']$, to get that, by the assumption, $g_3(k) = g_3(k')$. It follows that $\{g_i : i \neq 3\}$ separates points of K_0, a contradiction. This proves the claim. △

The recursion to define the finite sequence $K_0, K_1, \dots, K_n$ with the stated properties should now be clear.

Lemma 12.34 *Let K as in Proposition* 12.31. *Then, the mapping $F : K \to T$ is onto.*

Note that the mapping T is always one-to-one.

Proof of Lemma 12.34: Let $t = (t_n) \in T$ such that $|t_n| < 1$ for all $n \in \mathbb{N}$. The recursion in the construction preceding this lemma can be carried over indefinitely to define the (strictly) decreasing sequence $(K_n)_{n=0}^{\infty}$ of nonempty compact sets. It is enough to take $x \in \bigcap_{n=0}^{\infty} K_n$. This element satisfies $F(x) = t$.

Assume now that, for some $n \in \mathbb{N}$, $|t_n| = 1$. Let $n \in \mathbb{N}$ be the first index with this property. If $n = 1$, since $a_1 < b_1$ it is enough to choose an element in K_0 $(= K)$ where g_1 attains the supremum or the infimum. If $n > 1$, we can construct

$K_0 \supset K_1 \supset \ldots \supset K_{n-1}$ as above and still we have $a_n < b_n$, so $x \in K_{n-1}$ where g_n attains the supremum or the infimum is the sought element satisfying $F(x) = t$. □

Lemma 12.35 *Let K be a nonempty compact and elliptically convex subset of a normed space X. Let $f \in X^*$. For $x_0 \in K$, put $K(x_0) = \{x \in K : f(x) = f(x_0)\}$. Let (x_n) be a sequence in K such that $x_n \to x_0$. Then, for every $y_0 \in K(x_0)$ we can find elements $y_n \in K(x_n)$, $n \in \mathbb{N}$, such that $y_n \to y_0$.*

Proof: We may assume, without loss of generality, that $x_0 = 0$. Then $f(x_n) \to 0$. Put $\mathbb{N}^0 = \{n \in \mathbb{N} : f(x_n) = 0\}$, $\mathbb{N}^+ = \{n \in \mathbb{N} : f(x_n) > 0\}$, $\mathbb{N}^- = \{n \in \mathbb{N} : f(x_n) < 0\}$. If $\mathbb{N}^+ \cup \mathbb{N}^-$ is finite there is nothing to prove. Assume that $\mathbb{N}^+$ is infinite. Let n_0 be its first element; find $n_1 \in \mathbb{N}^+$, $n_1 > n_0$, such that $(0 <)\ f(x_n) < f(x_{n_0})$ for all $n \in \mathbb{N}^+, n \geq n_1$. For $n \in \mathbb{N}^+$ and $n \geq n_1$ we can find $y_n \in (y_0, x_{n_0})$ such that $f(y_n) = f(x_n)$. Put, for those indices, $y_n = \lambda_n y_0 + (1 - \lambda_n) x_{n_0}$, where $\lambda_n \in [0, 1]$. Since $f(y_n) = (1 - \lambda_n) f(x_{n_0}) \to 0$ whenever $n \in N^+, n \geq n_1$ and $n \to +\infty$, we get $\lambda_n \to 1$, so $y_n \to y_0$. If $\mathbb{N}^-$ is infinite, proceed similarly. □

Lemma 12.36 *Let K, f and $K(x)$, for $x \in K$, be as in Lemma* 12.35. *Let g be a continuous function on K. For $x \in K$, put $a(x) = \inf_{x \in K(x)} g(x)$, $b(x) = \sup_{x \in K(x)} g(x)$. Then, the mappings a and b are continuous.*

Proof: Let (x_i) be a converging sequence in K, and let x_0 be its limit. For $i \in \mathbb{N} \cup \{0\}$, let $z_i \in K(x_i)$ such that $g(z_i) = b(x_i)$. There exists a subsequence (z_{i_j}) of (z_i) that converges to some $u_0 \in K$. Since $f(z_{i_j}) = f(x_{i_j}) \to_j f(x_0)$, we get $u_0 \in K(x_0)$. Then we have

$$b(x_0) \geq g(u_0) = \lim_j g(z_{i_j}) = \lim_j b(x_{i_j}). \tag{12.17}$$

Let $z_0 \in K(x_0)$ such that $g(z_0) = b(x_0)$. By Lemma 12.35 we can find $y_i \in K(x_i)$, $i \in \mathbb{N}$, such that $y_i \to z_0$. Then $g(y_i) \leq b(y_i) = b(x_i) = b(z_i)$ for all $i \in \mathbb{N}$. We have

$$b(x_0) = g(z_0) = \lim_j g(y_{i_j}) \leq \lim_j b(y_{i_j}) = \lim_j b(x_{i_j}). \tag{12.18}$$

From (12.17) and (12.18) we get that for every sequence (x_i) that converges to some x_0 in K we can extract a subsequence (x_{i_j}) such that $b(x_{i_j}) \to_j b(x_0)$. This proves that b is continuous. A similar argument proves that a is continuous, too. □

Proof of Proposition 12.31: Let (x_i) be a sequence in K that converges to some $x_0 \in K$. Let $F(x_i)$ $(= (f_n(x_i))_{n=1}^{\infty}) = (t_n^i)_{n=1}^{\infty}$ for $i = 0, 1, 2, \ldots$. Assume first that $|t_n^0| < 1$ for every $n \in \mathbb{N}$. Fix $n \in \mathbb{N}$. We proved in the paragraph right before Lemma 12.34 that $b_n(x_0) - a_n(x_0) \neq 0$. Lemma 12.36 implies that $b_n(x_i) - a_n(x_i) \to_i b_n(x_0) - a_n(x_0)$, so we may find $i(n) \in \mathbb{N}$ such that $b_n(x_i) - a_n(x_i) > 0$ for all $i \geq i(n)$. The continuity of g_n, a_n, and b_n implies that

$f_n(x_i) \to_i f_n(x_0)$. Since this is true for every $n \in \mathbb{N}$, we proved the pointwise convergence of $(F(x_i))$ to $F(x_0)$.

Assume now that, for some $m \in \mathbb{N}$, $|t_m^0| = 1$. Repeat the former argument to get that $t_n^i \to_i t_n^0$ for indices $n = 1, 2, \ldots, m$. The description of the topology on T given above concludes that $F(x_i) \to F(x_0)$.

We proved that F is continuous from K into T. From Lemma 12.34, the injectivity of T and the fact that K is compact, we conclude that F is a homeomorphism from K onto T. □

12.3.5 Keller Theorem

Now we can formulate and prove the following result. Recall that a *Keller space* is a compact convex subset of a topological vector space that is affinely homeomorphic to an infinite-dimensional compact convex subset of ℓ_2. In the Introduction to this section, we gave examples of Keller spaces. In particular, the Hilbert cube and every compact convex subset of a metrizable locally convex space are Keller spaces.

Theorem 12.37 (Keller, [Kelr]) *All Keller spaces are mutually homeomorphic.*

Proof: Let K be a Keller space. By Lemma 12.26 and Proposition 12.29, K is homeomorphic to an infinite-dimensional compact elliptically convex subset C of ℓ_2. By Proposition 12.31, C is homeomorphic to the compact space T introduced above. □

Corollary 12.38 *Every compact convex set in a Banach space in its norm topology, every weakly compact convex set in a separable Banach space in its weak topology, and every w^*-compact convex set in the dual of a Banach space in its w^*-topology are homeomorphic to $\mathbb{I}^{\mathbb{N}}$.*

Remark: Note that the result in Theorem 12.37 is not longer true in nonseparable situations. See, e.g., [Avil2].

12.3.6 Applications to Fixed Points

The following result, Proposition 12.39, henceforth its consequences, Corollaries 12.40 and 12.41, are less general than Theorem 12.19. However, they are a straightforward consequence of Theorem 12.37 and the fact that the Hilbert cube Q has the fixed point property (see Exercise 12.22).

Proposition 12.39 *Every Keller space has the fixed point property. In particular, every compact convex subset of a metrizable locally convex space has the fixed point property.*

Proof: It is enough to apply Theorem 12.37, the fact that Q is a Keller space (see the Introduction to this section) and Exercise 12.22. For the last assertion, recall, too, that every such a set, endowed with induced topology, is a Keller space, see again the aforesaid Introduction. □

Corollary 12.40 (Schauder fixed point theorem) *Every nonempty compact convex subset of a Banach space has the fixed point property.*

Corollary 12.41 (Alternate version of Corollary 12.40) *Let M be a nonempty closed convex and bounded subset of a Banach space X, and let $\phi : M \to M$ be a continuous mapping with compact range. Then ϕ has a fixed point.*

Proof: The set $K := \overline{\text{conv}}\,\phi(M)$ is a compact convex subset of M, and ϕ maps K into K. It is enough to apply Proposition 12.39 to K and $\phi\big|_K$. □

Another simple consequence of Proposition 12.39 is the following result.

Corollary 12.42 (Schauder) *Let H be a separable Hilbert space. If f is a weakly sequentially continuous mapping from B_H into B_H, then f has a fixed point in B_H.*

Recall that f is *weakly sequentially continuous* if $f(x_n) \stackrel{w}{\to} f(x)$ whenever $x_n \stackrel{w}{\to} x$.

Proof: Let D be a countable and dense subset of H^*. The space $(H, w(H, D))$ (see Definition 3.4) is locally convex and metrizable, and the topologies $w(H, D)$ and w $(= w(H, H^*))$ coincide on B_H, so f is a continuous mapping from $(B_H, w(H, D))$ into itself. It is enough to apply now Proposition 12.39. □

12.4 Homeomorphisms: Kadec's Theorem

In this section we will first review some results in the finite-dimensional setting.

Theorem 12.43 (Brouwer, see, e.g., [Dugu2, Ch. 16]) *If $n \neq m$, then $\mathbb{R}^n$ is not homeomorphic to $\mathbb{R}^m$.*

Proof: (Main idea) This follows from Brouwer's theorem on invariance of domains in $\mathbb{R}^m$. This theorem asserts that if h is an homeomorphism from $\mathbb{R}^m$ onto a set $h(\mathbb{R}^m)$ in $\mathbb{R}^m$, then $h(\mathbb{R}^m)$ is an open set in $\mathbb{R}^m$ (see [Dugu2, p. 358]). Indeed, then, if $n < m$ consider $\mathbb{R}^n \subset \mathbb{R}^m$. If $\mathbb{R}^n$ were homeomorphic to $\mathbb{R}^m$, then $\mathbb{R}^n$ would be an open set in $\mathbb{R}^m$, which is not true as $\mathbb{R}^n$ is a proper subspace of $\mathbb{R}^m$. □

Theorem 12.44 (Brouwer, see, e.g., [Dugu2, Ch. 16]) *If $n \in \mathbb{N}$, then $B_{\mathbb{R}^n}$ is not homeomorphic to $S_{\mathbb{R}^n}$.*

Proof: Assume h is an homeomorphism of $B_{\mathbb{R}^n}$ onto $S_{\mathbb{R}^n}$ and put, for $x \in S_{\mathbb{R}^n}$, $f(x) = -x$. Then $h^{-1}(f(h))$ is a continuous mapping from $B_{\mathbb{R}^n}$ into itself. By Brouwer's Theorem 12.14, there is $x_0 \in B_{\mathbb{R}^n}$ such that $h^{-1}f(h(x_0)) = x_0$, i.e., $f(h(x_0)) = h(x_0)$, i.e., $-h(x_0) = h(x_0)$, which is impossible as $h(x_0) \in S_{\mathbb{R}^n}$. □

Theorem 12.45 *If X is an infinite-dimensional Banach space, then both B_X and S_X are homeomorphic to X.*

Proof: For the proof we refer to [BePe2, p.190]. □

Since only finite-dimensional Banach spaces have compact balls, it follows that an infinite-dimensional Banach space cannot be homeomorphic to a finite-dimensional one. A Banach space cannot be homeomorphic to a non-complete normed space (Exercise 1.79).

If a locally convex topological vector space E is uniformly homeomorphic to a normed space, then the topology of E is defined by a norm (Exercise 12.31).

Troyanski proved in [Troy0] that $c_0(\Gamma)$ and $\ell_1(\Gamma)$ are homeomorphic for all sets Γ. Then Bessaga showed in [Bess3] that all reflexive Banach spaces are homeomorphic to a Hilbert space.

Toruńczyk proved by topological methods that all infinite-dimensional Banach spaces of the same density character are mutually homeomorphic ([Toru2]). It would be interesting to have a "Banach space" proof of Torunczyk's result for, say, WCG spaces.

Theorem 12.46 (Anderson [Ander], Kadec [Kade4a], Klee, [Klee2]) *All separable infinite-dimensional Fréchet spaces are homeomorphic to $\mathbb{R}^{\mathbb{N}}$.*

We refer to [BePe2, Chapter 6] for the proof of Theorem 12.46. Here we prove only its special case that all infinite-dimensional separable reflexive Banach spaces are mutually homeomorphic.

First we will do some preparatory work. Let X be a separable reflexive infinite-dimensional Banach space. By Theorem 8.7 we may assume that the norm of X is locally uniformly rotund. Let $\{x_i; f_i\}$ be a Markushevich basis of X (Theorem 4.59). For $n \in \mathbb{N}_0$ we put $X_n = \overline{\operatorname{span}}\{x_{n+1}, \dots\}$. Note that $\bigcap_{n=1}^{\infty} X_n = \{0\}$. Indeed, given $i \in \mathbb{N}$, we have $f_i(x) = 0$ for every $x \in X_i$ and thus $f_i(x) = 0$ for every $x \in \bigcap X_n$ and every i. Since $\{f_i\}$ separates points of X, we get that $\bigcap X_n = \{0\}$. Also, $X_n + \operatorname{span}(x_n)$ is closed in X_{n-1} and $X_n + \operatorname{span}(x_n)$ contains $\{x_n, x_{n+1}, \dots\}$, hence $X_n + \operatorname{span}(x_n) = X_{n-1}$. Thus X_n is a hyperplane in X_{n-1} generated by f_n that splits X_{n-1} into three mutually disjoint parts: X_n and the open half-spaces $X_{n-1}^{+} = f_n^{-1}(0, \infty)$, $X_{n-1}^{-} = f_n^{-1}(-\infty, 0)$.

By induction, the codimension of X_n in X is n. Note that the distance function to each X_n is weakly continuous. Indeed, if $z_k \xrightarrow{w} z$ and $\hat{z}_k, \hat{z}$ denotes their cosets in X/X_n, then $\hat{z}_k \xrightarrow{w} \hat{z}$ in X/X_n. Since X/X_n is finite-dimensional, $\hat{z}_k \to \hat{z}$ in X/X_n and thus $\operatorname{dist}(z_k, X_n) = \|\hat{z}_k\| \to \|\hat{z}\| = \operatorname{dist}(z, X_n)$. Note that given n and $x \in X$, there is a unique $z_n \in X_n$ such that $\|x - z_n\| = \operatorname{dist}(x, X_n)$ (Exercise 7.46).

Finally, for every $x \in X$ we have $0 \le \operatorname{dist}(x, X_n) \le \operatorname{dist}(x, X_{n+1}) \le \|x\|$ and $\lim_{n\to\infty} \operatorname{dist}(x, X_n) = \|x\|$. To see the latter, let $z_n \in X_n$ be such that $\|x - z_n\| = \operatorname{dist}(x, X_n)$ for every n. Then $z_n \xrightarrow{w} 0$. Indeed, given $i \in \mathbb{N}$, $f_i(z_n) = 0$ for $n \ge i$ as $f_i = 0$ on X_n. Since B_X is weakly compact and the topology on B_X of the pointwise convergence on $\{f_i\}$ is Hausdorff, the weak and the $\{f_i\}$-topology coincide on B_X. As $\{z_n\}$ is bounded, $z_n \xrightarrow{w} 0$.

From the weak lower semicontinuity of the norm we thus get $\|x\| \leq \liminf_{n\to\infty} \|x - z_n\|$. Therefore $\|x\| \leq \liminf_{n\to\infty} \|x - z_n\| \leq \limsup_{n\to\infty} \|x - z_n\| \leq \|x\|$ as $\|x - z_n\| = \mathrm{dist}(x, X_n) \leq \|x\|$ for all n. Hence $\lim_{n\to\infty} \big(\mathrm{dist}(x, X_n)\big) = \|x\|$.

For $n \in \mathbb{N}$ and $x \in X$ define $H_n(x) = \mathrm{dist}(x, X_n) = \|x - z_n\|$ for some unique $z_n \in X_n$. Furthermore, to every $x \in X$ we assign a sequence $\{h_n(x)\}$ defined inductively by

$$h_1(x) = \begin{cases} H_1(x) & \text{if } x \in X_1^+ \cup X_1 \\ -H_1(x) & \text{if } x \in X_1^- \end{cases}$$

$$h_{n+1}(x) = \begin{cases} H_{n+1}(x) & \text{if } z_n \in X_n^+ \\ h_n(x) & \text{if } z_n \in X_{n+1}, \text{ i.e., if } z_{n+1} = z_n \\ -H_{n+1}(x) & \text{if } z_n \in X_n^-. \end{cases}$$

We now have the following:

Proposition 12.47 *In the above notation, let $\{h_n\}$ be a sequence of real numbers such that $|h_{n+1}| \geq |h_n|$ for every n, $\sup |h_n| = 1$, and $h_{n+1} = h_n$ if $|h_{n+1}| = |h_n|$. Then there is a unique $x \in S_X$ such that $h_n(x) = h_n$ for every n.*

The proof of Proposition 12.47 is based on the following lemma.

Lemma 12.48 *Let $h_1, \ldots, h_n$ be a sequence of real numbers such that $|h_{i+1}| \geq |h_i|$ for every $i < n$ and $h_{i+1} = h_i$ if $|h_{i+1}| = |h_i|$. Let Q_n be the set of all $x \in X$ such that $h_i(x) = h_i$ for $i = 1, \ldots, n$. Then $Q_n = a_n + X_n$ for some $a_n \in X$.*

Proof: For $n = 1$, the set Q_1 is a shift of X_1 and the result holds. Assume the assertion holds for $h_1, \ldots, h_{n-1}$ and denote the resulting affine subspace by Q_{n-1}. The set Q_n is obviously a subset of Q_{n-1}. Let x be an element of Q_{n-1} with minimal norm, in particular $\|x\| = |h_{n-1}|$, and let y be an arbitrary element in X_{n-1}^+.

Given $\lambda \in \mathbb{R}$, for every element x_λ of the affine subspace $Q_{n,\lambda} = X_n + x + \lambda y$ we have $H_i(x_\lambda) = |h_i|$ for $i = 1, \ldots, n-1$, since $Q_{n,\lambda} \subset Q_{n-1}$ and $H_n(x_\lambda) = \psi(\lambda)$. If we vary λ, the function $\psi(\lambda)$ is a strictly convex function that attains its minimum at 0 with $\psi(0) = |h_{n-1}|$. The equation $\psi(\lambda) = |h_n|$ for $|h_n| > |h_{n-1}|$ has exactly one positive and one negative solution.

Each of these solutions gives us one of the affine subspaces $Q_{n,0}$, Q_{n,λ_1} and Q_{n,λ_2} whose elements satisfy $h_n(x) = h_{n-1}(x)$, $h_n(x_{\lambda_1}) = |h_n|$ and $h_n(x_{\lambda_2}) = -|h_n|$, respectively, by the definition of $h_n(x)$. This completes the proof of Lemma 12.48. □

Proof of Proposition 12.47: The set S of all elements $x \in X$ such that $h_n(x) = h_n$ for every n is the intersection of the sets Q_n from Lemma 12.48. Since Q_n is a shift of X_n, if the intersection of all Q_n contained at least two different points, then shifting back to n we would have that $\bigcap X_n$ would contain at least two different points, which is not the case. Therefore the set S consists at most of a singleton. To see that S is nonempty, consider the intersection of Q_n with S_X. Since $|h_n| \leq 1$ and the distance of Q_n to X_n is $|h_n|$, we have $S_X \cap Q_n \neq \emptyset$ (here we used the reflexivity of X). The weakly compact sets $Q_n \cap B_X$ form a nested family of weakly compact sets, and thus their intersection is nonempty and contained in S_X. □

Proof of the special case of Theorem 12.46: Let X, Y be infinite-dimensional separable reflexive Banach spaces. We assume that their norms are locally uniformly rotund and that $\{x_i; f_i\}$ and $\{y_i; g_i\}$ are Markushevich bases of X and Y, respectively. We construct closed subspaces X_n of X and Y_n of Y as above, and define H_n^X, h_n^X on X and H_n^Y, h_n^Y on Y accordingly. We construct a mapping φ of S_X into S_Y as follows: Given $x \in S_X$, let $\varphi(x)$ be the unique element of S_Y such that $h_n^Y(\varphi(x)) = h_n^X(x)$ for all n. Then φ is one-to-one and onto S_Y by Proposition 12.47. Since the role of X and Y is interchangeable, to prove that φ is a homeomorphism of S_X onto S_Y we only need to show that φ is continuous.

To this end, assume that $x_k, x \in S_X$ and $x_k \to x$. Since $\varphi(x_k), \varphi(x) \in S_Y$ and since on S_Y the norm and weak topology coincide, we only need to show that $\varphi(x_k) \overset{w}{\to} \varphi(x)$. Assume the contrary. Then for some subsequence $\{k_l\}$ of $\mathbb{N}$, $\{\varphi(x_{k_l})\} \overset{w}{\to} y \in B_Y$ (using reflexivity and separability), and $y \neq \varphi(x)$. Let $n_0 = \min\{n \in \mathbb{N} : h_n^Y(y) \neq h_n^X(x)\}$. Then either $|h_{n_0}^Y(y)| > |h_{n_0}^X(x)| \geq |h_{n_0-1}^Y(y)|$ or $|h_{n_0}^X(x)| > |h_{n_0-1}^X(x)|$.

Assume without loss of generality that $|h_{n_0}^Y(y)| > |h_{n_0-1}^Y(y)|$. We know that $h_{n_0}^X$ is continuous in the weak topology and thus in the $\{f_i\}$ topology at points $p \in B_X$ where $|h_{n_0-1}^X(p)| < |h_{n_0}^X(p)|$. This is due to the fact that $\operatorname{sign}(h_{n_0}^X)$ is continuous in the $\{f_i\}$-topology at points where $x_{n_0-1} \notin X_{n_0}$. Consequently, $\lim_{l\to\infty} h_{n_0}^Y(\varphi(x_{k_l})) = h_{n_0}^Y(y)$. By the assumption, however, $\lim_{l\to\infty} |h_{n_0}^Y(\varphi(x_{k_l}))| = \lim_{l\to\infty} |h_{n_0}^X(x_{k_l})| = |h_{n_0}^X(x)|$, a contradiction. Therefore φ is continuous and φ is a homeomorphism of S_X and S_Y. The homeomorphism of X onto Y can then be constructed as in Exercise 12.29. □

12.5 Lipschitz Homeomorphisms

Definition 12.49 *We say that a metric space (P, ρ) is* Lipschitz equivalent *to a metric space (Q, σ) if there is a one-to-one mapping φ of P onto Q such that φ and φ^{-1} are both Lipschitz, i.e., there is $K > 0$ such that $K^{-1}\rho(x, y) \leq \sigma(\varphi(x), \varphi(y)) \leq K\rho(x, y)$, for all $x, y \in P$. Such a mapping φ is then called a* Lipschitz homeomorphism *of P onto Q.*

Theorem 12.50 (Mazur, [Mazu]) *If $1 \leq p < \infty$, $1 \leq q < \infty$, then S_{L_p} is uniformly homeomorphic to S_{L_q}. The same result holds for S_{ℓ_p} and S_{ℓ_q}.*

Proof: Assume $p > q$. For $f \in L_p$ put $\varphi(f) = |f|^{p/q} \operatorname{sign} f$. Then φ (called the *Mazur mapping*) takes S_{L_p} onto S_{L_q}.

If $f, h \in L_p$, then $\big(\varphi(f+th) - \varphi(f)\big)/t$ converges pointwise to $\big((p/q)|f| \operatorname{sign} f\big)h \operatorname{sign} f$.

Thus, by Hölder's inequality used for $\widetilde{p} := p/(p-q)$ and $\widetilde{q} := p/q$,

$$\|\varphi'(f)(h)\|_q^q \leq \left(\frac{p}{q}\right)^q \int |f|^{p-q}|h|^q \leq \left(\frac{p}{q}\right)^q \left(\int |f|^p\right)^{\frac{p-q}{p}} \left(\int |h|^p\right)^{q/p}.$$

Therefore, the Gâteaux derivative $\varphi'(f)$ satisfies

$$\|\varphi'(f)\| \leq \sup_{\|h\|\leq 1} \|\varphi'(f)(h)\|_{L_q} \leq \frac{p}{q}\|f\|^{(p-q)/q} \leq \frac{p}{q}, \text{ if } \|f\| \leq 1,$$

and this gives the upper estimate needed for the uniform homeomorphism. We will now estimate

$$\|\varphi(f) - \varphi(g)\|_q^q \left(= \int \big||f|^{p/q} \operatorname{sign} f - |g|^{p/q} \operatorname{sign} g\big|^q\right)$$

from below. To that end, we will use the following inequalities: If $p > q$, then there is a constant C which depends on p/q such that, if $a \geq b \geq 0$, then

$$|a^{p/q} - b^{p/q}| \geq C|a-b|^{p/q}, \text{ and } |a^{p/q} + b^{p/q}| \geq C|a+b|^{p/q}.$$

By integrating these inequalities we get

$$\|\varphi(f) - \varphi(g)\|_q^q \geq C^q \int |f-g|^p = C^q\|f-g\|_p^p.$$

This gives the lower estimate needed. □

Theorem 12.51 (Aharoni [Ahar]) *Let X be a Banach space. If X is separable then it is Lipschitz equivalent to a subset of c_0.*

Using Corollary 5.9 we immediately have the following result:

Corollary 12.52 *Every separable metric space is Lipschitz equivalent to a subset of c_0.*

In the proof of Theorem 12.51, we will need the following assertion. Denote by $\mathbb{N}_0$ the set $\{0, 1, 2, \dots\}$.

Lemma 12.53 *Let X_0 be a subset of a separable metric space X. There exists a sequence $\{M_i\}_{i=0}^{\infty}$ of subsets of X_0 such that*
(1) *For every $x \in X_0$ there is $i \in \mathbb{N}_0$ such that* $\operatorname{dist}(x, M_i) < 1$,
(2) *For every $x \in X$,* $\operatorname{card}\{i \in \mathbb{N}_0 : \operatorname{dist}(x, M_i) < 3\} < \infty$,
(3) $\operatorname{diam} M_i \leq 8$ *for every $i \in \mathbb{N}_0$.*

Proof: Let Z be a maximal 1-separated family in X_0 and let $Y \supset Z$ be a maximal 1-separated family in X with respect to inclusion. Enumerate $Y = \{y_j\}_{j \in \mathbb{N}_0}$, assuming without loss of generality that Y is infinite. We put $M_0 = B_X^O(y_0, 4) \cap Z$, where $B_X^O(y_0, 4)$ is the open ball centered at y_0 with radius 4, and for $i \geq 1$ we put

$$M_i = \big(B_X^O(y_i, 4) \cap Z\big) \setminus \bigcup_{j=0}^{i-1} M_j.$$

Since $\{M_i\}$ covers Z, part (1) follows, as any maximal 1-separated family F in a metric space W has the property that $\operatorname{dist}(w, F) < 1$ for every $w \in W$. Part (3) follows from the fact that $M_i \subset B_X^O(y_i, 4)$ for every i.

To show part (2), let $x \in X$ be given. There is $j_0 \geq 1$ such that $\|x - y_{j_0}\| < 1$. If i is such that $\operatorname{dist}(x, M_i) < 3$, then $\operatorname{dist}(y_{j_0}, M_i) < 4$. If $i > j_0$ and $y \in M_i$ is such that $\operatorname{dist}(y_{j_0}, y) < 4$, then $y \in \bigcup_{j=0}^{j_0} M_j$, a contradiction. Therefore $i \leq j_0$ whenever $\operatorname{dist}(y_{j_0}, M_i) < 4$. This gives that

$$\operatorname{card}\big\{i \in \mathbb{N}_0 : \operatorname{dist}(x, M_i) < 3\big\} \leq j_0 + 1.$$

$\square$

Proof of Theorem 12.51: (Assouad) Fix $e \in X$ and for $n \in \mathbb{Z}$ consider the subset X_n of X defined by $X_n = X \setminus B_X^O\big(e, 6(\frac{2}{3})^n\big)$. For $n \in \mathbb{N}$ consider on X the metric $\rho_n = \big(\frac{3}{2}\big)^n \rho$, where ρ is the canonical metric of X. For $n \in \mathbb{Z}$ construct the sequence $\{M_{i,n}\}_{i \in \mathbb{N}_0}$ applying Lemma 12.53 to the subset X_n of the metric space (X, ρ_n).

For $n \in \mathbb{Z}$, $i \in \mathbb{N}_0$ and $x \in X$, put $f_{i,n}(x) = \big(3(\frac{2}{3})^n - \rho(x, M_{i,n})\big)^+$, where $u^+ := \max\{u, 0\}$. Let $e_{i,n}$ denote the standard unit vectors in $\ell_\infty(\mathbb{N}_0 \times \mathbb{Z})$ and define for $x \in X$ formally

$$f(x) = \sum_{(i,n) \in \mathbb{N}_0 \times \mathbb{Z}} f_{i,n}(x) e_{i,n}.$$

We claim that $f(x) \in c_0(\mathbb{N}_0 \times \mathbb{Z})$ for $x \in X$ and

$$\frac{\rho(x, y)}{9} \leq \|f(x) - f(y)\|_{c_0(\mathbb{N}_0 \times \mathbb{Z})} \leq \rho(x, y)$$

for $x, y \in X$. Having the claim proven, it suffices to note that $c_0(\mathbb{N}_0 \times \mathbb{Z})$ is isometric to $c_0(\mathbb{N})$.

To prove the claim, given $n_0 \in \mathbb{Z}$, consider $i, m \in \mathbb{N}_0 \times \mathbb{Z}$ such that $f_{i,m}(x) \geq 3\big(\frac{2}{3}\big)^{n_0}$. Then from the definition of $f_{i,m}(x)$ we have $m \leq n_0$, so $f_{i,m}(x) > 3\big(\frac{2}{3}\big)^{n_0}$ for only finitely many integers m.

Now, for any such m, whenever $i \in \mathbb{N}$ satisfies $f_{i,m}(x) > 3\big(\frac{2}{3}\big)^{n_0}$, we get in particular $f_{i,m}(x) > 0$, and thus $\rho(x, M_{i,m}) < 3\big(\frac{2}{3}\big)^m$ from the definition of $f_{i,m}(x)$. So by the definition of the distance ρ_m we have $\rho_m(x, M_{i,m}) < 3$, and by the definition of $M_{i,m}$, there are only finitely many such i. This proves that

$$\operatorname{card}\left\{(i, n) \in \mathbb{N}_0 \times \mathbb{Z} : f_{i,n}(x) > 3\left(\tfrac{2}{3}\right)^{n_0}\right\} < \infty,$$

showing that $f(x) \in c_0(\mathbb{N}_0 \times \mathbb{Z})$ for every $x \in X$.

Let $x, y \in X$, $x \neq y$. There is $n \in \mathbb{Z}$ such that

$$3 \cdot 4\left(\tfrac{2}{3}\right)^n \leq \rho(x, y) \leq 3 \cdot 4\left(\tfrac{3}{2}\right) \cdot \left(\tfrac{2}{3}\right)^n = 18 \cdot \left(\tfrac{2}{3}\right)^n.$$

Thus at least one of x, y is further from e then $\frac{3}{2}4\left(\frac{2}{3}\right)^n = 6 \cdot \left(\frac{2}{3}\right)^n$. We may assume that $\rho(x, e) \geq 6 \cdot \left(\frac{2}{3}\right)^n$, so $x \in X_n$. By the definition of $M_{i,n}$ we have that there is $i \in \mathbb{N}$ such that $\rho(x, M_{i,n}) < \left(\frac{2}{3}\right)^n$. Therefore $f_{i,n}(x) \geq 3 \cdot \left(\frac{2}{3}\right)^n - \left(\frac{2}{3}\right)^n = 2 \cdot \left(\frac{2}{3}\right)^n$. On the other hand, if $f_{i,n}(z) > 0$, then from the definition of $f_{i,n}$ we have $\rho(z, M_{i,n}) < 3\left(\frac{2}{3}\right)^n$. As the diameter of $M_{i,n}$ in the canonical metric of X is at most $8 \cdot \left(\frac{2}{3}\right)^n$ and $\rho(x, M_{i,n}) < \left(\frac{2}{3}\right)^n$, assuming that $f_{i,n}(z) > 0$ we obtain an estimate for the diameter of $M_{i,n}$ in ρ:

$$\begin{aligned}\rho(x, z) &\leq \rho(x, M_{i,n}) + \operatorname{diam} M_{i,n} + \rho(z, M_{i,n}) \\ &< \left(\tfrac{2}{3}\right)^n + 8 \cdot \left(\tfrac{2}{3}\right)^n + 3 \cdot \left(\tfrac{2}{3}\right)^n = 12 \cdot \left(\tfrac{2}{3}\right)^n.\end{aligned}$$

Since $\rho(x, y) \geq 12 \cdot \left(\frac{2}{3}\right)^n$, necessarily $f_{i,n}(y) = 0$. Consequently

$$\|f(x) - f(y)\|_{c_0(\mathbb{N}_0 \times \mathbb{Z})} \geq |f_{i,n}(x) - f_{i,n}(y)| \geq 2\left(\tfrac{2}{3}\right)^n \geq \frac{\rho(x, y)}{9}.$$

This finishes the proof of the Claim. $\triangle$

On the other hand, from the definition of $f_{i,n}$ we have that all $f_{i,n}$ are 1-Lipschitz. We conclude that $\|f(x) - f(y)\|_{c_0(\mathbb{N}_0 \times \mathbb{Z})} \leq \rho(x, y)$. $\square$

Pelant, Holický, and Kalenda proved that $C([0, \omega_1])$ cannot be homeomorphically embedded into any $c_0(\Gamma)$, [PHK].

Theorem 12.54 (Mazur, Ulam, see, e.g., [BeLi, p.341]) *Every isometry from a Banach space X onto a Banach space Y that takes 0 to 0 is necessarily linear.*

Proof: Let f be an isometry of X onto Y such that $f(0) = 0$. Take $x, y \in X$. Put for $n \in \mathbb{N}$,

$$K_0 = \left\{u \in X : \|u - x\| = \|u - y\| = \left\|\frac{x + y}{2}\right\|\right\}$$

$$K_{n+1} = \left\{u \in K_n : K_n \subset B\left(u, \frac{d_n}{2}\right)\right\},$$

where $d_n := \operatorname{diam} K_n$, $n \in \mathbb{N}$. It suffices to show that $\bigcap K_n = \{\frac{x+y}{2}\}$. Indeed, using then the fact that $f(B_r(x)) = B_r(f(x))$ for every $x \in X$ and every $r > 0$, it follows that if we define similarly the sets K_n for $f(x)$ and $f(y)$, then $\frac{f(x)+f(y)}{2}$ is their intersection. Hence $f\left(\frac{x+y}{2}\right) = \frac{f(x)+f(y)}{2}$.

Using the continuity of f and the fact that $f(0) = 0$ we get, by a standard argument, that f is a (linear) operator. So we will now show that $\bigcap K_n = \{\frac{x+y}{2}\}$. For it first translate so that $y = -x$. We Claim that $0 \in K_n$ and that $\operatorname{diam} K_{n+1} \leq \frac{d_n}{2}$ for each n.

To prove the Claim, we will first show by induction that $0 \in K_n$ for each n. Note that each K_n is symmetric, i.e., $u \in K_n$ implies $-u \in K_n$. Clearly, $0 \in K_0$. Pick $u \in K_n$. Then $-u \in K_n$ and $2\|u\| = \|u - (-u)\| \leq d_n$. Thus $K_n \subset B(0, \frac{d_n}{2})$ and therefore $0 \in K_{n+1}$. To show that $\operatorname{diam} K_{n+1} \leq \frac{d_n}{2}$, pick $u, v \in K_{n+1}$. Then, by the definition of K_{n+1}, $v \in B(u, \frac{d_n}{2})$. Therefore $\|u - v\| \leq \frac{d_n}{2}$, which gives that $\operatorname{diam} K_{n+1} \leq \frac{d_n}{2}$. This proves the Claim. $\triangle$

Thus $\operatorname{diam} K_n \to 0$. Hence $\{0\} = \bigcap K_n$ and the proof of the theorem is finished. $\square$

Remark: It is proven in [GoKa1b] that if a separable Banach space X isometrically (non-linearly, in general) embeds into a Banach space Y, then Y contains a linear subspace that is linearly isometric to X.

Theorem 12.55 (Heinrich and Mankiewicz, [HeMa]) *Assume that a separable Banach space X is Lipschitz equivalent to a subset of a Banach space Y that has the RNP property. Then X is isomorphic to a subspace of Y. If Y is reflexive, and X is Lipschitz equivalent to Y, then X is isomorphic to a complemented subspace of Y. If X and Y are reflexive separable Banach spaces that are Lipschitz equivalent and both satisfy the Pełczyński decomposition method, then X and Y are isomorphic.*

Proof: Let f be a Lipschitz homeomorphism of X onto a subset of Y such that, for some $k > 0$,

$$k^{-1}\|x - y\| \leq \|f(x) - f(y)\| \leq k\|x - y\|$$

for all $x, y \in X$. Assume, without loss of generality, that f is Gâteaux differentiable at $x_0 \in X$ and let T be its derivative (Theorem 11.21). Then, for $x \in X$,

$$Tx = \lim_{t \to 0} \frac{f(x_0 + tx) - f(x_0)}{t},$$

and T is a (linear) operator from X into Y, and we have, for all $x \in X$,

$$K^{-1}\|x - x_0\| \leq \|T(x - x_0)\| \leq K\|x - x_0\|.$$

This proves the first part of the statement in the theorem.

Let f be now a Lipschitz homeomorphism from a separable Banach space onto a reflexive Banach space Y. Assume, without loss of generality, that f is Gâteaux differentiable at $x_0 := 0$ and that $f(0) = 0$.

Put $T = f'(0)$. For $y \in Y$, put $g(y) = f^{-1}(y)$ and $g_n(y) = ng(y/n)$. Then the functions g_n are uniformly Lipschitz mappings from Y (with the same Lipschitz constant as f^{-1}, i.e., K^{-1}) and, since Y is reflexive, X is reflexive by the first part of the statement in this theorem. Therefore there is a subnet $\{g_{n_\alpha}\}$ of $\{g_n\}$ such that $\{g_{n_\alpha}(y)\}$ weakly converges to some $h(y)$ for all $y \in Y$, where h is a Lipschitz function from Y into X. This follows from the weak lower semicontinuity of the norm on X.

We Claim that $h(Tx) = x$ for all $x \in X$. Indeed, for $x \in X$ put $y = Tx$. Since $x_0 = 0$, $f(0) = 0$ and $T = f'(0)$, we have $y = Tx = \lim_n nf(\frac{x}{n})$. Then

$$g_n\left(nf\left(\frac{x}{n}\right)\right) = ng\left(f\left(\frac{x}{n}\right)\right) = x, \text{ for all } x \in X.$$

Thus we have, for $x \in X$,

$$\|g_n(y) - x\| = \left\|g_n(y) - g_n\left(nf\left(\frac{x}{n}\right)\right)\right\| \leq K^{-1}\left\|y - nf\left(\frac{x}{n}\right)\right\| \to 0$$

as $n \to \infty$.

From this, from the weak lower semicontinuity and from the definition of h we get $\|h(Tx) - x\| = \|h(y) - x\| = 0$, and the Claim is proved. $\triangle$

Put now for $y \in Y$, $r(y) = T(h(y))$. Then r is a Lipschitz mapping from Y into $T(X)$. Moreover, $r(r(y)) = r(y)$ for every $y \in Y$. Indeed, given $y \in Y$, by the above,

$$r(r(y)) = r\Big(T(h(y))\Big) = T\Big(h\Big(T(h(y))\Big)\Big) = T(h(y)) = r(y).$$

Therefore r is a Lipschitz retraction from Y onto $T(X)$. As Y is reflexive, $T(X)$ is complemented in Y by Theorem 12.57 below.

The last statement in the theorem then follows from the statement in the Pełczyński decomposition method (Theorem 4.47). □

Corollary 12.56 *If a Banach space X is Lipschitz equivalent to a subset of a Banach space Y and Y is (super)reflexive, respectively has the RNP, respectively is isomorphic to a Hilbert space, then X has the same property.*

Proof: If X is separable, the statement follows from Theorem 12.55. The general case then follows because a Banach is (super)reflexive, respectively has the RNP, respectively is isomorphic to a Hilbert space, whenever all its separable subspaces are. For reflexivity see Exercise 3.115, for superreflexivity see Exercise 9.23, for RNP see Exercises 11.30 and 11.37, and, finally, for being isomorphic to a Hilbert space see Exercise 5.2. □

Theorem 12.57 (Lindenstrauss [Lind1]) *Assume that Y is a subspace of a reflexive Banach space X and that there is a Lipschitz retraction of X onto Y. Then Y is complemented in X.*

Proof: We will first assume that X is finite-dimensional and prove an estimate on the norm of a projection from X onto Y:

Let $X = Y \oplus Y_1$, and let f be a Lipschitz retraction from X onto Y with Lipschitz constant $K > 0$. We will first "smooth up" f in the direction of Y. For it, let $\varphi \geq 0$ be a C^∞-smooth function with compact support on Y such that $\int_Y \varphi = 1$ and $\varphi(y) = \varphi(-y)$ for all $y \in Y$. Put for $x \in X$,

$$\int_Y f(x+y)\varphi(y)\,\mathrm{d}y.$$

Then the Lipschitz constant of g is less than or equal to K. If $x_0 \in X$, $y_0 \in Y$, and t is a real number, then

$$g(x_0+ty_0) = \int_Y f(x_0+ty_0+y)\varphi(y)\,\mathrm{d}y = \int_Y f(x_0+u)\varphi(u-ty_0)\,\mathrm{d}u,$$

g is differentiable at x_0 in the direction of Y, and its derivative in the directions of Y is a (linear) operator $g'(x_0)(y)$. It follows that $g'(x_0)(y)$ is a continuous function in the variable x_0 on X.

Moreover, if $x_0 \in Y$, then

$$\begin{aligned} g(x_0) &= \int_Y f(x_0+y)\varphi(y)\,\mathrm{d}y = \int_Y (x_0+y)\varphi(y)\,\mathrm{d}y \\ &= \int_Y x_0\varphi(y)\,\mathrm{d}y + \int_Y y\varphi(y)\,\mathrm{d}y = x_0\int_Y \varphi(y)\,\mathrm{d}y = x_0, \end{aligned}$$

since $\varphi(y) = \varphi(-y)$ for all $y \in Y$.

Now we will "smooth up" g in the direction of Y_1. For it, let $\psi \geq 0$ be a C^∞-function with compact support on Y_1 such that $\int_{Y_1} \psi = 1$.

Put for $x_0 \in X$ and $n \in \mathbb{N}$,

$$f_n(x_0) = \int_{Y_1} g\left(x_0 + \frac{u}{n}\right)\psi(u)\,\mathrm{d}u.$$

Since g is differentiable in the directions of Y and ψ is a C^∞-function, f_n is differentiable at x_0 as a function on X, and its derivative $f_n'(x_0)$ is a (linear) operator from X into Y with norm less than or equal to K. Let P be a limit point of $\{f_n'(0)\}$ in the pointwise topology on X. Then P is a bounded (linear) operator from X into Y with norm less than or equal to K. We claim that P is the identity on Y. For it, let $y \in Y$. Since $g'(u)(y)$ exists for every $u \in Y_1$, we have

$$f_n'(0)(y) = \int_{Y_1} g'\left(\frac{u}{n}\right)(y)\psi(u)\,du.$$

Since g is the identity on Y, we have $g'(0)(y) = y$. Therefore, by passing to the limit of a subnet $\{f_{n_\alpha}'(0)\}$ of $\{f_n'(0)\}$ used in the definition of P, and by using the fact that ψ has compact support and $g'(u)(y)$ is continuous on Y_1, we get that $\left\{\int_{Y_1} g'\left(\frac{u}{n_\alpha}\right)(y)\psi(u)\,du\right\}_\alpha$ converges to $\int_{Y_1} g'(0)(y)\psi(u)\,du = y\int_{Y_1}\psi(u)\,du = y$. Since, by the definition of P, $Py = \lim \int_{Y_1} g'(0)(y)\psi(u)\,du$, we have that $Py = y$ for $y \in Y$.

This completes the proof for a finite-dimensional space X. For the general case of a reflexive space X, we follow a standard compactness argument: For each finite-dimensional subspace G of X we put $G_0 = G \bigcap Y$ and find a linear projection P_G from G onto G_0 with $\|P_G\| \leq K$. Then we use the weak compactness of the unit ball of X to pass to the limit of a convergent subnet of the net $\{P_G\}_G$. This completes the proof of the theorem. □

Theorem 12.58 (Godefroy, Kalton, [GoKa1b]) *Assume that X is a Banach space and Y is a subspace of X such that X/Y is separable. Then Y is complemented in X if and only if there is a Lipschitz lifting of the quotient mapping from X onto Y.*

Proof: We give the proof only in the case that X has the RNP property, and refer to [GoKa1b] for the general case. Let φ be a Lipschitz lifting of q and assume that φ is Gâteaux differentiable at some $\hat{x_0}$ with Gâteaux derivative D (Theorem 11.21). Then D is a lifting of q as well. Indeed, given $\hat{h} \in X/Y$,

$$\begin{aligned} q\big(D(\hat{h})\big) &= q\left(\lim_{t\to 0}\frac{1}{t}(\varphi(\hat{x}_0 + t\hat{h}) - \varphi(\hat{x}_0))\right) \\ &= \lim_{t\to 0}\frac{1}{t}\big(q(\varphi(\hat{x}_0 + t\hat{h})) - q(\varphi(\hat{x}_0))\big) = \lim_{t\to 0}\frac{1}{t}(\hat{x}_0 + t\hat{h} - \hat{x}_0) = \hat{h}. \end{aligned}$$

Define a bounded linear mapping P from X into X by $Px = x - D(\hat{x})$ for $x \in X$. Then $\widehat{Px} = \hat{x} - \widehat{D(\hat{x})} = 0$ and if $y \in Y$, then $Py = y - D(\hat{y}) = y$. Thus P is a bounded linear projection from X onto Y.

If P is a bounded linear projection from X onto Y, and $Q = I - P$, and $\widehat{Q}\hat{x} := Qx$ where $x \in \hat{x}$, then $\widehat{Q}$ is a bounded linear lifting of the quotient mapping from X onto X/Y. □

Corollary 12.59 (Godefroy, Kalton, [GoKa1b]) *Assume that X is a Banach space and Y is a subspace of X such that X/Y is separable. If there is a positive homogeneous* uniformly continuous *lifting of the quotient mapping $q : X \to X/Y$, then Y is complemented in X.*

Proof: It suffices to note that a positively homogeneous continuous mapping φ from a Banach space W into a Banach space Z is Lipschitz as soon as it is uniformly continuous: Indeed, then for some $\delta > 0$, $\|\varphi(x) - \varphi(y)\| < 1$ as soon as $\|x - y\| \leq \delta$. Given $x, y \in Z$, $x \neq y$, $\|\delta\frac{x}{\|x-y\|} - \delta\frac{y}{\|x-y\|}\| = \delta$ and thus, by the positive homogeneity, $\|\varphi(x) - \varphi(y)\| \leq \frac{1}{\delta}\|x - y\|$. □

Corollary 12.59 should be compared with Corollary 7.56.

Definition 12.60 *Let $(X, \|\cdot\|)$ be a separable Banach space. We say that, for some $c \in (0, 1]$, the norm is* c-Lipschitz Kadec–Klee smooth *if, for every $\varepsilon > 0$, $\limsup_n \|x^* + x_n^*\| \geq 1 + c\varepsilon$ for every $x^* \in S_{X^*}$ and every sequence $\{x_n^*\}$ in X^* such that $\|x_n^*\| \geq \varepsilon$ and $x_n^* \xrightarrow{w^*} 0$. If it is c-Lipschitz Kadec–Klee smooth for some $c \in (0, 1]$, then the norm is called* Lipschitz Kadec–Klee smooth.

Theorem 12.61 ([GKL1]) *If a Banach X is Lipschitz homeomorphic to a subspace of c_0, then X is isomorphic to a subspace of c_0.*

Proof: We will first show that X admits an equivalent norm $|||\cdot|||$ that is, for some $c > 0$, c-Lipschitz Kadec–Klee smooth. The statement then follows from Theorem 12.65 below.

The required norm is constructed as follows. Assume that φ is a Lipschitz homeomorphism from a subspace E of c_0 onto X such that $\|x - y\| \leq \|\varphi(x) - \varphi(y)\| \leq \|x - y\|$ for all $x, y \in E$. Define a new norm $|||\cdot|||$ on X^* by

$$|||x^*||| = \sup\left\{\frac{\langle x^*, \varphi(u) - \varphi(v)\rangle}{\|u - v\|} : u, v \in E, u \neq v\right\}$$

for $x^* \in X^*$. Then $\|x^*\| \leq |||x^*||| \leq c\|x^*\|$ for every $x^* \in X^*$, where $\|\cdot\|$ denotes the original norm of X^*. As $|||\cdot|||$ is moreover w^*-lower semicontinuous, $|||\cdot|||$ is a dual norm to an equivalent norm that will also be denoted by $|||\cdot|||$.

In order to show that $|||\cdot|||$ has the required property for $c := \frac{1}{10C}$, choose $\delta > 0$ such that $(1 - \delta)^{-1}(1 - 4\delta + \frac{\varepsilon}{5C}) \geq 1 + \frac{\varepsilon}{10C}$. As $|||x^*||| = 1$, pick vectors $u, v \in E$ such that $x^*(\varphi(u) - \varphi(v)) \geq (1 - \delta)\|u - v\|$. By a translation in E and then another in X, and by a rescaling in E, we may assume that $v = -u$, $\varphi(v) = -\varphi(u)$ and $\|u\| = 1$. Then $x^*(\varphi(u)) \geq 1 - \delta$.

Since E is a subspace of c_0, there is a subspace $E_1 \subset E$ of finite codimension in E such that $\|u + w\| = 1 + \delta$ for every $w \in B_{E_1}$.

Since φ^{-1} is 1-Lipschitz, from the version of Gorelik's principle (Theorem 12.62) it follows, for $d = 1$ and $b = \frac{1}{10}$, that there is a compact set $K \subset X$ such that

$$B_X \subset K + 5\varphi(B_{E_1}).$$

Since $x_n^* \to 0$ uniformly on the compact set K, by discarding finitely many x_n^*'s we may assume that for each n, there is $w_n \in B_{E_1}$ such that

$$\begin{aligned}
&(x_n^*, -\varphi(w_n)) = (-x_n^*, \varphi(w_n)) \\
&\geq 5^{-1}\| - x_n^*\| - \delta = 5^{-1}\|x_n^*\| - \delta \geq (5C)^{-1}|||x_n^*||| - \delta \geq (5C)^{-1}\varepsilon - \delta
\end{aligned}$$

Note that

$$(x^*, \varphi(w_n)) \leq 2\delta.$$

Indeed, $\varphi(-u) = -\varphi(u)$, $\|u + w_n\| \leq 1 + \delta$ and $x^*(\varphi(u)) \geq (1 - \delta)$ imply that

$$1 = |||x^*||| \geq \frac{(x^*, \varphi(w_n) + \varphi(u))}{\|u + w_n\|} \geq (1+\delta)^{-1}\big((x^*, \varphi(w_n)) + 1 - \delta\big).$$

Therefore $(x^*, \varphi(w_n)) + (1 - \delta) \leq 1 + \delta$, giving $(x^*, \varphi(w_n)) \leq 2\delta$.

Finally, $\|u - w_n\| \leq 1 + \delta$ and $x_n^*(\varphi(u)) \leq \delta$ for n large enough give

$$\begin{aligned}
|||x^* + x_n^*||| &\geq \frac{(x^* + x_n^*, \varphi(u) - \varphi(w_n))}{\|u - w_n\|} \\
&\geq (1+\delta)^{-1}\Big(\big(x^*, \varphi(u)\big) + \big(x_n^*, -\varphi(w_n)\big) + \big(x_n^*, \varphi(n)\big) - \big(x^*, \varphi(w_n)\big)\Big) \\
&\geq (1+\delta)^{-1}(1 - \delta + (5C)^{-1}\varepsilon - \delta - \delta - 2\delta) \geq 1 + \frac{\varepsilon}{10C}
\end{aligned}$$

for n large enough. This finishes the proof of the fact that X admits an equivalent norm that is Lipschitz Kadec–Klee smooth. □

To show that X is isomorphic to a subspace of c_0 we need the following couple of results. We recall that the *modulus of continuity* of a function f from a normed space X into a normed space Y is a function $\omega(f, \cdot) : [0, +\infty) \to [0, +\infty)$ that satisfies $\|f(x_1) - f(x_2)\| \leq \omega(\|x_1 - x_2\|)$ for every $x_1, x_2 \in X$, and $\omega(t) \to 0$ as $t \to 0$.

Theorem 12.62 (Gorelik's principle, [Gore], see, e.g., [JLS] and [GKL1]) *Let E and X be two Banach spaces and U be a homeomorphism from E onto X with uniformly continuous inverse. Let b and d be two positive constants such that $\frac{d}{4} > \omega(U^{-1}, 2b)$, where $\omega(\cdot, \cdot)$ denotes the modulus of continuity of U^{-1}. Then for every finite-codimensional subspace E_0 of E, there exists a compact set $K \subset X$ such that*

$$2bB_X \subset K + U(dB_{E_0}).$$

Proof: We will need the following Gorelik's lemma.

Lemma 12.63 (Gorelik, see, e.g., [JLS] and [GKL1]) *Let X_0 be a finite-codimensional subspace of a Banach space X. Then, for every $\tau > 0$ there is a compact set $A \subset \tau B_X$ such that if φ is a continuous mapping from A into X for which $\|\varphi(x) - x\| < \tau/2$ for all $x \in A$, then $\varphi(A) \cap X_0 \neq \emptyset$.*

Proof: Put $A = f(\frac{3}{4}\tau B_{X/X_0})$, where $f : \frac{3}{4}\tau B_{X/X_0} \to \tau B_X$ is the Bartle–Graves selector, see Corollary 7.56. Consider the mapping $x \mapsto x - \pi\varphi(f(x))$ from

$\frac{3}{4}\tau B_{X/X_0}$ into $\frac{3}{4}\tau B_{X/X_0}$, where π is the quotient mapping. It has a fixed point x_0 by Brouwer's fixed point theorem (Theorem 12.14). Then $x_0 - \pi\varphi(f(x_0)) = x_0$ and thus $\varphi(f(x_0)) \in X_0$. □

Now we continue with the proof of Theorem 12.62.

Let $A \subset \frac{d}{2}B_E$ be a compact set from Lemma 12.63 for $X = E$, $X_0 = E_0$, and $\tau = \frac{d}{2}$. Put $K = -U(A)$. Fix an arbitrary $y \in 2bB_X$. Put for $x \in A$, $\varphi(x) = U^{-1}(U(x) + y)$.

Then, for $x \in A$,

$$\begin{aligned}\|\varphi(x) - x\| &= \|U^{-1}(U(x) + y) - U^{-1}(U(x))\| \\ &\leq \omega(U^{-1}, \|y\|) \leq \omega(U^{-1}, 2b) < \frac{d}{4}.\end{aligned}$$

Therefore, by Lemma 12.63, $\varphi(A) \cap E_0 \neq \emptyset$. Since $A \subset \frac{d}{2}B_E$, $\varphi(A) \subset dB_E$. It follows that $\varphi(A) \cap dB_{E_0} \neq \emptyset$. This means that there is $x_0 \in A$ such that $U^{-1}(U(x_0) + y) \in dB_{E_0}$. This in turn means that $(U(x_0) + y) \in U(dB_{E_0})$, i.e., $y \in -U(x_0) + U(dB_{E_0}) \subset K + U(dB_{E_0})$. This completes the proof of Theorem 12.62. □

Proposition 12.64 *The canonical norm of $c_0(\Gamma)$ is 1-Lipschitz Kadec–Klee smooth for each Γ.*

Proof: Let X be the space $c_0(\Gamma)$ in its canonical norm. Let $f_\gamma \in S_{X^*}$, $f \in X^*$ be such that the net f_γ converges to f in the w^* topology of X^*, and $\|f_\gamma - f\| \geq \varepsilon$. We need to show that $\|f\| \leq 1 - \varepsilon$.

Given $\delta > 0$, let $F \subset \Gamma$ be a finite set such that $\sum_{i \notin F} |f^i| < \delta$, where f^i is the ith coordinate of f in $l_1(\Gamma)$. Let γ_0 be such that $\sum_{i \in F} |f^i_\gamma - f^i| < \delta$ for every $\gamma \geq \gamma_0$. Then $\sum_{i \notin F} |f^i_\gamma - f^i| \geq \varepsilon - \delta$ for every $\gamma \geq \gamma_0$.

Since $\sum_{i \notin F} |f^i| \leq \delta$, we have $\sum_{i \notin F} |f^i_\gamma| \geq 2 - 2\delta$ for every $\gamma \geq \gamma_0$. Since $\|f_\gamma\| = 1$ for all γ, we get $\sum_{i \in F} |f^i_\gamma| \leq 1 - \varepsilon + 2\delta$. By the w^*-lower semicontinuity of the dual norm, we get that $\sum_{i \in F} |f^i| \leq 1 - \varepsilon + 2\delta$. Thus

$$\|f\| = \sum_{i \in F} |f^i| + \sum_{i \notin F} |f^i| \leq 1 - \varepsilon + 2\delta + \delta = 1 - \varepsilon + 3\delta.$$

As $\delta > 0$ was arbitrary, we get $\|f\| \leq 1 - \varepsilon$. This finishes the proof of the proposition. □

Theorem 12.65 ([GKL1]) *A separable Banach space is isomorphic to a subspace of c_0 if and only if it admits an equivalent Lipschitz Kadec–Klee smooth norm.*

Proof: Assume that X is a subspace of c_0, and $\|\cdot\|$ be the canonical norm on c_0. If $f_n \in S_{X^*}$ are such that $f_n \to f$ in the w^*-topology of X^* and $\|f_n - f\| \geq \varepsilon$, then extend f_n in the Hahn–Banach way to $\widetilde{f_n}$ on c_0.

Let $\widetilde{f}_{n_k} \to g \in B_{c_0^*}$ in the w^*-topology of c_0^*, for some subsequence $n_1 < n_2 < \dots$ Then the restriction of g to X equals to f. We have $\|\widetilde{f}_n - g\|_{c_0^*} \geq \|f_{n_k} - f\|_{X^*} \geq \varepsilon$, and thus, by Proposition 12.64, $1 - \varepsilon \geq \|g\|_{c_0^*} \geq \|f\|_{X^*}$.

This proves the "only if" part of the statement. We will prove the "if" part only in case that X has a shrinking Schauder basis (note that any separable space with uniformly Kadec–Klee smooth norm has separable dual), and leave the proof of the general situation to Exercise 12.51, where the notion of a FDD is used. We will first prove the following.

Lemma 12.66 *Assume that the norm of a separable Banach space is c-Lipschitz Kadec–Klee smooth. Then for every $x \in X$,*

$$\max\left(\|x\|, \frac{1}{2-c} \limsup \|x_n\|\right) \leq \limsup \|x + x_n\|$$
$$\leq \max\left(\|x\|, \frac{1}{c} \limsup \|x_n\|\right) \quad (12.19)$$

wherever $\{x_n\}$ is a sequence in X such that $x_n \xrightarrow{w} 0$.

Proof: Let x be an element in X, and let $\{x_n\}$ be a sequence in X such that $x_n \xrightarrow{w} 0$. Without loss of generality, assume that $\lim \|x_n\|$ and $\lim \|x+x_n\|$ both exist and that $\lim \|x_n\| > 0$. We will first prove the right-hand-side inequality in (12.19).

For it, for $n \in \mathbb{N}$, pick $y_n^* \in X^*$ such that $\|y_n^*\| = 1$ and $y_n^*(x+x_n) = \|x+x_n\|$. By passing to a subsequence if needed, we assume that there exists $y^* \in X^*$ such that $y_n^* \xrightarrow{w^*} y^*$ and that $\lim \|y_n^* + y^*\|$ exists. Since the norm is c-Lipschitz Kadec–Klee smooth, we get

$$c \lim \|y_n^* - y^*\| \leq 1 - \|y^*\|. \quad (12.20)$$

Indeed, (12.20) obviously holds if $\|y_n^* - y^*\| \to 0$ or if $y^* = 0$. If not, it is enough to put $x^* = y^*/\|y^*\|$ and $x_n^* = (y_n^* - y^*)/\|y^*\|$ in the definition of a c-Lipschitz Kadec–Klee smooth norm.

Note that

$$\|x + x_n\| = y_n^*(x + x_n) = y_n^*(x) + y^*(x_n) + (y_n^* - y)(x_n).$$

Thus

$$\lim \|x + x_n\| \leq \|y^*\| \cdot \|x\| + \frac{(1 - \|y^*\|)}{c} \lim \|x_n\| \leq \max\left(\|x\|, \frac{1}{c} \lim \|x_n\|\right).$$

Here we used that $y_n^*(x) \to y^*(x)$, that $y^*(x_n) \to 0$, Equation (12.20), that $0 \leq \|y^*\| \leq 1$, and that the term in the middle is a convex combination of $\|x\|$ and $\frac{1}{c} \lim \|x_n\|$. This shows the right-hand-side inequality in (12.19). By using the weak lower semicontinuity of the norm, for the left-hand-side inequality we only need to show that

$$\lim \|x + x_n\| \geq \frac{1}{2-c} \lim \|x_n\|.$$

For it, for $n \in \mathbb{N}$, pick $x_n^* \in X^*$ such that $\|x_n^*\| = 1$ and $x_n^*(x_n) = \|x_n\|$. Assume without loss of generality that there exists $x^* \in X^*$ such that $x_n^* \xrightarrow{w^*} x^*$ and $\lim \|x_n^* - x^*\|$ exists. As before we have, by the c-Lipschitz Kadec–Klee smoothness property of the norm, that

$$c \lim \|x_n^* - x^*\| \leq 1 - \|x^*\|.$$

Since $(x_n^* - x^*)(x_n) \to \lim \|x_n\|$ as $\{x_n\}$ is weakly null, we have

$$\lim \|x_n\| = \lim (x_n^* - x^*)(x_n) \leq \lim \|x_n^* - x^*\| \lim \|x_n\|.$$

This gives $\lim \|x_n^* - x^*\| \geq 1$. Therefore $c \cdot 1 \leq c \lim \|x_n^* - x\| \leq 1 - \|x^*\|$, which gives $\|x^*\| \leq 1 - c$. We have

$$\begin{aligned} \|x + x_n\| \geq x_n^*(x + x_n) &= \|x_n\| + (x_n^* - x^*)(x) + x^*(x) \\ &\geq \|x_n\| - \|x^*\| \cdot \|x\| + (x_n^* - x^*)(x). \end{aligned}$$

Thus

$$\|x + x_n\| + \|x^*\| \|x\| \geq \|x_n\| + (x_n^* - x^*)(x).$$

Therefore

$$\|x + x_n\| + (1 - c)\|x\| \geq \|x_n\| + (x_n^* - x^*)(x).$$

By passing to the limit, we get

$$\lim \|x + x_n\| + (1 - c)\|x\| \geq \lim \|x_n\|.$$

Since, by the weak lower semicontinuity of the norm, we have $\|x\| \leq \lim \|x + x_n\|$, we thus get

$$\lim \|x + x_n\| + (1 - c) \lim \|x + x_n\| \geq \lim \|x_n\|.$$

So, $(2 - c) \lim \|x + x_n\| \geq \lim \|x_n\|$, and this finishes the proof of the lemma. □

As announced, we will only sketch the proof of Theorem 12.65 in the case that X has a shrinking Schauder basis $\{e_n\}$. For the proof of the general case see Exercise 12.51.

We use Lemma 12.66 to have that

$$\limsup \|x + x_n\| \leq \max\left(\|x\|, \frac{1}{c} \limsup \|x_n\| \right)$$

for all $x \in X$ and all weakly null sequences $\{x_n\}$ in X.

We note that the fact that $\{e_n\}$ is shrinking gives that the convergence in $\limsup \|x + x_n\|$ is uniform for x in the unit ball of any finite-dimensional space when $\{x_n\}$ have supports in the basis $\{e_n\}$ gliding to infinity.

Thus, given a summable sequence $\{\varepsilon_n\} \downarrow 0$, we can choose $1 =: n_1 < n_2 < \dots$ so that for each $k \geq 1$, if x is in the unit sphere of E_{1,n_k} and y is in $E_{n_{k+1},\infty}$ with $\|y\| \leq c$, then $\|x + y\| \leq 1 + \varepsilon_k$, where $E_{n,k}$ denotes the linear span of $\{e_j : n \leq j \leq k\}$. Put $F_k = E_{n_{k-1},n_k}$ for $k = 1, 2, \dots$

It is enough to show that $(F_{2n})_{n=1}^{\infty}$ and $(F_{2n-1})_{n=1}^{\infty}$ are both c_0-decompositions, i.e., for $x_n \in F_{2n}$, the series $\sum x_n$ converges if and only if $\{\|x_n\|\} \in c_0$.

Indeed, then X is the direct sum of the spaces $(\sum F_{2n})_{c_0}$ and $(\sum F_{2n-1})_{c_0}$. Since F_{2n} and F_{2n-1} are all finite-dimensional, the direct sum almost isometrically embeds into c_0.

To check that, for example, $(F_{2n})_{n=1}^{\infty}$ is a c_0-decomposition, it is enough to observe that, if x_k is in F_{2k} and $\sup \|x_k\| \leq c$, for $k \in \mathbb{N}$, then for each $m \in \mathbb{N}$ for which $\| \sum_{k=1}^{m} x_k \| \geq 1$, the inequality $\| \sum_{k=1}^{m+1} x_k \| \leq (1 + \varepsilon_{2m+1}) \left\| \sum_{k=1}^{m} x_k \right\|$ holds true. This finishes the proof of Theorem 12.65, and therefore also the proof of Theorem 12.61. □

Note that it is proved in [GKL1] that if X is Lipschitz equivalent to c_0, then it is linearly isomorphic to c_0. This is in contrast with Gul'ko result in [Gulk4] that all $C_p(K)$ spaces for K a countable compact space are uniformly homeomorphic to c_0 in its pointwise topology. Gul'ko showed also in [Gulk5], in particular, that $C_p(\Delta)$ is not uniformly homeomorphic to $C_p[0, 1]$.

Theorem 12.67 (Ribe, [Ribe0]) *If a Banach space X is uniformly homeomorphic to a Banach space Y, then X and Y are crudely finitely represented in each other.*

Proof: We will sketch the proof of the theorem only in case that both X and Y are separable superreflexive Banach spaces.

Let X and Y be two separable superreflexive Banach spaces and φ be a uniform homeomorhism from X onto Y. Then φ and φ^{-1} satisfy a *Lipschitz condition for large distances* (see Exercise 12.32). That is, there is a constant $K > 0$ such that

$$\frac{\|x - y\|}{K} \leq \|\varphi(x) - \varphi(y)\| \leq K\|x - y\| \tag{12.21}$$

for every $x, y \in X$ with $\|x - y\| \geq 1$. Assume that $\varphi(0) = 0$ and put, for $n \in \mathbb{N}$ and $x \in X$, $\varphi_n(x) = (1/n)\varphi(nx)$. Then $\varphi_n^{-1}(y) = (1/n)\varphi^{-1}(ny)$ for each $n \in \mathbb{N}$ and each $y \in Y$. Indeed, if $y = \varphi_n(x)$, we have

$$\varphi_n^{-1}(y) = \frac{1}{n}\varphi^{-1}(n\varphi_n(x)) = \frac{1}{n}\varphi^{-1}(\varphi(nx)) = \frac{1}{n}(nx) = x.$$

Moreover, φ_n satisfies (12.21) whenever $\|x - y\| \geq 1/n$. If $\{x_n\}$ is a bounded sequence in X, then $\{\varphi_n(x_n)\}$ is bounded in Y. Indeed, let ω be the modulus of continuity of φ. Then

$$\|\varphi_n(x_n)\| \leq \frac{1}{n}\omega(n\|x_n\|) \leq \omega(\|x_n\|)$$

because ω is subadditive. Let $\mathcal{U}$ be a free ultrafilter on $\mathbb{N}$ (see Section 17.2). Then, the mapping $\Phi := (\varphi_n)$ from the ultraproduct (see Section 17.4) $X_{\mathcal{U}}$ into the ultraproduct $Y_{\mathcal{U}}$ defined by

$$\Phi((x_n)) = (\varphi(x_n))$$

is a Lipschitz homeomorphism.

Indeed, fix two different points $x = (x_1, x_2, \ldots)$ and $y = (y_1, y_2, \ldots)$ in $X_{\mathcal{U}}$. Then $\{n \in \mathbb{N} : \|x_n - y_n\| \leq (1/n)\} \notin \mathcal{U}$ as $0 < \|x - y\|_{X_{\mathcal{U}}} = \lim_{\mathcal{U}} \|x_n - y_n\|$. Hence Φ satisfies (12.21) for all $x, y \in X_{\mathcal{U}}$. The mapping Φ^{-1} from $Y_{\mathcal{U}}$ into $X_{\mathcal{U}}$ is defined similarly by using the φ_n^{-1}s. Therefore $X_{\mathcal{U}}$ and $Y_{\mathcal{U}}$ are Lipschitz equivalent.

Since Y is superreflexive and $Y_{\mathcal{U}}$ is (always) finitely represented in Y, we have by the definition of superreflexivity (for Y) that $Y_{\mathcal{U}}$ is reflexive.

The space X is isometrically embedded into $X_{\mathcal{U}}$ by the mapping $x \mapsto (x)_{\mathcal{U}}$. Therefore X is Lipschitz homeomorphic to a subset of $Y_{\mathcal{U}}$. By Theorem 12.55, X is isomorphic to a subspace of $Y_{\mathcal{U}}$. Therefore X is crudely finitely represented in Y. Similarly we get that Y is crudely finitely represented in X.

For the general case we refer to [BeLi, Chapter 10]. □

Corollary 12.68 *If X is a Banach space that is uniformly homeomorphic to ℓ_2 (respectively to a superreflexive space), then X is isomorphic to ℓ_2 (respectively to a superreflexive space).*

Proof: Use Proposition 9.12. □

12.6 Remarks and Open Problems

Remarks

1. We refer the reader to the books [BeLi], [BePe2], [vMill], and [Kalt4] for more in this area.
2. Cauty proved in [Caut2] that every continuous mapping from a compact convex set C into itself has a fixed point whenever C is in an F-space, i.e., a complete metric topological vector space, not necessarily locally convex.

3. We refer to [Kalt4] and references therein for the theory of nonlinear type and cotype.
4. Tsarkov proved in [Tsa] that there is a Lipschitz mapping of B_{ℓ_2} into itself that cannot be uniformly approximated by mappings with uniformly Fréchet derivatives, see [BeLi, p. 38]. On the other hand, any Lipschitz mapping from B_{ℓ_2} into the real line can be uniformly approximated by mappings with Lipschitz derivative ([NeSe], [LasLi], [Cepe2]). There is a norm on ℓ_2 that cannot be approximated on the ball by functions with uniformly continuous second derivative ([NeSe], [Vand2]). Every Lipschitz mapping from a separable Banach space X into a Banach space Y can be uniformly approximated by uniformly Gâteaux differentiable Lipschitz mappings, [Joha]. It is shown in [DFH1] and [DFH2] that if X is either a separable Banach space with countable James boundary or an ℓ_p, space, p even, then any equivalent norm on X can be approximated uniformly on the ball by C^∞ smooth norms.
5. Concerning iterative procedures for fixed points we mention [GeLi].
6. If $p_n = \frac{n+1}{n}$, $n = 1, 2, \dots$, then the space $X := \ell_1 \oplus (\sum \ell_{p_n})_2$ is uniformly homeomorphic to $Y := (\sum \ell_{p_n})_2$ but not isomorphic to it [Ribe1]. Note that X is not Asplund while Y is even reflexive. In connection with this check Exercise 12.44.
7. If $1 < p < \infty$, and X is uniformly homeomorphic to ℓ_p, then X is isomorphic to ℓ_p [JLS].
8. There are two nonseparable WCG Banach spaces that are Lipschitz homeomorphic but not isomorphic, see Remark 14.5.
9. (Kalton, [Kalt5]), There is a Banach space X that is not a Lipschitz retract of X^{**}. This solves a question raised by Benyamini and Lindenstrauss (see [BeLi, p. 33]).
10. It is shown in [Kalt3b] that c_0 is not uniformly homeomorphic to any subset of a reflexive space.

Open Problems

1. Does there exist in a separable Hilbert space a nonconvex Chebyshev set? (A set C in a Banach space X is called a *Chebyshev set* if every $x \in X$ has a unique nearest point in C.) See also Exercise 12.16.
2. Is every separable Banach space X isomorphic to Y if X is Lipschitz equivalent to Y?
3. Let T be a nonexpansive mapping of the unit ball of a superreflexive Banach space X into itself. Does T have a fixed point?
4. Is the unit sphere of a superreflexive Banach space uniformly homeomorphic to the unit sphere of a Hilbert space? (See [BeLi, p. 218].)
5. Assume that X is a Banach space. Does the norm topology of X have a σ-discrete base formed by convex sets? Compare with Exercise 12.42.
6. Is a hyperplane uniformly homeomorphic to the whole space in every Banach space?

7. Assume that a Banach space X is uniformly homeomorphic to c_0. Is X isomorphic to c_0? (See [GKL2].)
8. Is every separable quasi-Banach space homeomorphic to a Hilbert space? (See [Kalt1, p. 1127].) Note that there is a separable F-space that is not homeomorphic to a Hilbert space [Caut1].
9. If X is an infinite-dimensional Banach space, it is not known if B_X is uniformly homeomorphic to S_X (unknown even in the Hilbert case) see [BeLi, p. 80]. On the other hand, in infinite-dimensional Banach space there is always a Lipschitz retraction from B_X onto S_X [BeSte].
10. If c_0 Lipschitz embeds into a Banach space X, does c_0 linearly embed into X? See [Kalt4, p. 21].
11. If B_X uniformly embeds into a reflexive space, is X weakly sequentially complete? See [Kalt4, p. 54].
12. If X is Lipschitz homeomorphic to ℓ_1, is X isomorphic to ℓ_1? (see [Kalt4, p. 28]).
13. Assume that Banach spaces X and Y are Lipschitz homeomorphic. Assume that X admits a C^k-smooth norm, resp. LUR norm. Does Y admit C^k-smooth norm, resp. LUR norm?
14. Assume that a nonseparable reflexive Banach space X is Lipschitz homeomorphic to a Banach space Y. Is X isomorphic to a subspace of Y? For separable Banach spaces the answer is affirmative, see Theorem 12.55.
15. It is apparently unknown whether Keller's Theorem 12.37 holds for compact convex sets in complete metric linear spaces (see, e.g., [Kalt4, p.13]).

Exercises for Chapter 12

12.1 Can a compact metric space be isometric to a proper subspace? Can a weakly compact set in a Banach space be isometric to a proper subset?
Hint. No for the first part (see, e.g., [Dugu2, p. 314]). Yes for the second one: let $T : B_{\ell_2} \to B_{\ell_2}$ be defined by $T(x_1, x_2 \ldots) = (0, x_1, x_2, \ldots)$.

12.2 Let T be a contraction from a Banach space X into X. Show that $F = I_X - T$ is an open one-to-one mapping from X onto X.
Hint. Let $y_0 \in X$ be given. Consider the mapping $\varphi(x) = T(x) + y_0$. Note that $x \in X$ is a solution of $F(x) = y_0$ if and only if x is a fixed point of the mapping φ. Observe that φ is a contraction on X. Thus F is onto.

Next show that $B^O(F(z), (1-q)r) \subset F\big(B^O(z,r)\big)$, where q is the contraction constant of T and $B^O(z,r)$ is the open ball centered at z and having radius r: Let $y \in B^O_{F(z)}((1-q)r)$. Put $f(x) = x - F(x) + y = y + T(x)$ for $x \in B^O(z,r)$. Then $\|f(x) - z\| = \|T(x) + y - z\| \le \|T(x) - T(z)\| + \|T(z) + y - z\| \le q\|x - z\| + \|y - F(z)\| < qr + (1-q)r = r$. Therefore $f\colon B^O(z,r) \to B^O(z,r)$ and F is open. By continuity, $f : B(z,r) \to B(z,r)$, and f is a q-contraction. Thus there is $w \in B(z,r)$ such that $f(w) = w$. This means $w = w - F(w) + y$, i.e., $F(w) = y$.

12.3 Define $\varphi\colon \mathbb{R} \to \mathbb{R}$ by $\varphi(x) = x - \arctan(x) + \pi/2$. Show that $|\varphi(x) - \varphi(y)| < |x - y|$ for $x, y \in \mathbb{R}$ and there is no fixed point for φ on $\mathbb{R}$.
Hint. The mean value theorem.

12.4 Find an example of a non-complete metric space and a contraction on it without a fixed point.
Hint. Think of open interval (0,1) and linear functions.

12.5 (Alspach [Alsp]) Consider $K = \{f \in L_1[0, 1] : 0 \le f \le 2, \int_0^1 f \, dt = 1\}$ and $T : K \to K$ defined as

$$T(f)(t) = \begin{cases} \min(2f(2t), 2) & \text{if } 0 \le t \le \frac{1}{2}, \\ \max(0, 2f(2t-1) - 2) & \text{if } \frac{1}{2} < t \le 1. \end{cases}$$

Show that T is an isometry (in particular a nonexpansive mapping) on the weakly compact set K without any fixed point. See also the open problem 12.3
Hint. The weak compactness of K follows from the following criterion: A set $M \subset L_1$ is weakly relatively compact if for every $\varepsilon > 0$ there is $\delta > 0$ such that for every $A \subset [0, 1]$ satisfying $\lambda(A) < \delta$ and every $f \in M$ we have $\int_A |f| \, d\lambda < \varepsilon$ (where λ is the Lebesgue measure, see, e.g., [DuSc]).

Assume by contradiction that $f \in K$ is a fixed point of T. Prove by induction that $\lambda\big(\{t : 2^{-k} \le f(t) < 2^{-k+1}\}\big) = 0$ for $k = 0, 1, \ldots$. Thus $\lambda(\{t : f(t) = 0\}) = \lambda(\{t : f(t) = 2\}) = \frac{1}{2}$. Next show that $\lambda\big(\{t : f(t) = 0\} \cap \big[\frac{n}{2^k}, \frac{n+1}{2^k}\big]\big) = \frac{1}{2^k}$ for arbitrary $k \in \mathbb{N}$, $0 \le n \le 2^k - 1$. This is a contradiction with the Lebesgue density point theorem for the set $\{t : f(t) = 0\}$.

12.6 Let K be a nonempty weakly compact subset of a Banach space X and let $f : K \to K$ be a mapping. Prove that K contains an f-invariant subset K_0 that is w-compact-minimal.
Hint. Let $\mathcal{K}$ be the (nonempty) family of all weakly compact subsets of K that are f-invariant, partially ordered by reverse inclusion. This family is inductive; indeed, if $\{K_i\}$ is a chain in $\mathcal{K}$, the set $\bigcap K_i$ is nonempty and obviously f-invariant. Apply now Zorn's lemma.

12.7 Let C be a closed convex bounded subset of a Banach space X. Show that if $T : C \to C$ is a nonexpansive mapping, then $\inf\{\|x - T(x)\| : x \in C\} = 0$.
Hint. Pick $z \in C$, $\varepsilon > 0$, and consider $T_\varepsilon(x) = \varepsilon z + (1 - \varepsilon)T(x)$. Then $T_\varepsilon : C \to C$ as C is convex. Moreover, T_ε is a contraction. Therefore there is $x_\varepsilon \in C$ such that $x_\varepsilon = T_\varepsilon(x_\varepsilon)$. Then we have $\|x_\varepsilon - T(x_\varepsilon)\| = \|\varepsilon z + (1 - \varepsilon)T(x_\varepsilon) - T(x_\varepsilon)\| = \varepsilon\|z - T(x_\varepsilon)\| \le \varepsilon \operatorname{diam} C$.

12.8 Use Exercise 12.7 to show that a nonexpansive mapping on a convex compact set in a Banach space has a fixed point.
Hint. The infimum is attained.

12.9 Define a mapping $T: S_{c_0} \to S_{c_0}$ by $T(x_1, x_2, \dots) = (1, x_1, x_2 \dots)$. Show that T is an isometry without a fixed point.

12.10 Let $A = \{x(t) \in C[0, 1] : 0 = x(0) \le x(t) \le x(1) = 1\}$. Define a mapping $T: A \to A$ by $T(x)(t) = tx(t)$. Show that A is a closed bounded convex set in $C[0, 1]$ and T is a nonexpansive mapping (in fact, $\|T(x) - T(y)\| < \|x - y\|$ for every $x \ne y$). Yet, T has no fixed point.
Hint. Solve $tx(t) = x(t)$ for $t \in [0, 1)$.

12.11 Let A be a subset of a metric space (X, ρ). The definition of a diametral point was given right before Lemma 12.8. Define $A = \{x(t) \in C[0, 1] : 0 = x(0) \le x(t) \le x(1) = 1\}$. Show that every point of A is a diametral point.
Hint. Draw a picture.

12.12 Show that every separable Banach space X can be renormed by an equivalent norm so that every convex set with more than one point has a non-diametral point.
Hint. Let $\{f_i\} \subset S_{X^*}$ be a separating set. Define $|||\cdot|||$ by $|||x|||^2 = \|x\|^2 + \sum 2^{-i} f_i^2(x)$. Then, from the convexity, whenever $2|||x_n|||^2 + 2|||y_n|||^2 - |||x_n + y_n|||^2 \to 0$ and $\{x_n\}$ is bounded, then $f_i(x_n - y_n) \to 0$. Let C be convex, $\operatorname{diam} C = 1$, and $x, y \in C$. Let $|||\frac{1}{2}(x + y) - x_n||| \to 1$. Then since $|||x - x_n|||, |||y - x_n||| \le 1$, we have that $f_i(x - x_n - (y - x_n)) = f_i(x - y) \to 0$ as $n \to \infty$. Since $\{f_i\}$ is separating, we have $x = y$.

12.13 The definition of a closed convex bounded subset C of a Banach space having normal structure was given right before Proposition 12.11. Show that every compact convex set in a Banach space has normal structure.
Hint. Assume the contrary. Let $d = \operatorname{diam} K$. Choose $x_1 \in K$. Find $x_2 \in K$ with $\|x_1 - x_2\| = d$. Pick $x_3 \in K$ such that $\left\|\frac{x_1+x_2}{2} - x_3\right\| = d$ etc. Then

$$d = \left\|x_{n+1} - \frac{x_1 + \dots + x_n}{n}\right\| = \left\|\sum_{i=1}^{n} \frac{x_{n+1} - x_i}{n}\right\| \le \frac{1}{n}\sum_{i=1}^{n} \|x_{n+1} - x_i\| \le d.$$

So $\sum_{i=1}^{n}(d - \|x_{n+1} - x_i\|) = 0$, and since $\|x_{n+1} - x_i\| \le d$ for every i, we get $\|x_{n+1} - x_i\| = d$. Hence $\{x_n\}$ has no convergent subsequence.

12.14 Show that there is a continuous mapping from B_{ℓ_2} into B_{ℓ_2} that has no fixed point. This means that the requirement of finite dimension cannot in general be dropped in Brouwer's Theorem 12.14.
Hint. (Kakutani) Define $\varphi(x_1, x_2, \dots) = \left(\frac{1}{2}(1 - \|x\|_2), x_1, x_2, \dots\right)$.

12.15 A *retraction* of A onto $B \subset A$ is a mapping from A onto B whose restriction to B is the identity mapping on B.

Show that the following two statements are equivalent for every n:

(i) There exists a continuous retraction of $B_{\ell_2^n}$ onto $S_{\ell_2^n}$.

(ii) There is a continuous mapping of $B_{\ell_2^n}$ into $B_{\ell_2^n}$ without a fixed point.

By Brouwer's theorem both statements fail. The situation is different if X is an infinite-dimensional Banach space. Then S_X is even a Lipschitz retract of B_X (see, e.g., [BeLi, p. 64]).

Hint. If r is such a retraction, consider the mapping $x \mapsto -r(x)$ and note that any fixed point of it would be necessarily on $S_{\ell_2^n}$. If f is a mapping of $B_{\ell_2^n}$ into $B_{\ell_2^n}$ without a fixed point, consider the mapping that maps $x \in B_{\ell_2^n}$ onto the intersection of the ray from x to $f(x)$ with $S_{\ell_2^n}$.

12.16 Let K be a compact convex set in a Banach space X and assume that each point of X has a unique farthest point in K. Show that K then necessarily consists of a single point. The same holds for general sets K in finite dimensions but for infinite-dimensional spaces in general is an open problem which is equivalent to a problem on Chebyshev sets mentioned in Open Problem 12.1 in the case of a Hilbert space.

Hint. Brouwer Theorem 12.14.

12.17 Klee showed that $\ell_1(c)$ contains a pairwise disjoint family of translates of the closed unit ball that covers the whole space [Klee3], [Klee4]. Show that the result of Klee mentioned above gives in $\ell_1(c)$ a nonconvex Chebyshev set. See also the open problem 12.1 and the Exercise 12.16.

Hint. The centers of the balls. For more in this direction see, e.g., [FLP].

12.18 Define $U_r\colon L_2(\mathbb{R}) \to L_2(\mathbb{R})$ by $U_r(f)\colon t \mapsto f(t-r)$. Show that $\{U_r : r \in \mathbb{R}\}$ is a commutative group of unitary operators.

12.19 (Markov–Kakutani) Let K be a compact convex set in a locally convex space E. Show that if $T\colon K \to K$ is a continuous affine mapping, then T has a fixed point.

Hint. (Werner [Wern]): By contradiction, assume that the intersection of the diagonal $\Delta := \{(x,x) : x \in K\}$ of $K \times K$ with the graph Γ of T is empty. By the Hahn–Banach theorem there exists continuous linear functionals ℓ_1 and ℓ_2 on E and numbers $\alpha < \beta$ such that

$$l_1(x) + l_2(x) \le \alpha < \beta \le l_1(y) + l_2\big(T(y)\big)$$

for all $x, y \in K$. Consequently, $l_2\big(T(x)\big) - l_2(x) \ge \beta - \alpha$ for all $x \in K$. Iterating this inequality yields $l_2\big(T^n(x)\big) - l_2(x) \ge n(\beta - \alpha) \to \infty$ for arbitrary $x \in K$. Thus the sequence $\{l_2\big(T^n(x)\big)\}$ is unbounded. This contradicts the compactness of $l_2(K)$.

The full statement of the Markov–Kakutani theorem follows from this by considering the sets of fixed points of mappings T_i, $i \in I$. Indeed, if K_i is the set of all fixed points of T_i, we have $K_i \neq \emptyset$ by the above and K_i is compact and convex. From the commutativity we get $T_i(K_j) \subset K_j$. Hence $T_i\big|_{K_j}$ has a fixed point by the above and $K_i \cap K_j \neq \emptyset$. Thus $\bigcap_{i \in F} K_i \neq \emptyset$ for every finite $F \subset I$. Therefore $\bigcap_{i \in I} K_i \neq \emptyset$, which proves the result.

12.20 (i) Let $(X, \mathcal{T})$ be a metrizable locally convex space, and let $S \subset X$ be a separable subset of X. Prove that there exists a countable subset of X^* that separates points of S.

(ii) Let C be a compact convex and metrizable subset of a locally convex space $(X, \mathcal{T})$. Prove that there exists a countable subset of X^* that separates points of C.

Hint. (i) By taking the closed span of S, we may assume that X is separable. Let $\{U_n : n \in \mathbb{N}\}$ be a fundamental system of neighborhoods of 0 in X. Then, by Alaoglu's theorem, each $U_n^\circ \subset X^*$ (the *polar* set of U_n) is a w^*-compact (and w^*-metrizable) set. It follows that (X^*, w^*) is separable.

(ii) Let ρ be a metric on C that induces its topology. Put, for $x_0 \in C$ and $r > 0$, $B(x_0, r) = \{x \in C : \rho(x, x_0) < r\}$. The set $B(x_0, r)$ is a neighborhood of x_0 in C, hence we can find an open closed and convex neighborhood $U_{x_0,r}$ of x_0 in X such that $\overline{U}_{x_0,r} \cap C \subset B(x_0, r)$. Given $n \in \mathbb{N}$, the family $\mathcal{U}_n := \{U_{x,1/n} : x \in C\}$ is an open cover of C, so we can find a finite subcover $\mathcal{V}_n$ of C. Note that if $U \in \mathcal{U}_n$, then ρ-diam$(\overline{U} \cap C) < 2/n$. For any couple U, V of elements in $\mathcal{V}_n$ such that $(\overline{U} \cap C) \cap (\overline{V} \cap C) = \emptyset$, we can find, by the separation theorem, an element $f_{U,V}$ in X^* that separates strongly the two compact convex sets $\overline{U} \cap C$ and $\overline{V} \cap C$. This is done for $n \in \mathbb{N}$, and we obtain in this way a countable subset M of X^*. Let us prove that M separates points of C. Indeed, given x and y in C such that $x \neq y$, find $n \in \mathbb{N}$ such that $\rho(x, y) > 4/n$. Since $\mathcal{V}_n$ is a covering of C, we can find V, W in $\mathcal{V}_n$ such that $x \in V$ and $y \in W$. The previous observation on the ρ-diameter of these two sets ensures the existence of $f \in M$ such that $f(x) \neq f(y)$.

12.21 Let K be an convex balanced subset of ℓ_2. Prove that there exists an orthogonal set $V := \{v_n : n \in \mathbb{N}\} \subset K$ such that V (as a family of functionals on ℓ_2) separates points of K, i.e., if $k, k' \in K$ satisfy $\langle k, v_n \rangle = \langle k', v_n \rangle$ for every $n \in \mathbb{N}$, then $k = k'$), and no proper subset nor superset enjoy the same property (orthogonality and separation).

Hint. The family of all orthogonal systems consisting of non-zero vectors in K has a maximal element $\{v_n : n \in \mathbb{N}\}$. Maximality implies that $\{v_n : n \in \mathbb{N}\}$, as a family of functionals on ℓ_2, separates points of K. It is also clear that no proper subset of $\{v_n : n \in \mathbb{N}\}$ separates points of K.

12.22 Prove, by a direct method using Brouwer's Fixed Point Theorem, that the Hilbert cube Q has the fixed point property.

Hint. We identify the Hilbert cube, as usual, with the subset $Q := \{x = (x_n) \in \ell_2 : |x_n| \leq (1/n), n \in \mathbb{N}\}$, endowed with the norm topology (see the Introduction to Section 12.3). For $n \in \mathbb{N}$, let $P_n : \ell_2 \to \ell_2$ be the canonical projection $P_n(x_1, x_2, \ldots) = (x_1, \ldots, x_n, 0, \ldots)$, and let $Q_n = P_n Q$ $(\subset Q)$, a set clearly homeomorphic to the closed unit ball of $\mathbb{R}^n$ (so, thanks to Brouwer's fixed point Theorem 12.14, having the fixed point property). A sequence $\{\varepsilon_n\}$ of positive numbers exists such that $\varepsilon_n \to 0$ and $\|x - P_n x\|_2 \leq \varepsilon_n$ for every $x \in Q$ and every $n \in \mathbb{N}$.

Let $\phi : Q \to Q$ be a continuous function. For $n \in \mathbb{N}$, put $\phi_n = P_n \circ (\phi|_{Q_n})$ for all n. Then ϕ_n is a continuous function from Q_n into Q_n, so there exists a fixed point x_n for ϕ_n in Q_n. This defines a sequence $\{x_n\}$ in Q. By compactness, there exists a convergent subsequence $\{x_{n_k}\}$. We shall prove that its limit x_0 is a fixed point for ϕ in Q. Indeed,

$$\begin{aligned}\|x_{n_k} - \phi(x_0)\|_2 &= \|(P_{n_k} \circ \phi)x_{n_k} - \phi(x_0)\|_2 \\ &\leq \|(P_{n_k} \circ \phi)x_{n_k} - \phi(x_{n_k})\|_2 + \|\phi(x_{n_k}) - \phi(x_0)\|_2 \leq \varepsilon_{n_k} + \|\phi(x_{n_k}) - \phi(x_0)\|_2 \to_k 0.\end{aligned}$$

This proves that $x_{n_k} \to \phi(x_0)$. It follows that $\phi(x_0) = x_0$.

12.23 Prove the assertion in the text after Definition 12.27: a nonempty subset C of a normed space X is elliptically convex if it is convex and every proper support functional $f \in X^*$ attains its supremum on C precisely at a single point.
Hint. Note that if some $f \in X^*$ attains the supremum on a convex set $C \subset X$ at a point $c \in (x, y)$, where $x, y \in C$, then f is constant on $[x, y]$.

12.24 Prove that the topology $\mathcal{T}$ defined on T by the metric d is described by the convergence of sequences as mentioned in Section 12.3.
Hint. Note first that $x_k \xrightarrow{\mathcal{T}} x_0$ in T if and only if $\lim_k \varphi_n(x_k) = \varphi_n(x_0)$ for every $n \in \mathbb{N}$. Second, observe that $|\varphi(x)| \leq |x(n)|$ for every $x \in T$ and $n \in \mathbb{N}$. Consider separately the case $x_0 = 0$ and the case $x_0 \neq 0$. In this last situation, distinguish between the case $\operatorname{ord} x_0 = \infty$ and the case $\operatorname{ord} x_0 < \infty$.

12.25 Prove that the metric space $(T, \mathcal{T})$ defined in Section 12.3 is compact.
Hint. Extract from every sequence (x_k) in T, by a diagonal procedure, a pointwise convergent subsequence (denoted again (x_k)) to some $y_0 \in \mathbb{I}^{\mathbb{N}}$. If $|y_0(n)| < 1$ for all $n \in \mathbb{N}$, put $x_0 = y_0$ ($\in T$). Otherwise, let n be the first $m \in \mathbb{N}$ such that $|y_0(m)| = 1$; put $x_0(j) = y_0(j)$ for $j \leq n$, $x_0(j) = 0$ for $j > n$. In both cases $x_0 \in T$ and, from the description of the topology $\mathcal{T}$ made in Exercise 12.24, we get $x_k \xrightarrow{\mathcal{T}} x_0$.

12.26 Show that the closed unit ball of ℓ_2 in its weak topology (a Keller space in the space (ℓ_2, w)) is not affinely homeomorphic to the Hilbert cube Q (a convex $\|\cdot\|$-compact subset of ℓ_2, see Exercise 1.51).
Hint. Affine homeomorphisms keep extreme points. All points of the sphere S_{ℓ_2} are extreme. Extreme points of Q are precisely the vertex. The first set (in the weak topology) is connected. Not the second one (in the norm topology).

12.27 Let X be a finite-dimensional Banach space. Show that every nonempty compact and convex subset C of X is homeomorphic to the closed unit ball of a certain subspace of X.
Hint. We consider the associated real Banach space $X_{\mathbb{R}}$, isomorphic to some $\mathbb{R}^n$. Let C be a nonempty convex subset of X. Put $L(C) = \operatorname{span}(C - C)$, isomorphic to $\mathbb{R}^m$ for some $m \leq n$, and $M(C) = x_0 + L(C)$, where $x_0 \in C$ is arbitrary. The set

C contains an m-dimensional simplex, whose centroid c_0 is an interior point of C relatively to $M(C)$. With respect to the metric space $M(C)$, $B(c_0, \varepsilon) \subset C$ for some $\varepsilon > 0$. If C is, moreover, compact, a homomorphism from $B(c_0, \varepsilon)$ onto C can be defined by noticing that every half-ray in $M(C)$ from c_0 meets the boundary of C at a single point.

12.28 Let X be a Banach space. Show that $h(x) = \frac{x}{1+\|x\|}$ defines a homeomorphism of B_X^O onto X.

12.29 Let X, Y be Banach spaces. Show that if φ is a homeomorphism of S_X onto S_Y, then $x \mapsto \|x\|\varphi(\frac{x}{\|x\|})$, $0 \mapsto 0$, is a homeomorphism of X onto Y.

12.30 Let S_1 and S_2 be spheres of two equivalent norms on an X. Show that the radial mapping $x \mapsto \frac{x}{\|x\|_2}$ is a Lipschitz homeomorphism of S_1 onto S_2. Is it a homeomorphism in the weak topologies?
Hint. About the w-topology: No. Try norm with the Kadec–Klee property.

12.31 A mapping $\varphi\colon E \to F$, where E, F are topological vector spaces, is called a *uniform homeomorphism* of E and F if φ is a bijection onto F and both φ and φ^{-1} are uniformly continuous. If such φ exists, the spaces E, F are called *uniformly homeomorphic*.

Prove that if a locally convex space E is uniformly homeomorphic to a Banach space X, then its topology is given by a norm (see [Bess2]).
Hint. Given a convex neighborhood V of 0 in E, find $n \in \mathbb{N}$ such that $\varphi(x) - \varphi(y) \in V$ whenever $x - y \in B_X^O\left(\frac{1}{n}\right)$, where $B_X^O\left(\frac{1}{n}\right)$ is the open ball in X centered at 0 with radius $\frac{1}{n}$. Note that by convexity of $B_X^O\left(\frac{1}{n}\right)$ we have $B_X^O\left(\frac{1}{n}\right) + \cdots + B_X^O\left(\frac{1}{n}\right) = nB_X^O\left(\frac{1}{n}\right) = B_X^O(1)$ and thus, by the telescopic argument,

$$\varphi(B_X^O) = \varphi\Big(B_X^O\left(\tfrac{1}{n}\right) + \cdots + B_X^O\left(\tfrac{1}{n}\right)\Big) \subset V + \cdots + V = nV.$$

Thus $\varphi(B_X^O)$ is a bounded neighborhood in E. Since E has a bounded neighborhood of zero, it is normable by Proposition 3.30.

12.32 Show that every uniformly continuous mapping f of a Banach space X into a Banach space Y is Lipschitz for large distances, i.e., for every $\delta > 0$ there is $K = K(\delta)$ such that $\|f(x) - f(y)\| \leq K\|x - y\|$ for all $x, y \in X$ satisfying $\|x - y\| \geq \delta$.
Hint. Given $\delta > 0$, first get M such that $\|f(a) - f(b)\| \leq M$ for $a, b \in X$ satisfying $\|a - b\| < \delta$ as follows: Using the uniform continuity for $\varepsilon = 1$, find $\Delta > 0$ such that $x, y \in X$, $\|x - y\| < \Delta$ implies $\|f(x) - f(y)\| < 1$. Now if $\|a - b\| < \delta$, we can get from a to b in $[\delta/\Delta]$ steps ($[w]$ denotes the integer part of w) not longer than Δ, so $\|f(a) - f(b)\| \leq \delta/\Delta = M$.

Put $K = 2M/\delta$. Given $x, y \in X$ with $\|x-y\| \geq \delta$, let $x = a_0, a_1, \dots, a_m = y$ be points on the line segment $[x, y]$ such that $\|a_{j+1} - a_j\| < \delta$ and $m = [2\|x-y\|/\delta]$. Then

$$\|f(x) - f(y)\| \leq \sum_{i=1}^{m} \|f(a_i) - f(a_{i-1})\| \leq mM \leq \tfrac{2\|x-y\|}{\delta} K \tfrac{\delta}{2} = K\|x - y\|.$$

12.33 Let X, Y be Banach spaces. Assume that S_X is Lipschitz equivalent to S_Y via a mapping φ. Show that X, Y are Lipschitz equivalent. (See also Exercise 12.30.)
Hint. $x \mapsto \|x\|\varphi\big(\frac{x}{\|x\|}\big)$.

12.34 ([BeLi, 170]) Define $\tau(t) = 2t$ for $t > 0$ and $\tau(t) = t$ for $t \leq 0$. Define $\varphi\colon \ell_2 \to \ell_2$ by $\varphi\big((x_i)\big) = \big(\tau(x_i)\big)$. Show that φ is a Lipschitz homeomorphism from ℓ_2 onto ℓ_2 which is nowhere Fréchet differentiable.
Hint. Direct calculation.

12.35 Let φ be a Lipschitz homeomorphism of X onto Y that has a Fréchet derivative at some point x_0. Show that $\varphi'(x_0)$ is a linear isomorphism of X onto Y.
Hint. Assume that φ and φ^{-1} are K-Lipschitz, $x_0 = 0$, and $\varphi(0) = 0$. Let $T = \varphi'(0)$. Find $\delta > 0$ so that $\|\varphi(h) - T(h)\| < \frac{1}{8}\|h\|$ whenever $\|h\| < \delta$. Then T is an isomorphism of X into Y. Assume $T(X) \neq Y$. As $T(X)$ is closed, find $y \in Y$ such that $\|y\| < \frac{\delta}{K}$ and $\operatorname{dist}\big(y, T(X)\big) \geq \frac{\delta}{4}$. Let $x = \varphi^{-1}(x)$. Then $\|x\| < \delta$, so we get a contradiction:

$$\tfrac{\delta}{4} \leq \operatorname{dist}\big(y, T(X)\big) \leq \|y - T(x)\| = \|\varphi(x) - T(x)\| \leq \tfrac{\delta}{8}.$$

12.36 Let P be Lipschitz (non-linear) projection of a Banach space X onto a subspace Y of X such that it is Gâteaux differentiable at a point of Y. Then Y is complemented in X.
Hint. The derivative is a linear projection.

12.37 Show that if Y is a reflexive subspace of a Banach space X and there is a uniformly continuous non-linear projection P of X onto Y, then there is a Lipschitz non-linear projection of X onto Y.
Hint. Since P is Lipschitz on large distances (Exercise 12.32), there is $\lambda > 0$ such that $\|P(x) - P(y)\| \leq \lambda\|x - y\|$ for all $x, y \in X$ with $\|x - y\| \geq 1$. Consider projections P_n of X onto Y defined by $P_n(x) = \frac{1}{n}P(nx)$ for $n \in \mathbb{N}$. Note that for a fixed $x \in X$,

$$\|P_n(x)\| = \|P_n(x) - P_n(0)\| = \|\tfrac{1}{n}P(nx) - \tfrac{1}{n}P(0)\| \leq \tfrac{1}{n}\lambda\|nx\| = \lambda\|x\|$$

for $\|nx\| \geq 1$, i.e., for $n \geq \|x\|^{-1}$. As Y is reflexive, B_Y is weakly compact and thus by Tychonoff's theorem, $\{P_n\}$ has a cluster point P_0 in the topology of the pointwise convergence for mappings into Y in its weak topology.

As $P_n(y) = y$ for all $y \in Y$, we have $P_0(y) = y$ for all $y \in Y$. If $x, y \in X$, then $\|P_n(x) - P_n(y)\| \leq \lambda\|x - y\|$ for all $n > \|x - y\|^{-1}$. Therefore $\|P_0(x) - P_0(y)\| \leq \lambda\|x - y\|$ for all $x, y \in X$. Hence P_0 is a Lipschitz projection of X onto Y.

Note that this is the first step in the proof of the Lindenstrauss' result (see, e.g., [BeLi, p. 173]) that if Y is a reflexive subspace of a Banach space X and there is a uniformly continuous non-linear projection P from X onto Y, then Y is complemented in X.

12.38 Let X, Y be Banach spaces and $T \in \mathcal{B}(X, Y)$, T is onto. One of the Bartle–Graves results, see Corollary 7.56, asserts that there is a continuous (in general non-linear) mapping φ from Y into X such that $T(\varphi(y)) = y$ for every $y \in Y$. Show that if $\mathrm{Ker}(T)$ is not complemented in X, then φ is nowhere Gâteaux differentiable.

By the differentiability results for Lipschitz maps, it therefore follows that in general, there is a little hope to have Lipschitz selectors of the Bartle–Graves type.
Hint. Otherwise $I_X - \varphi'(y)T$ is a projection of X onto $\mathrm{Ker}(T)$.

12.39 (Gorin) Show that for every n, ℓ_2^n is uniformly equivalent to a bounded subset of ℓ_2, that is, there is a mapping φ from ℓ_2^n onto a bounded subset of ℓ_2 such that both φ and φ^{-1} are uniformly continuous.
Hint. First, for $n = 1$, map $[-1, 1] \subset \mathbb{R}$ onto the line segment connecting $-e_1$ and e_1 (the standard unit vectors), then $[1, 1 + \sqrt{2}]$ onto the line segment connecting e_1 to e_2, then $[1 + \sqrt{2}, 1 + 2\sqrt{2}]$ onto the line segment between e_2 and e_3, etc., then map $(-1, -1 - \sqrt{2}]$ onto the line segment connecting $-e_1$ to $-e_2$, etc. Since the distance of the disjoint line segments in ℓ_2 we consider is at least $\frac{1}{\sqrt{2}}$ and angles between the adjacent line segments are either $\frac{\pi}{4}$ or $\frac{\pi}{3}$, we calculate that the inverse mapping is uniformly continuous. The mapping itself is easily seen to be uniformly continuous.

For $n > 1$ we map ℓ_2^n into the product of ℓ_2's, using the above mapping for coordinates.

12.40 Let H_n be defined as before Proposition 12.47. Show that a set $M \subset S_X$ is relatively compact if and only if $\lim\limits_{n\to\infty} H_n(x) = 1$ uniformly for $x \in M$.
Hint. Assume that $\lim\limits_{n\to\infty} H_n(x) = 1$ is not uniform on M. Then there are $\varepsilon > 0$, a subsequence $\{n_k\}$ of $\mathbb{N}$ and $x_{n_k} \in M$ such that $H_{n_k}(x_{n_k}) < 1 - \varepsilon$. Since $\lim\limits_{n\to\infty} H_n(x) = 1$ for every $x \in S_X$, we get $n_{k_2} > n_{k_1}$ such that $H_{n_{k_2}}(x_{n_{k_1}}) > 1-\varepsilon/2$, then get $n_{k_3} > n_{k_2}$ such that $H_{n_{k_3}}(x_{n_{k_1}}) > 1 - \varepsilon/2$ and $H_{n_{k_3}}(x_{n_{k_2}}) > 1 - \varepsilon/2$ etc.

Using the fact that the distance to a subspace is a subadditive function, we have for $i > j$,

$$\|x_{n_{k_i}} - x_{n_{k_j}}\| \geq H_{n_{k_i}}(x_{n_{k_j}} - x_{n_{k_i}}) \geq H_{n_{k_i}}(x_{n_{k_j}}) - H_{n_{k_i}}(x_{n_{k_i}}) \geq 1 - \tfrac{\varepsilon}{2} - (1 - \varepsilon) = \tfrac{\varepsilon}{2}.$$

Therefore M is not relatively compact. If M is relatively compact, then $\overline{M}$ is compact and $H_n(x)$ is a sequence of continuous functions that converges to the constant function 1 on $\overline{M}$. Hence, the uniformity of the convergence follows from the classical Dini theorem.

12.41 (Hájek, [BeLi, p. 277]) Show that if Y is a separable infinite-dimensional Banach space, then there exists a C^∞-smooth mapping from ℓ_2 onto Y.

Note that in general, if X, Y are separable infinite-dimensional Banach spaces then there exists a C^1-smooth Lipschitz mapping from X onto Y (Bates).

Hint. Define $K = \{x = (x_i) \in B_{\ell_1} : x_i \geq 0, \sum \sqrt{x_i} \leq 1\}$. Construct a bounded linear mapping T from ℓ_1 into X such that $\frac{1}{2}B_X \subset T(K)$ as follows: Let $\{x^i\}$ be dense in B_X. Define $T(e_i) = x^i$ and extend it on ℓ_1. Then T is a bounded operator. Given $x \in B_X$, find inductively $i_n \in \mathbb{N}$ such that $\|T(e_{i_1}) - x\| < 2^{-2}$ and $\left\|\sum_{n=1}^{k} 2^{-2(n-1)} T(e_{i_n}) - x\right\| < 2^{-2k}$. Put $y = \sum_{n=1}^{\infty} 2^{-2(n-1)} e_{i_n}$. Then $T(y) = x$ and $\frac{1}{2}y \in K$, so $\frac{1}{2}B_X \subset T(K)$.

Define a mapping P from ℓ_2 into ℓ_1 by $P(a_1, a_2, \dots) = (|a_1|^2, |a_2|^2, \dots)$. Then $K \subset P(B_{\ell_2})$. The desired operator is $Q = T \circ P$.

12.42 (Raja [Raja1]) Let X be a Banach space whose norm is LUR. Show that the norm has a σ-discrete base consisting of convex sets. A family of subsets of X is *discrete* if every point of X has a neighborhood that intersects at most one member of the family. A family of sets is σ*-discrete* if it is a union of a countable number of discrete families. See Open Problem 12.5.

Hint. Fix $\varepsilon > 0$ and define by transfinite induction a family of convex sets $\{B_\alpha\}$ as follows: $B_0 = B_X$, $B_\alpha = \bigcap_{\beta<\alpha} B_\beta$ if α is a limit ordinal and $B_{\alpha+1} = B_\alpha \setminus \{x \in X : x^*_\alpha(x) > a_\alpha\}$, where $x^*_\alpha \in S_{X^*}$ and $a_\alpha \in \mathbb{R}$ are such that $\operatorname{diam}(B_X \cap \{x \in X : x^*_\alpha(x) > a_\alpha\}) < \varepsilon$. The process ends when $B_\gamma \subset \operatorname{Int}(B_X)$. Take $\delta > 0$. Define convex sets $C(\alpha, \varepsilon, \delta) = B_\alpha \cap \{x \in X : x^*_\alpha(x) \geq a_\alpha + \delta\}$. Then $S_X \subset \bigcup_{\delta>0} \bigcup_{\alpha<\gamma} C(\alpha, \varepsilon, \delta)$. The family $\{C(\alpha, \varepsilon, \delta) : \alpha < \gamma\}$ is δ-discrete: If $\alpha < \beta$, then for every $x \in C(\alpha, \varepsilon, \delta)$ and every $y \in C(\beta, \varepsilon, \delta)$ we have $x^*_\alpha(x) \geq a_\alpha + \delta$ and $x^*_\alpha(y) \leq a_\alpha$ as $y \in B_{\alpha+1}$. Thus $x^*_\alpha(x - y) \geq \delta$ and hence $\|x - y\| \geq \delta$.

For $n > 3m$ define $U(\alpha, m, n) = C(\alpha, \frac{1}{m}, \frac{3}{n}) + \operatorname{Int}(\frac{1}{n}B_X)$. We obtain a $\frac{1}{n}$-discrete family $\{U(\alpha, m, n) : \alpha < \gamma\}$ of open convex sets of diameter less than $\frac{2}{m}$. Prove that

$$\{rU(\alpha, m, n) : \alpha < \gamma, r > 0 \text{ rational}, m, n \in \mathbb{N}\} \cup \{\operatorname{Int}(\tfrac{1}{n}B_X) : n \in \mathbb{N}\}$$

is a σ-discrete base of the norm topology of X: Fix $x \in X \setminus \{0\}$. Take $m > \|x\|/\varepsilon$. For some α and n big enough, $x/\|x\| \in U(\alpha, m, n)$. There is a positive rational r with $0 < r < \|x\|$ such that $x/r \in U(\alpha, m, n)$. Thus $x \in rU(\alpha, m, n)$ and $\operatorname{diam} rU(\alpha, m, n) < r/m \leq \varepsilon$.

12.43 Use the hint to prove the following version of a result of Kirszbraun (see, e.g., [BeLi, p. 18]): *If A is a subset of a Hilbert space H_1 and $f : A \to H_2$ is* *a Lipschitz mapping from A into a Hilbert space H_2, then f can be extended to a Lipschitz (with the same Lipschitz constant) mapping F from H_1 into H_2.*

Hint. Use the following lemma: *Assume that $\{B(x_i, r_i) : i \in I\}$ and $\{B(y_i, r_i) : i \in I\}$ are two systems of closed balls in Hilbert spaces H_1 and H_2, respectively, and such that $\|y_i - y_j\| \leq \|x_i - x_j\|$ for all $i, j \in I$. If $\bigcap_I B(x_i, r_i) \neq \emptyset$, then also $\bigcap_I B(y_i, r_i) \neq \emptyset$.* Assume from the beginning that the Lipschitz constant is 1. Take $z \in H_1 \backslash A$. The families $\{B(x; \|x - z\|) : x \in A\}$ and $\{B(f(x); \|x - z\|) : x \in A\}$ satisfy the assumptions of the lemma. Moreover, $z \in \bigcap_{x\in A} B(x; \|x-z\|)$. Therefore there is some $y \in \bigcap_{x\in A} B(f(x); \|x - z\|)$. Put $F(z) = y$. Use Zorn's lemma to finish the proof.

12.44 [BJLPS] Let X be a Banach space such that X^* is separable. Let Y be a Banach space Lipschitz homeomorphic to X. Prove that Y^* is separable. (Use the following theorem of Preiss [Prei2]: *Real-valued Lipschitz functions on Asplund spaces are Fréchet differentiable at points of a dense subset.*) In connection with this exercise see Remark 12.6.

Hint. [BJLPS] Let the homeomorphism f satisfy $C\|x - y\| \leq \|f(x) - f(y)\| \leq \|x - y\|$ for all $x, y \in X$. Assume that Y^* is nonseparable. Then there is an uncountable set $\{y_\gamma^*\}_{\gamma\in\Gamma}$ of norm one functionals in Y^* so that $\|y_\gamma^* - y_{\gamma'}^*\| \geq \frac{1}{2}$ for each $\gamma \neq \gamma'$. Set $f_\gamma = y_\gamma^* \circ f$. Then f_γ are 1-Lipschitz. By Preiss' theorem quoted in the statement of the exercise, each f_γ has a Fréchet derivative at some point x_γ in the unit ball of X.

Let $\varepsilon = (10C)^{-1}$ and for each γ choose $\delta_\gamma > 0$ so that whenever $\|z\| \leq \delta_\gamma$,

$$|f_\gamma(x_\gamma + z) - f_\gamma(x_\gamma) - f_\gamma'(x_\gamma)(z)| \leq \varepsilon\|z\|.$$

By passing to a suitable uncountable subset of Γ, we may assume that $\delta := \inf_{\gamma\in\Gamma} \delta_\gamma > 0$ and also (since X^* and X are separable) for all $\gamma, \gamma' \in \Gamma$,

$$\|x_\gamma - x_{\gamma'}\| < \varepsilon\delta, \quad \|f_\gamma'(x_\gamma) - f_{\gamma'}'(x_{\gamma'})\| < \varepsilon, \quad |f_\gamma(x_\gamma) - f_{\gamma'}(x_{\gamma'})| < \varepsilon\delta.$$

If $\|z\| \leq \delta$, we have

$$\begin{aligned}
&|f_\gamma(x_\gamma + z) - f_{\gamma'}(x_\gamma + z)| = |(f_\gamma(x_\gamma + z) - f_\gamma(x_\gamma) - f_\gamma'(x_\gamma)(z)) \\
&\quad +(f_\gamma(x_\gamma) - f_{\gamma'}(x_{\gamma'})) + (f_\gamma'(x_\gamma)(z) - f_{\gamma'}'(x_{\gamma'})(z)) + (-f_{\gamma'}(x_{\gamma'} + z) \\
&\qquad +f_{\gamma'}(x_{\gamma'}) + f_{\gamma'}'(x_{\gamma'})(z)) + (f_{\gamma'}(x_{\gamma'} + z) - f_{\gamma'}(x_\gamma + z))| \\
&\leq \varepsilon\|z\| + \varepsilon\delta + \|f_\gamma'(x_\gamma) - f_{\gamma'}'(x_{\gamma'})\| \cdot \|z\| + \varepsilon\|z\| + \|x_\gamma - x_{\gamma'}\| < 5\varepsilon\delta = \delta/(2C).
\end{aligned}$$

On the other hand,

$$\begin{aligned}
&\sup_{\|z\|\leq\delta} |f_\gamma(x_\gamma + z) - f_{\gamma'}(x_\gamma + z)| = \sup_{\|z\|\leq\delta} |(y_\gamma^* - y_{\gamma'}^*)f(x_\gamma + z)| \\
&\geq \sup_{\|y\|\leq(\delta/C)} |(y_\gamma^* - y_{\gamma'}^*)(f(x_\gamma) + y)| \geq \|y_\gamma^* - y_{\gamma'}^*\|(\delta/C) > \delta/(2C).
\end{aligned}$$

12.45 Let X and Y be nonseparable Banach spaces and φ be a uniform homeomorphism from X onto Y. Show that given separable subspaces W of X and Z of Y, there are separable subspaces $\widetilde{W}$ of X, $\widetilde{W} \supset W$ and $\widetilde{Z}$ of Y with $\widetilde{Z} \supset Z$ such that the restriction of φ to $\widetilde{W}$ is a uniform homeomorphism of $\widetilde{W}$ onto $\widetilde{Z}$.
Hint. Let $[S]$ denote the closed linear hull of S. Consider the following sequence of subsets: $S_1 := \varphi(W)$, $S_2 := \varphi^{-1}([[\varphi(W)] \cup Z])$, $S_3 := \varphi([S_2])$, $S_4 = \varphi^{-1}([S_3])$, etc. Put $\widetilde{W} = [\bigcup_k [S_{2k}]]$, $\widetilde{Z} = [\bigcup_k [S_{2k+1}]]$. Then it suffices to show that $\varphi(\widetilde{W}) = \widetilde{Z}$. To show this equality, let $z \in \widetilde{Z}$ be given. Choose $z_k \in [S_{2k+1}]$ be such that $z_n \to z$ in Y. Then $z_k = \varphi(x_k)$, where $x_k \in [S_{2k+2}]$. As φ^{-1} is uniformly continuous, $x_k \to x$ in X for some $x \in X$. Thus $x \in \widetilde{W}$ and $\varphi(x) = \lim \varphi(x_k) = \lim z_k = z$.

12.46 (Benyamini, Lindenstrauss [BeLi, p. 65]) Let $\varphi : C[0, 1] \to C[0, 1]$ be defined by

$$\varphi(x)(t) = |x(t) + G(x, t)| - G(x, t)$$

where $G(x, t) = 1 - 2(1 - \|x\|)t$, where $\|\cdot\|$ is the standard norm of $C[0, 1]$.

Then φ is 5-Lipschitz, $\varphi(x) = x$ when $\|x\| = 1$ and $\inf\{\|\varphi(x)\| : x \in C[0, 1]\} > 0$. Thus $f(x) := \varphi(x)/\|\varphi(x)\|$ is a Lipschitz retraction from the unit ball of $C[0, 1]$ onto the unit sphere.
Hint. Standard.

12.47 Show that $\left(\sum C[0, 1]\right)_{c_0}$ is isomorphic to $C[0, 1]$.
Hint. Use Pełczyński's decomposition method. Represent $\left(\sum C[0, 1]\right)_{c_0}$ as a complemented subspace of $C[0, 1]$ using an infinite number of nods in $[0, 1]$ and the subspace of $C[0, 1]$ of functions that vanish at these nodes.

12.48 Show that $C[0, 1]$ is uniformly equivalent to a bounded subset of $C[0, 1]$.
Hint. Consider the mapping $\varphi\ C[0, 1] \mapsto C[0, 1]$, $\varphi(f(x)) = \max\{-2, \min\{0, f(x)\}\}$; define

$$T : C[0, 1] \mapsto \left(\sum_{n=0}^{\infty} C[0, 1]\right)_{c_0} \oplus \left(\sum_{n=0}^{\infty} C[0, 1]\right)_{c_0}$$

by

$$Tf = \left(\sum_{n=0}^{\infty} \varphi(n + f)\right)_{c_0} \oplus \left(\sum_{n=0}^{\infty} \varphi(n - f)\right)_{c_0}.$$

We have for $f, g \in C[0, 1]$:

$$\min\{1, \|f - g\|\} \le \|T(f) - T(g)\| \le \|f - g\|$$

and $\|T(f)\| \le 3$ for all $f \in C[0, 1]$. Then use Exercise 12.47.

12.49 Let X be a Banach space and Y be a subspace of X. Show directly (not using the Kadec–Toruńczyk theorem) that X is homeomorphic to $Y \oplus (X/Y)$.
Hint. Let ψ be a Bartle–Graves selector, see Corollary 7.56. Define a mapping $\varphi\, X \mapsto Y \oplus X/Y$ for $x \in X$ by

$$\varphi(x) = (x - \psi(\hat{x}), \hat{x}).$$

Note that given $(y, \hat{x}) \in Y \oplus X/Y$, we have $\varphi(y + x) = (y, \hat{x})$.

12.50 There is a separable Banach space X that is not isomorphic to a Hilbert space and has a subspace Y such that both Y and X/Y are isomorphic to Hilbert spaces [ELP]. Show that for this space there is no Lipschitz selector $\psi : X/Y \mapsto X$ with the property $\pi(\psi(\hat{x})) = \hat{x}$ for every $\hat{x} \in X/Y$.
Hint. If such ψ existed, by the proof of Exercise 12.49, X would be Lipschitz equivalent to $Y \oplus X/Y$, which is isomorphic to a Hilbert space. Then X would be isomorphic to a Hilbert space by Corollary 12.68.

12.51 Theorem 12.65 claims in particular that every separable Banach space that admits a Lipschitz Kadec–Klee smooth norm is isomorphic to a subspace of c_0. We gave a proof of this theorem in case of X having a shrinking Schauder basis. Use the notion of FDDs (see Definition 4.31) to prove Theorem 12.65 in full generality.
Hint. First note that X^* is separable as the dual norm has the w^*-Kadec–Klee property. By Theorem 4.34 there is a subspace Y of X such that both Y and X/Y have a shrinking FDD. Extend the proof of the theorem for FDDs instead of Schauder bases. Then use Exercise 5.58.

12.52 (Godefroy, Kalton, and Lancien, [GKL1]) Use Theorem 12.65 to give an alternative proof of the result of Johnson and Zippin [JoZi0], see also, e.g., [LiTz3, p. 107], that every quotient space of c_0 is isomorphic to a subspace of c_0.
Hint. Every quotient of c_0 admits an equivalent Lipschitz Kadec–Klee norm as this property is easily seen to be carried to quotients.

12.53 A topological space T is called *totally disconnected* if the only connected subsets are the empty set and the singletons. A topological space T is called *zero-dimensional* if it has a base for the topology consisting on *clopen* (i.e., simultaneously closed and open) sets. Both concepts coincide in the class of compact topological spaces. Show that for every compact metric space K there is a zero-dimensional compact space L so that $C(K)$ is isomorphic to $C(L)$.
Hint. Use Miljutin theorem (see, e.g., [AlKa, p. 94]) and Mazurkiewicz–Sierpiński theorem (see, e.g., [HMVZ, p. 73]), plus the fact that the Cantor set Δ is a compact metrizable and zero-dimensional space.

12.54 Let $f(x, y)$ be a continuous bounded function on an open set Ω in $\mathbb{R}^2$ that is uniformly Lipschitz in y on Ω (there is a constant K such that $|f(x, y_1) - f(x, y_2)| \leq K|y_1 - y_2|$ for every $(x, y_1), (x, y_2) \in \Omega$).

Let $(x_0, y_0) \in \Omega$. Show that there is $\delta > 0$ such that on $[x_0 - \delta, x_0 + \delta]$ there is a unique continuously differentiable solution to $\frac{dy}{dx} = f(x, y)$ with the initial condition $y(x_0) = y_0$.
Hint. Observe that $y(x)$ is a solution if and only if $y(x) = y_0 + \int_{x_0}^{x} f\big(t, y(t)\big)\, dt$ for $x \in [x_0 - \delta, x_0 + \delta]$.

There is $M > 0$ such that $|f(x, y)| \le M$ for every $(x, y) \in \Omega$. Choose $\delta > 0$ such that $K\delta < 1$, set $I = [x_0 - \delta, x_0 + \delta]$ and

$$R = \{(x, y) \in \mathbb{R}^2 : \ |x - x_0| \le \delta, |y - y_0| \le M\delta\} \subset \Omega.$$

Consider the closed subset Y of $C(I)$ formed by all functions $y(x)$ for which $y(x_0) = y_0$ and $|y(x) - y_0| \le M\delta$. Then Y is a complete metric space in the metric induced from $C(I)$. Finally, define a mapping T on Y by

$$T(y) \colon x \mapsto y_0 + \int_{x_0}^{x} f\big(t, y(t)\big)\, dt$$

for $y(x) \in Y$. Check that indeed T maps Y into Y.

Using the Lipschitz property of f we get that T is a contraction, hence it has a fixed point. But $T(y)(x) = y(x)$ is equivalent to y being a solution on I.

12.55 (The Peano theorem) Let $f(x, y)$ be a continuous function on $R = \{(x, y) \in \mathbb{R}^2 : \ |x - x_0| \le a, |y - y_0| \le b\}$. Let $M = \max_{(x,y)\in R} |f(x, y)|$ and $\delta = \min\{a, \frac{b}{M}\}$. Show that on the interval $[x_0 - \delta, x_0 + \delta]$ there exists at least one solution to the equation $\frac{dy}{dx} = f(x, y)$ that satisfies $y(x_0) = y_0$.
Hint. A function y is a solution if and only if $y(x) = y_0 + \int_{x_0}^{x} f\big(t, y(t)\big)\, dt$. Set $I = [x_0 - \delta, x_0 + \delta]$ and let B be the closed ball in $C(I)$ centered at the constant function y_0 with radius b. Define a mapping T on B by

$$T(y) = y_0 + \int_{x_0}^{x} f\big(t, y(t)\big)\, dt.$$

T is continuous: If $y_n(x)$ converges to $y(x)$ in B, then by the uniform continuity of $f(x, y)$ on the rectangle R we have that $f(x, y_n(x)) \to f(x, y(x))$ uniformly on I. Therefore $T(y_n)(x) \to T(y)(x)$ uniformly on I.

The total boundedness of $T(B)$ is proved directly using the Arzelà–Ascoli theorem. Thus $\overline{T(B)}$ is compact and by Schauder's theorem, T has a fixed point $y(t) \in B$.

Chapter 13
Weakly Compactly Generated Spaces

In this chapter we study weakly compact operators and the related class of Banach spaces that are generated by weakly compact sets (i.e., weakly compactly generated spaces, in short WCG spaces). We focus on their decomposition properties, renormings, and on the topological properties of their dual spaces. We prove that WCG spaces are generated by reflexive spaces. Then we study absolutely summing operators and the Dunford–Pettis property.

13.1 Introduction

Definition 13.1 *A Banach space X is called* weakly compactly generated (WCG) *if there is a weakly compact set K in X such that $X = \overline{\text{span}}(K)$.*

We say that K *generates* X if $X = \overline{\text{span}}(K)$.

If K is weakly compact, then so is $K \cup (-K)$. Thus $\overline{\text{conv}}\{K \cup (-K)\}$ is convex, symmetric, and weakly compact by Theorem 3.133. Therefore we may assume in the definition of WCG spaces that the set K is weakly compact, convex, and symmetric.

Let X, Y be Banach spaces, let T be a bounded operator from X onto a dense set in Y. If X is weakly compactly generated by a weakly compact set K, then Y is also weakly compactly generated, namely by the weakly compact set $T(K)$. In particular, if Y is a closed subspace of a WCG space X, then X/Y is a WCG space. A subspace of WCG space need not be WCG ([Rose5], see, e.g., [Fabi1, p. 31] and Exercise 13.56).

Examples:

(i) Every reflexive Banach space is weakly compactly generated by its unit ball.

(ii) If X is a separable Banach space and $\{x_n\}$ is a dense set in S_X, then $K = \left\{\frac{1}{n}x_n\right\} \cup \{0\}$ is a compact set that generates X. Therefore every separable Banach space is $\|\cdot\|$- (then weakly-) compactly generated. Conversely, every $\|\cdot\|$-compactly generated Banach space is clearly separable.

(iii) Let Γ be a nonempty set, and let $\{e_\gamma : \gamma \in \Gamma\}$ be the family of the standard unit vectors in $c_0(\Gamma)$. Let $\mathcal{U}$ be a covering of $\{e_\gamma : \gamma \in \Gamma\} \cup \{0\}$ consisting of w-open sets, and let $U \in \mathcal{U}$ be such that $0 \in U$. There exists a finite subset F of

DOI 10.1007/978-1-4419-7515-7_13,

$\ell_1(\Gamma)$ and an $\varepsilon > 0$ such that $\{x \in c_0(\Gamma) : |\langle f, x\rangle| < \varepsilon, f \in F\} \subset U$. It is obvious that a finite set $\Gamma_0 \subset \Gamma$ exists such that $|\langle f, e_\gamma\rangle| < \varepsilon$ for every $f \in F$ and $\gamma \in \Gamma\backslash\Gamma_0$. This implies that $\{e_\gamma : \gamma \in \Gamma\backslash\Gamma_0\} \subset U$, hence $\mathcal{U}$ has a finite subcovering of $\{e_\gamma : \gamma \in \Gamma\} \cup \{0\}$ and so $\{e_\gamma : \gamma \in \Gamma\}$ is weakly compact. Therefore, $c_0(\Gamma)$ is a WCG space. Alternatively, the formal identity maps $\ell_2(\Gamma)$ onto a dense set in $c_0(\Gamma)$.

(iv) Since every weakly compact set in ℓ_∞ is norm-separable (Exercise 3.104), the space ℓ_∞ is not WCG. Every weakly compact set in $\ell_1(\Gamma)$ is compact (Exercise 5.47) and thus separable. Therefore the space $\ell_1(\Gamma)$ is not WCG if Γ is uncountable, as $\ell_1(\Gamma)$ is then nonseparable.

(v) Hölder's inequality (1.1) implies that the identity operator T from $L_2(\mu)$ into $L_1(\mu)$, where μ is a finite measure, is a bounded operator. From the theory of the Lebesgue integral, we know that T maps $L_2(\mu)$ onto a dense set in $L_1(\mu)$. Therefore $L_1(\mu)$ is weakly compactly generated by the set $T(B_{L_2(\mu)})$. More generally, we obtain that $L_1(\mu)$ is weakly compactly generated if μ is σ-finite.

If μ is not σ-finite, then $L_1(\mu)$ contains a complemented copy of $\ell_1(\Gamma)$ for some uncountable Γ. By (iv), $L_1(\mu)$ is not WCG.

Definition 13.2 *Let X be a Banach space. The* density character *of X (or just the* density *of X), denoted* dens(X), *is the smallest cardinal number of the form* card(A), *where A is a dense subset of X.*
The weak* density character of X^* *(or just the w^**-density of X^**), denoted* w^*-dens(X^*), *is the smallest cardinal number of the form* card(A), *where A is a w^*-dense subset of X^*.*

We have dens$(X) \le$ dens(X^*) (see the proof of Proposition 2.8).

Note also that w^*-dens$(X^*) \le$ dens(X). Indeed, assume that $\{x_\alpha\}_{\alpha\in A}$ is norm-dense in S_X and $f_\alpha \in S_{X^*}$ are such that $f_\alpha(x_\alpha) = 1$ for each $\alpha \in A$. If $f_\alpha(x) = 0$ for all α and $\alpha_0 \in A$ is such that $\|x - x_{\alpha_0}\| < 1/2$, we get $f_{\alpha_0}(x) = f_{\alpha_0}(x_{\alpha_0}) - f_{\alpha_0}(x_{\alpha_0} - x) \ge 1 - 1/2 = 1/2$, a contradiction (see Exercise 3.95). Then $\overline{\text{span}}^{w^*}\{f_\alpha : \alpha \in A\} = X^*$. The conclusion follows since the set of all rational linear combinations of $\{f_\alpha : \alpha \in A\}$ is still w^*-dense in X^*, and it has cardinality card(A).

In general we do not have equality. For instance, ℓ_1 is w^*-dense in ℓ_∞^* and thus w^*-dens$(\ell_\infty^*) \le \aleph_0 <$ dens(ℓ_∞). However, in the case of weakly compactly generated spaces we have the following statement.

Theorem 13.3 (Amir, Lindenstrauss [AmLi]) *Let X be a weakly compactly generated Banach space. Then* dens$(X) = w^*$-dens(X^*).

Proof: Let K be a weakly compact convex subset of X that generates X and let D be a set of cardinality w^*-dens(X^*) that is w^*-dense in X^*. Define an operator $T\colon X^* \to C(K)$ by $T(f)\colon k \mapsto f(k)$. Since $T(D)$ separates points of K, the algebra generated by $T(D)$ and the constant function contains a dense set N of cardinality w^*-dens$(X^*) = \text{card}\big(T(D)\big)$ that is, by the Stone–Weierstrass theorem, dense in $C(K)$. For every $g \in N$, choose $k_g \in K$ such that $g(k_g) = \sup_{k\in K}\big(g(k)\big)$. We claim that $S = \{k_g : g \in N\}$ is w-dense in K. Indeed, we have $\sup_K(\tilde{g}) = \sup_{g\in N}\big(\tilde{g}(k_g)\big)$

for every $\tilde{g} \in C(K)$ as otherwise for $g \in N$ sufficiently close in $C(K)$ to $\tilde{g}$ we cannot have that $\sup_K(g) = g(k_g)$. Therefore S is weakly dense in K by Urysohn's theorem, so $\text{span}(S)$ is w-dense in X. The set $\text{span}(S)$ has a norm dense subset of cardinality $\text{card}(S) = w^*\text{-dens}(X^*)$ (use rational combinations). Since $\text{span}(S)$ is weakly and thus norm dense in X, we have $\text{dens}(X) \leq w^*\text{-dens}(X^*)$. As noted above, it is always true that $w^*\text{-dens}(X^*) \leq \text{dens}(X)$.

Alternatively, let a weakly compact set K generate X and let D be a set of cardinality $w^*\text{-dens}(X^*)$ that is w^*-dense in X^*. The topology $w(X, D)$ of pointwise convergence on D coincides with the w-topology of K as K is weakly compact. The topology $w(X, D)$ has a base of cardinality $\text{card}(D)$ and thus there is a set S of cardinality $\text{card}(D)$ that is w-dense in K. Therefore $\text{dens}(X) \leq \text{card}(D)$. □

Corollary 13.4 *Let X be a Banach space such that X^* is weakly compactly generated. Then* $\text{dens}(X) = \text{dens}(X^*)$.

Proof: From the Goldstine's theorem we have $w^*\text{-dens}(X^{**}) \leq \text{dens}(X)$. From Theorem 13.3 we have $\text{dens}(X^*) = w^*\text{-dens}(X^{**})$. Therefore $\text{dens}(X^*) \leq \text{dens}(X)$. As noted above, $\text{dens}(X) \leq \text{dens}(X^*)$. □

13.2 Projectional Resolutions of the Identity

Recall that ω_0 is the first infinite ordinal.

Definition 13.5 *Let X be a Banach space with infinite density character $\aleph$, and let μ be the first ordinal with cardinality $\aleph$. A transfinite sequence of bounded linear projections $\{P_\alpha\}_{\omega_0 \leq \alpha \leq \mu}$ on X is called a* projectional resolution of identity (PRI) *if*
(i) $P_{\omega_0} = 0$, $P_\mu = I_X$,
and for all $\omega_0 < \alpha, \beta \leq \mu$ we have
(ii) $\|P_\alpha\| = 1$,
(iii) $\text{dens}\big(P_\alpha(X)\big) \leq \text{card}(\alpha)$,
(iv) $P_\alpha P_\beta = P_\beta P_\alpha = P_{\min(\alpha,\beta)}$, *and*
(v) *for every $x \in X$ the mapping $\alpha \mapsto P_\alpha(x)$ is a continuous mapping from the ordinal segment $[\omega_0, \mu]$ in its standard topology into X.*

For information on the standard topology of $[\omega_0, \mu]$ we refer to [Dugu2].

Theorem 13.6 (Amir, Lindenstrauss [AmLi]) *Let X be a Banach space weakly compactly generated by a convex symmetric and weakly compact set K. Let μ be the first ordinal of cardinality* $\text{dens}(X)$. *Then there is a PRI $\{P_\alpha\}_{\omega_0 \leq \alpha \leq \mu}$ such that $P_\alpha(K) \subset K$ for every $\alpha \in [\omega_0, \mu]$.*

The proof of Theorem 13.6 will be presented after a series of preliminary results. In order to illustrate the ideas behind the construction it is worth to see how a single norm-one projection whose range contains an a priori chosen separable subspace of a reflexive Banach space is constructed in Exercise 13.2.

For the aim of constructing a norm-one projection in a Banach space, it is useful to have a look at Exercise 4.19.

Given a subset M of a linear space X, put $\operatorname{span}_{\mathbb{Q}}(M)$ for the set of all rational linear combinations of elements in M. We say that M is a $\mathbb{Q}$*-linear* subset of X if $M = \operatorname{span}_{\mathbb{Q}}(M)$.

Lemma 13.7 *Let X be a Banach space, $W \subset X^*$ be $\mathbb{Q}$-linear, and $\Phi : W \to 2^X$ and $\Psi : X \to 2^W$ be two at most countably valued mappings. Let $A_0 \subset X$, $B_0 \subset W$ be such that* $\operatorname{card}(A_0) \leq \Gamma$ *and* $\operatorname{card}(B_0) \leq \Gamma$, *for some infinite cardinal Γ. Then there exist $\mathbb{Q}$-linear sets A and B such that $A_0 \subset A \subset X$, $B_0 \subset B \subset W$,* $\operatorname{card}(A) \leq \Gamma$, $\operatorname{card}(B) \leq \Gamma$, *$\Phi(B) \subset A$, and $\Psi(A) \subset B$.*

Proof: We will construct by induction two sequences of sets

$$(A_0 \subset)\ A_1 \subset A_2 \subset \ldots \subset X, \quad (B_0 \subset)\ B_1 \subset B_2 \subset \ldots \subset W$$

as follows. Having constructed $(A_0,)\ A_1, \ldots, A_n$, $(B_0,)\ B_1, \ldots, B_n$, we put

$$A_{n+1} = \operatorname{span}_{\mathbb{Q}}\big(A_n \cup \Phi(B_n)\big),$$
$$B_{n+1} = \operatorname{span}_{\mathbb{Q}}\big(B_n \cup \Psi(A_n)\big).$$

Finally, we set $A = \bigcup_{n=0}^{\infty} A_n$, $B = \bigcup_{n=0}^{\infty} B_n$. That A and B satisfy the required properties is obvious. □

Lemma 13.8 *Let X, W, Φ, and Ψ be as in Lemma* 13.7. *Let μ be the first ordinal with cardinal* $\operatorname{dens}(X)$ $(> \omega_0)$. *Then there exist families $\{A_\alpha : \omega_0 \leq \alpha \leq \mu\}$ and $\{B_\alpha : \omega_0 \leq \alpha \leq \mu\}$ of $\mathbb{Q}$-linear subsets of X and W, respectively, such that the following holds:*

(i) $A_{\omega_0} = \{0\}$, $B_{\omega_0} = \{0\}$.

(ii) $\operatorname{card}(A_\alpha) \leq \alpha$, $\operatorname{card}(B_\alpha) \leq \alpha$, *if* $\omega_0 \leq \alpha \leq \mu$, $\overline{A_\mu} = X$,

(iii) $\Phi(B_\alpha) \subset A_\alpha$, $\Psi(A_\alpha) \subset B_\alpha$, *if* $\omega_0 < \alpha \leq \mu$

(iv) $A_\alpha \subset A_\beta$, $B_\alpha \subset B_\beta$, *if* $\omega_0 \leq \alpha < \beta \leq \mu$,

(v) $A_\alpha = \bigcup_{\beta<\alpha} A_{\beta+1}$, $B_\alpha = \bigcup_{\beta<\alpha} B_{\beta+1}$, *if* $\omega_0 < \alpha \leq \mu$.

Proof: Suppose that $\{x_\alpha\}_{\omega_0 \leq \alpha < \mu}$ is a dense sequence in X. We proceed by transfinite induction. Put $A_{\omega_0} = \{0\}$ and $B_{\omega_0} = \{0\}$. Assume that for some γ such that $\omega_0 < \gamma \leq \mu$ and for every α such that $\omega_0 \leq \alpha < \gamma$, we already constructed sets $A_\alpha \subset X$ and $B_\alpha \subset W$ satisfying the required properties. If γ is a limit ordinal, put $A_\gamma = \bigcup_{\alpha<\gamma} A_\alpha$ and $B_\gamma = \bigcup_{\alpha<\gamma} B_\alpha$. If γ is nonlimit, apply Lemma 13.7 to the pair $A_{\gamma-1} \cup \{x_{\gamma-1}\}$, $B_{\gamma-1}$, and the cardinality $\operatorname{card}(\gamma)$, in order to obtain the sets A_γ and B_γ. □

Definition 13.9 *Let X be a Banach space, W be a* 1*-norming $\mathbb{Q}$-linear subset of X^*, and $\Phi : W \to 2^X$ be an at most countably valued mapping such that for every nonempty $\mathbb{Q}$-linear subset B of W we have $\Phi(B)^\perp \cap \overline{B}^{w^*} = \{0\}$. The couple (W, Φ) is called a* projectional generator *(PG) on X.*

In Proposition 13.13 and in Exercise 13.31, we shall provide examples of projectional generators in certain classes of Banach spaces.

Definition 13.10 *A Banach space X has the separable complementation property* (SCP) *if every separable subspace of X is contained in a separable complemented subspace Z of X. The space X has the 1-separable complementation property* (1-SCP) *if it has the* SCP *and for all Z as above, the projection $P : X \to Z$ satisfies $\|P\| = 1$.*

The notion of SCP is quite useful especially for constructing counterexamples. We will show in Theorem 13.11 that all spaces with a projectional generator have 1-SCP and a PRI.

Theorem 13.11 *Let X be a nonseparable Banach space X, and let μ be the first ordinal with cardinal dens X. If X has a projectional generator (W, Φ), then X has the 1-SCP and admits a PRI $\{P_\alpha : \omega_0 \leq \alpha \leq \mu\}$.*

Proof: We can define a countable valued mapping $\Psi : X \to 2^W$, such that $\|x\| = \sup_{f \in \Psi(x) \cap B_{X^*}} f(x)$ for all $x \in X$. Lemma 13.8 gives two families $\{A_\alpha : \omega_0 \leq \alpha \leq \mu\}$ and $\{B_\alpha : \omega_0 \leq \alpha \leq \mu\}$ of subsets of X and W, respectively, with the properties listed therein. Observe that, for each $\alpha \in (\omega_0, \mu]$, $\overline{A}_\alpha$ is linear and it follows from (iii) in that lemma that B_α 1-norms A_α and that A_α separates points of $\overline{B}_\alpha^{w^*}$. By Exercise 4.19 there exist norm-one projections $P_\alpha : X \to X$, such that $P_\alpha(X) = \overline{A_\alpha}$ and Ker $P_\alpha = (B_\alpha)_\perp$. Put $P_{\omega_0} = 0$. The properties of $\{A_\alpha\}$ and $\{B_\alpha\}$ listed in Lemma 13.8 yield that $\{P_\alpha : \omega_0 \leq \alpha \leq \mu\}$ is a PRI on X.

To prove the 1-SCP property, it is enough, given a separable subspace Y of X, to choose in the proof of Lemma 13.8 a dense subset $\{x_\alpha\}_{\omega_0 \leq \alpha < \mu}$ of X such that $\{x_\alpha\}_{\omega_0 \leq \alpha < 2\omega_0}$ is dense in Y. In this way Y is a subspace of the separable and 1-complemented subspace $\overline{A}_{2\omega_0}$. □

The following result is a version of a theorem that can be found in [FGMZ]. It strengths Theorem 13.11 for the case that the 1-norming subset in the projectional generator is the whole dual space. In this situation, the projectional resolution of the identity can be constructed in such a way that it fixes a countable family of subsets of X, in particular a closed subspace of X. Given a set Γ in a Banach space X, we say that it *countably supports* X^* if the set $\{\gamma \in \Gamma;\ \langle x^*, \gamma\rangle \neq 0\}$ is at most countable for every $x^* \in X^*$.

Theorem 13.12 *Let $(X, \|\cdot\|)$ be a nonseparable Banach space admitting a projectional generator (X^*, Φ_0). Let $M_1, M_2, \ldots$ be an at most countable family of bounded closed convex and symmetric subsets in X. Let $\Gamma \subset B_X$ be a set which countably supports X^*. Then there exists a PRI $(P_\alpha : \omega_0 \leq \alpha \leq \mu)$ on X such that $P_\alpha(M_n) \subset M_n$ and $P_\alpha(\gamma) \in \{\gamma, 0\}$ for every $\alpha \in [\omega_0, \mu]$, every $n \in \mathbb{N}$, and every $\gamma \in \Gamma$.*

Proof: Denote $M_0 = B_X$. Put

$$\Phi(x^*) = \Phi_0(x^*) \cup \{\gamma \in \Gamma;\ \langle x^*, \gamma\rangle \neq 0\}, \quad x^* \in X^*.$$

Clearly, Φ is also a projectional generator. For $n \in \mathbb{N} \cup \{0\}$ and $m \in \mathbb{N}$, let $\|\cdot\|_{n,m}$ be the Minkowski functional of the set $M_n + \frac{1}{m} B_X$; this will be an equivalent norm on X. For every $x \in X$ we find a countable set $\Psi(x) \subset X^*$ such that

$$\|x\|_{n,m} = \sup\{\langle x^*, x\rangle;\ x^* \in \Psi(x) \text{ and } \|x^*\|_{n,m}^* \leq 1\}$$

for every $n \in \mathbb{N} \cup \{0\}$ and $m \in \mathbb{N}$. Thus we defined $\Psi : X \to 2^{X^*}$.

Lemma 13.8 provides two families $\{A_\alpha : \omega_0 \leq \alpha \leq \mu\}$ and $\{B_\alpha : \omega_0 \leq \alpha \leq \mu\}$ with the properties listed there, and so, the proof of Theorem 13.11 gives an associated projectional resolution of the identity $\{P_\alpha : \omega_0 \leq \alpha \leq \mu\}$ in X. Each P_α satisfies, by the construction, that $P_\alpha X = \overline{A_\alpha}$, $P_\alpha^{-1}(0) = (B_\alpha)_\perp$, and $P_\alpha^* X^* = \overline{B_\alpha}^{w^*}$, and, moreover, $\|P_\alpha\|_{n,m} = 1$ for every $n \in \mathbb{N} \cup \{0\}$ and $m \in \mathbb{N}$. Then

$$P_\alpha M_n \subset \bigcap_{m=1}^{\infty} P_\alpha\big(M_n + \tfrac{1}{m} B_X\big) \subset \bigcap_{m=1}^{\infty} \overline{M_n + \tfrac{1}{m} B_X} \subset \bigcap_{m=1}^{\infty} \big(M_n + \tfrac{2}{m} B_X\big) = M_n$$

for every $n \in \mathbb{N} \cup \{0\}$, and in particular, $\|P_\alpha\| = 1$.

Fix any $\gamma \in \Gamma$ and $\alpha \in [\omega_0, \mu]$. It remains to prove that $P_\alpha \gamma \in \{\gamma, 0\}$. If $\gamma \in P_\alpha X$, then, trivially, $P_\alpha \gamma = \gamma$. Second, assume that $\gamma \notin P_\alpha X$ $(= \overline{A_\alpha})$. Then $\gamma \notin \Phi(B_\alpha)$, which implies that $\langle b, \gamma\rangle = 0$ for every $b \in B_\alpha$, that is, that $\gamma \in (B_\alpha)_\perp$. But $(B_\alpha)_\perp = P_\alpha^{-1}(0)$. Hence $P_\alpha \gamma = 0$. □

The following proposition provides a natural example of a PG for the class of WCG Banach spaces. As a consequence of it and of Theorem 13.11, we obtain the Amir–Lindenstrauss Theorem 13.6 on the existence of a PRI in every WCG Banach space and the 1-SCP of those spaces.

Proposition 13.13 *If $(X, \|\cdot\|)$ is a WCG Banach space, and K is an absolutely convex and weakly compact set that generates X, then X admits a projectional generator (X^*, Φ) such that $\Phi : X^* \to X$ is single-valued and $\Phi(x^*) \in K$ for every $x^* \in X^*$. Therefore, a PRI $\{P_\alpha : \omega_0 \leq \alpha \leq \mu\}$ exists on X and it can be constructed in such a way that $P_\alpha(K) \subset K$ for all $\omega_0 \leq \alpha \leq \mu$.*

Proof: Given $x^* \in X^*$, let $\Phi(x^*)$ be an element in K such that $\langle x^*, \Phi(x^*)\rangle = \sup_{x \in K} |\langle x^*, x\rangle|$. We shall show that (X^*, Φ) is a projectional generator. To this end, let W be a $\mathbb{Q}$-linear subset of X^*. The set $\overline{W}^{w^*}$ is a linear subspace of X^* (note that $\overline{W}^{w^*} = \overline{\overline{W}^{\|\cdot\|}}^{w^*}$ and that $\overline{W}^{\|\cdot\|}$ is already a linear subspace; see also Exercise 13.1). By the Mackey–Arens Theorem 3.41, $\overline{W}^{w^*} = \overline{W}^{\mu(X^*,X)}$, where $\mu(X^*, X)$ is the Mackey topology on X^* associated to the dual pair $\langle X, X^*\rangle$. Note that $\overline{W}^{\mu(X^*,X)} \subset \overline{W}^{\mathcal{T}_K}$, where $\mathcal{T}_K$ is the (metrizable) topology on X^* of the uniform convergence on K. Let $x^* \in \Phi(W)^\perp \cap \overline{W}^{w^*}$. Therefore we can find a sequence $\{x_n^*\}$ in W such that $x_n^* \xrightarrow{\mathcal{T}_K} x^*$. Fix $\varepsilon > 0$ and find $n_0 \in \mathbb{N}$ such that $\sup_{x \in K} |\langle x^* - x_n^*, x\rangle| < \varepsilon$ for all $n \geq n_0$. Then, in particular, $\sup_{x \in K} |\langle x_n^*, x\rangle| = |\langle x_n^*, \Phi(x_n^*)\rangle| = |\langle x^* -$

$x_n^*, \Phi(x_n^*))| < \varepsilon$ for all $n \geq n_0$. This implies that $\sup_{x \in K} |\langle x^*, x\rangle| \leq \varepsilon$. As $\varepsilon > 0$ is arbitrary, we get $x^*|_K \equiv 0$, and so $x^* = 0$. This shows that (X^*, Φ) is a projectional generator.

To prove the last assertion, define for each $n \in \mathbb{N}$ an equivalent norm $\|\cdot\|_n$ in X by letting $\overline{K + (1/n)B_{(X,\|\cdot\|)}}$ be its closed unit ball. Given $x \in X$, let us take a countable subset $\Psi(x)$ of X^* such that, for all $n \in \mathbb{N}$,

$$\|x\| = \sup\{\langle x^*, x\rangle : x^* \in \Psi(x) \cap B_{(X^*,\|\cdot\|)}\},$$
$$\|x\|_n = \sup\{\langle x^*, x\rangle : x^* \in \Psi(x) \cap B_{(X^*,\|\cdot\|_n)}\}.$$

Follow the proof of Theorem 13.11 to obtain a PRI $\{P_\alpha\}_{\omega_0 \leq \alpha \leq \mu}$ in X with $\|P_\alpha\| = 1$ and $\|P_\alpha\|_n = 1$ for all $n \in \mathbb{N}$ and for all $\omega_0 < \alpha \leq \mu$. This last assertion implies that

$$P_\alpha\big(K + (1/n)B_{(X,\|\cdot\|)}\big) \subset \overline{K + (1/n)B_{(X,\|\cdot\|)}}$$

for all $n \in \mathbb{N}$, hence

$$P_\alpha(K) \subset \bigcap_{n \in \mathbb{N}} \overline{K + (1/n)B_{(X,\|\cdot\|)}} \subset K,$$

for all $\alpha \in [\omega_0, \mu]$. □

Remark: A similar technique can be used to prove that every weakly countably determined Banach space has a projectional generator, hence it has the 1-SCP and a projectional resolution of the identity (see Exercise 13.31). Here we remark that using Theorem 13.12 and Proposition 13.13, we readily get that every closed subspace of a WCG Banach space—a particular example of a weakly countably determined Banach space—has a projectional resolution of the identity.

13.3 Consequences of the Existence of a Projectional Resolution

Proposition 13.14 *Let X be a Banach space with a PRI $(P_\alpha)_{\omega_0 \leq \alpha \leq \mu}$. Then*
(i) *For every $x \in X$ and for every $\varepsilon > 0$ the set $\{\alpha \in [\omega_0, \mu) : \|(P_{\alpha+1} - P_\alpha)x\| > \varepsilon\}$ is finite. As a consequence, the set $\{\alpha \in [\omega_0, \mu) : (P_{\alpha+1} - P_\alpha)x \neq 0\}$ is countable.*
(ii) *For every $x \in X$ we have $x = \sum_{\omega_0 \leq \alpha < \mu}(P_{\alpha+1} - P_\alpha)(x)$, where only a countable number of summands are non-zero (so this sum should be understood as a usual series). In particular, $X = \overline{\bigcup_{\omega_0 \leq \alpha < \mu} P_\alpha(X)}$.*

Proof: (i) Assume that for some $x \in X$ and some $\varepsilon > 0$ the set $A(x, \varepsilon) = \{\alpha \in [\omega_0, \mu) : \|(P_{\alpha+1} - P_\alpha)(x)\| > \varepsilon\}$ is infinite. Let $\alpha_1 < \alpha_2 < \ldots$ be a sequence in $A(x, \varepsilon)$ and let $\alpha_\infty = \lim_n \alpha_n$. Since $P_\alpha(x) \to P_{\alpha_\infty}(x)$ as $\alpha \uparrow \alpha_\infty$, the net $\{P_\alpha(x)\}_{\omega_0 \leq \alpha < \alpha_\infty}$ is Cauchy. It follows that, for some $\alpha_0 < \alpha_\infty$, we have $\|P_\alpha(x) -$

$P_\beta(x)\| < \varepsilon$ for every α, β in $[\alpha_0, \alpha_\infty)$. Find $n \in \mathbb{N}$ such that $\alpha_n \geq \alpha_0$. Then, in particular, $\|(P_{\alpha_n+1} - P_{\alpha_n})(x)\| < \varepsilon$, a contradiction. Letting $\varepsilon = 1/n$, $n \in \mathbb{N}$, we get the consequence in (i).

(ii) follows from the fact that, if $\{\alpha \in [\omega_0, \mu) : (P_{\alpha+1} - P_\alpha)x \neq 0\} = \{\alpha_n : n \in \mathbb{N}\}$, where $\alpha_1 < \alpha_2 < \ldots$, then $\sum_{\omega_0 \leq \alpha < \mu}(P_{\alpha+1} - P_\alpha)(x) = P_{\alpha_1}(x) + (P_{\alpha_1+1} - P_{\alpha_1})(x) + (P_{\alpha_2+1} - P_{\alpha_2})(x) + \ldots$, and from (v) in the definition of a PRI. □

Definition 13.15 *Let X be a Banach space. We say that a Markushevich basis $\{x_\alpha; f_\alpha\}_{\alpha\in\Gamma}$ of X is* weakly compact *if $\{x_\alpha : \alpha \in \Gamma\} \cup \{0\}$ is a weakly compact set in X.*

Theorem 13.16 (Amir, Lindenstrauss [AmLi], Corson, see [Lind13]) *Every weakly compactly generated Banach space admits a weakly compact Markushevich basis.*

Proof: By transfinite induction on $\text{dens}(X)$ we prove that given a w-compact convex symmetric set K that generates X, there is a Markushevich basis $\{x_\alpha; f_\alpha\}$ of X such that $\{x_\alpha\} \subset K$. First assume that $\text{dens}(X) = \aleph_0$. Since K is norm-separable, choose $\{z_n\}$ dense in K. By Theorem 4.59 there is a Markushevich basis $\{x_n; f_n\}$ of X such that $\text{span}\{x_n\} = \text{span}\{z_n\}$. Since $\text{span}\{z_n\} \subset \bigcup_{n\in\mathbb{N}} nK$, by scaling we can assume that $x_n \in K$ for every n. Then $\{x_n\} \cup \{0\}$ is weakly compact. Indeed, any infinite sequence $\{x'_n\}$ of distinct elements chosen from $\{x_n\}$ converges to zero in the topology of pointwise convergence on $\{f_n\}$, which is a Hausdorff topology. Since $\{x_n\} \cup \{0\} \subset K$ and K is weakly compact, we have $x'_n \overset{w}{\to} 0$. Thus $\{x_n\} \cup \{0\}$ is weakly compact by the Eberlein–Šmulyan theorem.

Assume that the theorem was proven for every WCG space of density character less than $\aleph$. Let X be a WCG space of density character $\aleph$. Let $\{P_\alpha\}_{\omega_0\leq\alpha\leq\mu}$ be a PRI in X. Then, for every $\alpha \in [\omega_0, \mu)$, $(P_{\alpha+1} - P_\alpha)(X)$ is weakly compactly generated by $(P_{\alpha+1} - P_\alpha)(K)$ $(\subset 2K)$. Using the induction hypothesis, for every $\alpha \in [\omega_0, \mu)$ we find a Markushevich basis $\{x^\alpha_\beta; f^\alpha_\beta\}_{\beta\in\Lambda_\alpha}$ of $(P_{\alpha+1} - P_\alpha)(X)$ such that $\{x^\alpha_\beta\}_{\beta\in\Lambda_\alpha} \subset (P_{\alpha+1} - P_\alpha)(K)$ $(\subset 2K)$. We consider $P_{\alpha+1} - P_\alpha$ as a mapping from X into $(P_{\alpha+1} - P_\alpha)(X)$, so $(P_{\alpha+1} - P_\alpha)^*$ maps $(P_{\alpha+1} - P_\alpha)(X)^*$ into X^*. Certainly, $\{x^\alpha_\beta; (P_{\alpha+1} - P_\alpha)^*(f^\alpha_\beta)\}_{\alpha<\mu,\beta\in\Lambda_\alpha}$ is a biorthogonal system in $X \times X^*$. By Proposition 13.14, $\{x^\alpha_\beta\}_{\alpha<\mu,\beta\in\Lambda_\alpha}$ is a linearly dense subset of X. If $(P_{\alpha+1} - P_\alpha)^*(f^\alpha_\beta)(x) = 0$ for every $\alpha < \mu$ and every $\beta \in \Lambda_\alpha$, then $(P_{\alpha+1} - P_\alpha)x = 0$ for all $\alpha < \mu$. Again by Proposition 13.14, we get $x = 0$. Thus $\{x^\alpha_\beta; (P_{\alpha+1} - P_\alpha)^*(f^\alpha_\beta)\}_{\alpha<\mu,\beta\in\Lambda_\alpha}$ is a Markushevich basis of X with $\{x^\alpha_\beta\}_{\alpha<\mu,\beta\in\Lambda_\alpha} \subset 2K$, so $\{\frac{1}{2}x^\alpha_\beta; 2(P_{\alpha+1} - P_\alpha)^*(f^\alpha_\beta)\}_{\alpha<\mu,\beta\in\Lambda_\alpha}$ is a weakly compact Markushevich basis with $\frac{1}{2}\{x^\alpha_\beta\}_{\alpha<\mu,\beta\in\Lambda_\alpha} \subset K$. □

Related to Theorem 13.16, we note that there exists a *convex non-symmetric non-metrizable compact* subset C of $(c_0(\Gamma), w)$ (hence Eberlein compact, see Definition 13.18), for Γ uncountable, so that every point of C is a G_δ-point. This set cannot contain a one-point-compactification of an uncountable discrete set. In particular, it cannot contain a shift of an (uncountable) Markushevich basis [Lind13].

Corollary 13.17 (Amir, Lindenstrauss [AmLi]) *Let X be a weakly compactly generated Banach space. Then there exists a bounded one-to-one operator from X into $c_0(\Gamma)$.*

Proof: Let $\{x_\alpha; f_\alpha\}_{\alpha\in\Gamma}$ be a Markushevich basis of X. By scaling we may assume that $\|f_\alpha\| \leq 1$ for every $\alpha \in \Gamma$. Define a bounded operator T from X into $\ell_\infty(\Gamma)$ by $T(x) = \big(f_\alpha(x)\big)$. Then $\|T\| \leq 1$ and T maps $\operatorname{span}\{x_\alpha\}$ into the set of finitely supported vectors in $\ell_\infty(\Gamma)$. In particular, it maps $\operatorname{span}\{x_\alpha\}$ into $c_0(\Gamma)$. Since T is continuous and $c_0(\Gamma)$ is closed in $\ell_\infty(\Gamma)$, we obtain $T(X) \subset c_0(\Gamma.)$ Since $\{f_\alpha\}$ separates points of X, T must be one-to-one. □

Definition 13.18 *A compact space K is said to be* Eberlein *if K is homeomorphic to a weakly compact set in $c_0(\Gamma)$, in its weak topology.*

Remark: It follows from Corollary 13.19 below that a compact space is Eberlein if, and only if, it is homeomorphic to a weakly compact subset of a Banach space, equipped with the restriction of its weak topology. Corollary 13.23 below shows that it is enough to restrict here the class of Banach spaces to the reflexive Banach spaces.

Example: Every compact metric space is Eberlein. Indeed, recall that if K is a compact metric space, then K is a w^*-compact subset of $C(K)^*$ (Lemma 3.55); moreover, $C(K)$ is separable (Lemma 3.102). If $\{f_i\}_{i=1}^{\infty}$ is dense in $S_{C(K)}$, then the mapping $T(k) := \big(\frac{1}{i} f_i(k)\big)$ gives a homeomorphism of K onto a compact set in c_0.

Proposition 3.108 shows that every Eberlein compact space is angelic. An example of a compact space that is not Eberlein is the ordinal segment $[0, \omega_1]$. Indeed, since ω_1 is not a limit of a sequence of ordinals smaller than ω_1, this space is not angelic and thus not Eberlein compact.

Corollary 13.19 (Amir, Lindenstrauss [AmLi]) *Every weakly compact set in a Banach space in its weak topology is Eberlein.*

Proof: Let K be a weakly compact set in a Banach space X. Let $Y = \overline{\operatorname{span}}(K)$. Then Y is weakly compactly generated and, by Corollary 13.17, there is a bounded linear one-to-one mapping T from Y into $c_0(\Gamma)$. Since T is w–w-continuous and since the restriction to Y of the weak topology on X is the weak topology on Y, T is a homeomorphism from (K, w_1) onto $(T(K), w_2)$, where w_1 (w_2) denotes the restriction of the weak topology of X (respectively, of $c_0(\Gamma)$) to K (respectively, to $T(K)$). □

Theorem 13.20 (Amir, Lindenstrauss [AmLi]) *Let X be a Banach space. Then X is weakly compactly generated if and only if there is a w^*-w-continuous one-to-one bounded operator from X^* into some $c_0(\Gamma)$.*
In particular, if X is weakly compactly generated, then B_{X^*} in its w^*-topology is Eberlein compact.

Proof: Let X be a weakly compactly generated Banach space. By Theorem 13.16, there exists a weakly compact Markushevich basis $\{x_\alpha; f_\alpha\}_{\alpha\in\Gamma}$ of X. Assume that $\|f_\alpha\| \leq 1$ and define a bounded operator T from X^* into $\ell_\infty(\Gamma)$ by $T(f) = \big(f(x_\alpha)\big)$.

We claim that $T(X^*) \subset c_0(\Gamma)$. If $\{x_n\}$ is an arbitrary sequence of distinct elements in $\{x_\alpha\}_{\alpha\in\Gamma}$, then $f_\alpha(x_n) \to 0$ as $n \to \infty$ for every $\alpha \in \Gamma$. Since $\{x_\alpha\}_{\alpha\in\Gamma}\cup\{0\}$ is weakly compact and $\{f_\alpha\}$ separates points of X, we have $x_n \overset{w}{\to} 0$ in X. Thus if $f \in X^*$ and $\varepsilon > 0$, $|f(x_\alpha)| > \varepsilon$ for only a finite number of α, so $T(f) \in c_0(\Gamma)$.

We will show that T is continuous from the w^*-topology of X^* into the w-topology of $c_0(\Gamma)$. By Theorem 3.92 it is enough to show that T is w^*-w-continuous on B_{X^*}. Let $f_\nu \overset{w^*}{\to} f$ in B_{X^*} and let e_α be a standard unit vector in $\ell_1(\Gamma)$. Then $e_\alpha\big(T(f_\nu)\big) = f_\nu(x_\alpha) \to f(x_\alpha) = e_\alpha\big(T(f)\big)$. Since $\{T(f_\nu)\}$ is bounded in $c_0(\Gamma)$ and $\overline{\operatorname{span}}\{e_\alpha : \alpha \in \Gamma\} = \ell_1(\Gamma)$, we have $T(f_\nu) \overset{w}{\to} T(f)$ in $c_0(\Gamma)$ and so T is w^*-w-continuous.

If $T(f) = 0$ for some $f \in X^*$, then $f(x_\alpha) = 0$ for all $\alpha \in \Gamma$ and $f = 0$ on X as $\overline{\operatorname{span}}\{x_\alpha\} = X$. Therefore T is one-to-one.

On the other hand, if such a mapping T exists, T^* is a w^*-w-continuous mapping from $\ell_1(\Gamma)$ onto a linearly dense subspace of X. Therefore, $T^*(B_{\ell_1(\Gamma)})$ is a weakly compact set generating X. □

Corollary 13.21 *Let X be a Banach space. Then X^* is weakly compactly generated if and only if there is a weakly compact operator $T : X \to c_0(\Gamma)$ such that $T^*(\ell_1(\Gamma))$ is norm-dense in X^*.*

Proof: If X^* is weakly compactly generated, Theorem 13.20 gives a mapping $T : X^{**} \to c_0(\Gamma)$ that is linear, w^*-w-continuous and one-to-one. The restriction $T\big|_X$ is clearly weakly compact and $T^*(\ell_1(\Gamma))$ is norm-dense.

If $T : X \to c_0(\Gamma)$ is a weakly compact operator, then T^{**} maps X^{**} into $c_0(\Gamma)$, and it is clearly w^*-w-continuous and one-to-one. The result follows again from Theorem 13.20. □

Theorem 13.22 (Davis, Figiel, Johnson, Pełczyński [DFJP]) *Let X be a WCG space generated by a weakly compact set K. Then there is a reflexive space Y and a bounded one-to-one operator T from Y into X such that $K \subset T(B_Y)$.*

Proof: According to Theorem 3.133, we may assume that K is a weakly compact, convex, and symmetric subset of B_X. For $n \in \mathbb{N}$, set $U_n = 2^n K + 2^{-n} B_X$, let $\|\cdot\|_n$ be the Minkowski functional of U_n. For $x \in X$ let

$$\|\|x\|\| = \Big(\sum \|x\|_n^2\Big)^{\frac{1}{2}}.$$

Let $Y = \{x \in X : \|\|x\|\| < \infty\}$, denote $B_Y = \{x \in Y : \|\|x\|\| \leq 1\}$. Let T be the inclusion mapping of the normed space Y into X. If $x \in K$, then $\|x\|_n \leq 2^{-n}$ and thus $|||x||| \leq 1$. Hence $K \subset T(B_Y)$. We claim that Y is a reflexive Banach space, T is continuous and $K \subset T(B_Y)$, where B_Y is the unit ball of Y in the norm $|||\cdot|||$.

Since $U_n \subset 2^{n+1}B_X$ for every $n \in \mathbb{N}$, we have $\|\cdot\|_n \geq 2^{-(n+1)}\|\cdot\|$ and thus Y is a normed space. To show that Y is a Banach space, put $X_n = (X, \|\cdot\|_n)$ for $n \in \mathbb{N}$. Then X_n is a Banach space isomorphic to X. Set $Z = \left(\sum_{n=1}^{\infty} X_n\right)_2$ and define $\varphi\colon Y \to Z$ by $\varphi(y) = \big(T(y), T(y), \ldots\big)$. Then $\varphi(Y) = \{z \in Z : z = (x_n), x_n = x_1 \text{ for all } n\}$ is a closed subspace of Z and φ is a linear isometry of Y onto the Banach space $\varphi(Y)$. Thus Y is a Banach space. T can be viewed as a composition of φ and the projection on the first coordinate in Z, so T is continuous, it is obviously one-to-one.

We claim that $T^{**}\colon Y^{**} \to X^{**}$ is one-to-one and that $(T^{**})^{-1}(X) = Y$. Indeed, $\varphi^{**}\colon Y^{**} \to \left(\sum X_n^{**}\right)_2$ takes $F \in Y^{**}$ to $(T^{**}(F), T^{**}(F), \ldots) \in \left(\sum X_n^{**}\right)_2$. Since φ is an isometry into, so is φ^{**} (see Exercise 2.49). In particular, T^{**} is one-to-one. From $(\varphi^{**})^{-1}\varphi(Y) = Y$ we get $(T^{**})^{-1}(X) = Y$.

We will now show that Y is reflexive. Observe that the w^*-closure of $T(B_Y)$ in X^{**} is $T^{**}(B_{Y^{**}})$. If $x \in T(B_Y)$, then $\|x\|_n \leq 1$ for every n and thus $x \in U_n$ for every n. Therefore $\overline{T(B_Y)}^{w^*} \subset \overline{U}^{w^*}$ in X^{**}. Thus for every n we have

$$\overline{T(B_Y)}^{w^*} \subset \overline{2^n K + 2^{-n} B_X}^{w^*} \subset \overline{2^n K + 2^{-n} B_{X^{**}}}^{w^*} = 2^n K + 2^{-n} B_{X^{**}}$$

as K and $B_{X^{**}}$ are both w^*-compact in X^{**}. Using the remark above we get

$$T^{**}(B_{Y^{**}}) \subset \bigcap_n (2^n K + 2^{-n} B_{X^{**}}) \subset \bigcap_n (X + 2^{-n} B_{X^{**}}) = X$$

as X is closed in X^{**}. Thus $T^{**}(Y^{**}) \subset X$ and so $Y^{**} \subset (T^{**})^{-1}(X) = Y$, hence Y is reflexive. □

Observe that, if $T : Y \to X$ is the operator in Theorem 13.22, then $T^* : X^* \to Y^*$ is again one-to-one and has dense range. Indeed, assume $T^*(x^*) = 0$. Then, for all $y \in Y$, $0 = \langle T^*x^*, y\rangle$ $(= \langle x^*, Ty\rangle)$. In particular, $\langle x^*, x\rangle = 0$ for all $x \in K$, hence $x^* = 0$. This proves that T^* is one-to-one. Assume now that $0 = \langle T^*x^*, y\rangle$ $(= \langle x^*, Ty\rangle)$ for all $x^* \in X^*$. Then $Ty = 0$, hence $y = 0$ by the injectivity of T. This proves that $T^*(X^*)$ is w^*-dense. The conclusion follows since Y is reflexive.

Corollary 13.23 *Every weakly compact set in a Banach space in its weak topology is linearly homeomorphic to a weakly compact set in a reflexive Banach space in its weak topology.*

In particular, every Eberlein compact space is homeomorphic to a weakly compact set in a reflexive Banach space (in its weak topology).

Proof: Let K be a weakly compact set in a Banach space Z. Set $X = \overline{\text{span}}(K)$. By Theorem 13.22, there is a reflexive Banach space Y and a bounded one-to-one operator T from Y into X such that $K \subset T(B_Y)$. Then $T^{-1}(K)$ is a weakly closed subset of the weakly compact set B_Y, thus $T^{-1}(K)$ is weakly compact. Consequently $T^{-1}\colon K \to T^{-1}(K)$ is a homeomorphism in the weak topologies of Z and Y. □

Corollary 13.24 *Let X be a Banach space. Then X is a WCG space if and only if there is a reflexive Banach space Y and a bounded (one-to-one) operator T from Y into X such that $T(Y)$ is dense in X.*

Proof: If Y is reflexive and T is a bounded operator from Y onto a dense subset of X, then X is weakly compactly generated by the weakly compact set $T(B_Y)$.

The other direction follows from Theorem 13.22. □

We shall continue the study of Eberlein compact spaces in Section 14.1.

13.4 Renormings of Weakly Compactly Generated Banach Spaces

Theorem 13.25 *Let X be a Banach space.*
(i) (Troyanski, [Troy1]) *If X is WCG, then X admits an equivalent norm that is simultaneously LUR and Gâteaux differentiable.*
(ii) ([GTWZ]) *If X^* is WCG, then X admits an equivalent norm $\|\cdot\|$ the dual norm of which is LUR. In particular, $\|\cdot\|$ is Fréchet differentiable.*

In the proof we will use the following two results.

Lemma 13.26 *Define a norm $\|\cdot\|$ on $\ell_1(\Gamma)$ by $\|x\|^2 = \|x\|_1^2 + \|x\|_2^2$, where $\|\cdot\|_i$ denotes the canonical norm on $\ell_i(\Gamma)$, $i = 1, 2$. Then $\|\cdot\|$ is an equivalent LUR norm on $\ell_1(\Gamma)$ that is pointwise-lower semicontinuous.*

Proof: It is easy to check that $\|\cdot\|$ is an equivalent norm on $\ell_1(\Gamma)$ that is pointwise-lower semicontinuous. Let $f, f_1, f_2, \dots \in \ell_1(\Gamma)$ satisfy

$$\lim_{n\to\infty} (2\|f\|^2 + 2\|f_n\|^2 - \|f + f_n\|^2) = 0.$$

Then $\lim(2\|f\|_i^2 + 2\|f_n\|_i^2 - \|f + f_n\|_i^2) = 0$ for $i = 1, 2$ (see the proof of Theorem 8.1). Thus

$$\lim \|f_n\|_1 = \|f\|_1. \tag{13.1}$$

Since $\|\cdot\|_2$ is uniformly convex, we get $\|f - f_n\|_2 \to 0$, also

$$\lim(f - f_n)(\gamma) = 0 \text{ for each } \gamma \in \Gamma. \tag{13.2}$$

For a subset A of Γ, we denote by χ_A the characteristic function of A. For $g \in \ell_1(\Gamma)$ define $g^A = g \cdot \chi_A \in \ell_1(\Gamma)$, then $\|g\|_1 = \|g^A\|_1 + \|g^{\Gamma\setminus A}\|_1$. If A is a finite set in Γ, then from (13.2) we get $\lim \|f_n^A\|_1 = \|f^A\|$ and consequently (13.1) implies

$$\lim \|f_n^{\Gamma\setminus A}\|_1 = \|f^{\Gamma\setminus A}\|_1. \tag{13.3}$$

For $n \in \mathbb{N}$ and a finite $A \subset \Gamma$ we also have

$$\begin{aligned}\|f - f_n\|_1 &= \|(f - f_n)^A\|_1 + \|(f - f_n)^{\Gamma\setminus A}\|_1\\ &\leq \|(f - f_n)^A\|_1 + \|f^{\Gamma\setminus A}\|_1 + \|f_n^{\Gamma\setminus A}\|_1\\ &= \|(f - f_n)^A\|_1 + 2\|f^{\Gamma\setminus A}\|_1 + (\|f_n^{\Gamma\setminus A}\|_1 - \|f^{\Gamma\setminus A}\|_1). \quad (13.4)\end{aligned}$$

Given $\varepsilon > 0$, choose a finite set $A \subset \Gamma$ such that $\|f^{\Gamma\setminus A}\|_1 < \varepsilon$. Then using (13.2) and (13.3) choose $n_0 \in \mathbb{N}$ such that for $n \geq n_0$ we have $\|(f - f_n)^A\|_1 < \varepsilon$ and $\big|\|f_n^{\Gamma\setminus A}\|_1 - \|f^{\Gamma\setminus A}\|_1\big| < \varepsilon$. From (13.4) it follows that $\|f - f_n\|_1 < 4\varepsilon$ for $n \geq n_0$. □

Theorem 13.27 (Troyanski [Troy1]) *If a Banach space X is WCG, then it admits an equivalent LUR norm.*

We give two proofs to Theorem 13.27. First we give a detailed one that uses the transfer technique for LUR renormings, a principle found in [Gode1b], and further developed in [GTWZ]. This proof follows [Fabi0]. Then we sketch the original proof of Troyanski in his seminal work, that influenced a large part of this area and related areas of Banach spaces. For example, it motivated most of the results in [MOTV2].

Proof of Theorem 13.27: Let U be a w^*-w-continuous one-to-one operator from X^* into $c_0(\Gamma)$ for some Γ (Theorem 13.20). Put $T = U^*$. Then T maps $c_0^*(\Gamma)$ onto a norm dense set in X as U is one-to-one and w^*-w-continuous. Moreover, T is w^*-w-continuous.

Let $\|\cdot\|$ be the original norm of X and let $|\cdot|$ be an equivalent dual LUR norm on $c_0^*(\Gamma)$ (Lemma 13.26). For $n \in \mathbb{N}$ and $x \in X$ put

$$|x|_n^2 = \inf\Big\{\|x - T(g)\|^2 + \tfrac{1}{n}|g|^2 : g \in c_0^*(\Gamma)\Big\}$$

and $|||x|||^2 = \sum_{n=1}^{\infty} 2^{-n}|x|_n^2$. This is a norm on X such that $|||\cdot||| \leq \|\cdot\|$. If $\|x\| = 1$, then $\|x\|_n \geq \min\{\frac{1}{2\sqrt{n}\|T\|}, \frac{1}{2}\}$. This follows by considering separately cases when $|g| \geq \frac{1}{2\|T\|}$ and $|g| \leq \frac{1}{2\|T\|}$. Hence $|\cdot|_n$ is an equivalent norm on X and so is $|||\cdot|||$.

As $T(c_0^*(\Gamma))$ is norm dense in X, it follows that $|x|_n \to 0$ for each $x \in X$. Indeed, given $x \in X$ and $\varepsilon > 0$, find $g \in c_0(\Gamma)^*$ such that $\|x - T(g)\| < \varepsilon$. Then for n large enough, $|x|_n^2 \leq \|x - T(g)\|^2 + \frac{1}{n}|g|^2 < \varepsilon^2$.

Furthermore, note that for each $x \in X$ and each n, the infimum in the definition of $|x|_n^2$ is attained at some $g \in c_0(\Gamma)^*$. Indeed, given $x \in X$, the mapping

$\phi_x : c_0(\Gamma)^* \to \mathbb{R}$ defined by $\phi(g) = \|x - T(g)\|^2 + \frac{1}{n}|g|^2$ is w^*-lower semicontinuous and satisfies $\lim_{|g|\to\infty} \phi_x(g) = \infty$. Hence ϕ_x attains its minimum on $c_0(\Gamma)^*$.

We will now show that $|||\cdot|||$ is LUR. Assume that $x, x_1, x_2, \dots \in X^*$ are such that $2|||x_j|||^2 + 2|||x|||^2 - |||x + x_j|||^2 \to 0$. Then for every n, $2|x|_n^2 + |x_j|_n^2 - |x + x_j|_n^2 \to 0$ as $j \to \infty$. Find $g, g_j \in c_0^*(\Gamma)$ such that $|x|_n^2 = \|x - T(g)\|^2 + \frac{1}{n}|g|^2$, $|x_j|_n^2 = \|x_j - T(g_j)\|^2 + \frac{1}{n}|g_j|^2$. Then

$$\begin{aligned} &2|x|_n^2 + 2|x_j|_n^2 - |x + x_j|_n^2 \\ &\geq 2\|x - T(g)\|^2 + \tfrac{2}{n}|g|^2 + 2\|x_j - T(g_j)\|^2 + \tfrac{2}{n}|g_j|^2 \\ &\quad -\|x + x_j - T(g + g_j)\|^2 - \tfrac{1}{n}|g + g_j|^2 \\ &\geq (\|x - T(g)\| - \|x_j - T(g_j)\|)^2 + \tfrac{1}{n}(2|g|^2 + 2|g_j|^2 - |g + g_j|^2). \end{aligned}$$

This implies that $\|x_j - T(g_j)\| \to \|x - T(g)\|$ and $2|g|^2 + 2|g_j|^2 - |g + g_j|^2 \to 0$. As $|\cdot|$ is LUR, we have $|g - g_j| \to 0$. Thus for all n, $\limsup \|x - x_j\| \leq \limsup(\|x - T(g)\| + \|T(g - g_j)\| + \|x_j - T(g_j)\|) = 2\|x - T(g)\| \leq 2|x|_n$. Since $|x|_n \to 0$, we get $\|x - x_j\| \to 0$ and thus $|||\cdot|||$ is LUR. □

Now we sketch Troyanski's original proof in [Troy1]—actually more flexible then the previous one. We will show that *if a Banach space X admits a PRI $\{P_\alpha\}_{\alpha\leq\mu}$, and for each $\alpha < \mu$ the space $(P_{\alpha+1} - P_\alpha)(X)$ admits an LUR norm, then so does X*. The key idea of the proof is shown by proving a typical particular case, namely when the Banach space X has a transfinite Schauder basis $\{e_\alpha, f_\alpha\}$, i.e., when X admits a PRI $\{P_\alpha\}_{\alpha\leq\mu}$, such that $\dim (P_{\alpha+1} - P_\alpha)(X) = 1$ for all $\alpha < \mu$.

Proof of Theorem 13.27: (Sketch) Put $\Gamma = [0, \mu)$. Let $\mathcal{A}_n$ be the family of all finite subsets of Γ with no more than n elements. For $x \in X$ and $A \in \mathcal{A}_n$, let $E_n^A(x) = \operatorname{dist}(x, \operatorname{span}\{e_\alpha\}_{\alpha\in A})$ and $F_n^A(x) = \sum_{\alpha\in A}|f_\alpha(x)|$. Put $G_n(x) = \sup\{E^A(x) + nF^A(x) : A \in \mathcal{A}_n\}$ for $n \in \mathbb{N}$ and $x \in X$. Finally, put $G_0(x) = \|x\|_0$ for $x \in X$, where $\|\cdot\|_0$ is the original norm of X. Let $\Omega = \{0, -1, -2, \dots\} \cup \Gamma$ and define $\Phi : X \to c_0(\Omega)$ by $\Phi(x)(-n) = 2^{-n}G_n(x)$ for $-n \in \{0, -1, -2, \dots\}$, $\Phi(x)(\alpha) = |f_\alpha(x)|$ for $\alpha \in \Gamma$. Define an equivalent norm $\|x\|$ on X by $\|x\| = \|\Phi(x)\|_D$, where $\|\cdot\|_D$ is Day's norm on $c_0(\Omega)$. Let us briefly outline the main idea of the proof that $\|\cdot\|$ is an LUR norm on X. Let $x_n, x \in X$ be such that $2\|x_n\|^2 + 2\|x\|^2 - \|x + x_n\|^2 \to 0$. Let $\varepsilon > 0$. Find $n \in \mathbb{N}$ and $A' \in \mathcal{A}_n$ such that $E_n^{A'}(x) < \varepsilon$. Assume without loss of generality that

$$\sup\{|f_\alpha(x)| : \alpha \notin A'\} < \min\{|f_\alpha(x)| : \alpha \in A'\}. \tag{13.5}$$

Due to the term n in the definition of G_n's, if m is big enough and $A \in \mathcal{A}_m$ is so chosen that $G_m(x) - (E_m^A(x) + mF_m^A(x)) < \varepsilon$ with $A \in \mathcal{A}_m$, then, *necessarily*, $A \supset A'$ and thus $E_m^A(x) < \varepsilon$. This is because $\{f_\alpha(x)\} \in c_0(\Gamma)$. From the LUR property of Day's norm on $c_0(\Omega)$ it follows that $F_n^A(x_k) \to F_n^A(x)$ for all $A \in \bigcup \mathcal{A}_n$ and that $G_n(x_k) \to G_n(x)$ for all n. As the topology of the coordinatewise convergence in X is Hausdorff, in order to prove that $\|x_k - x\| \to 0$ it suffices to show that $\{x_k\}$

is relatively norm compact in X. The latter is seen from the fact that for large k, $E_m^A(x_k) \le G_m(x_k) - mF_m^A(x_k) \le G_m(x) + \varepsilon - mF_m^A(x) + \varepsilon \le E_m^A(x) + 3\varepsilon \le 4\varepsilon$. □

Remark: The fact that the set A in Troyanski's proof above is enlarged to ensure that the relation in (13.5) holds true can be avoided by adding more parameters [Zizl1c]. Overall, the key point in the proof above is that the set A in the definition of E_n^A was chosen so that $E_n^A(x)$ is such that $F_n^A(x) \neq 0$, and the supremum in the definition of G_n is "uniquely located" (see the "rigidity" condition in [Hayd3]). This phenomenon in Troyanski's construction, explicitly or implicitly appears again in many results in this area, including results on smooth partitions of unity or results of Haydon, Talagrand and others on higher order smoothness.

In particular, it shows that a Banach space X admits an LUR norm if it has a Markushevich basis $\{x_\alpha; f_\alpha\}$ such that $x \in \overline{\operatorname{span}}\{x_\alpha : f_\alpha(x) \neq 0\}$ for every $x \in X$ (i.e., a *strong Markushevich basis*, see, e.g., [HMVZ, Ch. 1]).

Proof of Theorem 13.25:

(i): Let X be WCG and let T be a bounded one-to-one operator from X^* into $c_0(\Gamma)$ that is w^*-w-continuous (Theorem 13.20).

Since $c_0(\Gamma)$ is WCG for every Γ, it admits an equivalent LUR norm $|||\cdot|||$ by Theorem 13.27. Let $\|\cdot\|$ denote the original dual norm of X^* and define a norm $|\cdot|$ on X^* by $|f|^2 = \|f\|^2 + |||T(x)|||^2$. Then $|\cdot|$ is an equivalent w^*-lower semicontinuous norm on X^* that is strictly convex. Indeed, if for some $f, g \in X^*$ we have $2|f|^2 + 2|g|^2 - |f+g|^2 = 0$, then by convexity $2|||T(x)|||^2 + 2|||T(g)|||^2 - |||T(f) + T(g)|||^2 = 0$. Since $|||\cdot|||$ is strictly convex on $c_0(\Gamma)$, we get $T(f) = T(g)$ and thus $f = g$ as T is one-to-one.

It remains to prove that in the case (ii), $|||\cdot|||$ is a dual norm. It is enough to show that for each n, $|\cdot|_n$ is a dual norm. Let $\{x_i\}$ be a net in X w^*-converging to some $x \in X$. For each i, we can choose $g_i \in c_0(\Gamma)^*$ such that

$$|x_i|_n^2 = \|x_i - Tg_i\|^2 + \frac{1}{n}|g_i|^2.$$

Let g be a w^*-cluster point of $\{g_i\}$. Without loss of generality, we may assume that the net $\{g_i\}$ w^*-converges to g. The norms $|\cdot|$ and $\|\cdot\|$ are w^*-lower semicontinuous. Thus

$$|x|_n^2 \le \|x - Tg\|^2 + \frac{1}{n}|g|^2 \le \liminf_i(\|x_i - Tg_i\|^2 + \frac{1}{n}\|g_i\|^2) = \liminf_i |x_i|_n^2.$$

Hence $|\cdot|_n$ is w^*-lower semicontinuous and thus it is a dual norm.

The combination of norms needed to get the rest of the proof follows the proof of Theorem 8.2. □

Corollary 13.28 *Let X be a Banach space. Then, if X^* is WCG, the space X is Asplund.*

Proof: If X^* is WCG then, by (ii) in Theorem 13.25, X has an equivalent Fréchet differentiable norm. In particular, so does every separable subspace Y of X; this implies that Y^* is separable, by Theorem 8.6. It follows that X is an Asplund space by Theorem 11.8. □

Theorem 13.29 ([JoZi1]) *Let X be a Banach space. The following are equivalent:*
(i) X admits a shrinking Markushevich basis.
(ii) X is WCG and admits an equivalent Fréchet differentiable norm.

Proof: (Sketch) (i)⟹(ii): Let $\{x_\alpha; f_\alpha\}_{\alpha\in\Gamma}$ be a shrinking Markushevich basis of X, $\{x_\alpha\}$ bounded. Let $\{y_n\}$ be a sequence of distinct elements of $\{x_\alpha\}$. Then $f_\alpha(y_n) \to 0$ for every α and thus $y_n \overset{w}{\to} 0$. Hence $\{x_\alpha\} \cup \{0\}$ is a weakly compact set in X by the Eberlein–Šmulyan theorem. Therefore X is weakly compactly generated.

Assume that $\{x_\alpha; f_\alpha\}$ is a shrinking Markushevich basis of X such that $\{f_\alpha\}$ is bounded. Define a bounded operator $T\colon X \to c_0(\Gamma)$ by $T(x) = \big(f_\alpha(x)\big)$. Let e_α, $\alpha \in \Gamma$, denote the standard unit vector in $\ell_1(\Gamma)$. Then $T^*(e_\alpha) = f_\alpha$ for each $\alpha \in \Gamma$ and thus T^* maps $\ell_1(\Gamma)$ onto a norm dense set in X^*. By the proof of Theorem 13.27, X^* admits an equivalent dual LUR norm and thus X admits an equivalent Fréchet differentiable norm by Corollary 7.25.

(ii)⟹(i): Assume for simplicity that $\operatorname{dens}(X) = \aleph_1$ and that the dual norm $\|\cdot\|^*$ is LUR. Let $\{P_\alpha\}_{\alpha<\omega_1}$ be a PRI for the norm $\|\cdot\|$ (Theorem 13.6). Then by the proof of Proposition 8.34, $\{P_\alpha^*\}$ is a PRI for $(X^*, \|\cdot\|^*)$. As all $P_\alpha(X)$ are separable and thus $P_\alpha^*(X^*)$ is separable, we get that each $(P_{\alpha+1} - P_\alpha)(X)$ admits a shrinking Markushevich basis by Exercise 4.61. By a standard "gluing together" argument (see Theorem 13.16), we obtain a shrinking Markushevich basis of X.

For the general case we follow [JoZi1], see also [HMVZ, Chapter 6]. □

Remark: Let us mention that (i) —and then (ii)— in the previous theorem is equivalent to X being Asplund and simultaneously WCG, see, e.g., [HMVZ, Theorem 6.2].

Corollary 13.30 ([JoZi1]) *Let X be a Banach space. If X admits a shrinking Markushevich basis, then so does every closed subspace of X.*

Proof: Every closed subspace of a weakly compactly generated Banach space is weakly countably determined (see Exercise 13.28), hence it admits a projectional resolution of identity (see Exercise 13.31). Then we can proceed as in the proof of (ii)⟹(i) in the theorem above. Alternatively, one can build a projectional resolution of the identity on a closed subspace of a weakly compactly generated Banach space by Theorem 13.6. □

Corollary 13.31 *If X is a WCG Banach space, then every convex continuous function on X is Gâteaux differentiable at the points of a dense subset of X.*

Proof: It follows from Theorem 13.25 and Corollary 7.44. □

Remark: We will show in Exercise 13.45 that every WCG space is even a weak Asplund space.

13.5 Weakly Compact Operators

Let X, Y be Banach spaces. Recall that an operator $T \in \mathcal{B}(X, Y)$ is called *weakly compact* if $\overline{T(B_X)}$ is a weakly compact set in Y.

If X is reflexive then every bounded operator T from X into Y is weakly compact. Indeed, since T is w-w-continuous (Exercise 3.59) and B_X is weakly compact, $T(B_X)$ is weakly compact in Y. On the other hand, the identity operator I_X on a nonreflexive Banach space X is not weakly compact as B_X is not weakly compact.

Lemma 13.32 (Grothendieck) *Let X be a Banach space, $A \subset X$. If for every $\varepsilon > 0$ there is a w-compact set $A_\varepsilon \subset X$ such that $A \subset A_\varepsilon + \varepsilon B_X$, then A is relatively w-compact.*

Proof: Since A is bounded, it is enough to show that the w^*-closure of A in X^{**} is actually in X. Since A_ε and $\varepsilon B_{X^{**}}$ are both w^*-compact, $\overline{A_\varepsilon + \varepsilon B_X}^{w^*} = A_\varepsilon + \varepsilon B_{X^{**}}$. Moreover, X is closed in X^{**}, so $\overline{A}^{w^*} \subset \bigcap_{\varepsilon>0}(A_\varepsilon + \varepsilon B_{X^{**}}) \subset \bigcap_{\varepsilon>0}(X \cap \varepsilon B_{X^{**}}) = X$. □

Note that the space of all weakly compact operators from a Banach space X into a Banach space Y is a closed subspace of the Banach space $\mathcal{B}(X, Y)$. Indeed, assume that T_n are weakly compact and $T_n \to T$ in $\mathcal{B}(X, Y)$. Given $\varepsilon > 0$, there is n_0 such that $T(B_X) \subset T_{n_0}(B_X) + \varepsilon B_Y$. Therefore $T(B_X) \subset \overline{T_{n_0}(B_X)} + \varepsilon B_Y$ and the latter set is closed as $\overline{T_{n_0}(B_X)}$ is w-compact and εB_Y is closed. Thus $\overline{T(B_X)} \subset \overline{T_{n_0}(B_X)} + \varepsilon B_Y$ and $\overline{T(B_X)}$ is w-compact.

Note also that a composition of a weakly compact operator with a bounded operator in any order is weakly compact. In particular, consider a bounded operator $T \in \mathcal{B}(X, Y)$. If there is a reflexive space Z and operators $R \in \mathcal{B}(X, Z)$, $S \in \mathcal{B}(Z, Y)$ such that $T = SR$, then T is weakly compact.

The following theorem shows that factorization through reflexive spaces characterizes weakly compact operators.

Theorem 13.33 (Davis, Figiel, Johnson, Pełczyński [DFJP]) *Let X, Y be Banach spaces. If T is a weakly compact operator from X into Y, then there is a reflexive Banach space Z and bounded operators $R\colon X \to Z$, $S\colon Z \to Y$ such that $T = SR$.*

Proof: Assume without loss of generality that $\overline{T(X)} = Y$, otherwise replace Y with $\overline{T(X)}$. Then $\overline{T(B_X)}$ is a weakly compact, convex and symmetric set that generates Y. From the construction of a reflexive space Z in Theorem 13.22 it follows that T maps X into Z, as $T(B_X) \subset B_Z$. Define $S(x) = T(x) \in Z$ and for R take the operator denoted by T in the proof of Theorem 13.22. □

As a corollary we have the following result:

Theorem 13.34 (Gantmacher) *Let X, Y be Banach spaces, $T \in \mathcal{B}(X, Y)$. T is weakly compact if and only if T^* is weakly compact.*

Proof: If T is weakly compact, by Theorem 13.33 there is a reflexive Banach space Z and $R \in \mathcal{B}(X, Z)$, $S \in \mathcal{B}(Z, Y)$ such that $T = SR$. Then $T^* = R^*S^*$ and $R^*: Z^* \to X^*$ is weakly compact as Z^* is reflexive. Thus T^* is weakly compact.

If T^* is weakly compact, then $T^{**}: X^{**} \to Y^{**}$ is weakly compact by the first part of the proof. Since $T^{**}|_X = T$, we have $T(B_X) = T^{**}(B_X) \subset T^{**}(B_{X^{**}})$. Since $\overline{T^{**}(B_{X^{**}})}^w$ is w-compact in Y^{**}, we have $\overline{T(B_X)}^w \subset \overline{T^{**}(B_{X^{**}})}^w$ and $\overline{T(B_X)}^w$ is w-compact in Y^{**}, thus w-compact in Y. □

13.6 Absolutely Summing Operators

Definition 13.35 *Let X, Y be Banach spaces, $T \in \mathcal{B}(X, Y)$. T is called* absolutely summing *if $\sum \|T(x_i)\|_Y < \infty$ whenever $\sum x_i$ is unconditionally convergent in X.*

In other words, T carries every unconditionally convergent series in X to an absolutely convergent series in Y.

Examples:

(i) The identity operator I on ℓ_2 is not absolutely summing. Indeed, let $\{e_i\}$ be the canonical basis of ℓ_2. Then $\sum \frac{1}{i}e_i$ is unconditionally convergent but $\sum \|\frac{1}{i}e_i\|_2 = \sum \frac{1}{i} = \infty$.

(ii) The formal identity operator I from $C[0, 1]$ into $L_1[0, 1]$ is absolutely summing. Indeed, recall that if $x_1, \ldots, x_n$ are vectors in a Banach space $\in X$, then $\sup\Big\{\Big\|\sum_{i=1}^n \varepsilon_i x_i\Big\| : \varepsilon_i = \pm 1\Big\} = \sup\Big\{\sum_{i=1}^n |x^*(x_i)| : x^* \in S_{X^*}\Big\}$ (see Exercise 1.37). Using this, for $f_1, \ldots, f_n \in C[0, 1]$ we get

$$\begin{aligned}\sum_{i=1}^n \|f_i\|_1 &= \sum_{i=1}^n \int_0^1 |f_i(t)|\,\mathrm{d}t = \int_0^1 \sum_{i=1}^n |f_i(t)|\,\mathrm{d}t \le \sup_{t\in[0,1]} \sum_{i=1}^n |f_i(t)| \\ &\le \sup_{F\in S_{C[0,1]^*}} \sum_{i=1}^n |F(f_i)| = \sup\Big\{\Big\|\sum_{i=1}^n \varepsilon_i f_i\Big\|_\infty : \varepsilon_i = \pm 1\Big\}.\end{aligned}$$

If $\sum f_i$ is unconditionally convergent in $C[0, 1]$, then

$$\sup_n\Big\{\sup\Big\|\sum_{i=1}^n \varepsilon_i f_i\Big\|_\infty : \varepsilon_i = \pm 1\Big\} = M$$

is finite (Exercise 1.44). Therefore $\sum_{i=1}^n \|f_i\|_1 \le M$ for all n, which means that I is absolutely summing.

Theorem 13.36 (Grothendieck [Groth2]) *Every bounded operator from ℓ_1 into ℓ_2 is absolutely summing.*

Proof: Assume that $u_i = \sum_{j=1}^{m} \alpha_{i,j} e_j$, $i = 1, 2, \ldots, n$ are vectors in ℓ_1^m for some m such that $\sum_{i=1}^{n} |x^*(u_i)| \le \|x^*\|$ for every $x^* \in \ell_1^*$ (where (e_i) are the standard unit vectors in ℓ_1).

Let $(s_j)_{j=1}^m$ be real numbers of absolute value ≤ 1 and let $x_s^* \in \ell_1^*$ be defined by $x_s^*(e_j) = s_j$ if $1 \le j \le m$ and $x_s^*(e_j) = 0$ otherwise. For every choice of $(t_i)_{i=1}^n$,

$$\left|\sum_{j=1}^{m}\sum_{i=1}^{n} \alpha_{i,j} t_i s_j\right| \le \sum_{i=1}^{n} |t_i| \left|\sum_{j=1}^{m} \alpha_{i,j} s_j\right| \le \sum_{i=1}^{n} |x_s^*(u_i)| \le 1.$$

For every $1 \le i \le n$ let $y_i \in S_{\ell_2}$ be such that $(Tu_i, y_i) = \|Tu_i\|$. From the Grothendieck inequality (6.67), we get

$$\sum_{i=1}^{n} \|Tu_i\| = \sum_{i=1}^{n} (Tu_i, y_i) = \sum_{i=1}^{n}\sum_{j=1}^{m} \alpha_{i,j} (Te_j, y_i) \le K_G \|T\|,$$

where K_G is the Grothendieck constant. Therefore T is absolutely summing. □

Remark: It is shown in [LiPe] that *if X and Y are infinite-dimensional separable Banach spaces such that X has an unconditional basis and that every bounded operator $T : X \to Y$ is absolutely summing, then X is isomorphic to ℓ_1 and Y is isomorphic to ℓ_2.*

Theorem 13.37 *Let X, Y be Banach spaces, $T \in \mathcal{B}(X, Y)$. If T is absolutely summing, then T is weakly compact.*
If moreover X is reflexive, then T is a compact operator.

Before proving Theorem 13.37, we state one consequence. It shows that finite-dimensional spaces are characterized by Riemann's result that all unconditionally convergent series converge absolutely.

Theorem 13.38 (Dvoretzky, Rogers, see, e.g., [LiTz3]) *In every infinite-dimensional Banach space there is a series that is unconditionally convergent but not absolutely convergent.*

Proof: If in some Banach space X every unconditionally convergent series is absolutely convergent, then the identity mapping I_X is an absolutely summing operator. By Theorem 13.37, I_X is weakly compact, thus B_X is weakly compact and X is reflexive. By the second part of Theorem 13.37, I_X is a compact operator, which means that B_X is compact and X must be finite-dimensional. □

To prove Theorem 13.37, we will need the following lemmas:

Lemma 13.39 *Let X, Y be Banach spaces, $T \in \mathcal{B}(X, Y)$. If T is absolutely summing, then there is a constant $K > 0$ such that for all $n \in \mathbb{N}$ and vectors $\{x_i\}_{i=1}^n$ in X we have*

$$\sum_{i=1}^{n} \|T(x_i)\|_Y \le K \sup_{x \in S_{X^*}} \Big\{ \sum_{i=1}^{n} |x^*(x_i)| \Big\}.$$

Proof: Let Z be the space of all sequences $\{x_i\}_{i=1}^{\infty}$ in X for which the series $\sum x_i$ is unconditionally convergent, with the norm $\sup_{x^* \in S_{X^*}} \Big\{ \sum_{i=1}^{\infty} |x^*(x_i)| \Big\}$. Similarly as in the proof of Proposition 1.16, we find that Z is a Banach space. If $\{x_i\} \in Z$, then $\sum \|T(x_i)\|_Y < \infty$. The Baire category theorem gives that there are $K' \in \mathbb{N}$ and $c > 0$ such that $\sum \|T(x_i)\|_Y \le K'$ whenever $\{x_i\} \in Z$ is such that $\sup_{x^* \in S_{X^*}} \Big\{ \sum_{i=1}^{\infty} |x^*(x_i)| \Big\} \le c$. A standard homogeneity argument then gives the existence of $K > 0$ such that

$$\sum_{i=1}^{\infty} \|T(x_i)\|_Y \le K \sup_{x \in S_{X^*}} \Big\{ \sum_{i=1}^{\infty} |x^*(x_i)| \Big\}.$$

Now given $x_1, \dots, x_n$, we can choose $x_{n+1} = x_{n+1} = \dots = 0$ and use the above estimate. □

A bounded operator is absolutely summing if and only if it satisfies the conclusion of Lemma 13.39 for some $K > 0$. The smallest possible $K > 0$ is called the *absolutely summing norm* π_1 *of* T. It is standard to show that the space of all absolutely summing operators from X into Y, denoted $\Pi_1(X, Y)$, is a Banach space.

Lemma 13.40 (Pietsch, see, e.g., [LiTz3]) *Let* X, Y *be Banach spaces,* $T \in \mathcal{B}(X, Y)$. *If* T *is absolutely summing, then there is a regular probability measure* μ *on* (B_{X^*}, w^*) *and a constant* $K > 0$ *such that for every* x *we have*

$$\|T(x)\|_Y \le K \int_{B_{X^*}} |x^*(x)| \,\mathrm{d}\mu(x^*).$$

Proof: Consider (B_{X^*}, w^*). We treat X as a subspace of $C(B_{X^*})$; in particular, for $x \in X$ we have the function $|x| \colon x^* \mapsto |x(x^*)|$ for $x^* \in B_{X^*}$. We may assume that for every $n \in \mathbb{N}$ and $x_1, \dots, x_n \in X$ we have $\sum_{i=1}^{n} \|T(x_i)\|_Y \le \sup_{x^* \in S_{X^*}} \Big\{ \sum_{i=1}^{n} |x^*(x_i)| \Big\}$. Define $F_1, F_2 \subset C(B_{X^*})$ by

$$F_1 = \big\{ f \in C(B_{X^*}) : \sup_{x^* \in B_{X^*}} \big(f(x^*) \big) < 1 \big\}$$

and

$$F_2 = \operatorname{conv}\big\{ |x| : x \in X, \|T(x)\|_Y = 1 \big\}.$$

Then F_1 and F_2 are convex subsets of $C(B_{X^*})$ and F_1 is open.

We claim that $F_1 \cap F_2 = \emptyset$. Indeed, take $f \in F_2$. By the definition of F_2, there are $x_i \in X$, $i = 1, \dots, n$, with $\|T(x_i)\|_Y = 1$ and $\lambda_i \in [0, 1]$ with $\sum_{i=1}^n \lambda_i = 1$ such that $f(x^*) = \sum_{i=1}^n \lambda_i |x^*(x_i)|$ for $x^* \in B_{X^*}$. Then

$$\sup_{x^* \in B_{X^*}} \big(f(x^*)\big) = \sup_{x^* \in B_{X^*}} \Big\{\sum_{i=1}^n \lambda_i |x^*(x_i)|\Big\} = \sup_{x^* \in B_{X^*}} \Big\{\sum_{i=1}^n |x^*(\lambda_i x_i)|\Big\}$$
$$\geq \sum_{i=1}^n \|T(\lambda_i x_i)\|_Y = \sum_{i=1}^n \lambda_i \|T(x_i)\| = 1.$$

Therefore $f \notin F_1$.

By the separation theorem (and using $0 \in F_1$), there is $\Phi \in C(K)^*$ such that $\Phi(f) < 1$ for every $f \in F_1$ and $\Phi(f) \geq 1$ for every $f \in F_2$. We claim that if $g \in C(B_{X^*})$, $g \geq 0$, then $\Phi(g) \geq 0$. Assume the contrary. Then for β negative but large in absolute value we would have $\Phi(\beta g) > 1$, but also $\beta g \leq 0$, hence clearly $\beta g \in F_1$ and so $\Phi(\beta g) < 1$, a contradiction.

Therefore Φ is a non-negative functional and by the Riesz representation theorem there is a positive regular measure $\tilde{\mu}$ on B_{X^*} which represents Φ. Since F_1 contains the open unit ball of $C(B_{X^*})$, we have $\|\Phi\| \leq 1$ and thus $\tilde{\mu}(B_{X^*}) = \Phi(\chi_{B_{X^*}}) = \Phi\big(\lim(1 - \frac{1}{n})\chi_{B_{X^*}}\big) \leq 1$. Moreover, for every $f \in F_2$ we have $\int_{B_{X^*}} f \, d\tilde{\mu} = \Phi(f) \geq 1$ and thus the measure $\mu = \tilde{\mu}/\tilde{\mu}(B_{X^*})$ is a probability measure on B_{X^*} such that $\mu(f) \geq 1$ for every $f \in F_2$. Thus if $x \in X$ is such that $\|T(x)\| = 1$, then $|x| \in F_2$ and hence

$$\|T(x)\|_Y = 1 \leq \int_{B_{X^*}} |x| \, d\mu = \int_{B_{X^*}} |x^*(x)| \, d\mu(x^*).$$

A standard homogeneity argument concludes the proof. □

Proof of Theorem 13.37: Let $x \in X$. Consider x as an element of $L_2(B_{X^*}, \mu)$, where μ is the Pietsch measure from Lemma 13.40. By Hölder's inequality (1.1) we then have

$$\|T(x)\|_Y \leq K \int_{B_{X^*}} |x(x^*)| \, d\mu(x^*)$$
$$\leq K \Big(\int_{B_{X^*}} |x(x^*)|^2 \, d\mu(x^*)\Big)^{\frac{1}{2}} = K \|x\|_{L_2(B_{X^*}, \mu)}.$$

Hence we may consider T a bounded operator (also w–w-continuous) from the reflexive space $\overline{\operatorname{span}}^{\|\cdot\|_2}(X) \subset L_2(B_{X^*}, \mu)$ into Y. Thus $\overline{T(B_X)}$ is a weakly compact subset of Y.

Now assume that X is a reflexive Banach space. Given a sequence $\{x_n\}$ in B_X, by w-compactness we may assume that $x_n \xrightarrow{w} x \in B_X$ in X. Then x_n and x considered as continuous functions on (B_{X^*}, w^*) are all bounded in absolute value by 1 and

$x_n(x^*) \to x(x^*)$ for every $x^* \in B_{X^*}$. By the Lebesgue dominated convergence theorem used for the Pietsch measure μ from Lemma 13.40, we have that $x_n \to x$ in $L_1(B_{X^*}, \mu)$. By the Pietsch inequality, $\|T(x_n - x)\|_Y \leq \int_{B_{X^*}} |x_n - x| \, \mathrm{d}\mu \to 0$. Therefore T is a compact operator. □

Consider an operator $T \in \mathcal{B}(X, Y)$. Let i be the natural embedding of X into the space $C(B_{X^*}, w^*)$. Denote $X_\infty = i(X)$. Let μ be a probability measure on (B_{X^*}, w^*). Consider a norm-one inclusion i_1 of $C(B_{X^*}, w^*)$ into $L_1(B_{X^*}, \mu)$ and denote $X_1 = \overline{i_1(X_\infty)}^{L_1}$. Using Lemma 13.40 one can show that the operator T is absolutely summing if and only if there exists an operator $S\colon X_1 \to Y$ such that $T = S i_1 i$.

Thus absolutely summing operators are characterized by this factorization. This is why Lemma 13.40 is often called the *Pietsch factorization lemma*.

For more information on the theory of absolutely summing operators and the related field of the local theory of Banach spaces we refer to [Pisi3], [Tomc], and [AlKa, Chapter 8].

13.7 The Dunford–Pettis Property

Definition 13.41 *Let X be a Banach space. We say that X has the* Dunford–Pettis property *if $x_n^*(x_n) \to 0$ whenever $x_n \in X$ and $x_n^* \in X^*$, $n \in \mathbb{N}$, satisfy $x_n \xrightarrow{w} 0$ in X and $x_n^* \xrightarrow{w} 0$ in X^*.*

An example of a Banach space with the Dunford–Pettis property is ℓ_1. Indeed, if $x_n \xrightarrow{w} 0$ in ℓ_1, then $x_n \to 0$ by the Schur property of ℓ_1. If, moreover, $x_n^* \in \ell_1^*$ are such that $x_n^* \xrightarrow{w} 0$ in ℓ_1^*, then $\sup \|x_n^*\| \leq C < \infty$ for some $C > 0$ and thus $|x_n^*(x_n)| \leq \|x_n^*\| \, \|x_n\| \leq C\|x_n\| \to 0$.

It is easy to see that X has the Dunford–Pettis property if X^* does. Thus c_0 has the Dunford–Pettis property.

Proposition 13.42 *Let X be a Banach space. Then the following are equivalent.*
(i) *X has the Dunford–Pettis property.*
(ii) *Every weakly compact operator from X into any Banach space maps weakly compact sets to norm compact sets.*

In view of Exercise 3.173, (ii) can be formulated by saying that *every weakly compact operator from X into any Banach space is completely continuous.*

Proof of Proposition 13.42:. (i)$\Longrightarrow$(ii): Assume that for some $\delta > 0$ and $x_n \xrightarrow{w} 0$ we have $\|T(x_n)\| \geq \delta$ for all n. Let $y_n^* \in S_{Y^*}$ be such that $y_n^*\big(T(x_n)\big) = \|T(x_n)\|$ for all n. Since T^* is weakly compact, by eventual passing to a subsequence we may assume that for some $x^* \in X^*$, $T^*(y_n^*) \xrightarrow{w} x^* \in X^*$ in X^*. Since X has the Dunford–Pettis property, we have

$$0 = \lim\big(T^*(y_n^*) - x^*\big)(x_n) = \lim\big(y_n^*\big(T(x_n)\big) - x^*(x_n)\big) = \lim \|T(x_n)\|$$

as $\lim x^*(x_n) = 0$. This contradicts $\|T(x_n)\| \geq \delta > 0$ for all n.

(ii)$\Longrightarrow$(i): Assume $x_n \to 0$ weakly in X and $x_n^* \to 0$ weakly in X^*. Define an operator $T : X \to c_0$ by $Tx - (x_1^*(x), x_2(x)^*, \cdots)$. If e_n denotes the unit vector in ℓ_1, then $T^*e_n(x) = e_n(Tx) = x_n^*(x)$ for every n and every $x \in X$. Thus $T^*(e_n)$ is contained in the closed convex hull C of (x_n^*) and so is $T^*B_{\ell_1}$. Since $x_n^* \to 0$ weakly, C is weakly compact by Krein's theorem. Thus T^* is a weakly compact operator and so is T by Gantmacher's Theorem 13.34. Since $x_n \to 0$ weakly, by (ii), $\|Tx_n\| \to 0$ and since (x_n^*) is a bounded set in X^*, $|x_n^*(x_n)| \leq sup_k|x_k(x_n)| \to 0$ as $n \to \infty$. Therefore (i) holds. □

One consequence is that *an infinite-dimensional reflexive Banach space cannot have the Dunford–Pettis property*. Indeed, the identity mapping I_X on a reflexive Banach space X is weakly compact, so if X had the Dunford–Pettis property then B_X will be $\|\cdot\|$-compact, by Proposition 13.42, and so $\dim(X) < \infty$, a contradiction.

Theorem 13.43 (Dunford, Pettis, see, e.g., [DiUh]) *Let K be a compact space. Then $C(K)$ has the Dunford–Pettis property.*

Proof: Assume that $f_n \in C(K)$, $f_n \overset{w}{\to} 0$ in $C(K)$ and $F_n \in C(K)^*$, $F_n \overset{w}{\to} 0$ in $C(K)^*$. Set $A = \sup\|f_n\| < \infty$ and $B = \sup\|F_n\| < \infty$.

We identify F_n with regular measures on K by the Riesz representation theorem, and by the Hahn decomposition theorem we assume without loss of generality that F_n are non-negative measures. Since $F_n \overset{w}{\to} 0$, there is (see, e.g., [DuSc]) a positive measure μ on K such that the sequence $\{F_n\}$ of measures is equiabsolutely continuous with respect to μ, i.e., for every $\varepsilon > 0$ there is $\delta_\varepsilon > 0$ such that $|F_n(U)| < \varepsilon$ for every n whenever $U \subset K$ satisfies $\mu(U) < \delta_\varepsilon$.

We will show that $F_n(f_n) \to 0$. Given $\varepsilon > 0$, get $\delta_\varepsilon > 0$ from the equiabsolute continuity of F_n as above. By Egorov's theorem, there is $U \subset K$ such that $\mu(U) < \delta_\varepsilon$ and $\lim\limits_{n\to\infty} f_n = 0$ uniformly on $K\backslash U$. Thus there is n_0 such that for $n \geq n_0$ and every $t \in K\backslash U$ we have $|f_n(t)| \leq \varepsilon$. Then for $n \geq n_0$ we have

$$|F_n(f_n)| = \left|\int_K f_n \, \mathrm{d}F_n\right| \leq \int_U |f_n| \, \mathrm{d}F_n + \int_{K\backslash U} |f_n| \, \mathrm{d}F_n \leq \varepsilon A + \varepsilon B.$$

□

Proposition 13.44 *Let X be a Banach space with the Dunford–Pettis property. If Y is a complemented subspace of X, then Y has the Dunford–Pettis property.*

Proof: Let P be a bounded linear projection of X onto Y. Let $y_n \in Y$, $y_n \overset{w}{\to} 0$ in Y and $y_n^* \in Y^*$, $y_n^* \overset{w}{\to} 0$ in Y^*. Then $y_n \overset{w}{\to} 0$ in X and $P^*(y_n^*) \overset{w}{\to} 0$ in X^* as P^* is w-w-continuous. Since X has the Dunford–Pettis property, $P_n^*(y_n^*)(y_n) \to 0$. Hence $y_n^*(y_n) = y_n^*\big(P(y_n)\big) = P_n^*(y_n^*)(y_n) \to 0$. □

Corollary 13.45 *If K is a compact space and if Y is a reflexive complemented subspace of $C(K)$, then Y is finite-dimensional.*

Proof: By the preceding proposition, Y has the Dunford–Pettis property. By the remark following Proposition 13.42, every reflexive Banach space with the Dunford–Pettis property is finite-dimensional. □

13.8 Applications

Definition 13.46 *Let Y be a closed subspace of a Banach space X. A closed subspace Z of X is called a* quasicomplement *of Y in X if $Y \cap Z = \{0\}$ and $Y + Z$ is dense in X.*
If such a quasicomplement exists, Y is called quasicomplemented *in X.*

Theorem 13.47 (Murray [Murr], Mackey [Mack]) *Let X be a separable Banach space. Then every closed subspace of X is quasicomplemented in X.*

Proof: (Gurarii, Kadec) Let Y be a closed subspace of X. Let $\{x_i; f_i\}$ be a Markushevich basis of Y and $\big\{\{x_i\} \cup \{z_j\}; \{\psi_i\} \cup \{\varphi_j\}\big\}$ its extension to a Markushevich basis of X (Theorem 4.60). Put $Z = \overline{\operatorname{span}}\{z_i\}$. Let $w \in Y \cap Z$. Since $w \in \overline{\operatorname{span}}\{z_i\}$, we get $\psi_i(w) = 0$ for every i. Since $w \in \overline{\operatorname{span}}\{x_i\}$, we also get $\varphi_i(w) = 0$ for every i. Then $w = 0$ as $\{\psi_i\} \cup \{\varphi_i\}$ separates points in X. Moreover, $\overline{Y + Z} \supset \overline{\operatorname{span}}\{\{x_i\} \cup \{z_i\}\} = X$. □

In fact, the more general result below is true. For its proof we refer to [JoZi1] or [HMVZ, Section 5.7], where more results on quasicomplements can be found.

Theorem 13.48 (Lindenstrauss [Lind13], [JoZi1]) *Let X be a weakly compactly generated Banach space. Then every closed subspace of X is quasicomplemented in X.*

Theorem 13.49 (Rosenthal [Rose2]) *The space c_0 is quasicomplemented in ℓ_∞.*

Proof: (Sketch) As $\beta\mathbb{N} \setminus \mathbb{N}$ is perfect, there is a continuous mapping φ of $\beta\mathbb{N}\setminus\mathbb{N}$ onto $[0, 1]$ ([Lace2]). Therefore $C[0, 1]$ is isomorphic to a subspace of $C(\beta\mathbb{N}\setminus\mathbb{N})^* = c_0^\perp \subset \ell_\infty^*$. As $L_1[0, 1]$ is isomorphic to a subspace of $C[0, 1]^*$, it is isomorphic to a quotient of $C(\beta\mathbb{N}\setminus\mathbb{N})^*$ and thus by Rosenthal's result in [Rose2], $L_1[0, 1]$ is isomorphic to a subspace of $C(\beta\mathbb{N}\setminus\mathbb{N})^*$. Since ℓ_2 is isomorphic to a subspace of $L_1[0, 1]$ by Theorem 4.53, ℓ_2 is isomorphic to a subspace of $c_0^\perp \subset \ell_\infty^*$; call this subspace H_1.

For $n \in \mathbb{N}$, define $\delta_n \in \ell_\infty^*$ by $\delta_n(f) = f(n)$. Let $H = \overline{\operatorname{span}}\big\{\frac{\delta_n}{n} + \mu_n : n \in \mathbb{N}\big\}$, where $\{\mu_n\}$ is a Schauder basis of H_1 equivalent to the canonical basis of ℓ_2. It follows that there are constants $K_1, K_2 > 0$ such that

$$K_1\Big(\sum_{n=1}^{m} |\alpha_n|^2\Big)^{\frac{1}{2}} \leq \Big\|\sum_{n=1}^{m} \alpha_n\big(\mu_n + \tfrac{\delta_n}{n}\big)\Big\| \leq K_2\Big(\sum_{n=1}^{m} |\alpha_n|^2\Big)^{\frac{1}{2}}$$

for all $m \in \mathbb{N}$. Thus H is isomorphic to ℓ_2 and therefore H is w^*-closed in ℓ_∞^* (Lemma 4.62). We will show that $H_\perp$ is a quasicomplement of c_0 in ℓ_∞. By duality it suffices to show that $H \cap c_0^\perp = \{0\}$ and $H_\perp \cap c_0 = \{0\}$.

Take $f \in c_0 \cap H_\perp$. Then $(\mu_n + \frac{\delta_n}{n})(f) = 0$ for all n. As $\mu_n \in c_0^\perp$, we get $(\frac{\delta_n}{n})(f) = \frac{1}{n} f(n) = 0$ for all n and thus $f = 0$.

Now take $y \in H \cap c_0^\perp$. There are real numbers α_i such that $\sum |\alpha_i|^2 < \infty$ and $y = \sum_{n=1}^\infty \alpha_n(\mu_n + \frac{\delta_n}{n})$. Since $\mu_n \in c_0^\perp$ for all n, we have $y - \sum_{n=1}^\infty \alpha_n \mu_n \in c_0^\perp$ and thus $\sum_{n=1}^\infty \alpha_n(\frac{\delta_n}{n}) \in c_0^\perp$. For $n \in \mathbb{N}$, define a function e_n on $\mathbb{N}$ by $e_n(m) = \delta_{nm}$ (Kronecker delta). Then $\sum_{k=1}^\infty \alpha_k(\frac{\delta_k}{k}(e_n)) = \frac{\alpha_n}{n}$. Hence $\alpha_n = 0$ for all n, so $y = 0$. □

Theorem 13.50 (Lindenstrauss [Lind12]) *If Γ is uncountable, then $c_0(\Gamma)$ is not quasicomplemented in $\ell_\infty(\Gamma)$.*

Proof: By contradiction, assume that Z is a quasicomplement of $c_0(\Gamma)$ in $\ell_\infty(\Gamma)$. Let q be the quotient mapping of $\ell_\infty(\Gamma)$ onto $\ell_\infty(\Gamma)/Z$, consider $T = q|_{c_0(\Gamma)}$. $T(c_0(\Gamma))$ is dense in $\ell_\infty(\Gamma)/Z$ and thus $\ell_\infty(\Gamma)/Z$ is weakly compactly generated. Hence $B_{(\ell_\infty(\Gamma)/Z)^*}$ is w^*-sequentially compact (Theorems 13.20 and 3.109 in $c_0(\Gamma)$). We claim that q^* is weakly compact.

Let $y_n = q^*(x_n)$, $x_n \in B_{(\ell_\infty(\Gamma)/Z)^*}$ and x_{n_k} be a w^*-convergent subsequence of $\{x_n\}$. Then $\{y_{n_k}\}$ is a w^*-convergent sequence in $\ell_\infty^*(\Gamma)$. From the Grothendieck property of $\ell_\infty(\Gamma)$, we have that $\{y_{n_k}\}$ is w-convergent. Hence q^* is a weakly compact operator and so is q, the same is then true for T.

Therefore $T^*\colon (\ell_\infty(\Gamma)/Z)^* \to \ell_1(\Gamma)$ is a weakly compact operator and thus a compact operator as $\ell_1(\Gamma)$ has the Schur property. Hence $T^*\big((\ell_\infty(\Gamma)/Z)^*\big)$ is a separable subset of $\ell_1(\Gamma)$. Since T is one-to-one (from the definition of the quasicomplement), T^* maps $\ell_\infty(\Gamma)/Z$ onto a w^*-dense set in $c_0^*(\Gamma)$. This means that $c_0^*(\Gamma)$ is w^*-separable, a contradiction (see Proposition 13.3). □

Theorem 13.49 was later generalized:

Theorem 13.51 (Johnson [Johns2]) *Let Y be a closed subspace of a Banach space X. If Y^* is w^*-separable and X/Y has a separable and infinite-dimensional quotient, then Y is quasicomplemented in X.*

We remark that the problem of quasicomplementation is closely related to the problem of the existence of nontrivial separable quotients: Indeed, as shown by Rosenthal in [Rose2], X admits an infinite-dimensional separable quotient if and only if X has an infinite-dimensional separable quasicomplemented subspace. We refer the reader to [Muji] for a survey in this area.

Theorem 13.52 (Valdivia, [Vald1]) *Let E be a Banach space. Assume that E^{**}/E is separable. Then E is isomorphic to $R \oplus S$, where R is a reflexive Banach space and S a separable Banach space. Moreover, for every separable subspace X of E, X^{**} is separable.*

In the proof we shall use the following three lemmas.

Lemma 13.53 *Let F be a subspace of a Banach space X. Then X/F is reflexive if the following condition is satisfied:*
Whenever $x^{**} \in X^{**}$ belongs to the w^*-closure of a countable bounded subset of X, then x^{**} is necessarily of the form $x^{**} = x + f^{\perp\perp}$ for some $x \in X$ and $f^{\perp\perp} \in F^{\perp\perp}$ $(\subset X^{**})$.

Proof: Let $\{\hat{x}_n\}$ be a bounded sequence in X/F. Let q denote the quotient mapping from X onto X/F and let $\{x_n\}$ be a bounded sequence in X such that $q(x_n) = \hat{x}_n$ for $n \in \mathbb{N}$. Let x^{**} be a w^*-accumulation point of $\{x_n\}$ in X^{**}. Put

$$x^{**} = x + f^{\perp\perp}, \quad x \in X, \quad f^{\perp\perp} \in F^{\perp\perp}.$$

There exists a subnet $\{x_\gamma\}$ of $\{x_n\}$ such that $x_\gamma \overset{w^*}{\to} x^{**}$. If $u \in F^\perp$ $(= (X/F)^*)$ then

$$u(\hat{x}_\gamma) = u(x_\gamma) \to_\gamma u(x^{**}) = u(x) = u(\hat{x})$$

Thus $\hat{x}_\gamma \to \hat{x}$ in the weak topology of X/F, and X/F is thus reflexive by the Eberlein–Šmulyan theorem. □

In fact, the implication in Lemma 13.53 is an equivalence (see Exercise 13.53).

We will also use the following folklore lemma. We include the proof for the reader's convenience.

Lemma 13.54 *Let Y be a subspace of X such that X/Y is separable. Then there is a separable subspace $F \subset X$ such that $Y + F = X$.*

Proof: Let q be the canonical quotient mapping from X onto X/Y. Choose a countable dense set $\{\hat{x}_1, \hat{x}_2, \dots\}$ in X/Y. For every $n \in \mathbb{N}$, choose an element x_n of X such that

$$q(x_n) = \hat{x}_n, \quad \|x_n\| \le \|\hat{x}_n\| + \frac{1}{n^2}.$$

Let F be the closed linear hull of $\{x_n : n \in \mathbb{N}\}$.

Given $\hat{x} \in X$, choose a strictly increasing sequence (n_p) of natural numbers so that

$$\|\hat{x} - (\hat{x}_{n_1} + \hat{x}_{n_2} + \dots \hat{x}_{n_p})\| \le \frac{1}{p^2}, \quad p = 1, 2, \dots$$

We have

$$\begin{aligned}\|x_{n_{p+1}}\| &\le \|\hat{x}_{n_{p+1}}\| + \frac{1}{n_{p+1}^2}\\ &\le \|(\hat{x} - (\hat{x}_{n_1} + \hat{x}_{n_2} + \ldots \hat{x}_{n_{p+1}})) - (\hat{x} - (\hat{x}_{n_1} + \hat{x}_{n_2} + \cdots + \hat{x}_{n_p}))\| + \frac{1}{n_{p+1}^2}\\ &\le \|\hat{x} - (\hat{x}_{n_1} + \hat{x}_{n_2} + \cdots + \hat{x}_{n_{p+1}})\| + \|\hat{x} - (\hat{x}_{n_1} + \hat{x}_{n_2} + \cdots + \hat{x}_{n_p})\| + \frac{1}{n_{p+1}^2}\\ &\le \frac{1}{(p+1)^2} + \frac{1}{p^2} + \frac{1}{n_{p+1}^2} < \frac{3}{p^2}, \quad p = 1, 2, \ldots\end{aligned}$$

The series $\sum_{p=1}^{\infty} x_{n_p}$ converges to an element $x \in F$ and $q(x) = \hat{x}$. Thus $Y + F = X$. □

Lemma 13.55 *If E is a Banach space such that E^{**}/E is separable, then there is a separable subspace F of E^* such that E^*/F is reflexive and F is w^*-closed in E^*.*

Proof: Let $E^\perp$ be the annihilator of E in E^{***}. By Lemma 13.54 there is a separable subspace H of E^{**} such that $E^{**} = E + H$. Observe that $(E^{**}/E)^* = E^\perp$, and that E^{**}/E is, by hypothesis, separable. In particular, the set $\big(B_{E^\perp}, w(E^\perp, E^{**}/E)\big)$ is a compact metric space, hence separable. Let

$$D = \big\{d^{***} \in B_{E^\perp} : \text{ there exists a countable set } N := N(d^{***}) \subset E^* \text{ such that } d^{***} \in \overline{N}^{w(E^{***}, E^{**})}\big\}.$$

The set D has a $w(E^{***}, E^{**})$-dense and countable subset D_0. Put $A_0 = \{N(d_0^{***}) : d_0^{***} \in D_0\}$; it is a countable subset of E^*. Let $F_0 = \overline{\text{span}}^{\|\cdot\|}(A_0)$, a separable subspace of E^*. Given $e^{***} \in E^{***}$ in the $w(E^{***}, E^{**})$-closure of a bounded and countable subset of E^*, put $e^{***} = e^* + e^\perp$ according to the fact that $E^{***} = E^* \oplus E^\perp$. Then $e^\perp$ is also in the $w(E^{***}, E^{**})$-closure of a bounded and countable subset of E^*. It follows that $e^\perp \in \overline{F_0}^{w(E^{***}, E^{**})}$ $(= F_0^{\perp\perp})$. An application of Lemma 13.53 shows that E^*/F_0 is a reflexive space. Clearly, $(E^*/F_0)^{**} = E^{***}/F_0^{\perp\perp}$. Since E^*/F_0 is reflexive, $E^{***}/F_0^{\perp\perp} = E^*/F_0$, hence

$$E^{***} = E^* + F_0^{\perp\perp}. \tag{13.6}$$

The topological space $(B_{E^\perp}, w(E^\perp, E^{**}/E)$ is separable, so it has a dense subset $\{e_n^\perp : n \in \mathbb{N}\}$. According to (13.6), each $e_n^\perp$ can be written $e_n^\perp = e_n^* + f_n^{\perp\perp}$, where $e_n^* \in E^*$ and $f_n^{\perp\perp} \in F^{\perp\perp}$. Let $F = \overline{\text{span}}^{\|\cdot\|}\{F_0 \bigcup \{e_n^* : n \in \mathbb{N}\}\}$ (a separable subspace of E^*). The space E^*/F is still reflexive. Moreover, $E^\perp \subset F^{\perp\perp}$. This implies, obviously, that $F^\perp \subset E$, hence that F is $w(E^*, E)$-closed. □

Proof of Theorem 13.52: Let F be the separable subspace of E^* constructed in Lemma 13.55. Recall that $F^\perp = F_\perp$ $(\subset E)$. Then $F_\perp \subset E$ is reflexive. Moreover

$E/F_{\perp}$ is separable, since its dual space is isometric to F, a separable space. By Exercise 13.4, E is weakly compactly generated. Use Lemma 13.53 to write $E = F_{\perp} + G$, where G is a separable subspace of E. By Theorem 13.11, there exists a separable complemented subspace S of E such that $G \subset S$. Let $\hat{q} : F_{\perp} \to E/S$ be the restriction to $F_{\perp}$ of the canonical quotient mapping. Then, since $E = F_{\perp} + G$, $\hat{q}$ is onto. It follows that E/S is reflexive (and isomorphic to the complement R of S in E). Incidentally, if S is *any* separable subspace of E, notice that S^{**}/S can be identified with a subspace of E^{**}/E, hence S^{**}/S is separable. It follows from Lemma 13.53 that $S^{**} = S + H$, where H is a separable subspace of S^{**}. In particular, S^{**} is separable. □

We remark that for every WCG space Y, there is a Banach space X such that X^{**}/X is isomorphic to Y ([DFJP]).

13.9 Remarks and Open Problems

Remarks

1. The WCG spaces were introduced and their basic properties were established by Lindenstrauss and his collaborators in the 1960s. It becomes an important tool in the study of nonseparable Banach spaces. Lindenstrauss' idea for decompositions of WCG spaces crystalized in the following result in [Lind5]: *Let X be a Banach space and let Y be a separable subspace. Then there is a separable subspace $Z_0 \subset X$ such that for every $\varepsilon > 0$ and for every finite-dimensional subspace B of X, there is an operator $T_{B,\varepsilon} : B \to Z_0$ such that $\|T_{B,\varepsilon}\| \leq 1+\varepsilon$ and $T_{B,\varepsilon}y = y$ for every $y \in B \cap Y$.* If X is reflexive, then by Tychonoff's theorem, the net $\{T_{B,\varepsilon}\}$ has a limit point T in the topology of the pointwise convergence, putting in X the weak topology. This operator T is a norm-one operator from X into Z_0, and $Ty = y$ for $y \in Y$. By an iterative exhaustion argument, one can construct a separable subspace Z that contains Y and a norm-one projection from X onto Z ([Lind9]). For WCG spaces, this result allows for a construction of projectional resolution of the identity ([AmLi], see, e.g., [FHHMPZ, p. 360]). As a variant of Lindenstrauss method, in [JoZi1], the decomposition of WCG spaces was shown, alternatively, by working on their dual space and preserving X by dual projections (see Exercise 13.14). This allowed for some additional flexibility of the construction. Later on, Vašák [Vasa] constructed PRIs in WCD spaces by working on X^{**} and preserving X. A seminal paper in this area, that is not mentioned in the text, was also [Taco].
   The use of countable exhaustion arguments for Corson compact spaces (see Definition 14.40) was initiated by S.P. Gul'ko [Gulk3] and A.N. Plichko [Plic1]. The method of Gul'ko [Gulk3] (see the presentation of this by Namioka and Wheeler in [NaWh]) was purely topological.
   Independently, Valdivia [Vald2b] developed a technique that was fully adapted to the Banach space setting. In his method, the germ of a projectional generator was

already present. The present formulation of a projectional generator is due to Orihuela and Valdivia [OrVa], who presented this more general tool in order to unify previous constructions and to include very general classes of Banach spaces where the decomposition can be obtained. Valdivia produced similar structures in the class of spaces of continuous functions on certain general compact spaces (the so-called *Valdivia compact spaces*, see the definition in Exercise 14.64) endowed with the topology of the pointwise convergence [Vald3].
More information on this topic can be found in [DGZ3] and [Fabi1].
2. If X is a WCG space of density ω_1 and $c_0(\Gamma)$ is a subspace of X, then the space $c_0(\Gamma)$ is 4-complemented in X (see [GKL1]). This is no longer true for higher densities (see [ACGJM]).
3. There is a WCG Banach space with unconditional basis that has a non-WCG subspace having unconditional basis (see [ArMe2]).
4. There is a Banach space X so that X^* is a subspace of a WCG Banach space and there is no one-to-one bounded operator from X into any $c_0(\Gamma)$ space, see Remark 2 in Chapter 14. Note that X^* is not WCG as otherwise X^{**} would inject into $c_0(\Gamma)$.
5. It is proven in [FaGo], see, e.g., [DGZ3, p. 242], that if X is an Asplund space, then X^* admits a projectional resolution of the identity, and thus X^* admits an equivalent LUR norm, see, e.g., [DGZ3, p. 286].
6. We refer to [LiTz3] and references therein for the theory of p-absolutely summing operators.
7. We refer to, e.g., [HMVZ, Ch. 6], for classification of several classes of WCG spaces by using Markushevich bases.

Open Problems

1. It is not known whether a Banach space X is WCG whenever X^{**} is. See, in particular, Exercise 13.4.
2. (Godefroy) Assume an Asplund space X admits a Markusevich basis $\{x_\alpha; f_\alpha\}$ with $\text{span}\{f_\alpha\}$ norming in X^*. It is an open problem whether X is necessarily WCG.
3. [MOTV2, p. 120] Is it true that every space with the RNP property admits an equivalent LUR norm? A result of Plichko and Yost [PliYo] shows that the RNP does not imply the separable complementation property (see Definition 13.10).
4. [MOTV2, p. 122] Assume X admits an equivalent Fréchet differentiable norm. Does X admit an equivalent LUR norm? Haydon showed in [Hay4] that X admits an equivalent LUR norm if X^* admits a dual LUR norm.

Exercises for Chapter 13

13.1 Let E be a topological vector space. Let W be a $\mathbb{Q}$-linear subset of E. Prove that $\overline{W}$ is a linear subspace of E.

Hint. First, show that $\overline{W}$ is additive, i.e., $x+y \in \overline{W}$ whenever $x, y \in \overline{W}$. Prove, then, that $\overline{A}$ is $\mathbb{Q}$-linear. By using this two assertions, show that $\overline{A}$ is convex. It follows immediately that $\overline{A}$ is a linear subspace.

13.2 Let Y be a separable subspace of a reflexive Banach space X and let Z be a separable subspace of X^*. Follow the hint to show the existence of a norm-one projection on X such that $P(X)$ is separable and contains Y, and $Z \subset P^*(X^*)$.
Hint. (Valdivia) Let N_1 be a separable 1-norming subspace for Y, $N_1 \supset Z$, and let M_1 be a separable 1-norming subspace for N_1, $M_1 \supset Y$. Let $N_2 \supset N_1$ be a separable 1-norming subspace for M_1, $M_2 \supset M_1$ a separable 1-norming subspace for N_2, etc. Put $M = \overline{\bigcup M_i}$ and $N = \overline{\bigcup N_i}$. Use Exercise 4.19 to show that $X = M \oplus N_\perp$ and that $\|P\| = 1$, where P is the associated projection from X onto M parallel to $N_\perp$. To prove the last statement, let $z^* \in Z$. For $x \in X$ put $x = m + n_\perp$, where $m \in M$ and $n_\perp \in N_\perp$. Then

$$\langle P^*z^*, x\rangle = \langle P^*z^*, m+n_\perp\rangle = \langle z^*, P(m+n_\perp)\rangle = \langle z^*, m\rangle = \langle z^*, m+n_\perp\rangle = \langle z^*, x\rangle,$$

so $P^*z^* = z^*$.

13.3 Let C be a w-compact set in a Banach space X. Show that if X^* is w^*-separable, then C in its w-topology is metrizable.

In particular, C is separable. Thus we get an alternative proof that a WCG space X is separable if X^* is w^*-separable.
Hint. Proof of Propositions 3.105 and 3.106.

13.4 (Johnson–Lindenstrauss [JoLi1]) Let Y be a closed subspace of a Banach space X such that X/Y is separable. Show that X is WCG if and only if Y is WCG.

There are WCG Banach spaces with closed subspaces that are not WCG ([Rose5]). It is not known whether X is WCG if X^{**} is WCG.
Hint. Assume that Y is WCG. Let $\{\hat{x}_n\}$ be dense in $S_{X/Y}$. Pick $x_n \in \hat{x}_n$ with $\|x_n\| \le 2$. Let K be a w-compact set generating Y. Then $\left\{\frac{1}{n}x_n\right\} \cup K$ is a w-compact set generating X.

If X is WCG, we may assume that X is generated by a weakly compact Markushevich basis $\{x_\alpha\}_{\alpha\in\Gamma}$. Given $x \in X$, put $\hat{x} = q(x)$, where $q : X \to X/Y$ is the canonical quotient mapping. Then $\Gamma_0 = \{\alpha : \hat{x}_\alpha \neq 0\}$ is countable. Indeed, otherwise, since X/Y is separable, there would be non-zero condensation points of $\left\{\hat{x}_\alpha : \alpha \in \Gamma_0\right\}$ in the w-topology, which contradicts the fact that for every sequence of distinct points $\{\hat{x}_n\}$ in $\{\hat{x}_\alpha\}$ we have $\hat{x}_n \overset{w}{\to} 0$. Let $Z = \overline{\operatorname{span}}\{x_\alpha : \alpha \notin \Gamma_0\}$. Then Z is a WCG closed subspace of Y. Y/Z is separable since X/Z is separable. By the first part, Y is WCG.

13.5 (Johnson–Lindenstrauss [JoLi1]) Let Y be a closed subspace of a Banach space X. Show that if Y is reflexive and X/Y is WCG, then X is also WCG.
Hint. Let $\{\hat{x}_\alpha\}$ be a weakly compact Markushevich basis of X/Y. Choose $x_\alpha \in \hat{x}_\alpha$, $\|x_\alpha\| < 1$. We claim that $\{x_\alpha\} \cup \{0\} \cup B_Y$ is weakly compact. Indeed, if $y_\beta \in \{x_\alpha\}$

and $y_\beta \to y \in X^{**}$, then $\hat{y}_\beta \xrightarrow{w} 0$ in X/Y. If q denotes the quotient mapping $X \to X/Y$, then $q^{**}(y_\beta) \xrightarrow{w^*} q^{**}(y)$ in $X^{**}/\overline{Y}^{w^*}$. Since $\hat{y}_\beta \to 0$, we have $q^{**}(y) = 0$. Therefore $y \in \overline{Y}^{w^*} = Y$ as Y is reflexive. Thus $y \in B_Y$ and the rest is standard.

13.6 Prove that for every separable Banach space X there is a compact space $K \subset S_X$ such that $X = \overline{\text{span}}(K)$. What about such a weak-compact set for WCG spaces?
Hint. Let $X = \bigcup F_n$, where $F_1 \subset F_2 \subset \dots$ are finite-dimensional. By induction find a convergent sequence $\{x_k\}$ in S_X such that $F_n \subset \text{span}\{x_k\}$ for every n.

The answer for the second part is negative if the space is nonseparable, it is enough to take any space on the unit sphere of which the norm and weak topologies (and hence compact sets) coincide.

13.7 Show that ℓ_∞ is not a subspace of any WCG Banach space.
Hint. Assume that $\ell_\infty \subset X$ and X is weakly compactly generated by a weakly compact subset K. Let P be a projection of X onto ℓ_∞ (ℓ_∞ is injective; this follows from Proposition 5.10 and (ii) in Proposition 5.13). Then ℓ_∞ is weakly compactly generated by $P(K)$, a contradiction.

13.8 Does there exist a bounded operator from $c_0(c)$ onto a dense subset of ℓ_∞?
Hint. No, ℓ_∞ is not WCG while $c_0(c)$ is.

13.9 Is there a bounded operator from ℓ_∞ onto a dense set in $\ell_2(c)$? Is there a bounded operator from ℓ_∞ onto a dense set in c_0?
Hint. Yes, $\ell_2(c)$ is a quotient of ℓ_∞ (follows by reflexivity from $\ell_2(c) \subset \ell_\infty^*$, [Rose3]). Then use the formal identity mapping from $\ell_2(c)$ into $c_0(c)$.

13.10 It is known that $\ell_\infty(\Gamma)$ does not admit an equivalent strictly convex norm ([DGZ3]). Use this to show that if Γ is uncountable, then there is no bounded one-to-one operator from $\ell_\infty(\Gamma)$ into $c_0(\Gamma)$.
Hint. Theorem 13.27.

13.11 We proved that as a WCG space, $c_0(\Gamma)$ admits an equivalent LUR norm (see Theorem 13.27). Consider the following "Day's norm": For $x = (x_\gamma) \in c_0(\Gamma)$ we define $\|x\| = \sup\Big\{\Big(\sum_{k=1}^n x_{\gamma_k}^2/4^k\Big)^{\frac{1}{2}}\Big\}$, where the supremum is taken over all $n \in \mathbb{N}$ and all ordered n-tuples $(\gamma_1, \dots, \gamma_n)$ of distinct elements of Γ (see Fig. 13.1).

Show that this norm is strictly convex. In fact it is also LUR ([DGZ3]).

Hint. If $m > n$ and $|a| < |b|$, then $\frac{a^2}{n^2} + \frac{b^2}{m^2} < \frac{b^2}{n^2} + \frac{a^2}{m^2}$. Thus $\|x\| = \Big(\sum_{j=1}^\infty \frac{x_{\gamma_j}^2}{4^j}\Big)^{\frac{1}{2}}$, where γ_j are distinct and such that $|x_{\gamma_1}| \geq |x_{\gamma_2}| \geq \dots$. Calling such a sequence $\{\gamma_j\}$ an appropriate sequence for x, we have that if $\|x+y\| = \|x\|+\|y\|$, $\|x\| = \|y\| = 1$ and $\{\gamma_j\}$ is an appropriate sequence for $x + y$, then

$$2 = \|x + y\| = \Big(\sum_{j=1}^{\infty} \frac{(x+y)^2_{\gamma_j}}{4^j}\Big)^{\frac{1}{2}}$$

$$\leq \Big(\sum_{j=1}^{\infty} \frac{x^2_{\gamma_j}}{4^j}\Big)^{\frac{1}{2}} + \Big(\sum_{j=1}^{\infty} \frac{y^2_{\gamma_j}}{4^j}\Big)^{\frac{1}{2}} < \Big(\sum_{j=1}^{\infty} \frac{x^2_{\rho_j}}{4^j}\Big)^{\frac{1}{2}} + \Big(\sum_{j=1}^{\infty} \frac{y^2_{\gamma_j}}{4^j}\Big)^{\frac{1}{2}} = 2,$$

where $\{\rho_j\}$ is an appropriate sequence for x, a contradiction unless $\{\gamma_j\}$ is an appropriate sequence for x. Similarly we argue for y. Thus γ_j is an appropriate sequence for both x and y. By the parallelogram equality, we then get $x = y$.

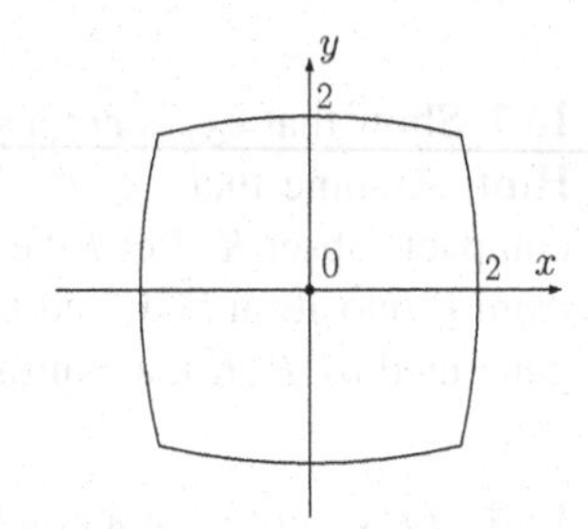

Fig. 13.1 The closed unit ball of Day's norm in $\mathbb{R}^2$

13.12 Let X be a WCG Banach space X. Show that if c_0 is a subspace of X, then c_0 is complemented in X.
Hint. There is a complemented separable $Z \subset X$ such that $c_0 \subset Z$. Use Sobczyk's theorem in Z.

13.13 Let $\{X_\mu\}_{\mu\in\Gamma}$ be WCG Banach spaces. Then $(\sum X_\mu)_p$ is WCG if $p \in (1, \infty)$ and $(\sum X_\mu)_{c_0(\Gamma)}$ is WCG. Also, $(\sum X_\mu)_{\ell_1(\Gamma)}$ is WCG if Γ is countable. Prove these statements.
Hint. Let $p \in (1, \infty)$ and K_μ be w-compact, symmetric, convex set generating X_μ, then the set $K = \{(x_\mu) \in (\sum X_\mu)_p : x_\mu \in K_\mu, \|(x_\mu)\| \leq 1\}$ is w-compact and generates $(\sum X_\mu)_p$. The space $(\sum X_\mu)_2$ is mapped onto a dense set in $(\sum X_\mu)_{c_0(\Gamma)}$ by the formal identity operator. Finally, if X_n are weakly compactly generated by w-compact symmetric convex sets K_n, then $K = \Big\{(x_n) \in \big(\sum X_n\big)_1 : x_n \in K_n, \|(x_n)\| < \frac{1}{n^2}\Big\}$ is a weakly compact set which generates $(\sum X_n)_1$.

13.14 Let K be a linearly dense weakly compact convex symmetric subset of a Banach space X. Consider a norm on X^* defined by $|f| = \sup_{x\in K} |f(x)|$. Note that $|\cdot| \leq c\|\cdot\|$ for some $c > 0$ by the boundedness of K. Let $\widetilde{B}$ be the closed unit ball of $|\cdot|$, note that $\widetilde{B} = K^0$. Show the following statement: *Let $T \in \mathcal{B}(X^*)$ satisfy $|T| \leq C$ for some $C > 0$. Then T is the dual operator to some $G \in \mathcal{B}(X)$.*

Hint. Let $T^*: X^{**} \to X^{**}$ be the dual operator of T. Then $T^*(\widetilde{B}^0) \subset C\widetilde{B}^0$. By the bipolar theorem, $C\widetilde{B}^0 = \overline{CK}^{w^*}$ in X^{**}, so $C\widetilde{B}^0 = CK$ as K is w-compact. Therefore $T^*(K) \subset CK$ and thus $T^*\big(\operatorname{span}(K)\big) \subset \operatorname{span}(K)$. Since T^* is continuous in $\|\cdot\|$, we have $T^*(X) = T^*\big(\overline{\operatorname{span}}(K)\big) \subset \overline{\operatorname{span}}(K) = X$. Therefore $T^*|_X$ preserves X and it is standard to check that $T = \big(T^*|_X\big)^*$.

13.15 Let X be a Banach space such that there is a WCG Banach space Y with $X^* \subset Y$. Show that there is an equivalent norm on X such that its dual norm on X^* is LUR.
Hint. Let F be the closed linear subspace of Y^* formed by all functionals whose restriction to X^* is w^*-lower semicontinuous (see Exercise 3.93). By the proof of (i) in Theorem 13.25, find an equivalent norm on Y that is LUR and $w(Y, F)$-lower semicontinuous. Its restriction to X^* is a dual LUR norm on X^*. For more details, see [GTWZ].

13.16 Show that if X is a separable Banach space, then $\operatorname{card}(X) = \operatorname{card}(X^*)$.
Hint. Continuous functions are determined on a countable set. For nonseparable X even density of the dual can be larger than the cardinality of X (ℓ_∞, see Exercise 14.34).

13.17 Let X be a Banach space of density character $\aleph_1$ that admits a PRI. Is it true that X^* is not w^*-separable?
Hint. No. $\ell_1(c)$ has PRI and its dual is w^*-separable. Indeed, $\ell_1(c) \subset \ell_\infty$ (Exercise 5.34) and ℓ_∞ is w^*-separable, then use the restriction mapping.

13.18 (Kadec) Let X be a Banach space. Show that if X has an equivalent Gâteaux differentiable norm, then $\operatorname{card}(X) \geq \operatorname{dens}(X^*)$. Thus ℓ_∞ has no Gâteaux differentiable norm.
Hint. Bishop–Phelps. $\operatorname{card}(\beta\mathbb{N}) > c$.

13.19 Prove that every WCG Banach space with the Schur property is separable.
Hint. Weakly compact sets are norm compact.

13.20 Let X be a WCG Banach space. Show that $C \subset B_{X^*}$ is w^*-compact if and only if C is w^*-sequentially compact.
Hint. (B_{X^*}, w^*) is Eberlein compact, Theorem 3.109 in $c_0(\Gamma)$.

13.21 Let $\{x_\alpha; f_\alpha\}_{\alpha\in\Gamma}$ be a Markushevich basis of a WCG Banach space X. Show that $\operatorname{card}\{\alpha \in \Gamma;\ f(x_\alpha) \neq 0\} \leq \aleph_0$ for every $f \in X^*$.
Hint. Let $H = \big\{f \in X^* :\ \operatorname{card}\{\alpha \in \Gamma :\ f(x_\alpha) \neq 0\} \leq \aleph_0\big\}$. Clearly H is a closed subset of X^* that contains all f_α, hence H is w^*-dense in X^*. By the Banach–Dieudonné theorem, we only need to show that $H \cap B_{X^*}$ is w^*-closed in B_{X^*}. Let $g \in \overline{H \cap B_{X^*}}^{w^*}$. Since (B_{X^*}, w^*) is angelic, there is a sequence $h_n \in H \cap B_{X^*}$

such that $h_n \xrightarrow{w^*} g$. Since each h_α has only countable support over x_αs, the same is true for g, meaning $g \in H \cap B_{X^*}$.

13.22 Let X be a nonseparable WCG space. Show that there is a sequence $\{x_n\} \subset S_X$ such that $x_n \xrightarrow{w} 0$.
Hint. Let $\{x_\gamma\}$ be a weakly compact Markushevich basis of X. There is $\delta > 0$ such that $\|x_\gamma\| \geq \delta$ for all $\gamma \in \Gamma'$, where Γ' is uncountable. Then 0 is a w-cluster point of $x_\gamma : \gamma \in \Gamma'$. The result then follows from the angelicity of w-compact sets.

13.23 Let X be a Banach space with $\text{dens}(X) = \aleph_1$. Assume that $\{P_\alpha : \alpha < \omega_1\}$ is a PRI in X. Show that given $x \in X$, there is $\alpha < \omega_1$ such that $x \in P_\alpha(X)$.
Hint. $\text{dist}(x, P_\alpha(x))$ is a non-decreasing function converging in α to 0, so it must be eventually zero. Otherwise, taking α_n such that $\text{dist}(\alpha, P_{\alpha_n}(X)) < \frac{1}{n}$ we would have a countable cofinal set in the segment $[0, \omega_1]$. As ω_1 is the first uncountable ordinal, this is a contradiction.

13.24 Let X be a WCG space and $\{P_\alpha : \alpha \leq \omega_1\}$ be a PRI on X. Show that given $f \in X^*$, there is $\alpha < \omega_1$ such that $f \in P_\alpha^*(X^*)$.
Hint. (B_{X^*}, w^*) is Eberlein compact and thus angelic. Given $f \in B_{X^*}$, there is a sequence $\alpha_n < \omega_1$ such that $P_{\alpha_n}^*(f) \xrightarrow{w^*} f$ in X^*. Let $\alpha < \omega_1$ be the supremum of $\{\alpha_n\}$. Then $P_{\alpha_n}^*(X^*) \subset P_\alpha^*(X^*)$ for all n, so $f \in P_\alpha^*(X^*)$

13.25 Let X be a Banach space of density character $\aleph_1$ that admits a PRI. Show that if c_0 is a subspace of X, then c_0 is complemented in X.
Hint. Each $x \in c_0$ lies in some $P_\alpha(X)$. Use the supremum of such $\alpha's$. Then use Sobczyk's theorem.

13.26 Let X be a Banach space of density character $\aleph_1$ with a PRI $\{P_\alpha\}$. Show that if the norm of X is Gâteaux differentiable, then $\bigcup P_\alpha(X) = X$ and $\bigcup P_\alpha^*(X^*) = X^*$.

Hint. Given $x \in X$, the function $\alpha \to \|x - P_\alpha(x)\|$ is continuous and equal to zero at ω_1. Thus it is zero starting from some countable ordinal.

Let $f \in S_{X^*}$ attain its norm at $x \in S_X \cap P_\alpha(X)$. Then $\|P_\alpha^*(f)\| = 1$ and $P_\alpha^*(f)(x) = f\big(P_\alpha(x)\big) = 1$, from the uniqueness of the support functional due to Gâteaux smoothness we get $P^*(f) = f$. If $f \in S_{X^*}$ is a general element, use the Bishop–Phelps theorem.

13.27 Let X be a Banach space. Show that if X^* is not w^*-separable, then there is a subspace Y of X of density character $\aleph_1$ that has a PRI.
Hint. Follow the proof of Lemma 4.20 with $\varepsilon = 0$, use transfinite induction.

13.28 A Banach space X is called *weakly countably determined (WCD)* or a *Vašák space* if there exists a countable collection $\{K_n\}$ of w^*-compact subsets of X^{**} such that for every $x \in X$ and $u \in X^{**}\setminus X$ there is n_0 for which $x \in K_{n_0}$ and $u \notin K_{n_0}$. Show that every WCG space is a Vašák space.

Show that every closed subspace of a WCD space is a WCD space.

Hint. Let K be a w-compact convex symmetric set generating X. Note that $X = \overline{\bigcup_n nK}$. For $n, m \in \mathbb{N}$ put $K_{n,m} = nK + \frac{1}{m}B_{X^{**}}$. Then $K_{n,m}$ are w^*-compact sets in X^{**}. Given $x \in X$ and $u \in X^{**}\setminus X$, choose $m \in \mathbb{N}$ such that $\text{dist}(u, X) > \frac{1}{m}$ and n such that $\text{dist}(x, nK) < \frac{1}{m}$.

If Y is a closed subspace of a WCD space X, to show that Y is WCD it suffices to consider the sets $(K_n \cap Y^{\perp\perp})$, where $Y^{\perp\perp} \subset X^{**}$ is identified with Y^{**}.

13.29 (Vašák [Vasa]) Prove that the sets K_n in the family $\{K_n\}_{n=1}^{\infty}$ from the definition of WCD Banach space in Exercise 13.28 can be taken to be convex symmetric and having nonempty norm-interior.

Hint. Given a subset S of a vector space, denote by $\Gamma(S)$ the symmetric convex hull of S. From the very beginning, we may assume that the family $\{K_n\}_{n=1}^{\infty}$ is closed under finite intersections. Given $x \in X$, let $\{K_{n_i}\}_{i=1}^{\infty}$ be the family of all sets in $\{K_n\}_{n=1}^{\infty}$ that contain x. Put $M_i = \bigcap_{j=1}^{i} K_i$. Then $\bigcap_{i=1}^{\infty} M_i = \bigcap_{i=1}^{\infty} K_{n_i}$. Certainly $\bigcap_{i=1}^{\infty} M_i \subset X$, and so $\bigcap_{i=1}^{\infty} M_i$ is w-compact in X. By Krein's theorem, $K := \overline{\Gamma(\bigcap_{i=1}^{\infty} M_i)}^{(X,w)}$ is also w-compact in X. Put $M = \bigcap_{i=1}^{\infty} \overline{\Gamma(M_i)}^{w^*}$. We claim that $K = M$. Certainly, $K \subset M$. If there exists $x^{**} \in M\setminus K$, use the separation theorem with respect to the dual pair $\langle X^{**}, X^* \rangle$ to get $x^* \in S_{X^*}$ and $r \in \mathbb{R}$ such that $\langle x^*, x \rangle < r < \langle x^{**}, x^* \rangle$ for all $x \in K$. Fix $i \in \mathbb{N}$. Since $x^{**} \in \overline{\Gamma(M_i)}^{w^*}$, we get that, for some $x_i^{**} \in M_i$, $\langle x_i^{**}, x^* \rangle \geq r$. This holds for every $i \in \mathbb{N}$, so we get a sequence $\{x_i^{**}\}$ in M_1; a w^*-cluster point x^{**} must be in $\bigcap_{i=1}^{\infty} M_i$, so in K, and this is a contradiction with $\langle x^{**}, x^* \rangle \geq r$. This proves $K = M$ and, in particular, $M \subset X$. This holds for every $x \in X$, so the family $\{\overline{\Gamma(K_n)}^{w^*}\}_{n=1}^{\infty}$ can be used in the definition of WCD. To obtain sets with nonempty norm-interior, consider the family $\{\overline{\Gamma(K_n)}^{w^*} + (1/n)B_{X^{**}}\}_{n=1}^{\infty}$.

13.30 (Preiss, Talagrand) Prove directly that every WCD Banach space X is Lindelöf in its weak topology, i.e., it is *weakly Lindelöf* (a more general result appears in Theorems 14.42 and 14.46).

Hint. (Vašák [Vasa]) Let $\{K_n\}_{n=1}^{\infty}$ be the sequence of w^*-compact subsets of X^{**} from the definition of WCD. Put $\Sigma = \{\pi \in \mathbb{N}^{\mathbb{N}} : \bigcap_{i=1}^{\infty} K_{\pi(i)} \subset X\}$, and let $\varphi : \Sigma \to \mathcal{P}(X)$ be the mapping defined by $\varphi(\pi) = \bigcap_{i=1}^{\infty} K_{\pi(i)}$ for $\pi \in \Sigma$. Show first that $\varphi : \Sigma \to (X, w)$ is upper semicontinuous by proving that $\varphi^{-1}(V) := \{\pi \in \Sigma : \varphi(\pi) \subset V\}$ is open for every open subset V of X. Let $\mathcal{U}$ be an open covering of (X, w). Let $\mathcal{U}'$ be the family of all finite unions of elements in $\mathcal{U}$. The family $\{\varphi^{-1}(U') : U' \in \mathcal{U}'\}$ in an open covering of Σ (a metrizable and separable space, hence Lindelöf), so it has a countable subcovering, say $\{\varphi^{-1}(U'_m) : m \in \mathbb{N}\}$, where $U'(m) = \bigcup_{i=1}^{r_m} U_i^m$, and $U_i^m \in \mathcal{U}$ for all $i = 1, 2, \ldots, r_m$, $m \in \mathbb{N}$. The family $\{U_i^m : i = 1, 2, \ldots, r_m, m \in \mathbb{N}\}$ is a countable subcovering of $\mathcal{U}$.

13.31 Prove that every weakly countably determined Banach space has a projectional generator, hence a projectional resolution of the identity. For the definition of such a space see Exercise 13.28. The construction of a projectional generator in a WCD space was done by Valdivia. A projectional resolution of the identity in WCD spaces was already present in [Vasa]. He worked on X^{**} preserving X, using the classical Lindenstrauss method (see, e.g., [FHHMPZ, Ch. 11]).
Hint. Let $\{K_n\}_{n=1}^{\infty}$ be the sequence of w^*-compact subsets of X^{**} associated to the definition of weak countable determinacy.

Let $\mathcal{P}_f(\mathbb{N})$ be the (countable) family of all finite subsets of $\mathbb{N}$. Put

$$L_s = \overline{X \cap \bigcap_{n\in s} K_n}^{w^*}, \quad s \in \mathcal{P}_f(\mathbb{N}).$$

Each L_s is a w^*-compact subset of X^{**}. For $x^* \in X^*$, let $\varphi(x^*)$ be a countable subset of X such that $\sup_{x\in L_s} |\langle x^*, x\rangle| = \sup_{x\in L_s\cap\varphi(x^*)} |\langle x^*, x\rangle|$ for every $s \in \mathcal{P}_f(\mathbb{N})$. This defines a set-valued mapping from X^* into the family of countable subsets of X. We claim that the couple (X^*, φ), where φ has been defined above, is a projectional generator. Indeed, let B be a $\mathbb{Q}$-linear subset of X^*. Assume that for some $x^* \in \varphi(B)^{\perp} \cap \overline{B}^{w^*}$ we have $x^* \neq 0$. Then we can find $x \in X$ such that $\langle x^*, x\rangle = 1$. Let $\{n_i : i \in \mathbb{N}\} = \{n : n \in \mathbb{N}, x \in K_n\}$. Put $K = \bigcap_{i=1}^{\infty} L_{n_1 n_2 \ldots n_i}$ and note that $K \subset X$, so K is a w-compact subset of X that contains x. The same argument used in the proof of Proposition 13.13 shows that $\overline{B}^{w^*} = \overline{B}^{\mu(X^*,X)}$. Thus we can find, for a fixed $\varepsilon < 1/3$, an element $y^* \in B$ such that $\sup_{z\in K} |\langle y^* - x^*, z\rangle| < \varepsilon$.

The set $U := \{x^{**} \in X^{**} : |\langle x^{**}, y^* - x^*\rangle| < \varepsilon\}$ is a w^*-open subset of X^{**} containing K, so there exists $i \in \mathbb{N}$ such that $(K \subset)\ L_{n_1,\ldots,n_i} \subset U$. Notice that $|\langle y^* - x^*, x\rangle| < \varepsilon$, hence $|\langle y^*, x\rangle| > 1 - \varepsilon$. Therefore

$$\sup\{|\langle y^*, z\rangle| : z \in L_{n_1\ldots n_i} \cap \varphi(y^*)\} \left(= \sup\{|\langle y^*, z\rangle| : z \in L_{n_1\ldots n_i}\}\right) > 1 - \varepsilon.$$

Since $L_{n_1\ldots n_i} \subset U$, we get

$$\sup\{|\langle x^*, z\rangle| : z \in L_{n_1\ldots n_i} \cap \varphi(y^*)\} > 1 - 2\varepsilon.$$

This is a contradiction, since $x^* \in (\varphi(B))^{\perp}$, and $y^* \in B$.

13.32 Let $\Sigma = \mathbb{N}^{\mathbb{N}}$ endowed with the product topology of the discrete topology. Let X be a Banach space, and Σ' be a subset of Σ. Recall that a set-valued mapping Φ of Σ' into X is *upper semicontinuous mapping with compact values* (usco for short) if

(i) $\Phi(\sigma)$ is weakly compact for every $\sigma \in \Sigma'$, and
(ii) For every weakly open subset U of X, $\{\sigma \in \Sigma' : \Phi(\sigma) \subset U\}$ is open in Σ'.

Prove the following result: *Let X be a Banach space. The following are equivalent. (i) X is weakly countably determined. (ii) There is a subset Σ' of Σ and an usco mapping Φ of Σ' into X such that $X = \bigcup\{\Phi(\sigma);\ \sigma \in \Sigma'\}$.*
Hint. (i)$\Longrightarrow$(ii): We use the notation from the definition of WCD in Exercise 13.28. First, by reindexing if necessary, we assume that any K_n repeats infinite number of times in the sequence $\{K_n;\ n \geq 1\}$ in Exercise 13.28. We denote by

$$\Sigma' = \{\sigma = (n_i)_{i\geq 1} \in \Sigma : \bigcap_{i\geq 1} K_{n_i} \subset X\}$$

and define the set-valued mapping Φ from Σ' into X by

$$\Phi(\sigma) = \bigcap_{i\geq 1} K_{n_i}.$$

By the w^*-compactness of the K_ns, Φ has w-compact values and is usco. By the definition of WCD Banach space (Exercise 13.28), we have for any $x \in X$,

$$\bigcap\{K_n :\ x \in K_n\} \subset X,$$

and this shows (ii).

(ii)$\Longrightarrow$(i): Let $\mathcal{G}$ be the set of all finite sequences in $\mathbb{N}$. We denote by $|s|$ the length of $s \in \mathcal{G}$. If $s \in \mathcal{G}$ and $\sigma \in \Sigma$, we write $s < \sigma$ whenever s is the sequence of the first $|s|$ terms of σ.

For $s \in \mathcal{G}$ and $n \geq 1$ we let

$$K_{s,n} = \overline{\bigcup\{\Phi(\sigma) :\ s < \sigma\} \cap nB_X}^{w^*}$$

where the closure is taken in (X^{**}, w^*). The countable collection of w^*-compact sets $\{K_{s,n} :\ s \in \mathcal{G}, n \geq 1\}$ is what we are looking for. Indeed, for $x \in X$ and $u \in X^{**}\setminus X$, choose $n \geq 1$ and $\sigma \in \Sigma'$ such that $\|x\| < n$ and $x \in \Phi(\sigma)$. Clearly $x \in K_{s,n}$ for any $s < \sigma$. Since Φ is usco, we have

$$\bigcap\{K_{s,n} :\ s < \sigma\} \subset \Phi(\sigma) \subset X.$$

Thus there is $s < \sigma$ such that $u \notin K_{s,n}$.

Recall that a Banach space is said to be weak $\mathcal{K}$-analytic if (ii) in the statement in Exercise 13.32 holds true for X with $\Sigma' = \Sigma$. There is a weak $\mathcal{K}$-analytic space X that is not $K_{\sigma\delta}$ in (X^{**}, w^*), i.e., it is not a countable intersection of sets, each of them is a countable union of compact sets in (X^{**}, w^*) [ArArMe].

13.33 Let X, Y be Banach spaces, $T \in \mathcal{B}(X, Y)$. Show that if X has the Grothendieck property and Y is WCG, then T is weakly compact.

Hint. B_{Y^*} is w^*-sequentially compact since Y is WCG. Thus $T^*(B_{Y^*})$ is w^*-sequentially compact and, by the Grothendieck property, w-compact. Therefore T^* is weakly compact. Use now Theorem 13.34.

13.34 A Banach space X is said to have the *Mazur property* if w^*-sequentially continuous linear functionals on X^* are w^*-continuous. Show that ℓ_∞ does not have the Mazur property. Show that every WCG Banach space has the Mazur property.
Hint. ℓ_∞ has the Grothendieck property. Every element of ℓ_∞^{**} is w-continuous and thus w^*-sequentially continuous. If ℓ_∞ had the Mazur property, every element from ℓ_∞^{**} would be w^*-continuous, hence from ℓ_∞. Thus ℓ_∞ would be reflexive, a contradiction. To show that a WCG Banach space has the Mazur property use Theorem 13.20, the fact that every Eberlein compact space is angelic (see Proposition 3.108) and the Banach–Dieudonné theorem, more precisely its Corollary 3.94.

13.35 ([FMZ1]) Prove that if X is a Banach space such that for every norming subspace Y of X^* there is an equivalent Fréchet differentiable norm which is Y-lower semicontinuous, then X is reflexive.
Hint. Use Exercise 3.88, the Šmulyan lemma 7.22 and the Bishop–Phelps theorem.

13.36 Define $T\colon L_1[0,1] \to C[0,1]$ by $T(f)\colon x \mapsto \int_0^x f(t)\,\mathrm{d}t$. Show that T is continuous but not weakly compact. Show that T is completely continuous.
Hint. Take $f_n = 2n\big(\chi_{[\frac{1}{2}-\frac{1}{n},\frac{1}{2}]} - \chi_{[\frac{1}{2},\frac{1}{2}+\frac{1}{n}]}\big)$ and show that $\{T(f_n)\}$ has no weakly convergent subsequence.

To show that T is completely continuous, it is enough to show that T takes w-compact subsets of $L_1[0,1]$ to compact sets in $C[0,1]$. Prove this using the Arzelà–Ascoli criterion in $C[0,1]$ and the Dunford criterion for weak compactness in $L_1[0,1]$: A subset K of $L_1[0,1]$ is relatively weakly compact if and only if K is bounded and uniformly integrable, that is, for every $\varepsilon > 0$ there is $\delta > 0$ such that $\int_M |f|\,\mathrm{d}\lambda < \varepsilon$ for every $f \in K$ and $M \subset [0,1]$ satisfying $\lambda(M) < \delta$.

13.37 Consider the identity injection I from $C[0,1]$ to $L_2[0,1]$. Show that that it is continuous but not weakly compact.
Hint. The space L_2 is reflexive and $C[0,1]$ is not.

13.38 Consider $\varphi \in C[0,1]^*$, $\varphi(f) = \int_0^{1/2} f(t)\,\mathrm{d}t - \int_{1/2}^1 f(t)\,\mathrm{d}t$. Show that φ as an operator from $C[0,1]$ to $\mathbb{R}$ is weakly compact but $\varphi(B_{C[0,1]})$ is not w-compact in $\mathbb{R}$.
Hint. Show that $\varphi(B_{C[0,1]}) = (-1,1)$.

13.39 Show that there is no one-to-one bounded operator from $c_0(\Gamma)$ into any reflexive space if Γ is uncountable.
Hint. Assume that T is a one-to-one bounded operator from $c_0(\Gamma)$ into a reflexive space X. Then T is weakly compact, hence so is T^*. Note that B_{X^*} is weakly compact, $T(B_{X^*})$ is compact in $\ell_1(\Gamma)$ and $\overline{\operatorname{span}}^{w^*}\big(T(B_{X^*})\big) = \ell_1(\Gamma)$. Hence $c_0(\Gamma)^*$

is w^*-separable, which is not the case as all elements of $c_0(\Gamma)^*$ have countable supports.

13.40 Does there exist an isomorphic copy Z of ℓ_2 in ℓ_∞ that is complemented in ℓ_∞?
Hint. No. $\ell_\infty = C(K)$, where K is the Stone–Čech compactification of $\mathbb{N}$, and thus ℓ_∞ has the Dunford–Pettis property. Then use Proposition 13.44.

13.41 Let $p \in [1, \infty)$. Show that $C[0, 1] \oplus \ell_p$ is not isomorphic to $C[0, 1]$.
Hint. If $p > 1$, then ℓ_p would be isomorphic to a complemented reflexive subspace of $C[0, 1]$ and as such it would have the Dunford–Pettis property. No infinite-dimensional reflexive space has the Dunford–Pettis property.

If $p = 1$, use the fact that $C[0, 1]^*$ has an equivalent LUR norm ([Troy1]) and ℓ_∞ does not have such a norm ([Lind13], [Troy1]).

13.42 Show that $L_1[0, 1] \oplus \ell_2$ is not isomorphic to $L_1[0, 1]$.
Hint. L_1^* is isometric to L_∞, which is in turn isomorphic to ℓ_∞ (Exercise 4.42), which is isomorphic to the space of continuous functions on the Stone–Čech compactification of $\mathbb{N}$ (see Section 17.1), which has the Dunford–Pettis property. Thus we get that L_1 has the Dunford–Pettis property. Hence ℓ_2 cannot be isomorphic to a complemented subspace of L_1 as it is reflexive and infinite-dimensional.

13.43 Let $p \in (1, \infty)$. Show that $L_1[0, 1]$ does not contain a complemented subspace isomorphic to ℓ_p.
Hint. $L_1[0, 1]$ has the Dunford–Pettis property. Use Proposition 13.44.

13.44 Let X be a Banach space, $C \subset X$ be a convex set in X and let $x^{**} \in \overline{C}^{w^*}$. Show that $\operatorname{dist}(x^{**}, C) \le 2 \operatorname{dist}(x^{**}, X)$.
Hint. Take any $\delta > 0$ such that $\operatorname{dist}(x^{**}, X) < \delta$ and find $x \in X$ such that $\|x^{**} - x\| < \delta$. Then $x \in \overline{C}^{w^*} + \delta B_{X^{**}} \subset \overline{C + \delta B_X}^{w^*}$. It follows that $x \in \overline{C + \delta B_X}^{\|\cdot\|}$, since C and B_X are convex sets in X, and $x \in X$. Therefore, given $\varepsilon > 0$, there exists $c \in C$ and $b \in B_X$ such that $\|x - c - \delta b\| < \varepsilon$. Thus $\|x^{**} - c\| = \|x^{**} - x + x - c\| < 2\delta + \varepsilon$. As $\varepsilon > 0$ was arbitrary, we get $\operatorname{dist}(x^{**}, C) < 2\delta$. Therefore $\operatorname{dist}(x^{**}, C) \le 2 \operatorname{dist}(x^{**}, X)$.

13.45 Show that any WCG Banach space is a weak Asplund space.
Hint. Let $(Z, \|\cdot\|)$ be a reflexive Banach space the dual norm of $\|\cdot\|$ being LUR, and let T be a bounded operator from Z onto a dense set in X. Let the dual norm $|||\cdot|||$ on X^* be defined by

$$|||f|||^2 = \|f\|_{X^*}^2 + \|T^* f\|_{Z^*}^2.$$

Then the predual norm to $|||\cdot|||$ on X is Gâteaux differentiable on X with uniformity in directions of $T B_Z$. The points of such differentiability automatically form a G_δ-

set S in X, and by the smooth variational principle for this kind of differentiability, the set S is dense in X.

Note that from this and the smooth variational principle, it follows that $C[0, 1]$ does not admit any equivalent weak Hadamard differentiable norm (Borwein, [Bor]).

13.46 Let f_1, f_2, f_3 be real-valued functions on the real line defined by $f_1(t) = |t|$ if $|t| \leq 1$ and $f_1(t) = +\infty$ otherwise, $f_2(t) = 0$ for $|t| \leq 1$ and $f_2(t) = +\infty$ otherwise, $f_3(t) = t^p$ if $|t| \leq 1$ and $f_3(t) = +\infty$ otherwise, where $p > 1$. Calculate the Fenchel dual functions to f_1, f_2, f_3 and note the connection to the duality between strict convexity and smoothness and the duality of ℓ_p and ℓ_q spaces.
Hint. Direct calculation.

13.47 ([ScWh]) A Banach space X is called β-*weakly compactly generated* (for short, βWCG) if there exists a weakly compact subset $K \subset X$ such that, for every weakly compact subset $W \subset X$, we can find $n \in \mathbb{N}$ such that $W \subset nK + \varepsilon B_X$ (we say, in this case, that K *strongly generates* X, or that X *is strongly generated by* K. Prove that X is βWCG if and only if $\big(B_{X^*}, \mu(X, X^*)\big)$ is metrizable, where $\mu(X^*, X)$ denotes the *dual Mackey topology*, i.e., the topology on X^* of the uniform convergence on the family of all convex, balanced and w-compact subsets of X.
Hint. It is simple to prove that $\mu(X^*, X)$ and $\mathcal{T}_K$, the topology of the uniform convergence on the set K, agree on B_{X^*}.

13.48 ([ScWh]) Prove that every βWCG Banach space (see Exercise 13.47) is weakly sequentially complete.
Hint. [FMZ4] Let (x_n) be a Cauchy sequence in X. Put $D_n = \overline{\text{aco}}\{x_p - x_q : p, q \geq n\}$, $n \in \mathbb{N}$, where $\text{aco}(S)$ denotes the absolutely convex (i.e., the convex and balanced) hull of a set $S \subset X$. Obviously, $X^* = \bigcup_{n\in\mathbb{N}} D_n{}^\circ$. In particular, $mB_{X^*} = \bigcup_{n\in\mathbb{N}}(D_n{}^\circ \cap mB_{X^*})$ for every $m \in \mathbb{N}$. It follows from Exercises 3.41 and 13.47 that $(B_{X^*}, \mu(X^*, X))$ is a complete metrizable space. Fix $m \in \mathbb{N}$. The sets $(D_n{}^\circ \cap mB_{X^*})$ are $\mu(X^*, X)$-closed, hence, by the Baire category theorem, there exists $n(m) \in \mathbb{N}$ and an absolutely convex weakly compact subset K_m of X such that

$$(K_m{}^\circ \cap mB_{X^*}) \subset (D_{n(m)}{}^\circ \cap mB_{X^*}).$$

By taking polars in X we get

$$(D_{n(m)} \subset)\ \overline{\text{conv}}\left(D_{n(m)} \cup \frac{1}{m}B_X\right) \subset \overline{\text{conv}}\left(K_m \cup \frac{1}{m}B_X\right)\ \left(\subset K_m + \frac{1}{m}B_X\right).$$

In particular, $x_p - x_q \in K_m + \frac{1}{m}B_X$ for every $p, q \geq n(m)$. Let x^{**} be the weak*-limit of the sequence (x_n) in X^{**}. Then $x^{**} - x_q \in K_m + \frac{1}{m}B_{X^{**}}$ for every $q \geq n(m)$ and we obtain $x^{**} \in X + \frac{1}{m}B_{X^{**}}$. This happens for every $m \in \mathbb{N}$, so $x^{**} \in X$.

13.49 Prove that if μ is a finite measure defined on a σ-algebra Σ of subsets of a certain set Ω, then $L_1(\mu)$ is βWCG.
Hint. Assume without loss of generality that μ is a probability measure. By using the identity operators, we have $B_{L_\infty(\mu)} \subset B_{L_2(\mu)} \subset B_{L_1(\mu)}$. Let K be a weakly compact set in the unit ball of $L_1(\mu)$. Then K is *uniformly integrable* in $L_1(\mu)$ ([DuSc, p. 292]), i.e., for every $\varepsilon > 0$ there is $\delta > 0$ such that for every $x \in K$, $\int_M |x| \mathrm{d}\mu < \varepsilon$ whenever $M \in \Sigma$ and $\mu(M) < \delta$.

For $k \in \mathbb{N}$ and for $x \in K$, put $M_k(x) = \{t \in \Omega : |x(t)| \geq k\}$, and write $x = x_1 + x_2$, where $x_1 := x.\chi(\Omega \setminus M_k(x))$ and $x_2 := x.\chi(M_k(x))$ (where $\chi(S)$ denotes the characteristic function of a set $S \subset \Omega$). Let $a_k(K) = \sup\{\|x_2\|_1 : x \in K\}$. Then

$$K \subset kB_{L_\infty(\mu)} + a_k(K)B_{L_1(\mu)} \subset kB_{L_2(\mu)} + a_k(K)B_{L_1(\mu)}.$$

We have $k\mu(M_k(x)) \leq \|x_2\|_1 \leq 1$, hence $\mu(M_k(x)) \leq 1/k$ for all $x \in K$. From the uniform integrability of K, we get that $a_k(K) \to 0$ when $k \to \infty$. This finishes the proof.

13.50 Find an example of a compact non-absolutely summing operator.
Hint. $T(x_i) = \big((1/\sqrt{i})x_i\big)$ in c_0.

13.51 Find an example of an absolutely summing operator that is not compact.
Hint. Id : $\ell_1 \to \ell_2$.

13.52 All Hilbert spaces of the same density are isomorphic. Compare this with the fact that $\ell_1(\omega_1)$ is not isomorphic to any subspace of $L_1(\mu)$ for a finite measure μ.
Hint. $(B_{\ell_\infty(\omega_1)}, w^*)$ is not angelic (see the example preceding Theorem 3.54)—hence not Eberlein compact—, hence $\ell_1(\omega_1)$ is not a subspace of a WCG Banach space.

13.53 Prove that the implication in Lemma 13.53 is in fact an equivalence.
Hint. Assume that X/F is reflexive. Let $\{x_n : n \in \mathbb{N}\}$ be a bounded set in X and let $x^{**} \in X^{**}$ be a w^*-accumulation point. Let $\{x_\gamma\}$ be a net in $\{x_n : n \in \mathbb{N}\}$ that w^*-converges to x^{**}. The net $\{q(x_\gamma)\}$ is w-convergent to some $\hat{x} \in X/F$, since X/F is reflexive. Find $x \in X$ such that $q(x) = \hat{x}$. Given $f^\perp \in F^\perp$, we have $\langle f^\perp, x_\gamma - x\rangle \to \langle x^{**} - x, f^\perp\rangle$. Observe that $\langle f^\perp, x_\gamma - x\rangle = \langle f^\perp, qx_\gamma - \hat{x}\rangle \to 0$; thus, $\langle x^{**} - x, f^\perp\rangle = 0$ for all $f^\perp \in F^\perp$.

13.54 Prove that the converse of Lemma 13.54 holds (i.e., if X is a Banach space and Y is a subspace of X such that $Y + H = X$ for some separable subspace H, then X/Y is separable).
Hint. Let $\{h_n : n \in \mathbb{N}\}$ be a dense subset of H. The set $\{y + h_n : n \in \mathbb{N}\}$ is obviously dense in $Y + H$ $(= X)$. Then, if $q : X \to X/Y$ denotes the canonical quotient mapping, the set $q(\{y + h_n : n \in \mathbb{N}\})$ $(= \{q(h_n) : n \in \mathbb{N}\})$ is dense in X/Y.

13.55 (Haar) Let G be a metrizable compact group. Then there is a probability measure on G that is invariant under left translations.
Hint. The following proof is due to Godefroy and Li [GoLi]: Let $\{f_k\} \subset C(G)$, $\|f_k\| \le 2^{-k}$, span a dense subspace of $C(G)$. On the set $\mathcal{P}(G)$ of probabilities on G define $\varphi(\mu) = \sum \mu(f_i)^2$. Note that φ is a strictly convex w^*-continuous function on $\mathcal{P}(G)$. For $\mu \in \mathcal{P}(G)$ put

$$\Phi(\mu) = \sup\{\varphi(\tau_g \mu) : g \in G\},$$

where τ_g is the left translation operator. Note that Φ is w^*-lower semicontinuous and convex. We claim that Φ is strictly convex.

Indeed, let $\mu, \lambda, \nu \in \mathcal{P}(G)$ be such that $\mu = \frac{1}{2}(\lambda + \nu)$ and $\Phi(\mu) = \frac{1}{2}(\Phi(\lambda) + \Phi(\nu))$. Since the mapping $g \mapsto \tau_g(\mu)$ is continuous from G into $(\mathcal{P}(G), w^*)$ and φ is w^*-continuous, by compactness there is $h \in G$ such that $\Phi(\mu) = \varphi(\tau_h \mu)$. Then from the definition of Φ we have

$$\varphi(\tau_h \mu) \ge \frac{1}{2}(\varphi(\tau_h \lambda) + \varphi(\tau_h \nu))$$

and from the strict convexity of φ we get $\tau_h \mu = \tau_h \lambda = \tau_h \nu$ and thus $\mu = \lambda = \nu$. Hence Φ is strictly convex.

The strictly convex and w^*-lower semicontinuous function Φ attains its minimum at a unique $m \in \mathcal{P}(G)$. Since $\Phi(\tau_g \lambda) = \Phi(\lambda)$ for all $g \in G$ and $\lambda \in \mathcal{P}(G)$, we have $\tau_g m = m$ for all $g \in G$ by the uniqueness of the minimum.

13.56 Modifying an example of Talagrand, Argyros constructed a uniform Eberlein compact space T such that the corresponding Banach space $C(T)$ (which is then WCG) contains a non-WCG subspace, X say. Note that T is a closed subset of the unit ball $B_{\ell_2(\mathbb{N}^{\mathbb{N}})} =: K$ provided with the weak topology; see [Fabi1, pp. 29–32]. Show that the (WCG, even Hilbert generated, space) $C(K)$ is not hereditarily WCG [AvKal, Problem 11]. We say that a Banach space X is *Hilbert generated* if there are a Hilbert space H and a bounded operator $T : H \to X$ with dense range.
Hint. Define $Q : C(K) \to C(T)$ by $C(K) \ni f \longmapsto Qf = f|_T$; this is a linear mapping with $\|Q\| \le 1$. According to Tietze's extension theorem (Corollary 7.55), Q is surjective. Then $Y := Q^{-1}X$ is not WCG. Indeed, if yes, then X $(= QY)$ would also be WCG, a contradiction.

Chapter 14
Topics in Weak Topologies on Banach Spaces

In this chapter we study the weak and weak* topologies of Banach spaces in more detail. We discuss several types of compacta (Eberlein, uniform Eberlein, scattered, Corson, and more), weakly Lindelöf determined spaces and properties of tightness in weak topologies. We discuss some applications in the structural properties of some Banach spaces.

14.1 Eberlein Compact Spaces

Recall that a compact space is said to be *Eberlein* if it is homeomorphic to a weakly compact set in some space $c_0(\Gamma)$, endowed with its weak topology, see Definition 13.18. By Corollary 13.19, the class of Eberlein compact spaces coincides with the class of weakly compact subset of Banach spaces, endowed with the restriction of their weak topology (and, by Corollary 13.23 it is enough to restrict ourselves to the class of reflexive Banach spaces). Let X be a Banach space. Endow B_{X^*} with the w^*-topology, and let $\mathcal{T}_p$ be the topology of the pointwise convergence in $C(B_{X^*})$. The mapping $\Phi : (X, w) \to (C(B_{X^*}), \mathcal{T}_p)$ is an isomorphism into. In particular, every weakly compact subset of a Banach space is (linearly) homeomorphic to a pointwise compact subset of a $C(K)$ space in its pointwise topology. This gives one of the implications in the following result.

Proposition 14.1 *A compact space L is Eberlein if and only if there is a compact space K such that L is homeomorphic to a pointwise compact subset of $C(K)$ considered in its pointwise topology.*

Proof: The "only if" part follows from the previous observation. To prove the "if" part, let L be homeomorphic to a pointwise compact set L_1 in $C(K)$ space for some compact space K. Let τ be a homeomorphism of $\mathbb{R}$ onto $(-1, +1)$ and define a mapping Φ from L_1 into $B_{C(K)}$ by $\Phi(f)\colon x \mapsto \tau(f(x))$. Then Φ is a homeomorphism in the pointwise topology and $\Phi(L_1)$ is thus a uniformly bounded pointwise compact set in $C(K)$. By Theorem 3.139, $\Phi(L_1)$ is weakly compact. By Corollary 13.19, $\Phi(L_1)$ is Eberlein compact, and so it is L. $\square$

M. Fabian et al., *Banach Space Theory*, CMS Books in Mathematics,

DOI 10.1007/978-1-4419-7515-7_14,

Proposition 14.2 *Let K be a Hausdorff space. If $M \subset C(K)$ is pointwise separable, then it is separable.*

Proof: ([Nami3]) Let D be a pointwise dense subset of M. Define $\varphi\colon K \to \mathbb{R}^D$ by $\varphi(t)\colon f \to f(t)$. Then φ is continuous and $\varphi(K)$ is compact and metrizable. For each $f \in M$ there is a unique $\hat{f} \in C(\varphi(K))$ such that $f = \hat{f} \circ \varphi$. Then $(M, \|\cdot\|)$ and $(\widehat{M}, \|\cdot\|)$ are isometric. Since $C(\varphi(K))$ is separable, so are $(\widehat{M}, \|\cdot\|)$ and $(M, \|\cdot\|)$. $\square$

Theorem 14.3 (Namioka, see, e.g., [Todo1]) *Let K be a compact set. Let L be a compact set in $C(K)$ considered in the pointwise topology. Then there is a dense G_δ subset M of L such that the pointwise and norm topologies of $C(K)$ coincide on M.*

Proof: Unless stated otherwise, we will consider $C(K)$ with its pointwise topology.

Consider the (non-linear) norm-to-norm-continuous mapping $\varphi\colon C(K) \to C(K)$ defined as $\varphi(f)\colon k \mapsto \frac{2}{\pi}\arctan\big(f(k)\big)$. Clearly $\varphi\big(C(K)\big) \subset B_{C(K)}$. For every $S \subset C(K)$, the set S with the pointwise topology is homeomorphic via φ with $\varphi(S) \subset B_{C(K)}$ in the pointwise topology of $C(K)$. φ is also a homeomorphism of S in the norm topology onto $\varphi(S)$ in the norm topology of $C(K)$. Thus we may assume that the set L is bounded.

For $f \in L$ define the oscillation of norm at f by

$$O(f) = \inf\big\{\sup\{\|g_1 - g_2\| : g_1, g_2 \in W\},\ W \subset L, W \text{ open}, f \in W\big\}.$$

For $n \in \mathbb{N}$, put $U_n = \big\{f \in L : O(f) < \frac{1}{n}\big\}$. From the definition, it follows that U_n is an open set in L.

Let $U = \bigcap U_n$. Then U is a G_δ set in L. By the Baire category theorem, the proof will be complete if we show that every U_n is dense in L. Supposing the contrary and replacing L by a nonempty set $L\backslash\overline{U_n}$ if needed, we can assume that $O(f) \geq \varepsilon > 0$ for all $f \in L$ and derive a contradiction. To this end we will construct a sequence $\{V_n\}$ of open sets in L and $f_n \in V_n$ such that $\overline{V_{n+1}} \subset V_n$ and $\operatorname{dist}\big(\overline{V_{n+1}}, \operatorname{conv}\{f_1, \dots, f_n\}\big) \geq \frac{\varepsilon}{3}$ for all n.

Assume that $f_1, \dots, f_n$ and $V_1, \dots, V_n$ have been constructed. Put $K_n = \operatorname{conv}\{f_1, \dots, f_n\}$. Then K_n is a compact set in the norm topology inherited from $C(K)$, so there is a finite $\frac{\varepsilon}{12}$-net $A_n \subset K_n$. For each $g \in A_n$, let $B(g)$ be the closed ball in $C(K)$ centered at g with radius $\frac{5\varepsilon}{12}$. Note that $B(g)$ is pointwise closed in $C(K)$.

Assume that $B(g)$ contains an open set W in L. Then for every g_1 and g_2 in W we have $\|g_1 - g_2\| \leq \|g_1 - g\| + \|g - g_2\| \leq \frac{10\varepsilon}{12} < \varepsilon$. Thus every point in W would have oscillation less than ε, contradicting our assumption.

Hence $B(g) \cap L$ is closed and nowhere dense in L for every $g \in A_n$. There is an open set V_{n+1} such that $\overline{V_{n+1}} \cap \Big(\bigcup_{g\in A_n} B(g)\Big) = \emptyset$ and $\overline{V_{n+1}} \subset V_n$. Assume that $\operatorname{dist}(f, K_n) < \varepsilon/3$ for some $f \in \overline{V_{n+1}}$. Then $\operatorname{dist}(f, A_n) < \frac{\varepsilon}{3} + \frac{\varepsilon}{12} = \frac{5\varepsilon}{12}$.

However, as $f \in \overline{V_{n+1}}$ and $\overline{V_{n+1}} \cap (\bigcup_{g \in A_n} B(g)) = \emptyset$, we get that $\operatorname{dist}(f, A_n) > \frac{5\varepsilon}{12}$, a contradiction. So we may choose f_{n+1} to be any point in V_{n+1}.

Let f_∞ be an accumulation point of $\{f_n\}$. Then $f_\infty \in \overline{V_n}$ for each n. As L is angelic, we get a subsequence $\{f_{n_k}\}$ of $\{f_n\}$ such that $f_\infty = \lim f_{n_k}$. Since L is bounded, we actually have $f_{n_k} \xrightarrow{w} f_\infty$, so by Mazur's theorem, $f_\infty \in \overline{\operatorname{conv}}\{f_{n_k}\}$. However, for every k, $\operatorname{dist}\big(f_\infty, \operatorname{conv}\{f_{n_1}, \ldots, f_{n_k}\}\big) \geq \frac{\varepsilon}{3}$, a contradiction. □

Corollary 14.4 *Let C be a weakly compact set in a Banach space X considered in its weak topology (i.e., an Eberlein compact space). Then there is a dense G_δ subset M of C on which the weak and norm topologies from X coincide.*

Proof: Put $K = B_{X^*}$ in its weak star topology. Then on C, the weak topology from X and the pointwise topology from $C(K)$ coincide. Moreover, the norm topology on $C \subset X$ coincides with the norm topology for $C \subset C(K)$. Then use Theorem 14.3. □

Before stating another corollary we need a definition:

Definition 14.5 *A topological space T is said to have* property CCC *(the* countable chain condition, *or the* Souslin property*) if T does not contain any uncountable family of nonempty open pairwise disjoint sets.*

Clearly, any separable space has property CCC. For metrizable spaces, property CCC is equivalent to separability. The same applies to Eberlein compact spaces:

Corollary 14.6 (Benyamini, Namioka, Rosenthal) *Let L be an Eberlein compact space. If L is nonseparable, then L does not have property CCC.*

Proof: Let M be a dense metrizable subset of L. Then M is not separable and thus there is an uncountable collection $\mathcal{C}$ of pairwise disjoint open sets in M. For every $C \in \mathcal{C}$ choose an open set $\tilde{C}$ in L such that $C = M \cap \tilde{C}$. We note that if C_1 and C_2 are distinct members of $\mathcal{C}$, then $\tilde{C}_1 \cap \tilde{C}_2 = \emptyset$. Indeed, if $x \in \tilde{C}_1 \cap \tilde{C}_2$ and U is a neighborhood of x in L such that $x \in U \subset \tilde{C}_1 \cap \tilde{C}_2$, then there is some $y \in M \cap U$ and so $y \in C_1 \cap C_2$. Thus $\{\tilde{C} : C \in \mathcal{C}\}$ forms an uncountable pairwise disjoint collection of open sets in L. □

Theorem 14.7 (see, e.g., [Enge, Theorem 2.3.18]) *Let X be a Banach space. Then X in the weak topology has property CCC.*

Proof: Let $S \subset X^*$ be an algebraic basis of X^*. Then (X, w) is canonically homeomorphic to a dense subspace of $\mathbb{R}^S$ endowed with the product topology. It is enough to prove that $\mathbb{R}^S$ has CCC. To this end, let $\{U_i : i \in I\}$ be a pairwise disjoint family of (basic) nonempty open sets in $\mathbb{R}^S$. Then for every $i \in I$ there exists a finite set $S_i \subset S$ and a family $\{W_s^i\}_{s \in S}$ of nonempty open subsets of $\mathbb{R}_s = \mathbb{R}$ such that $U_i = \prod_{s \in S} W_s^i$ and $W_s^i = \mathbb{R}$ for all $s \notin S_i$.

Let $I_0 \subset I$ be such that $\operatorname{card}(I_0) \le 2^{\aleph_0}$ and set $S_0 = \bigcup_{i\in I_0} S_i$. Then $\operatorname{card}(S_0) \le 2^{\aleph_0}$ and $U_i = \prod_{s\in S_0} W_s^i \times \prod_{s\in S\setminus S_0} \mathbb{R}_s$ for $i \in I_0$, so $\{\prod_{s\in S_0} W_s^i\}_{i\in I_0}$ is a pairwise disjoint family of nonempty open sets in $\mathbb{R}^{S_0}$.

We claim that $\mathbb{R}^{S_0}$ is separable, then I_0 is countable and the proof is finished. It is enough to prove that $\mathbb{Q}^{S_0}$ is separable and, as $\mathbb{Q}$ is countable, it suffices to prove that $\mathbb{N}^{S_0}$ is separable when $\mathbb{N}$ carries the discrete topology. $\mathbb{N}^{S_0}$ can be identified with the space F of all mappings $f: \Delta \to \mathbb{N}$ endowed with the pointwise topology, where Δ denotes the Cantor set. The set A of all mappings $f: \Delta \to \mathbb{N}$ which are constant on each of the 2^n dyadic subintervals of Δ at level n is countable and dense in F, so the separability follows. □

We quote one more result that will be used later.

Lemma 14.8 (Rosenthal [Rose3]) *Let T be topological space. If T has property CCC, then for every uncountable family $\mathcal{A}$ of distinct open sets in T there exists an infinite sequence $F_1, F_2, \ldots$ of distinct members of $\mathcal{A}$ with $\bigcap_{i=1}^{\infty} F_i \neq \emptyset$.*

We refer to [HMVZ, Lemma 7.21] for the proof.

Theorem 14.9 (Amir, Lindenstrauss [AmLi]) *Let K be a compact space. The following are equivalent:*
(i) *K is Eberlein.*
(ii) *$C(K)$ is weakly compactly generated.*
(iii) *$B_{C(K)^*}$ in its w^*-topology is Eberlein compact.*
(iv) *There is a weakly compact set $L \subset C(K)$ that separates points of K.*

Proof: (i)⟹(ii): Assume without loss of generality that K is a weakly compact set in $c_0(\Gamma)$ such that $K \subset \frac{1}{2}B_{c_0(\Gamma)}$. Let Φ be the family of all finite sequences $(\gamma_1, \ldots, \gamma_n)$ in Γ, note that γ_i need not be distinct. For $\phi = (\gamma_1, \ldots, \gamma_n) \in \Phi$ and $x \in K$ define $f_\phi(x) = \prod_{i=1}^{n} x(\gamma_i)$. Then f_ϕ is weakly continuous on K. Let $A = \{f_\phi : \phi \in \Phi\} \cup \{1\}$. For every $x \in K$ and $\varepsilon > 0$ there are only finitely many distinct $f \in A$ such that $|f(x)| > \varepsilon$. Indeed, if $|f_\phi(x)| > \frac{1}{2^n}$ then ϕ is necessarily a sequence of length smaller than n, and if $|f_\phi(x)| > \varepsilon$ then $|x(\gamma_i)| > \varepsilon$ for all $\gamma_i \in \phi$. Therefore every sequence of distinct elements in A converges pointwise and thus weakly to 0, so $A \cup \{0\}$ is weakly compact. Then $\operatorname{span}(A)$ is an algebra in $C(K)$ that separates points of K. By the Stone–Weierstrass theorem, $\operatorname{span}(A)$ is dense in $C(K)$, showing that $C(K)$ is weakly compactly generated.

(ii)⟹(iii): Theorem 13.20.

(iii)⟹(i): K is homeomorphic to a closed subset in $B_{C(K)^*}$ and thus K is Eberlein compact if $B_{C(K)^*}$ is.

(ii)⟹(iv): Any set $S \subset C(K)$ for which $\overline{\operatorname{span}}(S) = C(K)$ necessarily separates the points of K.

(iv)⟹(ii): Let L be a weakly compact set in $C(K)$ that separates points of K. For $n \in \mathbb{N}$, denote $L^n = \{f_1 \cdots f_n : f_i \in L\}$, where $f_1 \cdots f_n$ denotes the standard product of functions $f_1, \ldots, f_n$. Every sequence in L^n has a pointwise convergent subsequence in L^n. By Theorem 3.139, L^n is a weakly compact set in $C(K)$. For $n \in \mathbb{N}$, let $s_n = \sup\{\|f\| : f \in L^n\}$. Define $A = \left(\bigcup_n \frac{1}{n s_n} L^n\right) \cup \{1\}$.

It is easy to show that A is weakly sequentially compact in $C(K)$ (we distinguish between the cases when a sequence lies in infinitely many $\frac{1}{ns_n}L^n$'s and when it does not). The subspace span(A) is an algebra that separates points of K and by the Stone–Weierstrass theorem it is dense in $C(K)$. Thus $C(K)$ is weakly compactly generated. □

Corollary 14.10 (Corson and Lindenstrauss, [CoLi]) *Let K be an Eberlein compact space. Then the set of all G_δ-points of K is dense in K.*

Proof: The space $C(K)$ is weakly compactly generated by Theorem 14.9. By Theorem 13.25 (i), $C(K)$ admits an equivalent Gâteaux differentiable norm. By Corollary 7.44, the supremum norm of $C(K)$ is Gâteaux differentiable on a dense set in $C(K)$. By the Šmulyan Lemma 7.22 and the bipolar theorem, we have that the set E of all points of $B_{C(K)^*}$ (in its canonical dual norm) that are exposed by the elements of $C(K)$ has the property that $\overline{\text{conv}}^{w^*}(E) = B_{C(K)^*}$. By Milman's Theorem 3.66, we have $\text{Ext}(B_{C(K)^*}) \subset \overline{E}^{w^*}$. Since the extreme points of $B_{C(K)^*}$ are $\pm$ evaluation functionals $f \to f(k)$ for $k \in K$, the result follows. □

Recall that a subset S of a compact space K is called a *cozero set* if there is $f \in C(K)$ such that $S = \{k \in K : f(k) \neq 0\}$. A subset S of a compact space K is a cozero set if and only if S is open and F_σ.

A family of sets $\{F_\alpha\}$ is called *point-finite in a set A* if every point of A lies in at most finite number of sets F_α. It is called *σ-point-finite* if it is a countable union of point-finite families.

Theorem 14.11 (Rosenthal [Rose5]) *A compact space K is Eberlein if and only if it contains a σ-point-finite family $\mathcal{F}$ of cozero sets such that $\mathcal{F}$ weakly separates the points of K, i.e., given $x_1, x_2 \in K$, $x_1 \neq x_2$, there is $U \in \mathcal{F}$ such that either $x_1 \in U$ and $x_2 \notin U$ or $x_2 \in U$ and $x_1 \notin U$.*

Proof: Let $\mathcal{F} = \bigcup \mathcal{U}_n$ be a family of cozero sets such that each $\mathcal{U}_n$ is a point-finite family of sets. Given $n \in \mathbb{N}$, for $U \in \mathcal{U}_n$ take $f_U^n \in C(K)$ such that $U = \{x \in K : f_U^n(x) \neq 0\}$ and $0 \leq f_U^n \leq \frac{1}{n}$. Put $A = \{f_U^n : n \in \mathbb{N}, U \in \mathcal{U}_n\}$. It follows that every sequence of distinct elements in A converges pointwise and hence weakly to 0 (Lebesgue dominated convergence theorem). The set A is thus weakly compact in $C(K)$. Since A moreover separates the points of K, K is an Eberlein compact space by Theorem 14.9.

Conversely, assume that K is an Eberlein compact space. By Theorems 14.9 and 13.16, let $\{f_\alpha; F_\alpha\}_{\alpha\in\Gamma}$ be a weakly compact Markushevich basis of $C(K)$. Assume without loss of generality that $\|f_\alpha\| \leq 1$ for every $\alpha \in \Gamma$. Define cozero sets $U_{\alpha,j}^n = \left\{x \in K : \frac{j-1}{n} < f_\alpha(x) < \frac{j+1}{n}\right\}$ for $\alpha \in \Gamma$, $j \in \mathbb{Z}$, $|j| \geq 2$. For $n \in \mathbb{N}$, put

$$\mathcal{U}_n = \left\{U_{\alpha,j}^n : \alpha \in \Gamma, j \in \mathbb{Z}, 2 \leq |j| \leq n\right\}.$$

Note that for every infinite sequence of distinct α_i we have $f_{\alpha_i} \to 0$ pointwise and that $\bigcup_{2\leq|j|\leq n} U_{\alpha,j}^n = \left\{x \in K : |f_\alpha(x)| > \frac{1}{n}\right\}$. Thus $\mathcal{U}_n$ is point-finite. If $x_1, x_2 \in K$,

$x_1 \neq x_2$, then $f_\alpha(x_1) \neq f_\alpha(x_2)$ for some $\alpha \in \Gamma$ and then $U^n_{\alpha,j}$ separates x_1 and x_2 for some j and n. □

To compare Eberlein compact spaces with metrizable compact spaces, we state the following result.

Theorem 14.12 (see [Rose5]) *A compact space K is metrizable if and only if it contains a σ-point-finite family $\mathcal{F}$ of cozero sets that separates the points of K in the sense that for every $x_1, x_2 \in K$, $x_1 \neq x_2$ there is $G \in \mathcal{F}$ such that $x_1 \in G$ and $x_2 \notin G$.*

We omit the proof.

14.2 Uniform Eberlein Compact Spaces

Definition 14.13 *A compact space K is said to be* uniform Eberlein *if K is homeomorphic to a weakly compact subset of a Hilbert space in its weak topology.*

Theorem 14.14 (Benyamini, Starbird [BeSta]) *Let K be a compact space. The following are equivalent:*
(i) *K is uniform Eberlein.*
(ii) *K is homeomorphic to a weakly compact set $\tilde{K}$ in $c_0(\Gamma)$ which has the property that for every $\varepsilon > 0$ there is $N(\varepsilon) \in \mathbb{N}$ such that* $\operatorname{card}\{\gamma \in \Gamma : |k(\gamma)| > \varepsilon\} < N(\varepsilon)$ *for all $k \in \tilde{K}$.*

Proof: (i)⟹(ii): Let K be homeomorphic to a weakly compact set $\tilde{K}$ in $\ell_2(\Gamma)$ for some Γ. Let $\tilde{K} \subset mB_{\ell_2(\Gamma)}$ for some $m \in \mathbb{N}$. Given $\varepsilon > 0$ and $k \in \tilde{K}$, if S denotes all those $\gamma \in \Gamma$ for which $|k(\gamma)| > \varepsilon$, then we have

$$m^2 \geq \sum_S |k(\gamma)|^2 \geq \operatorname{card}(S)\varepsilon^2.$$

The implication now follows as the formal identity mapping from $\ell_2(\Gamma)$ into $c_0(\Gamma)$ restricted to K is weak–weak continuous.

(ii)⟹(i): Assume without loss of generality that $\|k\| \leq 1$ for every $k \in \tilde{K}$. Let $f\colon [-1,1] \to [-1,1]$ be a continuous, strictly increasing and odd function such that $f\left(\frac{1}{n}\right) \leq \left(2^n N\left(\frac{1}{n+1}\right)\right)^{-\frac{1}{2}}$ for all $n \in \mathbb{N}$. Define a mapping φ from $\tilde{K}$ into $\ell_\infty(\Gamma)$ by $\varphi(k)\colon \gamma \mapsto f(k(\gamma))$. Given $k \in \tilde{K}$ and $n \in \mathbb{N}$, put $A_n = \left\{\gamma \in \Gamma : \frac{1}{n+1} < |k(\gamma)| \leq \frac{1}{n}\right\}$. Then

$$\begin{aligned}\|\varphi(k)\|^2_{\ell_2(\Gamma)} &= \sum_\gamma |f(k(\gamma)|^2 = \sum_n \sum_{\gamma\in A_n} |f(k(\gamma))|^2\\ &\le \sum_n \operatorname{card}(A_n)\big|f\big(\tfrac{1}{n}\big)\big|^2 \le \sum_n \operatorname{card}(A_n)\cdot 2^{-n}N\big(\tfrac{1}{n+1}\big)^{-1}\\ &\le \sum_n N\big(\tfrac{1}{n+1}\big)2^{-n}N\big(\tfrac{1}{n+1}\big)^{-1} = \sum_n 2^{-n} = 1.\end{aligned}$$

Thus φ maps $\tilde{K}$ into $B_{\ell_2(\Gamma)}$ and it is clearly one-to-one and continuous in pointwise topologies of $c_0(\Gamma)$ and $\ell_2(\Gamma)$, which coincide with weak topologies on bounded sets in these spaces. □

Theorem 14.15 (Benyamini, Rudin, Wage [BRW]) *Let K be a compact space. The following are equivalent:*
(i) *K is uniform Eberlein.*
(ii) *There is a Hilbert space H and a bounded operator T from H into $C(K)$ such that $T(H)$ is dense in $C(K)$.*
(iii) *$(B_{C(K)^*}, w^*)$ is uniform Eberlein compact.*

Proof: (i)$\Longrightarrow$(ii): Similarly as in the proof of (ii)$\Longrightarrow$(i) in Theorem 14.14 we show that K is homeomorphic to a weakly compact subset $\tilde{K}$ of $\ell_2(\Gamma)$ for some Γ such that $0 \notin \tilde{K}$ and $\sum_{\gamma\in\Gamma} |x(\gamma)| \le 1$ for every $x \in \tilde{K}$. Set $A_0 = \emptyset$ and for $n \in \mathbb{N}$ put $\mathcal{A}_n = \Gamma \times \cdots \times \Gamma$ (n times), let $\mathcal{A} = \bigcup_{n=0}^{\infty} A_n$ and $H = \left(\sum_{\mathcal{A}} \ell_2(\Gamma)\right)_2$. Define an operator T from H into $C(K)$ as follows: For $h = (h_A)_{A\in\mathcal{A}} \in H$ and $x \in K$ we put

$$T(h)(x) = \sum_n 2^{-n}\Big(\sum_{A=(\alpha_1,\dots,\alpha_n)\in\mathcal{A}_n} \prod_{i=1}^n x(e_{\alpha_i})x(h_A)\Big),$$

where x is considered an element of $\ell_2(\Gamma) = \ell_2(\Gamma)^*$ and $\{e_\gamma\}$ is the canonical basis of $\ell_2(\Gamma)$.

Then T is a bounded operator because for every $x \in K$ and $\|h\| \le 1$ we have $|x(h_A)| \le \|h_A\| \le \|h\| \le 1$ for all A and thus, since $\sum_{\gamma\in\Gamma} |x(\gamma)| \le 1$, we have

$$\begin{aligned}\Big|\sum_{A=(\alpha_1,\dots,\alpha_n)\in\mathcal{A}_n} \prod_{i=1}^n x(\alpha_i)x(h_A)\Big| &\le \sum_{A=(\alpha_1,\dots,\alpha_n)\in\mathcal{A}_n} \prod_{i=1}^n |x(e_{\alpha_i})|\\ &= \Big(\sum_{\gamma\in\Gamma} |x(e_\gamma)|\Big)^n \le 1.\end{aligned}$$

The set $T(H)$ contains all polynomials in the coordinate functionals $x(e_\alpha)$. These polynomials separate points in $\tilde{K}$ and do not have a common zero because $0 \notin \tilde{K}$. Therefore $T(H)$ is dense in $C(K)$ by the Stone–Weierstrass theorem.

(ii)$\Longrightarrow$(iii): The dual operator T^* of the operator from (ii) maps $C(K)^*$ one-to-one and w^*–w^*-continuously into $H^* = H$. Therefore $B_{C(K)^*}$ is homeomorphic to a weakly compact subset of a Hilbert space.

(iii)$\Longrightarrow$(i): This follows from the fact that K is w^*-closed in $B_{C(K)^*}$. □

The following result was motivated by [MMOT].

Theorem 14.16 ([FGZ], see, e.g., [HMVZ, Chapter 6]) (i) *A Banach space X admits an equivalent 'UG-smooth norm if and only if (B_{X^*}, w^*) is uniform Eberlein compact.*
(ii) *Let K be a compact space. $C(K)$ admits an equivalent UG-smooth norm if and only if K is uniform Eberlein.*

The proof is based on the following result:

Lemma 14.17 *Let X be a Banach space. If X admits an equivalent UG-smooth norm, then X is a WCD space.*

Proof: (see Fig. 14.1) By Exercise 7.30, we have that $f_n - g_n \xrightarrow{w^*} 0$ whenever $f_n, g_n \in S_{X^*}$ are such that $\|f_n + g_n\| \to 2$.

For $\varepsilon > 0$ and $n \in I\!N$, put

$$B_n^\varepsilon = \left\{x \in B_X : |(f - g)(x)| < \varepsilon, \text{ if } f, g \in B_{X^*} \text{ satisfy } \|f + g\| > 2 - \frac{1}{n}\right\}$$

We have for every $\varepsilon > 0$ that $\bigcup_n B_n^\varepsilon = B_X$. We claim that for each $\varepsilon > 0$ and each n,

$$\overline{B_n^\varepsilon}^{w^*} \subset X + 2\varepsilon B_{X^{**}}$$

Indeed, if not, take $x_0^{**} \in \overline{B_n^\varepsilon}^{w^*} \subset X^{**}$ with the distance greater than 2ε from X. Then take $F \in S_{X^{***}}$ such that F equals to 0 on X and $F(x_0^{**}) > 2\varepsilon$.

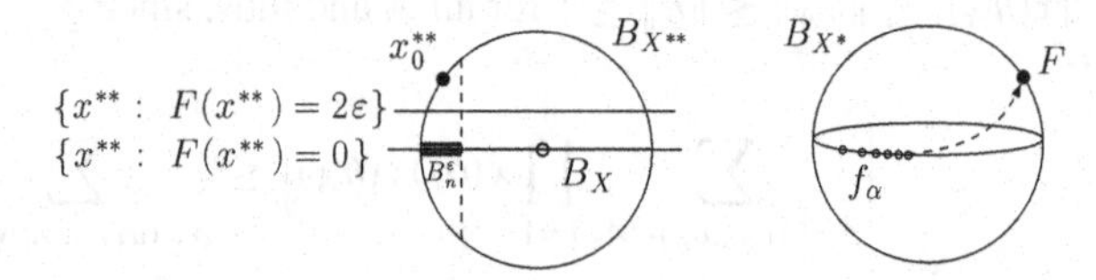

Fig. 14.1 The construction in Lemma 14.17

Let $f_\alpha \in S_{X^*}$ be such that $f_\alpha \to F$ in the w^*-topology of X^{***}.

Then $\|f_\alpha + f_\beta\| \to 2$ and thus $|(f_\alpha - f_\beta)(x)| < \varepsilon$ for all $x \in B_n^\varepsilon$ for large α, β. As f_α is w^*-convergent to F, we have $|(f_\alpha - F)(x)| \le \varepsilon$ for all $x \in B_n^\varepsilon$ for large α. Since $F = 0$ on X in particular on B_n^ε, we have $|f_\alpha(x)| \le \varepsilon$ for every $x \in B_n^\varepsilon$ and thus $|f_\alpha(x_0^{**})| \le \varepsilon$ for large α from the continuity of f_α in (X^{**}, w^*). Since

$f_\alpha \to F$ in (X^{***}, w^*) we get $|F(x_0^{**})| \leq \varepsilon$, which is a contradiction. The existence of the countable family $\{\overline{B_n^{1/m}}^{w^*}, n, m \in \mathbb{N}\}$ ensures that X is WCD. □

Proof of Theorem 14.16: (Sketch) If (B_{X^*}, w^*) is uniform Eberlein compact, then by Theorem 14.15 there is a bounded operator T from a Hilbert space H onto a dense set in $C\big((B_{X^*}, w^*)\big)$. By a standard transfer renorming technique, $C\big((B_{X^*}, w^*)\big)$ admits an equivalent uniformly Gâteaux differentiable norm and so does its subspace X.

If X admits an equivalent uniformly Gâteaux differentiable norm, then, by Lemma 14.17, X is WCD, and thus it admits a projectional resolution of the identity (see Exercise 13.31 and Theorem 13.11). Then, by [HMVZ, Theorem 6.30], (B_{X^*}, w^*) is uniform Eberlein compact. □

We remark that a Banach space X admits an equivalent UG-smooth norm if X admits a UG bump ([Tan]).

Proposition 14.18 *Let $M \subset X$ be a bounded set that satisfies that $\sup_{x\in M}(f_n - g_n)(x) \to 0$ whenever $f_n, g_n \in S_{X^*}$ are such that $\|f_n + g_n\| \to 2$. Then M is weakly relatively compact.*

Proof: Given $\varepsilon > 0$ build, according to the proof of Lemma 14.17, the sequence $\{B_n^\varepsilon\}_{n=1}^\infty$. Then, due to the hypothesis, there exists $n_0 \in \mathbb{N}$ such that $M \subset B_n^\varepsilon$ for all $n \geq n_0$. This implies that $\overline{M}^{w^*} \subset \bigcap_{\varepsilon>0}(X + \varepsilon B_{X^{**}})$ $(= X)$. Since M is bounded, it is w-relatively compact. □

Note that if $M = B_X$, then Proposition 14.18 gives a proof to the fact that X is reflexive if it has a uniformly Fréchet differentiable norm.

14.3 Scattered Compact Spaces

Definition 14.19 *A compact space K is said to be* scattered compact *if every closed subset $L \subset K$ has an isolated point in L.*

Recall that a point p is isolated in K if there is a neighborhood U of p in K such that $U \cap K = \{p\}$. Note that a compact space K is scattered if and only if every subset of K has a relatively isolated point.

Recall the definition of the *Cantor derivative* of a set K: $K^{(0)} = K$, $K^{(1)} = K'$, the set of all cluster points of K, i.e., the points that are not isolated. If α is an ordinal and $K^{(\beta)}$ are defined for all $\beta < \alpha$, then we put $K^{(\alpha)} = (K^{(\beta)})'$ for $\alpha = \beta + 1$ and $K^{(\alpha)} = \bigcap_{\beta<\alpha} K^{(\beta)}$ for a limit ordinal α. The *height* $\eta(K)$ of K is the least ordinal β for which the Cantor derivative $K^{(\beta)}$ is empty.

A compact set K is scattered if and only if $K^{(\gamma)} = \emptyset$ for some ordinal γ. Indeed, if K is not scattered, there is a nonempty perfect set $L \subset K$, i.e., a closed set whose all points are cluster points of L. Then $L \subset \bigcap_\alpha K^{(\alpha)}$ and thus $K^{(\alpha)} \neq \emptyset$ for all

α. If $L = \bigcap_\alpha K^{(\alpha)} \neq \emptyset$, then L is a perfect set in K, so K is not scattered by the definition.

If K is a scattered compact space and an ordinal λ is the first ordinal such that $K^{(\lambda)} = \emptyset$, then λ is not a limit ordinal. Indeed, by the compactness of K we have that $K^{(\lambda)} = \bigcap_{\beta<\lambda} K^{(\beta)} \neq \emptyset$ for all limit ordinals. Thus $\lambda = \beta + 1$ and β is the last ordinal such that $K^{(\beta)} \neq \emptyset$. From the compactness of $K^{(\beta)}$ we get that then $K^{(\beta)}$ is a finite set in K.

Lemma 14.20 *A continuous image of a scattered compact space is scattered.*

Proof: Let K be a scattered compact space and let f be a continuous mapping of K onto a compact space L. Assume that P is a perfect subset of L. Consider the family $\mathcal{A}$ of all compact subsets A of K such that $f(A) = P$ ordered by inclusion. If $\{A_\alpha\}$ is a chain in $\mathcal{A}$, then by the compactness, $\bigcap A_\alpha \neq \emptyset$, and $f(\bigcap A_\alpha) = P$. Let B be a minimal set in $\mathcal{A}$. Since K is scattered and B is compact, B contains a point q that is isolated in B. Put $B' = B\backslash\{q\}$. Then B' is compact and the minimality of B gives that $f(B')$ is a proper subset of P. Note that $f(q) \notin f(B')$ as otherwise $f(B') = f(B) = P$. Since $f(B')$ is compact, $f(q)$ is not a cluster point of $f(B')$. Since $P\backslash f(B') = f(q)$, we get that $f(q)$ is not a cluster point of P either. Therefore the set P contains a point $f(q)$ that is isolated in P. This shows that P is not perfect, a contradiction. □

Lemma 14.21 *A compact space K is countable if and only if K is metrizable and scattered.*

Proof: Assume that K is countable. For every $(x, y) \in K \times K$ with $x \neq y$, the set $H_{x,y} = \{f \in C(K) : f(x) = f(y)\}$ is a closed hyperplane in $C(K)$. By the Baire category theorem, there exists $f \in C(K)$ that belongs to no $H_{x,y}$, hence f is a one-to-one mapping of K into $\mathbb{R}$. Therefore K is metrizable. Any closed subset L of K is countable and has isolated points in L by the Baire category theorem. Hence K is scattered.

Conversely, assume that K is metrizable and scattered. Because the topology of K has a countable basis, it follows that there is a countable ordinal α_0 such that $K^{(\alpha_0)} = \emptyset$. Every set in K is separable and thus $K^{(\alpha)}\backslash K^{(\alpha+1)}$ is countable for any α. Hence K is countable. □

Corollary 14.22 *Let K be a scattered compact space. If f is a continuous mapping from K into $\mathbb{R}$, then $f(K)$ is countable.*

Proof: $f(K)$ is metrizable and scattered. □

Lemma 14.23 (Rudin, see, e.g., [DGZ3]) *Let μ be a non-negative regular finite Borel measure on a scattered compact space K. If $\mu(\{p\}) = 0$ for every $p \in K$, then μ vanishes identically on K.*

Proof: Suppose $\mu(K) > 0$. Let an ordinal λ be the last ordinal such that $K^{(\lambda)} \neq \emptyset$. Then $K^{(\lambda)}$ is a finite set. Thus $\mu(K^{(\lambda)}) = 0$ by the assumption on μ. Let α be the first ordinal such that $\mu(K^{(\alpha)}) < \mu(K)$. If $\alpha = \beta + 1$ for some β, then $\mu(K^{(\beta)}) = \mu(K^{(\alpha)}) + \mu(K^{(\beta)} \setminus K^{(\alpha)})$. The set $K^{(\beta)} \setminus K^{(\alpha)}$ contains no infinite compact set by the definition of the Cantor derivative. Hence by the regularity of μ and the property that μ vanishes on all singletons we get that $\mu(K^{(\beta)} \setminus K^{(\alpha)}) = 0$. Thus $\mu(K^{(\alpha)}) = \mu(K^{(\beta)})$, contradicting our choice of α. Hence α is a limit ordinal. By a simple compactness argument, for every open set G which contains $K^{(\alpha)}$ there then exists an ordinal $\beta < \alpha$ such that $K^{(\beta)} \subset G$. Since $\beta < \alpha$, we have $\mu(K^{(\beta)}) = \mu(K)$. Hence $\mu(G) = \mu(K)$ for every open set G containing $K^{(\alpha)}$. From the regularity of μ we get that $\mu(K^{(\alpha)}) = \mu(K)$, a contradiction. □

Theorem 14.24 (Rudin [Rudi1]) *If K is a scattered compact space then $C(K)^*$ is isometric to $\ell_1(\Gamma)$ for some Γ.*

Proof: Let μ be a non-negative regular finite Borel measure on a scattered compact space K. Let S be a collection of all points p of K satisfying $\mu(\{p\}) > 0$. Note that $\{p \in K : \mu(\{p\}) > \varepsilon\}$ is finite for every $\varepsilon > 0$ as μ is a finite measure. Thus S is countable. Define a measure ν on Borel subsets A of K by $\nu(A) = \mu(A \cap S)$. By Lemma 14.23, the measure $\mu - \nu$ vanishes on K and thus $\mu = \nu$. Since $S = \{q_n\}$ is countable, by the general measure theory, $\int_K f \,\mathrm{d}\mu = \int_K f \,\mathrm{d}\nu = \sum_{i=1}^{\infty} c_n f(q_n)$ for every $f \in C(K)$, where $\sum |c_n| < \infty$.

Conversely, given $\{c_n\}$ with $\sum |c_n| < \infty$ and $\{q_n\} \subset K$, the functional F defined for $f \in C(K)$ by $F(f) = \sum c_n f(q_n)$ is a continuous linear functional on $C(K)$. Clearly $\|F\| \leq \sum |c_n|$. To get the opposite inequality, for a finite set $\{q_1, \ldots, q_n\}$ we consider $f \in B_{C(K)}$ such that $f(q_n) = \operatorname{sign}(c_n)$. □

Note that the isometry in Theorem 14.24 is not in general a dual mapping. Indeed, c_0 is not isometric to c (Exercise 3.132), yet c_0^* is isometric to c^* (Exercise 2.31).

Theorem 14.25 *Let K be a compact space. Then K is scattered if and only if $C(K)$ is an Asplund space.*

Proof: Assume that K is a scattered compact space and X is a separable closed subspace of $C(K)$. Choose a dense sequence $\{g_n\}$ in $B_{C(K)}$ and define a mapping $G\colon K \to [-1, 1]^{\mathbb{N}}$ by $G(k) = \{g_n(k)\}$. The compact space $L = G(K)$ is scattered by Lemma 14.20 and metrizable. Thus L is countable by Lemma 14.21, hence $C(L)^*$ is isometric to ℓ_1 by the proof of Theorem 14.24. Therefore $C(L)^*$ is separable. Moreover, X is isometric to a closed subspace of $C(L)$ and X^* is thus separable.

Conversely, assume that K is not scattered and $C(K)$ is an Asplund space. Let P be a perfect subset of K. The restriction of continuous functions on K to P is a bounded operator from $C(K)$ into $C(P)$ which is onto due to the Tietze extension theorem. Therefore $C(P)$ is a quotient of an Asplund space. If Z is a separable closed subspace of $C(P)$, then there is a separable closed subspace W of $C(K)$ that is mapped onto Z by the quotient mapping (Exercise 2.56). Thus Z^* is isomorphic to a closed subspace of W^* (Exercise 2.49). Since W^* is separable, we have that Z^* is separable. Hence $C(P)$ is an Asplund space. In Exercise 7.38 we proved that the supremum norm in $C[0, 1]$ has no point of Fréchet smoothness. As P is perfect, we similarly prove that the supremum norm of $C(P)$ (and hence of Z) has no point of Fréchet smoothness. This is a contradiction. □

The following theorem lists some topological properties of compact sets and $C(K)$ spaces.

Theorem 14.26 ([Rose3]) *Let K be an infinite compact space.*
(i) *$C(K)$ contains an isomorphic copy of c_0.*
(ii) *K has property CCC if and only if every weakly compact set in $C(K)$ is separable if and only if $C(K)$ does not contain any isomorphic copy of $c_0(\Gamma)$, Γ uncountable.*
(iii) *If K contains a non-trivial infinite convergent sequence, then $C(K)$ has a quotient isomorphic to c_0.*
(iv) *If K is scattered, then $C(K)$ contains an isomorphic copy of c_0 that is complemented in $C(K)$.*
(v) *K is scattered if and only if $C(K)$ is saturated with isomorphic copies of c_0.*

Proof: (Sketch) (i) Let $\{U_n\}$ be an infinite sequence of open pairwise disjoint sets in K. For each n, let $f_n \in C(K)$ be such that $f_n(t_n) = 1$ for some $t_n \in U_n$, $0 \leq f_n \leq 1$ and $f_n = 0$ outside U_n. Then consider the operator $T: c_0 \to C(K)$ defined by $T((a_n)) = \sum_n a_n f_n$.

(ii) Assume that S is a nonseparable w-compact symmetric convex subset of $C(K)$. As in Theorem 13.16, we find an uncountable subset M of $S\backslash\{0\}$ such that every sequence of distinct members of M weakly converges to 0 in $C(K)$. Thus there is $\delta > 0$ such that $\widetilde{M} = \{g \in M : \|g\| > \delta\}$ is uncountable. For every $g \in \widetilde{M}$, let $U_g = \{t \in K : \|g(t)\| > \delta/2\}$.

Let $g_1, g_2, \ldots$ be an infinite sequence in $\widetilde{M}$. Then $\{U_{g_i}\}$ is an infinite sequence of open nonempty sets in K. If $t_0 \in \bigcap_{i=1}^{\infty} U_{g_i}$, then $|g_i(t_0)| > \frac{\delta}{2}$ for all i, a contradiction with $g_i \xrightarrow{w} 0$. Hence $\bigcap_{i=1}^{\infty} U_{g_i} = \emptyset$. Then K does not have property CCC due to Lemma 14.8

Thus we showed that K does not have property CCC if $C(K)$ contains a w-compact nonseparable subset. To finish the proof of (ii) we now assume that for some uncountable Γ, $c_0(\Gamma)$ is isomorphic to a subspace of $C(K)$. Let $\{e_\gamma\}$ be the canonical basis of $c_0(\Gamma)$. Then $\{e_\gamma\} \cup \{0\}$ is a nonseparable and w-compact set in $C(K)$.

Finally, if K does not have property CCC, then similarly to (i), $C(K)$ has a subspace isomorphic to $c_0(\Gamma)$ for some uncountable Γ.

(iii) Let $a_n \in K$ be distinct with $\lim a_n = a \in K$. Then $\{a_n\} \cup \{0\}$ is a compact subset of K and $C(\{a_n\} \cup \{a\})$ is isomorphic to c which is in turn isomorphic to c_0 (Exercise 5.16). Then we can use Tietze's extension theorem to show that the restriction of elements of $C(K)$ to $\{a_n\} \cup \{a\}$ is a bounded operator of $C(K)$ onto $C(\{a_n\} \cup \{a\})$.

(iv) Let $\{a_n\}$ be an infinite sequence of isolated points of K. As K is scattered, $C(K)$ is an Asplund space (Theorem 14.25) and thus $B_{C(K)^*}$ is w^*-sequentially compact (see Exercise 14.3). Hence we can assume that $a_n \to a \in K$ for some a. Given $f \in C(K)$, put $\tilde{f}(t) = f(a)$ if $t \neq a_n$ for all n and $\tilde{f}(t) = f(a_n)$ if $t = a_n$. Then $\tilde{f} \in C(K)$, the operator $P: C(K) \to C(K)$ defined by $P(f) = \tilde{f}$ is a projection, and $P(C(K))$ is isomorphic to c which is in turn isomorphic to c_0. As c_0 is not isomorphic to any quotient of ℓ_∞ (Exercise 3.44), we obtain as a corollary that $\beta\mathbb{N}$ does not contain any non-trivial convergent sequence.

(v) If K is not scattered, K contains a perfect set $\widetilde{K}$ and thus there is a continuous mapping $\tilde{\varphi}$ of $\widetilde{K}$ onto $[0, 1]$ (see, e.g., [Lace2]). From Tietze's extension theorem it follows that there is a continuous mapping φ of K onto $[0, 1]$. Using the composition mapping we see that $C[0, 1]$ is isomorphic to a subspace of $C(K)$ and $C[0, 1]$ contains for instance ℓ_2 which does not contain c_0. If K is scattered, we refer to [Lace1] for the proof that $C(K)$ is saturated with subspaces isomorphic to c_0. □

14.4 Weakly Lindelöf Spaces, Property C

Theorem 14.27 *Let A be a set. For $\alpha \in A$ let X_α be separable topological spaces, and let Y be a topological space all points of which are G_δ. For every continuous function $f: \prod_{\alpha\in A} X_\alpha \to Y$ there is a countable subset $S \subset A$ and a continuous function $f_S: \prod_{\alpha\in S} X_\alpha \to Y$ such that $f = f_S \circ p_S$, where p_S denotes the canonical projection of $\prod_{\alpha\in A} X_\alpha$ onto $\prod_{\alpha\in S} X_\alpha$.*

Proof: Let $x \in X = \prod_{\alpha\in A} X_\alpha$. Let $y = f(x)$. As $\{y\}$ is a G_δ set, there exists a sequence $\{V(y, n)\}_{n=1}^\infty$ of open subsets of Y such that $\{y\} = \bigcap_{n=1}^\infty V(y, n)$. Then $f^{-1}(y) = \bigcap_{n=1}^\infty f^{-1}[V(y, n)]$. Let $U(x, n)$ be a basic open subset of X such that $x \in U(x, n) \subset f^{-1}[V(y, n)]$, $n \in \mathbb{N}$. Then $x \in W_x = \bigcap_{n=1}^\infty U(x, n) \subset f^{-1}(y)$. Given a basic open set $U \subset X$, define $S[U] = \{\alpha \in A : \ p_\alpha(U) \neq X_\alpha\}$ (a finite subset of A) and let $S_x = \bigcup_{n=1}^\infty S[U(x, n)]$. Observe that $f(x) = f(x')$ if $x, x' \in X$ satisfy $p_{S_x}(x) = p_{S_x}(x')$.

Fix $a \in X$ and set $a_\alpha = p_\alpha(a)$ for $\alpha \in A$. Given a countable subset S of A, we define the mapping $e: \prod_{\alpha\in S} X_\alpha \to \prod_{\alpha\in A} X_\alpha$ so that for every $h \in \prod_{\alpha\in S} X_\alpha$, $e(h)$ is the following "extension" of h:

$$\begin{cases} p_S\big(e(h)\big) = h, \\ p_\alpha\big(e(h)\big) = a_\alpha, \alpha \in A\backslash S. \end{cases}$$

Set $S_0 = S_a$. By induction, assume that $S_0 \subset S_1 \subset \cdots \subset S_n$ have been defined. Let D_n be a countable dense subset of $\prod_{\alpha\in S_n} X_\alpha$. Define $S_{n+1} = S_n \cup \bigcup_{d_n\in D_n} S_{e(d_n)}$. Note that all S_n are countable subsets of A. Finally, put $S = \bigcup_{n=0}^\infty S_n$ and $D = \{a\} \cup \bigcup_{n=0}^\infty e(D_n)$. It is easy to see that $E = p_S(D)$ is dense in $\prod_{\alpha\in S} X_\alpha$.

Claim:
If $x, y \in X$ satisfy $p_S(x) = p_S(y)$, then $f(x) = f(y)$.

Proof: It is enough to check $f(x) = f\big(e(p_S(x))\big)$ for all $x \in X$. To this end, first note that $f\big(p_S^{-1}(h)\big) = f\big(e(h)\big)$ for all $h \in E$. This is a consequence of the following observation: if $x \in p_S^{-1}(h)$ then $p_{S_{e(h)}}\big(e(h)\big) = p_{S_{e(h)}}(x)$.

Now, given $h \in E$ and $x \in X$, define $x_h \in X$ so that

$$\begin{cases} p_S(x_h) = h, \\ p_{A\setminus S}(x_h) = x. \end{cases}$$

We have $f(x_h) = f\big(e(h)\big)$. As E is dense in $\prod_{\alpha\in S} X_\alpha$, there exists a net $\{d_i\}$ in D such that $p_S(d_i) \to p_S(x)$. It follows that $x_{p_S(d_i)} \to x$ in X. Since f is a continuous function on X, $f(x_{p_S(d_i)}) \to f(x)$. We also have $e(x_{p_S(d_i)}) \to e\big(p_S(x)\big)$, hence $f(x_{p_S(d_i)}) = f\big(e(x_{p_S(d_i)})\big) \to f\big(e(p_S(x))\big)$, which proves the Claim. $\triangle$

Thus we can define $f_S\colon \prod_{\alpha\in S} X_\alpha \to Y$ such that $f = f_S \circ p_S$. Clearly f_S is continuous. $\square$

Theorem 14.28 *Let X be a Banach space and let W be the family of all real-valued functions on X that are w-continuous. Then* $\mathrm{card}(W) = \mathrm{card}(X^*)$.

Proof: Let A be an algebraic basis of X^*. We may assume that X is infinite-dimensional, so $\mathrm{card}(A) \geq 2^{\mathbb{N}}$ (Exercise 4.26). As (X, w) can be canonically identified with a dense subspace of $\mathbb{R}^A$ endowed with the product topology, there are as many w-continuous functions on X as continuous functions on $\mathbb{R}^A$. By Theorem 14.27, given a real-valued continuous function f on $\mathbb{R}^A$ there exists a countable subset A_0 of A and a factorization $f = f_0 \circ p_{A_0}$, where $f_0\colon \mathbb{R}^{A_0} \to \mathbb{R}$ is also continuous. As $\mathbb{R}^{A_0}$ is separable, there are $2^{\aleph_0}$ real continuous functions on it. We have $\mathrm{card}(A) = \mathrm{card}(X^*)$, so the number of countable subsets of A is also $\mathrm{card}(X^*)$. The statement now follows. $\square$

Recall that a topological space T is called *Lindelöf* if every open cover of T has a countable subcover. We will show (Theorem 14.31) that every WCG space is Lindelöf in its weak topology (we say that the space is *weakly Lindelöf*).

Theorem 14.29 (Orihuela [Orih]) *Let X be a Banach space. Assume that for every mapping ϕ from X into finite subsets of X^*, X admits a norm-one projection P of X onto $P(X)$ such that $P(X)$ is separable and for some countable dense set $A \subset P(X)$, $P^*(f) = f$ for all $f \in \phi(A)$. Then X is weakly Lindelöf.*

Proof: Consider X in its weak topology. Assume that $\{V_\alpha\}_{\alpha\in\Gamma}$ is an open cover of X. For every $x \in X$, let $r_x > 0$ be the supremum of all positive numbers r such that $B_X^O(x, r)$, the open ball of radius r centered at x, lies in some V_α. Choose $V_x \in \{V_\alpha\}$ so that $B_X^O(x, \frac{r_x}{2}) \subset V_x$ and assume that V_x is formed by the intersection of half-spaces given by a finite set K_x of functionals. Define $\phi(x) = K_x$.

By our assumption, there is a projection P of X onto $P(X)$ and an appropriate set $A \subset P(X)$ constructed for the mapping ϕ. We claim that $\{V_z :\ z \in A\}$ is a cover of X, which will complete the proof.

Choose any $x \in X$ and let $B_X^O(P(x), r) \subset V_{P(x)}$. Find $z \in A$ such that $z \in B_X^O(P(x), \frac{1}{10}r)$. Then $r_z > \frac{9}{10}r$ and thus $B_X^O(z, \frac{2}{5}r) \subset V_z$. Hence $P(x) \in V_z$. As $\phi(z) \subset P^*(X^*)$, it follows that $x \in V_z$. Indeed, if, say $|f(Px - z)| < \varepsilon$ for all $f \in \phi(z)$, then $|f(x-z)| = |P^*(f)(x-z)| = |f(P(x) - P(z))| = |f(P(x) - z)| < \varepsilon$. This shows that $\{V_z\}$ is a cover of X. □

Theorem 14.30 (Amir, Lindenstrauss [AmLi]) *Let X be a weakly compactly generated space. If ϕ is a mapping that assigns to each $x \in X$ a finite set $\phi(x) \in X^*$, then there is a norm-one projection P of X onto $P(X)$ such that $P(X)$ is separable, and there is a countable dense set $A \subset P(X)$ such that $P^*(f) = f$ for all $f \in \phi(A)$.*

Proof: Similar to the beginning of the proof of Theorem 13.11, by using Proposition 13.13. The only adjustment needed is that we play the "exhaustion game" also in X. □

From Theorems 14.29 and 14.30 we obtain (see also Exercise 13.30 and Theorems 14.42 and 14.46):

Theorem 14.31 (Preiss, Talagrand [Tala3]) *Every weakly compactly generated space is weakly Lindelöf.*

A Banach space in its weak topology is Lindelöf if and only if it is paracompact if and only if it is normal ([Batu], [Rezn]).

We will see (Theorem 14.39) that the Lindelöf property is not a three-space property. This is not the case with the following (weaker) property which was introduced by Corson in [Cors1]:

Definition 14.32 *Let X be a Banach space. A closed convex subset M of X is said to have* property C *if for every family $\mathcal{A}$ of closed convex subsets of M with empty intersection there is a countable subfamily $\mathcal{B}$ of $\mathcal{A}$ with empty intersection.*
We say that X has property C if the set X has property C.

In other words, X has property C if and only if every family $\mathcal{F}$ of complements of closed convex sets in X that covers X has a countable subfamily that covers X. Hence every X which is weakly Lindelöf has property C.

Note that property C passes to closed subspaces and quotients.

Theorem 14.33 (Pol [Pol3]) *Let Y be a closed subspace of a Banach space X. If both Y and X/Y have property C, then X has property C.*

In the proof we will use the following statements:

Lemma 14.34 *A Banach space X has property C if the following condition is true:*
($*$) *If a family $\mathcal{K}$ of nonempty closed convex sets in X is closed under countable intersections, then for every $\sigma > 0$ there is $a \in X$ with* $\operatorname{dist}(a, C) < \sigma$ *for every $C \in \mathcal{K}$.*

Proof: Let $\mathcal{C}$ be a family of nonempty closed convex subsets of X closed under countable intersections. We need to show that $\bigcap \mathcal{C} \neq \emptyset$.

To this end define inductively a sequence $a_i \in X$ and nonempty closed convex sets C^i for every $C \in \mathcal{C}$ in such a way that

(1) $C = C^0 \supset C^1 \supset C^2 \dots$, $\operatorname{diam} C^{i+1} \le 2^{-i}$,

(2) every collection $\mathcal{C}^i = \{C^i : C \in \mathcal{C}\}$ is closed under countable intersections,

(3) $\operatorname{dist}(a_i, C^i) < 2^{-i-1}$, $i = 0, 1, 2, \dots$

We choose a_0 by using $(*)$ with $\mathcal{K} = \mathcal{C}$ and $\sigma = 2^{-1}$. Assume that C^j and a_j are chosen, $j = 0, 1, 2, \dots, i$. For every $C \in \mathcal{C}$ put $C^{i+1} = C^i \cap (a_i + 2^{-i-1} B_X)$. The sets C^{i+1} are closed, convex, and nonempty by (3), the condition (1) is satisfied and the family $\mathcal{C}^{i+1} = \{C^{i+1} : C \in \mathcal{C}\}$ satisfies (2) as $\mathcal{C}^i$ does. We complete the inductive step by choosing a_{i+1} by applying $(*)$ to $\mathcal{K} = \mathcal{C}^{i+1}$ and $\sigma = 2^{-i-2}$.

The points a_i form a Cauchy sequence, as by (3) and (1) we have

$$\|a_i - a_{i+1}\| \le 2^{-i-1} + 2^{-i+1} + 2^{-i-2} \le 3 \cdot 2^{-i+1}$$

for $i \ge 1$. The limit point of $\{a_i\}$ yields the conclusion. □

Lemma 14.35 *If a Banach space X does not have property C, then there exists $\varepsilon > 0$ and a family $\mathcal{C}$ of nonempty closed convex subsets of B_X closed under countable intersections such that for every closed convex subset M of X with property C there is a $C_M \in \mathcal{C}$ with* $\operatorname{dist}(M, C_M) \ge \varepsilon$.

Proof: As X does not have property C, by Lemma 14.34 the property $(*)$ fails for some $\mathcal{K}$ and some $\sigma > 0$. Since $\mathcal{K}$ is closed under countable intersections, there is a natural number n such that each member of $\mathcal{K}$ intersects the ball nB_X. Put $\mathcal{C} = \left\{\frac{1}{n}(C \cap nB_X) : C \in \mathcal{K}\right\}$ and $\varepsilon = \frac{\sigma}{n}$. The family $\mathcal{C}$ consists of nonempty closed convex subsets of B_X, it is closed under countable intersections and for every $x \in X$ there is $C_x \in \mathcal{C}$ with $\operatorname{dist}(x, C_x) > \varepsilon$.

Let M be a closed convex subset of X with property C and suppose by contradiction that $\operatorname{dist}(M, C) < \varepsilon$ for $C \in \mathcal{C}$. Put $C' = (C + \varepsilon B_X) \cap M$ for every $C \in \mathcal{C}$. The sets C' are closed convex nonempty.

Given a countable collection $\mathcal{A} \subset \mathcal{C}$ we have $(\bigcap \mathcal{A})' \subset \bigcap\{C' : C \in \mathcal{A}\}$ (note that $\bigcap \mathcal{A} \in \mathcal{C}$). Hence by property C of M there is $x \in \bigcap\{C' : C \in \mathcal{C}\}$. But then $\operatorname{dist}(x, C) \le \varepsilon$ for every $C \in \mathcal{C}$, a contradiction. □

Proof of Theorem 14.33: Let q be the quotient mapping of X onto X/Y. Assume that X does not have property C. Find $\varepsilon > 0$ and $\mathcal{C}$ by Lemma 14.35. Since for every $z \in X/Y$, $q^{-1}(z)$ has property C (as Y has property C), there is a set $C_z \in \mathcal{C}$ with $\operatorname{dist}(q^{-1}(z), C_z) \ge \varepsilon$. Thus $z \notin \overline{q(C_z)}$, hence $\bigcap\{\overline{q(C)} : C \in \mathcal{C}\} = \emptyset$. This is a contradiction with property C of X/Y. □

Theorem 14.36 (Corson [Cors1]) *The space $C[0, \omega_1]$ does not have property C.*

Proof: Observe that the hyperplane $L = \{x \in C[0, \omega_1] : x(0) = 0\}$ of $C[0, \omega_1]$ is isomorphic to $C[0, \omega_1]$. Since all hyperplanes of a given Banach space are isomorphic (Exercise 2.9), the hyperplane $C_0[0, \omega_1] = \{x \in C[0, \omega_1] : x(\omega_1) = 0\}$ is isomorphic to $C[0, \omega_1]$. Therefore it suffices to show that $C_0[0, \omega_1]$ does not have property C. For $\alpha < \omega_1$, define $x_\alpha \in C_0[0, \omega_1]$ by

$$x_\alpha(\beta) = \begin{cases} 0 & \text{if } \beta > \alpha, \\ 1 & \text{if } \beta \le \alpha. \end{cases}$$

For $\alpha \in [0, \omega_1)$, let $K_\alpha = \{x \in C_0[0, \omega_1] : \|x - x_\alpha\| \le \frac{1}{2}\}$. If $\alpha_i < \omega_1$ for $i \in \mathbb{N}$ and $\sup(\alpha_i) < \beta < \omega_1$, then $\frac{1}{2} x_\beta \in \bigcap_{i=1}^{\infty} K_{\alpha_i}$. Hence $\bigcap_{i=1}^{\infty} K_{\alpha_i} \neq \emptyset$. If $x \in \bigcap_{\alpha < \omega_1} K_\alpha$, then $x(\alpha) = \frac{1}{2}$ for all $0 < \alpha < \omega_1$, which is impossible by the definition of $C_0[0, \omega_1]$. Thus $\bigcap_{\alpha < \omega_1} K_\alpha = \emptyset$. □

Theorem 14.37 (Pol [Pol3]) *A Banach space has property C if and only if for every $A \subset B_{X^*}$ and $f \in \overline{A}^{w^*}$ there is a countable subset B of A such that $f \in \overline{\operatorname{conv}}^{w^*}(B)$.*

In the proof we will use the following lemma:

Lemma 14.38 *Assume that a Banach space X does not have property C. Then there is $A \subset B_{X^*}$ and $\varepsilon > 0$ such that*
(1) for every closed subspace M of X with property C there is $x^* \in A$ vanishing on M,
(2) for every countable $B \subset A$ there is $x \in B_X$ with $x^*(x) \ge \varepsilon$ for all $x^* \in B$.

Proof: Find $\mathcal{C}$ and $\varepsilon > 0$ by Lemma 14.35. Let $\mathcal{M}$ be the family of all closed subspaces of X with property C. Given $M \in \mathcal{M}$, by Lemma 14.35 there is $C_M \in \mathcal{C}$ such that $\operatorname{dist}(M, C_M) \ge \varepsilon$. Let B_X^O be the open unit ball of X. Since $(M + \varepsilon B_X^O) \cap C_M = \emptyset$, by the separation theorem there is $x_M^* \in S_{X^*}$ such that

$$\sup\{x_M^*(x) : x \in M + \varepsilon B_X^O\} \le \inf\{x_M^*(x) : x \in C_M\}.$$

As M is a closed subspace of X and $x_M^* \in S_{X^*}$, we have

$$x_M^*\big|_M = 0 \text{ and } x_M^*(x) \ge \varepsilon \text{ for every } x \in C_M. \tag{14.1}$$

Put $A = \{x_M^* : M \in \mathcal{M}\}$. Clearly (1) is satisfied. If $B = \{x_{M_1}^*, x_{M_2}^*, \dots\}$ is a countable subset of A, we take $x \in \bigcap_i C_{M_i}$ and (2) follows from the second part of (14.1). □

Proof of Theorem 14.37: Assume that X does not have property C. Find A by Lemma 14.38. Then by considering finite-dimensional subspaces of X we can see that $0 \in \overline{A}^{w^*}$ by (1) in Lemma 14.38. However, by (2) in Lemma 14.38, $0 \notin \overline{\operatorname{conv}}^{w^*}(B)$ for any $B \subset A$, B countable.

Assume now that X has property C. Let $A \subset B_{X^*}$ and $f \in \overline{A'}^{w^*}$. For $x^* \in A$ put $C_{x^*} = \{x \in X : x^*(x) \ge f(x) + 1\}$. The sets C_{x^*} are closed convex and $\bigcap\{C_{x^*} :$

$x^* \in A\} = \emptyset$. Hence there is a countable set $B \subset A$ with $\bigcap\{C_{x^*} : x^* \in B\} = \emptyset$. Then $f \in \overline{\text{conv}}^{w^*}(B)$. Indeed, otherwise, by the separation theorem, there is $x \in X$ such that $g(x) \geq f(x) + 1$ for all $g \in \overline{\text{conv}}^{w^*}(B)$. Thus $x \in \bigcap\{C_{x^*} : x^* \in B\}$, a contradiction. □

By a "two-arrow space" we mean the compact space $K = \{x \in [0, 1]\} \cup \{x^+ : x \in [0, 1]\}$ ordered by the lexicographic order, i.e., $x < x^+ < y$ whenever $x < y$ in $[0, 1]$. The topology of K is the order topology, that is, a basis is given by the intervals $[x^+, y)$ and $(z, x]$ (see, e.g., [DGZ3] or [Fabi1]).

Let D be the subspace of $\ell_\infty[0, 1]$ formed by all bounded real-valued functions on $[0, 1]$ that are right continuous and have finite left limits.

Theorem 14.39 (Corson [Cors1]) (i) *D is isomorphic to a $C(K)$ space, where K it the two-arrow space. D is not weakly Lindelöf but has property C.*
(ii) *The quotient $D/C[0, 1]$ is isomorphic to $c_0[0, 1]$.*
(iii) *The dual space D^* is w^*-separable.*

Since both $C[0, 1]$ and $c_0[0, 1]$ are WCG, it follows that the WCG property and the Lindelöf property are not three-space properties.

Proof: (ii) Define $\phi\colon D/C[0, 1] \to c_0[0, 1]$ by $\phi(\hat{x})\colon t \mapsto x(t) - x(t^-)$. ϕ is well defined and $\|\phi(\hat{x})\| = 2\|\hat{x}\|$ for all $x \in D$, hence ϕ is an isomorphism of $D/C[0, 1]$ onto $c_0[0, 1]$ as it has dense range.

(i) For $t \in [0, 1]$ define $x_t \in D$ by $x_t = \chi_{[t,1]}$, where $\chi_{[t,1]}$ denotes the characteristic function of the interval $[t, 1]$. Then $S = \{x_t : t \in [0, 1]\}$ is a closed discrete subset of D in its w-topology. Note that $\text{card}(S) = 2^{\mathbb{N}}$ and thus there are $2^{2^{\mathbb{N}}}$ continuous functions on S. If D were normal in its weak topology, by Tietze's extension theorem we would obtain $2^{2^{\mathbb{N}}}$ distinct continuous functions on D in its w-topology. This is impossible for the following reason:

$D^*/C[0, 1]^\perp$ is isomorphic to $C[0, 1]^*$ and that $C[0, 1]^\perp$ is isomorphic to $c_0[0, 1]^*$. The cardinality of $C[0, 1]^*$ is $2^{\mathbb{N}}$ as $C[0, 1]$ is separable, and the cardinality of $c_0[0, 1]^*$ is $2^{\mathbb{N}}$ as every element of $c_0[0, 1]^*$ has a countable support. Thus the cardinality of D^* is $2^{\mathbb{N}}$. Hence the cardinality of all functions on D that are continuous in the w-topology of D is $2^{\mathbb{N}}$ due to Theorem 14.28. Thus D in its w-topology is not normal and therefore not weakly Lindelöf ([Batu], [Rezn]).

D has property C because it is a three-space property and both $C[0, 1]$ and $c_0[0, 1]$ have it (as WCG spaces).

(iii) To see that D^* is w^*-separable, consider rational points in $[0, 1]$. □

14.5 Weak* Topology of the Dual Unit Ball

Definition 14.40 *A compact space K is said to be* Corson compact *if K is homeomorphic to a subset S of $[-1, +1]^\Gamma$ that is compact in the pointwise topology, and such that for every $x \in S$ we have* $\text{card}\{\gamma \in \Gamma : x(\gamma) \neq 0\} \leq \aleph_0$.

Every Eberlein compact space is clearly Corson. In particular, the dual unit ball of every WCG space is Corson compact in its w^*-topology (Theorem 13.20). We refer to [Fabi1] and [ArMe1] for examples of Corson compact spaces that are not Eberlein.

The space $[0, \omega_1]$ in its usual order topology is not Corson compact as every Corson compact space is angelic (Exercise 14.57) and ω_1 is not a limit of any sequence of smaller ordinals.

Under the Continuum Hypothesis, there are nonseparable Corson compact spaces with property CCC. Such compact spaces do not contain any dense metrizable subset by the proof of Corollary 14.6. The Continuum Hypothesis (CH) asserts that the first uncountable cardinal is the cardinal of the continuum, i.e., $2^{\aleph_0} = \aleph_1$, where $\aleph_0$ is the cardinal of $\mathbb{N}$ and $\aleph_1$ is the first uncountable cardinal.

However, we have the following result:

Theorem 14.41 (Shapirovskii [Shap]) *Every Corson compact space K contains a dense set formed by G_δ-points of K.*

Proof: (Kalenda [Kale1]) We first show that K contains at least one G_δ-point. To this end, let $K \subset [-1, 1]^\Gamma$ and define a partial order on K by $x < y$ if $y = x$ on the support of x. Then use the Zorn lemma to find a maximal element x_0 in K. By the definition of the partial order, $\{x_0\} = \{z \in K : z = x_0 \text{ on } \operatorname{supp}(x_0)\}$. As $\operatorname{supp}(x_0)$ is countable, it follows that the right-hand-side set is a G_δ set in K. This shows that x_0 is a G_δ-point of K.

Note that if in general $G_1 \supset F_1 \supset G_2 \supset F_2 \supset \dots$, where G_i is open and F_i is closed for all i, then $\bigcap G_i = \bigcap F_i$ is a closed G_δ set.

Let U be an open set in K. By the preceding remark, there is a G_δ closed set H in U. As H is Corson compact, it has a G_δ-point x_1. As H is a G_δ set in K, it follows that x_1 is a G_δ-point of K. □

Theorem 14.42 (Alster, Pol [AlPo], Gul'ko [Gulk1]) *Let X be a Banach space. If (B_{X^*}, w^*) is Corson compact, then X is weakly Lindelöf.*

Proof: If (B_{X^*}, w^*) is Corson compact, then X satisfies the assumption of Theorem 14.29 ([Fabi1]). □

Lemma 14.43 *Let $K \subset [0, 1]^\Gamma$ be a compact set such that* $\operatorname{card}\{\gamma \in \Gamma : x(\gamma) \neq 0\} \leq \aleph_0$ *for every $x \in K$. For every infinite countable subset J of Γ there is a countable subset $\tilde{J}$ of Γ such that $J \subset \tilde{J}$ and $R_{\tilde{J}} \subset K$, where*

$$R_{\tilde{J}}(x)(\gamma) = \begin{cases} x(\gamma) & \text{for } \gamma \in \tilde{J}, \\ 0 & \text{for } \gamma \notin \tilde{J}. \end{cases}$$

Denote $K_{\tilde{J}} = R_{\tilde{J}}(K)$. Then the operator defined on $C(K)$ by $P_{\tilde{J}}(f) = f \circ R_{\tilde{J}}$ is a norm-one projection on $C(K)$ with separable range.

Proof: (Sketch) Put $J_0 = J$. Let D_0 be a countable set in K such that $R_{J_0}(D_0)$ is dense in $R_{J_0}(K)$. Put $J_1 = \bigcup_{x \in D_0} \operatorname{supp}(x)$. Then J_1 is countable, we may assume that $J_1 \supset J_0$. Furthermore, $R_{J_0}(K) = R_{J_0}(\{x \in K : \operatorname{supp}(x) \subset J_1\})$.

We continue by induction and obtain an increasing sequence $\{J_i\}$ of countable subsets of Γ such that $R_{J_i}(K) = R_{J_i}(\{x \in K : \operatorname{supp}(x) \subset J_{i+1}\})$ for every i. Put $\tilde{J} = \bigcup J_n$. Note that $R_{\tilde{J}}(x) = \lim R_{J_n}(x)$ for every $x \in K$. Now for every n and every $x \in K$, $R_{J_n}(x) \in R_{J_n}(\{x \in K : \operatorname{supp}(x) \subset \tilde{J}\})$.

Denote $A = \{x \in K : \operatorname{supp}(x) \subset \tilde{J}\}$. Then given $x \in K$, for every n there is $x_n \in A$ such that $R_{J_n}(x) = R_{J_n}(x_n)$. Let y be a cluster point of $\{x_n\}$ in A. Then $R_{J_n}(x) = R_{J_n}(y)$ for every n. Hence $R_{\tilde{J}}(x) = R_{\tilde{J}}(y) = y$ and thus $R_{\tilde{J}}(x) \subset A \subset K$. Therefore $R_{\tilde{J}}(K) \subset K$.

The separability of $P\big(C(K)\big)$ follows from the fact that $[0, 1]^I$ has a base of its pointwise topology of the cardinality $\operatorname{card}(I)$.

Theorem 14.44 ([Orih], [VWZ]) *Let a Banach space X admit a Markushevich basis $\{x_\alpha; f_\alpha\}$. Then the following are equivalent:*
(i) *X is weakly Lindelöf.*
(ii) *X has property C.*
(iii) *(B_{X^*}, w^*) is Corson compact.*
If the conditions (i)–(iii) *are true for a nonseparable Banach space X with a Markushevich basis, then X^* is not w^*-separable.*

Proof: (i)$\Longrightarrow$(ii) is immediate and (iii)$\Longrightarrow$(i) is by Theorem 14.42.

(ii)$\Longrightarrow$(iii): It is enough to show that $\operatorname{card}\{\alpha : f(x_\alpha) \neq 0\} \leq \aleph_0$ for every $f \in B_{X^*}$. Assume by contradiction that there are $f \in B_{X^*}$, $\varepsilon > 0$ and x_β so that $f(x_\beta) > \varepsilon$ for every $\beta < \omega_1$. Define

$$C_\beta = \overline{\operatorname{span}}\{x_\gamma : \beta \leq \gamma < \omega_1\} \cap \{x \in X : f(x) \geq \varepsilon\}.$$

Then C_β is closed and convex for every β and if $x \in \bigcap_{\beta<\omega_1} C_\beta$, then $f_\alpha(x) = 0$ for all $\alpha \in I$ by the definition of a Markushevich basis. Therefore $x = 0$ as $\{f_\alpha\}$ is separating. This contradicts $f(x) \geq \varepsilon$. Hence $\bigcap_{\beta<\omega_1} C_\beta = \emptyset$. On the other hand, if $\beta_i < \omega_1$ for all $i \in \mathbb{N}$ and $\beta < \omega_1$, $\beta > \sup \beta_i$, then $x_\beta \in \bigcap_i C_{\beta_i}$. Thus X does not have property C. □

Definition 14.45 *We will say that a Markushevich basis $\{x_\alpha; f_\alpha\}$ of X is* weakly Lindelöf *if $\{x_\alpha\} \cup \{0\}$ is Lindelöf in its relative weak topology.*

Recall that $\ell_\infty^c(\Gamma)$ denotes the closed subspace of $\ell_\infty(\Gamma)$ formed by all functions that have countable support.

Theorem 14.46 ([Orih], [VWZ]) *Let X be a Banach space. The following are equivalent:*
(i) *(B_{X^*}, w^*) is Corson compact.*
(ii) *There is a set Γ and a bounded one-to-one operator from X^* into $\ell_\infty^c(\Gamma)$ that is w^*-to-pointwise continuous.*

(iii) *X is weakly Lindelöf and admits a Markushevich basis.*
(iv) *X has property C and admits a Markushevich basis.*
(v) *X admits a weakly Lindelöf Markushevich basis.*

Proof: We will sketch a proof under the assumption $\text{dens}(X) = \aleph_1$.

(i)$\Longrightarrow$(iii): If (B_{X^*}, w^*) is Corson then X is weakly Lindelöf by the preceding theorem. X also has a PRI ([Fabi1]), so it has a Markushevich basis by the proof of Theorem 13.16.

(iii)$\Longrightarrow$(iv) is trivial.

(iv)$\Longrightarrow$(ii): Assume that $\{x_\alpha; f_\alpha\}_{\alpha\in\Gamma}$ is a Markushevich basis with $\{x_\alpha\}$ bounded. Define a bounded operator $T\colon X^* \to \ell_\infty(\Gamma)$ by $T(f) = \big(f(x_\alpha)\big)_\alpha$. We claim that T maps X^* into $\ell_\infty^c(\Gamma)$: Let S be the collection of all elements of X^* that have countable support on $\{x_\alpha\}$. Note that S is a closed subspace of X^*. From Theorem 14.37 it follows that $S \cap B_{X^*}$ is w^*-closed in X^*. By the Banach–Dieudonné theorem, S is w^*-closed in X^*. As S contains all f_α and $\text{span}\{f_\alpha\}$ is w^*-dense in X^*, we get that $S = X^*$.

(ii)$\Longrightarrow$(i) is trivial.

Thus (i)–(iv) are equivalent.

(iii)$\Longrightarrow$(v): Note that if $\{x_\alpha; f_\alpha\}$ is a Markushevich basis of a Banach space X, then $\{x_\alpha\}\cup\{0\}$ is weakly closed in X. (v) follows, as a closed subspace of a Lindelöf space is Lindelöf.

(v)$\Longrightarrow$(ii): Let $\{x_\alpha; f_\alpha\}_{\alpha\in\Gamma}$ be a weakly Lindelöf Markushevich basis of X. Given $\varepsilon > 0$ and $f \in X^*$, let $U(f,\varepsilon) = \{x \in X : |f(x)| < \varepsilon\}$ and for $\alpha \in \Gamma$, let $U_\alpha = \{x \in X : |f_\alpha(x)| > 0\}$. By the hypothesis, the cover $\{U_\alpha, U\}$ of $\{x_\alpha\}\cup\{0\}$ has a countable subcover. Since $x_\alpha \notin U_\beta$ if $\alpha \neq \beta$, it follows that all but countably many x_α are in $U(f,\varepsilon)$. Thus $\text{card}\{\alpha \in \Gamma : f(x_\alpha) \neq 0\} \leq \aleph_0$. Normalize $\{x_\alpha\}$ so that $\{x_\alpha\} \subset B_X$ and define an operator T from X^* into $\ell_\infty^c(\Gamma)$ by $T(f) = \big(f(x_\alpha)\big)_\alpha$. □

The proof of (iv)$\Longrightarrow$(ii) also gives the following:

Proposition 14.47 *Let $\{x_\alpha; f_\alpha\}_{\alpha\in\Gamma}$ be a Markushevich basis of a Banach space X. If (B_{X^*}, w^*) is Corson compact, then* $\text{card}\{\alpha \in \Gamma : f(x_\alpha) \neq 0\} \leq \aleph_0$ *for all $f \in X^*$.*

Corollary 14.48 *Let X be a Banach space. If X^* is w^*-separable and (B_{X^*}, w^*) is Corson compact, then X is separable.*

Proof: By Theorem 14.46, assume that $\{x_\alpha; f_\alpha\}_{\alpha\in\Gamma}$ is a Markushevich basis of X. Let S be a countable set that is w^*-dense in X^*. Then S separates points of X and therefore $\Gamma = \bigcup_{f\in S}\{\alpha \in \Gamma : f(x_\alpha) \neq 0\}$.

Since $\text{card}\{\alpha \in \Gamma : f(x_\alpha) \neq 0\} \leq \aleph_0$ for all $f \in X^*$ by the above proposition, we have that Γ is countable, which proves the corollary. □

Remarks:

(i) Let Y be a nonseparable closed subspace of the space D from Theorem 14.39. Then Y does not have any Markushevich basis as its dual is w^*-separable and Y has property C.

(ii) Under the Continuum Hypothesis, Kunen constructed an uncountable separable scattered compact space K such that every subset of $C(K)$ is weakly Lindelöf and every subset of $C(K)^*$ is w^*-separable ([HSZ], [JiMo], [Negr]). By Theorem 14.46, no nonseparable closed subspace Y of this $C(K)$ admits a Markushevich basis since Y^* is w^*-separable (use the restriction mapping).

Recall that a topological space T is said to be *pseudocompact* if every continuous real-valued function on T is bounded, equivalently, if every continuous real-valued function on T attains its supremum over T.

Theorem 14.49 (Preiss, Simon [PrSi], Pták [Ptak1], Valdivia [Vald2]) *Let X be a Banach space. If $K \subset X$ is pseudocompact in the weak topology of X, then K is weakly compact in X.*

In the proof the following lemma is used:

Lemma 14.50 (Preiss, Simon) *Let x_0 be a non-isolated point in a weakly compact subset K (in its weak topology) of a Banach space X. Then there is a sequence $\{U_n\}$ of open sets in K that converges to x_0 (i.e., for every neighborhood U of x_0 in K there is $n_0 \in \mathbb{N}$ such that $U_n \subset U$ for $n \geq n_0$).*

Proof: By Corollary 13.19, K is homeomorphic to a weakly compact subset of $c_0(\Gamma)$ for some set Γ, considered in its weak topology. We will thus assume that $K \subset c_0(\Gamma)$ and $x_\gamma \in [0, 1]$ for all $x \in K$ and $\gamma \in \Gamma$. Moreover, we may assume that $x_0 = 0$. This is seen by replacing Γ by $\Delta = \Gamma \times \{0, 1\}$ and defining for $x \in K$: $\hat{x}(\gamma, 0) = (x_\gamma - (x_0)_\gamma)_+$, $\hat{x}(\gamma, 1) = ((x_0)_\gamma - x_\gamma)_+$. We use again Γ for the index set.

We will inductively define finite subsets Γ_i of Γ (possibly empty) and open subsets U_n and V_n of K as follows:

Put $\Gamma_1 = \emptyset$, $U_1 = V_1 = K$. If Γ_i, U_i and V_i have been defined for $i = 1, \ldots, n-1$, put $V_n = \left\{x \in K : x_\gamma < \frac{1}{n} \text{ for all } \gamma \in \bigcup_{i=1}^{n-1} \Gamma_i\right\}$.

Consider the two alternatives: either

(1) $x_\gamma \leq \frac{1}{n}$ for all $x \in V_n$, $\gamma \in \Gamma$, or

(2) there exists $x_1 \in V_n$ and $\gamma_1 \in \Gamma$ such that $(x_1)_{\gamma_1} > \frac{1}{n}$.

In case (1), set $\Gamma_n = \emptyset$. In case (2), let $F_1 = \{\gamma_1\}$ and suppose that an infinite strictly increasing sequence $F_1 = \{\gamma_1\} \subset F_2 = \{\gamma_1, \gamma_2\} \subset \ldots$ of finite subsets of Γ and a sequence $x_1, x_2, \ldots$ of elements in V_n can be defined in such a way that $x_m(\gamma_i) > \frac{1}{n}$, $m \in \mathbb{N}$, $i = 1, \ldots, m$. This is clearly impossible, as (x_m) has cluster points in $c_0(\Gamma)$. It follows that in case (2), $m \in \mathbb{N}$ can be found such that if $x \in V_n$ satisfies $x_\gamma > \frac{1}{n}$ for all $\gamma \in F_m$, then $x_\gamma \leq \frac{1}{n}$ for all $\gamma \notin F_m$. Define $\Gamma_n = F_m$.

In both cases, $U_n = \left\{x \in V_n : x_\gamma > \frac{1}{n} \text{ for all } \gamma \in \Gamma_n\right\}$ is a nonempty open subset of V_n such that $x_\gamma \leq \frac{1}{n}$ for all $\gamma \notin \Gamma_n$.

We claim that $\{U_n\}$ converges to 0. Let U be a neighborhood of $x_0 = 0$ in K of the form $U = \{x \in K : x_\gamma < \frac{1}{n_0}$ for all $\gamma \in \Gamma_0\}$ for some finite set Γ_0 in Γ and $n_0 \in \mathbb{N}$. Since $\{\Gamma_i\}$ are disjoint, there is $m_0 > n_0$ such that $\Gamma_m \cap \Gamma_0 = \emptyset$ for all $m \geq m_0$. Then given $m \geq m_0$, from the definition of U_m we get $x(\gamma) \leq \frac{1}{m} \leq \frac{1}{m_0} < \frac{1}{n_0}$ for all $x \in U_m$ and $\gamma \notin \Gamma_m$. It follows that $U_m \subset U$ if $m \geq m_0$. □

Proof of Theorem 14.49: Let K be a pseudocompact set in the weak topology of a Banach space X and $f \in X^*$. Then $\sup_K(f) = \sup_{\overline{\text{conv}}(K)}(f)$ and there is $x_0 \in K$ such that $f(x_0) = \sup_K(f)$. Hence every $f \in X^*$ attains its supremum over $\overline{\text{conv}}(K)$ and $\overline{\text{conv}}(K)$ is thus weakly compact by Theorem 3.130. Thus it suffices to show that K is weakly closed.

Let $x_0 \in \overline{K}^w \setminus K$. By Lemma 14.50, there is a sequence $\{U_n\}$ of open sets in $\overline{K}^w$ that converges to x_0. For $n \in \mathbb{N}$, let $x_n \in U_n \cap K$. Since the weak topology on X is completely regular (Proposition 3.27), there is $f \in C(K)$ such that $f(K \setminus U_n) = 0$, $0 \leq f \leq n$ on K and $f(x_n) = n$. Note that for every $x \in K$ there is a neighborhood V of x in K and $n_0 \in \mathbb{N}$ such that $V \cap U_n = \emptyset$ for all $n \geq n_0$. Therefore $\sum f_n$ is well defined, continuous and unbounded on K, a contradiction with the pseudocompactness of K. □

We already saw in Chapter 3 that (B_X, w) is a compact space if and only if X is reflexive, and (B_X, w) is metrizable if and only if X^* is separable (Theorem 3.111 and Proposition 3.106).

Let X^* be separable, let $\{x_i^*\}$ be norm dense in S_{X^*} and consider on $B_{X^{**}}$ the metric $\rho(x^{**}, y^{**}) = \sum 2^{-i}|x^{**}(x_i^*) - y^{**}(x_i^*)|$. This metric is compatible with the w^*-topology on X^{**}. (B_X, ρ) is a subspace of the compact metric space $(B_{X^{**}}, \rho)$ and is thus totally bounded. So if (B_X, ρ) is complete, then it is compact and the space X is reflexive. However, (B_X, w) can be metrizable by some other complete metric than ρ without X being reflexive. In fact we have the following result.

Theorem 14.51 (Godefroy [Gode1]) *Let X be a Banach space. If X^{**} is separable then (B_X, w) is a Polish space.*

Polish spaces were introduced in Section 5.2 (see also Section 17.9). Note that the unit ball of a separable space X can be a Polish space in its weak topology without X^{**} being separable. An example of this is the predual of the James tree space ([EdWh]).

Proof: Let $X^\perp = \{x^{***} \in X^{***} : x^{***}(x) = 0$ for every $x \in X\}$ and consider $X^\perp$ in the w^*-topology from X^{***}. Since $(B_{X^{***}}, w^*)$ is a metrizable compact space, it is separable and we can find a dense sequence $\{y_n\}$ in $(B_{X^\perp}, w^*)$. By Goldstine's theorem, for every n there is a sequence $\{x_{n,k}^*\}_k \subset B_{X^*}$ such that $x_{n,k}^* \xrightarrow{w^*} y_n$ in X^{***}. By Exercise 2.28, $X = (X^\perp)_\perp$, where $(.)_\perp$ denotes the annihilator in X^{**} of a set in X^{***}. Therefore

$$\begin{aligned} B_X = (B_{X^{**}}) \cap X &= B_{X^{**}} \cap (X^{\perp})_{\perp} \\ &= \{x^{**} \in B_{X^{**}} : \; y_n^*(x^{**}) = 0 \text{ for every } n\} \\ &= \bigcap_{n=1}^{\infty} \bigcap_{m=1}^{\infty} \bigcup_{k=m}^{\infty} \{x^{**} \in B_{X^{**}} : |x^{**}(x_{n,k}^*)| < \tfrac{1}{m}\}. \end{aligned}$$

Hence B_X is a G_δ set in (B_X^{**}, w^*). Since (B_X^{**}, w^*) is a compact metric space, (B_X, w) is metrizable by a complete metric by the Mazurkiewicz theorem ([Royd]). Finally, (B_X, w) is separable as X is a separable Banach space. □

Therefore (B_X, w) is a Baire space if X^{**} is separable. Concerning the Baire property for (B_X, w) we also have the following result.

Proposition 14.52 *Let $(X, \|\cdot\|)$ be a Banach space. If $\|\cdot\|$ has the Kadec–Klee property then (B_X, w) is a Baire space.*

Proof: Assume without loss of generality that X is infinite-dimensional. By Exercise 3.108, S_X is a dense G_δ set in (B_X, w). It is easy to see that a topological space is a Baire space if it contains a dense set which is Baire in the induced topology. In our case this is satisfied as on S_X the weak and norm topology coincide and S_X in the norm topology is a complete metric space. □

Since the canonical norm of ℓ_1 has the Kadec–Klee property (Exercise 5.46), we have that B_{ℓ_1} is a Baire space in its weak topology. The situation in c_0 is different:

Proposition 14.53 *B_{c_0} is not a Baire space in its weak topology.*

Proof: For $n \in \mathbb{N}$, let $B_n = \{x = (x_i) \in B_{c_0} : |x_i| \leq \frac{1}{2} \text{ for } i \geq n\}$. Then B_n is closed in B_{c_0} in its weak topology, which coincides there with the pointwise topology. Moreover, B_n is nowhere dense in this topology. Indeed, if $x \in B_n$ and $p \in \mathbb{N}$, $\varepsilon > 0$ are given, there is $\tilde{x} \in B_{c_0}$ such that $|\tilde{x}_i - x_i| < \varepsilon$ for $i \leq p$ and $|\tilde{x}_{\max\{p+1,n\}}| > \frac{1}{2}$, so $\tilde{x} \notin B_n$. Thus $B_{c_0} = \bigcup_{n=1}^{\infty} B_n$ is of the first category in itself in its weak topology, hence not a Baire space. □

As any separable space, c_0 can be equivalently renormed by a locally uniformly rotund norm (Theorem 13.27). Let B_1 be the unit ball of such a norm on c_0. Assume without loss of generality that $B_{c_0} \subset B_1$. By Proposition 14.52, (B_1, w) is a Baire space. However, (B_1, w) is not metrizable by any complete metric as otherwise (B_{c_0}, w), being a closed subspace of (B_1, w), would be metrizable by a complete metric and thus it would be a Baire space.

Theorem 14.54 (Aharoni, Johnson, Lindenstrauss, [AhLi1], [JoLi1]) *Let $\{N_\gamma\}_{\gamma \in \Gamma}$ be an uncountable collection of infinite subsets of natural numbers such that $N_\gamma \cap N_\beta$ is finite for $\gamma \neq \beta$ (see Lemma 5.7). Let X (called JL_0) be the closed subspace of ℓ_∞ (in the sup-norm) spanned by c_0 and the characteristic functions χ_{N_γ} of the sets N_γ. Then,*

(i) X/c_0 is isometric to $c_0(\Gamma)$, and X is isomorphic to $C(K)$, where K is a separable scattered compact space such that $K^{(3)} = \emptyset$.

(ii) *X is Lipschitz equivalent to $c_0(\Gamma)$.*
(iii) *X admits a C^∞-smooth norm, has property C, and is not weakly Lindelöf.*
(iv) *X does not contain any nonseparable closed subspace with a Markushevich basis. In particular, X does not contain any subspace isomorphic to $c_0(\Gamma)$ or $\ell_p(\Gamma)$ for Γ uncountable.*
(v) *B_{X^*} is w^*-separable. However, there is an equivalent norm on X whose dual ball is not w^*-separable.*

Remark: Since X is not weakly Lindelöf, X and $c_0(\Gamma)$ in their weak topologies are not homeomorphic.

Proof of Theorem 14.54: (Sketch) (i) follows form the fact that for any $\gamma_1, \dots, \gamma_n \in \Gamma$ and $a_1, \dots, a_n$ we have $\left\|\sum_{j=1}^n a_j \chi_{N_{\gamma_j}}\right\|_{X/c_0} = \max |a_j|$. The space K is the disjoint union of $\mathbb{N} \cup \Gamma \cup \{\infty\}$ topologized by letting all points of $\mathbb{N}$ be open sets, and the neighborhoods of $\gamma \in \Gamma$ being the sets that contain γ and sets $N_\gamma \backslash S$ for finite $S \subset N_\gamma$. The point ∞ is then the compactification point of $\mathbb{N} \cup \Gamma$, and K carries the topology of this compactification. It is easy to see that X is isomorphic to $C_0(K)$, which in turn is isomorphic to $C(K)$.

(ii) First we prove that there is a Lipschitz selector of X/c_0 to X, i.e., a mapping $\psi\colon X/c_0 \to X$ such that $q(\psi(\hat{x})) = \hat{x}$ for $\hat{x} \in X/c_0$, where q is the quotient mapping.

To obtain such a selector, we identify $q(\chi_{N_\gamma})$ with e_γ in $c_0(\Gamma)$ and define first such a selector f on $c_0^+(\Gamma)$, the positive functions in $c_0(\Gamma)$, as follows: If $y \in c_0^+(\Gamma)$, write $y = \sum_{j=1}^\infty a_j e_{\gamma_j}$, where $a_1 \ge a_2 \ge \dots$. Put $M_1 = N_{\gamma_1}$ and inductively $M_n = N_{\gamma_n} \backslash \bigcup_{j<n} N_{\gamma_j}$. Note that $\chi_{M_j} \in X$ as it differs from $\chi_{N_{\gamma_j}}$ by an element of c_0. Let $f(y) = \sum a_j \chi_{M_j}$. Note that $q(\chi_{M_j}) = q(\chi_{N_{\gamma_j}}) = e_{\gamma_j}$ and thus $q\big(f(y)\big) = y$.

To see that f is Lipschitz, note that $f(y)_n = a_i$ if and only if $n \in N_{\gamma_i} \backslash \big(N_{\gamma_1} \cup \dots \cup N_{\gamma_{i-1}}\big)$. By the monotonicity of $\{a_i\}$, if $A_n = \overline{\operatorname{span}}\{e_\gamma : n \in N_\mu\}$ then $(f(y))_n = \operatorname{dist}(y, A_n)$ and f is thus Lipschitz. Define a Lipschitz selector for $y \in c_0(\Gamma)$ by $f(y) = f(y^+) - f(y^-)$, where $y = y^+ - y^-$ is the canonical representation of y as a difference of two disjointly supported non-negative terms. We have

$$\begin{aligned}\|f(y) - f(z)\| &\le \|f(y^+) - f(z^+)\| + \|f(z^-) - f(y^-)\| \\ &\le 2\max\big(\|y^+ - z^+\|, \|y^- - z^-\|\big) \le 2\|y - z\|.\end{aligned}$$

Now define a mapping $\varphi\colon X \to c_0 \oplus X/c_0$ by $\varphi(x) = \big(x - \varphi(\hat{x}), \hat{x}\big)$. If $(y, \hat{x}) \in c_0 \oplus X/c_0$ then $\varphi(x + y) = (y, \hat{x})$, so φ is onto. Thus φ is a Lipschitz homeomorphism of X and $c_0 \oplus X/c_0$, which is isomorphic to $c_0 \oplus c_0(\Gamma)$, which is in turn isomorphic to $c_0(\Gamma)$.

(iii) Property C follows from Theorem 14.33. The existence of a C^∞-smooth equivalent renorming of X was proved in [DGZ3]. Pol showed that if K is a separable compact space such that $K^{(\omega_0)} = \emptyset$ and $C(K)$ is weakly Lindelöf, then $C(K)$ is separable ([Pol1]).

(iv) If a nonseparable closed subspace Y had a Markushevich basis, then Y^* could not be w^*-separable as it has property C. However, since K is separable, X^* is w^*-separable and thus, because of the restriction, so is Y^*.

(v) Since K is separable, B_{X^*} is w^*-separable by the Krein–Milman theorem. Let $\|\cdot\|$ be an equivalent norm on X such that its dual norm $\|\cdot\|^*$ is LUR ([DGZ3]). Let B be the unit ball and S the unit sphere for $\|\cdot\|^*$. Assume that there is a countable set $C \subset B$ such that $\overline{C}^{w^*} = B$. By the LUR property, it follows that $S \subset \overline{C}$ and thus X^* would be separable, a contradiction. □

Remarks:

(i) Note that if X is Lipschitz equivalent to c_0, then X is linearly isomorphic to c_0 (Godefroy, Kalton, and Lancien [GKL1]).

(ii) There is a Banach space X such that X^* is w^*-separable and yet there is no equivalent norm on X whose dual unit ball is w^*-separable ([JoLi1]).

(iii) Theorem 14.54 (iv) should be compared to Deville's result that every separable infinite-dimensional Banach space with a C^∞-smooth norm contains an isomorphic copy of c_0 or some ℓ_p, p an even integer (see, e.g., [DGZ3, Ch. V].

(iv) Since the dual unit ball of the space in Theorem 14.54 is weak*-separable, the space X there is isomorphic to a subspace of ℓ_∞.

(v) Note that it follows from (v) in Theorem 14.26 that X is saturated with c_0; however, it does not contain any copy of $c_0(\Gamma)$ for Γ uncountable.

(vi) Similarly to the statement in Theorem 14.54, there is a Banach space $X := JL_2$ such that X/c_0 is isometric to $\ell_2(\Gamma)$, X^* is weak*-separable but is not isomorphic to any subspace of ℓ_∞ ([JoLi1]). Let us mention in this direction that for a separable Banach space X, X^* is separable if and only if, for every equivalent norm in X, the dual ball of X^{**} is weak*-separable [HMVZ, Corollary 8.20].

(vii) We remark that Talagrand proved in [Tala3b] that, under CH, there is a non-separable compact space K so that $(B_{C(K)^*}, w^*)$ is separable.

(viii) We refer to [Ryc] for a characterization of compact spaces K that carry a strictly positive Radon probability measure μ, i.e. such that $\mu(O) > 0$ for every nonempty open set O in K, and for a characterization of compact spaces K such that $L_1(\mu)$ is separable for every Radon probability measure μ on K, in terms of renormings of $C(K)$, respectively $C(K)^*$, spaces.

14.6 Remarks and Open Problems

Remarks

1. Koszmider proved in [Kosz] that there is a zero-dimensional compact space K such that $C(K)$ is not isomorphic to any of its proper subspaces or to proper quotients. He also showed that there is a compact space L such that $C(L)$ is not isomorphic to any $C(S)$ for S a zero-dimensional compact space. (See also Exercise 12.53, the paper by Plebanek [Pleb], and [HMVZ].)

2. There is a WUR Banach space X that does not admit any bounded, one-to-one operator into some space $c_0(\Gamma)$ [ArMe3]. Then the dual norm in X^* is UG smooth (see Exercise 7.30) and thus $(B_{X^{**}}, w^*)$ is uniform Eberlein compact (see Theorem 14.16 (i)), hence X^* is a subspace of the WCG space $C((B_{X^{**}}, w^*))$ (see Theorem 14.9). Note that X^* is not a WCG space, since otherwise X^{**} will inject into $c_0(\Gamma)$ by Theorem 13.20.
3. A Corson compact space is Eberlein if and only if it is homeomorphic to a w^*-compact subset of the dual space of an Asplund space in its w^*-topology, [Steg5], [OSV].
4. If K is a compact space then the following are equivalent. (i) $C(K)$ is Lipschitz equivalent to $c_0(\Gamma)$ for some Γ. (ii) $C(K)$ is uniformly homeomorphic to $c_0(\Gamma)$ for some Γ. (iii) $K^{(\omega_0)} = \emptyset$ (see, e.g., [DGZ3, p. 264] and [BeLi, p. 256]). However, there is a compact space K such that $K^{(\omega_0+1)} = \emptyset$ and $C(K)$ does not homeomorphically embed into any $c_0(\Gamma)$ space, [PHK].
5. If a WCG Banach space of density character ω_1 is Lipschitz equivalent to a subspace of $c_0(\Gamma)$ for some Γ, then it is linearly isomorphic to a subspace of $c_0(\Gamma)$, see [GKL1]. However, there is a WCG space $C(K)$ with $K^{(3)} = \emptyset$ such that $C(K)$ is not isomorphic to a subspace of $c_0(\Gamma)$, see [BelMar] and [Mar].
6. Any scattered Eberlein compact space of height less than or equal to $\omega_0 + 1$ is uniform Eberlein, [BelMar].
7. We refer to, e.g., [HMVZ, Ch. 4 and 8] for the use of biorthogonal systems in descriptive topology and set theory.

Open Problems

1. It is an open problem whether a space $C(K)$ admits a C^∞-norm as soon as it admits a C^1-norm.
2. It is not known whether, for the compact space K defined in Exercise 14.63 (the so-called *Kunen compact space*), $C(K)$ admits a Fréchet differentiable bump.
3. Does there exist a Baire space E, a compact set K and a separately continuous function $f : E \times K \to \mathbb{R}$ with no points of joint continuity?
4. Assume that K is a *Radon–Nikodým compact space*, i.e., homeomorphic to a w^*-compact set in the dual of an Asplund space. Is the continuous image of K Radon–Nikodým compact? For information about this question see, e.g., [Fabi2].

Exercises for Chapter 14

14.1 A compact space K is said to have the *Namioka property* (see [Nami1]) if for every Baire space E and every separately continuous function $f : E \times K \to \mathbb{R}$, there is a dense G_δ subset Ω of E such that f is jointly continuous at each point of $\Omega \times K$. Let $\mathcal{T}_p$ denote the topology of the pointwise convergence on K in the space $C(K)$. Show that the following are equivalent for a compact space K.

(i) K has the Namioka property.
(ii) For any Baire space E and any continuous mapping $\varphi : E \to (C(K), \mathcal{T}_p)$, there exists a dense G_δ subset Ω of E such that $\varphi : E \to (C(K), \|\cdot\|_\infty)$ is continuous at every point of Ω.
(iii) For any Baire space E, for any continuous mapping $\varphi : E \to (C(K), \mathcal{T}_p)$, and for any $\varepsilon > 0$, there exists a nonempty open subset U of E such that $\|\cdot\|_\infty$- diam $\varphi(U) \leq \varepsilon$.
Hint. (i)$\Longleftrightarrow$(ii) The formula $f(x, k) = \varphi(x)(k)$ relates separately continuous functions f on $E \times K$ to continuous mappings $\varphi : E \to (C(K), \mathcal{T}_p)$. Given $x \in E$, we have, by compactness, that f is jointly continuous at each point of $\{x\} \times K$ if and only if $\varphi : E \to (C(K), \|\cdot\|_\infty)$ is continuous at x.
(ii)$\Longrightarrow$(iii) is obvious.
(iii)$\Longrightarrow$(ii): For $x \in E$ we put $w(x) = \|\cdot\|_\infty$-Osc $\varphi(x)$, where $\varphi : E \to (C(K), \|\cdot\|_\infty)$, i.e.,

$$w(x) := \{\inf_U \sup_{y,z \in U} \|\varphi(y) - \varphi(z)\|_\infty : U \text{ a neighborhood of } x \text{ in } E\}.$$

Since any open set of a Baire space is a Baire space, (iii) implies that for every $n \geq 1$, $\mathcal{O}_n := \{x \in E : w(x) < 1/n\}$ is an open and dense set in E, and then $\Omega := \bigcap_n \mathcal{O}_n$ satisfies (ii).

14.2 Show that every metrizable compact space has the Namioka property.
Hint. We use (iii) in Exercise 14.1. Since $C(K)$ is separable, given $\varepsilon > 0$ we can write $C(K) = \bigcup_n B_n$, where B_n are closed balls of radius ε. Since closed balls are $\mathcal{T}_p$-closed, $F_n := \varphi^{-1}(B_n)$ is closed in E for every n and $E = \bigcup_n F_n$. Since E is a Baire space, (iii) follows.

14.3 ([HaSu], [Steg4]) Let X be a Banach space. If X admits a Lipschitz Gâteaux differentiable bump function. Prove then that (B_{X^*}, w^*) is sequentially compact.
Hint. Let $\{f_n\}$ be a bounded sequence in X^*. For $n \in \mathbb{N}$ put $A_n = \overline{\{f_j\}_{j \geq n}}^{w^*}$ and $A = \bigcap A_n$. Define a function on X by $p(x) = \sup\{f(x) : f \in A\}$. By Corollary 7.44, p is Gâteaux differentiable at some point $x_0 \in X$. From Šmulyan's Lemma 7.20, we obtain that there is a unique $f_0 \in A$ such that $f_0(x_0) = p(x_0)$. As f_0 is in all A_n, for each j there is $f_{n_j} \in \{f_i\}$ such that $|f_{n_j}(x_0) - f_0(x_0)| < \frac{1}{j}$. Let f_1 be a w^*-cluster point of $\{f_{n_j}\}$. Then $f_1(x_0) = f(x_0) = p(x_0)$. Moreover, $f_1 \in A$. From the uniqueness of the element of A such that $f_0(x_0) = p(x_0)$ we have that $f_0 = f_1$. By a standard argument, $f_{n_j} \overset{w^*}{\to} f_0$.

14.4 Let K be an Eberlein compact space. Show that K is separable if and only if K is metrizable.
Hint. If K is metrizable, then $C(K)$ is separable and thus $(B_{C(K)^*}, w^*)$ is a metrizable compact space. Since K is homeomorphic to a subspace of $(B_{C(K)^*}, w^*)$, K is separable. If K is separable, then by the Krein–Milman theorem, $(B_{C(K)^*}, w^*)$

is separable as the extreme points of $B_{C(K)^*}$ are points of K, hence w^*-dens$\big(C(K)^*\big) \leq \aleph_0$. Since $C(K)$ is WCG, $C(K)$ is separable by Proposition 13.3. Then $(B_{C(K)}, w^*)$ and thus also K are metrizable.

14.5 Illustrate Theorem 14.3 on $(B_{\ell_2(\Gamma)}, w)$.
Hint. S_X is metrizable by the norm metric as on $S_{\ell_2(\Gamma)}$ the norm and weak topologies coincide.

14.6 A point in a compact space is a G_δ-point if and only if it has a countable base of neighborhoods.
Hint. One direction is obvious. If x_0 is a G_δ-point, say $x_0 = \bigcap_n V_n$, where $\{V_n\}$ is a decreasing sequence of open sets, and U is a (open) neighborhood of x_0, then it must contain some V_n. Otherwise there will exist a decreasing sequence of compact sets with empty intersection.

14.7 Show that the G_δ-points of $(B_{\ell_2(\Gamma)}, w)$, for Γ uncountable, are exactly those in $S_{\ell_2(\Gamma)}$.
Hint. Follows similarly as the proof of Proposition 3.23 for infinite-dimensional spaces. Use Exercise 14.6. Points on the sphere are G_δ since they are exposed.

14.8 Let $K = [-1, 1]^\Gamma$ in its pointwise topology, Γ uncountable. Does K contain a G_δ-point?
Hint. No, use the definition of the topology in K and Exercise 14.6.

14.9 Show that every compact metric space K is uniform Eberlein.
Hint. $C(K)$ is separable; if $\{x_n\} \subset S_{C(K)}$ is dense, the mapping $T\colon C(K)^* \to \ell_2$ defined by $T(f) = (f(x_i)/2^i)$ is one-to-one and w^*-w continuous. K is homeomorphic to a subset of $B_{C(K)^*}$ in its w^*-topology.

14.10 ([BRW]) Let K be a uniform Eberlein compact space, φ a continuous mapping from K onto a compact space L. Show that L is uniform Eberlein.
Hint. $C(L)$ is isomorphic to a closed subspace of $C(K)$. Use Theorem 14.16. We refer to [HMVZ, Ch. 6] for a "Banach spac" proof of the corresponding result on Eberlein compact spaces [BRW], and to [MiRu] and [Vald4] for the proof of the corresponding result for Corson compact spaces.

14.11 Show that every WCG Banach space X is generated by a set that is uniform Eberlein compact in the weak topology.
Hint. Let $T\colon X^* \to c_0(\Gamma)$ be a bounded one-to-one w^*–w-continuous operator onto a dense set in $c_0(\Gamma)$. Then X is generated by $T^*\big(B_{\ell_1(\Gamma)}\big)$. The formal identity mapping of $\ell_1(\Gamma)$ into $\ell_2(\Gamma)$ shows that $B_{\ell_1(\Gamma)}$ is uniform Eberlein compact in its w^*-topology.

14.12 Show that any weakly compact subset of a superreflexive Banach space is uniform Eberlein.

Hint. Theorems 9.14 and 14.16.

14.13 Following the hint, show that there exists a scattered Eberlein compact space K which is not uniform Eberlein. The first example of this kind was found in [BeSta].

Hint. Let $\Gamma = \prod_{n=1}^{\infty}\{1,\dots,n\}$. A subset A of Γ is said to be *admissible* if it satisfies the following condition: There exists a natural number $n = n(A)$ so that for any $x, y \in A$ with $x \neq y$ we have $x(j) = y(j)$ for $j < n-1$ and $x(n) \neq y(n)$.

Consider $K = \{\chi_A :\ A \subset \Gamma \text{ admissible}\}$ with the topology of pointwise convergence. The space K is Eberlein compact: We can naturally embed K into $c_0(\Gamma)$ with the w-topology. Hence we only need to show that K is w^*-closed in $\ell_\infty(\Gamma)$. This is true, because if $A \subset \Gamma$ is not admissible then there exist $\gamma_i \in A$, $i = 1, 2, 3$, such that the set $\{\gamma_1, \gamma_2, \gamma_3\}$ is not admissible. Hence $G = \{f \in \ell_\infty(\Gamma) :\ f(\gamma_i) > \frac{1}{2}, i = 1, 2, 3\}$ is an open neighborhood of χ_A which does not intersect K.

Assume that K is uniform Eberlein compact. From Theorem 1.8 in [ArFa], where we put $X = c_0(\Gamma)$, $x_\gamma = e_\gamma$, $\varepsilon = \frac{1}{2}$, we obtain a decomposition $\{\Gamma_n\}_{n=1}^{\infty}$ of Γ and numbers $\{k(n)\}_{n=1}^{\infty}$ so that $\operatorname{card}(A \cap \Gamma_n) < k(n)$ for any admissible set A and $n \in \mathbb{N}$. Because Γ is a complete metric space and $\Gamma = \bigcup_{n=1}^{\infty}\Gamma_n$, by the Baire category theorem there exists $n_0 \in \mathbb{N}$ and a basic open subset $V = (n_1,\dots,n_k) \times \prod_{n=k+1}^{\infty}\{1,\dots,n\}$ of Γ such that $\Gamma_{n_0} \cap V$ is dense in V. Hence Γ_{n_0} contains an arbitrary large admissible set, which is a contradiction. This proves that K is not uniform Eberlein compact.

K is scattered: Let $A = \{\gamma_1,\dots,\gamma_n\} \subset \Gamma$ be an admissible set. Consider $G = \{x \in K :\ x(\gamma_i) = 1, i = 1,\dots,n\} \subset K \subset c_0(\Gamma)$. If $B \subset \Gamma$ is such that $\chi_B \in G$, then $\operatorname{card}(B) \geq n$ and $\{\gamma_i :\ i = 1,\dots,n\} \subset B$. Hence if $n(A) = n$, then the set G defined above contains no other characteristic function of an admissible subset of Γ. Consequently, A is an isolated point of K. By induction,

$$K^{(n)} \setminus K^{(n+1)} = \{\chi_A :\ A \text{ admissible}, \operatorname{card}(A) > 1, n(A) - \operatorname{card}(A) = n\},$$

therefore $K^{(\omega_0)} \setminus K^{(\omega_0+1)} = \{\chi_A :\ A \text{ admissible}, \operatorname{card}(A) = 1\}$, $K^{(\omega_0+1)} = \{\chi_\emptyset\}$ and $K^{(\omega_0+2)} = \emptyset$.

14.14 ([KutTro]) Show that there is a reflexive Banach space X that does not admit any equivalent uniformly Gâteaux differentiable norm.

Hint. Use Exercise 14.13 to get such WCG space and then use Theorem 13.33.

14.15 Prove that every scattered compact space is sequentially compact.

Hint. Theorem 14.25 and Exercise 14.3, which is also valid for Asplund spaces.

14.16 Let K be an infinite scattered compact space. Show that $C(K)$ is isomorphic to its hyperplanes.

Hint. Theorem 14.26.

14.17 Let K be a scattered compact space. Show that every weakly compact operator from $C(K)$ into a Banach space X is a compact operator.
Hint. Theorems 13.34, 14.24, and Schur's property of $\ell_1(\Gamma)$.

14.18 Let Γ be an infinite set. Let $K = \Gamma \cup \{\infty\}$ be an Alexandrov compactification of the discrete space Γ. Show that K is a scattered Eberlein compact space and $C(K)$ is isomorphic to $c_0(\Gamma)$.

Note that K is uniform Eberlein compact by Theorem 14.15.
Hint. K is Eberlein since $c_0(\Gamma)$ is weakly compactly generated (Theorem 14.9). From the definition of scattered compact space, it follows that K is scattered.

If $\gamma \in \Gamma$, then $H = \{f \in C(K) : f(\gamma) = 0\}$ is a hyperplane in $C(K)$ which is isomorphic to $C(K)$. This is seen similarly as in c_0, use the mapping $(x_1, x_2, \dots) \mapsto (0, x_1, \dots)$. Since all hyperplanes of a given Banach space are mutually isomorphic (Exercise 2.9), $C(K)$ is isomorphic to the hyperplane $\{f \in C(K) : f(\infty) = 0\}$, which is in turn isomorphic to $c_0(\Gamma)$.

14.19 Let K be a compact metric space. Show that K is countable if and only if $C(K)^*$ is separable.
Hint. First note that $C(K)$ is separable as K is a compact metric space (Lemma 3.102). If K is countable, then K is scattered by Lemma 14.21. Therefore $C(K)^*$ is separable by Theorem 14.25. If $C(K)^*$ is separable, then K is scattered by Theorem 14.25. Since K is moreover metrizable, K is countable by Lemma 14.21.

14.20 Let K_1, K_2 be compact spaces. Assume that $C(K_1)$ and $C(K_2)$ are isomorphic.

(i) If K_1 is Eberlein, is K_2 necessarily Eberlein?
(ii) If K_1 is scattered, is K_2 necessarily scattered?
(iii) If K_1 is countable, is K_2 necessarily countable?

Hint. (i) Yes, $C(K_2)$ is weakly compactly generated.
(ii) Yes, Theorem 14.25.
(iii) Yes. K_1 is metrizable by Lemma 14.21. Therefore $C(K_1)$ is separable by Lemma 3.102. Thus $C(K_2)$ is separable. Since K_1 is scattered, $C(K_1)$ is an Asplund space by Theorem 14.25. Thus K_2 is scattered by Theorem 14.25. Since $C(K_2)$ is separable, K_2 is metrizable by Lemma 3.102. Thus K_2 is metrizable and scattered, hence countable by Lemma 14.21.

14.21 For a scattered compact space K, let $\alpha(K)$ be the smallest ordinal such that $K^{(\alpha(K))} = \emptyset$. The Bessaga–Pełczyński theorem [BePe1] asserts that if K_1, K_2 are infinite countable compact spaces such that $\alpha(K_1) \leq \alpha(K_2)$, then $C(K_1)$ is isomorphic to $C(K_2)$ if and only if there is $n \in \mathbb{N}$ such that $\alpha(K_2) \leq (\alpha(K_1))^n$. This theorem compares with the result by Milyutin (see, e.g., [Wojt]) that if K is an uncountable compact metric space, then $C(K)$ is isomorphic to $C[0, 1]$.

Use the quoted Bessaga–Pełczyński theorem to derive that for a compact space K, $C(K)$ is isomorphic to c_0 if and only if $K^{(\omega_0)} = \emptyset$.

Hint. c_0 is isomorphic to $C(K_0)$, where K_0 is an Alexandrov compactification of $\mathbb{N}$. So $C(K)$ is isomorphic to $C(K_0)$ and by the previous exercise, K is countable, it is also infinite. Since K is compact, we have $\alpha(K) \geq 2 = \alpha(K_0)$. Thus we can use the Bessaga–Pełczyński theorem. To finish the proof it is enough to observe that, by compactness, for a compact set L we have $L^{(\omega_0)} = \emptyset$ if and only $L^{(n)} = \emptyset$ for some positive integer n.

14.22 Let K be an infinite metrizable compact space. Prove that:

(i) $C(K)^*$ is separable if and only if $C(K)$ does not contain a subspace isomorphic to ℓ_1.

(ii) w^*-exposed points of $B_{C(K)^*}$ form a James boundary of $C(K)$.

Recall that $f \in B_{X^*}$ is a w^*-exposed point of B_{X^*} if there is $x \in S_X$ such that $f(x) = 1$ and x is a point of Gâteaux smoothness of the canonical sup-norm of $C(K)$.

Hint. (i) If $C(K)^*$ is not separable, K is not scattered (Theorem 14.25) and thus $C(K)$ contains a subspace isomorphic to $C[0, 1]$. The space $C[0, 1]$ contains an isomorphic copy of ℓ_1 (Theorem 5.8).

(ii) Given $f \in S_{C(K)}$, find $t_0 \in K$ is such that $f(t_0) = 1$. Define a function φ on K by $\varphi(t) = 1 - \text{dist}(t, t_0)$. Then φ is a point of Gâteaux differentiability of the sup-norm on $C(K)$ by the Šmulyan Lemma 7.22. Hence t_0 is a w^*-exposed point of $B_{C(K)^*}$. Thus w^*-exposed points of B_{X^*} form a James boundary of $C(K)$.

14.23 Find an example of a compact set K such that $C(K)$ is nonseparable but does not contain $c_0(\Gamma)$ for any Γ uncountable.

Hint. $\beta\mathbb{N}$, as ℓ_∞ has a w^*-separable dual unlike $c_0(\Gamma)$, Γ uncountable.

14.24 Show that if $C(K)$ is reflexive, then K is finite.

Hint. c_0 in $C(K)$?

14.25 Show that a continuous image of a metrizable compact space is metrizable, by using $C(K)$ spaces, subspaces, and their separability.

Hint. If $L = \varphi(K)$ then $C(L)$ is a subspace of $C(K)$ and $C(K)$ is separable.

14.26 Show that $\beta\mathbb{N}$ does not contain any infinite metrizable subset.

Hint. Otherwise $C(\beta\mathbb{N})$ would have a non-trivial convergent sequence and by Tietze's extension theorem, $\ell_\infty = C(\beta\mathbb{N})$ would factor to c_0, which is not the case (Exercise 3.44).

14.27 Let K be a metrizable compact space. Show that $C(K)$ admits an equivalent C^∞-smooth norm if it admits a C^1-smooth bump.

For non-metrizable spaces this result no longer holds in general (Haydon, see, e.g., [DGZ3]). However, it is not known whether a separable Banach space X admits an equivalent C^∞-smooth norm if it admits a C^∞-smooth bump.

Hint. $C(K)^*$ is then separable, so K is scattered. As K is metrizable, this means that K is countable. Then use Theorem 10.12.

14.28 A collection $\mathcal{F}$ of subsets of $\mathbb{N}$ is *hereditary* if $G \subset F$, $F \in \mathcal{F}$ implies $G \in \mathcal{F}$.

Prove the following Pták combinatorial lemma ([BHO], [Ptak1]):

Let $\mathcal{F}$ be a hereditary collection of finite subsets of $\mathbb{N}$ and let $\delta > 0$ be given. Assume that for all $a_1, \ldots, a_n \geq 0$ with $\sum_{i=1}^{n} a_i = 1$ there exists $F \in \mathcal{F}$ such that $\sum_{i \in F} a_i \geq \delta$. Then there is an infinite subsequence M of $\mathbb{N}$ such that $F \in \mathcal{F}$ for all finite $F \subset M$.

Hint. Define a norm on c_{00} by $|\sum a_i e_i| = \sup\Big\{\Big\|\sum_{i \in F} a_i\Big\|_\infty : F \in \mathcal{F}\Big\}$. Let X be the completion of c_{00} in this norm. Then $\{e_i\}$ is a Schauder basis of X that is equivalent to the canonical basis of ℓ_1. Note that $\mathcal{F}$ can be identified with a closed subspace of X^* by using the mapping $\Phi(\sum a_i e_i) = \sum_{i \in F} a_i$. In this way $\mathcal{F}$ is a 1-norming set in X^*.

Denote by K the w^*-closure of $\mathcal{F}$ in X^*. ℓ_1 is isomorphic to X, which is in turn isometric to a subspace of $C(K)$. Thus K is not countable, for otherwise $C(K)$ would be an Asplund space, which is not true for ℓ_1. As K can be identified with the closure of $\mathcal{F}$ in $2^{\mathbb{N}}$ in its pointwise topology, K contains an infinite sequence M. As $\mathcal{F}$ is hereditary, every finite subset of M is in $\mathcal{F}$.

14.29 Using Pták combinatorial lemma, prove the following variant of Mazur's theorem:

Let K be a compact space and $f, f_1, f_2, \cdots \in C(K)$ be such that $\{f_k\}$ is bounded in $C(K)$ and $f_n \to f$ pointwise. Then for every $\varepsilon > 0$ there exist $n_1, \ldots, n_k \in \mathbb{N}$ and $\lambda_1, \ldots, \lambda_k \in \mathbb{R}$ such that $0 \leq \lambda_i \leq 1$ for every i, $\sum \lambda_i = 1$ and $\Big\| f - \sum_{i=1}^{k} \lambda_i f_{n_i} \Big\|_\infty \leq \varepsilon$.

Hint. Assume that $f = 0$ and $\|f_n\|_\infty \leq 1$ for every n. Given $\varepsilon > 0$, for each $x \in K$ put $G_x = \{n \in \mathbb{N} : |f_n(x)| \geq \varepsilon/2\}$. Note that G_x is finite for all $x \in K$. Set $F_x = \{G : G \subset G_x\}$, let $\mathcal{F} = \{F_x : x \in K\}$. Assume that $\mathcal{F}$ satisfies the conclusion in Pták combinatorial lemma. Then there is an increasing sequence $\{n_i\}$ of natural numbers such that $\{n_1, \ldots, n_k\} \subset F_{x_k}$ for each k. Let x_0 be an accumulation point of $\{x_k\}$. Fix $k \in \mathbb{N}$. Since $n_k \in F_{x_i}$ for all $i > k$, $|f_{n_k}(x_i)| \geq \varepsilon/2$ for all $i > k$. From the continuity of f_{n_k} we thus have $|f_{n_k}(x_0)| \geq \varepsilon/2$. Hence $n_k \in F_{x_0}$ for all $k \in \mathbb{N}$. This shows that F_{x_0} is infinite, a contradiction.

Thus by Pták combinatorial lemma there exist $a_1, \ldots, a_k$, $a_i \geq 0$ and $\sum a_i = 1$ such that $\sum_{i \in F_x} a_i \leq \varepsilon/2$ for all $x \in K$. Let $x \in K$. Then

$$\begin{aligned}\Big|\sum_{i=1}^{k} a_i f_i(x)\Big| &= \Big|\sum_{i \in \{1,\ldots,k\} \cap F_x} a_i f_i(x) + \sum_{i \in \{1,\ldots k\} \setminus F_x} a_i f_i(x)\Big| \\ &\leq \sum_{i \in \{1,\ldots,k\} \cap F_x} a_i + \sum_{i \in \{1,\ldots,k\} \setminus F_x} a_i \varepsilon/2 \leq \varepsilon.\end{aligned}$$

14.30 (Root lemma) Let $\mathcal{A}$ be an uncountable family of finite sets. Show that there is an uncountable subfamily $\mathcal{B}$ of $\mathcal{A}$ and a finite (possibly empty) set S such that $A \cap B = S$ for every pair of distinct elements A, B of $\mathcal{B}$.

The family $\mathcal{B}$ is called a Δ-system and S is called a root of $\mathcal{B}$.

Hint. Assume that $\operatorname{card}(\mathcal{A}) = \aleph_1$ and that $\mathcal{A}$ is formed by finite sets in $[0, \omega_1]$. For $n \in \mathbb{N}$ set $\mathcal{A}_n = \{A \in \mathcal{A} : \operatorname{card}(A) = n\}$. There is n such that $\operatorname{card}(\mathcal{A}_n) = \aleph_1$. Fix this n. If $A \in \mathcal{A}_n$, write $A = \{A(1), \ldots, A(n)\}$ with $A(1) < A(2) < \cdots < A(n)$.

For every $\alpha < \omega_1$, the set $\{A \in \mathcal{A}_n : A \subset [0, \alpha]\}$ is countable. Hence $\sup\left(\bigcup_{A \in \mathcal{A}_n} A\right) = \omega_1$. Let $p \in \{1, \ldots, n\}$ be the least integer such that $\sup\{A(p) : A \in \mathcal{A}_n\} = \omega_1$. Put $\alpha_0 = \sup\{A(p-1) + 1 : A \in \mathcal{A}_n\}$ (if $p = 1$, we put $\alpha_0 = 0$). By transfinite induction on $\mu < \omega_1$, choose $A_\mu \in \mathcal{A}_n$ such that $A_\mu(p) > \max\{\alpha_0, \sup\{A_\nu(n) : \nu < \mu\}\}$ and set $\mathcal{B}_1 = \{A_\mu : \mu \in [0, \omega_1)\}$. Then $\operatorname{card}(\mathcal{B}_1) = \aleph_1$ and $A \cap B \subset [0, \alpha_0]$ whenever A, B are distinct elements of $\mathcal{B}_1$. Since the family of all finite sets in $[0, \alpha_0]$ is countable, there is an uncountable family $\mathcal{B} \subset \mathcal{B}_1$ and a finite set $S \subset [0, \alpha_0]$ such that $A \cap [0, \alpha_0] = S$ for every $A \in \mathcal{B}$. Clearly, $\mathcal{B}$ is a Δ-system with root S.

14.31 Let X be a nonseparable Banach space equipped with its weak topology. Does X contain a dense metrizable subset?

Hint. No, X would not have property CCC.

14.32 Show that if a Banach space X admits a Lipschitz Gâteaux differentiable bump, then ℓ_∞ is not a quotient of X.

Hint. Otherwise $\beta\mathbb{N}$ is in X^* and has no non-trivial convergent subsequence, a contradiction with Exercise 14.3.

14.33 ([Pol3]): Let μ be a Radon measure (see Section 17.13.1) on a compact space K such that $C(K)$ has property C. Show that μ has separable support. (The *support of a measure* is the complement of the set of points that have neighborhoods with measure 0.)

Hint. Assume that $\operatorname{supp}(\mu) = K$. We will show that K is separable.

Fix $i \in \mathbb{N}$. For $x \in K$ put

$$C_x = \left\{f \in C(K) : \int_K f \,\mathrm{d}\mu \geq \frac{1}{i} \text{ and } f(x) = 0\right\}.$$

If $z \in \bigcap_{x \in K} C_x$ then $f = 0$ identically on K and thus $\int_K f \,\mathrm{d}\mu = 0$, a contradiction. Hence $\bigcap_{x \in K} C_x = \emptyset$. As C_x are all closed and convex, and $C(K)$ has property C, $\bigcap_{x \in A_i} C_x = \emptyset$ for some countable set $A_i \subset K$. We claim that $K = \overline{\bigcup_{i=1}^{\infty} A_i}$: Assume that there is $f \in C(K)$ such that $f = 0$ on $\overline{\bigcup A_i}$ and $\int_K f \,\mathrm{d}\mu > 0$. Find $i \in \mathbb{N}$ such that $\int_K f \,\mathrm{d}\mu > \frac{1}{i}$. Then $f \in \bigcap_{x \in A_i} C_x$, a contradiction.

14.34 Show that $\operatorname{dens}(\ell_\infty^*) = 2^c$, hence $\operatorname{dens}(\ell_\infty^*) = 2^c > \operatorname{card}(\ell_\infty) = c$.

Hint. The cardinality of $\beta\mathbb{N}$.

14.35 Let X be a separable Banach space and assume that the dual norm of X^* is Gâteaux differentiable. Show that every element of X^{**} is a first Baire class function when considered as a function on (B_{X^*}, w^*).
Hint. Let $F \in S_{X^{**}}$ attain its norm at $f \in S_{X^*}$. Let $x_n \in S_X$ be such that $f(x_n) \to 1$. Then x_n as functions on (B_{X^*}, w^*) are continuous and by Šmulyan's Lemma 7.22, $x_n \to F$ pointwise on B_{X^*}. Thus F is a first Baire class function on (B_{X^*}, w^*).
By the Bishop–Phelps theorem, every $G \in S_{X^{**}}$ is a uniform limit (on B_{X^*}) of elements of $S_{X^{**}}$ that attain their norms.
For further use of Baire methods we refer to [Nata].

14.36 Let T be a topological space. We define the *weight* of T, $\mathrm{w}(T)$, as the minimal cardinal $\aleph$ such that there is a basis $\mathcal{F}$ of topology of T with $\mathrm{card}(\mathcal{F}) \leq \aleph$.
Show that if K is a compact space, then $\mathrm{w}(K) = \mathrm{dens}(C(K))$.
Hint. $\mathrm{dens}(C(K)) \leq \mathrm{w}(K)$: follow the proof of Lemma 3.102.
$\mathrm{w}(K) \leq \mathrm{dens}(C(K))$: Assume that $\mathcal{F}$ is dense in $C(K)$ and $\mathrm{card}(\mathcal{F}) = \mathrm{dens}(C(K))$. The topology on K of pointwise convergence on elements of $\mathcal{F}$ has a basis of cardinality $\mathrm{card}(\mathcal{F})$.

14.37 Let X be a separable Banach space, assume that F is a closed subspace of X^* that is w^*-dense in X^*. Is F w^*-sequentially dense in X^*? Is it true that every element of X^* is a w^*-limit of a w^*-bounded net in F?
Hint. No. Let $G_n = \overline{F \cap nB_{X^*}}^{w^*}$. Then assuming $\bigcup G_n = X^*$, by the Baire category theorem and the symmetry and convexity of G_n, at least one G_n contains a ball δB_{X^*} for some $\delta > 0$. Then F is a norming subset. Each separable space such that $\dim(X^{**}/X)$ is $\geq \aleph_0$ contains a w^*-dense non-norming closed subspace $F \subset X^*$ ([DaLi]).

14.38 Assume that X is a separable Banach space and $\{f_\alpha\} \subset X^*$ is a net that converges to 0 in the topology of uniform convergence on sequences $\{x_i\}$ in X that converge to zero. Is $\{f_\alpha\}$ necessarily bounded?
Hint. No. Take w^*-dense non-norming closed subspace F in X^* (see the previous exercise). Then the closure of F in the topology of uniform convergence on sequences converging to zero is X^* ([DuSc]). Take a point f in X^* that is not in $G_n = \overline{F \cap nB_{X^*}}^{w^*}$ as above for any n. Then f can be reached by a net $\{f_\alpha\}$ converging to f in the topology of uniform convergence on converging to zero sequences. This $\{f_\alpha\}$ is not bounded, as f is not in any G_n.

14.39 Let X be a separable space. Is X^* w^*-sequentially separable, i.e., does there exist a countable $C \subset X^*$ such that each element of X^* is a w^*-limit of a sequence in C?
Hint. Yes. nB_{X^*} is metrizable w^*-separable and $X^* = \bigcup nB_{X^*}$.

14.40 Show that ℓ_∞^* is not w^*-sequentially separable.
Note that by Goldstine's theorem, it is w^*-separable.

Hint. Use the Grothendieck property.

14.41 Show that $(\ell_\infty/c_0)^* = c_0^\perp \subset \ell_\infty^*$ is not w^*-separable.
Hint. $c_0(c) \subset \ell_\infty/c_0$ (Exercise 5.36).

14.42 Show that there is an equivalent norm on ℓ_∞ such that its dual ball is not w^*-separable, although the standard unit ball of $\ell_\infty = c_0^{**}$ is w^*-separable by Goldstine's theorem.

Note that there is a Banach space X such that its dual is w^*-separable and the dual ball of no equivalent norm on X is w^*-separable ([JoLi1]). Note also that the dual space $C(K)^*$, where K is Kunen's compact space, has the property that every subset of $C(K)^*$ is w^*-separable.
Hint. Extend on ℓ_∞ the norm in Theorem 14.54.

14.43 Let X be a separable Banach space. Is every subspace of X^* w^*-separable?
Hint. Yes. Let Y be a subspace of X^*. (nB_{X^*}, w^*) is separable metrizable space, so $A_n = Y \cap nB_{X^*}$ is a separable space in the w^*-topology. Thus $Y = \bigcup A_n$ is a w^*-separable space.

14.44 Let $f_n \in X^*$ be such that $\|f_n\| = \frac{1}{n}$ for each n. Write each f_n as $w^*\text{-}\lim\limits_k f_k^n$, where $\|f_n - f_k^n\| = n$ (Josefson–Nissenzweig, see Exercise 3.39). Show that the origin of X^* is not a w^*-limit of any sequence in $\{f_k^n\}_{n,k}$.
Hint. Banach–Steinhaus.

14.45 Prove that $C[0, \omega_1]^*$ is not w^*-separable.
Hint. $c_0[0, \omega_1]$ can be isomorphically embedded into $C[0, \omega_1]$ and $c_0[0, \omega_1]^*$ is not w^*-separable.

14.46 Show that $\ell_1(\Gamma)$ does not have property C if Γ is uncountable. Show that ℓ_∞ does not have property C.
Hint. The space $\ell_1(\Gamma)$ does not have property C as it has a Markushevich basis (the standard basis), and its dual ball in the w^*-topology is not Corson (not even angelic, use Goldstine's theorem and the fact that each element of $c_0(c)$ is countably supported). For the second question, $\ell_1(c) \subset C[0, 1]^*$ (Exercise 3.143) and hence $\ell_1(c) \subset \ell_\infty$.

14.47 Let $f \in C[0, \omega_1]$. Show that there is $\alpha < \omega_1$ such that f is constant on $[\alpha, \omega_1]$.
Hint. For $n \in \mathbb{N}$, let $\alpha_n < \omega_1$ be such that $|f(\beta) - f(\omega_1)| < \frac{1}{n}$ for all $\beta \geq \alpha_n$. Consider $\sup(\alpha_n)$.

14.48 Prove that the Stone–Čech compactification of the ordinal segment $[0, \alpha)$ is homeomorphic to its one-point compactification, see Section 17.1 and Exercise 14.66.

Hint. Continuous functions are eventually constant.

14.49 For $\omega_0 \le \alpha \le \omega_1$ define the projections in $C[0, \omega_1]$ by

$$P_\alpha(f)(\beta) = \begin{cases} f(\beta) & \text{for } \beta \le \alpha, \\ f(\alpha) & \text{for } \beta \ge \alpha. \end{cases}$$

Show that $\{P_\alpha\}$ is a norm-one projectional resolution of the identity on $C[0, \omega_1]$.

14.50 For $\alpha < \omega_1$, define a projection P_α of $C_0[0, \omega_1]$ by $P_\alpha(x)(\beta) = x(\beta)$ for $\beta < \alpha + 1$ and $P_\alpha(x)(\beta) = 0$ if $\beta \ge \alpha + 1$. Show that the projections P_α together with the identity operator satisfy all the properties needed to form a PRI but one, namely the continuity of the mappings $\alpha \mapsto P_\alpha(x)$ on the ordinal segment (use the characteristic functions of $[0, \alpha]$). However, $\bigcup_{\alpha<\omega_1} P_\alpha(C_0[0, \omega_1]) = C_0[0, \omega_1]$ and $\bigcup_{\alpha<\omega_1} P^*(C_0[0, \omega_1]^*) = C_0[0, \omega_1]^*$.
Hint. If $x \in C_0[0, \omega_1]$, then $x(\alpha) = 0$ for all $\alpha > \alpha_0$. From this argument, the first part in the statement follows. To see the latter part, let $f \in S_{C_0[0,\omega_1]^*}$ be such that $f(x) = 1$ for some $x \in S_{C_0[0,\omega_1]}$. Let $x(\beta) = 0$ for all $\beta \ge \beta_0$. If $h \in C_0[0, \omega_1]$, $\|h\| \le \frac{1}{2}$, then $\|x \pm h\| \le 1$ and thus $|f(x \pm h)| \le 1$. By a convexity argument, $f(x \pm h) = 1$, which means that $f(h) = 0$. From this we get $P^*(f) = f$.

14.51 (Semadeni) Show that $C[0, \omega_1] \oplus C[0, \omega_1]$ is not isomorphic to $C[0, \omega_1]$.
Hint. Show that the codimension of $C[0, \omega_1]$ in the space of all sequentially w^*-continuous functions on $C[0, \omega_1]^*$ is 1.

14.52 Prove that $C[0, \omega_1]$ does not have an unconditional basis.
Hint. Such a basis would be shrinking as the space is Asplund. Thus the space would be WCG, a contradiction.

14.53 (i) Is ℓ_∞ isomorphic to a subspace of $C[0, \omega_1]$?
(ii) Is $C[0, \omega_1]$ isomorphic to a subspace of ℓ_∞?
Hint. (i) No: Asplund property is hereditary ([Phelps]).
(ii) $C[0, \omega_1]^*$ is not w^*-separable.

14.54 A bounded operator T from a subspace X of a Banach space $C(K)$ into $c_0(K)$ is called a Talagrand operator ([Tala6], [Hayd3]) if for every $x \in S_X$ there is $k \in K$ such that $|x(k)| = 1$ and $T(x)(k) \ne 0$. Construct a Talagrand operator on $C_0[0, \omega_1]$.
Hint. For $x \in C_0[0, \omega_1]$ put

$$T(x)(\alpha) = \begin{cases} x(\alpha) - x(\alpha + 1) & \text{if } \alpha < \omega_1, \\ 0 & \text{if } \alpha = \omega_1. \end{cases}$$

Then $T(x) \in c_0[0, \omega_1]$ as $x \in C[0, \omega_1]$. Let $\|x\| = 1$. Choose $\alpha < \omega_1$ maximal such that $|x(\alpha)| = 1$. Then check that $T(x)(\alpha) \ne 0$.

14.55 Assume that a subspace X of $C(K)$ admits a Talagrand operator T into $c_0(K)$. Show that X admits an equivalent norm that locally depends on finitely many coordinates.
Hint. $|||x||| = \sup\{|x(k)| + |T(x)(k)| : k \in K\}$.

14.56 Show that $C[0, \omega_1]$ admits a norm that locally depends on finitely many coordinates.
Hint. $C_0[0, \omega_1]$ is isomorphic to $C[0, \omega_1]$.

14.57 Show that every Corson compact space K is angelic.
Hint. Let $K \subset [0, 1]^\Gamma$ be a compact space such that for every $f \in K$, the set $\{\mu \in \Gamma : f(\mu) \neq 0\}$ is countable. Let $H \subset K$ and $h \in \overline{H}$. Let $\{\mu_i^0\}$ be the support of h. Find $h_1 \in H$ such that $|(h - h_1)(\mu_1^0)| < 1$. Let $\{\mu_i^1\}$ be the support of h_1. Find $h_2 \in H$ such that $|(h - h_2)(\mu_j^i)| < \frac{1}{2}$, $i = 0, 1$, $j = 1, 2$, etc. The sequence h_i converges to h at every point of $\bigcup_{i=0}^\infty \operatorname{supp}(\mu^i)$. Outside this set, h and h_j are zero.

14.58 Show that every separable Corson compact space is metrizable. Thus for Corson compact spaces, metrizability is equivalent to separability.
Hint. Let $K \subset [0, 1]^\Gamma$ be a compact space such that for every $k \in K$, the support of k on Γ is countable. The set $[0, 1]^\Gamma$ is equipped with its natural product topology. Let $\{k_n\}$ be dense in K. Let Γ_1 be the union of all supports of $\{k_n\}$. Use the fact that $[0, 1]^{\Gamma_1}$ is metrizable in its product topology.

14.59 Assume that X is a separable Banach space such that $B_{X^{**}}$ in its w^*-topology is Corson compact. Show that X^* is separable.
Hint. Since X is separable, $B_{X^{**}}$ is w^*-separable by Goldstine's theorem. Thus $B_{X^{**}}$ in its w^*-topology is metrizable by the previous exercise. Hence X^* is separable by Proposition 3.103.

14.60 Show that the ordinal segment $[0, \omega_1]$ in its standard topology is a scattered compact space which is not Corson compact.
Hint. The point ω_1 is in the closure of the set of smaller ordinals, but it is not a limit of any sequence of them. Thus the space is not angelic. Every Corson compact space is angelic (Exercise 14.57).
From the definition it follows that $[0, \omega_1]$ is scattered.

14.61 Show that the space $C[0, \omega_1]$ is not weakly compactly generated.
Hint. By the previous exercise, $[0, \omega_1]$ is not an angelic space and thus not Eberlein compact, use Theorem 14.9.

14.62 Verify that the alternative definitions of pseudocompact spaces (before Theorem 14.49) are equivalent.
Hint. $f \mapsto \frac{1}{f}$.

14.63 Let K be the Kunen compact space mentioned in the remarks after Corollary 14.48.

(i) Show that $C(K)$ admits no locally uniformly rotund norm.

(ii) Show that $C(K)$ admits no Fréchet differentiable norm.

(iii) Show that every closed subspace of $C(K)$ is an intersection of a countable family of hyperplanes.

It is not known whether this $C(K)$ admits a Fréchet differentiable bump or a Gâteaux differentiable norm.

Hint. (i) If a norm on $C(K)$ were LUR, then its unit sphere would be norm Lindelöf, which it cannot be as it is not separable.

(ii) In such a norm, the set S of $x^* \in S_{X^*}$ that attain their norm is a w^*-separable set and on S the norm and w^*-topology coincide. Since S is dense in S_{X^*} by the Bishop–Phelps theorem, S_{X^*} would then be separable, a contradiction.

(iii) If $Y \subset C(K)$, then $Y^\perp$ is w^*-separable.

14.64 A compact space K is said to be *Valdivia compact* if K is homeomorphic to a compact set K_0 in some $[0, 1]^\Gamma$ in its pointwise topology such that the set of all points in K_0 that have countable support is dense in K_0. Note that every Corson compact space is Valdivia compact.

Prove that $[0, \omega_1]$ and $[0, 1]^\Gamma$ for every Γ are Valdivia compact spaces. For more information on Valdivia compacta we refer for instance to [Vald3], [AMN], [DeGo], [Kale2], [Kale1], and Chapter VI of [DGZ3].

Hint. Consider characteristic functions of intervals $(\alpha, \omega_1]$ and with every ordinal associate evaluations on those functions. For the second example, observe that finitely supported vectors are dense in $[0, 1]^\Gamma$.

14.65 Let f be a continuous real-valued function on $C[0, \omega_1)$ Show that f is eventually constant, i.e., there is c and a countable ordinal α_0 such that $f(\alpha) = c$ for all $\alpha \geq \alpha_0$.

Hint. First show a similar statement for the oscillation of f.

14.66 Show that the Stone–Čech compactification of $[0, \omega_1)$ is $[0, \omega_1]$.

Hint. Use the preceding exercise. See also Exercise 14.48.

14.67 For a set X and $n \in \mathbb{N}$, denote by $\sigma_n(2^X)$ the set $\{\chi_A : A \subset X, \text{ card}(A) \leq n\}$. Show that $\sigma_n(X)$ is a uniform Eberlein compact space of height $n + 1$. Note that Argyros and Godefroy proved that every Eberlein compact space of weight $< \omega_{\omega_0}$ can be embedded into $\sigma_0(2^X)$ for some set X and some $n \in \mathbb{N}$. The restriction on the weight is necessary, see [BelMar].

Hint. Check the cardinality of supports of the elements that are in the Cantor $(\sigma_n(2^X))'$.

14.68 Show that spaces JL_0 defined in Theorem 14.54 and JL_2 in Remark (vi) after the proof of Theorem 14.54 are not WCG.

Hint. Since X^* are weak*-separable, weak-compact sets in them are metrizable and thus separable.

14.69 Show the following version of the "pressing down lemma": Assume $f : [0, \omega_1) \to [0, \omega_1)$ be such a mapping that $f(\alpha) < \alpha$ for any α. Then

$$\exists \beta_0; \forall \beta \, \exists \alpha; f(\alpha) \le \beta_0.$$

Thus there is β_0 so that $f(\alpha) = f(\beta_0)$ for uncountably many α.
Hint. Assuming the contrary, we find a countable set of ordinals α_n such that $f(\alpha) > \alpha_n$ whenever $\alpha \ge \alpha_{n+1}$. Let α be the least upper bound of $\{\alpha_n\}$. Then $f(\alpha) > \alpha_n$ for each n, as $\alpha \ge \alpha_{n+1}$. Therefore $f(\alpha) \ge \alpha$, a contradiction.

Chapter 15
Compact Operators on Banach Spaces

In this chapter we study basic properties of compact operators on Banach spaces. We present the elementary spectral theory of compact operators in Banach spaces, including the spectral radius and properties of eigenvalues. Then we discus basic spectral properties of selfadjoint operators on Hilbert spaces, their spectral decomposition, and show some of the applications of these topics.

Unless stated otherwise, the word "space" in this chapter means a *complex* Banach space.

15.1 Compact Operators

Let X and Y be Banach spaces. Recall that an operator $T \in \mathcal{B}(X, Y)$ is called a *compact operator* if $\overline{T(B_X)}$ is compact in Y, and that the class of compact operators between X and Y is denoted by $\mathcal{K}(X, Y)$ ($\mathcal{K}(X)$ if $X = Y$). Recall, too, that an operator $T \in \mathcal{B}(X, Y)$ is called a *finite rank operator* or a *finite-dimensional operator* if $\dim\big(T(X)\big) < \infty$. By $\mathcal{F}(X, Y)$ ($\mathcal{F}(X)$ if $X = Y$) we denote the space of all finite rank operators from X into Y (see Definition 1.31, Proposition 1.40 and Exercise 1.77).

Unless stated otherwise, the closure operation in $\mathcal{B}(X, Y)$ is meant in the norm operator topology, see Definition 1.26.

Proposition 15.1 *Let X be a Banach space with a Schauder basis. Then $\overline{\mathcal{F}(X)} = \mathcal{K}(X)$.*

Proof: Let P_n be the canonical projection associated with a Schauder basis. For every $x \in X$ we have $\lim P_n(x) = x = I_X(x)$, where I_X is the identity operator on X. Given $T \in \mathcal{K}(X)$, we claim that the finite-dimensional operators $P_n \circ T$ converge to T in $\mathcal{B}(X)$. To see this we need to show that $(P_n - I_X)\big(T(x)\big)$ converges uniformly to zero on B_X, that is, $(P_n - I_X)$ converges uniformly to zero on $T(B_X)$. This follows from Corollary 3.87 as $\overline{T(B_X)}$ is compact. □

Proposition 15.2 *Let X be a Banach space with a Schauder basis. If X^* is separable, then $\mathcal{K}(X)$ is separable.*

M. Fabian et al., *Banach Space Theory*, CMS Books in Mathematics,

DOI 10.1007/978-1-4419-7515-7_15, © Springer Science+Business Media, LLC 2011

Proof: First we will show that one-dimensional operators form a separable subset in $\mathcal{K}(X)$: Choose a countable dense set $\{f_i\}$ in X^* and a countable dense set $\{x_n\}$ in X. Then the sequence of operators $T_{i,n}: x \mapsto f_i(x)x_n$ is dense in the set of one-dimensional operators on X. Indeed, let T be a non-trivial one-dimensional operator on X of the form $T(x) = f(x)e$, where $f \in X^*$, $e \in X$. Given $\varepsilon > 0$, choose f_i such that $\|f - f_i\| < \varepsilon/\|e\|$ and x_n such that $\|e - x_n\| < \varepsilon/(\|f\| + \varepsilon/\|e\|)$. For $\|x\| \leq 1$ we have

$$\begin{aligned}\|f(x)e - f_i(x)x_n\| &\leq \|e\| \cdot \|f - f_i\| \cdot \|x\| + \|f_i\| \cdot \|e - x_n\| \cdot \|x\| \\ &\leq \|e\| \frac{\varepsilon}{\|e\|} + \left(\|f\| + \tfrac{\varepsilon}{\|e\|}\right) \frac{\varepsilon}{\|f\| + \varepsilon/\|e\|} < 2\varepsilon.\end{aligned}$$

Thus $\|T_{i,n} - T\| < 2\varepsilon$.

Since the span of one-dimensional operators is $\mathcal{F}(X)$, this space is separable. From $\mathcal{K}(X) = \overline{\mathcal{F}(X)}$ we get that $\mathcal{K}(X)$ is separable. □

Recall that $\mathcal{B}(\ell_2)$ is not separable (Proposition 1.44).

Theorem 15.3 (Schauder) *Let X, Y be Banach spaces and $T \in \mathcal{B}(X, Y)$. Then $T^* \in \mathcal{K}(Y^*, X^*)$ if and only if $T \in \mathcal{K}(X, Y)$.*

Proof: Let $T \in \mathcal{K}(X, Y)$. We need to show that $T^*(B_{Y^*})$ is totally bounded in X^*. Let $\{f_n\} \subset B_{Y^*}$ be an arbitrary sequence.

Consider f_n restricted to $\overline{T(B_X)}$, which is compact in Y. Then $\{f_n\}$ are uniformly bounded and equicontinuous. By Arzelà–Ascoli, the restrictions of f_n to $\overline{T(B_X)}$ form a totally bounded set in $C\big(\overline{T(B_X)}\big)$. Hence there is a subsequence f_{n_k} such that $\sup\limits_{x \in B_X} |f_{n_k}(T(x)) - f_{n_l}(T(x))| \to 0$ as $k, l \to \infty$. Consequently

$$\begin{aligned}\lim_{k,l\to\infty} \|T^*(f_{n_k}) - T^*(f_{n_l})\| &= \lim_{k,l\to\infty} \sup_{x\in B_X} |(T^*(f_{n_k}) - T^*(f_{n_l}))(x)| \\ &= \lim_{k,l\to\infty} \sup_{x\in B_X} |f_{n_k}(T(x)) - f_{n_l}(T(x))| \to 0.\end{aligned}$$

Thus $T^*(f_{n_k})$ is Cauchy in X^* and $\overline{T^*(B_{Y^*})}$ is compact.

To prove the opposite implication, recall that $T^{**}\big|_X = T$. By the first part, $T^* \in \mathcal{K}(Y^*, X^*)$ implies that $\overline{T^{**}(B_{X^{**}})}$ is compact. Since $\overline{T^{**}(B_X)}$ is a closed subset of $\overline{T^{**}(B_{X^{**}})}$, we get that $\overline{T^{**}(B_X)}$ is compact in X^{**} and hence in X. Thus $T \in \mathcal{K}(X, Y)$. □

Corollary 15.4 *Let X and Y be Banach spaces and $T : X \to Y$ be a compact operator. For every $\varepsilon > 0$, there is a closed subspace Z of X of finite codimension such that for the restriction of T to Z we have $\|T_{\restriction Z}\| < \varepsilon$.*

Proof: Denote by i^* the canonical quotient mapping from X^* onto $X^*/Z^\perp = Z^*$. We have

$$\|T_{\restriction Z}\| = \|i^* T^*\|,$$

since the norm of an operator is equal to the norm of its conjugate. The operator T^* is compact (Theorem 15.3), hence there is a finite subset $\{f_1, f_2, \dots, f_n\}$ of X^* such that for any $g \in B_{Y^*}$,

$$\inf\{\|f_i - T^*(g)\| : 1 \le i \le n\} \le \varepsilon.$$

If $Z^\perp := \operatorname{sp}\{f_1, f_2, \dots, f_n\}$, then Z has the desired property. □

Lemma 15.5 *Let X be a Banach space. Let $T \in \mathcal{B}(X)$, denote $S = I_X - T$ and $Y = S(X)$. If Y is a proper closed subspace of X, then for every $\varepsilon > 0$ there is $x_0 \in B_X$ such that* $\operatorname{dist}\big(T(x_0), T(Y)\big) > 1 - \varepsilon$.

Proof: By Riesz's lemma (Lemma 1.37) there is $x_0 \in S_X$ such that $\operatorname{dist}(x_0, Y) > 1 - \varepsilon$. We have $S(x_0) \in Y$ and $T(Y) = (I_X - S)(Y) \subset Y$. Therefore $\operatorname{dist}\big(T(x_0), T(Y)\big) \ge \operatorname{dist}\big(T(x_0) + S(x_0), Y\big) = \operatorname{dist}(x_0, Y) > 1 - \varepsilon$. □

Theorem 15.6 *Let X be a Banach space. Assume that $T \in \mathcal{K}(X)$ and $\lambda \neq 0$. Then* $\operatorname{Ker}(\lambda I_X - T)$ *is finite-dimensional, and* $(\lambda I_X - T)(X)$ *is closed and finite-codimensional.*

Proof: We may assume that $\lambda = 1$. Let $N_\lambda = \operatorname{Ker}(I_X - T)$. For every $x \in N_\lambda$ we have $T(x) = x$, hence $T\big|_{N_\lambda}$ is an isomorphism into and also compact, so N_λ is finite-dimensional.

By Theorem 4.5 and Proposition 4.2, there is a closed subspace X_1 of X such that $X = N_\lambda \oplus X_1$. Denote $S = I_X - T$, $S_1 = S\big|_{X_1}$, and note that $S(X) = S(X_1) = S_1(X_1)$. Since $\operatorname{Ker}(S_1) = N_\lambda \cap X_1 = \{0\}$, we get that S_1 is one-to-one. We will show that $\inf\limits_{x \in S_{X_1}} \|S_1(x)\| > 0$.

By contradiction, assume that there are $x_n \in S_{X_1}$ such that $\|S_1(x_n)\| \to 0$. Since T is compact, we may assume that $T(x_n) \to y$. Then $x_n = (S_1 + T)(x_n) \to y$. Therefore $\|y\| = 1$ and also $S_1(x_n) \to S_1(y)$, so $S_1(y) = 0$. This contradicts S_1 being one-to-one.

Thus there is $c > 0$ such that $\|S_1(x)\| \ge c\|x\|$ for all $x \in X_1$ and by Exercise 1.73, $S_1(X_1) = S(X)$ is closed.

We will now prove that $S(X)$ is finite-codimensional. For $k \in \mathbb{N}_0$ define S^k as $S^0 = I_X$, $S^1 = S$, $S^{k+1} = S \circ S^k$. Let $N_k = \operatorname{Ker}(S^k)$. Since $S^k = (I_X - T)^k = I_X - T_k$ for some compact operator T_k (powers of T are again compact operators), we have $\dim(N_k) < \infty$ for every k. Denote $M_k = S^k(X) = S^k(X_1)$. We have $N_0 \subset N_1 \subset N_2 \cdots$, and $M_0 \supset M_1 \cdots$. We claim that there is n such that $M_n = M_{n+1}$. By contradiction, if all the inclusions $M_0 \supset M_1 \supset \dots$ were strict, we could find by Lemma 15.5 applied to $S\big|_{M_n} : M_n \to M_n$ elements $y_n \in B_{M_n}$ such that $\operatorname{dist}\big(T(y_n), T(M_{n+1})\big) \ge \frac{1}{2}$. This would in particular mean that $\|T(y_n) - T(y_m)\| \ge \frac{1}{2}$ for $n \neq m$, a contradiction with the compactness of T.

Similarly, there is m such that $N_m = N_{m+1}$. Indeed, if $x \in N_k$, i.e., $S^k(x) = 0$, then $S^{k-1}\big(S(x)\big) = 0$ and thus $S(x) \in N_{k-1} \subset N_k$. Therefore we again use Lemma 15.5 for $S\big|_{N_k} : N_k \to N_k$ to obtain the claim. Consequently, $M_n = M_{n'}$ for any $n' \geq n$ and $N_m = N_{m'}$ for any $m' \geq m$.

Finally, we claim that for $p = \max\{n, m\}$ we have $X = N_p \oplus M_p$. For arbitrary $x \in X$ we have $S^p(x) \in M_p$. However, $S^p(M_p) = S^p(S^p(X)) = S^{2p}(X) = S^p(X) = M_p$. Therefore there exists some $y \in M_p$ such that $S^p(y) = S^p(x)$, and so $S^p(y - x) = 0$. Thus $y - x \in N_p$ and $x = (x - y) + y$. From $X = N_p \oplus M_p$ we see that the codimension of M_p (and hence of $M_1 \supset M_p$) is finite and the proof is complete. □

In particular, $(\lambda I_X - T)(X) = \mathrm{Ker}(\lambda I_{X^*} - T^*)_\perp$ (see Exercise 2.44). It can be proved that $(\lambda I_{X^*} - T^*)(X^*)$ is w^*-closed, hence $(\lambda I_{X^*} - T^*)(X^*) = \mathrm{Ker}(\lambda I_X - T)^\perp$ (see Exercise 3.85).

Definition 15.7 *Let X, Y be Banach spaces. An operator $T \in \mathcal{B}(X, Y)$ is called a* Fredholm operator *if $\mathrm{Ker}(T)$ is finite-dimensional and $T(X)$ is finite-codimensional. The number $i(T) = \dim\big(\mathrm{Ker}(T)\big) - \mathrm{codim}\big(T(X)\big)$ is called the* index *of T.*

Note that $T(X)$ in the previous definition is necessarily closed, see Exercise 5.10. As in the proof of Theorem 15.6, if T is a Fredholm operator, we can write $X = \mathrm{Ker}(T) \oplus X_1$ and $T\big|_{X_1}$ is an isomorphism of X_1 onto $T(X)$.

From Theorem 15.6 we immediately obtain:

Proposition 15.8 *Let X be a Banach space and $T \in \mathcal{K}(X)$. Then $\lambda I_X - T$ is a Fredholm operator for every $\lambda \neq 0$.*

Example: Let $X = Y = \ell_2$. For $k \in \mathbb{N}$ define $T(x_i) = (x_{i+k}) \in \ell_2$. Then $\mathrm{Ker}(T) = \mathrm{span}\{e_1, \ldots, e_k\}$, $X_1 = \mathrm{span}\{e_{k+1}, \ldots\}$, $T(X) = X$ and T is a Fredholm operator with $i(T) = k$.

Theorem 15.9 (Fredholm alternative) *Let X be a Banach space, let $T \in \mathcal{K}(X)$ and $\lambda \neq 0$. Then the equation $T(x) - \lambda x = y$ has a solution for every $y \in X$ if and only if the equation $T(x) - \lambda x = 0$ has only the trivial solution $x = 0$.*

In other words, $\mathrm{Ker}(\lambda I_X - T) = \{0\}$ if and only if $(\lambda I_X - T)(X) = X$. In fact, a more general result is true: If T is a compact operator on X and $\lambda \neq 0$, then $i(\lambda I_X - T) = 0$ ([Murp]).

Recall that in general, for $S \in \mathcal{B}(X)$ we have $\mathrm{codim}\big(S(X)\big) = \dim\big(\mathrm{Ker}(S^*)\big)$ if they are both finite.

Proof: We can assume that $\lambda = 1$, denote $S = I_X - T$. If $T(x) - x = 0$ has only the trivial solution $x = 0$, then $N_\lambda = \mathrm{Ker}(S) = \{0\}$ and thus S is an isomorphism into by the proof of Theorem 15.6. We need to show that it is in fact onto.

Set $M_k = S^k(X)$ for $k = 0, 1, \ldots$. In Theorem 15.6 we proved that there is n so that $M_m = M_n$ for all $m \geq n$. We claim that $M_1 = M_0 = X$. If this is not the case, let m be the smallest integer such that $M_{m-1} \neq M_m = M_{m+1}$. Pick $u \in M_{m-1} \backslash M_m$.

Then $S(u) \in M_m = M_{m+1}$. Therefore there is $v \in M_m$ such that $S(v) = S(u)$ and $u \neq v$ since $u \notin M_m$. Hence $S(u - v) = 0$ and $u \neq v$, a contradiction with $\mathrm{Ker}(S) = \{0\}$.

Now assume that S maps X onto X. Define $N_k = \mathrm{Ker}(S^k)$ for $k \in \mathbb{N}$. We need to show that $N_1 = \mathrm{Ker}(S) = \{0\}$. Clearly, $N_k \subset N_{k+1}$ for every k. Assume by contradiction that there is $x_1 \neq 0$ such that $x_1 \in N_1$. By induction we will construct a sequence x_k such that $S(x_{k+1}) = x_k$ and $x_k \in N_k \backslash N_{k-1}$. This will complete the proof, since we know from the proof of Theorem 15.6 that $N_m = N_{m+1}$ for some m.

Assume that $x_1, \dots, x_k$ were constructed. Since S is onto, there is x_{k+1} such that $S(x_{k+1}) = x_k$. Then $S^k(x_{k+1}) = S^{k-1}(x_k) = \cdots = x_1 \neq 0$ and $S^{k+1}(x_{k+1}) = S(x_1) = 0$, which concludes the proof. □

15.2 Spectral Theory

Recall that if X and Y are Banach spaces, an operator $T \in \mathcal{B}(X, Y)$ is called *invertible* if T is an isomorphism from X onto Y (see the paragraphs preceding Definition 1.30).

An operator $T \in \mathcal{B}(X, Y)$ is invertible if and only if there is a bounded operator $T^{-1} \in \mathcal{B}(Y, X)$ such that $T^{-1}T = I_X$ (the identity on X) and $TT^{-1} = I_Y$. By the open mapping theorem, this is equivalent to T being one-to-one and onto.

Thus $T \in \mathcal{B}(X, Y)$ is invertible if and only if T^* is invertible, and $(T^*)^{-1} = (T^{-1})^*$. Also, if $T \in \mathcal{B}(X, Y)$ and $S \in \mathcal{B}(Y, Z)$ are invertible, then ST is invertible and $(ST)^{-1} = T^{-1}S^{-1}$.

Lemma 15.10 *Let X be a Banach space and $T \in \mathcal{B}(X)$. If $\|T\| < 1$ then $(I_X - T)$ is invertible and $(I_X - T)^{-1} = \sum_{k=0}^{\infty} T^k$, where the series converges absolutely in $\mathcal{B}(X)$.*

Proof: First note that $\sum_{k=0}^{\infty} \|T^k\| \leq \sum_{k=0}^{\infty} \|T\|^k = \frac{1}{1-\|T\|}$ and so $\sum_{k=0}^{\infty} T^k$ is absolutely convergent in $\mathcal{B}(X)$. Hence

$$(I_X - T)\sum_{k=0}^{\infty} T^k = (I_X - T) + (T - T^2) + \cdots = I_X.$$

Similarly $\left(\sum_{k=0}^{\infty} T^k\right)(I_X - T) = I_X$. □

Lemma 15.11 *Let X be a Banach space and $S, T \in \mathcal{B}(X)$. If T is invertible and $\|S - T\| < \|T^{-1}\|^{-1}$, then S is invertible and*

$$\|S^{-1} - T^{-1}\| \leq \frac{\|T^{-1}\|^2 \|S - T\|}{1 - \|T^{-1}\| \, \|S - T\|}.$$

Proof: We have $\|T^{-1}(T-S)\| \leq \|T^{-1}\| \cdot \|T-S\| < 1$. Therefore by Lemma 15.10, $I - T^{-1}(T-S) = T^{-1}S$ is invertible, hence S is invertible. We also have $[I_X - T^{-1}(T-S)]^{-1} = \sum_{i=0}^{\infty}(T^{-1}(T-S))^n$. Hence

$$S^{-1} = (T-(T-S))^{-1} = (T(I_X - T^{-1}(T-S)))^{-1} = \sum_{n=0}^{\infty}(T^{-1}(T-S))^n T^{-1}$$

and thus

$$\|S^{-1} - T^{-1}\| \leq \sum_{n=1}^{\infty} \|(T^{-1}(T-S))^n T^{-1}\|$$
$$\leq \|T^{-1}\| \sum_{n=1}^{\infty} (\|T-S\| \cdot \|T^{-1}\|)^n = \frac{\|T^{-1}\|^2 \|T-S\|}{1 - \|T^{-1}\| \, \|T-S\|}.$$

□

Corollary 15.12 *Let X be a Banach space. The set $\mathcal{C}$ of all invertible operators on X is an open set in $\mathcal{B}(X)$ and the mapping $T \mapsto T^{-1}$ is a homeomorphism of $\mathcal{C}$ onto $\mathcal{C}$.*

Definition 15.13 *Let X be a Banach space over $\mathbb{K}$, $T \in \mathcal{B}(X)$. The* spectrum $\sigma(T)$ *of T is defined by*

$$\sigma(T) = \{\lambda \in \mathbb{K} : \lambda I_X - T \text{ is not invertible}\}.$$

The resolvent set $\rho(T)$ *is defined by $\rho(T) = \mathbb{K} \backslash \sigma(T)$. The points of $\rho(T)$ are called* regular values *of T.*
If $\lambda \in \rho(T)$, then $R(\lambda) = (\lambda I_X - T)^{-1}$ is called the resolvent *of T at λ.*

There are precisely three (mutually exclusive) reasons for the operator $(\lambda I_X - T)$ not being invertible, i.e., for λ being in $\sigma(T)$:

(i) $\text{Ker}(\lambda I_X - T) \neq \{0\}$, i.e., the operator $(\lambda I_X - T)$ is not one-to-one. The set of scalars λ with this property is called the *point spectrum* of the operator T, and is denoted by $\sigma_p(T)$. Observe that $\lambda \in \sigma_p(T)$ if and only if there exists $x \neq 0$ such that $Tx = \lambda x$. Such a vector x is called an *eigenvector* of the operator T, and the corresponding λ is called an *eigenvalue* of T. Thus $\sigma_p(T)$ is the set of all eigenvalues of T. The subspace $\text{Ker}(\lambda I_X - T)$ is called the *eigenspace* corresponding to the eigenvalue λ.

(ii) $\text{Ker}(\lambda I_X - T) = \{0\}$ and $(\lambda I_X - T)X$ dense in X. The set of scalars $\lambda \in \sigma(T)$ with this property is called the *continuous spectrum*, and is denoted by $\sigma_c(T)$.

(iii) $\text{Ker}(\lambda I_X - T) = \{0\}$ and $(\lambda I_X - T)X$ not dense in X. The set of scalars λ with this property is called the *residual spectrum*, and is denoted by $\sigma_r(T)$.

It follows that

$$\sigma(T) = \sigma_p(T) \cup \sigma_c(T) \cup \sigma_r(T), \tag{15.1}$$

and the three sets to the right of the equality in (15.1) are pairwise disjoint.

Another way to classify the points of $\sigma(T)$ is by noticing that if X is a Banach space and Y is a normed space, an operator $S \in \mathcal{B}(X, Y)$ has a continuous inverse if and only if $S(X)$ is dense in Y and S is *bounded below* (i.e., there exists $c > 0$ such that $\|S(x)\| \geq c\|x\|$ for all $x \in X$). Indeed, the necessary condition is obvious. On the other side, if S is bounded below, then clearly S is one-to-one, and $S : X \to S(X)$ is an isomorphism. In particular $S(X)$ is a Banach space. Since $S(X)$ is dense in Y it follows that $S(X) = Y$.

Thus, if $(\lambda I_X - T)$ is not invertible, i.e., if $\lambda \in \sigma(T)$, either the space $(\lambda I_X - T)X$ is not dense in X or the operator $(\lambda I_X - T)$ is not bounded below (or both, of course). Elements $\lambda \in \sigma(T)$ for which $(\lambda I_X - T)X$ is not dense in X form the so-called *compression spectrum* $\sigma_{\text{com}}(T)$, and elements $\lambda \in \sigma(T)$ for which $(\lambda I_X - T)$ is not bounded below form the so-called *approximate spectrum* $\sigma_{\text{ap}}(T)$. Observe that $\lambda \in \sigma_{\text{ap}}(T)$ if and only if there exists a sequence $\{x_n\}$ in S_X such that $(\lambda I_X - T)x_n \to 0$, i.e., $Tx_n - \lambda x_n \to 0$. Such a sequence is called an *approximate eigenvector* with *approximate eigenvalue* λ.

So we have

$$\sigma(T) = \sigma_{\text{ap}}(T) \cup \sigma_{\text{com}}(T). \tag{15.2}$$

Observe that, from the very definition, we have

$$\sigma_{\text{r}}(T) = \sigma_{\text{com}}(T) \backslash \sigma_{\text{p}}(T), \tag{15.3}$$

and

$$\sigma_{\text{c}}(T) = \sigma(T) \backslash \big(\sigma_{\text{com}}(T) \cup \sigma_{p}(T)\big). \tag{15.4}$$

Remarks: (i) The spectrum of an operator T in a *complex* Banach space X is never empty, see Theorem 15.16 below. However, if X is a *real* space, then it may happen that $\sigma(T) = \emptyset$, even in the case that X is finite-dimensional (and so T is compact), as a rotation of angle $\pi/2$ in $\mathbb{R}^2$ shows.

(ii) Due to the fact that an operator from a finite-dimensional space into itself is one-to-one if and only if it is onto, it turns out that, if X is finite-dimensional and T is an operator from X into X, then $\sigma(T) = \sigma_{\text{p}}(T)$.

(iii) In general, $\sigma_{\text{p}}(T) \subsetneq \sigma(T)$. The example preceding Proposition 15.22 shows an operator T from a complex Hilbert space into itself that has no eigenvalues. However, Theorem 15.16 shows that $\sigma(T) \neq \emptyset$. Even more, there is a compact operator on a complex Banach space such that its spectrum is $\{0\}$, and it has no eigenvalues, see Exercise 15.29.

(iv) Note that $0 \in \sigma_{\text{p}}(T)$ if and only if $\operatorname{Ker} T \neq \{0\}$, i.e., if and only if T is not one-to-one.

(v) Note that if X is infinite-dimensional and $T \in \mathcal{K}(X)$, then $0 \in \sigma(T)$ as otherwise T would be a compact isomorphism.

Fact 15.14 *Let X be a Banach space and $T \in \mathcal{B}(X)$. Then $\sigma(T) = \sigma(T^*)$.*

Proof: $(\lambda I - T)$ is invertible if and only if $(\lambda I - T)^*$ is invertible, and $(\lambda I_X - T)^* = \lambda I_{X^*} - T^*$. □

Proposition 15.15 *Let X be a Banach space over $\mathbb{K}$ and $T \in \mathcal{B}(X)$. The spectrum $\sigma(T)$ of T is a compact set in $\mathbb{K}$ bounded by $\|T\|$.*

Proof: From Corollary 15.12 it follows that $\rho(T)$ is an open set in $\mathbb{C}$ and thus $\sigma(T)$ is a closed set in $\mathbb{K}$. If $|\lambda| > \|T\|$, then $(\lambda I_X - T)^{-1} = \lambda^{-1}\big(I_X - \frac{T}{\lambda}\big)^{-1} = \sum_{n=0}^{\infty} \frac{T^n}{\lambda^{n+1}}$ exists by Lemma 15.10 and $\lambda \notin \sigma(T)$. Thus $\sigma(T)$ is bounded (by $\|T\|$) and closed, hence compact. □

The next two statements are valid only in the complex case. We will use several results from the theory of complex functions. Recall that if Z is a complex Banach space and D is an open set in the complex plane $\mathbb{C}$, a function $f : D \to Z$ is said to be *analytic* if for every $z_0 \in D$ there is $r = r(z_0)$ and $a_n \in Z$ such that $f(z) = \sum_{n=0}^{\infty} a_n(z - z_0)^n$ for $z \in D(z_0, r) = \{z \in \mathbb{C} : |z - z_0| < r\} \subset D$ and the series is absolutely convergent in $D(z_0, r)$.

We will now show that Liouville's theorem remains valid for complex Banach space valued analytic functions. Precisely, if Z is a complex Banach space and $f : \mathbb{C} \to Z$ is an entire function such that $\sup_{z\in\mathbb{C}} \|f(z)\| < \infty$, then f is a constant function. Indeed, if $h \in X^*$, then $g(z) = h(f(z))$ is an entire function on $\mathbb{C}$ since $h(\sum a_n(z - z_0)^n) = \sum h(a_n)(z - z_0)^n$ and $\sum |h(a_n)|(z - z_0)^n \le \|h\| \sum \|a_n\|(z - z_0)^n < \infty$. Therefore by the standard Liouville theorem, g is constant on $\mathbb{C}$, i.e., $g(z) = g(0)$ for every $z \in \mathbb{C}$. Hence $h(f(z) - f(0)) = 0$ for all $h \in X^*$ and $f(z) = f(0)$.

For other results used in the proofs of the following theorems we refer to [Rudi2].

Theorem 15.16 *If X is a complex Banach space, then, for every $T \in \mathcal{B}(X)$, we have $\sigma(T) \neq \emptyset$.*

Proof: Fix $\lambda_0 \in \rho(T)$ and choose λ satisfying $|\lambda_0 - \lambda| < \|(\lambda_0 I_X - T)^{-1}\|^{-1}$. By Lemma 15.11 applied to $\lambda_0 I_X - T$ and $\lambda I_X - T$ we get

$$R(\lambda) = (\lambda I_X - T)^{-1} = \sum_{n=0}^{\infty}[(\lambda_0 I_X - T)^{-1}(\lambda_0 - \lambda)I_X]^n(\lambda_0 I_X - T)^{-1}$$

$$= \sum_{n=0}^{\infty}(\lambda_0 - \lambda)^n(\lambda_0 I_X - T)^{-n-1} = \sum_{n=0}^{\infty}(\lambda_0 - \lambda)^n R(\lambda_0)^{n+1},$$

where the series is absolutely convergent. We have just proved that the resolvent function R is an analytic function on $\rho(T)$ with values in the Banach space $\mathcal{B}(X)$.

In the proof of Proposition 15.15 we observed that for $|\lambda| > \|T\|$ we have $R(\lambda) = \sum \frac{T^k}{\lambda^{k+1}}$, hence $\|R(\lambda)\| \le \frac{1}{|\lambda| - \|T\|}$. In particular, $R(\lambda) \to 0$ as $|\lambda| \to \infty$.

Therefore, assuming $\rho(T) = \mathbb{C}$, by Liouville's theorem we would have that $R = 0$ on $\mathbb{C}$, which is impossible as $R(\lambda)$ is an inverse operator. Hence $\sigma(T) \neq \emptyset$. □

Definition 15.17 *Let X be a Banach space and $T \in \mathcal{B}(X)$. The* spectral radius $r(T)$ *of T is defined by $r(T) = \sup\{|\lambda| : \lambda \in \sigma(T)\}$.*

From Proposition 15.15 we have $r(T) \leq \|T\|$.

Theorem 15.18 (Gelfand) *Let X be a complex Banach space. For $T \in \mathcal{B}(X)$ we have*

$$r(T) = \lim\big(\|T^n\|^{\frac{1}{n}}\big).$$

In the proof we will use the following statements:

Fact 15.19 *Let X be a Banach space. Assume that $T, S \in \mathcal{B}(X)$ commute, that is, $TS = ST$. Then ST is invertible if and only if both T and S are invertible.*

Proof: We already observed that if T and S are invertible, then so is ST.

On the other hand, assume that ST is invertible. We claim that T and S are both one-to-one. Indeed, if for $x \neq 0$ we have $T(x) = 0$, then $(ST)(x) = S(0) = 0$ and ST is not one-to-one. If for some $x \neq 0$ we have $S(x) = 0$, then TS is not one-to-one again. If $T(X) \neq X$ then TS is not onto and similarly we proceed if $S(X) \neq X$, using the commutability of S and T. Thus, T and S are one-to-one and onto, hence invertible. □

Fact 15.20 *Let X be a complex Banach space. If T is an operator in $\mathcal{B}(X)$ and $n \in \mathbb{N}$, then $\sigma(T^n) = \{\mu^n : \mu \in \sigma(T)\}$.*

Proof: For $\lambda \in \mathbb{C}$, factor the complex polynomial $t^n - \lambda$ as $(t-\lambda_1)\cdot(t-\lambda_2)\cdots(t-\lambda_n)$ for every $t \in \mathbb{C}$. Then clearly $(T^n - \lambda I_X) = (T - \lambda_1 I_X)(T - \lambda_2 I_X)\cdots(T - \lambda_n I_X)$. By the inductive use of Fact 15.19 we get that $T^n - \lambda I_X$ is invertible if and only if $(T - \lambda_i I_X)$ are invertible for all i. This means that $\lambda \in \sigma(T^n)$ if and only if at least one nth root λ_i of λ is in $\sigma(T)$. Thus $\lambda \in \sigma(T^n)$ if and only if $\lambda = \mu^n$ for some $\mu \in \sigma(T)$. □

Proof of Theorem 15.18: (Sketch) By Fact 15.20, $\sigma(T^n) = \{t^n : t \in \sigma(T)\}$, so $r(T^n) = r(T)^n$. By Proposition 15.15 we have $r(T^n) \leq \|T^n\|$, therefore $r(T)^n \leq \|T^n\|$ for every $n \in \mathbb{N}$. Hence $r(T) \leq \|T^n\|^{\frac{1}{n}}$ for all n and thus $r(T) \leq \liminf \|T^n\|^{\frac{1}{n}}$.

On the other hand, we proved that $R(\lambda)$ is an analytic function on $\rho(T)$. For $|\lambda| > \|T\|$ we have an expansion $R(\lambda) = \frac{1}{\lambda}\sum \frac{T^k}{\lambda^k}$. By the properties of Laurent series, this expansion converges for all λ such that $|\lambda| > r(T)$, so there is C such that $\sum_n \|\frac{T^k}{\lambda^k}\| \leq C$. These facts follow from the general theory of Banach space valued analytic functions ([Rudi2]) and are by no means obvious.

Thus $\|T^n\|^{\frac{1}{n}} \leq |\lambda| C^{\frac{1}{n}} \to |\lambda|$. Since $|\lambda| > r(T)$ was arbitrary, we get $r(t) \leq \liminf \|T^n\|^{\frac{1}{n}} \leq \limsup \|T^n\|^{\frac{1}{n}} \leq r(t)$, so $\lim \|T^n\|^{\frac{1}{n}} = r(t)$. □

Remark: The example of a rotation by angle $\pi/2$ in $\mathbb{R}^2$ mentioned in Remark (i) preceding Fact 15.14 shows that Theorems 15.16 and 15.18 are not valid in the real case.

Lemma 15.21 *Let X be a Banach space, $T \in \mathcal{B}(X)$. Assume that $\lambda_1, \ldots, \lambda_n$ are distinct eigenvalues of T. If e_i is an eigenvector corresponding to λ_i for $i = 1, \ldots, n$, then the set $\{e_1, \ldots, e_n\}$ is linearly independent.*

Proof: By induction. Assume that $e_1, \ldots, e_{n-1}$ are linearly independent and let $e_n = \sum_{i=1}^{n-1} \alpha_i e_i$. Then $\sum_{i=1}^{n-1} \lambda_n \alpha_i e_i = \lambda_n e_n = T(e_n) = \sum_{i=1}^{n-1} \lambda_i \alpha_i e_i$, that is, $\sum_{i=1}^{n-1} (\lambda_n - \lambda_i)\alpha_i e_i = 0$. Since $e_1, \ldots, e_{n-1}$ are linearly independent and $\lambda_n - \lambda_i \neq 0$, we get $\alpha_i = 0$ for all i. □

Let X be a Banach space and $T \in \mathcal{B}(X)$. A closed subspace Y of X is called *invariant* for T if $T(Y) \subset Y$. Obviously, $\{0\}$, X, and all eigenspaces of T are invariant for T.

It is not known whether every bounded operator on a Hilbert space has an invariant subspace other than $\{0\}$ and the whole space (a *non-trivial* invariant subspace). However, Enflo constructed the first example of a Banach space X and an operator from $\mathcal{B}(X)$ without a non-trivial invariant subspace. Note that there exist compact operators in complex Hilbert spaces without any eigenvalue (see Exercise 15.29), and so with no eigenspaces. It is known, in contrast, that every compact operator has a non-trivial invariant subspace, see Theorem 12.23.

Example: The following example shows an operator T from the complex Hilbert space $L_2[0, 1]$ into itself with the following properties:

(i) $\sigma_p(T) = \emptyset$.

(ii) $\sigma(T) = \sigma_{ap}(T) = \sigma_c(T) = [0, 1]$. Consider the operator $T \in \mathcal{B}(L_2[0, 1])$ (over the complex scalars) defined by $T(x)\colon t \mapsto tx(t)$. Clearly $\|T\| \leq 1$.

(i) If $\lambda \in \mathbb{C}$ and for some $x(t)$ from L_2 we have $\lambda x(t) - tx(t) = 0$, then $(\lambda - t)x(t) = 0$ almost everywhere on $[0, 1]$, so considering $t \neq \lambda$ we get $x(t) = 0$ a.e. This shows that $\sigma_p(T) = \emptyset$

(iia) We shall prove first that $[0, 1] \subset \sigma_{ap}(T)$ $\big(\subset \sigma(T)\big)$, (so in particular $\|T\| = 1$). To this end, fix $\lambda \in [0, 1]$ and choose $\varepsilon > 0$ such that $[\lambda, \lambda + \varepsilon] \subset [0, 1]$ or $[\lambda - \varepsilon, \lambda] \subset [0, 1]$. Assume that $[\lambda, \lambda + \varepsilon] \subset [0, 1]$ is the case and put

$$x_\varepsilon = \begin{cases} \frac{1}{\sqrt{\varepsilon}} & \text{for } t \in [\lambda, \lambda + \varepsilon], \\ 0 & \text{for } t \notin [\lambda, \lambda + \varepsilon] \end{cases}$$

(the definition of x_ε in the case $[\lambda-\varepsilon,\lambda]\subset[0,1]$ is analogous). We easily observe that $\int_0^1 x_\varepsilon^2(t)\,dt=\int_\lambda^{\lambda+\varepsilon}\frac{1}{\varepsilon}=1$, so $x_\varepsilon\in S_{L_2}$. On the other hand, $(\lambda I_X-T)(x_\varepsilon)\colon t\mapsto(\lambda-t)x_\varepsilon(t)$, therefore

$$\begin{aligned}\|(\lambda I_X-T)(x_\varepsilon)\|^2&=\int_\lambda^{\lambda+\varepsilon}\frac{1}{\varepsilon}(\lambda-t)^2\,dt=\frac{1}{\varepsilon}\int_\lambda^{\lambda+\varepsilon}(\lambda-t)^2\,dt\\&=\frac{1}{\varepsilon}\Big[-\frac{(\lambda-t)^3}{3}\Big]_{t=\lambda}^{t=\lambda+\varepsilon}=\frac{1}{3\varepsilon}\cdot(\varepsilon^3)=\frac{\varepsilon^2}{3},\end{aligned}$$

so $(\lambda I_X-T)(x_\varepsilon)\to0$ as $\varepsilon\to0$. This proves that $\lambda\in\sigma_{\mathrm{ap}}(T)$. It follows that $[0,1]\subset\sigma_{\mathrm{ap}}(T)$.

(iib) Let $\lambda\in\mathbb{C}\setminus[0,1]$. Since $(t-\lambda)$ is bounded away from 0 for $t\in[0,1]$, given $x\in L_2$ the mapping $y(t):=(t-\lambda)^{-1}x(t)$ belongs to L_2; moreover, $(\lambda I-T)y=-x$. This shows that $(\lambda I-T)$ is onto. Since $\sigma_{\mathrm{p}}(T)=\emptyset$, the operator $(\lambda I-T)$ is one-to-one. The open mapping theorem concludes that $(\lambda I-T)$ is invertible, so $\lambda\notin\sigma(T)$.

This proves $\sigma(T)=\sigma_{\mathrm{ap}}(T)=[0,1]$.

(iic) If $\lambda\in[0,1]$, the range of $(\lambda I-T)$ is dense in L_2. Indeed, given $x\in L_2$ and $n\in\mathbb{N}$, put

$$x_n(t)=\begin{cases}x(t), & \text{if } |t-\lambda|\ge 1/n,\\ 0, & \text{if } |t-\lambda|<1/n.\end{cases}$$

Then $x_n\to x$ in L_2, and $(\lambda I-T)y_n=-x_n$, where $y_n(t):=(t-\lambda)^{-1}x_n(t)$ for all $t\in[0,1]$ and $n\in\mathbb{N}$. This proves (iic).

Putting together (i) and (iic) we get $\sigma_{\mathrm{c}}(T)=[0,1]\ \big(=\sigma(T)\big)$.

It is easy to observe that the operator T has a rich structure of non-trivial invariant subspaces. For example, the subspace $L_2[0,r]$ of $L_2[0,1]$ is invariant for every $r\in[0,1]$.

Proposition 15.22 *If X is a Banach space and $T\in\mathcal{K}(X)$, then $\sigma(T)=\sigma_p(T)\cup\{0\}$.*

Proof: If $\lambda\neq0$ is not an eigenvalue, then $\mathrm{Ker}(\lambda I_X-T)=\{0\}$. By Theorem 15.9, λI_X-T is also onto, hence invertible. Thus $\lambda\notin\sigma(T)$. □

Proposition 15.23 (Riesz, Schauder) *Let X be a Banach space, $T\in\mathcal{K}(X)$. For every $\varepsilon>0$, T has only finitely many eigenvalues with absolute value larger than ε.*

Proof: Assume that there is an infinite sequence $\{\lambda_i\}$ of distinct eigenvalues such that $|\lambda_i|\ge\varepsilon$ for every i. For every λ_i choose an eigenvector x_i. For $n\in\mathbb{N}$ define $X_n=\mathrm{span}\{x_1,\dots,x_n\}$, note that $T(X_n)=X_n$ and $X_{n-1}\neq X_n$ by Lemma 15.21.

By Lemma 1.37 we obtain $y_n\in X_n$ such that $\mathrm{dist}(y_n,X_{n-1})\ge\frac{1}{2}$ and $\|y_n\|=1$ for every n. Put $z_n=y_n/\lambda_n$ and note that $\|z_n\|\le\frac{1}{\varepsilon}$. We have $T(z_n)\in X_n$ and also $y_n-T(z_n)\in X_{n-1}$, since for $y_n=\sum_{k=1}^n c_kx_k$ we have

$$y_n - T(z_n) = \sum_{k=1}^{n} \big(1 - \tfrac{\lambda_k}{\lambda_n}\big) c_k x_k = \sum_{k=1}^{n-1} \big(1 - \tfrac{\lambda_k}{\lambda_n}\big) c_k x_k \in X_{n-1}.$$

If $n > m$ then $T(z_m) \in X_m \subset X_{n-1}$ and $y_n - T(z_n) \in X_{n-1}$ and thus we have

$$\begin{aligned} \|T(z_n) - T(z_m)\| &\geq \operatorname{dist}\big(T(z_n), X_{n-1}\big) \\ &= \operatorname{dist}\big(T(z_n) + y_n - T(z_n), X_{n-1}\big) = \operatorname{dist}(y_n, X_{n-1}) \geq \tfrac{1}{2}. \end{aligned}$$

We obtained an infinite bounded sequence $\{z_n\}$ such that $\|T(z_n) - T(z_m)\| \geq \frac{1}{2}$ for $m \neq n$, a contradiction with the compactness of $\overline{T(B_X)}$. □

Theorem 15.6 together with Proposition 15.23 yield the following:

Corollary 15.24 *Let X be a Banach space and $T \in \mathcal{K}(X)$. Then $\sigma(T) = \{0, \lambda_1, \lambda_2, \dots\}$, where $\{\lambda_i\}$ is either a finite set (possibly empty) or a sequence tending to zero, formed by non-zero eigenvalues, each of λ_i having a finite-dimensional eigenspace.*

15.3 Self-Adjoint Operators

Let H be a Hilbert space and $T \in \mathcal{B}(H)$. Using Theorem 2.22 we observe that there exists a unique operator $Q \in \mathcal{B}(H)$ satisfying $\big(T(x), y\big) = \big(x, Q(y)\big)$ for every $x, y \in H$.

Definition 15.25 *Let T be an operator on a Hilbert space H. The* adjoint operator *to T, denoted T^*, is defined by*

$$\big(T(x), y\big) = \big(x, T^*(y)\big) \text{ for all } x, y \in H.$$

In order to preserve the customary notation, for the rest of this section the symbol T^* is reserved for the adjoint operator rather than for the usual dual operator.

Let H be a Hilbert space, $T \in \mathcal{B}(H)$ and let $T^* \in \mathcal{B}(H)$ be the adjoint operator. We easily observe that $T^{**} = T$. Note also that $I_H^* = I_H$, $(\lambda T)^* = \bar{\lambda} T^*$, $(ST)^* = T^* S^*$, and $(T + S)^* = T^* + S^*$ for $T, S \in \mathcal{B}(H)$.

Denote for the moment by T^d the dual operator $T^d \in \mathcal{B}(H^*)$. Let $f, g \in H^*$, let $a, b \in H$ be vectors assigned to f, g by the Riesz identification (Theorem 2.22), that is, $f(x) = (x, a)$ and $g(x) = (x, b)$ for all $x \in H$. It follows from our definitions that $T^d(f) = g$ if and only if $T^*(a) = b$. Similarly to Fact 15.14 we prove the following:

Fact 15.26 *Let H be a Hilbert space, $T \in \mathcal{B}(H)$. $\lambda \in \sigma(T)$ if and only if $\bar{\lambda} \in \sigma(T^*)$.*

Proposition 15.27 *Let H be a Hilbert space and $T \in \mathcal{B}(H)$. Then* $\mathrm{Ker}(T) = T^*(H)^\perp$.

Proof:

$$\begin{aligned}\mathrm{Ker}(T) &= \{x \in H : T(x) = 0\} = \{x \in H : \big(T(x), y\big) = 0 \text{ for all } y \in H\} \\ &= \{x \in H : \big(x, T^*(y)\big) = 0 \text{ for all } y \in H\} = T^*(H)^\perp.\end{aligned}$$

□

In particular we have $H = \mathrm{Ker}(T) \oplus T^*(H)$.

Definition 15.28 *Let H be a Hilbert space. An operator $T \in \mathcal{B}(H)$ is called* self-adjoint *or* hermitian *if $T^* = T$, that is, $\big(T(x), y\big) = \big(x, T(y)\big)$ for all $x, y \in H$.*

Note that TT^* and T^*T are self-adjoint for any $T \in \mathcal{B}(H)$.

Proposition 15.29 *Let H be a complex Hilbert space. For every $T \in \mathcal{B}(H)$ there are self-adjoint operators T_1, T_2 on H such that $T = T_1 + iT_2$. Moreover, this decomposition is unique.*

Proof: Put $T_1 = \frac{1}{2}(T + T^*)$, $T_2 = -\frac{1}{2}i(T - T^*)$. Then clearly $T_{1,2}$ are self-adjoint and $T_1 + iT_2 = T$.

If T_1, T_2 are self-adjoint such that $T_1 + iT_2 = 0$, then $T_1 - iT_2 = (T_1 + iT_2)^* = 0$. Adding and subtracting these two equations we get $T_2 = 0$ and $T_1 = 0$, from which the uniqueness follows. □

Similarly to the polarization identities for a norm of a Hilbert space, in a complex Hilbert space H we have for every $T \in \mathcal{B}(H)$:

$$\begin{aligned}4\big(T(x), y\big) = \big(T(x+y), x+y\big) - \big(T(x-y), x-y\big) + i\big(T(x+iy), x+iy\big) \\ -i\big(T(x-iy), x-iy\big).\end{aligned}$$

If H is a real Hilbert space and T is a self-adjoint operator on H, then

$$4\big(T(x), y\big) = \big(T(x+y), x+y\big) - \big(T(x-y), x-y\big).$$

Proposition 15.30 *Let H be a complex Hilbert space and $T \in \mathcal{B}(H)$. If $\big(T(x), x\big) = 0$ for every $x \in H$, then $T = 0$.*
Let H be a real Hilbert space. If T is self-adjoint and $\big(T(x), x\big) = 0$ for every $x \in H$, then $T = 0$.

Proof: By an appropriate polarization identity above we have $\big(T(x), y\big) = 0$ for every x and $y \in H$ and in particular $\big(T(x), T(x)\big) = 0$ for every $x \in H$. Therefore $T = 0$. □

For self-adjoint operators, this is a consequence of a more general result below.

Example: Let T be the operator on $\mathbb{R}^2$ given by $T\big((x_1, x_2)\big) = (-x_2, x_1)$ (rotation by $\pi/2$). Then $\big(T(x), x\big) = (-x_2x_1 + x_1x_2) = 0$ for $x \in \mathbb{R}^2$. So Proposition 15.30 does not hold in general in the real case for operators that are not self-adjoint.

Proposition 15.31 *Let T be an operator on a Hilbert space H. If T is self-adjoint then*

$$\|T\| = \sup_{x \in B_H} |\big(T(x), x\big)|.$$

Proof: If $\|x\| \leq 1$, then $|\big(T(x), x\big)| \leq \|T(x)\| \cdot \|x\| \leq \|T\| \cdot \|x\|^2 \leq \|T\|$. Thus $\sup_{x \in B_H} |\big(T(x), x\big)| \leq \|T\|$. Put $\sup_{x \in B_H} |\big(T(x), x\big)| = C$. Note that $|\big(T(z), z\big)| \leq C\|z\|^2$ for every $z \in H$.

If $z \in H$, $z \neq 0$, put $\lambda = \big(\frac{\|T(z)\|}{\|z\|}\big)^{\frac{1}{2}}$ and $u = \frac{1}{\lambda}T(z)$. Using the fact that $\big(T(\lambda z), u\big)$ is a real number, and the parallelogram equality we get

$$\begin{aligned} \|T(z)\|^2 &= \big(T(z), T(z)\big) = \big(T(\lambda z), \tfrac{1}{\lambda}T(z)\big) = (T(\lambda z), u) \\ &= \tfrac{1}{4}[\big(T(\lambda z + u), \lambda z + u\big) - \big(T(\lambda z - u), \lambda z - u\big)] \\ &\leq \tfrac{1}{4}C(\|\lambda z + u\|^2 + \|\lambda z - u\|^2) = \tfrac{1}{2}C(\|\lambda z\|^2 + \|u\|^2) \\ &= \tfrac{1}{2}C(\lambda^2\|z\|^2 + \tfrac{1}{\lambda^2}\|T(z)\|^2) = C\|z\| \cdot \|T(z)\|. \end{aligned}$$

Therefore $\|T(z)\| \leq C\|z\|$ for every $z \in H$, hence $\|T\| \leq C$. □

Lemma 15.32 *Let H be a Hilbert space H, $T \in \mathcal{B}(H)$. If T is self-adjoint, then $\big(T(x), x\big)$ is a real number for all $x \in H$ and all eigenvalues of T are real number.*

Proof: Let $x \in H$. Then $\big(T(x), x\big) = \big(x, T(x)\big) = \overline{\big(T(x), x\big)}$.

If λ is an eigenvalue of T with an eigenvector x, then $\big(T(x), x\big) = (\lambda x, x) = \lambda(x, x)$, hence $\lambda = \frac{(T(x), x)}{\|x\|^2}$ is real. □

Proposition 15.33 *Let H be a Hilbert space and $T \in \mathcal{B}(H)$. If T is self-adjoint then $\|T^n\| = \|T\|^n$ for every $n \geq 1$.*

Proof: We can estimate

$$\|T\|^2 = \sup_{x \in B_H} \big(T(x), T(x)\big) = \sup_{x \in B_H} \big(T^*T(x), x\big) \leq \|T^*T\| \leq \|T^*\| \cdot \|T\| = \|T\|^2.$$

Therefore $\|T^*T\| = \|TT^*\| = \|T\|^2$. Since T is self-adjoint, $\|T\|^2 = \|T^2\|$. The operator T^k is also self-adjoint, so we have $\|T^{2^k}\| = \|T\|^{2^k}$ for every k. If $1 \leq n \leq 2^k$ then

$$\|T\|^{2^k} = \|T^{2^k}\| = \|T^nT^{2^k - n}\| \leq \|T^n\| \cdot \|T\|^{2^k - n} \leq \|T\|^n\|T\|^{2^k - n} = \|T\|^{2^k},$$

hence $\|T^n\| \cdot \|T\|^{2^k-n} = \|T\|^{2k}$. Thus $\|T^n\| = \|T\|^n$. □

Proposition 15.34 *Let T be a self-adjoint operator on a Hilbert space H and let λ be a scalar. Then $\lambda \in \sigma(T)$ if and only if* $\inf_{x \in S_H} \|(\lambda I_H - T)(x)\| = 0$. *In other words, $\sigma(T) = \sigma_{\text{ap}}(T)$.*

Proof: If $\lambda \in \rho(T)$, then $(\lambda I_H - T)^{-1} \in \mathcal{B}(H)$ and for $x \in S_H$ we have

$$1 = \|x\| = \|(\lambda I_H - T)^{-1}(\lambda I_H - T)(x)\| \le \|(\lambda I_H - T)^{-1}\| \cdot \|(\lambda I_H - T)(x)\|.$$

Hence $\inf_{\|x\|=1} \|(\lambda I_H - T)(x)\| \ge \|(\lambda I_H - T)^{-1}\|^{-1}$.

Now assume that $\inf_{x \in S_H} \|(\lambda I_H - T)(x)\| - C > 0$ for some $C > 0$. We then have $\|(\lambda I_H - T)(x)\| \ge C\|x\|$ for every $x \in X$, hence $\lambda I_H - T$ is an isomorphism into (Exercise 1.73). In particular, its range is closed in H. We will prove that $\lambda I_H - T$ is also dense in H, thus showing that $\lambda I_H - T$ is invertible.

By contradiction, assume that $(\lambda I_H - T)(H)$ is not dense in H. Then by Proposition 2.7 and Theorem 2.22 there is $y_0 \in H$, $y_0 \neq 0$, such that $\big((\lambda I_H - T)(x), y_0\big) = 0$ for every $x \in H$. Since $\big((\lambda I_H - T)(x), y_0\big) = \big(x, (\bar{\lambda} I_H - T)(y_0)\big)$ for every $x \in H$ as T is a self-adjoint operator, we get that $\big(x, (\bar{\lambda} I_H - T)(y_0)\big) = 0$ for every $x \in H$, hence $(\bar{\lambda} I_H - T)(y_0) = 0$ and $y_0 \neq 0$. This means that $\bar{\lambda}$ is an eigenvalue of T. Since all eigenvalues of T are real, $(\lambda I_H - T)(y_0) = 0$ and $y_0 \neq 0$, contradicting $\|(\lambda I_H - T)(y_0)\| \ge C\|y_0\|$. Therefore $(\lambda I_H - T)(H)$ is dense in H. □

Remark: The operator T in the example preceding Proposition 15.22 is easily seen to be self-adjoint.

Theorem 15.35 *Let T be a self-adjoint operator on a Hilbert space H. Define numbers $m_T = \inf_{x \in S_H}\big(T(x), x\big)$ and $M_T = \sup_{x \in S_H}\big(T(x), x\big)$. Then $\sigma(T) \subset [m_T, M_T]$ (closed interval on the real line) and $m_T, M_T \in \sigma(T)$.*

Proof: First we will show that $\sigma(T)$ lies on the real line. Assume that H is a complex Hilbert space and take any $\lambda = \alpha + i\beta$. For every $x \in S_H$ we write

$$\begin{aligned}\big(\lambda x - T(x), x\big) - \big(x, \lambda x - T(x)\big) &= \lambda\|x\|^2 - \big(T(x), x\big) - \bar{\lambda}\|x\|^2 + \big(x, T(x)\big)\\ &= (\lambda - \bar{\lambda})(x, x) = 2i\beta,\end{aligned}$$

hence

$$\begin{aligned}2|\beta| &= |\big(\lambda x - T(x), x\big) - \big(x, \lambda x - T(x)\big)| \le |\big(\lambda x - T(x), x\big)| + |\big(x, \lambda x - T(x)\big)|\\ &\le \|\lambda x - T(x)\| \cdot \|x\| + \|x\| \cdot \|\lambda x - T(x)\| = 2\|(\lambda I_H - T)(x)\|.\end{aligned}$$

This shows that $\inf\limits_{x \in S_H} \|(\lambda I_H - T)(x)\| \ge |\beta|$, so $\lambda \in \sigma(T)$ implies $|\beta| = 0$ by Proposition 15.34, that is, λ is a real number.

Considering $S = T + \mu I_H$ we have $\sigma(S) = \sigma(T) + \mu$, $m_S = m_T + \mu$, and $M_S = M_T + \mu$. So we may assume that $0 \le m_T \le M_T$.

Since then $\|T\| = M_T$ by Proposition 15.31 and $\sigma(T) \subset \mathbb{R}$, we have that $\sigma(T) \subset [-M_T, M_T]$ by Proposition 15.15. We will show $\lambda = m_T - d \notin \sigma(T)$ for every $d > 0$. For $x \in S_H$ we estimate

$$\big((T - \lambda I_H)(x), x\big) = \big(T(x), x\big) - (\lambda x, x) \ge m_T - \lambda\|x\|^2 = m_T - \lambda = d.$$

Since also $|\big((T - \lambda I_H)(x), x\big)| \le \|(\lambda I_H - T)(x)\|\,\|x\| = \|(\lambda I_X - T)(x)\|$, we get $\inf_{x \in S_H} \|(\lambda I_H - T)(x)\| \ge d > 0$. By Proposition 15.34, $\lambda \notin \sigma(T)$ and we established $\sigma(T) \subset [m_T, M_T]$.

Now we will show that $M_T \in \sigma(T)$. Again we will use Proposition 15.34. Let $x_n \in S_H$ be such that $\big(T(x_n), x_n\big) \to M_T$. Recall that we are assuming $0 \le m_T \le M_T$, hence $M_T = \|T\|$, in particular $\|T(x_n)\| \le M_T$. Using the fact that $\big(T(x_n), x_n\big)$ is real we get

$$\begin{aligned}0 &\le \|(M_T I_H - T)(x_n)\|^2 = \|M_T x_n - T(x_n)\|^2 \\ &= M_T^2\|x_n\|^2 + \|T(x_n)\|^2 - 2M_T\big(T(x_n), x_n\big) \le M_T^2 + M_T^2 - 2M_T\big(T(x_n), x_n\big) \to 0.\end{aligned}$$

Therefore $\|(M_T I_H - T)(x_n)\| \to 0$ and hence $M_T \in \sigma(T)$.

By considering $S = T - M_T I_H$ we get $m_S \le M_S = 0$, so $|m_S| = \|S\|$ and we prove in the same way that $m_S \in \sigma(S)$, that is, $m_T \in \sigma(T)$. □

In particular, $\|T\| \in \sigma(T)$ if T is self-adjoint.

Lemma 15.36 *Let T be an operator on a Hilbert space H. If T is self-adjoint then eigenvectors corresponding to different eigenvalues are orthogonal.*

Proof: If $T(x_1) = \lambda x_1$ and $T(x_2) = \mu x_2$ for $x_1 \ne 0$ and $x_2 \ne 0$, $\lambda \ne \mu$, then (recalling that λ, μ must be real)

$$\begin{aligned}\big(T(x_1), x_2\big) &= (\lambda x_1, x_2) = \lambda(x_1, x_2)\\ \big(T(x_1), x_2\big) &= \big(x_1, T(x_2)\big) = (x_1, \mu x_2) = \mu(x_1, x_2).\end{aligned}$$

Therefore $(\lambda - \mu)(x_1, x_2) = 0$. Since $(\lambda - \mu) \ne 0$, we have $(x_1, x_2) = 0$. □

This implies that eigenspaces corresponding to distinct eigenvalues of a self-adjoint operator are mutually orthogonal. We observed that eigenspaces are invariant subspaces for the given operator. Recall that given a closed subspace M of a Hilbert space H, its orthogonal complement $M^\perp$ in H satisfies $H = M \oplus M^\perp$.

Proposition 15.37 *Let T be a self-adjoint operator on a Hilbert space H. Let M be a closed subspace of H that is invariant under T. Then $N = M^\perp$ is invariant under T. Denote $T_1 = T\big|_M$ and $T_2 = T\big|_N$. Then T_1 is a self-adjoint operator on M, T_2 is a self-adjoint operator on N, $T(H) = T_1(M) \oplus T_2(N)$ and $\sigma(T) = \sigma(T_1) \cup \sigma(T_2)$.*

Proof: Let $y \in N$. Since M is invariant under T, for every $x \in M$ we have $0 = \big(T(x), y\big) = \big(x, T(y)\big)$ and thus $T(y) \in N$. Therefore N is invariant under T.

Since both M and N are invariant, the restrictions are self-adjoint operators on the corresponding subspaces.

Let $\lambda \in \sigma(T_1)$. By Proposition 15.34, there are $x_n \in S_M$ such that $\|\lambda x_n - T_1(x_n)\| \to 0$. Therefore $\|\lambda x_n - T(x_n)\| \to 0$ showing $\lambda \in \sigma(T)$. Similarly we show that $\sigma(T_2) \subset \sigma(T)$.

Assume that $\lambda \notin \sigma(T_1) \cup \sigma(T_2)$. Then there is $C > 0$ such that for every $x \in M$ and for every $y \in N$ we have $\|\lambda x - T(x)\| \geq C\|x\|$ and $\|\lambda y - Ty\| \geq C\|y\|$. Write $z \in H$ as $z = x + y$, $x \in M$ and $y \in N$. Since $\lambda x - T(x) \in M$ and $\lambda y - T(y) \in N$, we get

$$\|\lambda z - T(z)\|^2 = \|\lambda x - T(x)\|^2 + \|\lambda y - T(y)\|^2 \geq C^2(\|x\|^2 + \|y\|^2) = C^2\|z\|^2.$$

Thus $\lambda \notin \sigma(T)$ by Proposition 15.34. □

We will now put together the results of the previous two sections.

Proposition 15.38 *For every compact self-adjoint operator T on a Hilbert space we have $\sigma_{\mathrm{p}}(T) \neq \emptyset$.*

Proof: If $T = 0$ then 0 is an eigenvalue. If $T \neq 0$, then by Theorem 15.35, $\|T\| \in \sigma(T)$, and $\|T\| \neq 0$ is an eigenvalue by Proposition 15.22. □

Theorem 15.39 *Let $T \neq 0$ be a compact self-adjoint operator on an infinite-dimensional Hilbert space H. Then $\sigma(T) = \{0\} \cup \{\lambda_i\}$, where λ_i are distinct real non-zero eigenvalues of T. The set $\{\lambda_i\}$ contains $\|T\|$ and is either finite or a countable sequence convergent to zero.*
Moreover, the space H has an orthonormal basis formed by eigenvectors corresponding to eigenvalues of T.

Proof: Since H is infinite-dimensional and T is compact, $0 \in \sigma(T)$. Since also $\|T\| \in \sigma(T)$, there is at least one non-zero eigenvalue and by Proposition 15.23 and Theorem 15.35 there is at most countably many real eigenvalues of T convergent to zero.

It remains to show that we can form an orthonormal basis out of eigenvectors of T. For an eigenvalue λ denote $N_\lambda = \operatorname{Ker}(\lambda I_H - T)$. We form an orthonormal basis B_λ of each N_λ. By Lemma 15.36, $B = \bigcup B_\lambda$ is an orthonormal set in H, clearly $\overline{\operatorname{span}}(B)$ contains all eigenvectors of T. If $\overline{\operatorname{span}}(B) \neq H$, consider $G = \overline{\operatorname{span}}(B)^\perp$. Since all eigenspaces are invariant for T, so is $\overline{\operatorname{span}}(B)$ and hence by Proposition 15.37, G is also an invariant subspace for G. Moreover, $\sigma(T) = \sigma(T\big|_{\overline{\operatorname{span}}(B)}) + \sigma(T\big|_G)$. However, $T\big|_G$ has an eigenvalue, hence a non-zero eigenvector v. It must also be an eigenvector of T and thus $v \in G \cap \overline{\operatorname{span}}(B)$, a contradiction. This shows that $\overline{\operatorname{span}}(B) = H$, hence B is an orthonormal basis of H. □

Corollary 15.40 *Let T be a compact self-adjoint operator on a Hilbert space H. Then $\overline{\sigma_p(T)} = \sigma(T)$.*

Proof: If H is finite-dimensional, then every $\lambda \in \sigma(T)$ is an eigenvalue. If H is infinite-dimensional, then the only point of $\sigma(T)$ that need not be an eigenvalue is 0. If the set of non-zero eigenvalues is countable, then it converges to 0 and we are done.

The last case is that the set of non-zero eigenvalues is finite and H is infinite-dimensional. Since eigenspaces of non-zero eigenvalues are finite-dimensional (Theorem 15.6), the only possibility for eigenvectors to form an orthonormal basis is that 0 is also an eigenvalue. □

Theorem 15.41 (Spectral decomposition) *Let T be a compact self-adjoint operator on an infinite-dimensional separable Hilbert space H. Then there is an orthonormal basis $\{e_i\}$ of H such that each e_i is an eigenvector corresponding to some real eigenvalue λ_i of T, and for all $x \in H$ we have*

$$T(x) = \sum_{i=1}^{\infty} \lambda_i (x, e_i) e_i.$$

Moreover, for every $\lambda \notin \sigma(T)$ and $x \in H$ we have

$$R(\lambda)(x) = \sum_{i=1}^{\infty} \frac{(x, e_i)}{\lambda - \lambda_i} e_i.$$

Proof: Let $\{e_i\}$ be the (countable) orthonormal basis of H from Theorem 15.39. Note that for $x \in H$, the series $\sum \lambda_i (x, e_i) e_i$ is convergent, as

$$\left\| \sum_{i=n}^{m} \lambda_i (x, e_i) e_i \right\|^2 = \sum_{i=n}^{m} |\lambda_i (x, e_i)|^2 \leq \|T\| \sum_{i=n}^{m} |(x, e_i)|^2 \to 0$$

as $n, m \to \infty$, since $\{e_i\}$ is an orthonormal basis. Also, if $\|x\| \leq 1$, then for every $n \in \mathbb{N}$ we have

$$\left\| \sum_{i=1}^{n} \lambda_i (x, e_i) e_i \right\|^2 = \sum_{i=1}^{n} \lambda_i^2 |(x, e_i)|^2 \leq \|T\|^2 \sum_{i=1}^{n} |(x, e_i)|^2$$
$$\leq \|T\|^2 \cdot \sum_{i=1}^{\infty} |(x, e_i)|^2 = \|T\|^2 \cdot \|x\|^2.$$

Thus the operator G defined by $G(x) = \sum_{i=1}^{\infty} \lambda_i (x, e_i) e_i \in \mathcal{B}(H)$ is continuous. From $T(e_i) = \lambda_i e_i$ we have that $T(e_i) = G(e_i)$, so by linearity and continuity, $T = G$ on H.

Now consider some $\lambda \in \rho(T)$. Since $\sigma(T)$ is closed, there is $\delta > 0$ such that $\operatorname{dist}(\lambda, \sigma(T)) > \delta$. Then $|\lambda_i - \lambda| \geq \delta$ for every i. Therefore $\left\| \sum_{i=n}^m \frac{(x, e_i)}{\lambda - \lambda_i} e_i \right\|^2 = \sum_{i=n}^m \frac{|(x, e_i)|^2}{|\lambda - \lambda_i|^2} \leq \delta^{-2} \sum_{i=n}^m |(x, e_i)|^2$ due to the orthonormality of $\{e_i\}$. Thus the series is convergent for every $x \in H$ and we can define an operator on H by $G(x) = \sum \frac{(x, e_i)}{\lambda - \lambda_i} e_i$. For $\|x\| \leq 1$ we have

$$\left\| \sum_{i=1}^{n} \frac{(x, e_i)}{(\lambda - \lambda_i)} e_i \right\|^2 = \sum_{i=1}^{n} \frac{|(x, e_i)|^2}{|\lambda - \lambda_i|^2} \leq \delta^{-2} \sum_{i=1}^{n} |(x, e_i)|^2$$
$$\leq \delta^{-2} \sum_{i=1}^{\infty} |(x, e_i)|^2 = \delta^{-2} \|x\|^2 \leq \delta^{-2},$$

in particular G is a bounded operator on H. For $x = \sum (x, e_j) e_j$ we have $T(x) = \sum_j \lambda_j (x, e_j) e_j$ and $(\lambda I_H - T)(x) = \sum_j (\lambda - \lambda_j)(x, e_j) e_j$, hence using $(e_i, e_j) = \delta_{i,j}$ we get

$$(\lambda I_H - T)(G(x)) = \sum_j (\lambda - \lambda_j) \Big(\sum_i \frac{(x, e_i)}{\lambda - \lambda_i} e_i, e_j \Big) e_j$$
$$= \sum_{i,j} (\lambda - \lambda_j) \frac{(x, e_i)}{\lambda - \lambda_i} (e_i, e_j) e_j = \sum_j (x, e_j) e_j = x.$$

Similarly we show that $G(\lambda I_H - T) = I_H$, hence $G = R(\lambda)$. □

Definition 15.42 *Let H be a Hilbert space and let T be an operator on H.*
T is called normal *if $TT^* = T^*T$.*
T is called unitary *if it is invertible and $T^{-1} = T^*$.*

Clearly, every unitary operator is normal and every self-adjoint operator is normal.

Proposition 15.43 *Let H be a Hilbert space, $T \in \mathcal{B}(H)$. T is normal if and only if $\|T(x)\| = \|T^*(x)\|$ for every $x \in H$.*

Proof: For $x \in H$ we have

$$\|T(x)\|^2 - \|T^*(x)\|^2 = \big(T(x), T(x)\big) - \big(T^*(x), T^*(x)\big)$$
$$= \big(T^*T(x), x\big) - \big(TT^*(x), x\big) = \big((T^*T - TT^*)(x), x\big).$$

If T is normal, then the latter quantity is zero for every $x \in H$, so $\|T(x)\| = \|T^*(x)\|$ for every $x \in H$. If $\big((T^*T - TT^*)(x), x\big) = 0$, then since $T^*T - TT^*$ is always self-adjoint, we have that $T^*T - TT^* = 0$ by Proposition 15.31 and thus T is normal. □

Proposition 15.44 *Let H be a Hilbert space. If $T \in \mathcal{B}(H)$ is onto, then the following are equivalent:*
(i) *T is unitary.*
(ii) *T is an isometry.*
(iii) $\big(T(x), T(y)\big) = (x, y)$ *for every $x, y \in H$.*

If the condition (iii) is satisfied for an operator $T \in \mathcal{B}(H)$, we say that T *preserves the inner product.*

Proof: From the polarization identities for a real or complex Hilbert space it follows that T preserves the inner product if and only if T is an isometry. Therefore (ii) and (iii) are equivalent. If U is unitary, then $\big(T(x), T(y)\big) = \big(T^*T(x), y\big) = (x, y)$ for every (x, y), so, T satisfies (iii). If T satisfies (iii), then it is an isometry and onto, hence T^{-1} exists. By (iii) also $\big(T^*T(x), y\big) = (x, y)$ for all $x, y \in H$, thus $T^*T = I_X$ and $T^{-1} = T^*$ follows. □

Recall that if P is a bounded linear projection of a Banach space X onto $P(X)$, then $X = P(X) \oplus \mathrm{Ker}(P)$.

Definition 15.45 *Let H be a Hilbert space and P be a bounded linear projection of H onto $P(H)$. P is called an* orthogonal projection *if* $\mathrm{Ker}(P) \perp P(H)$.

Since we always have $\mathrm{Ker}(P) \oplus P(H) = H$ and the two subspaces are closed, we can equivalently write $P(H) = \mathrm{Ker}(P)^{\perp}$ and $\mathrm{Ker}(P) = P(H)^{\perp}$. Note that in Chapter 1 we proved that every closed subspace of a Hilbert space H is complemented by an orthogonal projection.

Lemma 15.46 *Let H be a Hilbert space and $x, y \in H$. If $\|x + \alpha y\| \geq \|x\|$ for all scalars α, then $(x, y) = 0$.*

Proof: We have $\|x\|^2 \leq \|x + \alpha y\|^2 = (x, x) + |\alpha|^2(y, y) + 2\,\mathrm{Re}\big(\alpha(y, x)\big)$. Write $\alpha = r_1 e^{it}$ and $(y, x) = r_2 e^{i\xi}$. Then $|\alpha|^2(y, y) + 2\,\mathrm{Re}(\alpha(y, x)) = r_1^2(y, y) + 2\,\mathrm{Re}(r_1 r_2 e^{i(t+\xi)})$. If we choose t such that $t + \xi = \pi$, then $2\,\mathrm{Re}\big(r_1 r_2 e^{i(t+\xi)}\big) = -2r_1 r_2$ and thus we have $-2r_1 r_2 + r_1^2\|y\| \geq 0$ for all $r_1 > 0$.

Then $-2r_2 + r_1 > 0$ for all $r_1 > 0$, which is possible only if $r_2 = 0$. Hence $(y, x) = 0$. □

Proposition 15.47 *Let P be a bounded linear projection of a Hilbert space H onto $P(H)$. Then the following are equivalent:*
(i) *P is an orthogonal projection.*
(ii) *P is a self-adjoint operator.*
(iii) *P is a normal operator.*
(iv) $\big(x - P(x), P(x)\big) = 0$ *for all $x \in H$.*
(v) $\|P\| = 1$.

Proof: (i)$\Longrightarrow$(ii): Follows easily from $\mathrm{Ker}(P) \perp P(H)$ and $x - P(x) \in \mathrm{Ker}(P)$.

Clearly (ii)$\Longrightarrow$(iii).

(iii)$\Longrightarrow$(i): By Proposition 15.43 we get $\text{Ker}(P) = \text{Ker}(P^*)$, also $\text{Ker}(P^*) = P(H)^\perp$ by Proposition 15.27, hence $\text{Ker}(P) = P(H)^\perp$.

(i)$\Longrightarrow$(iv): If $P(H)$ and $\text{Ker}(P)$ are orthogonal complemented subspaces of H, write $x = y + z$, $y \in P(H)$ and $z \in \text{Ker}(P)$. Then $P(x) = y$ and $\big(x - P(x), P(x)\big) = (z, y) = 0$.

(iv)$\Longrightarrow$(i): For $y \in P(H)$ and $z \in \text{Ker}(P)$ set $x = y + z$. Then $P(x) = y$, $x - P(x) = z$, and we have $(z, y) = \big(x - P(x), P(x)\big) = 0$, that is, $y \perp z$. Consequently $\text{Ker}(P) \perp P(H)$.

(i)$\Longrightarrow$(v): Write $x \in H$ as $x = y + z$, where $y \in P(H)$, $z \in \text{Ker}(P)$. Then $y \perp z$, so $\|P(x)\|^2 = \|y\|^2 \le \|y\|^2 + \|z\|^2 = \|y + z\|^2$. Therefore $\|P\| \le 1$. Since $P(x) = x$ for $x \in P(X)$, we have $\|P\| \ge 1$. Thus $\|P\| = 1$.

(v)$\Longrightarrow$(i): Write $x = y + z$ for $y \in P(H)$, $z \in \text{Ker}(P)$. Let α be an arbitrary scalar. Then $\|y\| = \|P(y+\alpha z)\| \le \|y+\alpha z\|$. By Lemma 15.46 we have $(y, z) = 0$, hence $\text{Ker}(P) \perp P(H)$. □

Theorem 15.48 (Spectral decomposition) *Let H be a complex infinite-dimensional separable Hilbert space. If T is a compact normal operator on H, then there exists an orthonormal basis $\{e_i\}$ of H, where each e_i is an eigenvector corresponding to an eigenvalue λ_i of T, such that for all $x \in H$ we have*

$$T(x) = \sum_i \lambda_i (x, e_i) e_i.$$

In particular, if $\{\lambda_n\}$ denotes the set of all distinct eigenvalues of T and P_n are the orthogonal projections of H onto the eigenspaces $\text{Ker}(\lambda_n I_H - T)$, *then $T = \sum_n \lambda_n P_n$, where the sum converges in $\mathcal{B}(H)$.*

Proof: Consider the self-adjoint operator $U = TT^* = T^*T$. By Theorem 15.39 there are (mutually orthogonal) eigenspaces H_n corresponding to distinct eigenvalues λ_n of U such that $H = \overline{\text{span}}\big(\bigcup H_n\big)$; in particular, orthonormal bases of H_n (formed by eigenvectors of U) ordered into a sequence form an orthonormal basis of H.

We claim that every H_n is an invariant subspace of T. Since $UT = TT^*T = TTT^* = TU$, for $h \in H_n$ we get $(\lambda_n I_H - U)\big(T(h)\big) = T(\lambda_n I_H - U)(h) = T(0) = 0$, that is, $T(h) \in \text{Ker}(\lambda_n I_H - U) = H_n$.

Each H_n has an orthonormal basis formed by eigenvectors of T. Indeed, if $\lambda_n = 0$, then $U = 0$ on H_n and also $\|T(x)\|^2 = \big(T(x), T(x)\big) = \big(T^*T(x), x\big) = \big(U(x), x\big) = 0$, that is, $T = 0$ on H_n. Thus every non-zero vector of H_n is an eigenvector. If $\lambda_n \neq 0$, then H_n is a finite-dimensional complex vector space and the existence of such a basis follows from linear algebra and the fact that $T\big|_{H_n}$ is normal.

Collecting the bases of all H_n we get an orthogonal basis of H. The convergence of $\sum \lambda_i(x, e_i)e_i$ is proved as in Theorem 15.41 and similar proof shows that $\sum_n \lambda_n P_n$ converges to T. □

Conversely, one can prove that if $\{e_i\}$ is an orthonormal basis of a separable Hilbert space and $\lambda_i \to 0$, then the operator $x \mapsto \sum_i \lambda_i(x, e_i)e_i$ is a compact normal operator.

Theorem 15.49 (Polar decomposition) *Let H be a separable complex Hilbert space. Let $A \in \mathcal{K}(H)$. Then there exist a unitary operator U and a positive operator $C \geq 0$, such that $A = UC$.*

Proof: (Sketch) Given $A \in \mathcal{K}(\ell_2)$, we have that $D = A^*A$ is self-adjoint and positive, thus diagonalizable with nonnegative terms on diagonal by Theorem 15.48. By taking square roots on the diagonal, we obtain a positive operator C, such that $C^2 = D$. Thus $\langle Ax, Ax\rangle = \langle A^*Ax, x\rangle = \langle C^2x, x\rangle = \langle Cx, Cx\rangle$. It follows that there is some unitary U such that $A = UC$. □

15.4 Remarks and Open Problems

Remarks

1. Recently, Argyros and Haydon [ArHay] proved that there is an infinite-dimensional Banach space X such that every bounded operator from X into X has the form $T + \rho I$, where T is a compact operator, I is the identity operator, and ρ is a real number.

Open Problems

1. Enflo showed in [Enfl3] (submitted in 1981) that there is a Banach space X and a bounded operator on X without nontrivial invariant subspace, i.e., different from zero and from the whole space. Later on, Read gave in [Read] another solution to the invariant subspace problem, this time on the space ℓ_1. The existence of such an example for reflexive spaces, in particular, for Hilbert spaces is still unknown. Note that there exist compact operators on complex spaces without any eigenvalue (see Exercise 15.29). It is known, in contrast, that every compact operator has a non-trivial invariant subspace, see Theorem 12.23.

Exercises for Chapter 15

15.1 Show that if T, S are bounded operators on a Banach space X and one of them is compact, then TS and ST are compact.
Hint. Standard.

15.2 Let X be a Banach space and $T \in \mathcal{K}(X)$. Show that $0 \in \overline{T(S_X)}$.
Hint. Since X is infinite-dimensional, $0 \in \overline{S_X}^w$. T is w-w-continuous, so $0 \in \overline{T(S_X)}^w$. The set $\overline{T(B_X)}^{\|\cdot\|}$ is $\|\cdot\|$-compact, so both topologies w and the induced by the norm agree on $\overline{T(B_X)}^{\|\cdot\|}$. In particular, $\overline{T(S_X)}^{\|\cdot\|} = \overline{T(S_X)}^w$, and the conclusion follows.

15.3 Let X, Y be Banach spaces, $T \in \mathcal{B}(X, Y)$. Show that if T is continuous from the weak topology of X into the norm topology of Y, then T is a finite rank operator.
Hint. Let $W = \{x : |f_i(x)| < 1\}$ for some $f_1, \dots, f_n \in X^*$ be such that $T(W) \subset B_Y$. Then T is 0 on $\bigcap f_i^{-1}(0)$. Write $X = \bigcap f_i^{-1}(0) \oplus Z$, where $\dim(Z) < \infty$. Note that $T(X) = T(Z)$.

15.4 Show that if T is a compact operator from X into Y, then $T(X)$ is separable.
Hint. $T(B_X)$ is totally bounded, hence separable.

15.5 Let X, Y be Banach spaces, $T \in \mathcal{B}(X, Y)$. Show that $T \in \mathcal{K}(X, Y)$ if and only if there is $\{x_n^*\} \subset X^*$ such that $\|x_n^*\| \to 0$ and $\|T(x)\| \le \sup_n |x_n^*(x)|$ for every $x \in X$.
Hint. The condition implies that $T(B_X) \subset C(\{x_n^*\} \cup \{0\})$ is relatively compact, using the Arzelà–Ascoli theorem ($\{x_n^*\} \cup \{0\}$ is compact). Assume that T is a compact operator. Then T^* is compact and by Exercise 1.69 there is a sequence $\{x_n^*\} \subset X^*$ such that $x_n^* \to 0$ and $T^*(B_{X^*}) \subset \overline{\operatorname{conv}}\{x_n^*\}$. Then for all $x \in X$,

$$\|T(x)\| = \sup_{f \in B_{X^*}} f\big(T(x)\big) = \sup_{T^*(B_{X^*})} (x) \le \sup_n |x_n^*(x)|.$$

15.6 Let X, Y be Banach spaces, $T \in \mathcal{B}(X, Y)$. Show that $K \in \mathcal{K}(X, Y)$ if and only if there is $\lambda \in c_0$ and a bounded sequence $\{y_n^*\}$ in X^* such that $\|T(x)\| \le \sup_n |\lambda_n|\,|y_n^*(x)|$ for all $x \in X$.
Hint. In the preceding exercise write $x_n^* = \frac{x_n^*}{\|x_n^*\|}\|x_n^*\|$.

15.7 (Grothendieck) Let X, Y be Banach spaces. Show that if $T \in \mathcal{K}(X, Y)$ then there is a closed subspace Z of c_0 and $R \in \mathcal{K}(X, Z)$, $S \in \mathcal{K}(Z, Y)$ such that $T = SR$ (i.e., T factors compactly through a subspace of c_0).
Hint. In the notation from the previous exercise, $\|T(x)\| \le \sup_n |\lambda_n|\,|y_n^*(x)|$ for all $x \in X$. Put $R(x) = \big(\lambda_i y_i^*(x)\big) \in c_0$ for $x \in X$. Define $Z = \overline{R(X)}$. Then $R \in \mathcal{K}(X, c_0)$ by the previous exercise (we use the coordinate functionals). Define an operator $S\colon R(X) \to Y$ by $S\big(\lambda_i y_i^*(x)\big) = T(x)$ for $x \in X$. Clearly S is continuous, so we can extend it to a bounded operator $Z \to Y$. Then $T = S \circ R$ and since $S(B_Z) \subset \overline{T(B_Z)}$, S is compact.

15.8 Let X, Y be Banach spaces, X reflexive, and $T \in \mathcal{B}(X, Y)$. Show that if T is completely continuous then it is compact.

Show that if T is not compact, then there is $x_n \in X$ and $\varepsilon > 0$ such that $x_n \xrightarrow{w} 0$ and $\|T(x_n)\| \geq \varepsilon$ for all n.
Hint. If $T(x_n) \in T(B_X)$, let $x_{n_k} \xrightarrow{w} x \in B_X$ in X by the weak sequential compactness of B_X. Then $T(x_{n_k}) \to T(x)$, showing that every sequence in $T(B_X)$ has a convergent subsequence.

If T is not compact, by the first part get $y_n \in X$ with $y_n \xrightarrow{w} y$ and $T(y_n) \not\to T(y)$. Then use $x_n = y_n - y$.

15.9 Let K be a compact space and T be a weakly compact operator from $C(K)$ into $C(K)$. Show that T^2 is a compact operator.
Hint. Since T is weakly compact, $C := \overline{TB_{C(K)}} \subset C(K)$ is weakly compact. Thus by Theorem 13.43 and (ii) in Proposition 13.42, $T(C)$ is norm compact. Therefore T^2 is a compact operator.

15.10 (Grothendieck) Show that there is no infinite-dimensional separable injective space.
Hint. If such a space existed, denote it by X and assume that $X \subset \ell_\infty$. Let P be a projection from ℓ_∞ onto X. Since B_{X^*} is w^*-sequentially compact, so is $P^*(B_{X^*}) \subset \ell_\infty^*$. Because ℓ_∞ has the Grothendieck property, $P^*(B_{X^*}) \subset \ell_\infty^*$ is w-sequentially compact. Thus P^* is a weakly compact operator and so is P by the Gantmacher Theorem 13.34. By Exercise 15.9, $P = P^2$ is a compact operator, which gives that X is finite-dimensional.

15.11 Let X, Y be Banach spaces and $T \in \mathcal{B}(X, Y)$. If Z is a closed subspace of X, is $T(Z)$ necessarily closed in Y? Would it help if T were open?
Hint. No, no: Consider $X = \ell_2 \oplus \ell_2$, $Y = \ell_2$, $Z = 0 \oplus \ell_2$ and the operator T from X onto Y defined by $T(x, y) = x + L(y)$, where L is a compact operator with $\|L\| = \frac{1}{2}$. Then $T(Z)$ is not closed in Y.

15.12 Let X_k be the space $C^k[0, 1]$ endowed with the norm

$$\|f\| = \sum_{j=1}^{k} \max\{|f^{(j)}(t)| : t \in [0, 1]\}.$$

Show that the identity mapping from X_k into X_{k-1} is a compact operator.
Hint. The Arzelà–Ascoli theorem.

15.13 Let $T \in \mathcal{B}(X, \ell_1)$. Show that if X is reflexive then T is a compact operator.
Hint. $T(B_X)$ is weakly compact. In ℓ_1 the weak and the norm compactness coincide (Schur).

15.14 Let $T \in \mathcal{B}(c_0, X)$. Show that if X is reflexive then T is a compact operator.
Hint. Check T^* using the previous exercise.

15.15 Find a non-compact operator $T \in \mathcal{B}(\ell_2)$ such that $T^2 = 0$.
Hint. Put $T(e_{2n}) = e_{2n+1}$ and $T(e_{2n+1}) = 0$.

15.16 Let X be an infinite-dimensional Banach space. Show that there is a bounded linear non-compact operator from X into c_0.
Hint. Let $f_n \in S_{X^*}$ be such that $f_n \overset{w^*}{\to} 0$ in X^* (the Josefson–Nissenzweig theorem, see Exercise 3.39). Define $T(x) = \big(f_i(x)\big)_i \in c_0$ for $x \in X$. $\overline{T(B_X)}$ is not compact by Exercise 1.50.

15.17 Show that the space c_0 is isometric to a subspace of $\mathcal{K}(\ell_2)$.
Hint. If $(a_i) \in c_0$, then the operator $T(x) = (a_i x_i)$ is in $\mathcal{K}(\ell_2)$. Use the proof of Proposition 1.44.

15.18 Show that $\mathcal{K}(\ell_2)$ is not reflexive.
Hint. Use Exercise 15.17.

15.19 Show that $\mathcal{K}(\ell_2)$ is not complemented in $\mathcal{B}(\ell_2)$.
Hint. Assume there is a projection P from $\mathcal{B}(\ell_2)$ onto $\mathcal{K}(\ell_2)$. Using Exercise 15.17 and the Sobczyk theorem, we obtain a projection Q from $\mathcal{K}(\ell_2)$ onto c_0. Then the restriction of QP to the subspace of $\mathcal{B}(\ell_2)$ of diagonal operators, which is isomorphic to ℓ_∞, would yield a projection from ℓ_∞ onto c_0, a contradiction.

15.20 Show that $\mathcal{K}(\ell_2)$ contains an isometric copy of ℓ_2.
Hint. For $y \in \ell_2$, define $T_y \in \mathcal{B}(\ell_2)$ by $T_y(x) = \left(\sum_{i=1}^{\infty} x_i y_i\right) e_1$. Clearly, $T_y \in \mathcal{K}(\ell_2)$, and $y \mapsto T_y$ is a linear mapping. Using $\left|\sum_{i=1}^{\infty} x_i y_i\right| \le \|x\|_2 \|y\|_2$ we get $\|T_y\| \le \|y\|_2$. To get the opposite inequality, use x defined by $x_i = \overline{y_i}$.

15.21 Let X be a separable Banach space and let $\{x_i\}$ be dense in B_X. Define an operator T from X^* into ℓ_2 by $T(f) = \big(2^{-i} f(x_i)\big)$. Show that T is a homeomorphism of (B_{X^*}, w^*) onto the compact set $T(B_{X^*})$ in ℓ_2 taken in its norm topology.
Hint. T is a compact operator. In norm compact sets in ℓ_2 the pointwise and norm topologies coincide.

15.22 Show that every compact metric space K is homeomorphic to a norm compact subset of ℓ_2 taken in its norm topology.
Hint. Put $X = C(K)$ in the previous exercise. Note that $C(K)$ is separable, also $x_n \to x$ in K if and only if $x_n \overset{w^*}{\to} x$ in X^*.

15.23 Show that Theorem 15.9 fails for $\lambda = 0$.
Hint. $T\big((x_i)\big) = (2^{-i} x_i)$ in ℓ_2. It is compact, see Exercise 1.51.

15.24 Let P be a bounded linear projection of a Banach space X onto Y. Show that if Y is non-trivial, then $\sigma(P) = \sigma_p(P) = \{0, 1\}$, and for $\lambda \neq 0, 1$ we have $(\lambda I_X - P)^{-1} = \frac{1}{\lambda} I_X + \frac{1}{\lambda(\lambda-1)} P$.

15.25 Let T be a diagonal operator on ℓ_2 associated with a bounded sequence of complex numbers $\{c_i\}$, that is, $T\big((x_i)\big) = (c_i x_i)$. Find $\sigma_p(T)$ and $\sigma(T)$.
Hint. Show that λ is an eigenvalue if and only if λ equals to one of the numbers c_i. Since $\sigma(T)$ is closed, we have that $\overline{\{c_i\}} \subset \sigma(T)$. If $\lambda \notin \overline{\{c_i\}}$ then $\inf |\lambda - c_i| > 0$ and the diagonal operator with diagonal $\frac{1}{\lambda - c_i}$ (as an inverse operator) shows that such λ is not in $\sigma(T)$.

15.26 Let K be a compact set in the scalar field. Show that there is an operator $T \in \mathcal{B}(\ell_2)$ such that $\sigma(T) = K$.
Hint. Consider a dense sequence $\{c_i\}$ in K and the previous exercise.

15.27 Is there $T \in \mathcal{K}(\ell_2)$ with $\sigma(T) = \{1 + \frac{1}{n}\}_{n=1}^{\infty} \cup \{0\}$?
Hint. Corollary 15.24.

15.28 (i) Let L be a left-shift operator in ℓ_2, $L(x_1, x_2, \ldots) := (x_2, x_3, \ldots)$. Show that the set of all eigenvalues of L is the open unit disc.

(ii) Let R be a right-shift operator on ℓ_2, $R(x_1, x_2, \ldots) := (0, x_1, x_2, \ldots)$. Show that the set of all eigenvalues of R is empty.

(iii) Show that $\sigma(L)$ and $\sigma(R)$ are both equal to the closed unit disc.

Hint. (i) The vector $(1, \lambda, \lambda^2, \ldots)$ is an eigenvector for λ if $|\lambda| < 1$. If $|\lambda| \geq 1$ then the equation for the eigenvalue is not solvable in ℓ_2.

(ii) R is one-to-one and thus 0 is not an eigenvalue. If $\lambda \neq 0$, then by solving $(0, x_1, x_2, \ldots) = (\lambda x_1, \lambda x_2, \ldots)$ we get $x_i = 0$ for all i.

(iii) Note that $L = R^*$, so L and R have the same spectrum (Fact 15.14). Since $\|L\| = \|R\| = 1$, $\sigma(L) \subset B_{\mathbb{C}}$, and $\sigma(L)$ is a closed set (Proposition 15.15). We conclude that $\sigma(L) = \sigma(R) = B_{\mathbb{C}}$.

15.29 Consider the right shift R on ℓ_2 and the diagonal operator D associated with $d_i = 2^{-i}$. Define a weighted shift operator T on the *complex* space ℓ_2 by $T = R \circ D$. Show that T is a compact operator with spectral radius 0 and T is one-to-one. Thus T has no eigenvalues and $\sigma(T) = \{0\}$.
Hint. Show first that D is a compact operator, hence T is compact. Write down the explicit formula for T^n to see that $\|T^n\| \to 0$. Since the spectrum must be nonempty, $\sigma(T) = \{0\}$.

15.30 Find an operator $T \in \mathcal{B}(\ell_2)$ such that its spectrum $\sigma(T)$ consists exactly of the points 0 and 1 and such that neither 0 nor 1 is an eigenvalue for T.
Hint. Recall that $\sigma(S + \mu I_X) = \sigma(S) + \mu$. Split ℓ_2 into $\overline{\operatorname{span}}\{e_{2i}\} \oplus \overline{\operatorname{span}}\{e_{2i+1}\}$ and use the previous exercise.

15.31 Let T be a Banach space and $T \in \mathcal{B}(X)$. Prove that $\exp(T) = \sum \frac{T^n}{n!}$ exists, it is an invertible operator, and $\sigma(\exp(T)) = \exp(\sigma(T))$.
Hint. $\exp(T)\exp(-T) = I_X$. Show first that $\sigma(p(T)) = p(\sigma(T))$ for every polynomial p.

15.32 Assume that T is a bounded operator from a Hilbert space H into H such that $\big(T(x), x\big) \geq (x, x)$ for every $x \in H$. Show that T is an invertible operator on H.
Hint. By assumption, $\|T(x)\|\|x\| \geq \|x\|^2$, that is, $\|T(x)\| \geq \|x\|$ for every $x \in H$, so T is an isomorphism into (Exercise 1.73). We also have $\big(x, T^*(x)\big) \geq (x, x)$, so $\|T^*(x)\| \geq \|x\|$ and T^* is isomorphism into. By an analog of Exercise 2.49, T is onto.

15.33 Let T be an operator on a Hilbert space H with $\big(T(x), y\big) = \big(x, T(y)\big)$ for all x, y in H. Show that T is a bounded operator (the Hellinger–Töplitz theorem).
Hint. Let $x_n \to 0$ in H. Then $|\big(x_n, T(y)\big)| \leq \|x_n\|\|T(y)\| \to 0$ for every $y \in H$. Thus for every $y \in H$ we have $\big(T(x_n), y\big) = \big(x_n, T(y)\big) \to 0$. This means that by the Riesz representation theorem, the assumptions in Exercise 3.97 are satisfied, hence T is continuous.

15.34 Let $T \in \mathcal{B}(H)$ be a self-adjoint isomorphism of a Hilbert space H onto H. Show that if T is *positive*, i.e., $\big(T(x), x\big) \geq 0$ for all $x \in X$, then $[x, y] = \big(T(x), y\big)$ defines a new inner product on H and $|||x||| = [x, x]^{\frac{1}{2}}$ is an equivalent norm on H.

15.35 Let H be a finite-dimensional Hilbert space. Follow the hint to show that $\operatorname{Ext}(B_{\mathcal{B}(H)})$ consists exactly of all unitary operators.

The statement is in fact true for all Hilbert spaces.

In the hint we use several facts from the operator theory. First, for every $T \in \mathcal{B}(H)$ there exists a polar decomposition $T = US$, where U is a unitary operator and S is a positive operator, that is, $\big(S(x), x\big) \geq 0$ for every $x \in H$.

Second, if H is a finite-dimensional Hilbert space and S is a positive operator on H, we can find an orthogonal basis of H so that the matrix of S with respect to this basis is diagonal. Note that elements on the diagonal have magnitudes at most $\|S\|$.
Hint. Let $T \in \operatorname{Ext}(B_{\mathcal{B}(H)})$, and $T = US$ be its polar decomposition. Then $\|S\| \leq \|T\|$, $\|U\| = 1$. We claim that $S = I_H$. Indeed, otherwise we find $S_1 \neq S_2$ with $\|S_1\|, \|S_2\| \leq 1$ such that $S = \frac{S_1+S_2}{2}$. Thus T would not be an extreme point.

On the other hand, let U be unitary. By the parallelogram equality

$$\|(U+V)(x)\|^2 + \|(U-V)(x)\|^2 = 2\|U(x)\|^2 + 2\|V(x)\|^2.$$

Assuming $\|U+V\|, \|U-V\| \leq 1$ we get $V(x) = 0$ for every $x \in H$.

15.36 Let H be a separable Hilbert space. An operator $T \in \mathcal{B}(H)$ is called a *Hilbert–Schmidt operator* if there is an orthonormal basis $\{e_i\}$ of H such that

$\sum \|T(e_i)\|^2 < \infty$. Show that if $\{f_i\}$ is another orthonormal basis of H, then $\sum \|T(f_i)\|^2 = \sum \|T(e_i)\|^2$.

The number $\|T\|_{HS} = (\sum \|T(e_i)\|^2)^{\frac{1}{2}}$ is called the Hilbert–Schmidt norm of T. Show that $\|T\|_{HS} \geq \|T\|$.

Hint.

$$\sum_i \|T(e_i)\|^2 = \sum_i \sum_j |(T(e_i), e_j)|^2 = \sum_i \sum_j |(e_i, T^*(e_j))|^2 = \sum_j \|T^*(e_j)\|^2.$$

Thus

$$\sum_i \|T(f_i)\|^2 = \sum_i \sum_j |(T(f_i), e_j)|^2 = \sum_j \|T^*(e_j)\|^2 = \sum_i \|T(e_i)\|^2.$$

By Hölder's inequality (1.1), we have

$$\|T(x)\| = \|\sum (x, e_i) T(e_i)\| \leq \sum |(x, e_i)| \, \|T(e_i)\| \leq \|x\| (\sum \|T(e_i)\|^2)^{\frac{1}{2}}.$$

15.37 Show that every Hilbert–Schmidt operator T on a Hilbert space H is compact. Find a compact operator that is not a Hilbert–Schmidt operator.

Hint. Let $\{e_i\}$ be an orthonormal basis of H, denote $a_{ij} = \big(T(e_i), e_j\big)$. Then $T(\sum x_i e_i) = \sum_j (\sum_i a_{ij} x_i) e_j$, so $\|T\|_{HS} = (\sum_i \sum_j |a_{ij}|^2)^{\frac{1}{2}}$. Set $a_j = \sum_i \bar{a}_{ij} e_i$. Then $a_j \in H$ for every j, $\|T\|_{HS} = (\sum_j \|a_j\|^2)^{\frac{1}{2}}$, and $T(x) = \sum_j (x, a_j) e_j$. Clearly $T(B_H) \subset \{x = \sum x_i e_i : |x_i| \leq \|a_i\| \text{ for every } i\}$ and the set if compact in H.

For the second question, consider a diagonal operator on ℓ_2 with diagonal $\{c_i\}$ such that $\lim c_i = 0$ and $\sum |c_i|^2$ is not convergent.

15.38 Let X be a Banach space, $T \in \mathcal{B}(X)$, and $x_0 \in X \backslash \{0\}$. Define a subspace Y of X as $Y = \overline{\operatorname{span}}\{x_0, T(x_0), T^2(x_0), \dots\}$. Show that Y is an invariant subspace for T.

The vector x_0 is called *cyclic* for T if $Y = X$. Show that T has no non-trivial invariant subspace if and only if every $x_0 \in X \backslash \{0\}$ is a cyclic vector for T.

15.39 Show that every operator on a non-separable Banach space has a non-trivial invariant subspace.

Hint. Use the previous exercise

15.40 (Korovkin) Let $\{T_n\}$ be a sequence of bounded operators on $C[0, 1]$ that are positive, i.e., $T_n(f) \geq 0$ if $f \geq 0$ on $[0, 1]$. Assume that $T_n(1) \to 1$, $T_n(x) \to x$, and $T_n(x^2) \to x^2$ in $C[0, 1]$. Show that $T_n(f) \to f$ in $C[0, 1]$ for every $f \in C[0, 1]$. Is it true that $\|T_n - I_{C[0,1]}\| \to 0$?

Hint. To prove $T_n(f) \to f$, it is enough to find for every $t \in [0, 1]$ and $\varepsilon > 0$ some $n_0 \in \mathbb{N}$ and $\delta > 0$ satisfying $|T_n(f)(\tau) - f(\tau)| < \varepsilon$ whenever $|\tau - t| < \delta, n > n_0$.

Since f is continuous, there are quadratic functions $\varphi_1, \varphi_2 \in C[0,1]$ satisfying $\varphi_2 < f < \varphi_1$ on $[0,1]$ and $|\varphi_1(\tau) - \varphi_2(\tau)| < \varepsilon$ for $|\tau - t| < \delta$. Thus, $T_n(\varphi_1 - f)(\tau) \geq 0$, $T_n(f - \varphi_2)(\tau) \geq 0$, and $T_n(\varphi_1)(\tau) \to \varphi_1(\tau)$, $T_n(\varphi_2)(\tau) \to \varphi_2(\tau)$ as $n \to \infty$ for $|\tau - t| < \delta$. Thus for n large enough, $\varphi_1(\tau) \geq T_n(f)(\tau) \geq \varphi_2(\tau)$ on $[0,1]$, and so $|T_n(f)(\tau) - f(\tau)| \leq \varepsilon$.

$\|T_n - I_{C[0,1]}\| \to 0$ may fail, consider $T_n\colon f \mapsto f^{(n)}$, where $f^{(n)}$ is the broken line agreeing with f at its nodes at points $\frac{i}{n}$, $0 \leq i \leq n$.

Chapter 16
Tensor Products

This chapter is an introduction to the topological theory of tensor products of Banach spaces. The focus lies on the applications of tensors in the duality theory for spaces of operators, and their structure as Banach spaces. We discuss the role of the approximation property and Enflo's example of a Banach space without the approximation property.

16.1 Tensor Products and Their Topologies

Let E and F be linear spaces over a real or complex scalar field $\mathbb{K}$. By $\Lambda(E \times F)$ we denote the set of all formal finite linear combinations $\sum_{i=1}^{n} a_i(e_i, f_i)$, where $a_i \in \mathbb{K}$, $e_i \in E$, $f_i \in F$. (We identify $\sum_{i=1}^{n} a_i(e_i, f_i)$ and $\sum_{i=1}^{n} a_{\pi(i)}(e_{\pi(i)}, f_{\pi(i)})$ for any permutation π of $\{1, 2, \ldots, n\}$, and likewise $\sum_{i=1}^{n+1} a_i(e_i, f_i)$ and $\sum_{i=1}^{n} a_i(e_i, f_i)$ for $a_{n+1} = 0$.) This set has a natural structure of linear space over $\mathbb{K}$, defined by the relations

$$a \sum_{i=1}^{n} a_i(e_i, f_i) = \sum_{i=1}^{n} a a_i(e_i, f_i), \text{ and}$$

$$\sum_{i=1}^{n} a_i(e_i, f_i) + \sum_{i=1}^{n} b_i(e_i, f_i) = \sum_{i=1}^{n} (a_i + b_i)(e_i, f_i).$$

By $\Lambda_0(E \times F)$ we denote the linear subspace generated by all elements of the form

$$(a_1 e_1 + a_2 e_2, b_1 f_1 + b_2 f_2) - \sum_{1 \le i, j \le 2} a_i b_j (e_i, f_j). \tag{16.1}$$

Definition 16.1 *Let E and F be linear spaces over the real or complex scalar field $\mathbb{K}$. We define their* algebraic tensor product *$E \otimes F$ as the linear quotient space $\Lambda(E \times F)/\Lambda_0(E \times F)$. Elements (cosets) of $E \otimes F$ will be called* tensors. *We denote by $e \otimes f \in E \otimes F$ the tensor (coset) containing (e, f). A tensor allowing such expression is called an* elementary tensor.

M. Fabian et al., *Banach Space Theory*, CMS Books in Mathematics,
DOI 10.1007/978-1-4419-7515-7_16, © Springer Science+Business Media, LLC 2011

Informally, $E \otimes F$ is the linear space consisting of all expressions $\sum_{i=1}^{n} a_i e_i \otimes f_i$, where $a_i \in K$, $e_i \in E$, $f_i \in F$, that are subject to an equivalence relation $\sim$:

$$\sum_{i=1}^{n} a_i e_i \otimes f_i \sim \sum_{j=1}^{m} a'_j e'_j \otimes f'_j \tag{16.2}$$

if and only if $\sum_{i=1}^{n} a_i (e_i, f_i) - \sum_{j=1}^{m} a'_j (e'_j, f'_j) \in \Lambda_0(E, F)$. Naturally, the zero element in $E \otimes F$ is represented by $0 \otimes 0 \sim 0 \otimes f \sim e \otimes 0$.

We refer to Exercise 5.52 for an alternative definition of a tensor product.

Lemma 16.2 *Let E, F be linear spaces, $A \subset E^{\#}$, $B \subset F^{\#}$ be linear subspaces of their algebraic duals that separate points in E, respectively F. Then* (16.2) *occurs if and only if*

$$\sum_{i=1}^{n} a_i \phi(e_i) \psi(f_i) - \sum_{j=1}^{m} a'_j \phi(e'_j) \psi(f'_j) = 0 \tag{16.3}$$

for all $\phi \in A$, $\psi \in B$.

Proof: It is clear from the definition of $\Lambda_0(E \times F)$ using (16.1), that (16.2) implies (16.3). On the other hand, given expressions as in (16.3), by using (16.1) repeatedly, we may assume without loss of generality that both $\{e_i\}_{i=1}^{n} \cup \{e'_j\}_{j=1}^{m}$ and $\{f_i\}_{i=1}^{n} \cup \{f'_j\}_{j=1}^{m}$ form a linearly independent set in E, F and $a_1 \neq 0$, or else the left-hand side is a combination of elementary tensors equal to 0 (which implies (16.2)). As A, B separate points, a simple linear algebra argument gives that there exist $\phi \in A$, $\psi \in B$ such that

$$\phi(e_i) = \begin{cases} 1, & \text{if } i = 1, \\ 0, & \text{otherwise,} \end{cases}$$

and moreover $\phi(e'_j) = 0$ for all j.

$$\psi(f_i) = \begin{cases} 1, & \text{if } i = 1, \\ 0, & \text{otherwise,} \end{cases}$$

and moreover $\psi(f'_j) = 0$ for all j. Using these functionals in (16.3) finishes the proof. □

Proposition 16.3 *Let X, Y be Banach spaces. Then $X \otimes Y$ is canonically isomorphic to the linear space of all finite rank and w^*-w-continuous operators $\mathcal{F}_{w^*}(X^*, Y)$ from X^* into Y. The isomorphism $i : X \otimes Y \to \mathcal{F}_{w^*}(X^*, Y)$ is given by*

$$i\left(\sum_{i=1}^{n} x_i \otimes y_i\right)(x^*) = \sum_{i=1}^{n} \langle x^*, x_i \rangle y_i.$$

Proof: The mapping i is clearly well defined. It remains to show that it is injective and onto. Lemma 16.2 with $A = X^*$, $B = Y^*$, implies that i is injective. If $T \in \mathcal{F}_{w^*}(X^*, Y)$, then clearly $T = \sum_{i=1}^{n} \phi_i(x^*)y_i$ for some $\phi_i \in X^{**}$, $y_i \in Y$. Using Proposition 1.36 we get that $\phi_i \in X$ for $i = 1, 2, \ldots, n$, which proves the claim. □

Since $X \otimes Y$ is linearly isomorphic to $Y \otimes X$, the above proposition implies that $\mathcal{F}_{w^*}(X^*, Y)$, $\mathcal{F}_{w^*}(Y^*, X)$, and $X \otimes Y$ are all canonically linearly isomorphic via the tensor representation of their elements as $\sum_{i=1}^{n} x_i \otimes y_i$. The identification of tensors and finite rank operators will be exploited constantly.

Definition 16.4 *We introduce the* injective norm $\varepsilon(\cdot)$ *on* $X \otimes Y$ *as follows:*

$$\varepsilon\left(\sum_{i=1}^{n} x_i \otimes y_i\right) = \left\| i\left(\sum_{i=1}^{n} x_i \otimes y_i\right)\right\|_{\mathcal{F}_{w^*}(X^*,Y)} = \sup_{\|x^*\|,\|y^*\|\le 1} \left|\sum_{i=1}^{n} x^*(x_i)y^*(y_i)\right|.$$

By $X \otimes_\varepsilon Y$ *we denote the* injective tensor product, *that is the completion of the normed space* $(X \otimes Y, \varepsilon)$.

Unless stated otherwise, the closure operation in $\mathcal{B}(X, Y)$ is meant in the norm operator topology, see Definition 1.26.

The following is immediate.

Proposition 16.5 $X \otimes_\varepsilon Y$ *is canonically isometric to* $(\overline{\mathcal{F}_{w^*}(X^*, Y)}, \|\cdot\|)$.

Definition 16.6 *We say that the mapping* $t : E \otimes F \to F \otimes E$ *given by* $t : \sum_{i=1}^{n} e_i \otimes f_i \mapsto \sum_{i=1}^{n} f_i \otimes e_i$ *is a* transposition mapping.

It is immediate that the transposition mapping is a linear isomorphism. Moreover, the next proposition follows readily from Definition 16.4.

Proposition 16.7 *The transposition mapping* $t : (E \otimes F, \varepsilon) \to (F \otimes E, \varepsilon)$ *is an isometric linear isomorphism. By* t *we will denote its unique isometric extension* $t : E \otimes_\varepsilon F \to F \otimes_\varepsilon E$.

Similarly, we obtain the identification:

Proposition 16.8 *Let* X, Y *be Banach spaces. Then* $X^* \otimes Y$ *is linearly isomorphic to the space of all finite rank bounded operators* $\mathcal{F}(X, Y)$. *The canonical isomorphism* $i : X^* \otimes Y \to \mathcal{F}(X, Y)$ *is given by*

$$i\left(\sum_{i=1}^{n} x_i^* \otimes y_i\right)(x) = \sum_{i=1}^{n} \langle x_i^*, x\rangle y_i.$$

The completion $X^* \otimes_\varepsilon Y$ *of the normed space* $(X^* \otimes Y, \varepsilon)$ *is isometric to* $(\overline{\mathcal{F}(X, Y)}, \|\cdot\|)$.

The concept of bilinear form was introduced after Definition 3.1. In Exercise 5.53 we listed several equivalent conditions for a bilinear form to be *bounded*, and in

Exercise 5.54 we introduced the space of bounded bilinear forms and we defined a norm on it that turned this space into a Banach space. Now we extend this to bilinear mappings from a product $E \times F$ of linear spaces into a linear space G.

Definition 16.9 *Let E, F, G be linear spaces. We say that a mapping $B : E \times F \to G$ is* bilinear, *if*

$$B(a_1x_1 + a_2x_2, b_1y_1 + b_2y_2) = \sum_{1\le i,j\le 2} a_i b_j B(x_i, y_j)$$

for all scalars a_i, b_i, $x_i \in E$, and $y_i \in F$.

Definition 16.10 *Let X, Y, Z be Banach spaces. We say that a bilinear mapping $B : X \times Y \to Z$ is* bounded, *if there exists $K > 0$ such that*

$$\sup_{\|x\|,\|y\|\le 1} \|B(x, y)\|_Z \le K. \tag{16.4}$$

The space of all bounded bilinear mappings from $X \times Y$ into Z is denoted by $\mathcal{B}il(X \times Y, Z)$. We define a norm on $\mathcal{B}il(X \times Y, Z)$ by $\|B\| = \inf\{K : K$ as in (16.4)$\}$.

If Z is the space of scalars $\mathbb{K}$, we use the abbreviation $\mathcal{B}il(X \times Y)$ for $\mathcal{B}il(X \times Y, \mathbb{K})$.

We leave as an exercise to the reader the proof that the space $(\mathcal{B}il(X \times Y, Z), \|\cdot\|)$ defined above is indeed a Banach space (see Exercise 5.54 for the case of bilinear forms). The bilinearity of B means, in particular, that for a fixed $x \in X$, we have $B(x, \cdot) \in \mathcal{B}(Y, Z)$. A simple, but key, fact for us is the following correspondence.

Proposition 16.11 *Let X, Y be Banach spaces. The two Banach spaces $\mathcal{B}\mathcal{B}il(XtimesY)$ and $\mathcal{B}(X, Y^*)$ are canonically isometric, via the correspondence $B \to T$ given by,*

$$\langle T(x), \cdot\rangle = B(x, \cdot) \text{for all} x \in X.$$

Proof: The bilinearity of B implies that $T(x)$ is linear in x, and $T(x) \in Y^*$ for every $x \in X$. Clearly,

$$\|T\| = \sup_{\|x\|,\|y\|\le 1} |\langle T(x), y\rangle| = \sup_{\|x\|,\|y\|\le 1} \|B(x, y)\| = \|B\|.$$

□

Definition 16.12 *Let E, F be linear spaces. We denote by $j : E \times F \to E \otimes F$ the canonical mapping $j(e, f) = e \otimes f$.*

It is easy to see that j is bilinear. Moreover, it has the following universality property.

Proposition 16.13 *(Universal property of tensor product) Let E, F, G be linear spaces. Then there is a one-to-one correspondence between bilinear mappings $B : E \times F \to G$ and linear mappings $T : E \otimes F \to G$, given by the relation $B = T \circ j$.*

Proof: Let $T : E \otimes F \to G$ be a linear mapping. Then $B = T \circ j$ is clearly bilinear. On the other hand, let $B : E \times F \to G$ be a bilinear mapping. Define a mapping T from $j(E \times F) \subset E \otimes F$ to G canonically, i.e., $T(e \otimes f) = B(e, f)$, and extend T onto $E \otimes F$ by linearity.

$$T\left(\sum_{i=1}^{n} e_i \otimes f_i\right) = \sum_{i=1}^{n} T(e_i \otimes f_i).$$

Such an extension exists and is unique. It is clear that the correspondences described above are inverse to each other. □

Theorem 16.14 *Let X, Y be Banach spaces. Then $\langle \mathcal{B}il(X \times Y), X \otimes Y\rangle$ forms a dual pair, where the bilinear form $\langle \cdot, \cdot \rangle$ associated to the dual pair is given, for $B \in \mathcal{B}il(X \times Y)$ and $z = \sum_{i=1}^{n} x_i \otimes y_i \in X \otimes Y$, by*

$$\langle B, z\rangle = \sum_{i=1}^{n} B(x_i, y_i). \tag{16.5}$$

Moreover, $X \otimes Y \hookrightarrow (\mathcal{B}il(X \times Y), \|\cdot\|)^*$.

Proof: Proposition 16.13 gives the identification of $\mathcal{B}il(X \times Y)$ with a subspace of the linear dual $(X \otimes Y)^{\#}$. By Lemma 16.2 the pair is separating, thus forming a dual pair (see Definition 3.2). The formula (16.5) is obvious from Proposition 16.13. Finally,

$$\sup_{\|B\| \leq 1} \left\langle B, \sum_{i=1}^{n} x_i \otimes y_i \right\rangle \leq \sum_{i=1}^{n} \|x_i\| \|y_i\| < \infty,$$

so we have $\sum_{i=1}^{n} x_i \otimes y_i \in \mathcal{B}il(X \times Y)^*$. □

By Proposition 16.11 we have an alternative description of the above duality pair as $\langle \mathcal{B}(X, Y^*), X \otimes Y\rangle$. For $T \in \mathcal{B}(X, Y^*)$, $z = \sum_{i=1}^{n} x_i \otimes y_i \in X \otimes Y$ we have $\langle T, z\rangle = \sum_{i=1}^{n} \langle T(x_i), y_i\rangle$. In particular, by Proposition 16.8, $X^* \otimes_\varepsilon Y^*$ is canonically isometric to a subspace of $\mathcal{B}(X, Y^*)$, and so we also have a canonical duality pair $\langle X^* \otimes Y^*, X \otimes Y\rangle$.

Definition 16.15 *Let X, Y be Banach spaces. We define the* projective norm $\pi(\cdot)$ *on $X \otimes Y$ for $z \in X \otimes Y$ as follows:*

$$\pi(z) = \sup\{\langle B, z\rangle : \|B\| \leq 1,\ B \in \mathcal{B}il(X \times Y)\}$$
$$= \sup\{\langle T, z\rangle : \|T\| \leq 1,\ T \in \mathcal{B}(X, Y^*)\}. \quad (16.6)$$

We denote by $X \otimes_\pi Y$ the projective tensor product, *that is the completion of* $(X \otimes Y, \pi)$.

The fact that π is indeed a norm, rather than just a seminorm, follows from Theorem 16.14, namely $\langle \mathcal{B}il(X \times Y), X \otimes Y\rangle$ is in particular a separating pair.

Proposition 16.16 *Let X, Y be Banach spaces. Then we have*

$$(X \otimes_\pi Y)^* = \mathcal{B}il(X \times Y) = \mathcal{B}(X, Y^*). \quad (16.7)$$

Proof: Given $x \in X$, $y \in Y$, there exist Hahn–Banach functionals $x^* \in B_{X^*}$, $y^* \in B_{Y^*}$ such that $x^*(x) = \|x\|$, $y^*(y) = \|y\|$. Thus $x^*(\cdot)y^*(\cdot) \in \mathcal{B}il(X \times Y)$ witnesses the fact that $\|x \otimes y\|_\pi = \|x\|\|y\|$. Thus for any $\phi \in (X \otimes_\pi Y)^*$, $\|\phi\|_\pi = 1$, we have $\sup_{\|x\|,\|y\|\leq 1} |\phi(x \otimes y)| \leq 1$. Therefore, ϕ is a bilinear mapping from $X \times Y$ that is also bounded. So (16.7) follows using Theorem 16.14. □

Proposition 16.17 *Let X, Y be Banach spaces. Then for $z \in X \otimes Y$*

$$\pi(z) = \inf\left\{\sum_{i=1}^{n} \|x_i\|\|y_i\| : z = \sum_{i=1}^{n} x_i \otimes y_i\right\} \quad (16.8)$$

Proof: We let

$$\lambda(z) = \inf\left\{\sum_{i=1}^{n} \|x_i\|\|y_i\| : z = \sum_{i=1}^{n} x_i \otimes y_i\right\}, \quad (16.9)$$

and we denote $S = \{z \in X \otimes Y : \lambda(z) \leq 1\}$. The set S is clearly a closed and convex set in $(X \otimes Y, \pi)$, which contains all $x \otimes y$, $\|x\| = \|y\| = 1$. So it is a 1-norming set and $\|B\| = \sup_{z\in S}\langle B, z\rangle$. The triangle inequality implies that $\pi(z) \leq \lambda(z)$ for every $z \in X \otimes Y$.

To prove the opposite inequality, assume by contradiction that $z \in X \otimes Y$ be such that $\pi(z) < 1 < \lambda(z)$. Applying the Hahn–Banach separation theorem to S (and using Proposition 16.16), there exists some $B \in \mathcal{B}il(X \times Y)$, $\|B\| = 1$, such that

$$\pi(z) \geq \langle B, z\rangle > 1 \geq \sup\{\langle B, z'\rangle : z' \in S\}.$$

This is a contradiction. □

From Lemma 3.100 and (16.8) we obtain immediately the following result.

Proposition 16.18 *Every element $z \in X \otimes_\pi Y$ admits a representation $z = \sum_{i=1}^\infty x_i \otimes y_i$, such that $\sum_{i=1}^\infty \|x_i\| \|y_i\| < \infty$. Moreover,*

$$\pi(z) = \inf \left\{ \sum_{i=1}^\infty \|x_i\| \|y_i\| : z = \sum_{i=1}^\infty x_i \otimes y_i, \sum_{i=1}^\infty \|x_i\| \|y_i\| < \infty \right\}. \quad (16.10)$$

The dual pairing satisfies the formulas

$$\langle B, z \rangle = \sum_{i=1}^\infty B(x_i, y_i), \text{ for all } B \in \mathcal{B}il(X \times Y), \text{ respectively} \quad (16.11)$$

$$\langle T, z \rangle = \sum_{i=1}^\infty \langle T(x_i), y_i \rangle, \text{ for all } T \in \mathcal{B}(X, Y^*). \quad (16.12)$$

Moreover, we may assume, without loss of generality, that $(\|x_i\|)_{i=1}^\infty \in c_0$ and $(\|y_i\|)_{i=1}^\infty \in \ell_1$.

Not that, clearly, for $z \in X \otimes Y$, we have $\varepsilon(z) \leq \pi(z)$.

Proposition 16.19 *The transposition mapping $t : X \otimes_\pi Y \to Y \otimes_\pi X$,*

$$t : \sum_{i=1}^\infty x_i \otimes y_i \to \sum_{i=1}^\infty y_i \otimes x_i, \qquad \sum_{i=1}^\infty \|x_i\| \|y_i\| < \infty$$

is an isometric isomorphism.

Proof: This follows easily from the duality pairing $(X \otimes_\pi Y)^* = \mathcal{B}il(X\times, Y)$, and the fact that $t^* : \mathcal{B}il(X \times Y) \to \mathcal{B}il(Y \times X)$ is an isometry. □

The alternative description of the dual to $X \otimes_\pi Y$, using the operators, leads to an alternative description of the conjugate transposition.

Proposition 16.20 *Let X, Y be Banach spaces. Let $t : X^* \otimes_\pi Y \to Y \otimes_\pi X^*$ be the transposition isometry. Then $t^* : \mathcal{B}(X^*, Y^*) \to \mathcal{B}(Y, X^{**})$ is an isometric isomorphism. Given $T \in \mathcal{B}(X^*, Y^*)$, we have $t^*(T) = T^*\big|_Y$. Given $S \in \mathcal{B}(Y, X^{**})$, we have $(t^*)^{-1}(S) = S^*\big|_{X^*}$.*

Proof: Combining Proposition 16.19 and Proposition 16.16, t^* is clearly an isometric isomorphism. Given $T \in \mathcal{B}(X^*, Y^*)$, we have

$$\langle T, y \otimes x^* \rangle = \langle T(x^*), y \rangle = \langle T^*(y), x^* \rangle = \langle t^*(T), x^* \otimes y \rangle.$$

The formula for $(t^*)^{-1}$ is proved similarly. □

It would be nice to represent elements of the projective tensor product as operators, similarly to the case of the injective tensor product. As we will see later, this is not always possible. However, the natural candidate for such representation is the following.

Definition 16.21 *A bounded operator $T : X \to Y$ is called* nuclear *if*

$$T(x) = \sum_{i=1}^{\infty} \langle x_i^*, x\rangle y_i$$

for some $\{x_i^*\}_{i=1}^{\infty} \in X^*$, $\{y_i\}_{i=1}^{\infty} \in Y$ such that $\sum_{i=1}^{\infty} \|x_i^*\| \|y_i\| < \infty$. We introduce the nuclear norm *of T as*

$$N(T) = \inf\left\{\sum_{i=1}^{\infty} \|x_i^*\| \|y_i\| \,:\, T(x) = \sum_{i=1}^{\infty} \langle x_i^*, x\rangle y_i\right\}. \tag{16.13}$$

By $\mathcal{N}(X, Y)$ we denote the space of all nuclear operators, with the nuclear norm.

Clearly, finite rank operators $\mathcal{F}(X, Y)$ are N-dense in the space of nuclear operators $\mathcal{N}(X, Y)$. It is easy to see that if $T \in \mathcal{N}(X, Y)$, $S \in \mathcal{B}(Y, Z)$, and $R \in \mathcal{B}(W, X)$, then $T \circ R \in \mathcal{N}(W, Y)$, $S \circ T \in \mathcal{N}(X, Z)$, and $T^* \in \mathcal{N}(Y^*, X^*)$.

Proposition 16.22 *The formal identity*

$$I : \mathcal{N}(X, Y) \hookrightarrow \mathcal{K}(X, Y) \tag{16.14}$$

is a continuous injection. Moreover, $(\mathcal{N}(X, Y), N)$ is a Banach space.

Proof: It is immediate that $N(\cdot) \geq \|\cdot\|$ holds on $X^* \otimes Y$. Thus a N-Cauchy sequence from $X^* \otimes Y$ is convergent in the operator norm, and (16.14) follows. To prove that $\mathcal{N}(X, Y)$ is a Banach space it suffices to show by Lemma 3.100 that whenever $T_j \in \mathcal{N}(X, Y)$, and $\sum_{j=1}^{n} N(T_j) < \infty$, we have $T(x) = \sum_{j=1}^{\infty} T_j(x) \in \mathcal{N}(X, Y)$. Choose representations $T_j(x) = \sum_{i=1}^{\infty} \langle f_i^j, x\rangle y_i^j$, such that $\sum_{i=1}^{\infty} \|f_i^j\| \|y_i^j\| < 2N(T_j)$. Then clearly

$$T(x) = \sum_{j=1}^{\infty} T_j(x) = \sum_{j=1}^{\infty}\sum_{i=1}^{\infty} \langle f_i^j, x\rangle y_i^j, \text{ where } \sum_{j=1}^{\infty}\sum_{i=1}^{\infty} \|f_i^j\| \|y_i^j\| < \infty.$$

□

Definition 16.23 *Let $J : \sum_{i=1}^{\infty} x_i^* \otimes y_i \to \sum_{i=1}^{\infty} x_i^* \otimes y_i$ be the formal identity mapping from $(X^* \otimes_\pi Y)$ into $\mathcal{N}(X, Y)$ defined for all pairs of sequences $\{x_i^*\}_{i=1}^{\infty} \in X^*$, $\{y_i\}_{i=1}^{\infty} \in Y$ such that $\sum_{i=1}^{\infty} \|x_i^*\| \|y_i\| < \infty$, as $(\sum_{i=1}^{\infty} x_i^* \otimes y_i)(x) := \sum_{i=1}^{\infty} x_i^*(x) y_i$ for $x \in X$.*

As we have seen above, such series may be used to represent all elements of $X^* \otimes_\pi Y$ and $\mathcal{N}(X, Y)$.

Proposition 16.24 *The formal identity J is a well-defined quotient mapping $J : X^* \otimes_\pi Y \to \mathcal{N}(X, Y)$. More precisely, let $z \in X^* \otimes_\pi Y$ have a representation $z = \sum_{i=1}^\infty x_i^* \otimes y_i$. Then, the nuclear operator T represented by $J(z)$,*

$$T = J\left(\sum_{i=1}^\infty x_i^* \otimes y_i\right), \quad T(x) = \sum_{i=1}^\infty x_i^*(x) y_i, \tag{16.15}$$

is independent of the concrete representation of the tensor z.

Proof: By (16.12), two absolutely convergent series $\sum_{i=1}^\infty x_i^* \otimes y_i$, $\sum_{i=1}^\infty u_i^* \otimes v_i$ represent a single tensor $z \in X^* \otimes_\pi Y$ if and only if

$$\langle S, z\rangle = \sum_{i=1}^\infty \langle S(x_i^*,)y_i\rangle = \sum_{i=1}^\infty \langle S(u_i^*), v_i\rangle, \quad \text{for all } S \in \mathcal{B}(X^*, Y^*). \tag{16.16}$$

By Definition 16.21 two series $\sum_{i=1}^\infty x_i^* \otimes y_i$, $\sum_{i=1}^\infty u_i^* \otimes v_i$ represent a single nuclear operator $T \in \mathcal{N}(X, Y)$ if and only if

$$\langle S, T\rangle = \sum_{i=1}^\infty \langle S(x_i^*), y_i\rangle = \sum_{i=1}^\infty \langle S(u_i^*), v_i\rangle = \sum_{i=1}^\infty x_i^*(x) y^*(y) = \sum_{i=1}^\infty u_i^*(x) y^*(y_i) \tag{16.17}$$

for all $S = x \otimes y^* \in \mathcal{F}_{w^*}(X^*, Y^*)$. Since (16.16) is formally stronger than (16.17), the result follows. □

Proposition 16.25 *Let X, Y be Banach spaces. Then the two spaces $\mathcal{N}(Y, X^{**})$ and $\mathcal{N}(X^*, Y^*)$ are canonically isometric, via the transposition of their elements $z = \sum_{i=1}^\infty y_i^* \otimes x_i^{**} \leftrightarrow z' = \sum_{i=1}^\infty x_i^{**} \otimes y_i^*$.*

Proof: By Proposition 16.19, the transposition mapping $t : Y^* \otimes_\pi X^{**} \to X^{**} \otimes_\pi Y^*$ is an isometric isomorphism. Next, $\mathcal{N}(Y, X^{**})$ is a quotient (via J) of $Y^* \otimes_\pi X^{**}$, while $\mathcal{N}(X^*, Y^*)$ is a quotient (via J') of the isometric transpose $t(Y^* \otimes_\pi X^{**}) = X^{**} \otimes_\pi Y^*$. The kernels are described as follows.

$$\mathrm{Ker}(J) = \left\{ z = \sum_{i=1}^\infty y_i^* \otimes x_i^{**} : \sum_{i=1}^\infty y_i^*(y) x_i^{**} = 0 \text{ for all } y \in Y \right\}.$$

$$\mathrm{Ker}(J') = \left\{ z' = \sum_{i=1}^\infty x_i^{**} \otimes y_i^* : \sum_{i=1}^\infty x_i^{**}(x^*) y_i^* = 0 \text{ for all } x^* \in X^* \right\}.$$

Both of these conditions are indeed equivalent to the single condition $z \in \mathrm{Ker}(J) \iff t(z) \in \mathrm{Ker}(J')$, which is to say $\sum_{i=1}^\infty x_i^{**}(x^*) y_i^*(y) = 0$ for all $y \in Y$, $x^* \in X^*$. □

16.2 Duality of Injective Tensor Products

In this section we are going to investigate the Banach space dual to the injective tensor product space $X \otimes_\varepsilon Y$. The results rely on the presence of the Radon–Nikodým property. The dual balls B_{X^*}, B_{Y^*} are assumed to be equipped with the w^*-topology, unless specified otherwise. A key to our approach is the next simple embedding result.

Lemma 16.26 *There is a canonical isometric embedding*

$$I : \overline{\mathcal{F}_{w^*}(X^*, Y)} = X \otimes_\varepsilon Y \hookrightarrow C(B_{X^*} \times B_{Y^*})$$

given by

$$I(S)(x^*, y^*) = \langle y^*, S(x^*)\rangle \text{ for } S \in X \oplus_\varepsilon Y.$$

Proof: Indeed, if $S = \sum_{i=1}^n x_i \otimes y_i$, then clearly

$$I(S)(x^*, y^*) = \sum_{i=1}^n \langle y^*, y\rangle\langle x^*, x\rangle \in C(B_{X^*} \times B_{Y^*})$$

so I is an isometry on $\mathcal{F}_{w^*}(X^*, Y)$. The rest follows as $X \otimes Y$ is dense in $X \otimes_\varepsilon Y$. □

Theorem 16.27 (Grothendieck) *Let $I : X \otimes_\varepsilon Y \hookrightarrow C(B_{X^*} \times B_{Y^*})$ be the isometric embedding from Lemma* 16.26. *Then every $\phi \in (X \otimes_\varepsilon Y)^*$ has a representation as a positive w^*-Radon measure μ on $(B_{X^*} \times B_{Y^*}, w^* \times w^*)$, so that for $z \in X \otimes_\varepsilon Y$*

$$\langle \phi, z\rangle = \int_{B_{X^*} \times B_{Y^*}} I(z)(x^*, y^*)\, d\mu = \int_{B_{X^*} \times B_{Y^*}} \langle x^* \otimes y^*, z\rangle\, d\mu. \tag{16.18}$$

Moreover, $\|\phi\| = |\mu|(B_{X^*} \times B_{Y^*})$.

Proof: Denote again by ϕ a Hahn–Banach extension of ϕ on the whole of $C(B_{X^*} \times B_{Y^*})$. By the Riesz representation theorem, ϕ is represented by a Radon measure μ on $B_{X^*} \times B_{Y^*}$, $\|\phi\| = |\mu|$. The positivity of μ is achieved easily. Indeed, let $\mu = \mu^+ - \mu^-$, where μ^+, μ^- are positive. We may replace μ by the positive measure $\rho = \mu^+ + \eta$, where $\eta = (-Id)(\mu^-)$, $Id : X \times Y \to X \times Y$ is the identity mapping. Clearly, $|\rho| \le |\mu|$. □

The role of RNP lies in the next theorem.

Theorem 16.28 (Schwartz) *Let X^* be a RNP space. Then for every w^*-Radon measure μ on B_{X^*}, $Id : B_{X^*} \to B_{X^*}$ is μ-Bochner integrable. More precisely, given* *any w^*-compact set $K \subset B_{X^*}$ and $\rho > 0$, there exists a compact set $K_\rho \subset K$, such that $|\mu|(K \backslash K_\rho) < \rho$ and $Id\big|_{K_\rho}$ is w^*-$\|\cdot\|$ continuous.*

For a proof see Theorem 11.16. Therefore, under the assumptions of the theorem, we are in a position to apply the theory of measure and Bochner integration (see Section 17.13.1).

In particular, note that $Id : B_{X^*} \to B_{X^*}$ is a w^*-$\|\cdot\|$-Borel mapping up to a set of μ-measure zero. In fact, in the important special case when X^* is a separable dual space, Id is in fact a w^*-Borel mapping. Indeed, for every $\lambda > 0$, $\lambda B_{X^*} = \bigcap_{i=1}^{\infty} x_i^{-1}[-\lambda, \lambda]$, where $\{x_i\}_{i=1}^{\infty}$ is norm dense in B_X. Being a countable intersection of w^*-open sets with a w^*-compact set, λB_{X^*} is w^*-Borel. Consequently, every norm open subset of B_{X^*} is w^*-Borel, and finally every norm Borel subset of B_{X^*} is w^*-Borel.

Lemma 16.29 *Let (S, Σ, μ) be a finite positive measure space and $f : S \to X$ be Bochner integrable. For each $\varepsilon > 0$ there are sequences $\{x_n\}_{n=1}^{\infty}$ in X and $\{E_n\}_{n=1}^{\infty}$ (not necessarily disjoint) in Σ, such that*

$$\sum_{n=1}^{\infty} \chi_{E_n} x_n \text{ converges to } f \text{ absolutely } \mu\text{-a.e.} \tag{16.19}$$

$$\int_S \|f\|\,\mathrm{d}\mu - \varepsilon \le \sum_{n=1}^{\infty} \|x_n\|\mu(E_i) \le \int_S \|f\|\,\mathrm{d}\mu + \varepsilon. \tag{16.20}$$

Proof: We may assume for simplicity that $|\mu| = 1$. Let $f_1 = f$. Fix a positive sequence $\{\delta_i\}_{i=1}^{\infty}$, with $\sum_{i=1}^{\infty} \delta_i < \frac{\varepsilon}{2}$. Given $\delta > 0$ and a point x in X, we set

$$S(f, x, \delta) = \{s \in S : \|f(s) - x\| < \delta\}.$$

Given $\delta > 0$ and a sequence of points $\{y_i\}_{i=1}^{\infty}$, we let $R(f, y_i, \delta) = S(f, y_i, \delta) \setminus \bigcup_{j<i} S(f, y_j, \delta)$. Since f_1 is Bochner integrable, there exists a sequence of points $\{x_j^1\}_{j=1}^{\infty}$ in X such that

$$\int_{S \setminus \bigcup_{j=1}^{\infty} R(f_1, x_j^1, \delta_1)} \|f_1\|\,\mathrm{d}\mu < \delta_1.$$

We now proceed inductively as follows. Having found f_n, a sequence $\{x_j^n\}_{j=1}^{\infty}$ in X and a disjoint system $\{R(f_n, x_j^n, \delta_n)\}_{j=1}^{\infty}$ of sets in Σ, such that

$$\int_{S \setminus \bigcup_{j=1}^{\infty} R(f_n, x_j^n, \delta_n)} \|f_n\|\,\mathrm{d}\mu < \delta_n, \tag{16.21}$$

we let $f_{n+1} = f_n - \sum_{j=1}^{\infty} \chi_{R(f_n, x_j^n, \delta_n)} x_j^n$ (note that f_{n+1} is again Bochner integrable). Note that $\| f_n - \sum_{j=1}^{\infty} \chi_{R(f_n, x_j^n, \delta_n)} x_j^n \| \leq \delta_n$ on $\bigcup_{j=1}^{\infty} R(f_n, x_j^n, \delta_n)$, so combining with (16.21) we get $\int \| f_{n+1} \| \mathrm{d}\mu < 2\delta_n$. We repeat the inductive argument. Finally, consider the systems $\{x_j^n\}_{n,j=1}^{\infty}$ in X and $\{R(f_n, x_j^n, \delta_n)\}_{n,j=1}^{\infty}$ in Σ.

$$\int_S \Big\| f - \sum_{n,j \in \mathbb{N}} \chi_{R(f_n, x_j^n, \delta_n)} x_j^n \Big\| \mathrm{d}\mu \leq \sum_{n=1}^{\infty} \Big\| f_n - \sum_{j \in \mathbb{N}} \chi_{R(f_n, x_j^n, \delta_n)} x_j^n \Big\| d\mu \leq 2 \sum_{n=1}^{\infty} \delta_n < \varepsilon.$$

Upon re-indexing, our systems satisfy both (16.19) and (16.20). □

Theorem 16.30 (Grothendieck) *Let Y^* be an RNP space. Then there is an isometry*

$$(X \otimes_{\varepsilon} Y)^* = \mathcal{N}(X, Y^*). \tag{16.22}$$

More precisely, every $\phi \in (X \otimes_{\varepsilon} Y)^*$, $\|\phi\| < 1$, is represented by a nuclear operator $T \in \mathcal{N}(X, Y^*)$, $T = \sum_{n=1}^{\infty} x_n^* \otimes y_n^*$, $\sum_{n=1}^{\infty} \|x_n^*\| \|y_n^*\| < 1$ so that for every $S \in \overline{\mathcal{F}_{w^*}(X^*, Y)} = X \otimes_{\varepsilon} Y$ we have

$$\langle T, S \rangle = \sum_{n=1}^{\infty} \langle y_n^*, S(x_n^*) \rangle \tag{16.23}$$

Proof: By Theorem 16.27, every $\phi \in (X \otimes_{\varepsilon} Y)^*$, $\|\phi\| < 1$ is represented by a positive w^*-Radon measure μ on $B_{X^*} \times B_{Y^*}$, $|\mu| < 1$. Our goal is to represent ϕ as a nuclear operator $T \in \mathcal{N}(X, Y^*)$. We are going to define T by using the commutative diagram:

$$\begin{array}{ccc} X & \xrightarrow{\;T\;} & Y^* \\ \downarrow{\scriptstyle i_1} & & \uparrow{\scriptstyle i_3} \\ C(B_{X^*} \times B_{Y^*}) & \xrightarrow{\;i_2\;} & L_1(\mu) \end{array}$$

where the mappings are defined as follows:

$i_1(x)(x^*, y^*) = x^*(x)$. Clearly, $\|i_1\| = 1$.

i_2 is the formal identity mapping from $C(B_{X^*} \times B_{Y^*})$ to $L_1(\mu)$. Thus $\|i_2\| = |\mu| < 1$.

$i_3 : L_1(\mu) \to Y^*$ is defined by the formula

$$i_3(f) = \int_{B_{X^*} \times B_{Y^*}} f(x^*, y^*) y^* \mathrm{d}\mu.$$

The integrated function is a product of a μ-integrable scalar function with the mapping $(x^*, y^*) \to y^*$. Due to Theorem 16.28, the latter is μ-Bochner integrable. Indeed, as it depends only on the second variable y^*, its Bochner integrability is equivalent to the Bochner integrability of the vector function $Id : (B_{Y^*}, w^*) \to (B_{Y^*}, \|\cdot\|)$ under the w^*-continuous image $P_{Y^*}(\mu|_U)$ of the measure μ, restricted to any w^*-Borel set in $B_{X^*} \times B_{Y^*}$. Clearly, $P_{Y^*}(\mu|_U)$ is again a w^*-Radon measure on B_{Y^*}.

Again, we have $\|i_3\| < 1$. Thus $T = i_3 \circ i_2 \circ i_1$ is well defined. Next we claim that the operator $i_3 \circ i_2 : C(B_{X^*} \times B_{Y^*}) \to Y^*$ is nuclear. Using Lemma 16.29, for $\varepsilon > 0$ small enough, there are sequences $\{y_n\}_{n=1}^{\infty}$ in Y^* and $\{E_n\}_{n=1}^{\infty}$ of w^*-Borel subsets of $B_{X^*} \times B_{Y^*}$, so that

$$\int_{B_{X^*} \times B_{Y^*}} \|y^*\| \mathrm{d}\mu - \varepsilon \le \sum_{n=1}^{\infty} \|y_n^*\| \mu(E_i) \le \int_{B_{X^*} \times B_{Y^*}} \|y^*\| \mathrm{d}\mu + \varepsilon < 1. \quad (16.24)$$

and moreover

$$\begin{aligned} i_3 \circ i_2(f) &= \int_{B_{X^*} \times B_{Y^*}} f(x^*, y^*) y^* \mathrm{d}\mu \\ &= \int_{B_{X^*} \times B_{Y^*}} f(x^*, y^*) \sum_{n=1}^{\infty} \chi_{E_n} y_n^* \mathrm{d}\mu = \sum_{n=1}^{\infty} \left(\int_{E_n} f \mathrm{d}\mu \right) y_n^*. \end{aligned} \quad (16.25)$$

Note that $l_n(f) = \int_{E_n} f \, \mathrm{d}\mu \in C(B_{X^*} \times B_{Y^*})^*$, $\|l_n\| = \mu(E_n)$. By (16.24), we see that $i_3 \circ i_2 = \sum_{n=1}^{\infty} l_n \otimes y_n^*$ is a nuclear operator with $N(i_3 \circ i_2) < 1$. Therefore, putting $x_n^* = i_1^*(l_n)$, we get that $T = \sum_{n=1}^{\infty} x_n^* \otimes y_n^*$ is a nuclear operator of norm less than one. Equation (16.25) yields

$$T(x) = \int_{B_{X^*} \times B_{Y^*}} x(x^*, y^*) y^* \mathrm{d}\mu = \int_{B_{X^*} \times B_{Y^*}} x^*(x) y^* \mathrm{d}\mu = \sum_{n=1}^{\infty} x_n^*(x) y_n^*. \quad (16.26)$$

Given $z = \sum_{i=1}^{k} u_i \otimes v_i \in X \otimes_{\varepsilon} Y$, by (16.18) and (16.26)

$$\begin{aligned} \langle \phi, z \rangle &= \int_{B_{X^*} \times B_{Y^*}} \sum_{i=1}^{k} y^*(v_i) x^*(u_i) \, \mathrm{d}\mu \\ &= \sum_{i=1}^{k} \left\langle \int_{B_{X^*} \times B_{Y^*}} x^*(u_i) y^* \mathrm{d}\mu, v_i \right\rangle = \sum_{i=1}^{k} \langle T(u_i), v_i \rangle \\ &= \sum_{n=1}^{\infty} \sum_{i=1}^{k} x_n^*(u_i) y_n^*(v_i) = \sum_{n=1}^{\infty} \left\langle y_n^*, \sum_{i=1}^{k} (u_i \otimes v_i)(x_n^*) \right\rangle = \langle T, z \rangle, \end{aligned}$$

and (16.23) follows. □

16.3 Approximation Property and Duality of Spaces of Operators

In the present section, we are going to introduce the approximation property of a Banach space, and show its consequences for various duality relations among the operator spaces.

Definition 16.31 *Let X, Y be Banach spaces. By τ we denote the locally convex topology on $\mathcal{B}(X, Y)$ of uniform convergence on compact sets in X, i.e., the topology on $\mathcal{B}(X, Y)$ generated by the family of seminorms $\{\|T\|_K$: K a norm-compact set in $X\}$, where $\|T\|_K = \sup\{\|T(x)\| : x \in K\}$.*

By Proposition 16.16 (and the transposition isometry $Y^* \otimes_\pi X = t(X \otimes_\pi Y^*)$), we have the duality of Banach spaces $(Y^* \otimes_\pi X)^* = \mathcal{B}(X, Y^{**})$. Denote by $i : \mathcal{B}(X, Y) \to \mathcal{B}(X, Y^{**})$ the formal identity embedding.

Lemma 16.32 *The mapping*

$$i : (\mathcal{B}(X, Y), \tau) \to (\mathcal{B}(X, Y^{**}), w^*) \tag{16.27}$$

is continuous. Hence, its adjoint mapping i^* is w-w^* continuous:

$$i^* : Y^* \otimes_\pi X \to (\mathcal{B}(X, Y), \tau)^* \tag{16.28}$$

Proof: By Proposition 16.18, every $z \in Y^* \otimes_\pi X$ admits a representation $z = \sum_{i=1}^\infty y_i^* \otimes x_i$, such that $(\|x_i\|)_{i=1}^\infty \in c_0$ and $(\|y_i^*\|)_{i=1}^\infty \in \ell_1$. Then $K := \overline{\text{conv}}\{x_i\}_{i=1}^\infty$ is a compact and convex set in X. Let U be a τ-open set in $\mathcal{B}(X, Y)$ defined as $U = \{T : \sup_{x \in K} \|T(x)\| < 1\}$. Clearly, $T \in U$ implies that $|\langle y^*, T(x)\rangle| < \|y^*\|$ for all $y^* \in Y^*, x \in K$. Thus $|\langle T, \sum_{i=1}^\infty y_i^* \otimes x_i\rangle| \le \sum_{i=1}^\infty \|y_i^*\| < \infty$ for all $T \in U$. Since every w^*-neighborhood of zero in $\mathcal{B}(X, Y^{**})$ is determined by finitely many vectors from its predual $Y^* \otimes_\pi X$, the result follows. The second result follows by duality, using Proposition 16.16. $\square$

Theorem 16.33 (Grothendieck) *The mapping $i^* : Y^* \otimes_\pi X \to (\mathcal{B}(X, Y), \tau)^*$ from* (16.28) *is surjective. In particular, every $\phi \in (\mathcal{B}(X, Y), \tau)^*$ can be represented as*

$$\phi(T) = \sum_{i=1}^\infty \langle y_i^*, T x_i\rangle, \; x_i \in X, \; y_i^* \in Y^*, \; \sum_{i=1}^\infty \|x_i\| \|y_i^*\| < \infty. \tag{16.29}$$

Proof: Given $\phi \in (\mathcal{B}(X, Y), \tau)^*$, there is some $C > 0$ and a compact set $K \subset X$, so that $|\phi(T)| \le C\|T\|_K$ holds for all $T \in \mathcal{B}(X, Y)$. By Exercise 1.69 (Grothendieck) we may, without loss of generality, assume that $K = \overline{\text{conv}}\,\{x_i\}_{i=1}^\infty$, $x_i \to 0$. We set $S : \mathcal{B}(X, Y) \to (Y \oplus Y \oplus \dots)_{c_0}$, $S(T) = (T(x_1), T(x_2), \dots)$. Since $|\phi(T)| \le C\|S(T)\|$, we may use the Hahn–Banach theorem in order to extend ϕ from $S(\mathcal{B}(X, Y))$, preserving the notation, into a functional

$$\phi \in (Y \oplus Y \oplus \dots)_{c_0}^* = (Y^* \oplus Y^* \oplus \dots)_{\ell_1}.$$

It follows that there exists a sequence $\{y_i^*\}_{i=1}^\infty$, $\sum_{i=1}^\infty \|y_i^*\| < \infty$ such that $\phi(T) = \sum_{i=1}^\infty \langle y_i^*, T(x_i)\rangle$. Clearly, $z = \sum_{i=1}^\infty y_i^* \otimes x_i \in Y^* \otimes_\pi X$, and $i^*(z) = \phi$. □

Definition 16.34 *We say that a Banach space X has the approximation property (AP in short), if*

$$Id \in \overline{\mathcal{F}(X)}^\tau,$$

where τ is the topology introduced in Definition 16.31.

Theorem 16.35 (Grothendieck) *Let X be a Banach space. The following are equivalent:*
(1) *X has the AP.*
(2) *For every pair of sequences $\{x_n\}_{n=1}^\infty$ from X and $\{x_n^*\}_{n=1}^\infty$ from X^*, such that $\sum_{n=1}^\infty \|x_n^*\|\|x_n\| < \infty$ and $\sum_{n=1}^\infty x_n^*(x)x_n = 0$ for all $x \in X$, we have $\sum_{n=1}^\infty x_n^*(x_n) = 0$.*
(3) *For every Banach space Y, $\overline{\mathcal{F}(X,Y)}^\tau = \mathcal{B}(X,Y)$.*
(4) *For every Banach space Y, $\overline{\mathcal{F}(Y,X)}^\tau = \mathcal{B}(Y,X)$.*
(5) *For every Banach space Y, $\overline{\mathcal{F}(Y,X)} = \mathcal{K}(Y,X)$.*
(6) *For every Banach space Y, $Y^* \otimes_\varepsilon X = \mathcal{K}(Y,X)$.*
(7) *$J : X^* \otimes_\pi X \to \mathcal{N}(X)$ is injective, or equivalently it is an isometry.*
(7') *For every Banach space Y, $J : Y^* \otimes_\pi X \to \mathcal{N}(Y,X)$ is injective, or equivalently it is an isometry.*
(8) *$j : X^* \otimes_\pi X \to X^* \otimes_\varepsilon X$ is injective.*
(8') *For every Banach space Y, $j : Y^* \otimes_\pi X \to Y^* \otimes_\varepsilon X$ is injective.*

Proof: (1)$\Longleftrightarrow$(2): Let $z = \sum_{n=1}^\infty x_n^* \otimes x_n \in X^* \otimes_\pi X$. The condition $\sum x_n^*(x)x_n = 0$ for all $x \in X$ is equivalent to $\langle z, T\rangle = 0$ for all $T \in \mathcal{F}(X)$. Indeed, if $T = \sum_{i=1}^k u_i^* \otimes u_i$, then

$$\langle z, T\rangle = \sum_{i=1}^k \left\langle u_i^*, \sum_{n=1}^\infty x_n^*(u_i)x_n \right\rangle = 0.$$

Also, the condition $\sum_{n=1}^\infty x_n^*(x_n) = 0$ is equivalent to

$$\langle z, Id\rangle = \sum_{n=1}^\infty x_n^*(Id(x_n)) = 0.$$

By Theorem 16.33 and the Hahn–Banach separation theorem, we obtain the conclusion.

Putting $X = Y$ verifies that (3)$\Longrightarrow$(1). The opposite implication is immediate. Indeed, let $T \in \mathcal{B}(X,Y)$. Then $T = T \circ Id_X$, so it remains to note that if a net $\{L_\alpha\}_\alpha$ of elements from $\mathcal{F}(X)$ τ-converges to Id_X, then the net $\{T \circ L_\alpha\}_\alpha$ τ-converges to T.

The equivalence (1)$\Longleftrightarrow$(4) is similar.

(1)$\Longrightarrow$(5) is easy.

To prove (5)$\Longrightarrow$(1), let $K \subset X$ be a compact convex and symmetric set, $\varepsilon > 0$. We may assume, without loss of generality, that $K = \overline{\text{conv}}\{x_i\}_{i=1}^{\infty}$, for some sequence in X, $x_i \to 0$. Choose a sequence $\lambda_i \to \infty$ such that $L = \overline{\text{conv}}\{\lambda_i x_i\}_{i=1}^{\infty}$ is a compact, convex and symmetric set in X. We know from Exercise 2.22, that the linear space span L, equipped with the Minkowski functional of L, is a Banach space, that we denote by Y.

The formal identity mapping $I : Y \to X$ is compact and injective. Therefore $I^* : X^* \to Y^*$ is compact and has a w^*-dense range.

Let (Y^*, τ_M) be a topology of uniform convergence on norm compact sets in Y. By Mackey's Theorem 3.41, this topology is compatible with the dual pair $\langle Y, Y^*\rangle$, so that $(Y^*, \tau_M)^* = Y$. Consider $S = \overline{I^*(X^*)}^{\tau_M}$. Since $I^*(X^*)$ is w^*-dense in Y^*, an appeal to the Hahn–Banach separation theorem yields that $S = Y^*$. By our assumption, there exists $T = \sum_{i=1}^{n} y_i^* \otimes x_i \in \mathcal{F}(Y, X)$ such that $\|T - I\| < \varepsilon$. Choose $0 < \delta < \varepsilon(n \max\{\|x_i\|\}_{i=1}^{n} \max_{x\in K}\|x\|)^{-1}$. Then there exist $x_i^* \in X^*$, such that $\|I^*(x_i^*) - y_i^*\|_K < \delta$, $i = 1, \dots, n$. Thus

$$\sup_{x\in K}\left\|\left(\sum_{i=1}^{n} I^*(x_i^*) \otimes x_i - I\right)(x)\right\| < \varepsilon + \delta n \max\{\|x_i\|\}_{i=1}^{n} \max_{x\in K}\|x\|.$$

Thus

$$\sup_{x\in K}\left\|\left(\sum_{i=1}^{n} x_i^* \otimes x_i - Id\right)(x)\right\| < 2\varepsilon.$$

The condition (1) has been verified.

By Proposition 16.8, (6) is a reformulation of (5). The equivalence of (1)–(6) has been established.

(1)$\Longrightarrow$(7'): Let $0 \neq z = \sum_{i=1}^{\infty} y_i^* \otimes x_i \in Y^* \otimes_\pi X$, where $\sum_{i=1}^{\infty}\|y_i\| < \infty$ and $\lim\limits_{i\to\infty} x_i = 0$. We proceed by contradiction, assuming that $J(z)(y) = \sum_{i=1}^{\infty} y_i^*(y)x_i = 0$ for all $y \in Y$. Given $\varepsilon > 0$, by condition (5), there is some

$$F = \sum_{k=1}^{n} u_k^* \otimes u_k \in \mathcal{F}(X, X), \text{ such that } \sup_{i\in\mathbb{N}}\|F(x_i) - x_i\| < \varepsilon.$$

We let $z' = \sum_{i=1}^{\infty} y_i^* \otimes F(x_i) \in Y^* \otimes_\pi X$. Note the important fact that $z' \in Y^* \otimes X$ is actually a finite tensor. Indeed,

$$z' = \sum_{i=1}^{\infty} y_i^* \otimes \left(\sum_{k=1}^{n} u_k^*(x_i)u_k\right) = \sum_{k=1}^{n}\left(\sum_{i=1}^{\infty} u_k^*(x_i)y_i\right) \otimes u_k.$$

Next, $J(z')$ satisfies the following:

$$J(z')(y) = \sum_{i=1}^{\infty} y_i^*(y)F(x_i) = F\left(\sum_{i=1}^{\infty} y_i^*(y)x_i\right) = 0, \text{ for every } y \in Y.$$

Hence $J(z') = 0$, and since z' is also a finite tensor we conclude that $z' = 0$ as an element of $Y^* \otimes_\pi X$. Hence we have an estimate

$$\pi(z) = \pi(z - z') = \pi\left(\sum_{i=1}^{\infty} y_i^* \otimes x_i - \sum_{i=1}^{\infty} y_i^* \otimes F(x_i)\right) \le \varepsilon \sum_{i=1}^{\infty} \|y_i^*\|.$$

Since ε was arbitrarily small, we conclude that $\pi(z) = 0$ as desired. It is clear by the Banach open mapping theorem that J is an isometry.

Clearly, (7')$\Longrightarrow$(7).

Let us show that (7)$\Longrightarrow$(1): We first identify the kernel of the quotient operator $J : X^* \otimes_\pi X \to \mathcal{N}(X)$ from Proposition 16.24. By (16.17), $z = \sum_{i=1}^{\infty} x_i^* \otimes x_i \in \operatorname{Ker}(J)$ if and only if

$$\langle S, z\rangle = \sum_{i=1}^{\infty} \langle x_i^*, S(x_i)\rangle = \sum_{i=1}^{\infty} x_i^*(x)x^*(x_i) = 0$$

for all $S = x^* \otimes x \in \mathcal{F}(X)$ $(= \mathcal{F}_{w^*}(X^*))$. Next let us consider the kernel of the quotient mapping $i^* : X^* \otimes_\pi X \to (\mathcal{B}(X), \tau)^*$, from (16.28). Clearly, $z = \sum_{i=1}^{\infty} x_i^* \otimes x_i \in \operatorname{Ker}(i^*)$ if and only if

$$\langle S, z\rangle = \sum_{i=1}^{\infty} \langle x_i^*, S(x_i)\rangle = \sum_{i=1}^{\infty} x_i^*(x)x^*(x_i) = 0$$

for all $S \in \mathcal{B}(X)$. If J is injective, then for every $z \in X^* \otimes_\pi X$, we have that $i^*(z)\big|_{\mathcal{F}(X)} = 0$ implies $i^*(z) = 0$. By the Hahn–Banach separation theorem and duality, this fact is equivalent to X having the AP.

To obtain (7)$\Longleftrightarrow$(8), it suffices to apply Exercise 16.6.

To obtain (7')$\Longleftrightarrow$(8'), it suffices to apply Exercise 16.6. □

Theorem 16.36 (Grothendieck) *Let X be a Banach space. The following are equivalent:*

(1) *X^* has the AP.*

(2) *For every Banach space Y, $\overline{\mathcal{F}(X, Y)} = \mathcal{K}(X, Y)$.*

(3) *$J : X^* \otimes_\pi X^{**} \to \mathcal{N}(X, X^{**})$ is an isometry.*

(4) *For every Banach space Y, $J : X^* \otimes_\pi Y \to \mathcal{N}(X, Y)$ is an isometry.*

Proof: (1)$\Longrightarrow$(2): If $T \in \mathcal{K}(X, Y)$, then $T^* \in \mathcal{K}(Y^*, X^*)$. By the AP of X^*, there is $S \in \mathcal{F}(X^*)$ such that $\|T^* - S \circ T^*\| \le \varepsilon$. Hence also $\|T^{**} - T^{**} \circ S^*\| \le \varepsilon$. If

$S = \sum_{i=1}^n x_i^{**} \otimes x_i^*$, then

$$T^{**} \circ S^* = \sum_{i=1}^n x_i^* \otimes T^{**}(x_i^{**}).$$

Since T is a compact operator, $T^{**}(X^{**}) \subset Y$. Thus $T^{**} \circ S^*\big|_X \in \mathcal{F}(X, Y)$ and the implication follows.

(2)$\Longrightarrow$(1): Let $T \in \mathcal{K}(Y, X^*)$, $\varepsilon > 0$. Then $T_1 = T^*\big|_X \in \mathcal{K}(X, Y^*)$. By assumption, there exists $S \in \mathcal{F}(X, Y^*)$, such that it holds $\|S - T_1\| < \varepsilon$. Thus $\|S^* - T_1^*\| < \varepsilon$. However, $T_1^*\big|_Y = T$, so $S^*\big|_Y \in \mathcal{F}(Y, X^*)$ verifies the condition (5) of Theorem 16.35.

(3)$\Longrightarrow$(1): By Proposition 16.25, $\mathcal{N}(X, X^{**})$ and $\mathcal{N}(X^*, X^*)$ are canonically isometric, via the transposition of their elements $z = \sum_{i=1}^\infty x_i^* \otimes x_i^{**} \leftrightarrow z' = \sum_{i=1}^\infty x_i^{**} \otimes x_i^*$. Using the transposition, we may transform (3) of Theorem 16.36 into the equivalent statement that $J' : X^{**} \otimes_\pi X^* \to \mathcal{N}(X^*, X^*)$ is an isometry. By condition (7) of Theorem 16.35 we conclude that X^* has the AP.

(4)$\Longrightarrow$(3) is immediate.

It remains to show (1)$\Longrightarrow$(4). Let $0 \neq z = \sum_{i=1}^\infty x_i^* \otimes y_i \in X^* \otimes_\pi Y$; our goal is to show that $J(z) \neq 0$. Without loss of generality, we may assume that $\sum_{i=1}^\infty \|y_i\| < \infty$ and $\lim\limits_{i\to\infty} \|x_i^*\| = 0$. We proceed by contradiction, assuming that $J(z)(y) = \sum_{i=1}^\infty x_i^*(x) y_i = 0$ for all $x \in X$. Given $\varepsilon > 0$, there is a

$$F = \sum_{k=1}^n u_k^{**} \otimes u_k^* \in \mathcal{F}(X^*), \text{ such that } \sup_i \|F(x_i^*) - x_i^*\| < \varepsilon.$$

We let $z' = \sum_{i=1}^\infty F(x_i^*) \otimes y_i \in X^* \otimes_\pi Y$. Note the important fact that $z' \in X^* \otimes Y$ is actually a finite tensor. Indeed,

$$z' = \sum_{i=1}^\infty \left(\sum_{k=1}^n \langle u_k^{**}, x_i^* \rangle u_k^* \right) \otimes y_i = \sum_{k=1}^n u_k^* \otimes \left(\sum_{i=1}^\infty \langle u_k^{**}, x_i^* \rangle y_i \right).$$

Next, $J(z')$ satisfies the following:

$$J(z')(x) = \sum_{i=1}^\infty \langle F(x_i^*), x \rangle y_i = \sum_{i=1}^\infty \langle x_i^*, F^*(x) \rangle y_i = 0, \text{ for every } x \in X.$$

Hence $J(z') = 0$, as an element of $\mathcal{K}(X, Y)$, and since z' is also a finite tensor we conclude that $z' = 0$ as an element of $X^* \otimes_\pi Y$. Hence we have an estimate

$$\pi(z) = \pi(z - z') = \pi\left(\sum_{i=1}^\infty x_i^* \otimes y_i - \sum_{i=1}^\infty F(x_i^*) \otimes y_i \right) \leq \varepsilon \sum_{i=1}^\infty \|y_i\|.$$

Since ε was arbitrarily small, we conclude that $\pi(z) = 0$ as desired. It is clear by the Banach open mapping theorem that J is an isometry. □

We are now going to present several important applications of the AP.

Theorem 16.37 (Grothendieck) *Let X be a Banach space, such that X^* has the AP. Then X also has the AP.*

Proof: By condition (2) of Theorem 16.35, the AP for X^* is equivalent to the following condition. For every pair of sequences $\{x_n^{**}\}_{n=1}^{\infty}$ from X^{**} and $\{x_n^*\}_{n=1}^{\infty}$ from X^*, such that $\sum_{n=1}^{\infty} \|x_n^*\|\|x_n^{**}\| < \infty$ and $\sum_{n=1}^{\infty} x_n^*(x)x_n^{**}(x^*) = 0$ for all $x \in X, x^* \in X^*$, we have $\sum_{n=1}^{\infty}\langle x_n^{**}, x_n^*\rangle = 0$. Specializing to the case when $x_n^{**} \in X$ verifies that X has the AP by Theorem 16.35. □

Theorem 16.38 *Let X, Y be Banach spaces. Suppose that either X or Y^* has the AP. Then the mapping $i^* : Y^* \otimes_\pi X \to (\mathcal{B}(X, Y), \tau)^*$ from* (16.28) *is injective. In particular, we may write $(\mathcal{B}(X, Y), \tau)^* = Y^* \otimes_\pi X$. The pairing is canonical,*

$$\langle z, T\rangle = \sum_{i=1}^{\infty}\langle y_i^*, Tx_i\rangle, \quad T \in \mathcal{B}(X, Y), \quad z = \sum_{i=1}^{\infty} y_i^* \otimes x_i \in Y^* \otimes_\pi X. \tag{16.30}$$

Proof: It follows from Theorems 16.35 and 16.36, that under these assumptions the mapping J is injective. So if $z \neq 0$, then $\langle z, x^* \otimes y\rangle \neq 0$ for some $y \in Y^*$, $x^* \in X$, and the injectivity of i^* is obvious. The surjectivity of i^* and the dual pairing formula have already been shown in Theorem 16.33. □

It is immediate from the definition that an adjoint to a nuclear operator is again nuclear. The opposite implication holds under the following assumptions.

Proposition 16.39 (Grothendieck) *Let X be a Banach space such that X^* has the AP. Let $T \in \mathcal{B}(X, Y)$ be such that $T^* \in \mathcal{B}(Y^*, X^*)$ is nuclear. Then T is also nuclear.*

Proof: Recall that $X \otimes_\pi Y \hookrightarrow X \otimes_\pi Y^{**}$ is an isometry (Exercise 16.10). Since T^* is nuclear, there exists a unique (by (7') in Theorem 16.35) $z = \sum_{i=1}^{\infty} y_i^{**} \otimes x_i^* \in Y^{**} \otimes_\pi X^*$ with $J(z) = T^*$. The claim of the proposition is equivalent (upon transposition) to the statement $z \in Y \otimes_\pi X^*$. Recall the duality $(Y^{**} \otimes_\pi X^*)^* = \mathcal{B}(Y^{**}, X^{**})$. By the Hahn–Banach theorem, our claim means that $\langle L, z\rangle = 0$ for every $L \in \mathcal{B}(Y^{**}, X^{**})$ which vanishes on $Y \otimes_\pi X^*$. So let us suppose that $\langle L(y), x^*\rangle = 0$ for all $y \in Y$, $x^* \in X^*$, so $L(y) = 0$ for all $y \in Y$. Since T^* is nuclear, it is also a compact operator, and so $T^{**}(X^{**}) \subset Y$. Combining the last statements, the nuclear operator $L \circ T^{**} = 0$. This means that $\sum_{i=1}^{\infty}\langle x_i^{**}, x^*\rangle L(y_i^{**}) = 0 \in X^{**}$ for all $x^{**} \in X^{**}$. In particular, we have $\sum_{i=1}^{\infty}\langle x^{**}, x_i^*\rangle\langle L(y_i^{**}), x^*\rangle = 0$ for all $x^* \in X^*$, and $x^{**} \in X^{**}$. Using the fact that X^* has the AP, and applying the condition (2) of Theorem 16.35, we obtain $\sum_{i=1}^{\infty}\langle L(y_i^{**}), x_i^*\rangle = 0$. However, the last formula is exactly $\langle L, z\rangle = 0$, which finishes the argument. □

Theorem 16.40 (Grothendieck) *Let Y be a Banach space such that* Y^* *has the RNP and the AP. Then there is an isometry* $(X \otimes_\varepsilon Y)^* = X^* \otimes_\pi Y^*$*, which results from the canonical pairing of the participating spaces. More precisely, for every* $z = \sum_{i=1}^\infty x_i^* \otimes y_i^* \in X^* \otimes_\pi Y^*$*, and every* $S \in \overline{\mathcal{F}_{w^*}(X^*, Y)} = X \otimes_\varepsilon Y$ *the dual pairing satisfies the formula*

$$\langle z, S\rangle = \sum_{i=1}^{\infty} \langle y_i^*, S(x_i^*)\rangle$$

Proof: By Theorem 16.30, we have $(X \otimes_\varepsilon Y)^* = \mathcal{N}(X, Y^*)$. More precisely, for every nuclear operator $T \in \mathcal{N}(X, Y^*)$, $T = \sum_{i=1}^\infty x_i^* \otimes y_i^*$, and every $S \in \overline{\mathcal{F}_{w^*}(X^*, Y)} = X \otimes_\varepsilon Y$ the dual pairing satisfies the formula

$$\langle T, S\rangle = \sum_{i=1}^{\infty} \langle y_i^*, S(x_i^*)\rangle$$

On the other hand, by Theorem 16.35, $J : X^* \otimes_\pi Y^* \to \mathcal{N}(X, Y^*)$ is an isometry. This finishes the proof. □

Theorem 16.41 (Grothendieck) *Let X, Y be Banach spaces, such that either* X^* *or Y has the AP, and either* X^{**} *or* Y^* *has the RNP. Then:*

$$\mathcal{K}(X, Y)^* = \mathcal{N}(X^*, Y^*). \tag{16.31}$$

The duality pairing for $S \in \mathcal{K}(X, Y)$, $T = \sum_{i=1}^\infty x_i^{**} \otimes y_i^* \in \mathcal{N}(X^*, Y^*)$ *is given by the formula*

$$\langle T, S\rangle = \sum_{i=1}^{\infty} \langle x_i^{**}, S^*(y_i^*)\rangle. \tag{16.32}$$

Proof: Our first step is the isometric identification.

$$X^* \otimes_\varepsilon Y = (\overline{\mathcal{F}(X, Y)}, \|\cdot\|) = \mathcal{K}(X, Y).$$

In the case where X^* has the AP it follows from condition (2) of Theorem 16.36. Otherwise, if Y has the AP, then it follows similarly from condition (5) of Theorem 16.35. It remains to show that

$$(X^* \otimes_\varepsilon Y)^* = \mathcal{N}(X^*, Y^*). \tag{16.33}$$

First assume that Y^* is a RNP space. Then (16.33) follows immediately from Theorem 16.30. More precisely, for every nuclear operator $T \in \mathcal{N}(X^*, Y^*)$,

$T = \sum_{i=1}^{\infty} x_i^{**} \otimes y_i^*$, and every $S \in \mathcal{K}(X, Y) = X^* \otimes_\varepsilon Y$, the dual pairing satisfies the formula (the first equality is the identification of S and S^{**} using the compactness of S and the w^*-continuity of S^{**}).

$$\langle T, S\rangle = \sum_{i=1}^{\infty} \langle y_i^*, S^{**}(x_i^{**})\rangle = \sum_{i=1}^{\infty} \langle S^*(y_i^*), x_i^{**}\rangle \tag{16.34}$$

Assuming instead that X^{**} has the RNP, we obtain from Theorem 16.30 that $(X^* \otimes_\varepsilon Y)^* = \mathcal{N}(Y, X^{**})$. In order to get (16.33), it remains to use Proposition 16.25. This finishes the proof. □

Theorem 16.42 (Grothendieck) *Let X, Y be Banach spaces, such that either X^{**} or Y^* has the AP. Then:*

$$\mathcal{N}(X^*, Y^*)^* = \mathcal{B}(X^{**}, Y^{**}). \tag{16.35}$$

The duality pairing for $S \in \mathcal{B}(X^{**}, Y^{**})$, $T = \sum_{i=1}^{\infty} x_i^{**} \otimes y_i^* \in \mathcal{N}(X^*, Y^*)$ is given by the formula

$$\langle S, T\rangle = \sum_{i=1}^{\infty} \langle S(x_i^{**}), y_i^*\rangle. \tag{16.36}$$

Proof: Assume first that X^{**} has the AP, and apply condition (4) of Theorem 16.36 in order to obtain the canonical isometry

$$X^{**} \otimes_\pi Y^* = \mathcal{N}(X^*, Y^*). \tag{16.37}$$

Composing this with the canonical dual pairing in Theorem 16.16 gives the desired relation

$$(X^{**} \otimes_\pi Y^*)^* = \mathcal{B}(X^{**}, Y^{**}). \tag{16.38}$$

The formula (16.36) is just (16.12). In the remaining case when Y^* has the AP, (16.37) follows from condition (7') of Theorem 16.35, and the rest of the argument is the same. □

Corollary 16.43 *Let X, Y be Banach spaces, such that either X^{**} or Y^* has the AP, and either X^{**} or Y^* has the RNP. Then:*

$$\mathcal{K}(X, Y)^* = \mathcal{N}(X^*, Y^*),\ \mathcal{K}(X, Y)^{**} = \mathcal{N}(X^*, Y^*)^* = \mathcal{B}(X^{**}, Y^{**}). \tag{16.39}$$

Proof: It suffices to combine Theorems 16.41 and 16.42 using the fact that the AP passes down from the duals. □

Corollary 16.44 (Holub [Holu]) *Let X, Y be Banach spaces, with at least one of them having the AP. If $\mathcal{B}(X, Y)$ is reflexive, then $\mathcal{B}(X, Y) = \mathcal{K}(X, Y)$.*

Proof: By Exercise 2.42, we have that $\mathcal{B}(X, Y)$ contains isomorphic copies of both Y and X^*. Thus both X, Y must be reflexive (in particular RNP spaces) and one of them, together with its dual, has the AP. Thus by Theorem 16.43, $\mathcal{K}(X, Y)^{**} = \mathcal{B}(X, Y)$ and the conclusion follows by the reflexivity of $\mathcal{B}(X, Y)$. □

16.4 The Trace

The concept of trace and the corresponding trace duality relation between spaces of operators is in some sense a reformulation of the duality pairing used so far. We are going to introduce the elementary facts concerning the trace, and study its basic properties in the Hilbert space.

Our starting point is the duality relation $(X^* \otimes_\pi X)^* = \mathcal{B}(X^*, X^*)$, and the natural identification $\mathcal{F}(X) = X^* \otimes X$. Using these facts, we see that every operator from $\mathcal{B}(X^*)$ acts naturally as a linear functional on the linear space $\mathcal{F}(X)$. Of course, we know that every such functional is bounded, if we consider $\mathcal{F}(X)$ as a subspace of $X^* \otimes_\pi X$.

Definition 16.45 *Let X be a Banach space. The linear functional on the linear space $\mathcal{F}(X, X)$, corresponding to $Id_{X^*} \in \mathcal{B}(X^*, X^*)$, is called the* trace tr_X on X. *We have the formula:*

$$tr_X\left(\sum_{i=1}^{n} x_i^* \otimes x_i\right) = \left\langle Id, \sum_{i=1}^{n} x_i^* \otimes x_i\right\rangle = \sum_{i=1}^{n} \langle x_i^*, x_i\rangle. \tag{16.40}$$

Let us compare tr_X with the usual concept from linear algebra. Let E be a finite-dimensional linear space with a vector space basis $\{e_i\}_{i=1}^n$. Let L be an operator on E, and let $A = (a_{i,j})_{1\leq i,j\leq n}$ be the matrix which represents L with respect to the basis $\{e_i\}_{i=1}^n$. Then we have the following simple fact.

Proposition 16.46 *The trace of the operator L is equal to the "usual trace" of its representing matrix A, i.e., $tr_E(L) = \sum_{i=1}^n a_{i,i}$ holds.*

Proof: By the matrix formalism, we have that

$$L\left(\sum_{i=1}^{n} \alpha_i e_i\right) = \sum_{j=1}^{n}\left(\sum_{i=1}^{n} a_{j,i}\alpha_i\right) e_j.$$

Denote by $\{e_i^*\}_{i=1}^n$ the dual basis of E^*, biorthogonal to $\{e_i\}_{i=1}^n$, and let $y_j^* = \sum_{i=1}^n a_{j,i} e_i^*$. Thus $L(x) = \sum_{j=1}^n y_j^* \otimes e_j$, and so $tr_E(L) = \sum_{j=1}^n y_j^*(e_j) = \sum_{j=1}^n a_{j,j}$ as claimed. □

The formula (16.40) suggests that the trace functional can be naturally extended to all $\mathcal{N}(X)$. However, this is false in general.

Theorem 16.47 *Let X be a Banach space. Then, the following are equivalent:*
(1) X has the AP.
(2) There exists a (necessarily unique) continuous extension of tr_X from $\mathcal{F}(X)$ onto $\mathcal{N}(X)$. We have the formula $tr_X(\sum_{i=1}^\infty x_i^* \otimes x_i) = \sum_{i=1}^\infty \langle x_i^*, x_i \rangle$.

Proof: (1)⟹(2) follows from condition (7) in Theorem 16.35 and the fact that tr_X has a unique continuous extension onto $X^* \otimes_\pi X$.

(2)⟹(1): Since $\mathcal{F}(X)$ is dense in both spaces $X^* \otimes_\pi X$ and $\mathcal{N}(X)$, we see that tr_X may exist on the latter space only if $tr_X(z) = 0$ for all $z \in \operatorname{Ker}(J)$. In other words, for every $z \in X^* \otimes_\pi X$, $\langle Id_{X^*}, z \rangle = 0$ whenever z annihilates $\mathcal{F}(X)$. By Proposition 16.20, we have that $t^* : \mathcal{B}(X^*, X^*) \to \mathcal{B}(X, X^{**})$ is an isometric isomorphism, such that $t^*(Id_{X^*}) = Id_{X \to X^{**}}$. Therefore we obtain $\langle Id_X, z \rangle = 0$ in the sense of duality in Theorem 16.33. By condition (2) of Theorem 16.35, we see that this is equivalent to X having the AP. □

Proposition 16.48 *Let X be a Banach space. Let $T \in \mathcal{B}(X)$ be a finite rank operator. Alternatively, let T be a nuclear operator, and X^* have the AP. Then $tr_{X^*}(T^*) = tr_X(T)$.*

Proof: Immediate from the representation formula of the adjoint operator. □

Proposition 16.49 *Let X, Y be a Banach spaces. Let $T \in \mathcal{B}(X, Y)$, $S \in \mathcal{B}(Y, X)$. Assume that T has finite rank. Alternatively suppose that T is a nuclear operator and X, Y have the AP. Then $tr_Y(T \circ S) = tr_X(S \circ T)$.*

Proof: Let $T = \sum_{i=1}^\infty x_i^* \otimes y_i$. We have

$$tr_X(S \circ T) = \sum_{i=1}^\infty \langle x_i^*, S \circ y_i \rangle = \sum_{i=1}^\infty \langle S^*(x_i^*), y_i \rangle = \sum_{i=1}^\infty \langle x_i^* \circ S, y_i \rangle = tr_Y(T \circ S).$$

□

The trace functional allows an elegant alternative description of the duality pairing of spaces of operators.

Theorem 16.50 *Let X be a reflexive Banach space with the AP. Then we have the isometric isomorphisms $\mathcal{K}(X)^* = \mathcal{N}(X)$, $\mathcal{N}(X)^* = \mathcal{B}(X)$. The pairing is given by formula*

$$\langle T, S\rangle = tr_X(T \circ S) = tr_X(S \circ T),$$

whenever $T \in \mathcal{B}(X)$ (including the case of compact operators), and $S \in \mathcal{N}(X)$.

Proof: We have by Theorem 16.43 that $\mathcal{K}(X)^* = \mathcal{N}(X^*)$, and $\mathcal{N}(X^*)^* = \mathcal{B}(X^{**})$. However, by Proposition 16.25, $\mathcal{N}(X^*, X^*) = \mathcal{N}(X, X^{**})$. By the reflexivity of X we obtain the first identification. The second one results from the duality relation $(X \otimes_\pi X^*)^* = \mathcal{B}(X, X^{**})$, using the AP and reflexivity of X. Finally, the pairing formula using the trace follows from (16.12) by inspection of the aforementioned identifications. □

Next we are going to investigate in some detail the nuclear operators on a separable complex Hilbert space. In the remaining part of this section, we are using the conjugation of operators in the sense usual in operator theory, namely for a $T \in \mathcal{B}(H)$, the adjoint T^* is again an element from $\mathcal{B}(H)$. This convention again implies that $tr_H(T) = tr_H(T^*)$ for every nuclear operator T. The bracket $\langle\cdot,\cdot\rangle$ is used in the sense of the dot product on H.

Lemma 16.51 *Let H be a separable complex Hilbert space, $T \in \mathcal{N}(H)$. Then*

$$tr_H(T) = \sum_{i=1}^{\infty} \langle T(e_i), e_i\rangle \text{ for any orthonormal basis of } H. \tag{16.41}$$

Proof: Since nuclear operators are compact, we have $T = UC$, where U is a unitary operator and C is diagonalizable (and also nuclear, as $C = U^*T$) with respect to some orthonormal basis $\{e_i\}_{i=1}^{\infty}$ by Theorem 15.49. Let $\{\lambda_i\}_{i=1}^{\infty}$ be its eigenvalues. Thus

$$C(x) = \sum_{i=1}^{\infty} \lambda_i \langle e_i, x\rangle e_i = \sum_{i=1}^{\infty} \langle C(e_i), x\rangle e_i = \sum_{i=1}^{\infty} (C(e_i) \otimes e_i)(x).$$

It follows that $tr_H(C) = \sum_{i=1}^{\infty} \langle C(e_i), e_i\rangle = \sum_{i=1}^{\infty} \lambda_i$. Using the fact that C is self-adjoint, we have

$$T^*(x) = CU^*(x) = \sum_{i=1}^{\infty} \langle C(e_i), U^*(x)\rangle e_i = \sum_{i=1}^{\infty} \langle UC(e_i), x\rangle e_i = \sum_{i=1}^{\infty} \big(T(e_i) \otimes e_i\big)(x).$$

Thus

$$tr_H(T) = tr_H(T^*) = \sum_{i=1}^{\infty} \langle T(e_i), e_i\rangle.$$

It remains to show that this expression is independent of the orthonormal basis. Let V be an arbitrary unitary operator. Using Proposition 16.49,

$$\sum_{i=1}^{\infty}\langle TV(e_i), V(e_i)\rangle = \sum_{i=1}^{\infty}\langle V^*TV(e_i), e_i\rangle = tr_H(V^*TV) = tr_H(TVV^*) = tr_H(T).$$

□

It follows from Lemma 16.51, that if a nuclear operator on H is represented by an infinite matrix with respect to some orthonormal basis, then its trace equals the sum of the entries on the main diagonal. This is a direct generalization of the finite-dimensional case.

Theorem 16.52 *Let $T \in \mathcal{N}(H)$ have a polar decomposition $T = UC$. Then*
(i) $N(T) = N(C) = tr_H(C)$.
(ii) $N(T) = \sup\{tr_H(TU) : U \in \mathcal{B}(H)$ *is a unitary operator*$\}$.
(iii) $N(T) = sup\left\{\sum_{i=1}^{\infty}\langle T(e_i), f_i\rangle :\right.$
over all orthonormal bases $\left.\{e_i\}_{i=1}^{\infty}, \{f_i\}_{i=1}^{\infty}\ of\, H\right\}$.
(iv) $N(T) = sup\{\sum_{i=1}^{\infty}\|T(e_i)\| :$ *over all orthonormal bases* $\{e_i\}_{i=1}^{\infty}$ *in* $H\}$.

Proof: Let $\{e_i\}_{i=1}^{\infty}$ be the orthonormal basis diagonalizing C, and $\{\lambda_i\}_{i=1}^{\infty}$ be the positive eigenvalues of C. By Theorem 16.50 we obtain

$$N(T) = N(T^*) = \sup\{tr_H(CU^*S) :\ S \in \mathcal{B}(H),\ \|S\| \le 1\}.$$

Thus by Lemma 16.51

$$N(T) = N(T^*) = \sup\left\{\sum_{i=1}^{\infty}\langle C(e_i), S^*U(e_i)\rangle :\ \|S\| \le 1\right\}.$$

As $C(e_i) = \lambda_i e_i$, and $\|S^*U\| \le 1$, we see that the supremum is attained precisely for $S = U$, and its value is equal to $\sum_{i=1}^{\infty}\lambda_i$. This proves (i) and also (ii), as $tr_H(T^*U) = tr_H(U^*T) = tr_H(TU^*)$. The last equality easily implies the remaining conditions (iii) and (iv). □

16.5 Banach Spaces Without the Approximation Property

The AP is easily verified for all Banach spaces with a Schauder basis. Indeed, the sequence of canonical projections P_n associated with the Schauder basis τ-converges to Id on X. Banach spaces with a Schauder basis are the main examples of spaces with AP. By a similar argument, the AP passes to complemented subspaces. The first example of a Banach space failing the AP, a subspace of c_0, was constructed by Enflo [Enfl1]. The construction was subsequently simplified and generalized by many authors. Davie's method of proof [Davi], [Davi2], works for c_0 and ℓ_p, $p > 2$ (see Exercise 16.14), giving subspaces that fail the AP; Szankowski [Szan1b] did it for ℓ_p, $1 \le p < 2$. For other proofs of Enflo's theorem and related results see, e.g., [FiPe] and [Kwap2]. Kwapień result is quoted in Remark 16.3. In this section we are going to follow Davie's method of construction for producing subspaces of c_0 without the AP, modified by Lindenstrauss and Tzafriri, see [LiTz3, p. 86]. The basic idea is to look for a subspace of c_0 failing the AP, by using (2) in Theorem 16.35. Thus we are searching for a sequence $\{x_i\}_{i=1}^{\infty}$ of vectors from c_0, and using $x_i^* = e_i \in \ell_1$, we hope to achieve that $X = \overline{\text{span}}\{x_i\}_{i=1}^{\infty}$ is a Banach space failing (2) in Theorem 16.35. The construction can be conveniently described using an infinite matrix A, whose rows consist of vectors $x_i \in c_0$.

Lemma 16.53 (Grothendieck) *Let $A = (a_{i,j})_{i,j=1}^{\infty}$ be an infinite matrix satisfying the following conditions:*
1. *For every $i \in \mathbb{N}$, $a_{i,j} \neq 0$ for only finitely many $j \in \mathbb{N}$.*
2. $\sum_{i=1}^{\infty} \max_j |a_{i,j}| < \infty$.
3. $A^2 = 0$ *and* $tr(A) = \sum_{i=1}^{\infty} a_{i,i} \neq 0$.

Then the Banach space $X = \overline{\text{span}}\{(a_{i,1}, a_{i,2}, \dots), i \in \mathbb{N}\} \hookrightarrow c_0$ fails the AP.

Proof: We let $x_i = (a_{i,1}, a_{i,2}, \dots) \in c_0$, and $x_i^* = e_i$ be the coefficient functionals from ℓ_1. Condition 2 is equivalent to $\sum_{i=1}^{\infty} \|x_i\| \|x_i^*\| < \infty$. Condition $A^2 = 0$ means that

$$\sum_{i=1}^{\infty} a_{j,i} a_{i,k} = 0 \text{ for all } j, k \in \mathbb{N}. \tag{16.42}$$

It follows that

$$\sum_{i=1}^{\infty} x_i^*(x_j) x_i = \sum_{i=1}^{\infty} a_{j,i} x_i = \left(\sum_{i=1}^{\infty} a_{j,i} a_{i,k} \right)_{k \in \mathbb{N}} = 0 \in c_0$$

for every $j \in \mathbb{N}$. Since $X = \overline{\text{span}}\{x_i : i \in \mathbb{N}\}$, we conclude that

$$\sum_{i=1}^{\infty} x_i^*(x) x_i = 0 \in c_0 \text{ for all } x \in X.$$

Finally, $tr(A) = \sum_{i=1}^{\infty} x_i^*(x_i) \neq 0$. We have thus verified that X fails the condition 2 of Theorem 16.35, so it fails to have the AP. □

The following result is implicitly in [Enfl1], although we are presenting a simplified proof due to Davie, see, e.g., [LiTz3, p. 86].

Theorem 16.54 (Enflo) *There exists an infinite matrix A satisfying conditions* 1 *to* 3 *of Lemma* 16.53. *In particular, c_0 has a subspace without the AP.*

Proof: For convenience, we work with the complex scalars. For $k = 0, 1, 2, \dots$, we denote by I_k the identity matrix of order k. The infinite matrix A will be constructed from finite blocks, originating from certain unitary matrices U_k of order $3 \cdot 2^k$. Let us first describe the general form and properties of A. We write

$$U_k = \begin{pmatrix} 2^{\frac{k+1}{2}} P_k \\ 2^{\frac{k}{2}} Q_k \end{pmatrix},$$

where $2^{\frac{k+1}{2}} P_k$ is the top $2^{k+1} \times 3 \cdot 2^k$ matrix and $2^{\frac{k}{2}} Q_k$ is the bottom $2^k \times 3 \cdot 2^k$ matrix. Since $U_k U_k^* = I_{3 \cdot 2^k}$, we get

$$P_k P_k^* = 2^{-(k+1)} I_{2^{k+1}}, \ Q_k Q_k^* = 2^{-k} I_{2^k}, \ P_k Q_k^* = Q_k P_k^* = 0. \tag{16.43}$$

We let A to be the following infinite matrix.

$$\begin{pmatrix} P_0^* P_0 & P_0^* Q_1 & 0 & 0 & 0 & \dots \\ -Q_1^* P_0 & P_1^* P_1 - Q_1^* Q_1 & P_1^* Q_2 & 0 & 0 & \dots \\ 0 & -Q_2^* P_1 & P_2^* P_2 - Q_2^* Q_2 & P_2^* Q_3 & 0 & \dots \\ 0 & 0 & -Q_3^* P_2 & P_3^* P_3 - Q_3^* Q_3 & P_3^* Q_4 & \dots \\ \dots & \dots & \dots & \dots & \dots & \end{pmatrix}.$$

It is easily verified that $A^2 = 0$. By (16.43), $tr(P_k^* P_k - Q_k^* Q_k) = 0$ so $tr(A) = tr(P_0^* P_0) = 1$. Thus A satisfies conditions 1 and 3 from Lemma 16.53. In order to guarantee condition 2, we need to choose U_k with additional properties. To this end, let us denote by $G_k = (\mathbb{Z}_{3 \cdot 2^k}, +)$ the additive Abelian group of integers modulo $m = 3 \cdot 2^k$. Consider the system of complex functions (known as the *group characters*) on G_k

$$\gamma_j : G_k \to \mathbb{C}, \ \gamma_j(l) = e^{-2\pi i \frac{jl}{m}}, \ 0 \le j < m.$$

The characters have the important property of being translation invariant, that is to say $\gamma_j(s + t) = \gamma_j(s)\gamma_j(t)$ for every $s, t \in G_k$, and every j. Moreover, the system $\{\gamma_j\}_{j=0}^{m-1}$ is orthogonal in the complex Hilbert space $L_2(\mathbb{Z}_{3 \cdot 2^k})$, i.e., $\sum_{l=0}^{m-1} \overline{\gamma_{j_1}(l)} \gamma_{j_2}(l) = 0$ whenever $j_1 \neq j_2$. This means that the matrix formed by $(\gamma_j(l))_{0 \le j,l < m}$ is an orthogonal matrix, but this is not yet the matrix we are looking for. We also need to carefully choose the particular order (and sign) of the rows of this matrix. This task poses the main technical obstacle in the construction.

Lemma 16.55 *The set* $\{\gamma_j\}_{j=0}^{m-1}$, $m = 3 \cdot 2^k$, *can be split into disjoint sets denoted by* $\{\tau_j^k\}_{j=1}^{2^{k+1}}$ *and* $\{\sigma_j^k\}_{j=1}^{2^k}$, *so that*

$$\left| 2 \sum_{j=1}^{2^k} \sigma_j^k(l) - \sum_{j=1}^{2^{k+1}} \tau_j^k(l) \right| \le L(k+1)^{\frac{1}{2}} 2^{\frac{k}{2}}, \ \text{for every } l \in G_k,$$

where L is a constant independent of k.

Proof: Let (S, Σ, μ) be a probability space, and let $\{\theta_j\}_{j=0}^{m-1}$ be independent random variables on S such that $\mu(\theta_j^{-1}(2)) = \frac{1}{3}$ and $\mu(\theta_j^{-1}(-1)) = \frac{2}{3}$. By Exercise 16.13, it follows easily that there exists a constant L independent of k so that for all k large enough,

$$\mu\left\{\left|\sum_{j=0}^{m-1} \gamma_j(l)\theta_j\right| > L(k+1)^{\frac{1}{2}} 2^{\frac{k}{2}}\right\} < \frac{L}{m^3} \tag{16.44}$$

is valid for every $l \in G_k$. If $m < \frac{m^3}{L}$, then there exists a set of nonzero measure $R \in \Sigma$, such that

$$\left|\sum_{j=0}^{m-1} \gamma_j(l)\theta_j\right| \leq L(k+1)^{\frac{1}{2}} 2^{\frac{k}{2}} \text{ holds on } R \text{ for every } l \in G_k. \tag{16.45}$$

So there exists a sequence $\{\theta'_j\}_{j=0}^{m-1}$ of numbers from $\{2, -1\}$, such that

$$\left|\sum_{j=0}^{m-1} \gamma_j(l)\theta'_j\right| \leq L(k+1)^{\frac{1}{2}} 2^{\frac{k}{2}} \text{ holds on } R \text{ for every } l \in G_k. \tag{16.46}$$

For $l = 0$ (the neutral element of G_k), we get $|\sum_{j=0}^{m-1} \theta'_j| \leq L(k+1)^{\frac{1}{2}} 2^{\frac{k}{2}}$. Thus changing at most $2L(k+1)^{\frac{1}{2}} 2^{\frac{k}{2}}$ of the values θ'_j and replacing L by $2L$, we obtain (16.46) along with the additional fact that $\sum_{j=0}^{m-1} \theta'_j = 0$. Finally, we set $\{\tau_j^k\}_{j=1}^{2^{k+1}} = \{\gamma_i : \theta'_i = -1\}$ and $\{\sigma_j^k\}_{j=1}^{2^k} = \{\gamma_i : \theta'_i = 2\}$. □

We may now define the top $2^{k+1} \times 3 \cdot 2^k$ matrix.

$$P_k = (3^{-\frac{1}{2}} 2^{-\frac{2k+1}{2}} \tau_j^k(l))_{1 \leq j \leq 2^{k+1},\ 0 \leq l \leq m-1}. \tag{16.47}$$

The bottom $2^k \times 3 \cdot 2^k$ matrix will be

$$Q_k = (3^{-\frac{1}{2}} 2^{-k} \theta_j^k \sigma_j^k(l))_{1 \leq j \leq 2^k,\ 0 \leq l \leq m-1}, \tag{16.48}$$

where $\theta_j^k \in \{1, -1\}$ are selected so that for some constant C, independent of k, and all k large enough,

$$\left|\sum_{j=1}^{2^k} \theta_j^k \overline{\sigma_j^k(h)} \tau_j^{k-1}(g)\right| \leq C(k+1)^{\frac{1}{2}} 2^{\frac{k}{2}}, \text{ for all } h \in G_k,\ g \in G_{k-1}. \tag{16.49}$$

Such a choice is again possible. Indeed, let (S, Σ, μ) be a probability space, and let $\{\theta_j\}_{j=1}^{2^k}$ be independent random variables on S such that $\mu(\theta_j^{-1}(1)) = \frac{1}{2}$ and $\mu(\theta_j^{-1}(-1)) = \frac{1}{2}$. By Exercise 16.13, it follows easily that there exists a constant C independent of k so that

$$\mu\left\{\left|\sum_{j=1}^{2^k} \theta_j \overline{\sigma_j^k(h)\tau_j^{k-1}(g)}\right| > C(k+1)^{\frac{1}{2}}2^{\frac{k}{2}}\right\} < \frac{C}{2^{3k}} \tag{16.50}$$

is valid for every $h \in G_k$, $g \in G_{k-1}$. If $|G_k| \cdot |G_{k-1}| = \frac{9}{2} \cdot 2^{2k} < 2^{3k}/C$, then there exists a sequence $\{\theta_j^k\}_{j=1}^{2^k}$ of numbers from $\{1, -1\}$, such that (16.49) holds. We have finished the description of matrices P_k, Q_k. It is clear that the corresponding U_k are unitary matrices. It remains to verify the condition 2 in Lemma 16.53 for the matrix A. We claim that that each scalar entry in the block

$$\left(0 \ldots 0 \; -Q_k^* P_{k-1} \;\; P_k^* P_k - Q_k^* Q_k \;\; P_k^* Q_{k+1} \; 0 \ldots\right)$$

has absolute value less than $D(k+1)^{1/2}2^{-3k/2}$, where $D > 0$ is a constant independent of k. Since the block contains only $3 \cdot 2^k$ nontrivial rows, this will imply that

$$\sum_{i=1}^{\infty}(\max_j |a_{i,j}|) \leq \sum_{k=0}^{\infty} 3 \cdot 2^k D(k+1)^{\frac{1}{2}}2^{-\frac{3k}{2}} < \infty, \tag{16.51}$$

and the proof will be finished. Observe that since $P_k^* Q_{k+1} = (Q_{k+1}^* P_k)^*$, it suffices to examine only the entries in the matrices $Q_k^* P_{k-1}$ and $P_k^* P_k - Q_k^* Q_k$. The entries of $Q_k^* P_{k-1}$ and $P_k^* P_k - Q_k^* Q_k$ are precisely the following:

$$3^{-1}2^{\frac{1}{2}-2k}\sum_{j=1}^{2^k} \theta_j^k \overline{\sigma_j^k(h)\tau_j^{k-1}(g)}, \; h \in G_k, \; g \in G_{k-1}, \tag{16.52}$$

$$3^{-1}2^{-2k}\left(\frac{1}{2}\sum_{j=1}^{2^{k+1}} \tau_j^k(h) - \sum_{j=1}^{2^k} \sigma_j^k(h)\right), \; h \in G_k. \tag{16.53}$$

In the derivation of (16.53) we used the translation invariance of characters, i.e., $\tau_j^k(h_1+h_2) = \tau_j^k(h_1)\tau_j^k(h_2)$. The sought estimate now follows directly from (16.46) and (16.49). □

Finally, let us see that the complex space X fails AP also as a real Banach space. Indeed, if T is a real operator on the real Banach space X, then $\tilde{T} = \frac{1}{2}\big(T(x) - iT(ix)\big)$ is a complex operator on the complex Banach space X. $T = \tilde{T}$ if and only if T is a complex operator. It is now easy to see that if X has the AP, as a real space, and the finite rank real operators $T_\alpha \to Id$ in the τ-topology, then $\tilde{T}_\alpha \to Id$ in the complex τ-topology. A small modification of the construction leads to examples of subspaces of ℓ_p, $p > 2$ without the AP (see Exercise 16.14). General arguments allow to further strengthen the properties of counterexamples to the AP. We are going to investigate in this direction, and show the consequences of such examples for the general theory. As a first example, let us see that the result in Theorem 16.37 cannot be reversed.

Theorem 16.56 *There exists a space with a Schauder basis, whose dual is separable and fails the AP.*

Proof: Using Exercise 4.80, there exists a Banach space Z, such that Z^{**} has a boundedly complete basis, and $Z^{**}/Z = X$, where X is a Banach space without the AP. Thus X^* also fails the AP. We have $Z^{***} = Z^* \oplus X^*$. The separability of Z^{***} is clear. As the AP passes to complemented subspaces, Z^{***} fails the AP (since X^* does). Thus Z^{**} satisfies the statement of the theorem. □

Theorem 16.57 ([HajSm]) *Let Y be a Banach space with the AP. Then the mapping $i^* : Y^* \otimes_\pi X \to (\mathcal{B}(X, Y), \tau)^*$ from* (16.28) *is injective for every Banach space X if and only if Y^* has the AP.*

Proof: We first assume the injectivity of i^* for every Banach space X. In fact, the case $X = Y^{**}$ is sufficient in order to prove the direct implication of our theorem. Our goal is to establish that Y^* has the AP. By condition 3 in Theorem 16.36, it suffices to show that $J : Y^* \otimes_\pi X \to \mathcal{N}(Y, X)$ is an isometry. Recall that

$$\operatorname{Ker}(i^*) = \left\{ z = \sum_{i=1}^{\infty} y_i^* \otimes x_i : \langle z, S \rangle = \sum_{i=1}^{\infty} \langle y_i^*, S(x_i) \rangle = 0, \text{ for all } S \in \mathcal{B}(X, Y) \right\}. \tag{16.54}$$

As Y is assumed to have the AP, we have by condition 3 in Theorem 16.35 that for every X, $\overline{\mathcal{F}(X, Y)}^{\tau} = \mathcal{B}(X, Y)$. Thus by the bipolar and Hahn–Banach theorem, (16.54) is equivalent to the next condition.

$$\operatorname{Ker}(i^*) = \{ z = \sum_{i=1}^{\infty} y_i^* \otimes x_i : \langle z, S \rangle = \sum_{i=1}^{\infty} \langle y_i^*, S(x_i) \rangle = 0, \text{ for all } S \in \mathcal{F}(X, Y) \}. \tag{16.55}$$

Next, compare this condition with the condition describing the kernel of J:

$$\operatorname{Ker}(J) = \left\{ z = \sum_{i=1}^{\infty} y_i^* \otimes x_i : \langle T, z \rangle = \sum_{i=1}^{\infty} \langle T(y_i^*), x_i \rangle = 0, \text{ for all } T \in \mathcal{F}_{w^*}(Y^*, X^*) \right\}. \tag{16.56}$$

We claim that (16.55) and (16.56) are equivalent conditions. Indeed, it suffices to note that taking the adjoints $S \to S^*$ makes an isometry from $\mathcal{F}(X, Y)$ onto $\mathcal{F}_{w^*}(Y^*, X^*)$, and thus a reformulation of (16.55)

$$\mathrm{Ker}(i^*) = \{z = \sum_{i=1}^{\infty} y_i^* \otimes x_i : \langle z, S\rangle = \sum_{i=1}^{\infty} \langle S^*(y_i^*), x_i\rangle = 0, \text{ for all } S \in \mathcal{F}(X, Y)\}$$

is precisely (16.56). Since i^* is assumed to be injective, so is J. It is clear by the Banach open mapping theorem that J is an isometry. This proves that Y^* indeed has the AP. The opposite implication follows from Theorem 16.38. □

Checking the case $X = Y^{**}$ in the previous proof, we obtain the following corollary.

Corollary 16.58 ([HajSm]) *Let Y be a Banach space with the AP, whose dual Y^* fails the AP. Then $i^* : Y^* \otimes_\pi Y^{**} \to (\mathcal{B}(Y^{**}, Y), \tau)^*$ is not injective.*

16.6 The Bounded Approximation Property

Let us start by introducing a stronger variant of the AP.

Definition 16.59 *A Banach space X is said to have the* λ-bounded approximation property *(λ-BAP for short), if*

$$Id \in \overline{\lambda B_{\mathcal{F}(X)}}^{\tau}.$$

This is equivalent to $B_{\mathcal{B}(X)} \subset \overline{\lambda B_{\mathcal{F}(X)}}^{\tau}$. We say that X has the bounded approximation property (BAP) *if it has λ-BAP for some $\lambda > 0$.*

It is immediate that BAP$\Longrightarrow$AP. The BAP is easily verified for all Banach spaces with a Schauder basis. Indeed, the sequence of canonical projections P_n associated with the Schauder basis is uniformly bounded and it τ-converge to Id on X. By a similar argument, the BAP passes to complemented subspaces. In fact, every Banach space with the BAP can be complementably embedded into a Banach space with a Schauder basis. This is the content of the next important theorem whose proof can be found, e.g., in [LiTz3, p. 38].

Theorem 16.60 ([Pelc6b], [JRZ]) *A separable Banach space X has the BAP if and only if it is isomorphic to a complemented subspace of a space with a Schauder basis.*

An example of a separable Banach space with the BAP, but without a Schauder basis, was constructed by Szarek [Szar]. The proof of the next theorem is similar to that of Theorem 16.35, and it is left to the reader as an exercise.

Theorem 16.61 (Grothendieck) *Let X be a Banach space. The following are equivalent:*
(1) *X has the λ-BAP.*
(2) *For every pair of sequences $\{x_n\}_{n=1}^{\infty}$ from X and $\{x_n^*\}_{n=1}^{\infty}$ from X^*, such that*

$$\sum_{n=1}^{\infty} \|x_n^*\| \|x_n\| < \infty \text{ and } \left|\sum_{n=1}^{\infty} x_n^*(Tx_n)\right| \leq \|T\| \text{ for all } T \in \mathcal{F}(X),$$

we have $\sum_{n=1}^{\infty} x_n^*(x_n) \leq \lambda$.
(3) *For every Banach space* Y, $\overline{\lambda B_{\mathcal{F}(X,Y)}}^{\tau} \supset B_{\mathcal{B}(X,Y)}$.
(4) *For every Banach space* Y, $\overline{\lambda B_{\mathcal{F}(Y,X)}}^{\tau} \supset B_{\mathcal{B}(Y,X)}$.

We now focus on a construction of a separable Banach space with the AP, but failing the BAP.

Theorem 16.62 (Figiel, Johnson [FiJo]) *Let* X *be a Banach space. Take* $\lambda > 0$. *Assume that* X *has* λ*-BAP under every equivalent norm. Then* X^* *has* $2\lambda(1+4\lambda)$*-BAP.*

Proof: Let $\varepsilon, \lambda > 0$. We introduce the auxiliary notion of (ε, λ)-AP as follows. A Banach space X is said to have (ε, λ)-AP if for every finite-dimensional subspace F of X and $\delta > 0$, there is $T \in \mathcal{F}(X)$, $\|T\| \leq \lambda + \delta$, so that $\|Tx - x\| \leq (\varepsilon + \delta)\|x\|$ for all $x \in F$.

In order to complete the proof, we need the following lemma.

Lemma 16.63 *Let* $0 < \varepsilon < 1$, $\lambda > 0$, *and* X *be a Banach space with the* (ε, λ)*-AP. Then* X *has the* $(\frac{\lambda}{1-\varepsilon})$*-BAP.*

Proof: Choose $\delta > 0$ such that $\varepsilon + \delta < 1$, and let $F \hookrightarrow X$ be a finite-dimensional subspace. By assumption, we can inductively find a sequence $\{T_n\}_{n=1}^{\infty}$ of finite rank operators from $(\lambda + \delta)B_{\mathcal{F}(X)}$ so that $\|T_1(x) - x\| \leq (\varepsilon + \delta)\|x\|$ for $x \in F$, and

$$\|T_{n+1}(x) - x\| \leq (\varepsilon + \delta)\|x\|, \text{ for } x \in \operatorname{span}\{F \cup \bigcup_{i=1}^{n} T_i(X)\}.$$

For $n \geq 1$, let $S_n \in \mathcal{B}(X)$ be defined by the relation

$$(Id - S_n) = (Id - T_n)(Id - T_{n-1}) \ldots (Id - T_1).$$

Then for $x \in F$, $\|(Id - S_n)(x)\| \leq (\varepsilon + \delta)^n \|x\|$. Also,

$$S_n = (Id - T_n) \ldots (Id - T_2)T_1 + (Id - T_n) \ldots (Id - T_3)T_2 + \cdots + T_n.$$

Hence,

$$\|S_n\| \leq (\lambda + \delta)((\varepsilon + \delta)^{n-1} + (\varepsilon + \delta)^{n-2} + \cdots + (\varepsilon + \delta) + 1) \leq \frac{(\lambda + \delta)}{(1 - \varepsilon - \delta)}.$$

Since $\delta > 0$ can be taken arbitrarily small, this proves the assertion by Exercise 4.36. $\square$

We continue with the proof of Theorem 16.62.

Let F^* be a finite-dimensional subspace of X^*, $\delta > 0$ and $\beta = \lambda + \delta$. Let F be a finite-dimensional subspace of X such that, for every $x^* \in F^*$, $\|x^*\| \leq (1+\delta)\sup\{|x^*(x)| : x \in F, \|x\| = 1\}$. We fix $\varepsilon > 0$ and introduce a new norm $|||\cdot|||$ on X^* by

$$|||x^*||| = \|x^*\| + \frac{2\beta}{\varepsilon}\operatorname{dist}(x^*, F^*).$$

Claim: $|||\cdot|||$ is a dual norm to some equivalent norm on X, which we denote again by $|||\cdot|||$.

The claim is equivalent to the statement that $B = \{x^* : |||x^*||| \leq 1\}$ is w^*-compact. So suppose that $2 \geq |||x^*||| = 1 + 5\xi > 1$. Since F^* is finite-dimensional, there exists a finite set $S \subset F^*$, such that for every $z^* \in X^*$ satisfying $\|z^*\| \leq 2$, we have

$$M(z^*) = \|z^*\| + \frac{2\beta}{\varepsilon}\operatorname{dist}(z^*, S) \geq |||z^*||| \geq M(z^*) - \xi. \tag{16.57}$$

Note that the function $M(z^*) = \inf_{s\in S}\{\|z^*\| + \frac{2\beta}{\varepsilon}\|z^* - s\|\}$, being an infimum over a finite family of w^*-lsc functions, is w^*-lsc. So the set $M^{-1}([0, 1+\xi])$ is w^*-compact. By (16.57), $B \subset M^{-1}([0, 1+\xi])$, and $x^{**} \notin M^{-1}([0, 1+\xi])$. This proves the claim.

By assumption, $(X, |||\cdot|||)$ has the λ-BAP. Thus there is $T \in \mathcal{F}(X)$, $|||T||| \leq \beta$, and so that $\|T(x) - x\| \leq \delta\|x\|$ for $x \in F$. Passing to the dual, we get for $x^* \in X^*$

$$\|T^*(x^*)\| + \frac{2\beta}{\varepsilon}\operatorname{dist}\big(T^*(x^*), F^*\big) \leq \beta\left(\|x^*\| + \frac{2\beta}{\varepsilon}\operatorname{dist}(x^*, F^*)\right). \tag{16.58}$$

It follows that $\|T^*(x^*)\| \leq \beta(1 + \frac{2\beta}{\varepsilon})\|x^*\|$, and hence $\|T\| \leq \beta(1 + \frac{2\beta}{\varepsilon})$. For $x^* \in F^*$ we get from (16.58) that $\operatorname{dist}\big(T^*(x^*), F^*\big) \leq \frac{\varepsilon\|x^*\|}{2}$. Thus there is $y^* \in F^*$ such that $\|T^*(x^*) - y^*\| \leq \frac{\varepsilon\|x^*\|}{2}$. Let $y \in F$, $\|y\| = 1$. Then $|\langle T^*(x^*), y\rangle - \langle x^*, y\rangle| = |x^*(T(y) - y)| \leq \delta\|x^*\|$. Hence

$$\|x^* - y^*\| \leq (1+\delta)\sup\{|x^*(y) - y^*(y)| : y \in F, \|y\| = 1\} \leq (1+\delta)\big(\delta + \frac{\varepsilon}{2}\big)\|x^*\|.$$

and consequently $\|T^*(x^*) - x^*\| \leq (1+\delta)\big(\delta + \frac{\varepsilon}{2}\big)\|x^*\|$. Since $\delta > 0$ is arbitrary, we get that X^* has the $(\varepsilon, \lambda(1 + 2\varepsilon^{-1}\lambda))$-AP. Clearly, the λ-BAP implies the $(0, \lambda)$-AP. By Lemma 16.63, the theorem is proved. □

Theorem 16.64 (Figiel, Johnson [FiJo]) *There exists a separable Banach space X, with a separable dual, such that X has the AP, but it fails to have the BAP.*

Proof: Consider the Banach space X from Theorem 16.56. By applying Theorem 16.62, we see that for every $\lambda > 0$ there exists a renorming $\|\cdot\|_\lambda$ of X, such that $(X, \|\cdot\|_\lambda)$ fails to have the λ-BAP. Let $Y = \sum_{\ell_2}(X, \|\cdot\|_n)$. This space clearly has the AP but fails the BAP. The separability conditions are easily verified. □

Proposition 16.65 (Figiel, Johnson [FiJo]) *Let X be a Banach space with a separable dual, such that X has the AP, but fails the BAP. Then there is a non-nuclear operator $T \in \mathcal{B}(X)$ such that T^* is nuclear.*

Proof: As X has the AP, we have the isometry $X^* \otimes_\pi X = \mathcal{N}(X)$. Since X has the AP, and X^* has the RNP, by Theorems 16.38 and 16.41 we have that $(\mathcal{B}(X), \tau)^* = \mathcal{N}(X)$, and $\mathcal{K}(X)^* = \mathcal{N}(X^*)$. Assume, by contradiction, that T is nuclear whenever T^* is nuclear. Thus, by the Banach open mapping theorem, the mapping $T \to T^*$ is an isomorphism from $\mathcal{N}(X)$ onto $\mathcal{N}_{w^*}(X^*)$ the Banach space of all nuclear operators on X^*, that are simultaneously dual operators. For every $T = \sum_{i=1}^\infty x_i^* \otimes x_i \in \mathcal{N}(X)$, we have $N(T^*) = \sup\{\langle T^*, L\rangle : L \in \mathcal{F}(X), \|L\| \leq 1\}$. Due to the isomorphism there is some $C > 0$, so that we also get $N(T) \leq C \sup\{\langle T, L\rangle : L \in \mathcal{F}(X), \|L\| \leq 1\}$. Thus $\langle T, L\rangle \leq \|L\|$ for all $L \in \mathcal{F}(X)$ implies that $N(T) \leq C$. On the other hand, $N(T) \geq |tr(T)| = |\sum_{n=1}^\infty x_n^*(x_n)|$. By condition 2 of Theorem 16.61 we conclude that X has the BAP, a contradiction. □

The following result applies, in particular, to all reflexive spaces and all separable dual spaces.

Theorem 16.66 (Grothendieck) *Let X be a dual Banach space with the RNP. Then X has the 1-BAP whenever X has the AP.*

Proof: Let Y be a Banach space, $X = Y^*$ be its dual with the AP, and $z \in X^* \otimes_\pi X$. By Corollary 16.43 and Theorem 16.35 we have $(X^* \otimes_\pi X)^* = \mathcal{B}(X^*)$, so

$$\pi(z) = \sup_{\|T\|\leq 1,\ T\in\mathcal{B}(X^*)} \langle T, z\rangle \geq \sup_{\|T\|\leq 1,\ T\in\mathcal{B}(X)} \langle T^*, z\rangle. \tag{16.59}$$

On the other hand, by Corollary 16.43 and Theorem 16.35 we have $\mathcal{K}(Y)^* = X^* \otimes_\pi X$, so:

$$\pi(z) = \sup_{\|T\|\leq 1,\ T\in\mathcal{K}(Y)} \langle z, T\rangle \leq \sup_{\|T\|\leq 1,\ T\in\mathcal{K}(X)} \langle T^*, z\rangle = \sup_{\|T\|\leq 1,\ T\in\mathcal{F}(X)} \langle T^*, z\rangle. \tag{16.60}$$

The last equality follows from condition 5 of Theorem 16.35, since X has the AP. Combining (16.59) with (16.60), we obtain

$$\pi(z) = \sup_{\|T\|\leq 1,\ T\in\mathcal{B}(X)} \langle T^*, z\rangle = \sup_{\|T\|\leq 1,\ T\in\mathcal{F}(X)} \langle T^*, z\rangle. \tag{16.61}$$

Given $z = \sum_{i=1}^\infty x_i^* \otimes x_i \in X^* \otimes_\pi X$ and $T \in \mathcal{B}(X)$, we have the equality

$$\langle T^*, z\rangle = \sum_{i=1}^\infty \langle T^*(x_i^*), x_i\rangle = \sum_{i=1}^\infty \langle x_i^*, T(x_i)\rangle.$$

By Theorem 16.33, the mapping $i^* : X^* \otimes_\pi X \to (\mathcal{B}(X), \tau)^*$ is surjective. Thus by applying the Hahn–Banach theorem to the set $\{T \in \mathcal{F}(X) : \|T\| \le 1\} \subset (\mathcal{B}(X), \tau)$, and using (16.61), we see that no operator $T \in \mathcal{B}(X)$, $\|T\| < 1$, can be separated by a τ-continuous hyperplane. This is clearly a reformulation of the 1-BAP. □

16.7 Schauder Bases in Tensor Products

Consider a linear ordering $\prec$ of $\mathbb{N}^2$ such that $(1,1) \prec (2,1) \prec (2,2) \prec (1,2) \prec (3,1) \prec (3,2) \prec (3,3) \prec (2,3) \prec (1,3) \prec (4,1) \prec \dots$. Clearly $(\mathbb{N}^2, \prec)$, as a linearly ordered set, is isomorphic to the usual integers $(\mathbb{N}, <)$. Let X, be a Banach space with a normalized Schauder basis $\{e_i\}_{i=1}^\infty$, and Y be a Banach space with a normalized Schauder basis $\{f_i\}_{i=1}^\infty$. We denote by P_n the initial projections associated with a Schauder basis. Our intention is to investigate the basis properties of the linearly ordered sequence of vectors $\{e_i \otimes f_j\}_{(i,j)\in(\mathbb{N}^2,\prec)}$ from $X \otimes Y$. We already know that the set $\{x \otimes y : x \in X, y \in Y\}$ is dense in both $X \otimes_\pi Y$ and $X \otimes_\varepsilon Y$. It is therefore clear that the linear span of $\{e_i \otimes f_j\}_{(i,j)\in(\mathbb{N}^2,\prec)}$ is also dense in these spaces. We will use the notation P_A, where $A \subset \mathbb{N}^2$, as follows:

$$P_A\left(\sum_{(i,j)\in\mathbb{N}^2} a_{(i,j)} e_i \otimes f_j\right) = \sum_{(i,j)\in A} a_{(i,j)} e_i \otimes f_j.$$

For $A = \{(i,j) : (i,j) \preceq (n,m)\}$ we abbreviate P_A as $P_{(n,m)}$. In order to prove that $\{e_i \otimes f_j\}_{(i,j)\in(\mathbb{N}^2,\prec)}$ is a Schauder basis for $X \otimes_\pi Y$ and $X \otimes_\varepsilon Y$, it suffices to establish the uniform boundedness of the initial projections $P_{(n,m)}$, $n, m \in \mathbb{N}$.

Let $D \subset \mathbb{N}^2$ be a subset satisfying

$$D = \{(i,j) : i_1 \le i < i_2,\ j_1 \le j < j_2 \text{ for some } i_1, i_2, j_1, j_2 \in \mathbb{N}\}.$$

We will call such sets *rectangular sets* in $\mathbb{N}^2$. In particular, if $i_1 = j_1 = 1$, then the rectangular set $D \subset \mathbb{N}^2$ is an initial segment in the ordering $\prec$.

Lemma 16.67 *Let X be a Banach spaces with a normalized monotone Schauder basis $\{e_i\}_{i=1}^\infty$, and Y be a Banach space with a normalized monotone Schauder basis $\{f_i\}_{i=1}^\infty$. Let D be a rectangular set in $\mathbb{N}^2$. Then $P_D : X \otimes Y \to X \otimes Y$ is a 4-bounded projection with respect to either π or ε. Thus, if $S \subset \mathbb{N}^2$ can be written as a disjoint union of at most k rectangular sets, then P_S is $4k$-bounded with respect to either π or ε.*

Proof: Let $D = A \times B$, where $A = [i_1, i_2]$, $B = [j_1, j_2] \subset \mathbb{N}$ are intervals of integers. Consider first the π-norm case. Let $\delta > 0$, and $z = \sum_{j=1}^n x_j \otimes y_j$, $x_j \in X$ and $y_j \in Y$, be a representation such that $\pi(z) \le \sum_{j=1}^n \|x_j\|\|y_j\| + \delta$. Note that

$$P_D(z) = \sum_{j=1}^{n} P_A(x_j) \otimes P_B(y_j).$$

Since our bases are assumed to be monotone, we have $\|P_A\|, \|P_B\| \leq 2$. Thus $\pi(P_D(z)) \leq \sum_{j=1}^{n} \|P_A(x_j)\|\|P_B(y_j)\| \leq 4(\pi(z)+\delta)$. As δ can be chosen arbitrarily small, the case of π is proved.

The ε norm case. Given $z = \sum_{j=1}^{n} x_j \otimes y_j$

$$\varepsilon\left(\sum_{j=1}^{n} x_j \otimes y_j\right) = \sup_{\|x^*\|,\ \|y^*\| \leq 1} \left|\sum_{j=1}^{n} \langle x^*, x_j\rangle\langle y^*, y_j\rangle\right|.$$

Similarly,

$$\begin{aligned} \varepsilon(P_D(z)) &= \sup_{\|x^*\|,\ \|y^*\| \leq 1} \left|\sum_{j=1}^{n} \langle x^*, P_A(x_j)\rangle\langle y^*, P_B(y_j)\rangle\right| \\ &= \sup_{\|x^*\|,\ \|y^*\| \leq 1} \left|\sum_{j=1}^{n} \langle P_A^*(x^*), x_j\rangle\langle P_B^*(y^*), y_j\rangle\right|. \end{aligned}$$

Again, since $\|P_A^*\|, \|P_B^*\| \leq 2$, we obtain $\varepsilon(P_D) \leq 4$. The estimate for unions of rectangular sets is immediate. □

Theorem 16.68 (Gelbaum, de la Madrid, Holub) *Let X be a Banach spaces with a normalized monotone Schauder basis $\{e_i\}_{i=1}^{\infty}$, and Y be a Banach space with a normalized monotone Schauder basis $\{f_i\}_{i=1}^{\infty}$. Then $\{e_i \otimes f_j\}_{(i,j)\in(\mathbb{N}^2,\prec)}$ is a Schauder basis for both $X \otimes_\pi Y$, and $X \otimes_\varepsilon Y$. Moreover, if both $\{e_i\}_{i=1}^{\infty}$ and $\{f_i\}_{i=1}^{\infty}$ are shrinking, then $\{e_i \otimes f_j\}_{(i,j)\in(\mathbb{N}^2,\prec)}$ is a shrinking basis of $X \otimes_\varepsilon Y$. If both $\{e_i\}_{i=1}^{\infty}$ and $\{f_i\}_{i=1}^{\infty}$ are boundedly complete, then $\{e_i \otimes f_j\}_{(i,j)\in(\mathbb{N}^2,\prec)}$ is a boundedly complete basis of $X \otimes_\pi Y$.*

Proof: By Lemma 4.7 and Fact 4.8, it suffices to show that the initial projections $P_{(n,m)}$, $n, m \in \mathbb{N}$, are uniformly bounded in both $X \otimes_\pi Y$ and $X \otimes_\varepsilon Y$. It is immediate to observe that every initial set $S = \{(i, j) : (i, j) \preceq (n, m)\}$ in $(\mathbb{N}^2, \prec)$ is a disjoint union of at most two rectangular sets. Thus by Lemma 16.67, we have a uniform estimate 8 for the norm of the initial projections, and the sequence is a Schauder basis.

Next assume that $\{e_i, e_i^*\}_{i=1}^{\infty}$ and $\{f_i, f_i^*\}_{i=1}^{\infty}$ are shrinking bases of X, Y. Then $\{e_i \otimes f_j\}_{(i,j)\in(\mathbb{N}^2,\prec)}$ is a Schauder basis of $X \otimes_\varepsilon Y$, and $\{e_i^* \otimes f_j^*\}_{(i,j)\in(\mathbb{N}^2,\prec)}$ is a Schauder basis of $X^* \otimes_\pi Y$. Note that Y^* is a RNP space with the AP. By Theorem 16.40 we have $(X \otimes_\varepsilon Y)^* = X^* \otimes_\pi Y^*$, so the previous bases are dual to each other. Therefore, $\{e_i \otimes f_j\}_{(i,j)\in(\mathbb{N}^2,\prec)}$ is a shrinking basis of $X \otimes_\varepsilon Y$.

Finally, let $\{e_i, e_i^*\}_{i=1}^\infty$ and $\{f_i, f_i^*\}_{i=1}^\infty$ the respective boundedly complete bases of X, Y. Set $X_* = \overline{\operatorname{span}}\{e_i^*\}_{i=1}^\infty$, $Y_* = \overline{\operatorname{span}}\{f_i^*\}_{i=1}^\infty$. Then $\{e_i^*\}_{i=1}^\infty$ is a Schauder basis of X_* and $\{f_i^*\}_{i=1}^\infty$ is a Schauder basis of Y_*. Moreover, these bases are shrinking and their dual bases are precisely $\{e_i\}_{i=1}^\infty$ and $\{f_i\}_{i=1}^\infty$. By applying the previous result, we obtain the conclusion that $\{e_i \otimes f_j\}_{(i,j)\in(\mathbb{N}^2,\prec)}$ is a boundedly complete basis of $X \otimes_\pi Y$. □

Corollary 16.69 *Let X be a Banach spaces with a shrinking Schauder basis $\{e_i\}_{i=1}^\infty$, and Y be a Banach space with a Schauder basis $\{f_i\}_{i=1}^\infty$. Then $\{e_i^* \otimes f_j\}_{(i,j)\in(\mathbb{N}^2,\prec)}$ is a Schauder basis of $\mathcal{K}(X, Y)$. Moreover, if both $\{e_i^*\}_{i=1}^\infty$ and $\{f_i\}_{i=1}^\infty$ are shrinking, then $\{e_i^* \otimes f_j\}_{(i,j)\in(\mathbb{N}^2,\prec)}$ is a shrinking basis of $\mathcal{K}(X, Y)$.*

Proof: It suffices to note that, under the assumptions, $\mathcal{K}(X, Y) = X^* \otimes_\varepsilon Y$. □

Corollary 16.69 applies in particular to all classical Banach spaces ℓ_p, $L_p[0, 1]$, c_0. In particular, $\mathcal{K}(\ell_2)$ has a shrinking Schauder basis. However, its bidual $\mathcal{B}(\ell_2)$ fails to have the AP, by a result of Szankowski [Szan2]. The basis $\{e_i \otimes f_j\}_{(i,j)\in(\mathbb{N}^2,\prec)}$ is seldom unconditional, even when both $\{e_i\}_{i=1}^\infty$ and $\{f_i\}_{i=1}^\infty$ are unconditional.

Example: The basis $\{e_i^* \otimes e_j\}_{(i,j)\in(\mathbb{N}^2,\prec)}$ of $\mathcal{K}(\ell_2)$ is not unconditional. Indeed, let W_n be a $2^n \times 2^n$-Walsh matrix, i.e., a unitary matrix whose all entries belong to $\{-2^{-\frac{n}{2}}, 2^{-\frac{n}{2}}\}$. Then W_n defines a compact operator of norm one on ℓ_2 (by using only the first 2^n coordinates). However, the operator $2^{-\frac{n}{2}} \sum_{i,j\le 2^n} e_i^* \otimes e_j$ has a norm at least $2^{\frac{n}{2}}$.

In [Pisi3, p. 112] it is proved that the space $\mathcal{K}(\ell_2)$ has no unconditional basis.

Proposition 16.70 (Holub [Holu]) *Let X be a Banach space with an unconditional basis $\{e_i\}_{i=1}^\infty$, and $\{f_i\}_{i=1}^\infty$ be the unit basis of c_0. Then $\{e_i \otimes f_j\}_{(i,j)\in(\mathbb{N}^2,\prec)}$ is an unconditional basis of $X \otimes_\varepsilon c_0$.*

Proof: Assume, without loss of generality, that $\{e_i\}_{i=1}^\infty$ is 1-unconditional. We have

$$\Big\| \sum_{(i,j)\in\mathbb{N}^2} a_{(i,j)} e_i \otimes f_j \Big\| = \sup_{\|x^*\|\le 1} \Big\| \sum_{i=1}^\infty a_{(i,j)} \langle x^*, e_i\rangle f_j \Big\|_{c_0}$$
$$= \sup_{j\in\mathbb{N},\ \|x^*\|\le 1} \Big| \sum_{i=1}^\infty a_{(i,j)} \langle x^*, e_i\rangle \Big| = \sup_{j\in\mathbb{N}} \Big\| \sum_{i=1}^\infty a_{(i,j)} e_i \Big\|_X .$$

The last expression depends only on the absolute values of $a_{(i,j)}$, so the unconditionality follows. □

Corollary 16.71 *Let X be a Banach space with a shrinking unconditional basis. Then $\mathcal{K}(X, c_0)$ has an unconditional basis (consisting of $\{e_i^* \otimes f_j\}_{(i,j)\in(\mathbb{N}^2,\prec)}$).*

Proposition 16.72 (Holub [Holu]) *Let X be a Banach space with an unconditional basis $\{e_i\}_{i=1}^{\infty}$, and Y have a Schauder basis $\{f_i\}_{i=1}^{\infty}$. Then $\{e_i \otimes f_i\}_{i=1}^{\infty}$ is an unconditional basic sequence in of $X \otimes_{\varepsilon} Y$.*

Proof: We already know that $\{e_i \otimes f_i\}_{i=1}^{\infty}$ is a basic sequence. Assume without loss of generality that $\{e_i\}_{i=1}^{\infty}$ is 1-unconditional. We have

$$\left\|\sum_{i=1}^{\infty} a_{(i,i)} e_i \otimes f_i\right\| = \sup_{\|y^*\| \leq 1} \left\|\sum_{i=1}^{\infty} a_{(i,i)} \langle y^*, f_i\rangle e_i\right\|_X.$$

The last expression depends only on the absolute values of $a_{(i,j)}$, so the unconditionality follows. □

Theorem 16.73 (Holub [Holu]) *Let $1 < p, q < \infty$. Let $\{e_i\}_{i=1}^{\infty}$ be the canonical basis in ℓ_p, and $\{f_i\}_{i=1}^{\infty}$ be the canonical basis in ℓ_q. Then the basic sequence $\{e_i^* \otimes f_i\}_{i=1}^{\infty}$ in $\mathcal{K}(\ell_p, \ell_q)$ has the following properties:*
1. If $p \leq q$ then it is equivalent to the canonical basis of c_0.
2. If $p > q$, then it is equivalent to the canonical basis of ℓ_s, where $s = \frac{pq}{p-q}$.

Proof: Case 1:

$$\left\|\sum_{i=1}^{\infty} a_{(i,i)} e_i^* \otimes f_i\right\| = \sup_{\|x\|_{\ell_p} \leq 1} \left\|\sum_{i=1}^{\infty} a_{(i,i)} \langle e_i^*, x\rangle f_i\right\|_{\ell_q} \leq \sup_{i \in \mathbb{N}}\{|a_{(i,i)}|\}\|x\|_{\ell_q}.$$

Since $p \leq q$, the result follows.
Case 2: Suppose that $\sum_{i=1}^{\infty} a_{(i,i)} e_i^* \otimes f_i$ is convergent in $\mathcal{K}(\ell_p, \ell_q)$. Recall that it is an unconditional basic sequence. Then

$$\lim_{m \to \infty} \sup_{\|x\|_{\ell_p} \leq 1} \sum_{i=m}^{n} |a_{(i,i)}|^q |\langle e_i^*, x\rangle|^q \to 0, \tag{16.62}$$

which implies that $\sum_{i=1}^{\infty} |a_{(i,i)}|^q |\langle e_i^*, x\rangle|^q$ converges for all $x \in \ell_p$. For any sequence $(b_i) \in \ell_{\frac{p}{q}}$, let $x = (|b_i|^{\frac{1}{q}}) \in \ell_p$. Thus we have that

$$\sum_{i=1}^{\infty} |a_{(i,i)}|^q (|b_i|^{\frac{1}{q}})^q = \sum_{i=1}^{\infty} |a_{(i,i)}|^q |b_i|$$

converges for all $(b_i) \in \ell_{\frac{p}{q}}$, and therefore $(|a_{(i,i)}|^q) \in \ell_{\frac{p}{q}}^* = \ell_{\frac{p}{p-q}}$. Thus $(a_{(i,i)}) \in \ell_{\frac{pq}{p-q}}$. Conversely, if $(a_{(i,i)}) \in \ell_{\frac{pq}{p-q}}$, then for any m, n:

$$\sup_{\|x\|_{\ell_p}\le 1}\left(\sum_{i=m}^{n}|a_{(i,i)}|^q|\langle e_i^*,x\rangle|^q\right)^{\frac{1}{q}}$$

$$\le \sup_{\|x\|_{\ell_p}\le 1}\left[\left(\sum_{i=m}^{n}(|a_{(i,i)}|^q)^{\frac{p}{p-q}}\right)^{\frac{p-q}{p}}\left(\sum_{i=m}^{n}(|\langle e_i^*,x\rangle|^q)^{\frac{p}{q}}\right)^{\frac{q}{p}}\right]^{\frac{1}{q}},$$

by Holder's inequality. Therefore

$$\sup_{\|x\|_{\ell_p}\le 1}\left(\sum_{i=m}^{n}|a_{(i,i)}|^{\frac{pq}{p-q}}\right)^{\frac{p-q}{pq}}\left(\sum_{i=m}^{n}|\langle e_i^*,x\rangle|^p\right)^{\frac{1}{p}}\le \sup_{\|x\|_{\ell_p}\le 1}\left(\sum_{i=m}^{n}|a_{(i,i)}|^{\frac{pq}{p-q}}\right)^{\frac{p-q}{pq}}\xrightarrow{m,n}0.$$

Hence, $\sum_{i=1}^{\infty}a_{(i,i)}e_i^*\otimes f_i$ converges in $\mathcal{K}(\ell_p,\ell_q)$, and the conclusion follows. □

Theorem 16.74 (Holub [Holu]) *Let X be a closed subspace of $\mathcal{K}(\ell_2)$. Then either X is isomorphic to ℓ_2 and complemented in $\mathcal{K}(\ell_2)$, or X contains a subspace isomorphic to c_0.*

Proof: Fix an orthonormal basis $\{e_i\}_{i=1}^{\infty}$ of ℓ_2, and represent $\mathcal{K}(\ell_2)$ by using the basis $\{e_i^*\otimes e_j\}_{(i,j)\in(\mathbb{N}^2,\prec)}$. Given $n\in\mathbb{N}$, denote by $R_n=\{(i,j)\in\mathbb{N}^2: i\ge n \text{ and } j\ge n\}$ the rectangular set, and $S_n=\mathbb{N}^2\setminus R_n$. It is easy to see that S_n is a disjoint union of two rectangular sets S_n^1 and S_n^2. As the basis $\{e_i\}_{i=1}^{\infty}$ is bimonotone, we have that $\|P_{R_n}\|\le 1$, and $\|S_n\|\le 2$ in $\mathbb{K}$.

We have the following two possibilities. Either there exists a $\delta>0$, and an integer $n\in\mathbb{N}$, such that for every $x\in X$, $\|P_{S_n}(x)\|\ge\delta\|x\|$. In this case, clearly, $P_{S_n}(X)$ is isomorphic to X. Note that $P_{S_n}=\sum_{i=1}^{2n}P_{D_i}$, where $D_i=\{(i,j): j\ge i\}$ for $i\le n$, and $D_i=\{(j,i-n): j>i-n\}$ for $i>n$. Next note that for every $i\le 2n$, $P_{D_i}(\mathcal{K}(\ell_2))$ is a Hilbert space. Indeed, if $i\le n$, then

$$\left\|\sum_{j=i}^{\infty}a_{(i,j)}e_i^*\otimes e_j\right\|=\sup_{\|x\|\le 1}\sum_{j=i}^{\infty}a_{(i,j)}x_j.$$

By Holder's inequality,

$$\left\|\sum_{j=i}^{\infty}a_{(i,j)}e_i^*\otimes e_j\right\|=\left(\sum_{j=i}^{\infty}a_{(i,j)}^2\right)^{\frac{1}{2}}.$$

If $n<i\le 2n$ a similar argument works. Thus we obtain that X is isomorphic to a closed subspace of a finite direct sum of Hilbert spaces $P_{S_n}(\mathcal{K}(\ell_2))=\sum_{i=1}^{2n}P_{D_i}(\mathcal{K}(\ell_2))$. Thus in this case X is a Hilbert space. Therefore there is a bounded projection $Q: P_{S_n}(\mathcal{K}(\ell_2))\to X$. Since $P_{S_n}:\mathcal{K}(\ell_2)\to\sum_{i=1}^{2n}P_{D_i}(\mathcal{K}(\ell_2))$

is a bounded projection, we finally obtain that $Q \circ P_{S_n} : \mathcal{K}(\ell_2) \to X$ is a bounded projection, and so X is a complemented subspace.

It remains to deal with the case when for every $\delta > 0$ and every $n \in \mathbb{N}$ there is some $x \in B_X$ such that $\|P_{S_n}(x)\| \leq \delta$. Choose a sequence $\delta_i > 0$, $\sum_{i=1}^{\infty} \delta_i < 1$. By assumption, there exists an increasing sequence of integers n_i and vectors $x_i \in B_X$, such that $\|P_{R_i}(x_i) - x_i\| < \varepsilon_i$, where $R_i = \{(i, j) : n_i \leq i, j < n_{i+1}\}$. Put $y_i = P_{R_i}(x_i)$. It is easy to see that $\{y_i\}_{i=1}^{\infty} \in \mathbb{K}$ is equivalent to the canonical basis in c_0. Indeed, $\sum_{i=1}^{\infty} a_i y_i$ correspond to block diagonal operators from $\mathcal{K}(\ell_2)$, where the blocks are supported by rectangular sets R_i. By the Mazur perturbed basis theorem, $\{x_i\}_{i=1}^{\infty}$ contains a subsequence equivalent to the canonical basis of c_0. $\square$

16.8 Remarks and Open Problems

Remarks

1. The classical part of the theory of tensor products was initiated by Schatten, von Neumann and their co-authors, as it is summarized in [Scha]. Most of the material in Section 16.1 originates from this period. The central part of the theory was completed by Grothendieck, to whom belong most of the results in Sections 16.2, 16.3, and 16.4. The outline of Grothendieck's monumental contribution to the field is contained in his Resume [Groth2]. The more recent development of this field was sparked by the paper of Lindenstrauss and Pełczyński [LiPe], and by the negative solution of the approximation problem by Enflo [Enfl2]. Enflo's result and its offsprings are studied in Sections 16.5 and 16.6. Finally, in the last Section 16.7 we present an investigation of Schauder bases in tensor products, due mostly to Holub [Holu]. We refer to [Pisi3], where contributions to the theory of tensor products made in the direction opened by Grothendieck in [Groth2] since 1968 are presented. In particular, this work contains Pisier's solution to the sixth problem in [Groth2] (published in [Pisi2], see Remark 16.1). As our book is an introductory text, we have not attempted to compile an exhaustive reference list, and our attribution of theorems conforms with the earlier books on the subject. We refer to [DiUh] for a more detailed historical account. For more on tensor norms we refer to [DeFl].
2. We refer to the articles [Casa, pp. 271–316] and [DJP, pp. 437–496] for more in this area.
3. The result of Kwapień mentioned in the introduction to Section 16.5 is the following [Kwap2], see, e.g., [PeBe, p. 242]: *For each p with $2 < p < \infty$, there exist increasing sequences $\{n_k\}$ and $\{m_k\}$ of positive integers such that the closed linear subspace of L_p spanned by the functions $f_k(t) := e^{in_k 2\pi t} + e^{im_k 2\pi t}$, $k = 1, 2, \ldots$, fails the AP.*
4. The space $\mathcal{B}(H)$ of bounded operators on a Hilbert space H does not have the approximation property (Szankowski, [Szan2]) and neither does the Calkin space $\mathcal{B}(H)/\mathcal{K}(H)$ (Godefroy and Saphar [Gosa2]).

5. A locally convex space E is nuclear if and only if $E \otimes_\epsilon E \otimes_\epsilon E = E \otimes_\pi E \otimes_\pi E$ isomorphically [Johnk1b].

Open Problems

1. Pisier proved in [Pisi2] and [Pisi3] that there is an infinite-dimensional Banach space X such that $X \oplus_\pi X = X \oplus_\varepsilon X$ algebraically and topologically. This was a solution to Grothendieck's problem in [Groth2]. K. John [Johnk2] proved that Pisier's space satisfies moreover that $\mathcal{N}(X, X^*) = \mathcal{K}(X, X^*)$. Note that Pisier's space does not have the approximation property. It seems to be unknown if for some infinite-dimensional Banach space one can have $\mathcal{K}(X) = \mathcal{N}(X)$, see [DJP, p. 489]. Very recently, K. John proved that no infinite-dimensional reflexive Banach space may satisfy the relation $X \otimes_\pi X = X \otimes_\varepsilon X$.
   A negative solution to the aforesaid Grothendieck's problem in the field of locally convex spaces was independently found in [JohnK1].
2. It seems to be an open problem whether X has the approximation property if $\overline{\mathcal{F}(X)} = \mathcal{K}(X)$, see [Casa, p. 282].
3. It is an open problem whether X^* has BAP if X is a Banach space such that X^* has AP.

Exercises for Chapter 16

16.1 Denote by $\mathcal{P}(\mathbb{R}^n)$ the space of real polynomials on $\mathbb{R}^n$. Then $\mathcal{P}(\mathbb{R}^n) \otimes \mathcal{P}(\mathbb{R}^m) = \mathcal{P}(\mathbb{R}^{n+m})$.
Hint. Standard.

16.2 We denote by $L_1(\mu, X)$ the Banach space of all Bochner integrable functions on a σ-additive finite measure space (S, Σ, μ), equipped with the L_1-norm. Then $L_1(\mu) \otimes_\pi X = L_1(\mu, X)$.
Hint. Since $L_1(\mu) \times X \to L_1(\mu, X)$ is a 1-bounded bilinear mapping, we get an isometric embedding $L_1(\mu) \otimes_\pi X \hookrightarrow L_1(\mu, X)$. It remains to check that $\sum_{i=1}^n \chi_{E_i} \otimes x_i$ are dense in $L_1(\mu, X)$.

16.3 Let K be a compact space and X be a Banach space. By $C(K, X)$ we denote the Banach space of continuous X-valued functions on K with the supremum norm. Then $C(K) \otimes_\varepsilon X = C(K, X)$.
Hint. Let $j : C(K) \otimes X \to C(K, X)$ be defined by $j(\sum_{i=1}^n f_i \otimes x_i) = \sum_{i=1}^n f_i(\cdot)x_i$. It remains to prove that this mapping has norm one, and a dense range. See [DiUh, p. 244].

16.4 Let K, L be a compact spaces and $X = C(L)$ be a Banach space. Then $C(K) \otimes_\varepsilon C(L) = C(K \times L)$.

Hint. [DeFl, p. 48].

16.5 Let X, Y be Banach spaces. Then the formal identity mapping $j : X \otimes Y \to X \otimes Y$ has a (unique) continuous extension $j : X \otimes_\pi Y \to X \otimes_\varepsilon Y$.
Hint. Standard.

16.6 Let X, Y be Banach spaces. Then the formal identity mapping $j : X^* \otimes_\pi Y \to X^* \otimes_\varepsilon Y$ has a factorization $j = I \circ J$, where $J : X^* \otimes_\pi Y \to \mathcal{N}(X, Y)$ from 16.24 and $I : (\mathcal{N}(X, Y), N) \to X^* \otimes_\varepsilon Y$ is from Proposition 16.22.
Hint. It suffices to note that each operator j, J, I is just a unique continuous extension (in the proper spaces) of the formal identity mapping among finite tensors.

16.7 Give an example of a compact operator on ℓ_2 that is not nuclear.
Hint. $Tx = (\frac{1}{i}x_i)$. Use Theorem 16.52.

16.8 Is $\mathcal{K}(\ell_2)$ a separable Asplund space?
Hint. Yes, Corollary 16.69.

16.9 Let $B \in \mathcal{B}il(X \times Y)$ be a bounded bilinear mapping. Then there exists a canonical bilinear extension $\hat{B} \in \mathcal{B}il(X \times Y^{**})$, defined by $\hat{B}(x, \cdot) = B(x, \cdot)^{**}$. We have $\|\hat{B}\| = \|B\|$.
Hint. Standard.

16.10 Let X, Y be Banach spaces. Then the natural embedding $X \otimes_\pi Y \hookrightarrow X \otimes_\pi Y^{**}$ is an isometry.
Hint. Clearly, given $z \in X \otimes Y$, $\pi_{X\otimes_\pi Y^{**}}(z) \le \pi_{X\otimes_\pi Y}(z)$. Let $B \in \mathcal{B}il(X \times Y) = (X \otimes_\pi Y)^*$ be the Hahn–Banach functional for z, and let $\hat{B} \in \mathcal{B}il(X \times Y^{**})$ be its canonical isometric extension (16.9). Then $\langle \hat{B}, z\rangle = \langle B, z\rangle$, so $\pi_{X\otimes_\pi Y^{**}}(z) \ge \pi_{X\otimes_\pi Y}(z)$.

16.11 Let X, Y be Banach spaces. The canonical image $\mathcal{B}(X, Y) \hookrightarrow \mathcal{B}(X, Y^{**})$, respectively $\mathcal{T}(\mathcal{B}(X, Y)) \hookrightarrow \mathcal{B}(Y^*, X^*)$, is w^*-dense if and only if i^* from (16.28) is injective.
Hint. $\mathcal{B}(X, Y)$ is w^*-dense in $\mathcal{B}(X, Y^{**})$ if and only if for every $z \in Y^* \otimes_\pi X$ such that $z \in \mathcal{B}(X, Y)_\perp$ it holds that $z = 0$. Alternatively, $i^*(z) = 0$ implies $z = 0$ which is clearly equivalent to the injectivity of i^*. The respective case follows by standard transposition.

16.12 If Y is reflexive, then $Y^* \otimes_\pi X = (\mathcal{B}(X, Y), \tau)^*$.
Hint. Standard.

16.13 Let (S, Σ, μ) be a probability space, and let $\{\theta_j\}_{j=1}^n$ be independent random variables such that $\mu(\theta_j^{-1}(1)) = \mu(\theta_j^{-1}(-1)) = \frac{1}{2}$, (respectively $\mu(\theta_j^{-1}(2)) = \frac{1}{3}$ and $\mu(\theta_j^{-1}(-1)) = \frac{2}{3}$). Then there exists a constant K independent of n so that

$$\mu\{|\sum_{j=1}^{n}\alpha_j\theta_j| > K(\sum_{j=1}^{n}|\alpha_j|^2\log n)^{\frac{1}{2}}\} < \frac{K}{n^3}$$

for every choice of complex numbers $\{\alpha_j\}_{j=1}^n$.

Hint. Without loss of generality, α_j are real numbers such that $\sum_{j=1}^n \alpha_j^2 = 1$. Denote $f = |\sum_{j=1}^n \alpha_j\theta_j|$ a random variable on S. Then for every $\lambda > 0$:

$$\int e^{\lambda f}\,\mathrm{d}\mu \le \int (e^{\lambda\sum_{j=1}^n \alpha_j\theta_j} + e^{-\lambda\sum_{j=1}^n \alpha_j\theta_j})\,\mathrm{d}\mu = 2\prod_{j=1}^{n}\frac{(e^{\lambda\alpha_j}+e^{-\lambda\alpha_j})}{2}.$$

Since $\frac{e^x+e^{-x}}{2} \le e^{x^2}$ (in the respective case we use that $\frac{1}{3}(e^{2x}+2e^{-x}) \le e^{2x^2}$), we get that

$$\mu\{\lambda f - \lambda^2 - 3\log n > 0\} \le \int (e^{\lambda f-\lambda^2-3\log n})\,\mathrm{d}\mu \le \frac{2}{n^3}.$$

The result follows by taking $\lambda = (3\log n)^{\frac{1}{2}}$.

16.14 (Davie [Davi]) Let $p > 2$. Then ℓ_p contains a subspace without the AP.

Hint. First, by a standard modification of the proof, Lemma 16.53 can be strengthened to yield a matrix A satisfying a stronger condition than 2, namely $\sum_{i=1}^\infty(\max_j|a_{i,j}|)^r < \infty$, where $r = \frac{p}{p+1} > \frac{2}{3}$. The calculations have to be adjusted so as to strengthen (16.51) into

$$\sum_{i=1}^{\infty}(\max_j|a_{i,j}|)^r \le \sum_{k=0}^{\infty}3\cdot 2^k D^r(k+1)^{\frac{r}{2}}2^{-\frac{3rk}{2}} < \infty.$$

Then, put $\lambda_i = \max_j|a_{i,j}|$, $b_{i,j} = (\frac{\lambda_j}{\lambda_i})^{\frac{1}{p+1}}$, and form a new infinite matrix $B = (b_{i,j})_{i,j\in\mathbb{N}}$. It holds $B^2 = 0$, $tr(B) = tr(A) \ne 0$. Finally, the rows $y_i = (b_{i,1}, b_{i,2},\dots)$ of B belong to ℓ_p, so the argument of Lemma 16.53 yields a subspace of ℓ_p without the AP.

16.15 (Willis, [Willis]) We say that a Banach space X has the *compact approximation property* (CAP for short), if $Id \in \overline{\mathcal{K}(X)}^{\tau}$. CAP is a strictly weaker condition than AP, due to a result of Willis. Show that a Banach space X fails CAP, provided that there exist bounded sequences $\{x_i^*\}_{i=1}^\infty \in X^*$, $\{x_i\}_{i=1}^\infty \in X$, $I_n = \{2^n+1, 2^n+2,\dots,2^{n+1}\}$, and finite subsets $\{F_i\}_{i=1}^\infty$ of X, so that

1. $x_i^*(x_i) = 1$ for all $i \in \mathbb{N}$.
2. $x_i^* \xrightarrow{w^*} 0$.
3. For every $T : X \to X$,

$$\left| 2^{-n} \sum_{i \in I_n} \langle x_i^*, T(x_i) \rangle - 2^{-n-1} \sum_{i \in I_{n+1}} \langle x_i^*, T(x_i) \rangle \right| \leq \max\{\|T(f)\| : f \in F_n\}.$$

4. $\sum_{i=1}^{\infty} \max\{\|f\| : f \in F_n\} < \infty$.

Hint. Let $\beta_n \in (\mathcal{B}(X), \tau)^*$ be defined by the formula $\beta_n(T) = 2^{-n} \sum_{i \in I_n} \langle x_i^*, T(x_i) \rangle$, and $\alpha_n = \max\{\|f\| : f \in F_n\}$. By 3 and 4, $\beta_n(T)$ is convergent for every $T \in \mathcal{B}(X)$, so we put $\beta(T) = \lim_{n \to \infty} \beta_n(T)$. In particular, $\beta = w^*\text{-}\lim \beta_n \in (\mathcal{B}(X), \tau)^*$. By 1, $\beta(Id) = 1$. If $T \in \mathcal{K}(X)$, then $\{T(x_i)\}_{i=1}^{\infty}$ is compact, so $\lim_{i \to \infty} \langle x_i^*, T(x_i) \rangle = 0$, and $\beta(T) = \lim_{n \to \infty} \beta_n(T) = 0$. Thus β τ-separates Id from $\mathcal{K}(X)$.

16.16 (Davie, [Davi]) Let $p > 2$. Then ℓ_p has a subspace without the CAP.
Hint. See [Davi].

For the exercises below we need some definitions. The *weak operator topology* (WOT for short) on $\mathcal{B}(X, Y)$ is a locally convex topology generated by functionals: $T \to \langle y^*, T(x) \rangle$, $y^* \in Y^*$, $x \in X$.

The *strong operator topology* (SOT for short) on $\mathcal{B}(X, Y)$ is a locally convex topology generated by: $T \to \|T(x)\|$, $x \in X$.

The *dual weak operator topology* (WOT* for short) is defined by (T^*) $T \to \langle x^{**}, T^*(y^*) \rangle$, $y^* \in Y^*$, $x^{**} \in X^{**}$.

16.17 Show that WOT* is stronger than WOT.

16.18 Let X, Y be Banach spaces, Y be reflexive. By Theorem 16.16, $(X \otimes_\pi Y)^* = \mathcal{B}(X, Y^*)$. Show that WOT and w^*-topology coincide on bounded sets in $K \subset \mathcal{B}(X, Y^*)$. In particular, $B_{\mathcal{B}(X,Y^*)}$ is WOT-compact.
Hint. $X \otimes Y$ is dense in $X \otimes_\pi Y$.

16.19 Let X, Y be Banach spaces. Show that a bounded convex set $B \subset \mathcal{B}(X, Y)$ is WOT-closed if and only if it is SOT-closed.
Hint. Let B be bounded convex and SOT-closed, $T \notin B$. There are $\{x_i\}_{i=1}^n \in X$, such that

$$\tilde{T} = (T(x_1), \ldots, T(x_n)) \notin \tilde{B} = \overline{\{(S(x_1), \ldots, S(x_n)) : S \in B\}}^{\|\cdot\|} \subset Y \times \cdots \times Y.$$

By the Hahn–Banach theorem, there exists $y^* = (y_1^*, \ldots, y_n^*) \in Y^* \times \cdots \times Y^*$, separating $\tilde{T}$ from $\tilde{B}$. The finite collection of pairs $\{(x_i, y_j^*)\}_{1 \leq i,j \leq n}$ determines a WOT-neighborhood of T disjoint from B.

16.20 Let X be a Banach space. Then $B_{\mathcal{B}(X)}$ is SOT-compact if and only if X is finite-dimensional.

Hint. Choose a sequence from B_X, $x_n \overset{w}{\to} 0$. Then $U_n = \{T \in B_{\mathcal{B}(X)} : \|T(x_n)\| < \frac{1}{2}\}$ forms an open cover of $B_{\mathcal{B}(X)}$.

16.21 Is $\mathcal{K}(\ell_2)$ isomorphic to a subspace of a separable dual?
Hint. No, c_0 in it.

16.22 (Kalton, [Kalt2]) Let X, Y be Banach spaces. The mapping $i : \mathcal{K}(X, Y) \to C((B_{X^{**}}, w^*) \times (B_{Y^*}, w^*))$, $i(T) = \langle x^{**}, T^* y^* \rangle$ is an isometric embedding.
Hint. Let $T \in \mathcal{K}(X, Y)$, and convergent nets $u_\alpha \overset{w^*}{\to} u \in B_{X^{**}}$, $v_\alpha \overset{w}{\to} v \in B_{Y^*}$. As T^* is compact, $\|T^* v_\alpha - T^* v\| \to 0$. Thus

$$|u_\alpha(T^* v_\alpha) - u(T^* v)| \le |u_\alpha(T^* v_\alpha - T^* v)| + |(u_\alpha - u)(T^* v)| \to 0.$$

16.23 (Kalton, [Kalt2]) Let X, Y be Banach spaces. Let A be a subset of $\mathcal{K}(X, Y)$. Then A is weakly compact if and only if is WOT*-compact.
Hint. WOT*-compactness is equivalent to the pointwise compactness of $i(A) \subset C(B_{X^{**}} \times B_{Y^*})$. By Grothendieck's Theorem 3.52, this is equivalent to the weak compactness of $i(A)$.

16.24 If X is reflexive, then a subset $A \subset \mathcal{K}(X, Y)$ is weakly compact if and only if it is WOT-compact.
Hint. Use Exercise 16.23 and reflexivity.

16.25 (Kalton, [Kalt2]) If X and Y are reflexive, and $\mathcal{K}(X, Y) = \mathcal{B}(X, Y)$, then $\mathcal{K}(X, Y)$ is reflexive.
Hint. It suffices to show that $B_{\mathcal{K}(X,Y)}$ is WOT-compact. Suppose that $\{T_\alpha\}$ is a WOT-Cauchy net in $B_{\mathcal{K}(X,Y)}$. For every $x \in X$, $T_\alpha(x)$ is weakly Cauchy (and so weakly convergent) in Y, hence there exists $T \in B_{\mathcal{B}(X,Y)} = B_{\mathcal{K}(X,Y)}$, such that $T_\alpha \to T$ in WOT.

16.26 Let $p > q \ge 1$. Then $\mathcal{B}(\ell_p, \ell_q)$ is reflexive.
Hint. By Pitt's theorem, $\mathcal{K}(\ell_p, \ell_q) = \mathcal{B}(\ell_p, \ell_q)$.

16.27 (Ruess and Stegall, [RuSt]) Let X, Y be Banach spaces. Then $\text{Ext}(B_{\mathcal{K}(X,Y)^*}) = \text{Ext}(B_{X^{**}}) \otimes \text{Ext}(B_{Y^*})$.
Hint. As $i : \mathcal{K}(X, Y) \hookrightarrow C((B_{X^{**}}, w^*) \times (B_{Y^*}, w^*)$ is an isometric embedding, the extreme points of the dual ball of $C((B_{X^{**}}, w^*) \times (B_{Y^*}, w^*))$ are the Dirac functionals $\pm\delta_p$, where $p = (x^{**}, y^*)$, $\|x^{**}\|, \|y^*\| \le 1$. By the Krein–Milman theorem, i^* maps the extreme points in $C((B_{X^{**}}, w^*) \times (B_{Y^*}, w^*)^*$ onto the set $\text{Ext}(B_{\mathcal{K}(X,Y)^*})$.

16.28 (Feder and Saphar, [FeSa]) Let X^{**} or Y^* be an RNP space. Then $\mathcal{K}(X, Y)^*$ is a quotient space of $Y^* \otimes_\pi X^{**}$. In particular, $\mathcal{K}(X, Y)^{**} \hookrightarrow \mathcal{B}(X^{**}, Y^{**})$. The embedding is canonical, i.e., $T \to T^{**}$.

Hint. Use the embedding $i : \mathcal{K}(X, Y) \to C((B_{X^{**}}, w^*) \times (B_{Y^*}, w^*))$ and follow the line of proof of Theorem 16.30.

16.29 (Godefroy and Saphar, [GoSa1]) Let X and Y be reflexive spaces. Then $\mathcal{K}(X, Y)^{**} = \overline{\mathcal{K}(X, Y)}^{\tau} \hookrightarrow \mathcal{B}(X, Y)$. In particular, if X or Y has the CAP, then $\mathcal{K}(X, Y)^{**} = \mathcal{B}(X, Y)$.
Hint. Let X^{**} or Y^* be an RNP space. Then the canonical embedding $T \to T^{**}$ represents $\mathcal{K}(X, Y)$ as a subspace of $\mathcal{B}(X^{**}, Y^{**}) = (X^{**} \otimes_\pi Y^*)^*$. Then $\mathcal{K}(X, Y)^{**} = \overline{\mathcal{K}(X, Y)}^{w^*} \hookrightarrow \mathcal{B}(X^{**}, Y^{**})$. Use Theorem 16.33 and Exercise 16.28.

16.30 (Feder and Saphar, [FeSa]) Let X be a separable reflexive space without the CAP. Then $\mathcal{K}(X)^{**} \neq \mathcal{B}(X)$.
Hint. By contradiction, assume that $\mathcal{K}(X)^{**} = \mathcal{B}(X)$. Then $\overline{B_{\mathcal{K}(X)}}^{w^*} = B_{\mathcal{B}(X)}$. Thus $\overline{B_{\mathcal{K}(X)}}^{WOT} = B_{\mathcal{B}(X)}$. Thus $\overline{B_{\mathcal{K}(X)}}^{SOT} = B_{\mathcal{B}(X)}$. The last statement can be rephrased as 1-bounded CAP.

16.31 If X is finite-dimensional, then $\mathcal{B}(X, Y)^{**} = \mathcal{B}(X, Y^{**})$.
Hint. Standard.

16.32 (Johnson, [JJohn]) Let X, Y be Banach spaces, such that Y has the BAP. Then there is an isomorphism $\mathcal{B}(X, Y) \hookrightarrow \mathcal{K}(X, Y)^{**}$, whose restriction to $\mathcal{K}(X, Y)$ is the identity.
Hint. Let (T_α) be a λ-bounded net from $\mathcal{F}(Y)$, that is τ-convergent to Id_Y. For every $\phi \in \mathcal{K}(X, Y)^*$, there is a subnet of α, such that $\phi \circ T_\alpha$ converges to ϕ when restricted to $\mathcal{K}(X, Y)$.

16.33 Assume that, for some Banach space X, the space $\mathcal{K}(X)$ admits an equivalent C^2-smooth norm. Show that then X is isomorphic to a Hilbert space.
Hint. Exercise 2.42 and Meshkov Theorem 11.7.

16.34 Is $\mathcal{K}(c_0)$ an Asplund space?
Hint. No. See Exercise 2.42.

Chapter 17
Appendix

In this short chapter we collect, for the reader's convenience, some basic definitions and results that are used in the book. A list of sources for them is provided at the end of each section here.

17.1 Basics in Topology

A *topology* on a set S is a family $\mathcal{T}$ of subsets (called *open sets*) of S with the following properties:

(i) $\emptyset \in \mathcal{T}$, $S \in \mathcal{T}$.

(ii) If $\{O_i : i \in I\}$ is a subfamily of $\mathcal{T}$, then $\bigcup_{i \in I} O_i \in \mathcal{T}$.

(iii) If $\{O_j : j \in J\}$ is a *finite* subfamily of $\mathcal{T}$, then $\bigcap_{j \in J} O_j \in \mathcal{T}$.

A *topological space* $(S, \mathcal{T})$ is a pair consisting of a set S and a topology $\mathcal{T}$ on S (we shall write just S for a topological space if the topology $\mathcal{T}$ is understood). A *neighborhood* of a point $x \in S$ is a set $U \subset S$ such that $x \in O \subset U$ for some $O \in \mathcal{T}$. A subset F of S is called *closed* if its complement $S \setminus F$ in S is open. The *interior* $\operatorname{Int} A$ of a set $A \subset S$ is the largest (in the sense of inclusion) open set contained in A, and its *closure* $\overline{A}$ is the smallest (again in the sense of inclusion) closed set containing A. The *boundary* ∂A of A is the set $\overline{A} \setminus \operatorname{Int} A$.

If $(S, \mathcal{T})$ is a topological space and Z is a nonempty subset of S, Z becomes a topological space when endowed with the topology $\{O \cap Z : O \in \mathcal{T}\}$. It is easy to check that this is indeed a topology. We shall denote this topology, if no confusion arises, again by $\mathcal{T}$.

The *discrete topology* on a nonempty set S consists of all subsets of S. Of course, in this topology every subset is, simultaneously, open and closed.

Let $\mathcal{R}$ be an *equivalence relation* in S, i.e., a set $\mathcal{R}$ of couples from S such that, for all $x, y, z \in S$, (i) $(x, x) \in \mathcal{R}$, (ii) if $(x, y) \in \mathcal{R}$, then $(y, x) \in \mathcal{R}$, and (iii) if $(x, y) \in \mathcal{R}$ and $(y, z) \in \mathcal{R}$, then $(x, z) \in \mathcal{R}$. If $(S, \mathcal{T})$ is a topological space, then the set $X/\mathcal{R}$ of all equivalence classes becomes a topological space (called a *quotient topological space*) when endowed with the topology $\widehat{\mathcal{T}} := \{O \subset X/\mathcal{R} : q^{-1}(O) \in \mathcal{T}\}$, where $q : S \to S/\mathcal{R}$ is the canonical quotient mapping, i.e., the mapping that associates to each $x \in S$ the class containing it.

M. Fabian et al., *Banach Space Theory*, CMS Books in Mathematics,
DOI 10.1007/978-1-4419-7515-7_17,

A *metric* (also called a *distance*) d on a set S is a mapping $d : S \times S \to \mathbb{R}$ such that

(i) $d(x, y) \geq 0$ for all $x, y \in S$.
(ii) $d(x, y) = 0$ if and only if $x = y$, $x, y \in S$.
(iii) $d(x, y) = d(y, x)$ for all $x, y \in S$.
(iv) $d(x, z) \leq d(x, y) + d(y, z)$ for all $x, y, z \in S$.

A *metric space* is a pair (S, d), where S is a set and d is a metric on S.

If (S, d) is a metric space, $x_0 \in S$ and $r > 0$, the *closed (open) ball* centered at x_0 of radius r is the set $B(x_0, r) := \{x \in S : d(x, x_0) \leq r\}$ ($B^O(x_0, r) := \{x \in S : d(x, x_0) < r\}$). A metric space is a particular instance of a topological space. The topology $\mathcal{T}$ consists of all sets $O \subset X$ such that for every $x \in O$ there exists $r > 0$ such that $B(x, r) \subset O$. A sequence $\{x_n\}$ in S *converges* to $x \in S$ in the metric d (we also say that x *is the limit of the sequence* $\{x_n\}$, and we denote this fact by $x_n \to x$ or by $\lim_n x_n = x$) if for every $\varepsilon > 0$ there exists $n_0 \in \mathbb{N}$ such that $x_n \in B(x, \varepsilon)$ for all $n \geq n_0$. All topological notions in a metric space (S, d) can be described by using convergence of sequences. For example, a set $A \subset S$ is closed if and only if, for every sequence in A that converges to some $x \in S$, then $x \in A$.

If $(X, \|\cdot\|)$ is a normed space, the function $\rho(x, y) := \|x - y\|$ in Definition 1.1 is indeed a metric on X. To check the triangle inequality, we write

$$\rho(x, z) = \|x - z\| = \|x - y + y - z\| \leq \|x - y\| + \|y - z\| = \rho(x, y) + \rho(y, z).$$

In this way, a normed space $(X, \|\cdot\|)$ becomes a metric space if endowed with ρ (called the *canonical metric associated to* $\|\cdot\|$). All topological notions in a normed space refer to this metric, unless stated otherwise.

A topological space $(S, \mathcal{T})$ is called *Hausdorff* if given any two different points x and y in S there exist neighborhoods $U(x)$ of x and $U(y)$ of y such that $U(x) \cap U(y) = \emptyset$. It is called *regular* if it is Hausdorff and, for each x in S and each closed subset F of S such that $x \notin F$, there are open sets O_1 and O_2 in S such that $O_1 \cap O_2 = \emptyset$, $x \in O_1$, and $F \subset O_2$. The space $(S, \mathcal{T})$ is called *completely regular* if it is Hausdorff and for each $x \in S$ and each closed subset F of S such that $x \notin F$, there exists a continuous function $\varphi : S \to [0, 1]$ such that $\varphi(x) = 1$ and $\varphi(y) = 0$ for all $y \in F$. A space $(S, \mathcal{T})$ is called *normal* if it is Hausdorff and for any two disjoint closed subsets F_1 and F_2 of S, there exist disjoint open subsets A_1 and A_2 of S such that $F_i \subset A_i$ for $i = 1, 2$. A topological space $(S, \mathcal{T})$ is said to be *compact* if it is Hausdorff and every covering of S by open sets has a finite subcovering. A space is *locally compact* if it is Hausdorff and every point has a compact neighborhood.

For any locally compact Hausdorff topological space X, the *(Alexandroff) one-point compactification* of X is obtained by adding one extra point (often called a *point at infinity* and denoted by ∞) and defining the open sets of $X \cup \{\infty\}$ to be the open sets of X together with the sets of the form $G \cup \{\infty\}$, where G is a subset of X such that $X \backslash G$ is compact. The one-point compactification is a compact Hausdorff space.

If X is a completely regular topological space, the space $CB(X)$ of all bounded continuous real functions on X, endowed with the supremum norm, is a Banach space. Naturally, X is homeomorphic to a subset (denoted again X) of $(B_{CB(X)^*}, w^*)$. The closure βX of X in this space is called the *Stone–Čech compactification of* X. Every bounded, continuous real-valued function on X can be extended to a continuous real-valued function on βX. The spaces $C(\beta X)$ and $CB(X)$ in their supremum norm are linearly isometric.

A family $\mathcal{F}$ of subsets of a certain nonempty set S is called a Σ*-algebra* if $\mathcal{F}$ contains the empty set and it is closed under the two operations of taking complements and taking countable unions. If $(S, \mathcal{T})$ is a topological space, the family of *Borel sets* is the smallest Σ-algebra containing $\mathcal{T}$.

For information on this area see, e.g., [Kell], [Enge], [Dugu2], and [Jmsn].

17.2 Nets and Filters

Properties of a general topological space cannot always be described by using only sequences (see the example after Definition 3.3). The more general concept of net is needed for dealing with convergence. Another possibility is to use filters. We will describe here the required definitions and some simple facts about nets and filters.

Definition 17.1 *By a* directed set *we mean any nonempty preordered set I directed upwards, that is, a nonempty set I with a binary relation $\leq$ on I satisfying for all $\alpha, \beta, \gamma \in I$:*
(N1) $\alpha \leq \alpha$,
(N2) *if* $\alpha \leq \beta$ *and* $\beta \leq \gamma$, *then* $\alpha \leq \gamma$,
(N3) *for every α and β in I there exists γ in I such that $\alpha \leq \gamma$ and $\beta \leq \gamma$.*

Note that we do not assume that two arbitrary members of I are always related.

Definition 17.2 *A* net *in a nonempty set X is a mapping N from a directed set I into X. Instead of $N(\alpha)$ we will often write x_α, and denote the net as $\{x_\alpha\}_{\alpha \in I}$. Let $\{x_\alpha\}_{\alpha \in I}$ be a net. Let J be a directed set and $S\colon J \to I$ be a mapping with the following property: given $\alpha_0 \in I$, there exists $\beta_0 \in J$ such that $\alpha_0 \leq S(\beta)$ whenever $\beta_0 \leq \beta$. Then the net $\{x_{S(\beta)}\}_{\beta \in J}$ is called a* subnet *of the net $\{x_\alpha\}_{\alpha \in I}$.*

A *sequence* is a net where the directed set is $\mathbb{N}$ endowed with its natural order.

Definition 17.3 *A* filter *$\mathcal{F}$ on a nonempty set S is a nonempty family of subsets of S with the three following properties:*
(F1) $\emptyset \notin \mathcal{F}$.
(F2) *If* $F_1, F_2 \in \mathcal{F}$ *then* $F_1 \cap F_2 \in \mathcal{F}$.
(F3) *If* $F \subset G \subset S$ *and* $F \in \mathcal{F}$ *then* $G \in \mathcal{F}$.
A filter base *$\mathcal{B}$ on S is a nonempty family of subsets of S such that*
(FB1) *the empty set does not belong to $\mathcal{B}$ and*
(FB2) *the intersection of two elements of $\mathcal{B}$ contains an element of $\mathcal{B}$.*
The family of all supersets of a filter base is a filter (called the filter generated by *$\mathcal{B}$).*
A filter subbase *is a nonempty family $\mathcal{S}$ of subsets of S such that*

(FS1) *every finite intersection of its members is nonempty.*
The family of all these finite intersections is a filter base (that generates a filter also called the filter generated by $\mathcal{S}$*).*

Some simple filters in a nonempty set S can be easily described: Fix a nonempty subset C of S. The family $\mathcal{F}(C)$ of all subsets of S containing C is clearly a filter, called the *principal* (or *fixed*) *filter generated by* C. A filter $\mathcal{F}$ is *free* if $\bigcap_{F\in\mathcal{F}} F = \emptyset$. The *Fréchet filter* $\mathcal{FF}$ on an infinite set S is the family of all subsets of S that have finite complement. The Fréchet filter is free (hence non-principal), and it is contained in every free filter on S.

An *ultrafilter* on S is a filter that is maximal (in the sense of inclusion) in the family of all filters on S. Given a filter on S, it is always included in an ultrafilter (just use Zorn's lemma). There is a simple characterization of ultrafilters among filters: a filter $\mathcal{F}$ on a set S is an ultrafilter if and only if given an arbitrary subset S_0 of S we have only two (exclusive) possibilities: either $S_0 \in \mathcal{F}$ or $S\backslash S_0 \in \mathcal{F}$.

Given an element $s \in S$, $\mathcal{F}(\{s\})$ is seen to be a (*principal*) ultrafilter by looking at the previous characterization. If S is infinite, there are non-principal ultrafilters (also called *free* ultrafilters). Take, for example, the (non-principal) filter $\mathcal{FF}$ and an ultrafilter containing $\mathcal{FF}$.

For information on this area see, e.g., [Kell], [Enge], and [Dugu2].

17.3 Nets and Filters in Topological Spaces

Assume that X is a topological space. We will describe convergence of nets defined in X.

Definition 17.4 *We say that a net* $\{x_\alpha\}_{\alpha\in I}$ *in a topological space* (X,τ) converges *to some point* $x \in X$ *if for every neighborhood* $U(x)$ *of* x *there exists* $\alpha_0 \in I$ *such that* $x_\alpha \in U(x)$ *whenever* $\alpha_0 \leq \alpha$. *We then say that* x *is the* limit *of* $\{x_\alpha\}_{\alpha\in I}$. *We shall use indistinctly any of the following notation for this fact:* $x_\alpha \overset{\tau}{\rightarrow}_\alpha x$, $x_\alpha \overset{\tau}{\rightarrow} x$, $\tau\text{-}\lim_{\alpha\in I} x_\alpha = x$ *or simply* $\tau\text{-}\lim x_\alpha = x$, *dropping the reference to* τ *if no misunderstanding arises about the topology involved.*
We say that $x \in X$ *is a* cluster point *of a net* $\{x_\alpha\}_{\alpha\in I}$ *if for every neighborhood* $U(x)$ *of* x *and* $\alpha_0 \in I$ *there exists* $\alpha \in I$ *such that* $\alpha_0 \leq \alpha$ *and* $x_\alpha \in U(x)$.

Note that if $x_\alpha \rightarrow x$, then every subnet of $\{x_\alpha\}$ also converges to x. A net converges to x if and only if every subnet has x as a cluster point. Note, also, that x is a cluster point of a net $\{x_\alpha\}_{\alpha\in I}$ if and only if there is a subnet of $\{x_\alpha\}_{\alpha\in I}$ that converges to x.

The topology of a topological space X can be specified by describing the convergence of nets. Hence, all topological concepts can be defined in terms of nets. For example, given a subset A of a topological space X, its closure $\overline{A}$ is the set of all limits of nets in A that converge in X. Thus a set A is closed if and only if it contains the limits of all convergent nets with elements in A.

Note that in a metric space (S,d) it is enough to consider sequences for describing the topology. Indeed, it is enough to prove that an arbitrary closed subset A of

S can be characterized by sequences in the following way: *A is closed if and only if for every sequence $\{a_n\}_{n\in\mathbb{N}}$ in A that converges to some $x \in S$, then $x \in A$.* Indeed, since a sequence is a particular case of a net, the condition is necessary. Assume now that A is not closed. Then its complement C is not open, so there exists $c \in C$ such that $B(c, 1/n) \cap A \neq \emptyset$ for each $n \in \mathbb{N}$. Choose $a_n \in B(c, 1/n) \cap A$ for $n \in \mathbb{N}$. Then we have $a_n \to c$ and $c \notin A$.

If S is a topological space, the family of all neighborhoods of a point is a filter. A filter $\mathcal{F}$ on a topological space S is said to *converge to an element $s \in S$* if it contains the filter of all neighborhoods of s. All topological concepts described by using nets can be equivalently described by using (ultra)filters, and conversely. For example, a topological space K is compact if and only if every ultrafilter on K converges to some element in K.

Nets and filters are closely related. For example, given a net $\{s_i\}_{i\in I}$ in a set S, the family $\mathcal{B} := \{S_i : i \in I\}$, where $S_i := \{s_j : i \leq j\}$, $i \in I$, is a filter basis. The corresponding filter is called the *filter generated by the net* $\{s_i\}_{i\in I}$. Conversely, observe that a filter $\mathcal{F}$ is a directed set (ordered by the reverse set inclusion). The nets obtained by choosing an element in each of the sets F of the filter $\mathcal{F}$ are called *nets corresponding to the filter $\mathcal{F}$*.

For more information on this area we refer to, e.g., [Kell], [Enge], and [Dugu2].

17.4 Ultraproducts

Let $\mathcal{X} = \{(X_i, \|\cdot\|_{X_i}) : i \in I\}$ be a family of Banach spaces, and $\mathcal{U}$ be an ultrafilter on the set I. Equip $\prod_{i\in I} X_i$ with the natural vector structure, and consider the subspace $E := \{(x_i)_{i\in I} : \sup_{i\in I} \|x_i\|_{X_i} < \infty\}$. On E we can define a seminorm $p_{\mathcal{U}}$ by $p_{\mathcal{U}}((x_i)) = \lim_{\mathcal{U}} \|x_i\|_{X_i}$. An equivalence relation on E is defined by $(x_i) \sim (y_i) \Leftrightarrow p_{\mathcal{U}}((x_i - y_i)) = 0$. The quotient space $E/\sim$ becomes a normed space (called the *$\mathcal{U}$-ultraproduct of the family $\mathcal{X}$*, and denoted by $\prod_{\mathcal{U}} X_i$) when endowed with the norm (denoted $\|\cdot\|_{\mathcal{U}}$) induced by the seminorm $p_{\mathcal{U}}$. The equivalence class where an element $(x_i) \in \prod_{\mathcal{U}} X_i$ belongs is denoted by $[(x_i)]$. The space $(\prod_{\mathcal{U}} X_i, \|\cdot\|_{\mathcal{U}})$ is, in fact, a Banach space.

When all elements of the family $\mathcal{X}$ coincide with a given Banach space $(X, \|\cdot\|)$, the ultraproduct $\prod_{\mathcal{U}} X_i$ is called an *ultrapower* of $(X, \|\cdot\|)$, and is denoted by $X_{\mathcal{U}}$.

Every Banach space embeds (linearly and isometrically) in any of its ultrapowers via the "diagonal" mapping $D : X \to X_{\mathcal{U}}$ given by $D(x) = [(x_i)]$, where $x_i = x$ for all $i \in I$.

We refer to, e.g., [Heinr] and [Pisi3] and references therein for more on this topic.

17.5 The Order Topology on the Ordinals

Let Γ be an ordinal number. It can be identified with the segment $[0, \Gamma)$ of all the ordinal numbers that are greater than or equal to 0 and less than Γ. This set can be endowed with a topology $\mathcal{O}$, called the *order topology* of the segment $[0, \Gamma)$.

Precisely, the topology $\mathcal{O}$ is generated by the family of all the sets $\{x \in [0, \Gamma) : x < \alpha\}$ (denoted by $\{x < \alpha\}$) and $\{x \in [0, \Gamma) : x > \beta\}$ (denoted by $\{x > \beta\}$), where $\alpha, \beta \in [0, \Gamma]$ (i.e., a set is open in $\mathcal{O}$ whenever it is a union of sets of the former family).

A base of the topology is given by the family of all sets $(\alpha, \beta]$, where $0 \le \alpha < \beta < \Gamma$. In particular, a base of neighborhoods of any $\beta \in [0, \Gamma)$ is given by the family $\{(\alpha, \beta] : \alpha < \beta\}$. It follows then that a net $\{\alpha_i\}$ of elements in $[0, \Gamma)$ converges to some element $\beta \in [0, \Gamma)$ if and only if for every $\alpha < \beta$ there exists i_0 such that $\alpha < \alpha_i \le \beta$ for all $i \ge i_0$.

The space $([0, \Gamma), \mathcal{O})$ is Hausdorff (indeed, given $\alpha < \beta$ in $[0, \Gamma)$, (i) if $\beta = \alpha + 1$, then $\{x < \beta\}$ and $\{x > \alpha\}$ are disjoint open sets that separate α and β; (ii) if there exists γ such that $\alpha < \gamma < \beta$, then $\{x < \gamma\}$ and $\{x > \gamma\}$ are again two disjoint open sets that separate α and β. Indeed, the space is even normal. It is totally disconnected, i.e., the only connected components are the points (use an argument similar to the previous one), zero-dimensional (it has a base of clopen, i.e., simultaneously open and closed) subsets of $[0, \Gamma)$; indeed, $(\alpha, \beta] = (\alpha, \beta + 1) = [\alpha + 1, \beta]$), and it is scattered (i.e., every nonempty subset S has an isolated point; just take the first element in S).

17.6 Continuity of Set-Valued Mappings

Given a function $f : S \to T$, where S and T are topological spaces, and a filter $\mathcal{F}$ on S, we say that $t \in T$ is the *limit of f along the filter $\mathcal{F}$* if the family $\{f(F) : F \in \mathcal{F}\}$ (a filter basis) generates a filter on T that converges to t.

If S and T are topological spaces, a function $f : S \to T$ is *continuous at $s \in S$* if, for every net $\{s_i\}_{i \in I}$ in S that converges to s, the net $\{f(s_i)\}_{i \in I}$ converges to $f(s)$. This is equivalent to say that for every filter $\mathcal{F}$ in S that converges to s, then $f(s)$ is the limit of f along the filter $\mathcal{F}$.

The function f is said to be *continuous* if it is continuous at every point of S.

Theorem 17.5 (Tietze–Urysohn) *Let $(S, \mathcal{T})$ be a normal topological space. Then, given a point $x_0 \in S$ and an open neighborhood $U(x_0)$ of x_0, there exists a continuous function $f : S \to [0, 1]$ such that $f(x_0) = 1$ and $f(x) = 0$ for all $x \in S \setminus U(x_0)$.*

For paracompact spaces, this is a consequence of Corollary 7.55. Note that every paracompact space is normal (see, e.g., [Enge, Theorem 5.1.5]).

Definition 17.6 *A real-valued function f defined on a topological space X is* lower (upper) semicontinuous *if for every $x_0 \in X$ and every real number r satisfying the inequality $f(x_0) > r$ (the inequality $f(x_0) < r$) there exists a neighborhood $U \subset X$ of x such that $f(x) > r$ (that $f(x) < r$) for every $x \in U$.*

A real-valued function defined on a topological space X is lower (upper) semicontinuous if and only if for every real number r the set $\{x \in X : f(x) \le r\}$ (the set $\{x \in X : f(x) \ge r\}$) is closed, if and only if the epigraph (the subgraph) of f is closed. The definition extends to functions $f : X \to \mathbb{R} \cup \{+\infty\}$ in the same terms, and the characterizations applies as well.

If f and g are lower (upper) semicontinuous, then $\min(f, g)$, $\max(f, g)$ and $f+g$ are lower (upper) semicontinuous, as well as $f.g$ provided that $f(x) \geq 0$ and $g(x) \geq 0$ for all $x \in X$. If f is lower (upper) semicontinuous then $-f$ is upper (lower) semicontinuous. Given a family $\{f_i\}_{i\in I}$ of lower (upper) semicontinuous functions, the function $\sup_{i\in I} f_i$ (the function $\inf_{i\in I} f_i$) is lower (upper) semicontinuous.

Observe that a lower semicontinuous real-valued function defined on a compact topological space K is bounded below and attains its infimum: the boundedness assertion follows from the fact that a sequence $\{x_n\}$ in K such that $f(x_n) \leq -n$ for all $n \in \mathbb{N}$ has a cluster point x_0, and that $f(x_0) \leq -n$ for all $n \in \mathbb{N}$, something impossible. If m is the infimum on K, f attains the infimum at a cluster point of a sequence $\{x_n\}$ in K such that $f(x_n) \leq m + 1/n$ for all $n \in \mathbb{N}$.

Definition 17.7 *A set-valued mapping Φ from a topological space X into a topological space Y is said to be* lower semicontinuous at a point $x_0 \in X$ *if, for every open subset V of Y that intersects $\Phi(x_0)$, there exists a neighborhood U of x_0 such that $\Phi(x) \cap V \neq \emptyset$ for all $x \in U$. The mapping Φ is said to be* upper semicontinuous at x_0 *if, for every open subset V of Y such that $\Phi(x_0) \subset V$, there exists a neighborhood U of x_0 such that $\Phi(x) \subset V$ for all $x \in U$. The mapping Φ is said to be* lower *(respectively* upper*) semicontinuous at X if it is lower (respectively upper) semicontinuous at every point of X.*

Another equivalent way to formulate the lower (upper) semicontinuity of a set-valued mapping Φ from X into Y is to say that *for every open subset G of Y, the set $\{x \in X : \Phi(x) \cap G \neq \emptyset\}$ (respectively, the set $\{x \in X : \Phi(x) \subset G\}$) is open in X.*

For information on this area see, e.g., [Enge], [vMill], and [Dugu2].

17.7 The Cantor Space

In this section, the symbol 2 stands for the set $\{0, 1\}$ endowed with the discrete topology, while $[0, 1]$ carries the restriction of the usual topology of $\mathbb{R}$.

Definition 17.8 *The* Cantor space*, denoted by Δ, is the space $2^{\mathbb{N}}$ endowed with the product topology $\mathcal{T}_p$.*

The space Δ is compact. It is easy to see that Δ is zero-dimensional. Since it is a countable product of metrizable spaces, it is itself metrizable. A metric d generating its topology is given explicitly by the formula $d(a, b) = \sum_{n=1}^{\infty} 2^{-n}|a_n - b_n|$, where $a := (a_n)$ and $b := (b_n)$ are elements in Δ. The Cantor space is a perfect (i.e., without isolated points) space. It can be embedded homeomorphically in every uncountable separable complete metric space M—more generally, in the closure of every uncountable subset of M. It can be embedded homeomorphically in every uncountable G_δ subset of a separable complete metric space (Kuratowski). It can be embedded homeomorphically in every perfect complete metric space. The space

Δ is homeomorphic to Δ^n for every $n \in \mathbb{N}$, even to $\Delta^{\mathbb{N}}$ (see the proof of Theorem 17.11).

Proposition 17.9 *The Cantor space is homeomorphic to the subspace of* [0, 1], *called the* Cantor ternary set, *consisting of all numbers having a ternary expansion that uses only digits* 0 *and* 2.

Proof: Consider the mapping $\varphi : \Delta \to [0,1]$ given by $\varphi(\alpha) = 2\sum_{n=1}^{\infty} \frac{\alpha_n}{3^n}$, for $\alpha = (\alpha_n) \in \Delta$. It is easy to check that φ is, indeed, a homeomorphism from Δ onto the Cantor ternary set. □

Recall (Exercise 12.15) that a subset R of a topological space S is called a *retract* if there exists a mapping $r : S \to R$ that is a *retraction*, i.e., $r(x) = x$ for every $x \in R$. If the retraction is continuous, the set R is called a *continuous retract.*

Proposition 17.10 *Every nonempty closed subspace of Δ is a continuous retract.*

Proof: Let A be a nonempty closed subspace of Δ. We shall define the retraction r in the following way. Let $x = (x_n) \in \Delta$. If there exists $a \in A$ such that $x_1 = a_1$, then put $r(x)_1 = x_1$; otherwise, put $r(x)_1 = 1 - x_1$. Assume that for some $n \in \mathbb{N}$ the coordinates $r(x)_i$, $i = 1, 2, \ldots, n$ have been already defined. If there exists $a \in A$ such that $a_i = x_i$ for $i = 1, 2, \ldots, n+1$, put $r(x)_{n+1} = x_{n+1}$; otherwise put $r(x)_{n+1} = 1 - x_{n+1}$.

This defines a mapping $r : \Delta \to \Delta$. Given $x \in \Delta$, there exists by construction a sequence of elements in A that converges to $r(x)$. Since A is closed, $r(x) \in A$. Obviously, $r(a) = a$ for every $a \in A$. It is clear, too, that r is continuous. □

One of the most striking (and most useful) properties of Δ is given in the following result.

Theorem 17.11 (Alexandrov and Urysohn) *Every compact metrizable space is a continuous image of Δ.*

Proof: Let K be a compact metrizable space. By Lemma 3.102, the space $C(K)$ is separable. Let $D \subset B_{C(K)}$ be a countable dense subset. The space K is homeomorphic to the subset $\{\delta_k;\ k \in K\}$ of the space $(B_{C(K)^*}, w^*)$ (see the proof of Lemma 3.102). By evaluating each δ_k on the elements of D we can see that K is indeed homeomorphic to a (closed) subset (called again K) of $[-1, 1]^{\mathbb{N}}$, hence to a subset of $[0, 1]^{\mathbb{N}}$.

There is a continuous function g mapping Δ *onto* [0, 1] (a simple description of such a mapping can be provided readily: if $a := (a_n) \in \Delta$, put $g(a) = \sum_{k=1}^{\infty} a_k 2^k$). Therefore, $G : \Delta^{\mathbb{N}} \to [0, 1]^{\mathbb{N}}$ given by $G((c^n)) = (g(c^n))$ if $(c^n) \in \Delta^{\mathbb{N}}$, maps continuously $\Delta^{\mathbb{N}}$ *onto* $[0, 1]^{\mathbb{N}}$. The spaces Δ and $\Delta^{\mathbb{N}}$ are homeomorphic. A simple way to see this is to write $a := (a_n) \in \Delta$ as an infinite two-dimensional matrix. Let $h : \Delta \to \Delta^{\mathbb{N}}$ be a homeomorphism. Then $f := G \circ h$ maps homeomorphically Δ onto $[0, 1]^{\mathbb{N}}$. Since $f^{-1}(K)$ is a closed subspace of Δ, there is, according to Lemma 17.10 a continuous retraction r from Δ onto $f^{-1}(K)$. Finally we get $K = (f \circ r)(\Delta)$. □

For more on this subject see, e.g., [Kech].

17.8 Baire's Great Theorem

Theorem 17.12 (Baire's great theorem, see, e.g., [DGZ3, Theorem I.4.1]) *Let Z be a complete metric space, X be a normed space, and $f : Z \to X$ be a function. Then the following are equivalent.*
(i) *For every nonempty closed subset F of Z, $f|_Z$ has a point of continuity.*
(ii) *The mapping f is the pointwise limit of a sequence $\{f_n\}$ of continuous mapping from Z into X.*

For information on this area see, e.g., [DGZ3].

17.9 Polish Spaces

A topological space $(S, \mathcal{T})$ is *Polish* if it is separable and completely metrizable, i.e., if it is separable and homeomorphic to a complete metric space.

Theorem 17.13 (Alexandrov) *If X is Polish then so is any G_δ subset of X.*

The converse of Alexandrov's theorem is true as well: if a subspace S of a Polish space X is Polish, then it is a G_δ subset of X.

Theorem 17.14 (Cantor–Bendixson) *If X is Polish then any closed subset of X can be written as the disjoint union of a perfect subset and a countable subset.*

Proposition 17.15 *Every Polish space is a continuous image of $\mathbb{N}^{\mathbb{N}}$.*

The following result will be used in the proof of Lemma 5.51.

Theorem 17.16 (see, e.g., [Kech, p. 83]) *Let (P, ρ) be a Polish space, and $\{U_n\}_{n=1}^{\infty}$ be a collection of Borel subsets of P. Then there is a metric $\widetilde{\rho}$ which turns $(P, \widetilde{\rho})$ into a Polish space such that* (i) *the topology defined by $\widetilde{\rho}$ on P is stronger that the topology defined by ρ,* (ii) *the spaces (P, ρ) and $(P, \widetilde{\rho})$ have the same family of Borel sets, and* (iii) *all sets U_n, $n \in \mathbb{N}$ are clopen in $(P, \widetilde{\rho})$. The metric $\widetilde{\rho}$ can even be chosen so that $(P, \widetilde{\rho})$ is zero-dimensional.*

For information on this area see, e.g., [Kech].

17.10 Uniform Spaces

Let S be a nonempty set. Given a subset U of $S \times S$, denote $U^{-1} := \{(y, x) : (x, y) \in U\}$, and if V is another subset of $S \times S$, let $VU = \{(x, z) :$ there exists $y \in S$ such that $(x, y) \in V,\ (y, z) \in U\}$. Put $U^2 = UU$.

A *uniformity* $\mathcal{U}$ in S is a filter of subsets (called *vicinities*) of $S \times S$ that satisfy
(U1) $\Delta \subset U$ for all $U \in \mathcal{U}$, where $\Delta := \{(x, x) : x \in S\}$.
(U2) If $U \in \mathcal{U}$ then $U^{-1} \in \mathcal{U}$.
(U2) For each $U \in \mathcal{U}$ there exists $V \in \mathcal{U}$ such that $V^2 \subset U$.

The pair $(S, \mathcal{U})$ is called a *uniform space*. Every uniform space $(S, \mathcal{U})$ becomes a topological space. It is enough to describe the family $\mathcal{U}(x)$ of neighborhoods of any

$x \in S$; precisely $\mathcal{U}(x) := \{U(x) : U \in \mathcal{U}\}$, where $U(x) := \{y \in S : (y, x) \in U\}$ if $U \in \mathcal{U}$. The topology so defined is called the *topology associated to the uniformity*.

Every metric space (S, d) is a uniform space. Indeed, it suffices to consider the family $\mathcal{U}$ of all supersets of sets of the form $\{(x, y) \in S \times S : d(x, y) < 1/n\}, n \in \mathbb{N}$. Then $\mathcal{U}$ is a uniformity on S whose associated topology is the topology defined by the metric. Every topological vector space $(E, \mathcal{T})$ has a unique translation-invariant uniformity defined on E whose associated topology is $\mathcal{T}$. Precisely the uniformity $\mathcal{U}$ is defined this way: $U \subset E \times E$ belongs to $\mathcal{U}$ if and only if $U = \{(x, y) \in E \times E : x - y \in B\}$, where B is a neighborhood of 0.

If M is a subset of a uniform space $(S, \mathcal{U})$ and $U \in \mathcal{U}$, M is said to be *U-small* if $M \times M \subset U$. The set M is *totally bounded* if, for every $U \in \mathcal{U}$, there exists a finite covering of M with U-small sets.

A function f from a uniform space $(T, \mathcal{U})$ into another uniform space $(S, \mathcal{V})$ is called *uniformly continuous* if, for every $V \in \mathcal{V}$, there exists $U \in \mathcal{U}$ such that $\{(f(t_1), f(t_2)) : (t_1, t_2) \in U\} \subset V$. Obviously, every uniformly continuous function is continuous when T and S are endowed with their associated topologies.

A family of functions $\mathcal{F}$ from a topological space $(T, \mathcal{T})$ into a uniform space $(S, \mathcal{V})$ is said to be *equicontinuous at $t_0 \in T$* if, given a vicinity $V \in \mathcal{V}$, there exists a neighborhood $U(t_0)$ of t_0 in T such that $\{(f(t), f(t_0)) : t \in U(t_0), f \in \mathcal{F}\} \subset V$. The family is called *equicontinuous* if it is equicontinuous at every point $t \in T$. If $(T, \mathcal{T})$ is, moreover, a uniform space and $\mathcal{U}$ is its system of vicinities, the family $\mathcal{F}$ is said to be *uniformly equicontinuous* if for each vicinity $V \in \mathcal{V}$ there exists a vicinity $U \in \mathcal{U}$ such that $\{(f(t_1), f(t_2)) : (t_1, t_2) \in U, f \in \mathcal{F}\} \subset V$.

For information on this area see, e.g., [Kell], [Dugu2], and [Enge].

17.11 Nets and Filters in Uniform Spaces

A net $\{x_\alpha\}_{\alpha \in I}$ in a uniform space $(S, \mathcal{U})$ is said to be *Cauchy* if, given $U \in \mathcal{U}$, there exists $\alpha_0 \in I$ such that $(x_\alpha, x_\beta) \in U$ for all $\alpha, \beta \in I, \alpha_0 \leq \alpha, \alpha_0 \leq \beta$, i.e., the set $\{x_\alpha : \alpha_0 \leq \alpha\}$ is U-small.

A filter $\mathcal{F}$ on a uniform space $(S, \mathcal{U})$ is said to be *Cauchy* if for every $U \in U$ there exists $F \in \mathcal{F}$ such that F is U-small.

A uniform space $(S, \mathcal{U})$ is said to be *complete* if every Cauchy net in it (or, alternatively, if every Cauchy filter on it) converges. It is easy to prove that *if a uniform space $(S, \mathcal{U})$ is complete then every totally bounded and closed subset of S is compact, and that the converse trivially holds in metric spaces.*

The following simple result holds:

Proposition 17.17 *Let $(S, \mathcal{U})$ be a uniform space, and let $\mathcal{T}$ be its associated topology. Assume that a second (Hausdorff) topology $\mathcal{T}'$ is given on S having the two following properties:* (i) *$\mathcal{U}$ has a base of vicinities that are closed in $(S, \mathcal{T}') \times (S, \mathcal{T}')$, and* (ii) *$\mathcal{T}'$ is coarser than $\mathcal{T}$. Then a net $\{s_\alpha\}$ (a filter $\mathcal{F}$) in S is $\mathcal{T}$-convergent to some $s_0 \in S$ whenever is $\mathcal{U}$-Cauchy and $\mathcal{T}'$-converges to s_0.*

Proof: Given a $\mathcal{T}'$-closed vicinity $U \in \mathcal{U}$, there exists α_0 such that $(s_\alpha, s_\beta) \in U$ for every $\alpha_0 \le \alpha, \beta$. Taking $\mathcal{T}'$-limit with respect to β we get $(s_\alpha, s_0) \in U$ for every α such that $\alpha_0 \le \alpha$. This proves that the net $\{s_\alpha\}$ is $\mathcal{T}$-convergent to s_0. The proof for filters is similar. □

Note that a uniformly continuous function from a uniform space into another takes Cauchy nets (filters) into Cauchy nets (respectively, filters).

A normed space $(X, \|\cdot\|)$ is an example of a uniform space. The description of a Cauchy sequence in X adopts the following more familiar aspect: a sequence $\{x_n\}$ in X is Cauchy if, for every $\varepsilon > 0$ there exists n_0 such that $\|x_n - x_m\| < \varepsilon$ for every $n, m \ge n_0$.

For information on this area see, e.g., [Kell], [Dugu2], and [Enge].

17.12 Partitions of Unity

This section follows the presentation of this topic made in [BeLi, Appendix B].

Lemma 17.18 *Let $\mathcal{F}$ be a locally finite family of subsets of a topological space X. Then $\overline{\bigcup_{F\in\mathcal{F}} F} = \bigcup_{F\in\mathcal{F}} \overline{F}$.*

Proof: Given $F_0 \in \mathcal{F}$, we have $\overline{F_0} \subset \overline{\bigcup_{F\in\mathcal{F}} F}$, hence $\bigcup_{F\in\mathcal{F}} \overline{F} \subset \overline{\bigcup_{F\in\mathcal{F}} F}$. To prove the reverse inclusion, take $x \in \overline{\bigcup_{F\in\mathcal{F}} F}$; then x is the limit of a net $\{x_i\}$ in $\bigcup_{F\in\mathcal{F}} F$. There exists a neighborhood $U(x)$ of x such that $\mathcal{F}_0 := \{F \in \mathcal{F} : U(x)\cap F \neq \emptyset\}$ is finite. We may assume $x_i \in U(x)$ for all i, hence $x \in \overline{\bigcup_{F\in\mathcal{F}_0} F} = \bigcup_{F\in\mathcal{F}_0} \overline{F} \subset \bigcup_{F\in\mathcal{F}} \overline{F}$. □

Proposition 17.19 *Let X be a paracompact space and A, B be a pair of closed subsets of X. If for every $b \in B$ there exist open sets U_b such that $A \subset U_b$, $b \in V_b$ and $U_b \cap V_b = \emptyset$, then there also exist open sets U, V such that $A \subset U$, $B \subset V$, and $U \cap V = \emptyset$. In particular, every paracompact space is normal.*

Proof: The family $\{X\backslash B, V_b : b \in B\}$ is an open cover of X. Let $\mathcal{W}$ be a locally finite open refinement. Put $\mathcal{W}_0 = \{W \in \mathcal{W} : W \cap B \neq \emptyset\}$. Observe that $\overline{W} \cap A = \emptyset$ for every $W \in \mathcal{W}_0$. We have

$$B \subset \bigcup_{W\in\mathcal{W}_0} W \subset \overline{\bigcup_{W\in\mathcal{W}_0} W} = \bigcup_{W\in\mathcal{W}_0} \overline{W},$$

where the equality follows from Lemma 17.18. It is enough then to put $V = \bigcup_{W\in\mathcal{W}_0} W$ and $U = X \setminus \bigcup_{W\in\mathcal{W}_0} \overline{W}$. By letting A to be a singleton, we obtain that every paracompact space is regular. Normality follows now by applying the first part again. □

Lemma 17.20 *Let X be a paracompact topological space. Let $\mathcal{V}$ be an open cover of X. Then there exists a locally finite open refinement $\mathcal{W}$ with the property that for each $W \in \mathcal{W}$ there is a closed set $F_W \subset W$ such that $\bigcup_{W \in \mathcal{W}} F_W = X$.*

Proof: We may assume that $\mathcal{V}$ is already locally finite. The regularity of X (see Proposition 17.19) implies that each $x \in X$ has an open neighborhood U_x such that $\overline{U_x} \subset V$ for some $V \in \mathcal{V}$. This gives an open refinement $\{U_x : x \in X\}$ of $\mathcal{V}$. In turn, $\{U_x : x \in X\}$ has a locally finite refinement $\mathcal{U}$. By the construction, given $U \in \mathcal{U}$ there exists $V \in \mathcal{V}$ such that $\overline{U} \subset V_U$ for some $V_U \in \mathcal{V}$. Put $\mathcal{W} = \{V_U : U \in \mathcal{U}\}$; since $\mathcal{W} \subset \mathcal{V}$, it follows that $\mathcal{W}$ is locally finite, too. For each $W \in \mathcal{W}$, put $F_W = \bigcup\{\overline{U} : U \in \mathcal{U}, V_U = W\}$. Due to the fact that $\mathcal{U}$ is locally finite, F_W is closed (see Lemma 17.18). □

Now we are ready to prove the main result of this section.

Theorem 17.21 *Let X be a paracompact topological space and let $\mathcal{V}$ be an open cover of X. Then there exists a locally finite partition of unity subordinated to $\mathcal{V}$.*

Proof: Let $\mathcal{W}$ be the locally finite open refinement constructed in Lemma 17.20. By Urysohn's lemma we can find, for each $W \in \mathcal{W}$, a continuous function $g_W : X \to [0, 1]$ such that $g_W(x) = 1$ for $x \in F_W$ and $g_W(x) = 0$ for $x \notin W$. Since $\mathcal{W}$ is a locally finite cover, the function $g := \sum_{W \in \mathcal{W}} g_W$ is well defined, strictly positive, and continuous on X. Put $f_W = g_W/g$ for $W \in \mathcal{W}$. The required partition of unity is $\{f_W : W \in \mathcal{W}\}$. □

For information on this area see, e.g., [BeLi], [Dugu2], and [Enge].

17.13 Measure and Integral

In this section we follow [BeLi, Appendix D].

17.13.1 Measure

A *measurable space* is a couple (Ω, Σ), where Ω is a nonempty set and Σ is a σ-algebra of subsets of Ω. A (extended-valued) *measure* τ on Ω is a $[0, +\infty]$-valued function on a σ-algebra Σ of subsets of Ω that has the value 0 on $\emptyset$ and is *countably additive*, i.e., given a sequence $\{S_n\}$ of pairwise disjoint sets in Σ, then $\tau(\bigcup_{n=1}^{\infty} S_n) = \sum_{n=1}^{\infty} \tau(S_n)$. The triple (Ω, Σ, τ) is called a *measure space*. The measure τ is called σ-finite if $\Omega = \bigcup_{n=1}^{\infty} \Omega_n$, where $\Omega_n \in \Sigma$ and $\tau(\Omega_n) < \infty$ for all $n \in \mathbb{N}$.

Let T be a (Hausdorff) topological space, and let $\mathcal{B}$ be the σ-algebra of all Borel sets in T. Let τ be a measure on $\mathcal{B}$ (in this case, we say that τ is a *Borel measure*). The measure τ is called *inner regular* or *tight* if, for all $B \in \mathcal{B}$, $\tau(B)$ is the supremum of $\tau(K)$ for K a compact set contained in B. The measure τ is called *locally finite* if every point has a neighborhood of finite measure. The measure τ is called a *Radon measure* if it is inner regular and locally finite.

Let X be a Banach space, and (Ω, Σ) be a measurable space. A function $\mu : \Sigma \to X$ is called a *vector-valued measure with values in* X (or just an X*-valued measure*) if $\mu(\emptyset) = 0$ and μ is countably additive, i.e., given a pairwise disjoint sequence $\{S_n\}$ of sets in Σ, we must have $\mu(\bigcup_{n=1}^{\infty} S_n) = \sum_{n=1}^{\infty} \mu(S_n)$. (Observe that, for the consistency of this definition, the aforesaid series must converge unconditionally.) Again, the triple (Ω, Σ, μ) is called a *measure space*. The *variation* (sometimes also called *total variation*) of an X-valued measure μ is the nonnegative measure $|\mu|$ on Ω defined for $S \in \Sigma$ by $|\mu|(S) = \sup\{\sum \|\mu(S_i)\|\}$, where the supremum is taken over all *finite* partitions of S with $S_i \in \Sigma$. An X-valued measure μ is said to be *of bounded variation* if $|\mu|(\Omega) < \infty$.

A measure space (Ω, Σ, μ) is said to be *complete* if, for every set $N \in \Sigma$ with $\mu(N) = 0$, then the family of subsets of N is in Σ.

A measure μ of bounded variation is said to be *absolutely continuous* with respect to a scalar-valued σ-finite positive measure τ (and is denoted $\mu \prec \tau$) if $\mu(S) = 0$ whenever $S \in \Sigma$ and $\tau(S) = 0$. This is equivalent to say that for every $\varepsilon > 0$ there exists $\delta > 0$ such that $|\mu(A)| < \varepsilon$ whenever $A \in \Sigma$ and $\tau(A) < \delta$. In this direction see also Exercise 11.23.

17.13.2 Integral

Let (Ω, Σ, τ) be a measure space, where τ is a scalar-valued finite measure. A function $s : \Omega \to \mathbb{R}$ is called *simple* if $s = \sum_{i=1}^{n} a_i \chi_{S_i}$, where $a_i \in \mathbb{R}$, $S_i \in \Sigma$ for $i = 1, 2, \ldots, n$, and $n \in \mathbb{N}$. For $S \in \Sigma$, we define $\int_S s\,\mathrm{d}\tau = \sum_{i=1}^{n} a_i \tau(S_i \cap S)$. A function $f : \Omega \to \mathbb{R}$ is said to be *measurable* if it is the limit almost everywhere (a.e.) of a sequence $\{s_n\}$ of simple functions. We consider first a nonnegative measurable function $f : \Omega \to \mathbb{R}$. We say that f is *integrable* (with respect to the measure space (Ω, Σ, τ)) if there exists a nondecreasing sequence $\{s_n\}$ of nonnegative simple functions that converges to f (a.e.) and such that $\lim \int_\Omega s_n \mathrm{d}\tau < \infty$. In this case we define $\int_S f\,\mathrm{d}\tau = \lim_n \int_S s_n \mathrm{d}\tau$. It is routine to prove that the fact that f is integrable and the value of the integral does not depend on the particular sequence $\{s_n\}$ chosen. Now, if $f : \Omega \to \mathbb{R}$ is an arbitrary measurable function, we say that f is *integrable* if it can be written as a difference $f = u - v$ of two nonnegative integrable functions u, v. We define then $\int_S f\,\mathrm{d}\tau = \int_S u\,\mathrm{d}\tau - \int_S v\,\mathrm{d}\tau$. Again, the fact that f would be integrable and the value of the integral is independent of the particular representation $f = u - v$. The space of τ-integrable functions is denoted by $L_1(\tau)$ (rather, the space of classes of functions that are equal (a.e.)) and, equipped with the norm $\int_\Omega |f|\mathrm{d}\tau$, becomes a Banach space.

Let (Ω, Σ, μ) be a measure space, where μ is an X-valued measure with bounded variation and X is a Banach space. For $S \in \Sigma$ and a simple function $s : \Omega \to \mathbb{R}$ as above, we define $\int_S s\,\mathrm{d}\mu = \sum a_i \mu(S \cap S_i)$ $(\in X)$. We obviously have $\| \int_\Omega s\,\mathrm{d}\mu\| \leq \int_\Omega |s|\mathrm{d}|\mu|$. Thus, there exists an extension of the operator $s \mapsto \int_\Omega s\,\mathrm{d}\tau$ to the space of functions f such that $|f| \in L_1(|\mu|)$ (denoted $\mathcal{L}_1(\mu)$), such that

$\| \int_\Omega f \mathrm{d}\mu \| \leq \int_\Omega |f| \mathrm{d}|\mu|$. It defines an integral on the space $L_1(\mu)$. Again, $L_1(\mu)$ equipped with the norm $\int_\Omega |f| \mathrm{d}|\mu|$ is a Banach space.

We discuss now integration of vector-valued functions with respect to a scalar-valued measure. Let X be a Banach space and let (Ω, Σ, τ) be a complete measure space, where τ is a σ-finite scalar-valued measure. A *simple X-valued function* is a function $s : \Omega \to X$ that can be written as $s = \sum_{i=1}^n x_i \chi_{S_i}$, where $x_i \in X$, $S_i \in \Sigma$, for $i = 1, 2, \ldots, n$, and $n \in \mathbb{N}$. For $S \in \Sigma$ we define $\int_S s \mathrm{d}\tau = \sum_{i=1}^n \tau(S_i \cap S) x_i$ ($\in X$). A function $f : \Omega \to X$ is said to be *measurable* if there is a sequence $\{s_n\}$ of simple functions from Ω into X that converges to f (a.e.). If, moreover, we have $\int_\Omega \|s_n - s_m\| \mathrm{d}\tau \to 0$ whenever n and m tend to ∞, the measurable function f is called *Bochner integrable*. If this is the case, the *Bochner integral* $\int_S f \mathrm{d}\tau$ of f with respect to the measure m is defined as the limit of the sequence $\int_S s_n \mathrm{d}\tau$; again, this limit is found to be independent of the particular approximate sequence of simple functions. The space of equivalence classes of Bochner integrable functions $f : \Omega \to X$ equipped with the norm $\|f\|_1 := \int_\Omega \|f\| \mathrm{d}\tau$ is denoted $L_1(\tau, X)$. It is a Banach space.

The two following results reduce the concept of measurability and the Bochner integral to the (scalar-valued) concepts of measurability and integrability, respectively, that have been considered above:

Theorem 17.22 (Pettis) *A function $f : \Omega \to X$ is measurable if and only if the two following assertions hold:*
(i) For every $x^* \in X^*$, the scalar-valued function $\omega \mapsto \langle x^*, f(w) \rangle$ is measurable.
(ii) The function f is almost separably valued, *i.e., there exists $N \in \Sigma$ such that $\mu(N) = 0$ and $\overline{\mathrm{span}}\{f(\omega) : \omega \in \Omega \setminus N\}$ is separable.*

Proposition 17.23 *A function $f : \Omega \to X$ is Bochner integrable if and only if it is measurable and $\int_\Omega \|f\| \mathrm{d}\tau < +\infty$.*

For more on this topic see, e.g., [BeLi, Appendix D], [DiUh], [Fede], [LuMa], [Rudi2], and [WhZy].

17.14 Continued Fractions and the Representation of the Irrational Numbers

Let $x \in \mathbb{R}$. The symbol $\lfloor x \rfloor$ denotes the *floor function* (sometimes called the *integer part*) of x. We describe the following procedure (in (ii) below we follow the use of assigning to the variable preceding the symbol := the numerical value succeeding it). We start by setting $n := 0$, $r := x$. Do, in order,

$$\begin{aligned} &\text{(i) } a_n := \lfloor r \rfloor, \ x_n := r - \lfloor r \rfloor. \\ &\text{(ii) If } x_n = 0, \text{ then stop. Otherwise, } r := \tfrac{1}{x_n}, \ n := n + 1. \\ &\text{(iii) Go to (i).} \end{aligned} \tag{17.1}$$

This generates a finite or infinite list $\{a_0; a_1, a_2, \ldots\}$ of integers called the *(simple) continued fraction associated to x* (it is customary to write the integer a_0, called the *root*, separated by a semicolon from the subsequent natural numbers; the word *simple* is added sometimes to emphasize that the numerator at each fraction in (17.2) is 1). It also defines a (finite or infinite) sequence $\{x_0, x_1, x_2, \ldots\}$. Note that only a_0 may be 0, and that

$$x = a_0 + x_0 = a_0 + \frac{1}{a_1 + x_1} = a_0 + \frac{1}{a_1 + \frac{1}{a_2 + x_2}} = a_0 + \frac{1}{a_1 + \frac{1}{a_2 + \frac{1}{a_3 + x_3}}} \ldots \quad (17.2)$$

We consider also the sequence $\{f_n\}_{n \geq 0}$ of the *convergents of the continued fraction*, i.e.,

$$f_0 := a_0, \; f_1 := a_0 + \frac{1}{a_1}, \; f_2 := a_0 + \frac{1}{a_1 + \frac{1}{a_2}}, \; f_3 := a_0 + \frac{1}{a_1 + \frac{1}{a_2 + \frac{1}{a_3}}}, \; \ldots. \quad (17.3)$$

The following lemma collects some easy facts about those sequences.

Lemma 17.24 *Let $\{a_n\}_{n=0}^{\infty}$ and $\{f_n\}_{n=0}^{\infty}$ be the sequences associated to some $x \in \mathbb{R}$ defined above. Define by recurrence the sequences $\{b_n\}_{n=0}^{\infty}$ and $\{c_n\}_{n=0}^{\infty}$ letting $b_0 = a_0$, $b_1 = a_0 a_1 + 1$, $b_n = a_n b_{n-1} + b_{n-2}$, $c_0 = 1$, $c_1 = a_1$, and $c_n = a_n c_{n-1} + c_{n-2}$, for $n = 2, 3, \ldots$. Then*

(i) $\frac{b_n}{c_n} = f_n$, *for* $n \geq 0$.
(ii) $b_n c_{n-1} - b_{n-1} c_n = (-1)^{n-1}$ *for* $n \geq 1$.
(iii) The two numbers b_n and c_n are relatively primes, for $n = 0, 1, 2, \ldots$.
(iv) $c_n \geq n$ *for* $n = 0, 1, 2, \ldots$.

Proof: (i) is proved by induction. Indeed, the assertion is obviously true for $n = 0, 1, 2$. Assume now that we already proved that, for some $n \geq 2$, we have $(b_i/c_i) = f_i$, $i = 0, 1, 2, \ldots, n$. Note that the $(n+1)$-th convergent is obtained by replacing a_n by $a_n + 1/a_{n+1}$ in the formula of the n-th convergent (i.e., b_n/c_n by the assumption). So, the $(n+1)$-th convergent is

$$\frac{(a_n + \frac{1}{a_{n+1}})b_{n-1} + b_{n-2}}{(a_n + \frac{1}{a_{n+1}})c_{n-1} + c_{n-2}} = \frac{a_{n+1}(a_n b_{n-1} + b_{n-2}) + b_{n-1}}{a_{n+1}(a_n c_{n-1} + c_{n-2}) + c_{n-1}} = \frac{a_{n+1} b_n + b_{n-1}}{a_{n+1} c_n + c_{n-1}}.$$

This completes the induction, and proves (i).

Part (ii) is also proved by induction. It is obviously true for $n = 1$. Assume that it holds for $i \leq n$, where n is some natural number. Then, by (i),

$$\begin{aligned} b_{n+1} c_n - b_n c_{n+1} &= (a_{n+1} b_n + b_{n-1}) c_n - b_n (a_{n+1} c_n + c_{n-1}) \\ &= b_{n-1} c_n - b_n c_{n-1} = -(-1)^{n-1} = (-1)^n, \end{aligned}$$

by the induction hypothesis. This proves (ii) for all $n \in \mathbb{N}$.

(iii) is a consequence of (ii). Indeed, a common factor of b_n and c_n will be a factor of 1.

(iv) Note that $a_n \geq 1$ for $n \in \mathbb{N}$. The assertion follows again by induction. □

Some straightforward consequences of Lemma 17.24 follow. For example,

$$\left| \frac{b_n}{c_n} - \frac{b_{n-1}}{c_{n-1}} \right| = \frac{1}{c_n c_{n-1}}, \text{ for } n \in \mathbb{N}, \tag{17.4}$$

Therefore, by (iv) in Lemma 17.24, the sequence $\{f_n\}_{n\geq 0}$ converges. Moreover,

$$f_0 < f_2 < f_4 < \ldots < f_5 < f_3 < f_1. \tag{17.5}$$

The sequence $\{f_n\}_{n\geq 0}$ converges to x. This can be seen by noticing that x is obtained from the expression of the n-th convergent f_n by replacing a_{n+1} by $1/x_n$ (see (17.2) and (17.3)). Then

$$x = \frac{\frac{b_n}{x_n} + b_{n-1}}{\frac{c_n}{x_n} + c_{n-1}} = \frac{b_n + x_n b_{n-1}}{c_n + x_n c_{n-1}}.$$

Since $x_n \in (0, 1)$, we obtain that x is strictly between f_{n-1} and f_n (geometrically, x is an intermediate slope in between the slopes of the two-dimensional vectors (c_{n-1}, b_{n-1}) and (c_n, b_n)). The oscillating behavior of the sequence $\{f_n\}$ (see (17.5)), and its convergence, proves the assertion.

Lemma 17.25 *Let $x \in \mathbb{R}$.*
(i) *The continued fraction $\{a_0; a_1, a_2, \ldots\}$ associated to x is finite if and only if $x \in \mathbb{Q}$.*
(ii) *Put $\neg\mathbb{Q}$ for the set of all irrational numbers. Then, the mapping $\phi : \neg\mathbb{Q} \cap (0, 1) \to \mathbb{N}^{\mathbb{N}}$ given by $\phi(x) := (a_1, a_2, \ldots)$ is one-to-one and onto.*

Proof: (i) Obviously, a finite continued fraction is associated to a rational number. Conversely, if $x \in \mathbb{Q}$ then performing (17.1) above we get a sequence $\{x_n\}_{n\geq 0}$ of rational numbers in $[0, 1)$ with strictly decreasing denominators; hence it must be finite.

(ii) The injectivity of ϕ follows from the fact that the sequence $\{f_n\}_{n\geq 0}$ of convergents defined by the continued fraction $\{a_0; a_1, a_2, \ldots\}$ associated to x converges to x. The surjectivity follows from the observation that Lemma 17.24 applies to the sequence $\{0, a_1, a_2, \ldots\}$ and the corresponding sequence $\{f_n\}_{n=0}^{\infty}$ defined in (17.3), for any $(a_1, a_2, \ldots) \in \mathbb{N}^{\mathbb{N}}$. □

We endow $\mathbb{N}^{\mathbb{N}}$ with the product topology, i.e., the topology of the coordinatewise convergence, which makes $\mathbb{N}^{\mathbb{N}}$ a complete metric space. A complete metric defining this topology is given by d, where $d(a, b) := 1/n$ if n is the first natural number k with $a_k \neq b_k$, for $a := (a_k) \in \mathbb{N}^{\mathbb{N}}$ and $b := (b_k) \in \mathbb{N}^{\mathbb{N}}$. The set $\neg\mathbb{Q} \cap (0, 1) \subset \mathbb{R}$ is endowed with the restriction of the usual topology in $\mathbb{R}$; it is homeomorphic to the set $\neg\mathbb{Q}$ of all irrational numbers.

Proposition 17.26 *The mapping* $\phi : \neg\mathbb{Q} \cap (0, 1) \to \mathbb{N}^{\mathbb{N}}$ *defined in Lemma 17.25 is a homeomorphism.*

Proof: It was proved in Lemma 17.25 that ϕ maps $\neg\mathbb{Q} \cap (0, 1)$ onto $\mathbb{N}^{\mathbb{N}}$ in a one-to-one way. Let $x \in \neg\mathbb{Q} \cap (0, 1)$. Fixing $n \in \mathbb{N} \cup \{0\}$, it is clear that for $y \in \neg\mathbb{Q} \cap (0, 1)$ close enough to x, the process (17.1) applied to x and to y gives the same integer parts up to n (i.e., $a_0, a_1, \ldots, a_n$). This proves the continuity of ϕ. The continuity of ϕ^{-1} follows from the fact that x is in between the two convergents b_{n-1}/c_{n-1} and b_n/c_n, whose mutual distance in $\mathbb{R}$ is $(c_{n-1}c_n)^{-1}$ $(\leq (n-1)^{-2})$, according to (17.4). □

The set $\neg\mathbb{Q}$, endowed with the restriction of the usual metric in $\mathbb{R}$, is not complete. However, it is topologically homeomorphic to $\neg\mathbb{Q} \cap (0, 1)$, in turn topologically homeomorphic to $\mathbb{N}^{\mathbb{N}}$. The complete metric d on this last space introduced in the paragraph preceding Proposition 17.26 induces, via the homeomorphism, a (complete) metric on $\neg\mathbb{Q}$ generating the usual topology. The existence of a complete metric on $\neg\mathbb{Q}$ compatible with the usual topology is a consequence of the fact that $\neg\mathbb{Q}$ is a G_δ subset of $\mathbb{R}$ (see, e.g., [Enge, Theorem 4.3.23]). Here we gave an explicit construction of such a metric.

References

[AaLu] J. Aarts and D. Lutzer, *Completeness properties designed for recognizing Baire space*, Dissert. Math. **116** (1974).

[AAB] Y.A. Abramovich, C.D. Aliprantis, and O. Burkinshaw, *The Daugavet equation in uniformly convex Banach spaces*, J. Funct. Anal. **97** (1991), 215–230.

[Ahar] I. Aharoni, *Uniform embeddings of Banach spaces*, Israel J. Math. **27** (1977), 174–179.

[AhLi1] I. Aharoni and J. Lindenstrauss, *Uniform equivalence between Banach spaces*, Bull. Amer. Math. Soc. **84** (1978), 281–283.

[AhLi2] I. Aharoni and J. Lindenstrauss, *An extension of a result of Ribe*, Israel J. Math. **52** (1985), 50–64.

[AlKa] F. Albiac and N.J. Kalton, *Topics in Banach space theory*, Graduate Text in Mathematics **233**, Springer, 2006.

[AlArDi] R. Alencar, R.M. Aron, and S. Dineen, *A reflexive space of holomorphic functions in infinitely many variables*, Proc. Amer. Math. Soc. **90** (1984), 407–411.

[Alex] A.D. Alexandrov, *Almost everywhere existence of the second differential of a convex function and some properties of convex surfaces connected with it*, Učenye Zapiski Leningrad. Univ. **6** (1939), 3–35.

[Alsp] D. Alspach, *A fixed point free nonexpansive map*, Proc. Amer. Math. Soc. **82** (1981), 423–424.

[AlPo] K. Alster and R. Pol, *On function spaces of compact subspaces of σ-products of the real line*, Fund. Math. **107** (1980), 135–143.

[Amir] D. Amir, *Banach spaces*, Lecture Notes (unpublished), Edmonton, Canada, 1975.

[AmLi] D. Amir and J. Lindenstrauss, *The structure of weakly compact sets in Banach spaces*, Ann. Math. **88** (1968), 35–44.

[Ander] R.D. Anderson, *Hilbert space is homeomorphic to the countable infinite product of lines*, Bull. Amer. Math. Soc. **72** (1966), 515–519.

[Arch] A.V. Archangelskij, *Topological function spaces*, Mathematics and Its Applications **78**, Kluwer Academic Publishers, 1992.

[ArArMe] S. Argyros, A.D. Arvanitakis, and S. Mercourakis, *Talagrand's $K_{\sigma\delta}$ problem*, Topology Appl. **155** (2008), 1737–1755.

[ACGJM] S. Argyros, J.M.F. Castillo, A.S. Granero, M. Jiménez, and J.P. Moreno, *Complementation and embeddings of $c_0(I)$ in Banach spaces*, Proc. London Math. Soc. **85**(3) (2002), 742–768.

[ArFa] S. Argyros and V. Farmaki, *On the structure of weakly compact subsets of Hilbert spaces and applications to the geometry of Banach spaces*, Trans. Amer. Math. Soc. **289** (1985), 409–427.

[ArHay] S. Argyros and R. Haydon, *A hereditarily indecomposable L_∞-space that solves the scalar-plus-compact problem*, submitted March 23rd, 2009. arXiv:0903.3921v2 [math.FA]

M. Fabian et al., *Banach Space Theory*, CMS Books in Mathematics,
DOI 10.1007/978-1-4419-7515-7,

[ArMe1] S. Argyros and S. Mercourakis, *On weakly Lindelöf Banach spaces*, Rocky Mount. J. Math. **23** (1993), 395–446.

[ArMe2] S. Argyros and S. Mercourakis, *Examples concerning heredity problems of WCG Banach spaces*, Proc. Amer. Math. Soc. **133** (2005), 773–785.

[ArMe3] S. Argyros and S. Mercourakis, *A note on the structure of WUR Banach spaces*, Comment. Math. Univ. Carolinae **46** (2005), 399–408.

[AMN] S. Argyros, S. Mercourakis, and S. Negrepontis, *Functional analytic properties of Corson compact spaces*, Studia Math. **89** (1988), 197–229.

[ArTo] S. Argyros and A. Tolias, *Methods in the theory of hereditarily indecomposable Banach spaces*, Mem. Amer. Math. Soc. **170** (2004).

[Aro] N. Aronszajn, *Differentiability of Lipschitz mappings between Banach spaces*, Studia Math. **57** (1976), 147–190.

[ArSm] N. Aronszajn and K.T. Smith, *Invariant subspaces of completely continuous operators*, Ann. Math. **60** (1954), 345–350.

[Arva] A.D. Arvanitakis, *Some remarks on Radon–Nikodým compact spaces*, Studia Math. **172** (2002), 41–60.

[Aspl] E. Asplund, *Fréchet differentiability of convex functions*, Acta Math. **121** (1968), 31–47.

[Aspl2] E. Asplund, *Boundedly Krein-compact Banach spaces*, Proceedings of the Functional Analysis Week, Aarhus, 1969, 1–4, Matematisk Institute Aarhus University

[Avil] A. Avilés, *Radon–Nikodým compact spaces of low weight and Banach spaces*, Studia Math. **166** (2005), 71–82.

[Avil2] A. Avilés, *Compact space that do not map onto finite products*, Fund. Math. **202** (2009), 81–96.

[AvKal] A. Avilés and O. Kalenda, *Compactness in Banach space theory – selected problems*, Revista de la Real Academia de Ciencias, to appear.

[AzDe1] D. Azagra and R. Deville, *Starlike bodies and bump functions in Banach spaces*, Prépublication no. 116, Mathématiques Pures de Bordeaux, C.N.R.S., 1999.

[AzDe2] D. Azagra and R. Deville, *James' theorem fails for starlike bodies*, J. Funct. Anal. **180**(2) (2001), 328–346.

[AzFe] D. Azagra and J. Ferrera, *Every closed convex set is the set of minimizers of some C^1-smooth convex function*, Proc. Amer. Math. Soc. **130** (2002), 3687–3892.

[Ball] K.M. Ball, *Ellipsoids of maximal volume in convex bodies*, Geom. Dedicata **41** (1992), 241–250.

[Barr] J.A. Barroso, *Introduction to holomorphy*, Mathematical Studies **106**, North Holland 1985.

[BarGra] R.G. Bartle and L.M. Graves, *Mappings between function spaces*, Trans. Amer. Math. Soc. **72** (1952), 400–413.

[Bate] S.M. Bates, *On smooth, nonlinear surjections of Banach spaces*, Israel J. Math. **100** (1997), 209–220.

[BJLPS] S.M. Bates, W.B. Johnson, J. Lindenstrauss, D. Preiss, and G. Schechtman, *Affine approximation of Lipschitz functions and nonlinear quotients*, GAFA, Geom. Funct. Anal. **9** (1999), 1092–1127.

[Bato] E. Bator, *Unconditionally converging and compact operators on c_0*, Rocky Mount. J. Math. **22** (1992), 417–422.

[Batu] D.P. Baturov, *On subspaces of function spaces*, Vestnik Moskov. Univ. Ser. Mat. no. 4 (1987), 66–69. English transl. in Moscow Univ. Math. Bull. **42** (1987). MR **89a**:54018.

[Beau] B. Beauzamy, *Introduction to Banach spaces and their geometry*, Mathematics Studies **68**, North Holland, 1982.

[BelMar] M. Bell and W. Marciszewski, *On scattered Eberlein compact spaces*, Israel J. Math. **158** (2007), 217–224.

[BHO] S. Bellenot, R. Haydon, and E.W. Odell, *Quasireflexive and tree spaces constructed in the spirit of R.C. James*, Cont. Math. **85** (1989), 19–43.

[BeMo] J. Benítez and V. Montesinos, *Restricted weak upper semicontinuous differentials of convex functions*. Bull. Austral. Math. Soc. **63** (2001), 93-100.

[BDGJN] G. Bennett, L.E. Dor, V. Goodman, W.B. Johnson, and C.M. Newman, *On uncomplemented subspaces of L_p, $1 < p < 2$*, Israel J. Math. **26** (1997), 178–187.

[Beny] Y. Benyamini, *The uniform classification of Banach spaces*, Longhorn Notes, The University of Texas at Austin, Functional Analysis Seminar, 1984–1985.

[BeLi0] Y. Benyamini and J. Lindenstrauss, *A predual of ℓ_1 which is not isomorphic to a $C(K)$ space*, Israel J. Math. **13** (1972), 246–259.

[BeLi] Y. Benyamini and J. Lindenstrauss, *Geometric nonlinear functional analysis*, Vol. 1, Colloquium Publications **48**, American Mathematical Society, 2000.

[BRW] Y. Benyamini, M.E. Rudin, and M. Wage, *Continuous images of weakly compact subsets of Banach spaces*, Pacific J. Math. **70** (1977), 309–324.

[BeSta] Y. Benyamini and T. Starbird, *Embedding weakly compact sets into Hilbert space*, Israel J. Math. **23** (1976), 137–141.

[BeSte] Y. Benyamini and Y. Sternfeld, *Spheres in infinite-dimensional normed spaces are Lipschitz contractible*, Proc. Amer. Math. Soc. **88** (1983), 439–445.

[Bess1] C. Bessaga, *A note on universal Banach spaces of finite dimension*, Bull. Pol. Acad. **6** (1958), 97–101.

[Bess2] C. Bessaga, *On topological classification of complete linear metric spaces*, Fund. Math. **56** (1965), 251–288.

[Bess3] C. Bessaga, *Topological equivalence of nonseparable reflexive Banach spaces, ordinal resolutions of identity and monotone bases*, Ann. Math. Studies **69** (1972), 3–14.

[BePe0] C. Bessaga and A. Pełczyński, *On bases and unconditional convergence of series in Banach spaces*, Studia Math. **17** (1958), 151–164.

[BePe1] C. Bessaga and A. Pełczyński, *Spaces of continuous functions IV*, Studia Math. **19** (1960), 53–62.

[BePe1a] C. Bessaga and A. Pełczyński, *On extreme points in separable conjugate spaces*, Israel J. Math. **4** (1966), 262–264.

[BePe2] C. Bessaga and A. Pełczyński, *Selected topics in infinite-dimensional topology*, Polish Scientific Publishers, Warszawa, 1975.

[BiPh] E. Bishop and R.R. Phelps, *A proof that every Banach space is subreflexive*, Bull. Amer. Math. Soc. **67** (1961), 97–98.

[Bohn] F. Bohnenblust, *Subspaces of $\ell_{p,n}$ spaces*, Amer. J. Math. **63** (1941), 64–72.

[Boll1] B. Bollobás, *Linear analysis, an introductory course*, Cambridge Mathematical Textbooks, 1990.

[Boll2] B. Bollobás, *Linear analysis, an introductory course* (2nd. Ed.), Cambridge Mathematical Textbooks, Cambridge University Press, 1999.

[BoFr] R. Bonic and J. Frampton, *Smooth functions on Banach manifolds*, J. Math. Mech. **15** (1966), 877–898.

[Bors] K. Borsuk, *Drei Sätze über die n-dimensionale euklidische Sphäre*, Fund. Math. **20** (1933), 177–190.

[Bor] J. Borwein, *Asplund spaces are "sequentially reflexive"*, University of Waterloo Technical Report, 1991.

[BoFa] J. Borwein and M. Fabian, *On convex functions having points of Gateaux differentiability which are not points of Fréchet differentiability*, Can. J. Math. **45** (6) (1993), 1121–1134.

[BoNo] J. Borwein and D. Noll, *Second order differentiability of convex functions: A.D. Alexandrov's theorem in Hilbert space*, Trans. Amer. Math. Soc. **342** (1994), 43–81.

[BoPr] J. Borwein and D. Preiss, *A smooth variational principle with applications to subdifferentiability and differentiability of convex functions*, Trans. Amer. Math. Soc. **303** (1987), 517–527.

[BoVa] J. Borwein and J. Vanderwerff, *Convex functions: Constructions, characterizations and counterexamples*. Encyclopedia of Mathematics and Its Applications **109**, Cambridge University Press, 2009.

[BGK] B. Bossard, G. Godefroy, and R. Kaufman, *Hurewicz theorem and renormings of Banach spaces*, J. Funct. Anal. **140** (1996), 142–150.

[Bou] J. Bourgain, *On dentability and the Bishop–Phelps property*, Israel J. Math. **28** (1976), 265–271.

[Bou2] J. Bourgain, *ℓ_∞/c_0 has no equivalent strictly convex norm*, Proc. Amer. Math. Soc, **78** (1985), 225–226.

[BFT] J. Bourgain, D.H. Fremlin, and M. Talagrand, *Pointwise compact sets in Baire-measurable functions*, Amer. J. Math. **100** (1978), 845–886.

[BoRo] J. Bourgain and H.P. Rosenthal, *Martingales-valued in certain subspaces of L_1*, Israel J. Math. **37** (1980), 55–76.

[BoTa] J. Bourgain and M. Talagrand, *Compacité extrémale*, Proc. Amer. Math. Soc. **80** (1980), 68–70.

[Bour] R. Bourgin, *Geometric aspects of convex sets with the Radon–Nikodým property*, Lecture Notes in Mathematics **93**, Springer, 1983.

[BrSu1] A. Brunel and L. Sucheston, *On B-convex Banach spaces*, Math. Syst. Theory **7** (1973).

[BrSu2] A. Brunel and L. Sucheston, *On G-convexity and some ergodic superproperties of Banach spaces*, Trans. Amer. Math. Soc. **204** (1975), 79–90.

[Casa] P. Casazza, *Approximation properties*, Handbook of Banach Spaces I, Editors W.B. Johnson and J.Lindenstrauss, Elsevier, 2001, 271–316.

[CasGod] B. Cascales and G. Godefroy, *Angelicity and the boundary problem*, Mathematika **45**(1) (1998), 105–112.

[CasVe] B. Cascales and G. Vera, *Norming sets and compactness*, Rocky Mountain J. Math. **25**(3) (1995), 919–925.

[CasGon] J.M.F. Castillo and M. González, *Three space problems in Banach space theory*, Lecture Notes in Mathematics **1667**, Springer, 1997.

[Caut1] R. Cauty, *Un espace métrique linéaire qui n'est pas un rétracte absolu*, Fund. Math. **146** (1994), 85–99.

[Caut2] R. Cauty, *Solution du problème de point fixe de Schauder*, Fund. Math. **170** (2001), 231–246.

[Cepe] M. Cepedello, *Approximation of Lipschitz functions by Δ-convex functions in Banach spaces*, Israel J. Math. **106** (1998), 269–284.

[Cepe2] M. Cepedello, *On regularization in superreflexive Banach spaces by infimal convolution formulas*, Studia Math. **129** (1998), 265–284.

[Choq] G. Choquet, *Lectures on analysis I, II, III*, W.A. Benjamin, New York, 1969.

[Chri] P.R. Christensen, *Measure theoretic zero sets in infinite-dimensional spaces and applications to differentiability of Lipschitz mappings II*, Publ. Dep. Math. Lyon **109** (1973), 29–39.

[Cla] J.A. Clarkson, *Uniformly convex spaces*. Trans. Amer. Math. Soc. **40** (1936), 396–414.

[Cohe] J.S. Cohen, *Absolutely p-summing, p-nuclear operators and their conjugates*, Math. Ann. **201** (1973), 177–200.

[CoEd] J. Collier and M. Edelstein, *On strongly exposed points and Fréchet differentiability*, Israel J. Math. **17** (1974), 66–68.

[Cors1] H.H. Corson, *The weak topology of a Banach space*, Trans. Amer. Math. Soc. **101** (1961), 1–15.

[Cors2] H.H. Corson, *A compact convex set in E^3 whose exposed points are of the first category*, Proc. Amer. Math. Soc. **16** (1965), 1015–1021.

[CoLi] H.H. Corson and J. Lindenstrauss, *On weakly compact subsets of Banach spaces*, Proc. Amer. Math. Soc. **17** (1966), 407–412.

[Dane] J. Daneš, *Equivalence of some geometric and related results of nonlinear functional analysis*, Comm. Math. Univ. Carolinae **26** (1985), 443–454.
[DGK] L. Danzer, B. Grünbaum, and V. Klee, *Helly's theorem and its relatives*, Proc. Sym. Pure Math. VII, Convexity, Amer. Math. Soc. (1963), 101–181.
[Davd] K.R. Davidson, *Nest algebras*, Research Notes in Mathematics, Pitman **191**, 1988.
[Davi] A.M. Davie, *The approximation problem for Banach spaces*, Bull. London Math. Soc. **5** (1973), 261–266.
[Davi2] A.M. Davie, *The Banach approximation problem*, J. Approx. Theory **13** (1975), 392–394.
[DFJP] W.J. Davis, T. Figiel, W.B. Johnson, and A. Pełczyński, *Factoring weakly compact operators*, J. Funct. Anal. **17** (1974), 311–327.
[DaJo] W.J. Davis and W.B. Johnson, *Renorming of nonreflexive Banach spaces*, Proc. Amer. Math. Soc. **37** (1973), 486–488.
[DaLi] W.J. Davis and J. Lindenstrauss, *On total non-norming subspaces*, Proc. Amer. Math. Soc. **31** (1972), 109–111.
[DaPh] W.J. Davis and R.R. Phelps, *The Radon–Nikodým property and dentable sets in Banach spaces*, Proc. Amer. Math. Soc. **45** (1974), 119–122.
[Day0] M.M. Day, *Uniform convexity in factor and conjugate spaces*, Ann. Math. **45** (1944), 375–385.
[Day1] M.M. Day, *Strict convexity and smoothness of normed spaces*, Trans. Amer. Math. Soc. **78** (1955), 516–528.
[Day] M.M. Day, *Normed linear spaces*, Springer, 1973.
[DJS] M.M. Day, R.C. James, and S. Swaminathan, *Normed linear spaces that are uniformly convex in every direction*, Canadian J. Math. **83** (1972), 1051–1059.
[DJP] F. Delbaen, H. Jarchow, and A. Pełczyński, *Subspaces of L_p isometric to subspaces of ℓ_p*, Positivity **2** (1998), 339–367.
[DeFl] A. Defant and K. Floret, *Tensor Norms and Operator Ideals*, North-Holland Math. Studies **176** (1993).
[Devi] R. Deville, *Geometrical implications of the existence of very smooth bump functions in Banach spaces*, Israel J. Math. **6** (1989), 1–22.
[DFH1] R. Deville, V.P. Fonf, and P. Hájek, *Analytic and C^k approximations of norms in separable Banach spaces*, Studia Math. **120** (1996), 61–74.
[DFH2] R. Deville, V.P. Fonf, and P. Hájek, *Analytic and polyhedral approximation of convex bodies in separable polyhedral Banach spaces*, Israel J. Math. **105** (1998), 139–154.
[DeGh] R. Deville and N. Ghoussoub, *Perturbed minimization principles and applications*, Handbook of Banach Spaces I, Editors W.B. Johnson and J. Lindenstrauss, Elsevier, 2001, 393–435.
[DeGo] R. Deville and G. Godefroy, *Some applications of projectional resolutions of identity*, Proc. London Math. Soc. **67** (1993), 183–199.
[DGZ1] R. Deville, G. Godefroy, and V. Zizler, *Un principle variationnel utilisant des fonctions bosses*, C.R. Acad. Sci. Paris **312** Serie I (1991), 281–286.
[DGZ2] R. Deville, G. Godefroy, and V. Zizler, *A smooth variational principle with applications to Hamilton–Jacobi equations in infinite dimensions*, J. Funct. Anal. **111** (1993), 197–212.
[DGZ3] R. Deville, G. Godefroy, and V. Zizler, *Smoothness and renormings in Banach spaces*, Pitman Monographs **64**, London, Logman, 1993.
[DGZ4] R. Deville, G. Godefroy, and V. Zizler, *Smooth bump functions and geometry of Banach spaces*, Mathematika **40** (1993), 305–321.
[Died] J. Dieudonné, *Foundations of modern analysis*, Academic Press, 1969.
[Dies1] J. Diestel, *Geometry of Banach spaces: Selected topics*, Lecture Notes in Mathematics **485**, Springer, 1975.
[Dies2] J. Diestel, *Sequences and series in Banach spaces*, Graduate text in Mathematics **92**, Springer, 1984.

[DJP] J. Diestel, H. Jarchow, and A. Pietsch, *Operator ideals*, Handbook of Banach Spaces I, Editors W.B. Johnson and J. Lindenstrauss, Elsevier, 2001, 437–496.

[DJT] J. Diestel, H. Jarchow, and A. Tonge, *Absolutely summing operators*, Cambridge Studies in Advanced Mathematics **43**, Cambridge University Press, 1995.

[DiUh] J. Diestel and J. Uhl, *Vector measures*, Mathematical Surveys **15**, American Mathematical Society, 1977.

[D-B] T. Domínguez-Benavides, *A renorming of some nonseparable Banach spaces with the fixed point property*, J. Math. Anal. Appl. **350**(2) (2009), 525–530.

[Dugu1] J. Dugundji, *An extension of Tietze's theorem*, Pacific J. Math. **1** (1951), 353–367.

[Dugu2] J. Dugundji, *Topology*, Allyn and Bacon, Inc., Boston, 1966.

[DuNa] D. van Dulst and I. Namioka, *A note on trees in conjugate Banach spaces*, Indag. Math. **87** (1984), 7–10.

[DuSc] N. Dunford and J.T. Schwartz, *Linear operators, Part I*, Interscience, New York, 1958.

[Dvor] A. Dvoretzky, *Some results on convex bodies and Banach spaces*, Proceedings of the Symposium on Linear Spaces, Jerusalem, 1961, 123–160.

[EdWh] G.A. Edgar and R.F. Wheeler, *Topological properties of Banach spaces*, Pacific J. Math. **115** (1984), 317–350.

[Ekel] I. Ekeland, *On the variational principle*, J. Math. Anal. Appl. **47** (1974), 324–353.

[Enfl1] P. Enflo, *Banach spaces which can be given an equivalent uniformly convex norm*, Israel J. Math. **13** (1972), 281–288.

[Enfl2] P. Enflo, *A counterexample to the approximation problem in Banach spaces*, Acta Math. **130** (1973), 309–317.

[Enfl3] P. Enflo, *On the invariant subspace problem for Banach spaces*, Acta Math. **158**, (1987), 213–313 (submitted in 1981).

[ELP] P. Enflo, J. Lindenstrauss, and G. Pisier, *On the three space problem*, Math. Scand. **36** (1975), 189–210.

[Enge] R. Engelking, *General Topology*, PWN Warszawa, 1977.

[Fab0] M. Fabian, *On projectional resolution of identity on the duals of certain Banach spaces*, Bull. Australian Math. Soc. **35** (1987), 363–372.

[Fab1] M. Fabian, *Each weakly countably determined Asplund space admits a Fréchet differentiable norm*, Bull. Australian Math. Soc. **36** (1987), 367–374.

[Fabi0] M. Fabian, *On a dual locally uniformly rotund norm on a dual Vašák space*, Studia Math. **101** (1991), 69–81.

[Fabi1] M. Fabian, *Differentiability of convex functions and topology: Weak Asplund spaces*, Wiley, 1997.

[Fabi2] M. Fabian, *Overclasses of the class of Radon–Nikodým compact spaces*, Methods in Banach Spaces, Editons J.M.F. Castillo and W.B. Johnson, London Math. Soc. Lect. Notes No. **337**, 2006, 197–214.

[FabFin] M. Fabian and C. Finet, *On Stegall's smooth variational principle*, Nonlinear Anal., Theory, Methods, Appl. **66** (2007), 565–570.

[FaGo] M. Fabian and G. Godefroy, *The dual of every Asplund space admits a projectional resolution of identity*, Studia Math. **91** (1988), 141–151.

[FGMZ] M. Fabian, G. Godefroy, V. Montesinos, and V. Zizler, *Inner characterizations of weakly compactly generated Banach spaces and their relatives*, J. Math. Anal. Appl. **297** (2004), 419–455.

[FGZ] M. Fabian, G. Godefroy, and V. Zizler, *The structure of uniformly Gâteaux smooth Banach spaces*, Israel J. Math. **124** (2001), 243–252.

[FGM] M. Fabian, A. González, and V. Montesinos, *A note on Markushevich bases in weakly compactly generated Banach spaces*. Ann. Acad. Sci. Fenn. **34** (2009), 555–564.

[FHHMPZ] M. Fabian, P. Habala, P. Hájek, J. Pelant, V. Montesinos, and V. Zizler, *Functional analysis and infinite-dimensional geometry*, Canadian Mathematical Society Books in Mathematics, No. **8**, Springer, New York, 2001.

[FHMZ] M. Fabian, P. Hájek, V. Montesinos, and V. Zizler, *A quantitative version of Krein's theorem*, Rev. Mat. Iberoamericana, **21** (2005), 237–248.

[FHZ] M. Fabian, P. Hájek, and V. Zizler, *Uniformly Eberlein compacta and uniformly Gâteaux smooth norms*, Serdica Math. J. **23** (1977), 1001–1010.

[FHM] M. Fabian, M. Heisler, and E. Matoušková, *Remarks on continuous images of Radon–Nikodým compacta*, Comment. Math. Univ. Carolinae **39** (1998), 59–69.

[FMZ1] M. Fabian, V. Montesinos, and V. Zizler, *Pointwise semicontinuous smooth norms*, Arch. Math. **78** (2002), 459–464.

[FMZ2] M. Fabian. V. Montesinos, and V. Zizler, *Weakly compact sets and smooth norms in Banach spaces*, Bull. Australian Math. Soc. **65** (2002), 223–230.

[FMZ3] M. Fabian, V. Montesinos, and V. Zizler, *A characterization of subspaces of weakly compactly generated spaces*, J. London Math. Soc. **69** (2004), 457–464.

[FMZ4] M. Fabian, V. Montesinos, and V. Zizler, *On weak compactness in L_1 spaces*, Rocky Mountains J. Math. **39** (2009), 1885–1893.

[FMZ5] M. Fabian, V. Montesinos, and V. Zizler, *Smoothness in Banach spaces: Selected problems*, Rev. Real Acad. Cien. Serie A. Mat. **100**(1–2) (2006), 101–125.

[FaWh] M. Fabian and J.H.M. Whitfield, *On equivalent characterization of weakly compactly generated Banach spaces*, Rocky Mount. J. Math. **24** (1994), 1363–1378.

[FWZ] M. Fabian, J.H.M. Whitfield, and V. Zizler, *Norms with locally Lipschitz derivatives*, Israel J. Math. **44** (1983), 262–276.

[FZZ] M. Fabian, L. Zajíček, and V. Zizler, *On residuality of the set of rotund norms on Banach spaces*, Math. Ann. **258** (1981/1982), 349–351.

[FaZi1] M. Fabian and V. Zizler, *A note on bump functions that locally depend on finitely many coordinates*, Bull. Austral. Math. Soc. **56** (1997), 447–451.

[FaZi2] M. Fabian and V. Zizler, *An elementary approach to some problems in higher order smoothness in Banach spaces*, Extracta Math. **14** (1999), 295–327.

[Farm] V. Farmaki, *The structure of Eberlein, uniformly Eberlein and Talagrand compact spaces in $\Sigma(\mathbb{R}^{\Gamma})$*, Fund. Math. **128** (1987), 15–28.

[FeSa] M. Feder and P.D. Saphar, *Spaces of compact operators and their dual spaces*, Israel J. Math. **21** (1975), 38–49.

[Fede] H. Federer, *Geometric measure theory*, Classics in Mathematics, Springer, 1996.

[Fere] V. Ferenczi, *A uniformly convex hereditarily indecomposable Banach space*, Israel J. Math. **102** (1997), 199–225.

[FeGa] H. Fetter and B. Gamboa de Buen, *The James forest*, London Mathematical Society Lecture Notes Series **236**, Cambridge University Press, 1997.

[Figi1] T. Figiel, *On the moduli of convexity and smoothness*, Studia Math. **66** (1976), 121–155.

[Figi2] T. Figiel, *An example of an infinite-dimensional Banach space not isomorphic to its Cartesian square*, Studia Math. **42** (1972), 295–306.

[FiJo] T. Figiel and W.B. Johnson, *The approximation property does not imply the bounded approximation property*, Proc. Amer. Math. Soc. **41** (1973), 197–200.

[FiJo2] T. Figiel and W.B. Johnson, *A uniformly convex Banach space which contains no ℓ_p*, Compos. Math. **29** (1974), 179–190.

[FLM] T. Figiel, J. Lindenstrauss, and V.D. Milman, *The dimension of almost spherical sections of convex bodies*, Acta Math. **139** (1977), 53–94.

[FiPe] T. Figiel and A. Pełczyński, *On Enflo's method of construction of Banach spaces without the approximation property*, Uspehi Mat. Nauk **28** (1973), 95–108 (Russian).

[Fitz] S. Fitzpatrick, *Differentiability of real-valued functions and continuity of metric projections*, Proc. Amer. Math. Soc. **91** (1984), 544–548.

[Flor] K. Floret, *Weakly compact sets*, Lecture Notes in Mathematics **801**, Springer, 1980.

[Fonf0] V.P. Fonf, *Three characterizations of polyhedral Banach spaces*, Ukrainian Math. J. **42**(3) (1990), 1145–1148.

[Fonf] V.P. Fonf, *On supportless convex sets in incomplete normed spaces*, Proc. Amer. Math. Soc. **120** (1994), 1173–1176.

[FoKa] V.P. Fonf and M.I. Kadec, *Subspaces of ℓ_1 with strictly convex norm*, Math. Notes Ukrainsk. Academy of Sciences **33** (1983), 213–215.

[FLP] V.P. Fonf, J. Lindenstrauss, and R.R. Phelps, *Infinite-dimensional convexity*, Handbook of Banach Spaces I, Editors W.B. Johnson and J.Lindenstrauss, Elsevier, 2001, 599–670.

[FoRo] J.H. Fourie and I.M. Rontgen, *Banach space sequences and projective tensor products*, J. Math. Anal. Appl. **277** (2003), 629–644.

[FrSe] D.H. Fremlin and A. Sersouri, *On ω-independence in separable Banach spaces*, Quarterly J. Math. **39** (1988), 323–331.

[GeLi] A. Genel and J. Lindenstrauss, *An example concerning fixed points*, Israel J. Math. **22** (1975), 81–86.

[GhMa1] N. Ghoussoub and B. Maurey, *G_δ-embeddings in Hilbert space*, J. Funct. Anal. **61** (1985), 72–97.

[GhMa2] N. Ghoussoub and B. Maurey, *A nonlinear method for constructing certain basic sequences in Banach spaces*, Illnois J. Math. **34** (1990), 607–613.

[Gile] J.R. Giles, *Convex analysis with applications in differentiation of convex functions*, Research Notes in Mathematics **58**, Pitman, 1982.

[GGS] J.R. Giles, D.A. Gregory, and B. Sims, *Characterization of normed linear spaces with Mazur's intersection property*, Bull. Aust. Math. Soc. **18** (1978), 105–123.

[GiJe] L. Gillman and M. Jerison, *Rings of continuous functions*, Princeton University Press, Princeton, NJ, 1960.

[Glus] E. Gluskin, *The diameter of the Minkowski compactum is roughly equal to n*, Funct. Anal. Appl. **15** (1981), 72–73.

[Gode1] G. Godefroy, *Espace de Banach, existence et unicité de certains préduaux*, Ann. Inst. Fourier (Grenoble) **28** (1978), 87–105.

[Gode1b] G. Godefroy, *Existence de normes très lisess sur certains espaces de Banach*, Bull. Sci. Math. **106**(2) (1982), 63–68.

[Gode2] G. Godefroy, *Boundaries of convex sets and interpolation sets,* Math. Ann. **277** (1987), 173–184.

[Gode3] G. Godefroy, *Five lectures in geometry of Banach spaces*, Notas de Matemática, Vol. **1**, Seminar on Functional Analysis, 1987, Universidad de Murcia, 1988, 9–67.

[Gode4] G. Godefroy, *Renormings of Banach spaces*, Handbook of Banach Spaces I, Editors W.B. Johnson and J. Lindenstrauss, Elsevier, 2001, 781–835.

[Gode5] G. Godefroy, *Some applications of Simons inequality*, Serdica Math. J. **26** (2000), 59–78.

[Gode6] G. Godefroy, *On the diameter of the Banach–Mazur set*, Czechoslovak Math. J. **60**(1) (2010), 95–100.

[GoKa1] G. Godefroy and N.J. Kalton, *The ball topology and its applications*, Cont. Math. **85** (1989), 195–238.

[GoKa1b] G. Godefroy and N.J. Kalton, *Lipschitz-free Banach spaces*, Studia Math. **159** (2003), 121–141.

[GoKa2] G. Godefroy and N.J. Kalton, *Isometric embegddings and universal spaces*, Extracta Math. **22** (2007), 179–189.

[GKL1] G. Godefroy, N.J. Kalton, and G. Lancien, *Lipschitz isomorphisms and subspaces of $c_0(\mathbb{N})$*, Geom. Funct. Anal. **10** (2000), 798–820.

[GKL2] G. Godefroy, N.J. Kalton, and G. Lancien, *Szlenk indices and uniform homeomorphisms*, Trans. Amer. Math. Soc. **353** (2001), 3895–3918.

[GoLi] G. Godefroy and D. Li, *Strictly convex functions on compact convex sets and their use*, Functional Analysis, 182–192, Narosa, New Delhi, 1998.

[GoSa1] G. Godefroy and P.D. Saphar, *Duality in spaces of operators and smooth norms on Banach spaces*, Illinois J. Math. **32** (1988), 672–695.

[Gosa2] G. Godefroy and P.D. Saphar, *Three-space problems for the approximation properties*, Proc. Amer. Math. Soc. **185** (1989), 70–75.

[GTWZ] G. Godefroy, S. Troyanski, J.H.M. Whitfield, and V. Zizler, *Smoothness in weakly compactly generated Banach spaces*, J. Funct. Anal. **52** (1983), 344–352.

[GTWZ1] G. Godefroy, S. Troyanski, J.H.M. Whitfield, and V. Zizler, *Locally uniformly rotund renorming and injections into $c_0(\Gamma)$*, Canad. Math. Bull. **27** (1984), 494–500.

[GTWZ2] G. Godefroy, S. Troyanski, J.H.M. Whitfield, and V. Zizler *Three space problem for locally uniformly rotund renormings of Banach spaces*, Proc. Amer. Math. Soc. **94** (1985), 647–652.

[Godu] B.V. Godun, *Points of smoothness of convex bodies in separable Banach spaces*, Matem. Zametki, **38** (1985), 713–716.

[Godu2] B.V. Godun, *Preserved extreme points*, Functional Anal. i Prilozhen. **19** (1985), 75–76.

[Goeb] K. Goebel, *On a fixed point theorem for multivalued nonexpansive mappings*, Annal. Univ. Mariae Curie-Skłodowska **29** (1975), 70–72.

[Gon] A. González, *Compacta in Banach spaces*, PHD Dissertation. Universidad Politécnica de Valencia, Valencia, Spain, 2008.

[GoLe] Y. Gordon and D.R. Lewis, *Absolutely summing operators and local unconditional structure*, Acta Math. **133** (1974), 27–48.

[Gore] E. Gorelik, *The uniform non-equivalence of $\mathcal{L}_p$ and ℓ_p*, Israel J. Math. **87** (1994), 1–8.

[Gowe1] W.T. Gowers, *Lipschitz functions on classical spaces*, European J. Combin. **13** (1992), 141–151.

[Gowe2] W.T. Gowers, *Recent results in the theory of infinite-dimensional Banach spaces*, Proceedings of the International Congress of Mathematicians, Zürich, Switzerland, 1994.

[Gowe3] W.T. Gowers, *A solution to Banach's hyperplane problem*, Bull. London Math. Soc. **26** (1994), 523–530.

[Gowe4] W.T. Gowers, *A space not containing c_0, ℓ_1 or a reflexive subspace*, Trans. Amer. Math. Soc. **344** (1994), 407–420.

[Gowe5] W.T. Gowers, *A solution to the Schroeder–Bernstein problem for Banach spaces*, Bull. London Math. Soc. **28** (1996), 297–304.

[Gowe6] W.T. Gowers, *A new dichotomy for Banach spaces*, Geom. Funct. Anal. **6** (1996), 1083–1093.

[GoMa] W.T. Gowers and B. Maurey, *The unconditional basic sequence problem*, J. Amer. Math. Soc. **6** (1993), 851–874.

[GHM] A.S. Granero, P. Hájek, and V. Montesinos, *Convexity and w^*-compactness in Banach spaces*, Math. Ann. **328** (2004), 625–631.

[Groth1] A. Grothendieck, *Produits tensoriels topologiques et espaces nucléaires*, Mem. Amer. Math. Soc. **16** (1955).

[Groth2] A. Grothendieck, *Résumé de la théorie métrique des produits tensoriels topologiques*, Bol. Soc. Mat. São Paulo **8** (1953), 1–79.

[Groth3] A. Grothendieck, *Sur les applications lineaires faiblement compactes d'espace du type $C(K)$*, Canad. J. Math. **5**, (1953), 129–173.

[GrSc] P.M. Gruber and F.E. Schuster, *An arithmetic proof of John's ellipsoid theorem*, Arch. Math. **85** (2005), 82–88.

[Grun] B. Grünbaum, *Projection constants*, Trans. Amer. Math. Soc. **95** (1960), 451–465.

[Guer] S. Guerre-Delabrière, *Classical sequences in Banach spaces*, Monographs and textbooks in pure and applied mathematics, Marcel Dekker, Inc., 1992.

[GuiHa1] A.J. Guirao and P. Hájek, *On the moduli of convexity*, Proc. Amer. Math. Soc. **10**, (2007), 3233–3240.

[GuiHa2] A.J. Guirao and P. Hájek, *Schauder bases under uniform renormings*, Positivity **11**, (2007), 627–638.

[Gulk1] S.P. Gul'ko, *On properties of subsets of Σ-products*, Dokl. Akad. Nauk SSSR **237** (1977), 505–507.

[Gulk2] S.P. Gul'ko, *On the properties of some function spaces*, Soviet Math. Dokl. **19** (1978), 1420–1424.

[Gulk3] S.P. Gul'ko, *On the structure of spaces of continuous functions and their complete paracompactness*, Russian Math. Surveys **34** (1979), 36–44.

[Gulk4] S.P. Gul'ko, *The space $C_p(X)$ for countable compact X is uniformly homeomorphic to c_0*, Bull. Pol. Acad. Sci. **36** (1988), 391–396.

[Gulk5] S.P. Gulko, *Uniformly homeomorphic spaces of functions*, Proc. Steklov Inst. Math. **3** (1993), 87–93.

[GuGu] V.I. Gurarii and N.I. Gurarii, *On bases in uniformly convex and uniformly smooth spaces*, Izv. Akad. Nauk. SSSR Ser. Mat. **35** (1971), 210–215.

[GuKa] V.I. Gurarii and M.I Kadec, *Minimal systems and quasicomplements*, Soviet Math. Dokl. **3** (1962), 966–968.

[GKM1] V.I. Gurarii, M.I. Kadec, and V.I. Macaev, *On Banach–Mazur distance between certain Minkowski spaces*, Bull. Acad. Polon. Sci. Sér. Sci. Math. Astronom. Phys. **13** (1965), 719–722.

[GKM2] V.I. Gurarii, M.I. Kadec, and V.I. Macaev, *Distances between finite-dimensional analogs of the L_p-spaces*, Mat. Sb. **70**(112) (1966), 24–29.

[HHZ] P. Habala, P. Hájek, and V. Zizler, *Introduction to Banach spaces I, II*, Matfyzpress, Prague, 1996.

[Hag] J. Hagler, *A counterexample to several questions about Banach spaces*, Studia Math. **60** (1977), 289–308.

[Hag] J. Hagler, *A note on separable Banach spaces with non-separable dual*, Proc. Amer. Math. Soc. **99** (1987), 452–454.

[HaSu] J. Hagler and F.E. Sullivan, *Smoothness and weak star sequential compactness*, Proc. Amer. Math. Soc. **78** (1980), 497–503.

[Haje1] P. Hájek, *Smooth norms that depend locally on finitely many coordinates*, Proc. Amer. Math. Soc. **123** (1995), 3817–3821.

[Haje2] P. Hájek, *Dual renormings of Banach spaces*, Comment. Math. Univ. Carolinae **37** (1996), 241–253.

[Haje2b] P. Hájek, *Analytic renormings of $C(K)$ spaces*, Serdica Math. J. **22** (1996), 25–28.

[Haje3] P. Hájek, *Smooth functions on c_0*, Israel J. Math. **104** (1998), 89–96.

[Haje4] P. Hájek, *Smooth functions on $C(K)$*, Israel J. Math. **107** (1998), 237–252.

[Haje5] P. Hájek, *Smooth norms on certain $C(K)$ spaces*, Proc. Amer. Math. Soc. **131** (2003), 2049–2051.

[HaLa] P. Hájek and G. Lancien, *Various slicing indices on Banach spaces*, Mediterranean J. Math. **7** (2007), 2031–2035.

[HMZ] P. Hájek, V. Montesinos, and V. Zizler, *Qualitative Gâteaux differentiability of norms and geometry of separable Banach spaces. A short survey with open problems*, to appear.

[HMVZ] P. Hájek, V. Montesinos, J. Vanderwerff, and V. Zizler, *Biorthogonal systems in Banach spaces*, CMS Books in Mathematics, Canadian Mathematical Society, Springer, 2007.

[HajSm] P. Hájek and R. Smith, *On some duality relations in the theory of tensor products*, preprint.

[HaTr] P. Hájek and S. Troyanski, *Analytic norms in Orlicz spaces*, Proc. Amer. Math. Soc. **129** (2000), 713–717.

[HaZi] P. Hájek and V. Zizler, *Functions locally dependent on finitely many coordinates*, Rev. Real Acad. Cien. Serie A. Mat. **100**(1–2) (2006), 147–154.

[Hann] O. Hanner, *On the uniform convexity of L_p and ℓ_p*, Ark. Mat. **3** (1956), 239–244.

[Hans] R.W. Hansell, *Descriptive sets and the topology of nonseparable Banach spaces*, preprint, 1989.

[HWW] P. Harmand, D. Werner, and W. Werner, *M-ideals in Banach spaces and Banach algebras*, Lecture Notes in Mathematics **1547**, Springer, 1993.

[Hayd1] R. Haydon, *A counterexample to several questions about scattered compact spaces*, Bull. London Math. Soc. **22** (1990), 261–268.

[Hayd2] R. Haydon, *Smooth functions and partitions of unity on certain Banach spaces*, Quart. J. Math. **47** (1996), 455–468.

[Hayd3] R. Haydon, *Trees in renorming theory*, Proc. London Math. Soc. **78** (1999), 541–584.

[Hay4] R. Haydon, *Locally uniformly rotund norms in Banach spaces and their duals*, J. Funct. Anal. **254** (2008), 2023–2039.

[Heinr] S. Heinrich, *Ultraproducts in Banach space theory*, Journal für die reine und angewandte Mathematik (Crelles Journal) **313** (1980), 72–104.

[HeMa] S. Heinrich and P. Mankiewicz, *Applications of ultrapowers to the uniform and Lipschitz classification of Banach spaces*, Studia Math. **73** (1982), 225–251.

[HKO] C.W. Henson, A.S. Kechris, and E.W. Odell, *Analysis and logic*, preprint, Universite de Mons-Hinaut, Mons, Belgium, 1997

[HSZ] P. Holický, M. Šmídek, and L. Zajíček, *Convex functions with non Borel set of Gâteaux differentiability points*, Comment. Math. Univ. Carolinae **39** (1998), 469–482.

[Holu] J.R. Holub, *Tensor product bases and tensor diagonals*, Trans. Amer. Math. Soc. **151** (1970), 563–579.

[Hu] R.E. Huff, *Dentability and the Radon Nikodým property*, Duke Math. J. **41** (1974), 111–114.

[HuMo1] R.E. Huff and P.D. Morris, *Dual spaces with the Krein–Milman property have the Radon–Nikodým property*, Proc. Amer. Math. Soc. **49** (1975), 104–108.

[HuMo2] R.E. Huff and P.D. Morris, *Geometric characterizations of the Radon–Nikodým property in Banach spaces*, Studia Math. **56** (1976), 157–164.

[Jame1] R.C. James, *Bases and reflexivity of Banach spaces*, Ann. Math. **52** (1950), 518–527.

[Jame1b] R.C. James, *On nonreflexive Banach space isometric with its second conjugate*, Proc. Nat. Acad. Sci. USA **37** (1951), 174–177.

[Jame2] R.C. James, *Weak compactness and reflexivity*, Israel J. Math. **2** (1964), 101–119.

[Jame3] R.C. James, *Uniformly non-square Banach spaces*. Ann. Math. **80** (1964), 542–550.

[Jame4] R.C. James, *A counterexample for a sup theorem in normed spaces*, Israel J. Math. **9** (1971), 511–512.

[Jame5] R.C. James, *Reflexivity and the sup of linear functionals*, Israel J. Math. **13** (1972), 289–300.

[Jame6] R.C. James, *Some self-dual properties of normed linear spaces*, Symposium on Infinite Dimentional Topolology, Editor D. Anderson, Annals of Mathematical Studies **69** (1972), 159–175.

[Jame7] R.C. James, *Superreflexive Banach spaces*, Canadian J. Math. **24** (1972), 896–904.

[Jame8] R.C. James, *Superreflexive spaces with bases*, Pacific J. Math. **41–42** (1972), 409–419.

[Jame9] R.C. James, *A separable somewhat reflexive Banach space with nonseparable dual*, Bull. Amer. Math. Soc. **80** (1974), 738–743.

[Jame9b] R.C. James, *A nonreflexive Banach space that is uniformly nonoctahedral*, Israel J. Math. **18** (1974), 145–155.

[James10] R.C. James, *Nonreflexive spaces of type* 2, Israel J. Math. **30** (1978), 1–13.

[JamSch] R.C. James and J.J. Schaffer, *Superreflexivity and the girth of spheres*, Israel J. Math. **11** (1972), 398–404.

[Jmsn] G.J.O. Jameson, *Topology and normed spaces*, Chapman and Hall, London, 1974.

[Jarc] H. Jarchow, *Locally convex spaces*, Teubner, 1981.

[JOPV] J.E. Jayne, J. Orihuela, A.J. Pallarés, and G. Vera, *σ-fragmentability of multivalued mappings and selection theorems*, J. Funct. Anal. **117** (1993), 243–273.

[JaRo] J.E. Jayne and C.A. Rogers, *Borel selectors for upper semicontinuous set valued maps*, Acta Math. **155** (1985), 41–79.

[JiMo] M. Jiménez and J.P. Moreno, *Renorming Banach spaces with the Mazur intersection property*, J. Funct. Analysis **144** (1997), 486–504.

[Joha] M. Johanis, *Approximation of Lipschitz mappings*, Serdica Math. **29** (2003), 141–148.

[JohaRy] M. Johanis and J. Rychtář, *On uniformly Gâteaux smooth norms and normal structure*, Proc. Amer. Math. Soc. **135** (2007), 1511–1514.

[JohnF] F. John, *Extremum problems with inequalities as subsidiary conditions, studies and essays presented to R. Courant on his 60th birthday, January 8, 1948*, Interscience Publishers, New York, 1948, 187–204.

[JohnK1] K. John *Counterexample to a conjecture of Grothendieck*, Math. Ann. **265** (1983), 169–179.

[Johnk1b] K. John, *Tensor products of several spaces and nuclearity*, Math. Ann. **256** (1983), 169–179.

[Johnk2] K. John, *On the compact non-nuclear problem*, Math. Ann. **287** (1990), 509–514.

[JTZ] K. John, H. Toruńczyk, and V. Zizler, *Uniformly smooth partitions of unity on superreflexive Banach spaces*, Studia Math. **70** (1981), 129–137.

[JoZi1] K. John and V. Zizler, *Smoothness and its equivalents in weakly compactly generated Banach spaces*, J. Funct. Anal. **15** (1974), 161–166.

[JoZi1b] K. John and V. Zizler, *Duals of Banach space which admit nontrivial smooth functions*, Bull. Australian Math. Soc. **11** (1974), 161–166.

[JoZi2] K. John and V. Zizler, *Topological linear spaces*, Lecture Notes, Charles University, Prague, 1978 (in Czech).

[JoZi3] K. John and V. Zizler, *A short proof of a version of Asplund averaging theorem*, Proc. Amer. Math. Soc. **73** (1979), 277–278.

[JJohn] J. Johnson, *Remarks on Banach spaces of compact operators*, J. Funct. Anal. **32** (1979), 304–311.

[Johns1] W.B. Johnson, *No infinite-dimensional P space admits a Markushevich basis*, Proc. Amer. Math. Soc. **28** (1970), 467–468.

[Johns2] W.B. Johnson, *On quasicomplements*, Pacific J. Math. **48** (1973), 113–118.

[JoLi1] W.B. Johnson and J. Lindenstrauss, *Some remarks on weakly compactly generated Banach spaces*, Israel J. Math. **17** (1974), 219–230.

[JoLi2] W.B. Johnson and J. Lindenstrauss, *Basic concepts in the geometry of Banach spaces*, Handbook of Banach Spaces I, Editors W.B. Johnson and J. Lindenstrauss, Elsevier, 2001, 1–84.

[JoLi3] W.B. Johnson and J. Lindenstrauss (Editors), *Handbook of Banach spaces I and II*, Elsevier, 2001, 2003.

[JLS] W.B. Johnson, J. Lindenstrauss, and G. Schechtman, *Banach spaces determined by their uniform structures*, Geom. Funct. Anal. **6** (1996), 430–470.

[JoOd] W.B. Johnson and E.W. Odell, *The diameter of the isomorphism class of a Banach space*, Annals Math. **162**(2) (2005), 423–437.

[JoRo] W.B. Johnson and H.P. Rosenthal, *On w^*-basic sequences and their applications to the study of Banach spaces*, Studia Math. **43** (1972), 77–92.

[JRZ] W.B. Johnson, H.P. Rosenthal, and M. Zippin, *On bases, finite-dimensional decompositions and weaker structures in Banach spaces*, Israel J. Math. **9** (1971), 488–506.

[JoZi0] W.B. Johnson and M. Zippin, *Subspaces and quotient spaces of $(\sum G_n)_{\ell_p}$ and $(\sum G_n)_0$*, Israel J. Math. **17** (1974), 50–55.

[JoZi] W.B. Johnson and M. Zippin, *Extension of operators from subspaces of $c_0(\Gamma)$ into $C(K)$ spaces*. Proc. Amer. Math. Soc. **107** (1989), 751–754.

[Jose] B. Josefson, *Weak sequential convergence in the dual of a Banach space does not imply norm convergence*, Ark. Math. **13** (1975), 79–89.

[Kade1] M.I. Kadec, *Unconditional convergence of series in uniformly convex spaces*, Uspechi Mat. Nauk SSSR **11** (1956), 185–190.

[Kade2] M.I. Kadec, *On linear dimension of the spaces L_p*, Uspechi Mat. Nauk SSSR **13** (1958), 95–98.

[Kade3] M.I. Kadec, *On spaces isomorphic to locally uniformly rotund spaces*, Izv. Vysš. Uč. Zav. Matem. **1** (1959), 51–57, and **1** (1961), 186–187.

[Kade4] M.I. Kadec, *Conditions on the differentiability of the norm of a Banach space*, Uspechi Mat. Nauk SSSR **20** (1965), 183–187.

[Kade4a] M.I. Kadec, *Proof of topological equivalence of separable infinite-dimensional Banach spaces*, Funct. Anal. Appl. **1** (1967), 53–62.

[KaPe] M.I. Kadec and A. Pełczyński, *Bases, lacunary sequences and complemented subspaces in the space $\mathcal{L}_p$*, Studia Math. **21** (1961/1962), 161–176.

[KaSn] M.I. Kadec and M.G. Snobar, *Certain functionals on the Minkowski compactum*, Mat. Zametki **10** (1971), 453–457.

[KaTo] M.I. Kadec and E.V. Tokarev, *On the dual spaces with countably many extreme points of their balls*, Izvestija Vysh. Ucz. Zavedenij **4** (1975), 98–99.

[KaKuLP] J. Kakol, W. Kubiś, and M. López-Pellicer, *Descriptive topology in selected topics of functional analysis*, to appear.

[Kaku1] S. Kakutani, *Weak convergence in uniformly convex spaces*, Tôhoku Math. J. **45** (1938), 188–193.

[Kaku2] S. Kakutani, *Simultaneous extensions of continuous functions considered as a positive linear operator*, Japanese J. Math. **17** (1940), 1–4.

[Kale1] O. Kalenda, *Valdivia compacta and subspaces of $C(K)$ spaces*, Extracta Math. **14** (1999), 355–371.

[Kale2] O. Kalenda, *An example concerning Valdivia compact spaces*, Serdica Math. J. **25**(2) (1999), 131–140.

[Kalt1] N.J. Kalton, *Quasi-Banach spaces*, Handbook of Banach Spaces II, Editors W.B. Johnson and J.Lindenstrauss, Elsevier, 2003, 1099–1130.

[Kalt2] N.J. Kalton, *Spaces of compact operators*, Math. Ann. **208** (1974), 267–278.

[Kalt3] N.J. Kalton, *Independence in separable Banach spaces*, Contemp. Math. **85** (1989), 319–324.

[Kalt3b] N.J. Kalton, *Coarse and uniform embeddings into reflexive spaces*, Q. J. Math. **58** (2007), 393–414.

[Kalt4] N.J. Kalton, *The nonlinear geometry of Banach spaces*, Rev. Mat. Complutense **21** (2008), 7–60.

[Kalt5] N.J. Kalton, *Lipschitz and uniform embeddings into ℓ_∞*, to appear.

[Kalt6] N.J. Kalton, *Uniform homeomorphisms of Banach spaces and asymptotic structure*, to appear.

[Kalt7] N.J. Kalton, *The uniform structure of Banach spaces*, to appear.

[Kalt8] N.J. Kalton, *Examples of uniformly homeomorphic Banach spaces*, to appear.

[KPR] N.J. Kalton, N.T. Peck, and J.W. Roberts, *An F-space sampler*, London Mathematical Society Lecture Notes **89**, Cambridge University Press, Cambridge (1985).

[KaRo] N.J. Kalton and J.W. Roberts, *A rigid subspace of L_0*, Trans. Amer. Math. soc. **266** (1981), 645–654.

[Karl] S. Karlin, *Bases in Banach spaces*, Duke Math. J. **15** (1948), 971–985.

[Karv] L.A. Karlovitz, *On nonexpansive mappings*, Proc. Amer. Math. Soc. **55** (1976), 321–325.

[Katz] Y. Katznelson, *An introduction to harmonic analysis*, Dover Publications, Inc., 1968.

[Kech] A.S. Kechris, *Classical descriptive set theory*, Graduate Texts in Mathematics **156**, Springer, 1994.

[Kelr] O.H. Keller, *Die Homoiomorphie der kompakten konvexen Mengen in Hilbertschen Raum*, Math. Ann. **105** (1931), 748–758.

[Kell] J.L. Kelley, *General topology*, D. van Nostrand, Princeton, 1957.

[Ken] P. Kenderov, *Monotone operators in Asplund spaces*, C.R. Acad. Bulgare Sci. **30** (1977), 963–964.

[Kirk] W.A. Kirk, *A fixed point theorem for mappings which do not increase distances*, Amer. Math. Monthly **72** (1965), 1004–1006.

[Klee1] V.L. Klee, *Some new results on smoothness and rotundity in normed linear spaces*, Math. Annalen **139** (1959), 51–63.

[Klee2] V.L. Klee, *Mappings into normed linear spaces*, Fund. Math. **49** (1960/1961), 25–34.

[Klee3] V.L. Klee, *Dispersed Chebyshev sets and coverings by balls*, Math. Ann. **257** (1981), 251–260.

[Klee4] V.L. Klee, *Do infinite-dimensional Banach spaces admit nice tilings?* Studia Scientiarum Mathematicarum Hungarica **21** (1986), 415–427.

[KoKo] A. Koldobsky and H. König, *Aspects of the isometric theory of Banach spaces*, Handbook of Banach Spaces I, Editors W.B. Johnson and J.Lindenstrauss, Elsevier, 2001, 899–939.

[KoFo] A.N. Kolmogorov and S.V. Fomin, *Introductory real analysis*, Dover Books on Advanced Mathematics, New York, 1970.

[KoTo] R. Komorowski and N. Tomczak-Jaegermann, *Banach spaces without local unconditional structure*, Israel J. Math. **89** (1995), 205–226, and ibidem **105** (1998), 85–92.

[Kosz] P. Koszmider, *Banach spaces of continuous functionals with few operators*, Math. Ann. **330** (2004), 151–183.

[Koth] G. Köthe, *Topological vector spaces I*, Springer, 1969.

[Kri] J.L. Krivine, *Sous-espaces de dimension finie des espaces de Banach réticulés*, Annals Math. **104** (1976), 1–29.

[KuRo] K. Kunen and H.P. Rosenthal, *Martingale proofs of some geometrical results in Banach space theory*, Pacific J. Math **100** (1982), 153–175.

[Kurz] J. Kurzweil, *On approximation in real Banach spaces*, Studia Math. **14** (1954), 213–231.

[KutTro] D. Kutzarova and S. Troyanski, *Reflexive Banach spaces without equivalent norms which are uniformly convex or uniformly differentiable in every direction*, Studia Math. **72** (1982), 91–95.

[Kwap] S. Kwapień, *Isomorphic characterization of inner product space by orthogonal series with vector valued coefficients*, Studia Math. **44** (1972), 583–595.

[Kwap2] S. Kwapień, *On Enflo's example of a Banach space without the approximation property*, Séminaire Goulaonic–Schwartz, 1972–1973, École Polytechnique, Paris.

[Lace1] H.E. Lacey, *Separable quotients of Banach spaces*, An. Acad. Brasil Ci. **44** (1972), 185–189.

[Lace2] H.E. Lacey, *The isometric theory of classical Banach spaces*, Springer, 1974.

[Lanc] G. Lancien, *On uniformly convex and uniformly Kadec–Klee renormings*, Serdica Math. J. **21** (1995), 1–18.

[Lanc2] G. Lancien, *A survey on the Szlenk index and some of its applications*, Rev. Real Acad. Cien. Serie A. Mat. **100**(1–2) (2006), 209–235.

[LasLi] J.M. Lasry and P.L. Lions, *A remark on regularization in Hilbert spaces*, Israel J. Math. **55** (1986), 257–266.

[Lau] K.S. Lau, *Farthest points in weakly compact sets*, Israel J. Math. **22** (1975), 168–174.

[LeWh] E.B. Leach and J.H.M. Whitfield, *Differentiable functions and rough norms on Banach spaces*, Proc. Amer. Math. Soc. **33** (1972), 120–126.

[Lind1] J. Lindenstrauss, *On nonlinear projections in Banach spaces*, Michigan Math. J. **11** (1954), 263–287.

[Lind2] J. Lindenstrauss, *On some subspaces of ℓ_1 and c_0*, Bull. Res. Council Israel. **10** (1961), 74–80.

[Lind3] J. Lindenstrauss, *On the modulus of smoothness and divergent series in Banach spaces*, Michigan J. Math. **10** (1963), 241–252.

[Lind4] J. Lindenstrauss, *On operators which attain their norms*, Israel J. Math. **3** (1963), 139–148.

[Lind5] J. Lindenstrauss, *Extension of compact operators*, Mem. Amer. Math. Soc. **48** (1964).

[Lind5b] J. Lindenstrauss, *On a certain subspace of ℓ_1*, Bull. Acad. Polon. Sci. **12** (1964), 539–542.

[Lind6] J. Lindenstrauss, *On the extension of operators with a finite-dimensional range*, Illinois J. Math. **8**, (1964), 488–499.

[Lind7] J. Lindenstrauss, *On reflexive spaces having the metric approximation property*, Israel J. Math. **3** (1965), 199–204.

[Lind8] J. Lindenstrauss, *Notes on Klee's paper "Polyhedral sections of convex bodies"*, Israel J. Math. **4** (1966), 235–242.

[Lind9] J. Lindenstrauss, *On nonseparable reflexive Banach spaces*, Bull. Amer. Math. Soc. **72** (1966), 967–970.

[Lind9b] J. Lindenstrauss, *On extreme points in ℓ_1*, Israel J. Math. **4** (1966), 59–61.

[Lind10] J. Lindenstrauss, *On a theorem of Murray and Mackey*, An. Acad. Brasil. Ci. **39** (1967), 1–6.

[Lind11] J. Lindenstrauss, *On complemented subspaces of m*. Israel J. Math. **5** (1967), 153–156.

[Lind12] J. Lindenstrauss, *On subspaces of Banach spaces without quasicomplements*, Israel J. Math. **6** (1968), 36–38.

[Lind12b] J. Lindenstrauss, *On James' paper "Separable conjugate spaces"*, Israel J. Math. **9** (1971), 279–284.

[Lind13] J. Lindenstrauss, *Weakly compact sets, their topological properties and spaces they generate*, Ann. Math. Studies **69** (1972).

[Lind14] J. Lindenstrauss, *Uniform embeddings, homeomorphisms and quotient maps between Banach spaces (A short survey)*, Topology Appl. **85** (1998), 265–279.

[LiPe] J. Lindenstrauss and A. Pełczyński, *Absolutely summing operators in $\mathcal{L}_p$ spaces and their applications*, Studia Math. **29** (1968), 275–326.

[LiPe2] J. Lindenstrauss and A. Pełczyński, *Contributions to the theory of the classical Banach spaces*, J. Funct. Analysis **8** (1971), 225–249.

[LiPh] J. Lindenstrauss and R.R. Phelps, *Extreme point properties of convex bodies in reflexive Banach spaces*, Israel J. Math. **6** (1968), 39–48.

[LPT] J. Lindenstrauss, D. Preiss, and J. Tišer, *Fréchet differentiability of Lipschitz functions and porous sets in Banach spaces*, to appear.

[LiRo1] J. Lindenstrauss and H.P. Rosenthal, *Automorphisms in c_0, ℓ_1 and m*, Israel J. Math. **7** (1969), 227–239.

[LiRo2] J. Lindenstrauss and H.P. Rosenthal, *The $\mathcal{L}_p$ spaces*, Israel J. Math. **7** (1969), 325–349.

[LiSt] J. Lindenstrauss and C. Stegall, *Examples of separable spaces which do not contain ℓ_1 and whose duals are nonseparable*, Studia Math. **54** (1975), 81–105.

[LiTz1] J. Lindenstrauss and L. Tzafriri, *On the complemented subspaces problem*, Israel J. Math. **9** (1971), 263–269.

[LiTz2] J. Lindenstrauss and L. Tzafriri, *Classical Banach spaces*, Lecture Notes in Mathematics **338**, Springer, 1973.

[LiTz3] J. Lindenstrauss and L. Tzafriri, *Classical Banach spaces I, Sequence spaces*, Springer, 1977.

[LiTz4] J. Lindenstrauss and L. Tzafriri, *Classical Banach spaces II, Function spaces*, Springer, 1979.

[LjSo] L.A. Ljusternik and B.I. Sobolev, *Elements of functional analysis*, New Delhi, 1961.

[Lom] V. Lomonosov, *A counterexample to the Bishop–Phelps theorem in complex spaces*, Israel J. Math. **115** (2000), 25–28.

[LoMo] M. López-Pellicer and V. Montesinos, *Cantor sets in the dual of a separable Banach space: Applications*, General Topology in Banach Spaces, Editor T. Banakh (QA322.2 .G45 2001), Nova Science Publishers.

[Lov] A. Lovaglia, *Locally uniformly convex spaces*, Trans. Amer. Math. Soc. **78** (1955), 225–238.

[LuMa] J. Lukeš and J. Malý, *Measure and integral*, Matfyzpress, Prague 1994.

[LyVa] Y. Lyubich and L. N. Vaserstein, *Isometric embeddings between classical Banach spaces, cubature formulas, and spherical designs*, Geom. Dedicata **47** (1993), 327–362.

[Mack] G. Mackey, *Note on a theorem of Murray*, Bull. Amer. Math. Soc. **52** (1946), 322–325.

[MaTr] R.P. Maleev and S. Troyanski, *Smooth norms in Orlicz spaces*, Canad. J. Math. **34** (1991), 74–82.

[Maka] B.M. Makarov, *One characterization of Hilbert space*, Mat. Zametki **26** (1979), 739–746.

[Mank] P. Mankiewicz, *On the differentiability of Lipschitz mappings in Fréchet spaces*, Studia Math. **XLV**, (1973), 15–29.

[Mar] W. Marciszewski, *On Banach spaces $C(K)$ isomorphic to $c_0(\Gamma)$*, Studia Math. **156** (2003), 295–302.

[Maur0] B. Maurey, *Un théorème de prolonguement*. C.R. Acad. Sci. Paris A **279** (1974), 329–332.

[Maur] B. Maurey, *Banach spaces with few operators*, Handbook of Banach Spaces II, Editors W.B. Johnson and J.Lindenstrauss, Elsevier, 2003, 1247–1332.

[May] H.B. Maynard, *A geometrical characterization of Banach spaces with the Radon-Nikodým property*, Trans. Amer. Math. Soc. **185** (1973), 493–500.

[Mazu] S. Mazur, *Une remarque sur l'homémorphic des champs fonctionnels*, Studia Math. **1** (1929), 83–85.

[Mazu2] S. Mazur, *Uber konvexe Mengen in linearen normierten Raumen*, Studia Math. **4** (1933), 70–84.

[McOb] P.W. McCartney and R.C. O'Brian, *A separable space with the Radon–Nikodým property that is not isomorphic to a subspace of a separable dual*, Proc. Amer. Math. Soc. **78** (1980), 40–42.

[MPVZ] D. McLaughlin, R. Poliquin, J. Vanderwerff, and V. Zizler, *Second-order Gâteaux differentiable bump functions and approximations in Banach spaces*, Can. J. Math. Vol. **45**(3) (1993), 612–625.

[Megg] R.E. Megginson, *An introduction to Banach space theory*, GTM **183**, Springer, 1998.

[Meir1] A. Meir, *On the moduli of smoothness of L_p spaces*, Rendiconti di Math. **4** (1984), 215–219.

[Meir2] A. Meir, *On the uniform convexity of L_p spaces*, Illinois J. Math. **28** (1984), 420–424.

[Merc] S. Mercourakis, *On weakly countably determined Banach spaces*, Trans. Amer. Math. Soc. **300** (1987), 307–327.

[MeNe] S. Mercourakis and S. Negrepontis, *Banach spaces and topology II*, Recent Progress in General Topology, Editors M. Hušek and J. van Mill, Elsevier Science Publishers BV, 1992, 493–536.

[Mesh] V.Z. Meshkov, *Smoothness properties in Banach spaces*, Studia Math. **63** (1978), 111–123.

[Mich] E. Michael, *Continuous selections I*, Ann. Math. **63** (1956), 361–382.

[MiRu] E. Michael and M.E. Rudin, *A note on Eberlein compacts*, Pacific J. Math. **72** (1972), 487–495.

[Milm1] V.D. Milman, *Geometric theory of Banach spaces. Part I. Theory of basic and minimal systems*, Uspechi Mat. Nauk SSSR **25** (1970), 113–173.

[Milm2] V.D. Milman, *Geometric theory of Banach spaces II, geometry of the unit sphere*, Russian Math. Surveys **26** (1971), 79–163.

[MiSc] V.D. Milman and G. Schechtman, *Asymptotic theory of finite-dimensional normed spaces*, Lecture Notes in Mathematics **1200**, Springer, 1986.

[Miln] J. Milnor, *Analytic proofs of the "Hairy Ball Theorem" and the Brouwer fixed point theorem*. Ameri. Math. Monthly, **85**(7) (1978), 521–524.

[MMOT] A. Moltó, V. Montesinos, J. Orihuela, and S. Troyanski, *Weakly uniformly rotund Banach spaces*, Comment. Math. Univ. Carolinae **39** (1998), 749–753.

[MOTV1] A. Moltó, J. Orihuela, S. Troyanski, and M. Valdivia, *On weakly locally uniformly rotund Banach spaces*, J. Funct. Anal. **163** (1999), 252–271.

[MOTV2] A. Moltó, J. Orihuela, S. Troyanski, and M. Valdivia, *A non-linear transfer technique for renorming*, Springer LNM **1951**, 2009.

[Mtr] I. Monterde, *Some compactness criteria in locally convex and Banach spaces*, PHD Dissertation, Universidad Politécnica de Valencia, Valencia, Spain, 2009.

[Mont1] V. Montesinos, *Solution to a problem of S. Rolewicz*, Studia Math. **81** (1985), 65–69.

[Mont2] V. Montesinos, *Drop property equals reflexivity*, Studia Math. **87** (1987), 93–100.

[Mont3] V. Montesinos, *On the drop property*, Notas de Matemática, Vol. **1**, Seminar on Functional Analysis 1987, Universidad de Murcia, 1988, 69–123.

[MoTo] V. Montesinos and J.R. Torregrosa, *Sobre espacios de Banach localmente uniformemente rotundos*, Revista de la Real Academia de Ciencias **86**, (1992), 263–277.

[MoSo] W.B. Moors and S. Somasundaram, *A Gâteaux differentiability space that is not weak Asplund*, Proc. Amer. Math. Soc. **134** (2006), 2745–2754.

[Morr] P.D. Morris, *Disappearance of extreme points*, Proc. Amer. Math. Soc. **88** (1983), 244–246.

[Muño] M. Muñoz, *Índice de K-determinación de espacios topológicos y σ-fragmentabilidad de aplicaciones*, PhD Thesis, Universidad de Murcia, 2004.

[Muji] J. Mujica, *Separable quotients of Banach spaces*, Rev. Math. **10** (1997), 299–330.

[Murp] G.J. Murphy, *C^*-algebras and operator theory*, Academic Press, 1996.

[Murr] F.J. Murray, *Quasi complements and closed projections in reflexive Banach spaces*, Trans. Amer. Math. Soc. **58** (1945), 77–95.

[Nami1] I. Namioka, *Separate continuity and joint continuity*, Pacific J. Math. **51** (1974), 515–531.

[Nami2] I. Namioka, *Radon–Nikodým compact spaces and fragmentability*, Mathematika **34** (1987), 258–281.

[Nami3] I. Namioka, *Fragmentability in Banach spaces: Interaction of topologies*, Lecture Notes Paseky, 1999.

[NaPh] I. Namioka and R.R. Phelps, *Banach spaces which are Asplund spaces*, Duke Math. J. **42** (1968), 735–750.

[NaWh] I. Namioka and R.F. Wheeler, *Gul'ko's proof of the Amir–Lindenstrauss theorem*, Contemp. Math. **52** (1986), 113–120.

[Nata] I.P. Natanson, *Theory of functions of real variable*, Moscow, 1950.

[Negr] S. Negrepontis, *Banach spaces and topology*, Handbook of Set-Theoretic Topology, Editors K. Kunen and J.E. Vaughan, Elsevier Science Publishers BV, 1984.

[NeTs] S. Negrepontis and A. Tsarpalias, *A nonlinear version of the Amir–Lindenstrauss method*, Israel J. Math. **38** (1981), 82–94.

[NeZa] A. Nekvinda and L. Zajíček, *A simple proof of Rademacher theorem*, Časopis pro pěstování matematiky **113** (1988), 337–341.

[NeSe] A.M. Nemirovski and S.M. Semenov, *On polynomial approximation in function spaces*, Mat. Sbornik **21** (1973), 255–277.

[Niss] A. Nissenzweig, *w^*-sequential convergence*, Israel J. Math. **22** (1975), 266–272.

[OdRo] E.W. Odell and H.P. Rosenthal, *A double dual characterization of separable Banach spaces containing ℓ_1*, Israel J. Math. **20** (1975), 375–384.
[OdSc1] E.W. Odell and T. Schlumprecht, *The distortion problem*, Acta Math. **173** (1994), 259–281.
[OdSc2] E.W. Odell and T. Schlumprecht, *Distortion and stabilized structure in Banach spaces, New geometric phenomenon for Banach and Hilbert spaces*, Proceedings of the International Congress of Math. Zürich, Switzerland, 1994.
[OdSc3] E.W. Odell and T. Schlumprecht, *On asymptotic properties of Banach spaces under renormings*, J. Amer. Math. Soc. **11** (1998), 175–188.
[OdSc4] E.W. Odell and T. Schlumprecht, *Distortion and asymptotic structure*, Handbook of Banach Spaces II, Editors W.B. Johnson and J.Lindenstrauss, Elsevier, 2003, 1333–1360.
[Oja1] E. Oja, *A proof of the Simons inequality*, Acta Comment. Univ. Tartu. Math. **2** (1998), 27–28.
[Oja2] E. Oja, *A short proof of a characterization of reflexivity of James*, Proc. Amer. Math. Soc. **126** (1998), 2507–2508.
[Orih] J. Orihuela, *On weakly Lindelöf Banach spaces*, Progress in Funct. Anal. K.D. Bierstedt, Editors J. Bonet, J. Horváth, and M. Maestre, Elsevier Science Publishers B.V., 1992.
[OSV] J. Orihuela, W. Schachermayer, and M. Valdivia, *Every Radon–Nikodým Corson compact is an Eberlein compact*, Studia Math. **98** (1991), 157–174.
[OrVa] J. Orihuela and M. Valdivia, *Projective generators and resolutions of identity in Banach spaces*, Rev. Mat. Univ. Complutense, Madrid **2**, Suppl. Issue (1990), 179–199.
[Orn91] P. Orno, *On J. Borwein's concept of sequentially reflexive Banach spaces*, Banach Space Bulletin Board, 1991.
[Oxto] J.C. Oxtoby, *Measure and category*, Graduate Texts in Mathematics, Springer, 1980.
[Part] J.R. Partington, *Equivalent norms on spaces of bounded functions*, Israel J. Math. **35** (1980), 205–209.
[PWZ] J. Pechanec, J.H.M. Whitfield, and V. Zizler, *Norms locally dependent on finitely many coordinates*, An. Acad. Brasil Ci. **53** (1981), 415–417.
[PHK] J. Pelant, P. Holický, and O. Kalenda, *$C(K)$ spaces which cannot be uniformly embedded into $c_0(\Gamma)$*, Fund. Math. **192** (2006), 245–254.
[Pelc1] A. Pełczyński, *A property of multilinear operations*, Studia Math. **16** (1957), 173–182.
[Pelc1b] A. Pełczyński, *A connection between weakly unconditional convergence and weakly completeness of Banach spaces*, Bull. Acad. Pol. Sci. Ser. Sci. Math. Astron. Phys. **6** (1958), 251–253 (unbound insert). (English, with Russian summary.)
[Pelc2] A. Pełczyński, *On the isomorphism of the spaces m and M*, Bull. Acad. Polon. **VI** (1958), 695–696.
[Pelc3] A. Pełczyński, *Projections in certain Banach spaces*, Studia Math. **19** (1960), 209–228.
[Pelc3b] A. Pełczyński, *Banach spaces on which every unconditionally converging operator is weakly compact*, Bull. Acad. Pol. Sci. **10** (1962), 641–648.
[Pelc3c] A. Pełczyński, *A note on the paper of I. Singer "Basic sequences and reflexivity of Banach spaces"*, Studia Math. **21** (1962), 371–374.
[Pelc3d] A. Pełczyński, *A proof of Eberlein–Šmulyan theorem by an application of basic sequences*, Bull. Acad. Pol. Sci. Ser. Math. Astron. Phys. **12** (1964), 543–548.
[Pelc4] A. Pełczyński, *On Banach spaces containing $L_1(\mu)$*, Studia Math. **30** (1968), 231–246.

[Pelc5] A. Pełczyński, *Linear extensions, linear averagings and their applications to linear topological calssification of spaces of continuous functions*, Dissert. Math. **58** (1968).

[Pelc6] A. Pełczyński, *On $C(S)$ subspaces of separable Banach spaces*, Studia Math. **31** (1968), 231–246.

[Pelc6b] A. Pełczyński, *Any separable Banach space with the bounded approximation property is a complemented subspace of a Banach space with a basis*, Studia Math. **40** (1971), 239–242.

[Pelc7] A. Pełczyński, *Remarks on John's theorem on the ellipsoid of maximal volume inscribed into a convex body in* $\mathbb{R}^n$, Note Mat. **10** (1990), Supp. 2, 395–410.

[Pel8] A. Pełczyński, *Selected problems on the structure of complemented subspaces of Banach spaces*, Methods in Banach Space Theory, Editors J.M.F. Castillo and W.B. Johnson, London Math. Soc. Lecture Notes Series **337**, 2006, 341–354.

[PeBe] A. Pełczyński and C. Bessaga, *Some aspects of the present theory of Banach spaces*, in 1979 Edition of S. Banach: Travaux sur L'Analyse Fonctionnelle, Warszaw, 1979.

[PeSi] A. Pełczyński and I. Singer, *On non-equivalent bases and conditional bases in Banach spaces*, Studia Math. **25** (1964/1965), 5–25.

[PeSz] A. Pełczyński and W. Szlenk, *An example of a nonshrinking basis*, Rev. Roumaine Math. Pures et Appl. **10** (1965), 961–966.

[PeWo] A. Pełczyński and M. Wojciechowski, *Sobolev spaces*, Handbook of Banach Spaces II, Editors W.B. Johnson and J.Lindenstrauss, Elsevier, 2003, 1361–1423.

[Phel0] R.R. Phelps, *A representation theorem for bounded convex sets*, Proc. Amer. Math. Soc. **11** (1960), 876–983.

[Phel1] R.R. Phelps, *Lectures on Choquet's theorem*, Van Nostrand, Princeton, 1966.

[Phel1b] R.R. Phelps, *Dentability and extreme points in Banach spaces*, J. Funct. Anal. **17** (1974), 78–90.

[Phelps] R.R. Phelps, *Convex functions, monotone operators and differentiability*, Lecture Notes in Mathematics **1364**, Springer, 1989.

[Phil] R.S. Phillips, *On linear transformations*, Trans. Amer. Math. Soc. **48** (1940), 516–541.

[Pfi] H. Pfitzner, *Boundaries for Banach spaces determine weak compactness*, to appear, electronic http://hal.archives-ouvertes.fr/hal-00300244/en/.

[Pisi1] G. Pisier, *Martingales with values in uniformly convex spaces*, Israel J. Math. **20** (1975), 326–350.

[Pisi2] G. Pisier, *Counterexamples to a conjecture of Grothendieck*, Acta Math. **151** (1983), 181–208.

[Pisi3] G. Pisier, *Factorization of linear operators and geometry of Banach spaces*, CBMS AMS **60** (1985).

[Pisi4] G. Pisier, *The volume of convex bodies and Banach space geometry*, Cambridge University Press, 1989.

[Pleb] G. Plebanek, *Banach spaces of continuous functions with few operators*, Math. Annalen **328**(1) (2004), 151–183.

[Plic1] A.N. Plichko, *On projectional resolution of the identity operator and Markushevich bases*, Soviet Mat. Dokl. **25** (1982), 386–389.

[Plic2] A.N. Plichko, *Projectional decompositions, Markushevich bases and equivalent norms*, Mat. Zametki **34** (1983), 719–726.

[Plic3] A.N. Plichko, *Bases and complements in nonseparable Banach spaces*, Sibir. Mat. J. **25** (1984), 155–162.

[Plic4] A.N. Plichko, *On bounded biorthogonal systems in some function spaces*, Studia Math. **84** (1986), 25–37.

[PliYo] A.N. Plichko and D. Yost, *The Radon–Nikodým property does not imply the separable complementation property*, J. Funct. Anal. **180** (2001), 481–487.

[Pol1] R. Pol, *Concerning function spaces on separable compact spaces*, Bull. de L'Acad. Polon. Serie des Sci., Math., Astr., et Phys. **25** (1977), 993–997.

[Pol2] R. Pol, *A function space $C(X)$ which is weakly Lindelöf but not weakly compactly generated*, Studia Math. **64** (1979), 279–285.

[Pol3] R. Pol, *On a question of H.H. Corson and some related problems*, Fund. Math. **109** (1980), 143–154.

[Pol4] R. Pol, *On pointwise and weak topology in function spaces*, Warszaw University preprint **4/84**, 1984.

[Prei1] D. Preiss, *Almost differentiability of convex functions in Banach spaces and determination of measures by their values on balls*, London Mathematical Society Lecture Notes Series **158**, Geometry of Banach Spaces, Proceedings of Conference held in Strobl, Austria, Editors P.F.X. Müller and W. Schachermayer, 1989.

[Prei2] D. Preiss, *Differentiability of Lipschitz functions on Banach spaces*, J. Funct. Anal. **91** (1990), 312–345.

[PPN] D. Preiss, R.R. Phelps, and I. Namioka, *Smooth Banach spaces, weak Asplund spaces and monotone or usco mappings*, Israel J. Math. **72** (1990), 257–279.

[PrSi] D. Preiss and P. Simon, *A weakly pseudocompact subspace of a Banach space is weakly compact*, Comment. Math. Univ. Carolinae **15** (1974), 603–610.

[PrZa] D. Preiss and L. Zajíček, *Fréchet differentiation of convex functions in a Banach space with a separable dual*, Proc. Amer. Math. Soc. **91** (1984), 202–204.

[Ptak1] V. Pták, *On a theorem of W.F. Eberlein*, Studia Math. **14** (1954), 276–287.

[Ptak2] V. Pták, *Biorthogonal systems and reflexivity of Banach spaces*, Czechoslovak Math. J. **9**(84) (1959), 319–326.

[Qu] S. Quian, *Nowhere differentiable Lipschitz maps and the Radon–Nikodým property*, J. Math. Anal. Appl. **185** (1994), 613–616.

[Rain1] J. Rainwater, *Weak convergence of bounded sequences*, Proc. Amer. Math. Soc. **14** (1963), 999.

[Rain2] J. Rainwater, *Day's norm on $c_0(\Gamma)$*, Proc. Amer. Math. Soc. **22** (1969), 335–339.

[Raja1] M. Raja, *Measurabilité de Borel et renormages dans les espaces de Banach*, PhD Thesis, Université de Bourdeaux, 1998.

[Raja2] M. Raja, *Kadec norms and Borel sets in Banach spaces*, Studia Math. **136** (1999), 1–16.

[Raja3] M. Raja, *On locally uniformly rotund norms*, Mathematika **46** (1999), 343–358.

[Raja4] M. Raja, *Weak* locally uniformly rotund norms and descriptive compact spaces*, J. Funct. Anal. **197** (2003), 1–13.

[Rams] F.P. Ramsey, *On a problem of formal logic*, Proc. London Math. Soc. **30** (1930), 264–286.

[Read] C. Read, *A solution to the invariant subspace problem on the space ℓ_1*. Bull. London Math. Soc. **17** (1985), 305–317.

[Rest] G. Restrepo, *Differentiable norms in Banach spaces*, Bull. Amer. Math. Soc. **70** (1964), 413–414.

[Rezn] E.A. Rezniczenko, *Normality and collectionwise normality of function spaces*, Vestnik. Mosk. Univ. Ser. Mat. (1990), 56–58. English trans. Moscow Univ. Math. Bull. **45**, no. 6 (1990), 25–26. MR **92b**:46003.

[Ribe0] M. Ribe, *On uniformly homeomorphic normed spaces*, Ark. Math. **14** (1976), 237–244.

[Ribe1] M. Ribe, *Existence of separable uniformly homeomorphic nonisomorphic Banach spaces*, Israel J. Math. **48** (1984), 139–147.

[Rieff] M.A. Rieffel, *The Radon–Nikodým theorem for the Bochner integral*, Trans. Amer. Math. Soc. **131** (1968), 466–487.

[Robe1] J.W. Roberts, *A compact convex set with no extreme points*, Studia Math. **60** (1977), 255–266.

[Robe2] J.W. Roberts, *Pathological compact convex sets in $L_p[0,1]$, $0 \leq p < 1$*, The Altgeld Book, University of Illinois Functional Analysis Seminar, 1975, 1976.

[Rock] R.T. Rockafellar, *Maximal monotone relations and the second derivatives of nonsmooth functions*, Ann. Inst. H. Poincare, Anal. Non Lin. **2** (1985), 167–184.

[Rode] G. Rodé, *Superkonvexitat und schwache Kompaktheit*, Arch. Math. **36** (1981), 62–72.

[Roge] C.A. Rogers, *A less strange version of Milnor's proof of Brouwer's fixed-point theorem*, Amer. Math. Monthly, **87**(7) (1980), 525–527.

[RoJa] C.A. Rogers and J.E. Jayne, *$\mathcal{K}$-analytic sets*, Academic Press, 1980.

[Rose1] H.P. Rosenthal, *Projections onto translation-invariant subspaces of $L_p(G)$*, Mem. Amer. Math. Soc. **63** (1966).

[Rose2] H.P. Rosenthal, *On quasi complemented subspaces of Banach spaces with an appendix on compactness of operators from $L^p(\mu)$ to $L^r(\mu)$*, J. Funct. Anal. **4** (1969), 176–214.

[Rose3] H.P. Rosenthal, *On injective Banach spaces and the spaces $L^\infty(\mu)$ for finite measures μ*, Acta Math. **124** (1970), 205–247.

[Rose4] H.P. Rosenthal, *On the subspaces of L_p spanned by sequences of independent random variables*, Israel J. Math. **8** (1970), 273–303.

[Rose5] H.P. Rosenthal, *The heredity problem for weakly compactly generated Banach spaces*, Comp. Math. **28** (1974), 83–111.

[Rose6] H.P. Rosenthal, *A characterization of Banach space containing ℓ_1*, Proc. Nat. Acad. Sci. USA **71** (1974), 2411–2413.

[Rose7] H.P. Rosenthal, *Pointwise compact subsets of the first Baire class*, Amer. J. Math. **99** (1977), 362–378.

[Rose8] H.P. Rosenthal, *Some recent discoveries in the isomorphic theory of Banach spaces*, Bull. Amer. Math. Soc. **84** (1978), 803–831.

[Rose9] H.P. Rosenthal, *Weak*-Polish Banach spaces*, J. Funct. Anal. **76** (1988), 267–316.

[Rose10] H.P. Rosenthal, *The Banach spaces $C(K)$*, Handbook of Banach Spaces II, Editors W.B. Johnson and J.Lindenstrauss, Elsevier, 2003, 1549–1602.

[Royd] H.L. Royden, *Real analysis*, Third Edition, Macmillan, 1988.

[Rudi1] W. Rudin, *Continuous functions on compact spaces without perfect subsets*, Proc. Amer. Math. Soc. **8** (1957), 39–42.

[Rudi2] W. Rudin, *Real and complex analysis*, McGraw Hill, New York, 1974.

[Rudi3] W. Rudin, *Functional analysis*, McGraw Hill, 1973.

[RuSt] W.M. Ruess and C. Stegall, *Extreme points in duals of operator spaces*, Math. Ann. **261** (1982), 535–546.

[Ruto] D. Rutovitz, *Some parameters associated with finite-dimensional Banach spaces*, J. London Math. Soc. **40** (1965), 241–255.

[Ryc] J. Rychtář, *Pointwise uniformly rotund norms*, Proc. Amer. Math Soc. **133** (2005), 2259–2266.

[Schc] W. Schachermayer, *Some more remarkable properties of the James tree space*, Cont. Math. **85** (1987), 465–496.

[SchSerWer] W. Schachermayer, A. Sersouri, and E. Werner, *Moduli of non-dentability and the Radon-Nikodým property in Banach spaces*, Israel J. Math. **65** (1989), 225–257.

[Scha] R. Schatten, *A theory of cross spaces*, Annals of Mathematics Studies No. **26**, Princeton University Press (1950).

[Schau] J. Schauder, *Der Fizpunktsatz in Funktionalräumen*, Studia Math. **2** (1930), 171–180.

[ScWh] G. Schlüchtermann and R.F. Wheeler, *On strongly WCG Banach spaces*, Math. Z. **199** (1988), 387–398.

[Sema1] Z. Semadeni, *Banach spaces non-isomorphic to their cartesian squares. II*, Bull. Acad. Polon. Ser. Sci. Math. Astron. et Phys. **8** (1960), 81–84.

[Sema2] Z. Semadeni, *Banach spaces of continuous functions*, Polish Scientific Publishers, Warszawa, 1971.

[Shap] B.E. Shapirovskii, *Special types of embeddings in Tychonoff cubes: Subspaces of Σ-products and cardinal invariants*, Collection, Topology **VII**, Colloq. Math. Soc. Janos Bolyai 23, North-Holland, 1980, 1055–1086.

[ShSt1] S. Shelah and J. Steprans, *A Banach space on which there are few operators*, Proc. Amer. Math. Soc. **104** (1988), 101-105.

[Simo] S. Simons, *A convergence theorem with boundary*, Pacific J. Math. **40** (1972), 703–708.

[Sing1] I. Singer, *Bases in Banach spaces I*, Springer, 1970.

[Sing2] I. Singer, *On the problem of nonsmoothness of nonreflexive second conjugate spaces*, Bull. Austral. Math. Soc. **12** (1975), 407–416.

[Sing3] I. Singer, *Bases in Banach spaces II*, Springer, 1981.

[Smul1] V.L. Šmulyan, *On the principle of inclusion in the space of type (B)*, Mat. Sbornik (N. S.), **5** (1939), 317–328 (in Russian).

[Smul2] V.L. Šmulyan, *Sur la dérivabilité de la norme dans l'espace de Banach*, C.R. Acad. Sci. URSS (Doklady) N.S. **27**, 1940, 643–648.

[Sobc] A. Sobczyk, *Projections of the space m onto its subspace c_0*, Bull. Amer. Math. Soc. **47** (1941), 938–947.

[Steg1] C. Stegall, *The Radon–Nikodým property in conjugate spaces*, Trans. Amer. Math. Soc. **206** (1975), 213–223.

[Steg2] C. Stegall, *Optimization of functions on certain subsets of Banach spaces*, Math. Ann. **236** (1978), 171–176.

[Steg3] C. Stegall, *A proof of the principle of local reflexivity*, Proc. Amer. Math. Soc. **78** (1980), 154–156.

[Steg4] C. Stegall, *The Radon–Nikodým property in conjugate Banach spaces II*, Trans. Amer. Math. Soc. **264** (1981), 507–519.

[Steg4b] C. Stegall, *Optimization and differentiability in Banach spaces*, J. Linear Algebra Appl. **84** (1986), 191–211.

[Steg5] C. Stegall, *More facts about conjugate Banach spaces eith the Radon–Nikodým property II*, Acta Univ. Carolinae–Math. Phys. **32** (1991), 47–54.

[Stern] S. Sternberg, *Lectures on differential geometry*, Prentice-Hall, Englewood Cliffs, NJ, 1964.

[Stra] S. Straszewicz, *Über exponierte Punkte abgeschlossener Punktmengen*, Fund. Math. **24** (1935), 139–143.

[Stro] K. R. Stromberg, *Introduction to classical real analysis*, Wadsworth International Group, Belmont, CA, 1981.

[Szan1] A. Szankowski, *An example of a universal Banach space*, Israel J. Math. **11** (1972), 292–296.

[Szan1b] A. Szankowski, *Subspaces without the approximation property*, Israel J. Math. **30** (1978), 123–129.

[Szan2] A. Szankowski, *$B(H)$ does not have the approximation property*, Acta Math. **146** (1981), 89–108.

[Szar] S.J. Szarek, *The finite-dimensional basis problem with an appendix on nets of Grassmann manifolds*, Acta Math. **151** (1983), 153–179.

[Szle] W. Szlenk, *The nonexistence of a separable reflexive Banach space universal for all separable reflexive Banach spaces*, Studia Math. **30** (1968), 53–61.

[Taco] D.G. Tacon, *The conjugate of a smooth Banach space*, Bull. Austral. Math. Soc. **2** (1970), 415–425.

[Tala1] M. Talagrand, *Sur la structure borelienne des espaces analytiques*, Bull. Sc. Math. **101** (1977), 415–422.

[Tala2] M. Talagrand, *Comparison des boréliens d'un espace de Banach pour topologies faibles et fortes*, Indiana Math. J. **27** (1978), 1001–1004

[Tala3] M. Talagrand, *Espaces de Banach faiblement $\mathbb{K}$-analytiques*, Ann. Math. **119** (1979), 407–438.

[Tala3a] M. Talagrand, *Sur la propiété de Radon–Nikodým dans les espaces de Banach réticulés*, C.R. Acad. Sci. Paris, **288**, 1979, no. saer A, 907–910.

[Tala3b] M. Talagrand, *Serabilité vague dans l'espace des measures sur un compact*, Israel J. Math. **37** (1980), 171–180.

[Tala4] M. Talagrand, *Sur les espaces de Banach contenant $\ell^1(\tau)$*, Israel J. Math. **40** (1981), 324–330.

[Tala5] M. Talagrand, *Pettis integral and measure theory*, Memoirs Amer. Math. Soc. **307** (1984).

[Tala6] M. Talagrand, *Renormages de quelques $C(K)$*, Israel J. Math. **54** (1986), 327–334.

[Tan0] W.-K. Tang, *On Fréchet differentiability of convex functions on Banach spaces*, Comment. Math. Univ. Carolinae **36** (1995), 249–253.

[Tan1] W.-K. Tang, *A note on preserved smoothness*, Serdica Math. J. **22** (1996), 29–32

[Tan] W.-K. Tang, *Uniformly differentiable bump functions*, Arch. Math. **68** (1997), 55–59.

[Todo1] S. Todorcevic, *Topics in topology*, Lecture Notes in Mathematics **1652**, Springer, 1997.

[Todo2] S. Todorcevic, *Compact subsets of the first Baire class*, J. Amer. Math. Soc. **12** (1999), 1179–1212.

[Tomc0] N. Tomczak-Jaegermann, *The moduli of smoothness and convexity and the Rademacher averages of trace classes S_p $(1 \le p < \infty)$*, Studia Math. **50** (1974), 163–182.

[Tomc] N. Tomczak-Jaegermann, *Banach–Mazur distances and finite-dimensional operator ideals*, Pitman Monographs and Surveys in Pure and Applied Mathematics **38**, 1989.

[Toru1] H. Toruńczyk, *Smooth partitions of unity on some nonseparable Banach spaces*, Studia Math. **46** (1973), 43–51.

[Toru2] H. Toruńczyk, *Characterizing Hilbert space topology*, Fund. Math. **111** (1981), 247–262.

[Troy0] S. Troyanski, *On topological equivalence of spaces $c_0(\aleph)$ and $\ell_1(\aleph)$*, Bull. Acad. Polon. Sci. Ser. Sci. Math. Astr. et Phys. **15** (1967), 389–396.

[Troy1] S. Troyanski, *On locally uniformly convex and differentiable norms in certain nonseparable Banach spaces*, Studia Math. **37** (1971), 173–180.

[Troy2] S. Troyanski, *On equivalent norms and minimal systems in nonseparable Banach spaces*, Studia Math. **43** (1972), 125–138.

[Troy3] S. Troyanski, *On nonseparable Banach spaces with a symmetric basis*, Studia Math. **53** (1975), 253–263.

[Troy4] S. Troyanski, *On uniform rotundity and smoothness in every direction in nonseparable Banach spaces with an unconditional basis*, Compt. Rend. Acad. Bulg. Sci. **30** (1977), 1243–1246.

[Troy5] S. Troyanski, *Locally uniformly convex norms*, Compt. Rend. Acad. Bulg. Sci. **32** (1979), 1167–1169.

[Troy6] S. Troyanski, *Construction of equivalent norms for certain local characteristics with rotundity and smoothness by means of martingales*, Proceedings of the 14th Spring conference of the Union of Bulgarian mathematicians, 129–156, 1985.

[Troy7] S. Troyanski, *Gâteaux differentiable norms in L_p*, Math. Ann. **287** (1990), 221–227.

[Tsa] I.G. Tsarkov, *On global existence of an implicit function*, Russian Acad. Sci. Sbornik, **79** (1994), 287–313.

[Tsir] B.S. Tsirelson, *Not every Banach space contains an embedding of ℓ_p or c_0*, Funct. Anal. Appl. **8** (1974), 138–141.

[Tur] B. Turett, *A dual view of a theorem of Baillon*, Lect. Notes Pure Appl. Mat. **80** (1982), Marcel Dekker 1982.

[Tych] A. Tychonoff, *Ein Fixpunktsatz*, Math. Annalen **111** (1935), 767–776.

[Tzaf] L. Tzafriri, *On Banach spaces with unconditional bases*, Israel J. Math. **17** (1974), 84–93.
[Vakh] F.S. Vakher, *On the basis on the space of continuous functions on a compact space*, Dokl. Ak. N. SSSR **4** (1955), 589–592
[Vald1] M. Valdivia, *On a class of Banach spaces*, Studia Math. **60** (1977), 11–30.
[Vald2] M. Valdivia, *Some more results on weak compactness*, J. Funct. Anal. **24** (1977), 1–10.
[Vald2b] M. Valdivia, *Resolution of the identity in certain Banach spaces*, Collect. Math. **39** (1988), 127–140.
[Vald3] M. Valdivia, *Projective resolutions of identity in $C(K)$ spaces*, Arch. Math. **54** (1990), 493–498.
[Vald4] M. Valdivia, *Simultaneous resolutions of the identity operator in normed spaces*, Collect. Math. **42** (1991), 265–284.
[Vald5] M. Valdivia, *On basic sequences in Banach spaces*, Note di Matematica **12** (1992), 245–258.
[Vald6] M. Valdivia, *Fréchet spaces with no subspaces isomorphic to ℓ_1*, Math. Japonica **38** (1993), 397–411.
[Vale] F.A. Valentine, *Convex sets*, McGraw Hill, New York, 1964.
[vMill] J. van Mill, *Infinite-dimensional topology*, North-Holland, 1989.
[Vand1] J. Vanderwerff, *Fréchet differentiable norms on spaces of countable dimension*, Arch. Math. **58** (1992), 471–476.
[Vand2] J. Vanderwerff, *Smooth approximations in Banach spaces*, Proc. Amer. Math. Soc. **115** (1992), 113–120.
[Vand3] J. Vanderwerff, *Second-order Gâteaux differentiability and an isomorphic characterization of Hilbert spaces*, Quart. J. Math. Oxford **44** (1993), 249–255.
[Vand4] J. Vanderwerff, *Extensions of Markushevich bases*, Math. Z. **219** (1995), 21–30.
[VWZ] J. Vanderwerff, J.H.M. Whitfield, and V. Zizler, *Markushevich bases and Corson compacta in duality*, Canad. J. Math. **46** (1994), 200–211.
[Vasa] L. Vašák, *On a generalization of weakly compactly generated Banach spaces*, Studia Math. **70** (1981), 11–19.
[Veec] W.A. Veech, *Short proof of Sobczyk's theorem*, Proc. Amer. Math. Soc. **28** (1971), 626–628.
[Wern] D. Werner, *A proof of the Markov–Kakutani fixed point theorem via the Hahn–Banach theorem*, Extracta Math. **8** (1993), 37–38.
[WhZy] R.L. Wheeden and A. Zygmund, *Measure and integral: An introduction to real analysis*, Marcel Dekker, 1977.
[WhZi] J.H.M. Whitfield and V. Zizler, *Extremal structure of convex sets in spaces not containing c_0*, Math. Z. **197** (1988), 219–221.
[Whitl] R.J. Whitley, *Projecting m onto c_0*, Amer. Math. Monthly **73** (1966), 285–286.
[Will] S. Willard, *General topology*, Addison-Wesley, 1968.
[Willis] G. Willis, *The compact approximation property does not imply the approximation property*, Studia Math. **103**(1) (1992), 99–108.
[Wojt] P. Wojtaszczyk, *Banach spaces for analysts*, Cambridge Studies in Advanced Mathematics **25**, 1991.
[Yost0] D. Yost, *M-ideals, the strong 2-property and some renorming theorems*, Proc. Amer. Math. Soc. **81** (1981), 299-303.
[Yost1] D. Yost, *Asplund spaces for beginners*, Acta Univ. Carolinae **34** (1993), 159–177.
[Yost2] D. Yost, *The Johnson–Lindenstrauss space*, Extracta Math. **12** (1997), 185–192.
[Zaho] Z. Zahorski, *Sur l'ensemble des points de non-derivabilite d'une fonction continue*, Bull. Soc. Math. France **74** (1946), 147–178.
[Zhen] X.Y. Zheng, *Measure of non-Radon–Nikodým property and differentiability of convex functions on Banach spaces*, Set-valued Analysis **131** (2005), 181–196.

[Zipp0] M. Zippin, *A remark on bases and reflexivity in Banach Spaces*, Israel J. Math. **6** (1968), 74–79.

[Zipp1] M. Zippin, *The separable extension problem*, Israel J. Math. **26** (1977), 372–387.

[Zipp2] M. Zippin, *Extension of bounded linear operators*, Handbook of Banach spaces II, Editors W.B. Johnson and J.Lindenstrauss, Elsevier, 2003, 1703–1741.

[Zipp3] M. Zippin, *Banach spaces with separable duals*, Trans. Amer. Math. Soc. **310** (1988), 371–379.

[Zizl1] V. Zizler, *On some rotundity and smoothness properties of Banach spaces*, Dissert. Math. **87** (1971), 1–33.

[Zizl1b] V. Zizler, *On some extremal problems in Banach spaces*, Math. Scandinavica **32** (1973), 214–224.

[Zizl1c] V. Zizler, *Locally uniformly rotund renorming and decomposition of Banach spaces*, Bull. Austr. Math. Soc. **29** (1984), 259–265.

[Zizl2] V. Zizler, *Smooth extension of norms and complementability of subspaces*, Arch. Math. **53** (1989), 585–589.

[Zizl3] V. Zizler, *Nonseparable Banach spaces*, Handbook of Banach Spaces II, Editors W.B. Johnson and J.Lindenstrauss, Elsevier, 2003, 1743–1816.

Symbol Index

Subject Index

Entries in bold typeface correspond to the pages where the corresponding concepts are defined.

M

O

Z

Author Index

The structure is Name [citation], pages where the citation appears.

K

Printed in the United States
By Bookmasters